7th International Conference on Applied Electrostatics

AA003190

(ICAES-2012)

Journal of Physics: Conference Series Volume 418

Dalian, China
17-19 September 2012

ISBN: 978-1-62748-296-7
ISSN: 1742-6588

Printed from e-media with permission by:

Curran Associates, Inc.
57 Morehouse Lane
Red Hook, NY 12571

proceedings.com

Some format issues inherent in the e-media version may also appear in this print version.

Printed by Curran Associates, Inc. (2013)

Institute of Physics
Dirac House, Temple Back
Bristol BS1 6BE UK

Phone: 44 1 17 929 7481
Fax: 44 1 17 920 0979

techtracking@iop.org

Additional copies of this publication are available from:

Curran Associates, Inc.
57 Morehouse Lane
Red Hook, NY 12571 USA
Phone: 845-758-0400
Fax: 845-758-2634
Email: curran@proceedings.com
Web: www.proceedings.com

Numerical Simulation and Calculation

ESD, EMC and Electric Device

Measurment and Analysis

Discharges and Plasma

Diagnosis

Inactivation and Purification of Pollutants in Gas, Water and Soil

Journal of Physics: Conference Series **418** (2013) 011002 doi:10.1088/1742-6596/418/1/011002

Committees

Honorary Chair

Shanghe Liu, Professor, Academician of the Chinese Academy of Engineering, National Key Lab. of EM and EM Protection, China

International Advisory Committee

Chair:

M. N. Horenstein, Professor, Boston University, USA

Co-Chair:

M. Rea, Professor, University of Padova, Italy
T. Yamamoto, Professor, Tokyo City University, Japan

Members:

J. O. Chae, Professor, Inha University, Korea
K. Ikezaki, Professor, Keio University, Japan
M. Sato, Professor, Gunma University, Japan

Local Advisory Committee

Chair:

Keping Sun, Professor, Shanghai Maritime University, China

Co-Chair:

Keping Yan, Professor, Zhejiang University, China

Members:

Zhenghao He, Professor, Huazhong University of Science & Technology, China
Ninghui Wang, Professor, Dalian University of Technology, China
Daiqi Ye, Professor, South China University of Technology, China
Jingchang Zhang, Professor, Zhongyuan University of Technology, China
Tianle Zhu, Professor, Beihang University, China
Yimin Zhu, Professor, Dalian Maritime University, China

International Organizing Committee

Chair:

T. B. Jones, Professor, University of Rochester, USA

Co-Chair:

T. Oda, Professor, the University of Tokyo, Japan
Yan Wu, Professor, Dalian University of Technology, China

Members:

R. H. Amirov, Dr., Joint Institute for High Temperatures of the Russian Academy of Sciences, Russia
J. B. Gajewski, Professor, Wroclaw University of Technology, Poland
S. Kanazawa, Professor, Oita University, Japan
B. R. Locke, Professor, Florida State University, USA
P. Lukeš, Dr., Academy of Sciences of the Czech Republic, Czech Republic
A. Mizuno, Professor, Toyohashi University of Technology, Japan
J. D. Moon, Professor, Kyungpook National University, Korea
J. Mizeraczyk, Professor, Polish Academy of Sciences, Poland
A. Ohsawa, Dr., National Institute of Industrial Safety, Japan
T. Ohshima, Professor, Gunma University, Japan
Keping Sun, Professor, Shanghai Maritime University, China

K. Takaki, Professor, Iwate University, Japan
X. Tu, Dr., University of Liverpool, UK
K. Urashima, Dr., MEXT, Japan
M. Washizu, Professor, the University of Tokyo, Japan
Qingguo Wang, Professor, National Key Lab of EM and EM Protection, China

Local Organizing Committee

Chair:

Yan Wu, Professor, Dalian University of Technology, China

Co-Chair:

Keping Sun, Professor, Shanghai Maritime University, China
Qingguo Wang, Professor, National Key Lab of EM and EM protection, China
Guofeng Li, Professor, Dalian University of Technology, China

Members:

Tianhai Chang, Dr., South China University of Technology, China
Lili Cui, Professor, Second Military Medical University, China
Mingjun He, President, Safety Technology Academy of CNPC, China
Jian Jiang, Professor, Second Military Medical University, China
Jie Li, Professor, Dalian University of Technology, China
Jiwu Li, Professor, Hangzhou University of Commerce, China
Qing Li, Professor, Hebei University, China
Quanzhen Liu, President, China Petroleum and Chemical Corporation, China
Zhiqiang Liu, Professor, Hebei University, China
Yanming Kang, Professor, Donghua University, China
Jiting Ouyang, Professor, Beijing Institute of Technology, China
Zhenhua Pei, President, TA&A Ultra Clean Technology Co. Ltd., China
Bing Sun, Professor, Dalian Maritime University, China
Ming Sun, Professor, Shanghai Maritime University, China
Guanghui Wei, Professor, Nation Key Lab. Of EM and EM Protection, China
Jianping Xiong, Professor, Maoming University, China
Xiaoying Xu, Professor, Wuhan University of Technology, China
Linfeng Yang, Professor, Zhongyuan University of Technology, China
Benmou Zhou, Professor, Nanjing University of Science &Technology, China
Yuanxiang Zhou, Professor, Tsinghua University, China
Zhitao Zhang, Professor, Dalian Maritime University, China
Bingyan Dong, Professor, Jiangxi University of Science and Technology, China

Secretary:

Jie Li, Professor, Dalian University of Technology, China

Co-Secretaries:

Dr. Na Lu, Dalian University of Technology, China
Dr. Kefeng Shang, Dalian University of Technology, China
Mr. Changhai Sun, Dalian University of Technology, China
Mr. Shoufeng Tang, Dalian University of Technology, China
Ms. Nan Jiang, Dalian University of Technology, China
Mr. Jiutao An, Dalian University of Technology, China

Preface

ICAES is an important conference organized every four years by the Committee on Electrostatics of the Chinese Physical Society, which serves as a forum for scientists, educators and engineers interested in the fundamentals, applications, disasters and safety of electrostatics, etc.

In recent years, new techniques, applications and fundamental theories on electrostatics have developed considerably. ICAES-7, held in Dalian, China, from 17–19 September 2012, aimed to provide a forum for all scholars to report the newest developments in electrostatics, to probe the questions that scholars faced and to discuss fresh ideas related to electrostatics.

ICAES-7 was co-organized and hosted by Dalian University of Technology, and was sponsored by the Ministry of Education of China, the National Natural Science Foundation of China, Dalian University of Technology, Nanjing Suman Electronics Co. Ltd (Suman, China), Shekonic (Yangzhou Shuanghong, China) Electric/Mechanical Co. Ltd, and Suzhou TA&A Ultra Clean Technology Co. Ltd. (China). On behalf of the organizing committee of ICAES-7, I express my great appreciation for their support of the conference.

Over 160 scholars and engineers from many countries including Croatia, The Czech Republic, D.P.R. Korea, Germany, Japan, Malaysia, Poland, Russia, the United States of America, China attended ICAES-7, and the conference collected and selected 149 papers for publication. The subjects of those papers cover the fundamentals of electrostatics, electrostatic disaster and safety, and electrostatic application (e.g. precipitation, pollutant control, biological treatment, mixture separation and food processing, etc). I cordially thank all authors and attendees for their support, and my appreciation is also given to the conference honorary chair, the organizing committee and advisory committee, and the conference secretaries for their hard work.

ICAES-7 is dedicated to the memory of Professor Jen-Shih Chang (professor emeritus in the Faculty of Engineering, McMaster University, Canada), Haitian Scholar of Dalian University of Technology (China), who passed away on 27 February 2011. Professor Chang was active in research fields including the applications of electrostatics, electromagnetic hydrodynamics, plasma environmental pollution control technologies, etc and he contributed much to the development of these fields. Professor Chang was the visiting professor at some Key Universities in China and was the friend of Chinese scholars engaged in electrostatics. Professor Chang was also active in joining and supporting the previous ICAES. We will cherish the memory of Professor Jen-Shih Chang forever.

Professor Jie Li
Proceedings Editor
Dalian, September 2012

The 7th International Conference on Applied Electrostatics

Peer review statement

All papers published in this volume of *Journal of Physics: Conference Series* have been peer reviewed through processes administered by the proceedings Editors. Reviews were conducted by expert referees to the professional and scientific standards expected of a proceedings journal published by IOP Publishing.

Study on electromagnetic field distribution of impulse current wave

Xiao-lei Bi[1,2*] , Quan-zhen Liu[1,2], Bao-quan Liu[1,2], Xin Gao[1,2], Hai-yan Hu[1,2], Yun-peng Zhang[1,2]
1. SINOPEC Safety Engineering Institute, Qingdao, China
2. State Key Laboratory of Chemical Safety, Qingdao, China

E-mail:bxlluck@163.com

Abstract. The 10/350 μs impulse current wave which is generated by impulse current generator in the laboratory is simulated by Heidler function current model. Based on dipole theory, electromagnetic field equations are calculated. Combined with Heidler function current model, electromagnetic field time-space distribution of 10/350 μs impulse current wave is simulated. The results of simulations show that with an increase in the length of the current element, the electromagnetic field intensity, caused by the current element, increases correspondingly. The electromagnetic field caused by the same current element decreases gradually with an increase in the distance. Based on the above conclusions, some electromagnetic protection measures of the laboratory are presented, which have specific meaning to the shield design of the lab and the health of workers in the lab.

1. Introduction

With the fast development of information industry, electronic devices are becoming highly integrated, highly accurate and low energy-consuming. Meanwhile, lighting electromagnetic pulse (LEMP) following lighting discharge has caused more and more hazards to these electronic devices. According to protection against lighting standard of GB/T 21714 and IEC 62305, 10/350 μs impulse current waveform is chosen to simulate the first lighting stroke. The 10/350 μs impulse current generator running in the laboratory can generate an impulse current of 100 kA. This powerful impulse current must cause very intensive electromagnetic field. As a result, studying the electromagnetic pulse of impulse current in the laboratory is of vital importance for the electromagnetic pulse shielding of electronic devices in the lab, the shield design of a lab and the health of people working in the lab. Based on the dipole

2. Theoretical Calculation

The output terminal of the impulse current generator is located in the sample case, which is composed of a high voltage terminal and low voltage terminal as is shown in figure 1. Ideally, when a sample is put in the case, set the conductor with a length of L between the two terminals as current element $i(t)L$. According to the Maxwell equations, the electric field $E(R,\ t)$ and the magnetic field $B(R,\ t)$ in any position can be determined. The Maxwell equations of differential form are as follows:

[*] To whom any correspondence should be addressed.

Journal of Physics: Conference Series **418** (2013) 012001 doi:10.1088/1742-6596/418/1/012001

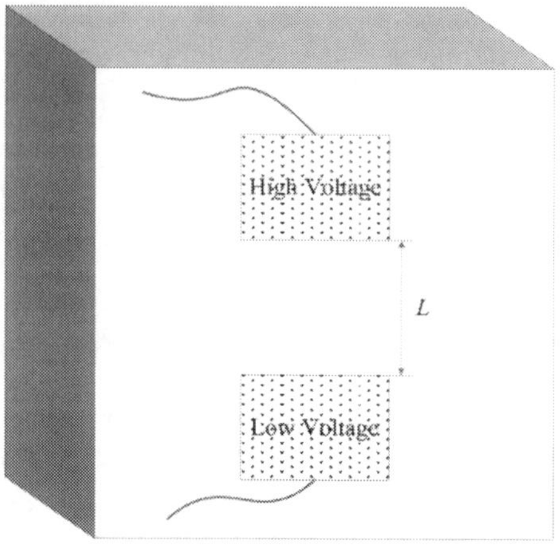

Figure 1. Sample case of the impulse current generator.

$$\nabla \times \boldsymbol{B} = \mu \boldsymbol{J}_e + \frac{1}{c^2}\frac{\partial \boldsymbol{E}}{\partial t} \tag{1}$$

$$\nabla \times \boldsymbol{E} = -\frac{\partial \boldsymbol{B}}{\partial t} \tag{2}$$

$$\nabla \cdot \boldsymbol{B} = 0 \tag{3}$$

$$\varepsilon_0 \nabla \cdot \boldsymbol{E} = \rho \tag{4}$$

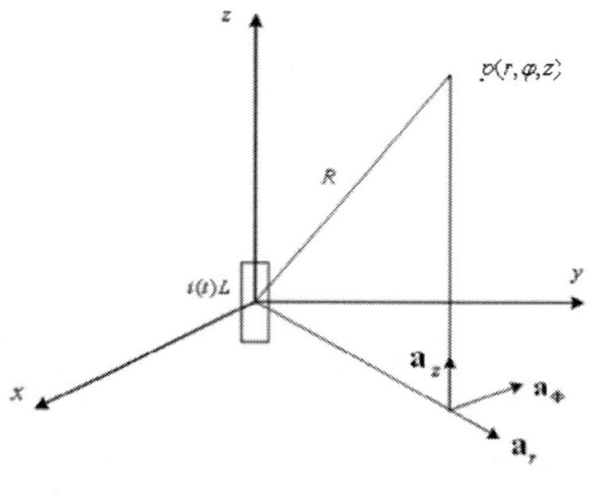

Figure 2. Calculation for the electromagnetic field generated by current element at the zero point.

When current element is located at the zero point and is in free space, the electromagnetic component, which includes magnetic field component H_ϕ, electric field radial component E_r and electric field axial component E_z can be determined by solving the Maxwell equations:

$$H_\phi(r,\varphi,z,t) = \frac{L}{4\pi R^2}\left[\frac{r}{c}\frac{\partial i(t-R/c)}{\partial t} + \frac{r}{R} i(t-R/c)\right] \tag{5}$$

$$E_r(r,\varphi,z,t) = \frac{Lrz}{4\pi\varepsilon_0 R^3}\left[\frac{3}{R^2}\int_0^t i(\tau - R/c)d\tau \frac{1}{c^2}\frac{\partial i(t-R/c)}{\partial t} + \frac{3}{cR} i(t-R/c)\right] \tag{6}$$

$$E_z(r,\varphi,z,t) = \frac{L}{4\pi\varepsilon_0 R^3}\left[\frac{2z^2-r^2}{R^2}\int_0^t i(\tau - R/c)d\tau + \frac{2z^2-r^2}{cR} i(t-R/c) - \frac{r^2}{c^2}\frac{\partial i(t-R/c)}{\partial t}\right] \tag{7}$$

Where: ε_0 (the free-space dielectric constant) is 0.854×10^{-12} F m^{-1}; μ_0 (the free-space magnetic conductivity) is $4\pi\times10^{-7}$; R is the distance between current element and observation point.

By applying the Heidler function current model proposed by Heidler, Current element $i(t)$ can be shown as follows:

$$i(t) = \frac{I_0}{h}\frac{\left(\frac{t}{\tau_1}\right)^n}{1+\left(\frac{t}{\tau_1}\right)^n} e^{-\frac{t}{\tau_2}} \tag{8}$$

Where: I_0 is the peak current, h is the coefficient of correction for the peak current, τ_1 is the time constant of the wave head, τ_2 is the time constant of the wave tail. The derivative about time of the Heidler function can be described as follows:

$$\frac{\partial i(t)}{\partial t} = \frac{I_0}{h}\frac{\left(\frac{t}{\tau_1}\right)^n}{1+\left(\frac{t}{\tau_1}\right)^n} e^{-\frac{t}{\tau_2}}\left\{\frac{n}{t(1+t/\tau_1)^n} - \frac{1}{\tau_2}\right\} \tag{9}$$

3. Calculation and simulation

3.1. Simulation for the 10/350 μs impulse current

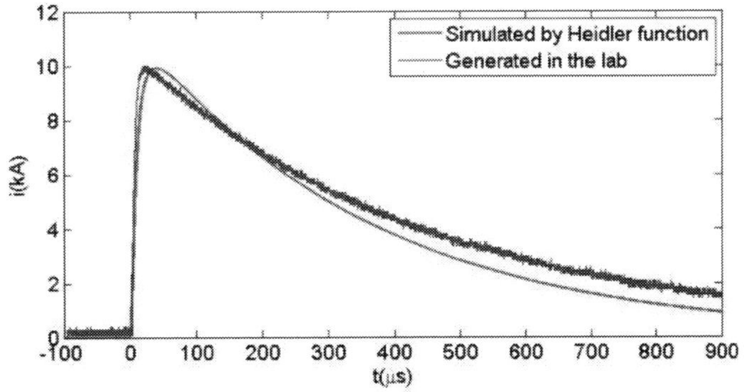

Figure 3. The comparison between two impulse currents separately simulated by Heidler function and generated by the impulse current generator in the lab.

The Heidler function is applied to simulate the 10/350 μs impulse current wave, and this current wave is compared with the impulse current generated by the impulse current generator in the lab, as is shown in figure 3, in which the waveform parameter I_0 (the peak current) is 10 kA and τ_1 (the time constant of the wave head) is 10 μs and τ_2 (the time constant of the wave tail) is 350 μs. As can be seen from the comparison result, the only difference between the simulated and the generated impulse current waveform is located at the wane tail, while in other locations such as the wave head and the peak, the two forms match very well. As a result, it can be reasonable to apply the Heidler function to simulate the impulse current generated by the impulse current generator in the lab, and calculate the electromagnetic field caused by this impulse current.

3.2. Simulation for the electromagnetic field distribution of the impulse current

As the distance between the two terminals of the impulse current generator output terminal is very short, it can be supposed that a current element with the length of L is caused between the two terminals. In this paper, the electromagnetic field distribution of the impulse current is simulated by combining the Heidler function and the current element electromagnetic field equations.

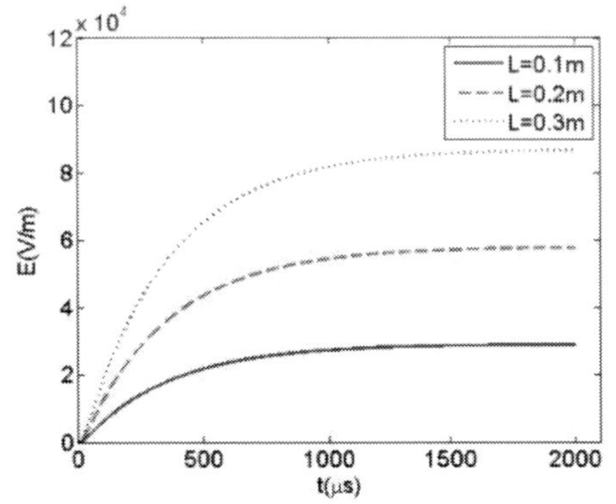

Figure 4. The electric-field intensity generated by different current elements at a radial distance (r=50 m).

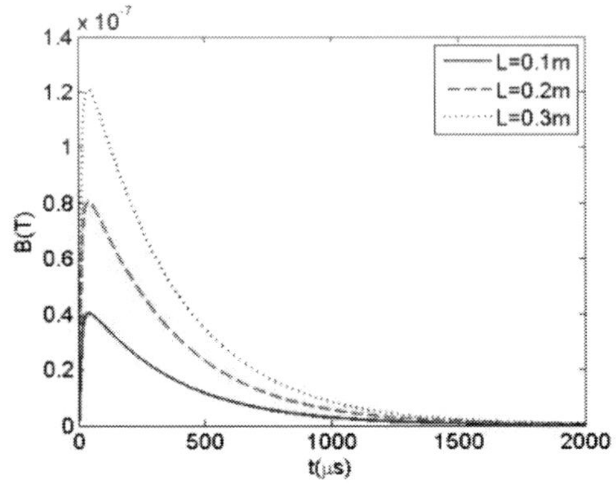

Figure 5. The magnetic induction intensity generated by different current elements at a radial distance (r=50 m).

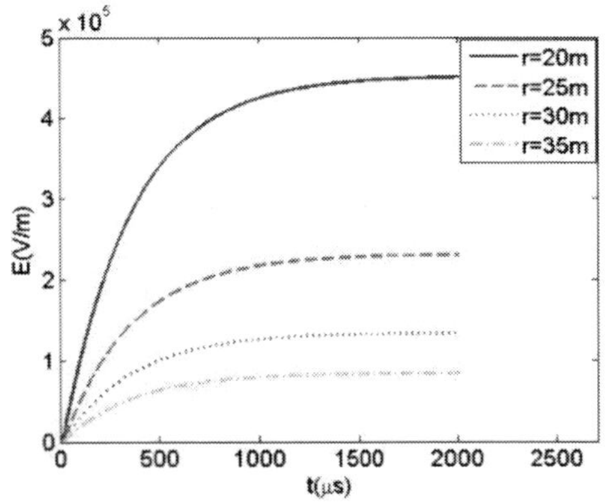

Figure 6. The electric-field intensity generated by the same current element at different distances.

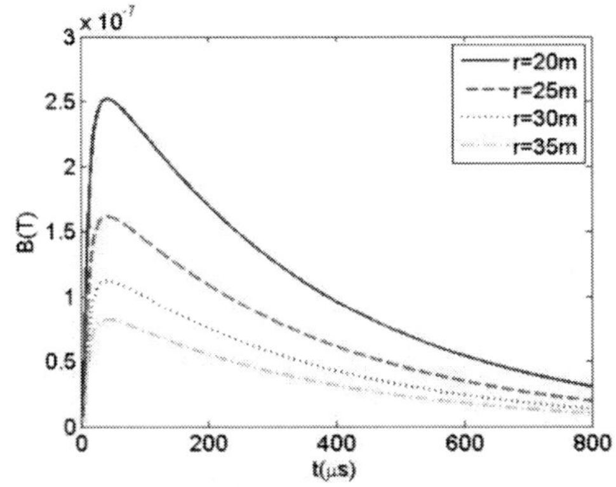

Figure 7. The magnetic induction intensity generated by the same current element at different distances.

Figure 4 and figure 5 separately shows the electric field and magnetic field generated by different current elements at a radial distance (r=50 m). As can be seen from figure 4 and figure 5, as the increase of the current element, the electric-field intensity and the magnetic induction intensity increase at the same time, and the relationship between the current element and field intensity is linear. The electric-field intensity increases fast at first and slow after that. While the magnetic induction intensity increases very fast to a peak value at first and then decreases fast and later gradually decreases to zero.

Figure 6 and figure7 separately shows the electric field and magnetic field generated by the same current element at different distances. As the distance goes further, the electric-field intensity and the magnetic induction intensity decrease gradually.

4. Conclusions

Based on the dipole theory and combing the Heidler function current model, the electromagnetic field of the 10/350 μs impulse current, generated by the impulse current generator, is calculated in this paper. By analyzing and comparing the simulation results, some conclusions and suggestions are proposed as follows:

- With an increase in the length of the current element, the electromagnetic field intensity, caused by the current element, increases correspondingly. Under the condition of meeting the experimental demands, the distance between the high voltage terminal and the low voltage terminal should be decreased as short as possible.
- The electromagnetic field caused by the same current element decreases gradually with an increase in the distance. So it's noticeable to keep an enough long distance between the impulse current generator and other electric devices and workers in the lab during the construction of the lab.
- Because the electromagnetic field intensity generated by the 10/350 impulse current is very powerful, it's very essential to improve the electromagnetic shielding in the lab, and set apart the experiment area and the working area.

Acknowledgments

This project is supported by National Key Technology R&D Program (Grant 2012BAK03B03).

References

[1] Shi G Q and Fang H H 2004 *Journal of Hunan Institute of Engineering* **14** 19
[2] Wei G H, Sun W C and Tian M H 2003 *Safety & EMC* **3** 37
[3] Chen N N 2010 *Electric Switchgear* **3** 82
[4] Zhou X G and Yang X H 2006 *Microcomputer Information* **22** 217
[5] GB/T 21714.1 *Protection against lightning- Part 1:General principles*(Beijing)
[6] IEC 62305-1 *Protection Against Lighting- Part1: General principles*(Beijing)
[7] Heidler F, Cvetic J M and Stanic B V 1999 *IEEE Trans. Power Delivery* **14** 399
[8] Yang C S and Zhou B H 2004 *IEEE Trans. Electromagn. Compat.* **46** 133

Journal of Physics: Conference Series **418** (2013) 012002 doi:10.1088/1742-6596/418/1/012002

Numerical calculations of human-body capacitance in mining tunnel environment

Linfeng Yang*, Hu Liu, Lin Xing, Haizhen Ren, Jingchang Zhang
Zhongyuan University of Technology, Zhengzhou, China

E-mail: laodizhi@yahoo.com.cn

Abstract. The Electrostatic Discharge (ESD) is the main cause of some disasters in the mining environments. The ESD depends mainly on the capacitance and resistance of human-body and the capacitance is very sensitive to the environments. In this paper, the figure capacitance of human-body in mining tunnel environment was studied with numerical simulation. It was found that the human-body capacitance in mining tunnel environment depended mainly on the human figure and the shape of the tunnel, only slightly related to the position. The results implied the ESD model was significant in the mining environments; moreover, the human-body capacitance in mining tunnel environment was also approximately linear to the figure size and its value was around 250 pF, which was coincident with Human Body Model (HBM); The figure capacitance was dependent on the constituent of coal layer as well, the typical value of capacitance was 20% larger than that at free space, which meant the mining tunnel was an electrostatic sensitive environment.

1. Introduction

One of the most common causes of electrostatic accidents is the direct transfer of electrostatic charge through a significant series resistor from the human body or from a charged material to the electrostatic discharge sensitive (ESDS) device. Human-body can be regarded as a mobile conductor, when one touches or is near a charged object or even just walks across a floor in dry air, electrostatic charge will accumulate on the body and the electrostatic voltage can be high to thousands even ten thousands volts. As the local electric field strength is higher than disruptive field intensity threshold, the air dielectric medium will be broken down and instantaneous discharge takes place and gives out electric spark, which possibly results in device damage or electrostatic accident in the environment of inflammable and explosive gases or particulate matter such as gasoline or mash gases.

The mineral environment is a place where the electrostatic accidents are easy and frequent to happen and result in great casualties: firstly, the rapid body movement during working process can generate electrostatic charges on the body; secondly, a relative dry room is formed around the body inside the thick working clothes; thirdly, the mineral tunnel is abundant of firedamp (methane) gases. It is significant to eliminate the potential safety hazard and prevent electrostatic losses.

* To whom any correspondence should be addressed.

2. Model and formulism

Most of static electricity protection standards are based on human-body electrostatic phenomena because human-body electrostatic is the main and frequent cause of electrostatic accidents [1-3]. To eliminate the potential safety hazards, we need to study the human-body discharging process mechanism and to setup human-body discharging models in different environments. The model used to simulate this event is the Human Body Model (HBM). The HBM testing model represents the discharge from the fingertip of a standing individual delivered to the device. It is simulated by a 100 pF capacitor discharged through a switching component and a 1.5 kΩ series resistor into the component. This model, which was derived from the nineteenth century, was developed for investigating explosions of gas mixtures in mines. A typical Human Body Model circuit is presented in figure 1.

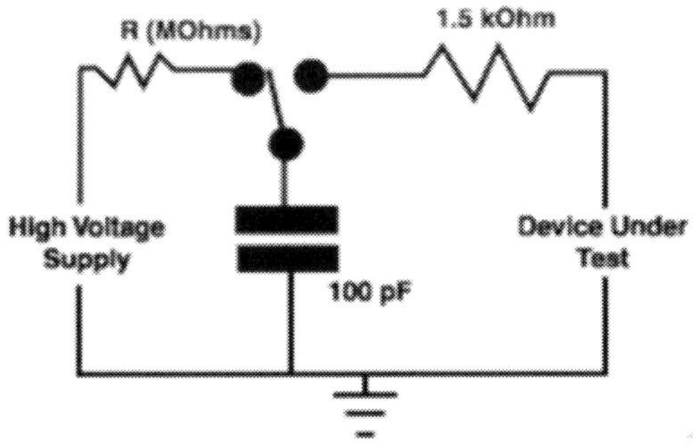

Figure 1. A typical human-body electrostatic discharging model.

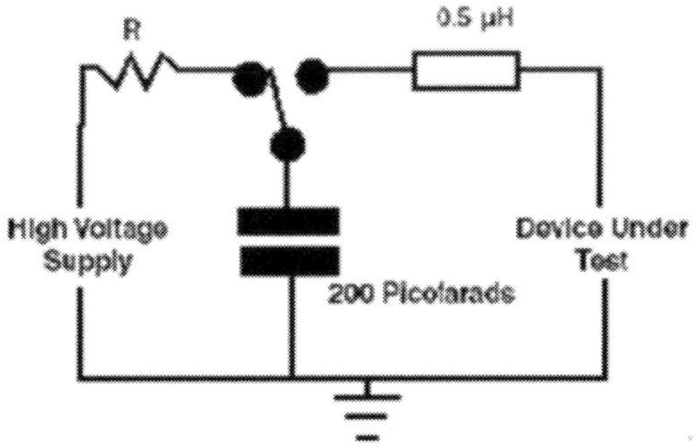

Figure 2. Typical machine model circuit.

A discharging similar to the HBM event also can occur from a charged conductive object [4, 5], such as a metallic tool, the model circuit can be seen in figure 2. The model is called machine model(MM), as a result of investigating worst-case scenarios of the HBM, the Machine Model simulates a more rapid and severe electrostatic discharge from a charged machine, fixture, or tool. The MM test circuit charges a 200 pF capacitor to a certain voltage and then discharges this capacitor directly into the device being tested through a 500 nH inductor with no series resistor.

To establish the human-body electrostatic discharging model in coal mineral tunnel environment we need to determine the parameters of human-body resister and figure capacitance in tunnel environment. The resister is easy to measure and is independent of the special environment, but the human figure capacitance is much more complicated. The human-body capacitance is dependent not only on the figure of human-body and the relative position in the mineral tunnel, but also is sensitive to the shape of the coal mineral tunnel and the dielectric constant ε of the coal layer. So it is a hard task to calculate the human-body figure capacitance precisely.

The simplest way to compute the human body capacitance is to consider the human-body in coal mineral tunnel as a hollow coaxial cylindrical conductors system, by using electrostatic charge image method of a dielectric cylindrical wire. We can replace the mineral tunnel environment equivalence to a parallel cylindrical wire (cylindrical wire electric image of human-body), then calculate the capacitance of the two parallel wires system. But this kind of treatment has three shortcomings: First, the section shape of most coal mineral tunnels are not circle but rectangle instead; second, the human-body is finite, we can't treat it as a point or a infinite long wire; last, the human-body always is not parallel to the axis of the tunnel.

To calculate the human body figure capacitance in coal mineral environment more accurately, we developed a multi-fold electrical images method in this paper, we treat the section shape of mineral tunnels as rectangle with width W and height L, treat the human-body as a cylinder shape with human height h and human waist circumference (radius r) standing on ground parallel to side tunnel walls; treat the coal layers as four infinite dielectric matter (dielectric constant ε) with plane surfaces enclose the human-body; by using the electric image method to every one of the four mirror surfaces, four first-order image human-body will be formed symmetry to the mirror surfaces. Apply the new formed "image human-body" to the four mirror surfaces, every new formed image human-body will then form four higher-order human-body images; and so forth, a plain "woods" of human-body images generated with different orders. The effects of the mineral tunnel environment can be equivalent to a human-body electric images woods conductor system. For the N conductors system, we can obtain the equation groups as follows:

$$\begin{cases} V_1 = p_{11}q_1 + p_{12}q_2 + p_{13}q_3 + \cdots + p_{1N}q_N \\ V_2 = p_{21}q_1 + p_{22}q_2 + p_{23}q_3 + \cdots + p_{2N}q_N \\ \cdots\cdots\cdots\cdots\cdots\cdots\cdots\cdots\cdots\cdots\cdots\cdots\cdots\cdots \\ V_N = p_{N1}q_1 + p_{N2}q_2 + p_{N3}q_3 + \ldots + p_{NN}q_N \end{cases}$$

Where p_{ij} is the potential coefficient, we can also express q_i with C_{ij} and V_i , we can also express q_i with C_{ij} and V_i as follows:

$$\begin{cases} q_1 = C_{11}V_1 + C_{12}(V_2 - V1) + C_{13}(V_3 - V_1) + \ldots + C_{1N}(V_N - V_1) \\ q_2 = C_{21}(V_1 - V_2) + C_{22}V_2 + C_{23}(V_3 - V_2) + \ldots + C_{2N}(V_N - V_2) \\ \cdots\cdots\cdots\cdots\cdots\cdots\cdots\cdots\cdots\cdots\cdots\cdots\cdots\cdots\cdots\cdots\cdots\cdots\cdots\cdots \\ q_N = C_{N1}(V_1 - V_N) + C_{N2}(V_2 - V_N) + C_{N3}(V_3 - V_N) + \ldots + C_{NN}V_N \end{cases}$$

Where C_{ii} is the capacitance coefficients and C_{ij} ($i \neq j$) is inducing coefficient.

3. Results and discussions

The calculation of human-body figure capacitance can be equivalent to calculate the human-body capacitance of human-body in electric images woods conductor system as illustrated in previous section. To calculate the figure capacitance, we assume that the human-body carries electrostatic charge q Coulomb, the *nth*-order electric image then has image charge as follows:

$$q_n = \left(-\frac{\varepsilon - \varepsilon_0}{\varepsilon - \varepsilon_0}\right)^n \cdot q \qquad (1)$$

Where ε is the dielectric constant of coal layer and ε_0 is the dielectric constant in air.

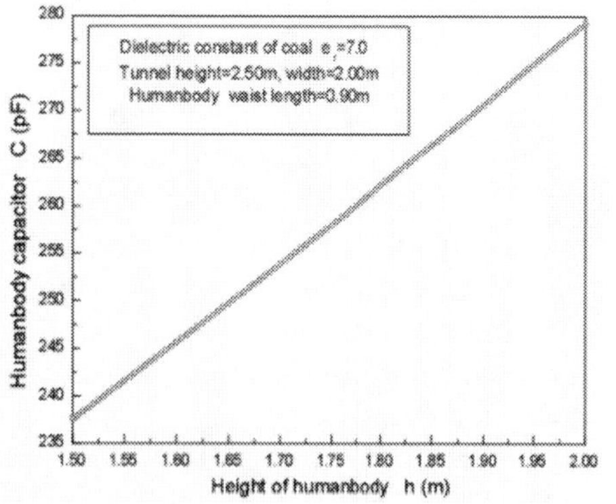

Figure 3. Human-body capacitance versus human height.

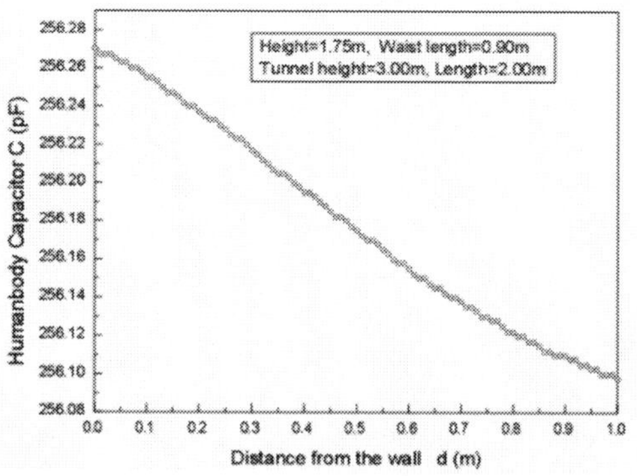

Figure 4. Human-body capacitance versus human position in tunnel.

Journal of Physics: Conference Series **418** (2013) 012002 doi:10.1088/1742-6596/418/1/012002

By calculating the total electrostatic potential of human-body in electric images woods, we can get the human-body figure capacitance in mineral environment. We first calculated the figure capacitance variation with human-body height; the result can be seen in figure 3. The human-body capacitance in mineral environment is also approximately linear to the man height, and the value, around 250 pF, is coincident well to HBM by Kirk [3, 4]. The relationship of human-body capacitance and the position in the tunnel is investigated as seen in figure 4, where *d* represents the distance between the human-body and the side wall of the tunnel, we can see from the figures that the human-body capacitance is mainly dependent on the human figure and the shape of the tunnel, and it is only slightly related to the position, which means the ESD model is significant in mineral environment.

The human-body capacitance is dependent on the shape of the tunnel. To study how the human-body capacitance change with the tunnel shape, we have calculated the human-body capacitance at different tunnel lengths with the section area keeping constant, the calculated result is seen in figure 5, it is clear that as the tunnel length rises, the human-body capacitance decreases, that is to say, the wider the tunnel, the smaller the figure capacitance. At last, we investigated the figure capacitance variation to the coal dielectric constant ε, the calculated result can be seen in figure 6.

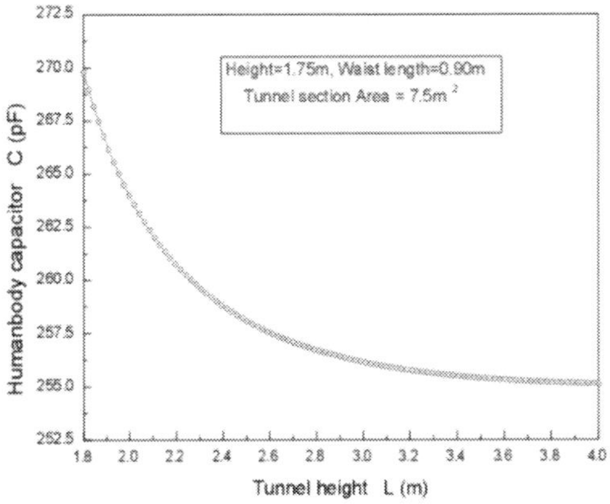

Figure 5. Human-body capacitance versus tunnel length.

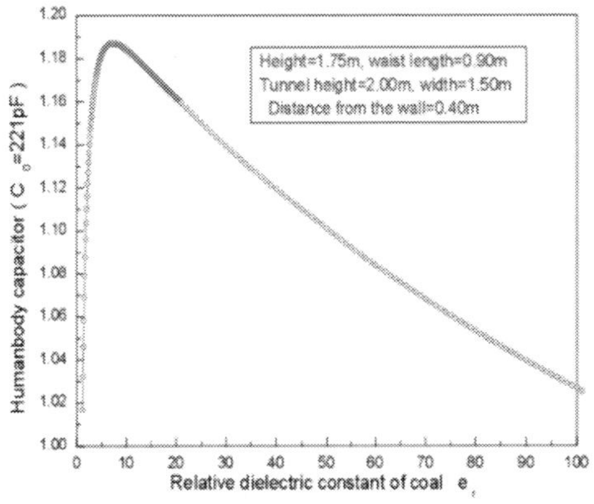

Figure 6. Human-body capacitance versus dielectric constant of the coal layer.

It is very interesting that the human-body capacitance is dependent on the constituent of coal layer as well, which is a peculiar property of human-body capacitance in coal mineral environment. As the dielectric constant ε approaching zero, the effect of tunnel environment can be neglected; as dielectric constant is nearly infinite, the effect is also small since the contribution of the equal image charge with opposite sign compensate mutually. The maximum peak around $\varepsilon_r = 7$ ($\varepsilon_r = 6.3$ for dry coal layer) explains why the electrostatic accidents occurred frequently in the coal mine environment.

4. Conclusions

The numerical modelling results indicate that the human-body capacitance mainly depends on the human figure and the shape of the tunnel, and it is approximately linear to the human-body figure size, and it is only slightly related to the position. So the human-body figure capacitance is a feature quantity of human-body and miner tunnel shape, which means the ESD model is significant in mineral environment. The human-body capacitance in all tunnels is related to the dielectric constant ε of the layer and becomes the largest in the coal mineral environment. Therefore, the coal mineral tunnel becomes the potential electrostatic disaster environment.

Acknowledgments
This work was supported by the Natural Science Foundation of China under Grant No. 51077134 and 50977093

References

[1] Avery L R 1987 *Electr. Overstress Electrost. Discharge Symp. Proc. (Orlando)* (NY, Rome: US/ESD Assn) p 186.
[2] Steven H V 2009 *ESD: Failure Mechanisms and Models* (New York: Wiley–Interscience)
[3] Renninger R G. 1991 *Electr. Overstress Electrost. Discharge Symp. Proc.(Las Vegas)* (Griffiss: US/AFB Reliability Analysis Cent) p 127
[4] Verhaege K, Russ C, Robinson-Hahn D, Farris M, Scanlon J, Lin D, Veltri J and Groeseneken G 1996 *Overstress Electrost. Discharge Symp. Proc. (Orlando)* (NY, Rome: US/ESD Assn) p 40
[5] Wada T 1995 *Electr. Overstress Electrost. Discharge Symp. Proc. (Phoenix)* (NY, Rome: US/ESD Assn) p 186

Analyses of characteristics of spatial lightning electromagnetic field at different horizontal distances and observation point heights

Xiaojia Wang*, Yazhou Chen, Haojiang Wan, Lin Wang
Institute of Electrostatic and Electromagnetic Protection, Ordnance Engineering College, Shijiazhuang, China

E-mail: wangxiaojia0@sina.com

Abstract. The analytic expressions of the spatial lightning electromagnetic field are obtained by solving Maxwell's Equations based on the dipole method. On this basis, the waveforms of the electric and magnetic fields at different heights and distances in the near-zone are presented using the transmission line (TL) model and the channel-base current was described by the impulse function. The result shows that at a same height to the ground, the rising edges of the electromagnetic field waveforms become subdued and the peak values decrease with the increase of the horizontal distance. For the electromagnetic field at the same distance, the rising edge of the waveform becomes steep gradually with the increasing height of the observation point. The peak value of the horizontal electric field increases with increasing height of the observation point and the peak value of the magnetic field basically remains unchanged. The study can provide useful information for protecting the electronic equipment from the destruction of LEMP.

1. Introduction

Lightning is a typical source of electromagnetic hazards in nature. When a lightning happens, the voltage of the lightning return-stroke channel may reach up to millions of volts, and the current in the channel may reach up to hundreds of kiloamperes, with a rise-rate of tens of kiloamperes per microsecond [1-2]. Such powerful transient current can produce a strong electromagnetic radiation around the lightning return stroke channel, which would generate a serious threat to the surrounding buildings and electronic equipments. Especially in the 20th century, with the rapid development of the electronic industry and information technology, the electromagnetic sensitivity of electronic equipment becomes very high and the capacity to resist the lightning electromagnetic interference gradually drops. Therefore, a very weak electromagnetic field is likely to cause hazard to the electronic equipment. The loss caused by lightning electromagnetic pulse (LEMP) gradually appears and increases year by year [3]. Since the spatial lightning electromagnetic field is difficult to be measured directly, we can utilize the lightning return-stroke engineering model [4-8] and conduct a theoretical calculation for the lightning electromagnetic field. M Rubinstein solved the Maxwell's equations of lightning electromagnetic field by using dipole method [9], and some scholars calculated the expressions of LEMP above the ground without considering the mirror current [10-13]. In this paper, the spatial lightning electromagnetic fields are calculated based on the TL model[14-15], and

* To whom any correspondence should be addressed.

the distribution laws of the electromagnetic fields at different heights and distances in near-zone are then discussed.

2. Theoretical derivation of spatial lightning electromagnetic field

For a linear, homogeneous and single media, Maxwell's equations can be written as:

$$\nabla\cdot\varepsilon_0\boldsymbol{E}=\rho_0,\ \ \nabla\cdot\mu_0\boldsymbol{H}=0,\ \ \nabla\times\boldsymbol{E}=-\frac{\partial\mu_0\boldsymbol{H}}{\partial t},\ \ \nabla\times\boldsymbol{H}=\boldsymbol{J}+\frac{\partial\varepsilon_0\boldsymbol{E}}{\partial t} \tag{1}$$

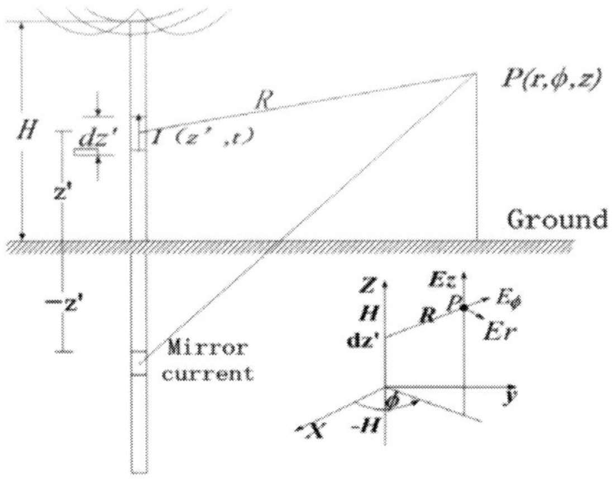

Figure 1. Computational model of lightning return stroke electromagnetic fields.

As shown in Figure 1, we simplify the discharge channel as a vertical antenna model with a height of *H*, assuming that the ground was an ideal conductor and the channel current above the return-stroke height is 0. Considering the current infinitesimal in the return-stroke channel as a current dipole, according to the dipole theory, we can obtain the integral expressions of lightning electromagnetic field in the cylindrical coordinate system as following:

$$E_r(r,\phi,z,t)=\frac{1}{4\pi\varepsilon_0}\int_{-H}^{H}\left[\frac{3r(z-z')}{R^5}\int_{-\infty}^{t}i(t-\frac{R}{c})dt+\frac{3r(z-z')}{cR^4}i(t-\frac{R}{c})+\frac{r(z-z')}{c^2R^3}\frac{\partial i(t-R/c)}{\partial t}\right]dz' \tag{2}$$

$$E_z(r,\phi,z,t)=\frac{1}{4\pi\varepsilon_0}\int_{-H}^{H}\left[\frac{2(z-z')^2-r^2}{R^5}\int_{-\infty}^{t}i(t-\frac{R}{c})dt+\frac{2(z-z')^2-r^2}{cR^4}i(t-\frac{R}{c})-\frac{r^2}{c^2R^3}\frac{\partial i(t-R/c)}{\partial t}\right]dz' \tag{3}$$

$$H_\phi(r,\phi,z,t)=\frac{1}{4\pi}\int_{-H}^{H}\left[\frac{r}{R^3}i(t-\frac{R}{c})-\frac{r}{cR^2}\frac{\partial i(t-R/c)}{\partial t}\right]dz' \tag{4}$$

Where $R=\left((z-z')^2+r^2\right)^{1/2}$ is the distance between the current dipole and the observation point $P(r,\phi,z)$, *c* is the velocity of light in free space, ε_0 and μ_0 represent the vacuum permittivity and permeability respectively, z' is the height of lightning current dipole. The first item, second item and third item at the right side of equations (2) and (3) are the electrostatic field, induction field and radiation field, respectively. The first item and the second item at right side of equation (4) are the induction field and radiation field, respectively.

Using the TL model to calculate electromagnetic field:

$$i(z',t)=u(t-z'/v)\cdot i(0,t-z'/v) \tag{5}$$

In process of solving equations (2)~(4), we will encounter the integral of function δ [16]. Considering that $g(x)$ and $q(x)$ are continuous functions and $g(x_0)=0$, we can obtain:

$$\int \delta[g(x)]q(x)dx = \int \delta(y)q\left[g^{-1}(y)\right]\frac{1}{|dg/dx|}dy = \left.\frac{q(x)}{|dg/dx|}\right|_{x=x_0} \tag{6}$$

Applying (6) in (2)~(4), we can get:

$$E_r(r,z,t) = \frac{I^+_0}{4\pi\varepsilon_0}\left\{\begin{array}{l} rt\left(\frac{1}{R^3_H}-\frac{1}{R^3_0}\right)-\frac{r}{v}\left[\frac{h}{R^3_H}+\frac{z-h}{r^2R_H}-\frac{z}{r^2R_0}\right] \\ +\frac{r(z-h)}{c^2R^3_H}\frac{1}{1/v-(z-h)/(cR_H)}\end{array}\right\} + \frac{I^-_0}{4\pi\varepsilon_0}\left\{\begin{array}{l} rt\left(\frac{1}{R^3_0}-\frac{1}{R^3_{-H}}\right)+\frac{r}{v}\left[\frac{h_2}{R^3_{-H}}-\frac{z+h_2}{r^2R_{-H}}+\frac{z}{r^2R_0}\right] \\ +\frac{r(z+h_2)}{c^2R^3_{-H}}\frac{1}{1/v+(z+h_2)/(cR_{-H})}\end{array}\right\} \tag{7}$$

$$E_z(r,z,t) = \frac{I^+_0}{4\pi\varepsilon_0}\left\{\begin{array}{l} t\left(\frac{z-h}{R^3_H}-\frac{z}{R^3_0}\right)-\frac{r^2}{c^2R^3_H}\frac{1}{1/v-(z-h)/(cR_H)} \\ -\frac{1}{v}\left[-2\left(\frac{1}{R_H}-\frac{1}{R_0}\right)+r^2\left(\frac{1}{R^3_H}-\frac{1}{R^3_0}\right)+\frac{z(z-h)}{R^3_H}-\frac{z^2}{R^3_0}\right]\end{array}\right\} + \frac{I^-_0}{4\pi\varepsilon_0}\left\{\begin{array}{l} t\left(\frac{z}{R^3_0}-\frac{z+h_2}{R^3_{-H}}\right)-\frac{r^2}{c^2R^3_{-H}}\frac{1}{1/v+(z+h_2)/(cR_{-H})} \\ +\frac{1}{v}\left[-2\left(\frac{1}{R_0}-\frac{1}{R_{-H}}\right)+r^2\left(\frac{1}{R^3_0}-\frac{1}{R^3_{-H}}\right)-\frac{z(z+h_2)}{R^3_{-H}}+\frac{z^2}{R^3_0}\right]\end{array}\right\} \tag{8}$$

$$H_\phi(r,z,t) = \frac{I^+_0}{4\pi}\left[\frac{z}{rR_0}-\frac{z-h}{rR_H}+\frac{r}{cR^2_H}\frac{1}{1/v-(z-h)/(cR_H)}\right] + \frac{I^-_0}{4\pi}\left[\frac{z+h_2}{rR_{-H}}-\frac{z}{rR_0}+\frac{r}{cR^2_{-H}}\frac{1}{1/v+(z+h_2)/(cR_{-H})}\right] \tag{9}$$

Where $R_0 = (z^2+r^2)^{1/2}$, $R_H = \left((z-h)^2+r^2\right)^{1/2}$, and $R_{-H} = \left((z+h_2)^2+r^2\right)^{1/2}$. I^+_0 and I^-_0 represent the value of the current at the bottom of the channel at a time of $t-RH/c$ and $t-Rh/c$, respectively.

The value of h and h_2 can be obtained by solving the following equation:

$$t-\left(r^2+(z-h)^2\right)^{1/2}/c-h/v=0, \quad t-\left(r^2+(z+h_2)^2\right)^{1/2}/c-h_2/v=0 \tag{10}$$

3. Distribution laws of lightning electromagnetic field at different horizontal distances

The distribution laws of lightning electromagnetic field at different horizontal distances with a same height to the ground are researched by using equations (7)~(9). The lightning current waveform of 8/20 μs is adopted to calculate the spatial lightning electromagnetic field, and the channel-base current is described by the pulse function [17] as following:

$$i(0,t) = I_0/\xi\left[1-\exp(-t/\tau_1)\right]^n\exp(-t/\tau_2) \tag{11}$$

Where, $\xi = \left[n\tau_2/(\tau_1+n\tau_2)\right]^n\left[\tau_1/(\tau_1+n\tau_2)\right]^{\tau_1/\tau_2}$

Hence, the lightning current parameters in the pulse function can be set as: I_0=30 kA, n=2, τ_1=4.05×10^{-7} s, τ_2=6.8×10^{-5} s.

Herein, the channel height H is 7.5 km and return stroke speed v is 1.3×10^8 m s^{-1}. At the height of z=50 m for the observation point, the spatial distribution of the electric and magnetic fields at different horizontal distances are shown in Figures 2~ 4.

As can be seen in Figure 2, there is a positive peak value for the vertical electric field when the height of the observation point is 50 m and the horizontal distance is smaller than 50 m. This peak value will decrease with the increase of the distance and forms a zero-crossing point with the time axis. When the horizontal distance is larger than 50 m, the positive peak value will disappear and the negative peak value will decrease with the increase of the horizontal distance. Moreover, the peak value of the vertical electric field is about 109 kV· m^{-1} at r =5 m, and the peak value is only 5.9 kV·m^{-1} when r=500 m, with an attenuation of almost 20 times.

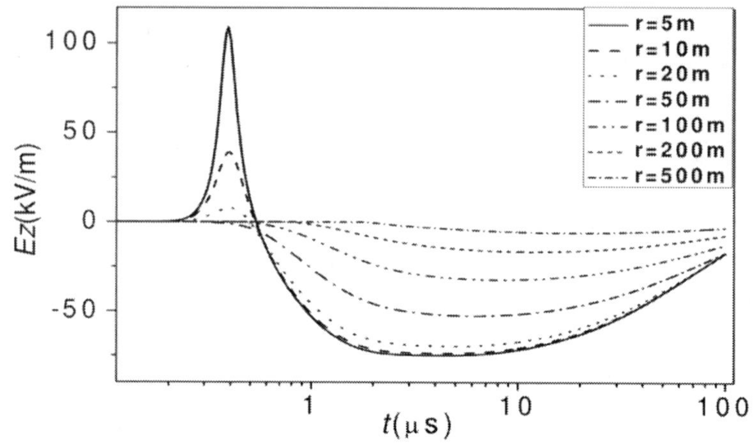

Figure 2. Vertical electric field *Ez* at different horizontal distances when z=50 m.

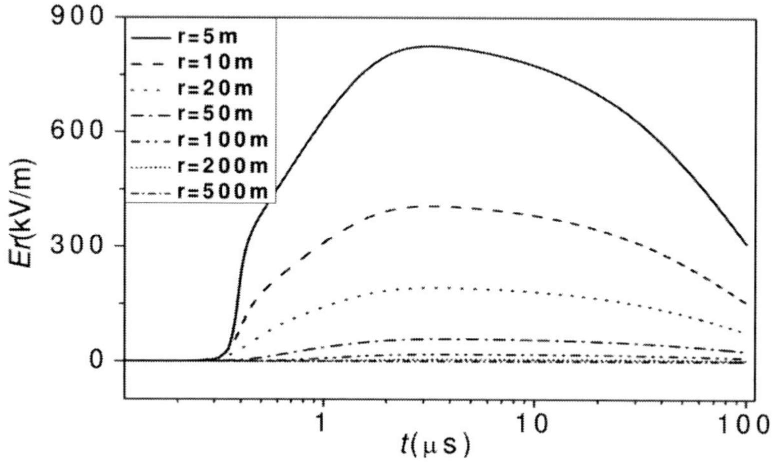

Figure 3. Horizontal electric field *Er* at different horizontal distances when z=50 m.

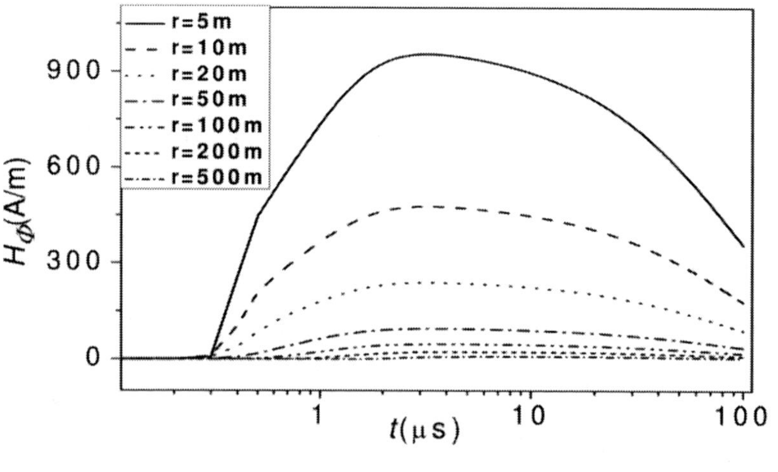

Figure 4. Magnetic field H_Φ at different horizontal distances when z= 50 m.

From Figures 3 and 4, we can find out that the rising edges of the horizontal electric field and magnetic field become gentle with the increase of the horizontal distance. Moreover, the peak value of the field decreases gradually with the increase of the horizontal distance. The peak value at the distance of r=500 m have attenuated about a hundred times than that at the distance of r=5 m.

4. Distribution laws of lightning electromagnetic field at different heights

We keep the parameters of the channel-base current unchanged. Assuming that the horizontal distance from the field point to the source point is r=500 m, the waveforms of lightning electromagnetic field at different heights with a same horizontal distance are shown in Figures 5~7 by using equations (7)~(9), respectively.

Figure 5 shows that when the height of the observation point is less than 80 m, the waveforms of the vertical electromagnetic field at different heights are basically consistent. When the height of the observation point is larger than 80 m, the peak value of the electric field will increase with the increase of height and the rising edge will become steep gradually. When the observation height is larger than 1,000 m, the polarity of the electric field will reverse and the value of electric field after the initial peak will decrease soon and decay to zero in tens of microseconds. As shown in Figure 6, the rising edge and falling edge of the horizontal electric field waveform will become steeper and steeper, and the peak value will increase gradually with the increase of height. As shown in Figure 7, we can conclude that the peak value and falling edge of the magnetic field waveform are nearly unchanged.

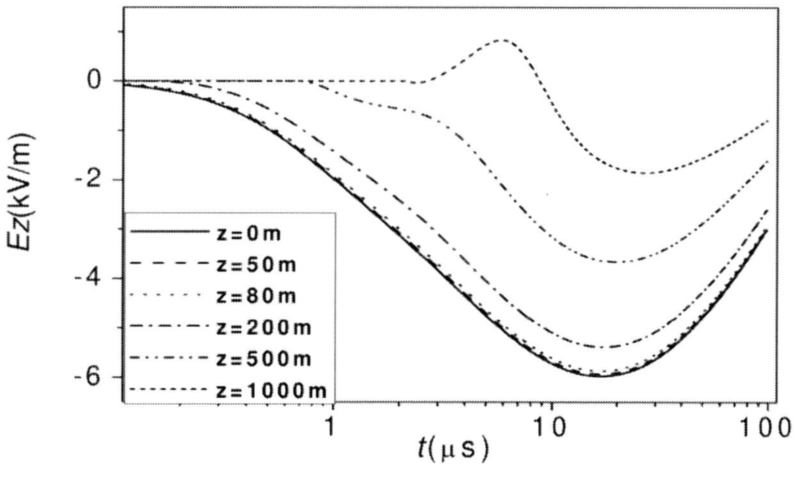

Figure 5. Vertical electric field Ez at different observation heights when r=500 m.

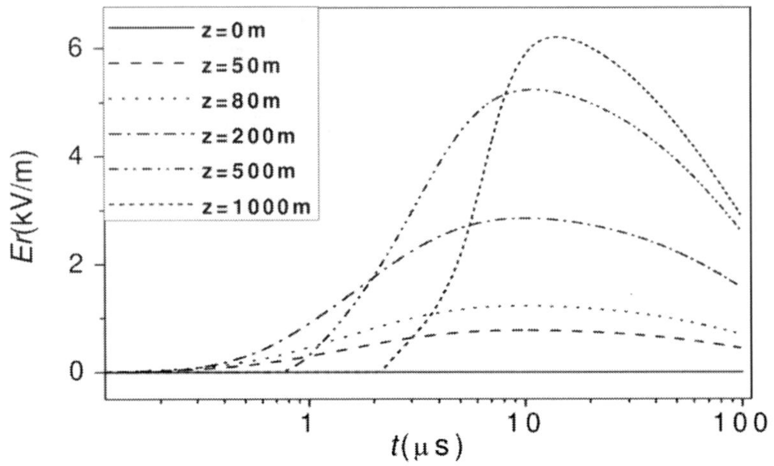

Figure 6. Horizontal electric field Er at different observation heights when r=500 m.

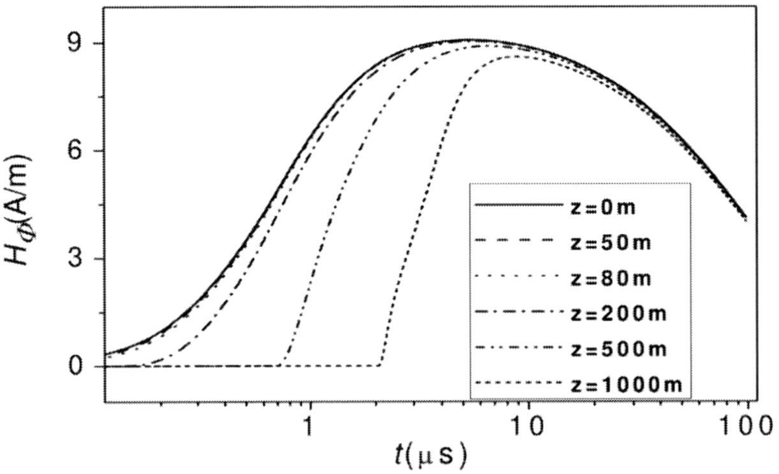

Figure 7. Magnetic field H_Φ at different observation heights when r=500 m.

5. Conclusions

In this paper, the analytic expressions of the spatial lightning electromagnetic field are obtained by using the transmission line (TL) model and the channel-base current is described by the impulse function based on dipole method. On this basis, the spatial lightning electromagnetic field at different observation heights and horizontal distances are calculated. The result shows that:

1) For the electric and magnetic fieldss at a same height above the ground, the rising edges of the electric and magnetic fields become gentle with the increase of the horizontal distance, and the peak value reduces sharply with the increase of the horizontal distance. These above result shows that the lightning electromagnetic field is much stronger for the observation point which is much closer to the

lightning channel. Thus, the threat of this field to sensitive electronic equipments and systems will also be more serious.

2) For the electric and magnetic fields at a same horizontal distance from the channel, the rising edges of the electric and magnetic fields become steep gradually with increasing height of the observation points. For the vertical electric field at a height of less than dozens of meters, the waveforms are basically consistent. When the height increases to more than dozens of meters, the peak value of the electric field will increases with increasing height of the observation point. When the observation height is larger than 1 km, the polarity of the electric field may reverse. Moreover, the peak value of the horizontal electric field increases with the increase of height, and the peak value and falling edge of the magnetic field waveform are nearly unchanged.

Acknowledgments
The research is supported by National Natural Science Foundation Project of China under Grant No. 51077132.

References

[1] Yu H 2005 *The basis of modern lightning protection technology* (Beijing, China)
[2] Wei M and Liu S H 2000 *High Voltage Engineering* **26** 28
[3] Bruce C E R and Golde R H 1941 *J. IEE* **88** 487
[4] Diendorfer G and Uman M A 1990 *J. Geophy. Res.* **95** 13621
[5] Uman M A and Mclain D K 1969 *J Geophy. Res* **74** 6899
[6] Nucci C A, Mazzetti C and Rachidi F 1988 *19th Int. Conf. on Lightning Protection* (Graz, Austria)
[7] Rakov V A and Dulzon A A 1987 *Tekh Elektrodinam* **1** 87
[8] Rubinstein M 1996 *IEEE Trans on EMC* **38** 531
[9] Vernon C 2000 *J. Atmos. Sol.-Terr. Phys.* **62** 169
[10] Rubinstein M A and Uman M A 1991 *IEEE Tran. Electromagn. Compat.* **33** 312
[11] Hu W and Cummer S A. 2003 *Antennas and Propagation Society Int. Symp.* **3** 1061
[12] Sartori C A F and Cardoso J R 2000 *IEEE Trans. Magnetics* **35** 1631
[13] Yang H 2008 *Electromagnetic fields calculation of lightning return stroke based on FDTD method* (Qingdao, China)
[14] Chen Y Z, Liu S H and Hou M S 2001 *Voltage Eng.* **27** 38
[15] Rakov V A, Uman M A 1998 *IEEE Trans. Electromagn. Compat.* **40** 403
[16] Chen Y Z 2002 *Lightning electromagnetic pulse field theory calculation and radiation effects on the electric fuze experiment* (Shijiazhuang, China)
[17] Chen Y Z, Liu S H, Zhang F Z, Wei M and WU X R 2002 *Chinese Journal of Radio Science* **17** 119

Effect of inclusions′ distribution on microwave absorbing properties of composites

Siliang Qin*, Qingguo Wang, Zhaoming Qu
Research Institute of Electrostatic and Electromagnetic Protection,
Mechanical Engineering College, Shijiazhuang, Hebei province, China

E-mail: chunqiu2006@126.com

Abstract. Effect of inclusions′ spatial distributions on the permeability and permittivity of composites is studied using the generalized Maxwell-Garnett equations. The result indicates that inclusions′ orientation distribution can increase the longitudinal electromagnetic parameters. For inclusions′ random and orientation distribution, single and three-layer absorbers are designed and optimized using genetic algorithm. The result shows that under a given absorbing requirement, absorber with inclusions′ orientation distribution is lighter and thinner than absorber with inclusions′ random distribution.

1. Introduction

With the fast improvement of electronic components, muti-phase composites are attracting more and more attentions in designing absorbers because of their lightweight broadband microwave absorbing properties [1, 2]. Many researchers have been focusing on using different types of inclusions to improve composites′ permeability and permittivity [3-5]. Maxwell-Garnett and Bruggeman equations are two famous equations of effective medium theory [6]. However both equations are applied to randomly distributed inclusions and will become invalid if the inclusions are nonrandomly distributed. Duan Huiling and other scholars have proposed a generalized M-G equation based on the variational principles by introducing the concept of distribution factor [7, 8]. The shape factor and distribution factor are brought into represent inclusions′ shapes and distribution [9]. The generalized M-G equations have advantages in estimating the effective permeability and permittivity of composites, especially composites with nonrandomly distributed inclusions of fiber, flake and other nonspherical shapes.

An optimized function between inclusions′ distribution and the reflection loss (RL) of composites can be built by combining the generalized M-G equation with transmission line approach. The main variables of the target function are concentration and thickness of composites. A certain algorithm is needed to find a suitable solution because the target function is not a linear equation. Genetic algorithm (GA) optimizers are robust and stochastic search methods modeled on the concepts of natural selection and evolution [10]. Single and three-layer absorbers considering inclusions′ random and orientational distribution are designed using GA. A new scheme of designing absorbers is proposed and discussed in this paper and it will be for reference to the following research in designing absorbers.

* To whom any correspondence should be addressed.

2. Generalized M-G equations

Generalized M-G equations are given below [7]:

$$\overline{\overline{\mu_e}} = \overline{\overline{\mu_1}} + \overline{\overline{\mu_l}}\left(\sum_{r=2}^{N}\overline{\overline{h_r}} - \overline{\overline{S_v}}\right)^{-1} \tag{1}$$

$$\overline{\overline{h_r}} = \frac{1}{3} f_r [2\left(\overline{\overline{\mu_l}} \cdot \left(\overline{\overline{\mu_r}} - \overline{\overline{\mu_l}}\right)^{-1} + \overline{\overline{S_r}}\right)^{-1} + \left(\overline{\overline{\mu_l}} \cdot \left(\overline{\overline{\mu_r}} - \overline{\overline{\mu_l}}\right)^{-1} + \mathbf{1} - \mathbf{2}\overline{\overline{S_r}}\right)^{-1}] \tag{2}$$

$\boldsymbol{\mu}_1$, $\boldsymbol{\mu}_r$ and $\boldsymbol{\mu}_e$ are the permeability tensors of the base, the r th-phase inclusions and the composites, respectively. $\mathbf{S}_r$ is the geometrical factor tensor of r th-phase and $\mathbf{S}_v$ is the inclusions′ distribution tensor. f_r is the concentration of r th-phase. As pointed out by Hashin and Strikman [11], the permittivity and permeability are mathematically analogous.

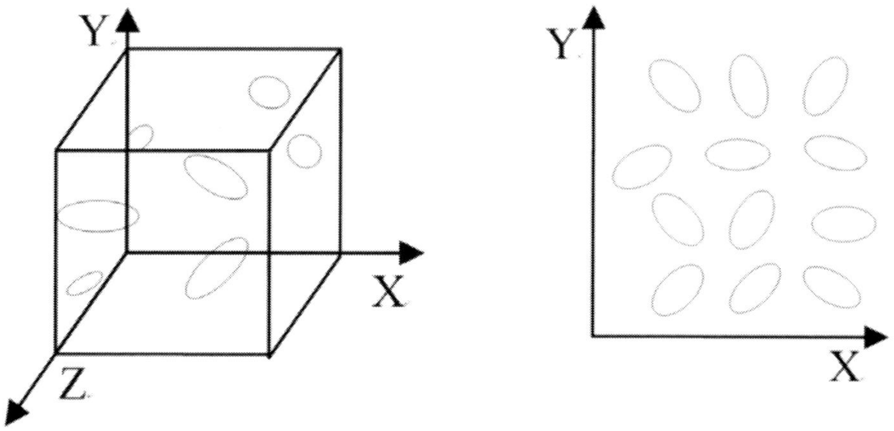

Figure 1. (a) Randomly distributed inclusions. (b) Orientationally distributed inclusions.

For two-phase composites whose inclusions are randomly distributed as shown in figure 1a, its effective permeability tensor $\overline{\overline{\mu_e}} = \mu_e \overline{\overline{I}}$, μ_e is given below:

$$\mu_e = \mu_1 + \mu_1\left(\frac{1}{h_{2\mu}} - S_v\right)^{-1} \tag{3}$$

$$h_{2\mu} = \frac{1}{3} f [2\left(\frac{\mu_1}{\mu_2 - \mu_1} + S_1\right)^{-1} + \left(\frac{\mu_1}{\mu_2 - \mu_1} + 1 - 2S_1\right)^{-1}] \tag{4}$$

If the inclusions are fibers with an aspect ratio larger than 1, S_1 is given as:

$$S_1 = \frac{1}{4\left(1-\gamma^2\right)^{3/2}}\left(2\sqrt{1-\gamma^2} - \gamma^2 \ln\frac{1+\sqrt{1-\gamma^2}}{1-\sqrt{1-\gamma^2}}\right) \tag{5}$$

$\gamma = \frac{a_1}{a_3} = \frac{a_2}{a_3}$, a_1 and a_2 denote the radial length and a_3 denotes semi-axial length. For randomly distributed inclusions $S_v = \frac{1}{3}$.

For two-phase composites with inclusions distributed orientationally, this paper focuses only on one situation for simplicity. The inclusions' axes are parallel to the XY plane as shown in figure 1b. The transverse and longitudinal permeability is as follows:

$$\mu_T = \mu_1 + \mu_1 \left(f \left(\frac{\mu_1}{\mu_2 - \mu_1} - S_1 \right)^{-1} - S_v \right)^{-1} \tag{6}$$

$$\mu_L = \mu_1 + \mu_1 \left(f \left(\frac{\mu_1}{\mu_2 - \mu_1} + 1 - 2S_1 \right)^{-1} - 1 + 2S_v \right)^{-1} \tag{7}$$

For orientationally distributed inclusions $S_v = S_1$.

Numerical calculation is carried out to observe the effect of inclusions' distribution on the EM parameters of composites. Let the base parameters be $\varepsilon_1 = 3, \mu_1 = 1$. Inclusions are iron fiber with 4mm length and 9um diameter. Inclusions' concentration is 10 % and its permeability is given by figure 2.

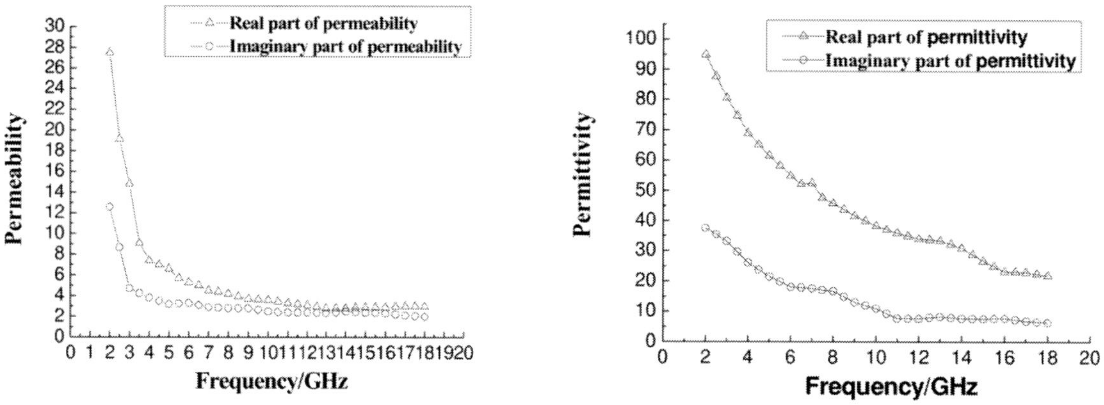

Figure 2. Permittivity and permeability of inclusions.

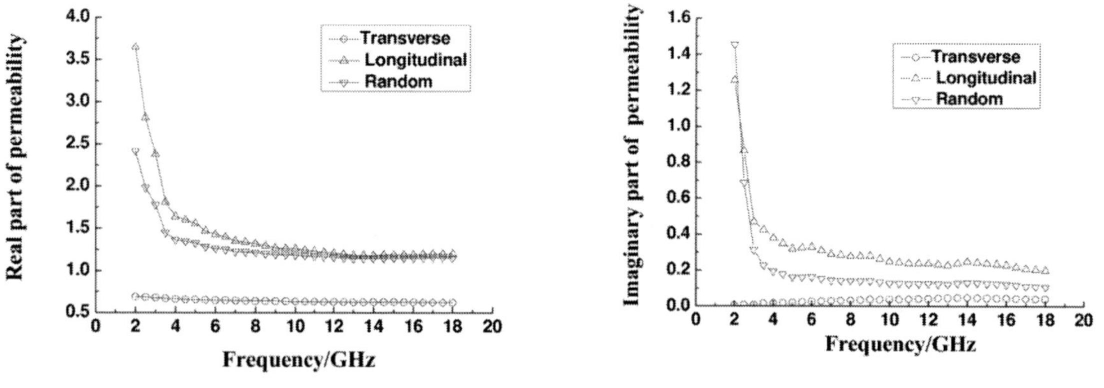

Figure 3. Permeability of composties with orientationally and randomly distributed inclusions.

As shown in figure 3, the longitudinal permeability of composites with orientational inclusions is larger than its transverse permeability. While the permeability of composites with randomly distributed inclusions is between the longitudinal and transverse one. If the plane wave incident along the Z axis, its electric and magnetic field will transmit in the XY plane, then the longitudinal parameters will mainly affect composites' absorbing properties. So comparing with inclusions randomly distributed, orientational inclusions can apparently increase composites' absorbing properties.

3. Reflection loss

For a plane wave normally incident on the air-composites-metal structure shown in figure 4, the reflection loss can be calculated using transmission-line approach 12]. The i th input impedance of N-layer composites with metal backing is given by:

$$Z_{in}^{i} = \eta_i \frac{Z_{in}^{i+1} + \eta_i \tanh(jk_i d_i)}{\eta_i + Z_{in}^{i+1} \tanh(jk_i d_i)} \tag{8}$$

d_i, k_i and η_i is the thickness, propagation constant and characteristic impedance of the i th layer, respectively:

$$k_i = \frac{2\pi f}{c}\sqrt{\mu_{ri}\varepsilon_{ri}}, \quad \eta_i = \frac{E_{ix}}{H_{iy}} = \eta_0\sqrt{\frac{\mu_{ir}}{\varepsilon_{ir}}} \tag{9}$$

η_0 is the characteristic impedance of the air. μ_{ir}, ε_{ir} are the relative permeability and permittivity of the i th layer. The reflection coefficient of the i th layer is calculated by equation (10).

$$R_i = \frac{Z_{in}^{i} - \eta_{i-1}}{Z_{in}^{i} + \eta_{i-1}} \tag{10}$$

For the $n+1$ th metal backing, $Z_{in}^{n+1} = 0$. For single layer with metal backing, $n = 1$

$$Z_{in}^{1} = \eta_1 \tanh(jk_1 d_1), \qquad R_1 = \frac{Z_{in}^{1} - 1}{Z_{in}^{1} + 1} \tag{11}$$

The refection loss (RL)

$$RL = 20\lg|R_1| \tag{12}$$

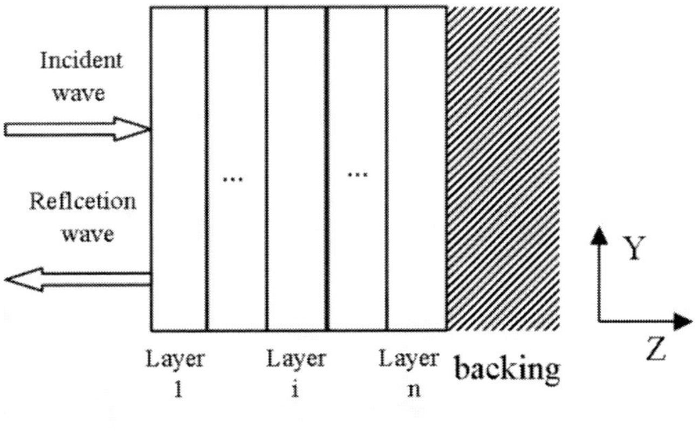

Figure 4. Geometry of multi-layer absorber with perfectly conducting backing.

Composites with randomly distributed inclusions of different concentrations are simulated to emphasize the effect of concentrations on reflection loss. Let the base parameters be $\varepsilon_1 = 3, \mu_1 = 1$. Inclusions are iron fiber with 4mm in length and 9um in diameter.

The calculated RL of five different concentrations of 5 %, 10 %, 15 %, 20 %, 25 % are shown in figure 5. It can be seen from figure 5 that composites' absorbing properties are related to EM frequency and inclusions′ concentrations. There is an absorption peak for single layer absorber, and the peak frequency goes down as the concentration increases. The absorption peak is not a constant but first increases then decreases as concentration goes up. However, the absorber will gain a broader absorbing band and more reflection loss as concentration increases.

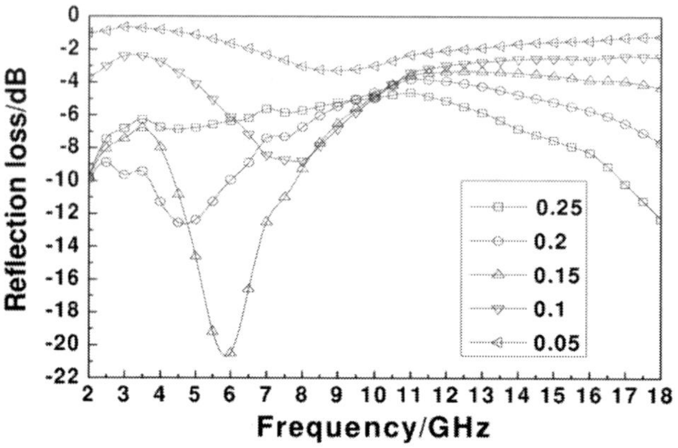

Figure 5. Reflection loss of composites of different concentrations.

4. Design and optimization of absorber

For randomly and orientationlly distributed inclusions, there is a relationship between its concentration and the reflection loss. In designing an absorber which meets certain requirement, its thickness and concentration are the main variables. This paper uses genetic algorithm for multi-variable optimization. Single and three-layer absorbers are designed using GA for random and orientational inclusions separately. The design and optimization flowchart is in figure 6:

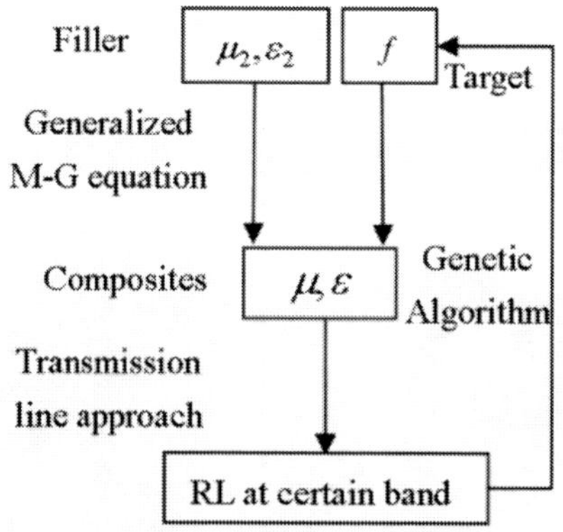

Figure 6. Design and optimization flowchart of absorber.

In the optimization process, the target absorber should be thinner than 4 mm and its RL should be less than -5 dB. For single layer absorbers, the optimization result using GA for random inclusions is 4 mm thickness and concentration of 27 %, and the result for orientational inclusions is 4 mm thickness and 20 % concentration. For three-layer absorbers, the results are showed in table 1 and table 2.

Table 1. Design parameters for three-layer absorber with randomly distributed inclusions.

	d (mm)	f
Layer 1	1.01	0.14941
Layer 2	1.81	0.23803
Layer 3	1	0.14849

Table 2. Design parameters for three-layer absorber with orientationally distributed inclusions.

	d (mm)	f
Layer 1	1.1	0.127
Layer 2	1.62	0.234
Layer 3	1.	0.13

The reflection loss of all four kinds of absorbers are shown in figure 7. For single layer absorber the optimized results of both kinds of inclusions have the same reflection loss above 5 GHz. However absorber with orientational inclusions is lighter and thinner than that with random inclusions, so absorber with orientational inclusions is better than the other one for designing single layer absorber.

For three-layer absorber the differences between two kinds of absorbers are more complicated. Absorber with orientational inclusions has a greater RL at the frequency range of 2-7 GHz and an absorbing peak at 5 GHz. At the range of 7-18 GHz absorber with random inclusions is better at absorbing than the other one. Though both kinds of absorbers can satisfy the design requirement the one with random inclusions is thicker and heavier than the other one. That is why the absorber with random inclusions is better than the other one at the range of 7-18 GHz.

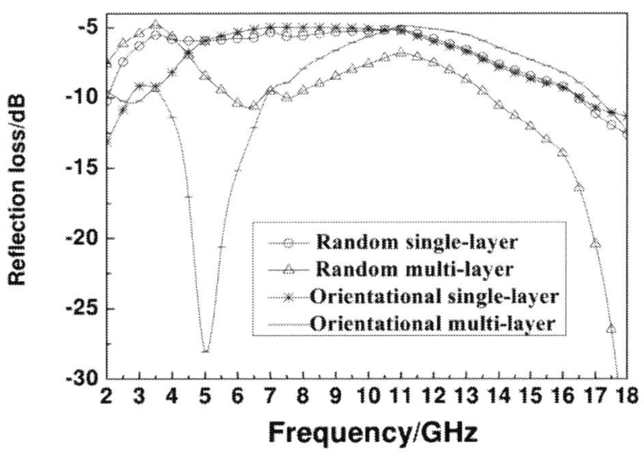

Figure 7. Reflection loss of optimized absorbers.

5. Conclusions

Effect of inlcusions' distribution on composites' absorbing properties is discussed by introducing the distribution tensor. The absorbing properties of composites are obviously affected by the distribution of inclusions. Composites with orientationally distributed inclusions have a greater absorbing ability than that with randomly distributed inclusions.

The relationship between inclusions' concentration and composites' reflection loss is brought in based on the transmission line approach. The absorber will have a lower peak frequency and broader absorbing frequency band as the inclusions' concentration increases.

Single and three-layer absorbers of randomly and orientationlly distributed inclusions are designed using genetic algorithm. The result shows that under a given absorbing requirement, absorber with inclusions' orientation distribution is lighter and thinner than absorber with inclusions' random distribution. For single layer absorber the two kinds of composites have almost the same absorbing ability. However for three-layer absorber composites with randomly distributed inclusions have a better absorbing ability at most of the frequencies.

References

[1] Enveniste Y and Milton G W 2011 *Inter. J. Eng. Sci.* **49** 2
[2] Bregar V B 2004 *IEEE Trans. Mag.* **40** 1679
[3] Merrill W M, Diaz R E, LoRe M M, et al. 1999 *IEEE Trans. Antennas Propag.* **47** 142
[4] Koledintseva M, Ravva P C, et al. 2006 *IEEE Inter. Symp. EMC* 598
[5] Youngs I J 2007 *IEEE Trans. Dielectr. Electr. Insul.* **14** 1145
[6] Milton G W 2002 *The Theory of Composites* (Cambridge: Cambridge University Press)
[7] Duan H L, Karihaloo B L, Wang J, et al. 2006 *Phys. Rev. B* **73** 4203
[8] Ponte P, Willis J R 1995 *J Mech Phys Solids* **43** 1919
[9] Mura T 1987 *Micromechanics of defects in solids* (Martinus Nijhoff: Dordrecht)
[10] Johnson J M, Rahmat S V 1997 *IEEE trans. on Antennas Propag.* **39** 7
[11] Hashin Z, Shtrikman S 1963 *J Mech Phys Solids* **11** 127
[12] Schulz R B, Plantz V C, Brush D R 1988 *IEEE Trans. EMC* **30** 187

Journal of Physics: Conference Series **418** (2013) 012005 doi:10.1088/1742-6596/418/1/012005

Approximation characteristic between LEMP on earth surface and channel-base current

Yazhou Chen*, Haojiang Wan, Xiaojia Wang
Key laboratory of electromagnetic environmental effects, Shijiazhuang Mechanical Engineering College, Shijiazhuang 050003, China

E-mail: chen_yazhou@sina.com

Abstract. Based on the dipole technique, the exact expressions and approximate expressions for LEMP on earth surface are derived from the TL model and specified channel-base current described by a pulse function. According to the comparisons between the exact results and approximate results, the waveforms of LEMP and its derivative are similar to the lightning channel-base current and its derivative, respectively. In near-field, the electric field and its derivative waveforms of the approximate results are essentially coincident with the corresponding waveforms of the exact results within 100 m. The magnetic field waveforms of approximate results are essentially coincident with the waveforms of the exact results within 500 m, and the magnetic field derivative waveforms of the approximate results are essentially coincident with the waveforms of the exact results within 200 m. In far-field, the waveforms of the approximate results are essentially coincident with the waveforms of the exact results in the rising edge before the initial peak value. At the distance beyond 100 km, the waveforms of the radiation field component and the scaled channel-base current appear identical. The difference between the derivative waveforms of the LEMP field and the scaled channel-base current in the first three microseconds is slight and can be ignored.

1. Introduction

The lightning electromagnetic pulse (LEMP) generated by the lightning return stroke current has caused wide concern in recent years due to the proliferation of sensitive loads [1-5]. A series of experiments were performed to measure the electric field, magnetic field, and time derivative of the electromagnetic field [6-16]. And lots of researches on the theoretical modeling and calculation methods of LEMP were also carried out by different researchers [17, 18]. Rakov *et al.* defined four classes of lightning return stroke models [17], including the gas dynamic models, the electromagnetic models, the distributed-circuit models, and the engineering models. Among them, the engineering models, which can be defined as equations relating the longitudinal current along the lightning channel at any height and any time to the current at the channel origin (e.g. at ground level), is widely used. The engineering models such as the Bruce-Golde model (BG) [19], the transmission line model (TL) [20], the traveling current source model (TCS) [21], the modified transmission-line model with linear current decay with height (MTLL) [22], the modified transmission line model with exponential current decay with height (MTLE) [23], and the Diendorfer-Uman model (DU) [24] were proposed one after the other to describe the lightning return stroke process. In addition to the propagation direction and production mode of the channel current, different engineering models are mainly distinguished by the

* To whom any correspondence should be addressed.

propagation characteristic (e.g. velocity, and attenuation law) of the return stroke current along the channel. Based on these engineering models, several analytic methods, such as variable separation method, conformal mapping method, the monopole and the dipole techniques, were adopted to calculate the electromagnetic field. Rubinstein presented a comparison between the two different techniques (monopole and dipole) for calculating the electric and magnetic fields from lightning, and showed that the monopole approach is applicable only to upward-traveling current waves and hence is not particularly useful in the realistic modeling of lightning return strokes [25, 26]. Cooray, who used the dipole technique, derived analytical expressions for LEMP fields from Maxwell's equations [27]. Uman, comparing the calculation results with experimental results, showed that the dipole technique has a high computing accuracy [28]. Henceforth, the dipole technique is widely used in the antenna theory and the calculation of electromagnetic fields.

With the increasing demand by customers for good quality in the electronic system, security testing is a necessary procedure for the sensitive electronic equipments. In order to meet the test requirements of LEMP environment effects for the sensitive loads, LEMP simulation technology is suggested to resolve this key issue. The above theoretical calculation methods provide the possibility for LEMP simulation. Moreover, Uman *et al.* demonstrated the approximation characteristics between the far-field of LEMP and channel-base current, and showed that the far-field within a certain distance can be approximated as a function of channel-base current [29]. Similar fine structure was also observed by Jerauld *et al.* in d*E*/d*t* and d*I*/d*t* waveforms produced by an unusual rocket-triggered lightning stroke, which involved a downward dart-stepped leader and a pronounced upward connecting leader [30]. These above researches connect LEMP simulation with the channel-base current of lightning return stroke closely. In order to establish the further relationship between LEMP fields and the channel-base current in theory, the analytical expressions of electric and magnetic fields based on the TL model are derived according to the dipole method. The approximation characteristics between LEMP fields and the channel-base current, and the approximation characteristics between the derivatives of LEMP fields and the derivative of channel-base current in near field and far field are analyzed, respectively. And the applicable ranges of the approximate expressions are also discussed.

2. Exact expressions of LEMP on earth surface

The return-stroke channel can be regarded as a vertical perfectly antenna, as shown in figure 1. The transient current propagates along the vertical perfectly conductor (the lightning channel) above a perfectly conducting plane (the ground).

According to the dipole method, the LEMP field due to an upward-moving return stroke for the case of a field point (in a cylindrical coordinate system) can be described as follows [28]:

$$E_r = \frac{1}{4\pi\varepsilon_0}\int_{-h}^{h}\left[\frac{3r(z-z')}{R^5}\int_{-\infty}^{t} i(z',t-R/c)\mathrm{d}t + \frac{3r(z-z')}{cR^4}i(z',t-R/c) + \frac{r(z-z')}{c^2R^3}\frac{\partial i(z',t-R/c)}{\partial t}\right]\mathrm{d}z' \quad (1)$$

$$E_z = \frac{1}{4\pi\varepsilon_0}\int_{-h}^{h}\left[\frac{2(z-z')^2-r^2}{R^5}\int_{-\infty}^{t} i(z',t-R/c)\mathrm{d}t + \frac{2(z-z')^2-r^2}{cR^4}i(z',t-R/c) - \frac{r^2}{c^2R^3}\frac{\partial i(z',t-R/c)}{\partial t}\right]\mathrm{d}z' \quad (2)$$

$$H_\phi = \frac{1}{4\pi}\int_{-h}^{h}\left[\frac{r}{R^3}i(z',t-R/c) + \frac{r}{cR^2}\frac{\partial i(z',t-R/c)}{\partial t}\right]\mathrm{d}z' \quad (3)$$

Where $R=\sqrt{(z-z')^2+r^2}$ is the distance between the return-stroke current wavefront and the field point, L is the length of the return-stroke channel, r is the horizontal distance between the return-stroke channel and the observation point, z is the height of the field point, z' is the position of the wavefront, h is the height of z', c is the speed of light, i is the return-stroke current, E_r is the radial electric field, E_z is the vertical electric field, H_ϕ is the poloidal magnetic field. In equations (1) and (2), the first term represents the electrostatic field, the second term is the induction electric field (i.e. intermediate-field), and the third term is the radiation electric field (i.e. far-field). In equation (3), the first term represents the induction magnetic field and the second term is the radiation magnetic field.

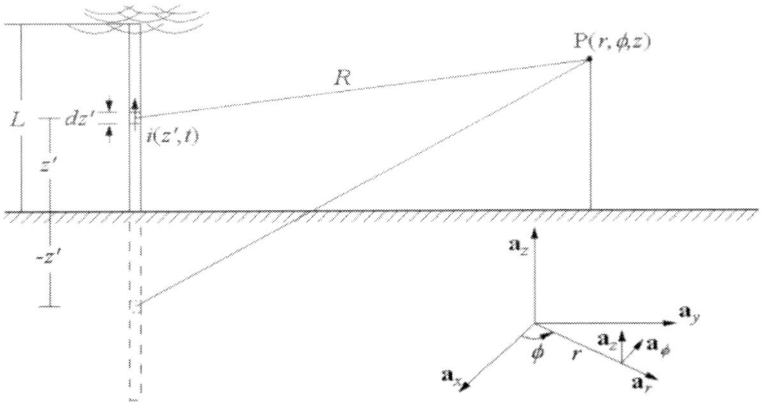

Figure 1. The antenna model of lightning return stroke.

In this paper, the emphasis is placed on the electromagnetic fields on earth surface (i.e. the height of the field point z=0). Thus, E_r=0, $R=\sqrt{h^2+r^2}$.

In the engineering models, the relationship between the channel current $i(z',t)$ and channel-base current $i(0,t)$ can be expressed by the following equation except DU model [17]:

$$i(z',t)=u\left(t-z'/v_f\right)P(z')i\left(0,t-z'/v\right) \tag{4}$$

Where $u(\xi)$ is the Heaviside function equal to unity for $\xi>0$ and zero otherwise, $P(z')$ is the height-dependent current attenuation factor, v_f is the upward-propagating front speed, and v is the current-wave propagation speed. Different $P(z')$ and v corresponds to different engineering model.

Here, the TL model is adopted to calculate the lightning electromagnetic field. Thus, $P(z')=1$, $v=v_f$. And the return stroke current can be written as:

$$i(z',t)=u(t-z'/v)\cdot i(0,t-z'/v) \tag{5}$$

Moreover, the pulse function is chosen to describe the channel-base current, expressed as follows:

$$i(0,t)=\frac{I_0}{\eta}[1-\exp(-t/\tau_1)]^n\exp(-t/\tau_2) \tag{6}$$

Where $\eta=[n\tau_2/(\tau_1+n\tau_2)]^n[\tau_1/(\tau_1+n\tau_2)]^{\tau_1/\tau_2}$. In this paper, we set n=2.

Thus, each component of equations (2) and (3), and the total electric and magnetic field can be described as follows:

Electrostatic field component:

$$E_z(\text{electrostatic})=\frac{i(z',t)}{2\pi\varepsilon_0}\left\{\frac{-th+\dfrac{2h^2}{v}+\dfrac{r^2}{v}}{\left(h^2+r^2\right)^{3/2}}-\frac{1}{rv}-\frac{1}{2cr}\left[\tan^{-1}\left(\frac{h}{r}\right)-\frac{3hr}{h^2+r^2}\right]\right\} \tag{7}$$

Induction electric field component:

$$E_z(\text{induction})=-\frac{i(z',t)}{4\pi\varepsilon_0 rc}\left[\tan^{-1}\left(\frac{h}{r}\right)-\frac{3hr}{h^2+r^2}\right] \tag{8}$$

Radiation electric field component:

$$E_z(\text{radiation}) = -\frac{i(z',t)}{2\pi\varepsilon_0}\frac{r^2}{\left(\frac{1}{v}+\frac{h}{c(h^2+r^2)^{1/2}}\right)c^2\left(h^2+r^2\right)^{3/2}} \tag{9}$$

Induction magnetic field component:

$$H_\phi(\text{induction}) = \frac{i(z',t)}{2\pi}\frac{h}{r\left(h^2+r^2\right)^{1/2}} \tag{10}$$

Radiation magnetic field component:

$$H_\phi(\text{radiation}) = \frac{i(z',t)}{2\pi}\frac{r}{\frac{c}{v}\left(h^2+r^2\right)+h\left(h^2+r^2\right)^{1/2}} \tag{11}$$

Total electric field on earth surface:

$$\vec{E}(r,0,t) = \frac{i(z',t)}{2\pi\varepsilon_0}\left\{\frac{-th+\frac{2h^2}{v}+\frac{r^2}{v}}{\left(h^2+r^2\right)^{3/2}}-\frac{1}{rv}-\frac{r^2}{c^2\left(h^2+r^2\right)^{3/2}\left(\frac{1}{v}+\frac{h}{c\sqrt{h^2+r^2}}\right)}\right\}\vec{\alpha}_z \tag{12}$$

Total magnetic field on earth surface:

$$\vec{H}(r,0,t) = \frac{i(z',t)}{2\pi}\left[\frac{h}{r\left(h^2+r^2\right)^{1/2}}+\frac{r}{\frac{c}{v}\left(h^2+r^2\right)+h\left(h^2+r^2\right)^{1/2}}\right]\vec{\alpha}_\phi \tag{13}$$

Where $h=\beta\frac{ct-\sqrt{(\beta ct)^2+r^2\left(1-\beta^2\right)}}{1-\beta^2}$, $\beta = v/c$.

3. Approximation characteristic of LEMP on earth surface and its derivative in near-field

3.1. Derivation of approximate expressions for LEMP in near-field

Let $\partial f(z',t)/\partial t = i(z',t)$, according to the TL model:

$$\int_{-\infty}^{t} i(z',\tau)\mathrm{d}\tau = \int_{-\infty}^{t} i(0,\tau - z'/v)\mathrm{d}\tau = \int_{z'/v}^{t} i(0,\tau - z'/v)\mathrm{d}\tau = \int_{0}^{t-z'/v} i(0,s)\mathrm{d}s$$
$$= f(0,t)\Big|_0^{t-z'/v} = f(0,t-z'/v) - f(0,0) = F(t-z'/v) \tag{14}$$

Thus, the current integration in the electrostatic field component of equation (2) is:

$$\int_{-\infty}^{t} i(z',\tau - R/c)\mathrm{d}\tau = \int_{-\infty}^{t} i(0,\tau - R/c - z'/v)\mathrm{d}\tau = F(t-R/c-z'/v) \tag{15}$$

For the near-field, we view $r \ll H$. $1/R$ attenuates to zero very rapidly with the increasing z'. Thus, using the Taylor series expansion, $F(t-R/c-z'/v)$ can be written as:

$$F(t-R/c-z'/v) = F(t-r/c) - F'(t-r/c)\left[(t-r/c)-(t-R/c-z'/v)\right]+o\left(c^{-2}\right)$$
$$\approx F(t-r/c) - i(0,t-r/c)\left[(R-r)/c+z'/v\right] \tag{16}$$

where $\partial F(t-r/c)/\partial t = i(0,t-r/c)$.

Thus, the electrostatic field component can be expressed approximately as:

$$\begin{aligned}E'_z(\text{electrostatic}) &= \frac{1}{2\pi\varepsilon_0}\int_0^h \frac{2R^2-3r^2}{R^5}\mathrm{d}z'\int_{-\infty}^{t} i(z',\tau-R/c)\mathrm{d}\tau \\ &= \frac{1}{2\pi\varepsilon_0}\int_0^h \frac{2R^2-3r^2}{R^5}\mathrm{d}z' F(t-R/c-z'/v)\mathrm{d}z' \\ &= \frac{F(t-r/c)}{2\pi\varepsilon_0}\left[\frac{h}{R(h)^3}\right]-\frac{i(0,t-r/c)}{2\pi\varepsilon_0} \\ &\times\left\{\frac{1}{c}\left(\frac{rh}{R(h)^3}-\frac{3h}{2R(h)^2}+\frac{\tan^{-1}(r/h)}{2r}\right)+\frac{1}{v}\left(\frac{r^2}{R(h)^3}-\frac{2}{R(h)}+\frac{1}{r}\right)\right\}\end{aligned} \tag{17}$$

$$\begin{aligned}E'_z(\text{induction}) &\approx \frac{i(0,t-r/c)}{2\pi\varepsilon_0 c}\int_0^h \frac{2R^2-3r^2}{R^4}\mathrm{d}z' \\ &= \frac{i(0,t-r/c)}{2\pi\varepsilon_0 c}\left[\frac{\tan^{-1}(h/v)}{2r}-\frac{3h}{2R(h)^2}\right]\end{aligned} \tag{18}$$

The radiation electric field component is extremely small, of the order of c^{-2}, and can be ignored. Thus, we obtain the first-level approximation of the electric field in near field as:

$$E'_z \approx -\frac{F(t-r/c)}{2\pi\varepsilon_0}\frac{h}{R(h)^3}-\frac{i(0,t-r/c)}{2\pi\varepsilon_0}\left[\frac{rh}{cR(h)^3}+\frac{1}{v}\left[\frac{r^2}{R(h)^3}-\frac{2}{R(h)}+\frac{1}{r}\right]\right] \tag{19}$$

Likewise, we can obtain the first-level approximation of the magnetic field in near field as:

$$H'_\phi \approx H'_\phi(\text{induction}) = \frac{1}{2\pi}i(0,t-r/c)\int_0^h \frac{r}{R^3}\mathrm{d}z' = \frac{1}{2\pi r}i(0,t-r/c)\frac{h}{R(h)} \tag{20}$$

As $r \ll H$ in near-field, $R(h)$ approaches h gradually along with the current propagating upward in the channel. Ignore the R^{-1} term and R^{-3} term in equation (19), we can obtain the second-level approximation of the electromagnetic field in near field as:

$$E''_z \approx -\frac{1}{2\pi\varepsilon_0 vr}i(0,t-r/c) \tag{21}$$

$$H''_\phi \approx \frac{1}{2\pi r}i(0,t-r/c) \tag{22}$$

Performing the derivation on both sides of equations (21) and (22), we obtain:

$$\frac{\mathrm{d}E_z(r,t)}{\mathrm{d}t} \approx \frac{\mathrm{d}E'(r,t)}{\mathrm{d}t} \approx \frac{\mathrm{d}E''_z(r,t)}{\mathrm{d}t} \approx -\frac{1}{2\pi\varepsilon_0 vr}\frac{\mathrm{d}i(0,t-r/c)}{\mathrm{d}t} \tag{23}$$

$$\frac{\mathrm{d}H_\phi(r,t)}{\mathrm{d}t} \approx \frac{\mathrm{d}H'_\phi(r,t)}{\mathrm{d}t} \approx \frac{\mathrm{d}H''_\phi(r,t)}{\mathrm{d}t} \approx \frac{1}{2\pi r}\frac{\mathrm{d}i(0,t-r/c)}{\mathrm{d}t} \tag{24}$$

3.2. Comparisons of LEMP in near-field between the exact results and approximate results

By comparing the waveforms of the first-level approximation, the second-level approximation in near-field with the waveforms of the exact solution, the LEMP field waveforms are similar to the channel-base current waveform of lightning return stroke, only with a coefficient difference. In order to verify the reliability of the approximate expressions in near-field, the lightning channel-base current with a typical waveform 8/20μs is adopted to calculate the ligthning electromagnetic fields. The height of the lightning channel is set to H=7.5 km, and the return stroke speed is v=1.3×10^8 m s^{-1}. The parameters of the pulse function are set as follows: $I_0 = 30\text{kA}$, $\tau_1 = 4.0\times10^{-6}\text{s}$, $\tau_2 = 6.25\times10^{-6}\text{s}$. Comparisons of the electric and magnetic field waveforms from the approximate results, including the first-level approximation and the second-level approximation, with the exact results at different distances are given in figures 2 to 9. And comparisons of the electric and magnetic field derivative waveforms from the approximate results, including the first-level approximation and the second-level approximation, with the exact results at different distances are given in figures 10 to 16.

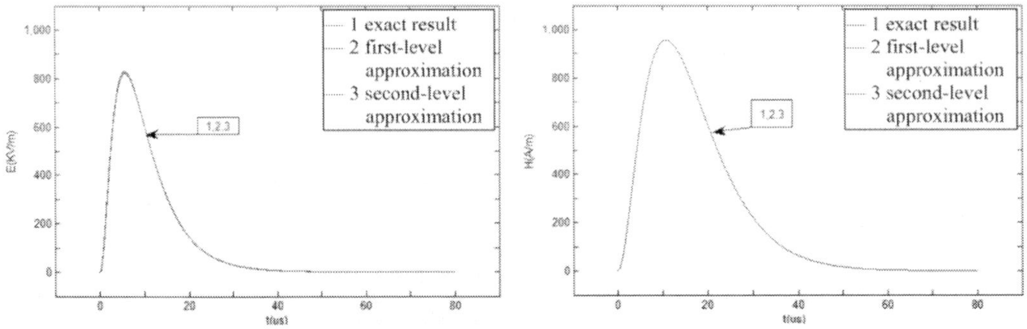

Figure 2. Comparison of the electric and magnetic field waveforms from the approximate results with the exact results at a distance of 5 m.

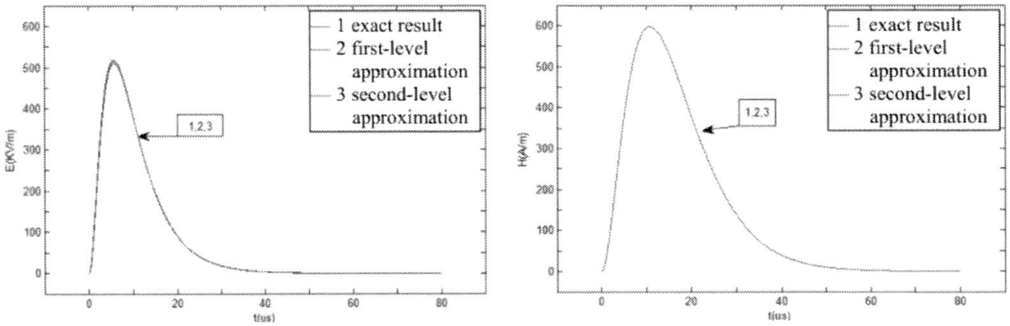

Figure 3. Comparison of the electric and magnetic field waveforms from the approximate results with the exact results at a distance of 8 m.

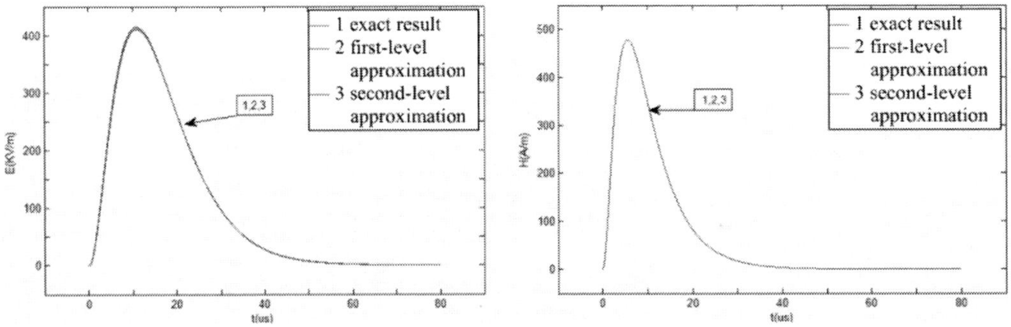

Figure 4. Comparison of the electric and magnetic field waveforms from the approximate results with the exact results at a distance of 10 m.

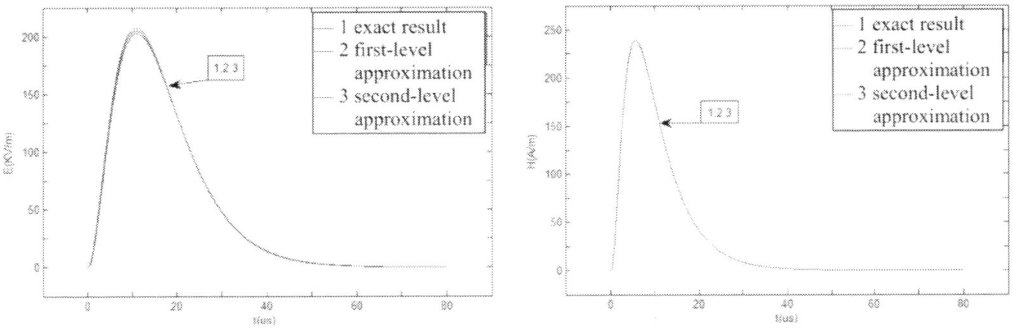

Figure 5. Comparison of the electric and magnetic field waveforms from the approximate results with the exact results at a distance of 20 m.

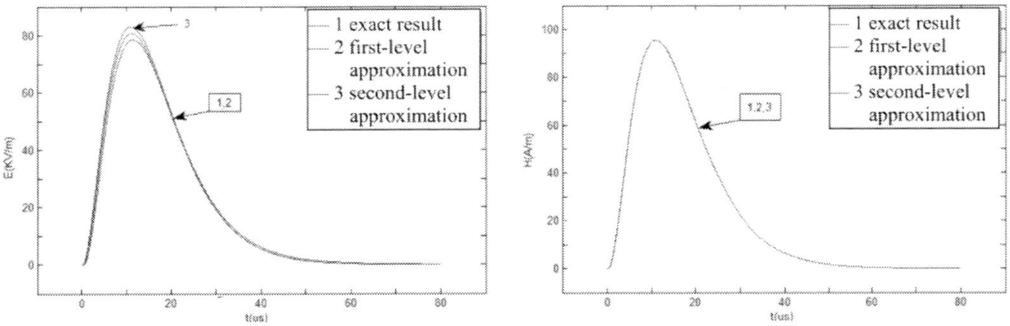

Figure 6. Comparison of the electric and magnetic field waveforms from the approximate results with the exact results at a distance of 50 m.

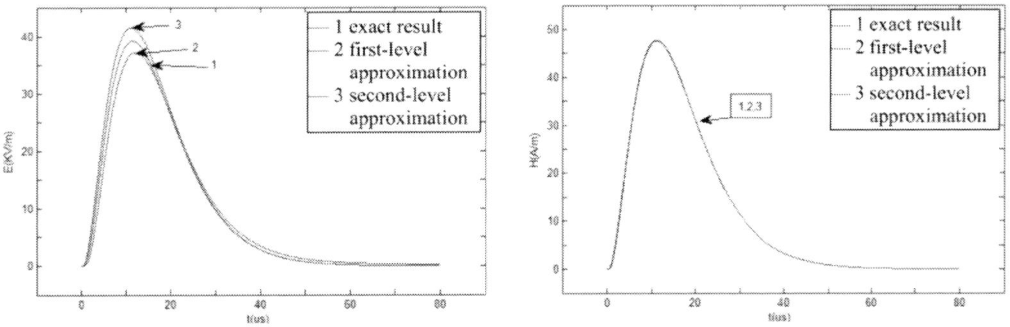

Figure 7. Comparison of the electric and magnetic field waveforms from the approximate results with the exact results at a distance of 100 m.

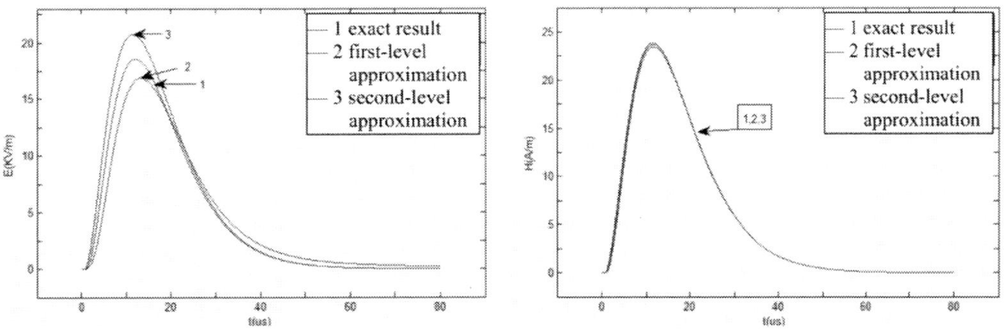

Figure 8. Comparison of the electric and magnetic field waveforms from the approximate results with the exact results at a distance of 200 m.

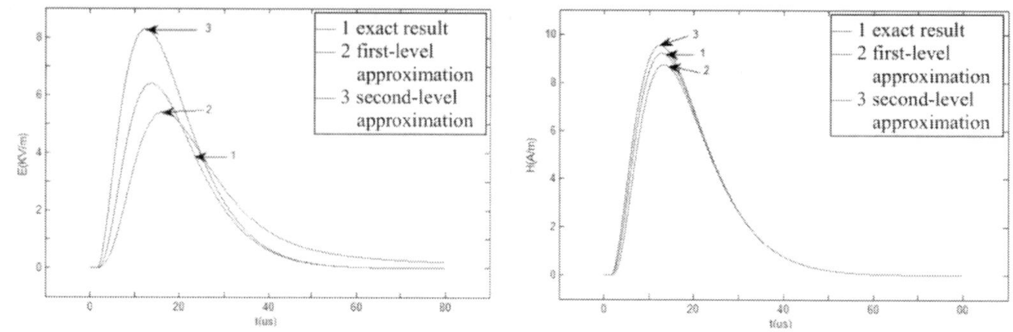

Figure 9. Comparison of the electric and magnetic field waveforms from the approximate results with the exact results at a distance of 500 m.

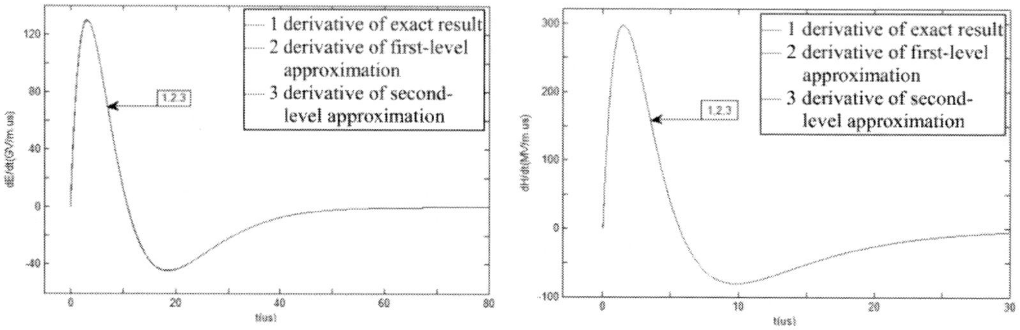

Figure 10. Comparison of the electric and magnetic field derivative waveforms from the approximate results with the exact results at a distance of 5 m.

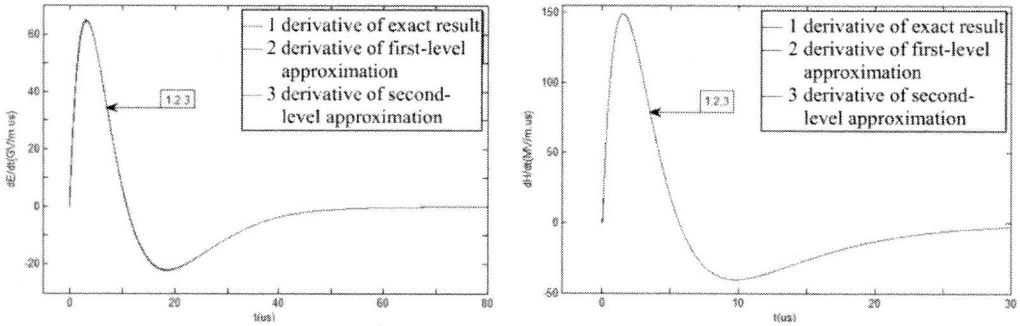

Figure 11. Comparison of the electric and magnetic field derivative waveforms from the approximate results with the exact results at a distance of 10 m.

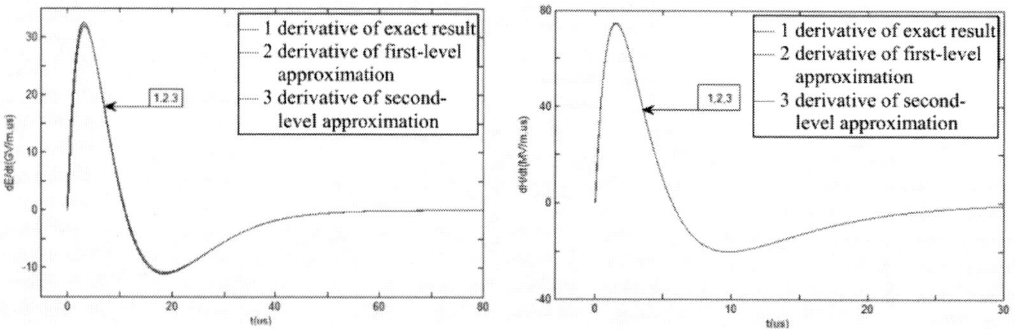

Figure 12. Comparison of the electric and magnetic field derivative waveforms from the approximate results with the exact results at a distance of 20 m.

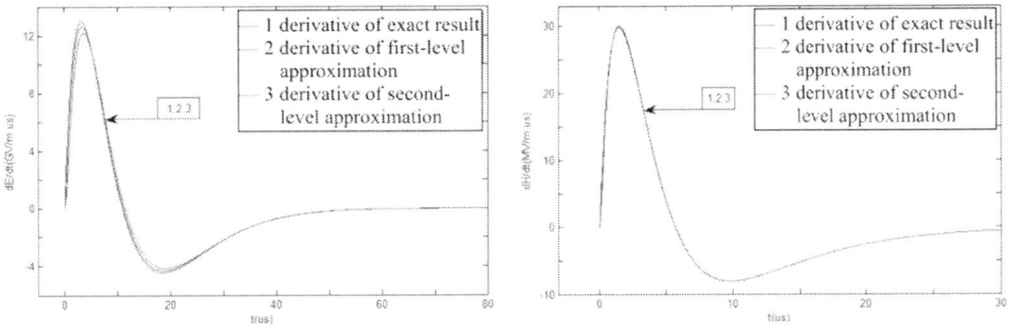

Figure 13. Comparison of the electric and magnetic field derivative waveforms from the approximate results with the exact results at a distance of 50 m.

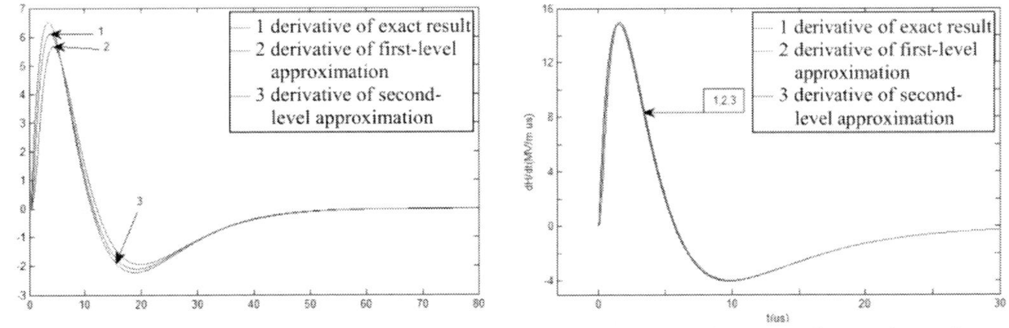

Figure 14. Comparison of the electric and magnetic field derivative waveforms from the approximate results with the exact results at a distance of 100 m.

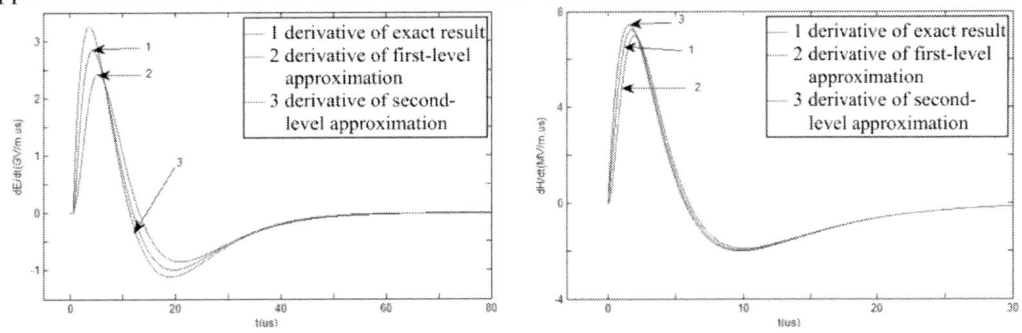

Figure 15. Comparison of the electric and magnetic field derivative waveforms from the approximate results with the exact results at a distance of 200 m.

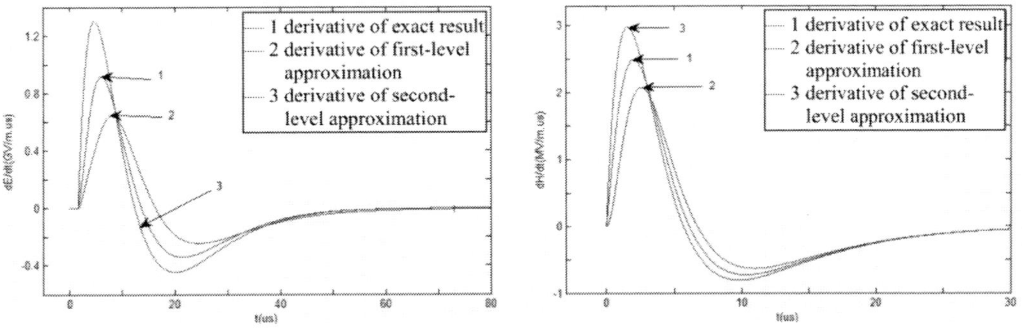

Figure 16. Comparison of the electric and magnetic field derivative waveforms from the approximate results with the exact results at a distance of 500 m.

As follows from figures 2 to 9, the LEMP waveforms of the approximate results are similar to the waveforms of the exact results. The deviations of the approximate results from the exact results increase with the distance r increases. The closer the distance is, the less the deviations are. The electric field waveforms of the approximate results are essentially coincident with the waveforms of the exact results within 100 m. When the distance exceeds 500 m, there is a great deal of difference between the electric field waveforms of the two approximate results and the exact results. By comparing the waveforms of electric and magnetic fields in figures 2 to 9, it can be found that the difference between the magnetic field and its two approximate results is smaller than the difference between the electric field and its two approximate results within near range. The magnetic field waveform of the two approximate results can not been distinguished from the exact results until the distance exceeds 500 m.

As follows from figures 10 to 16, the deviations of the approximate derivative waveforms from the exact derivative waveforms increase with the distance r increases. The closer the distance is, the less the deviations are. The electric field derivative waveforms of the approximate results are essentially coincident with the derivative waveforms of the exact results within 100 m. When the distance exceeds 500 m, there is a great deal of difference between the electric field derivative waveforms of the two approximate results and the exact results. By comparing the waveforms of electric and magnetic field derivatives in figures 2 to 9, it can be found that the difference between the magnetic field derivative and its two approximate results is smaller than the difference between the electric field derivative and its two approximate results within near range. The magnetic field derivative waveforms of the two approximate results can not been distinguished from the exact results until the distance exceeds 200 m.

4. Approximation characteristic of LEMP on earth surface and its derivative in far-field

4.1. Derivation of approximate expressions for LEMP in far-field

For the far-field, namely $R \approx r$ and $r \gg L$, the radiation field is the dominant component of the electromagnetic field produced by lightning return strokes, and the electrostatic, and induction terms tend to 0. For the transmission-line-type models, we can simplify the expressions as following because of the continuity of the upward-moving front [28]:

$$E_z \approx E_z(\text{radiation}) \approx -\frac{1}{2\pi\varepsilon_0 c^2 r}\int_0^h \frac{\partial i(0,t-r/c-z'/v)}{\partial t}\mathrm{d}z' \tag{25}$$

$$H_\phi \approx H_\phi(\text{radiation}) \approx \frac{1}{2\pi c r}\int_0^h \frac{\partial i(0,t-r/c-z'/v)}{\partial t}\mathrm{d}z' \tag{26}$$

Since v is constant, we obtain:

$$\frac{\partial i(0,t-r/c-z'/v)}{\partial t} = -v\frac{\partial i(0,t-r/c-z'/v)}{\partial z'} \tag{27}$$

Thus, the above equations (25) and (26) can be written as:

$$E_z(\text{radiation}) \approx -\frac{v}{2\pi\varepsilon_0 c^2 r}\int_0^h \frac{\partial i(0,t-r/c-z'/v)}{\partial z'}\mathrm{d}z' \tag{28}$$

$$H_\phi(\text{radiation}) \approx \frac{v}{2\pi c r}\int_0^h \frac{\partial i(0,t-r/c-z'/v)}{\partial z'}\mathrm{d}z' \tag{29}$$

Performing the integration, we get:

$$E_z(\text{radiation}) \approx -\frac{v}{2\pi\varepsilon_0 c^2 r}\left[i(0,t-r/c)-i(0,t-r/c-h/v)\right] \tag{30}$$

$$H_\phi(\text{radiation}) \approx \frac{v}{2\pi c r}\left[i(0,t-r/c)-i(0,t-r/c-h/v)\right] \tag{31}$$

Since $i(0,\tau)=0$, when $\tau \le 0$. If $t \le h/v + r/c$, the equations (30) and (31) can be expressed as:

$$E_z(r,t) \approx E_z(\text{radiation}) \approx -\frac{v}{2\pi\varepsilon_0 c^2 r} i(0, t - r/c) \tag{32}$$

$$H_\phi(r,t) \approx H_\phi(\text{radiation}) \approx \frac{v}{2\pi c r} i(0, t - r/c) \tag{33}$$

From equations (32) and (33), we can obtain that the electromagnetic field waveforms are essentially coincident with the waveform of the lightning return stroke channel-base current as long as the return stroke current has not reached the top of the channel. The negative sign in equation (32) represents that the direction of the electric field is opposite to the direction of the current propagation.

Performing the derivation on both sides of equations (32) and (33), we obtain:

$$\frac{\mathrm{d}E_z(r,t)}{\mathrm{d}t} \approx \frac{\mathrm{d}E_{\mathrm{rad}}(r,t)}{\mathrm{d}t} \approx \frac{v}{2\pi\varepsilon_0 c^2 r} \frac{\mathrm{d}i(0, t - r/c)}{\mathrm{d}t} \tag{34}$$

$$\frac{\mathrm{d}H_\phi(r,t)}{\mathrm{d}t} \approx \frac{dH_{\mathrm{rad}}(r,t)}{\mathrm{d}t} \approx \frac{v}{2\pi c r} \frac{\mathrm{d}i(0, t - r/c)}{\mathrm{d}t} \tag{35}$$

4.2. Comparisons of LEMP in far-field between the exact results and approximate results

The rising height of the return stroke current on the channel is generally several hundreds of meters while the electromagnetic field reaches the initial peak value. Thus, at the distance beyond 10km, the far-field approximation condition of the electromagnetic field derivative can be basically satisfied, and the approximate expressions of the far-field derivative have a wider applicable range. In order to verify the reliability of the approximate expressions in far-field calculation, the lightning channel-base current with a typical waveform 1.2/50 μs is adopted to calculate the ligthning electromagnetic fields. Accordingly, the parameters of the pulse function are set as follows: $I_0 = 30\mathrm{kA}$, $\tau_1 = 4.05\times10^{-7}\mathrm{s}$, $\tau_2 = 6.80\times10^{-5}\mathrm{s}$. And the height of the lightning channel is set to H=7.5km, and the return stroke speed is v=10^8 m s^{-1}. Comparisons of the waveforms from the radiation fields, and the scaled channel-base current with the total electromagnetic fields at different distances beyond 5 km are given in figures 17 to 20. And comparisons of derivative waveforms from the radiation fields, and the scaled channel-base current with the total electromagnetic fields at different distances are given in figures 21 to 23. Here, two different coefficients are introduced in the scaled channel-base current: While comparing with the electric field and its derivative, the multiplying factor of the current and its derivative is q_1. While comparing with the magnetic field and its derivative, the multiplying factor of the current and its derivative is q_2. The values of q_1 and q_2 are set as follows:

$$q_1 = \frac{v}{2\pi\varepsilon_0 c^2 r}, \quad q_2 = \frac{v}{2\pi c r}$$

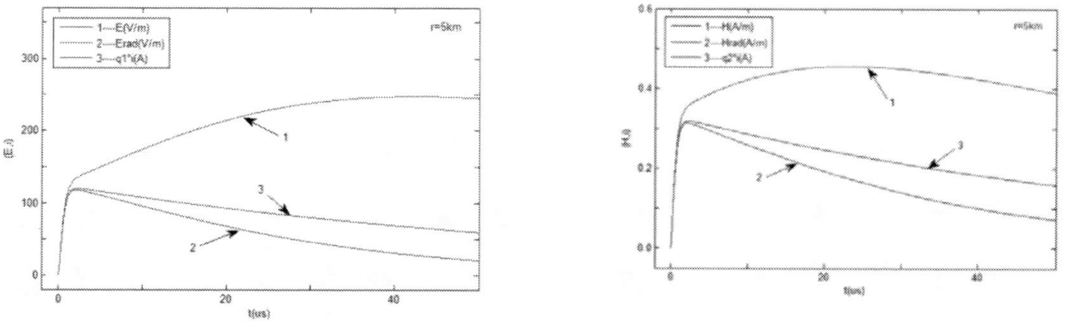

Figure 17. Comparison of the waveforms from the approximate results (2) radiation field component,and (3) scaled channel-base current with (1) the exact results at a distance of 5 km.

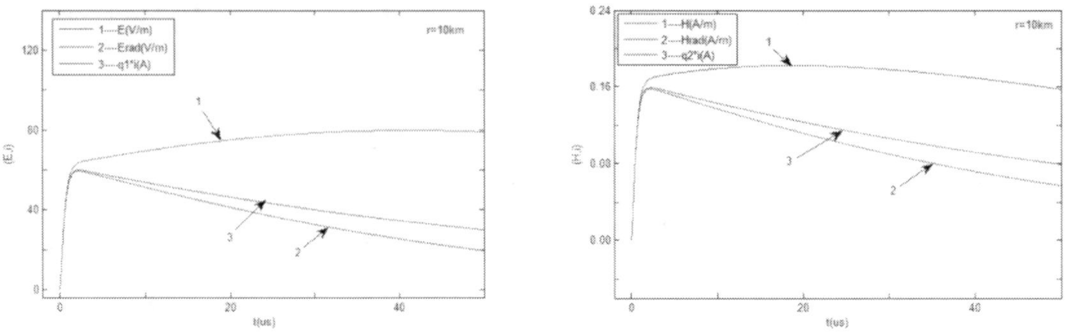

Figure 18. Comparison of the waveforms from the approximate results (2) radiation field component and (3) scaled channel-base current with (1) the exact results at a distance of 10 km.

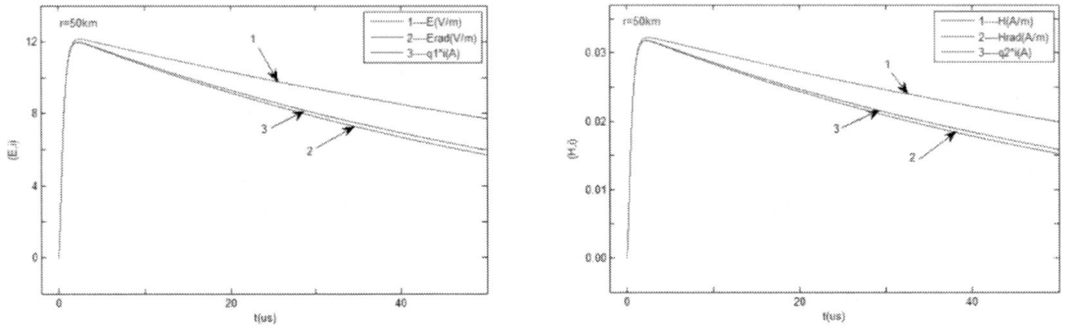

Figure 19. Comparison of the waveforms from the approximate results (2) radiation field component, and (3) scaled channel-base current with (1) the exact results at a distance of 50 km.

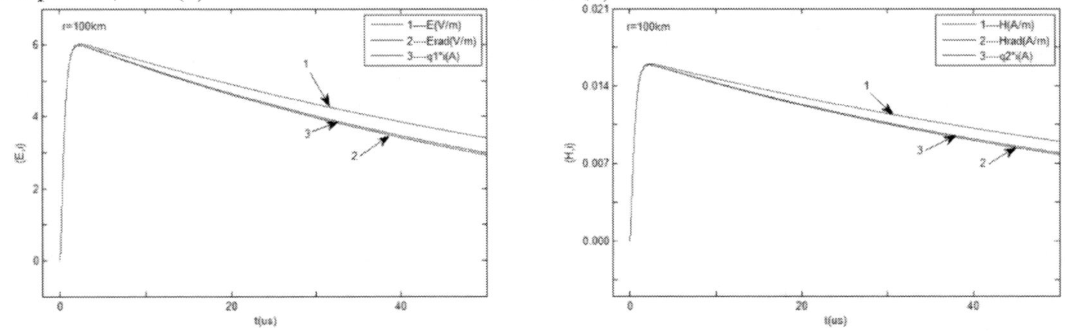

Figure 20. Comparison of the waveforms from the approximate results (2) radiation field component and (3) scaled channel-base current with (1) the exact results at a distance of 100 km.

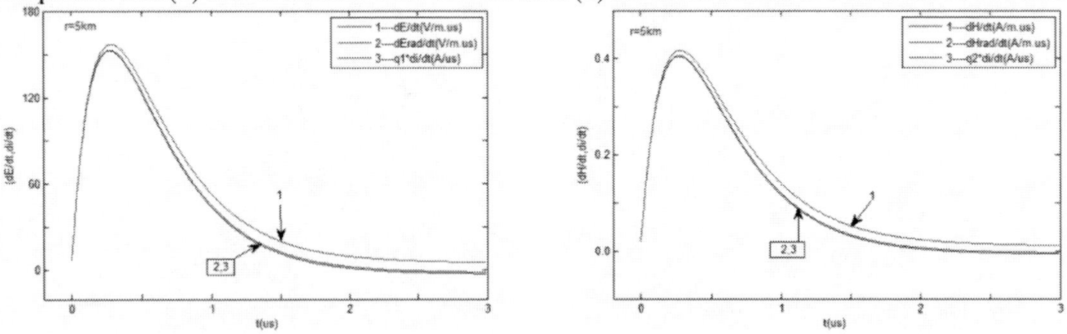

Figure 21. Comparison of the derivative waveforms from the approximate results (2) radiation field component and (3) scaled channel-base current with (1) the exact results at a distance of 5 km.

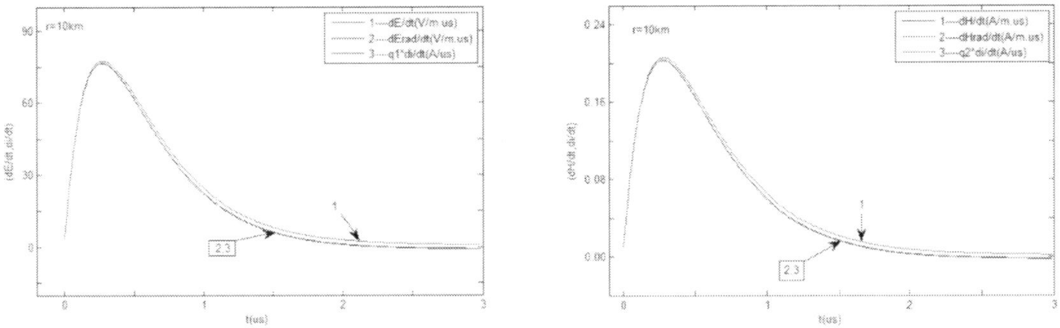

Figure 22. Comparison of the derivative waveforms from the approximate results (2) radiation field component and (3) scaled channel-base current with (1) the exact results at a distance of 10 km.

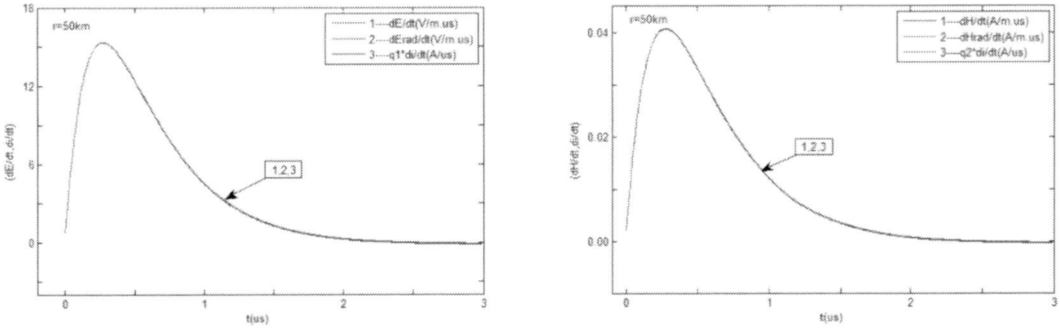

Figure 23. Comparison of the derivative waveforms from the approximate results (2) radiation field component and (3) scaled channel-base current with (1) the exact results at a distance of 50 km.

As follows from figures 17 to 20, the deviations of the radiation fields, and the scaled channel-base current from the total electromagnetic fields decrease with the distance r increases. The closer the distance is, the greater the deviations are. However, the rising edge of the radiation field waveforms, and the scaled channel-base current waveforms are essentially coincident with that of the total electromagnetic field waveforms. As shown in figures 17 and 18, at a distance of 5 to 10 km, the deviations of the radiation fields from the total electromagnetic fields increase gradually after the peak value, because the distance of 5 to 10 km located in the range of intermediate field, and the electrostatic, and induction components can not yet be ignored. Moreover, the deviations between the radiation fields and the scaled channel-base current are relatively small. These deviations mainly result from the approximate substitution of r for R in equations (32) and (33). As shown in figure 19, at a distance of 50 km, the electric and magnetic fields increase to a peak value then decay. Thus, the approximation characteristic between the total electromagnetic fields and the approximate results in the first several microseconds before reaching the initial peak value can basically represents the approximation characteristic of the full waveform. Furthermore, there is still some difference between the total electromagnetic fields, the radiation fields and the scaled channel-base current because the distance of 50km is only about 7 times larger than the height of the channel, which can not be viewed as $r \gg L$ (namely, the far-field approximation condition can not be met perfectly). When the distance exceeds 100 km, as shown in figure 20, the radiation field waveforms are essentially coincident with the scaled channel-base current waveforms. There is a little difference, which emerges after dozens of microseconds, between the total electromagnetic field and the radiation field component.

As follows from figures 21 to 23, the deviations of the radiation field derivatives, and the scaled channel-base current derivatives from the total electromagnetic field derivatives decrease with the distance r increases. The closer the distance is, the greater the deviations are. However, the radiation field waveforms, and the scaled channel-base current waveforms are essentially coincident with the total electromagnetic field waveforms in the first several microseconds. The difference between the scaled channel-base current derivatives and the total electromagnetic field derivatives in the first three microseconds is almost negligible at the distance beyond 5 km, because the rising height of the return stroke current on the channel is generally several hundreds of meters in the first several microseconds. At the distance beyond 5 km, the far-field approximation condition of the electromagnetic field derivative can be basically satisfied. With time going by and the further rising of the return stroke current on the channel, the distance requirement of the far-field approximation becomes farther gradually. Note that the electromagnetic field changes sharply only in the first several microseconds, then becomes slow finally. The far-field approximation characteristic in the first several microseconds between the total electromagnetic field derivatives and the approximate derivatives can basically represents the far-field approximation characteristic of the full waveform. It can be stated that the approximation condition of the far-field derivative is looser than the approximation condition of the far-field. The approximate expressions of the far-field derivative can be applied in the field derivative calculation at the distance beyond several kilometers, and the approximate expressions of the far-field can be applied only in the field calculation at the distance beyond 50 km.

Moreover, by comparing figures 17 to 20 with figures 21 to 23, it can be found that the difference among the total electromagnetic field derivatives, the radiation field derivatives, and the scaled channel-base current derivatives is smaller than the difference among the total electromagnetic fields, the radiation fields, and the scaled channel-base currents at the same distance.

5. Conclusions

According to the approximate derivation of the general expressions for LEMP on earth surface, the approximation characteristic between the near-field and far-field of LEMP and the channel-base current is obtained. The dipole method, based on the TL model, is used to calculate the near-field and far-field of LEMP on earth surface. The approximate results obtained from the approximate expressions are in a good agreement with the exact calculation results in the near-field and far-field, which indicates that the approximate derivation of the near-field and far-field expressions on earth surface is feasible, and the approximation characteristic between the near-field and far-field waveforms of LEMP and the channel-base current waveform is credible. This conclusion provides a theoretical basis for the LEMP wave simulation of near-field and far-field on earth surface.

Acknowledgments

The research is supported by National Natural Science Foundation Project of China under Grant No. 51077132.

References

[1] Rakov V A and Uman M A 2003 *Lightning: Physics and Effects* (Cambridge: Cambridge University Press)
[2] Rakov V A and Rachidi F 2009 *IEEE Trans. Electromagn. Compat.* **51** 428
[3] Miyazaki T, Ishii T and Okabe S 2010 *IEEE Trans. Electromagn. Compat.* **52** 921
[4] Samaras K, Sandberg C, Salmas C J and Koulaxouzidis A 2007 *IEEE Trans. Ind. Applicat.* **43** 150
[5] Thottappillil R 2002 *IEEE Trans. Electromagn. Compat.* **44** 203
[6] Jerauld J, Uman M A, Rakov V A, Rambo K J, Jordan D M and Schnetzer G H 2008 *J. Geophys. Res.* **113** D17111 doi: 10.1029/2008JD010171
[7] Barbosa C F, Paulino J O S, Miranda G C de, Boaventura W do C, Nalin F E, Person S and Zeddam A 2008 *IEEE Trans. Electromagn. Compat.* **50** 913
[8] Lee B H, Eom J H, Kang S M, Paek S K and Kawamura T 2004 *Jpn. J. Appl. Phys.* **43** 4379
[9] Hussein A M, Milewski M and Janischewskyj W 2008 *IEEE Trans. Electromagn. Compat.* **50** 642
[10] Cooray V, Fernando M, Gomes C and Sorenssen T 2004 *IEEE Trans. Electromagn. Compat.* **46** 87
[11] Uman M A, Rakov V A, Schnetzer G H, Rambo K J, Crawford D E and Fisher R J 2000 *J.*

Geophys. Res. **105** 15577
[12] Schoene J, Uman M A, Rakov V A, Rambo K J, Jerauld J and Schnetzer G H 2003 *J. Geophys. Res.* **108** D23, doi: 10.1029/2003JD003683
[13] Jerauld J E, Uman M A, Rakov V A, Rambo K J, Jordan D M and Schnetzer G H 2009 *J. Geophys. Res.* **114** D19115, doi:10.1029/2008JD011660
[14] Uman M A, Rakov V A, Versaggi J A, Thottappillil R, Eybert-Berard A, Barret L, Berlandis J P, Bador B, Barker P P, Hnat S P, Oravsky J P, Short T A, Warren C A and Bernstein R 1994 *Proc. EMC'94 ROMA, Int. Symp. Electromagn. Compat.* (Rome, Italy) p 33
[15] Uman M A, Rakov V A, Rambo K J, Vaught T W, Fernandez M I, Cordier D J, Chandler R M, Bernstein R and Golden C 1997 *Trans. Inst. Elect. Eng. Japan* **117-B** 446
[16] Uman M A, Swanberg C E, Tiller J A, Lin Y T and Krider E P 1976 *Radio Sci.* **11** 985
[17] Rakov V A and Uman M A 1998 *IEEE Trans. Electromagn. Compat.* **40** 403
[18] Cooray V 2009 *IEEE Trans. Electromagn. Compat.* **51** 748
[19] Bruce C E R and Golde R H 1941 *J. IEE.* **88** 487
[20] Uman M A and McLain D K 1969 *J. Geophys. Res.* **74** 6899
[21] Heidler F 1985 *Proc. 6th Int. Zurich Symp. Electromagn. Compat.* (Zurich, Switzerland) p 157
[22] Rakov V A and Dulzon A A 1987 *Tekh. Elektrodinam.* **1** 87
[23] Nucci C A, Mazzetti C, Rachidi F and Ianoz M 1988 *Proc. 19th Int. Conf. Lightning Protection* (Graz, Austria)
[24] Diendorfer G and Uman M A 1990 *J. Geophys. Res.* **95** 13621
[25] Rubinstein M and Uman M A 1989 *IEEE Trans. Electromagn. Compat.* **31** 183
[26] Rubinstein M and Uman M A 1991 *IEEE Trans. Electromagn. Compat.* **33** 312
[27] Cooray V 1993 *J. Electrostat.* **30** 343
[28] Uman M A 1987 *The Lightning Discharge* (San Diego, CA: Academic)
[29] Uman M A, Mclain D K and Krider E P 1975 *Amer. J. Phys.* **43** 33
[30] Jerauld J, Uman M A, Rakov V A, Rambo K J and Schnetzer G H 2007 *J. Geophys. Res.* **112** D13113-1

FDTD simulation of field performance in reverberation chamber excited by two excitation antennas

Song Wang*, **Zhan-cheng Wu, Yao-zhong Cui**
Research Institute of Electrostatic and Electromagnetic Protection, Mechanical Engineering College, Shijiazhuang, Hebei province, China

E-mail: wansonde@gmail.com

Abstract. The excitation source is one of the critical items that determine the electromagnetic fields in a reverberation chamber (RC). In order to optimize the electromagnetic fields performance, a new method of exciting RC with two antennas is proposed based on theoretical analysis. The full 3D simulation of RC is carried out by the finite difference time domain (FDTD) method on two excitation conditions of one antenna and two antennas. The broadband response of RC is obtained by fast Fourier transformation (FFT) after only one simulation. Numerical data show that the field uniformity in the test space is improved on the condition of two transmitting antennas while the normalized electric fields decreased slightly compared to the one antenna condition. It is straightforward to recognize that two antennas excitation can reduce the demands on power amplifier as the total input power is split among the two antennas, and consequently the cost of electromagnetic compatibility (EMC) test in large-scale RC can be reduced.

1. Introduction

Reverberation chamber is a new EMC test environment which has the evident advantages of higher quality factor (Q), lower test cost and bigger test space over the traditional test fields. The three-dimensional full simulation of reverberation chamber (RC) plays an important role in RC's design and its performance optimization. This simulation can be divided into two categories, frequency-domain and time-domain simulation. Since long time and big computational resource is called for the simulation running for one time, time-domain simulation is preferred because of wideband response of RC could be obtained using FFT in the post-stage. The finite difference time domain (FDTD) method has been used widely in numerical simulation of RC in recent years [1, 2].

The electric field (E-field) strength and field uniformity within test space are the key parameters in RC's performance validation. From theoretical analysis, one can see that the electromagnetic fields are related to the RC's geometry, working frequency and the feeding source. Towards the improvement of field uniformity, much work has been done focusing on the optimization of RC's geometry including the stirrer's structure and the RC's wall dimensions [3-5]. However, the reports on the effect of feeding source on field uniformity or field strength are found to be scarce, especially on the condition of two excitation antennas. To extend the previous work [6], the effect of two excitation antennas on the RC's field performance is investigated in wide frequency band using FDTD.

* To whom any correspondence should be addressed.

The full 3D simulation of RC is carried out by FDTD on both excitation conditions of one antenna and two antennas. The broadband response of RC is obtained after only one simulation by FFT. Numerical results show that using two excitation antennas could help improve the field uniformity and reduce the demands on power amplifier as the total input power is split. Consequently, the cost of EMC test in large-scale RC can be reduced.

2. Effect of excitation source on E-fields

Based on the Maxwell's equations, the expression of electric fields inside a RC can be deduced using eigenfunction superposition method [7].

$$\vec{E} = \sum_{m,n,p} (c_{xmnp} A\vec{e}_x + c_{ymnp} B\vec{e}_y + c_{zmnp} C\vec{e}_z) \tag{1}$$

where $\vec{E}$ is the E-field strength vector, $\vec{e}_x$, $\vec{e}_y$ and $\vec{e}_z$ are unit vector point to the x, y, z direction, mnp corresponds to a resonant mode of the RC, A, B, C are Eigen solutions related to the x, y, z component of E-field respectively and the weighting coefficients are given by

$$\begin{aligned} c_{xmnp} &= \frac{\mathrm{c}^2}{4\pi^2(f^2 - f_{mnp}^2)} \frac{8}{LWH} \iiint_{r\in D} \omega\mu_0 J_x A\mathrm{d}x\mathrm{d}y\mathrm{d}z \\ c_{ymnp} &= \frac{\mathrm{c}^2}{4\pi^2(f^2 - f_{mnp}^2)} \frac{8}{LWH} \iiint_{r\in D} \omega\mu_0 J_y B\mathrm{d}x\mathrm{d}y\mathrm{d}z \\ c_{zmnp} &= \frac{\mathrm{c}^2}{4\pi^2(f^2 - f_{mnp}^2)} \frac{8}{LWH} \iiint_{r\in D} \omega\mu_0 J_z C\mathrm{d}x\mathrm{d}y\mathrm{d}z \end{aligned} \tag{2}$$

where c is the speed of light in vacuum, ω is angular frequency, μ_0 is the permeability of vacuum, f is the RC's working frequency, f_{mnp} is the resonant frequency, L, W, H are dimensions of the RC, and J_x, J_y, J_z are the excitation currents density in the corresponding directions.

According to(1)(2), the E-fields mainly depend on three factors: the RC's geometry, the working frequency f and detailed configuration of the feeding source including its position in the RC and the specification of the driving current. It is well known that mechanical stirrers can help realize field uniformity by changing the boundary conditions which is equivalent to the first factor. So considering the third factor, it is assumed to be beneficial to field uniformity to apply two excitation antennas since the weighting coefficients are related to the excitation source. When putting the excitation antennas at different positions with typical directions, the electromagnetic fields from each transmitting antenna are superimposed randomly ending up with the field uniformity improvement within the test space. This assumption is supported by the following numerical simulation data.

3. RC's modelling and numerical simulation by FDTD

The studied mode-stirred RC has dimensions equal to $8\text{m} \times 10.5\text{m} \times 4.3\text{m}$ and the stirrers are similar to that in [8] but a little more irregular. An efficient discretization method [9] is applied to model the stirrers making the two stirrers can rotate independently with a random rotating degree. As the fundamental mode frequency is 23.6 MHz, the most interesting frequency band for this RC could be chosen as 60 MHz to 300MHz according to [10]. In accordance with the FDTD principles defined in [11], the space and time increment are set equal to $\Delta h = 0.1\text{m}$ (the space step is the same for the three dimensions) and $\Delta t = 1.67 \times 10^{-10}\text{s}$ with the Courant factor being $\sqrt{3}/2$. The RC is excited by dipole antennas which are modeled using the thin-wire technique [12, 13]. At the middle port, the dipole antenna is attached to a transmission line (TML) with the characteristic impedance being $Z_0 = 50\Omega$. A unit-amplitude modulated Gaussian pulse defined by the frequency range from 60MHz to 300MHz is applied into the TML as incident voltage source v_{inc} and the reflected voltage v_{ref} in time domain could be recorded as discussed in [14]. This kind of feeding method is convenient to calculate the input power p_{inp} of the excitation antenna.

$$p_{\text{inp}} = \frac{|\breve{v}_{inc}|^2}{Z_0} - \frac{|\breve{v}_{ref}|^2}{Z_0}. \tag{3}$$

where the $\breve{v}_{inc}$ and $\breve{v}_{ref}$ are the frequency components related to v_{inc} and v_{ref} respectively. The RC's geometry and the antennas' configuration are showed in Figure 1. The TML is in dotted lines indicating that its real structure is not included in the 3D full simulation but is solved by additional one-dimension FDTD codes.

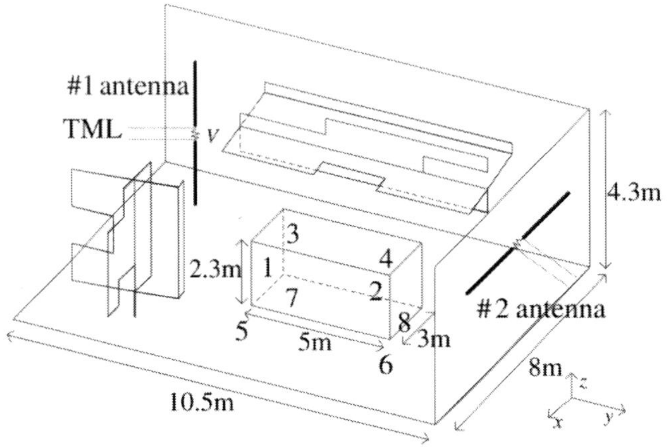

Figure 1. Schema of the RC's structure.

Disposal of loss is of great importance to the RC's simulation in time domain since the simulation should be running until the total energy inside decreases small enough. We adopt an approximation method by setting the electric conductivity of the air within the RC to 10^{-5}S/m while treating the RC as PEC structure according to [15]. In order to verify the conclusion from the end of section 2, the simulation is carried out on two excitation conditions of one antenna and two antennas, which is achieved by removing antenna #2 or not. A test volume ($5\times3\times2.3\text{m}^3$) is chosen as referred to the IEC-61000-4-21 standard. The eight vertices of this volume are chosen as the observation points for the field uniformity verification.

On each excitation condition, the stirrers are rotated in an interleaved way with the rotation step being 45 degrees. 64 different stirrers' configurations are obtained satisfying the specification in IEC-61000-4-21 standard. FDTD codes for every stirrer configuration are created and executed separately. Each execution is taken 50000 iterations and the E-field amplitude in RC decreases below $1/100$ of its peak value.

4. Simulation results and discussion

All three orthogonal components of E-field at each of the eight vertices are recorded during the execution of RC's FDTD code. The corresponding frequency values $\breve{E}_{i,j}$ $i = x, y, z,$ $j = 1,2...8.$ are obtained by FFT with the frequency resolution being $1/\Delta t/50000 = 120\,\text{KHz}$. The normalized E-field $\breve{E}_{i,j}^{\text{nor}}$ is computed by

$$\breve{E}_{i,j}^{\text{nor}} = \breve{E}_{i,j} / \sqrt{p_{\text{inp}}} \quad i = x, y, z, \; j = 1, 2...8. \tag{4}$$

At each observation point, for every orthogonal component of E-field, the maximal value occurred during a complete rotation of stirrers is selected and they are used to compute the standard deviations of the E-field in dB according to IEC-61000-4-21 standard.

4.1. Investigation into the field uniformity

There are four parameters $\sigma_x, \sigma_y, \sigma_z$ and σ_{24} embodying the standard deviations of E-fields which should not exceed a limit specified by IEC-61000-4-21 standard. From the theoretic analysis in section 2, it is possible to improve the field uniformity using two transmitting antennas.

Using the E-field vector data on eight vertices in frequency domain, the four parameter $\sigma_x, \sigma_y, \sigma_z$ and σ_{24} in dB on both conditions are computed by reference formulas in [16] . Glancing at the results showed from Figure 2 to Figure 5, we can see that the field uniformity satisfies the standard specification on both exciting conditions above 70MHz. These four figures are similar to each other, and the fourth exhibits less fluctuations because of its statistical components being 24. Comparing the two main lines in each figure, it is not hard to recognize that the two antennas excitation method can help improve the field uniformity at most of the studied frequency points.

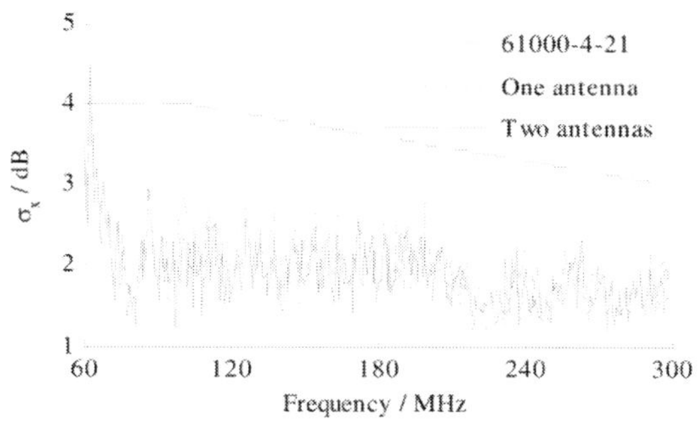

Figure 2. Comparison of X std. dev. Db.

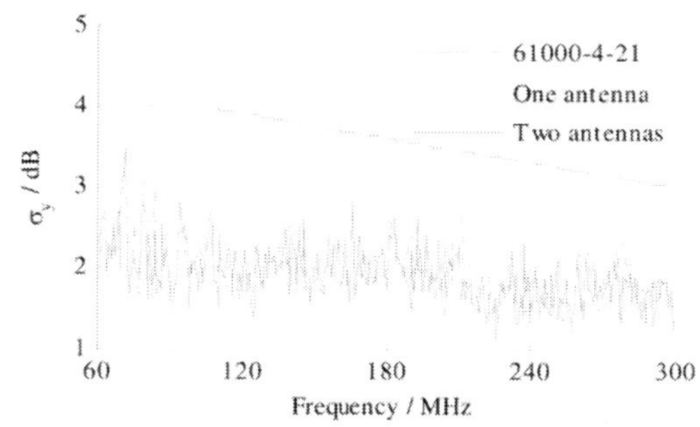

Figure 3. Comparison of Y std. dev. dB.

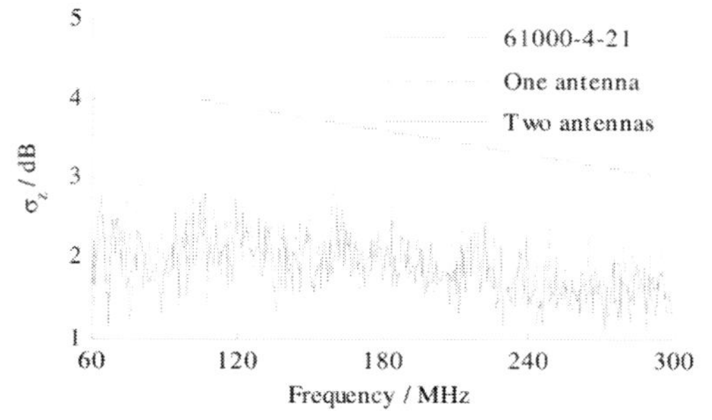

Figure 4. Comparison of Z std. dev. dB.

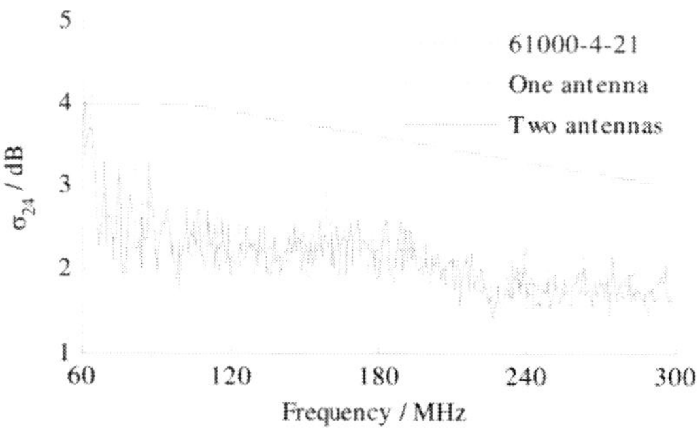

Figure 5. Comparison of X, Y, and Z std. dev. dB.

In order to see this improvement more explicitly, a statistical parameter η called improvement degree is proposed.

$$\eta = \frac{\langle \sigma' \rangle - \langle \sigma'' \rangle}{\langle \sigma' \rangle} \times 100\% \tag{5}$$

where σ can be any of $\sigma_x, \sigma_y, \sigma_z$ or σ_{24} , the superscripts ' and " stand for one antenna and two antennas excitation conditions, and $\langle\rangle$ denotes the mean value in a specified frequency band.

Table 1. Improvement degree of E-field standard deviation in separated frequency bands.

Std. dev.\freq. bands	60~120MHz	120~180MHz	180~240MHz	240~300MHz
σ_x	5.05 %	5.43 %	0.23 %	2.03 %
σ_y	2.94 %	3.87 %	3.63 %	1.23 %
σ_z	7.86 %	3.48 %	1.54 %	4.49 %
σ_{24}	3.36 %	1.88 %	-0.94 %	0.98 %

The studied frequency band is divided into four groups, and the corresponding η is displayed in Table 1. The results indicate that the two antennas excitation method can achieve an average field uniformity improvement in all the four frequency bands for all the uniformity parameters except σ_{24} within 180~240 MHz. This improvement is more obvious when the frequency is below 180 MHz, which can be explained by the well-known trend that the field uniformity gets better naturally as the frequency increases.

4.2. Comparison of normalized E-fields.

The normalized E-field modulus at vertex #1 is compute from the maximum E-field over the rotation of stirrers. Figure 6 shows that the normalized E-field modulus on condition of two transmitting antennas is slightly smaller than that of one antenna condition, especially when the frequency is below 150 MHz. Besides vertex #1, we can see from Figure that the E-field strength is also smaller at other vertices at the frequency of 200 MHz. There may be two reasons for this phenomenon, one is that the coupling between the two antennas adds additional losses to the RC's overall losses, the other is that the field uniformity's improvement on the two antennas condition may be realized by decreasing the maximal E-field strength and increasing the minimal E-field strength, whereas the reference quantity is the maximal E-field strength but not the average value.

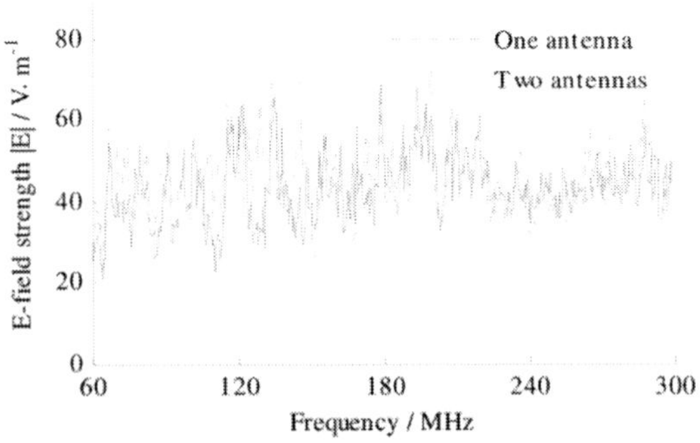

Figure 6. Comparison of normalized E-field strength versus frequency at vertex #1,

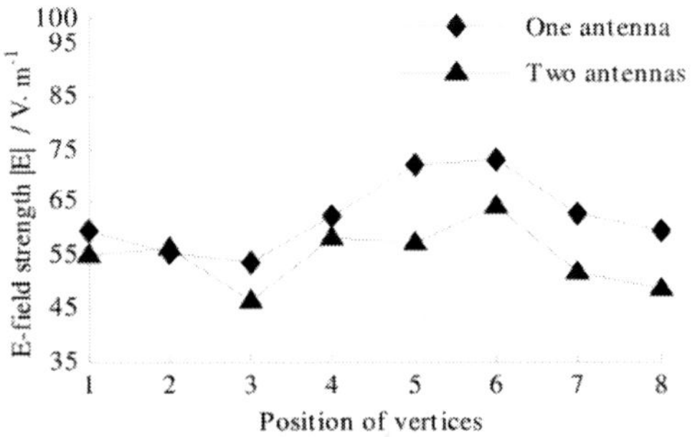

Figure 7. Comparison of E-field versus vertex's position at the frequency of 200 MHz.

4.3. Exploration of exciting a large RC by two antennas

In the engineer application, two advantages make it attractive to apply the new excitation method to large-scale RC. On one hand, it is instrumental in producing a reverberation environment in a large RC; on the other hand, it can save on the cost of power amplifier without the deterioration of field uniformity.

If the dimensions of the RC are very large, the electromagnetic field decreases obviously as its location moves away from the excitation antenna. Consequently, only one transmitting antenna can hardly generate statistical uniform electromagnetic field. Using two or more excitation antennas located at different positions with different orientations inside the RC will alleviate the non-uniformity distribution of electromagnetic energy. In this way, the new excitation method is helpful in producing a reverberation environment. In addition, the price of high-power amplifier is usually much larger than twice of the ordinary power amplifier's price. In this way, it is a good alternative to apply the new excitation method to generate high-intensity electromagnetic field if using one power amplifier is not economically acceptable.

5. Conclusion

A new excitation method is proposed using two transmitting antennas for reverberation chamber. The two antennas should be put at different position with different orientations. From the above comparison and analysis, this method can lead to a slight improvement on field uniformity and slightly lower normalized E-fields than the traditional method. More importantly, this method could reduce the demands on power amplifier without the deterioration of field uniformity. So it can be used to generate high E-field in RC for special equipments' EMC tests.

References

[1] Orjubin G, Petit F, Richalot E, Mengue S and Picon O 2006 *IEEE Trans. Electromagn. Compat.* **48** 429
[2] Moglie F and Pastore A P 2006 *IEEE Trans. Electromagn. Compat.* **48** 195
[3] Zhao H P and Shen Z X 2011 *IEEE Trans. Antennas Propag.* **59** 2943
[4] Liu X Q, Wei G H, Zhang Y Q and Zhang C H 2008 Effects of Stirrer on the Field Uniformity at Low Frequency in a Reverberation Chamber and its Simulation *Int. Sympo. on Computer Science and Computational Technology* pp 517-9
[5] Clegg J, Marvin A C, Dawson J F and Porter S J 2005 *IEEE Trans. Electromagn. Compat.* **47** 824
[6] Cui Y Z, Wei G H, Fan L S, Liu X Q, Pan X D and Chen Y Z 2011 *Journal of Microwaves* **27**
[7] Ding J J and Sha F 2005 *Chinese Journal of Radio Science* **20** 557
[8] Mengue S, Richalot E and Picon O 2008 *IEEE Trans. Electromagn. Compat.* **50** 237
[9] Haffar M E, Reineix A, Guiffaut C and Adardour A 2009 Reverberation chamber modeling using the FDTD method *Int. Conf. on Advances in Computational Tools for Engineering Applications* pp 151-6
[10] Bruns C and Vahldieck R 2005 *IEEE Trans. Electromagn. Compat.* **47** 612
[11] Elsherbeni A and Demir V 2009 *The Finite-Difference Time-Domain Method for Electromagnetics with MATLAB Simulations* (San francisco: peter classman)
[12] Umashankar K, Taflove A and Beker B 1987 *IEEE Transactions on Antennas and Propagation* **AP-35** 1248
[13] Boonzaaier J J and Pistorius C W I 1990 *ELECTRONICS LETTERS* **26** 1891
[14] Taflove A and Hagness S C 2000 *Computational Electrodynamics: The Finite-Difference Time-Domain Method* (Boston: MA: Artech House)
[15] Moglie F 2004 *IEEE Trans. Electromagn. Compat.* **46** 469
[16] IEC 61000-4-21: Electromagnetic Compatibility (EMC) - Part 4-21: Testing and Measurement Techniques - Reverberation Chamber Test Methods. pp 40-4

Journal of Physics: Conference Series **418** (2013) 012007 doi:10.1088/1742-6596/418/1/012007

Numerical calculations on the field emission of carbon nanotubes

Haizhen Ren*, **Linfeng Yang and Yuguang Zhang**
Department of Physics, College of Science, Zhongyuan University of Technology, Zhongyuan west road, Zhengzhou, China

E-mail: hzhren@gmail.com

Abstract. Carbon nanotubes (CNTs) have been proven experimentally to be well suited for field emission applications. However, the exact process of electron emission from CNTs is still not fully understood. In this paper, we numerically investigate the effects of electrostatic field screening at the CNT emitter tips by finite element method and show the impact on device performance. We further observe that to achieve maximum emission efficiency of CNTs in an infinite line, one need to have an intertube distance of three times the CNT height. These results can be used to develop an optimized CNT field emission array that fits within the constraints of currently available fabrication capabilities.

1. Introduction

Since Iijima published his seminal article in Nature identifying multi-walled carbon nanotubes in 1991, Carbon nanotubes (CNTs) have attracted tremendous interest from fundamental science and technological perspectives [1]. CNTs posses many unique properties due to their low dimensionality make them promising candidates for future technology applications. CNTs have been used in many laboratories to build prototype nanodevices. Understanding these novel properties and their impact on devices is crucial in the development and evolution of CNT applications.

CNTs have been proven experimentally to be well suited for field emission applications. This is related to that CNTs have many unique properties such as narrow diameters, high aspect ratios, high temperature stability, good conductivity and structural strength. Prototype devices using the superior field emission properties of CNT have been demonstrated. These devices include x-ray tubes [2], scanning x-ray sources [3], flat panel displays [4, 5] and lamps [6]. CNTs make excellent electron emitters not because of a low work function but due to the extremely high local electric field that forms at the small diameter tips.

The ability to grow aligned CNTs on a substrate has made it possible for designers to specify the height, radius, separation, etc. However, the exact process of electron emission from CNTs is still not fully understood. Much work has been done on the emission process through experimental and numerical investigations for further understanding [7, 8]. It has long been assumed that the field screening is negated when adjacent emitters are at a separation of twice their height [9]. In this paper, we investigate the effect of electric field screening on an infinite line of CNTs by means of two-dimensional simulation and show the impact on the device performance.

* To whom any correspondence should be addressed.

2. Numerical model and results

CNTs were simulated in 2D by solving the Laplace equation with finite element method (FEM). The simulation allows the extraction of potential and electric field at any point around the structure. The schematic diagram of simulated line of CNT is shown in the figure 1(a). The CNTs are treated as perfect conductors in our model. Vertically aligned CNTs of uniform height h and radius r are placed on a grounded cathode that is separated from the positive anode by distance D. The intertube distance is denoted as d. Field emission occurs primarily due to the physical dimensions of CNTs. CNTs can range from 0.6 nm to more than 100 nm in diameter and tens of nanometers to microns and even millimeters in length resulting in incredible aspect ratios. We assumed tubes of 1 μm length with a tip apex of 2 nm in our calculation the same as previous reported in Nilsson's work [10]. The resulting focus of the electric field at the tips of the CNTs is often referred to as the electric field enhancement factor β which depends on the geometry of the tip. The field enhancement will increase for smaller diameter CNTs and decrease for larger diameter CNTs.

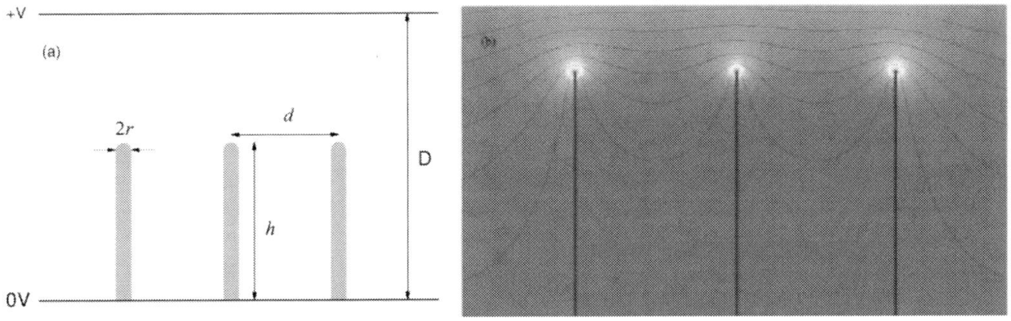

Figure 1. (a) Schematic of the simulation setup of a line of CNTs of height h and radius r placed on a grounded cathode and beneath an anode at a distance of D; (b) The simulation shows the equipotential lines of the electrostatic field with the background surface indicating the magnitude of the electric field. The CNTs used here have height of h=1 μm and radius of r=2 nm, the intertube distance is d=0.5 μm.

A simple 2D simulation of 3 element CNT line with intertube distance of 0.5 μm was first used and the calculation results were shown in the figure 1(b). The equipotential lines of the electric field with the background surface of the magnitude of the electric field are depicted. The anode is located relatively far away from the CNT emitters so its contribution to the electric field is assumed to be small and uniform. According to our calculation, the electric field at the central tube is lower than that of the outer tube. This is explainable by an electrostatic screening effect provoked by the proximity of neighbouring tubes. The CNTs are too close together for electrostatic field penetration which negates some of the expected field enhancement. These electrostatic screening effects are clearly evident in the simulation results shown in figure 1(b). Nilsson et al conducted a study of screening effects by performing 2D simulations of field penetration and the optimized separation between individual CNTs was determined to be twice the CNT height [10]. 3D simulations performed by Smith et al showed 2D simulations underestimated the effects of screening and determined an optimized separation of 3 times the height of the individual CNTs [11].

To investigate the electric field screening effect as a function of intertube distance, we simulate an infinite line of CNTs with periodic boundary condition. The intertube distance ranged from 0.1 μm to 6 μm in our calculation, which corresponds to the ratio d/h of 0.1 to 6. The local electric field was carefully extracted at the point immediately above the CNT tip. The percentage of screening was calculated by the difference in the local electric field to that of an identical but isolated CNT. Figure 2 shows the percentage of screening as a function d/h for CNT of radius 2 nm, height 1μm. We can see that the amount of screening decreases as d/h increases. The amount of screening at d=2 h indicates that there is a 4.3 % drop in the local electric field. When the intertube distance increased to d=3 h, the drop of screening falls to 1.1% which means the CNT is performing at 99 % of its optimum value. The percentage of screening was found to be 0.2 % when the intertube distance is further increased to d=4h. To achieve a fully unscreened line of CNT, one needs to have an intertube distance in excess of three times the height of CNT.

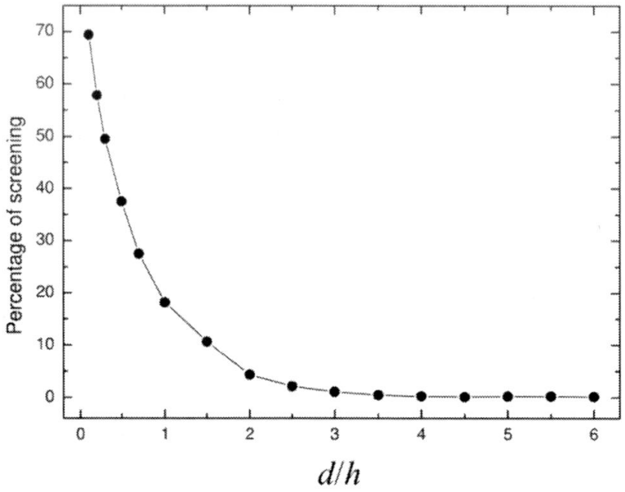

Figure 2. Percentage of electrostatic screening as a function of intertube distance over CNT height calculated by the difference in the local electric field between an isolated CNT.

It is important to consider the effect of screening in maximizing the current. Perhaps the most obvious approach to increasing the current density is to decrease the intertube distance. This will increase the number of emitters across the line resulting in a higher total current density, as long as the reduced intertube distance does not cause screening effect between neighbouring CNTs. The current density is calculated using the well known and widely accepted Fowler-Nordheim equation

$$J = \int \frac{aE_L^2}{\phi} \exp\left(\frac{-b\phi^{3/2}}{E_L}\right) ds \tag{1}$$

where a and b are constants with values 1.54×10^{-6} AeVV^{-2}and 6.83×10^{3} eV$^{-3/2}$Vμm^{-1}, respectively. E_L is local electric field in Vμm^{-1} at the surface of CNT and ϕ is the work function with value 4.5 eV. The integration is calculated over the CNT surface.

We calculated the emission current density for an infinite line of CNT and the result is plotted in figure 3. We observe a sharp increase in current density as the initially tightly packed CNTs become more separated. The current density reaches maximum when the intertube distance approaches three times the CNT height. Beyond this value, the current density decreases. This is because the far separated CNTs have very few efficient emitters while the closely packed CNTs suffer from electrostatic screening due to the neighbouring nanotubes. The nearly linear decrease beyond d=3 h would be expected, as at these intertube distance each emitter is unscreened, i.e., half the CNT equals half the current density. The most efficient arrangement of CNTs in a line, for maximizing the current density, is d=3 h, giving rise to a 1.1 % screening of the local field.

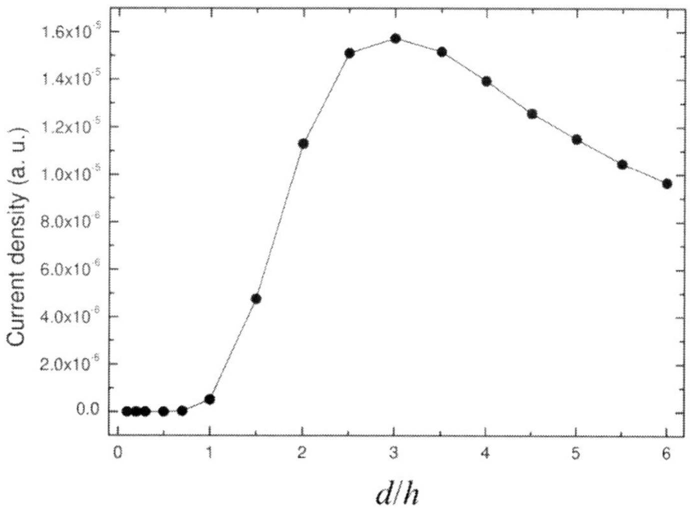

Figure 3. The calculated field emission current density of CNT with height 1 μm and radius 2 nm as a function of d/h.

3. Conclusions

In conclusion, we have simulated an infinite line of CNT emitters and investigated the effect of field screening in terms of intertube distance. We found that the optimum intertube distance needs to be in excess of three times the height of the CNT to achieve a fully unscreened line. Our calculations predicate that an intertube distance of three times the CNT height optimizes the emitted current density.

Acknowledgments

This work was supported by National Natural Science Foundation of China under grant No. 51077134.

References

[1] Sumio I 1991 *Nature* **354** 56
[2] Sugie H, Tanemura M, Filip V, Iwat K, Takahashi K and Okuyama F 2001 *Appl. Phys. Lett.* **78** 2578
[3] Zhang J, Yang G, Cheng Y, Gao B, Qiu Q, Lee Y Z, Lu J P and Zhou O 2005 *Appl. Phys. Lett.* **86** 184104
[4] Semet V, Binh V, Vincent P, Guillot D, Teo K, Chhowalla M, Amaratunga G, Milne W, Legagneux P and Pribat D 2002 Appl. Phys. Lett. 81 343
[5] Minoux E, Groening O, Teo K, Dalal S, Gangloff L, Schnell J, Hudanski L, Bu I, Vincent P, Legagneux P, Amaratunga G and Milne W 2005 Nano Lett. 5 2135
[6] Croci M, Arfaoui I, Stöckli T, Chatelain A and Bonard J-M 2004 Microelectron. J. 35 329
[7] De Heer W A, Chatelain A and Ugarte D 1995 *Science* **270** 1179
[8] Smith R, Cox D and Silva S 2005 *Appl. Phys. Lett.* **87** 103112
[9] Bonard J-M, Salvetat J-P, Stöckli T, Forró L and Ch^atelain A 1999 *Appl. Phys. A: Mater. Sci. Process.* **69** 245
[10] Nilsson L, Groening O, Emmenegger C, Kuettel O, Schaller E, Schlapbach L, Kind H, Bonard J-M and Kern K 2000 *Appl. Phys. Lett.* **76** 2071
[11] Smith R and Siliva S 2009 *Appl. Phys. Lett.* **94** 133104

Time-domain modeling for shielding effectiveness of materials against electromagnetic pulse based on system identification

Xiang Chen *, **Yong Guang Chen, Ming Wei, Xiao Feng Hu**
Electrostatic & Electromagnetic Protection Institute, Mechanical Engineering College, Heping West Road, No.97, Shijiazhuang, China

E-mail: chenxiang061101@163.com

Abstract. Shielding effectiveness (SE) of materials against electromagnetic pulse (EMP) cannot be well estimated by traditional test method of SE of materials which only consider the amplitude-frequency characteristic of materials, but ignore the phase-frequency ones. In order to solve this problem, the model of SE of materials against EMP was established based on system identification (SI) method with time-domain linear cosine frequency sweep signal. The feasibility of the method in this paper was examined depending on infinite planar material and the simulation research of coaxial test method and windowed semi-anechoic box of materials. The results show that the amplitude-frequency and phase-frequency information of each frequency can be fully extracted with this method. SE of materials against strong EMP can be evaluated with time-domain low field strength (voltage) of cosine frequency sweep signal. And SE of materials against a variety EMP will be predicted by the model.

1. Introduction

Testing standards of shielding effectiveness (SE) of materials are carried out through one by one measurement in frequency-domain by far. Then, the shielding performance of materials can be characterized by the amplitude-frequency curve. With in-depth study of high-power electromagnetic pulse (EMP) interference, it was discovered that only by frequency-domain SE of materials research cannot be completely characterized by shielding performance of time-domain pulsed-field [1, 2]. The results of frequency-domain measurement cannot directly determine time-domain shielding effectiveness of EMP mentioned in [3, 4]. And measurement equipment for EMP-field is not as continuous-wave measurement equipment as popular. People wish that time-domain response of the system can be estimated with the amplitude-frequency characteristics which were provided by manufacturers or experiment. The solution for this problem is to assume that the system is linear and minimum-phase system, estimate the phase frequency characteristics by amplitude-frequency characteristics through the Hilbert transform, and get the system impulse response discussed in [5, 6]. But some non-minimum phase systems is not well resolved. In addition, the vector fitting method can be used to solve this problem described in [7]. Overall, deconvolution problems which reconstruct time-domain data from frequency-domain are "sick", there are inherent deficiencies. In this paper, the shielding effect of materials against EMP was looked as a black-box system. Time-domain linear cosine swept signals were used to sufficiently excite the system. And time-domain waveform of unloaded and loaded materials was recorded. The model of SE of materials against EMP will be

* To whom any correspondence should be addressed.

established by SI. And 2 ns/50 ns double exponential EMP and square-wave EMP was used to validate the applicability of this method.

2. Model flow of SE of materials using SI

In this paper, linear cosine frequency sweep signal is an excitation signal. Electric field (or magnetic field), received in space, when shielded and unshielded was looked as an input / output (I/O) data of the system. SI was used to establish the model of SE of materials against EMP. Then, the time-domain waveform of EMP through shielding materials can be predicted by the model. Model flow chart of SE of materials against EMP was shown in figure 1.

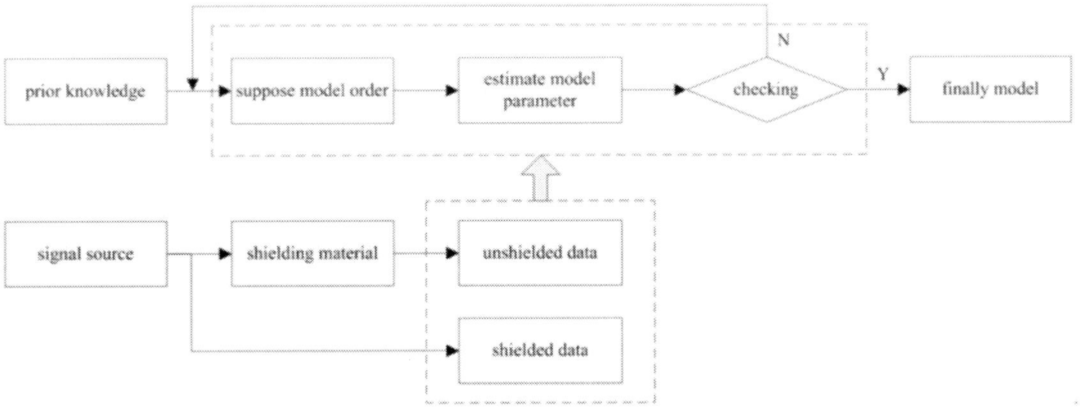

Figure 1. Modeling flow chart of SE of materials against EMP.

SI is to determine an equivalent system in the specified system on the basis of I/O data [8]. Differential equation is the most basic model for identification of linear discrete systems. The relationship between input ($u(k)$) and output ($y(k)$) sample value sequence of an dynamic system can be expressed as the following n-order linear differential equation (1):

$$y(k)+a_1y(k-1)+\cdots+a_ny(k-n)=b_0u(k)+b_1u(k-1)+\cdots+b_nu(k-n) \tag{1}$$

Equation (1) is the Auto-Regressive Moving Average model, referred to as the ARMA model. Equation (1) is operated by Z transform. In zero initial conditions, the ratio of the output z-transform and the input z-transform is the system transfer function.

$$G(z)=\frac{Z\{y(k)\}}{Z\{u(k)\}}=\frac{b_0+b_1z^{-1}+\cdots+b_nz^{-1}}{1+a_1z^{-1}+\cdots+a_nz^{-1}} \tag{2}$$

Equation (2) can be abbreviated as

$$A(z^{-1})y(k)=B(z^{-1})u(k) \tag{3}$$

$$A(z^{-1})=1+a_1z^{-1}+\cdots+a_nz^{-1}\text{, } B(z^{-1})=b_0+b_1z^{-1}+\cdots+b_nz^{-1}$$

When you consider the noise term $\varepsilon(k)$, equation (3) is changed to equation (4)

$$A(z^{-1})y(k)=B(z^{-1})u(k)+\varepsilon(k) \tag{4}$$

Equation (4) is called the ARX (auto-regressive model with exogenous input) model.
Equation (3) can become

$$y(k) = \frac{B(z^{-1})}{A(z^{-1})}u(k) + \varepsilon(k) \tag{5}$$

Equation (5) is the OE (Output Error) model.

Above all is the commonly used model in the linear system. And it is also used in this paper. The combination of various improvements or optimization algorithm can get better results. SI is transformed into estimation of model parameters used in the least squares method, through constant change model parameters, making it the closest to the actual system parameters, and the final system model is got. The selected error criterion is the basis of the model close to the actual system. And it is one of the important elements of the identification problem. In this paper, the fitting between the predict waveform and simulation waveform is finally used to judge the pros and cons of the model. In addition, the model order selection should be the appropriate. When precision of the model meets the requirements, it should be chosen by the lower order. The order of the model in this paper is less than 10.

3. Time-domain modeling for SE of materials against EMP

Time-domain linear cosine sweep signal is used as the modeling signal. And 2ns/50ns EMP and square wave EMP is used for validation signal. SE measurement of infinite planar materials and coaxial test and windowed semi-anechoic box is simulated by CST Microwave Studio software. The time-domain waveform shielded and unshielded of three different cases is recorded. And the data is used to validate the proposed method in this paper. Linear cosine sweep signal for the expression:

$$f_i(t) = f_0 + \beta t \quad (\beta = (f1 - f_0)/t_1) \tag{6}$$

2ns/50ns double exponential expression of EMP as:

$$y(t) = k(e^{-\alpha t} - e^{-\beta t}) \tag{7}$$

Detailed parameters of the three simulation signals are listed in table 1. The square wave EMP is determined by the rising time (t_r), falling time (t_f) and flat-topped to keep (t_h). The linear sweep frequency signal is described by sweep time range (t), end time (t_1), initial frequency (f_0) and end frequency (f_1).

Table 1. Parameters of three simulation signal.

	Modeling signal	Validation signal	
	Linear sweep frequency signal	2ns/50nsEMP	Square wave EMP(ns)
parameter of signal	t = 0-200 ns, f_0 = 0 Hz, t_1=200 ns,f_1=1 GHz	k=1.09,α =1.54×10^7, β =9×10^8	t_r=1, t_f=4, t_h=47

The thickness of 1 mm and the conductivity of 1000 s m^{-1} of isotropic material, is always used in the simulation. Most energy of EMP is concentrated in the sub-1 GHz, so the frequency range calculated in simulation is 0 to 1 GHz.

3.1. Method of coaxial test

Shielding of infinite planar material against vertical incidence plane wave is the classic problem of electromagnetic shielding. Through research of infinite planar materials, we can see that the proposed method in this paper is feasible in theory. This is not surprising that the minimum-phase method also succeeds in solving this problem. Only using 2-order Model can be achieved very good results, fitting up to 99.71 %. So we no longer display the results.

Simulation model of coaxial test device is established, in accordance with GJB6190-2008. One end is excitation port, the other end is load port, impedance is 50 Ω. Three signals have been given in table1, respectively, in setting the excitation port source required for modeling data and validation data. ARX model identification, take the model order $n_a = 1{:}10$, $n_b = 1{:}10$, the pure delay $n_k = 1{:}10$, found that when $n_a = 6$, $n_b = 4$, $n_k = 10$ and $n_a = 10$, $n_b = 5$, $n_k = 5$, the model output and the actual simulation output agreement well, the fitting of IEC waveform, respectively 88.56 % and 88.37 % (figure 2a), but still a poor prediction of square wave EMP. OE model is used to model, assuming no delay, ie, $n_a = n_b$. When the order is 4, the fitting of 2ns/50ns EMP and square wave EMP, respectively, achieve 97.67 % and 98.3 % (figure 2b). Transfer function of EMP SE is:

$$H(z^{-1}) = \frac{0.0002307z^{-1} + 0.0007235z^{-2} - 0.0007587z^{-3} + 0.0002661z^{-4}}{1 - 3.842z^{-1} + 5.546z^{-2} - 3.566z^{-3} + 0.8614z^{-4}} \tag{8}$$

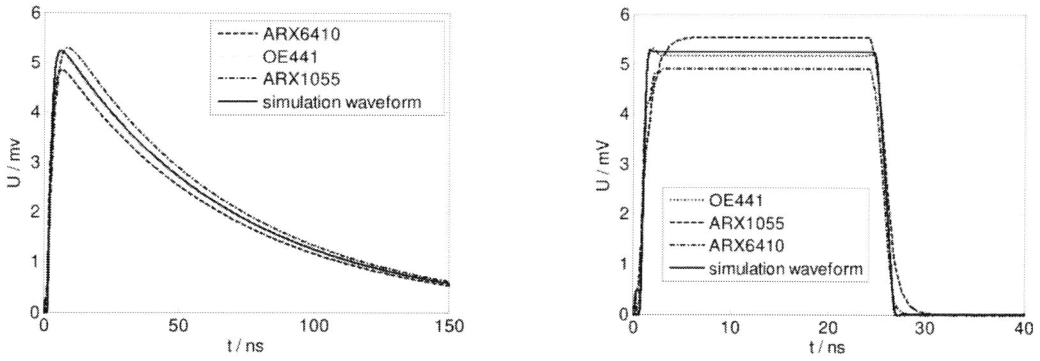

(a) (b)

Figure 2. Comparison of simulation shielded waveform and model prediction waveform using the method of coaxial test. (a) 2 ns/50 ns EMP; (b) square wave EMP.

3.2. Method of windowed semi-anechoic box

The method of windowed semi-anechoic box has been simulated based on IEC61587-3. Window size of 60 cm× 60 cm, box size of 1 m × 1 m × 1 m aluminium shielding box model was created by CST Microwave Studio. In order to restrain resonant caused by the box and control the space of absorber, hybrid absorber was mounted on the inner wall of the box according to the reference. The box is illuminated by a plane wave, plane wave along the window at normal incidence, the polarization direction of electric-field perpendicular to the ground. The proposed method in this paper is used to model the SE of materials against EMP using the windowed semi-anechoic box. It is found that the prediction waveform of other order of the ARX model is similar to the ARX441 model in figure 3a and figure 3b, but the proportion attenuation of unshielded materials. OE model can forecast better the 2ns/50ns EMP waveform unshielded materials, but is still large error for the forecast of square wave EMP, even if increased the model order. Although the box is loaded absorber using the windowed semi-anechoic box, there is still resonance. And taking into account the strong attenuation of the window for low-frequency electromagnetic waves coupled to box. There is much more complex than the two previous simulation examples. Maybe, the reason is that the model data cannot take enough information for the model of SE of materials against EMP.

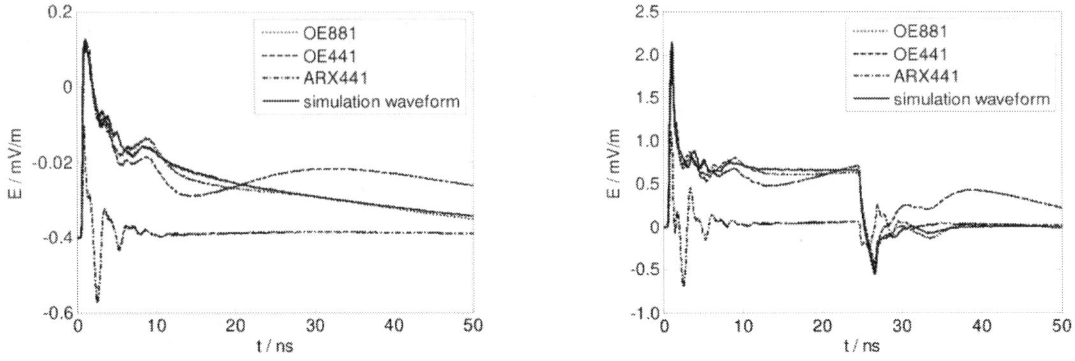

Figure 3. Comparison of simulation shielded waveform and model prediction waveform using the method of windowed semi-anechoic box. (a) 2 ns/50 ns EMP; (b) square wave EMP.

To this end, the phase 90°, 180°, 270° linear cosine swept signals are added to the original data. The combination signal is used for modelling the signal. 8-order OE model can well forecast the shielded time-domain waveform. The fitting degree respectively achieved 88.68 % and 89.66 % for 5ns/20ns EMP and square wave EMP. Transfer function of EMP SE is:

$$H(z^{-1}) = \frac{0.00249z^{-1} - 0.007811z^{-2} + 0.004685z^{-3} + 0.01377z^{-4} - 0.03254z^{-5}}{1 - 3.773z^{-1} + 5.269z^{-2} - 2.383z^{-3} - 2.371z^{-4} + 4.108z^{-5}} \quad \frac{+0.03307z^{-6} - 0.01757z^{-7} + 0.003913z^{-8}}{-2.07z^{-6} + 0.01251z^{-7} + 0.2079z^{-8}} \tag{9}$$

Overall, OE model is better than the ARX model, and the model order is lower. It is very important to select the appropriate modeling signal that can sufficiently excitation the system properties. Only in this way we can get a good generalization ability of the model.

4. Conclusion

In this paper, three kinds of simulation data of SE of materials are used to research the model of SE of materials against EMP based on method of SI. Model of SE of materials against EMP is successfully established through time-domain low field strength (voltage) of cosine frequency sweep signal. The time-domain waveform of a typical EMP through the shielding material is successfully predicted. It will lay the foundation to predict SE of materials against variety strong EMP

The next step will be to study the modeling method in engineering practice, such as the rational design of SE of materials against EMP in measured experiment, data preprocessing, and so on.

Acknowledgments

This work is supported by National Science Foundation of China under the grant number of No. 51177174 and 61001050.

References

[1] Andrew C. M, John F. D and Simon W 2004 *IEEE Trans EMC.* **46** 459
[2] Li Y X, Chen B and Shi L H 2006 *High Power Laser and Particle Beams (in Chinese).* **18** 972
[3] Zhou B H, Gao C and Ren H M 2003 *CEEM (Hangzhou)* vol 1 (IEEE) p562

[4] Zhou B H, Gao C and Chen B 2000 *CEEM (Shanghai)* vol 1 (IEEE) p296
[5] Shi Li-hua, Zhou Bi-hua and Chen Bin 2000 *Chinese journal of radio science.* **15** 467
[6] Xie Yan-zhao, Wang Zan-ji and Wang Qun-shu 2004 *High Power Laser and Particle Beams (in Chinese).* **16** 320
[7] Guo Jian, Zou Jun and He Jin-liang 2004 *Tsinghua Uni (Sci&Tech).* **44** 1302
[8] Ljung L 2002 *System identification: Theory for the user* (Beijing, Tsinghua University Press)

Journal of Physics: Conference Series **418** (2013) 012009 doi:10.1088/1742-6596/418/1/012009

Optimization design of electromagnetic shielding composites

Zhaoming Qu*, Qingguo Wang, Siliang Qin, Xiaofeng Hu
Research Institute of Electrostatic and Electromagnetic Protection,
Mechanical Engineering College, Shijiazhuang, Hebei province, China

E-mail: iamqzm@gmail.com

Abstract. The effective electromagnetic parameters physical model of composites and prediction formulas of composites' shielding effectiveness and reflectivity were derived based on micromechanics, variational principle and electromagnetic wave transmission theory. The multi-objective optimization design of multilayer composites was carried out using genetic algorithm. The optimized results indicate that material parameter proportioning of biggest absorption ability can be acquired under the condition of the minimum shielding effectiveness can be satisfied in certain frequency band. The validity of optimization design model was verified and the scheme has certain theoretical value and directive significance to the design of high efficiency shielding composites.

1. Introduction

The prediction of the physical properties (e.g., the dielectric constant, the thermal or electrical conductivity, and the effective elastic constants, etc.) of heterogeneous media has been the subject of scientific and engineering interest spanning more than a century. It has attracted the attention of such luminaries as Maxwell, Rayleigh, and Einstein [1–3]. Many predictive schemes have been well documented. Among these schemes, the Maxwell formula1 for the conductivity of a heterogeneous medium containing a very dilute suspension of spheres is the earliest and most well-known one. Since then, the corresponding formulas for the conductivities of media containing ellipsoids [4] and parallel circular cylinders [5] have been developed. In the current literature, the Maxwell formula for spheres and Fricke's expressions for ellipsoids are the results of the so-called Maxwell-Garnett effective medium theory [6]. These results hold for very low volume fractions of the dispersed constituents. Recently, Giordano [7] predicted the effective permittivity of two-phase heterogeneous media containing aligned or randomly oriented ellipsoidal inclusions by using the Maxwell formula for ellipsoids and Bruggeman's scheme.

However, the effective electromagnetic parameters of composite materials for predicting work focused mainly on parameter prediction itself, not with the shielding performance together, maked effects of composite microstructure on macroscopical shielding property not intuitive enough. Therefore, the physical model of shielding composite material was proposed and the prediction equation of effective electromagnetic parameters can be obtained based on micromechanics and variational principle, and then using the plane wave transmission theory identifies multilayer shielding effectiveness of materials and reflectivity and equivalent electromagnetic parameters of quantitative relationship and for multilayer composite shielding for multi-objective optimization design.

* To whom any correspondence should be addressed.

Optimization results show that the proposed model is feasible. The scheme proposed here can guide material optimization design, has important theoretical and practical significance.

2. Effective electromagnetic parameters prediction of composite material

Considering a macroscopically homogeneous isotropic material, is filled with ellipsoidal particle filler. Hypothesis of matrix material permittivity is ε^1, the permittivity of filling particles is ε^2, the effective permittivity of two-phase composites is $\bar{\varepsilon}$. In order to determine the effective permittivity of composite materials, a homogeneous material ε^0 was introduced as reference material. Hypothesis of the existence of the polarization field τ, there are:

$$\tau = \bar{\varepsilon} \cdot E^0 - \varepsilon^0 \cdot E^0 \quad (1)$$

$$\tau^2 = \varepsilon^2 \cdot E^2 - \varepsilon^0 \cdot E^2 \quad (2)$$

E^0 is Electric field intensity outside, E^2 is Electric field intensity of filling phase, and τ^2 is polarization field of filling phase. Draw lessons from the mechanics calculation of elastic modulus method [8], in volume V Introduce Green function G^0, can be:

$$E^2 = E^0 - \int_V \mathbf{G}^0(\mathbf{x}-\mathbf{x}')\cdot[\boldsymbol{\tau} - \langle\boldsymbol{\tau}\rangle]dV' \quad (3)$$

$\langle\boldsymbol{\tau}\rangle$ is average of $\boldsymbol{\tau}$ in the volume V. By using statistical average method, the polarization field is expressed as:

$$\boldsymbol{\tau}(\mathbf{x}) = \lambda^1 \cdot \boldsymbol{\tau}^1 + \lambda^2 \cdot \boldsymbol{\tau}^2 \quad (4)$$

$\lambda^2(x)$ is characteristic function of filling particles, When x in the filling phase is equal to 1, or equal to 0, and the filling phase concentration is $f^2 = \langle\lambda^2\rangle$. The ensemble average of formula (3) is:

$$\langle E\rangle^2 = E^0 - \frac{1}{f^r}\sum_{s=1}^{2}\mathbf{A}^{rs}\cdot\boldsymbol{\tau}^s \quad (5)$$

The interaction parameter of micro structure $\mathbf{A}^{rs}$ of inter-particle [9] is:

$$\mathbf{A}^{rs} = \left\langle \int_V \lambda^r(x)[\lambda^s(x') - f^s]\mathbf{G}^0(\mathbf{x}-\mathbf{x}')dV' \right\rangle \; (r,s=1,2) \quad (6)$$

Arranging formula (5) is:

$$E^0 = \frac{\boldsymbol{\tau}^r}{\varepsilon^r - \varepsilon^0} + \frac{1}{f^r}\sum_{s=1}^{2}\mathbf{A}^{rs}\cdot\boldsymbol{\tau}^s \; (r,s=1,2) \quad (7)$$

In order to simplify the calculation, we consider only the two-point interaction of material filler particles, ignoring the multi-point effect. Here assuming that the particles in the composite do not lap joint. According to Hashin-Shtrikman [10] variational theory, when the composite filler concentration is not very high, can choose the matrix as a uniform reference material, i.e.:

$$\varepsilon^0 = \varepsilon^1 \quad (8)$$

This can be:

$$\mathbf{A}^{22} = f^2\left(\frac{\mathbf{S}^2}{\varepsilon^1} - f^2\frac{\mathbf{S}^v}{\varepsilon^1}\right) \quad (9)$$

$\mathbf{S}^2$ is particle shape factor, which depends on its structure and size, $\mathbf{S}^v$ is particle distribution factor, when the filling phase distributes randomly, $\mathbf{S}^v = 1/3$, when the filling phase in the matrix distributes directional, its shape and orientation consistent is same as particles. Finally the Prediction formula can be obtained:

$$\bar{\varepsilon} = \varepsilon^1 + \varepsilon^1 \cdot [\frac{\varepsilon^1}{(\varepsilon^2 - \varepsilon^1)f^2} + \frac{\mathbf{S}^2}{f^2} - \mathbf{S}^{\mathrm{V}}]^{-1} \tag{10}$$

Based on the electromagnetic theory of the principle of reciprocity, the effective permeability of composite material is:

$$\bar{\mu} = \mu^1 + \mu^1 \cdot [\frac{\mu^1}{(\mu^2 - \mu^1)f^2} + \frac{\mathbf{S}^2}{f^2} - \mathbf{S}^{\mathrm{V}}]^{-1} \tag{11}$$

From the derivation process above-mentioned, in known filler microstructure information (concentration, size, shape and distribution) under the premise, formula (13), (14) can be used in predicting composite equivalent electromagnetic parameters (permittivity and permeability), and then to theoretically predict composite shielding effectiveness and reflectivity.

3. Shielding effectiveness calculation of composites

Figure 1 shows the physical model of composite electromagnetic shielding material, according to the electromagnetic wave theory, when the uniform plane electromagnetic wave vertical incidence the material surface, the input impedance of i layer of material is given:

$$Z_{in}^i = \eta_i \frac{Z_{in}^{i+1} + \eta_i \tanh(jk_i d_i)}{\eta_i + Z_{in}^{i+1} \tanh(jk_i d_i)} \tag{12}$$

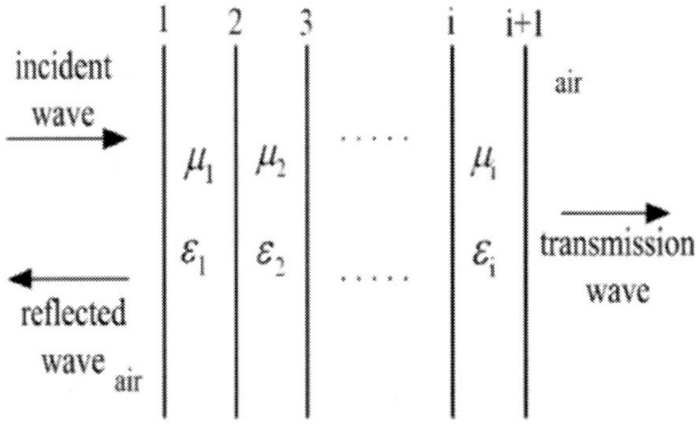

Figure 1. Model of multi-player shielding.

The characteristic impedance of i layer of material is $\eta_i = \sqrt{\bar{\mu}/\bar{\varepsilon}}$, the Propagation constant of i layer material is $k_i = 2\pi f \sqrt{\bar{\mu}\bar{\varepsilon}}$: $\eta_0 = \sqrt{\mu_0 / \varepsilon_0}$ is vacuum wave impedance, d_i is thickness of i layer material, $\bar{\varepsilon}$, $\bar{\mu}$ is equivalent complex permittivity and complex permeability of i layer material, respectively, c is the speed of light in vacuum. When the $n+1$ layer of material is air, $Z_{in}^{n+1} = \eta_{n+1} = \eta_0$, therefore, the reflection coefficient of i layer interface is $R_i = \frac{Z_{in}^i - \eta_{i-1}}{Z_{in}^i + \eta_{i-1}}$, transmission coefficient is $T_i = \frac{2Z_{in}^i}{Z_{in}^i + \eta_{i-1}}$, then the total transmission coefficient of multilayer materials is given:

$$T = \prod_{m=1}^{i+1} T_m \Bigg/ \prod_{m=1}^{i} (e^{-jk_m d_m} + R_{m+1} e^{jk_m d_m}) \tag{13}$$

The shielding effectiveness of multilayer materials is given:

$$SE = -20\lg\left|\prod_{m=1}^{i+1} T_m \Big/ \prod_{m=1}^{i}(e^{-jk_m d_m} + R_{m+1}e^{jk_m d_m})\right| \tag{14}$$

The reflectivity of multilayer materials is given:

$$R = \frac{Z_{in}^1 - \eta_0}{Z_{in}^1 + \eta_0} \tag{15}$$

According to the above analysis, the application of multilayer material composite material's shielding effectiveness calculation model can establish the material equivalent electromagnetic parameters and shielding characteristics of the quantitative relationship, the shielding mechanism is more accurate. At the same time, combined with the composite material equivalent electromagnetic parameters prediction models, the procedure of the composite material microstructure to predict the macroscopic equivalent electromagnetic shielding performance design can be realized.

4. Optimization design of shielding composites

Genetic algorithm is a kind of genetics and survival of the fittest mechanism based on global optimization method, it has a robust, efficient features and suitable for processing multiple target optimization design problem of complex system. The relationship of equivalent electromagnetic parameters (ε, μ) of material and filler particle size and concentration is presented above, thus the genetic algorithm can be applied for optimization design the microstructure of composite shielding material, then getting the optimal fill size, concentration and the thickness of each layer, to avoid the blindness of material processing.

The optimizing target set for the given number, maximum thickness, bandwidth and minimum shielding effectiveness under conditions of maximum absorption loss, based on which the layer thickness and the size of the packing and the concentration parameters were optimized. The detail optimization objective is: layers respectively 1, 3 and 5, in determined frequency range $f_{\min} \leqslant f_{\max}$, the shielding effectiveness of $SE=F(\gamma, f, d)$ is not less than 70 dB. Requirements: $d = \sum_{i=1}^{5} d_i$ is less than a given maximum thickness of each layer, which γ is aspect ratio of fillers, f is concentration of each layers' filler, and d is given as thickness of each layer material.

Table 1. Optimization result of three layer material.

layer	thickness	concentration	aspect ratio
1	0.1 mm	24.6 %	5.321
2	0.6 mm	20.7 %	10.48
3	3.3 mm	30 %	444.4

Table 2. Optimization result of five layer material.

layer	thickness	concentration	aspect ratio
1	0.25 mm	22.97 %	11.13
2	0.1 mm	23.68 %	3.2133
3	0.25 mm	23.96 %	4.0683
4	0.1 mm	27.22 %	9.6339
5	3.3 mm	30 %	444.4

According to the above optimization strategy, set the initial population number is 20, the maximum algebra is 500, and mutation rate is 0.01, thickness of shielding material limit below 4 mm. The single layer material optimization results as follows: the thickness is 4 mm, the concentration is 0.3, and the aspect ratio is 444.4. The optimization results of multilayer composite as shown in Table 1and table 2.

Table 1 and table 2 give the optimized results of nickel plated carbon fibre aspect ratio, concentration and thickness of each layer under the condition of shielding effectiveness more than 70dB and the absorption maximum of the materials between the frequency range of 2 ~ 18 GHz. The shielding effectiveness and reflectivity of figure 2 and figure 3 indicate that this paper proposed the multi-objective optimization design method can optimize the electromagnetic shielding material in meet specific shielding characteristics and give optimal material parameters matching, then design and guide the preparation of broadband layer electromagnetic shielding composite materials.

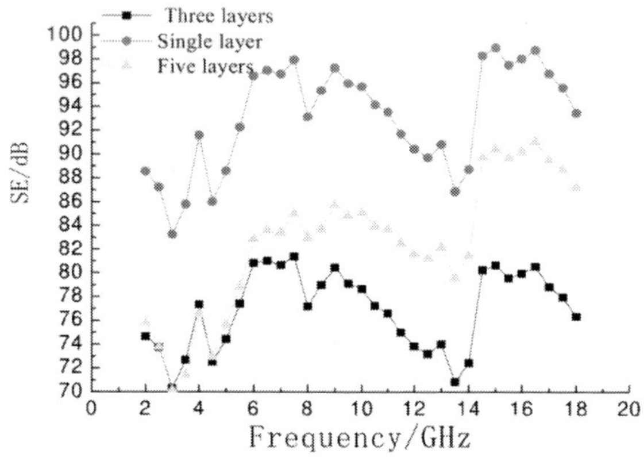

Figure 2. Comparison of shielding effectiveness.

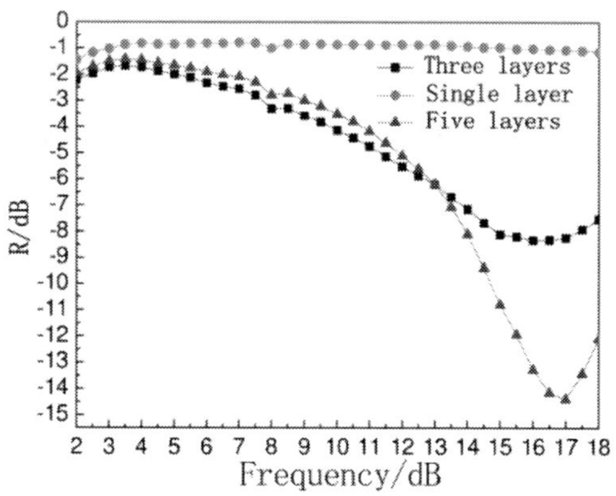

Figure 3. Comparison of reflectivity.

5. Conclusions

The effective electromagnetic parameters physical model of composites and prediction formulas of composites' shielding effectiveness and reflectivity were derived based on micromechanics, variational principle and electromagnetic wave transmission theory. The multi-objective optimization design of multilayer composites was carried out using genetic algorithm. The optimized results indicate that the results show that the proposed optimization method is feasible. The proposed scheme has certain theoretical value and directive significance to design low filler concentration, low cost and high efficiency shielding composites.

References

[1] Maxwell J C *Treatise on Electricity and Magnetism* (Clarendon, Oxford, 1873)
[2] Rayleigh J W 1892 *Philos. Mag.* **34** 481
[3] Einstein A 1905 Ann. *Phys.* **19** 289
[4] Fricke H 1924 *Phys. Rev.***24** 575
[5] Van Beek L K H , 1967 *Prog. Dielectr.* **7** 69
[6] Giordano S 2003 *J. Electrost.* **58** 59
[7] Bruggeman D A G 1935 *Ann. Phys.* **24** 636
[8] Mura T. *Micromechanics of defects in solids* (Martinus Nijhoff, Dordrecht, 1987)
[9] Ponte P and Willis J R 1995 *J. Mech. Phys. Solids* **43** 1919
[10] Hashin Z and Shtrikman S 1963 *J. Mech. Phys. Solids* **11** 127

The theory research and electric field simulation of the AHPC

Minkang Tang *, **Yanling Ma**, **Ping Yang**
Faculty of Resources and Environmental Engineering of Jiangxi University of Science and Technology

Email: tmkjxust@126.com

Abstract. Diameter of different corona electrode, distance of lines and distribution of the electric field of the porous dust collection plate of the AHPC electrostatic fabric filter were studied and analyzed on the basis of MATLAB PDE Toolbox. The simulation results of the AHPC electric-bag composite dust collector electrical fields show that when the corona polar diameter was 4 mm and distance between neighboring electrodes was 196 mm, the performance of the electric field was better. And the larger the aperture was, the higher the electric potential of the porous plate would be. Compared with the experimental research, it is simpler, visual and economical to use the MATLAB PDE to find the electric-field distribution of the electric-bag composite dust collector, it can provide theoretical basis for design and modification of corona wire diameter, line-spaces and precipitating plate aperture.

1. Introduction

In recent years, scholars at home and abroad have conducted lots of research work of the electric-bag dust collector [1]. The results show that electric-bag dust removal technology got the advantages of the electrostatic precipitator and the bag filter. Furthermore, new performance advantages will come into appearance when the advantages of the above two filters were put together. This made a complementary of the advantages of the dust removal come true, and has a good application prospect [2]. The research and development of the electric-bag composite will make an important contribution to precipitators in coal-fired power plants in China during the "energy conservation and emission reduction" environmental protection policy period.

The studies of this paper provided theoretical bases and the reference for the optimal design of the electric-bag composite precipitator and helped to improve its overall performance and dust removal efficiency and promote efficient filter design and application. It has a very important practical significance in reducing dust emission concentrations and atmospheric pollution.

2. The research of the AHPC theory

2.1. The AHPC

* To whom any correspondence should be addressed.

The AHPC was designed by the Energy and Environmental Research Center of the University of North Dakota ("EERC" for short) in 1999, the overhead view is shown in figure 1. In the AHPC dust removal, porous plates are arranged on both sides of the filter bag, with alternating electrodes and filter bags [3-4].

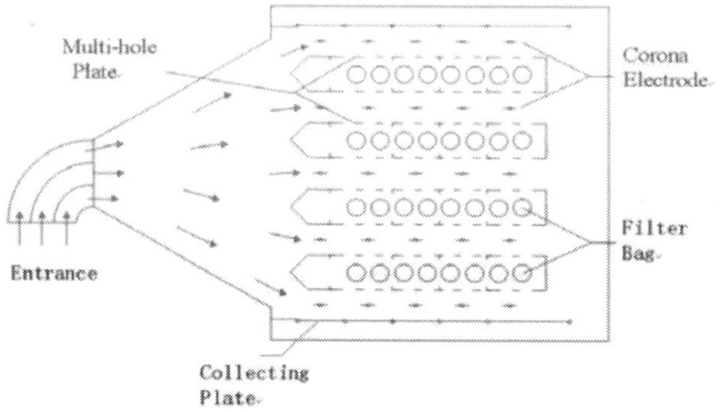

Figure 1. Schematic diagram of AHPC type electric-bag dust remover.

2.2. *Mechanism of dust removal*

As a new and effective dust collector, it owns the dust removal mechanisms and technics of electrostatic precipitator and bag filter (figure 2).

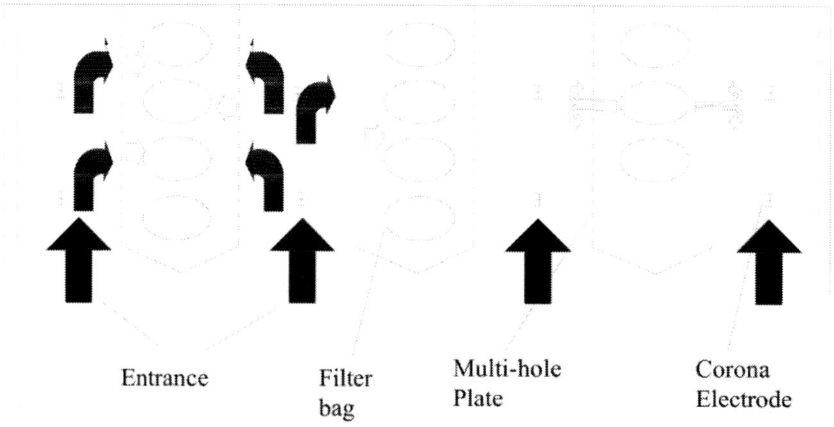

Figure 2. AHPC dust removal mechanism.

80%-90% of the amount of dust is collected when dust-containing gas flows through the ESP area, and then the remaining charged dust passes through the plate holes to the back filter bag by the electric field, and is collected by the filter bag to ensure a low concentration of dust emissions.

When dust is removed by the bag filters, there will be a backflow through the porous plate of the dust. Because there is a condensation of the dust collected on the surface of the filter bag, the size of the dust enlarged, the dust can be caught effectively by the collection plate in the electric field. And the probability that the dust will return to the filter bag decreased greatly by avoiding secondary adsorption of the filter bag. In the same way, dust which did not fall into the dust ash hopper may enter the filter area and be collected by the filter bag. Some shortcomings that might result in a decline of dust removal efficiency are overcome. The dust removal efficiency of the AHPC is improved. The porous boards of AHPC have two functions apart from capturing charged dusts, and another important role is that it can protect the filter bag from damage of corona discharge. The equipment has a long-term and stable operation, the dust removal efficiency can be higher than 99.99%, and the resistance can maintain at 2000 Pa or even less.

The AHPC mainly uses pre-charged dust to increase the performance of bag filtration. The technology which combines the pre-charged dust technology and the bag filter can improve the dust removal efficiency, reduce the loss of pressure and make dust removal easier. The filter pressure is small mainly because of the coagulation of the charged dust and the charged dust layer structure on the surface of the filter bag. The analysis is shown in figure 3.

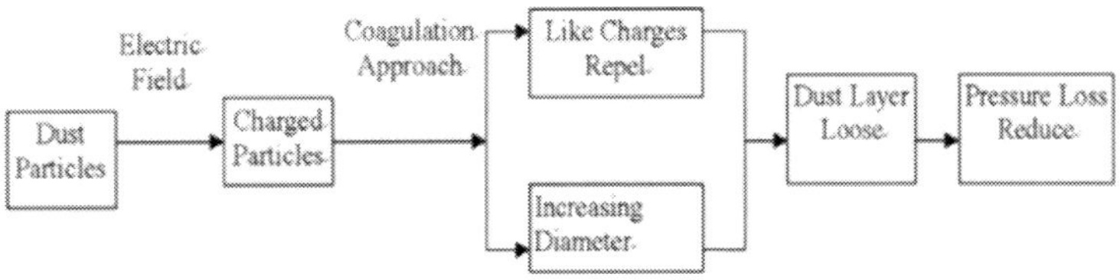

Figure 3. The influence of charged dust to pressure loss.

The charged dust layer structure on the surface of the filter bag is shown in figure 4. As shown in figure 4, the charged dust arrived the filter materials surface earlier repels the one arriving later, therefore, deposition on the surface of the filter bag forms a loose and uneven structure. This kind structure has good air permeability and great porosity. The uncharged dust on the surface of the filter bag is a close-grained flat structure (figure 5).

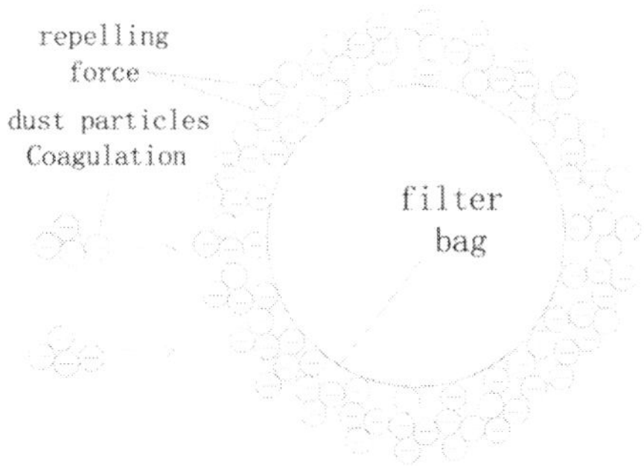

Figure 4. deposition of charged dust

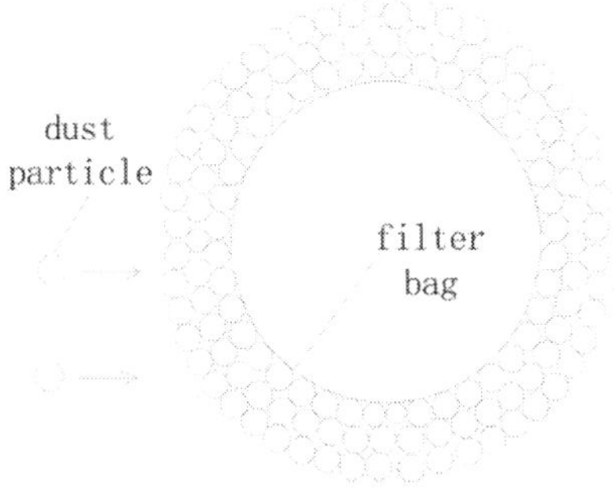

Figure 5. deposition of uncharged dust

3. Unit model of the AHPC

3.1 Model parameters of the bag filter unit

The four circular bags are used. The diameter is 130 mm and the length is 6 m with an area of 9.8 m^2. Filtering wind speed is 2.4 m min^{-1}; the dust capacity is 23.52 m^3 min^{-1}.

3.2. The model parameters of electrostatic precipitation unit

When filtering wind speed is 2.4 m min^{-1}, the plate-to-cloth ratio (the ratio of porous plate dust collecting area and the filter bag area) should be 1.151, the filtering area is 9.8 m^2, so the effective area of the dust collecting board is 11.28 m^2, the porous plate opening rate is 40%, and the dust collecting board length is 1.565 m.

(1) The channel width of the electric field (the distance between two adjacent plates) is 300 mm.

(2) The diameter of corona lines is 2 mm, 4 mm and 10 mm, respectively.

(3) According to the corona line number, line spacing is determined. The Corona line numbers were 6, 7 and 8, and the corresponding line spacing is separately 224 mm, 196 mm and 174 mm.

The porous plate is 2 mm thick, and the open hole is round. According to the dust collecting effective area and the plate opening rate, the total area of holes is 7.52 m^2. The hole diameter are 25 mm, 51 mm and 64 mm. According to the geometry of the dust collecting board, we calculate the number of holes with different diameter and place the holes equably.

The final geometry arrangement and the relevant parameters of the 2D model are shown in figure 6. The diameter of the corona line is 4 mm, line spacing is 196 mm, and aperture is 51 mm.

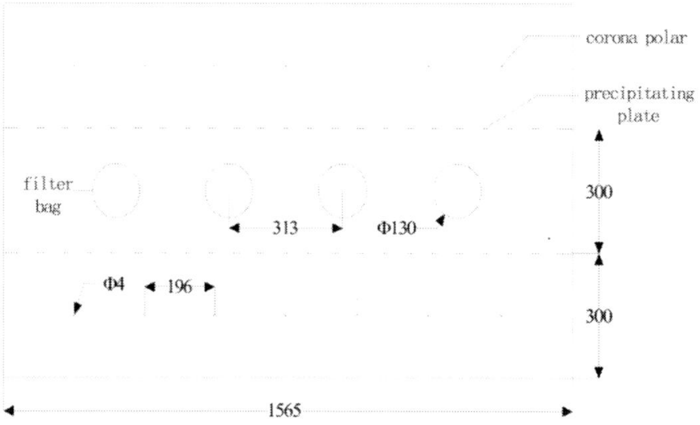

Figure 6. The 2D model.

4. Simulation of the AHPC electric field

Due to the structure complex of AHPC and the strong nonlinear process discharge of the corona, it is very difficult to obtain the analytical solution of the electric field. The PDE toolbox is applied to simulate the simplified 2D model of the electric field distribution of the AHPC and to analyze the theory. Compared with the electric field performance in different electrode structure configurations, certain bases and the references are provided to optimize and improve the AHPC.

4.1. Profile of PDE toolbox

PDE toolbox is a set of tools which is used to solve a partial differential equation based on MATLAB. Its core is a program which solves partial differential equations with the edge boundaries defined by the finite element method. The PDE toolbox has some professional communication modes, such as the electrostatic field, the static magnetic field and the electromagnetic field. The powerful numerical calculation functions made it is convenient for to solve the electric field distribution, and the result is graphic, intuitive audio-visual [5-7].

4.2. Simulation of the AHPC

The simulation models are built based on certain assumptions. The simulation process: build the calculation model in the GUI of the PDE toolbox, as shown in figure 7, choose electrostatics mode, set the dielectric constant to be 1, charge density of the body to be 0, and then set the boundary conditions, discrete grid, drawing solution. The simulation results is shown with graphics. 60 kV negative voltage is added as the boundary conditions of the electrode surface when the dust collector operates normally. The dust collector wall, the filter bag and orifice is grounded, and the potential of them is zero.

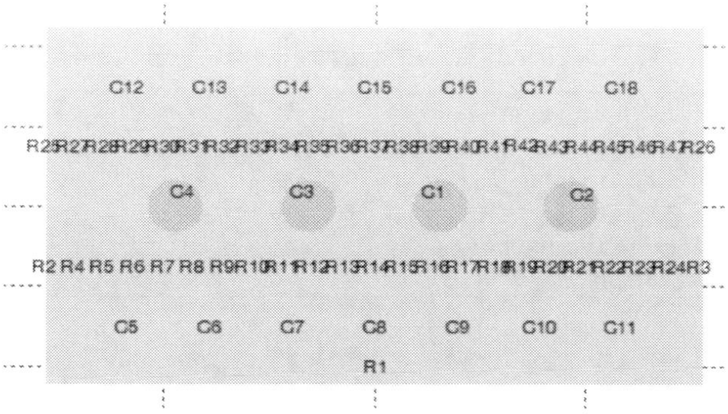

Figure 7. The simulation model.

4.3. The simulation result analysis

In the AHPC two-dimensional model, the electric field intensity of the negative corona discharge under different electrode parameters is derived from the isoline map of the field, as listed in table 1.

Table 1. Different electrode parameters of maximum value of the field.

Corona line diameter (mm)	Electrode distance (mm)	Aperture (mm)	Field Biggest ($kV\ cm^{-1}$)
2	196	51	143
4	196	51	75
10	196	51	27
4	174	51	70
4	196	51	75
4	224	51	68
4	196	25	69
4	196	51	75
4	196	64	71

(1) The influence of corona polar diameter on electric field distribution

When the corona line distance is 196 mm, pore diameter is 51 mm, with the increase of corona wire diameter, the electric field strength value near the electrode reduce and then the corona electrode diameter is smaller, and the corona discharge ability is stronger.

According to the Peek formula, the varying electric field strength of the electrode can be calculated [8]. The values under the different corona electrode diameters are calculated in table 2, and the electric field strength of the corona discharge is also listed.

It is known that the largest field strength of the corona discharge is less than the electric field intensity when the corona line is 10 mm in diameter, thus it cannot produce normal corona discharge. From the electric field non-uniform coefficient formula, it is observed that the average of the field is determined; the uneven electric field non-uniform coefficient is greater when the maximum value of the field is higher. In order to improve the breakdown voltage, we must minimize the non-uniform extent of the electric field. The maximum value of electric field strength is not 'the bigger the better', when achieve the varying electric field strength. Through analysis, a corona line of 4 mm in diameter is more appropriate under the model parameters, and has enough mechanical strength.

Table 2. The electric field strength under different diameter of the corona wire.

Corona Electrode Diameter (mm)	2	4	10
Electric Field Strength (kV cm^{-1})	143	75	27
Circulating Electric Intensity (kV cm^{-1})	48	42	36

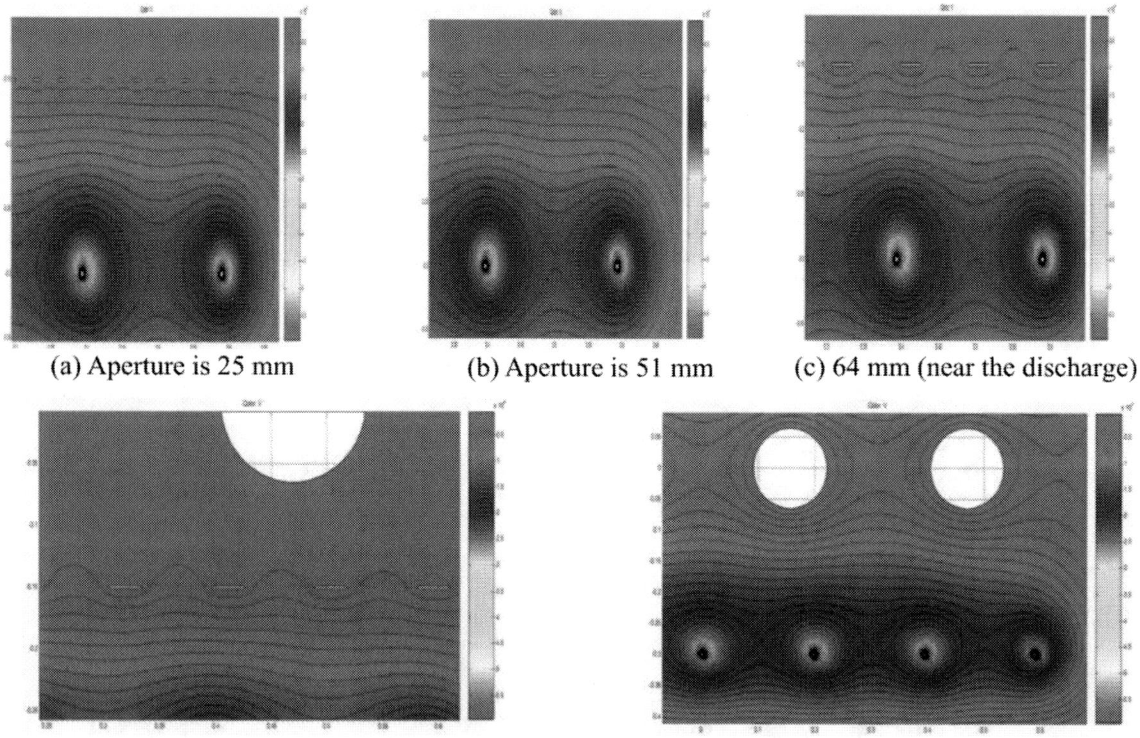

(a) Aperture is 25 mm (b) Aperture is 51 mm (c) 64 mm (near the discharge)

(d) 64 mm (near the filter bag) (e) no porous plate

Figure 8. The partial enlarged view of potential isoline (V).

(2) The influence of the electrode distance on electric field distribution

When the corona wire diameter is 4 mm and the aperture is 51 mm, with the increase of corona wire distance, the largest electric field strength increases, and then decreases, but with little change. The maximum values of electric field strength are higher than the circulating electric strength for 42 kV cm^{-1}. When the electrode distance is small, the electric field of the corona wire mutually influences, restrains and weakens the corona electric field, the corona current reduced. It is a kind of phenomenon of electric field shield, some attention should be paid to designing the electric dust removal unit. Along with the increase of corona wire distance, the electric field strength value increases a little, and then drops. The wire distance is too many, even though each corona wire can discharge fully, and the current value of the single wire is larger, but the corona wires reduce in the electric field causing the total corona current to reduce. Therefore, there exists an optimal wire distance to produce the maximum corona current. If corona wire distance is less than or greater than the best value, the corona current will be reduced.

(3) The influence of the aperture on electric field distribution

① the relationship between the aperture of the dust collecting plate and the greatest electric field strength

When the corona polar diameter is 4 mm, and the distance between corona wires is 196 mm, with pore diameter in the multi-hole plate changes, the change of the greatest corona field is not obvious. Visibly, the size of the aperture has little impact on corona discharge.

② the influence of the aperture on electric potential distribution

Pore plate whose orifice aperture is 25 mm, 51 mm or 64 mm are distributed on both sides of the bag, the electric potential distribution of the bags without perforated plate around the bag, see in figure 8.

It can be seen from figure 8 that the potential of the filter bag in the protection of the perforated plate is very low, what's more, it is much lower than the one without the perforated plate. It is indicated that the smaller the hole's size is, the lower of electric potential of the perforated plate is. The simulation results show that even if the hole's diameter takes 64 mm of the maximum one, the filter bag can also be protected well from corona discharge damage.

5. Conclusions

The electric-bag composite dust collector has the advantages of the electric and bag dust collector. It can make a reasonable assignment of dust loading during dedusting and deashing. Compared with the experimental research and to stimulate with a computer program, it is simpler, visual and economical to use the MATLAB PDE to find the electric-field distribution of the electric-bag composite dust collector, it can provide theory basis for the choices of corona wire diameter, line-spaces and precipitating plate aperture of the design and modification.

Acknowledgments

This project is supported by National Natural Science Foundation of China (51064008).

References

[1] Miller S, Schelkoph G, Dunham G, Walker K and Krigmont H 1997 *Advanced Hybrid Particulate Collector-a New Concept for Air Toxics and Fine–Particles Control* (Washington, D.C: EPRI–DOR–EPA Combined Utility Air Pollution Control Symposium) p 25
[2] Guo X, Tang X, Zhao H and Zhu Y 2008 *North China electric power technology (in Chinese)* **9** 50
[3] Zhang X 2006 *meishan science and technology (in Chinese)* **1** 21
[4] Wang L, Wang L, Zhang X and Zhang Y 2008 *Electric power environmental protection (in Chinese)* **5** 1
[5] Zhang B, Chu J and Xu X 2005 *High voltage technology (in Chinese)* **31** 1
[6] Zhang Y and Tian M 2005 *Journal of The China Railway Society (in Chinese)* **27** 124
[7] Lu J and Shang T 2001 *The solution of partial differential equation by MATLAB* (Wuhan: University Press) 1 p 3
[8] Xu X and Zhu D 1996 *Gas discharge physics* (Shanghai: Fudan University Press) p 246

Time-domain simulation and waveform reconstruction for shielding effectiveness of materials against electromagnetic pulse

Xiao-feng Hu *, xiang Chen, Ming Wei
Electrostatic & Electromagnetic Protection Institute, Mechanical Engineering College, Heping West Road, No.97, Shijiazhuang, China

E-mail: snowfox2270@163.com

Abstract. Shielding effectiveness (SE) of materials of current testing standards is often carried out by using continuous-wave measurement and amplitude-frequency characteristics curve is used to characterize the results. However, with in-depth study of high-power electromagnetic pulse (EMP) interference, it was discovered that only by frequency-domain SE of materials cannot be completely characterized by shielding performance of time-domain pulsed-field. And there is no uniform testing methods and standards of SE of materials against EMP. In this paper, the method of minimum phase transfer function is used to reconstruct shielded time-domain waveform based on the analysis of the waveform reconstruction method. Pulse of plane waves through an infinite planar material is simulated by using CST simulation software. The reconstructed waveform and simulation waveform is compared. The results show that the waveform reconstruction method based on the minimum phase can be well estimated EMP waveform through the infinite planar materials.

1. Introduction

Electromagnetic pulse (EMP) is a phenomena of transient electromagnetic, with a characteristics of transient, wideband, high field strength, wide effective scope and so on, the coupling path through the antenna through an aperture, cable and other electronic equipment system to cause a serious threat. EMP is an important part of electromagnetic environment, which is described in time-domain. EMP includes electrostatic discharge (ESD), lightning electromagnetic pulse (LEMP), nuclear electromagnetic pulse (NEMP), high power microwave (HPM), intends to electromagnetic interference (IEMI) and ultra-wide band (UWB). With the extensive application of microelectronic devices, the sensitivity and vulnerability of electronic device or system against EMP are gradually increasing. It seriously limits the electromagnetic viability of electronic device or system. Therefore, research and analysis of anti-jamming capability of electronic equipment or system in complex electromagnetic environment has an important theoretical and practical significance.

Shielding effectiveness (SE) of materials of existing testing standards is usually described by continuous-wave measurement in frequency domain. Amplitude-frequency characteristic curve is used to characterize SE by measuring the response amplitude at each frequency point one by one in a system. It can be more comprehensive understanding shielding ability of materials to the measured

* To whom any correspondence should be addressed.

spectrum of electromagnetic waves based on the SE in frequency-domain. But with the discovery of high-power EMP interference, only by the SE of materials in frequency-domain cannot be completely characterized by time-domain pulsed-field shielding effect [1-4]. The method of minimum-phase transfer function and flow chart of reconstruction waveform is proposed to reconstruct shielding time-domain waveform based on the analysis of the waveform reconstruction method. Pulse of plane waves through the infinite planar materials is simulated by using CST simulation software. In order to verify the applicability of minimum-phase method, the reconstructed waveform is compared with the simulation waveform.

2. Analysis of the waveform reconstruction method

There is no phase information in continuous-wave test. Researchers want to estimate time-domain response from the actual measurement of amplitude-frequency characteristics, and establish relationship between time-domain SE and frequency-domain SE. M. H. Hayes and other researchers have made a detailed discussion that reconstruct the phase information using amplitude-frequency characteristics. It is generally believed that the phase-frequency characteristics can be uniquely determined by its amplitude frequency characteristics when the system meets the minimum phase condition. Assumed that the system is linear and satisfy the minimum phase condition, the relationship between the real and imaginary parts of transfer function is established by Hilbert transform, and then shielding time-domain waveform of EMP is reconstructed.

For a stable, causal real sequence $x[n]$, the complex cepstrum $\hat{x}[n]$ is defined as the Fourier transform of $x[n]$ to take the logarithm after the inverse Fourier transform.

$$\hat{x}[n] = \frac{1}{2\pi}\int_{-\pi}^{\pi} \ln[X(e^{j\omega})]e^{j\omega} d\omega \tag{1}$$

$X(e^{j\omega})$ will be written in polar form

$$X(e^{j\omega}) = \left|X(e^{j\omega})\right| e^{\arg[X(e^{j\omega})]} \tag{2}$$

Real inverted spectrum $x_c[n]$ is defined as the Fourier transform $x[n]$ of taking the logarithm after inverse Fourier transform.

$$x_c[n] = \frac{1}{2\pi}\int_{-\pi}^{\pi} \ln\left|X(e^{j\omega})\right| e^{j\omega} d\omega \tag{3}$$

Obviously, $x_c[n]$ is the Fourier transform of the real part of the inverse Fourier transform of $\hat{X}(e^{j\omega})$, and $x_c[n]$ is the even part of $\hat{x}[n]$. Transfer function to meet amplitude-frequency characteristics of the system has infinitely many, because the phase-frequency characteristics can be arbitrarily chosen, but only one of which is the minimum-phase system. Minimum-phase system is the discrete transfer function of the system of poles and zeros outside the unit circle.

The complex cepstrum $\hat{x}[n]$ is causal discrete, so $x[n]$ is transfer function of minimum-phase system. Log and its phase function of frequency-domain amplitude spectrum of minimum phase system discrete transfer function is a mutual Hilbert Transform, which $\ln\left|X(e^{j\omega})\right|$ and $\arg[X(e^{j\omega})]$ satisfy the following relationship:

$$\arg[X(e^{j\omega})] = -\frac{1}{2\pi}\int_{-\pi}^{\pi} \ln\left|X(e^{j\theta})\right|]\cot(\frac{\omega-\theta}{2})d\theta \tag{4}$$

$$\ln\left|X(e^{j\omega})\right| = \frac{1}{2\pi}\int_{-\pi}^{\pi} \arg[X(e^{j\theta})]\cot(\frac{\omega-\theta}{2})d\theta + \hat{x}[0] \tag{5}$$

Where $\hat{x}[0] = -\frac{1}{2\pi}\int_{-\pi}^{\pi} \ln\left|X(e^{j\omega})\right| d\omega$

If $\left|X(e^{j\omega})\right|$ is known, the phase spectrum can be got by the equation (4). The processing of singular point should be noticed in the same.

The real cepstrum $x_c[n]$ can be got by logarithmic from frequency-domain amplitude spectrum $|X(e^{j\omega})|$. And $\hat{x}[n]$ can be recovered by adding window function form the real cepstrum $x_c[n]$.

$$\hat{x}[n] = x_c[n]l_{\min}[n] \tag{6}$$

Where $l_{\min}[n] = 2u[n] - \delta[n]$

If the frequency-domain SE of materials is known, unshielded EMP is looked as the input excitation of system and response of EMP penetrating the materials is looked as output of system, then frequency-domain transfer function H(W) can be calculated using the equation (8). Assuming minimum-phase system, the time-domain transfer function is calculated based on the frequency-domain transfer function, then it convolution with unshielded EMP time-domain waveform. Shielded time-domain waveform *E*m(t) is got, it can be estimate the time-domain SE of materials. The calculation flow chart is shown in figure 1.

$$SE = 20 * \log 10 \frac{E_{unshielidng}(\omega)}{E_{shielidng}(\omega)} \tag{7}$$

$$H(w) = \frac{E_{shielidng}(\omega)}{E_{unshielidng}(\omega)} = 10^{-\frac{SE}{20}} \tag{8}$$

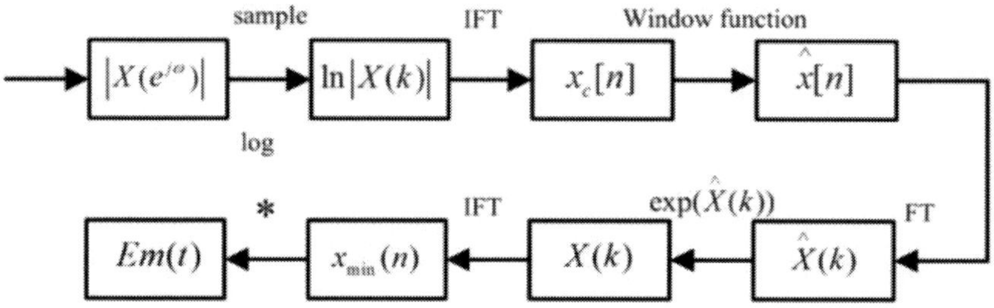

Figure 1. Modelling flow chart of SE of materials against EMP.

3. Simulation of SE of infinite planar materials against EMP

The waveform of plane wave through the infinite planar materials is calculated with the CST simulation software, the model shown in figure 2, the polarization direction of the plane wave + y, incident direction-z. Simulation with a uniform thickness of the material to take the 0.1 mm thick, the relative permittivity and relative permeability are 1, the conductivity of the material taken 10^3 s m^{-1}. Square wave EMP and IEC61000-4-4 standard EMP is used for excitation source signal. And the parameters of square wave EMP is rising edge 1ns, the falling edge 4 ns, flat top keeping 47 ns.

4. Comparison of reconstructed waveform and simulation waveform

The time-domain waveform of pulse through the shielding materials is calculated by the minimum-phase method, which was compared to simulation results. The results are shown in figure 4, in good agreement with the simulation results. It is known that waveform reconstruction method based on the minimum-phase can be well estimate the time-domain waveform of EMP through the planar materials.

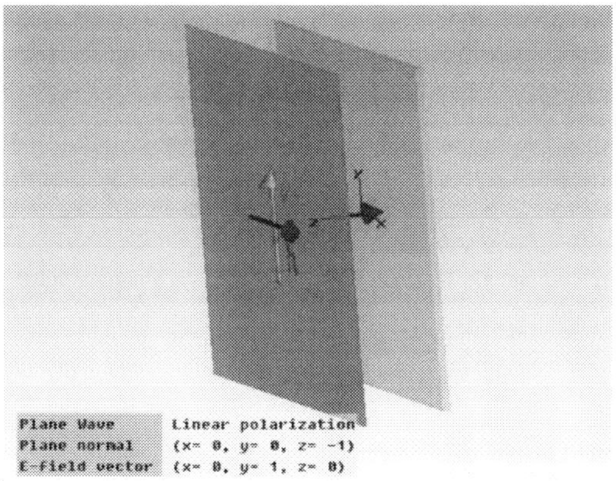

Figure 2. Infinite planar materials.

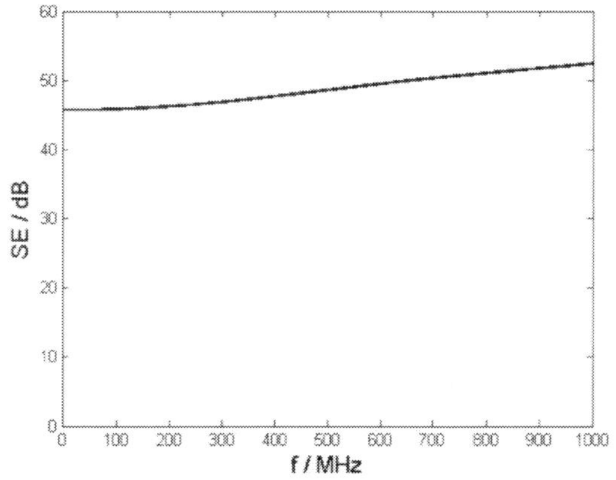

Figure 3. SE of infinite planar. Materials.

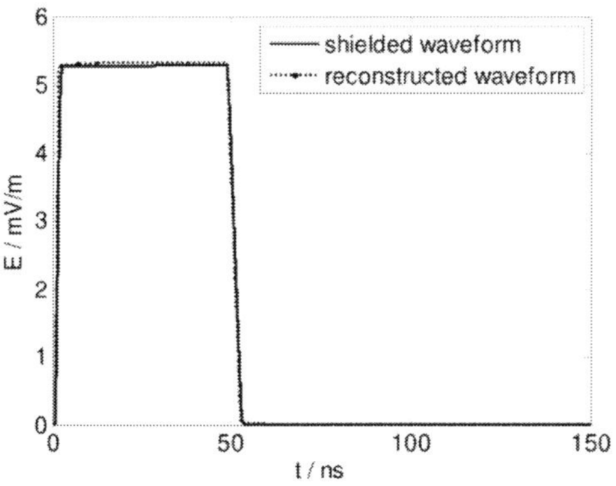

(a) square waveform EMP

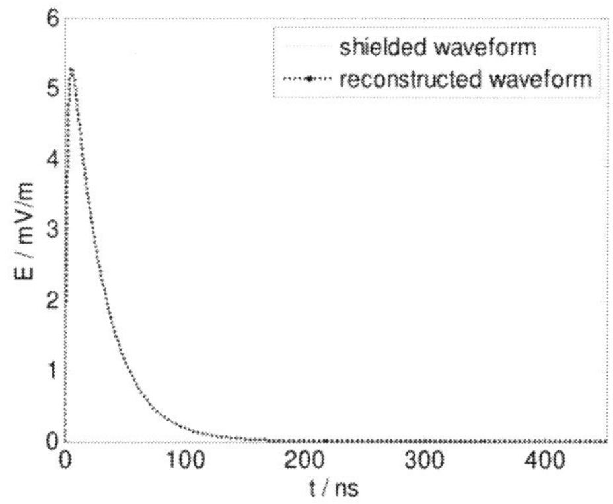

(b) IEC61000-4-4 EMP

Figure 4. Reconstructed waveform of two classical EMP through the shield.

Frequency-domain SE can provide a more comprehensive understanding of shielding ability of materials against each frequency band. But with the strong EMP interference research, it cannot fully characterize the SE of materials against EMP based on frequency-domain SE. Peak ratio SE is often used to characterize the SE of materials against EMP, and it cannot directly determine the peak ratio SE by frequency-domain SE. In addition, it cannot meet the needs of EMP protection design only by the time-domain peak ratio SE. Rise time, pulse width and other information of EMP is also important, except for the time-domain peak ratio SE. The use of minimum-phase method, you can get the materials to shield the pulse before and after the time-domain parameters change. Then, it can estimate the time-domain SE of materials, the results are listed in table 1.

Table 1. Time-domain parameter changes of EMP penetrate the planar materials.

	Peak of EMP(V m^{-1})			Rise time of EMP(s)	
	unshielded	shielded	SE(dB)	unshielded	shielded
Square EMP	1.0040	0.0053	45.5	7.9615e-010	8.2258e-010
IEC61000-4-4 EMP	1.0166	0.0053	45.7	4.9848e-009	4.9841e-009

5. Conclusions

In this paper, the method of minimum-phase transfer function is suggested to reconstruct shielding time-domain waveform based on the analysis of the waveform reconstruction method. Pulse of plane waves through the infinite planar material is simulated by using CST simulation software. And the reconstructed waveform is compared with the simulation waveform. The results show that the waveform reconstruction method can be well estimate the waveform of EMP through the infinite planar materials.

Acknowledgments

This work is supported by National Science Foundation of China under the grant number of No. 61001050 and 51177174.

References

[1] Ludger K 2005 *IEEE Trans. EMC.* **47** 589
[2] Bao Y F, Lu Y H, Zhang H X and Han C Y 2005 *Proc. Int. conf. on MAPE (Hangzhou)* vol 1 (IEEE) p 635
[3] William A R, Carl E B and Manuem W W 2004 *IEEE Trans. EMC.* **46** 314
[4] Andrew C M, John F D and Simon W 2004 *IEEE Trans. EMC.* **46** 459

Journal of Physics: Conference Series **418** (2013) 012013 doi:10.1088/1742-6596/418/1/012013

Numerical investigation of flow past a circular cylinder controlled by electromagnetic force

Huixing Liu, Benmou Zhou*, **Zongkai Liu, Jiansheng Dai**
Science and Technology on Transient Physics Laboratory, Nanjing University of Science and Technology, XiaoLingWei, Nanjing, China

E-mail: bmzhou@njust.edu.cn

Abstract. The flow past a circular cylinder controlled by Electromagnetic force is calculated by a finite volume method in order to investigate the control effect of the electromagnetic force on the three-dimensional vortical structures and the dynamics characteristics. The Reynolds number considered are from 200 to 1000, during which the wake transition occurs and the three-dimensional flow structures develop. The electromagnetic force (i.e. Lorentz force) is utilized to suppress the boundary layer separation and modify the vortical structures of the flow past the cylinder in low-conducting electrolyte. It is found that the spanwise wavelength of mode A and mode B instability are about 3.3 and 0.83 times cylinder diameter, respectively. The streamwise vortex becomes finer as the Reynolds number increase. The three dimensional flow structures during the period of vortex shedding may be suppressed effective by the Lorentz force, leading to drag reduction and vibration suppression.

1. Introduction

It is well known that vortex shedding occurs in the wake of a circular cylinder when the Reynolds number *Re* exceeds a critical value ($Re \approx 49$, *Re* based on the diameter of cylinder). The alternate shedding of vortices leads to fluctuation of drag and lift forces and may cause structure vibrations and acoustic noises, which in some cases can trigger damage. Therefore, control of the vortex shedding of a bluff body is of great importance from a practical engineering viewpoint.

The control methods developed in the past 20 years can be categorized into three groups: passive, active open-loop and active closed-loop controls [1]. In the case of low-conductivity electrolyte (such as seawater) the Lorentz force can be utilized to modify the boundary layer and control the vortex shedding, which is an active control strategy.

Few results have been reported on the three-dimensional wake of the cylinder controlled by Lorentz force. The objective of the present study is to numerically investigate the vortical structures and dynamic characteristics of the cylinder wake controlled by the Lorentz force.

2. Governing equations and numerical scheme

2.1 Governing equations

The incompressible Navier-Stokes equations with Lorentz force in dimensionless form read

---

* To whom any correspondence should be addressed.

$$\nabla \cdot \boldsymbol{U} = 0 \tag{1}$$

$$\frac{\partial \boldsymbol{U}}{\partial t} + \boldsymbol{U} \cdot \nabla \boldsymbol{U} = -\nabla p + \frac{1}{Re} \nabla^2 \boldsymbol{U} + N\boldsymbol{F} \tag{2}$$

where $\boldsymbol{U}$ and p are the velocity and pressure of the fluid, and $\boldsymbol{F}$ refers to the dimensionless Lorentz force. The Reynolds number Re is based on the diameter of cylinder D. The interaction parameter

$$N = \frac{j_0 B_0 D}{\rho u_\infty^2}$$

is the ratio of Lorentz force to the inertial force of the fluid (N=2 used in this article), where u_∞ is the free-stream velocity, j_0 and B_0 are the current density and magnetic field at the surface of the electrode and magnetic, respectively. The surface of the cylinder is flush mounted with alternating electrodes and magnets along the streamwise, therefore, the Lorentz force acts tangential to the surface of the cylinder along the streamwise. The distribution of the streamwise Lorentz force can be described as [2, 3]:

$$|F| = e^{-a((x^2+y^2)^{1/2}-r)} \tag{3}$$

where r is the radius of the cylinder, a is a constant showing the electromagnetic penetration into the flow which is mainly defined by the electrode spacing. It can be found that the Lorentz force decays exponentially in radial direction and is linear in azimuthal direction.

2.2 Spatial and temporal discretization

As shown in figure 1, the computational domain is a 40 D×10 D×10 D cuboid. Periodic boundary condition (b. c.) is applied on the spanwise domain boundary and inflow b.c. is applied on the side domain boundary. The cylinder surface is non-slip b.c and the outflow is pressure b.c.. The domain is spatially discretized using cubic finite volumes organise hierarchically as an octree [4-6]. The fundamental feature of tree type discretization is that each cell can be divided into eight sub-cells. As the cross-section mesh shown in figure 1, the mesh is adapted dynamically by vorticity criteria to follow the vortical evolution in the wake in order to save the computation time.

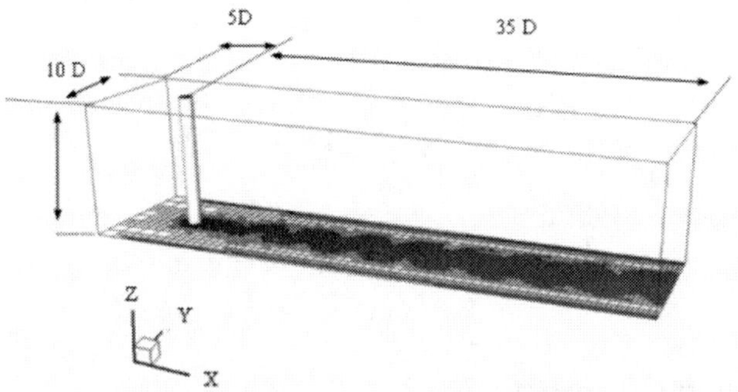

Figure 1. Computational details of flow past a circular cylinder.

The fractional-step projection method is used to implement temporal discretization [4, 5]. The advection term is discretized using the Bell-Colella-Glaz second-order upwind scheme. The advection term on the solid surface is treated using cell-merging technique. The diffusion term is discretized using an implicit Crank-Nicolson scheme. Second-order convergence in space and time is achieved.

3. Flow past a circular cylinder

Figure 2 shows the instantaneous vortical structures by λ_2 definition[7]. At *Re*=200, the primary Karman vortex becomes undulate and bifurcates to a three-dimensional flow (mode A). At *Re*=250, the primary vortex become more twisty and the streamwise vortex is clearly observed. At *Re*=300, finer-scale streamwise vortex pairs (mode B) are formed in the near wake. The spanwise wavelength of mode A and mode B is about 3.3 *D* and 0.83 *D*, respectively, which agrees well with [8]. The nonuniformity of the streamwise vortex may be caused by the interaction of the spanwise vortex and the streamwise vortex after shedding from the cylinder. At *Re*=1000, more streamwise vortex are generated and interact with the spanwise vortex, leading to an increasing disorder in the far wake. The scale of the streamwise vortex becomes finer as Reynolds number increase.

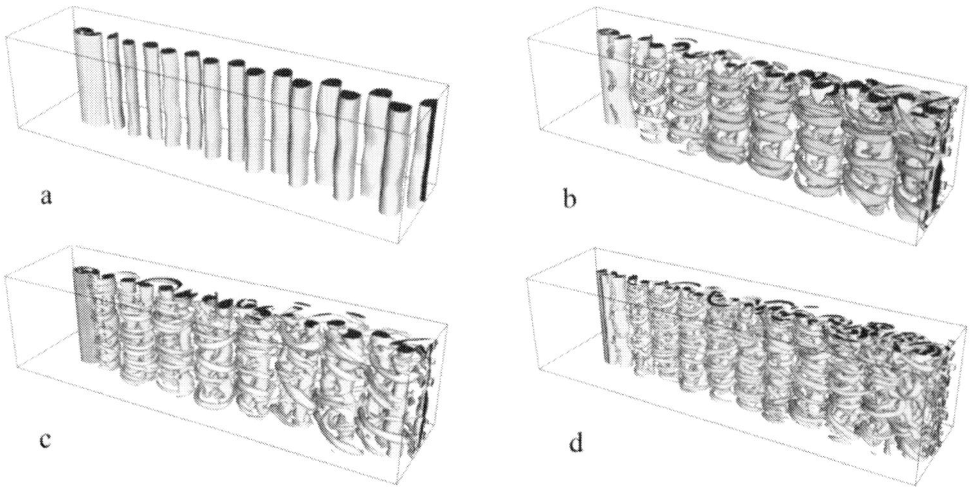

Figure 2. Vortical structures: (a)*Re*=200; (b)*Re*=250; (c)*Re*=300; (d)*Re*=1000.

The drag coefficient and lift coefficient are defined as

$$C_D = \frac{F_x}{\frac{1}{2}\rho u_\infty^2 S} \quad \text{and} \quad C_L = \frac{F_y}{\frac{1}{2}\rho u_\infty^2 S}$$

Where F_x and F_y are the drag and lift force, respectively. S is the projected area of the cylinder perpendicular to the direction of the free-stream, $S=D\times H$, where H is the spanwise length of the cylinder. $\overline{C_D}$ denotes the time-averaged drag coefficient and C_L' is the amplitude of lift coefficient. As figure 3 shows, the drag and lift force coefficients decrease as *Re* increase.

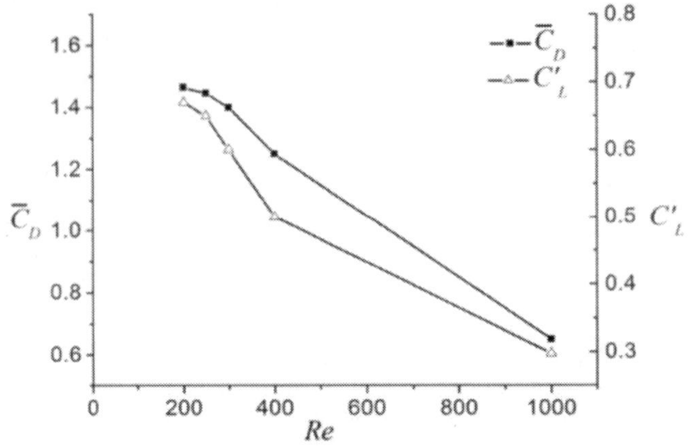

Figure 3. Variations of the drag and lift coefficients.

4. Control effect of Lorentz force

The boundary layer separation and vortical structures may be suppressed by the Lorentz force. The momentum is directly transported into the boundary layer by the Lorentz force without associated mass flux, making the Lorentz force an appealing approach for flow control.

Figure 4. Vortical structures controlled by Lorentz force (Re=300).

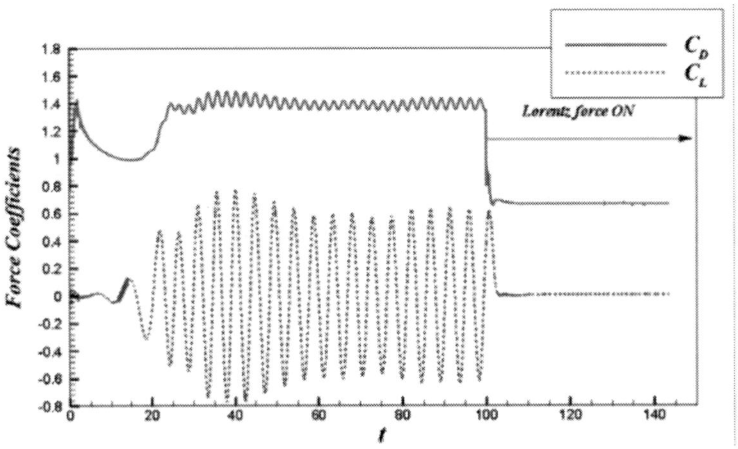

Figure 5. Drag and lift force coefficients (*Re*=300).

Figure 4 shows the vortical structures of flow past the cylinder at Reynolds number of 300 controlled by Lorentz force. The flow is from left to right and the Lorentz force is applied at nondimensional time t of 100. From the evolving vortical structures, it can be found that the shedding of spanwise primary vortex and the accompanying streamwise vortex are suppressed effectively by the Lorentz force. As time goes on, the vortex shed before the Lorentz force is applied is transported downstream out of the domain.

As figure 5 shows, the force coefficients become quasi-periodic at Reynolds number of 300 for the flow past a circular cylinder. The mean drag coefficient is about 1.396 without control and reduced to 0.67 controlled by the Lorentz force, leading to 28% of the drag reduction. The oscillation of the lift force is completely eliminated.

5. Discussion

Computations have been carried out for three-dimensional flow past a circular cylinder controlled by Lorentz force for Reynolds number varying from 200 to 1000. The spanwise wavelength of mode A and mode B are found to be about 3.3 and 0.83 times cylinder diameter, respectively. The boundary layer separation and vortex shedding is suppressed for the flow controlled by Lorentz force, leading to drag reduction and vibration suppression.

Acknowledgements

This study was supported by the Science and Technology on Transient Physics Laboratory Foundation of NJUST (KX 21373) and National Natural Science Foundation of China (Grant No.10572061).

References

[1] Choi H, Jeon W P and Kim J 2008 *Annu. Rev. Fluid Mech.* **40** 113
[2] Weier T, Gerbeth G, Mutschke G, Platacis E and Lielausis O 1998 *Exp. Therm. Fluid Sci.* **16** 84
[3] Zhou B M, Fan B C, Chen Z H, Ye J F, Ding H X and Jin J M 2004 *Acta. Mech. Sinica* **36** 472
[4] Popinet S 2003 *J. Comput. Phys.* **190** 572
[5] Popinet S 2009 *J. Comput. Phys.* **228** 5838
[6] Wang L and Wu C J 2010 *Sci. China Phys. Mech. Astron.* **53** 923
[7] Jeong J and Hussain F 1995 *J. Fluid Mech.* **285** 69
[8] Williamson C H K 1996 *J. Fluid Mech.* **328** 345

The influence of bulk conductivity on Lorentz force generated by coplanar waveguide

Zongkai Liu, Benmou Zhou*, Yanliang Ji, Jianwen Zhu
Science and Technology on Transient Physics Laboratory, Nanjing University of Science and Technology, Nanjing 210094, China

E-mail: bmzhou@njust.edu.cn

Abstract. Lorentz force has been used to modify the boundary layer around the controlled body. The Coplanar Waveguide (CPW) structure can support a quasi-TEM (Transverse Electromagnetic) mode of low frequency microwave propagation. Meanwhile the Lorentz force could be induced by electric and magnetic fields in the cross section. But the force is deeply influenced by the fluid bulk conductivity. Therefore, in this paper, particular emphasis is placed on exploring the effect of varying bulk conductivity on the Lorentz force. The results show that the microwave attenuation increases very rapidly with the increase of frequency and bulk conductivity. Moreover, the strength of Lorentz force is increased with the bulk conductivity over 0~1s m^{-1}.

1. Introduction

The researchers in different countries developed many new propulsion tools and fundamental theory of propulsion physics [1, 2]. And the Lorentz force was employed to boundary layer control and propulsion, successfully in the last few years, and attracted new attention again due to its potential engineering application background. With the use of the electromagnetic actuators covered on the object surface, Lorentz force distribution, noise suppression, drag force reduction, vibration absorption and noise suppression have been investigated widely [3-5].

In addition to the actuators, more new style methods have been discussed. The CPW structure can support a quasi-TEM mode of low frequency microwave propagation. Meanwhile electric and magnetic fields can be generated in the cross section. And the Lorentz force along the coplanar waveguide circuits can be induced by the mutual interaction of the double fields.

In this paper, we mainly discuss the distributions of electric, magnetic and Lorentz force fields around the CPW, immersed in flow field with different bulk conductivities [6]. In order to improve the propulsion efficiency more effectively, the influence of bulk conductivities on Lorentz force will be mainly discussed.

* To whom any correspondence should be addressed.

2. Theoretical Model

2.1. The design of CPW circuit

A common CPW on a dielectric substrate consisted of a center strip conductor with semi-infinite ground planes on either side is shown in figure 1. The L is the length of CPW, W is width of center strip conductor, $D1$ is the slot width, $D2$ is the width of ground planes, T and H is the thickness of double side and lower ground planes, respectively.

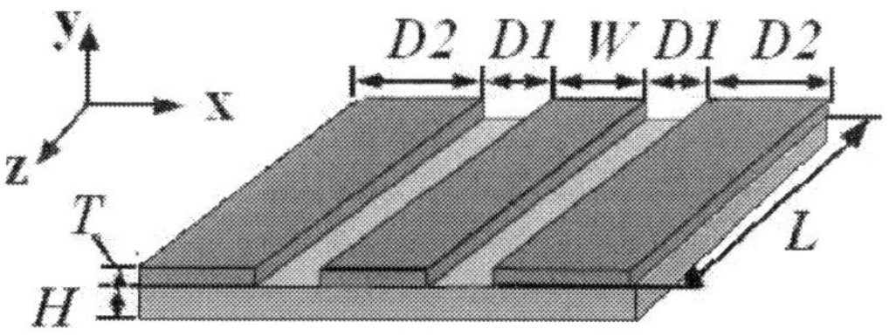

Figure 1. Schematic illustration of coplanar waveguide.

In this simulation we set W=1.5 mm, T=0.035 mm, H=7 mm, $D1$=1 mm, $D2$=13.28 mm and L=100 mm, respectively. During the microwave propagation, many losses will be generated, in which Ohmic loss is predominant, which is mainly induced by the unperfected medium material. So the linear uniformity materials of CPW should be used. In order to avoid the radiation loss and keep the quasi-TEM mode, silver (Ag) and GaAs are the excellent materials using as the material of strip and ground planes, respectively.

2.2. The propagation characters

The CPW structure can support a quasi-TEM mode of low frequency microwave propagation. Meanwhile electric and magnetic fields can be induced in the cross section, as shown in figure 2.

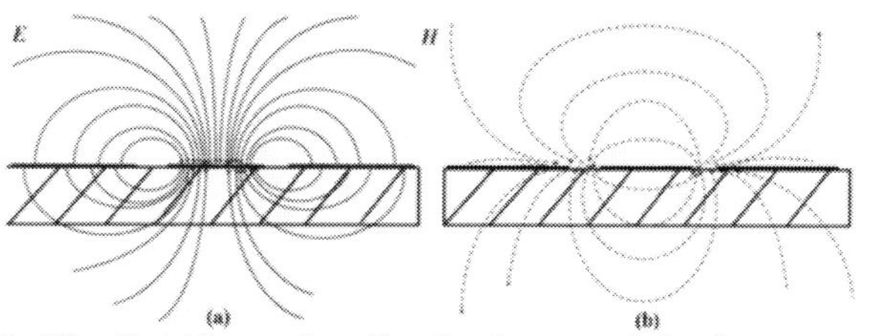

Figure2. The field intensity distribution graph in the cross section ((a) electric field, (b) magnetic field).

Figure 2(a) and (b) illustrate the distributions of electric field and magnetic field, respectively, in the cross section of coplanar waveguide. It is observed that the electric-field lines extend across the slot while the magnetic-field lines are perpendicular to the fluid-dielectric interface in the slot. The electric field structure in the figure 2(a) is in the opposite direction of the magnetic field structure in the figure 2b. So the Lorentz force is induced along the slot [7].

3. Theoretical Model

The vector wave equations of electromagnetic field are

$$\nabla \times \left(\frac{1}{\mu_r} \nabla \times \boldsymbol{E}\right) - k_0^2 \varepsilon_r \boldsymbol{E} = 0 \tag{1}$$

$$\nabla \times \left(\frac{1}{\varepsilon_r} \nabla \times \boldsymbol{H}\right) - k_0^2 \mu_r \boldsymbol{H} = 0 \tag{2}$$

where ε_r is relative dielectric constant, μ_r is relative permeability, k_0 is the wave number in free space, $\boldsymbol{E}$ is the electrical field strength and $\boldsymbol{H}$ is the magnetic field strength in equation (1) and equation (2).

Define the Poynting vector $\boldsymbol{S}$ (or instantaneous Poynting vector W m^{-2}) as the energy flows through unit vertical section, in the direction of energy flow, during the unit time. On the other hand, the Maxwell's equations conserve certain properties—the magnetic field strength, and the Poynting vector that describes the electric or magnetic flux of an electromagnetic field.

$$\boldsymbol{S} = \boldsymbol{E} \times \boldsymbol{H} \tag{3}$$

$$\boldsymbol{F} = \boldsymbol{J} \times \boldsymbol{B} \approx \sigma(\boldsymbol{E} \times \boldsymbol{B}) = \sigma(\boldsymbol{E} \times \mu\boldsymbol{H}) = \sigma\mu\boldsymbol{S} \tag{4}$$

Here μ and σ are relative permeability and electrical conductivity, $\boldsymbol{J}$ is current density, $\boldsymbol{B}$ is strength of magnetic flux, $\boldsymbol{F}$ is the strength of the Lorentz force. Due to the rapidly phase change of microwave, power density can be investigated by the average Poynting's vector $\boldsymbol{S}_{av}$ over one period T.

$$\boldsymbol{S}_{av} = \frac{1}{T}\int_0^T \boldsymbol{S}dt = \mathrm{Re}[\boldsymbol{E} \times \boldsymbol{H}^*] \tag{5}$$

$$\boldsymbol{F}_{av} = \sigma\mu\boldsymbol{S}_{av} \tag{6}$$

The $\boldsymbol{F}_{av}$ denotes the force average strength over one period based on equation (6).

4. Numerical investigation

4.1. The effect of frequency

The transmission character of microwave is deeply influenced by the surrounding medium, microwave frequency (*Freq*) and so on. And the quasi-TEM mode propagation of microwave on a CPW has low dispersion and hence offers the potential to construct the single direction force. So during the simulation, proper frequency should be set according to the different bulk conductivity.

The S-parameters are usually used to discuss the attenuation of CPW. Figure 3 (a-d) illustrate the distributions of the return loss *S11* and insertion loss *S21* with *Freq* for σ is equal to 0、0.2、0.5 and 0.8, respectively. Where *S11* is input reflection coefficient with matched output port. The *S21* is forward transmition gain with matched output port (the two S-parameters are analyzed in this paper).

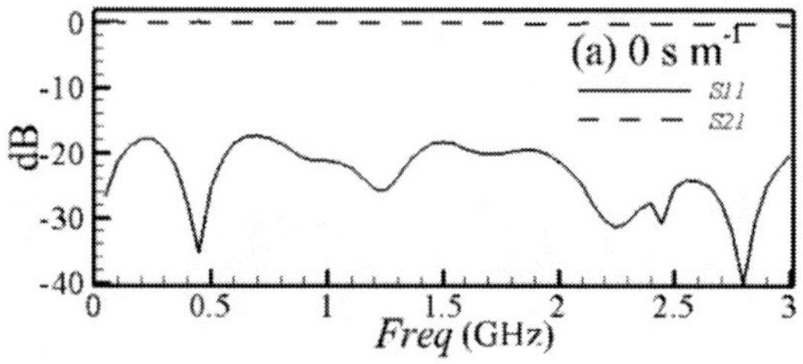

(a).0 s m^{-1}

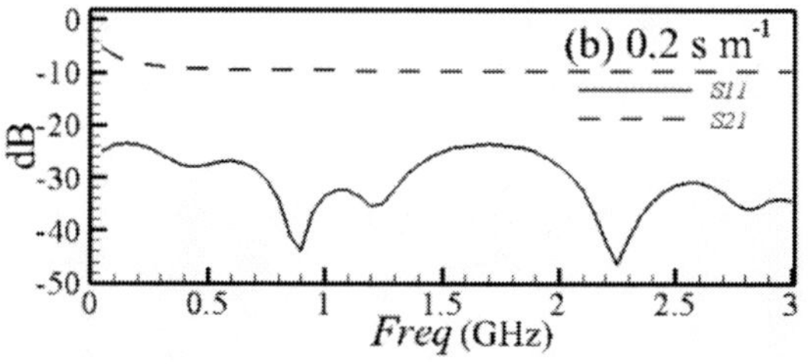

(b).0.2 s m^{-1}

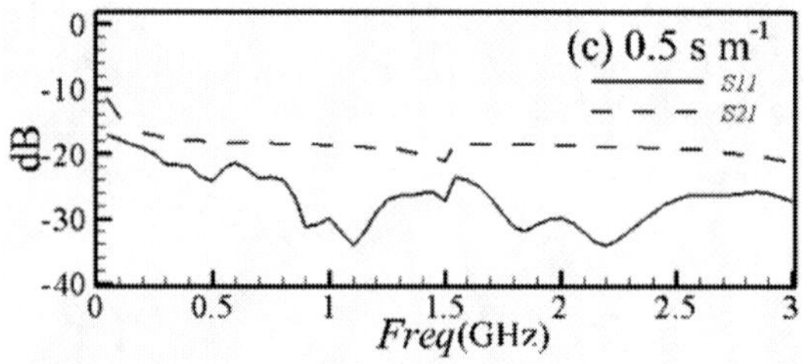

(c). 0.5 s m^{-1}

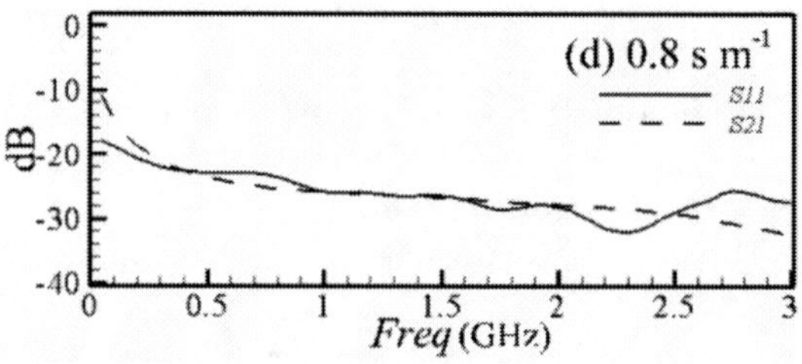

(d).0.8 s m^{-1}

Figure 3. The distribution curves of *S11* and *S21* with the frequency *Freq*.

These figures show that for σ equal to 0, the *S21* parameters keep almost to a parallel line, at about 0 dB. And the *S11* appears significantly reduction at several frequency points, such as 0.45 and 2.8 GHz. It is worth noting that these S-parameters at low frequencies decrease with the increase of frequency for the four cases. For CPW the larger *S11* and lower *S21* parameters denote the less loss of microwave. The analysis of frequency suggests that the low frequency microwave show less loss for CPW in the different bulk conductivity mediums and larger bulk conductivity could lead to much loss. Thus the low frequency *Freq*=0.05 GHz will be considered in the numerical simulation.

4.2. The influence of field strength distributions due to bulk conductivity

Figure 4 and 5 are the distribution of electric and magnetic fields for two different bulk conductivities σ=0.5 and 0.8, respectively.

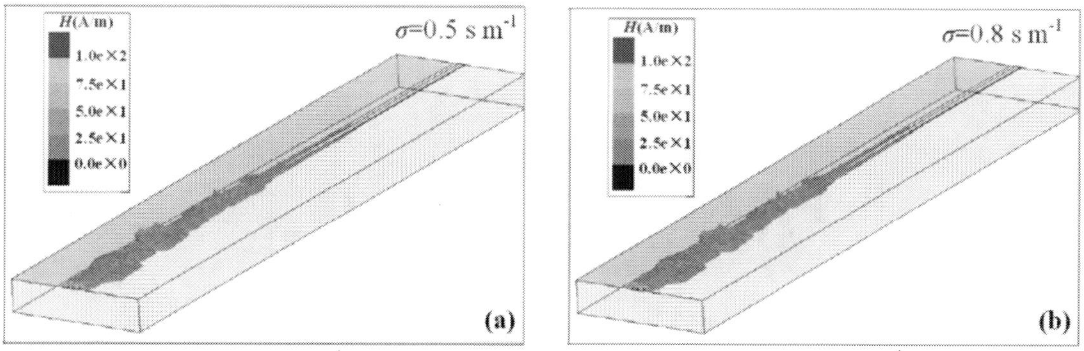

(a) σ=0.5 s m^{1} (b) σ=0.8 s m^{-1}

Figure 4. The distributions of electric field for zero phase.

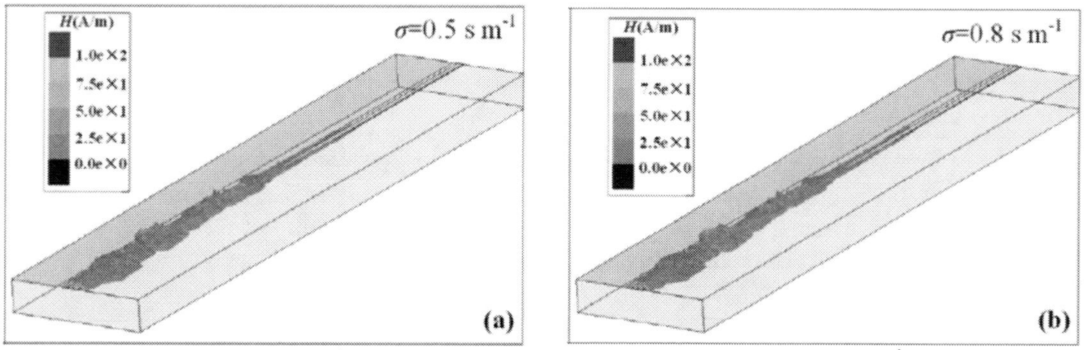

(a) σ=0.5 s m^{1} (b) σ=0.8 s m^{-1}

Figure 5. The distributions of magnetic field at zero phase.

The microwave travels with the light speed, which is about 300,000 kilometres per second. In this simulation, the rapid variation of the phase with time could be ignored, so we just consider the whole field strength is of the uniform distribution over the section (and the average value is about $1/\sqrt{2}$ of the maximum field strength). Note that, for larger bulk conductivity 0.8 s m^{-1}, the microwave is attenuation much more sharply over the surface of CPW and the distribution of strength field shapes like a circular truncated cone.

4.3. The math model of Lorentz force

The total force over the CPW can be obtained by the average energy density and the action volume. In order to simplify math model, in this numerical simulation, , the average electric field $\boldsymbol{E}_0$, magnetic field $\boldsymbol{H}_0$ and the Poynting's vector $\boldsymbol{S}_{\mathrm{av0}}$ in a vacuum, over the input port should be used as the benchmark to approximately calculate the effective energy density. And this simulation also ignores microwave heat and radiation loss. The figure 6 presents the distribution of $\boldsymbol{S}_{\mathrm{av0}}$ over the input port. And we should also assume that in microwave radiation zone of the same energy volume density as the face density of the input port $\boldsymbol{S}_{\mathrm{av0}}$. The figure 6 is the distribution of $\boldsymbol{S}_{\mathrm{av0}}$ over the input port and it presents that the strength of $\boldsymbol{S}_{\mathrm{av0}}$ is decreasing with the increase of the distance to the center strip conductor. The average radius R is 6 mm and S_{av0} is 0.8×10^4 (W m^{-2}).

$$S_{\mathrm{av0}} = \frac{1}{T}\int_0^T S dt = \mathrm{Re}\left[\boldsymbol{E}_0 \times \boldsymbol{H}_0^*\right] \tag{7}$$

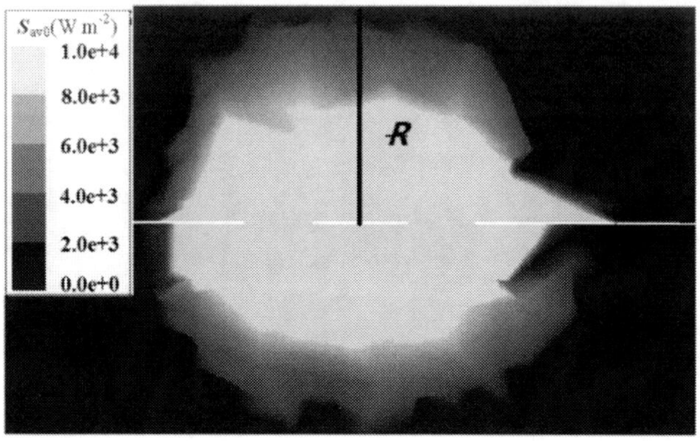

Figure 6. The distribution of $\boldsymbol{S}_{\mathrm{av0}}$ over the input port.

In figure 4 and 5, it is seen that strength of microwave decreases with the wall normal direction, but for larger σ and longer distance to the input port, the amplitude of electric and magnetic fields decreases significantly. Here a math model is used to investigate the propagation characteristics of Lorentz force. Based on the field structure characteristics, the input port over the CPW could be seemed as a semi-circle (the radius R is approximately equals to the average microwave amplitude of the input port) as well as the output port (the radius r). Here we only discuss the $S21$, based on its basis definition the relationships could be written as:

$$S21 = 10\log_{10}[(A_{\mathrm{out}})/(A_{\mathrm{in}})]^2 \approx 10\log_{10}(r/R)^2 \tag{8}$$

Where A_{out} and A_{in} denote the average microwave amplitude of output and input, respectively. And the volume of circular truncated cone could be calculated based on the long of CPW, the radius R and r. Figure 7 presents the body of circular truncated cone denotes the control zone of the force.

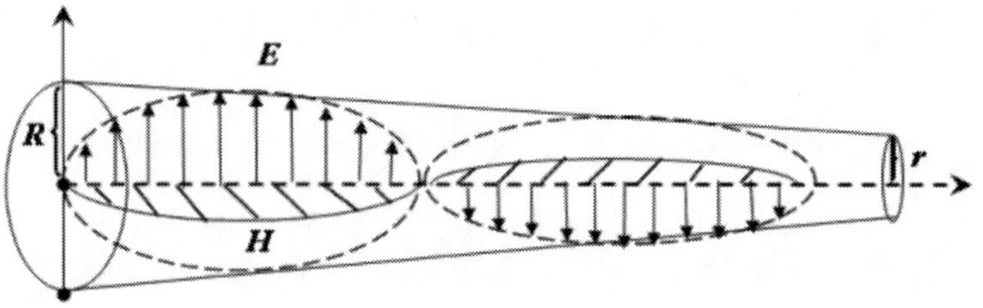

Figure 7. The diagram model of the microwave distribution.

The effects of bulk conductivity σ on the computed action range $V/2$ and strength $\boldsymbol{F}$ of Lorentz force at a fixed frequency of 0.05 GHz are shown in figure 8 and 9. As expected, the action region space of microwave attenuation increases very rapidly with the increase of the bulk conductivity σ.

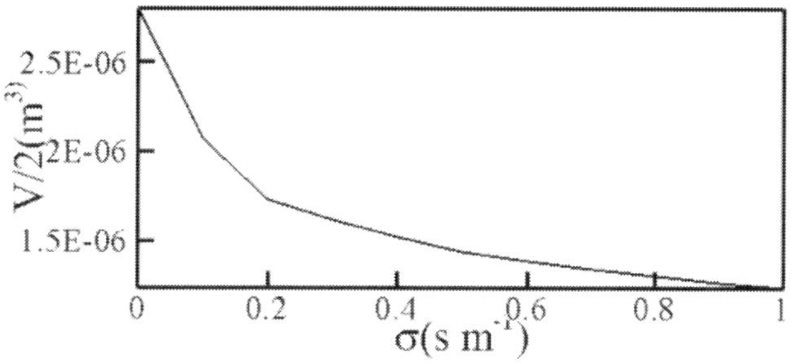

Figure 8. The half volume of circular truncated cone versus bulk conductivity.

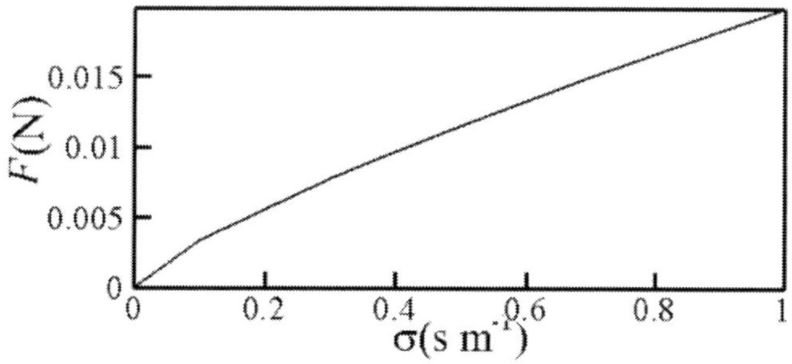

Figure 9. The Lorenz force versus bulk conductivity.

The figures 8 and 9 also show that $V/2$ is 2.7×10^{-6} m^3 and F is 0 N while σ is 0 s m^{-1} .And when σ is equal to 0.2 s m^{-1}, there is about 30 percent decrease in the microwave radiation space volume ($V/2$) comparing with the case of σ=0 s m^{-1}. It is important to note that the Lorentz force F is not depended on the action region space V. This is due to the fact that in addition to the loss introduced by the increasing of σ, the strength of Lorentz force is also proportional to σ in the equation 3. In general, for σ over the range of 0 and 1, the Lorentz force generated by CPW is increasing along with the bulk conductivity. It is clear from the above discussion that active control methods present an attractive alternative to the current methods used in micro-propulsion.

5. Conclusions

(1) The CPW structure can support a quasi-TEM mode of low frequency microwave propagation. And the Lorentz force along the CPW direction can be created by mutual coupling of landscape orientation electric and magnetic fields. (2) The attenuation increases very rapidly with the increase of frequency and bulk conductivity. (3) The strength of Lorentz force is increased with the bulk conductivity over 0~1 s m^{-1}.

Acknowledgements

This study was supported by the graduate student innovation fund of Jiangsu province ordinary college (CXZZ 11_0231) and the Science and Technology on Transient Physics Laboratory Foundation of NJUST (KX 21373).

References

[1] Sehra A K and Jr W W 2004 *Prog. Aerosp. Sci.* **40** 199
[2] Quarta A A and Mengali G 2011 *Aerosp. Sci. Technol.* **15** 381
[3] Lantzsch R *et al* V. 2007 *J. Crystal Growth* **305** 249

Journal of Physics: Conference Series **418** (2013) 012014 doi:10.1088/1742-6596/418/1/012014

[4] Shatrov V and Gerbeth G 2007 *Phys. Fluids* **19** 035109
[5] Dousset V and Potherat A 2008 *Phys. Fluids* **20** 017104
[6] Simons R N 2001 *Coplanar Waveguide Circuits, Compon., and Syst.* chap2 (New York: John Wiley & Sons) p 11
[7] Liu Z K and Zhou B M 2011 *Chin. Phys. Lett.* **28** 094703

The mechanism of propulsion effect of BLEVE based on discharge

Yanliang Ji, Benmou Zhou*, Zongkai Liu, Pengfei Sun
Science and Technology on Transient Physics Laboratory, Nanjing University of Science and Technology, XiaoLingWei, Nanjing, China

E-mail: bmzhou@njust.edu.cn

Abstract. Propellant could reach a high pressure and temperature situation in a very short time by the use of high voltage pulsed discharge. Once sharp depressurization occurs, BLEVE (boiling liquid expanding vapor explosion) happens immediately with a violent phase transition. A two-phase flow with high pressure and speed forms and often causes severe damage. It is also useful in other aspects such as propulsion. The numerical simulation of BLEVE for propulsion shows the pressure has a sharp increasing when phase transition begins. Then a two phase mixture flow moves forward with a high speed and compresses the gas in front.

1. Introduction
Since the sixties of last century, electric propulsion has been applied in areas such as aeronautics and astronautics. Traditional propellants mainly contain Teflon, xenon, mercury, etc. These propellants may lead to plume contamination. Several researchers try to find other alternative propellants. As we know, water is non-polluting and non-corrosive. It has higher specific impulse and larger explosion volume radio [1-4].Considering water is relatively ideal explosion propellant, it is a new attempt to use BLEVE as propulsion power. The BLEVE is a type of physical explosion and has been defined as an explosion resulting from the failure of a vessel containing a liquid at a temperature significantly above its boiling point to local pressure. There are a lot of experiments and theory researches about the process of BLEVE. But there is no accurate model which could describe the whole process. The high energy two-phase flow may have severe damage. But finding a way to use it is also valuable. In this paper, 2D numerical simulation is carried out as a preliminary exploration for the use of BLEVE.

The traditional external heat conduction is slow and non-uniform. The propellant could be directly discharged by a high-power pulsed power source [5, 6]. By the control of external conditions (i.e. voltage, charge power), the propellant could be heated up to 523 K or higher.

2. Model design

2.1 two-phase flow model
Actually the BLEVE is a process of violent phase transition. When depressurization occurs in liquid, the superheat situation leads to a violent phase transition from liquid to vapor and a large number of bubbles form in liquid. Energy transfer on the bubble surface makes bubble grow rapidly and two-phase mixture flow forms. The violent vapor generation causes rapid surge of the two-phase

---
* To whom any correspondence should be addressed.

mixture volume. This condition will increase the internal pressure. When bubbles burst, energy is released and this will contribute to the increase of pressure too. The high pressure has a propulsion effect on surroundings.

The physical model of two-phase flow is assumed incompressible. The momentum equation is [7]:

$$\frac{\partial}{\partial t}(\rho \vec{u})+(\nabla\cdot\rho\vec{u}\otimes\vec{u})=-\nabla p+\rho\vec{g}+\nabla\mu[(\nabla\vec{u})+(\nabla\vec{u})^{T}]+2\sigma\kappa\delta_{S}\vec{n} \tag{1}$$

$\vec{u}$, ρ, μ, p, $\vec{g}$ denote velocity, density, dynamic viscosity, pressure, volume force (gravity). σ is surface tension. κ is local interface curvature. δ_s is Dirac function on the interface of S. $\vec{n}$ is normal vector of interface. The energy and continuity equations are:

$$\rho C_p(\frac{\partial T}{\partial t}+\vec{u}\cdot\nabla T)=\nabla\cdot(k\nabla T)+\Phi \tag{2}$$

$$\nabla\cdot\vec{u}=\Gamma \tag{3}$$

Here the viscous dissipation in energy equation is neglected. T, C_p, k denote temperature, heat at constant pressure, thermal conductivity and Φ is latent heat energy source that comes from the phase transition on interface. Γ is the volume change caused by mass transfer on the interface between different phases.

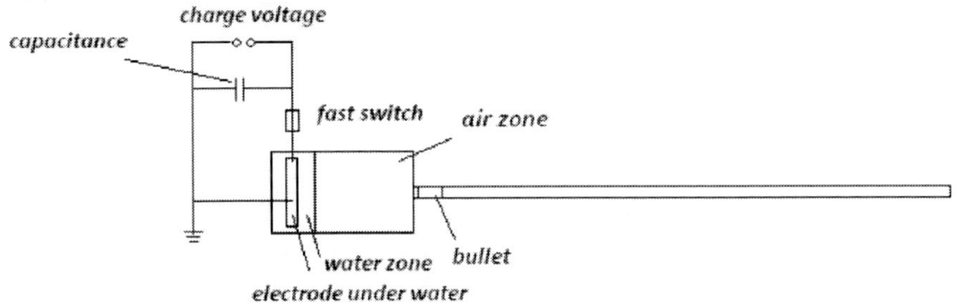

Figure 1. Propulsion model.

The phase transition rate is unsteady for the influence of some complex factors such as pressure, temperature and bubble generation. It is difficult to find a model to predict the rate accurately [8, 9]. Figure 1 shows the initial design of the propulsion model. The length and diameter of column vessel is 200 mm and 100 mm, which connect with a long barrel. The length of barrel is about 1000 mm with diameter of 20 mm. The bullet for propulsion is about 15 g. About one third of the vessel is filled with water.

2.2 Discharge model

By the way of high voltage pulsed discharge, we can make the water reach experimental temperature in a short time. In general the discharge includes corona discharge and arc discharge. When corona discharge occurs, plasma only generate around electrodes. Arc discharge requires more energy and plasma channel appears through electrodes. In general the charge and discharge process could be described by an equivalent circuit. The equivalent circuit is shown in figure 2.

The characteristic of discharge in water is determined by differential equation of RLC circuit as:

$$L\frac{d^2Q}{dt^2}+R(t)\frac{dQ}{dt}+\frac{Q}{C}=0 \tag{4}$$

Here L and C are inductance and capacitance. Q is the stored charge and $R(t)$ is the impedance mainly determined by the discharge channel.

The shape of electrode could be used is various (i.e. sphere-ring electrode) [10]. When discharge happens, the energy released from capacitance is divided into several parts such as circuit equivalent resistance, water energy, shock wave energy, radiation energy etc [11].

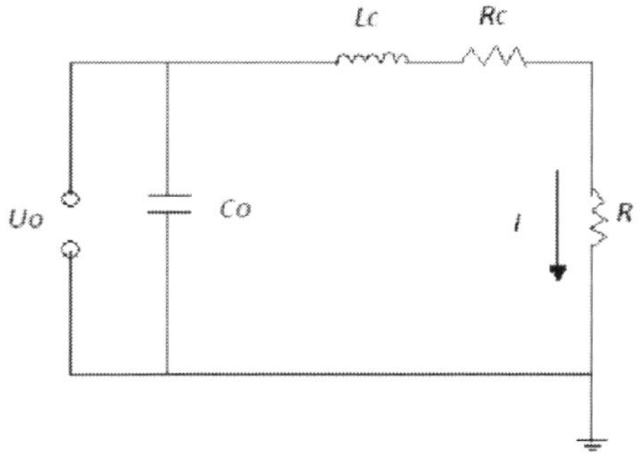

Figure 2. Equivalent circuit.

3. Numerical simulation method

Strictly speaking, phase transition happens after the spread of rarefaction wave in liquid. So it happens on the surface first. We assume that depressurization in liquid phase has finished so quickly that we can neglect it because the area of propellant is narrow. When the simulation starts, the phase transition occurs at once in the whole liquid.

At the beginning, the water has reached saturation condition of 523 K with pressure of approximate 3.9 MPa by the use of pulsed discharge. Triangle mesh and finite volume method is adopted. PISO (pressure implicit with splitting of operators) algorithm that is one of the pressure based schemes is used for pressure-velocity coupling. For spatial discretization, PRESTO (Pressure Staggering Option) is used for pressure terms and first order upwind is used for other terms such as volume fraction, energy, etc. The boundary conditions of wall are no slip and adiabatic. In the flow field, the relation between temperature and corresponding evaporation pressure can be obtained by Antoine equation. If the evaporation pressure is larger than local pressure, the mass transfer happens from liquid to gas. If the evaporation pressure is smaller than local pressure, the transfer happens in the opposite direction. Triangle mesh is used and the simulation stops when bullet move to the end of the barrel.

4. Flow field parameters analysis

4.1 pressure of monitoring point

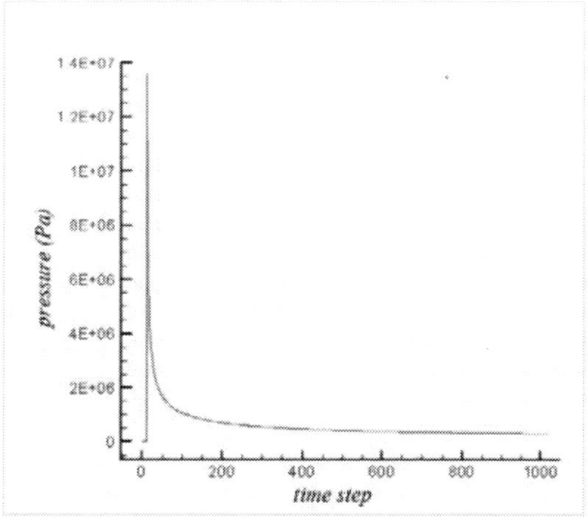

Figure 3. The pressure of monitoring point.

A monitoring point is set in the center of water area, so that we can record the change of pressure. Figure 3 shows pressure change of the point in initial iteration of about 1000 time steps. Each step equals $10e^{-6}$ s. When the calculation starts, depressurization accomplishes in a very short time. Then phase transition occurs in the whole liquid. The liquid phase's pressure increases suddenly because of violent evaporation. Internal pressure could rise to 3 or 4 times higher compared to the initial saturated pressure. Then a high speed two-phase mixture fluid erupts and impresses the gas in front. As the volume expands rapidly, pressure will decrease with timing process. So the flow field pressure is determined by both evaporation and expansion.

4.2 instantaneous flow field parameters

Figure 4 shows the vessel's internal states at about 1.5 ms. We consider the initial gas in air zone (shown in figure 1) as ideal gas, so the ideal gas equation is suitable for the compressed gas. When the gas is blocked by the front wall, the speed will slow down and both pressure and temperature rise up. As the evaporation pressure is higher than actual pressure in the whole field, there is no transition from gas to liquid. As shown in the third and fourth figures, the front speed of mixture flow is expanding rapidly with the speed of about 100 m s^{-1} at this moment. Mixture flow compresses front gas and the gas push bullet forward.

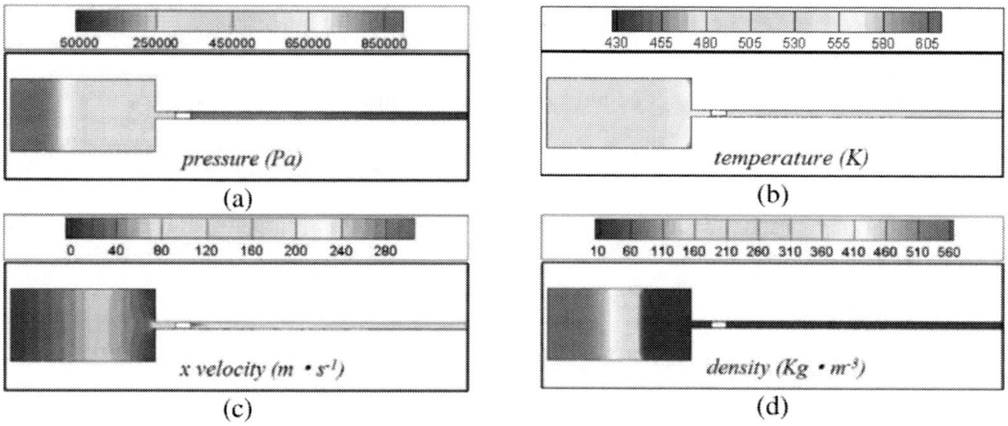

Figure 4. Pressure, temperature, velocity and density of flow field at 1.5 ms.

4.3 bullet velocity and tail pressure

The acceleration of bullet is mutative in the whole calculation process. The drag force increases when bullet speeds up. The final velocity keeps around 350 m s^{-1}. This indicates the forces role in bullet has come to a balance condition. Tail pressure curve is no larger than 500000 Pa, this may be a little small compared to the mixture flow pressure for the reason that the two-phase mixture flow just compresses the gas in front and the direct force that pushes bullet comes from the compressed gas. As a part of gas flow out through the gap between bullet and barrel, the tail pressure increases to about 500000 Pa and then decreases. At about 4 ms, as reflected shock wave reaches again, the tail pressure have a slightly ascension.

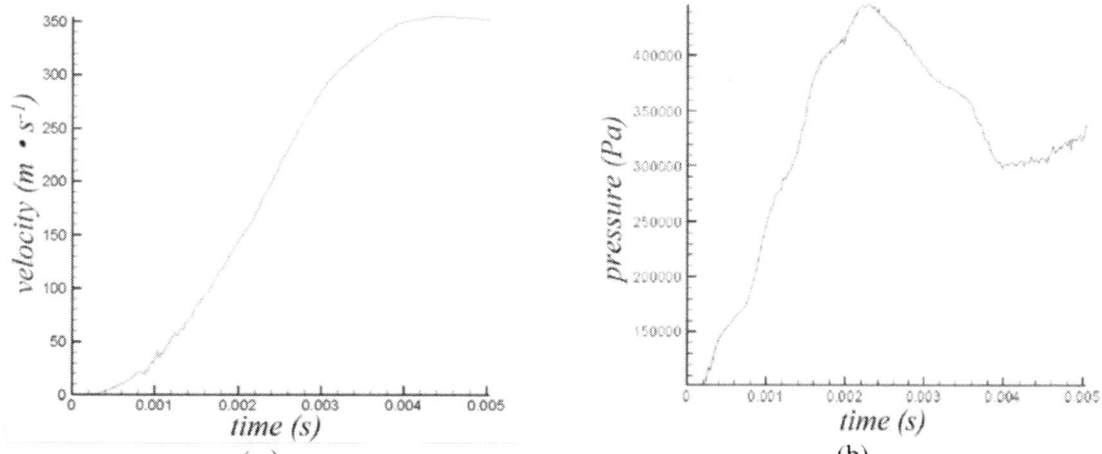

(a) (b)

Figure 5. Velocity and tail pressure of bullet.

5. Conclusions

Through the simulation, we could get some important flow parameters. The pressure increases several times compared to initial field pressure when phase transition begins. The propulsion effect of BLEVE is obvious for the violent two-phase mixture flow. As the shock wave has a repeated oscillation in vessel, this could disturb the propulsion and lead to energy loss. The BLEVE may have a better application prospect on propulsion area by further researches, such as fixing the model, improving simulation algorithm, carrying out more experiments and discussing its propulsion efficiency.

Acknowledgements

This study was supported by the Science and Technology on Transient Physics Laboratory Foundation of NJUST (KX 21373) and National Natural Science Foundation of China (Grant No.10572061).

References

[1] Mao G W, Han X W, Yang J and He H Q 2000 *Journal of Propulsion technology (in Chinese).* **21** 1
[2] Yang L, Li Z R, Yin L, Wu J J, Zhou J 2006 *Journal of Rocket propulsion (in Chinese).* **32** 32
[3] Scharlemann C A 2003 *Investigation of thruster mechanisms in a water fed pulsed plasma thruster* (UMI number: 3119256)
[4] Li J, Cao Y D, Yu L B and Yuan D S 2008 *Journal of Shenyang university of technology (in Chinese).* **30** 498
[5] Liu X C, Feng C G, Zhu Z L, Xu Y Z and Xu Z 1999 *Journal of Beijing institude of technology (in Chinese).* **19** 8
[6] Lu X P, Zhang H H, Pan Y, Liu K F and Liu M H 2001 *Explosion and shock waves. Explosion and shock waves (in Chinese).* **21** 282
[7] Yuan M H, Yang Y H, Li T S and Hu Z H 2007 *J. Eng. Thermophys.* **28** 962
[8] Chen S N, Sun J H and Wan W 2008 *J. Hazard. Mater.* **156** 530
[9] Pinhasi G A, Ullmann A and Dayan A 2007 *Int. J. Heat Mass Transfer* **50** 4780
[10] Zhu T Y, Wan X H, Zhang Q G and Yang L J 2008 *Journal of xi'an jiantong university (in Chinese).* **42** 723
[11] Li N, Huang J G, Chen J F, Lei K Z and Zhang Q F 2010 Journal of basic science and engineering(in Chinese). **18** 1010

Journal of Physics: Conference Series **418** (2013) 012016 doi:10.1088/1742-6596/418/1/012016

Finite-difference method on electrostatics field calculation in oil storage tank

Gefei Yu [*], **Xuewen Li**
SMU College of Arts and Sciences, Shanghai Maritime University, 1550 Hai Gang Dadao, Shanghai, P.R. China

E-mail: gfyu@shmtu.edu.cn

Abstract. This paper starts with the basic principle of Finite-Difference Method, combined with the practical application of the method, focusing on the difference scheme of Poisson equation and Laplace equation in 2D Electric Field and Axisymmetrical Field under the square grid partition. This study solves the Finite-Difference Method on electrostatics field calculation and introduces discretization on the Dirichlet Problem in oil storage tanker. The Finite-Difference Method gives the analytic solutions of the potential equations by changing the continuous field problems into discrete system. The numerical solutions of discrete points could infinitely approximate the real solution of the continuous field through discrete model.

1. Introduction

Of all the calculation methods of the electromagnetic field numerical analysis, the Finite-Difference Method, also named Grid Method, is applied the earliest for its simple and visual features. It involves aspects from linear field to nonlinear field, constant field to variable field. Finite-Difference Method is still one important numerical analysis method that could not be ignored even though the Finite Element Method, which is the combination of Finite-Difference Method and Calculus of Variations, is increasingly wide range of applications.

The Finite-Difference Method gives the analytic solutions of the potential equations by changing the continuous field problems into discrete system. The numerical solutions of discrete points could infinitely approximate the real solution of the continuous field through discrete model.

2. Concept of Finite-Difference Method

2.1. Basic concepts

We hypothesize one function $f(x)$, whose independent variable x increases tiny $\Delta x = h$, the corresponding increment of $f(x)$ would be

$$\Delta f(x) = f(x+h) - f(x) \tag{1}$$

Which is named as Finite-Difference (First order difference) of function $f(x)$ and is different from Differential. It is commonly known as Finite-Difference because it is limited Difference. However, the difference between Finite-Difference Δf and Differential df would be very tiny if h is tiny enough.

[*] To whom any correspondence should be addressed.

According to the definition of Finite-Difference, A second-order central difference is always used in the calculation of Finite-Difference

$$\Delta f(x) = f(x+\frac{h}{2}) - f(x-\frac{h}{2}) \tag{2}$$

And the First order difference quotient which is the division of Δf by h would close to the First order derivative $\frac{df}{dx}$

$$\frac{\Delta f(x)}{\Delta x} = \frac{f(x+h) - f(x)}{h} \tag{3}$$

The First order difference is still the function of the independent variable x, when calculating the Finite-Difference of the First order difference like the same method of equation (1), we could get $\Delta^2 f(x)$ which named Two order difference quotient of primitive function $f(x)$ and it would close to the Two order derivative $d^2 f$ as h goes to 0.

Obviously, the two order difference quotient closes to the two order derivative. So we could definite higher order difference quotient and higher order derivative. But in here, we wouldn't discuss the higher order difference quotient and higher order derivative, because they are not involve into the actual numerical calculation of Finite-Difference Method in the electromagnetic field.

2.2. Principle of Finite-Difference

The Finite-Difference Method is one numerical calculation on the basic of Difference Principle, which could transform the solving of boundary value problem into a corresponding group of finite difference equations by using the difference quotient instead of the partial derivative on the very discrete point. The numerical solution of the boundary value problem could get through the calculation of function value on every discrete point according to the group of finite difference equations.

The Finite-Difference Method, the following abbreviation is Difference Method, is one method of solving which would transform the continuous domain problem in the electromagnetic field into discrete system problem, which makes the numerical solution on every discrete point in the model of grid discretization system approximate infinitely to the real value in continuous field domain. It is an approximate calculation method and the computer could guarantee enough the calculation precision.

3. Mathematical model of electric field in oil tank

3.1. Basic equation of electric field

It is assumed that the oil charges uniform and there is no space charge in the above of oil in tank. Also we could consider the tank is good grounding according to actual situation, which means that the potential of the tank wall is 0. We should know that all of above assumptions could simplify the problem and would not affect the feasibility and reliability of the method.

After those appropriate assumptions, the electric field in oil tank could be treated as 2D field in no-uniform medium. It is difficult to solve the problem by Analytical method. But modern computer provides effective means in numerical calculation of electrostatic field. Its essence is to transform continuous field domain problems into discrete system problems, and makes the numerical solution on discrete point approximate infinitely to the real value in continuous field domain by discretization model. So we study the electric field in oil tank by Finite-Difference Method.

$$\nabla^2\varphi = \begin{cases} \dfrac{\partial^2\varphi}{\partial x^2} + \dfrac{\partial^2\varphi}{\partial y^2} = -\dfrac{\rho}{\varepsilon} \\ \dfrac{\partial^2\varphi}{\partial x^2} + \dfrac{\partial^2\varphi}{\partial y^2} = 0 \end{cases} \tag{4}$$

φ – Potential, V;

x, y – Coordinate, m;

ρ – Charge density, Cm^{-3};

ε – Dielectric constant, Fm^{-1}, $\varepsilon = \varepsilon_0 \varepsilon_r$.

The potential function in oil tank meets the first boundary value problem which is named Dirichlet problem. If there is free charge density area, the equation $\nabla^2 \varphi = -\frac{\rho}{\varepsilon}$ is named Poisson equation, and the oil in tank is belong to this situation; on the country, if there is not free charge density area, the equation $\nabla^2 \varphi = 0$ is named Laplasse equation, and the air space in tank is belong to this situation. Given surface potential of every conductor, that is

$$\varphi\big|_{\Gamma_i} = 0 \tag{5}$$

Γ is outer boundary, the surface potential of tank is 0 because of static grounding. $i = 1$ is the air space; $i = 2$ is the oil space. In interface, it meets join condition, that is

$$\begin{cases} D_{1n} = D_{2n} \\ E_{1r} = E_{2r} \end{cases} \text{ or } \begin{cases} \varepsilon_1 \dfrac{\partial \varphi_1}{\partial n}\Big|_{\Gamma_{12}} = \varepsilon_2 \dfrac{\partial \varphi_2}{\partial n}\Big|_{\Gamma_{12}} \\ \varphi_1\big|_{\Gamma_{12}} = \varphi_2\big|_{\Gamma_{12}} \end{cases} \tag{6}$$

D – Electric displacement, Cm^{-3};
E – Electric field strength, Vm^{-1}.

3.2. Numerical solution of Electrostatic Field

We adopt the Finite-Difference Method to solve the boundary problem of electrostatic field in tank. The distribution form of discrete points is governed by the grid division.

In order to export the solution of difference discretization from basic equations under irregular grid division, also meet the discussion of boundary conditions discretization, we start from the common situation of step distance ranging, as shown in figure 1. Assumed there are two groups of lines, which are no-equal but respectively parallel to x and y coordinate axis, to divide the field domain Γ. We take a arbitrarily point $o(x_0, y_0)$ which joints together with its adjacent node 1, 2, 3, 4 to combine to form a star type structure. The function of node o could be expressed as $\varphi(x_0, y_0)$ or φ_0, and the function of node 1, 2, 3, 4 also could be expressed as $\varphi_1, \varphi_2, \varphi_3, \varphi_4$, and so on.

So we could get following equations through Finite-Difference calculating.

$$\left(\frac{\partial^2 \varphi}{\partial x^2}\right)_0 \cong 2\frac{h_3(\varphi_1 - \varphi_0) + h_1(\varphi_3 - \varphi_0)}{h_1 h_3 (h_1 + h_3)} \tag{7}$$

$$\left(\frac{\partial^2 \varphi}{\partial y^2}\right)_0 \cong 2\frac{h_4(\varphi_2 - \varphi_0) + h_2(\varphi_4 - \varphi_0)}{h_2 h_4 (h_2 + h_4)} \tag{8}$$

After substituting equation (7) and (8) into Poisson equation in (4), we could get

$$\nabla^2 \varphi = 2\frac{h_3(\varphi_1 - \varphi_0) + h_1(\varphi_3 - \varphi_0)}{h_1 h_3 (h_1 + h_3)} + 2\frac{h_4(\varphi_2 - \varphi_0) + h_2(\varphi_4 - \varphi_0)}{h_2 h_4 (h_2 + h_4)} = P \tag{9}$$

$$\begin{aligned} \varphi_0 = {} & \frac{h_2 h_3 h_4 \varphi_1}{(h_1 + h_3)(h_1 h_3 + h_2 h_4)} + \frac{h_1 h_3 h_4 \varphi_2}{(h_2 + h_4)(h_1 h_3 + h_2 h_4)} \\ & + \frac{h_1 h_2 h_4 \varphi_3}{(h_1 + h_3)(h_1 h_3 + h_2 h_4)} + \frac{h_1 h_2 h_3 \varphi_4}{(h_2 + h_4)(h_1 h_3 + h_2 h_4)} - \frac{h_1 h_2 h_3 h_4 P}{2(h_1 h_3 + h_2 h_4)} \end{aligned} \tag{10}$$

The above two equations are suit to oil space in tank.

After substituting equation (7) and (8) into Laplasse equation in (4), we could get

$$\nabla^2\varphi = 2\frac{h_3(\varphi_1-\varphi_0)+h_1(\varphi_3-\varphi_0)}{h_1h_3(h_1+h_3)} + 2\frac{h_4(\varphi_2-\varphi_0)+h_2(\varphi_4-\varphi_0)}{h_2h_4(h_2+h_4)} = 0 \tag{11}$$

$$\begin{aligned}\varphi_0 = & \frac{h_2h_3h_4\varphi_1}{(h_1+h_3)(h_1h_3+h_2h_4)} + \frac{h_1h_3h_4\varphi_2}{(h_2+h_4)(h_1h_3+h_2h_4)} \\ & + \frac{h_1h_2h_4\varphi_3}{(h_1+h_3)(h_1h_3+h_2h_4)} + \frac{h_1h_2h_3\varphi_4}{(h_2+h_4)(h_1h_3+h_2h_4)}\end{aligned} \tag{12}$$

The above two equations are suit to air space in tank.

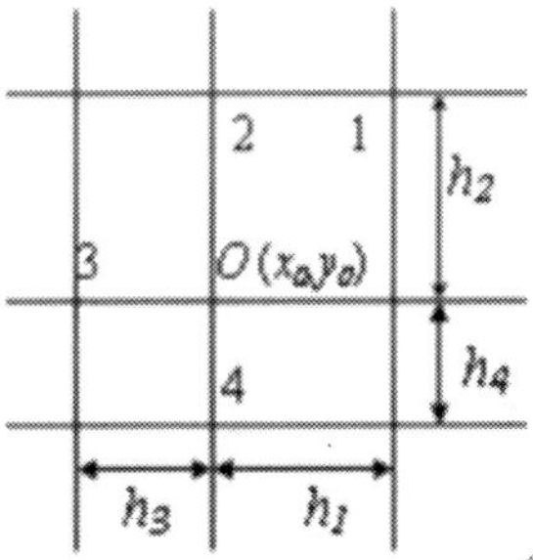

Figure 1. The asymmetric star type structure.

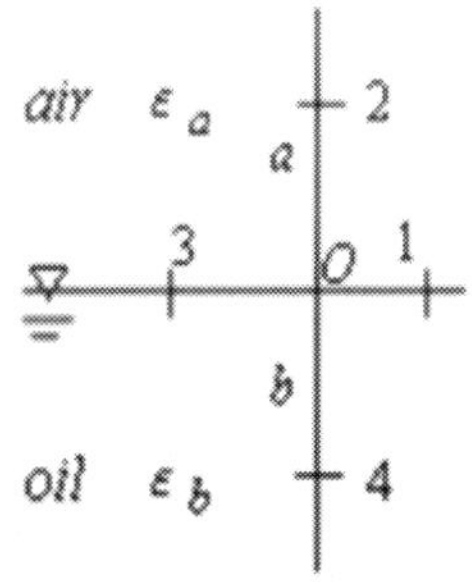

Figure 2. The interface of air-oil.

We could change (10) and (12) into followings:

$$\varphi_0 = K_1\varphi_1 + K_2\varphi_2 + K_3\varphi_3 + K_4\varphi_4 - KP \tag{13}$$

$$\varphi_0 = K_1\varphi_1 + K_2\varphi_2 + K_3\varphi_3 + K_4\varphi_4 \tag{14}$$

According to the air-oil interface, see figure 2, node o, 1, 3 are not on the same interface of same media. φ_a, φ_b is respectively expressed as the potential in media ε_a and ε_b, and there is charge density only in ε_a. That is, potential φ_a meets Poisson Equation (13) and φ_b meets Laplasse Equation (14). If media ε_b is changed into ε_a, Equation (13) on node o could be written as

$$\varphi_0^a = K_1\varphi_1^a + K_2\varphi_2^a + K_3\varphi_3^a + K_4\varphi_4^a - KP \tag{15}$$

And if media ε_a is changed into ε_b, Equation (14) on node o could be written as

$$\varphi_0^b = K_1\varphi_1^b + K_2\varphi_2^b + K_3\varphi_3^b + K_4\varphi_4^b \tag{16}$$

Actually, φ_2^a and φ_4^b are not existent in oil tank, they are virtual potentials just only in order to export the calculation format of Finite-Difference Method on the boundary node. So they should be deleted from the equations (15) and (16), and we could get following from equation (6)

$$\varepsilon_a(\varphi_2^a - \varphi_4^a) = \varepsilon_b(\varphi_2^b - \varphi_4^b)\text{, namely, } \varepsilon_a\varphi_2^a + \varepsilon_b\varphi_4^b = \varepsilon_a\varphi_4^a + \varepsilon_b\varphi_2^b \tag{17}$$

$$\varphi_0^a = \varphi_0^b = \varphi_0,\ \varphi_1^a = \varphi_1^b = \varphi_1,\ \varphi_3^a = \varphi_3^b = \varphi_3 \tag{18}$$

Equation (15) multiplied by ε_a, and (16) multiplied by ε_b, then plus the two results together and be substituted into (17) and (18), we could get the Finite-Difference equation

$$\varphi_0 = K_1\varphi_1 + K_2\varepsilon_b \frac{K_2 + K_4}{K_2\varepsilon_b + K_4\varepsilon_a}\varphi_2 + K_3\varphi_3 + K_4\varepsilon_a \frac{K_2 + K_4}{K_2\varepsilon_b + K_4\varepsilon_a}\varphi_4 - P\varepsilon_a \frac{KK_4}{K_2\varepsilon_b + K_4\varepsilon_a} \quad (19)$$

Equation (13) (14) (19) would be expressed by Over-relaxation iterative method

$$\varphi_{(i,j)}^{(n+1)} = \varphi_{(i,j)}^{(n)} + \alpha R_{(i,j)}^{(n)} = \varphi_{(i,j)}^{(n)} + \alpha\left(K_1\varphi_{(i+1,j)}^{(n)} + K_2\varphi_{(i,j+1)}^{(n)} + K_3\varphi_{(i-1,j)}^{(n+1)} + K_4\varphi_{(i,j-1)}^{(n+1)} - KP - \varphi_{(i,j)}^{(n)}\right) \quad (20)$$

$$\varphi_{(i,j)}^{(n+1)} = \varphi_{(i,j)}^{(n)} + \alpha R_{(i,j)}^{(n)} = \varphi_{(i,j)}^{(n)} + \alpha\left(K_1\varphi_{(i+1,j)}^{(n)} + K_2\varphi_{(i,j+1)}^{(n)} + K_3\varphi_{(i-1,j)}^{(n+1)} + K_4\varphi_{(i,j-1)}^{(n+1)} - \varphi_{(i,j)}^{(n)}\right) \quad (21)$$

$$\varphi_{(i,j)}^{(n+1)} = \varphi_{(i,j)}^{(n)} + \alpha R_{(i,j)}^{(n)} = \varphi_{(i,j)}^{(n)} + \alpha\left(K_1\varphi_{(i+1,j)}^{(n)} + K_2\varphi_{(i,j+1)}^{(n)} + K_3\varphi_{(i-1,j)}^{(n+1)} + K_4\varphi_{(i,j-1)}^{(n+1)} - KP - \varphi_{(i,j)}^{(n)}\right) \quad (22)$$

The coefficient α is commonly named as accelerating convergence factor. Its value would determine the degree of Over-relaxation and then affect the convergence velocity of iterative solutions. Our actual calculating indicates that the convergence velocity is the fastest when α is 1.8.

So the Linear equation group, whose coefficient is known, would be available according above given calculating equations. In the discrete model, the electric field strength would be

$$\vec{E} = \frac{\varphi_1 - \varphi_3}{h_1 + h_3}\vec{i} + \frac{\varphi_2 - \varphi_4}{h_2 + h_4}\vec{j} \quad (23)$$

Of course the following equation is suitable when by using field map

$$E = \left| -\frac{\Delta\varphi}{\Delta n} \right|$$

3.3. Mathematical model of oil tank electric field

The only one difference of mathematical models of electric field strength numerical calculation between rail tanker and road tanker is that the length of cross axle may not be the same as that of the longitudinal axle in road tanker, while the two lengths of that of rail tanker are similar. And the one difference could only reflect in calculating programming. All the others are the same in both types of tankers. Related computer simulation programming would be discussed in following paper.

Acknowledgments

My special thanks and gratitude to the promoter Fund Project (No.20120041) for continues support during my work and while preparing this work.

References

[1] Binns K J and Lawrenson P J 1973 *Analysis and computation of Electric and Magnetic Field Problems* (Oxford, Pergamon Press)

[2] Ming Wu 1999 *The study of Electrostatic Field Simulation Model on Railroad Tank* Proceedings of the 4th Beijing International Conference on System Simulation and Scintific Coputing p 214

[3] Stokll R L 1967 *IEEE Trans Magn* **144** 83

[4] Davies J B and Muilwyk C A 1966 *IEEE Trans Magn* **133** 157

The simulation of electrostatics elimination effect influenced by the cross-section shape of discharge needles

Wei Zheng, Youping Hu*, **Chengguo Liu, Xiaolei Wei, Limin Hao**
Physics Department, Wuhan University of Technology, Luoshi Road No.122, Wuhan, China

E-mail: hp@whut.edu.cn

Abstract. In this paper, the electrostatics elimination effect influenced by the cross-section shape of discharge needles in a tube was studied. Firstly, the electrostatics elimination effect was discussed according to the spatial distribution of the electric field . Secondly, the electric field generated by the needles with cross-section of triangle and quadrangle was simulated by finite element method and then compared with the electric field generated by rounded cross-section needles. The results show that the electric field intensity generated by needles of different cross-section decreases in the following sequence: triangle,quadrangle and rounded shape. The needle with triangle cross-section generated the strongest electric field in the direction of tube axis, and the average value of electric field intensity is 12.6% larger than that of the rounded cross-section. So the needles with triangle cross-section are best to the elimination of electrostatics in pipelines.

1. Introduction

In the process of light oil transmission in the pipe, the flowing oil will generate electrostatics. When the electrostatics accumulates to a certain value, it will cause transport security risk. In order to reduce and prevent electrostatic accidents happened in pipe oil transmission, we use a pipeline electrostatic-eliminator at present.

The traditional pipe electrostatic-eliminator included oil conveying inner tube and a grounded outer tube rounded cross-section discharge needles was fixed on the outertube , and got through the inner tube. When the charged oil flowed into the tube, the capacitance generated by inner and outer tubes raised the potential of oil. So, the high potential difference generated between oil and grounded discharge needles. In needle tip area nearby, under the function of the strong electric field generated by the high potential difference, the charge in oil would be neutralized by the heterocharge inducted out by grounding, so as to achieve the purpose of eliminate electrostatics in oil. However, when rounded cross-section needle cusp was abraded, it caused the weakening of point effects. Comparing to the cusp, the other part of needle have less effect on electrostatics elimination, so the whole electrostatics elimination effect of the eliminator would be diminished.

In order to solve the problem, this paper proposes several discharge needle shapes with different cross-section. The cusp and side blades of this needle form the cutting-edge inductive zones, which greatly increases the existence range of cutting-edge inductive zone. It can not only eliminate the

---

* To whom any correspondence should be addressed.

electrostatic charge near the needle tip, but also can eliminate the electrostatic charge near the needle side. Thus electrostatics elimination effect can be improved.

For purpose of studying the influence of the electrostatics eliminate effected by different discharge needle of several tip angle sections, using the finite element analysis method, this paper do simulations about electric field intensity, which caused by rounded cross-section, triangle cross-section and quadrangle cross-section with different cusp angle, and determine the optimal discharge needle shape.

2. The eliminate electrostatic principle of needles with several cusps

When the charged oil flows into pipe electrostatic-eliminator, due to electrostatic induction, the strong point effects generate by the new needle with several cusps. So the charges in the oil neutralize quickly with heterocharge inducted from ground, and achieve the purpose of electrostatic elimination.

In the pipe electrostatic-eliminator, electrostatic elimination speed is determined by the modules of space electric field intensity value. From the electric field force formula, combining with Newton's second law, we know in any position of space there is:

$$\vec{F}(\vec{s},t) = e\cdot\vec{E}(\vec{s},t) = m_e\vec{a}(\vec{s},t) = m_e\frac{d\vec{v}(\vec{s},t)}{dt} \tag{1}$$

In the formula, e is charge quantity of single charge in the oil, m_e the mass of charge, $\overline{E}$ the electric field in space. In the flowing oil, they are constants in steady state. Integrate the formula above, it can be expressed as:

$$e\cdot\vec{E}(\vec{s},t)t = m_e\left[\vec{v}(\vec{s}_1) - \vec{v}_0(\vec{s}_2)\right] \tag{2}$$

In the formula, $\vec{v}_0$ oil flow speed, it equates initial speed of charge. If time is constant, to make sure the charge can finish a distance as quickly as possible, we need a larger $\overline{E}$, which means a larger potential gradient. If a larger potential gradient exist in pipe electrostatic eliminator, it can eliminate charge faster and have better effects. Therefore, we can use the electric field strength in the pipe size as important evaluation parameter, which can evaluate the oil electrostatic eliminating effect in a pipeline electrostatic-eliminator.

3. The introduce of model

3.1. Geometric part of Model

For studying the electrostatics elimination effect on needle section, this paper constructs model is shown in figure 1. Considering the interval of discharge needle between each layer is 10-40 cm in eliminate charge tube, for simplicity the simulation space is set as a 400 mm×400 mm rectangle to simulate an axis direction section of tube. The lines that simulate out electric field value are expressed as lines r=0 and r=1/2R in the sketch.

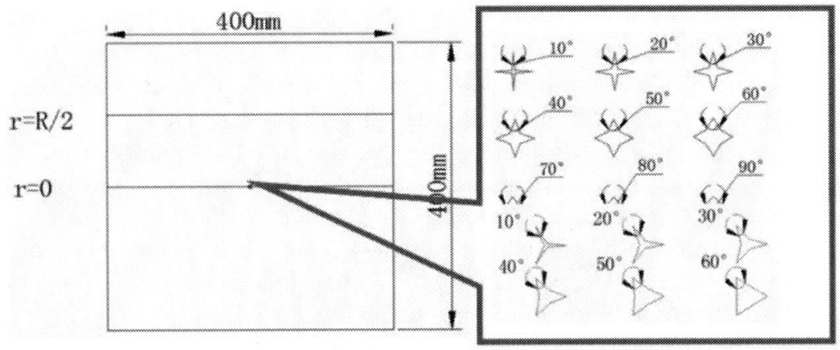

Figure 1. Structure of the model sketch and specific shape of the simulation samples.

In the axis of the text space, one rounded whose radius is 4 mm to express the discharge needle section is constructed. These sections can be replaced with triangle and square that inscribe the roundeds. From this, the other two basic types of simulation model can be construct. Meanwhile, based on two kinds of simulation models, through the method of changing interior angle of every section, several samples of needle section was simulated.

3.2. Boundary condition

The boundary conditions must be defined when solving the equations. Generally, speaking the charge density at the entrance of tube is known, and after charge generate in tube, if the tube is long enough the charge density will be fully developed. So the boundary conditions is $q\,|_{x=0}=0$, $\partial q/\partial x\,|_{x=L}=0$. L is the tube length. Boundary conditions in the tube wall involve complex electric mechanisms. We simplify mechanisms as "diffusion", according to the relative paper we have the following boundary conditions [1]:

$$C=\partial q/\partial r\,|_{r=R}=\frac{k\sigma_0 A_t (U_m)^n}{2FeD_0^2 T}\exp\left(-\frac{W_\sigma-2W_D}{kT}\right) \tag{3}$$

In this equation, R is radius of tube; F is Faraday constant; k is Boltzmann constant; e is electron charge; K is Kelvin temperature; W_σ and W_D are activation energy of conductivity and molecular diffusion; σ_0 and D_0 are coefficients in equations of oil conductivity and molecular diffusion coefficient in the form of activation energy; U_m is average flow velocity of oil.

According to M. J. Lee and J. K. Nelsons' theory [2], we make σ_0=2.35×10^{-6} S·m^{-1}, W_σ=5.37×10^{-20} J, W_b=W_D=5.22×10^{-20} J, D_0=1.4×10^{-8} m^2·s^{-1}. Meanwhile, the main purpose of adding At and n is fixing the oil aging, wall roughness and other factors that are difficult to quantify description, their values must be confirmed by experiment. Here, A_t=20.0, n=1 [3].

3.3. Initial condition

Abedian's research shows that [4] , the change of average charge density that in the direction of pipeline can be approximately considered as exponential change in the pipeline, when the charge density obtain full development in oil, the charge density change have nothing to do with x direction. So:

$$-\frac{1}{r}\frac{\partial}{\partial r}\left[r\left(D_m+D_T\right)\frac{\partial q}{\partial r}\right]+\frac{q}{\tau}=0 \tag{4}$$

Therefore, the charge density having different expressions in different layers, according to other literature, the simultaneous equation is expressed as:

$$\begin{cases} q_1(y)=\dfrac{\left[q_2(R)+C\lambda\sinh(\delta/\lambda)\right]}{\cosh(\delta/\lambda)}\cosh\left(\dfrac{y}{\lambda}\right)-C\lambda\sinh\left(\dfrac{y}{\lambda}\right) \\ q_2(r)=\dfrac{C\lambda^2\operatorname{sech}(\delta/\lambda)\mathrm{I}_0(r/\lambda_T)}{\lambda\tanh(\delta/\lambda)\mathrm{I}_0(R/\lambda_T)+\lambda_T\mathrm{I}_1(R/\lambda)} \end{cases} \tag{5}$$

In the simultaneous equation, $\lambda=\sqrt{D_m\tau}$ is Debye Length; $\lambda_T=\sqrt{D_T\tau}$ is turbulence Debye Length; r is distance to pipe center, y is distance to pipe wall; q_1 represent charge density distribution in the sticky bottom layer; q_2 represent charge density distribution in turbulence core area; I_0, I_1 is Bessel function; δ is the thickness of sticky bottom layer; it is changed by difference of oil fluid property, but the magnitude under 0.1 mm, the value is far less than the radius of the pipeline, here it have a same length as the Debye Length [5]. So when the electric field value in some positions were discussed, only the value of q_2 was considered.

The materials of the samples are set to high strength steel alloy and the relative dielectric constant

Journal of Physics: Conference Series **418** (2013) 012017 doi:10.1088/1742-6596/418/1/012017

is 1. The rectangular space around the sample is defined as limited size space, and fill it with the charge density that defined by equation. The material inside of the model is set to kerosene and the relative dielectric constant is set to 1.7, which refer to JP4 kerosene. The potential of needle is set to 0 V, and then the electric field and potential distribution in the pipeline section can be calculated.

Above all, in order to avoid the simulated result differences caused by the different parameters above, the test samples in the simulation are all in the same file and use the same kind of grid partition method and the same solver.

4. Result of simulation and discussion

The electric field intensity distribution in the tube was shown in figure 2. Electric field intensity distribution effected by several different needle shapes that with three cusps was extracted.

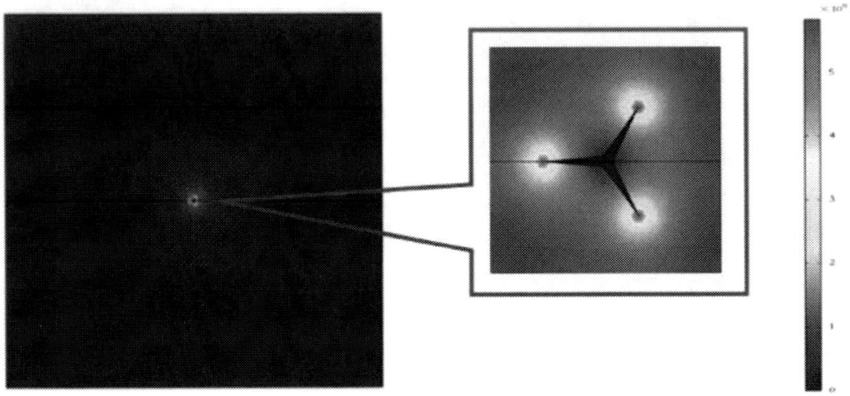

Figure 2. The electric field intensity distribution caused by three cusps needle section with 10° cusp angle.

In order to assess the electrostatics elimination effect of different needles with three cusps and four cusps, electric field intensity average in the space and particular directions need to be compared. In this paper, needles whose cross-sections have 10°-90° blade cusp angle were chosen, and then the effect of cross-sections on average field intensity was simulate. Similarly, needle cross-sections was replaced by a circumscribed rounded section, and then the electric field intensity average in corresponding place were simulated, all values above shown in figure 3-5.

In figure 3, x-coordinate represents the tip angle of cross-section, and y-coordinate represents the module value of electric field intensity. From the roughly trend of the curve in graph, it shows that when the tip angle getting smaller, the average value of electric field intensity module in r=0 position tend to increase. Meanwhile, comparing the maximum with minimum values that express the four cusps needle with 10° angle and three cusps needle in figure above, it turns out that the electric field intensity that formed by three cusps needle section with 10° cusp angle is 12.6% is higher than rounded section needle in this space.

The figure 4 expresses the average value of electric field intensity generated by the needle section with different tip angle in the $r=R/2$ position. In the image above, the electric field intensity have a minimum when the tip angle is around 40°, therefore this angle should be avoided for use in practice. Comparing the maximum with minimum values that express the three cusps needle section with 10° cusp angle and quadrangle needle in table above, it turns out that the electric field intensity that formed by three cusps needle section with 10°cusp angle is 0.05% is higher than rounded section needle in this space.

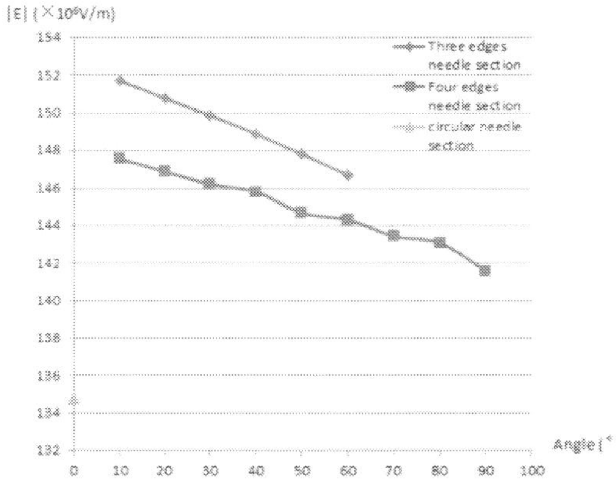

Figure 3. Average value of the electric field module when r=0 and with different needle section.

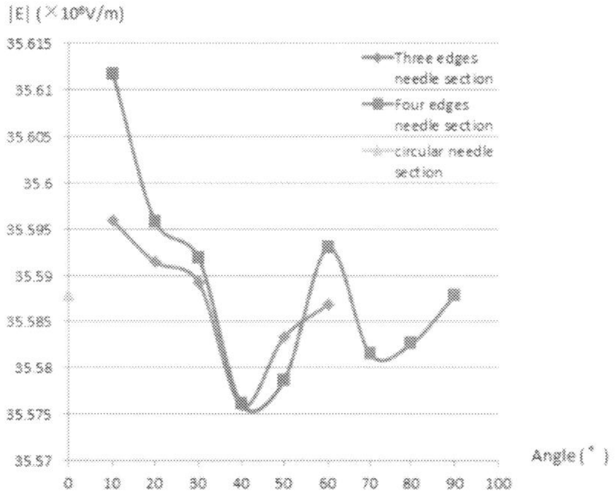

Figure 4. Average value of the electric field module when r=R/2 and with different needle section

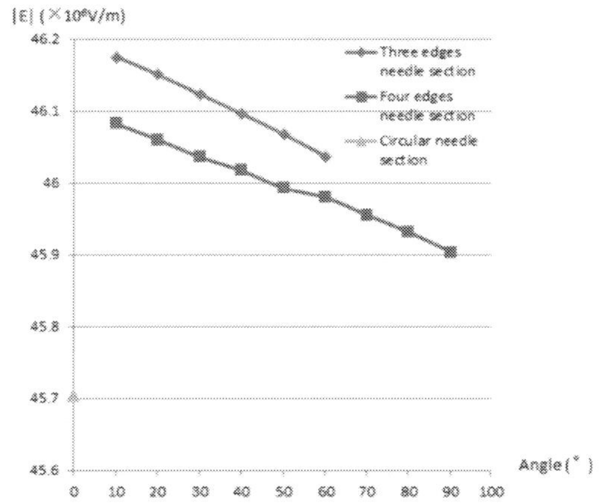

Figure 5. Average value of the electric field module with different needle section in the tube section.

The figure 5 expresses the average value of electric field intensity generated by the needle section with different tip angles in space. Similarly, comparing the maximum with the minimum values that express the three cusps needle section with 10° cusp angle in figure above, it turns out that the electric field intensity that formed by three cusps needle with 10° cusp angle is 1.02% is higher than rounded section needle in this space.

Although there may exist errors because of meshing is not much fine in the area nearby tip needle, but the three cusps section needle that have smaller cusp angle can form higher electric field intensity than common rounded section needle.

5. Conclusion

This paper gives a potential distribution diagram affected by different charge needle section on tube axial section, and simulates the electric field model average in tube section and particular direction. From the simulation results, the triangle needles that have smaller section cusp angle can form stronger electric field than common rounded section in tube, especially in the axial direction of tube section, and the electric field model average growth is 12.6%. In the whole space, average of the electric field model also can increase by 1.02%. Considering that charge needles are multilayer ring arranged in the electrostatics elimination tube, actually there will be more charge needles than that in the model for the static elimination tube. Therefore, in practice the point effects formed by more needles with smaller angle will increase electric field model average more than before, thus it is benefit for charge needle to collect the charge in the oil.

References

[1] Wang J F and Meng H L 2009 *IEEE Trans. Dielectr. Electr. Insul.* **16** 299
[2] Brubaker M A and Nelson J K 1997 *IEEE Trans. Dielectr. Electr. Insu.l* **4** 157
[3] Abedian B and Sonin A A. 1986 *Chem. Eng. Sci* **41** 3183.
[4] Abedian B and Sonin A A. 1982 *J. Fluid. Mech.* **120** 199
[5] Gerard T 2001 *Electrost.* **51** 440

7th International Conference on Applied Electrostatics (ICAES-2012) IOP Publishing
Journal of Physics: Conference Series **418** (2013) 012018 doi:10.1088/1742-6596/418/1/012018

Calculation of absorption cross section in a reverberation chamber

Rui Jia*, **Qing Guo Wang, Shu Qiao Wang**
Research Institute of Electrostatic & Electromagnetic Protection, Mechanical Engineering College, Shijiazhuang 050003, China

Email: piaoyang315@163.com

Abstract. In this paper, the mechanism of the power lost in the reverberation chamber is analyzed and a model to calculate the absorption cross section of loaded object is introduced. The cavity quality factor Q can be obtained from calculating the absorption cross sections of loaded objects. The cavity Q is proportional to the frequency selected, the relative permittivity and inversely proportional to the average absorption cross section of the lossy objects.

1. Introduction

Reverberation chambers are being used as an alternative test facility for a wide range of electromagnetic compatibility measurements, such as radiated immunity, radiated emissions, shielding effectiveness of composite materials, antenna efficiency and probe calibration. In practice, a typical reverberation chamber has lossy walls and is loaded with other objects, many of which absorb electromagnetic energy. This behavior may influence the chamber quality factor (Q) and the performance of the chamber. The key issue is whether these losses degrade the reverberation performance under an acceptable value, which needs further more research. Our concern, now, is what the relation between the quality factor and the loaded reverberation chambers.

A convenient way to model the losses of the loaded objects is by the absorption cross section which depends on the size, shape, and electrical parameters of the loaded objects. Especially when the losses are related to incident plane waves, the absorption cross section is an intuitively suitable parameter because it is an equivalent area over which an amount of the incident power falls [1]. The purpose of this paper is to describe how to calculate the absorption cross sections. The second part of this paper analyzes the path of the power lost in the reverberation chamber, and the third section introduces a model and investigates the absorption cross section deeply. Finally, in section fourth conclusions will be drawn.

---

* To whom any correspondence should be addressed.

2. The path of power lost

The power dissipated in reverberation chamber can be divided into two parts, one is dissipated in cavity walls, the other one is absorbed by loaded objects within the cavity. The Q of a reverberation chamber with finitely conducting walls and loaded objects is given by the following [2]:

$$Q^{-1} = Q_{wall}^{-1} + Q_{load}^{-1} \tag{1}$$

The Q is associated with the total power lost in cavity, and the Q_{wall} is associated with the wall loss, and the Q_{load} is associated with the loss due to the loaded objects. The Q_{wall} can be expressed as:

$$Q_{wall} = 3V/2S\delta\mu_r \tag{2}$$

Where S is the surface area of the walls of the cavity, V is the volume of the cavity, μ_r is the relative permeability of the walls, δ is the skin depth of the wall. And the Q_{load} can be expressed as:

$$Q_{load} = 2\pi V/\lambda\langle\sigma_a\rangle \tag{3}$$

Where $<\sigma_a>$ is the averaged absorption cross section.

From equations (2) and (3), it is not difficult to find the averaged absorption cross section $<\sigma_a>$ is the key factor of equation (1), for which can be used to determine the Q of the whole chamber with many of loaded spheres.

3. Absorption cross section

The absorption cross section of the loaded object is defined as the power dissipated in the object under planewave incidence [3]. The literature [4] gives the expression of absorption cross section, as follows:

$$\langle\sigma_o\rangle = \pi a^2(\eta_t - \eta_s) \tag{4}$$

Where:

$$\eta_t = \frac{2}{(ka)^2}\sum_{n=1}^{\infty}(2n+1)\mathrm{Re}\left[a_n + b_n\right] \tag{5}$$

$$\eta_s = \frac{2}{(ka)^2}\sum_{n=1}^{\infty}(2n+1)\left[|a_n|^2 + |b_n|^2\right] \tag{6}$$

The parameters a_n and b_n are given by:

$$a_n = \frac{\psi'_n(y)\psi_n(x) - m\psi_n(y)\psi'_n(x)}{\psi'_n(y)\xi_n(x) - m\psi_n(y)\xi'_n(x)} \tag{7}$$

$$b_n = \frac{m\psi'_n(y)\psi_n(x) - \psi_n(y)\psi'_n(x)}{m\psi'_n(y)\xi_n(x) - \psi_n(y)\xi'_n(x)} \tag{8}$$

Where $x=ka$, $y=mx$, $k=2\pi/\lambda$ and is the free-space wave number, and m is the refractive index. ξ_n, Ψ_n are the Riccati-Bessel functions defined in [5] and [6].

The equation (4) is based on the assumption that the loaded objects are all homogeneous spheres, and equation (4) expressed the absorption cross section of one single object. With this assumption, a is the radius of loaded object, and the total absorption cross section of all objects can be calculated by:

$$\langle\sigma_a\rangle=\sum_{i=1}^{N}\langle\sigma_{oi}\rangle \tag{9}$$

If the objects are identical, (9) can be simplified to

$$\langle\sigma_a\rangle=N\langle\sigma_{oi}\rangle \tag{10}$$

Actually, the incident electromagnetic wave to the surface of objects is oblique, as shown in figure 1. θ_i is the angle of incidence to the normal of the surface, and θ_m is the angle of the refracted wave (inside the material) to the normal, and θ_e is the angle of the refracted wave (outside the material) to the normal.

In practice, the absorption cross section of a homogeneous sphere is independent of both the incidence angle and polarization of the incident field [2]. This is identical to the reverberation chamber environment characteristics which are stated as statistically uniform, isotropic and randomly polarized. With the discussion above, obviously, the absorption cross sections of the loaded spheres have a closed relation with the relative permittivity of the loaded objects. A plot of the result can be seen in figure 2.

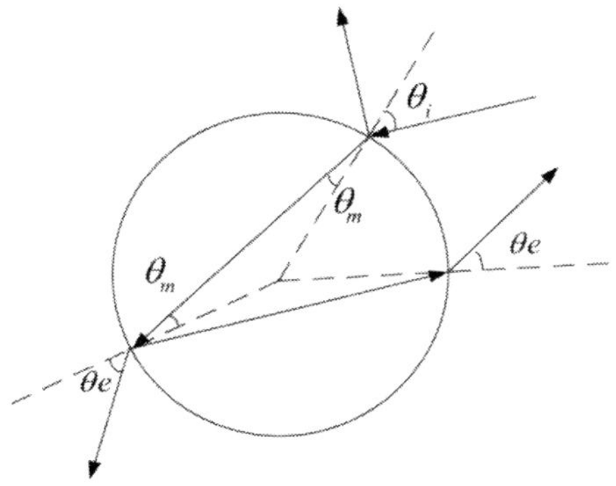

Figure 1. Example of planewave into sphere object.

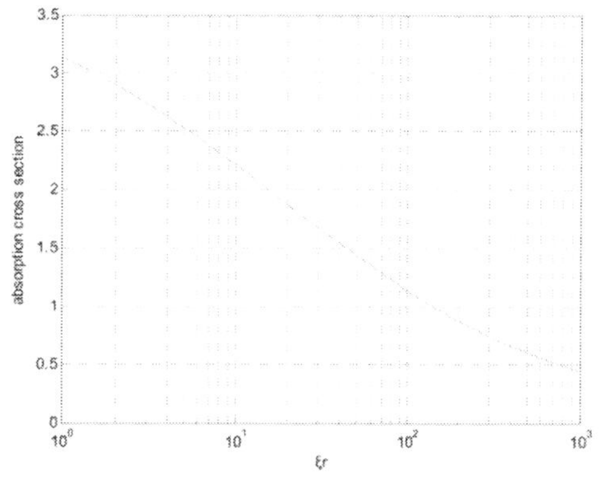

Figure 2. Absorption cross section for different relative permittivity.

From figure 2, it is not difficult to obtain the conclusion that the absorption cross section fall down with the increasing relative permittivity of the loaded objects, which indicated that the loaded dielectric object should be chosen with larger relative permittivity.

The curve in figure 3 is the relation between absorption cross section and the radius of sphere, which seems like quadratic curve.

Journal of Physics: Conference Series **418** (2013) 012018 doi:10.1088/1742-6596/418/1/012018

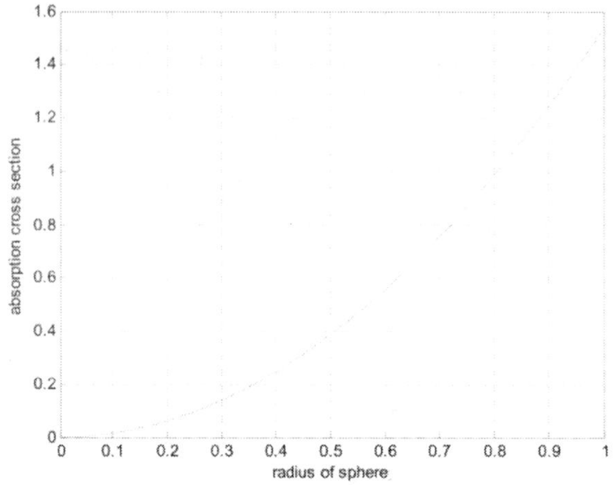

Figure 3. Absorption cross section for different radii of sphere objects.

The absorption cross section can be used to determine the Q of the chamber with lots of loaded objects, and figure 4 shows the Q of loaded chambers with various numbers of frequencies. The more absorption cross sections, the more electromagnetic energy lost. As expected, the Q of the loaded chamber decreases with the increasing absorption cross sections. IEC61000-4-21 provides an access to evaluate the reverberation chamber performance, while it is possible to investigate the degradation of the useable of a reverberation chamber use the quality of Q. But the problems are how much quality of Q lost can be seen as the chamber's effectiveness degrades and what is the expression for the threshold metric of Q that must be exceeded. Both of these are needed to investigate more in further time. Figure 5 gives the same trend that the decrease in the chamber with the absorption cross section as shown in figure 4.

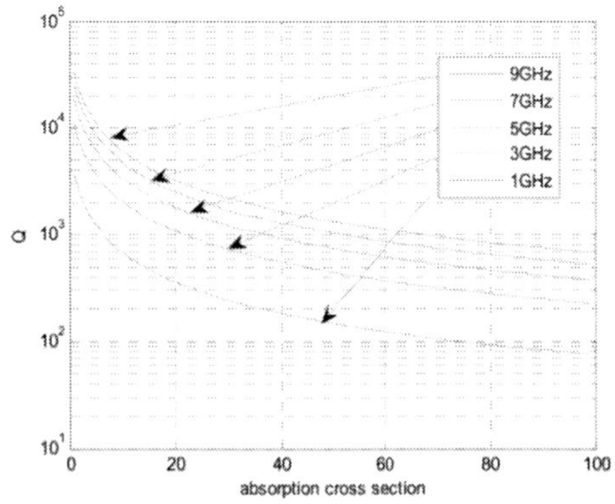

Figure 4. Quality factor Q for different absorption cross sections.

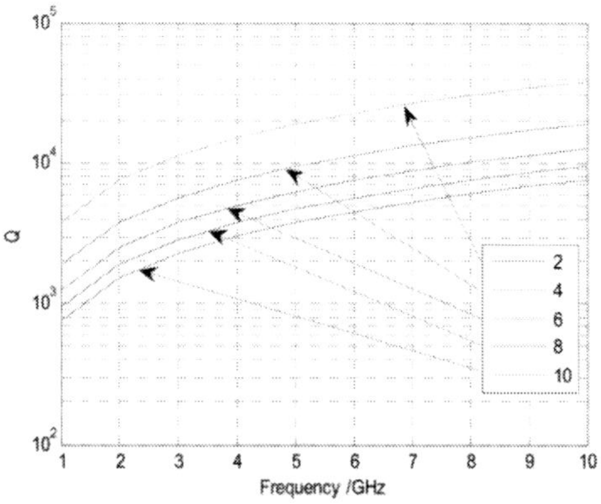

Figure 5. Quality factor Q for different frequencies.

According to figure 4 and figure 5, the Q of the loaded chamber increases with the increasing frequency. In figure 4, we can see that as the frequency increases, the curves are much closer to each other, which indicates that the quality of Q trend to be horizontal up to high frequencies. It means that the cavity Q should be within a certain range, and not be infinite. A completely lossless reverberation chamber results in poor performance. These calculations are all used the reverberation chamber model with the dimensions as 4.3 m×8 m×10.4 m.

After the discussion, the quality of Q can be relation with the relative permittivity of the loaded objects. Figure 6 presents the relation between the quality factor Q and he relative permittivity of the loaded objects. As expected, the Q of the loaded chamber increases with the increasing relative permittivity and frequency.

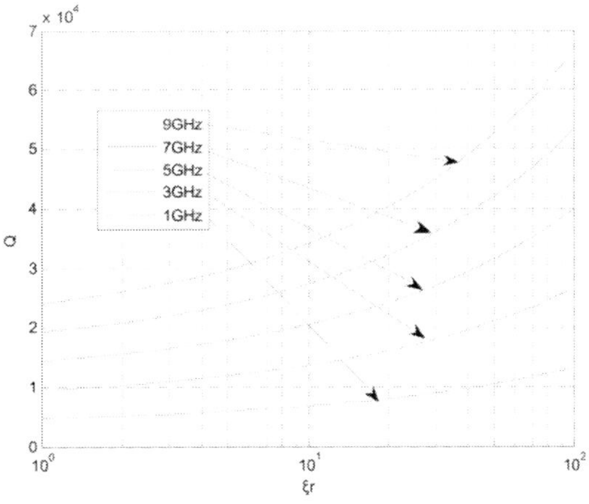

Figure 6. Quality of Q for different relative permittivity.

4. Conclusion

This paper presents a model to calculate the absorption cross sections of loaded objects within the reverberation chamber. This model simplified the loaded objects into a various number of identical lossy spheres. The cavity Q can be obtained from calculating the absorption cross sections of loaded objects. The analyses show that the absorption cross sections have closed relation with the radius, relative permittivity of the lossy spheres. The cavity Q is proportional to the frequency selected, the relative permittivity and inversely proportional to the average absorption cross section of the lossy objects. However, how to evaluate the chamber's effectiveness degrades and what is the expression for the threshold metric of Q that must be exceeded. These topics should be further investigated in the future research.

References

[1] Hallbjorner P, Carlberg U, Madsen K, et al. 2005 *IEEE Trans. on EMC* **47** 291
[2] Holloway L, Hill A, Ladbury J, et al. 2006 *IEEE Trans. on EMC* **48** 187
[3] Carlberg U, Simon P,et al. 2004 *IEEE Trans. on EMC* **46** 146
[4] Holloway, Christopher L, Hill A, Ladbury J, et al. 2003 *IEEE Symp. on EMC* Boston p 834
[5] Hill A, Ma M, Ondrejka R, et al. 1994 *IEEE Trans. on EMC* **36** 169
[6] Van de Hulst H C 1981 *Light scattering by small particles*. New York Dover

Journal of Physics: Conference Series **418** (2013) 012019 doi:10.1088/1742-6596/418/1/012019

Numerical simulation of oil charge density measured in pipeline by built-in capacitive coupling electrode

Xiaolei Wei, Youping Hu[*], Chengguo Liu, Wei Zheng
College of Science, Wuhan University of Technology, 122 Luoshi Road, Wuhan, China

E-mail: hp@whut.com.cn

Abstract. This paper presents a method of measuring charge density of oil in pipeline by built-in capacitive coupling inductive electrode. This new type of electrode solved the problem of charge leak in the contacting measurements, and effectively reduced the influence of external factors on the measured results. The three-dimensional model of the electrode in the pipeline was built and was calculated to give the amount of charge captured by the outer electrode which changed with the charge density of oil in the pipeline. The potential of the inner electrode was obtained on the basis of the relationship between the amount of charge and the capacitance, by which the charge density of oil in the pipeline was reflected. This research provided a more effective method for measuring charge density of oil in pipeline.

1. Introduction

Since the 1950s, it has never stopped that the electrostatic accidents of oil tanker caused by loading/ unloading oil too fast or other reasons. The accidents of fire and explosion in tankers have also occurred occasionally in our country. The oil products are high insulation and easily produce large amount of electrostatic charge in the process of loading/unloading and transportation. The electrostatic charge brings great risk of safety in the transportation of oil products. Therefore, it's particularly important to monitor the electrification situation in oil transportation. There are a variety of ways to measure oil charge density. The contacting measurement and contactless measurement are the most common. Since contacting electrodes leak charge in measurements and the re-measuring needs to be re-charged to steady state, the contacting electrodes are gradually substituted by contactless electrodes. Contactless electrodes mainly depend on the potential signal which is generated by the induced charge to reflect the charge density. Generally, the inductive portion of conventional electrodes for measuring oil charge density is placed outside of the pipe. Thus, the measuring results can be easily influenced by external factors. In this paper, the capacitive coupling electrode is placed inside the pipe, effectively reducing the impact of external factors.

2. Theory

Built-in capacitive coupling electrode was consisted of a coaxial cylindrical capacitor, the inside pillar was called the inner electrode, and the outside cylinder was called the outer electrode. While the outer electrode captured the charge in oil, the potential of inner electrode changes with the induced charge. For a certain structure of the capacitive coupling electrode, its capacitance value was determined.

[*] To whom any correspondence should be addressed.

Therefore, the quantity of charge captured by the electrodes was proportional to the potential differences. Once the reference potential of the outer electrode determined, the potential signal of inner electrode was given by $V_{inner}=V_{outer}-Q/C$. This reflected the relationship between the potential of inner electrode and the amount of charge captured by outer electrode. The built-in capacitive coupling electrode was placed in oil pipe, fixed on the pipe wall. In order to avoid the tip effect, the end of the electrode inserted into the pipe was designed to hemispherical. The other end was the fixing disk welded with the inner electrode. Both sides of the fixing disk use special insulation spacer whose diameter was bigger than that of the disk used to insulate the pipe. Simulating the relationship between the amount of charge captured by outer electrode and the charge density of oil in pipe, the relationship between charge density and the potential of inner electrode can be obtained.

3. Model

The three-dimensional model of the electrode in pipe was built. The outer electrode was a hollow cylinder of inner diameter 6 mm and outer diameter 8 mm. The inner electrode was a pillar of diameter 2 mm. It was filled with 2 mm dielectric medium between the inner electrode and the outer electrode. The diameter of the fixing disk and the spacers are 20 mm and 22 mm, respectively. The inner diameter of the pipe was 80 mm, the thickness of pipe wall was1.5 mm. A diameter of 10 mm circular hole (the outer diameter of electrode was 8 mm) was on the pipe wall, and the electrode was placed in pipe through the centre of the hole, the endpoint of the electrode at the axis of the pipe. In order to build the model, multi-layer drawing was used to build each part, so various parts of the model structure could be shown or hidden. For observing easily, the sleeve welded vertically in the pipe and the bolt for fixing the electrode was hidden, as shown in Figure 1.

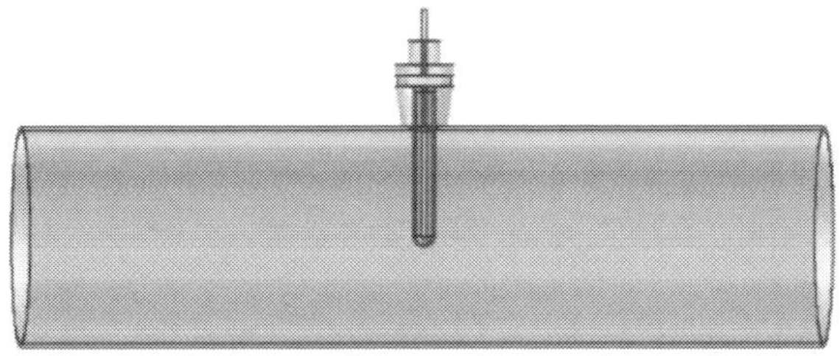

Figure 1. The model of electrode in pipe.

In calculations, the attachments for fixing the electrode have a little influence on the result of the simulation and the setting of their parameters and boundary conditions are complicated. To optimize the simulation process, all attachments are hidden. The Figure 2 was the diagram of the grid model using the finite element method.

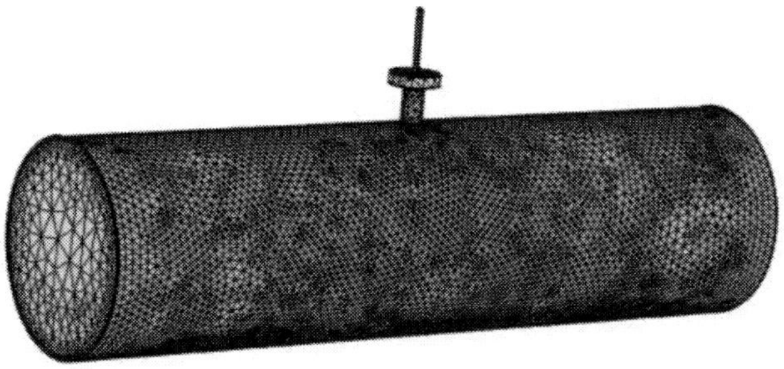

Figure 2. The grid model of electrode in pipe.

Journal of Physics: Conference Series **418** (2013) 012019 doi:10.1088/1742-6596/418/1/012019

4. Simulation process

4.1. Simulation conditions

When oil products flow through the pipe, electrical double layer (compact layer and diffuse layer) is formed at the pipe wall due to different adsorption capacity of various materials. According to the Debye length, the thickness of diffuse layer is less than 1 mm [1]. The rate of oil flow in pipeline is not uniform, viscous layer, transition layer and diffuse turbulent layer being defined from the pipe wall to its center. Since diffuse layer is thicker than viscous layer [2], charge in diffuse layer is brought to the transition layer or the turbulent layer. The oil carries charge and the charge density fits for the equation (1) [3]:

$$\vec{U} \bullet \nabla q = D_m \nabla^2 q - q/\tau \tag{1}$$

Here, q is the charge density, $\vec{U}$ is the speed vector of oil, D_m is the molecular diffusion coefficient, and τ is the relaxation time of the oil. This formula is applicable to laminar and turbulent flow. Under normal circumstances, the oil at the entrance of the pipe is not charged. When the oil charges sufficiently in pipeline to reach steady state, the charge density does not change. So,

$$q\,|_{x=0} = 0, \quad \partial q/\partial x\,|_{x=L} = 0,$$

Here, x is the pipe axial coordinate, L is the length of pipe. The boundary conditions at the pipe wall can be obtained as (2)[4]:

$$\partial q/\partial r\,|_{r=R} = \frac{k\sigma_0 A_t (U_m)^n}{2FeD_0^2 T}\exp\left(-\frac{W_\sigma - 2W_d}{kT}\right) = C \tag{2}$$

Where R is the pipe radius, k is the Boltzmann constant, F is the Faraday's constant, e is the electric charge, σ_0 and D_0 are coefficients of the expressions, W_σ and W_d are activation energies, and T is the absolute temperature. According to the above boundary conditions, the charge density of oil in the turbulent flow can be expressed as (3)[5]:

$$q(r) = \frac{C\lambda^2 \operatorname{sech}(\delta/\lambda) I_0(r/\lambda_T)}{\lambda \tanh(\delta/\lambda) I_0(R/\lambda_T) + \lambda_T I_1(R/\lambda)} \tag{3}$$

Which q(r) represents the distribution of charge density in turbulent zone; $\lambda = (D_m\tau)^{1/2}$ is the Debye length, $\lambda_T = (D_T\tau)^{1/2}$ is the turbulence Debye Length; r is the distance from the center of the pipe; I_0 and I_1 are the Bessel functions; δ is the thickness of the viscous sub-layer.

4.2. Parameter settings

Various parameters involved must be defined in the simulation process, all parameters' units use the international unit. In the expression of boundary conditions, R=0.004 m, $k=1.38\times10^{-23}$ J K^{-1}, $F=9.649\times10^4$ C mol^{-1}, $e=1.6\times10^{-19}$ C, $\sigma_0=2.35\times10^{-6}$ S m^{-1}, $D_0=1.4\times10^{-8}$ m^2 s^{-1}, $W_\sigma=5.37\times10^{-20}$ J [6]. A_t and n mainly amends the influence of oil ageing, the roughness of pipe wall or other factors which are difficult to be quantified, and their values are determined in experiment, here At=20.0, n =1 [6]. The materials of pipe wall, inner and outer electrode in the model are set to High-strength alloy steel. Dielectric material between the electrodes is the epoxy resin, the dielectric constant of 4, oil for Gasoline and dielectric constant of 1.7. In the expression of the charge density, the viscous sub-layer thickness δ changes according to the fluid nature of the oil, but the order of magnitude is less than 0.1 mm, and its value is much less than the pipe radius. It is assumed that the thickness δ is equal to Debye length in the simulation.

Journal of Physics: Conference Series **418** (2013) 012019 doi:10.1088/1742-6596/418/1/012019

5. Simulation results and discussion

5.1. Charge density distribution of oil in pipe

Three-dimensional model was imported into the electrostatic calculation module to simulate in the preset steady-state. In order to optimize the simulation process, the midpoint of the axis of pipe is set to the origin of coordinates, the axis of pipe and the x-axis coincided, the axis of electrode and the y-axis coincided. Therefore, the charge density was expressed as (4):

$$q(y,z)=\frac{C\lambda^2\operatorname{sech}(\delta/\lambda)I_0\left[\left(y^2+z^2\right)^{1/2}/\lambda_T\right]}{\lambda\tanh(\delta/\lambda)I_0(R/\lambda_T)+\lambda_T I_1(R/\lambda)} \tag{4}$$

In this condition, shielding the influence of the electrode on charge density, charge density distribution of oil in pipe was shown in Figure 3.

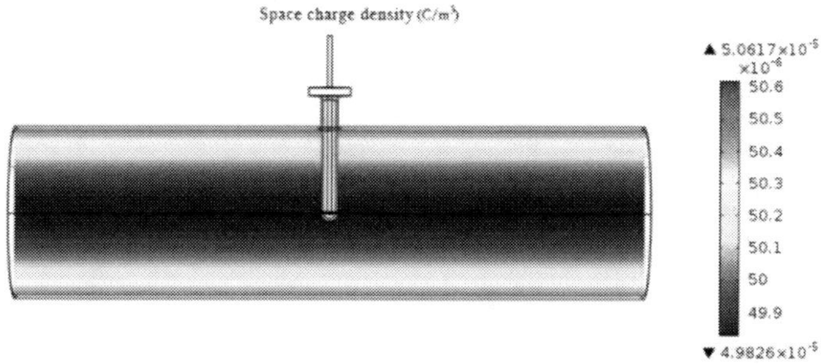

Figure 3. Charge density distribution of oil in pipe.

As shown in Figure 3, it could be found that low charge density at the center of pipe and high electrification was close to pipe wall. This indicated that the main reason for the electrification in oil pipe was the charge in diffuse layer taken away by oil flow. The simulation was built on the basis of no impurity contained in oil. However, in fact, the oil inevitably carried impurities so that charge density distribution might be different from the simulation results.

5.2. Potential distribution in pipeline and the calculation of the electrode potential

Potential distribution in the model could be obtained by the electrostatics calculation module. According to this module, the outer electrode was defined for the reference point of potential whose electric potential was 0V, and the potential distribution in the model was shown in Figure 4.

From the Figure 4, it could be seen that after the insertion of electrode, the potential distribution changes, the potential fell down around the electrode, which tended to balance after a distance equal to the diameter of the pipe. Under the drive of potential, the electrons moved to the electrode surface from the oil. The amount of charge captured by the outer electrode in the model could be obtained by global calculation. The calculation result was -7.553×10^{8}C. The electrostatic module contained the calculation of capacitance. The model of electrode was simulated separately to get a more accurate capacitance. The calculation result was 11.78 pF. The amount of charge and the capacitance value was known, and the potential of inner electrode was given by $V_{inner}=V_{outer}-Q/C=6.412\times10^{3}$ V.

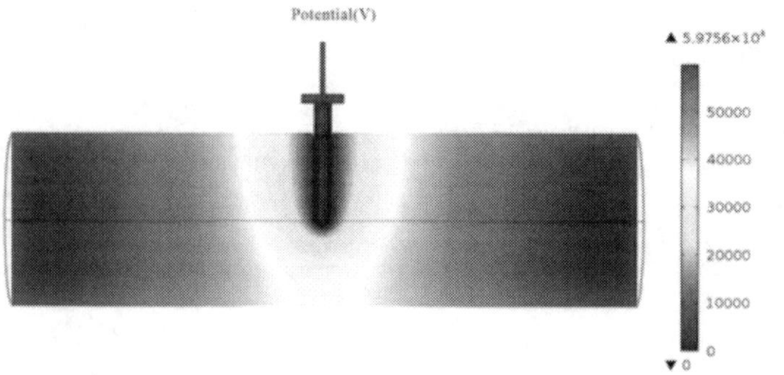

Figure 4. Potential distribution in the model.

6. Conclusion

This paper presents the model and simulation of a new type built-in capacitive coupling inductive electrode. As contactless electrode, the charge leak is avoided in measurements. The inductive portion of the electrode is placed inside the pipe, effectively reducing the influence of external factors on the measurement results. According to the law of electrification in oil pipe, the distribution of charge was simulated. In the case of the oil free of impurities, the charge density of oil at the centre of the pipe is lower than that of the pipe wall. Setting the outer electrode as the terminal, the amount of charge captured by outer electrode is calculated by the terminal charge calculation module. The amount of charge captured by the outer electrode changes according to the charge density of oil, the relationship between the potential of inner electrode and outer electrode is determined by the electrode capacitance. Therefore, the potential of inner electrode and charge density are interdependent, and the measurement of charge density in oil can be realized by this dependence.

References

[1] Martin G 2001 *J. Electrostat.* **51-52** 359
[2] Touchard G 2001 *J. Electrostat.* **51-52** 440
[3] Junfen W and Jianing P 2006 *Acta Petrolei Sinica* **27** 133
[4] Jufen W and Haolong M 2009 *IEEE Trans. Dielectr. Electr. Insul.* **16** 299
[5] Jufen W and Haolong M 2010 *J. China University of Petroleum (Edition of Social Sciences)* **34** 131
[6] Brubaker M and Nelson J 1997 *IEEE Trans. Dielectr. Electr. Insul.* **4** 157

Journal of Physics: Conference Series **418** (2013) 012020 doi:10.1088/1742-6596/418/1/012020

Numerical analysis on effective electric field penetration depth for interdigital impedance sensor

Chon-ung Kim*[1, 2], Guofeng Li[2], Jie Li[2], Hakchol Jong[1, 2], Cholwu Ro[1], Yunho Song[3], Gilhung Pak[1], Songil Im[1],

[1]Department of Electrical Engineering, Kim Chaek University of Technology, Yongguang Street, Pyongyang, D.P.R. Korea

[2]School of Electrical Engineering, Dalian University of Technology, Linggong Road 2, Dalian 116024, P.R. China

[3]Hamhung Institute of Water Resource and Power, 2.8 Street, Hamhung, D.P.R. Korea

E-mail: kcu1967@163.com

Abstract. Interdigital (finger-like) electrodes are widely used for electrical impedance and capacitance tomography of composite dielectric materials and complex insulating structures. Because of their advantages, they are now effectively introduced as capacitance sensors into a variety of industrial branches, agriculture, medical science, biological engineering, military branches, etc. In order to effectively apply the so-called interdigital impedance sensors in practice, of great importance is to optimize the sensor design parameters such as the electric field penetration depth, signal strength and so on. The general design principles of the interdigital capacitance sensor have been discussed for a long time by many researchers. However, there is no consensus on the definition of the effective electric field penetration depth of interdigital electrode. This paper discusses how to determine the effective electric field penetration depth of interdigital sensor on the basis of the refractive principle of electric field intensity and the FEM analyses of electric field distribution and capacitance for the sensor model.

1. Introduction

Dielectric permittivity and loss tangent are the important electrical properties of dielectrics and electrical insulating materials, which have been widely used as non-destructive diagnostic parameters in industrial applications, including the monitoring of curing degree for plastics and rubbers, the humidity measurement in oils, papers and gaseous media, the measurements of concentration for emulsions, suspensions and bio-cells, the measurements of thickness for dielectrics and the height level of liquids, the detection of freshness of foods and so on[1-13].

Traditional dielectrometry is based on measuring permittivity and loss tangent of dielectric samples placed in conventional parallel or cylindrical plate electrode cell, which is often not suitable to many practical cases, i.e. dielectric measurements for high-viscosity liquid, multi-layered and matrix composite materials, complex insulating structures, extra thin and thick liquids and solids, moving

* To whom any correspondence should be addressed.

bodies etc.

It is just the interdigital impedance sensor, also often called fringing electric field(FEF) sensor that has been suggested for the sake of overcoming the above-mentioned disadvantages of the conventional one, which is comprised of two interdigital (finger-like or comb-like) metal electrodes fixed on the surface of inert insulating substrate (e.g. quartz or Teflon). The dielectric material sample to be measured is placed on this surface in intimate contact with the pair of come-like electrodes, one electrode being excited with the sinusoidal voltage of fixed amplitude and frequency and the other electrode having the ground potential, as a result of which the fringing electric field (FEF) formed by the interdigital electrodes penetrates into the sample (figure 1).

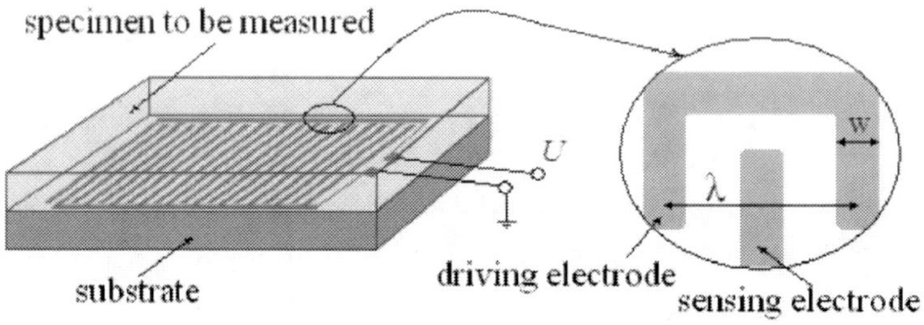

Figure 1. Interdigital impedance sensor.

Interdigital impedance sensors have the following advantages, compared with the conventional capacitance cells:

(i) The electrodes of the sensors are contacted only with one sides of the samples measured, making it possible to measure the impedance or capacitance of extra thin (i.e. nano- and micro-sized particles and films) and thick samples, complex insulating structures (i.e. power cable head insulation, etc.), the material samples difficult or inconvenient to be inserted into the conventional capacitance cells (i.e. bitmen, grease, heavy oil, etc.), the samples dimension-variable during electrode installation and measurements (i.e. powders, water converted into solid, liquid or gaseous state with the temperature changed, soft spongeous materials, etc.).

(ii) The electric field penetration depth of FEF sensor can be arbitrarily altered by changing the electrode geometry, allowing the measurements of the impedance or capacitance for individual layers of layered composite materials and the distribution of the defects inside the dielectric materials or structures considered (figure 2).

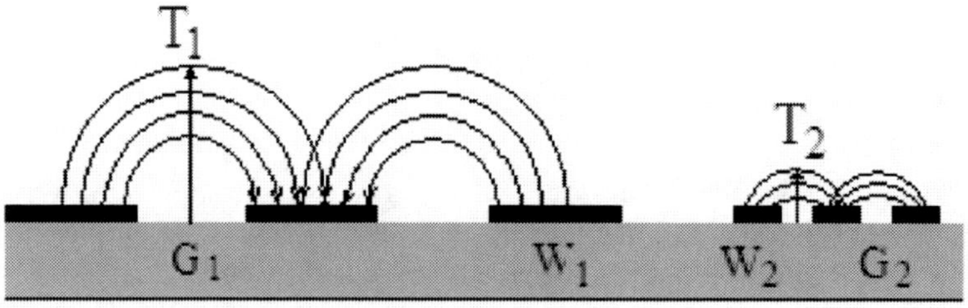

Figure 2. Dependence of electric field penetration depth on the spacing between the interdigital electrodes.

According to the previous studies, the longer the values of the spatial period of FEF sensors, the wider the width of interdigital electrodes, the greater the effective electric field penetration depth.

The general design principles of the interdigital impedance sensor have been discussed for a long time by many researchers since the early 1970s, and a variety of FEF sensors have been widely used in practical applications up to now [4-22], but some problems still require to be further investigated for the purpose of more effective applications of FEF sensors.

One of the most important problems is to optimize the geometries of the interdigital impedance sensors, including spatial period (λ) and width (W) of interdigital electrodes and so on, which play determinative roles in ensuring the effective electric field penetration depth of FEF sensors.

However, there is no consensus on the definition of the effective electric field penetration depth of the interdigital impedance sensor at present because the electric field between interdigital electrodes can be considered to show such an infinitely-spread distribution characteristic that it is nearly impossible to determinate the exact electric field penetration depth, and therefore the empirical relationship between the penetration depth and the spatial period of electrode system has been roughly accepted as

$$T = \frac{\lambda}{4} \sim \frac{\lambda}{3} \tag{1}$$

where T is the effective electric field penetration depth; λ is the spatial period

According to previous papers, one way of evaluating the effective penetration depth of EFE sensor is based on measuring the position at which the difference between the current value (for the sample tightly in contact with the sensor surface) and asymptotic value (for the sample infinitely far from the sensor surface) of sensor terminal impedance equals to 3% of the difference between the maximum and minimum of the terminal impedance [4], however, to our knowledge, few researchers have discussed what the meaning of the critical value 3% of the normalized impedance (or capacitance) for a given FEF sensor is and whether or not this critical value is absolutely accepted , and few users have not ever paid attention to them.

This paper aims to make the meaning of the critical value 3% of the normalized capacitance for a given interdigital impedance sensor clear, based on both the refractive principle of electric field intensity at the interface of two different dielectrics and the FEM analysis of electric field distribution and capacitance for FEF sensor.

2. Electric field behaviour at the interface between two different dielectrics

Let us consider the behaviours of the electrical potential and electric field intensity at the interface between two different dielectrics placed in a parallel plate capacitor (figure 3(a)).

If two kinds of dielectrics are placed between two electrodes as shown in figure 3(a), the electrical potential and electric field intensity are discontinuously varied along lines of PQR and KLMN between two electrodes, which are quite different from those in the single layer plate capacitor (right line PR and ST), as shown in figure 3(b) and (c).

In the case of the capacitor with double-layered dielectrics between two electrodes, the values of electric field intensity in the individual dielectric layers can be expressed respectively by

$$\left.\begin{aligned} E_1 &= \frac{\varepsilon_2}{\varepsilon_2 d_1 + \varepsilon_1 d_2} U \\ E_2 &= \frac{\varepsilon_1}{\varepsilon_2 d_1 + \varepsilon_1 d_2} U \end{aligned}\right\} \tag{1}$$

where E, d, ε and U mean the electric field intensity inside a dielectric, the thickness of dielectric layer, dielectric permittivity and the voltage across two electrodes, respectively; the numbers "1" and "2" denote the dielectric layers "1" and "2", respectively.

If the dielectric permittivity values of two individual dielectrics are the same ($\varepsilon_1 = \varepsilon_2$), the electric field intensity expressed by equation (1) is converted into

$$E = \frac{U}{d} \tag{2}$$

where d is the distance between two electrodes.

This hints us at that considering the degree of the discontinuity of electric field intensity may help us with defining the effective field penetration depth of the interdigital impedance sensor.

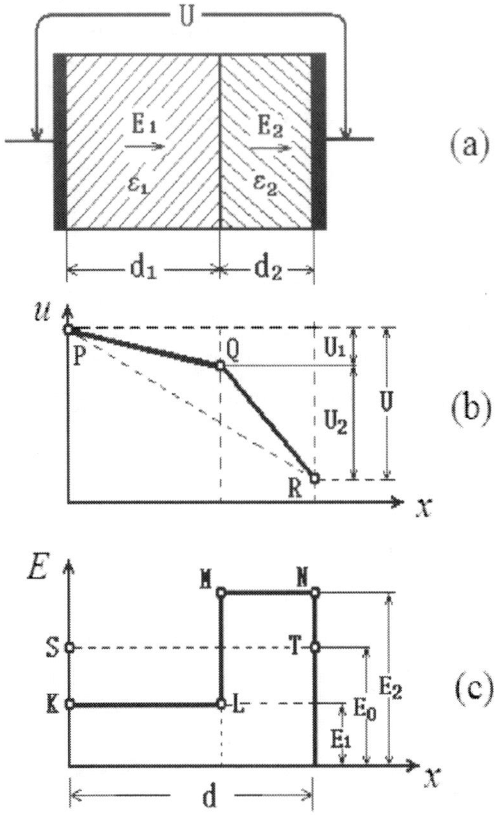

Figure 3. The changes of the electrical potential and electric field intensity at the interface between two different dielectrics inserted into two parallel plates: (a) the double-layered capacitor model; (b) the potential distribution; (c) the distribution of electric field.

3. FEM analysis for interdigital impedance sensor model

3.1 FEM analysis model

Since the electric field distribution of a FEF sensor is nonlinearly decreased along z-axis the direction vertical to the surface of the sensor, determining accurately the extent to which the electric field generated by one pair of FEF electrodes can spread out along the z-axis direction plays the key role in defining the E-field penetration depth of an interdigital impedance sensor.

For this, considering that a dielectric specimen ($\varepsilon_2 = 2.5$) on the sensor surface was going up farther and farther away from it to infinity as shown in figure (4), the electric field distribution and capacitance were analyzed using the software package Maxwell by Ansoft Corp. If the dielectric specimen is infinitely far away from the sensor, the sensor can be considered to be placed in air ($\varepsilon_1 = 1$).

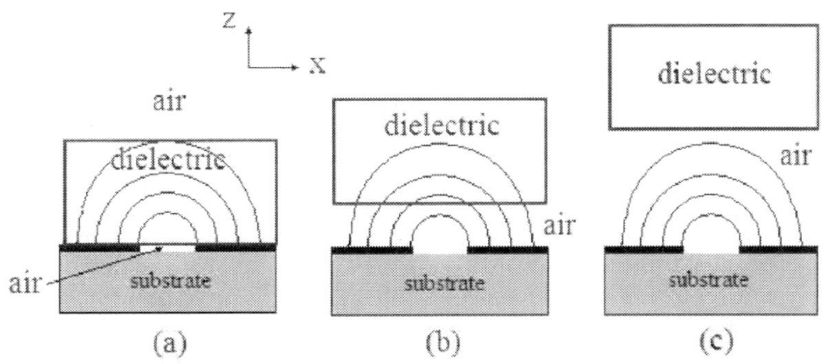

Figure 4. FEM model for a FEF sensor with the dielectric specimens attached on the sensor surface (a), some distance away from the sensor surface(b) and infinitely far away from the sensor surface(c), respectively.

In the FEM model shown in figure (4), the electrode-to-electrode-separation, the electrode width, electrode thickness and the thickness of the insulating substrate (ε_3=2.3) were taken as g=5 mm, w=5 mm, $d = 10$ μm, and h=2 mm, respectively.

The dielectric specimen on the sensor was considered as thick as the maximal value of the expected electric field penetration depth, and the very thin layer of air due to the thickness of metal electrode was also considered between the sensor surface and the dielectric specimen in FEM model (figure 4(a)). For the simplicity of analysis, the conductivity of the dielectric specimen considered was ignored, and the driving and the sensing electrodes were set to 5V and 0V, respectively.

3.2 Normalization of FEM analysis results

FEM analysis results were normalized in the range of 0 (minimum) to 100 (maximum) in order that it could be possible to compare an analysis result with another one.

Equation (2) and (3) show the normalized electric field intensity $E(\%)$ and capacitance $C(\%)$, respectively:

$$E(\%) = \frac{E - E_{\min}}{E_{\max} - E_{\min}} \times 100\ (\%) \tag{2}$$

$$C(\%) = \frac{C - C_{\min}}{C_{\max} - C_{\min}} \times 100\ (\%) \tag{3}$$

where E and C are arbitrary values of electric field intensity and capacitance between their maximum ($E_{\max}$ and $C_{\max}$) and the minimum ($E_{\min}$ and $C_{\min}$), respectively.

4. Results and discussion

Figure 5(a) and 5(b) represent FEM results for the change of the electric field intensity from the middle position between the driven and sensing electrodes along z axis when the dielectric specimen is attached to the electrode surfaces (figure 4(a)) and it is infinitely far away from the electrode surface (figure 4(c)), respectively, showing that the electric field intensity varies continuously with the distance z increasing inside the specimen and air, which is ascribed to that the single dielectric layers can be thought to exist on the sensor surface in these cases.

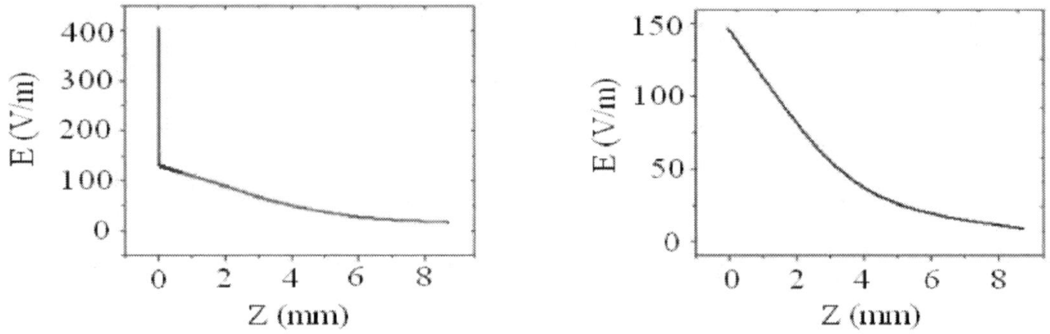

(a) (b)

Figure 5. The behaviours of E-field distribution along the z-axis direction for the dielectric specimens on the electrodes (a) and infinitely far away from the sensor surface(b).

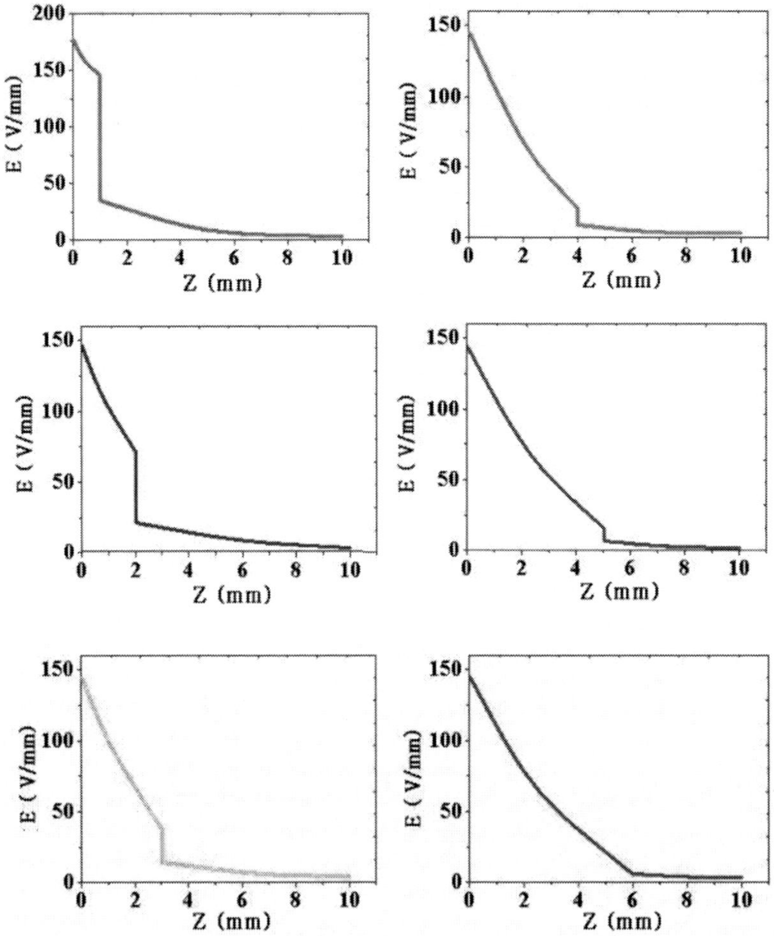

Figure 6. The dependences of the electric field intensity on the distance away from the sensor surface.

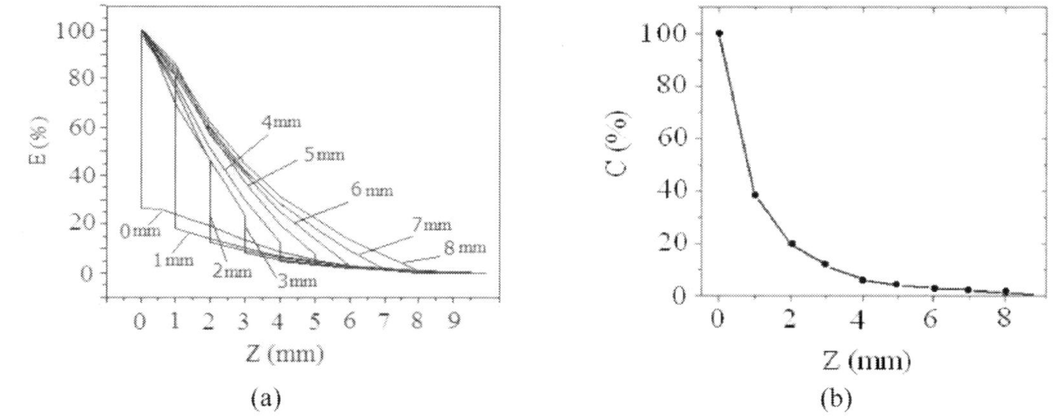

Figure 7. The normalized plots of electric field intensity (a) and capacitance (b) with the distance away from the FEF sensor surface.

However, when the dielectric specimen is a distance away from the sensor surface, an apparent jump-like change of electric field intensity is found at the air-dielectric interface, getting smaller and smaller as the dielectric specimen goes up farther and farther away from the sensor surface (figure 6), which may be ascribed both to the weakness of the electric field and to the conversion of the double-layered dielectric structure into the single one with the specimen-to-sensor distance increased.

It can be clearly seen from the results of FEM analyses in figure 6 and their normalization (figure 7(a)) that the jump-like change of the electric field intensity can hardly be observed when the dielectric specimen goes up more than 6 mm away from the surface of the sensor model considered in this study, but the refractive property of field intensity still exists at those positions, too, which means that the dielectric-to-air interface is still in the electric field formed by the FEF electrodes although its intensity is very weak.

To measure the electric field distribution in the dielectric layers of a capacitor is less realistic than to measure the capacitance, but the latter is not available to distinguish the interface between two dielectrics (figure 7(b)).

Expecting that too weak electric field may give little influence on the impedance values measured, the distance from the sensor surface to the position at which the jump-like change of electric field intensity may be nearly neglected can be accepted as the effective electric field penetration depth for an interdigital impedance sensor.

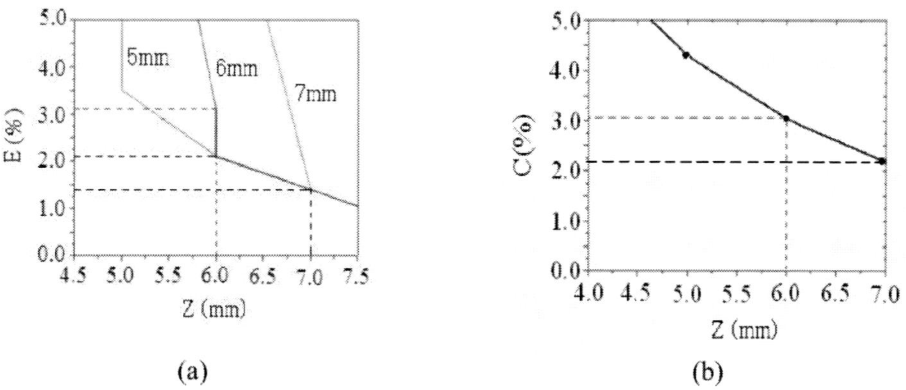

Figure 8. The normalized plots of electric field intensity (a) and capacitance (b) with the distance away from the FEF sensor surface.

As can be seen in figure 8, the normalized electric field intensity jumps from about 3% down to about 2% at the position of $z = 6$ mm, that is, the jump-like change of normalized field intensity is only 1%, which corresponds to 3% of the normalized capacitance for the FEF sensor considered in this study (figure 8(b)).

Let us consider in detail the meaning of the change of the normalized electric field intensity by 1%, based on the double-layered parallel plate capacitor model shown in figure 3(a).

The jump-like change of electric field intensity ΔE observed at the dielectric-to-air interface can be written as

$$\Delta E = E_2 - E_1 = \frac{(\varepsilon_2 - \varepsilon_1)}{\varepsilon_1 d_2 + \varepsilon_2 d_1} U \tag{4}$$

where E_1 and E_2 are the electric field intensities for the air and dielectric layers, respectively; ε_1 and ε_2 are the dielectric permittivity of air and the dielectric considered, respectively; d_1 and d_2 are the thickness of air and dielectric layers, respectively.

Assuming that $\varepsilon_1 = 1$, $d_1 = 6$ mm, $d_2 = 7$ mm and $U = 5$ V in figure 3, then the dielectric permittivity of the dielectric layer in the double-layered parallel plate capacitor model, required for the jump-like change of the normalized electric field intensity by $\Delta E = 0.01$ (or 1%) at the dielectric-to-air interface can be calculated from equation (4) to be $\varepsilon_2 = 1.026$, which is so close to the value of permittivity of air that it may be nearly identified with the permittivity of air ($\varepsilon_1 = 1$) from the viewpoint of practically-available measurement accuracy.

It can be clearly understood that at the position where the jump-like change of the normalized electric field intensity due to the difference in permittivity values at the dielectric-to-air interface is about 1%, the existence of the dielectric-to-air interface may be just as well neglected, which means that the electric field intensity at this position is too weak to distinguish the dielectric specimen from the air layer.

The results of FEM numerical analysis in this study suggest that the jump-like change of the normalized electric field intensity by about 1% is the premise for the previous definition of the effective electric field penetration depth which has been adopted without any doubt, and that for the sake of determination of more accurate penetration depth, less than 1% of the normalized electric field intensity or less than 3% of the normalized capacitance may be selected.

5. Conclusion

Numerical analysis on the effective electric field penetration depth has been performed for the interdigital impedance sensor, based both on the refractive characteristics of electric field at the interface between two different dielectrics and on the FEM analysis of electric field distribution and capacitance, through which it has been more clearly understood how to define the effective electric field penetration depth for FEF sensors, and what is the physical meaning of 3% of the sensor terminal capacitance which has been accepted as the critical value for defining the effective field penetration depth in the previous papers.

As discussed in this paper, as the dielectric specimen goes up farther and farther away from the sensor surface, the electric field distribution shows an apparent jump-like change at each position due to the existence of the dielectric-air interface, whereas the capacitance change is found to be monotonically continuous, which hints us at that the FEM analysis of electric field distribution makes it more convenient to grasp the extent to which the electric field generated in the FEF sensor spreads out.

According to the FEM analysis results, the jump-like change of the normalized electric field intensity by about 1% is the premise for accepting 3% of the normalized capacitance as the criteria for defining the effect electric field penetration depth for a FEF sensor, which may be reasonable from the viewpoint of the practical measurement accuracy.

However, for the sake of determination of more accurate penetration depth, less than 1% of the normalized electric field intensity or less than 3% of the normalized capacitance may be selected.

Acknowledgments

The authors would like to thank Mr. Seu-hon Hong, President of Kim Chaek University of Technology, D.P.R. Korea for his great support to our research group, and at the same time gratefully acknowledge Mr. Hong-jun Chen and Dalian University of Technology for providing with the opportunity and convenience for the international joint research. Mrs. Gyong-ok Jo from Measurement Centre, Pyongyang 326 Cable & Wire Factory is also thanked for helping and supporting during the measurements.

References

[1] Ganchev S I, Bakhtiari S and Zoughi R 1992 *IEEE trans on Instru. and Meas.* **41** 361
[2] Boifot A M 1991 *IEEE Trans. on Instrum. and Meas.* **40** 611
[3] van Roggen A 1990 *IEEE Transactions on Electrical Insulation* **25** 95
[4] X B Li, S D Larson, A S Zyuzin and A V Mamishev 2004 *Conf. rec. of the 2004 Int. Symp. on Electrical Insulation September (Indianapolis, USA, 19-22 September 2004)* p 406
[5] Nelson S O 2006 *IEEE Trans. on Diele. and Electr. Insul.* **13** 688
[6] M C Zaretsky, J R Melchor and C M Cooke 1989 *IEEE Trans. on Electr. Insul.* **24** 1167
[7] Fothergill J C, Dodd S J, Dissado L A, Liu T and Nilsson U H 2011 *IEEE Trans. on Diele. and Electr. Insul.* **18** 1544
[8] Blennow J, Ekanayake C, Walczak K, Garcia B and Gubanski S M 2006 *IEEE Trans. on Power Deliv.* **21** 681
[9] Sheiretov Y and Zahn M 1995 *IEEE Trans. on Diel. and Electr. Insul.* **2** 329
[10] García-Baños B, Canóñ A J, Peñaranda-Foix F L and Catalá-Civera J M 2011 *IEEE Sens. J.* **11** 62
[11] Dervos C T, Paraskevas C D, Skafidas P D, Stefanou N 2006 *IEEE Trans. on Diele. and Electr. Insul.* **13** 586
[12] Bertolini, D.; Cassettari, M.; Salvetti, G.; Tombari, E.; Veronesi, S. 1991 *Rev. of Scien. Instru.* **62** 450
[13] Saha T K and Purkait P 2008 *IEEE Trans. on Power Deliv.* **23** 252
[14] Taylor N and Edin H 2010 *IEEE Trans. on Diele. and Electr. Insul.* **17** 1489
[15] M W den Otter 2002 *Sens. and Actua.* A **96** 140
[16] Mamishev A V, Cantrell S R, Du Y, Lesieutre B C and Zahn M 2002 *IEEE Trans. on Instru. and Meas.* **51** 1192
[17] Sheiretov Y, Zahn M 2005 *IEEE Trans. on Diele. and Elec. Insul.* **12** 993
[18] Mamishev AV, Du Y, Bau J H, Lesieutre B C and Zahn M 2001 *IEEE Trans. on Dielec. and Electr. Insul.* **8** 785
[19] Li X B, Larson SD, Zyuzin AS, Mamishev AV 2006 *IEEE Sens. J.* **6** 434
[20] Zaretsky M C, Li P and Melcher J R 1989 *IEEE Trans. on Electr. Insul.* **24** 1159
[21] Zaretsky MC, Mouayad L and Melcher JR 1988 *IEEE Trans. on Electr. Insul.* **23** 897
[22] F Starzyk 2008 *Archiv. Of Mater. Scien. and Eng.* **34** 31

Molecular dynamics simulation of the electrical double layer in ionic liquids

S A Kislenko, R H Amirov *, I S Samoylov
Department for Gas-Discharge Physics and Plasma Technologies, Joint Institute for High Temperatures of the Russian Academy of Sciences, Izhorskaya 13/19, Moscow, 125412, Russia

E-mail: amirovravil@yandex.ru

Abstract. The structure of the electrical double layer in the strongly coupled ionic liquid 1-butyl-3-methylimidazolium hexafluorophosphate ([BMIM][PF_6]) near a basal plane of graphite was investigated by molecular dynamics simulation. It is found that near an uncharged surface the ionic liquid structure differs from its bulk structure and represents a well-ordered region, extending over ~2 nm from the surface. Interfacial layering is clearly observed at the surface. Ions adsorbed at the uncharged surface form the 2D molecular clusters of two types. In the first cluster type anions are self-assembled in a triangular lattice (containing ~5÷10 ions) while a cation subsystem is disordered. In the second one cations and anions self-assembled in a honeycomb lattice. The behavior of the screening potential in the ionic liquid [BMIM][PF_6] near the charged graphite surface with the charge density in the range $-1.7 \leq \sigma \leq 1.7$ μC/cm^2 was investigated. It was shown that the potential is a nonmonotonic function of distance. Asymmetric behavior of the screening potential at surface charge densities equal in magnitude and opposite in sign was detected. It was shown that the local self-diffusion coefficients of ions in the vicinity of the surface correlate with the local ion density. Finally, the influence of temperature on the screening potential in the vicinity of a charged graphite surface has been studied. It was shown that the increase of temperature from 300 K to 400 K induces the decrease of the potential drop across the interface that implies the increase of the capacitance of the electrical double layer.

1. Introduction

In modern technology there are numerous applications of ionic liquids (ILs) in particular in fuel cells, supercapacitors, dye-sensitized solar cells, heterogeneous catalysis, supported ionic liquid phase catalysis and electrodeposition of metals. The common feature of these systems is the presence of an interface between the ionic liquid and the other solid, liquid or gas phase through which a mass transfer is taking place during chemical and electrochemical reactions. The rates of the corresponding reactions are heavily related to structure and dynamics of the interface. Hence, the study of the interfacial properties of the ionic liquids is the key to understand their functional performance in technological applications.

In recent papers, various distinct structural features of the ILs at the surface have been observed. The formation of interfacial layers has been revealed by x-ray reflectivity [1], neutron reflectometry measurements [2] and atomic force microscopy [3, 4]. By means of sum frequency generation

* To whom any correspondence should be addressed.

vibrational spectroscopy it has been shown that 1-alkyl-3-methylimidazolium cations have specific orientation, which depends on alkyl chain length [5], anion type [6], material [7] and electric potential of the surface [8]. Also with the help of scanning tunneling microscopy it has been observed that ions of the ionic liquid can form a long range two-dimensional lattice structure at the surface [9, 10]. However, for obtaining more detailed information about ILs structure at the interface further investigations are needed. In particular, the development of adequate theoretical models of the electrical double layer (EDL) in ILs takes on great significance.

Owing to high ion concentration in ILs (~10^{21} cm^{-3}), the Debye-Huckel classical theory is no longer valid and the double layer model must take into account a finite size of ions [11]. Recently, based on the lattice-gas model, the double layer theory in ILs was developed and the novel analytical expression for the double layer capacitance obtained [11]. However, this model does not account for the asymmetry of molecules, intramolecular charge distribution and specific adsorption of ions at the surface [11]. Moreover, when developed in the framework of the mean-field approach, this theory does not describe the ion-correlation effect in high-concentrated electrolytes [11, 12].

A major goal of this paper is to investigate the double layer in the ionic liquid 1-butyl-3-methylimidazolium hexafluorophosphate [BMIM][PF_6] at the graphite surface by use of the molecular dynamics simulation (MD). The MD method allows for the above-listed characteristics of the medium and has already been successfully used on investigating the vacuum/IL [13, 14], liquid/IL [15] and solid/IL [16, 17] interfaces. [BMIM][PF_6] is chosen by reason of a high electrochemical stability which opens the door to its application as an electrolyte for supercapacitors. Account was also taken of a large body of experimental and theoretical data available for comparison with the simulation results. The basal plane of graphite emulates the surface of an activated carbon used as an electrode material for supercapacitors. In the present study we also investigated the influence of a surface charge ($-1.7 \leq \sigma \leq 1.7$ μC cm^{-2}) and temperature (T=300÷400 K) on the EDL structure as well as some dynamical properties of the IL near the surface.

2. Computational details

Molecular dynamics calculations were performed using the DL-POLY package. We used the modified AMBER force field for the ionic liquid [BMIM][PF_6] simulation [18].

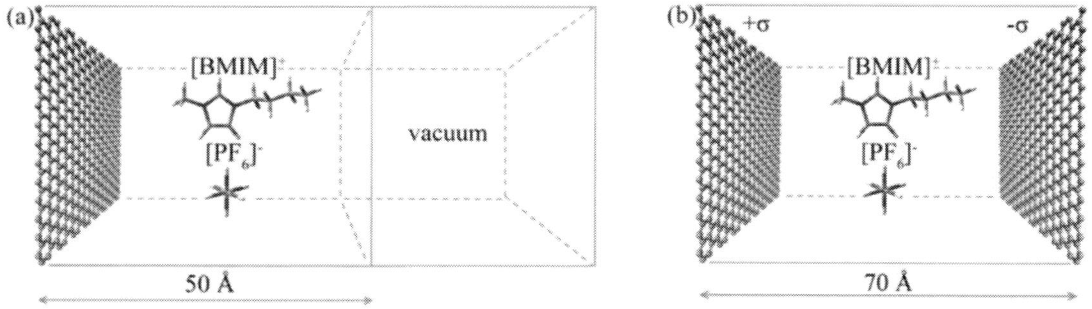

Figure 1. Schematic configurations of the simulation boxs and molecular structure of the cation $[BMIM]^+$ and anion $[PF_6]^-$ of the IL [BMIM][PF_6].

On investigation of the EDL structure at the uncharged graphite surface, the following simulation configuration has been used. Graphite surface was constructed of a single layer of immovable sp^2-hybridized carbon atoms arranged in a hexagonal pattern (basal plane of graphite) of 34.4x34 $Å^2$ in size. An approximately 50 Å wide ionic liquid layer adjoins the graphite wall. A vacuum region of ~100 Å is adjacent to the ionic liquid free surface (figure 1a).

On investigation of the EDL structure at the charged graphite surface we used the simulation box constructed of two parallel graphite surfaces, arranged at a distance of ~7 nm, the space between which was filled with ionic liquid [BMIM][PF_6] (figure 1b). Three-dimensional periodic boundary conditions were employed and the periodicity in the direction perpendicular to the surface was 200 Å. The surfaces were charged by surface charges σ equal in magnitude and opposite in sign, which induced a polarization voltage in the gap. (A single point charge was placed at the center of each surface carbon atom, its magnitude being determined by the overall surface charge density σ.)

The MD simulation was performed in the NVT ensemble. To maintain a constant temperature of the system the Nosé-Hoover thermostat was used. Ewald method with an accuracy of 10^{-5} and a real space cutoff value of 12.5 Å was used to calculate the electrostatic interaction. The equations of motion were solved using the Verlet leapfrog integration algorithm with a time step of 0.5 fs. The cutoff radius of the Van der Waals interaction was 7 Å.

To reduce statistical errors eight simulations starting from different initial configurations were performed under all conditions (temperature, surface charge). The final results were obtained by averaging over all eight configurations and over 1÷3 ns of dynamics per configuration. To generate an equilibrated system, 0.5 ns trajectory was initially calculated at 600 K. Then the temperature was reduced from 600 K to an appropriate temperature, and the systems were further equilibrated for 1 ns before collecting data.

All the simulations were performed on the cluster MVS-100K (140.16 TFlops peak performance) at the Joint Supercomputer Center of the Russian Academy of Sciences.

3. Results and discussions

3.1. Interfacial ordering of [BMIM][PF_6]

The variations of the relative mass density $\rho_m(x) = \rho(x)/\rho_{bulk}$ (where ρ_{bulk} – is a bulk density) and volume charge density ρ_q in [BMIM][PF_6] in the direction, normal to the graphite surface (along x-axis), were determined in the case of the uncharged wall (figure 2). The surface was placed at $x = 0$.

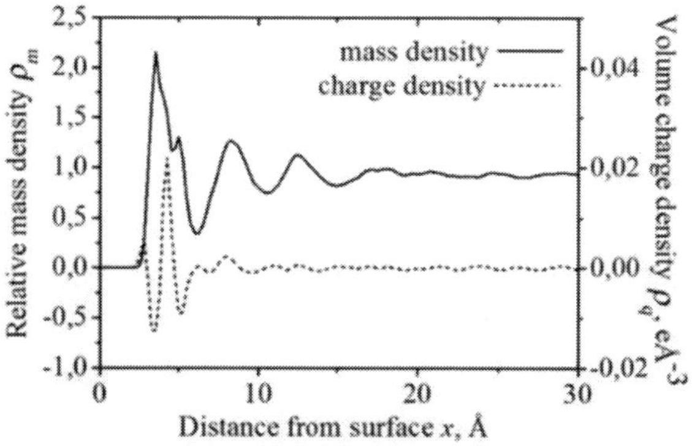

Figure 2. Relative mass density and volume charge density along the normal to the uncharged graphite surface.

A relative mass density near the wall exhibits damped oscillations with the period of ~5 Å. Three high-density peaks are clearly observed at the interface and density in first one reaches value two times larger than that of the bulk liquid. Such a behavior of ρ_m points to the existence of well-ordered region in ionic liquid, extending over ~2 nm from the surface. In the region x > 2 nm the density takes its bulk value.

The region of a spatial inhomogeneity of the volume charge density in figure 2 is less extended than the corresponding mass density one. The charge density is different from zero only in the area of the first and second oscillations of the mass density and shows a complex structure which arises from the most probable orientation of the adsorbed molecular ions and the associated intramolecular charge distribution.

Figure 3a shows the concentration profiles of imidazolium rings and anions on the distance to the uncharged surface. In the case of rings, the concentration was obtained from the position of its center of mass, and in the case of anions - from the position of its center of symmetry (phosphorus atom). In the distribution of rings and anions three concentration peaks were found (marked 1, 2 and 3 in figure 3a). The first peak corresponds to ions adsorbed at the surface. Since the first adsorption peak of anions higher than that of the rings, surface concentration of adsorbed anions is larger than the surface concentration of adsorbed cations.

The positions of concentration peaks in anion and ring distributions coincide with the regions of enhanced mass density in figure 2. Thus, the structure of $[BMIM][PF_6]$ near the graphite surface represents three dense layers composed of negative ions and positively charged rings.

Formation of layering structure in thin films of 1-alkyl-3-methylimidazolium ILs on silicon substrate was detected experimentally [19]; in addition authors assumed that high-density layers composed of associated 1-ethyl-3-methylimidazolium cation head groups and hexafluorophosphate anions. This assumption was confirmed by our simulation for graphite/$[BMIM][PF_6]$ interface.

Interfacial layering of $[BMIM][PF_6]$ has also been observed [4, 20]. Experimentally measured thickness of layers on mica and gold surface was 7-8 Å and 5.5 Å respectively. Calculated value of 5 Å is hence in a good agreement with experimental data. A small difference between experimental and calculated values can be explained by different $[BMIM]^+$ cation orientation in layers, that depends on a substrate material [5, 6].

The order parameter S as a function of the distance perpendicular to the surface was calculated to investigate the cation orientation according to the following formulation:

$$S = \left\langle \frac{3}{2}\cos^2\vartheta - \frac{1}{2} \right\rangle$$

where θ is the angle between the surface normal and a specific direction vector fixed in the cation frame. In completely disordered, isotropic phase the order parameter becomes zero ($S = 0$). If $\theta = 90^\circ$ then S = 1. Finally, if $\theta = 0^\circ$ then the order parameter $S = -0.5$.

Figure 3b shows the order parameters S_n and S_{NN} for a vector $\vec{n}$ normal to the plane of the imidazolium ring and a vector $\overrightarrow{NN}$ connecting the two nitrogen atoms. It can be seen that in the first concentration peak of rings at $x = 3.5$ Å the order parameters are equal to $S_n \sim 0.9$, $S_{NN} \sim -0.45$. Therefore adsorbed $[BMIM]^+$ cations is orientated in such a way that the ring tends to be arranged in parallel to the surface.

Figure 5 shows a snapshot taken from the MD simulation, demonstrating orientation and 2D ordering of the adsorbed ions at the uncharged surface. We found that ions adsorbed at the graphite surface can form the 2D molecular clusters of two types. In the first cluster type anions are self-assembled in a triangular lattice (containing ~5÷10 ions) while a cation subsystem is disordered (figure 5a). In the second one cations and anions self-assembled in a honeycomb lattice (figure 5b).

It can be deduced from the values of S_n and S_{NN} at $x = 8.5$ Å (figure 3b) that rings in the second layer have a small tilt angle and orient nearly parallel to the surface.

At $x = 6$ Å, in the local peak 2', $S_n \sim -0.23$, $S_{NN} \sim 0.4$. Thus, there is a small fraction of rings between first and second layers that demonstrates a tendency to be perpendicular to the surface.

Experimental studies of solid/liquid interfaces for 1-alkyl-3-methylimidasolium ILs shows that tilt angle of adsorbed cation rings depends on the alkyl chain length, anion and surface material [5, 6, 8]. It was found that the $[BMIM][PF_6]$ ionic liquid imidazolium ring had a tilt angle between 32° and 52° on quartz substrate [6]. A special case of the ring is lying in the plane of the surface, similar as in our work, was observed for 1-butylimidazole [6].

Currently there is no experimental data available for molecular orientation at the ILs/graphite interface. However, it was shown in a number of works that adsorbed on the graphite surface compounds, containing carbon cycles or heterocycles, orient parallel to the surface [21, 22]. This fact indirectly supports our simulation results.

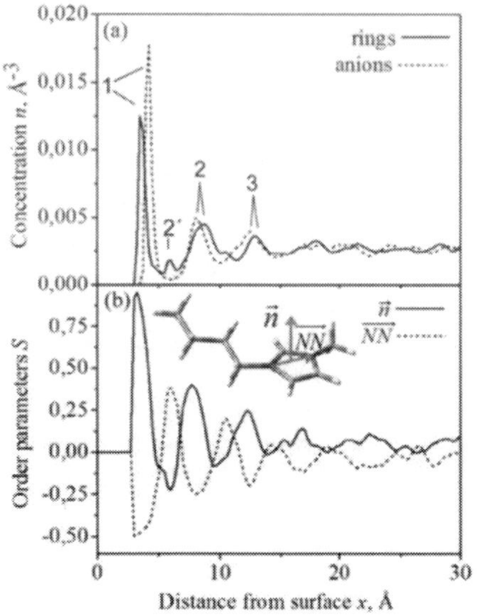

Figure 3. (a) – The concentration profiles of imidazolium rings and anions along the normal to the uncharged graphite surface. (b) – The orientational order parameters S_n and S_{NN} as a function of the distance x from the surface.

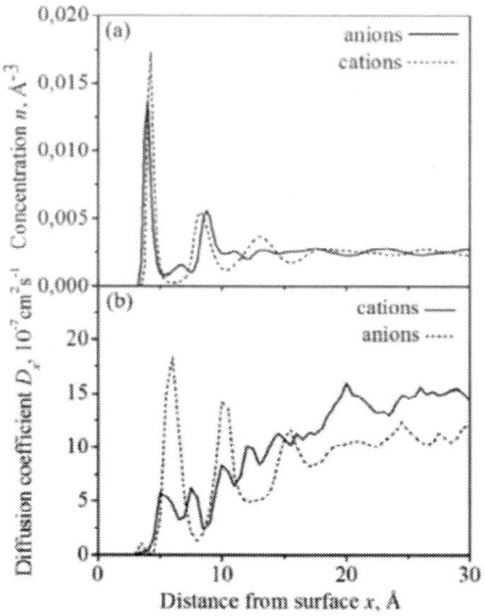

Figure 4. (a) – The anion and cation densities as a function of the distance x from the graphite surface at 400K. (b) – Local self-diffusion coefficient of ions in the direction normal to the surface at 400 K.

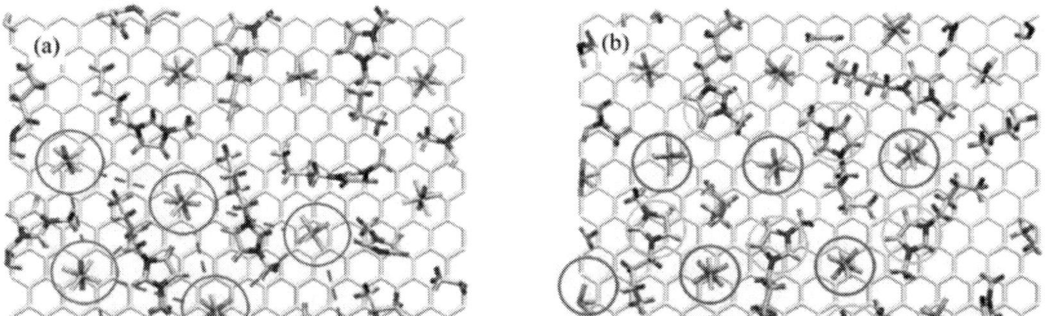

Figure 5. A snapshot taken from the MD simulation, demonstrating formation of 2D molecular clusters. (a) – Anions are self-assembled in a triangular lattice while a cation subsystem is disordered. (b) – Cations and anions self-assembled in a honeycomb lattice.

3.2. Ion dynamics at the interface

To investigate the influence of the graphite surface on the dynamical properties of the ionic liquid we derived the local self-diffusion coefficients $D_x(x)$ of ions as a function of a distance in the normal direction to the surface (figure 4b). The local self-diffusion coefficients were determined from the slope of a line approximating the mean-square displacement $\left\langle (x(t)-x(0))^2 \right\rangle$ of mass center of ions in the range 10-30 ps. It was assumed that the Einstein relation $\left\langle (x(t)-x(0))^2 \right\rangle \propto 2D_x t$ is satisfied within the analyzed time intervals. The lower boundary of the interval (t=10 ps) corresponds to an origin of a linear region of the function $\left\langle (x(t)-x(0))^2 \right\rangle$ [23]. The size of the interval is chosen to meet the requirement of small ion displacement during observation. Such approach allows us to obtain the local diffusion characteristics.

Figure 4b show the oscillatory behaviour of the local self-diffusion coefficients near the surface. The maxima of the function $D_x(x)$ for the cations (anions) coincide with the regions of decreased cation (anion) density and *vice versa* (figure 4a). As the ions move away from the surface the diffusion coefficient converge to a constant bulk value. Small statistical variation of the D_x in the region x>2 nm can be reduced by adding extra simulation time.

The main difference in the behavior of the local self-diffusion coefficients of $[BMIM]^+$ and $[PF_6]^-$ in the vicinity of the surface is that the maximum values of the function D_x for anions (in contrast to cations) exceed the bulk values. It can be accounted for a smaller size of the anion and its octahedral symmetry. These properties promote the $[PF_6]^-$ mobility in the region of reduced density.

3.3. Electric potential profiles

Figure 6 shows the dependences of the electric potential □*(x)* on the distance to the surface at various σ. The electric potential was calculated as a numerical solution of a one-dimensional Poison equation. We can see that the potential is a nonmonotonic function of distance, and screening occurs at a distance on the order of two molecular sizes. The screening feature is the asymmetry of the function □*(x)* with respect to the charge polarity of surfaces at charge densities equal in magnitude, which is explained by different sizes and atomic structures of cations and anions and their different Van der Waals interactions with the surface. Furthermore, at $\sigma = 0$, the potential difference between the surface and a remote point (potential of zero charge) is nonzero, U_0 = 0.10 V, and the function □*(x)* is characterized by a complex oscillatory behavior. This is explained by the formation of a quasicrystalline phase in an ionic liquid near the surface, which is characterized by low mobility and spatial and orientational ordering of ions.

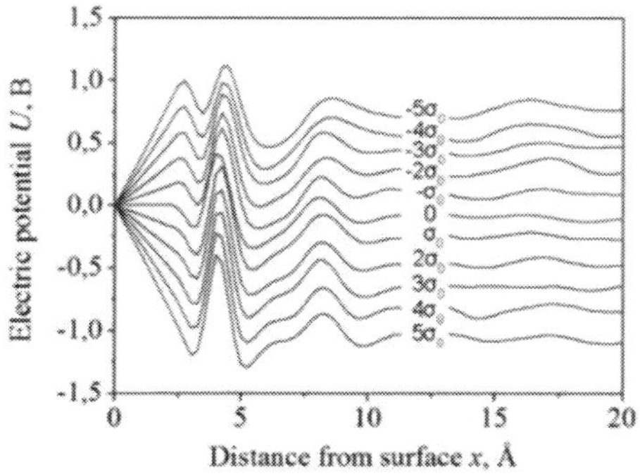

Figure 6. Dependences of the electric potential on the distance to the charged surface at the surface charge density in the range $-5\sigma_0 \le \sigma \le 5\sigma_0$ with the step σ_0 = 0.34 μC/cm^2.

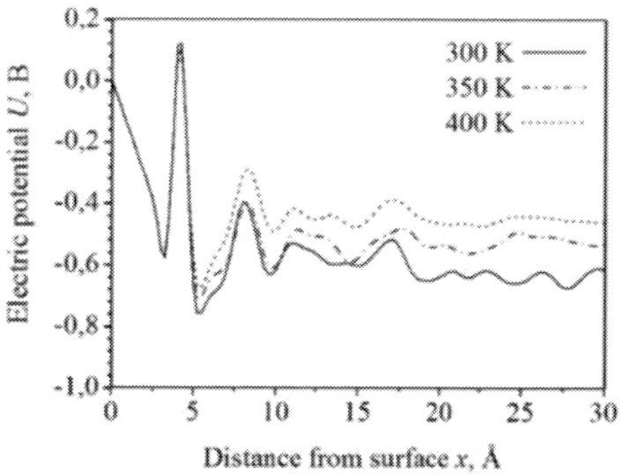

Figure 7. Influence of temperature on the electric potential in the vicinity of a positively charged surface.

Finally, we investigated the influence of temperature on the screening potential in the vicinity of a charged graphite surface. Figure 7 shows the electric potential profiles in the ionic liquid in the vicinity of the positively charged graphite surface at various temperature regimes. It can be easily seen that the potential difference ΔU between the graphite surface and a remote point in the IL decreases as temperature increases. This tendency implies the increase of the capacitance of the electrical double layer (EDL) $C=\sigma/\Delta U$, where σ is a surface charge density. For aqueous electrolytes the opposite tendency in temperature behaviour is detected. For example, for dilute electrolytes the Gouy-Chapman theory predicts the decrease of the electrical capacity as a function of temperature $1/T^{1/2}$ [24]. The anomalous temperature dependence of the capacity for ILs is confirmed experimentally [25, 26]. Similar behaviour was observed for high-temperature molten salts [27, 28]. One possible explanation can be the existence of ion pairs (consisting of electrostatically associated cation and anion) in molten salts that dissociate on separate ions when temperature increases [24]. This approach can possibly be applied to ionic liquids. In favor of this hypothesis one can argue the formation of ion pairs in ILs which was experimentally confirmed [29]. In addition, it should be taken into account that besides ion pairs the formation of domains due to aggregation of the chains is occurred in ILs [30, 31] that smear out with increasing temperature [32, 33].

4. Conclusion

The structure of the electrical double layer in the strongly coupled ionic liquid l-butyl-3-methylimidazolium hexafluorophosphate ([BMIM][PF_6]) near a basal plane of graphite was investigated by molecular dynamics simulation. It is found that near an uncharged surface the ionic liquid structure differs from its bulk structure and represents a well-ordered region, extending over ~2 nm from the surface. Interfacial layering is clearly observed at the surface. Ions adsorbed at the uncharged surface form the 2D molecular clusters of two types. In the first cluster type anions are self-assembled in a triangular lattice (containing ~5÷10 ions) while a cation subsystem is disordered. In the second one cations and anions self-assembled in a honeycomb lattice. The behavior of the screening potential in the ionic liquid [BMIM][PF_6] near the charged graphite surface with the charge density in the range $-1.7 \leq \sigma \leq 1.7$ μC/cm^2 was investigated. It was shown that the potential is a nonmonotonic function of distance. Asymmetric behavior of the screening potential at surface charge densities equal in magnitude and opposite in sign was detected. It was shown that the local self-diffusion coefficients of ions in the vicinity of the surface correlate with the local ion density. Finally, the influence of temperature on the screening potential in the vicinity of a charged graphite surface has been studied. It was shown that the increase of temperature from 300 K to 400 K induces the decrease of the potential drop across the interface that implies the increase of the capacitance of the electrical double layer.

References

[1] Mezger M, Schröder H, Reichert H, Schramm S, Okasinski J S, Schöder S and Honkimäki V 2008 *Science* **322** 424
[2] Bowers J, Vergara-Gutierrez M C and Webster J R P 2004 *Langmuir* **20** 309
[3] Atkin R and Warr G G 2007 *J. Phys. Chem. C* **111** 5162
[4] Liu Y, Zhang Y, Wu G and Hu J 2006 *J. Am. Chem. Soc.* **128** 7456
[5] Fitchett B D and Conboy J C 2004 *J. Phys. Chem. B* **108** 20255
[6] Romero C and Baldelli S 2006 *J. Phys. Chem. B* **110** 6213
[7] Aliaga C and Baldelli S 2008 *J. Phys. Chem. C* **112** 3064
[8] Rivera-Rubero S and Baldelli S 2004 *J. Phys. Chem. B* **108** 15133
[9] Ge-Bo Pan and Freyland W 2006 *Chem. Phys. Lett.* **427** 96
[10] Yu-Zhuan Su, Yong-Chun Fu, Jia-Wie Yan, Zhao-Bin Chen and Bing-Wei Mao 2009 *Angew. Chem.* **121** 5250
[11] Kornyshev A. 2007 *J. Phys. Chem. B* **111** 5545
[12] Keblinski P, Eggebrecht J, Wolf D and Phillpot S 2000 *J. Chem. Phys.* **113** 282
[13] Lynden-Bell R M and Del Pópolo M. 2006 *Phys. Chem. Chem. Phys.* **8** 949
[14] Pensado A S, Malfreyt P and Pádua A A H *J.* 2009 *Phys. Chem. B* 113 14708
[15] Lynden-Bell R M, Kohanoff J and Del Popolo M G 2005 *Faraday Discuss.*, **129** 57

[16] Liu L, Li S, Cao Z, Peng Y, Li G, Yan T and Gao X-P 2007 *J. Phys. Chem. C* **111** 12161
[17] Sieffert N and Wipff G *J.* 2008 *Phys. Chem. C* **112** 19590
[18] Kislenko S A, Samoylov I S and Amirov R H 2009 *Phys. Chem. Chem. Phys.* **11** 5584
[19] Carmichael A, Hardacre C, Holbrey J, Nieuwenhuyzen M and Seddon K 2001 *Mol. Phys.* **99** 795
[20] Xiao Zhang, Yun-Xin Zhong, Jia-Wei Yan, Yu-Zhuan Su, Meng Zhang and Bing-Wei Mao 2012 *Chem. Commun.* **48** 582
[21] Jin J, Yang W, Li Y, Li L, Zhao Y, Jiangb L and Li T 2003 *New J. Chem.* **27** 1463
[22] Martin D *Surf. Science* 2003 **536** 15
[23] Margulis C J, Stern H A and Berne B J 2002 *J. Phys. Chem. B* **106** 12017
[24] Holovko M, Kapko V, Henderson D and Boda D 2001 *Chem. Phys. Lett.* **341** 363
[25] Lockett V, Sedev R, Ralston J, Horne M and Rodopoulos T 2008 *J. Phys. Chem. C* **112** 7486
[26] Silva F, Gomes C, Figueiredo M, Costa R, Martins A and Pereira C M 2008 *J. Electroanal. Chem.* **622** 153
[27] Tumanova N K and Delimarskii Y K 1981 *Electrochim. Acta* **26** 1737
[28] Painter K R, Ballone P, Tosu M P, Grout P J and March N H 1983 *Surf. Sci.* **133** 89
[29] Weingärtner H, Knocks A, Schrader W and Kaatze U, 2001 *J. Phys. Chem. A* **105** 8646
[30] Wang Y and Voth G A 2005 *J. Am. Chem. Soc.* **127** 12192
[31] Canongia Lopes J N A and Pádua A A H 2006 *J. Phys. Chem. B* **110** 3330
[32] Wang Y and Voth G A 2006 *J. Phys. Chem. B* **110** 18601
[33] Triolo A, Russina O, Bleif H-J and Di Cola E 2007 *J. Phys. Chem. B* **111** 4641

Application of TEM horn antenna in radiating NEMP simulator

Yun Wang [1*], **Yongguang Chen** [2], **Qingguo Wang** [2]
1. Department of Postgraduate, the Academy of Equipment, Beijing 101416, China;
2. Institute of Electrostatic and Electromagnetic Protection, Mechanical Engineering College, Shijiazhuang, 050003, China

Email: wangyunjunxie@163.com

Abstract. In order to design a small radiating nuclear electromagnetic pulse simulator with fast rise time, we have investigated the application of TEM horn antenna in this kind of simulator. The TEM horn antenna is excitated by the integral double exponential pulse. Farfield response in time domain is analyzed based on the step response formula which is derived from the equivalent transmission line model of TEM horn antenna. Principle of antenna parameters selection is determined. The results of numerical calculation give a show of the radiation waveform and some problems. Then experiments are implemented and results are compared with the radiation waveforms generated by double exponential pulse excitation. It suggests that the TEM horn antenna can be used as radiating antenna in the NEMP simulator, which can radiate double-exponent-like pulse with the integral double exponential pulse as excitation. Meanwhile, impedance loading can improve farfield waveforms and enhance the low-frequency radiating ability of antenna.

1. Introduction

Nuclear electromagnetic pulse simulators (NEMPS) are usually used to produce nuclear electromagnetic pulse (NEMP) environment by generating double exponential pulse approximately when testing NEMP's electromagnetic effect. At present, international and domestic waveform standards have been made for electromagnetic pulse (EMP) simulators. In 1999, the U.S. standard "MIL-STD-461E" [1]. specified following standards: EMP waveform peak value is 50 kV m^{-1}, rise time is 1.8-2.8 ns, and half peak pulse width is 23±5 ns. Our country has released similar waveform standard "GJB1389A-2005" [2]. All these standards have shorter rise time and narrower pulse width than old standards (such as the 1986 U.S. standard "MIL-STD-461C" [3], 1993 U.S. standard "MIL-STD-461D" [4] etc.). Rise time of pulse generated by traditional NEMPS (such as bounded wave simulator, static simulator, etc. [5]) is not fast enough because of the transmission line structure. However, the radiating EMP simulators using antenna structure can get faster rise time and larger test space without special requirements, which benefit to the EMP effect experiments for large equipment.

Nowadays cage-like dipole antennas are commonly used [6] with the features of constant impedance, broad band and good field uniformity parallel to the axis. However, the volume of this kind of antenna is generally very huge, rise time is longer, directionality is poor, and the effective test area is small. In this paper, TEM horn antenna is tried for NEMPS, which has been widely adopted in ultra wideband fields [7] as a pulse radiating antenna with a certain directivity, wide bandwidth and

[*] To whom any correspondence should be addressed.

the phase centre of radiation invariance. Theoretic analysis and numerical simulation as well as experiment are implemented. Results show that TEM horn antenna can be used as the radiating antenna in EMP simulator if its low frequency radiation performance can be improved.

2. Basic theory of TEM horn antenna

As shown in figure 1, TEM horn antenna is composed of two triangular metallic plates. The intersection angle is β. Each plate's flare angle is α. Antenna length is expressed by l, height by h, width by w. The characteristic impedance of antenna is Z_c. For TEM horn antenna with small flare angle, current distribution on plates is mainly composed of the longitudinal component along with Z direction. Therefore, in literature [8], TEM horn antenna can be described by a transmission line circuit model with terminal open. Based on this equivalent model, a formula of step response for radiation on the antenna's axis is given:

$$\tilde{E}_y^{TOT}(r,t_r) = -\frac{V_0}{r}\frac{l}{h}E_y^{norm}\left[e^{-\omega_2 t_r}u(t_r) - \frac{c}{2l}\frac{1}{\omega_2}[(1-e^{-\omega_2 t_r})u(t_r) - (1-e^{-\omega_2(t_r-\frac{2l}{c})})u(t_r-\frac{2l}{c})]\right] \quad (1)$$

where $u(t)$ is the step pulse; $t_r = t - \dfrac{r+l}{c}$; c is light velocity; r is the far-field distance shift from the antenna aperture; $w_2 = 2\pi f_2$,which means high frequency cut-off point in antenna frequency response; V_0 is the input voltage peak, $E_y^{norm} = E_a/(V_0/h)$; for TEM horn antenna with small flare angle, $E_y^{norm} \approx 1$. From the frequency response asymptotes of the TEM horn with step excitation shown in figure 7 of literature [8], we get that:

$$f_1 = \frac{c}{2\pi l} \quad (2)$$

$$f_2 = \frac{2lcf_g E_y^{norm}}{h^2} \quad (3)$$

This frequency response can be viewed as the band pass filter. In the pass band, the curve is smoothing, which means that radiation waveforms keep up with the differential form of pulse excitation. So in order to get double-exponent-like pulse radiation field, we need to use the integral double exponential pulse excitation.

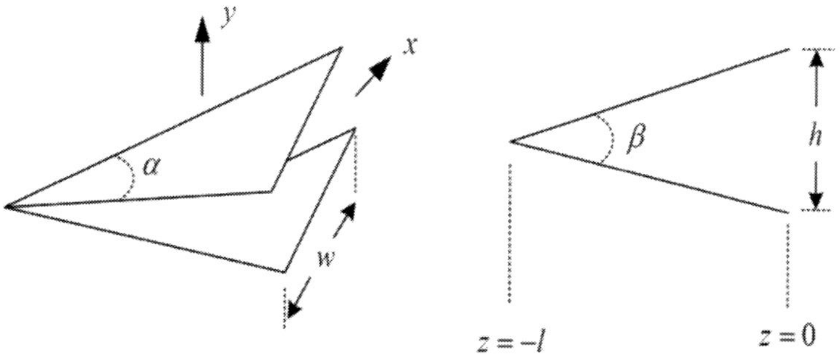

Figure 1. Diagram of the TEM horn antenna.

3. Farfield in time domain with integral double exponent pulse excitation.

The actual systemic response is the convolution of step response and the differential form of actual input voltage. Using formula (1), the formula of radiation farfield response in time domain can be deduced for TEM horn antenna excited by arbitrary pulse:

$$\tilde{E}_y^{TOT}(r,t) = formula(1) \otimes \frac{dV(t)}{dt}$$

$$= -\frac{V_0}{r}\frac{l}{h}E_y^{norm}\left[A - \frac{c}{2l}\frac{1}{\omega_2}[B-C]\right] = \begin{cases} 0 & t_r < 0 \\ -\frac{V_0}{r}\frac{l}{h}E_y^{norm}\left[A_1(t_r) - \frac{c}{2l}\frac{1}{\omega_2}B_1(t_r)\right] & 0 \le t_r < \frac{2l}{c} \\ -\frac{V_0}{r}\frac{l}{h}E_y^{norm}\left[A_1(t_r) - \frac{c}{2l}\frac{1}{\omega_2}[B_1(t_r) - B_1(t_r - \frac{2l}{c})]\right] & t_r > \frac{2l}{c} \end{cases} \quad (4)$$

Where $\frac{dV(t)}{dt} = V_1(t)u(t)$ indicates that $V_1(t)$ only has meaning at $t \ge 0$, and

$A = e^{-\omega_2 t_r}u(t_r) \otimes V_1(t)u(t) = A_1(t_r)u(t_r)$; $B = (1 - e^{-\omega_2 t_r})u(t_r) \otimes V_1(t)u(t) = B_1(t_r)u(t_r)$;

$A_1(t) = e^{-\omega_2 t}u(t) \otimes V_1(t)u(t)$ $B_1(t) = (1 - e^{-\omega_2 t})u(t) \otimes V_1(t)u(t)$

$$C = (1 - e^{-\omega_2 (t_r - \frac{2l}{c})})u(t_r - \frac{2l}{c}) \otimes V_1(t)u(t) = B_1(t_r - \frac{2l}{c})u(t_r - \frac{2l}{c})$$

According to IEC 61000-2-9 [1], we can get the integral double exponential pulse,

$$V(t) = \begin{cases} 0 & t \le 0 \\ k\frac{E_0}{\tau_1 \tau_2}\left(\tau_1 e^{-\tau_2 t} - \tau_2 e^{-\tau_1 t}\right) & t > 0 \end{cases} \quad (5)$$

Here E_0=50 kV m^{-1}, τ_1=4.0×10^7 s^{-1}, τ_2=6.0×10^8 s^{-1}, k=1.3, pulse duration time T=100 ns.

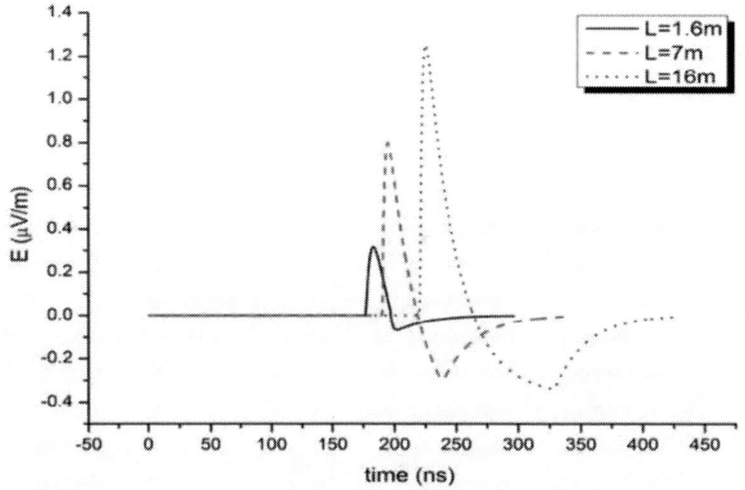

Figure 2. Radiated waveforms for different antenna length.

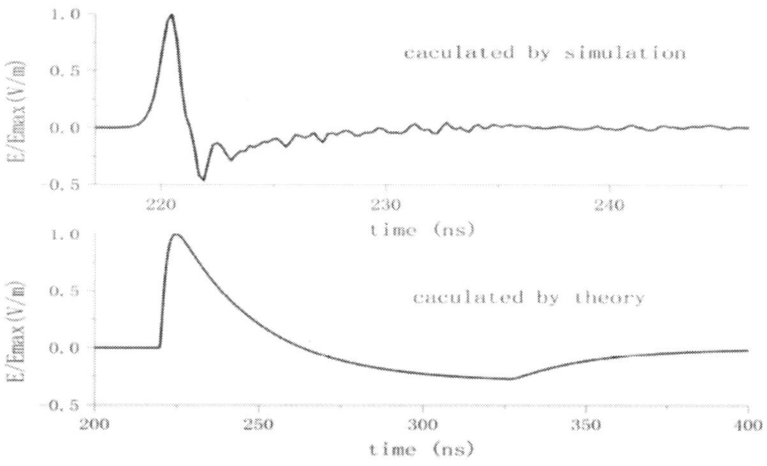

Figure 3. Comparison of farfield waveforms.

Set the antenna length as 1.6 m, 7 m, 16 m, respectively, the corresponding antenna height are 1.55 m, 3.24 m, 4.90 m, using integral double exponential pulse excitation, the far field radiation waveforms on the antenna's axis are shown in figure 2, which indicate that the amplitude of radiation field becomes large, the pulse width is broadening, and the pulse form is gradually close to double exponential pulse with the antenna length and height increasing. However, the zero-crossing phenomenon becomes serious, the negative peak value is larger and the waveform after zero comes to be poor.

4. Simulations and experiments

Now we execute some simulations and experiments to have a farther analysis for the TEM horn antenna.

4.1. Principle of antenna parameters selection

There are 5 parameters which are l、h、w、α and β. According to the geometric features, the relationship between them is shown as follows:

$$\begin{cases} \beta = 2\,\mathrm{atan}\left(\dfrac{h}{2l}\right) \\ \alpha = 2\,\mathrm{atan}\left(\dfrac{w}{h}\sin\left(\dfrac{\beta}{2}\right)\right) \end{cases} \tag{6}$$

In this formula, the ratio of $\dfrac{w}{h}$ is determined.by characteristic impedance Z_c, which can be calculated by the formula (21) and (22) in literature [9].

According to the above analysis, the principle of TEM horn antenna parameters selection can be summarized as: with Z_c known, E_y^{norm} is generally taken as one, we need firstly to determine the antenna length l and height h according to frequency range $f_1 \sim f_2$ by formula (2) and (3); and then get the values of aperture width w and structure angle (α, β) by the formulas in literature [10].

4.2. Numerical modeling and simulation results
According to the above principle, let $f_1 = 3$ MHz, $f_2 = 400$ MHz, $Z_c = Z_0 = 377\Omega$, then we can get that: $l \approx 16$ m, $h \approx 4.89$ m, $w \approx 0.85$ m, $\alpha \approx 3^\circ$, $\beta \approx 17^\circ$. With integral double exponential pulse as excitation, based on the finite integral in time domain (FITD) method [11], the farfield waveform of 50 m away from the antenna aperture, compared with the theoretical value (because of the difference between norm antenna and equivalent transmission line model, the farfield amplitudes are not comparable, so both the simulation and theoretical results are normalized by max values), is shown in figure 3.

It can be seen that there are negative peak in both numerical and theoretical calculations. Coincident front waveforms demonstrate that the high frequency radiation performance of actual antenna is good, so we can have lower requirement for antenna height in view of antenna parameter selection principle. However, the farfield pulse width calculated by simulation is far narrower than theoretical calculation. This is because that the actual antenna length is limited and then most of the low frequency components of the pulse excitation is not radiated out but returned from antenna terminal. Therefore, we have to consider methods of impedance loading, antenna deformation, filling medium, and low frequency compensation loop and antenna array and so on to improve low-frequency radiation performance for TEM horn antenna.

4.3. The experimental design and results analysis
TEM horn antenna for experiments has length 1.5 m and height 1 m, aperture width 2 m, two 180 ohms water resistances are placed at the antenna end. High voltage source is used for charging the capacitor, and then the spark gap switch breaks over, capacitor discharges to produce an integral double exponential pulse approximately. Electric field is measured by optical field tester, and displayed in a shielded room through the oscilloscope.

Do gradually increase the output voltage value to make switch breakdown. Because of the functional limitations of field tester, waveforms but not the field strength values are observed. E-field waveform at 5 meters shift from the aperture is shown in figure 4. In order to compare with radiation waveform generated by double exponential pulse excitation, we can use Marx generator to replace voltage source and capacitor, for which far field is shown in figure 5.

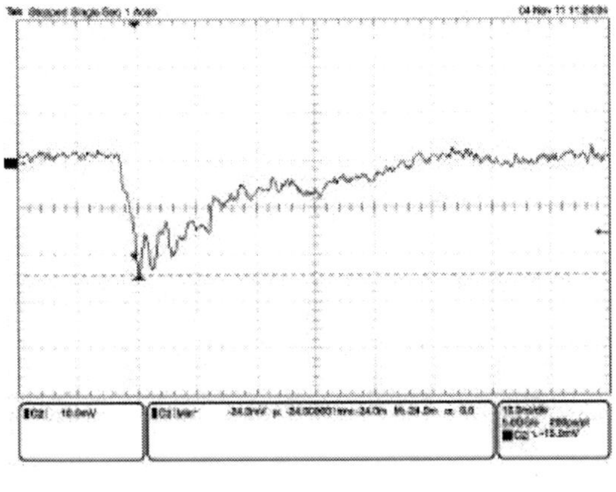

Figure 4. Farfield waveform with resistors at the antenna end.

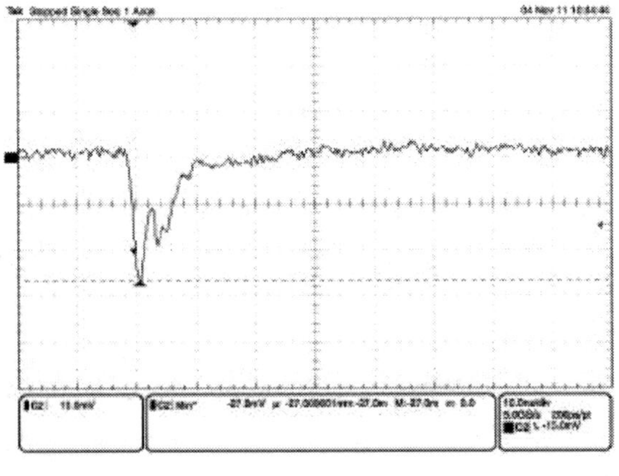

Figure 5. Farfield waveform with Mark generator excitation.

From experimental results it can be seen that when using integral double exponential pulse excitation, far field is approximately close to double exponential pulse form, and low frequency radiation performance can be improved through the method of resistance loading; but when using Marx generator as excitation, TEM horn antenna actually radiate the differential form of double exponential pulse, and pulse width is smaller, furthermore, radiation waveform is generally in some serious oscillation as the Marx generator itself output dithering double exponential pulse. Experiments indicate that TEM horn antenna can be used for radiating NEMPS, but the low frequency radiation ability needs to be enhanced through the way of resistance loading.

5. Conclusions

This paper tries to apply TEM horn antenna for NEMPS and get the following conclusions:

a) Radiation theory in time domain is analyzed based on the equivalent transmission line model of TEM horn antenna with small flare angle, which deduces far field response for the integral exponential pulse excitation, and indicates that with the antenna length and height increasing, amplitude of radiation field becomes larger, pulse width is broaden and waveform is gradually close to double exponential pulse form, and low frequency radiation performance is improved, but the zero-crossing phenomena becomes serious, the part of waveform after zero goes into poor;

b) Principle for antenna parameters selection is given as that: firstly, choose the proper antenna length and height with taking the frequency range into account; and then set the structure angles based on the characteristic impedance of antenna;

c) Numerical results and experimental results show that: the high frequency radiation performance of actual antenna is good; but the narrow half peak pulse width cannot satisfy the requirements. Therefore, when using TEM horn antenna as the radiating antenna of simulator, low frequency radiation performance need to be improved.

As limited by experimental conditions, the strength values of far field and field uniformity have not been analyzed. Further work will be concentrated on EMP waveform, field strength, field distribution and field uniformity and many other aspects. More importantly, we have to improve the low frequency radiation performance of antenna through antenna optimization design in depth.

References

[1] *MIL-STD-461E* 1999
[2] *GJB1389A-2005* 2005
[3] *MIL-STD-461C* 1986
[4] *MIL-STD-461D* 1993
[5] Baum C E 1978 *IEEE. Trans. EMC* **26** 35
[6] Bailey V, Carboni V and Eichenberger C 2010 *IEEE Trans. Plasma Sci.* **38** 2554
[7] Bassam S, Rashed-Mohassel J 2006 *Piers. Online.* **2** 706
[8] Everett G 1992 *In: Sens. and Simul. Notes* Note 340
[9] *IEC 61000-2-9* 1996
[10] Todd Lee R, Glenn Smith S 2004 *IEEE Trans. Antennas Propag.* **52** 315
[11] Weiland T *In: Int. J. Numer. model. electron. networks, devices fields* vol 9 p 295

A study on the equivalent electric circuit simulation model of DBD streamer and glow alternate discharge

J Yao[1], Z T Zhang[1,2*], Z Yu[2], S J Xu[1], Q X Yu[1], J S Zhao[2]
1 Institute of Marine Engineering, Dalian Maritime University, Dalian 116026, China
2 Department of Physics, Dalian Maritime University, Dalian 116026, China

E-mail: newzhangzhitao@163.com

Abstract. This paper presents a dynamic simulating model of the dielectric barrier discharge (DBD), structured as an equivalent electric circuit of the streamer and glow discharge generated alternately in DBD. The main parameters of DBD have been established by means of analysing the structural characteristics of a single discharge cell. An electrical comprehensive Simulink /MATLAB model was developed in order to reveal the interaction of the adjacent two discharge cell. A series of simulations was carried out in order to estimate the key structural parameters that affect the alternate streamer and glow discharge mode. The comparison results of experimental and simulate indicate that there exists a close similarity of the current waveforms graphic. Therefore, we can grasp a deep understanding mechanism of the dielectric barrier discharge and optimize the plasma reactor.

1. Introduction

Dielectric barrier discharge (DBD) is a typical non-equilibrium ac gas discharge. Its experimental device generally consists of two parallel electrodes, at least one of which is covered with a dielectric material [1, 2]. Recently, much attention has been paid to DBD because of its broad application prospect in numerous industrial areas. The DBD has been previously used in different fields such as plasma display, light emitting and environmental protection [3-6]. For these reasons, accurate and reliable information of the reactive species is critical for the understanding of plasma behavior and then optimizing the plasma generation process to improve the functioning of the plasma processing reactor. Furthermore, plasma discharge parameters can be selected to optimize the production of reactive species based on the optical measurement. Several researchers have been investigated the physical mechanism of the micro-discharge in pin-to-plane DBD [7-10], combined with a macro camera. When the excitation voltage large enough, it can be found that there are two different discharge models in a current cycle which were considered to be the streamer and glow alternately discharge.

This paper sets out to present a refined modeling of the DBD electrical elements so as to afford a better understanding of the current and voltage responses during the discharge, provided that these parameters are measurable throughout the simulation process. Furthermore, this DBD model takes into account the alternative discharge of the streamer and glow model with the applied voltage above 5 kV.

[*] To whom any correspondence should be addressed.

2. Modelling and analysis

2.1. The current-voltage characteristics of the streamer and glow discharge generated alternately in pin-to-plane DBD

The instrumental arrangement used for this study is shown in figure 1. It is mainly composed by the current and voltage detection and the DBD, which is supplied by a voltage source inverter coupled with a step-up high frequency. Measurements of the power source output voltage as well as of the DBD applied voltage and the discharge current is provided by a Tektronix oscilloscope. In all cases, the electric discharges are generated at atmospheric pressure.

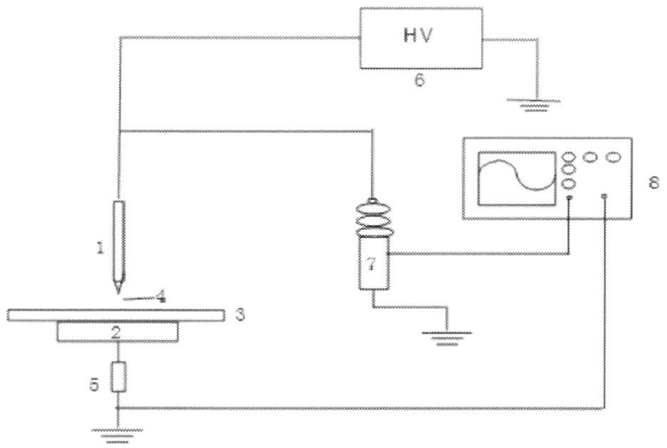

Figure 1. Dielectric barrier discharge experimental system.
1 High voltage electrode,
2 Plane electrode,
3 0.47 mm thick dielectric of aluminium,
4 Discharge gap,
5 Resistor,
6 Power source,
7 High-voltage probe,
8 Oscilloscope.

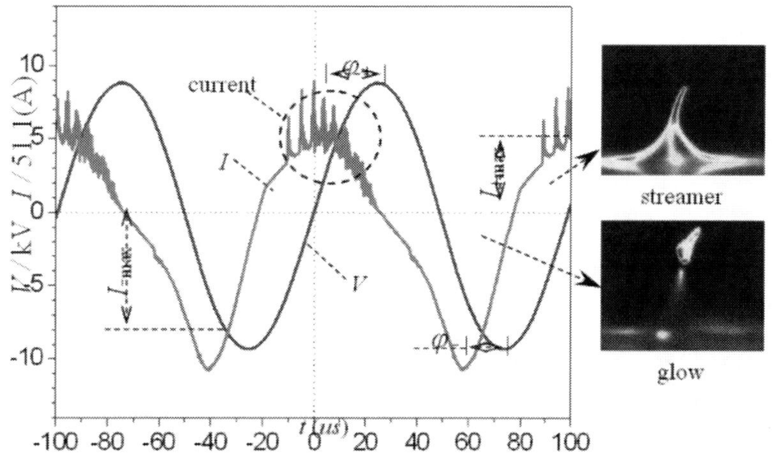

Figure 2. Current and voltage waveform of the pin-to-plane DBD (V >5 kV).

The pin electrode is made of brass and their dimensions are shown in figure 1.The gap clearance d is adjustable up to 0.9 mm and the excitation power frequency is 10 kHz. A resistor R is connected to the electrodes in series for us to measure the discharge current. A high voltage probe is used for us to measure the applied voltage between the electrodes at the same time. When the applied voltage is less than 5 kV the resulting is a typical streamer. However there will be a new type of discharge mode when the applied voltage above 5 kV. Although it still looks like a streamer discharge mode in our eyes, we can find the significant differences by a 1us macro high-speed CCD camera. According to the current and voltage waveform, as seen in figure 2, it would be the glow discharge mode.

2.2. *DBD discharge equivalent circuit*

An equivalent electric circuit of the pin-to-plane DBD is set up, as shown in figure 3, in order to understand the physical mechanism of the DBD and investigate the streamer and glow discharge generated alternately in pin-to-plane DBD. Thus, the electrical characteristics of such plasma discharges can be represented by a voltage-controlled current source. By this means the streamer discharge model will occur in the positive of the current waveform and the negative will be the glow. In this electric circuit the dielectric sheets between the electrodes can be represented by capacitors C_d. Both the gas capacitance value C_g and the equivalent resistance R_l depend on the applied gas composition.

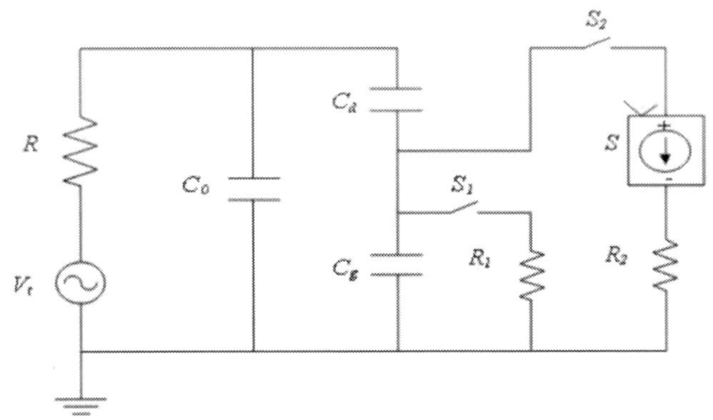

Figure 3. Equivalent circuit of single micro-discharge channel

2.3. *Micro-discharge dynamic simulation model*

According to the analysis of DBD discharge equivalent circuit, a dynamic simulation model is established by using the simulation software of MATLAB/ Simulink [11, 12], as shown in figure 4.

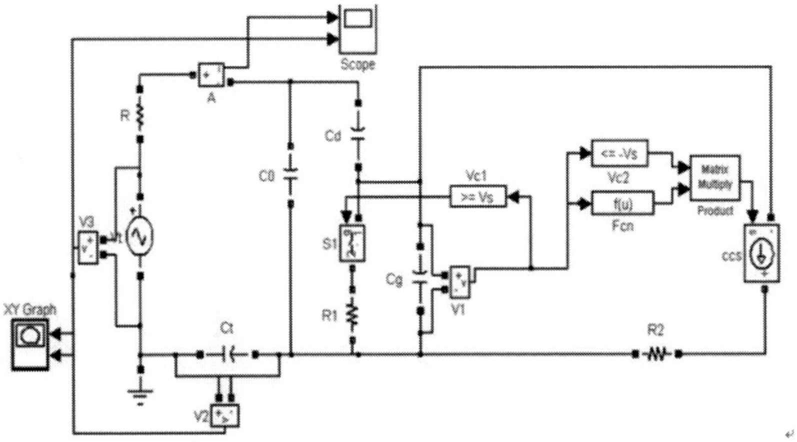

Figure 4. Dynamic simulation model of single micro-discharge channel.

In this model the initial state of the voltage-controlled switch S_l is disconnected. The streamer discharge model process are as follows, when the air-gap voltage V_g exceeds the value of the previously fixed block parameter *Vs*, then S_l is switched on, output "1" , R_l is connected to the circuit then the streamer discharge began. However when *Vg* is less than -*Vs*, *Vc2* output "1", the value of the function module Fcn control the output of current source CCS to start the glow discharge.

3. Simulation process

3.1. Current waveform simulation

In order to validate the proposed-model simulation, experimental tests have been carried out with the current waveform, as shown in figure 5. It can be seen, the discharge current waveform superimpose on a large number of current pulses in the positive half cycle, showing a capacitive load characteristics, so it is a typical streamer discharge model. In the negative half cycle, there is a wide current pulse. It can be determined that there are two completely different discharge models between the positive and the negative half cycle. It must be the streamer and glow discharge generated alternately in pin-to-plane DBD. In addition, the comparison between the experimental and simulated current waveform is shown in figure 6. There exists a considerable similarity between the experimentally measured values, those produced by Simulink /MATLAB.

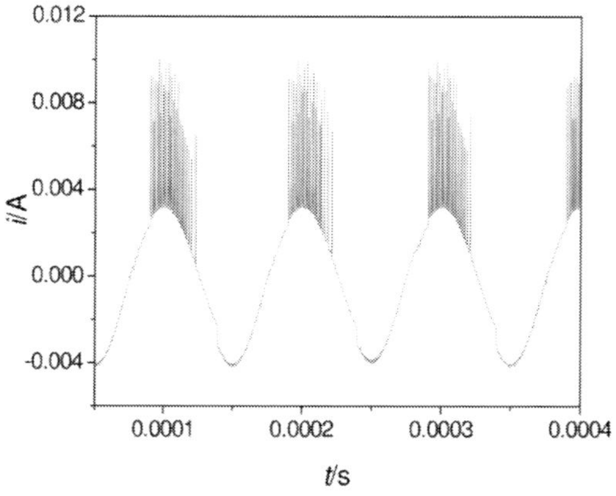

Figure 5. Current waveform of simulation result.

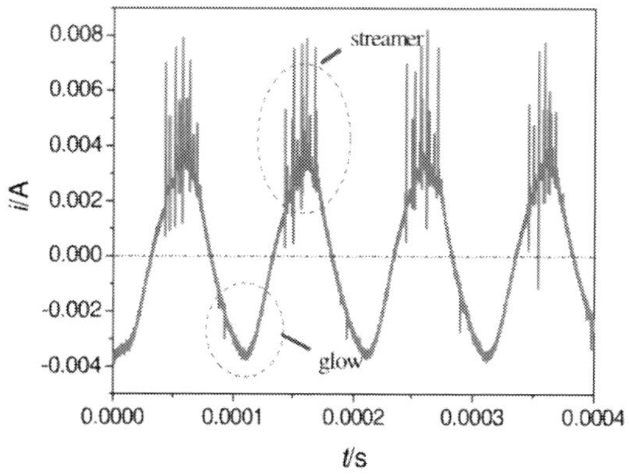

Figure 6. Current waveforms of single micro-discharge with two modes.

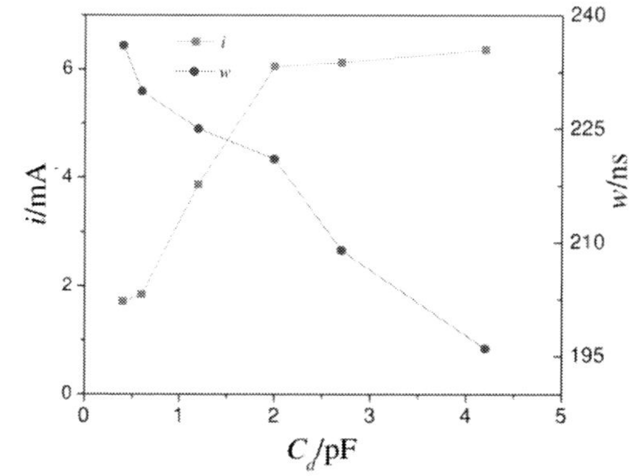

Figure 7. Micro-glow's amplitude with different C_d.

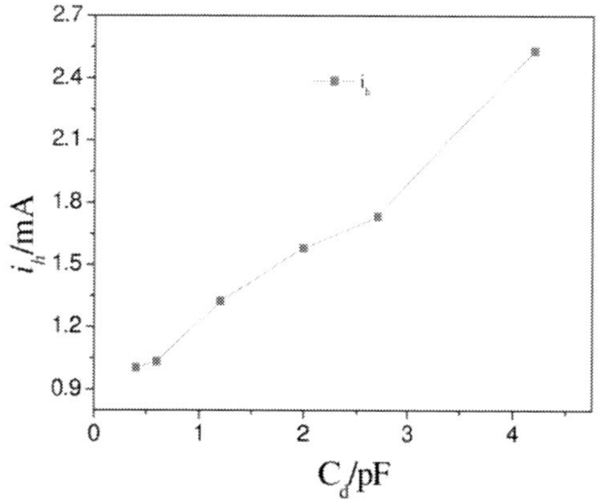

Figure 8. Streamer's amplitude and lifetime with different C_d.

3.2. The impact of different parameters on DBD micro-discharge

Departing from the obtained simulation results, the two influential electrical parameters C_d and C_g have been considered in figure 7 to figure 10. It indicates that the performance of the DBD discharge can be improved with the increasing of the gas capacitance C_d and dielectric capacitance C_g.

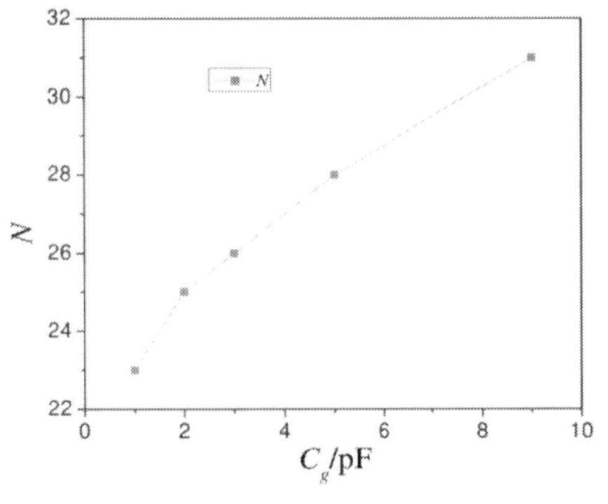

Figure 9. Discharge times with different C_g.

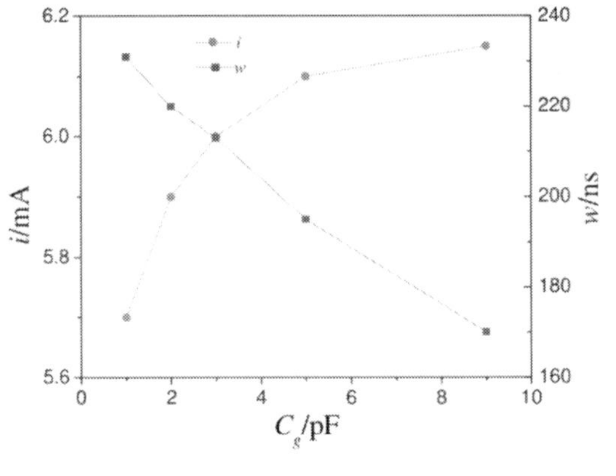

Figure 10. Streamer's amplitude and lifetime with different C_g.

4. Conclusion

There exists a close similarity between the experimental results and the predictions from the proposed DBD model. This can be considered relevant as the model goal is to provide an acceptable estimation of the electric parameters of the electrical characteristics of the high-power switches of the static converter, the intensity and frequency of the excitation voltage source, of course, the necessary power to be delivered to the charge.

A series of experiments with electric discharges in a DBD reactor has been carried out. The experimental results were satisfactorily compared with the simulation predictions of the proposed DBD reactor model. Thus, the model allows one to both predict with accuracy the discharge behaviour using the experimental parameters and verify the different discharge models of the streamer and glow discharge generated alternately in pin-to-plane DBD.

Acknowledgments

This work is supported by The National High Technology Research and Development Program of China under Grant No 2012AA062609, The International Science and Technology Cooperation Project of China under Grant No 2010DFA61470, The National Science Foundation for Distinguished Young Scholars of China under Grant No 61025001, The Fundamental Research Funds for the Central Universities under Grant No 2012QN067 and 2012TD061.

References

[1] Leys C 2004 *Contrib. Plasma Phys.* **44** 542
[2] Wei D D, Feng L, and Lan J Y 2009 *IEEE Trans. Plasma Sci.* **37** 2207
[3] Tenderoa C, Tixiera C, Tristanta P, Desmaisona J, and Leprinceb P 2006 *Spectrochim. Acta Part B* **61** 2
[4] Roth J R, Rahel J, and DAI X 2005 *J. Phys. D: Appl. Phys.* **38** 555
[5] Fang Z, Qiu Y, Luo Y 2003 *J. Phys. D: Appl. Phys.* **36** 2980
[6] Eden J G 2006 *Proc. IEEE* **94** 567
[7] Radu I, Bartnikas R and Wertheimer M R 2003 *J. Phys. D: Appl. Phys.* **36** 1284
[8] Yu. S. Akishev, A. V. Dem'yanov, V. B. Karal'nik, A. E. Monich, and N. I. Trushkin 2003 *Plasma Phys. Reports* **29** 82
[9] Kogelschatz U 2002 *IEEE Trans. Plasma Sci.* **30** 1400
[10] Hao Y P, Guan Z C, Wang L M, Wang X X and Li C R 2005 *High Voltage Apparatus* **31** 59
[11] Flores-Fuentes A, Pena-Eguiluz R, Lopez-Callejas R and Mercado-Cabrera A 2009 *IEEE Trans. Plasma Sci.* **37** 128
[12] Zhang C, Fang Z, Zhao L Z and Qiu Y C 2007 *High Voltage Apparatus* **43** 218

The 3-dimensional numerical simulation of artificially altitude-triggered negative lightning

Bo Zhang, Bin Chen*, Lihua Shi, Qiang Chen
National Key Laboratory on Electromagnetic Environmental Effects and Electro-optical Engineering, PLA University of Science and Technology, Nanjing 210007, China

E-mail: emcchen@163.com

Abstract. A 3-dimensional numerical model for artificially altitude-triggered negative lightning is developed based on an analytic thunderstorm model and the Dielectric Breakdown Model (DBM). Two major parameters are concerned, they are the thundercloud electric field and the length of the nylon wire which isolates the triggering wire from the ground. A few groups of contrast numerical experiments are done to study their effects on the success rates of altitude-triggered lightning. It is found that the success rates of altitude-triggered lightning increase when the thundercloud electric field enhances or the length of the nylon wire increases. Another interesting phenomenon is that the upward positive leader is always initiated earlier than the downward negative leader in either case.

1. Introduction

Artificially altitude-triggered lightning is an efficient way to study some unknown physical processes related to the lightning flashes. It is initiated by a small rocket towing an ungrounded triggering wire aloft under a thunderstorm. When the triggering wire is introduced rapidly into a strong-field region, it might actually initiate a bi-directional stepped leader, i.e., the upward positive leader and the downward negative leader. The positive leader propagates towards the thundercloud and might initiate a negative lightning, while the negative leader is an efficient way to imitate the lightning connection process to a grounded structure.

Since numerical work complements the experimental studies, much effort has been done to develop the lightning model. A numerical model on the stepped leader path in the earth atmosphere was developed by Odim Mended to study the effect of the tropospheric electric conductivity on the lightning behaviour [1], and a self-consistent upward leader model was proposed by Marley Becerra to study the initiation and propagation of an upward connecting leader in the presence of a downward moving leader [2], and so on. Nevertheless, much work is remained to be done. It is worthy to mention that P. Lalande once developed a model on the altitude-triggered lightning to study the lightning connection process [3]. In the model, the Charge Simulation Method (CSM) was applied to the computation of the electric field which was distorted by the introduction of the triggering wire, and the distribution of the induced charge on the surface of the triggering wire was assumed according to

* To whom any correspondence should be addressed.

experience, so it might be a good work but seems to be not very scientific. Besides, there are few systemic reports about the growth rule of the bi-directional stepped leader.

In this paper, a 3D numerical model on the altitude-triggered lightning is proposed and it contains 4 major parts:

- A background electric field model based on an analytic thunderstorm model.
- A new method to calculate the distorted electric field based on the combination of the Laplace equation and the Successive Over Relaxation (SOR) method.
- A stepped leader model based on the Dielectric Breakdown Model which ensures the lightning path to have enough branches.
- A space resolution conversion model to save the calculation resource.

A study of the relationship between the success rates of altitude-triggered lightning and the two parameters, the background electric field intensity and the length of the nylon wire which will be introduced in the following text, is carried out based on the model. And finally, some growth rules of the bi-directional stepped leader are indicated.

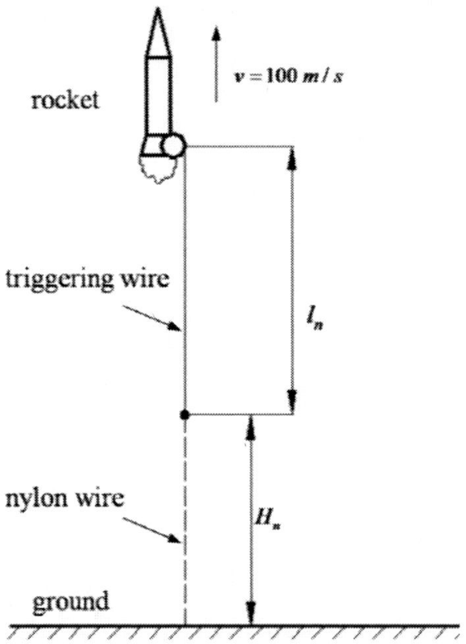

Figure 1. Altitude-triggered device.

2. Modelling

The experiment device of altitude-triggered lightning is shown in figure 1: the triggering wire is spooled to the end of a small rocket and introduced into the air. It has a constant total length of l_0 and is isolated from the ground by a nylon wire with a length of H_n. When the triggering wire has been trailed out over a sufficient length of l_n, the bi-directional leader is initiated from both of its two ends and propagates towards the cloud and the ground respectively. For simplification, the speed of the rocket is assumed to be constant and the value is 100 m s^{-1}.

2.1. Background electric field

An appropriate background electric field model is necessary to develop the numerical model of the altitude-triggered lightning. It should not only be able to express the main characteristics of the thundercloud electric field, but also not be too complex to calculate.

Williams, Uman and Rakov proposed the Three Layers Cylindrical Charged Pile Model (TLCCPM) to calculate the thundercloud electric field and it overcame the drawback of Dipole Model that the distribution of the charge was simplified too much. And then Amoruso and Lattarulo pointed out that when the charge density inside the thundercloud was symmetrical the TLCCPM could be simplified to the Charged Disks Model. It is convenient and reasonable to describe the thundercloud electric field with this model [4, 5]. And the final expression for the background electric field is obtained [6]

$$\vec{E}(x) \approx \frac{\rho_N}{4\pi\varepsilon}\int_{S_1} \vec{u}_1 \frac{dS'_1}{r_1} + \frac{\rho_N - \rho_L}{4\pi\varepsilon}\int_{S_2} \vec{u}_2 \frac{dS'_2}{r_2} + \frac{\rho_P}{4\pi\varepsilon}\int_{S_3} \vec{u}_3 \frac{dS'_3}{r_3} \tag{1}$$

It is an analytic expression, where ρ_N, ρ_L, ρ_P are the charge density of the main negative, lower positive, positive charge center respectively, and $\vec{u}_1$, $\vec{u}_2$, $\vec{u}_3$ are the respective outward unit normal vectors of the interface. The surface integral in equation (1) is

$$\int_S \frac{dS'}{r} = 2\left[-(1-\varepsilon')\frac{\pi}{2}h + r_d\,\mathrm{E}(p) + \frac{a^2 - R^2}{r_d}\mathrm{K}(p) + \frac{h^2}{r_d}\frac{a-R}{a+R}\Pi(m,p)\right] \tag{2}$$

Where K, E, Π are the first, second, and third kind of complete elliptic integral respectively, a is the radius of the thundercloud, R is the horizontal distance to the center of the thundercloud, h is the height of the charged disk, $r_d = \left[(a+R)^2 + h^2\right]^{1/2}$, $m = 2(aR)^{1/2}(a+R)^{-1}$, and $p = 2(aR)^{1/2} r_d^{-1}$. ε' is -1, 0 or 1 respectively when R is greater than, equal to or less than a. Taking account of the Mirror Effect of the ground, the total electric field is obtained.

The space concerned is cubic with a size of $4\times4\times5\text{km}$. Integrating the E-field, the electric potentials of the whole space will be obtained. The potentials of points which are on the surface of the space will be the constant outer boundary conditions of the Laplace Equation.

2.2. *The distorted electric field*

The electric field is distorted when the triggering wire is introduced into the air. Thus the E-field around the triggering wire is enhanced and might exceed the initiating criterion of the stepped leader. Since both the distorted E-field and the charge distribution on the wire are unknown, it is not easy to calculate the distribution of the space potential in fact. As the relaxation velocity of charge is much faster than the velocity of the rocket, the triggering wire could be treated as an equipotential conductor at each time step. Thus we have one more unknown parameter, the potential of the triggering wire, besides the solutions to the Laplace equations. On the other hand, the net charge on the surface of the triggering wire must be zero, so a new equation related to the potential of the triggering wire is obtained.

$$-\int_{\partial\Omega_1} \frac{\partial\phi}{\partial n} = 0$$

Where $\partial\Omega_1$ is the surface of the triggering wire. Adding this equation to the finite difference Laplace equations as an inner boundary condition, we obtain a definite problem

$$\begin{cases} \nabla^2\phi = 0 \\ -\int_{\partial\Omega_1} \frac{\partial\phi}{\partial n} = 0 \\ \phi|_{\partial\Omega_2} = \psi \end{cases} \tag{3}$$

Where $\partial\Omega_2$ is the outer boundary of the space.

The seven-point finite difference scheme of the Laplace equations [7] is

$$\frac{\varphi_{i+1,j,k}-2\varphi_{i,j,k}+\varphi_{i-1,j,k}}{\Delta x^2}+\frac{\varphi_{i,j+1,k}-2\varphi_{i,j,k}+\varphi_{i,j-1,k}}{\Delta y^2}+\frac{\varphi_{i,j,k+1}-2\varphi_{i,j,k}+\varphi_{i,j,k-1}}{\Delta z^2}=0 \quad (4)$$

Where Δx, Δy and Δz are the respective resolution of x, y and z axis. Applying the method of Successive Over Relaxation (SOR) to equation (4), we will obtain the iteration scheme of potential φ

$$\varphi^{t+1}_{i,j,k}=\varphi^{t}_{i,j,k}+\frac{\omega}{2(\Delta x^2+\Delta y^2+\Delta z^2)}\cdot[\Delta x^2\left(\varphi^{t}_{i+1,j,k}+\varphi^{t}_{i-1,j,k}\right)+\Delta y^2\left(\varphi^{t}_{i,j+1,k}+\varphi^{t}_{i,j-1,k}\right)$$

$$+\Delta z^2\left(\varphi^{t}_{i,j,k+1}+\varphi^{t}_{i,j,k-1}\right)-2(\Delta x^2+\Delta y^2+\Delta z^2)\varphi^{t}_{i,j,k}] \quad (5)$$

Where the superscript t represents the iteration, and ω is the relaxation factor

$$\omega=\frac{6}{3+[9-(\cos\frac{\pi}{l-1}+\cos\frac{\pi}{m-1}+\cos\frac{\pi}{n-1})^2]^{1/2}} \quad (6)$$

Where l, m, n are the respective number of lattices in the direction of x, y and z axis, and it is evident that $\omega\in[1, 2]$.

When the criterion is exceeded, the bi-directional leader will be initiated and propagate rapidly, the wire will be melted by the heat produced by the leader discharge. Both the channel of the wire and the developed leader will be treated as new constant boundary conditions for another iterative calculation.

2.3. *The stepped leader model*

The E-field can be determinated when the distribution of the space potential is known. And the Dielectric Breakdown Model [8] is applied to calculate the probability of each possible direction for the next stepped leader

$$P_i=\begin{cases}\left|E_{ij}-E_{crit}\right|^{\eta}/F & E_{ij}\ge E_{crit}\\ 0 & E_{ij}<E_{crit}\end{cases}$$

$$F=\sum_{j(E_{ij}>E_{crit})}\left|E_{ij}-E_{crit}\right|^{\eta} \quad (7)$$

Where $E_{ij}=\left(\varphi_i-\varphi_j\right)/\Delta l$ is the average electric field intensity between the developed and developing point, Δl is the distance of the two points, and E_{crit} is the threshold value. The Monte Carlo Method is used in the selection.

Considering the finite conductivity of the leader channel, we use E_{in} to describe the axial electric field intensity. Thus the potential of the newly chosen point becomes

$$\varphi_i=\varphi_0-E_{in}\sum\Delta l \quad (8)$$

Where φ_0 is the potential of the initiating point, and $\sum\Delta l$ is the length along the channel. Thus a new boundary condition is obtained.

2.4. *The resolution conversion*

In this paper, a resolution of 50 m is applied to the whole 3D space concerned, and it is enough to show the details of the stepped leader. But for some areas where the electric field is distorted by the triggering wire and has a greater gradient, it is necessary to convert the resolution into a finer one to ensure a precise solution. The 3-spline interpolation method is employed to do this conversion [9]. The procedure is shown in figure 2, and the finer resolution is 5 m.

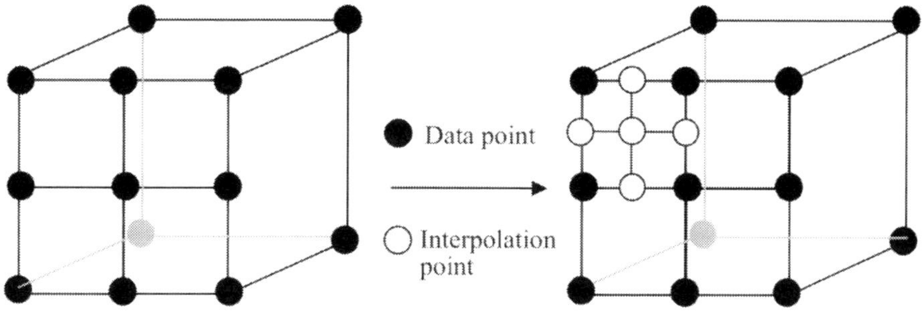

Figure 2. Sketch of the interpolation.

2.5. *The criterion of success rates*

An altitude-triggered lightning numerical experiment succeeds when the upward positive leader reaches the cloud and the downward negative leader reaches the ground respectively. But not every numerical experiment will succeed, because the total length of the triggering wire is limited. The numerical experiment will fail if the bi-directional leader is not initiated before the wire runs out. And the longer the triggering wire is used, more difficult the numerical experiment to trigger a lightning. So the success rate of the altitude-triggered lightning experiment will be

$$Rate = \alpha(l_0 - l_n)/l_0 \tag{9}$$

Where l_n is the length of the triggering wire, $(l_0 - l_n)$ is the remaining length, and α is an adjustable parameter which will be determined by the numerical experiment.

3. Numerical experiment

Numerical experiments are conducted while the parameters are: ρ_P=ρ_N=0.5 nCm^{-3}, ρ_L=0.1 nCm^{-3}, the respective height h_P=10 km, h_N=1.5 km, h_L=1.4 km, the radius of the thundercloud a=2 km, the radius of the lower positive center a_L=0.5 km, the total length of the triggering wire l_0=350 m, the length of the nylon wire H_0=350 m, the axial electric field intensity of the channel E_{in}=500 Vm^{-1}, the threshold value of the upward positive leader E_{critP}=150 Vm^{-1}, the threshold value of the downward negative leader E_{critN}=180 Vm^{-1}, the error threshold of the SOR method δ=0.1 V. The result of the simulation is shown in figure 3.

In this case, the electric field intensity E_0 at the center of the ground (0, 0, 0) is about -15 kVm^{-1}. An upward positive leader is initiated from the top of the triggering wire and propagates towards the cloud when the length is 110 m. And about 40 steps later, a downward negative leader is developed from the bottom of the triggering wire and propagates towards the ground. Both of them have abundant branches which make it look like a real lightning. And according to the criterion, the success rate in this case is obtained

$$Rate = \alpha(350 - 110)/350 = 68.6\alpha\%$$

If it is assumed that $\alpha = 1$, the success rate will be 68.6%.

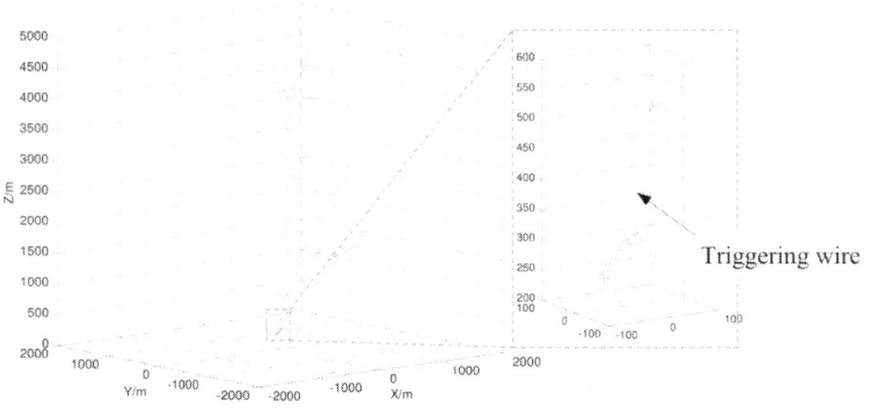

Figure 3. Diagram of an altitude-triggered lightning discharge and the resolution conversion.

In the following text two groups of contrast numerical experiments are conducted to study the effect of the thunderstorm electric field intensity E_n and the length of the nylon wire H_n on the success rates of altitude-triggered lightning. Other parameters are invariant unless mentioned.

3.1. The effect of the thunderstorm electric field intensity

To describe the problem quantitatively, the length of the nylon wire is assumed to be constant H_0=350 m in this case. And for convenience, the electric field intensity E_n at the center of the ground is employed to express the variation of the electric field instead of the thundercloud charge density, where $E_n = E_0 / n$ and E_0 is $-15\ \mathrm{kVm^{-1}}$. The results of the contrast numerical experiments are shown in table 1, where l_n is the length of the triggering wire.

Table 1. The success rates under different electric field intensities.

n	$E_n(\mathrm{kVm^{-1}})$	l_n (m)	*Rate*	n	$E_n(\mathrm{kVm^{-1}})$	l_n (m)	*Rate*
2	-7.500	130	62.9 %	10	-1.500	290	17.1 %
4	-3.750	165	52.9 %	12	-1.250	335	4.2 %
6	-2.500	215	38.6 %	14	-1.071	Failed	0
8	-1.875	260	25.7 %	16	-0.937	Failed	0

Evidently, l_n grows when the magnitude of the electric field intensity decreases, and the success rates decrease quickly at the same time. When E_n is less than $-1.071\ \mathrm{kVm^{-1}}$, there seems no possibility to trigger a lightning in the case. The relation between E_n and the success rates are shown in figure 4, where the success rates decrease in a parabola law with the variation of E_n.

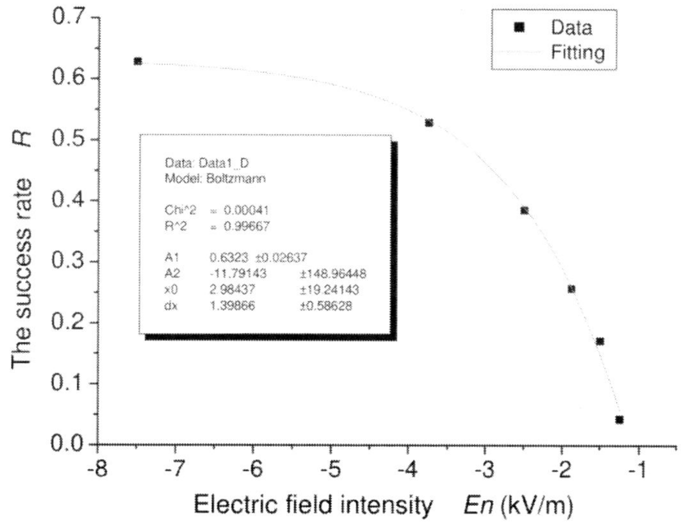

Figure 4. The parabola relation of E_n and the success rates.

3.2. The effect of the length of the nylon wire

In this case the length of the nylon wire H_n is variable, and E_n is constant. We assume $E_n = -7.5\,\mathrm{kVm^{-1}}$. Other parameters are invariant as mentioned above. The results of the contrast numerical experiments are shown in table 2, where l_n is the length of the triggering wire.

Table 2. The success rates beneath different lengths of the nylon wire.

H_n(m)	l_n (m)	*Rate*	H_n(m)	l_n (m)	*Rate*
250	140	60.0 %	450	120	65.7 %
300	140	60.0 %	500	120	65.7 %
350	130	62.3 %	550	110	68.8 %
400	130	62.3 %	600	110	68.6 %

The success rates increase when the nylon wire gets longer. A probable reason is that the electric field intensity increases along the altitude [10, 11]. But the variation is not as evident as the thundercloud electric field intensity does. The relation between H_n and the success rates are shown in figure 5, where the success rates form a ladder-type curve coursed by the discrete time step. A linear fitting has been done to the data, and it fits quit well.

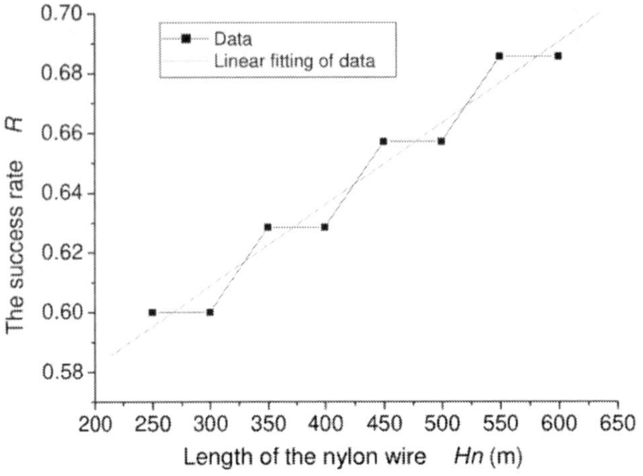

Figure 5. The relation of H_n and the success rates.

3.3. The bi-directional stepped leader

Another interesting phenomenon is found in all the simulations that the upward positive leader is always initiated earlier than the downward negative leader. And this phenomenon gets more obvious when the electric field intensity decreases. The initiating procedure is shown in figure 6.

The result is highly consistent with the experimental data proposed by P. Lalande [3]. Two probable reasons are responsible for the phenomenon. Firstly, the electric field intensity increases along the altitude, what has been mentioned in section 3.2. The top of the triggering wire where the upward positive leader is initiated has a higher electric field and it is easier to exceed the initiating criterion. Secondly, the threshold of the positive leader is lower than the negative leader [12]. It means that the positive leader is easier to be initiated.

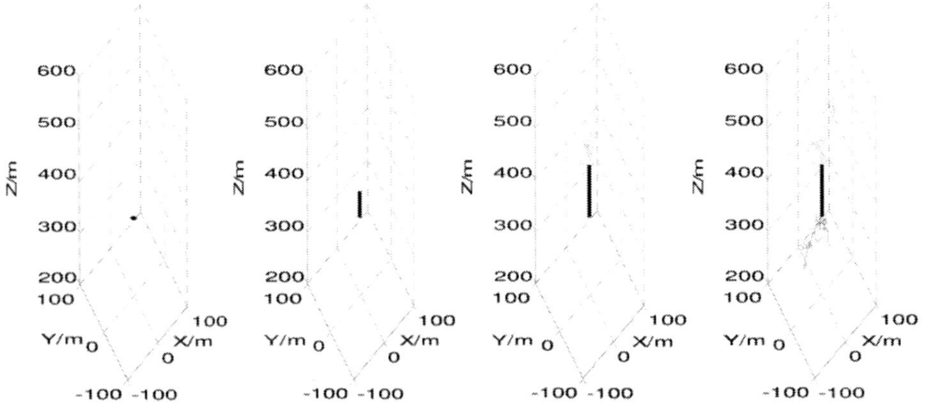

Figure 6. The initiating procedure of a bi-directional stepped leader.

4. Conclusion

A 3-dimensional numerical model on artificially altitude-triggered lightning has been developed in a new and more physical way. It combines the charge conservation equation with the Laplace equations and is solved with the application of the SOR method. A criterion is proposed to describe the success rates of artificially altitude-triggered lightning. Through groups of contrast numerical experiments, it is found that the success rates of artificially altitude-triggered lightning increase when the thundercloud electric field enhances or the length of the nylon wire increases. And the former one is the main influence factor. Another interesting phenomenon is found that the upward positive leader is always initiated earlier than the downward negative leader in either case. A typical time interval is about 40 steps and it becomes more obvious when the electric field decreases.

References

[1] Odim M and Margarete O 2005 *J. Atmos. Sol-terr. Phy.* **67** 1287
[2] Becerra M and Cooray V 2006 *J. Phys, D: Appl. Phys.* **39** 3708
[3] Philippe Lalande, Anne Bondio-Clergeria, Bacchiega G andGallimberti I 2002 *J. C. R. Physique.* **3** 1375
[4] Francesco L 2007 *Electromagnetic Compatibility in Power Systems* (Oxford: Elsevier Ltd)
[5] Amoruso V and Lattarulo F 2002 *J. Electrostatics.* **56**
[6] Jackson J 1999 *Classical Electrodynamics (3rd Edition)* (Hoboken: John Wiley & Sons Inc)
[7] Mathews J and Fink K 2004 *Numerical Methods Using MATLAB (4th Edition)* (Beijing: Electronics Industry Press)
[8] Wiesmann H and Zeller H 1986 *J. Appl. Phys.* **60** 1770
[9] Cheng Q, Wei G H, Chen Y Z and Wan H J 2011 *High Power Laser and Particle Beams.* **23** 721
[10] Qie X S, Liu X S and Yu Ye 1998 *J. Plateau Meteorology.* **17** 84
[11] Qie X S and Yan M H 1996 *J. Plateau Meteorology.* **15** 293
[12] Alessandro F 2003 *J. Electrostatics.* **58** 17

Journal of Physics: Conference Series **418** (2013) 012025 doi:10.1088/1742-6596/418/1/012025

Influence of relative humidity and temperature on quantity of electric charge of static protective clothing used in petrochemical industry

Yunpeng Zhang [1,2*], Quanzhen Liu [1,2], Baoquan Liu [1,2], Yipeng Li [1,2], Tingting Zhang [1,2]

1. SINOPEC Research Institute of Safety Engineering, Yan'an 3rd road, Qingdao 266071, China

2. State Key Laboratory for Safety Control of Chemicals, Yan'an 3rd road, Qingdao 266071, China

E-mail: colouredrain2009@hotmail.com

Abstract. In this paper, the working principle of static protective clothing and its testing method of quantity of electric charge are introduced, and the influence of temperature and relative humidity on the quantity of electric charge (q_e) of static protective clothing is studied by measuring q_e of different clothing samples. The result shows that temperature and relative humidity can influence q_e of static protective clothing to some extent and the influence of relative humidity is bigger than that of temperature. According to experimental results, the relationship of q_e and relative humidity and temperature was analysed, and the safety boundary of quantity of electric charge is discussed. In order to reduce the occurrence of electrostatic accidents and ensure safe production and operation of petrochemical industry, some suggestions on choosing and using of static protective clothing are given for guaranteeing its static protective performance.

1. Introduction

As there are plenty of inflammable and explosive sites in petrochemical industry, fire accidents caused by static electricity are very easy to take place. Static electricity is mainly caused by friction, which can be generated by any activity. Static electricity almost exists everywhere and its voltage can be very high. When a person moves, large amount of electrostatic charges would be generated on his body. For the people working at the inflammable and explosive sites, to reduce the electrostatic charges on human body and prevent the occurrence of fire accidents, static protective clothing is very essential, while the electrostatic charges generated by the clothing itself are of vital importance for its safety performance. If the quantity of electric charge of the clothing is large enough, when it's being used at the inflammable and explosive sites, fire accidents may take place very probably. As a result, it's very significant to intensify the testing work for quantity of electric charge of static protective clothing.

---

* To whom any correspondence should be addressed.

2. The working principle of static protective clothing and the testing method of its quantity of electric charge

Static protective clothing is a kind of working clothes widely used under inflammable and explosive circumstance, which is made of static protective textile. Static protective textile contains conductive fibre, which is mixed with other common fibre together symmetrically. The volume resistivity (ρ_v) of conductive fibre is 10^4~10^9 Ω cm^{-1}, which is made of conductive material or sub-conductive material using metal or organic substance. According to the distribution type of conductive substance in the fibre, conductive fibre can be divided into three different kinds: homogeneous, overlaying and complex. Among them, the third kind is mainly in use.

The working principle of static protective clothing, which contains conductive fibre, is based on the leakage and neutralization of electric charge. When the textile is connected to the ground, the electric charges on it may neutralize as a result of corona discharge or leak to the ground through conductive fibre; if not, the electric charges may disappear because of corona discharge on the conductive fibre.

So far, an important index to judge whether the static protective clothing is qualified or not is its quantity of electric charge (q_e), and the testing method of q_e is as follows: firstly the clothes should be washed as the national standard Static Protective Clothing requires; then the clothes are put into the roller friction machine to be treated for 15 minutes; thirdly the Faraday cup is used to measure the quantity of electric charge, which should meet the demands listed in table 1.

Table 1. Technical demands for the quantity of electric charge.

Testing items	Technical demands	
	Class A	Class B
Quantity of electric charge (μC / piece of clothes)	<0.2	0.2~0.6

Generally speaking, environmental element can influence the quantity of electric charge of static protective clothing, especially relative humidity and temperature. However, what is the relationship between them? How do relative humidity and temperature influence the quantity of electric charge? To answer these questions, a testing experiment is made to study this problem.

3. Testing experiment

Physical research proves that the electrical characteristics of objects can be directly influenced by humidity and temperature. Under the condition of low temperature and humidity, the resistance of an object will increase significantly, so that the release speed of electrostatic charge will slow down and the release time will increase; in high humidity conditions, water may condensate (or be adsorbed) on the object's surface to form a thin water layer. This hygroscopicity enhances the conductivity of the material, especially for the materials that can absorb water vapor. As the temperature rises, the mobility of the free charge is enhanced, so that the conductivity of the material is increased.

To study the relationship between the quantity of electric charge of static protective clothing and environmental elements, especially relative humidity and temperature, a series of testing experiments were carried out.

Three samples (Sample 1, Sample 2 and Sample 3) were chosen randomly from the static protective clothes produced by different manufacturers. The relationship between relative humidity and temperature and q_e of static protective clothing was studied. The corresponding variation range of relative humidity and temperature was set according to the actual using conditions of static protective clothing. Each sample was tested for five times and the average of these five results was taken as the final result.

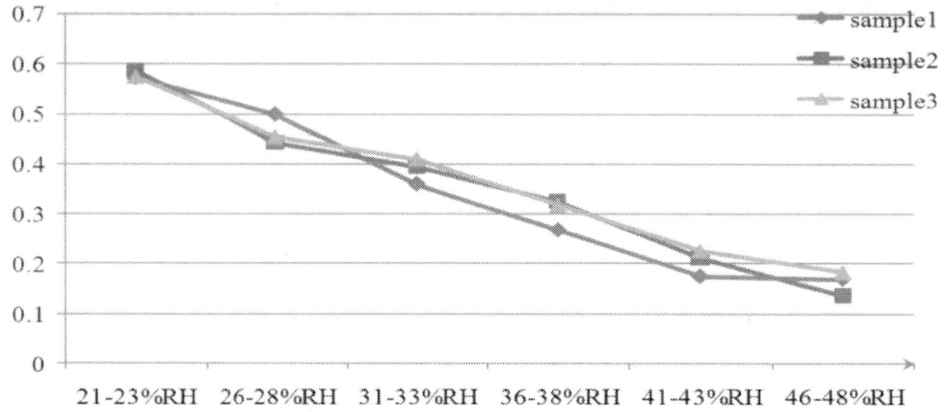

Figure 1. Quantity of electric charge under different relative humidity ranges (*T*: 16~17℃).

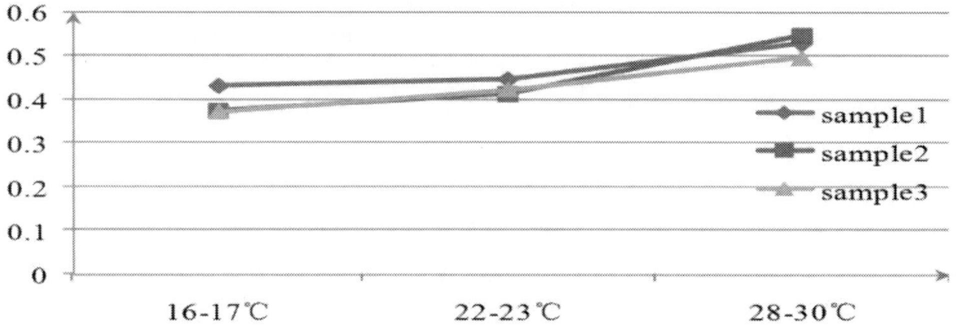

Figure 2. Quantity of electric charge under different temperature ranges (Relative humidity: 23~25%).

Figure 1 shows that when the temperature is constant, the quantity of electric charge tends to decrease with the relative humidity increasing gradually. Figure 2 shows that when the relative humidity is constant, the quantity of electric charge tends to increase with the temperature increasing gradually. What's more, the influence of relative humidity on the quantity of electric charge of static protective clothing is bigger than that of temperature on it.

To some extent, this can verify the phenomenon that in winter there is large quantity of electrostatic charges on human body while in summer it's the opposite. In winter, the temperature is very low and the air is very dry, when the clothes rub against with themselves or with the human body, many electrostatic charges will be generated. Under the condition of low relative humidity, corona discharge is difficult to go on. If the charges can't be led to the ground, they will gather up on the human body. In summer, the relative humidity is very high. Under this circumstance, corona discharge is easy to carry on, for which electrostatic charges will decrease gradually. As a result, electrostatic charges are difficult to gather up.

The technical demand for quantity of electric charge of static protective clothing is below "0.6 μC / piece of clothes" in the national standard (in China). This data is figured out from the minimum ignition energy data of most inflammable gas, about 0.25 mJ, so the data "0.6 μC / piece of clothes" could be a safety boundary for static protective clothing. If the quantity of electric charge of a piece of static protective clothing is more than 0.6 μC, it may be a dangerous source. While from the testing data it could be seen that under the condition of low relative humidity, the quantity of electric charge of a piece of static protective clothing could be very high, nearly 0.6 μC (the dangerous boundary). According to the data in figure 1, if the relative humidity gets lower, the quantity of electric charge may be larger than 0.6 μC, so it will be dangerous. While if it is tested under the circumstance as the national standard "Static protective clothing" demands (25±5℃, 35±5%RH), its quantity of electric charge is about 0.4 μC, less than 0.6 μC. So if the standard is taken as the only judging basis, this may be unreasonable. To judge the safety performance of static protective clothing, it may be more reasonable to test the clothing under the corresponding condition just as the clothing is used and the element of different places and seasons should be taken into consideration.

4. Conclusion

From the results of the testing experiments, it can be concluded that to some extent, relative humidity and temperature can really influence the quantity of electric charge of static protective clothing, and the influence of relative humidity is bigger than that of temperature. When the temperature is constant, with the relative humidity increasing gradually, the quantity of electric charge tends to decrease; when the relative humidity is constant, with the temperature increasing gradually, the quantity of electric charge tends to increase. As the quantity of electric charge of static protective clothing can be seriously influenced by relative humidity, it may be more reasonable to test the clothing in the corresponding condition just as the clothing is used.

Considering safety problem, petrochemical industry should choose static protective clothing as strictly as possible. Additionally, some suggestions about the using of static protective clothing are necessary for petrochemical industry:

(1) At the inflammable and explosive sites with dry weather, it is forbidden to put on or put off static protective clothing.

(2) When static protective clothing is in use, the upper garment and the lower garment should keep in touch as well as possible and they should be used with anti-static shoes together. Further more, strenuous exercise is not advisable.

(3) Static protective clothing is not anti-static forever. After it is be used for a period of time, an additional anti-static test is very essential. If it is not qualified, it should not be in use any longer.

Acknowledgments

The project was supported by National Key Technology R&D Program (2012BAK03B00), we thank for their support.

References

[1] Kodali W P 2001 *Engineering Electromagnetic Compatibility—Principles, Measurements, Technologies, and Computer Models* (Second Edition) (Wiley Publishing Inc)
[2] Hua J and Tao Z 1999 *Tech. Text.* **4** 24
[3] Gao Z, Wu J and Wang J 2011 *China Fibre Inspection* **2** 80
[4] Zhang Y and Shi M 2001 *Wool Text. J.* **4** 13
[5] Zhao Y and Teng J 2006 *Labor Protec.* **10** 90
[6] Fang S and Zhang Y 2010 *Modern Occupational Safety.* **11** 101
[7] Holme I, Mcintyre J E and Shen Z I 1998 *Text. Prog.* **28** 16

Journal of Physics: Conference Series **418** (2013) 012026 doi:10.1088/1742-6596/418/1/012026

Research on the static electricity of petrochemical powder pneumatic conveying and prevention of dust electrostatic explosion

Fenggui Tan, Wei Song, Hong Gong, Nanshan Xu *

School of Information Science and Technology, Beijing University of Chemical Technology, Beijing 100029, China

E-mail: xuns@mail.buct.edu.cn

Abstract. In this paper we present the problem of electrostatic phenomena in powder pneumatic conveying, which includes the dilute phase transport and dense phase conveying. It is certified that the dust explosion of polyolefin silo in the past is related to the electrostatic discharge in the silo. By analyzing the data from the field, it is revealed that the statics of dense phase conveying and dilute phase conveying are in the same order of magnitude. However, it is more difficult for dense phase to be controlled due to the absolute value of the amplitude fluctuation. The data of field survey and basic experiments show that polyolefin silo dust explosion is closely related to the failure of controlling gas and dust during the preparation. Finally, the result of experiments by the non-equilibrium bipolar ion wind electrostatic eliminator is introduced, including the effect when its application is expanded.

1. Introduction

In the 1970s to 1980s, as the rapid development of petrochemical powder production scales and the increasing number of silo explosion accidents, the research of powder electrostatic hazard was promoted. The warring of "a discharge of high insulation and large particles of dust in accumulation state is the igniting source of powder cloud" (M.G lor. 1998) [1] has been confirmed. It is warned that "when dealing with a product which produced high potential and was rounded in the midst of explosive atmosphere, the chance of explosion would be maximized". However, because of insufficient investigations on the electrostatic produced by pneumatic conveying powder and the controlling technology, domestic powder silo electrostatic explosion accident sporadically occurs. For instance, the powder silo of the LDPE device of a factory had exploded 13 times between 1987 and 1998. The powder silo of the PP device of a factory had exploded 14 times between 1989 and 1994. The powder silo of HDPE device of a factory had exploded 12 times between 1989 and 2002 [2]. The reason why there are so many accidents during that period can be gained from following factors: firstly, defective designed devices, inadequate ability of processing gas, and hidden danger of the silo (72% of 70 accidents are related to design defect); secondly, insufficient emergency treatment for

* To whom any correspondence should be addressed.

abnormal reaction, gas recovery and bunker ventilation equipment failure; finally, insufficient improvement of the gas recovery facilities and ventilations when promoting production capacity.

Although quantity of accidents caused by the petrochemical dust explosion has declined over the past decades, it still can be concluded from 11 cases [3] that the accidents are related to mistakes in complex process control. To avoid electrostatic explosion of the silo, it is necessary to investigate the risk of the electrostatic phenomenon caused by pneumatic conveying powder besides of strengthening regulatory of process control.

2. The electrostatic phenomenon of powder pneumatic conveying and the risk of blasting

2.1. The electrostatic phenomenon of the powder pneumatic conveying

The powder pneumatic conveying is influenced by many factors. Previous studies have indicated that the particle size is influenced by physical properties, such as the wind speed and sending capacity, air humidity, wall thickness, and roughness attachments. It is difficult to use quantitative characteristic parameters to forecast the lightning phenomenon, but the field statistic data may be significant. By simulation experiments, we collected electrostatic data of more than 200 silos conveying, including dilute phase conveying and dense phase conveying (bolt type). The major phenomena are as follows:

(1) The electrostatic phenomenon of powder silo conveying depends heavily on region and equipment. Typical data of six powder pneumatic conveying devices is listed in Table 1. These devices come from various regions of China. There are both positive and negative polarities in these data, most of which range from 0.8 to 5.8 μC kg^{-1}.

Table 1. The domestic field data collection and discharge limits of the calculated data.

Set	Bin number	Standard (m) diam×height	Specific charge (μC kg^{-1})	Tapered discharge critical value (μC kg^{-1})	D/2 Trace value (μC kg^{-1})
Yzbeta PP	TK2505	6×21	3.7	0.09	0.24
Fuxun PP	D902	2×15	5.8	0.28	0.72
KingSun 1PE	OT-401	4.4×18	2.0	0.14	0.36
YanHua LDPE	OH-251	4×18.5	0.81	0.14	0.36
Metallocene ethylene LDPE	V4264	3.6×17	2.56	0.15	0.40
Wide ethylene LLDPE	D6001	5×21	3.6	0.11	0.29

(2) Properties of the silo also have impact on the electrostatic phenomenon of powder silo conveying. As it is shown above, there are 16 silos for the LLDPE of a factory, 11 of which have their amplitude below 1.0 μC kg^{-1}, 2 in 1 ~ 2 μC $kg^{-1}$and 2 in 2 ~ 4 μC kg^{-1}.

(3) The amplitude of electrostatic phenomenon keeps stable. Table 2 lists a random testing data of a LDPE suction silo in 10 months. Static electricity of A silo is -0.51 ~ -2.16 μC kg^{-1}. Static electricity of B silo is -0.5 ~ -1.2 μC kg^{-1}.

Table 2. Electrostatic data on a factory's LDPE feeding tube of suction silo.

Time	Environment temperature (℃)	Environment temperature (%RH)	Electrostatic value（$\mu C\ kg^{-1}$）			
			A silo	B silo	C silo	D silo
1995.8.16	28.5	75	-1.22	-0.83	-1.05	0.27
9.1		76	-0.61	-0.95	-1.10	
9.16		72	-0.91	-0.5		
9.26		63	-2.16	-0.99		
9.27		65	-2.05	-1.07		
9.28		72	-1.77	-1.12		
9.29		76	-1.27	-1.20		
11.24		42	-0.51			
12.12	16	63	-1.09			
	14	43			-0.69	
1996.2.7	10	54			-0.39	0.27

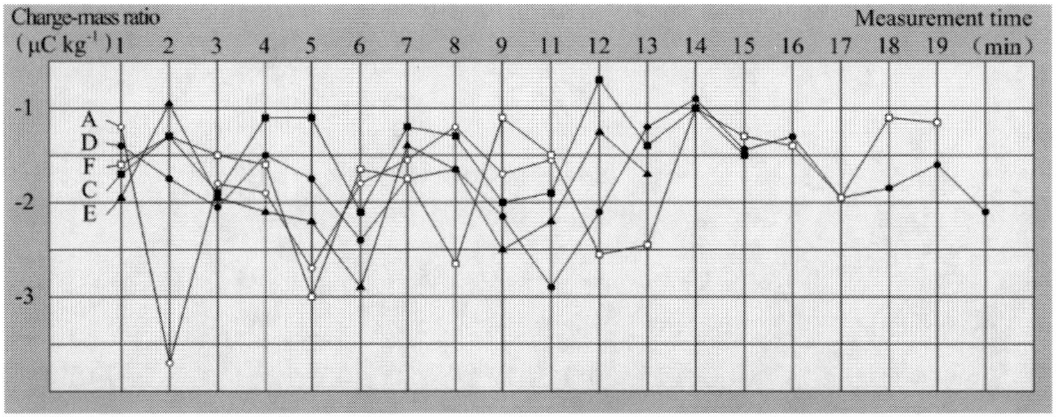

Figure 1. The data of conveying static electricity of a factory's HDPE set.

(4) Compared to dilute phase conveying electrostatic, density phase has higher amplitude. As dense phase transportation is a new technology which was introduced several years ago, experimental data are rare. In the past, it was speculated that their static electricity may be smaller. However, the field data shows that it is incorrect to forecast the electrification according to the dilute phase electrification phenomenon. If dilute phase pressure is small, and the wind speed is high (more than 40m s^{-1}), the pellets are "jump waft"; if dense phase pressure is big, and the wind speed is not very high (more than 10~ 15 m s^{-1}), it is a sand dune type waft. Figure 1 shows the measured data of conveying static electricity of 5 HDPE powder silos (bolt type) of a factory. The amplitudes fluctuate between -1 and -4 $\mu C\ kg^{-1}$. Table 3 lists the statistical data of powder silo conveying static electricity of 4 LDPE (bolt type). The conveying static electricity fluctuates more in blending operations, and it is more dangerous.

For comparison, Figure 2 plots the data of dilute phase conveying static electricity of 2 LLDPE devices. The data fluctuates relatively less.

Table 3. The data of static electricity of HDPE set powder silo conveying (bolt type).

Powder silo	Number of measurements	Charged range (μC kg^{-1})	Average (μC kg^{-1})
D5301A（Feed）	10	-0.31～1.04	-0.69
D5301B（Feed）	10	-0.33～-1.28	-0.627
D5301C（Feed）	10	-0.48～-2.43	-1.038
D5301D（Feed）	10	-0.47～-1.30	-0.635
D5301A（Small-doped）	10	-0.38～-1.17	-0.758
D5301B（Small-doped）	10	-0.25～-1.55	-0.572
D5301C（Small-doped）	10	-0.3～-1.90	-0.802
D5301D（Small-doped）	10	-0.49～-1.76	-0.953
D5301A（Large-doped）	3	-3.65～-4.5	-4.2
D5301B（Large-doped）	5	-0.45～-0.91	-0.754
D5301C（Large-doped）	10	-0.48～-2.52	-1.38
D5301D（Large-doped）	6	-0.67～-1.08	-0.855

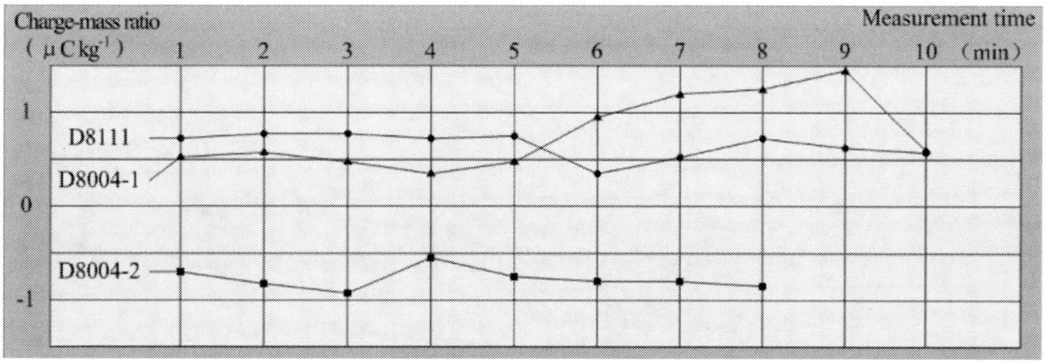

Figure 2. The data of static electricity of dilute phase conveying of 2 LLDPE sets.

2.2. *Powder silo electrostatic dangerous analysis*

Conical discharge on the surface of the stockpile is more possible amount all forms of discharge. Energy coming from discharge is a function of the powder silo diameter (*D*) and particle diameter (*d*). The upper limit value of 3m diameter silo can be expressed by $W = 5.22D^{3.36} \cdot d^{1.462}$. Such discharges can ignite miscellaneous mixed dust which contains gas, or combustible dust with the minimum ignition energy (*MIE*) ≤ 10 mJ. We can make use of a simple calculation （$E=\frac{\rho R}{2\varepsilon\varepsilon_0}$）of the silo farm or computer software (FDCS) to estimate the conditions of discharge. Assuming that there is a cone-shaped discharge conditions when the electric field strength is larger than 3 MV m^{-1}. The calculated data in Table 1 is from FDCS by different materials and different sizes of a silo in the tank wall and the center of the tank, with the charge to mass ratio critical value of E> 3MV m^{-1}. The result shows that in 3 ~ 6 m diameter polyolefin silo, if the material charge for quality is between -0.1 and -0.36 μC kg^{-1}, the surface of the stockpile can generate "linear" or "planar" conical discharge. The electric field distribution of the surface of the stockpile is illustrated in Figure 3. It is a factory 1PE silo.

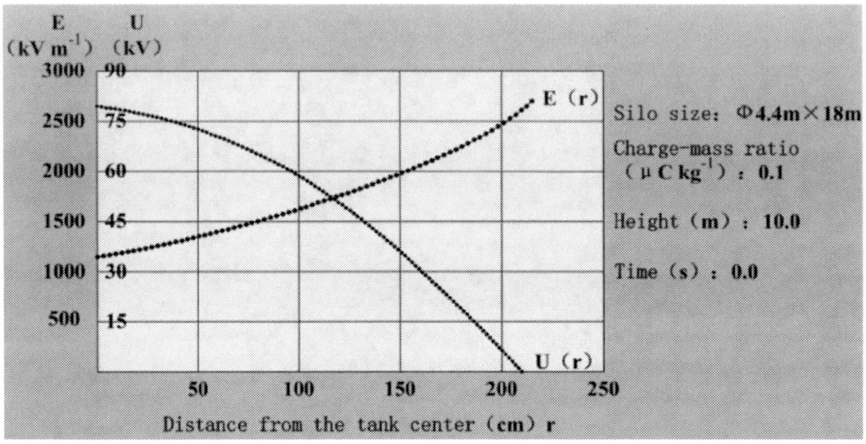

Figure 3. The surface of the stockpile electric field and potential calculation model.

If the charged material heap plane of the metal protrusions, including clean air damper mouth, level meter, the mixing tube or cutting tube mounting bracket, concentrated discharge can occur. This kind of discharge can ignite combustible dust with MIE ≤ 30 mJ. Figure 4 shows the data of potential distribution of simulated silo wall. The experiments show that when the material charge is more than 1.65 μC kg^{-1}, and from the silo wall 170 mm at the stockpile surface potential is equal to or lager than 40 kV, metal protrusions discharge can ignite the dust with MIE ≤ 30 mJ or miscellaneous mixed dust.

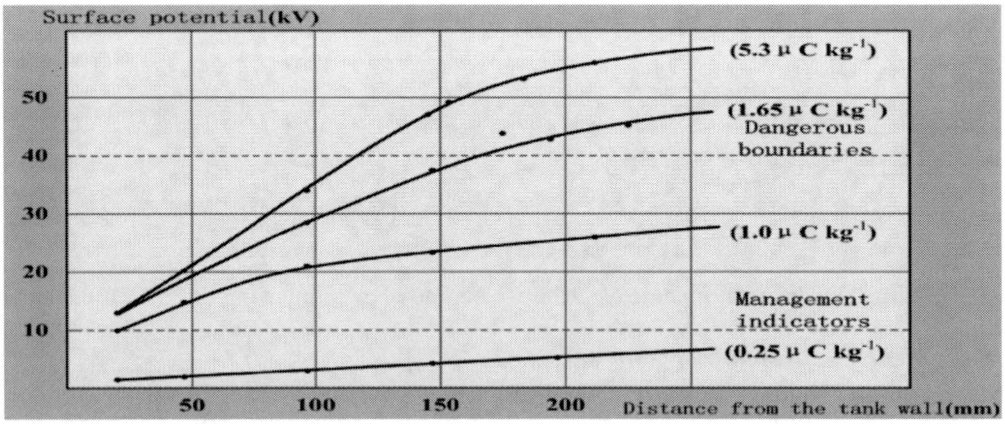

Figure 4. Surface potential distribution of the different charged powders in silos.

If the silo wall with sticky wall material or coating ($\delta = 4 \sim 8$ mm), or silo mixing tube stent, or purify the air damper mouth, is buried, complex material surface effects will accumulate because more charge of the back electrode can be induced by the distributed brush discharge. The discharge energy can be from several hundred to thousands mJ and can ignite most of polyolefin dust. If the insulated conductors of silo are not grounded, or there are ungrounded fasteners, broken pipe, or shedding parts caused by design shortcomings or improper maintenance in the silo, hundreds mJ of spark discharge may be induced. That can ignite most polyolefin dust.

3. Analysis of the combustible gas in powder hopper and its ignition danger

Every polyolefin powder processing device has a combustible gas processing unit which has two main functions. One is the process before granulation, in which the unreacted gas is extracted and recycled. The other one is the process after granulation, in which the gas is purged and emptied. When the reaction is abnormal, such as the gas processing equipment fails to work, or hopper's ventilation has something wrong etc., the overflowing gas in the hopper is extraordinary, gas concentration in dust space will increase and the minimum ignition energy of a mixture will decline. Finally, dust explosion will occur. For example, in the survey of 70 accidents, 60 accidents (about 86%) are caused by gas control failures.

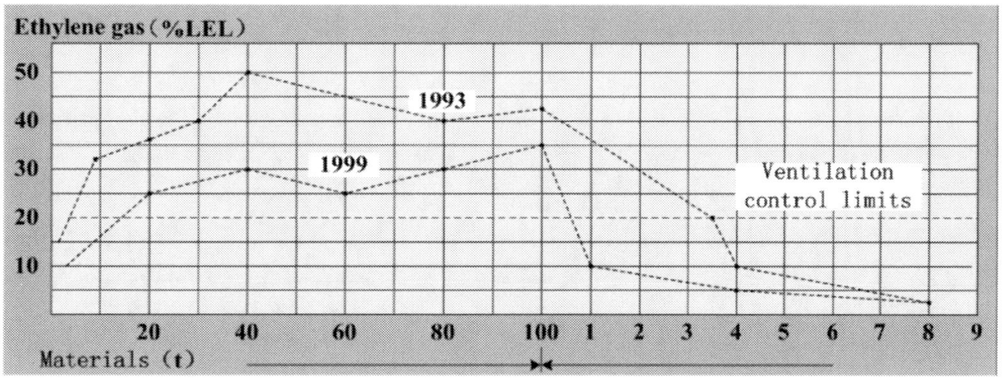

Figure 5. Analysis curves of combustible gases within the silo.

Figure 5 illustrates a detection data of LDPE exhaust hopper gas from a factory. Because of the bad design of back flush clean wind (24m^3 min^{-1}), the safe value is only 50% of theory. There are 14 explosions occurring in exhaust hopper, admixture hopper, and unqualified hopper in 4 years. As another example, a PP equipment in a factory was expanded and reconstructed. Dryer N2's pressure decreased from 0.6 MPa to 0.2~0.4 Mpa. Materials volatile increased to 2000~3000 ppm, which were larger than the design value of 1000 ppm. As a result, 11 explosions occurred in the admixture hopper in four years.

Figure 6 illustrates experimental data of LDPE dust and ethylene gas mixture's minimum ignition energy from a factory. The dust MIE collected in spot is 15.6 mJ. When the ethylene concentration in the mixture increases to 0.6 %Wt, the MIE of the mixture is 10 mJ. The experimental data is very close to the calculated value (curve1) recommended by CIBA. The experiment shows that with the increase of gas concentration, MIE of the mixture declines at an exponential rate. When the gas concentration is 0.5 %Wt, MIE of the mixture is 10mJ. Even if conical discharge occurs, the probability of ignition still exists. The higher the gas concentration is, the higher the chance for ignition will be. This is the main reason for the increasing hopper explosions when gas control fails in spot.

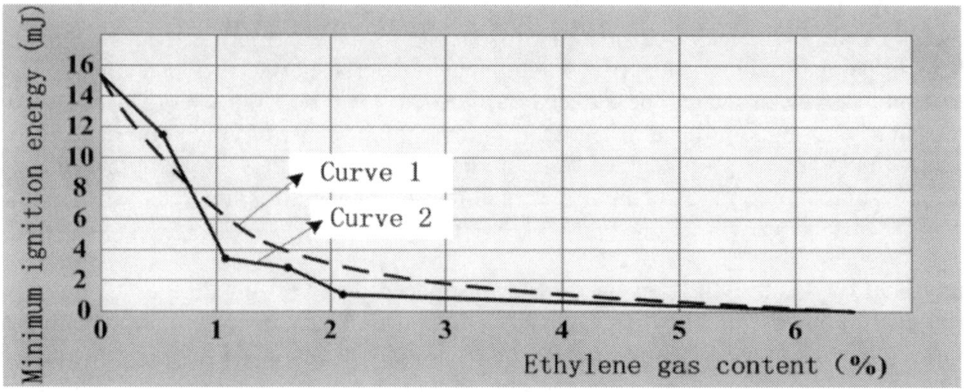

Figure 6. An experimental curve of minimum ignition energy of LDPE dust and ethylene gas mixture.

4. Research on the elimination of static electricity in powder pneumatic conveying

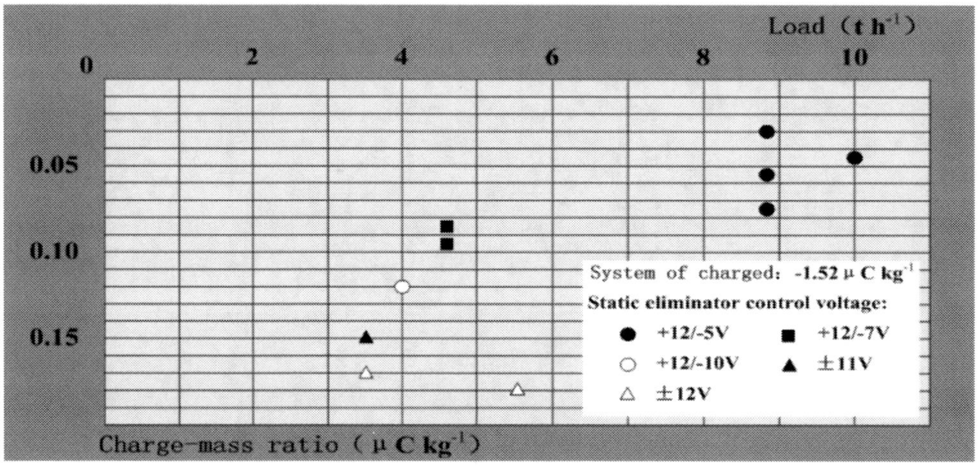

Figure 7. Eliminator eliminates the electrical effect under different control voltage.

In order to suppress the electrostatic explosion of powder hopper, relevant norms at home and abroad recommend the use of ion wind static eliminator. Relative to the gas control and dust control in spot, the practicality and operability of this measure are much better. Unbalanced bipolar ion wind static eliminator is a new technology and now has become more and more famous. Figure 7 shows the experimental data on the 50 mm pipeline simulator. The experiments show that the efficiency of electrical elimination of the unbalanced bipolar control is superior to the AC and DC mode. By the experiments, the conclusion of [4], which says that ion wind static eliminator must be supplemented by the online monitoring. That means this method avoids not only the risk operation of "reverse charged", but also the requirement of realizing the high eliminates electrical efficiency for the large hoppers demand. Since 2003, we have used the unbalanced bipolar ion wind static eliminator with online monitoring and feedback regulation. We tested it in 19 sets of devices and 167 hoppers, and we are very satisfied with the experience and effects.

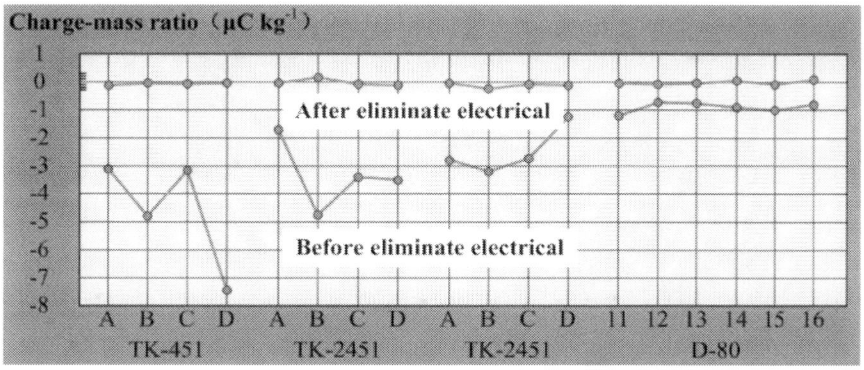

Figure 8. A corporate silo elimination electrical of two sets of PE usage.

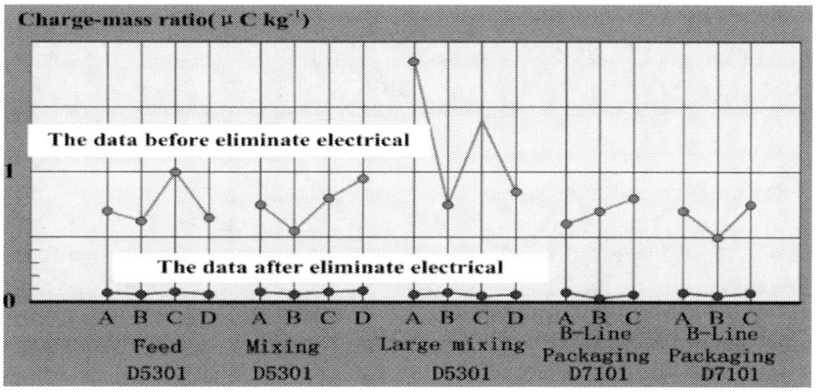

Figure 9. A corporate silo elimination electrical of a set of HDPE usage.

Table 4. Test data of electrostatic elimination of two sets of LDPE plant silo (dense phase) in an enterprise.

Powder silo		1PE451A/D	1PE2451A/D	1PE3451A/D	2PE8011/8016
Quantity		4	4	4	6
Before electrostatic elimination	Charged range (μC kg^{-1})	-2.62~-7.49	-1.67~-5.1	-1.04~-3.0	-0.6~-1.4
	Average (μC kg^{-1})	-4.585	-3.357	-2.14	-0.766
After electrostatic elimination	Charged range (μC kg^{-1})	-0.01~0.14	-0.01~-0.12	-0.02~-0.18	-0.03~-0.17
	Average (μC kg^{-1})	-0.07	-0.067	-0.097	-0.071
Electrostatic elimination efficiency (%)			98	95	91

Figure 8 is an application example of two PE plants (dilute phase conveying) of a factory in South China. Figure 9 is an application example of the HDPE plant of a factory in North China. Table 4 and

Table 5 list the statistics of them. It can be concluded that the technology can be used to eliminate non-equilibrium bipolar ion wind static electricity under electric charge online monitoring and feedback control. The efficiency of the elimination of electric charge is basically more than 90%. We can achieve management objectives of no-spark discharge and operate requirement of instrument management.

Table 5. Test data of electrostatic elimination of two sets of LDPE plant silo (dense phase) in an enterprise.

Powder silo		BL1801A	1801D	1801E	1801F
Before electrostatic elimination	Charged range (μC kg^{-1})	-1.06～-3.73	-0.93～-2.93	-0.76～-2.89	-1.04～-2.99
	Average (μC kg^{-1})	-1.83	-1.55	-1.89	-1.69
After electrostatic elimination	Charged range (μC kg^{-1})	-0.03～-0.25	-0.06～0.2	0.03～0.19	0.02～-0.15
	Average (μC kg^{-1})	-0.128	0.126	0.09	0.113
Electrostatic elimination efficiency (%)		93	92	95	93

5. Conclusions

(1) Petrochemical powder hopper dust explosion occurred in the past is confirmed by on-site investigation and testing. The vast majority of the powder pneumatic is related to the conveying of electric discharge. The failures of gas control and dust control in the manufacturing process are induced by external causes of the accident. However, it is difficult to control and easy to affect. (2) The conveying electrostatic phenomenon of powder dense phase pneumatic (bolt) fluctuates more easily than the dilute phase pneumatic. (3) In the dilute phase and dense phase pneumatic conveying process pipelines, online monitoring and timely control of non-equilibrium bipolar ion wind static eliminator help to achieve the ideal goal of discharging without spark. (4) The production process lost control or with a temporary failure is most dangerous for the dust electrostatic explosion frequency. That includes dealing with substandard material or transitional material, the production of high-melting grades of resigning material containing high-boiling gas-liquid materials (which includes raw material gas, catalyst, thinner, cold-induced, etc.), degassing equipment or ventilation equipment failure, granulator cutting knife or templates badly worn, metal off pieces or mixing tube breaking in the warehouse, and the temporary stop or start early etc.

Above period is not only prone to the flammable explosive atmospheres, its static electricity from powder is most unstable. The hopper is easy to produce an electrostatic discharge and dust explosion. The application of practice proved that the non-equilibrium bipolar ion wind static elimination is the final barrier to prevent dust explosions.

References

[1] Dr Martin Glor 2001 *Overview of the occurrence and incentives of cone discharges with case studies from industrial practice* Journal of Loss Prevention in the Process Industries pp 123-128
[2] Safety Engineering Research Institute of Sinopec 2001 *Polyolefin silo blasting risk assessment and control measures* Subject of the report
[3] Tan F 2008 *Analysis and Countermeasure of recent polyolefin silo dust explosion* Petroleum and Chemical Safety and Environmental Protection Technology p 48
[4] Er Y, Ling M and Zui S 2004 Suppression of Incendiary Electrostatic Discharge in a Silo Using a New Electrostatic Charge Eliminator for Powder Research Report of the National Institute of Industrial Safety NIIS-RR-2003

Journal of Physics: Conference Series 418 (2013) 012027 doi:10.1088/1742-6596/418/1/012027

Simulation study on radio frequency safety of electric explosive device

Peijie Yang[1*], Zhiliang Tan[1], Chaoyang Liu[2], Zhide Du[3]
1 Institute of Electrostatic and Electromagnetic Protection Mechanical Engineering College Shijiazhuang, China
2 The 3rd Department of Mechanical Engineering College, Shijiazhuang, China
3 Beijing International Studies University, Beijing, China

E-mail: yangpeijie0426@163.com

Abstract. Radio frequency (RF) is a great danger to the electric explosive device (EED) of typical ordnance. This paper introduced the RF firing mechanism of the EED and the measuring method of its RF impedance. Through the professional antenna simulation software CST, a dipole antenna model of the EED was set up, the gain coefficient of the antenna model was obtained, and the RF power penetrating into the EED was calculated. The multi-frequency analysis of the emulation indicates that in the certain frequency range of 0.5-2 GHz, the gain coefficient of the antenna model increases as the frequency does.

1. Introduction

Along with the rapid development of radio technology, RF electromagnetic field has widely spread among both human living space and cosmic space. The EED in ammunition can be affected by the broadcast on the ground, transmitter on the aircraft, and communication radars on the warship easily [1-2]. As a result, researches have been massively done around the world to study the RF safety of EED since 1950s. Although there are various EED, the most common one used in the weapon equipment is the bridge-wire EED. Statistics from the 213th Research Institute of China Ordnance Industry indicate that some 90% of EED are bridge-wire ones. The electrode leads of the bridge-wire EED, when put in particular environment, can be used as receiving antennas to absorb RF energy, which will result in malfunctions of the EED. Consequently, the bridge-wire EED can be taken as antenna model, the electrical parameter of which can be obtained through the professional simulation software. Then the RF power penetrating into the EED can be calculated and the coupling law of its electromagnetic energy will be summarized.

[*] To whom any correspondence should be addressed.

2. Basic Theory and the Antenna Model of EED

The basic structure of any bridge-wire EED is similar, mainly composed of electrode leads, bridge wire, explosive and shell, as shown in figure 1.

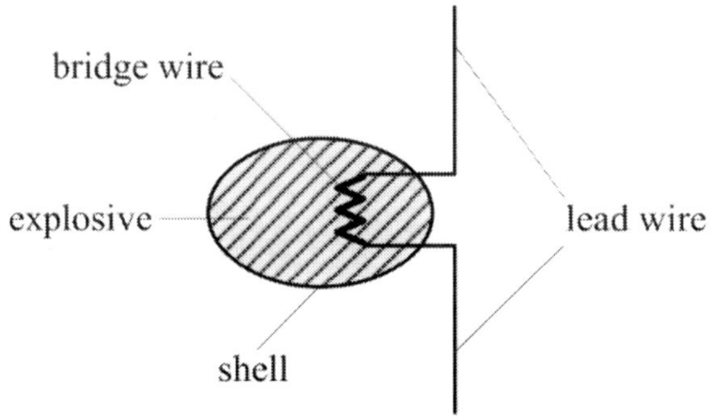

Figure 1. Structure diagram of bridge wire EED.

The EED in ammunition along with its connecting circuit and related parts can be used as a receiving antenna of radio waves to lead RF energy into the EED. This type of energy is generally too weak to induce accidental firing or ineffectiveness of the EED, but under certain conditions, an unexpected action of ammunition may be caused if RF energy gives birth to an earlier ignition of the EED. On the other hand, the reliability of the EED may be lowered due to its chronic exposure to RF energy insufficient for ignition. The RF energy it receives is mainly seen from the functions of electric current and voltage: the electric current, when transmitting from the feet of circuit to the bridge wire, will heat the bridge wire to ignition point, leading to an earlier action or ineffectiveness of the product; while the voltage can cause a breakdown between the feet and shell, the result of which will be a explosive detonation, similar to the static detonation.

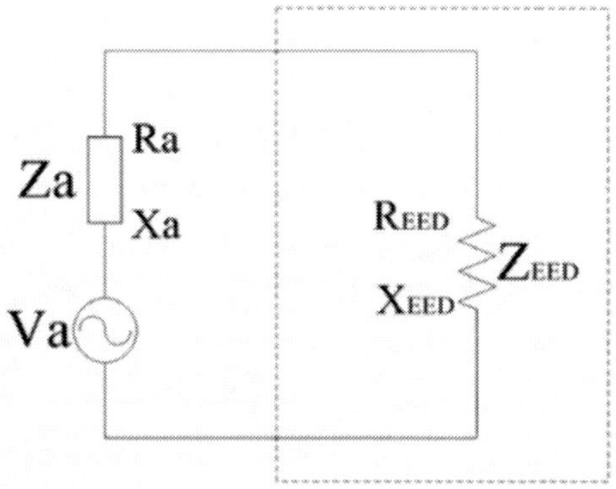

Figure 2. Equivalent circuit of the antenna model.

As for two-wire EED, the short-circulated EED lead functions as a loop antenna, while the one that is not short-circulated acts as a dipole antenna. Put the receiving antenna in external electromagnetic field, it will induce current and generate electromotive force (EMF) in the output terminal of antenna [3]. The EMF in the electric firing system will transmit electronic current to bridge-wire load through the firing circuit. The equivalent electrical circuit of antenna model has been shown in figure 2.

Where V_a is the EMF produced by the electromagnetic field in the antenna, Z_a is the equivalent impedance of antenna, Z_{EED} is the RF impedance of the EED.

The EMF produced by the electromagnetic field in the antenna V_a, can be expressed as follows:

$$V_a = \frac{E\lambda}{\pi}\left(\frac{R_a G}{120}\right)^{\frac{1}{2}} \tag{1}$$

Where E is the electromagnetic field intensity, λ is the wavelength of the radio wave, G is the gain of the equivalent antenna, R_a is the resistive component of the antenna equivalent impedance.

The RF power (P_{EED}) into the EED can be calculated by the following formula:

$$\begin{aligned} P_{EED} &= I^2 R_{EED} \\ &= \frac{V_a^{\,2} R_{EED}}{(R_a + R_{EED})^2 + (X_a + X_{EED})^2} \end{aligned} \tag{2}$$

Where I represents for Induced current in the load, Va represents for the Induced EMF in Ignition circuit, Xa represents for reactance component of the antenna equivalent impedance, R_{EED} represents for resistive component of the EED. X_{EED} represents for reactance component of the EED.

3. The RF Impedance of EED and Its Measurement

The RF impedance of EED is a vital parameter in analyzing the electromagnetic hazard of the weapon system [4]. It can be used to analyze and calculate the RF power into the EED, which is of importance to evaluate the safety of the EED and to the design the EED. The value of RF impedance is a function of the frequency. We can measure it under the help of RF impedance analyzer. The theory of measuring system is shown in figure 3.

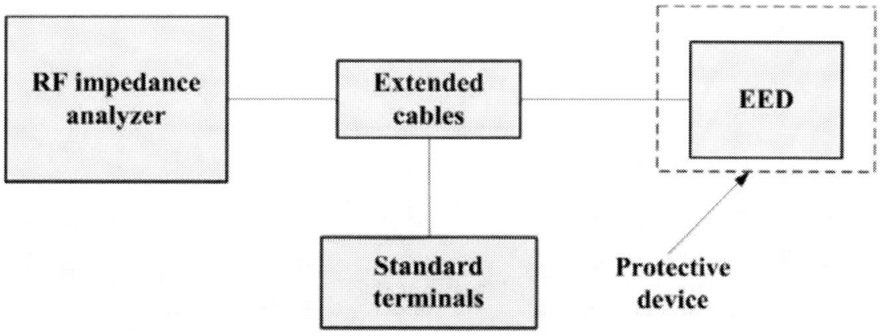

Figure 3. The RF impedance measurement principle diagram.

Connect the extended cables with three standard terminals. First, conduct automatic calibration with short circuit (0 Ω), circuit breaker (0 S) and 50 Ω in the frequency range from 1 MHz to 1300 MHz. Second, measure the RF impedance of the EED on the calibrated frequency point. The formula (3) will show the resistive value and reactance value of the RF impedance strictly.

$$Z_{EED} = R_{EED} + jX_{EED} \tag{3}$$

When the antenna is in a direction in which it can absorb radio wave by the largest margin, also, there is not power loss and the impedance between antenna and load is matching, the load can receive RF power by the largest margin, that is to say when $R_a = R_{EED}$ and $X_a = -X_{EED}$, the EED can absorb RF power by the largest margin.

4. The Establishment of Simulation Model and Results Analysis

CST MICROWAVE STUDIO is one of the software packages of CST studio product in CST Company and it is the flagship product of CST that is widely used in simulations of middle or high frequency passive devices. Another characteristic is that the CST MICROWAVE STUDIO is suitable for the whole electromagnetic wave, the electromagnet of the light wave band and the electromagnetic compatibility simulations. It can be used to conduct simulations in various types of antenna and antenna arrays, antenna layout, RCS, FSS, planar passive device. There exist lots of algorithmic methods such as integrated time-domain finite integral method, time domain transmission line matrix, finite integral, domain finite element [5]. So the writer has chosen this software to conduct the simulation research on dipole antenna model of EED and calculate the antenna model gain.

4.1 The Establishment of Simulation Model

The dipole antenna model is established for a type of EED of a particular artillery missile's off-engine, based on some relevant references. The arms of antenna dipole are solid cylinder, made of silver-plated copper wire which can be taken as an ideal conductor. The length is 100 mm and the diameter is 0.3 mm. The distance of the two arms is 0.6 mm which is used for antenna to feed. Only considering the characteristic of far field, set the feeding way as discrete ports feed. Regard the center of the cylinder as the origin of coordinates and regard the positive direction of the axis z as the upward direction. According to relevant regulations in GJB1389A – 2005 [6], set up the simulation frequency from 0.5 GHz to 2 GHz. It has been shown in figure 4.

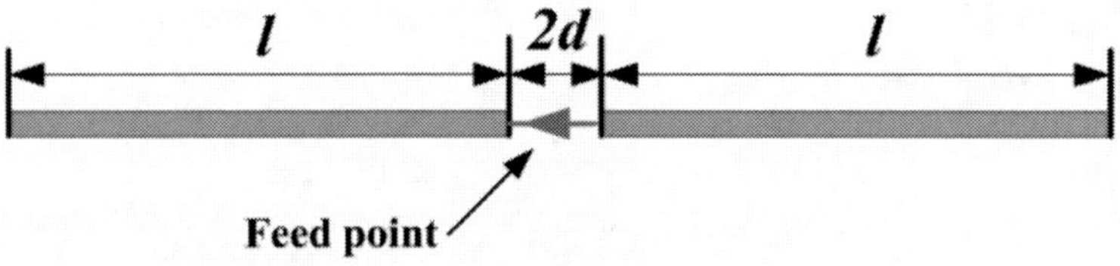

Figure 4. Simulation model figure of dipole antenna.

4.2 Analysis of the Simulation Results

Figure 5 is a dipole antenna far-field parameters on the condition that l = 100 mm and the frequency is 0.75 GHz, from which we can get the gain of the radiating antenna as 2.235 dBi. According to reciprocity principle, the functioning process of the receiving antenna and the radiating antenna is reversible, that is there is only physical meaning difference when the same antenna used as a receiving and a radiating antenna respectively, but the gain will not change. Therefore, when used as a receiving antenna, its gain is also 2.235 dBi, provided that the frequency is the same.

Type	Farfield
Approximation	enabled (kR >> 1)
Monitor	farfield (f=0.75) [1]
Component	Abs
Output	Directivity
Frequency	0.75
Rad. effic.	0.09679 dB
Tot. effic.	-0.7951 dB
Dir.	2.235 dBi

Figure 5. Dipole antenna far field parameters.

Take reference [4] as an example: The EED, its lead length being 10cm and the RF impedance being 1+j75 (Ω), when exposed to an electromagnetic environment of which test frequency is 0.75GHz and the field strength is 195 V m^{-1}, what will be the RF power the EED obtains?

The test frequency in the electromagnetic field is 750 MHz, so the wavelength is 0.4 m. At this moment, $2l = \lambda/2$. The antenna consisted of the lead line of EED is in a resonance condition. The equivalent impedance of antenna is pure resistance that is 73.1 Ω. From the simulated result, we can know the gain of dipole antenna is 2.235 dBi (G = 2.235 dBi). Put it into formula (2), then we can obtain the largest induced electromotive force the antenna may receive V_a = 25.06 V.

If the RF impedance of EED is matched with the antenna impedance, i.e. when $R_a = R_{EED}$ = 73.1 Ω, $X_a = -X_{EED}$, the maximum RF power penetrating into the EED can be got by formula (2) as P_{max}=2.147 W.

In fact, the RF impedance of EED is not matched with the antenna's impedance. In this example, the RF impedance of EED is 1+j75 (Ω). Put it into formula (2), we can obtain that the RF power into the EED is 0.056 W (P_{EED} = 0.056 W) in the unmatched condition.

Considering that there exist some errors caused by the parameter setting in the process of establishing the model and software algorithm. The error has so little influence on simulation that we consider it reasonable. Compare the result of the simulation about CST with that of reference [4]. Under the matched condition, the simulated calculation concludes that the largest RF power into the EED is 2.147 W; the reference's is 2.10 W. Under the unmatched condition, the simulated calculation concludes that the maximum RF power into the EED is 0.056 W; the reference's is 0.055 W. The fact mentioned above indicates that the result in the simulated calculation is almost the same.

In order to study the different effects that different frequencies have on the gain of the antenna, a far-field monitor is defined to calculate the gain of the antenna under different frequencies. When l = 100 mm, and the frequency is set as 0.5 GHz, 0.75 GHz, 1 GHz, 1.25 GHz, 1.5 GHz, 1.75 GHz and 2 GHz respectively, the simulation gains of the radiating antenna are shown in table 1.

Table 1. Relationship between dipole antenna gain and frequency.

Frequency/GHz	0.50	0.75	1.00	1.25	1.50	1.75	2.00
Gain/dBi	1.817	2.235	2.618	3.210	4.137	4.542	5.012

It can be seen from Table 1 that the gain of the antenna increases as the frequency does in the frequency range of 0.5 - 2 GHz. And changes of the gain will induce changes to the EMF of the antenna generated by RF waves, and further affect the RF power penetrating into the EED.

5. Conclusion

Through the professional antenna simulation software CST, a dipole antenna model of the EED was set up, the far-field characteristics of the radiation of the antenna model was analyzed, the gain coefficient of the antenna model was got, and then the RF powers penetrating into the EED was calculated when the impedance matches and does not match respectively. The simulating calculation results are basically consistent with the practical results. So it can be concluded that it is feasible to analyze the RF security of the EED with the assist of CST software. And the multi-frequency analysis of the emulation indicates that in certain frequency range, the gain of the antenna model will increase with the frequency.

References

[1] Li G M 1992 *Explosive Device* (China: East China Engineering Academy)
[2] Liu S H 2005 *Journal of the Academy of Equipment Command* **16** 1
[3] Zhang W X 2011*Antennas (Third Edition)* (Beijing: Publishing House of Electronics Industry)
[4] Li J R 1995 *Initiators & Pyrotechnics* **1** 11
[5] Zhang M 2004 *CST Microwave Studio* (Chengdu: Press of Electronic Science and Technology University)
[6] GJB1389A–2005. *System Compatibility Requirements*

A novel protection method against direct lightning strike based on the principle of electrostatic field counteraction

Guanghui Wei*, **Haojiang Wan***

Key laboratory of electromagnetic environmental effects, Shijiazhuang Mechanical Engineering College, Shijiazhuang 050003, China

E-mail: wei-guanghui@sohu.com, hbwhj1983@163.com

Abstract. Strong electromagnetic radiation around the conventional lightning rod system exists when it is struck by lightning, and serious electromagnetic interference on sensitive electronic equipments could be caused. A novel active protection method against direct lightning strike is proposed in this paper. The method, which is based on the principle of electrostatic field counteraction and shielding, is aimed at weakening the background electric field near the protection region to eliminate the condition for direct lightning strike there. The protection system is composed of three parts, including the monitoring equipments for atmospheric electric field, the active protection device, and the control system. Two working flow charts of the protection system are presented, and the feasibility of the protection method has been verified by scale-model experiment in the laboratory.

1. Introduction

Research on lightning has lasted over 250 years since lightning was validated as an electrical discharge by Benjamin Franklin. During these past years, the protection of structures and systems against lightning has experienced three stages: protection against direct lightning stroke, protection against induced effects of lightning, and protection against lightning electromagnetic pulse [1-3]. For the buildings, equipments and personnel inside, several technologies such as bonding, conducting, dividing, grounding, and shielding are adopted in lightning protection [4-5]. However, no protection measure is almighty. Although the conventional lightning rod system has a history of success in preventing or minimizing damage to structures, there is still potential safety hazard for sensitive electronic equipments [6-9]. Firstly, the conventional lightning rod systems for ground-based structures serve to provide lightning attachment points and paths for the lightning current to follow from the attachment points into the ground without harm to the protected structure. There is strong electromagnetic radiation around the lightning rod when it is stroked by lightning, which can generate serious electromagnetic interference on sensitive electronic equipments, or even damage them. Secondly, when the lightning rod is stroked by lightning, the touch voltage and step voltage caused by the lightning current can not be ignored. Moreover, air-termination system of the lightning rod is a huge grounded metallic conductor, which can affect the antenna characteristics of the devices near it (e.g. the antennas installed in the microwave stations, etc.) and thereby leads to technical performance degradation. Regardless of the above side effect, the latent dangers of being subjected to direct

---

* To whom any correspondence should be addressed.

lightning stroke are difficult to be avoided for mobile equipments and vehicles, since it is almost impossible to install a lightning rod system for them which can not always provide a good grounding. Hence, the conventional protection technology against direct lightning stroke can not meet the technical requirements of electromagnetic protection for sensitive electronic equipments. In this paper, an active direct lightning protection method based on the principle of electrostatic field counteraction is proposed to solve this problem.

2. Principle of the active lightning protection method

A typical cloud-to-ground lightning is composed of a downward-moving process termed a downward stepped leader, a lightning attachment process with one or more upward-connecting leaders and an upward-moving process termed a return stroke. The lightning stepped leader, beginning in the thundercloud, propagates toward the ground in a series of discrete steps without being affected by the presence of the ground-based structures. As the stepped leader approaches ground, the electric field at the ground surface or objects increases until it exceeds the critical value for the initiation of upward connecting leaders. Note that the initiation of an upward-connecting leader basically depends on the distribution of the electric field near the possible initiation points. Only the point with the stronger background field may initiate an upward-connecting leader. Thus, the basic idea of the active protection method against direct lightning stroke in this paper is to weaken the background field near the protection region based on the principle of electrostatic field counteraction and electrostatic shielding, and to destroy the condition for direct lightning stroke at the protection region. According to the principle of vector superposition of electrostatic field, the basic electrostatic shielding strategies tend to fall into three categories as following [10]:

(1) Interposition of a ''barrier'' with good electric conductivity between the source and the area where the electric field needs to be reduced.

(2) Introduction of a mean capable of diverging the electric field from the area of interest.

(3) Introduction of an additional source whose effect is the reduction of the electric field levels in the prescribed area with respect to a situation involving the original source or source system.

According to the physical characteristics of lightning discharge and lightning strike selectivity, a combination of the electrostatic shielding strategies (1) and (3) is applied for the protection against direct lightning stroke. In order to weaken the background field near the protection region and destroy the condition for direct lightning stroke, an active protection device composed of two polar plates with high conductivity is introduced. The upper plate of the active protection device is driven by a high voltage DC generator, and the lower plate of the active protection device is earthed. When the electric field near the protection region becomes large enough, an appropriate voltage excitation will be applied on the upper plate to excite a reverse electric field which can weaken the electric field above the protection region and the active protection device. Thus, the electric field near the protected objects and the active protection device can not reach the critical value for the initiation of upward-connecting leaders. And the lightning strike points can only be selected outside the protection region.

3. Structure and working flows of the protection system

The active protection system based on the above protection method is composed of three elements including the atmospheric electric field monitoring equipments, the active protection device, and the control system. The atmospheric electric field monitoring equipments is mainly utilized for real-time monitoring of the background electric field generated by the thundercloud, which serves as the criterion for turning on and turning off the active protection device. The active protection device is mainly employed for producing a reverse electric field to weaken or counteract the background field near the protected objects, and preventing the upward streamer from forming. The control system is used for coordinating the orderly working of the electric field monitoring equipments and the active protection device, in order to optimize the working hours of the active protection device and save power.

When a thundercloud appears over the protection region, the intensity and polarity of the electric field at the protection region can be detected by the atmospheric electric field monitoring equipment. If the absolute value of the electric field intensity is stronger than the preset safety threshold E_{S1}, a downward lightning stepped leader may occur. At this time, turn on the active protection device to excite an artificial reverse electric field above the protection device. According to the principle of vector superposition, the artificial reverse electric field will weaken the background field above the device and exert an inhibitory effect on the initiation of the upward connecting leader at the protection region. Moreover, as long as the reverse electric field excited by the active protection device is strong enough, the direction of the total electric field above the active protection device will be opposite to the direction of electric field excited by the charged thundercloud or stepped leader. So, the probability that the lightning strike point is chosen at the protection region is very low, then the purpose of lightning protection can be achieved at the protection region.

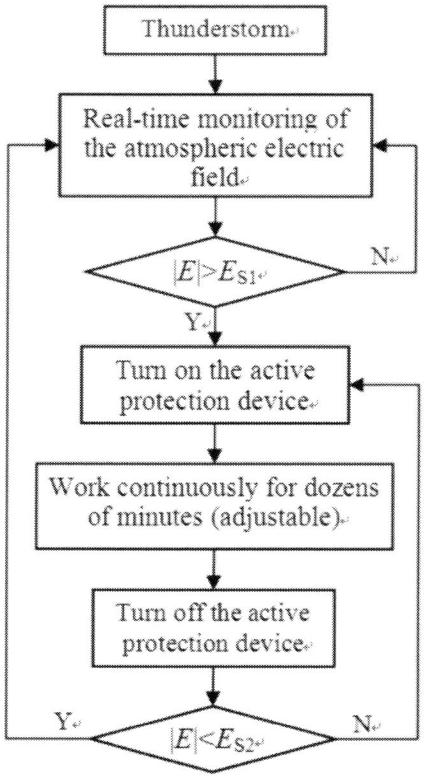

Figure 1. Working flow chart of the protection system using single sensor.

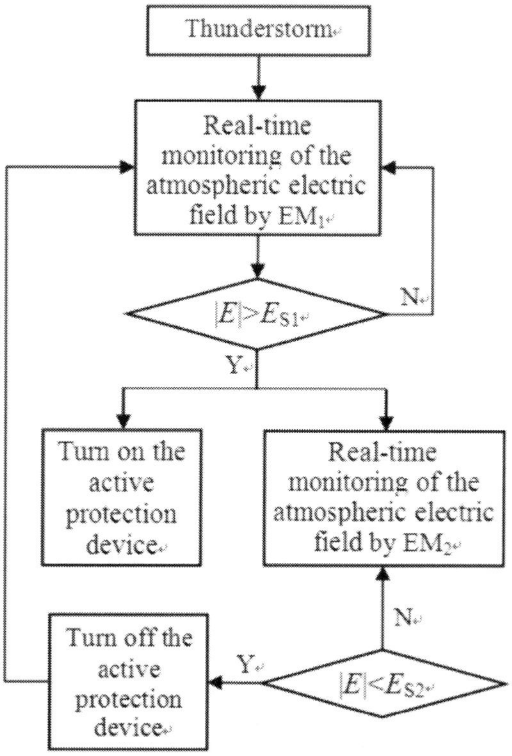

Figure 2. Working flow chart of the protection system using double sensors.

As the criterion for turning on the active protection device is the electric field above the active protection device, it is necessary for one electric field monitoring equipment to be located on the active protection device. However, once the active protection device is turned on, the electric field near the device will change sharply. Thus, the monitoring results above the device at this time can not be used as the criterion for turning off the active protection device. In order to resolve this conflict between electric field monitoring and active protection, two types of working flows are proposed: 1) alternative operation of the electric field monitoring equipment and the active protection device, as shown in figure 1. The working hours of the active protection device can be adjusted according to the research results. 2) Multi-sensors technology. As shown in figure 2, the electric field monitoring equipment EM_1 located on the active protection device is utilized to control the turn-on of the active protection device, while the other electric field monitoring equipment EM_2 located outside the protection region is utilized to control the turn-off of the active protection device. Note that the safety threshold for turning off the active protection device is set as E_{S2}, which is smaller than E_{S1}.

4. Experimental validation of the protection method

In order to verify the feasibility of electrostatic field counteraction using the active protection device, a scale-model experimental system is designed as shown in figure 3. In this experimental system, two metal planes (No.1 plane with a radius of 0.25 m at a height of 1.1 m and No.2 plane with a radius of 0.2 m at a height of 1 m) are employed to represent the active protection device. A metal plane (No.3 plane) with a radius of 0.52 m is suspended at a height of approximately 4 m to represent the charged thundercloud, and a grounding metal plane (No.4 plane) is located on the ground. The active protection device is located inside the plane-to-plane gap which enables to simulate the background field generated by the thundercloud. Electric field measurements are performed using the electrometer SIMCO FMX-002, above the No.1 plane at a height of 1.175 m.

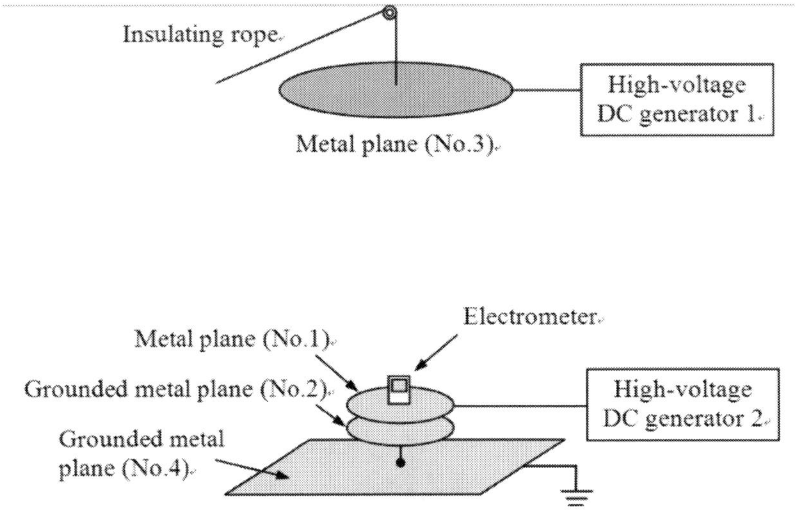

Figure 3. Experimental platform diagrammatic layout.

For charged thunderclouds of different polarities, submit the No.3 plane to a fixed voltage of 30kV and -30kV, respectively. By enhancing the applied voltage to the No.1 plane step by step (the same voltage polarity with that applied to the No.3 plane), we can obtain the relationship between the electric field monitoring results above the active protection device and the applied voltage to the No.1 plane, as shown in figures 4 and 5, respectively.

In figures 4 and 5, the absolute value of the total electric field above the active protection device decreases linearly with increasing applied voltage to the No.1 plane. When the applied voltage reaches a certain level (Here, its absolute value is approximately 2.5 to 3.0kV), the direction of the total electric field above the protection device will reverse; whereafter, the absolute value of the total electric field increases linearly with the increase of applied voltage. The above results show that the background electric field above the active protection device can be weakened or counteracted by the active protection device with appropriate applied voltage, which verify the feasibility of electrostatic field counteraction for the background field above the protection region.

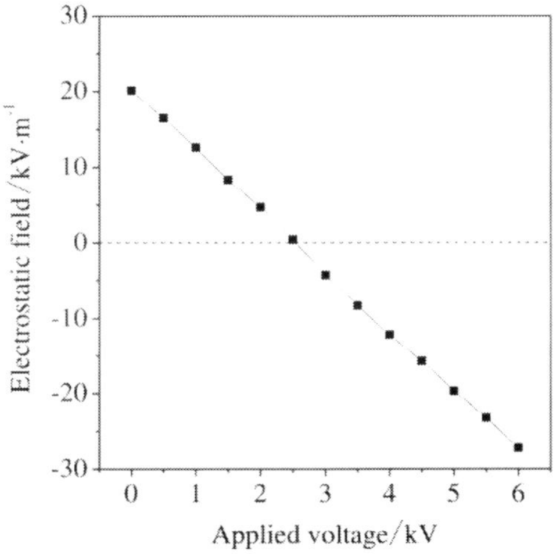

Figure 4. Relationship between the applied voltage to the No.1 plane and electrostatic field above the active protection device when the potential of simulated thundercloud is 30kV.

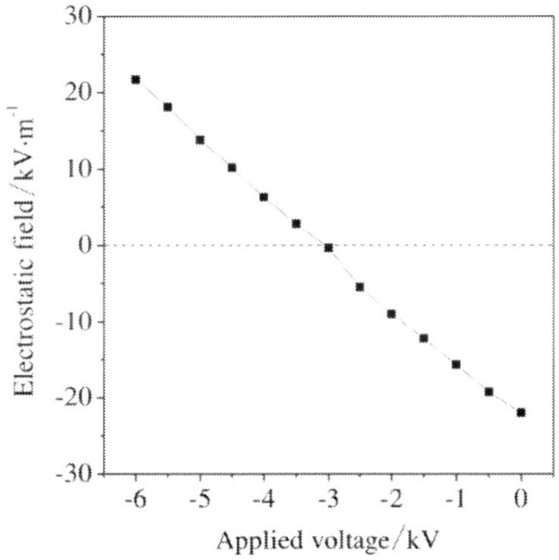

Figure 5. Relationship between the applied voltage to the No.1 plane and electrostatic field above the active protection device when the potential of simulated thundercloud is -30kV.

5. Conclusions

Based on the principle of electrostatic field counteraction and electrostatic shielding, a novel lightning protection method by destroying the condition for direct lightning stroke at the protection region is presented with the combination of the atmospheric electric field monitoring equipments and the active protection device. To resolve the conflict between electric field monitoring and active protection, two types of working flows for the protection system are proposed. The feasibility of this active lightning protection method is verified by scale-model experiment in the laboratory. Moreover, further researches on applied voltage control of the active protection device and forecast of the charge polarity deposited on the lightning downward stepped-leader channel are required.

Acknowledgments

The research is supported by Research Program of Army under Grant No. 7130933.

References

[1] Golde R H 1977 *Lightning* (London: Academy Press)
[2] Rakov V A and Uman M A 2003 *Lightning: Physics and Effects* (Cambridge: Cambridge University Press)
[3] Rakov V A and Rachidi F 2009 *IEEE Trans. Electromagn. Compat.* **51** 428
[4] IEC 62305 2006 *Protection against lightning*
[5] GB 50057-94 2000 *Design code for protection of structures against lightning*
[6] Liu S M 2005 *High Voltage Engine.* **31** 82
[7] Zhang S C, Yan P, Wang J, Yuan W Q, Ren C Y and Shao T 2009 *High Voltage Engine.* **35** 114
[8] Uman M A and Rakov V A 2002 *B. Am. Meteorol. Soc.* **12** 1809
[9] Mazzetti C and Flisowski Z 2007 *J. Lightning Res.* **1** 69
[10] Celozzi S, Araneo R and Lovat G 2009 *Electromagnetic Shielding* (Beijing: China Machine Press)

Journal of Physics: Conference Series **418** (2013) 012029 doi:10.1088/1742-6596/418/1/012029

Influence factor analysis of atmospheric electric field monitoring near ground under different weather conditions

Haojiang Wan*, Guanghui Wei, Yaozhong Cui, Yazhou Chen

Key laboratory of electromagnetic environmental effects, Shijiazhuang Mechanical Engineering College, Shijiazhuang 050003, China

E-mail: hbwhj1983@163.com

Abstract. Monitoring of atmospheric electric field near ground plays a critical role in atmospheric environment detecting and lightning warning. Different environmental conditions (e.g. buildings, plants, weather, etc.) have different influences on the data's coherence in an atmospheric electric field detection network. In order to study the main influence factors of atmospheric electric field monitoring under different weather conditions, with the combination of theoretical analysis and experiments, the electric field monitoring data on the ground and on the top of a building are compared in fair weather and thunderstorm weather respectively in this paper. The results show that: In fair weather, the field distortion due to the buildings is the main influence factor on the electric field monitoring. In thunderstorm weather, the corona ions produced from the ground, besides the field distortion due to the buildings, can also influence the electric field monitoring results.

1. Introduction

The atmospheric electric field is generated by the earth and the charged particles in atmosphere above the earth surface. Monitoring of the atmospheric electric field close to the ground surface is an essential part of atmospheric environment detecting and lightning warning [1-4]. Especially under thunderstorm conditions, the atmospheric electric field, as a diagnostic tool, can reflect the evolution of the charged centers in the thundercloud [5]. Moreover, the atmospheric electric field is generally monitored by a detection network, and the data's coherence among different detection positions is the key issue in the detection network. However, the electric field close to the ground surface is susceptible to the surface features, humidity, lightning, precipitation, etc. [6-7]. According to the instruction manual of the lightning warning system in the Cape Canaveral air force station, the installation location requirements of the field mill is too exacting to be fulfilled [7]. At present, in order to ensure the coherence of the monitoring data in the detection network, only a modification based on the surface features is performed [8]. The influence of different weather conditions on the modifying factor is usually ignored. In this paper, in order to study the main influence factors of atmospheric electric field monitoring under different weather conditions, the electric field monitoring data on the ground and on the top of a building are compared in fair weather and thunderstorm weather respectively. The research result can provide a basis and guidance for improving the accuracy and coherence of the electric field monitoring data from multiple sites.

* To whom any correspondence should be addressed.

2. Atmospheric electric field monitoring instruments and experimental setting

At present, for low-frequency or slowly varying electric field monitoring, the electric field sensor based on charge induction is widely used [9-12]. This type of sensor, with high stability, is suitable for electric field monitoring in the harsh environment such as low temperature, moisture and mechanical oscillations. In this paper, two electric field mills with a resolution close to 5 V m^{-1} in a range of ±50 kV m^{-1} are adopted to monitor the atmospheric electric field. Furthermore, in order to eliminate the zero drift and linearity error of the sensors, the two field mills are both calibrated before the comparative experiments. The calibration curves of the two field mills are presented in figure 1. As follows from figure 1, the two field mills show good accordance in the electric field measurement and exhibit good linearity in the measuring range.

According to the existing experimental conditions, the experimental setting of the electric field mills under different weather conditions is depicted in figure 2. The field mill EM1, located on the top of the building with a height of 15 m, is used to monitor the electric field on the top of the building. The field mill EM2, located on the ground away from the building with a distance of 25 m, is used to monitor the electric field on the ground. The surrounding regions of the field mills are flat, open and clear, which can eliminate the influence of environmental factor on the experimental results to the maximum extent.

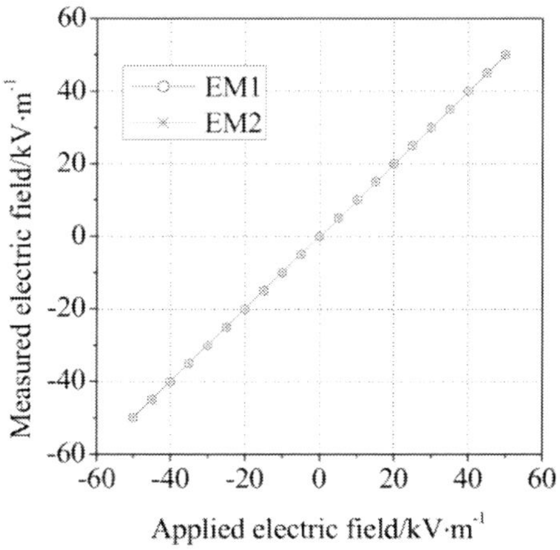

Figure 1. Calibration curves of the two electric field mills.

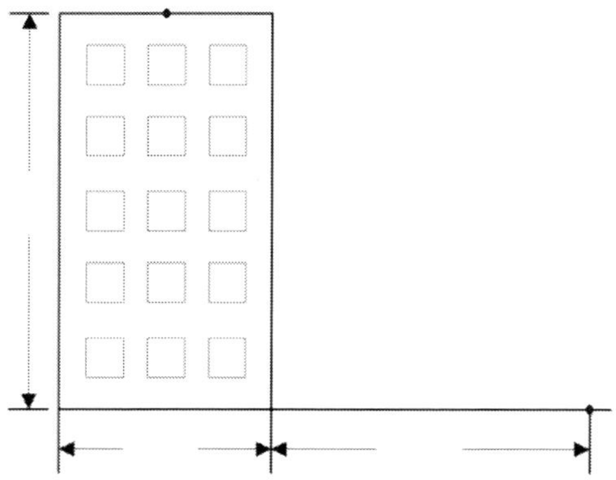

Figure 2. Experimental setting of the electric field mills.

3. Comparative experiments of electric field monitoring under different weather conditions

3.1. Comparative experiment in fair weather

In normal conditions, the atmospheric electric field is down directed in fair weather and in the pure air. The charged particles, such as light ions, heavy ions in the atmosphere, and the negative charge in the earth are the main origins of the atmospheric electric field. In the urban areas, the atmospheric electric field is a quasi-static field with statistical intensity, close to the ground surface, of 130 V m^{-1}. Moreover, the atmosphere electric field in fair weather always varies with the time and space, because of the influence of meteorological phenomena, geographical conditions, solar activity, etc. In general, the atmospheric electric field has a negative correlation with the atmospheric conductivity in fair weather. As the light ions concentration, which is closely related to the atmospheric conductivity, has a negative correlation with the atmosphere aerosol concentration which is in accordance with the atmospheric temperature inversion, the comparative experiment of electric field monitoring in fair weather is performed at the time with slight temperature fluctuation on a clear day. The ambient temperature during the experimental period is presented in figure 3. It can be seen from figure 3 that the average temperature during the experimental period is about 27.5 ℃, and the fluctuation range is no more than 1 ℃. Comparison of the electric field monitoring results between the two positions is shown in figure 4.

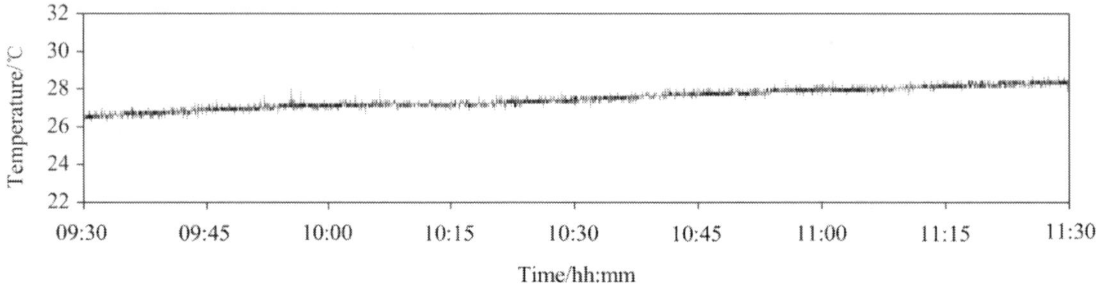

Figure 3. Ambient temperature during the experimental period.

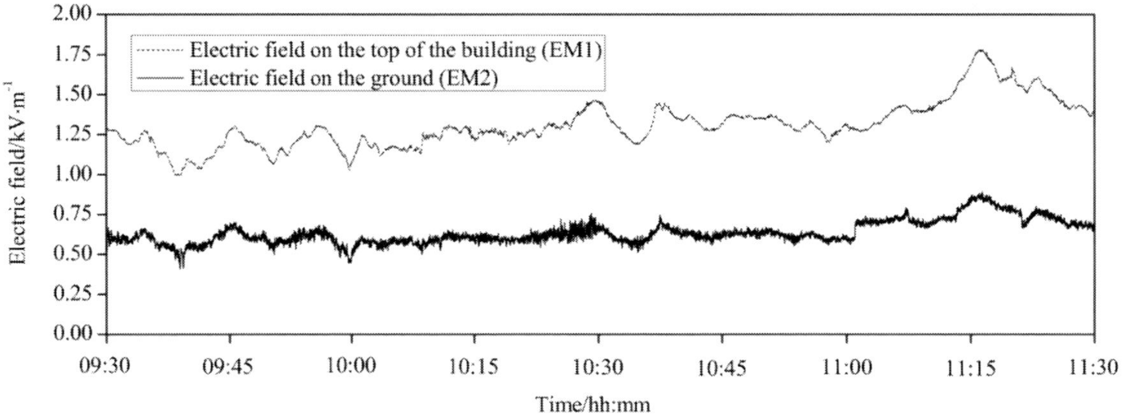

Figure 4. Comparison of the monitoring results between the two positions on a clear day.

According to the statistical calculation on the atmospheric electric field in figure 4, the average atmospheric electric field on the ground in fair weather is about 637 V m^{-1}, with a standard deviation of 73 V m^{-1}. This statistical value is much larger than the present statistical intensity of 130 V m^{-1} in the urban areas mainly for the following reasons: 1) there are obvious spatial-temporal differences for the atmospheric electric field distribution. 2) The statistical value of the atmospheric electric field in fair weather, with intensity of 130 V m^{-1}, is obtained in the case that the sensor surface of the field mill is located in the same surface with the ground surface. However, the electric field monitoring instrument EM2 in this experiment extends out from the ground, which will inevitably lead to the distortion of the electric field near the sensor.

By comparing the electric field on the top of the building with the electric field on the ground in figure 4, it can be seen that the electric field curve on the top of the building is similar to the electric field curve on the ground, only with a coefficient difference. The electric field on the top of the building is distinctly stronger than the electric field on the ground: The average atmospheric electric field on the top of the building during the experimental period is 1.306 kV m^{-1}, with a standard deviation of 0.145 kV m^{-1}, and the ratio of the electric field on the top of the building to the electric field on the ground is about 2. This is mainly due to the distortion effect of the metallic materials and steel mesh of the building on the atmospheric electric field. By employing the finite element analysis software ANSYS, the vertical electric field on the top of the building under a certain background electric field condition can be calculated. Relationship between the enhanced ratios of the electric field on the top of the building and the horizontal distance between the observation point and the center of the building at different heights *h* are shown in figure 5. It can be found that in the position of the electric field sensor with a height of about 1 m on the top of the building, the enhanced ratio is approximately 1.9, which is in accordance with the measurement result.

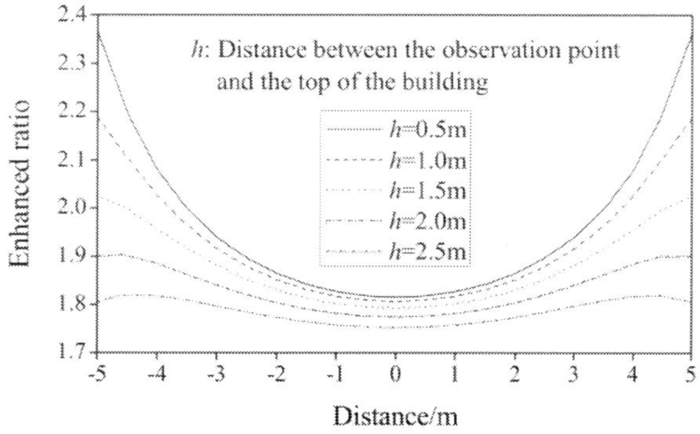

Figure 5. Influence of the building to the E-field at different heights.

3.2. Comparative experiment in thunderstorm weather

Monitoring of the electric field in thunderstorm weather is especially important to lighting warning. Figure 6 shows the comparison of the electric field monitoring results between the two positions on a thunderstorm day.

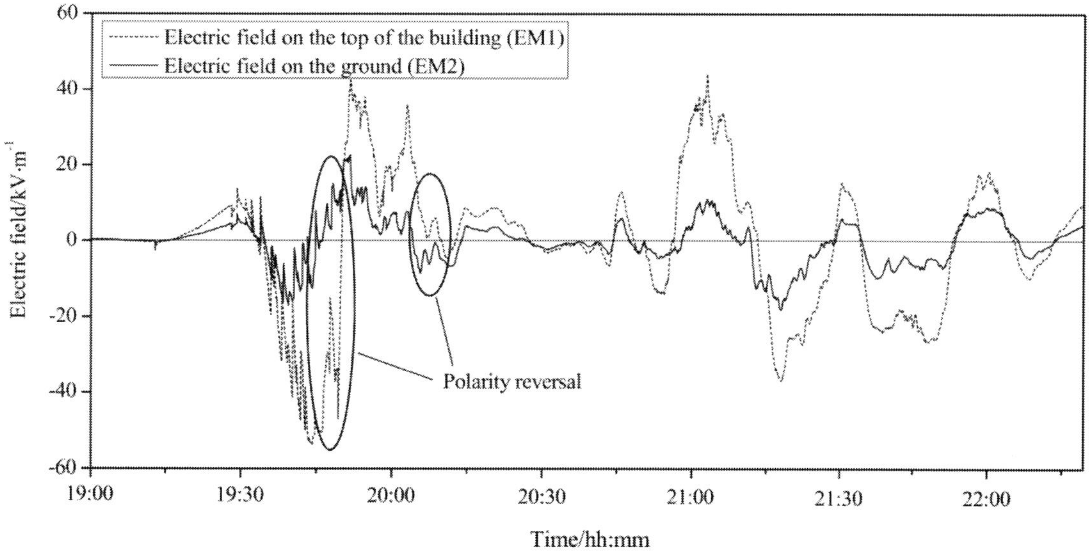

Figure 6. Comparison of the monitoring results between the two positions on a thunderstorm day.

As shown in figure 6, during the thunderstorm, the changing trends of the two electric field curves obtained from different monitoring positions are basically consistent. However, the degree of correlation between the two curves in thunderstorm weather is worse than that in fair weather. The fluctuation range of the electric field on the top of the building is significantly larger than that on the ground. By comparing the ratio of the electric field at the key feature points, it can be found that the ratio of the electric field on the top of the building to the electric field on the ground is not stable during the thunderstorm. Especially when the atmospheric electric field is strong enough, the ratio of the electric field in thunderstorm weather is much greater than the ratio in fair weather. Note that at the two time slices marked with the elliptical lines in figure 6, the electric fields on the top of the building are not directly proportional to the electric fields on the ground. There are obvious differences between each other not only in quantity, but also in the polarity of the electric fields. This is mainly due to that the corona ions produced from the ground, besides the field distortion due to the buildings, can also influence the electric field monitoring results in thunderstorm weather. When the absolute value of the electric field intensity close to the ground surface is stronger than 1.5 to 2 kV m^{-1}, the point discharge from the ground points with a small curve radius will produce large numbers of positive ions or negative ions. These corona ions have a suppressing action on the atmospheric electric field to a certain extent [13]. And this action decreases rapidly with the increasing altitude.

Furthermore, the electric field close to the ground surface will change suddenly while lightning occurs. The ratios of the sudden electric field change at the two positions are also calculated in this paper, as shown in figure 7. It can be seen from figure 7, the amplitude of the sudden electric field change on the top of the building ΔE_1 is about the double of that on the ground ΔE_2. Thus, the distortion effect of the building is the major factor influencing the amplitude of the sudden electric field change. The influence of charged ions from point discharge on the sudden electric field change is slight and can be ignored.

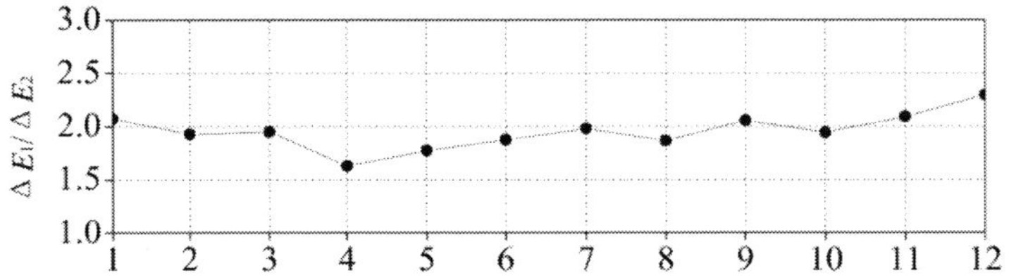

Figure 7. Ratios of the sudden electric field change at the two positions.

4. Conclusions

At present, monitoring of the atmospheric electric field is influenced by many complex factors. Aiming at this, two electric field mills located at different positions are employed to monitor the atmospheric electric field under different weather conditions. By comparing the experimental result with the simulation result from ANSYS, the main influence factors on the electric field monitoring in fair weather and thunderstorm weather are obtained: In fair weather, the field distortion effect due to the buildings is the main influence factor on the electric field monitoring. In thunderstorm weather, the corona ions produced from the point discharge, besides the field distortion due to the buildings, can also influence the electric field monitoring results. Therefore, traditional modification approach, only modifies the monitoring results based on the surface features, can hardly guarantee the consistency of the monitoring data from multiple sites. Moreover, further research is required on the influence laws of the corona ions produced from the point discharge on the monitoring of the atmospheric electric field.

Acknowledgments

The research is supported by Research Program of Army under Grant No. 7130933 and National Natural Science Foundation Project of China under Grant No. 51077132.

References

[1] Parker L W and Kasemir H W 1982 *IEEE Trans. Electromagn. Compat.* **EMC-24** 137
[2] Chen J H, Zhang Q, Feng W X and Fang Y H 2008 *High Voltage Engine.* **34** 425
[3] Singye J, Masugata K, Murai T, Kitamura I and Kontani K 2005 *IEEE Int. Symp. Circ. Syst.*

(Kobe, Japan: IEEE Press) p 244
[4] Yu S Q, Jiang X J, Liu C, Zhang H W and Luo F S 2010 *2010 Asia-Pacific Int. Symp. Electromagn. Compat.* (Beijing: IEEE Press) p 1586
[5] Golde R H 1977 *Lightning* (London: Academy Press)
[6] Guo F X, Zhang Y J, Yan M H and Zhao Y 2007 *Plateau Meteorol.* **26** 257
[7] Zhou B H, Jiang H, Yang B, Guo J M and Zhu K E 2010 *Chinese J. Radio Sci.* **25** 839
[8] Zhou J C, Wang Z H, Guo F X, Zeng Q F, Xu M Y, Yu J, Xing H Y and Ji X Y 2011 *J. Nanjing University Inf. Sci. Techn.: Nat. Sci. Edition* **3** 244
[9] Maruvada P S, Dallaire R D and Pedneault R 1983 *IEEE Trans. Power Apparatus and Systems* **PAS-102** 738
[10] Tantisattayakul T, Masugata K, Kitamura I and Kontani K 2006 *IEEE Trans. Electromagn. Compat.* **48** 435
[11] Jiang H, Zhou B H and Li Y X 2009 *5th Asia-Pacific Conf. Environ. Electromagnetics* (Xi'an: IEEE Press) p 159
[12] Fort A, Mugnaini M, Vignoli V, Rocchi S, Perini F, Monari J, Schiaffino M and Fiocchi F 2010 *2010 IEEE Sensors Appli. Symp.* (Limerick: IEEE Press) p 105
[13] Liu S J and Zhou X G 2009 *Chinese J. Geophys.* (in Chinese) **52** 1451

Research on spark discharge of floating roof tank shunt

Xiaolei Bi[1, 2*], Quanzhen Liu[1, 2], Baoquan Liu[1, 2], Xin Gao[1, 2], Haiyan Hu[1, 2], Juan Liu[1, 2]

1. SINOPEC Safety Engineering Institute, Yan'an 3rd RD, Qingdao, P. R. China
2. State Key Laboratory of Chemical Safety, Yan'an 3rd RD, Qingdao, P. R. China

E-mail: bxlluck@163.com

Abstract. In order to quantitatively analyze the spark discharge risk of floating roof tank shunts, the breakdown voltage of shunt has been calculated by Townsend theory, the shunt spark discharge experiment is carried out by using 1.2/50 μs impulse voltage wave, and the relationship between breakdown voltage of shunt spark discharge and air gap is analyzed. It has been indicated by theoretical analysis and experimental study that the small gap is more easily cause spark discharge than the big gap when the contact between shunt and tank shell is poor. When air gap distance is equal to 0.1 cm, average breakdown voltage is 5280 V. When the air gap distance is less than 0.3 cm, experiment data agree well with Townsend theory. Therefore, in the condition of small gap, Townsend theory can be used to calculated breakdown voltage of shunt. Finally, based on the above conclusions, improvements for avoiding the spark discharge risk of shunt of floating roof tanks have been proposed.

1. Introduction

With the development of world economy, consumption of oil is growing rapidly. The large floating roof tanks usually are used to reserve crude oil and the diameter of tanks is from 60 meters to 100 meters. Capacity is up to 150,000 cubic meters at the most. Because the area of floating roof is very large and floating roof is directly exposed in the air, probability of lightning strike increases obviously. If electrostatic charge in floating roof, which is induced by thundercloud, can not quickly discharge, the tank fire accidents could be caused by spark discharge of floating roof tanks. According to the incomplete statistics, 32.5 percent of 529 tank fire accidents are caused by lightning. The main discharge channel of lightning current and electrostatic charge in floating roof is shunts above the secondary seal [1]. In order to analyze the security and reliability of shunts, the risk of shunts is analyzed. Then, breakdown voltage of spark discharge between the shunt and the tank shell is calculated by the breakdown theory and the relationship between the breakdown voltage of spark discharge and the gap of the shunt is obtained by 1.2/50 μs impulse voltage test of shunt spark discharge, which provide the theoretical and experimental basis for the lightning protection design of the tank.

2. Risk analysis of the shunt

To make the floating roof move up and down, there is 200~300 mm gap between floating roof and tank shell. And in order to reduce the evaporation of the oil through the gap, the gap between floating roof and tank shell is equipped with both primary and secondary seals [2]. The shunts equipped on

* To whom any correspondence should be addressed.

Journal of Physics: Conference Series **418** (2013) 012030 doi:10.1088/1742-6596/418/1/012030

secondary seals are bonded to tank shell by own elasticity (Figure 1) or extrusion of rubber scraper (Figure 2), which are not reliable electrical connections [3]. But it is difficult to ensure the perfect connection between shunt and tank shell in actual operation because of the following three main reasons.

- The geometry size of tank can be changed because of influence of climate and temperature in actual operation and the floating roof will move up and down with elevation and subsidence of oil level. All of these factors will cause drift of floating roof, which would cause the air gap between shunt and tank shell (Figure 3).
- Aging and deformation, which cause the air gap between shunt and tank shell, could occur on rubber scraper because of insolation, wind and rain erosion, sulfidation of vapor, etc. (Figure 4).
- Crude oil contains much sticky wax, which is easy to curdle on the tank shell. This will make the shunts electrically insulate from the tank shell.

Therefore, it is difficult to ensure the perfect connection between shunt and tank shell by own elasticity or extrusion of rubber scraper in actual operation. It is easy to form air gap between shunt and tank shell. In this case, if an air-gap spark occurs in a location with a small gap where the lightning creates a voltage large enough to cause electrical breakdown of the air or vapor/air mixture in the gap, it is easy to cause fire disaster [4].

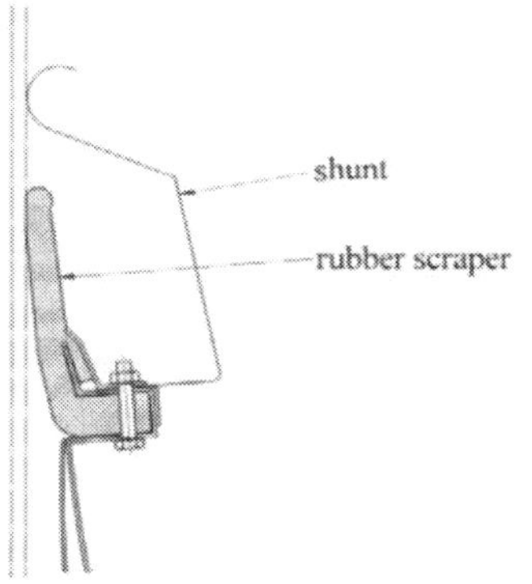

Figure 1. Shunt is bonded to tank shell by own elasticity.

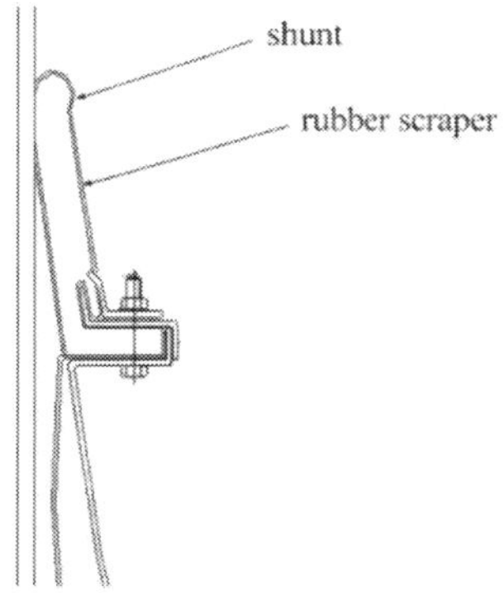

Figure 2. Shunt is bonded to tank shell by extrusion of rubber scraper.

Figure 3. The air gap between shunt and tank shell.

Figure 4. Aging and deformation of rubber scraper.

3. Theoretical calculation for spark discharge of the shunt

The breakdown voltage of spark discharge between shunt and tank shell can be calculated by gas discharge theory. There are two basic theories used to explaining the breakdown in gases. The two theories are popularly known as the Townsend theory and the streamer theory. Townsend theory is valid at low gas pressures and for small gap and the streamer theory developed in the contrary cases. According to the Townsend theory, the current that flows in a uniform field is expressed as follows [5]:

$$i = i_0 e^{\alpha d} / [1 - \gamma(e^{\alpha d} - 1)] \tag{1}$$

where α and γ are the Townsend's primary and secondary coefficient [6], respectively, and i_0 is the primary current, and d is gap distance between shunt and the tank shell.

The medium between shunt and tank shell is air, so relationship between α and electric field intensity E can be expressed as follows:

$$\alpha = Ape^{-Bp/E} \tag{2}$$

Where p=760 Torr for 1 atm, A=8 cm^{-1}Torr^{-1}, B=247 cm^{-1}Torr^{-1}.

According to Paschen's law, the breakdown criterion of air-gap between shunt and tank shell should satisfy the following equation [6-7]:

$$1 - \gamma(e^{\alpha d} - 1) = 0 \tag{3}$$

4. Spark discharge experiment of the shunt

4.1. Experimental waveform and environment

1.2/50 μs impulse voltage wave (Figure 4) is used to carry out spark discharge experiment. Wave head time of 1.2/50 μs impulse voltage wave generated in the lab is 1.2 μs±30% and semi-peak time is 50 μs±20%.

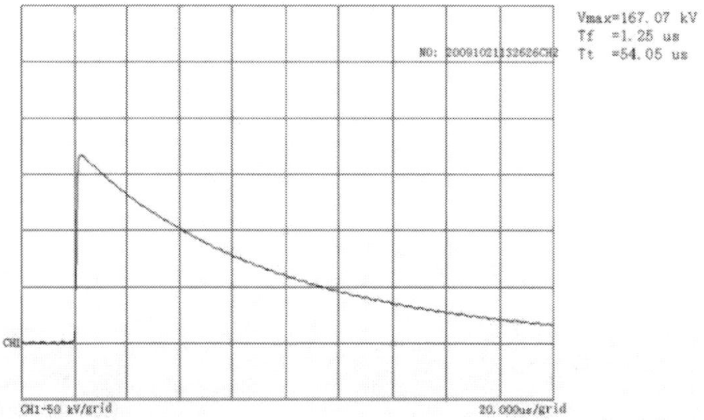

Figure 4. 1.2/50 μs impulse voltage wave.

During the experiment, air conditioner and dehumidifier are used to maintain stable lab environment. Temperature is 20~25 °C, relative humidity is 55%~60%, and atmospheric pressure is 101.8 kPa.

4.2. Experimental scheme

Figure 5 shows the setup for spark discharge of single shunt. Floating roof is insulated from the tank bottom by insulated material. Shunt is electrically bonded to floating roof through the bolt. 1.2/50 μs impulse voltage wave generated by impulse voltage generator is applied to the tank floating roof bonded to shunt. The breakdown voltage between shunt and the tank shell is measured by a Tektronix DPO4104 oscilloscope. The spark discharge process of the shunt is recorded by a high-speed camera.

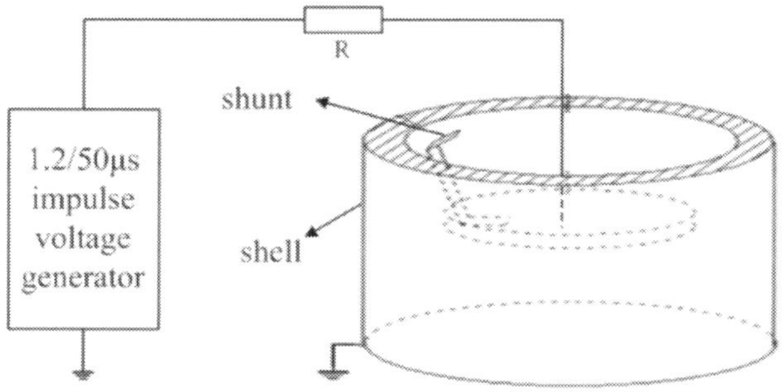

Figure 5. Schematic diagram of the spark discharge experimental.

5. Results and discussion

Figure 6 shows the spark discharge process recorded by the high-speed camera. Table 1 lists spark discharge breakdown voltage for the different gap distances. Table 1 shows that the smaller the gap, the smaller the breakdown voltage. The results indicate that the spark discharge in the small gap is easier to occur than in the big gap. When d is equal to 0.1 cm, the average breakdown voltage is only 5280 V. So spark discharge is easy to occur, when lightning stroke occurs on the tank. Therefore, we should try to avoid forming small gap between shunt and the tank shell and usually check and repair the shunts.

Figure 6. Spark discharge process recorded by the high-speed camera.

Table 1. Experimental air breakdown voltages of shunt.

Times	Air breakdown voltage U/V				
	d=0.1 cm	d=0.2 cm	d=0.3 cm	d=0.5 cm	d=0.8 cm
1	5439	8210	10227	13500	14500
2	5300	8190	10100	12900	15680
3	5100	7980	11600	12080	13500
Average	5280	8127	10642	12827	14560

Figure 8 shows the breakdown voltage as a function of gap distance. The air breakdown voltages calculated by Eq. (2) and Eq. (3) are compared with the corresponding experimental values in figure 1. The experimental data agree well with theoretical data for the air gap d=0.1 cm, 0.2 cm, 0.3 cm. But with an increase of air gap, the deviations between experimental data and theoretical data are increases correspondingly and experimental data are smaller than theoretical data. The reason of deviations is that Townsend theory is valid for small gap. Calculations and experiments indicate that the spark discharge breakdown voltage of small air gap for the shunt can be calculated by Townsend theory. The results are beneficial to the safe designs of tank, which can avoid lightning stroke spark discharge

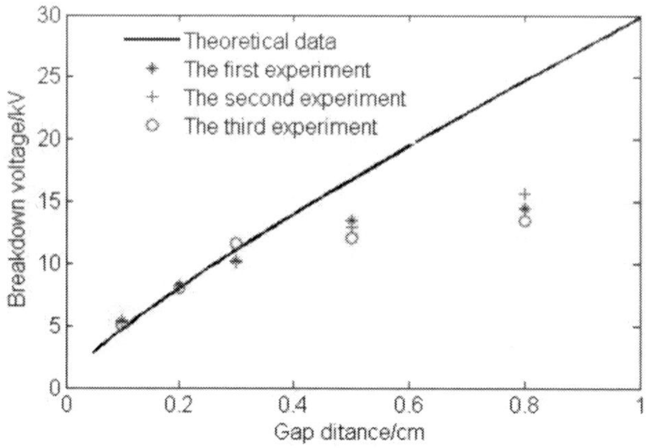

Figure 8. Theoretical and experimental air breakdown voltages.

6. Suggestions

In the actual operation, it is difficult to ensure reliable connection between shunt and tank shell and easy to form small air gap, because the tank is affected by the movement of the floating roof, mechanical stress, aging, oil, metal corrosion and other factors. In this case, the tank is struck by lightning, it would be easy to cause spark discharge. The following measures are recommended to avoid shunt causing the spark discharge.

- Detect oil-gas concentration around the shunt regularly.
- Strengthen the secondary sealing effect, replace the secondary seal of aging timely scraper, and control oil-gas concentration around the secondary seal.
- Cancel the shunt and install reliable electrical connection between floating roof and the tank shell if possible. The risk factors of the spark discharge will be eliminated virtually.

Acknowledgments

We thank the project supported by National Key Technology R&D Program (Grant 2012BAK03B03) for their support.

References

[1] Liu Q Z, Hu H Y, Liu B Q, et al. 2008 *The Proc. 6th Int. Conf. on Appl. Electrostat.* ed Sun K P (Shanghai: International Academic) p 206
[2] GB 50737-2011 *Code for design of petroleum storage depot* (Beijing)
[3] Hu H H, Liu Q Z, Liu B Q 2007 *Saf. Healthy& Environ.* **11** 17
[4] Liu B Q, Hu H Y, Liu Q Z 2012 *Oil &Gas Storage and Transp.* **31** 193
[5] Lin D L 1992 *J. Appl. Phys.* **71** 2580.
[6] Ward A L 1965 *Phys. Rev.* **138** 1357.
[7] Sun K P, Li X W 2005 *Trans. of Beijing Inst. of Technol.* **25** 24

Journal of Physics: Conference Series **418** (2013) 012031 doi:10.1088/1742-6596/418/1/012031

Application of electrostatic prevention technology on polyethylene silos

Hong Gong [1,2*], **Quanzhen Liu** [1,2], **Fenggui Tan** [1,2], **Yunpeng Zhang** [1,2]
1. SINOPEC Research Institute of Safety Engineering, Yan'an 3rd road, Qingdao 266071, China
2. State Key Laboratory for Safety Control of Chemicals, Yan'an 3rd road, Qingdao 266071, China

E-mail: gh64@163.com

Abstract. The main reasons of static electric explosion accidents in polyolefin plant silos were analyzed in this paper, and the study finds that the reasons include control failure of flammable gas content in the feed, high electrification caused by the wind supply, and frequent electrostatic discharge in silos. The electrostatic-reducing technologies of polyolefin powder were introduced, and its application performance in polyolefin plant silos was also clarified. In addition, the methods including FDCS and DGES for evaluation of electrostatic explosion in polyolefin plant silo were proposed. In the end, the risk of electrostatic explosion in PE plant blended silo was evaluated before and after application of electrostatic reducing technology.

1. Introduction

Electrostatic explosion accidents abroad reached a peak during 1970s and 1980s, while the peak delayed a little later in China. From 1985 to 2008, more than 70 explosion accidents occurred in China, nearly 90% of the accidents were related to polyolefin silos. For example, 13 flash explosion accidents occurred in the powder silo of PE (Polyethylene) facilities in 1987-1998 after the expansion and reconstruction of factory, and similar accidents happened many times in another factory's blending silo of PP(Polypropylene) facilities in 1989-1994. In other large-scale polyolefin production enterprises, these explosion accidents also took place occasionally. To solve these problems, the related administrative authorities have organized several security experts to investigate the explosion accidents in polyolefin powder silo for many times, and some common conclusions were acquired. Namely, most of these accidents were caused by the combination of powder, gas and static electricity; Inadequate ventilation design or process control errors including improper responses, devolatilization equipment failures or silo ventilation operational errors were also one of the important causes of explosion accidents; Large quantity of electrostatic charges caused by wind sending powder and frequent electrostatic discharge were produced frequently in polyolefin facilities silos, which plays a important role in explosion accidents. The investigation results of these accidents show that completely eliminating or avoiding the trigger sources related to processes was extremely difficult, which was affected by deficiencies or shortcomings of the current equipment design. For instance, according to a statistic of 68 accidents, there were 49 related to the deficiencies or shortcomings of equipment design. To enhance the prevention capability of electrostatic explosion in current silo

* To whom any correspondence should be addressed.

devices, the measures including elimination of static electricity caused by wind sending and enhancement of ventilation capacity in silos should be taken first. Therefore, a company cooperated with SINOPEC Research Institute of Safety Engineering to complete the program entitled "powder electrostatic prevention for mixing silo of PE devices" in 2011, and the expected results were acquired. The following will introduce the electrostatic elimination effect and silo explosion risk assessment.

2. Brief introduction of electrostatic prevention technology for wind sending powder in mixing silo of PE facilities

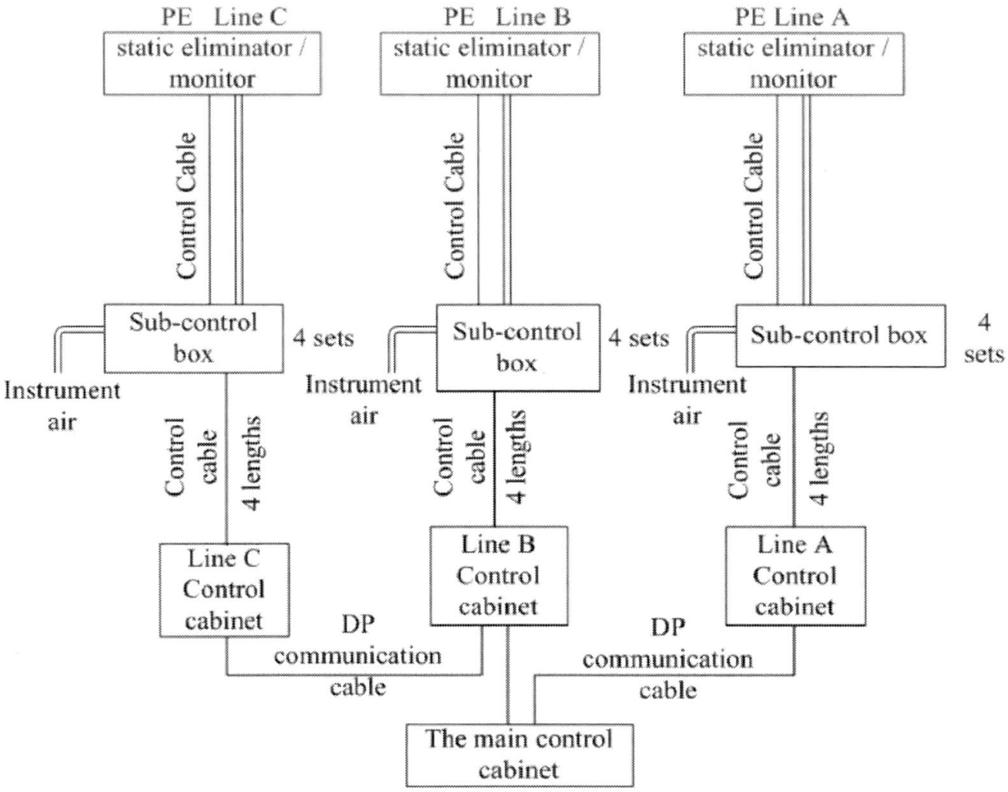

Figure 1. Schematic of system layout.

The PE plant employed density polyethylene reaction under low pressure gas-phase process, so the reaction period was short and the processing load was very heavy (30~33 t h^{-1}). But the content of volatile materials was also high and the devolatilization depended mainly on the degassing silo. Due to fast production, the reliability requirement for equipment maintenance and operating system was very strict. Once the system failure or inadequate devolatilization happened and the emergency was not handled properly, it will cause silo static flash explosion accidents (similar cases occurred many times domestically). For the above reasons , the company installed a set of ion wind powder static eliminator / monitor system in 12 blending silos of PE (A / B / C line) devices.

This system consisted of operation control cabinet, on-site electrical control box, sub-control box, eliminating electrical body and monitoring body. Using the most advanced non-equilibrium bipolar ion wind static elimination technology , electrostatic parameters (μc kg^{-1}) provided by electrostatic charge sensor on the monitor, the charge transmitter - the current controller of the electronic control box , the I / O module - master cabinet PLC signal processing unit , the touch screen system and others can be employed to randomly adjust the electrical positive and negative ion current of discharge needle in the electrostatic eliminator to neutralize material charge in the pipeline to achieve the designed effect. Positive and negative ions flow regulation was implemented by the following system, which was formed by master cabinet touch-screen communication unit , the PLC digital-analog processing unit - electronic control box I / O modules, the voltage controller-electrostatic eliminator body positive and negative high voltage control module, positive and negative discharge needle air current jet components. The schematic system layout was shown in figure 1.

3. Record of electrostatic prevention of wind sending powder in mixing silo of PE facilitie

The static electrification and static elimination record of wind sending material in silos of PE facilities were shown in Table1 - Table3.

Table 1. The static elimination record for PE(Line A) (unit: μc kg^{-1}).

Silo number	TK451A	TK451B	TK451C	TK451D
static electrification range of material	-2.62～-3.26	-4.4～-4.96	-2.97～-3.40	-7.19～-7.99
Average	-3.07	-4.78	-3.16	-7.34
Average after static elimination	0.11	-0.06	-0.09	0.03
Static elimination efficiency	>96%	>98%	>97%	>99%

Table 2. The static elimination record for PE(Line B) (unit: μc kg^{-1}).

Silo number	TK2451A	TK2451B	TK2451C	TK2451D
static electrification range of material	-1.67～-1.83	-4.41～-5.10	-3.28～-3.59	-3.24～-3.64
Average	-1.74	-4.77	-3.44	-3.48
Average after static elimination	-0.02	-0.13	0.04	-0.08
Static elimination efficiency	>98%	>97%	>98%	>97%

Table 3. The static elimination record for PE(Line C) (unit: μc kg^{-1}).

Silo number	TK2451A	TK2451B	TK2451C	TK2451D
Average	-2.77	-1.88	-2.64	-1.27
Average after static elimination	-0.05	-0.15	-0.11	-0.08
Static elimination efficiency	>98%	>92%	>95%	>93%

Static elimination records of PE facilities' (Line A/B/C) silos in table 1-table 3 show that the static eliminator reduced the static electricity of wind sending material from 1-8 μc kg^{-1} to less than 0.1 μc kg^{-1}. The probability of discharge on stockpile surface and the electrostatic risk in silo was significantly reduced.

4. Risk assessment of static electricity in polyolefin silos

To analyze the silo electrostatic explosion risk, two main methods were used. One was silo field diagnosis and counting for static electricity (FDCS), and the other was silo dust electrostatic explosion danger grading (DGES). The following described the practical effect employing the above assessment methods accordingly when the PE plant equipped with electrostatic eliminator.

4.1. Electrostatic field risk assessment for mixing silo of PE plant

Table 4. Basic data of PE plant mixing silo.

Silo number	Line A	Line B	Line C
Size of silo	Φ 5500×23450	Φ 5500×17500	Φ 6000×17000
Dielectric constant(ε_r) of material	2	2	2
Charge-mass ratio(μc/kg)	-1.74～-7.34	-1.27～-2.77	-0.7～-1.18
Resistivity of material(Ω.m)	10^{14}	10^{14}	10^{14}

Basic data of PE plant mixing silo was listed in Table 5 and data of silo electric field acquired by using FDCS method was analyzed as follows:

(1) When the charge-mass ratio "q" was -0.11μc kg^{-1}, the electric field distribution condition on the material surface was shown in figure 2. The electric field close to tank shell approached the critical discharge electric field(3 MV m^{-1}), which may cause brush discharge (discharge energy < 3.6 mJ).

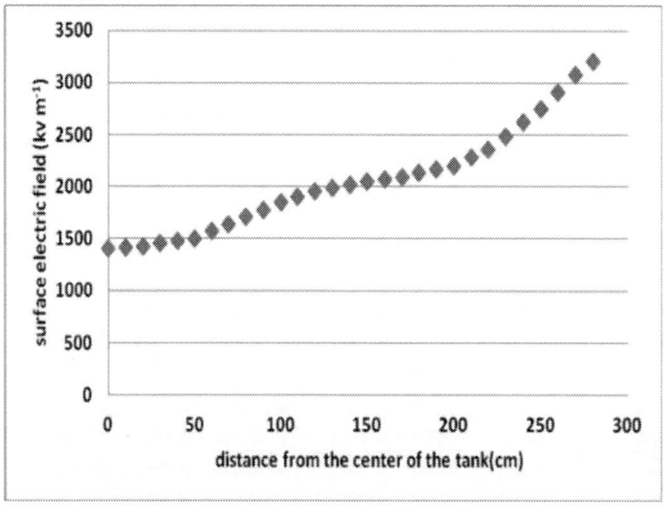

Figure 2. Silo of PE plant electric field distribution (-0.11 μc kg^{-1}).

(2) When the charge-mass ratio "q" was -0.11<q≤-0.24 μc kg^{-1}, the material surface electric field was 3~6.7 MV m^{-1} (as shown in figure 3). In this case, conical discharge may take place(discharge energy≤10 mJ).

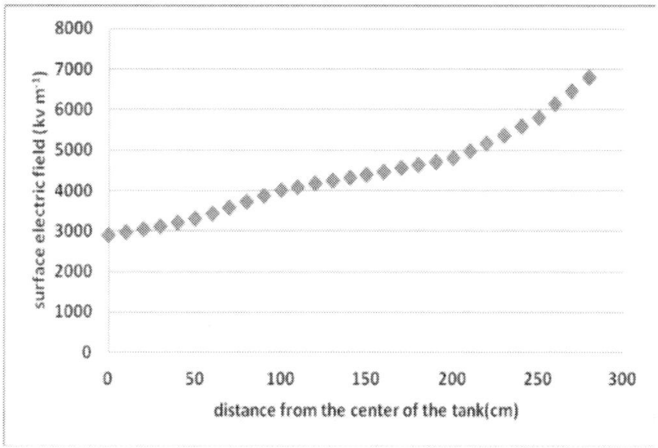

Figure 3. Silo of PE plant electric field distribution (-0.24 μc kg^{-1}).

(3) With simulating experiments (as shown in figure 4), it can be concluded that the surface of materials will be ⩾ 40 kV (powder ignition risk potential) when the charge-mass ratio "q" was ⩾ 1.65 μc kg^{-1}, 170 mm away from tank wall, the potential on the stockpile . If there was metal protrusion above the stockpile surface, igniting discharge may take place.

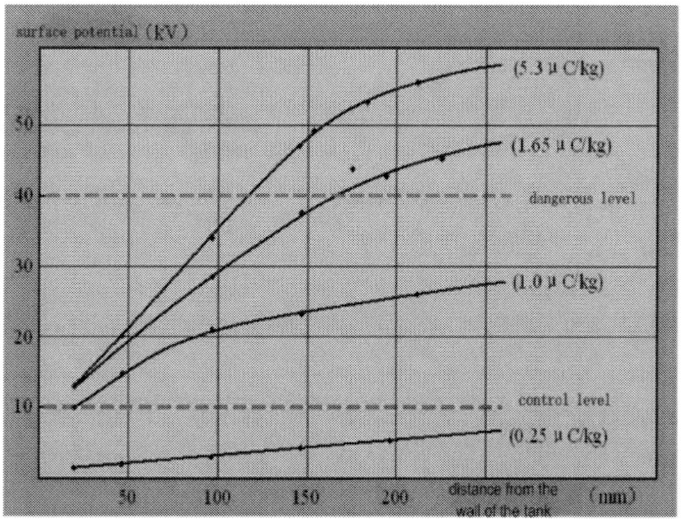

Figure 4. Potential distribution curve near the tank wall.

(4) When the charge-mass ratio "q" was ≤ 0.25 μc kg^{-1}, within 400 mm from the tank shell, the surface potential was <10 kV, which met the requirements of "<10 kV static electricity safety management target"[3].

From the above analysis, it can be conclude that:

① Wind sending systems of PE plants generated a great amount of static electricity, which was greatly beyond the charge-mass range of conical discharge (-0.11~-0.24 μc kg^{-1}) and the risk limit of metal protrusion discharge (1.65 μc kg^{-1}). When the Hexane gas volume in the silo was more than 0.5% wt, conical discharge may cause explosion of mixed miscellaneous dust; when material dust concentration exceeded the explosion limit (20 g m^{-3}), the discharge of material level sensors or mixing tube holder may ignite the silo dust and cause explosions.

② When the ion wind powder static eliminator was installed, the static electricity of wind sending material in PE lines was reduced to less than 0.1 μc kg^{-1}. As a result, conical discharge will not happen on the stockpile surface in the silo and ignitable discharge will not take place within 400 mm from tank shell, so the safety performance of material static electricity was improved dramatically.

4.2. Assessment of explosion risk grading for dust static electricity in mixing silos of PE facilities

The DGES method was used to calculate composite index of electrostatic explosion. The main data were shown in table 6 and table 7.

Table 5. Calculation factor

	Before elimination	After elimination
Dust factor D	55	55
Gas factor G	150	150
Electrostatic discharge factor E	370	10
Second blast factor S	70	70
Coefficient of correction K	1	1
Risk factor H	645	285

Table 6. Danger grading and recommended measures

Risk factor H	Recommended measures
Very serious（>350）	Comprehensive management of dust, gas, static electricity
Serious（300-350）	To focus on the management static electricity, taking dust and gas into account
Moderate（200-300）	Mainly to manage the dust and gas
Slight（<200）	Mainly to manage the technology of dust and gas

5. Conclusion

Before the application of the static elimination measures, the risk factor of PE facility blending silo was 645, which was a severe danger exposure frequency grade. When the measures were taken, the risk factor dropped to 285, which was a moderate danger exposure frequency grade. So the silo electrostatic risk was significantly reduced.

References

[1] Glor M, Maurer B and Rogers R 1995 *Loss Prevention and Safety Promotion* **1** 219
[2] Tan F G 2008 *Petrochemical Safety and Environmental Protection Technology (in Chinese)* **24** 48
[3] Sun L F, Liu Q Z and Liu B Q 2007 *Safety Health and Environment (in Chinese)* **12** 16
[4] Sun LF, Liu Q Z and Liu B Q 2006 *Safety Health and Environment (in Chinese)* **6** 21
[5] Liu Q Z , Sun L F, Zhang Y L and Gong H Transactions of Beijing Institute of Technology *(in Chinese)* **25** 99

Research on electrostatic electrification during jet kerosene spraying

Quanzhen Liu [1], Wentian Zhang [2], Yipeng Li [1], Keping Sun [2*]
1 Qingdao Safety Engineering Institute, 218 Yan'an 3rd RD, Qingdao, China
2 Shanghai Maritime University, 1550 Haigang RD, Shanghai, China

Email: kpsun@shmtu.edu.cn

Abstract. Multiple electrostatic electrifications during aircraft fuelling process may cause a fire disaster or explosion, so study on the protection measure for electrostatic electrification is very important for the security of aircraft fuelling. This paper investigated the electrostatic voltage and charge of the fuel nozzle and metal parts during the fuel spraying by self-designed jet kerosene spraying electrostatic electrification test system. The experimental results indicate that the voltage on the fuel nozzle and metal parts is very dangerous for electrostatic safety if they are not reliably grounded.

1. Introduction

There are multiple electrostatic electrifications during aircraft fuelling process, including fuelling streaming electrification in the oil pipeline caused by the pump, and the electrification also appears in contacting separation while fuel flows through the inner side of the pipeline and filter, and spraying electrification arises on the nozzle in a high speed [1-4]. The charge carried by the fuel is difficult to dissipate ,it also continues accumulating till an electrostatic discharge occurs to bring out a fire diaster or explosion [5]. So investigation of the electrification during jet kerosene spraying is very important for the security of aircraft fuelling.

A jet kerosene spraying electrostatic electrification test system was designed for studying the electrostatic electrifications of fuel nozzle (grounded or ungrounded, respectively) and metal parts (grounded or ungrounded) during the jet kerosene spraying.

2. Experimental system

Schematic of jet kerosene spraying electrostatic electrification test system is shown in figure 1. Fuel nozzle (3), an isolated metal fixture (2) and oil pipeline (5) are arranged in the airtight container (1) to ensure the security of this test. Three additional grounded fuel tanks are designed as a electrostatic charge relaxation devices, ensuring that the fuel is close to the uncharged state before the next cycle of the test starts. The photo of the actual experimental test system was shown in figure 2.

3.Results and discussion

3.1 Nozzle static voltage during spraying

Schematic diagram of nozzle voltage measurement is shown in figure 3. Figure 4 shows the

* To whom any correspondence should be addressed.

measurement results of nozzle voltage under the conditions of nozzle grounded and ungrounded. We should pay more attention to the maximum electrostatic voltage when analyzing and validating the safety situation of jet kerosene spraying. There is a dangerous threshold for ungrounded nozzle voltage up to 8 kV while the air humidity is 57% RH. And it is also found that the voltage reduced conspicuously when the nozzle was grounded. As we know, the minimum ignition energy of combustible hydrocarbon-air mixtures is 0.2 mJ and the minimum voltage that may cause ignition is 2.5 kV, which is in accordance with the formula:

$$E=1/2(CV^2)$$

Here, E is ignition energy; C is the capacitance of the isolated conductor; V is isolated conductor's electrostatic voltage.

When isolated conductor has a capacitance of 70 pF. It is concluded that the voltage on isolated conductors like metal nozzle is dangerous to electrostatic safety.

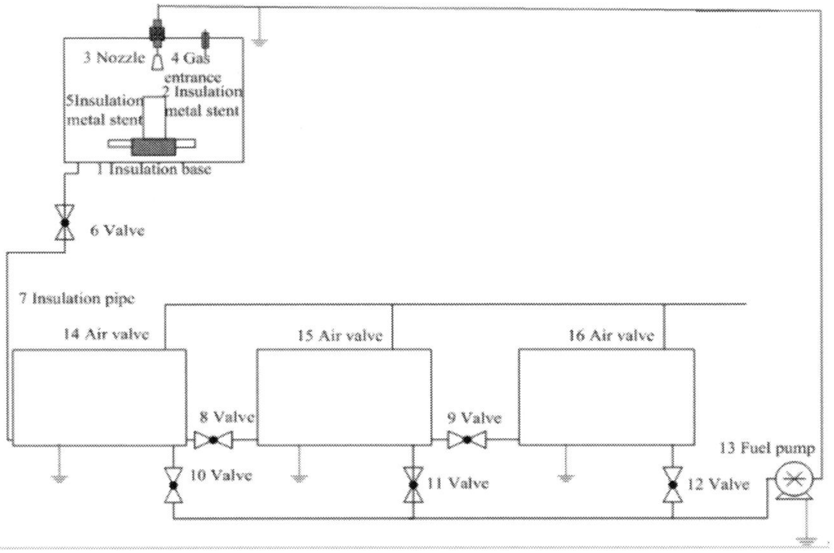

Figure 1. Schematic diagram of the experiment system.

Figure 2. Picture of experimental system.

3.2 Charge of jet kerosene

Theoretically, the metal nozzle eletrostatic charge and sprayed fuel charge should be equal and the polarity of them should be opposite. In order to make a comparison between them, we can directly measure the nozzle charge with digital microammeter (6517A).

As shown in figure 5, the charge of nozzle and faraday cage is almost equal, in fact the polarity is opposite, and in order to show the result in this figure, we change the polarity of faraday cage. The test results show that the theoretical analysis fits well with the test results.

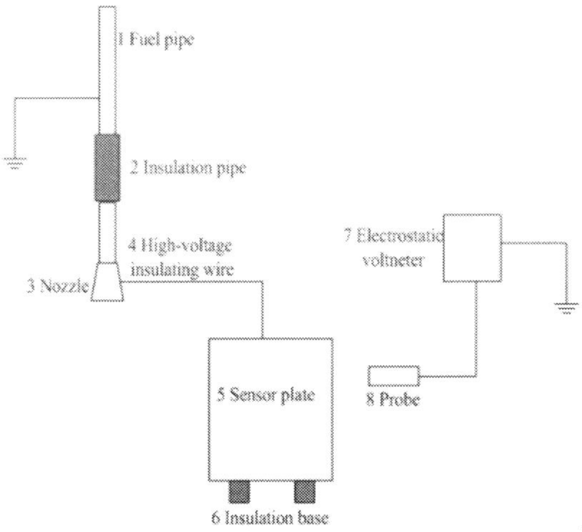

Figure 3. Nozzle measurement system.

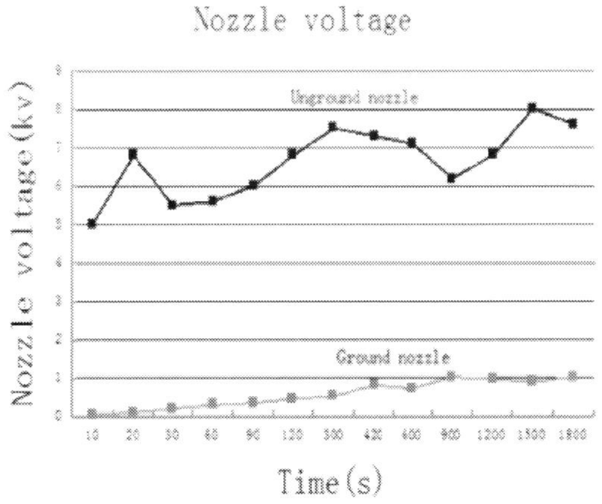

Figure 4. Nozzle voltage measurement result.

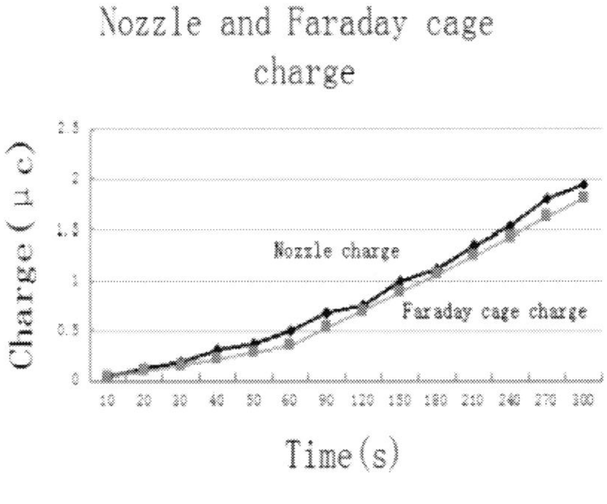

Figure 5. Charge of nozzle and Faraday cage.

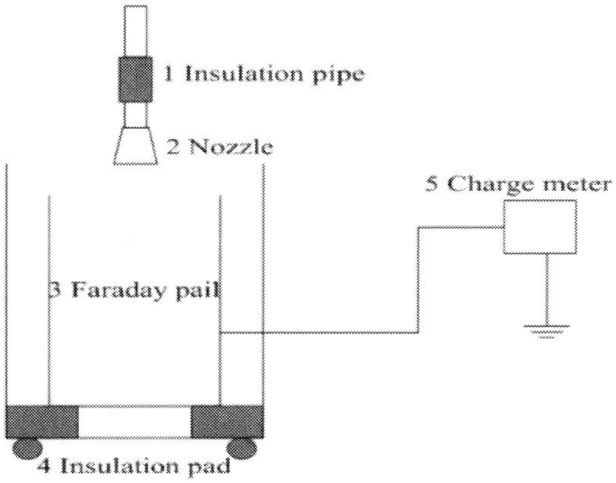

Figure 6. Faraday pail measurement diagram.

3.3 Special charge measurement

The experiments for measuring special charge can be conducted by using metal collecting tank that is a faraday pail instead of faraday cage metal. The schematic diagram of test devices using faraday pail to collect the fuel charge is shown in figure 6. Specific charge of fuel can be gotten through measuring total charge of the charged fuel droplets collected by faraday pail and mass of the JP 8 jet kerosene (its electric conducticity is 180 pS m^{-1}) . The results shown in figure 7 indicate that the fuel specific charge decreased obviously when test system was grounded. That is to say the ground technology is an effective measure to prevent electrostatic accidents.

3.4 Isolated conductor

Figure 8 shows the test system of evaluating the isolated conductor electrostatic safety performance, and the test system is mainly composed of the insulation metalic stent and a insulating material for insulation of the stent and pipeline.

The measuring results of voltage shown in figure 9 are conduced under the conditions of the isolated conductor grounded and ungrounded. From firgue 9 we can see the voltage has a sharp decline when the isolated conductor is grounded. And we can also know that the isolated conductor which immerses in the fuel has a higher voltage than the one above the fuel level.

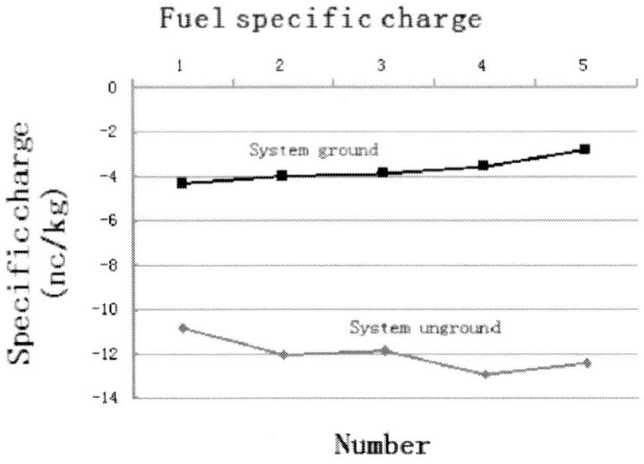

Figure 7. Results of fuel specific charge.

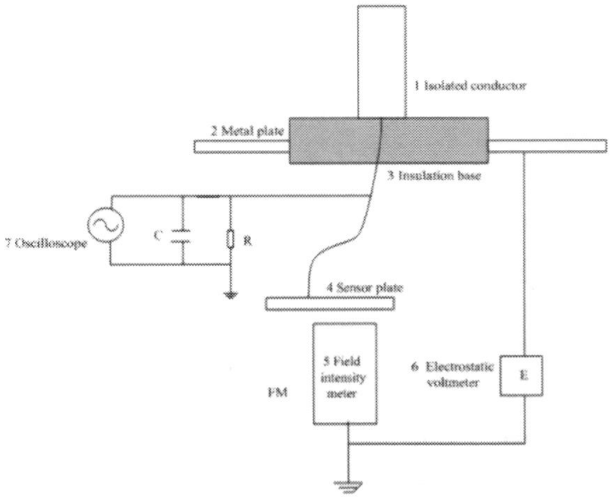

Figure 8. Measuring device for isolated conductor.

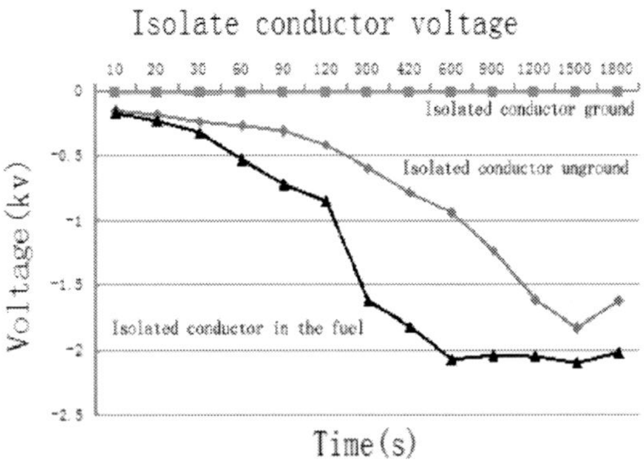

Figure 9. Measurement results of isolated conductor voltage.

It is observed that the electrostatic charge can be discharged quickly when the isolated conductor is grounded. The charge will be maintained in a stable value, and there is a slow discharge when we immersed isolated conductor in the fuel. That is to say electrostatic accident probability will be greatly reduced when the isolated conductors are thoroughly grounded in jet kerosene spraying environment.

4.Conclusions

1) The nozzle highest voltage can arise to 8 kV when the air humidity is 57% RH. It is concluded that the voltage on isolated conductors like metal nozzle is dangerous to electrostatic safety.

2) Voltage for the isolated conductor can reach only dozens of volts when isolated conductor is grounded and is immersed in the fuel. That is to say, they are safe in respect of electrostatic safety in our experimental conditions.

3) All metal parts on aircraft must be thoroughly grounded before refueling. Isolated conductor parts, such as metal float, should also be grounded. In addition, in order to improve the safety performance of airplane fuel tank, we suggest that the fuel tank should be filled with nitrogen, carbon dioxide and other inert gas during filling operation.

Acknowledgement

The study was supported by the National Key Technology R&D Program (Grant 2012BAK03B03).And the study was supported by State Key Laboratory Safety and Control for Chemicals (No.SKL-013)

References

[1] Davies K and Gerken G 2000 *J. Electrostatics* **49** 135
[2] Sun K P, Liu Q Z and Li X W 2009 *J. Electrostatics* **67** 340
[3] Choi S H and Huh C S 2012 *Trans. Korean Inst. Electri. Engineers* **61** 580
[4] Glor. M 2005 *J. Electrostatics* **63** 6
[5] Chang C K 1995 Handbook of Electrostatic Process. Marcel Dekker. New York

Journal of Physics: Conference Series **418** (2013) 012033 doi:10.1088/1742-6596/418/1/012033

Some electrostatic factors may cause gas explosion in coal mines

T T Wang, S M Liu, J C Zhang*, X Chen, Y G Zhang, M Zhang, Y Zhang, X X Ding, H Liu
Zhongyuan University of Technology, Zhengzhou, PR China, 450007

E-mail: zjc@zzti.edu.cn

Abstract. The human body capacitances vary greatly when miners leave a narrow tunnel. The human body electrostatic discharge energy is increased at the outlet of narrow tunnels. Thereby the possibility of detonating gas is greatly increased. Human body electrostatic half-life changes according to the dress of miner. Intense movements of human body will increase the static electricity produced in unit time. This can also increase the human body static electricity and cause gas explosion.

1. Introduction

Coal mine gas explosion is a serious threat to people's lives and property safety. Human body static is an important cause of coal mine gas explosion. The generation and discharge of human body static electricity is influenced by a number of factors. When there is too much human body static electricity, the discharge energy could cause gas explosion, resulting in significant harm.

2. Human body capacitance in narrow trench of coal mine

The human body capacitances of miners in uniform at different places of the narrow trench are tested with "the simulating device of mining narrow trench" (only parts of the data are listed in table 1).

Table 1. Human body capacitance at different positions of narrow trench [a].

No.	sex	height (cm)	weight (kg)	capacitance (pF)						surface resistivity of antistatic boots (Ω)
				outlet of the trench	inner	middle	back	left	right	
1	male	165	63	135	2720	2850	2730	2650	2690	2×10^{10}
2	male	173	72	150	2400	2810	2800	2500	2610	2×10^{10}
3	male	181	73	155	2710	3020	3000	2700	2700	2×10^{10}
4	male	177	72	160	1700	1550	1520	1620	1300	2×10^{10}

[a] Temperature: 19℃; Relative humidity: 45%

* To whom any correspondence should be addressed.

Results in table 1 show that the body capacitances of the miners at the outlet are less than one tenth of the inner.

"The simulating system of the human body electrostatic discharge detonating gas" consists of mixed gas chamber, the ignition device of human body ESD simulator, the control unit, gas measurement unit, oxygen measurement unit, temperature measurement unit, alarm screen, PC and gas supply devices. The schematic diagram is shown in figure 1. The gas concentration is measured with the system. The measured error of gas concentration is less than 0.1%.

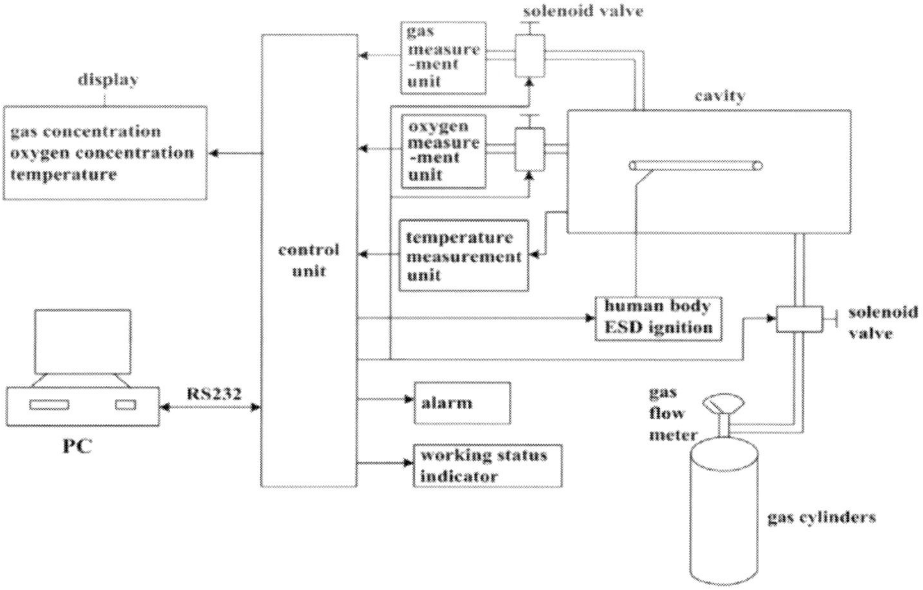

Figure 1. Schematic diagram.

With "the simulating system of the human body electrostatic discharge detonating gas", discharge experiments are made, and the human body capacitances range from 100 pF to 2700 pF (the human body resistance is chosen as 350 Ω).

In experiments, the gas of concentration from 5% to 16% is filled into the chamber. The gas mixture is detonated by "the simulating system of the human body electrostatic discharge detonating gas". The gas concentration, oxygen concentration and mixture temperature in time of the gas explosion are shown. In this way different human body ESD energy detonating gas and the easiest detonated gas concentration in different conditions of temperature and oxygen concentration can be verified. Human body leakage resistance, relative humidity, the discharge gap of the human body ESD model and the gas concentration are shown in table 2.

3. The human body discharge voltage detonating gas

Electrostatic discharge detonating gas must meet three conditions:

- Electrostatic discharge energy must be greater than the gas mixture's minimum ignition energy 0.28 mJ.
- Field strength of the discharge voltage must be greater than 22 kV/cm, which is the breakdown field strength of the gas mixture.
- Gas concentrations must be in the explosive concentration limits of 5% -16%.

That is to say, discharge voltage ensures that the human body discharge energy should be greater than 0.28 mJ, and the discharge field strength should be greater than 22 kV/cm. To detonate gas mixture, the human body voltage not only must be greater than $(2W/C)^{\frac{1}{2}}$ but also must be greater than

$E_0 d$. W is the minimum ignition energy. C is the human body capacitance. E_0 is the breakdown field strength of gas mixture, and d is the discharge gap.

Table 2. Some experimental parameters of gas mixture explosion with "the simulating system of the human body electrostatic discharge detonating gas".

No	Gas Concentration (%)	Oxygen concentration (%)	capacitance (pF)	Discharge gap (mm)	Leakage resistance (Ω)	Temperature (°C)	Relative humidity (%)
1	8.5	20.9	2700	0.2	350	22	45%
2	8.8	20.2	1800	0.3	350	22	45%
3	8.5	20	900	0.6	350	22	45%
4	8.5	20	500	0.6	350	22	45%
5	8.3	20	300	0.9	350	22	45%
6	8.5	20	280	1.2	350	22	45%
7	8.7	20	100	1.2	350	22	45%

4. The relationship between the human body discharge voltage and resistance

Take a Gaussian surface in the miner's dress; the current and current density vector $\vec{\delta}$ through the closed surface of the dress and the charge Q satisfy the following relationship.

$$I=\int_S \vec{\delta} \bullet \mathrm{d}\vec{S}=-\frac{\mathrm{d}Q}{\mathrm{d}t} \tag{1}$$

Where
$$\vec{\delta}=\vec{E}/\rho$$

$$\int_S \vec{E} \bullet \mathrm{d}\vec{S}=\frac{Q}{\varepsilon_0\varepsilon_r} \tag{2}$$

$$\int_S \rho\vec{\delta} \bullet \mathrm{d}\vec{S}=-\rho\frac{\mathrm{d}Q}{\mathrm{d}t} \tag{3}$$

$$-\rho\frac{\mathrm{d}Q}{\mathrm{d}t}=\frac{Q}{\varepsilon_0\varepsilon_r} \tag{4}$$

$$\frac{\mathrm{d}Q}{Q}=-\frac{\mathrm{d}t}{\varepsilon_0\varepsilon_r\rho} \tag{5}$$

The solution of the equation is
$$Q=Q_0 e^{-\frac{t}{\varepsilon_0\varepsilon_r\rho}} \tag{6}$$

When the human body electrostatic voltage attenuates to half of the original, the required time is:

$$\tau_{1/2}=0.69\varepsilon_0\varepsilon_r\rho \tag{7}$$

$\tau_{1/2}$ is known as the electrostatic half-life.

If the charge is Q at a moment, from the charge attenuation formula (6), the decrease of electricity from t to t + dt is:

$$-\mathrm{d}Q'=\frac{1}{\varepsilon_0\varepsilon_r\rho}Q\mathrm{d}t \tag{8}$$

If the electricity generated in unit time is P, then the total electricity increased from t to $t + \mathrm{d}t$ is:

$$\mathrm{d}Q = P\mathrm{d}t - \frac{1}{\varepsilon_0 \varepsilon_r \rho} Q\mathrm{d}t \tag{9}$$

$$\frac{\mathrm{d}Q}{\mathrm{d}t} = P - \frac{1}{\varepsilon_0 \varepsilon_r \rho} Q \tag{10}$$

$$Q = \varepsilon_0 \varepsilon_r \rho P (1 - e^{-\frac{t}{\varepsilon_0 \varepsilon_r \rho}}) \tag{11}$$

Based on Q = CU, if the surroundings around miners doesn't change, C is constant, therefore

$$U = \varepsilon_0 \varepsilon_r \rho C^{-1} P (1 - e^{-\frac{t}{\varepsilon_0 \varepsilon_r \rho}}) \tag{12}$$

It can be seen that there is a maximal limit of Q:

$$Q = \varepsilon_0 \varepsilon_r \rho P = 1.45 \tau_{1/2} P \tag{13}$$

Therefore, electrostatic charge has a maximum value, and the maximum value is proportional to the electrostatic half-life and the electricity generation rate. The larger the electrostatic half-life is, the more the maximal static electricity on human body is. The larger the electricity generation rate is, the more the maximal static electricity on human body is.

In this paper, the human body capacitances in different places of narrow mine tunnel is discussed. The human body capacitance at the outlet of narrow tunnel is less than one tenth of that in the middle of the tunnel. So it is dangerous for miners to move out of a narrow tunnel. The human body electrostatic discharge voltage detonating gas and the relationship between the human body discharge voltage and resistivity of miners' dress are also discussed. From the discussion it can be seen that the miners' dress and intense movements of miners influence the maximal static electricity on human body. So miners' dress should be selected strictly and intense movements of miners in coal mine should be forbidden.

Acknowledgments

This work is supported by the National Nature Science Foundation of China (NSFC) under contract No. 50977093 and 51077134.

References

[1] Mannan M S 1999 *Chem. Process Saf.* (AIChE, Chemical Process Safety). p 105
[2] Liu S and Wei G 1990 *Static Theor. and Prot.* (Beijing, Weapons Industry Press)
[3] Nabours R.E. 2003 *Ind.l and Commer. Power Syst.* (St. Louis, IEEE Technical Conference) p 66

An experimental study on anti-electrostatic gauge rulers

Renjie Lou[1*], Liping Dai[2], Hong Sun[3]
[1,2] CNPC Research Institute of Safety & Environment Technology, Dalian, China
[3] CNPC Dalian Petrochemical Company, Dalian, China

E-mail: lourj_dl@petrochina.com.cn

Abstract. The process of oil filling will produce electrostatic phenomena which may cause fire accidents. There were no reports about research on the danger of static electricity generation in the process of gauging operation to date. This paper presents an experiment on charge transferring quantity of gauge rulers, and calculates the charge transferring quantity of an anti-electrostatic gauge ruler and a metal one, respectively. The results indicate that the charge transferring quantity can be more than 0.1 μC for a metal gauge ruler, while it is less than 0.1 μC for an antistatic gauge ruler. Therefore, this experimental research proves that using an anti-electrostatic gauge ruler is safer than using a metal one. This study also provides some theoretical and experimental evidence for making anti-electrostatic gauge rulers.

1. Introduction

In the process of filling, loading and shipping the oil products or in the monthly inventory checking at the end of every month, the operators must perform the manual measurement. Because the oil products are highly insulated, the oil surface is apt to accumulate charge in the filling process. At this moment, if the operator uses a steel gauge to perform the manual measurement, it may produce electrostatic discharge when the metal copper weight of the grounded gauge touches the charged oil surface, which may lead to an oil tank fire accident. For example, the oil tank caught fire in the process of dynamic manual measurement in an oilfield company in September, 2003. The oil tank caught fire in the process of dynamic manual measurement in another oilfield company in February, 2005.

The electrostatic danger of the manual measurement manifests itself mainly when it produces brush discharge as soon as the metal copper weight touches the charged oil surface. Generally speaking, the energy is not more than 4.0 mJ, but charge transferring quantity is far beyond 0.1 μC [1], which is easy to a flash explosion in the oil tank in the process of manual measurement. In order to keep the process of manual measurement safe, the only way is to make sure the oil product standing time. Given that the first part to touch the oil surface is the metal weight in the manual measurement, it is necessary to limit the charge transferring quantity of the weight. Therefore, when we research on the material to be used in making the weight, we should make sure that the charge transferring quantity of this kind of material should be reduced to less than 0.1 μC, and it should also make the weight not keep a high potential because of oil product electrostatic reaction.

* To whom any correspondence should be addressed.

2. Selection of the material to make the anti-electrostatic weight

According to the requirements of the working situations which need to use an anti-electrostatic gauge weight, the following aspects should be considered.

2.1 Electrostatic conductivity

According to the rules defined in Section 6.3.7 in GB12158-2006 *General Guideline for preventing electrostatic accidents* [2], for the synthetic material for the sampler, thermoscope and gauge should be of electrostatic sub-conductive material, whose electrical resistance value per unit length should be $1\times10^5 \sim 1\times10^7\ \Omega\ m^{-1}$, or whose area specific resistivity and volume resistivity should be less than 1×10^{10} or $1\times10^8\ \Omega\ m^{-1}$ respectively.

2.2 Oil resistance

The oil-measuring gauge should be checked every half year, during which the time of using the gauge may be about 20 h. If the material is soaked into some common gases with stronger solubility for 96 h without changing its appearance and texture, we may say it has good oil resistance.

2.3 High-temperature resistance

The temperature of the oil product in the tank may be very high, so the material used must have a certain high-temperature resistance. The requirement in the process is that the material must resist the high temperature of more than 80 ℃.

2.4 Impact resistance

There may be some collisions in the manual measurement, so the material must have a certain impact resistance. If the finished weight is dropped freely from the height of 1.5 m and is not broken, we may say it has good impact resistance.

3. Research on the testing method of charge transferring quantity

Using a pump, the oil product is sent from the storage tank to the measuring tank by a filter. Use electrostatic potentiometer to test the surface potential of the oil product, and then test respectively the charge transferring quantity when using an anti-electrostatic gauge ruler or a common gauge ruler at different oil surface potentials and then the electrostatic dangers from the different kinds of gauge rulers can be determined. It is shown in figure 1.

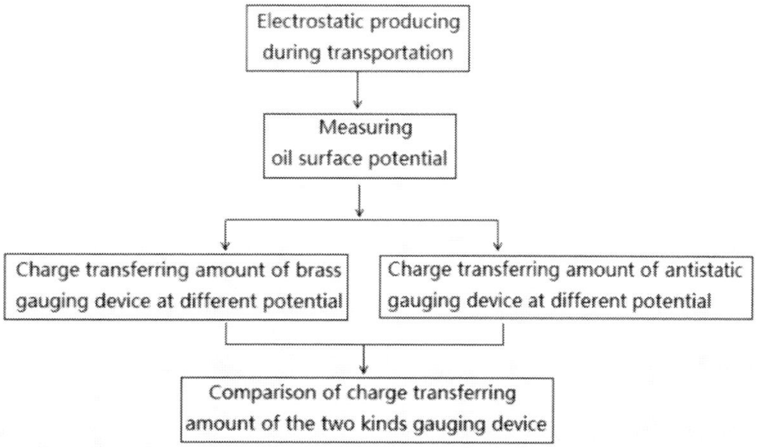

Figure 1. Block diagram of testing method of the charge transferring quantity.

3.1 Research purpose for the testing equipment of charge transferring quantity of oil product

Using this testing device, the oil product will be charged. After the potential of the oil product in the testing tank reaches to a certain value, the test of charge transferring quantity of the gauge ruler will be carried out to determine the dangerous degree of charge transferring quantity of the weights with different electrical resistance value.

3.2 Testing method of oil surface potential

Figure 2 shows how to test the potential according to GB6951-86 safe surface voltage of light fuel oil in tank filling operation [3]. When an isolated copper ball is placed on the tested surface as a collecting electrode, the ground insulation resistance of the copper ball and the tested surface is more than 10^{14} Ω, and the ground capacitance is less than 100 pF, the copper ball's potential is the potential of the oil surface where the ball is located.

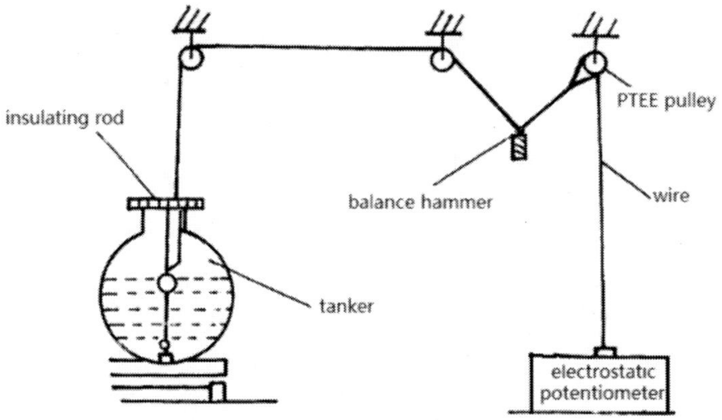

Figure 2. Diagram of the testing system of oil surface potential.

3.3 Testing method of charge transferring quantity

Figure 3 shows the testing method of charge transferring quantity. At different oil surface potentials, the weight lowers to the oil surface. By the RC circuit the maximum V_m of the RC discharge pulse is tested by an oscilloscope, based on $Q=C\ V_m$, we can get the max charge transferring quantity.

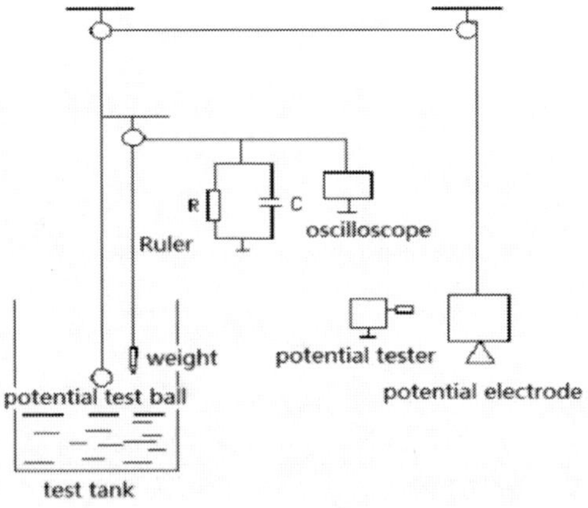

Figure 3. Schematic diagram of the testing method of charge transferring quantity.

3.4 Selection of oil surface potential in the testing tank

According to the definition in Section 7.2.5 of GB12158-2006 General guidelines for preventing electrostatic accidents, when loading the light oil product, the oil surface potential should be lower than 12 kV. So we should select the following oil surface potentials to carry out the test of charge transferring quantity:

$U<12$ kV; 12 kV$\leq U \leq$30 kV; 30 kV$< U \leq$40 kV; 40 kV$< U \leq$50 kV; 50 kV$< U \leq$60 kV

3.5 Testing device

Figure 4 shows the testing device used for testing charge transferring quantity. Oil storages in 1# tank, flowing through valve 1, valve 2, pump, valve 5, valve 6, filter, valve 8, insulation flange 1, valve 9, pumped into 2# tank. The electrostatic generated during transportation is collected in 2# tank, and then tests the charge transferring quantity of gauging ruler at different potentials. To begin another test the oil flows through flange 2, valve 10, valve 4, pump, valve 3, valve 1, back to 1# tank.

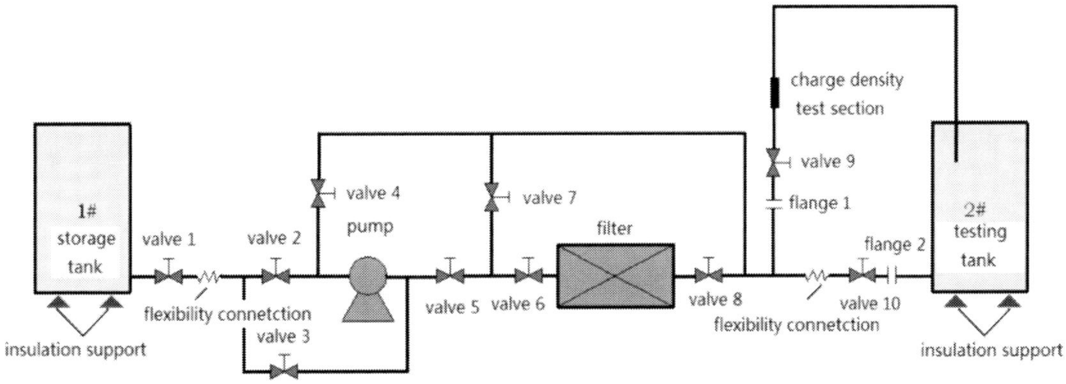

Figure 4. Schematic diagram of the testing device for charge transferring quantity.

4. Experimental research on charge transferring quantity between ruler weight and the charged oil surface

4.1 Metal weight

Table 1 shows the experimental data. Figure 5 is the data graph. From the table 1 and figure 5, we can see that the charge transferring quantity of metal weight is more than 0.1 μC when the oil surface potential reaches 30 kV, so it is dangerous for gauging operation at this condition.

Table 1. The data of the charge transferring quantity of the metal weight

No.	The surface potential of the discharge ball (kV)	Discharge potential (V)	Charge transferring quantity (μC)
1	10	0.93	0.0047
2	15	5.17	0.0259
3	20	9.56	0.0478
4	25	15.07	0.0754
5	30	22.13	0.1107

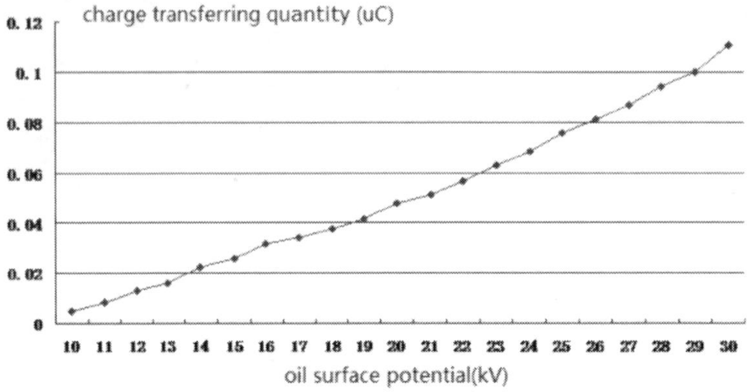

Figure 5. Data graph of the charge transferring quantity of the metal weight.

4.2 Anti-electrostatic weight

Table 2 shows the experimental data. Figure 6 is the data graph. From the table 2 and figure 6, we can see that the charge transferring quantity of anti-electrostatic weight is still very low until the oil surface discharges itself, so it is safer using anti-electrostatic weigh for gauging operation especially at high oil surface potential.

Table 2. Data of charge transferring quantity of anti-electrostatic weight.

No.	Surface potential of discharge ball(kV)	Discharge potential (V)	Charge transfer (μC)	Weight potential (V)
1	10	0.012	0.0001	0
2	15	0.059	0.0003	0
3	20	0.096	0.0005	0
4	25	0.188	0.0009	0
5	30	0.469	0.0023	0
6	35	0.513	0.0025	0
7	40	0.678	0.0034	0
8	45	0.821	0.0041	0
9	50	0.922	0.0046	0
10	55	1.332	0.0066	0
11	59	2.730	0.0137	0
12	60	The oil surface discharges itself.		

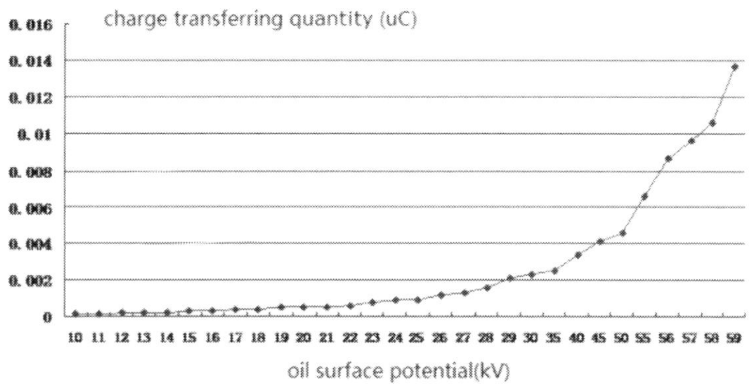

Figure 6. Data graph of the charge transferring quantity of the anti-electrostatic weight.

5. Conclusions

(1) The charge transferring quantity of the gauge ruler is related to its medium resistivity. The smaller the resistivity is, the larger the charge transferring quantity will be observed.
(2) The charge transferring quantity of the gauge weight of the same kind is related to the oil surface potential. The larger the oil surface potential is, the larger the charge transferring quantity will be.
(3) When the metal weight touches the oil product whose surface potential is 29.0 kV, the charge transferring quantity is 0.1001 μC, which is more than 0.1 μC. When the surface potential of charged oil product is more than 29.0 kV, all of the charge transferring quantities are more than 0.1 μC. So it is dangerous for the metal weight to touch the charged oil product.
(4) When anti-electrostatic weight touches the oil product whose surface potential is 59.0 kV, the charge transferring quantity is 0.0137 μC, which is much less than 0.1 μC. When the oil surface potential is up to 60 kV, the oil surface discharges itself and the oil surface potential descends rapidly. For this reason, the oil surface potential will not be over 60 kV. The experimental results indicate that it is very safe for the anti-electrostatic weight to touch the surface of the charged oil product and the weight will not be charged.

References

[1] *Electrostatic* Security Guidelines, Japan
[2] *General guideline for preventing electrostatic accidents* GB12158-2006
[3] *Safe surface voltage of light fuel oil in tank filling operation* GB6951-86

The risk analysis of dust electrostatic based on on-site survey of polypropylene plant

Xiumin Wu*, Mingjun He, Haibo Yu
CNPC Research Institute of Safety & Environment Technology, Dalian, China

E-mail: wuxm_dl@petrochina.com.cn
Abstract. The dust electrostatic explosion accidents in polypropylene plant are mainly caused by the interaction of combustible gas, dust and static electricity. This paper analyses the key parts easy to produce dust and the risks of dust electrostatic by on-site survey of polypropylene plant, and proposes corresponding safety protection measures. The analysis results indicate that any careless mistakes and deviation in every step of process control may lead to electrostatic explosion in the silo. And if the equipment has some inherent defects and there are some careless mistakes in the process control, it will be easier to cause dust electrostatic explosion accidents.

1. Introduction

The production of polyolefin powders in China has developed rapidly in recent two decades. However, the electrostatic explosion accidents have also increased. According to the statistics, there were more than 70 electrostatic explosion accidents in powder silo of China from 1986 to 2000. There were also more than 10 electrostatic explosion accidents after 2000. At the same time the ratio of electrostatic explosion accidents was also very high abroad. This paper analyzes the causes of the dust electrostatic explosion accidents in polypropylene production plant, and discusses the rules of dust electrostatic ignition and explosion, which is essential to propose related preventive countermeasures.

2. On-site survey of risk factors of dust in polypropylene plant

2.1. Process of polypropylene plant

The refined propylene is put into the loop reactor through measuring tank. The main catalyst, activator, molecular weight regulator and electron donor are also added to the loop reactor in a certain proportion in order to perform the bulk polymerization of liquid phase. There are mainly polypropylene solid particles and unreacted high pressure propylene gases in the reactor. Gas-phase propylene is cooled and then recycled by circulating gas compressor. After they are dried by the dryer, fed by the pneumatic handling system, polypropylene powder materials are treated by extruding granulation, homogenization in particle silo and packaging in the packing department.

2.2. Main causes of producing dust easily in polypropylene production plant

(1) Quality problem of granulation

When the melting index is high, it tends to produce the materials with tails and fiber drawing; when the melting index is low, it tends to produce broken particles caused by collision. As a result, the dust

* To whom any correspondence should be addressed.

is increased. After the reaction, if the oligomers are not treated properly, the increasing oligomers will make the melting point lower, and the particles tend to be soft and broken easily so that the dust quantity is increased.

(2) Malfunctioning of granulating process

It tends to produce materials with tails or irregular shapes in the following situation: tool breaking, template wearing, gap or rotational speed regulation.

(3) Mixing operation

In the process of internal or external mixing operation, the upper part of the mixing silo is easy to accumulate dust.

(4) Internal structure of the pipelines

When the surfaces of the pipelines are rubbed down, it tends to produce the materials with fiber drawing and broken particles so that the dust quantity is increased.

(5) Fast wind speed

If the wind speed is too fast, it tends to produce broken particles.

(6) Other causes

Emptying, filtration, and so on.

3. Analysis of electrostatic risk in the key parts of polypropylene production plant and suggestions

3.1. Mixing silo

The test of powder electrostatic mass-to-charge ratio is carried out at the entrance of the silo by faraday cup and the charge meter. The data is shown in table 1.

The data indicates, the average of powder electrostatic mass-to-charge ratio is -1.7 μC kg^{-1} (variation range: -0.4~3.24 μC kg^{-1}), which is far beyond the recommended safety values (<0.1~0.3 μC kg^{-1}).

The powder static electricity may cause the following discharging and ignition risks:

(1) It can produce conical discharging frequently at the surface of the material heap. The discharging energy ≥10 mJ, and the discharging period is about 26~71 seconds. When the concentration of combustible gas ≥0.5% wt, this conical discharging may cause flash burning or explosive accidents.

(2) When the materials approach the position of material level alarm, it is easy to produce long spark discharging [1], the discharging energy is large, and the ignition rate is high.

Suggestion: The material high level alarm should be changed into anti-electrostatic type [4]; the entrance of the silo should be equipped with electrostatic eliminator.

Table 1. Data of electrostatic test in the silo.

No.	Electric charge (μC)	Sample Mass (kg)	mass-to-charge ratio (μC kg^{-1})
1	-0.10	0.25	-0.40
2	-0.20	0.25	-0.80
3	-0.36	0.25	-1.44
4	-0.43	0.25	-1.72
5	-0.67	0.25	-2.68
6	-0.81	0.25	-3.24

3.2. Finished products silo

It also tends to produce electrostatic discharging in the finished products silo. The recent adopted protection measures include: clean wall-sticking materials regularly and so on.

3.3. Intermediate silo

The diameter of powder material silo is large, and the load of the pneumatic handling system is high. In the powder material simulation experiment, the charge of the material is up to 10-18 $\mu C\ kg^{-1}$ which can produce conical discharge frequently. If there is a tuning fork material level meter on the top of the material, it can cause long spark discharge. The discharging energy can be dozens of mJ. In the experiment, when the material charge $\geq 1.65\ \mu C\ kg^{-1}$, the potential of the material 200 mm away from the tank wall is up to 50 kV, which is far beyond the critical potential of igniting the dust (40 kV). The electrostatic safety in the silo mainly depends on the nitrogen protection system. Suggestions: monitor the oxygen content in the silo to prevent powder electrostatic explosion; increase the system safety by replacing tuning fork material level meter with the anti-electrostatic one.

3.4. Additive feeding platform

In the continuous feeding operation, the static electricity accumulated in the worker's body is 2-8 kV, and the discharging energy can be 0.4-6.4 mJ, which can cause the ignition of additive dust and electric shock. On the platform, intrinsically safe type of body electrostatic eliminator should be installed to eliminate or slow down the accumulation of static electricity in the worker's body. Suggestion: intrinsically safe type of body electrostatic eliminator should be installed on the platform to eliminate the accumulation of static electricity in the worker's body.

3.5. Vibrating screen

By multiple vibrating screen, the selected qualified materials will be sent to the downstream storage and transportation system and the unqualified materials will be sent to waste material tank to be retreated. When the vibrating screen is selecting the materials, it tends to produce dust as well as static electricity easily.

(1) When the powder material is treated, the nonmetallic material is not only the "electrostatic source" producing static electricity, but also the "hidden dangerous source" accumulating static electricity easily and producing surface discharging. For example, several flash burning happened to the DMT powder material suction machine in a company. Investigation showed that the accidental fire source is related with the electrostatic accumulation of soft connection rubber board. The similar accidents never happened after installment of an anti-electrostatic soft pipeline.

(2)The material potential in the bottom material tank >80 kV, which is far beyond the critical potential of dust ignition (40 kV)and controlling index (<10 kV) . Suggestion: take some reliable measures to solve the problem of electrostatic grounding and eliminate static electricity in this place.

3.6. Buffer hopper

The materials in the hopper may have high electrostatic charge, and produce a lot of dust. It tends to produce electrostatic discharging and flash explosion when the material is accumulated in the hopper. So it is necessary to limit strictly the height of material heap in the hopper.

3.7. Discharge and pack

In the process of discharge and packaging, it tends to produce dust and a lot of static electricity. Measures: make sure the function of the process of reaction and granulation; strengthen ventilation to reduce the dust; install electrostatic eliminator.

3.8. Bag filter

Bag filter tends to produce dust and a lot of static electricity. Measures: use anti- electrostatic bag filter; check the ventilation of the filter.

3.9. Temporary manual material receiving opening
Temporary manual material receiving opening tends to produce dust and static electricity. Measures: install powder electrostatic eliminator and body electrostatic releasing device for the temporary manual material receiving opening.

4. Risk analysis and suggestions of combustible gas in polypropylene production plant
The combustible gas may be out of control in the following situations: (1) When the reaction is not steady, there are unqualified materials or transitional material; (2) When devolatilization equipment is out of work or is out of control; (3) When something is wrong with the dryer, the concentration of the combustible gas increases. Suggestion: monitor the concentration of the combustible gas in the dryer. (4) When start or shut down the low pressure discharge tank, powder, gas and steam are discharged mixed, which tends to produce electrostatic discharge and explosive accidents. Suggestions: when we make sure the grounding of the equipment is safe, we should check and confirm $CH<0.2\%$ before the discharge. Install the electrostatic eliminator at the discharge port. (5) When receiving and discharging material are improperly operated, there will be a certain amount of combustible gas in the silo. At this moment, we should monitor the amount of combustible gas in the air conveying silo.

5. Typical case analysis of electrostatic flash explosion accident in the silo of polypropylene production plant

5.1. Brief introduction of the plant
The plant had three production lines which have a capacity of 100 thousand t/a respectively. The first two production lines, which were imported from Italy, used tubular process. The third production line used domestic tubular loop technology. All these three production lines could produce homopolymerization, random copolymerization and anti-impact copolymer products.

5.2. History of the accidents
(1) A flash explosion happened in 2D903B silo On February 12th, 2000,
(2) A flash explosion happened in 2D903E silo On September 16th, 2000.

5.3. Analysis of the accidents
(1) In both of the two flash explosion accidents, the plant was producing high melting index material with higher volatilization content.

(2) In accident of 9.16, the material level position was out of control in 2D501 steaming machine. The material level position was too low, and the devolatilization of the material was not enough. When steaming machine works properly, the volatile component of the feed is about 2~2.5% wt, and that of the discharging material <500 ppm. The concentration of the gas is about 20~50% LEL in 2D803. When tested after the accident, the concentration of the gas was up to 100% LEL in 2D803. When the production is normal, the concentration of the gas in the same products is only 57% LEL.

(3) The air-conveying PP particles tend to produce a lot of static electricity. The charge mass ratio in the feed pipes is about 0.43~2.72 $\mu C\ kg^{-1}$ (The measurement data of the charge mass ratio in the silo are seen in table 2), which is far beyond the critical value of conical discharging (0.1 $\mu C\ kg^{-1}$). When the concentration of the gas on the top of the material heap in the silo is beyond 0.5%wt, the dust may be ignited by the conical discharging on the surface of the material heap.

Table 2. Measurement data of electrification amount in the silo

2D903	Silo A	Silo B	Silo C	Silo D	Silo E
charge mass ratio, $\mu C\ kg^{-1}$	0.71	-1.12	-0.43	-2.72	-1.33

6. Conclusion

The dust electrostatic explosion accidents in polypropylene production plant are caused by the interaction of combustible gas, dust and static electricity. It is a common phenomenon in the silo that the air-conveying materials tend to produce a lot of static electricity and discharge, which is the inherent hidden danger; but whether the accident will happen mainly depends on the control of the combustible gas and the dust in the silo, which is the fuse or the inducement. A lot of factors influence the gases and the dust in the silo. Any careless mistakes and deviation appeared in any step of the process control can be the inducement of electrostatic explosion in the silo. If the plant has more inherent defects and there are more careless mistakes of the process control, it is much more likely to cause electrostatic explosion accidents in the silo. Therefore, it is our urgent task to find out the hidden dangers and defects in the production plant and define reasonable countermeasures.

References

[1] China Petrochemical Corporation (Sinopec Group) Safety and Environmental Protection Department 2007 *A Reader for Electrostat. Saf. Edu.* (China Petrochemical Press)
[2] Wang Z and Tan F 2005 *Invention patent, PR China*
[3] Liu Q Z and Tan F G 2002 *Saf. Health and Environ.(in Chinese)* **1** 9
[4] Tan F and Wang Z 2003 *Invention patent, PR China*

Journal of Physics: Conference Series **418** (2013) 012036 doi:10.1088/1742-6596/418/1/012036

Analysis on an electrostatic accident due to hydrogen gas ejection

Zhaohui Han, Renjie Lou*, Juya Shan
CNPC Research Institute of Safety & Environment Technology, Dalian, China

E-mail: lourj_dl@ petrochina.com.cn

Abstract. The article discusses a hydrogen flash explosion accident, and the mechanism of hydrogen charging and discharging; and analyses the stress condition at the anchoring points of the blowdown line. Moreover, the load-bearing capacity of the blowdown line and the maximum explosion pressure in flash explosion of hydrogen gas were calculated. It can be concluded that the electrostatic discharge of high pressure ejected hydrogen was the main reason for the accident.

1. Introduction

On January 5, 2012, the hydrogen compressor in the hydrotreating plant of a petrochemical company was in switching operation, and suddenly, hydrogen gas at a high pressure of 6.0 MPa was ejected out of the vent line with its relief valve not closed. The ejected hydrogen caused flash explosion in the space and pushed over the vent line. One operator was knocked and injured by the falling vent line and finally died although first-aid treatment was given.

There were two vent lines at the field which were located on the northern wall of the compressor building. One of the vent lines DN80 (Φ89×3.5, length: 12.3 m, weight: 91 kg) was provided for venting of the compressors Nos. 1 and 2, and the other one DN50 for venting of the connecting barrel of the compressors Nos. 1, 2, 3 and 4. The top of the vent line DN80 for the compressors Nos. 1 and 2 was 15 m above ground and its bottom 3.15 m above ground. There were 5 clamps totally for clamping the DN80 and each of them was fastened with 2 anchor bolts (Φ8 mm). The outlet of the line faced towards west and opposite to the upper platform with the inclined ladder. The vent line DN50 had the height equivalent to DN80. The top of the parapet wall, the northern wall of the building, was 15.6 m above ground. It is estimated that the top of the line DN50 and that of DN80 were 0.6 m lower than the top of the parapet wall of the building.

It was found at the field that 3 clamps were damaged and the anchor bolts for the clamp No. 1 (0.8 cm dia.) were corroded and pulled out completely. The bottom plate and anchor bolts for the clamp No. 2 were welded falsely and broken due to bad corrosion. Both anchor bolts for the clamp No. 3 were pulled and broken, as shown in figure 1.

2. Mechanism of electrostatic charging and discharging in hydrogen ejection

In theory, pure hydrogen would not cause electrostatic charging when it is ejected. However, if solid

* To whom any correspondence should be addressed.

particles, liquid drops or other foreign substances were entrapped in the gas, electrostatic charging may occur in case of high-speed impact, crashing or friction and such space charge would be built up leading to discharge, as described below: (1) If there are insulated or non-conductive objects in the space with gas ejection, such "charge collectors" would build up static leading to discharge; (2) With its dimensions and density developing over some extent, the "charge cloud" on scattering objects may cause brush discharge or lightning discharge to grounded objects around; (3)If the machine or nozzle is ungrounded or the nozzle is insulated, the machine or nozzle would cause spark discharge (to metal body) or propagating brush discharge (to insulated body).

3. Looking into the cause of electrostatic charging at the field

At the accident field, the compressor had been operated for 126s nitrogen replacement before hydrogen entered into the compressor. As nitrogen contained trace of moisture, water grains were entrapped in hydrogen after 6.0 MPa hydrogen was led in and electrostatic produced in case of high-speed impact and friction, as shown in figure 2.

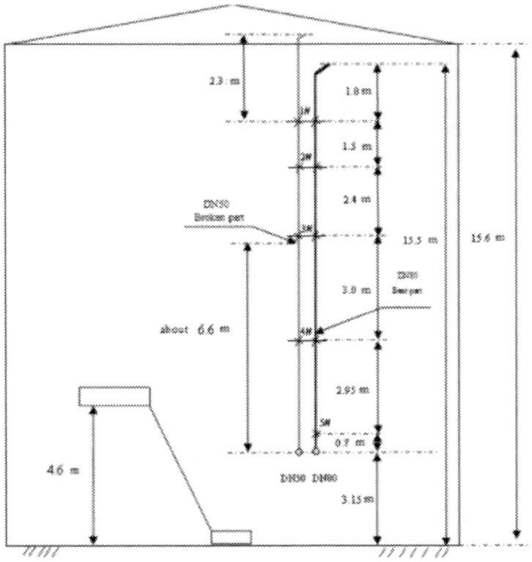

Figure1. Hydrogen vent line diagram.

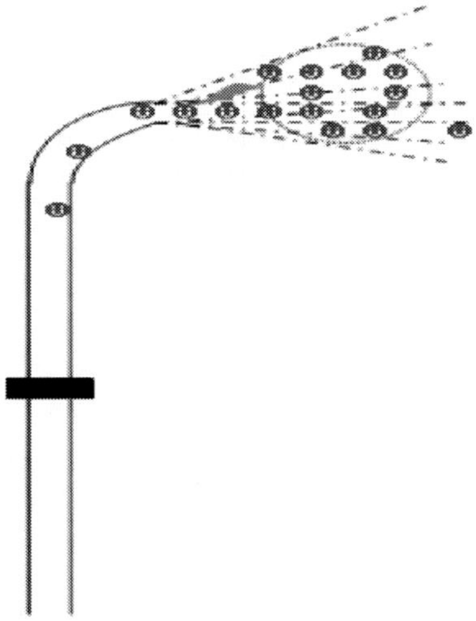

Figure 2. Diagram of electrostatic charging in hydrogen gas ejection.

Safety Guide for Static (Japan) [1] says that when its diameter is greater than 0.7 m and the average electrical field strength in space charge cloud is 1 $kVcm^{-1}$, the gas space charge cloud can cause brush discharge with the min. ignition energy of 0.1~1.0 mJ from the space charge cloud to a grounded conductor. The space charge density is calculated with the equation as follows:

$$\rho = \frac{3\varepsilon E}{a} = \frac{3\times 1\mathrm{kVcm}^{-1}\times 1.00026\times 8.85\times 10^{-12}\mathrm{Fm}^{-1}}{0.35\mathrm{m}} = 7.6\mu\mathrm{Cm}^{-3} \quad (1)$$

Where, E represented safety limit of field strength for the space charge cloud, ε represented dielectric constant of the medium and a represented a half of the diameter of ball charge cloud.

The quantity of electrostatic charge built up in the hydrogen space charge cloud with the diameter of 0.7 m is:

$$q = \rho v = 7.6\times\frac{4}{3}\times 3.14\times 0.35^3 = 1.36\mu\mathrm{C} \quad (2)$$

The grounded conductor with tip closest to the hydrogen space charge cloud at the field was the nozzle of vent line, to which the space charge cloud discharged. The hazard of discharge depended on the quantity of charge transfer in a single discharge.

Electrostatics - Code of Practice for the Avoidance of Hazards due to Static Electricity CLC/TR 50404-2000 (European Committee for Electrotechnics Standardization) [2] says that the quantity of charge transfer in discharge greater than 10 nC can ignite Class IIC substances such as hydrogen gas.

The charge quantity in the hydrogen space charge cloud with the diameter of 0.7 m at the field was 1.36 μC which is far greater than 10 nC so that the space charge cloud formed in hydrogen ejection discharged to the nozzle of the vent line causing hydrogen explosion in the space.

4. Stress analysis and calculation of load-bearing capacity of the anchoring points for vent pipeline

According to the process description, both lines were provided for venting and unused normally with low venting capacity and pressure. The clamp installation and fastening at the field showed that the vent line clamps were used mainly for clamping and supporting the lines to prevent from displacement without the function of load bearing. The vent lines DN80 and DN50 weighed 135.7 kg totally.

4.1 Calculation of the designed value of load-bearing capacity of the anchoring point for static load

For the stress condition with static loads and without alternate loads in consideration, the clamp bore permanent loads (dead weight of the structure, soil pressure, prestress and so on) according to Load Code for the Design of Building Structures (GB50009-2001) [3] and therefore it is unnecessary to consider variable loads (live load on floor, live load on roof, dust load, crane load, wind load, snow load and so on) and incidental loads (explosion force, collision force and so on).

The anchor bolts for clamp were in the stress condition of composite pull shear load according to Post-Anchoring Construction for Concrete (JGJ145—2004) [4]. The design value of load-bearing capacity of anchoring is:

$$R \geq r_A \times S = 1.1 \times 135.7 \times 1.35 = 201.6\text{kg} \tag{3}$$

where, R - the design value of load-bearing capacity of anchoring; r_A - the coefficient for importance of anchoring (taken as 1.1 as the anchoring are rated to Safety Class II); S - the design value of composite load effect for anchoring, $S = r_G S_{Gk}$ (where, r_G is the coefficient for permanent load taken as 1.35, and S_{Gk} is the standard value of permanent load.)

4.2 Calculation of the actual load-bearing capacity at a single anchoring point

It was found at the accident field that the anchor bolts for clamp were made of Φ8mm rebar. It can be considered that the load-bearing capacity at a single anchoring point means the tensile limit if they were installed and fastened properly. In general, anchor bolts are made of Class II steel (the tensile strength for Class I, II and III steels are 210, 300 and 360 N mm^{-2}, respectively).

As a result, the max load-bearing capacity at a single anchoring point is:

$$F = 300 \times (\frac{8}{2} \times \frac{8}{2} \times 3.14) \times 2 = 30144\text{N} = 3076\text{kgf} \tag{4}$$

4.3 Stress analysis on the anchoring points and calculation of the resistance to explosion force

The vent line was turned over due to the anchor bolts broken by the force of explosion applied on the vent line. The clamps Nos. 1, 2, 3, 4 and 5 were fixed onto the wall. The accident should occur such that the clamps were broken under the action of explosion force sequentially in the order of Nos. 1, 2 and 3 (It was impossible that No. 2 was destroyed and displaced with Nos. 1 and 3 fixed. In such a way, it was impossible that No. 3 was damaged before Nos. 1 and 2. Therefore, the clamp No. 1 should be damaged as the first one.)

We have a lever with the two points A and B fixed and applied with the force F1 at Point O. Calculate the stress at Point A. With Point B as the fulcrum, we have the following equation from the

Journal of Physics: Conference Series **418** (2013) 012036 doi:10.1088/1742-6596/418/1/012036

lever principle:

$$F_o \times (L_1 + L_2) = F_A \times L_2 \tag{5}$$

So, the stress at Point A is: F_1 or F_0

$$F_A = \frac{F_o \times (L_1 \times L_2)}{L_2} \tag{6}$$

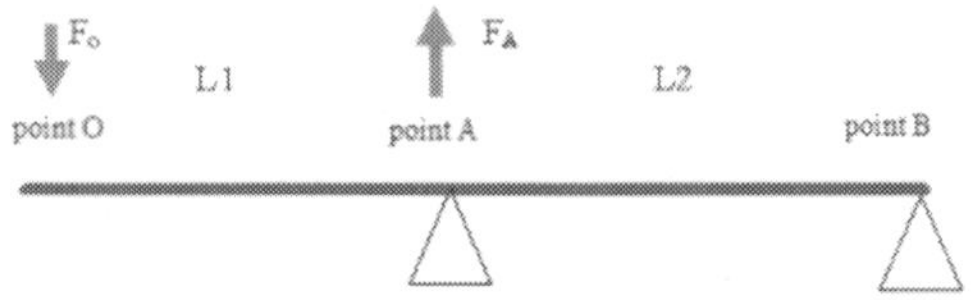

Figure 3. Lever stress diagram.

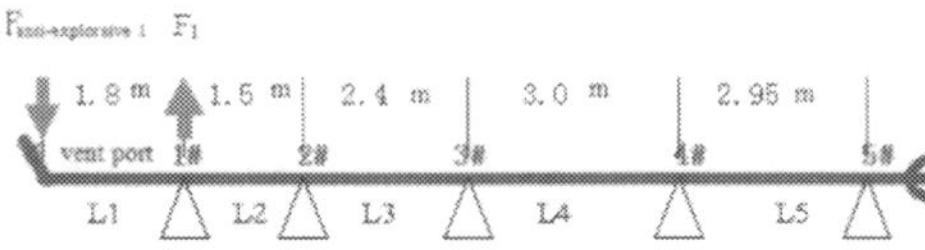

Figure 4. Stress diagram for the anchoring point No. 1.

4.3.1 Analysis on actual stress of the anchoring point No. 1 and calculation of the resistance to explosion force

According to the lever principle, the actual stress condition at the anchoring point No. 1 was the anchoring point No. 2 acting as the fulcrum. Therefore, the actual resistance to explosion force at the anchoring point No. 1 was:

$$F_{anti-explorsivel} = \frac{F_1 \times L_2}{L_1 + L_2} = \frac{30144 \times 1.5}{1.8 + 1.5} = 13702\text{N} = 1398\text{kgf} \tag{7}$$

F_1 is the max load-bearing capacity (30144 N/3076 kgf) at a single anchoring point.

4.3.2 Analysis on actual stress of the anchoring point No. 2 and calculation of the resistance to explosion force

According to the lever principle, the actual stress condition at the anchoring point No. 2 was the anchoring point No. 3 acting as the fulcrum. Therefore, the actual resistance to explosion force at the anchoring point No. 2 was:

$$F_{anti-explorsive2} = \frac{F_1 \times L_3}{L_1 + L_2 + L_3} = \frac{30144 \times 2.4}{1.8 + 1.5 + 2.4} = 12692\text{N} = 1295\text{kgf} \tag{8}$$

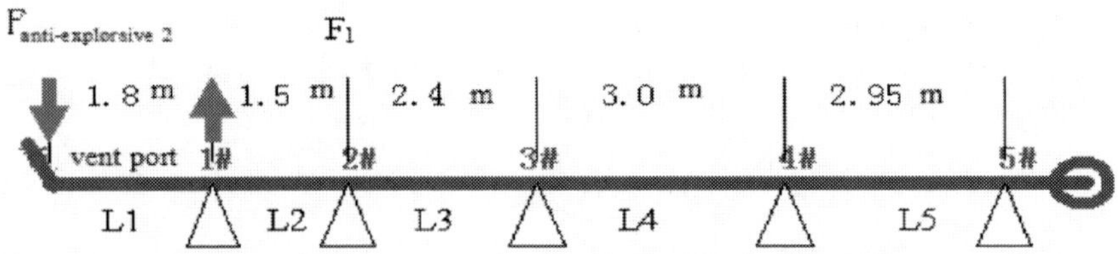

Figure 5. Stress diagram for the anchoring point No. 2.

4.3.3 Analysis on actual stress of the anchoring point No. 3 and calculation of the resistance to explosion force

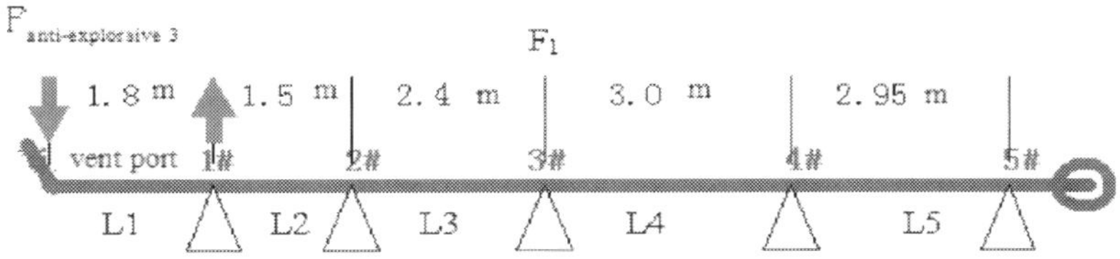

Figure 6. Stress diagram for the anchoring point No. 3.

According to the lever principle, the actual stress condition at the anchoring point No. 3 was the anchoring point No. 4 acting as the fulcrum. Therefore, the actual resistance to explosion force at the anchoring point No. 3 was:

$$F_{anti-ecploresive3} = \frac{F_1 \times L_4}{L_1 + L_2 + L_3 + L4} = \frac{30144 \times 3.0}{1.8 + 1.5 + 2.4 + 3.0} = 10394\text{N} = 1061\text{kgf} \quad (9)$$

5. Analysis and calculation of the explosion force of hydrogen on the vent line

The diameter of brush discharge from hydrogen cloud was 0.7 m. Considering that the cloud diameter was 0.7 m and the cloud had the ball shape for simplifying the calculation, the sectional area of the upper part of the gas pipe in the influence of explosive cloud should be 0.35 m×0.08 m. See Figure 2. According to the test made by the authority, the pressure of hydrogen gas (35%) was 1 atm initially, while the max. explosion pressure was 7.3 atm in explosion. The max explosion pressure went up at a rate of 2703 atm s^{-1} meaning that the hydrogen gas could reach the maximum explosion pressure in 0.027 s.

For the hydrogen cloud with initial pressure of 1 atm, the max explosion pressure was 7.3 atm in its explosion. In this case, the max thrust of explosion to the upper part of the hydrogen pipeline was:

$$F_{thrust} = PS = 7.3 \times 1.033 \times 10^4 \times 0.35 \times 0.08 = 2111\text{kgf} \quad (10)$$

It is concluded from the above analysis that the resistance to explosion force at the anchoring points Nos. 1, 2and 3 are 1398 kgf, 1295 kgf and 1061 kgf respectively. However, the field showed that both anchor bolts for the clamp No. 3 only were pulled to break, meaning that the force applied on the anchoring point No. 3 was greater than 1061 kgf in flash explosion. They should not be damaged as the composite force applied on the anchoring point Nos. 1 and 2 in flash explosion was not greater than 1050 kgf totally. However, they were damaged indeed. It was only one case that there were serious defects in the anchoring points Nos. 1 and 2 having the resistance to explosion force far lower than the designed value.

It was found in the field investigation that the anchor bolts for clamp No. 1 had been corroded and pulled out completely and the bottom plate and anchor bolt for the clamp No. 2 were welded falsely and broken due to serious corrosion. It is proved that the above analysis is correct.

6. Conclusion

(1) Ejected hydrogen at a high pressure (6.0 MPa) would form space electrostatic “charge cloud” causing brush discharge to grounded objects and ignited hydrogen in the space finally.

(2) There were defects in the clamps at the anchoring points Nos. 1 and 2 with poor quality. The anchor bolts and the bottom plate were not welded securely. The clamps were not treated for anti-corrosion properly, leading to serious corrosion and therefore pulled to break under the action of impact from flash explosion of hydrogen.

References

[1] *Safety Guide for Static* (Japan)

[2] *Electrostatics - Code of Practice for the Avoidance of Hazards due to Static Electricity* (CLC/TR 50404-2000)
[3] *Load Code for the Design of Building Structures* (GB 50009-2001)
[4] *Post-Anchoring Construction for Concrete* (GJ 145-2004)

A case study of electrostatic accidents in the process of oil-gas storage and transportation

Yuqin Hu, Diansheng Wang*, Jinyu Liu, Jianshen Gao
Key Laboratory of New Energy Physics & Materials Science in Universities of Shandong, College of Science, China University of Petroleum, Qingdao 266580, China

E-mail: dshw@upc.edu.cn

Abstract. Ninety nine electrostatic accidents were reviewed, based on information collected from published literature. All the accidents over the last 30 years occurred during the process of oil-gas storage and transportation. Statistical analysis of these accidents was performed based on the type of complex conditions where accidents occurred, type of tanks and contents, and type of accidents. It is shown that about 85% of the accidents occurred in tank farms, gas stations or petroleum refineries, and 96% of the accidents included fire or explosion. The fishbone diagram was used to summarize the effects and the causes of the effects. The results show that three major reasons were responsible for accidents, including improper operation during loading and unloading oil, poor grounding and static electricity on human bodies, which accounted for 29%, 24% and 13% of the accidents, respectively. Safety actions are suggested to help operating engineers to handle similar situations in the future.

1. Introduction

Oil and gas are flammable and explosive, respectively. A small accident occurred in the process of oil-gas storage and transportation may lead to million-dollar property loss and a few days of production interruption, and a large accident can result in lawsuits, stock devaluation, or company bankruptcy. There are many causes for the accident, an important one of which is involved in static electricity. In order to prevent electrostatic accidents and to mitigate their consequences, many strict engineering guidelines and standards for the control of static electricity in industry have been published by trade organizations and engineering societies [1]. Though most companies have followed those standards and guidelines, electrostatic accidents still occurred in oil and gas industry. Therefore, it is of great significance to learn from the history for the future safe operation of oil-gas storage and transportation.

The purpose of this paper is to categorize the causes that lead to 99 electrostatic accidents occurred in the process of oil-gas storage and transportation in last 30 years, and to reduce potential hazards associated with static electricity. The fishbone diagram [2] is used to summarize the effects and the causes that create or contribute to those electrostatic accidents.

2. Overall statistics

The 99 electrostatic accidents were reviewed on the information collected from published paper-based

* To whom any correspondence should be addressed.

literature [3-11] and web reports in this paper, which occurred in the process of oil-gas storage and transportation over last 30 years. As indicated in table 1, accidents occurred more frequently at gas stations with 45 cases (45.4%).The second most frequently involved place was tank farms (28 cases, 28.3%). About 27.3% of accidents happened in petroleum refineries (11.1%), petrochemical plants (6.1%), oil fields (1.0%), and other types of industrial facilities (8.1%) such as gas plants, pipelines. The atmospheric external floating roof tank was the most frequent type and the inner floating roof storage tank was the second most frequent type as could be seen in table 2. Both types were used extensively for the storage of crude oil and oil products such as gasoline, diesel, fuel oil, etc. Table 3 showed that fire was the most frequent type of loss with 51 cases, and explosion was the second most frequent type of loss with 44 cases. In other words, fire and explosion together accounted for 96% of total cases.

Table 1. Type of complex where accidents occurred.

Year	Refinery	Tank farm	Chemical plant	Oil field	Gas station	Misc	Total
1980-1989	7	15	1	1	9	1	34
1990-1999	3	2	2	0	14	4	25
2000-2009	1	7	2	0	21	3	34
2010-2012	0	4	1	0	1	0	6
Total	11	28	6	1	45	8	99

Table 2. Type of tanks and contents.

Content	External floating top	Cone top	Sphere	Cone roof internal floating top	Refrigerated tank	Submerged tank	Fiber glass	Subtotal
Crude oil	4	1	0	11	0	5	2	23
Oil products	10	3	0	4	0	4	1	22
Gasoline	13	0	0	3	0	0	0	16
LGP	0	0	7	0	1	0	0	8
Subtotal	27	4	7	18	1	9	3	69

Table 3. Type of accidents.

Year	Fire	Explosion	Spill	Misc	Total
1980-1989	18	14	1	1	34
1990-1999	14	10	1	0	25
2000-2009	16	17	1	0	34
2010-2012	3	3	0	0	6
Total	51	44	3	1	99

3. Causes of accidents

As indicated in table 4, improper operation during loading and unloading oil was the most frequent cause of accidents, and poor grounding and static electricity on human bodys were the second and the third most frequent causes. The rest was the materials of oil and gas containers, different mix of oils, storage tank fault, etc. To illustrate causes and effects, a fishbone diagram is shown in figure 1.

Table 4. Cause of electrostatic accidents.

Year	1980-1989	1990-1999	2000-2009	2010-2012	Total
Improper operation during loading and unloading oil	7	6	16	0	29
Poor grounding	6	8	10	0	24
Static electricity on human bodys	5	2	5	1	13
Materials of oil and gas containers	5	5	2	0	12
Different mix of oils	3	3	3	1	10
Storage tank fault	3	1	2	0	6
Misc	3	1	1	0	5
Total	32	26	39	2	99

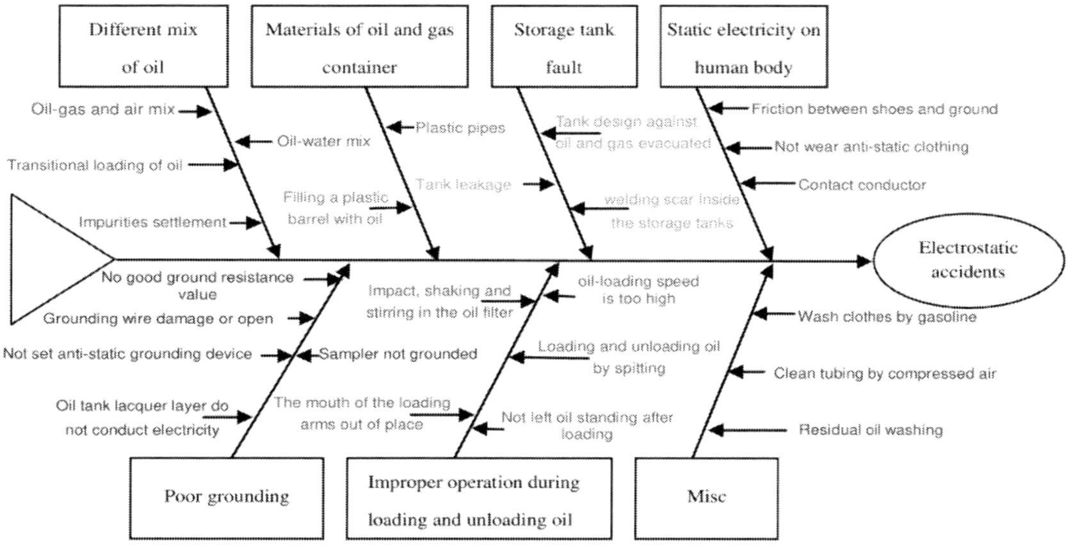

Figure 1. Fishbone diagram of accident causes.

3.1. Improper operation during loading and unloading oil

29 electrostatic accidents were caused by improper operation during loading and unloading oil. Among the 29 cases, loading and unloading oil by spitting is responsible for 15 accidents. In general, loading and unloading oil by spitting will make the liquid that ejects from the nozzle split into many small droplets. When the droplets are in contact with air, an electronic rearrangement occurs to minimize the energy at the interface. Since this process is generally not reversible, charge separation occurs when contact is lost between the surfaces. If the interface is disrupted at a rate faster than equilibrium conditions, additional charge separation will occur. Moreover, the position of the pipeline export will also affect the generation and accumulation of electrostatic charge when oil flows into the tanks or other containers. For example, on July 22, 1996 in a petroleum chemical plant, a fire happened and spread to other areas when a worker begun loading with putting the crane tube into the tank with good static grounding. Survey data showed that the vertical height from the mouth of crane tube to the bottom of tank was 1.19 m. Huge amounts of electrostatic charge was built up by the impact and splash during the oil falling, and then discharge detonated the oil vapor.

Too fast flow rate of oil is the second frequent cause in this category. Electrostatic hazards arise during oil and gas transporting when charge separation occurs and leads to an accumulation of one sign of charge within some defined boundary, such as inside a pipeline. When oil flow velocity is too fast, there also can be significant charge accumulation, even to the point where incendive discharge can occur. In 1982, Venezuela in the northern part of Latin America, the Caribbean coast, a group of tanks blew up to an oil fire while unloading crude oil to storage tanks. The fire lasted for 4 days. There were 145 workers died and more than 500 people injured in this incident. The investigation proved that this accident was caused by the discharge of static spark resulted from too fast flow rate of oil. In consideration of that, the faster liquid flow velocity is, the easier static electricity produces, the oil flow must keep within the speed limit. For example, if the pipe is less than 150 mm in diameter, the flow velocity of gasoline, kerosene and other light oil should be restricted below 3 to 4 ms^{-1}, and that of the heavy oil should be restricted below $7ms^{-1}$.

3.2. Poor grounding

There are two major causes of poor grounding related electrostatic accidents. The first one is no installation of static grounding wire, and the second is damage or open of grounding wire. Among 24 poor grounding accidents, some storage tanks are directly ignited by the discharge of static electricity. An explosion to a tank car containing gasoline on October 2, 2000 in Hubei resulted in 4 casualties and property damages of 3 million dollars. Investigations showed that this gas station was in violation of the norms of anti-static operation, and not installed static grounding device as required. In 1997 a accident happened during cargo oil transferring operations in Yangtze River, a tanker and three barges destroyed by fire, and nearly ten thousand tons of crude oil spilled. The cause of accident was that the rope of sampler did not conduct electricity, so the static charges gathered on the rope.

3.3. Static electricity on human bodys

The clothing of the staff is the most frequent cause in this category. Among the 13 cases, no use of anti-static clothing is responsible for 10 accidents. An accident in 1987 at a cave oil depot, in Liaoning of China, was caused by the ignition of static electricity on human bodys during instrument maintenance. 8 workers were killed, and 12 workers were injured in this explosion. In 2002, an explosion occurred while a worker wearing synthetic clothes was refueling a car in a gas station. The electrostatic discharge generated by human bodys is one of the important reasons for static ignition hazards, so operating personnel must wear anti-static clothing and must touch grounding body to remove static electricity when stepping into a potentially dangerous place with regard to explosions.

3.4. Materials of oil and gas containers

The cause of 12 out of 99 electrostatic accidents was the inappropriate materials of oil and gas containers. Loading oil with plastic barrel is the most frequent cause in this category. In 1999 an accident, a fire broke out at a gas station when a refueling worker was filling a plastic bucket with gasoline. A Jeep car, two computer tankers and many other auxiliary facilities burned down in the accident.

3.5. Different mix of oils

Transitional loading of oil and mixture of oil-gas and air are responsible for 40% and 30% of this category, respectively. Impurities settlement and oil-water mixture are in total responsible for the remaining 30%. When the oil contains water or different oil or gas with air, the amount of static electricity generated will increase [12]. The experiment proved that if water content in oil is 5%, it can make the electrostatic effects increase 10-50 times. In 1987, when the diesel fuel was filled into an empty gasoline tanker in an oil loading station, the tanker burst into flames. The accident survey indicated that the fire was caused by static electricity during the tanker transitional loading.

4. Conclusion

The information of 99 electrostatic accidents occurred in the process of oil-gas storage and transportation in last 30 years was reviewed. Based on this statistical data, this paper has expressed the causes and the contributing failures that led to these accidents with a fishbone diagram. Static electricity is a real and important cause of fires and explosions associated with oil and gas operations. Therefore, electrostatic hazards must be given proper consideration and effectively guarded against in the storage and transportation process of oil and gas. Thus, most of those electrostatic accidents could be avoided if good engineering in operation was practiced and safety management program was implemented effectively.

References

[1] Chang J I and Lin C 2006 *J. Loss. Prev. Proc.* **19** 51
[2] Ishikawa K and Lu D J 1985 *What Is Total Qual. Control* (Englewood Cliffs: Prentice-Hall Inc)
[3] Feng K, Wang F and Tang Q 2005 *Anal. and Prev. of Gas Station Accid.* (Beijing: China Petro-Chemical Publishing House)
[4] Fan J 2005 *Anal. of One Thousand Cases of Accid. Happened in Oil Depot* (Beijing: China Petro-Chemical Publishing House)
[5] Britton L G 1999 *Avoiding Static Ignition Hazard. in Chem. Oper.* (New York: Wiley-AIChE)
[6] Pratt T H 2000 *Electrostat. Ignitions of Fires and Explosions* (New York: Wiley-AIChE)
[7] Phil C, Mike J and Ben H 2007 *J. Hazard. Mater.* **140** 429
[8] Walmsley H L and Mills J S 1992 *J. Electrostat.* **28** 61
[9] Park A A and Fan L 2007 *Chem. Eng. Sci.* **62** 371
[10] Martin G 2005 *J. Electrostat.* **63** 447
[11] Pavey I D 2004 *Process. Saf. Environ.* **82** 132
[12] Li X M and Chen W B 2004 *OGST(Oil & Gas Storage and Transportation)* **23** 50

A simulation method for lightning surge response of switching power

Ming Wei*, **Xiang Chen**
Institute of Electrostatic & Electromagnetic Protection, Shijiazhuang Mechanical Engineering College, Number 97 of Peace west road, Shijiazhuang, China

E-mail: weiming1968@sohu.com

Abstract. In order to meet the need of protection design for lighting surge, a prediction method of lightning electromagnetic pulse (LEMP) response which is based on system identification is presented. Experiments of switching power's surge injection were conducted, and the input and output data were sampled, de-noised and de-trended. In addition, the model of energy coupling transfer function was obtained by system identification method. Simulation results show that the system identification method can predict the surge response of linear circuit well. The method proposed in the paper provided a convenient and effective technology for simulation of lightning effect.

1. Introduction

Surge poses an important LEMP (lightning electromagnetic pulse) effect on electric devices. According to the accident statistic data, 70% surge interferences come from power lines [1-2]. Effect of lightning is mainly presented in surge injection to power circuit. In order to promote the LEMP immunity of electric devices, it is very important to research on the lightning surge energy coupling to power.

LEMP energy coupling is the key for system level effect simulation. It can be divided into mechanism analysis method and data statistic method. Computation in the mechanism analysis method is complex, while data statistic method obtains energy coupling model from input and output data of a black box. Data statistic method includes many realized methods such as probability statistic method, spectrum estimating method, system identification method. As a comparatively mature method, system identification has been widely used in linear system modelling. The principle is to take the energy coupling way as a black box. Model parameters can be identified from experimental data in order to get energy transfer function model.

2. System identification

L. A. Zadeh defined the identification as follows: Identification is to choose a model equal to real system from a set of models based on input and output data. Identification contains experiment design, model structure identification, parameter identification, model verification [3]. Figure 1 shows a system identification procedure.

* To whom any correspondence should be addressed.

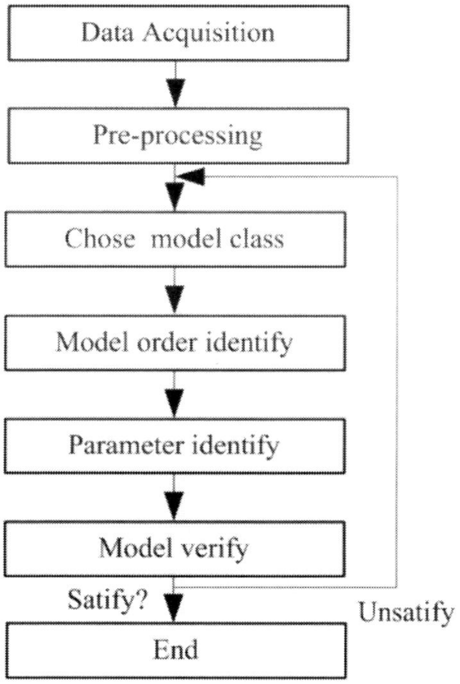

Figure 1. Block diagram of system identification.

Model structure identification includes model class choice and model structure parameter confirmation. MATLAB system identification tool-box has provided several models such as ARX model, ARMAX model, BJ model, etc [4]. Function of ARX model can be expressed as follows:

$$A(q)y(t) = q^{-nk} B(q)u(t) + e(t) \tag{1}$$

$$A(q) = 1 + a_1 q^{-1} + a_2 q^{-2} + \ldots + a_{na} q^{-na} \tag{2}$$

$$B(q) = b_1 + b_2 q^{-1} + \ldots + b_{nb} q^{-nb+1} \tag{3}$$

In equations (1)-(3), *na* and *nb* are the polynomial orders, *nk* is the pure delay time, *e*(*t*) is the zero mean noise.

The estimating criterion for unknown system includes residual error quadrum sum criterion, check criterion, final error prediction criterion and AIC (Akaika Information Criterion) [4]. AIC is a model order confirmation criterion which has an objective standard. Its order confirmation criterion is as follows:

$$AIC(n_\theta) = N \ln \hat{\sigma}_\varepsilon^2 + 2n_\theta \tag{4}$$

n_θ : Number of n order known parameters.

$\hat{\theta}$: Maximum likelihood estimating value.

$L(\hat{\theta}, n)$: Likelihood estimating function which reflects the accuracy. Akaika proved that $\hat{n}$ is a system order when $AIC(n_\theta)$ is the minimum.

3. Experiment design

Lightning surge injection test has been designed according to GB/T 17626.5-1999(surge immunity test) [5].Surge generator is a LSG-8015 made by Shanghai SANKI. Oscilloscope used in experiments is a Tektronix TDS7404B whose maximum sampling rate is 20 GSa/s (testing band width 4 GHz). EUT (Equipment under test, EUT) is a regulated power supply. The testing equipments for lightning surge injection are showed in figure 2.

Injection and response waveforms of EUT are collected by a digital oscilloscope. Three grades of surge peak voltages (600V, 700V, 800V) are chosen in order to avoid EUT works in the nonlinear region. Figure 3 shows the injection surge waveforms and EUT response voltage waveforms.

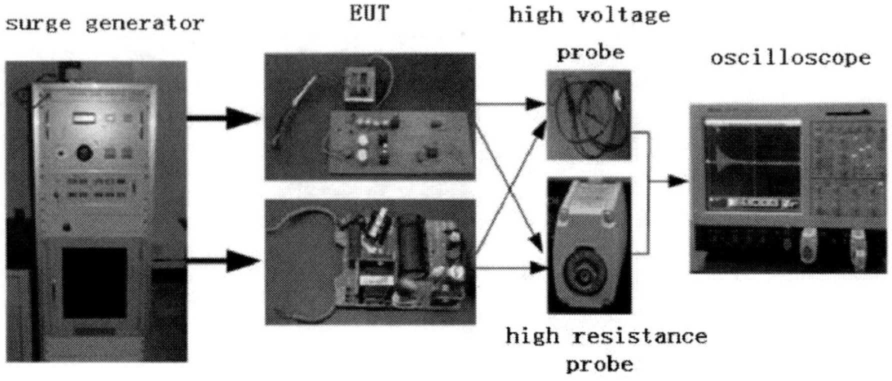

Figure 2. Lightning surge injection test.

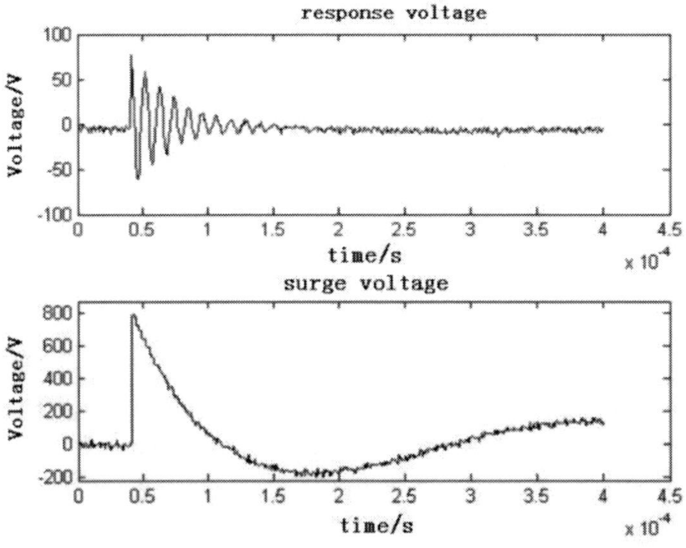

Figure 3. Waveforms of 800V injection test.

4. Data pre-process

Since noise burred in experimental data can affect the accuracy of model identification, it must be reduced before modelling. Wavelet can analyse signal in time domain, frequency domain and separate accident component and noise. Donoho-Johnstone shrinking de-noising method is used by db series of wavelet function decomposition [7-8]. It has been concluded from repeated tries that db3 wavelet function for surge signal and db9 for response signal can get a better result.

Identification demands that the mean value of data is zero and the statistic speciality has no relationship with start time. So data smoothing is mandatory to remove the tendency. In order to reduce the model order and increase the model accuracy, only middle frequency component switch has been excited thoroughly. Figure 4 is the waveforms after pre-processing. The peak and accident part of signals has been reserved well.

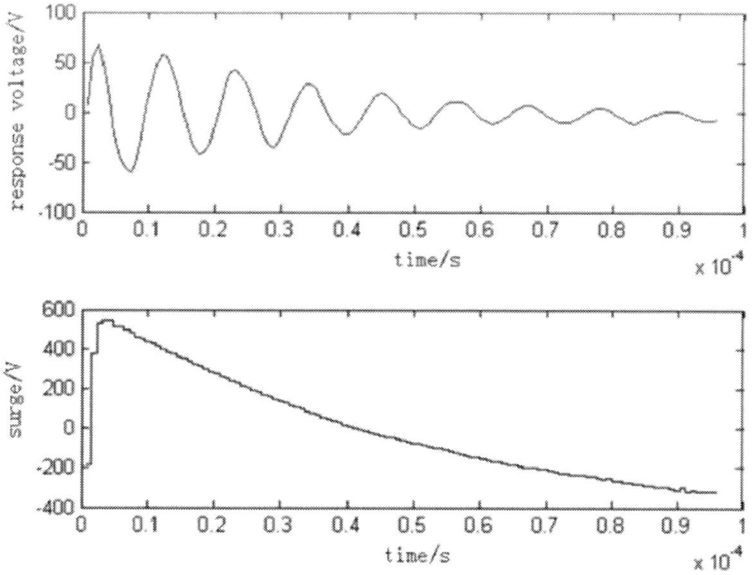

Figure 4. Waveform after pre-processing.

5. Modelling

In fact, a majority of systems are essentially non-linear. Linear model is only proximate in some characteristics. The energy between power circuit and lightning surge is assumed to be linear. Surge voltage is taken as the input signal of a black box. Response voltage is the output signal of a black box. Model is identified via ARX model based on the least square method. Experimental data of 800V surge is used for identification. Data of 600V and 700V are used to verify the model.

5.1Model identification

The order of model should be chosen as low as possible if the accuracy is satisfied. Program is written with MATLAB. Polynomial model can be identified as follows:

$$A(q) = 1 - 1.809q^{-1} + 1.816q^{-2} - 1.483q^{-3} + 0.8928q^{-4} - 0.0911q^{-5} \tag{5}$$

$$B(q) = 0.009799q^{-5} - 0.0067q^{-6} \tag{6}$$

Loss coefficient is 0.577032. Akaike final predictive error is 0.644352. Transfer function can be written as equation (7) if noise item $e(t)$ is ignored.

$$H(q\ ) = \frac{B(q)}{A(q)} \tag{7}$$

Figure 5 is a comparison of the model output and the real output with 800V surge injection data. The GOF (Goodness of Fit, GOF) is 90.12%.

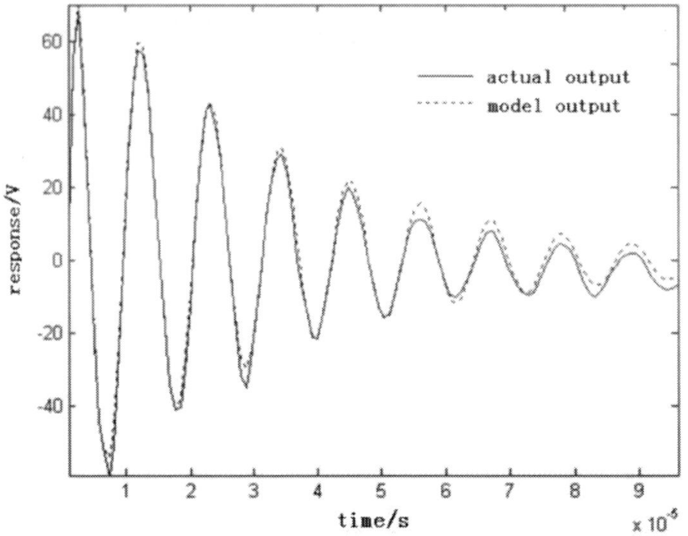

Figure 5. Comparison of model output and real output.

5.2 Model verifying

Model verification is necessary to system identification. In order to verify the generalization ability of model, estimating result is compared with the real output with a series of surge injection levels. It can be seen from figure 6 and figure 7 that the simulating results approximately match to the real output waveforms. The change trend is same. But the end part of waveforms is different.

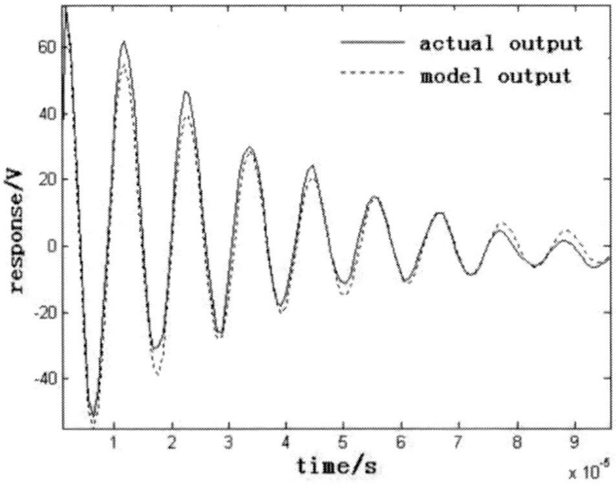

Figure 6. Model output compares with real output of 600V surge injection.

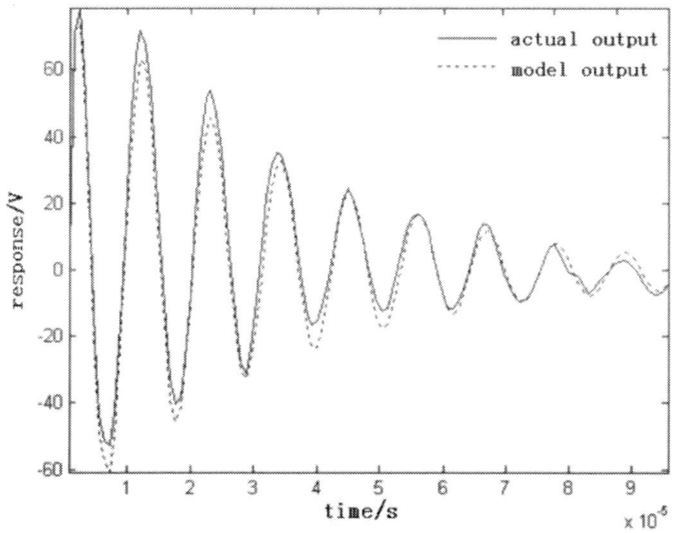

Figure 7. Comparison of model output and real output Model output compares with real output of 700V surge injection.

6. Conclusion

Modelling for lightning surge coupling based on system identification has been researched in this paper. It proves that the system identification is an efficient modelling and forecasting method for lightning surge simulation.

This modelling method is applicable to the linear system. When the surge level is high, the EUT may work in a nonlinear area where the coupling will be more complex. So a further study such as neural network will be needed to get nonlinear model.

Acknowledgements

This project has been supported by Natural Science Foundation of China (50877079) and National Key Lab Foundation (9140C8703021IJB34).

References

[1] Li F and Chen B 1999 *Transaction of Shanghai Railway University (in Chinese)* **20** 25
[2] Zhou B, Chen B and Shi L 2003 *EMP and Engineering Protection* (Beijing: National Defense Industry Press)
[3] Xiong R, Huang J and Deng W 2003 *Asia-Pacific Conference on Environmental Electromagnetic (Dalian)* (Amsterdam: North-Holland/American Elsevier) p 182
[4] Li Y and Zhang K 2003 *System Identification Theory and Application* (Beijing: National Defense Industry Press)
[5] Wei M 2012 *LEMP and Protection* (Beijing: National Defense Industry Press)

Lunar electrostatic effects and protection

Yongwei Sun*, Qingyun Yuan, and Jiuliang Xiong
Institute of Electrostatic & Electromagnetic Protection, Mechanical Engineering College, No 97 Heping West Road, 050003 Shijiazhuang, Hebei Province, China

E-mail: syw3369@sina.com

Abstract. The space environment and features on the moon surface are factors in strong electrostatic electrification. Static electricity will be produced in upon friction between lunar soil and detectors or astronauts on the lunar surface. Lunar electrostatic environment effects from lunar exploration equipment are very harmful. Lunar dust with electrostatic charge may enter the equipment or even cover the instruments. It can affect the normal performance of moon detectors. Owing to the huge environmental differences between the moon and the earth, the electrostatic protection technology on the earth can not be applied. In this paper, we review the electrostatic characteristics of lunar dust, its effects on aerospace equipment and moon static elimination technologies. It was concluded that the effect of charged lunar dust on detectors and astronauts should be completely researched as soon as possible.

1. Introduction

All of the moon detector activities, such as exploring the appearance of the moon, investigating the lunar resource material, detecting the moon's space environment and deepening the knowledge of the lunar soil, will face one of the major challenges that is the impact of lunar dust. Lunar surface is covered by a layer of lunar soil which contains a variety of lunar rock and mineral debris. The lunar soil records the lunar surface's impacted history and the solar activity. It is the main sources of studying the lunar resources, material composition and the evolution of the moon. And it is also one of the most important research content on lunar exploration. The lunar soil is a floating layer which consisted of a large number of fine particles, and it is about a few meters thick. It is the fission products by the rock and lunar crust in the role of meteorite impact, mainly composed by solidification or glass fragments and fragments. Below 100 microns, it is usually called lunar dust. The average diameter of the lunar dust is 20 microns, but the majority is less than 5 microns. The lunar dust has a high specific surface area and insulation. Due to the space environmental effects, the lunar dust is easy to get static electricity [1-3]. It is also static floating and has strong adsorption capacity due to the interaction among electrostatic lunar dust electrostatic. Floating lunar dust will block the probe line of sight, adsorb on the surface of the detection equipment or enter the equipment contained in the lunar probe, and even affect the working state of the moon detector optical system, power systems, thermal control system, and even the astronaut system, resulting in blurred vision, reading errors, sealing failure, materials wearing, and reducing the power efficiency of the system, astronauts inhalation and

* To whom any correspondence should be addressed.

allergy problems [4-6]. For example, a mirror of the "measurement on the 3rd (Surveyor3) lander dust pollution, 25 % of the area of the camera filters lunar dust pollution, resulting in decreasing the amount of light to reduce the image contrast. If the moon detector solar system is covered with lunar dust or leads to electrostatic discharge with electrostatic role of lunar dust, it will cause the performance degradation or the function loss, and most likely to make the detection task fail.

2. Electrostatic properties of the lunar dust

Dust particles usually have electrostatic. There are generally three factors accounting for the phenomenon: firstly, when the Moon goes through the Earth's magnetosphere, charged particles would generate static electricity; secondly, when solar storms occur, a high-speed stream of particles is formed by a large number of charged particles. The ray will be launched to the moon, the electrostatic charge of the lunar surface will be gathered to the highest when the solar activity reaches the highest peak; thirdly, when the moon detectors move on the lunar surface, the friction of the lunar surface will generate static electricity. The moon charges positively in the day, while the moon charges negatively in the night, which caused by solar ultraviolet and X-ray photoelectron charge dominant.

In 1968, NASA's Surveyor 7 (Surveyor7) lander photographed a lunar horizon after dark issued a strange light, which is shown in figure 1. A static dust particles floating on the surface of the moon, and the light is scattered when the sun light goes across the charged lunar dust surface. The phenomenon is consistent with the data sent back by NASA's Lunar Prospector "(Lunar Prospector) with this phenomenon lunar Prospector moon running around in 1998-1999. During the period of the moon and the Earth's magnetic tail phase cross, lunar Prospector recorded the significant voltage changes in the lunar surface.

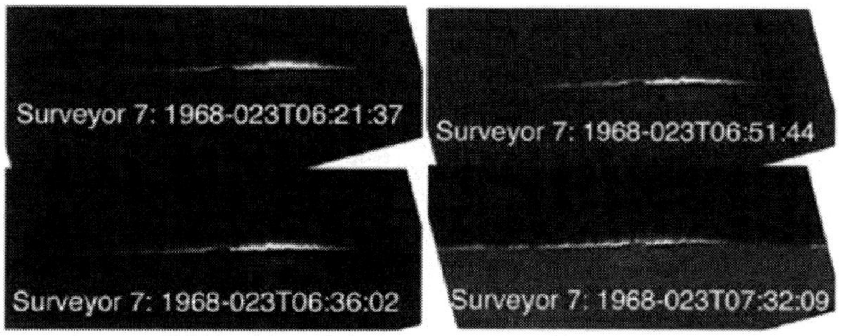

Figure 1. The moon dust scattering of Surveyor7.

Tim-Stubbs, working in Goddard Space Flight Centre, found that the earth's magnetic tail extend far, even cross the earth's orbit. The moon passes through the earth's magnetic tail once a month (full moon), forming a series of phenomena, from the lunar 'dust storms' to electrostatic discharge (Figure 2). The moon moves through the plasma which formed by electric particles during a full moon. The moon charged negative because of the lightest and the most mobile particles, which spreads over the surface of the moon. Positive side of the moon, this effect is offset by sunlight——the photon to hit the electronic bouncing back from the lunar surface, releasing the negative charge. On the other hand, in the back side of the moon, the effect will make the negative potential up to several kilovolts. After analysed the "lunar probe" probe's returned data and the lunar lander photographs for the period 1998-1999, Jasper Halekas, from University of California, found that the charge of the lunar surface is up to 4500 volts. These extreme differences in charge may make the negatively charged dust from the night side of the moon fly to the smooth side, and the strongest particle stream will be formed in the region of the sun rising or falling. The conductivity of the moon surface is low, so the lunar dust particles will maintain charged.

Journal of Physics: Conference Series **418** (2013) 012039 doi:10.1088/1742-6596/418/1/012039

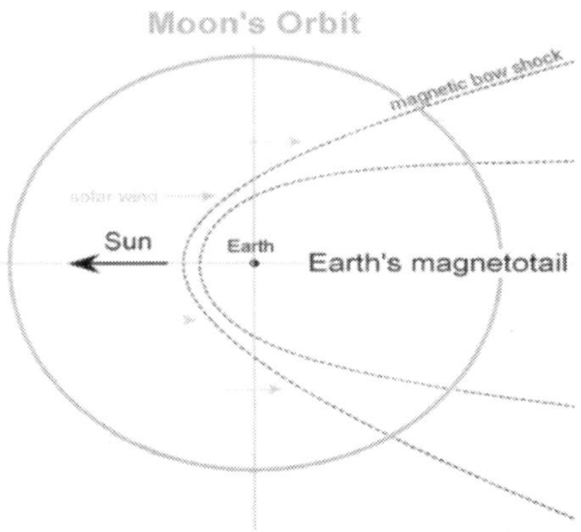

Figure 2. The moon through the Earth's magnetosphere in the full moon.

On the moon, except for the space environment, the environment of triboelectrification is also harsh to make the lunar soil charged. The dry environment of the lunar surface and the low conductivity make the electrical properties of lunar soil to be an insulator. When the moon probe or astronauts land on the lunar surface, lunar soil and the moon probe, as well as the most parts of the space suits material (such as aluminized polyester film, the diene coated nylon, polyester, polyurethane coated nylon, warp knitting fabrics, as well as stainless steel) will produce static electricity in the process of friction with grit and dust. As the lunar soil is insulated, there is no path leading to the ground. The lunar probe and spacesuits surface would accumulate huge amounts of electrostatic charges. Charged lunar dust has a strong adsorption friction which can easily adsorbs on any surface. It can enter into the equipment contained in the lunar rover, no matter the lander, rover or the suit, even in the aluminium surface and the coating surface. It would be no use to remove the lunar dust, even if surface drier, vibration and wiping were used. Large numbers of charged lunar dust may affect the normal working state of the circuit system, thermal control systems, and solar energy power supply system. All of these impacts would lead to communication failure, discharge breakdown, solar panels covering, and other electronic equipments failure and even form an irreversible damage. For example, the moon has no atmosphere, and the temperature varies greatly between days and nights. If the probe's thermal control system is impacted by the lunar dust, and the dust adsorbed on the surface would increase the absorption of radiation and the visible light, which would lead to the radiator deterioration. Studies found that the impact of the lunar dust on absorption ratio is nonlinear, 11 % of the lunar dust coverage area will lead to doubling solar energy absorption ratio.

Static lunar dust would also harmful to spacesuit and human health either. Since the Apollo 11 lunar mission, most of the astronauts considered that lunar dust is one of the main problems encountered after landing on the lunar surface. December 13, 1972, the Apollo 17 spacecraft Astronauts Eugene Cernan and Harrison Schmitt made an investigation of the Taurus-Littrow valley on the moon surface. The spacesuit surface covered with gray dust, and there were also numerous scratches appeared on the goggles. The dust also crept into the suit though the mechanical joints, and even the lunar module was covered with fine dust. After walking on the moon, Schmidt breathed some dust, appeared similar allergic reaction; Cernan reported that his eyes and throat were irritated by static lunar dust. The American space suits manufacturers Hamilton Standard and ILC-Dover engineers believed that the spacesuit providing for "Apollo" missions would make its wearer in serious danger because of these dust. Preliminary studies have shown that astronauts may cause health hazards, including iron poisoning through inhalation of lunar dust. For example, Taylor, a member of the NASA lunar air dust toxicity advisory group, carried out the study of lunar dust's influence on the respiratory system, and planed to develop the standard of touching lunar dust in 2010.

3. Lunar electrostatic environment simulation and static electricity protection research

The moon's atmosphere is very thin, about 12 to 14 orders of magnitude smaller than the density of the Earth's. The atmosphere of the moon's surface is made of neon, hydrogen, helium and argon. Neon, hydrogen, and most of the helium is brought from the solar wind, about 10 % of the helium is produced by the heavy nuclear radioactive decay of the moon itself. Because there is no atmosphere on the moon, the temperature of the lunar surface is determined by the sun directly, ranging from -150 ℃ to 130 ℃, thus high requirements on the efficiency of the lunar probe thermal control system is proposed. The moon rotates a cycle for 27.322 days; a night on the moon is as long as 14 days on the earth. If detect the moon on the moonlit night, the lunar probe has a great demand for solar power system. As the major equipment of the observation of the lunar environment, the lunar probe high-precision optical detection system must always remain in good condition, these equipments are most likely to be effected by the electrostatic lunar. There is no evidence for the existence of the water on the moon, so the grounding system for neutralizing the moon electrostatic would not work. The mature technology of static electricity protection on Earth can't be applied on the moon, because of the different environment between the Moon and the Earth. The lunar probe is completely exposed in a multi kind of cosmic rays, the acceleration of gravity on the moon is about 1.62 m s^{-2} , which is one sixth of the Earth's. These conditions limit the development of the lunar electrostatic protection technology significantly, and Moon static elimination technology is a new technology.

NASA has been very concerned about the effects of static electricity, the lunar dust on astronauts and the moon detectors. A large number of lunar electrostatic environment simulations and researches about the elimination technical are done. As part of the plan, NASA Glenn Research Center developed lunar dust simulation test equipment, which includes lunar dust, vacuum, temperature, solar electromagnetic irradiation and other environmental factors. Lunar dust simulation is one of the most important parts of the equipment. In order to prevent the expansion of the simulated lunar dust explosion, the pumping speed of the vacuum pump should be reasonably controlled; the lunar dust should be stirred constantly at the same time. The degree of vacuum should be achieved 10^{-6} Pa, and before the experiment, more than three times of temperature cycling treatment in the range of -150-200 ℃ should be done until the residual gas analyzer measurement data remained stable. For the interaction of dust particles with a payload in the vacuum environment, which will be inspected in the following ways: in a temperature-controlled vacuum experimental area, near-ultraviolet irradiation could be produced with the mercury-xenon lamp while far-ultraviolet irradiation could be produced by deuterium lamp. The analogy inputs are from the laboratory and the theoretical analysis and observation of the lunar surface. NASA also built a laboratory, which would be used to simulate the voltage change caused by the interaction of solar radiation and plasma. From these, we can understand how to control the surface voltage to resist or attract month dust, thereby reducing or eliminating the effect of static electricity lunar dust on the moon aerospace equipment. For the electrostatic problem, NASA Glenn Centre studied in the Pathfinder space vehicle. They made a tip in the bottom of the antenna, which is made of sharpened tungsten, one inch long and 2.5 microns in diameter. The tip can produce a strong field, and charges gathered on the device which is released into the thin air. Landis, university of California, believed that we can use a extremely thin air on the moon as a common electrode and put a small radioactive source connected to the space suit and the ground, the low-energy alpha particles will fly into the thin atmosphere and impact the gas molecules, then ionize them. The gas around the moon equipment and the astronauts will become a conductor, so any extra charge will be counteracted. NASA plans to build an ion gun, laser technology will be used to regulate the voltage of the conductive surface. This self-cleaning device is a combination of devices based on electronic or ion beams, which is used for cleaning the lunar dust. The quality and power consumption are limited to a few kilograms and a few watts level, respectively. The application of electric curtain to eliminate the lunar dust is also actively developing. In order to prevent electrostatic breakdown, transparent parallel electrodes are covered by a polyester film of electric curtain, which are connected to the multi-phase or single AC power, forming an alternating electric field. Lunar dust deposited on the electrode will be lifted and moved along the direction perpendicular to the electrode axis under the electric field, so as to achieve the purpose of lunar dust removal. This method has been used to prototype demonstration; removal efficiency can reach to 80-90 %. But there are lots of further researches should be done in the future, such as optimum conditions chosen including AC voltage, frequency, electrode spacing and electrode width.

The lunar dust seminar convened by NASA in 2007, discussed four issues of the lunar dust: the basic theory of the lunar dust, the effect of lunar dust on the mechanical systems, the impact of lunar dust on medicine/ health and the life support system. These studies cover the lunar plasma environmental studies, the lunar dust data, and the existing analogue lunar soil analysis to see if it is sufficient to describe the characteristics of the real lunar dust. According to the Apollo mission's knowledge, the lunar dust seminar studied the lunar dust reduction and cancellation technology, designed a mechanical device in the lunar dust environment guide, and developed the correct simulated lunar soil to design, and tested the performance of the lunar dust elimination devices. The standard testing techniques and improved technology were also used to study the sensitivity of the mechanical systems on the lunar dust.

4. Conclusion

To the long-term development of China's lunar exploration mission, the impact of lunar dust is a problem that can not be ignored. So the exploration of the moon must be carried on the basis of fully studying the problem. Considering the status and trends of the technology development of China's lunar exploration project, the effect of charged lunar dust on the detector and the astronauts should be completely researched as soon as possible.

Acknowledgments

This work was supported by National Natural Science Foundation of China under Grant 51177173.

References

[1] Wang J, He X M and Cao Y 2008 *Proc. Int. Conf. on Plasma Science* vol 36 p 2459
[2] Hine B, Spremo S, Turner M and Caffrey R 2010 *IEE Aerospace Conference* p 9
[3] Pirich B, Weir J, Leyble D, Chu S and DiGiuseppe M 2010 *IEE Long Island Applications and Technology Conference* p 6
[4] El-Shlbokshy M S, Mujahid A and Zakzouk A K M 1985 *IEE PROCEEDINGS I Communications Speech and Vision* vol 127 p5
[5] Hubbard G S, Binder A B and Feldman W 1998 *Proc. Int. Conf. on Nuclear Science* vol 45 p 880
[6] Wei E, Ren X D. 2010 *Proc. Int. Conf. on Multimedia Technology* p 3

The electrostatic properties of Fiber-Reinforced-Plastics double wall underground storage gasoline tanks

Yipeng Li*, **Quanzhen Liu, He Meng, Lifu Sun, Yunpeng Zhang**
SINOPEC Research Institute of Safety Engineering, Qingdao, China;
State Key Laboratory of Safety and Control for Chemicals, Qingdao, China

E-mail: liyp.qday@sinopec.com

Abstract. At present Fiber Reinforced Plastics (FRP) double wall underground storage gasoline tanks are wildly used. An FRP product with a resistance of more than 10^{11} Ω is a static non-conductor, so it is difficult for the static electricity in the FRP product to decay into the earth. In this paper an experimental system was built to simulate an automobile gasoline filling station. Some electrostatic parameters of the gasoline, including volume charge density, were tested when gasoline was unloaded into a FRP double wall underground storage tank. Measurements were taken to make sure the volume charge density in the oil-outlet was similar to the volume charge density in the tank. In most cases the volume charge density of the gasoline was more than 22.7 $\mu C\ m^{-3}$, which is likely to cause electrostatic discharge in FRP double wall underground storage gasoline tanks. On the other hand, it would be hard to ignite the vapor by electrostatic discharge since the vapor pressure in the tanks is over the explosion limit. But when the tank is repaired or re-used, the operators must pay attention to the static electricity and some measurements should be taken to avoid electrostatic accident. Besides the relaxation time of charge in the FRP double wall gasoline storage tanks should be longer.

1. Introduction

In the last few years, oil leakage accidents in the automobile gasoline filling station have taken place for many times. These accidents take pollution to the environment and sometimes may cause explosion accidents [1]. FRP double wall underground storage gasoline tanks have been used in foreign countries to avoid these accidents. This kind of storage gasoline tanks can alarm us when there is an oil leakage accident. So something can be done to avoid environmental pollution. But the FRP product is static non-conductor and its electrostatic characteristic should be noticed. In this paper an experimental system was built to simulate the automobile gasoline filling station. Gasoline in this system was unloaded into a FRP double wall underground storage gasoline tank, and at the same time some electrostatic parameters of the gasoline were tested.

* To whom any correspondence should be addressed.

2. The experiment

2.1. Introduction of the experimental system

According to the true automobile gasoline filling station an experimental system was built. This system was built by a FRP double wall underground storage gasoline tank, a tank truck, gasoline, an oil drum, gasoline dispenser, test equipment, and so on.

The tank was buried in an impermeable pool, and the diameter of the tank was 2.5 m, the length of the tank was 6.7 m. The diameter of sucker was 80 mm, and the distance between the terminal of the sucker and the bottom of the tank was 200 mm. The diameter of oil-outlet was 40 mm. In order to dissipate the electrostatic accumulation, there were two metal plates and each plate's acreage was 0.8 m^2, they located at the bottom of the tank and they were bonded to the external ground [2].

Because the existing liquid potential measures are severely influenced by the environment [3], the volume charge density was chosen to be tested and analyzed. The way of gathering volume charge density data was as follows. A meter was installed in the oil-outlet, the name of the meter was DQJ-4A and it was used for testing the volume charge density of the pipe. When the gasoline was unloaded into the tank, the operators run the gasoline dispenser and tested the volume charge density of the pipe. The meter was closed to the tank and the pipe between the meter and the tank didn't bond to the earth. So the volume charge density of the pipe was nearly equal to the volume charge density of the gasoline in the tank.

2.2. Some basic parameters of the experimental system

First some basic parameters of the experimental system were tested. The use of equipments were as follows, the ACL-800 surface resistivity meter, the MLA-900 conductivity meter, the BI-870 dielectric constant meter, the GEOX grounding resistance meter and so on. The relative conductivity of the gasoline was 16 pS • m^{-1}. The relative dielectric constant of the gasoline was 2.2. The surface resistivity of the tank was more than 10^{12} Ω, so the FRP was static non-conductor and the static electricity in FRP product was difficultly decayed to the earth [4]. The grounding resistance of the metal plate was lower than 0.72 Ω, so the metal plate was well connected to the earth.

2.3. The experimental process

The volume of the gasoline was 27000 L, before operation the gasoline had been reposed for 12 hours. The temperature of the environment was 26 (±2) ℃, and the relative humidity was 50 (±5) %RH. Under the gravitational effect gasoline was unloaded into the Double Wall FRP underground storage gasoline tank through the sucker and the data of the volume charge density was tested at the same time. The experiment had been done for many times and one group of the test data of the volume charge density of the gasoline is in table 1. Some data in other groups was close to 22.7 μC • m^{-3}, but all less than 22.7 μC • m^{-3} .

Table 1. The results of volume charge density

filling rate	volume charge density (μC • m^{-3})	filling rate	volume charge density (μC • m^{-3})
40%	17.4	60%	14.9
42%	22	65%	16.2
45%	22.7	70%	15.8
48%	18.9	75%	14.6
50%	16.9	78%	15.3
55%	13.4	80%	16.8

After the test mentioned above, the gasoline was pumped from the Double Wall FRP underground storage gasoline tank into the tank truck, until the filling rate of the Double Wall FRP underground storage gasoline tank was 40%. After this operates for 20 minutes, under the gravitational effect, the gasoline was unloaded into the Double Wall FRP underground storage gasoline tanks again. One group of the test data of the volume charge density of the gasoline is in table 2.

Table 2. The results of charge density

filling rate	volume charge density ($\mu C \cdot m^{-3}$)	filling rate	volume charge density ($\mu C \cdot m^{-3}$)
50%	25.3	65%	35.6
52%	27.1	68%	33.3
55%	29.8	70%	31.2
58%	37.1	72%	29.6
60%	36.4	76%	30.7
62%	33.9	80%	29.4

3. The electrostatic risk analysis of the Double Wall FRP underground storage gasoline tank

3.1. Oil surface potential

When the volume charge density of the tank is given, the oil surface potential of the tank can be calculated by special formula [6].

$$V = \frac{\rho \gamma^2}{4\varepsilon} \tag{1}$$

Where V is oil surface potential, ρ is volume charge density, ε is dielectric constant.

In this experiment, the diameter of the tank was 1.25 m, the relative dielectric constant of the gasoline was 2.2. When the relaxation time of charge was enough, the maximal number of the volume charge density was 22.7 $\mu C \cdot m^{-3}$. Assuming the charge in the tank was well distributed, it could be calculated that the maximal oil surface potential of the tank was 455 kV in theory. If the relaxation time of charge was not enough, the maximal number of the volume charge density was 37.1 $\mu C \cdot m^{-3}$, it can be calculated that the maximal oil surface potential of the tank was 705 kV in theory. The calculation of the oil surface potential shows that it has big probability to cause electrostatic discharge in the tank [7].

3.2. Time of repose of the gasoline

The relaxation time of charge of the gasoline can be calculated by special formula [8].

$$t = \frac{\varepsilon}{\sigma} \tag{2}$$

Where t is relaxation time of charge, σ is conductivity, ε is dielectric constant。

The conductivity of the gasoline was 16 $pS \cdot m^{-1}$, the relative dielectric constant was 2.2. The time of repose is three times of the relaxation time of charge, it could be calculated that the relaxation time of charge of the gasoline was more than 3.65 s. When the conductivity was 16 $pS \cdot m^{-1}$, the volume of the tank was 30 m^3, according to some standards, for metal tank the time of repose was more than 3 minutes [4].

The experiment had been done for many times, sometimes the time of repose was more than 12 hours and sometimes the time of repose was more than 20 minutes. If the tank was metallic, the times of repose were all satisfied. But from the data of the volume charge density it could be concluded that, after the operation the Double Wall FRP underground storage gasoline tanks need more time to relax the charge.

3.3. The risk analysis of explosion accident causing by electrostatic discharge

From the results of some experiments we know that the volume ratio of oil gas in a underground storage gasoline tank is more than 10% when unloading gasoline into the tank, and the vapor would be hard to be ignited by discharge even the energy is larger than 10 J. Under the general circumstance the electrostatic discharge is less than 10 J, so the vapor in the Double Wall FRP underground storage gasoline tank would be hardly to be ignited by electrostatic discharge. The Double Wall FRP underground storage gasoline tanks have been used in North America for thirty years, their operation experience can prove that. But when the tank is repaired or re-used, some appropriate measurements should be taken to avoid the occurrence of accident.

4. Conclusion

In this paper, some electrostatic parameters of the Double Wall FRP underground storage gasoline tank were tested in an experimental system. Some related data showed that it has big probability to cause electrostatic discharge in the tank. But in normal condition it has low risk to cause explosion accident, and some appropriate measurements should be taken when the tank is repaired or re-used. Besides the Double Wall FRP underground storage gasoline tanks need more rest time after operation.

Acknowledgments

The study was supported by the National Key Technology R&D Program (Grant 2012BAK03B03).

References

[1] Li J, Tao H, Zhang K and Li Y 2010 *Energy Conservation and Environment Protection* **9** 41
[2] API RP 2003-2008 *Protection against ignitions arising out of static, lightning, and stray currents* (Washington. D. C)
[3] GB 6951-1986 *Safe oil surface potential of the lightweight oil* (Beijing)
[4] GB 12158-2006 *Basic safety code of preventing electrostatic accident* (Beijing)
[5] GB 6950-2001 *Safe stationary conductivity of the lightweight oil* (Beijing)
[6] Sun K and Song G 1994 *Industry static electricity* (Shanghai)
[7] Yu G 2006 *The pivotal technology research on the electrostatic safety during the weight oil loading and unloading and transporting* (Shanghai)
[8] GB 13348-2009 *Safety code of the liquid petroleum product* (Beijing)

Journal of Physics: Conference Series **418** (2013) 012041 doi:10.1088/1742-6596/418/1/012041

A novel method for analyzing electrification in friction process

TianHai Chang, JinCheng Wu*
School of Electronic and Information Engineering, South China University of Technology, Guangzhou 510640, Guangdong, China

E-mail: wujinc@163.com

Abstract. This paper is trying to explore the electrification mechanism of solid material during the friction process in a new angle of secondary electron emission characteristics, and explain the course of electrostatic formation by using the SEE (Secondary electron emission) theory. It will demonstrate that there is a positive connection between SEE yield and the cause of electrostatic electrification by conducting the experiment with the SEE theory. We are going to show the part which SEE characteristics will take in electrification, and deduce a new formula model with SEE yield and electrostatic potential by using our experimental data, which demonstrates that there is a numerical relationship between them. Based on the conclusion from this experiment, it can provide a new thought for electrostatic theory development of solid material and analysis on some special electrification phenomena.

1. Introduction

Early about 600 years BC, materials electrification by friction was discovered by Thales from Greece. However, how to explain these phenomena scientifically, especially for micro-mechanism of electrification, is still one of knotty problems in the present field. As we know, there are various viewpoints and theory in academic circles [1]. One theory which is widely accepted is "contact-separate" theory that is based on material contacting and separating course, combining the energy band theory as a model. This theoretical system has provided a reasonable explanation for mechanism of electrification process of materials, including solid, liquid and gas. However, in variety of complicated electrification processes, many experiment phenomena can't be explained satisfyingly, which is needed us to study further .Based on above analysis, this paper is going to try to explain the course of electrostatic formation by using the SEE characteristics and considering SEE yield as a key factor in analyzing electrification process of solid materials. The theory of SEE should give a better explanation to some special electrification phenomenon and we plan to introduce the formula model of SEE to demonstrate the relationship between SEE yield and surface potential.

2. Secondary Electron Emission Theory

2.1 .SEE phenomenon

Electrons (or ion) will be forced to excite out from material when the primary incident electrons (or ion) possessed of some energy or speed have bombarded materials' surface, which are called SEE(Secondary Electron Emission).Early in 1899,Campbell had discovered this phenomenon. The electron bombarding material is usually called original electron (or primary electron), on the other

* To whom any correspondence should be addressed.

hand, what is emitted from material' surface is called secondary electron. The secondary electron with less than 50eV is usually called true secondary electron, besides, there are other electrons, including inelastic scattering, elastic scattering and auger electrons and so forth [2]. The true secondary electron takes a large part in this process. So, it is the main object we must study on. The model of secondary electron emission is showed in figure 1.

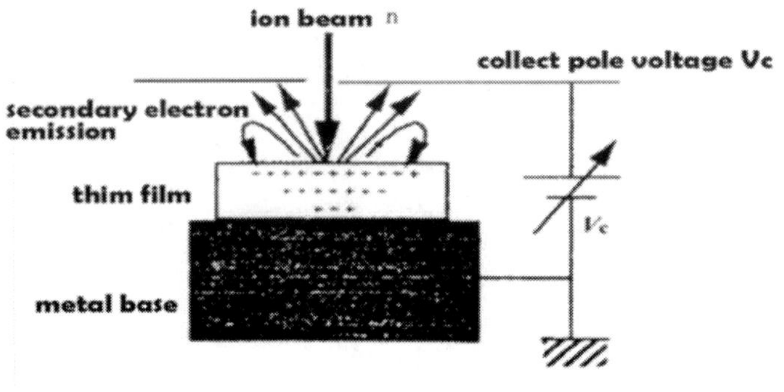

Figure 1. Secondary electron emission model.

2.2 .Secondary Electron Emission Yield
The capacity of SEE is usually presented by SEE yield δ. Simply, SEE yield is defined by the radio of SEE stream I_s and original incident electron stream I_p [3], this relationship can be described in formula 1.

$$\delta = \frac{Is}{Ip} \tag{1}$$

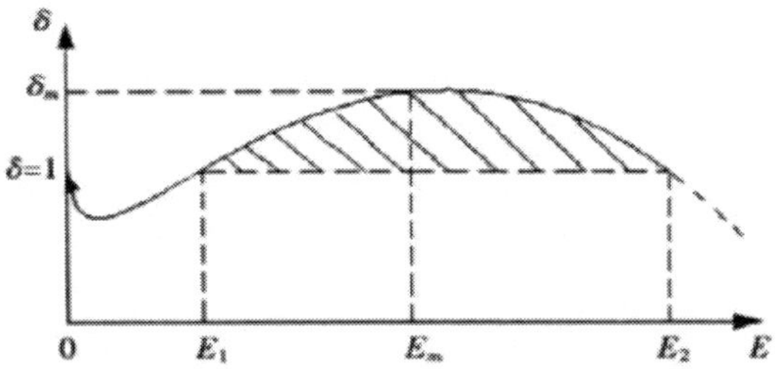

Figure 2. The classic relationship curve between SEE yield and incident energy.

Further, there is a close relationship between the amount of original electron and δ with the characteristic of bombarded material in the special environment conditions. The figure 2 is classic secondary electron emission characteristic curve of solid material [4].

3. Friction Electrification Experimental Scheme

3.1. Experimental Materials
Three solid materials are chosen for this experiment, including polyvinylchloride (PVC), high-density polyvinylchloride (HDPE) and polypropylene (PP).Their characteristics are listed in table 1 [3], and their shapes are showed in figure 3. We mainly choose the circle one as the analysis object of this experiment.

Table 1. Experimental material characteristic.

material	BRRR[a] /Ω. m	BRRM[b] /Ω. m	SRM[c] / Ω	MSY[d] /δ	COE[e] /eV
HDPE	$10^{13}\sim10^{14}$	$>10^{14}$	4.26×10^{9}	2.71	250
PP	10^{14}	$>10^{14}$	2.53×10^{11}	—	—
PVC	$10^{14}\sim10^{15}$	$>10^{14}$	1.53×10^{12}	2.12	400

[a] Body resistance ratio reference value
[b] Body resistance ratio measure value
[c] Surface resistance measure value
[d] Maximum SEE yield
[e] Corresponding original electron

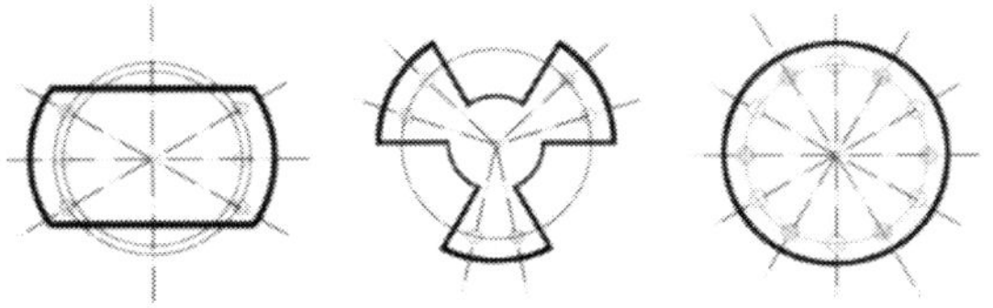

(1) Strip shape (2) trifocal shape (3) circle shape

Figure 3. Schematic diagram of materials in friction experiment.

3.2. Design of Experimental Scheme

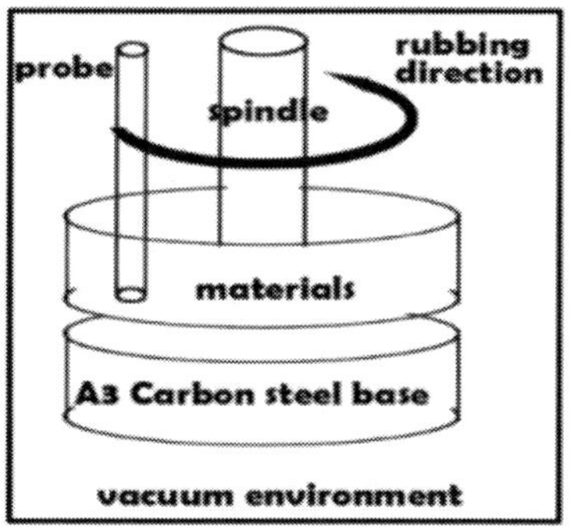

Figure 4. Friction experiment schematic diagram.

The experiment is conducted by using self-designed equipment "electrostatic dynamic measurement device", whose structure is showed in figure 5. A3 carbon steel is fixed on the base as a friction object, which rubs with above three materials by turns. From this experiment, we have gotten three groups of data by using the keen probe under different speeds which are showed in the table 3. Otherwise, some critical environment conditions are listed in table 2 (in the vacuum environment, the electric field and magnetic field are shielded.). To ensure the accuracy of this experiment, every material should be polished and cleaned carefully before experiment in order to eliminate electrostatic accumulated on the surface [5, 6].

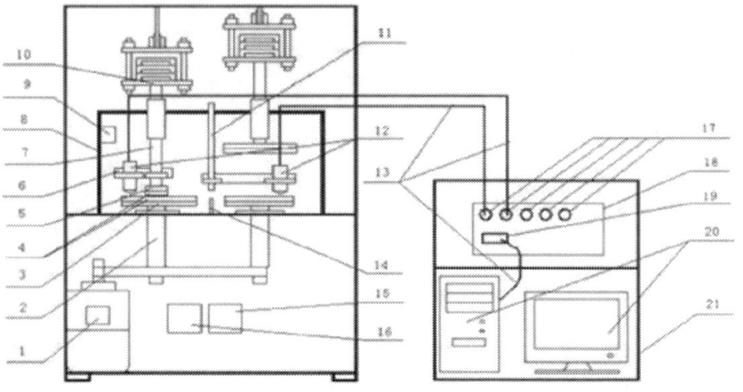

1. Variable-Frequency Motor 2. rotation bearing 3.rotation friction disc 4. test material 5.static friction disc 6. probe clamp 7. force application bearing 8. workshop 9. humidity probe 10. force weights 11. Manipulator 12. electrostatic probe 13. signal lines 14. temperature probe 15. humidity control panel 16. temperature control panel 17. probe interface 18. Acquisition Card 19. ESB interface 20. computer 21. Cabinet

Figure 5. Electrostatic dynamic potential measurement device.

Table 2. The experimental conditions of three materials.

RH[a]	E.T.[b] ℃	Spd.[c] r min^{-1}	FP[d] kg	PD[e] cm	S.T.[f] s	BM[g] /	RM[h] /	DM[i] /
(80 ±5) %	27.6	70-150	5	2	3	A3 carbon steel	PVC, HDPE, PP	PVC, HDPE, PP

[a] Relative Humidity; [b] Environment Temperature; [c] Rubbing Speed; [d] Friction Pressure
[e] Probe Distance; [f] Sampling Timing; [g] Base material; [h] Rotating materials; [i] Detected materials

Table 3. Electrostatic potential data of three materials in friction experiment.

Speed r min^{-1}	electrostatic potential on rotating circle shape material / kV								
	Maximum value			Minimum value			Average(absolute)		
	PVC	HDPE	PP	PVC	HDPE	PP	PVC	HDPE	PP
30	0.0782	0.6211	0.7328	-0.5018	-0.2924	-0.2687	0.0798	0.1585	0.0297
50	0.0391	0.6211	0.0977	-1.3288	-0.4805	-0.3175	0.2749	0.2548	0.0417
70	0.0391	0.5942	0.7328	-1.5246	-0.6417	-0.5129	0.3678	0.3841	0.1031
90	0.0391	0.5674	0.7084	-1.4851	-0.7223	-0.6106	0.3869	0.4736	0.1488
110	0.0391	0.7084	0.7084	-1.4851	-0.6595	-0.6595	0.3615	0.4973	0.1752
130	0.0391	0.5674	0.7328	-1.5633	-0.7761	-0.6839	0.4336	0.4897	0.1727
150	0.0391	0.5405	0.7084	-1.7192	-0.8835	-0.6351	0.4308	0.5298	0.1576

4. The Experimental Result and Phenomenon Analysis

We make a contrast among the three groups of electrostatic potential data from table 3, the curve is painted in figure 6.

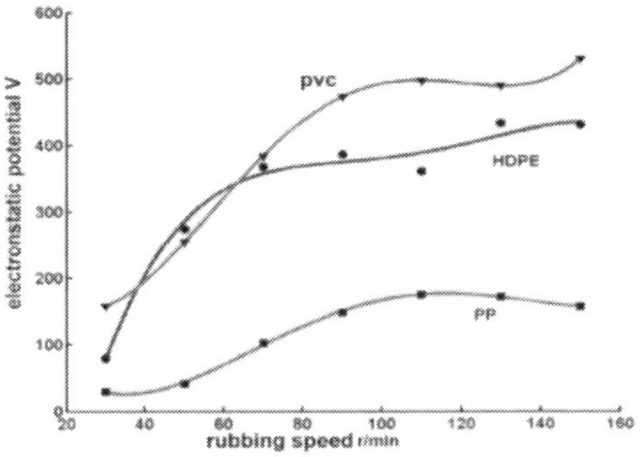

Figure 6. The relationship between friction speed and electrostatic potential of the three materials.

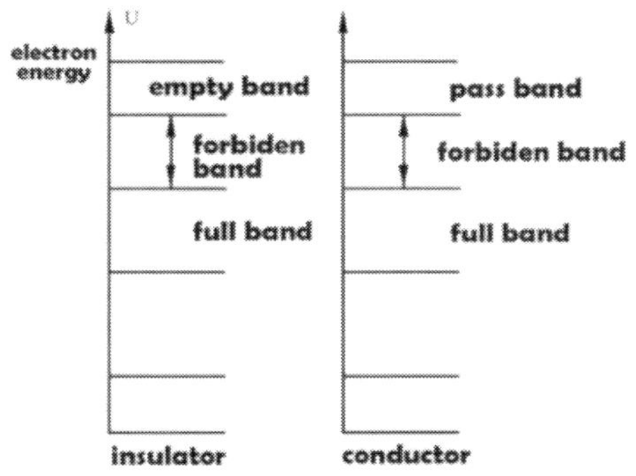

Figure 7. Energy band model of insulator and conductor.

From figure 6, we can find out a peculiar phenomenon that the electrostatic potential in Y axis does not increase with the rubbing speed in X axis all the time, especially when the speed reaches to some value, potential of PP and HDPE even has a trend of decrease at the speed value closing to 140 r min^{-1}. This phenomenon is not what we expect before. Because, based on the energy band theory model, rubbing process of materials is equivalent to countless contact point being the "contact-separate" status. At the same time, friction work will produce heat on material surface, which can be conducted into inner of material. As a result, some electrons will be emitted by heat to transit up to different energy band. Take the insulator for instance， electrons from the inner are excited to span forbidden band to arrive at empty band and finally become free ones. When energy they have is double to work function (minimum energy of electrons coming out of the solid inner) of materials they come from, free electrons will get rid of constraint and reach to the surface, at the end, electric double layer will be created gradually. Finally, two friction materials will get the inverse charge as soon as they separate.

In the experiment, according to energy band model in figure 7, with the increase of rubbing speed, more heat will be produced on the material surface which can evidently add the number of electrons coming out from materials. These electrons with enough energy will span over the energy band to become free ones. As a result, electrostatic potential goes up more when rubbing materials separate [1, 7]. However, the result from experiment does not accord with what we expected. So we are planning to introduce SEE characteristics to analyze it in this situation. We will simulate the course of electrification at the angle of SEE characteristics, just like figure 8 showed below.

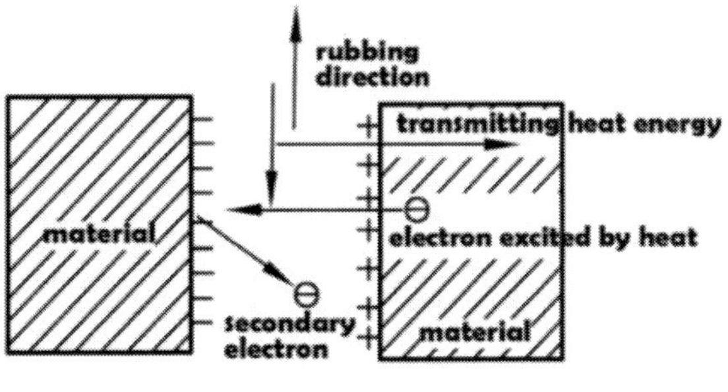

Figure 8. Secondary electron mission simulating graph in the friction process.

We try to take SEE yield consider as one of critical factors in the process of electrostatic electrification. At first, referring to figure 2 and table 1, SEE yield is not rising with the increase of original energy E on the whole. From the beginning of Y axis of figure 2, SEE yield δ climbs up rapidly, and then reaches the maximum value δ_m when E is equal to E_m, after that, δ turns to decline. According to the definition of SEE yield, the process of δ from up to down exactly reflects a trend of the number of electrons excited by incident electrons, which will affect the account of electrostatic potential when rubbing materials separate. We can analyze this experimental result as follows: materials rubbing with each other will produce more heat energy as well as inner electrons [7, 8, 9]. This result will lead to the increase of the number of incident electrons on the surface of materials. This changing process can be analyzed by dividing into two parts: part 1, for some specific SEE yield of material, when original energy $E< E_m$, the number of secondary electrons excited by rub heat energy will rise quickly ,which leads that surface electrostatic potential goes up much; part 2, when original energy $E> E_m$, from the curve showed in the figure 2, SEE yield δ is keeping a rapid decrease trend, which caused that the number of electrons aroused by heat rises slowly, or even goes down, as well as electrostatic potential, at the end it gradually reaches to a dynamic balance value [11, 12]. We can describe this change by using two simple flow charts below. In the case of $E< E_m$, δ goes up with a large slope, the whole changing process of potential will be showed in figure 9.

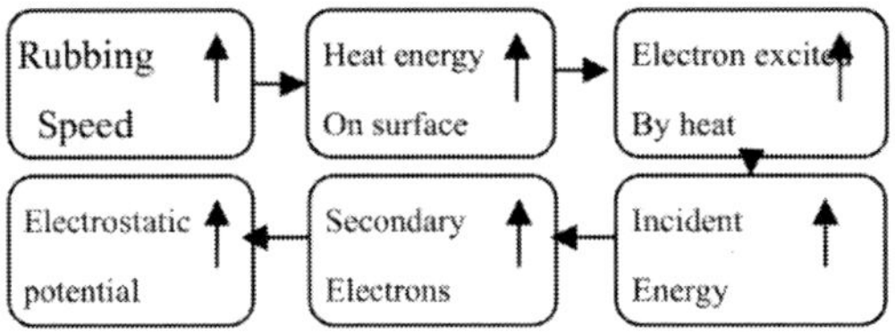

Figure 9. The verified process of surface potential when $E\leq E_m$.

On the other hand, when $E> E_m$, the value of δ goes down, the above changes will be showed in figure 10 below.

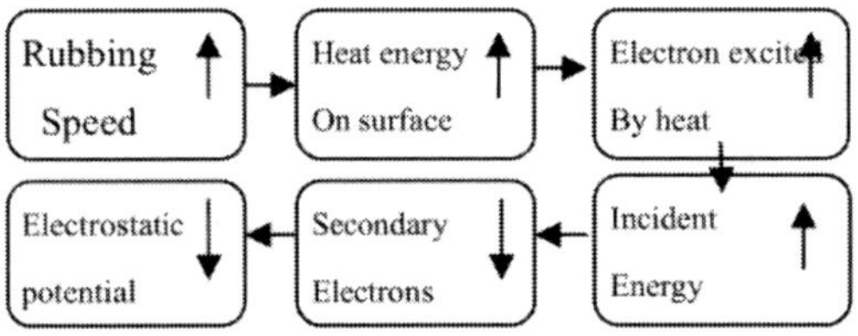

Figure 10. The verified process of surface potential when $E>E_m$.

5. Friction Electrification Formula Model with Secondary Electron Emission Yield.

From above analysis of experimental result, adding up the changing process description in figure 9 and figure 10, among some limited energy area of incident energy, we can probably get the special relationship between incident energy and electrostatic potential on the material surface which can be drafted in the figure 11.

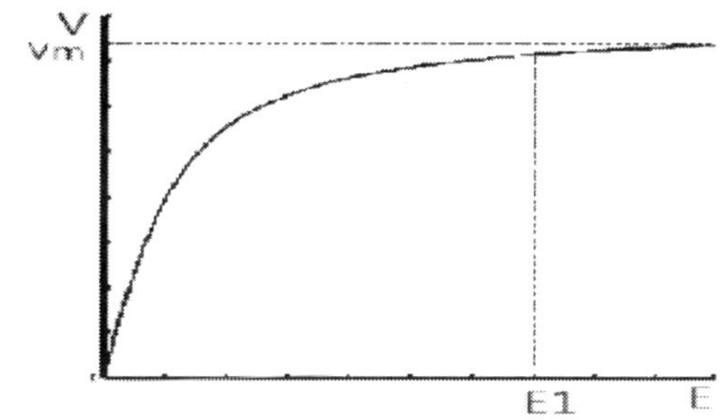

Figure 11. The relationship between original energy and surface electrostatic potential of rubbing material.

Electrostatic potential of materials during rubbing process begins to rise slowly when incident energy E reaches to E_1, and at the end it will reach to dynamic balance value V_m. According to the figure 11, we probably get a related formula:

$$V = k_1 \bullet \arctan(k_2 \bullet E) \tag{2}$$

V represents electrostatic potential of materials and E as the incident electron energy, k_1 and k_2 are relevant coefficients which are decided by characteristics of and SEE yield of materials. Formula 2 unites semi-empirical formula 3 [3, 13, 14], which gets:

$$\delta(E)/\delta_m = 1.526\,[1 - \exp(-z^{1.725})]/z^{0.725} \quad (z = 1.284E/E_m) \tag{3}$$

A new formula model will be established with electrostatic potential and secondary electron emission in this process of dynamic friction.

$$\begin{cases} \delta/\delta_m = 1.526[1-\exp(-z^{1.725})/z^{0.725}], z = 1.284E/E_m \\ V = k_1 \bullet \arctan(k_2 \bullet E) \end{cases} \tag{4}$$

Simply, the relationship model with secondary electron emission yield and electrostatic potential is showed by a formula 5 below.

$$\begin{aligned} &\delta/\delta_m = 1.526[1-\exp(-z^{1.725})/z^{0.725}], \\ &and, z = 1.284 \bullet \frac{1}{k_2} \bullet \tan(V/k_1)/E_m \end{aligned} \tag{5}$$

Therefore, the maximum value δ_m, incident energy E_m and related coefficient k_1 and k_2, which are relative stable for some special solid materials. So, based on the formula 5, we can conclude that the difference of SEE yield of material has a special relationship with electrostatic potential. As a result, we think that it is feasible to analyze the changing process of electrostatic potential on material surface at the angle of SEE yield and benefit to expand the thought of studding the electrostatic theory further.

6. Conclusion

It is a novel method to study the process of electrostatic electrification with SEE yield. And the following conclusions can be drawn:

(1) Based on the intrinsic "contact-separate" theory and energy band model, study of electrostatic theory with the characteristic of SEE contributes to comprehend understanding of electrification mechanism. The phenomenon occurred on this experiment reflects the characteristic presentation of SEE. According to this relationship between them, we can deduce the micro process of electrostatic generation and it also shows the changing trend of electrostatic potential value. Formula 5 also has

revealed some numerical connection between δ and V.

(2) It helps us to analyze electrostatic generation and electrostatic discharge by the formula model with δ and V. As long as SEE yield of materials can be measured, we could generally estimate the dynamic electrostatic potential during the process of friction, which is benefit to the prediction of high electrostatic voltage level and helps us to take a necessary measure to avoid an electrostatic accident ahead of time.

(3) By using the relationship between δ and V, we can change the characteristics of SEE of material in order to decrease electrostatic voltage efficiently during the process of practical production environment and lower the accident incidence in practice. In conclusion, it is useful to understand deeply the micromechanics of electrostatic formation in the inner of material and provide a full new method of electrostatic protection technology through introduction of SEE yield during the rubbing process between solid materials.

Acknowledgments
Grant from : The National Natural Science Funds (51047002)

References
[1] Liu S H, Wu Z C and Wei G H 1999 *Electrostatic Theory and Protection* (Beijing: Arms Industry Press) p 81
[2] Chang T H and Ma W, et al 2011 *J. South China Univ. Technol.* (in Chinese) **12** 117
[3] Burke E A 1980 *IEEE Trans. Nucl. Sci.* **27** 1759
[4] Dionne, Gerald F 1973 *J. Appl. Phys.* **44** 5364
[5] Chang T H and Yin J X 2008 *J. South China Univ. Technol.* (in Chinese) **36** 132
[6] Chen D X and Chang T H 2009 *Vac. Cryognics* (in Chinese) **15** 170
[7] Xie A G and Zhang J, et al 2011 *High Power Laser Part. Beams* (in Chinese) **23** 1663
[8] Richterova I, Pavlu J, Nemecek Z, Safrankova J and Beranek M 2007 *IEEE Trans. Plasma Sci.* **35** 286
[9] Ganachaud J P and Mokrani A. *Electrical Insulation and Dielectric Phenomena* (Pocono Manor, PA) p 233
[10] Watts C, Gilmore M and Schamiloglu E 2011 *IEEE Trans. Plasma Sci.* **39** 836
[11] Cheng H and Jiang J P 1986 *Cathode Electronics* (Xi'an: XiDian University Northwest Telecommunication Engineering College Press) p 90
[12] Mai S L and Wang L, et al 2006 *Vac. Cryognics* (in Chinese) **12** 87
[13] Yang C C and Huo A C 1979 *J. Yunnan Univ. Nat. Sci.*(in Chinese) **2** 34
[14] Abuelma'atti M T 1990 *IEEE Trans. Electr. Dev.* **37** 1590

Study on effect of space environment on surface charging of spacecraft

Bin Xu*, **Zhancheng Wu, Pei Zhang**
Institute of Electrostatic and Electromagnetic Protection, Ordnance Engineering College, 050003, Shijiazhuang, China

E-mail: tnbhandinhand@163.com

Abstract. The effects of space environments on the surface charging of spacecraft include space plasmas environment and space magnetic sub-storm environment. The investigation on these space environments is essential for analysing spacecraft surface charging mechanism, Electrostatic Discharge (ESD) safety and for research on deep dielectric charging and discharging in spacecraft. Furthermore, analysis of charging mechanism can provide basic guidelines to ESD protection for spacecrafts, which is important to the safety of spacecrafts.

1. Introduction

In modern times, the battlefield environment is the complex of sea, land, air, space, electromagnetism, in which the space means being preponderant in spaceflight domain. The development and martial application of spaceflight technology and equipment have impacted the modern wars deeply. As we know, 70 satellites were used in Gulf War, which reached to 50 respectively in Kosovo War and Afghanistan War. They offered all-sided information support and guaranteed to equipments in the sea, land and air.

In 1960s, along with the running of geostationary orbit satellites, many failures appeared in the spacecraft in and abroad, which interfered the functions. The unconventionality could not be explained correctly in the near 10 years. By analyzing the failures happened aboard, it is known that the space environment is the main causation rather than other reasons. And in the failures caused by space environment, 75% were relative with spacecraft's surface charging and discharging. The ratio reached to 83% in the failures of satellites of our country. The hazard can be seen clearly. The process was simulated and demonstrated in the laboratory by De Forest, proving that static was the main reason for the failures.

In such case, it is very necessary to research the spacecraft's surface charging in the space environment for military affairs. Furthermore, it is very important for spacecraft's anti-jamming ability, which means spacecrafts can play more important roles in modern wars.

2. The space environment influencing spacecraft's surface charging

Spacecraft runs on many kinds of orbits and faces many space environment basics, including electromagnetic radiation from the sun, neutral atmospheric globosity ionosphere, space plasmas globosity magnetic field, space electrification particle radiation and so on. These basics work on the

* To whom any correspondence should be addressed.

spacecraft running on the orbit by influencing the materials and electron devices of the spacecraft. And then the safety of the spacecraft is impacted.

In all basics, space plasmas environment and magnetospheres sub storm environment influence spacecraft's surface charging intensively. While running in such environment, the spacecrafts act with the plasmas, and surface charging/discharging effect come forth[1-2]: Potential differences between the near surfaces, surface and deep medium, surface and terra appear because of the differences in dielectric properties, light and geometry. Along with the rise of potential differences, charging by corona, flash-over, breakdown, and EMP appears. Besides, grounded system may immit the charging current into the electron system which causes the failures. Even the whole spacecraft would be in danger, as shown in figure 1.

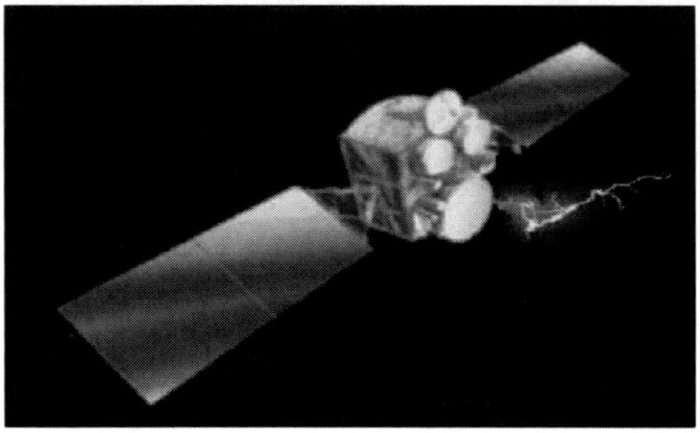

Figure 1. Charging and discharging on spacecraft's surface.

2.1 Space plasmas environment

Plasma sphere starts from about 60 km upside from the ground. The density, component, and energy of the plasmas change as the altitude changes. While sun activity is in the range of 60 km~3000 km, the relation is shown in table 1[3-4].

2.2 Space magnetic sub storm environment

Table 1. The variety of density, energy of electron along with altitude.

altitude (km)	density (cm^{-3})	energy (keV)
60	2×10^2	0.05
85	1×10^4	0.05
140	2×10^5	0.05
200	5×10^5	0.08
300	2×10^6	0.19
400	1.5×10^6	0.22
500	1×10^6	0.23
600	6×10^5	0.24
700	4×10^5	0.25
800	3×10^5	0.26
900	2×10^5	0.27
1000	1×10^5	0.28
2000	2.5×10^4	0.30
3000	1.5×10^4	0.35
10000	2×10^3	1.00

Earth's magnetosphere is the finite space area which is restricted and inhibited by solar wind and interplanetary magnetic field and controlled by geomagnetic field.

Magnetosphere sub storm is regularly disturbed which occurs in magnetosphere. The disturbance to magnetosphere occurs while stored energy about 1×10^{14}~1×10^{15}J is released by it in about 1000 s.

When magnetic sub storm occurs, the plasma of high density (10 cm^{-3} ~100 cm^{-3}), low energy (less than 1 eV) in geosynchronous orbits environment were replaced by that of low density (less than 1 cm^{-3}), high energy (1 keV~50 keV). Statistic shows that sub storm can be observed at 30% of the time, and 8%~10% of them can influence the spacecraft in geosynchronous orbits. With the environment data basic obtained by ATS-5, ATS-6, SCATHA (STP P78-20), the USA presents the average and extremely hard condition plasma environment in geosynchronous orbits for design and evaluation by computer [5].

Table 2. The average plasma environment in geosynchronous orbits while magnetic sub storm occurs.

model	ATS-5		ATS-6		P78-20	
particle	electron	ion	electron	ion	electron	ion
density (cm^{-3})	0.80	1.30	1.06	1.20	1.09	0.58
current density ($nA\cdot cm^{-2}$)	0.065	5.1	0.096	3.4	0.115	3.3
average energy (keV)	1.85	6.8	2.55	1.20	2.49	11.2

Table 3. The extremely hard condition plasma environment in geosynchronous orbits while magnetic sub storm occurs.

particle	electron	ion
density (cm^{-3})	1.12	0.236
average energy (keV)	12	29.5

3. The main influence factors and complexion of spacecraft's surface charging

3.1 While magnetism serenity without illumination

In magnetism serenity, the spacecraft is surrounded by plasma whose electron energy is about several eV. The high energy ion is so unwonted that it is out of consideration. Suppose that the speed distributing of the electron and ion abide Maxwell distributing, their average speed is [6~8]:

$$v_e = \left(\frac{8kT_e}{\pi m_e}\right)^{1/2}, \quad v_i = \left(\frac{8kT_i}{\pi m_i}\right)^{1/2} \qquad (1)$$

v_e, v_i stand for the average speed of the electron and ion; Te, Ti stand for the temperature of the electron and ion; m_e, m_i stand for the quality of the electron and ion; k stands for Boltzman constant, the cost of which is 8.617×10^5 ev $^{\circ}C^{-1}$.

While the ion stands for proton, $T_e \approx T_i$, because the quality of proton is about 1840 times of that of electron. So

$$\frac{v_e}{v_i} = \left(\frac{m_i}{m_e}\right)^{1/2} \approx 44 \qquad (2)$$

It means that as the densities of electron and ion are pretty much the same thing, the electron falling on the surface of the spacecraft is much more than that of the ion. The electron piles up at the surface of the spacecraft so that the negative potential is formed. When it is in magnetism serenity, the energy of electron on the GEO is only about 1 eV, the surface potential of spacecraft in the shadow of the globosity is only some volts and it does not work on the regular operation.

3.2 While magnetism serenity with illumination

While spacecraft gets in the illumination area, the photoelectron will be excited by photons. Although the energy of photons is quite low, its speed always exceeds the plasma. The photoelectron plays an important role because the amount is big. The photoelectron can get out of the surface when it is irradiated by the sun. When the summation of the ion current, electron current and photoelectron current is around zero, the surface in the irradiation gets in balance, namely

$$J_p + J_{ph} = J_e \tag{3}$$

There J_p, J_e and J_{ph} stand for the ion current, electron current and photoelectron current which arrive at the space of spacecraft.

Ignore the ion current J_p, the balance potential can be gotten as

$$\varphi_s = \frac{kT_e}{e} \ln \frac{J_e}{J_{ph}} \tag{4}$$

While magnetism serenity with illumination, on the GEO, $J_e \approx 1.5\times10^{-11}$ A (cm)$^{-2}$, $J_{ph} \approx 5.4\times10^{-10}$ A (cm)$^{-2}$, so the potential can be estimated as some positive volt.

3.3 While magnetic sub storm

While in magnetic sub storm, thermal plasma immits to the GEO, and the scope of the energy is from some thousand to some ten thousand. Je can reach to 1 nA cm^{-2}. $J_p \approx (1/20\text{-}1/50) J_e$, so the balance potential can be deduced as

$$\varphi_s = [(-3) \sim (-4)] E_e / e \tag{5}$$

As that, E_e can be 1 keV~ 30 keV, so the balance potential will be estimated to be some thousand votes to some ten thousand.

4. Conclusion

The paper summarized the investigations on spacecraft surface charging, in which the space environment resulting in surface charging of spacecraft was mainly studied. The investigation can provide the basic to spacecraft ESD protection, which is important to the safety of spacecraft.

References

[1] Wang L, Qin X G, Li K, Ma Y L and Huang L F 2004 *Chinese Space Technol.* **5** 32
[2] Shu T L 2007 *38th AIAA* p 98
[3] Huang B C, Ma Y L 2002 Industry of National Defense Publishing Company.
[4] Vayner B V, Galofaro J T and Ferguson D C 2006 *Coll. Tech. Pap. AIAA 37th Plasmadynamics Lasers Conf.* (San Francisco) vol 1 (American Institute of Aeronautics and Astronautics Inc.) p 178
[5] Mandell M J, Davist V A and Gardner B M 2007 *45th AIAA* p230
[6] Ding Y G 2005 *Graduate School of National University of Defense Technology degree of Master*
[7] Masui H, Toyoda K and Cho M 2009 *47th AIAA* p267
[8] Mandell M J, Davist V A and Gardner B M 2007 *45th AIAA* p328

Journal of Physics: Conference Series **418** (2013) 012043 doi:10.1088/1742-6596/418/1/012043

Harmful effects of lightning surge discharge on communications terminal equipments

Sisi Liang[1], Xiaoying Xu[1*], Zhigang Tao[2], Yanling Dai[3]
1. School of Sciences, Wuhan University of Technology, Wuhan, 430070, China
2. Neutron Engineering Inc, Dongguan, 523792, China
3. The Zhongshan Affiliated Middle School of South China Normal University Guangzhou, 528447, China

E-mail: xu_xiao_ying@126.com

Abstract. The interference problem of lightning surges on electronic and telecommunication products were examined, and a series of experiments were conducted to analyze the failure situations to find out the mechanisms of failures caused by the lightning surge. In addition, the ways in which lightning surges damaged equipment were deduced. It was found that failure positions were scattered and appeared in groups, and most of them were ground discharge. Internet access transformer had high withstand-voltage under the lightning pulse, and the lightning surge seldom passed through the internet access transformer. The lightning current can release to the ground via the computer network adapter of the terminal user. The study will help to improve the performance of lightning surge protection circuit and protection level.

1. Test method and equipment needed

With the increasing popularity of the electronic and telecom products, information communication is playing an increasingly significant role in our daily life and the development of the society. However, since the destruction of electromagnetic interference in the environment, especially the damage from lightning surge, which is of high voltage, large current, and instantaneity, the devices are in danger of communication interruption and blast fault which lead to serious accidents [1].

1.1 Test method

The lightning surge experiments can establish a common basis with the power lines, input/output lines and the telecommunication lines of the evaluation devices, when they are subjected to the high-energy pulse. As illustrated in the two figures, coupling and decoupling networks between the differential and the common mode tests are different. Coupling capacitance in the differential mode is 18 μF, while coupling capacitance in the common mode consists of a 9 μF capacitor and a 10 Ω resistor in series.

1.2 Experimental equipment and tested object

The YD/T993—2006 principle describes two different waveform generators, each has its own special applications according to the type of the tested ports. One kind of wave generators with ports connected to the symmetry of telecommunication lines is the 10/700 μS combination wave generator,

[*] To whom any correspondence should be addressed.

and for other conditions, especially for those connected to the power lines and short distance signal interconnect ports, are the 1.2/50 μS combination wave generators.

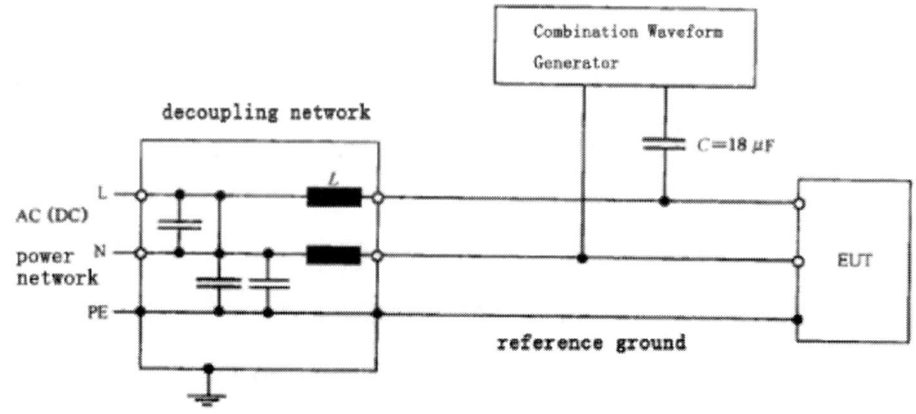

Figure 1. Lightning surge on the single-phase power line of the differential mode immunity test.

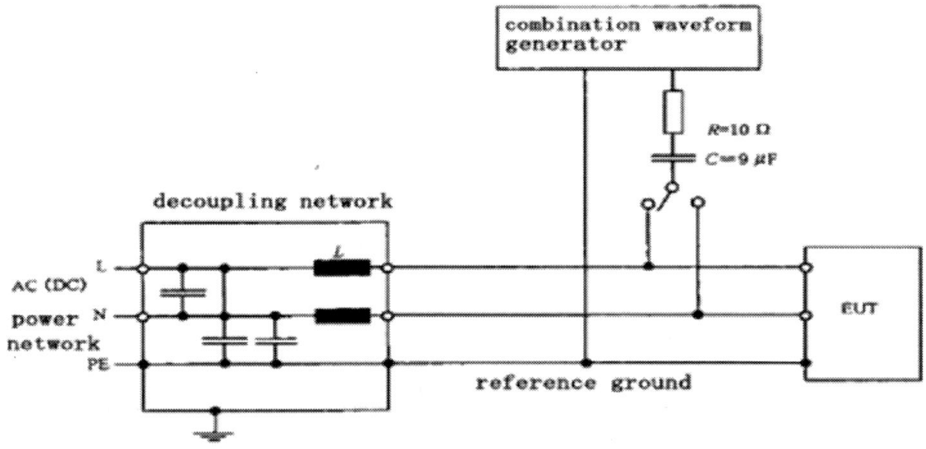

Figure 2. Lightning surge on the single-phase power line of the common mode immunity test.

DSL, which is short for Digital Subscriber Line, is adopted in the experiments. It is a peer-to-peer transmission technology with the transmission medium of copper phone lines, which is generally known as xDSL including HDSL, SDSL, VDSL, ADSL, and RADSL. In the experiment we selected ADSL, which is a kind of new data transmission mode [2], as the tested object. It is called as asymmetric digital subscriber line because of the asymmetry between ascending and descending bandwidth.

2. Research on lightning surge effect

2.1 Lightning failure phenomenon

The experiments mainly analyzed the situation of burnout in large areas caused by lightning surge, and showed preliminary analysis results for typical lightning failure. The distribution of the most serious damaged parts is shown in table 1. As can be seen, a large area of the PCB is burnout and one of the most typical cases is BCM6338 and BCM6301 chip failure.

Table 1. Statistical machine failure.

Failure parts	U203(FLASH)	T401(LAN Components)	PCB	U401(BCM6301)	U101(BCM6338)
Breakdown number	25	32	41	72	279

2.2 Withstand voltage test and current path analysis in the key components

2.2.1 Withstand voltage in ADSL transformer test and analysis.

Table 2 shows the withstand voltage test parameters of five ADSL transformers. As can be seen from the table, three samples have no significant changes in it, which indicates that high-voltage pulses pass into the lower-level circuit through ADSL transformer. We find that the value of the insulation resistance of the transformer in network port doesn't reduce in a damaged machine. However, for the fourth and fifth sample (l# sample, 2# sample retested respectively), when using a long rise time (25 s) in the test, the leakage current is less than 3.5 mA. The result demonstrates that the transformer doesn't breakdown under that voltage, and the leakage current in the first three samples is generated because the increasing high frequency components in voltage lead to distributed capacitance coupling with the coil resistance in a very short rise time.

Then we replaced the ADSL transformer with a 35 pF/3 kV capacitor connecting between the first level and the second level of ADSL transformer in TX and RX channel to simulate distributed capacitance of the ADSL transformer, respectively. The circuit worked normally when the voltage ranged from 5 kV to 9 kV. This indicates that the energy from equivalent capacitance (35 pF) between the first level and the second level of ADSL transformer coupling to lightning pulse can't damage TVS pipe or other circuits.

Table 2. Voltage test of the ADSL transformer.

Num / Type	Maximum voltage	Rise/hold time	Leakage current	Retest voltage	Resistance
1# / MT5422	5500 V	0.1 s/10 s	>3.5 mA	>5000 V	>10e+9 Ω
2# / MP9042	6000 V	0.1 s/10 s	>3.5 mA	>5800 V	>10e+9 Ω
3# / MP9042	6000 V	0.1 s/10 s	>3.5 mA	>5800 V	>10e+9 Ω
4# / MT5422	5500 V	25 s/10 s	<3.5 mA	——	>10e+9 Ω
5# / MP9042	6000 V	25 s/10 s	<3.5 mA	——	>10e+9 Ω

Tables 3 can be drawn when lightning tests are carried out in different ADSL transformers. Analyzing the phenomenon of table 3, we know that when lightning pulses pass through the ADSL transformer, the energy transfers through air discharge between the primary and the secondary transformer rather than capacitive coupling.

However, air discharge is a complex process. In a uniform electric field, breakdown voltage can be up to a maximum of 30 $kVcm^{-1}$ when air space is more than 1 cm; while in the non-uniform electric field, the breakdown voltage between the gas gaps drops [3], which is closely related to the two electrode surface shapes of air gap formation. Specifically, the smaller the curvature radius of the electrode is, the higher probability of the point discharge will be, and the lower the value of breakdown voltage will be [4].

Different point discharges for different discharging voltages, and the tests indicates that the minimum voltage 3.9 kV can generate discharge. The discharge of the transformer under the high voltage between first level and second level causes the explosion.

Table 3. Lightning current test of different ADSL transformers.

Test sample	ADSL transformer(2 kV)	ADSL transformer (5 kV)
Test conditions	All pins in the primary and the secondary transformer were short circuit, respectively. The 75 Ω resistance and the transformer were in series and the voltage increased from 3 kV to 6 kV.	All pins in the primary and the secondary transformer were short circuit, respectively. The 75 Ω resistance and the transformer were in series and the voltage increased from 5 kV to 9 kV.
Test phenomenon	The transformer worked normally as well as the resistance when the voltage was less than 6 kV, however, the transformer blasted when the voltage was larger than 6 kV. The value of the resistance increased. (>100Ω)	The transformer worked normally as well as the resistance when the voltage was less than 9 kV, however, the resistance got to an open circuit and blasted when the voltage was larger than 9 kV.

2.2.2 Internet access high voltage capacitor test and analysis.

Experiments about lightning tests on high voltage capacitor were performed. From the experiments we know, for a 1000 pF capacitor, it works well while transient current passes through and the internet access resistance blasts when the lightning surge voltage is about 2500 V. The value of the insulation resistance will reduce if the internet access transformer is breakdown. However, it never happens in the damaged devices, which indicates a low probability of the lightning surge going through internet access transformer. Table 4 and table 5 show the experiments about lightning test on high voltage capacitor with 0603 resistance in series connection. The test voltage ranges from 2 kV to 5 kV and the capacitor works well until the test voltage reaches 5000 V.

Table 4. Lightning test of the high voltage capacitor (1000 pF /3 kV).

Test voltage	2000V	2500V	3000V	3500V	4000V
Resistance value	normal	84Ω	106Ω	3.23KΩ	39.8KΩ
Capacitor value	normal	normal	normal	normal	normal

Table 5. Lightning test of the high voltage capacitor (1000 pF /2 kV).

Test voltage	2000 V	2500 V	3000 V	3500 V	4000 V	5000 V
Resistance value	normal	77.2Ω	78.4Ω	78.1Ω	84.4Ω	Burn out
Capacitor value	normal	normal	normal	normal	normal	Burn out

If we assume the rise time for the lightning pulse is 1.2 μs, and the frequency can be described as:

$$f = (\pi t_r)^{-1} \tag{1}$$

The derived frequency is about 265 KHz. And the equivalent impedance is:

$$Z_c = (2\pi f C)^{-1} \tag{2}$$

If we assume the capacity of the high-voltage capacitor is 1000 pF, and the derived impedance is 525 Ω. Therefore, the equivalent resistance from ground to internet access is:

$$Z_o = Z_C + Z_R \approx 600 \tag{3}$$

Capacitor will not have breakdown when high voltage surge is lower than it's withstand voltage capacity, but it couples with the 75 Ω matching resistance and discharges to ground through internet access. When the voltage is 2500 V, the transient current is estimated about 4 A. If this current is regarded as the resistance current, transient voltage can reach 300 V. It largely exceeds the rated voltage (50 V) of 0603 resistance, so the resistance is damaged.

The last test was the lightning surge in the combination model of the key components. The test circuit composed of a 2 kV ADSL transformer, a 1000 pF/2 kV capacitance and a 75 Ω resistance in series connection. The resistance worked normally when the voltage was less than 8.5 kV, however, the capacitance burned out when the voltage became even larger. We knew that there was no current in the circuit before the air discharge of the ADSL transformer [5]. The 1000 pF/2 kV capacitance broke down and the resistance burned out as long as the ADSL transformer discharged. And it never happened that the 75 Ω resistance burned out while capacitance worked normally in the process of the ADSL transformer discharge. There is no obvious relationship between the withstand voltage of the ADSL transformer and the resistant of the lightning surge effect [6]. But the withstand voltage of the internet access transformer is closely related to the resistant and higher withstand voltage in high voltage capacitance results in a better effect in resisting lightning surge.

3. Conclusions

One of the most significant tasks of researching lightning discharge failure is to determine the discharge route, which plays a key role in improving the design of the discharge protection circuit [7]. To analyze the influences of lightning surge to the electronic and telecom products, a series of experiments were performed. The following conclusions were drawn from the experiments:

(1) The failure positions were scattered and appeared in groups, and most of them were ground discharge;

(2) Lightning surge imported from telephone port and the current went through the way of air discharge into the ADSL transformer. Finally it coupled with the ADSL transformer of the subprime circuit;

(3) Internet access transformer had high withstand-voltage under the lightning pulse, and the lightning surge seldom passed through internet access transformer;

(4) After the current passed the AFE chip, the internet access matching resistance coupled with the large value capacitance. Because of the capacitance coupling, the lightning current passed through and then released to the ground via the computer network adapter of the terminal user.

According to the failure analysis, the paper deduced the discharge route of lightning surge which could provide some guiding significance for the design of lightning surge protection circuit and make a further contribution to improving the protection level.

References

[1] De Conti A, Silveira F H and Visacro S 2012 *30th Int. Conf. on Lightning Prot. (Cagliari, Italy)* vol 85 p 7

[2] Feng J H and Wang L M 2002 *ADSL Broadband Access Technol. Appl.* (Beijing: Press of People's Posts and Telecommunication)

[3] Caligaris C, Delfino E and Procopio R 2008 *IEEE Trans. Electromagn. Compat.* **50** 194

[4] Chang H C, Chen F H, Kuo C C and Chen T H 2011 *Energy Convers. Manage* **52** 693

[5] Aoki M and Oki E 2011 *IEICE Trans. Commun.* **E94-B** 3583

[6] Zamani M, Taghizadeh M, Moghadasi M N and Virdee B S 2012 *Analog. Integr. Circuits Signal Process* **71** 143

[7] Anderson D R, Gillberg J M, Torrey J W and Koneru J N 2012 *PACE* **35** 159

Journal of Physics: Conference Series **418** (2013) 012044 doi:10.1088/1742-6596/418/1/012044

Maximum capacity model of grid-connected multi-wind farms considering static security constraints in electrical grids

W Zhou[*1]**, G Y Qiu**[2]**, S O Oodo**[1]**, H He**[1,3]
1 College of Electrical Engineering, Dalian University of Technology, No.2 Linggong Road, Dalian City, China
2 State Key Laboratory of Structural Analysis for Industrial Equipment, School of Automotive Engineering, Dalian University of Technology, No.2 Linggong Road, Dalian City, China
3 East Inner Mongolia Electric Power Research Institute, Zhaowuda Road, Hohhot City, China

E-mail: zhouwei-ee@163.com

Abstract. An increasing interest in wind energy and the advance of related technologies have increased the connection of wind power generation into electrical grids. This paper proposes an optimization model for determining the maximum capacity of wind farms in a power system. In this model, generator power output limits, voltage limits and thermal limits of branches in the grid system were considered in order to limit the steady-state security influence of wind generators on the power system. The optimization model was solved by a nonlinear primal-dual interior-point method. An IEEE-30 bus system with two wind farms was tested through simulation studies, plus an analysis conducted to verify the effectiveness of the proposed model. The results indicated that the model is efficient and reasonable.

1. Introduction

Since wind energy is an environment friendly and renewable resource, installed wind power generation capacity is continuously increasing. However, as wind energy is increasing integrated into power systems, it has significant influence on the steady-state security of electrical grids, such as voltage stability problems and overloading of some transmission lines. For the power users, over-voltage will lead to a reduction in equipment lifetime, increase in energy consumed with no improvement in performance and even a puncture of insulation. While under-voltage will lead to the failure of premature equipment, failing to start up, and then increase temperature in the case of motor windings and loss of service. For the power system itself, it will suffer from a collapse of voltage and ultimately cascading blackouts. Therefore, the accepted maximum capacity issue of wind farms in power systems should be investigated in order to avoid system stability problems [1-5].

Similar studies for a given power system have been performed in literature [6-8], which present methods to find the maximum capacity of one grid-connected wind farm. The determination of the capacity in a wind farm is formulated as a maximization issue of the number of wind turbines.

* To whom any correspondence should be addressed.

However, there are generally more than one wind farm in an actual power system, and thus this paper investigates the maximum grid-connected capacity issue considering multi-wind farms.

The whole paper is structured as follows: firstly, the introduction is given; secondly, the power model in the wind farm is described, and then this paper introduces the proposed mathematics model of grid-connected wind capacity considering multi-wind farms in detail; finally, numerical simulation and analysis are elaborated to verify the feasibility of the proposed optimization model and method.

2. Power model in the wind farm

The power output of wind turbine is determined by the type of wind turbines. Usually, there are two kinds of wind turbine systems in current applications [8, 9]: the fixed-speed wind turbine system and the variable speed wind turbine system. The double-fed wind power generator regarded as a widely used variable speed wind turbine is chosen in this paper. So the active power for a variable speed wind turbine is as follows:

$$P_{wind} = \begin{cases} 0 & 0 \le v < v_i \\ a + bv^3 & v_i \le v < v_r \\ P_r & v_r \le v \le v_0 \\ 0 & v > v_0 \end{cases}, a = \frac{P_r v_i^3}{v_i^3 - v_r^3}, b = \frac{P_r}{v_r^3 - v_i^3} \tag{1}$$

where v is wind speed, v_i is cut-in wind speed, v_r is rated wind speed, v_0 is cut-out wind speed, P_r is rated power of wind turbine.

If all the wind turbines are identical in a wind farm, and then the total active power output of the wind farm is given by

$$P_{farm} = NP_{wind} \tag{2}$$

where N is the number of wind turbines in a wind farm.

The power factor or voltage of the bus which is connected to a wind farm will maintain constant by effective control, supposing the wind turbines operate at the constant power factor. Therefore, the reactive power output of the wind turbine can be expressed as

$$Q_{wind} = \frac{P_{wind}}{cos\varphi}\left(1 - cos^2\varphi\right)^{1/2} \tag{3}$$

where $cos\varphi$ is power factor. The total reactive power output of the wind farm is equal to

$$Q_{farm} = NQ_{wind} \tag{4}$$

3. Mathematics model of grid-connected wind capacity issue

As mentioned above, the determination of total capacity of all the grid-connected wind farms can be formulated as a maximization wind power accepted by systems, while satisfying the power flow equations and other several operation limits, including the restrictions of system stability, security and power quality, etc. In this paper, the objective function of optimization model is the sum of the power outputs of all the wind farms which can be expressed as

$$\max \sum_{m=1}^{M} N_m P_{wind,m} \tag{5}$$

where N_m and $P_{wind,m}$ are the number of wind turbines and power output per wind turbine with wind farm of m, respectively.

It can also be rewritten in the following form:

$$\min -\sum_{m=1}^{M} N_m P_{wind,m} \tag{6}$$

The constraints include equality and inequality constraints, and the power flow equations at each bus are treated as equality constraints, which must be satisfied. For the bus which is not connected to a wind farm, the active power and reactive power flow equations are

$$\begin{aligned} \Delta P &= P(x) - P_G + P_L = 0 \\ \Delta Q &= Q(x) - Q_G + Q_L = 0 \end{aligned} \tag{7}$$

where $x = [V \quad \theta \quad P_G \quad Q_G \quad N]^{\mathrm{T}}$, x is the optimization variable, V is the magnitude vector of bus voltages, θ is the angle vector of bus voltages, here N denotes the number vector of wind turbines, P_G and Q_G are active power and reactive power of conventional generators, P_L and Q_L are active power and reactive power demands. $P(x)$ and $Q(x)$ are nodal active and reactive power functions.

While the power flow equations at the bus connected to a wind farm are given by

$$\begin{aligned} \Delta P_{i_{wind}} &= P_{i_{wind}}(x) - P_{farm} - P_{G,i_{wind}} + P_{L,i_{wind}} = 0 \\ \Delta Q_{i_{wind}} &= Q_{i_{wind}}(x) - Q_{farm} - Q_{G,i_{wind}} + Q_{L,i_{wind}} = 0 \end{aligned} \tag{8}$$

where i_{wind} indicates the bus number at which the wind farm is connected to the grid.

The main concerns of this paper are operational limits of conventional generator, including the voltage limits, active power and reactive power limits of conventional generators, and the thermal limits of transmission lines in the grid system. They are given by

$$\begin{aligned} & V_{\min} \le V \le V_{\max} \\ & P_{G,\min} \le P_G \le P_{G,\max} \\ & Q_{G,\min} \le Q_G \le Q_{G,\max} \\ & -P_{L,\max} \le P_{ij} \le P_{L,\max} \\ & -P_{L,\max} \le P_{ji} \le P_{L,\max} \end{aligned} \tag{9}$$

4. Case study analysis

4.1. Setup of the test system

To examine the feasibility and merits of the proposed optimization model, the IEEE-30 test system consisting of two wind farms was studied. There are 2 generators and 4 synchronous condensers in the power system. Bus 1 is considered as a slack bus, and the synchronous condenser at bus 11 and 13 are replaced by wind farms. Double-fed wind turbine is selected maintaining a constant power factor, and the power factors in these two different wind farms are set to 0.95 and 0.98, respectively. All the maximum and minimum voltage limits are set to 1.06 (p.u.) and 0.94 (p.u.), respectively. In order to simplify the calculation, only active power constrains of main line are involved in the model. And the thermal limits of main line 7, 9 and 22 are equal to 150 MW, 130 MW and 110 MW, respectively. The rated power output of wind turbine in the wind farm connected at bus 11 is 2 MW, and that in the wind farm connected at bus 13 is 1.5 MW. The initial wind turbine numbers of these two wind farms are both set to 100.

4.2. Simulation results

In order to obtain the optimal solution of the proposed model, the nonlinear primal-dual interior point method is applied to solve the grid-connected wind capacity issue [10]. The related program was coded using Matlab language, and then the result was achieved by running the optimization program for 2.9 seconds. The variation process of total capacity of grid-connected wind farms is given in Figure 1. Simulation results demonstrate that the final capacity of the grid-connected wind farms is 375.9 MW, and the optimal wind turbine number of the wind farm connected to bus 11 is 88, the other one is 132.

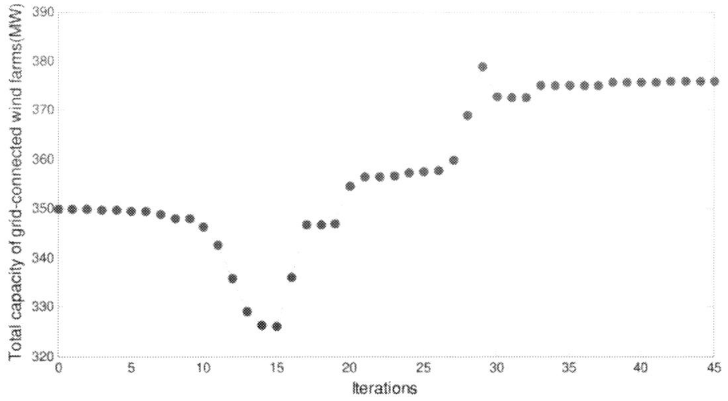

Figure 1. The iteration curve of total capacity of grid-connected wind farms.

Table 1. Active and reactive power of conventional generator and wind farm in the optimal solution.

Generator bus number	Active power (MW)	Generator bus number	Reactive power (Mvar)
1	-200	**1**	101.17
2	0	**2**	50
5	100	**5**	40
8	27.39	**8**	40
11	177.11	**11**	58.21
13	198.79	**13**	40.37

Table 2. Magnitude of bus voltage in the optimal solution.

Bus number	Magnitude of bus voltage (p.u.)	Bus number	Magnitude of bus voltage (p.u.)	Bus number	Magnitude of bus voltage (p.u.)
1	1.0600	**11**	1.0346	**21**	0.9669
2	1.0543	**12**	0.9927	**22**	0.9676
3	1.0253	**13**	1.0096	**23**	0.9582
4	1.0228	**14**	0.9779	**24**	0.9549
5	1.0522	**15**	0.9693	**25**	0.9588
6	1.0233	**16**	0.9748	**26**	0.9400
7	1.0276	**17**	0.9736	**27**	0.9754
8	1.0261	**18**	0.9593	**28**	1.0162
9	0.9842	**19**	0.9571	**29**	0.9545
10	0.9798	**20**	0.9618	**30**	0.9424

The specific optimal results are shown in Tables 1-3. It can be found that the active power output of the conventional generator connected to bus 5 reaches the upper boundary, the active power outputs of the conventional generators connected to bus 1 and bus 2 reach the lower boundary, the reactive power outputs of the conventional generators connected to bus 2, bus 5 and bus 8 reach the upper limit, the voltage magnitude of bus 1 reaches the upper boundary, and that of bus 26 reaches the lower boundary. Other optimal variables are between the upper and lower limits. That means we can get the acceptable maximum capacity of grid-connected wind farms while satisfying the operation constraints of power systems.

Table 3. Calculation results for the active power flow of transmission line.

Branch number	Active power flow of transmission line P_{ij} (MW)	Active power flow of transmission line P_{ji} (MW)
7	-34.74	34.88
9	44.74	-44.14
22	12.98	-12.79

5. Conclusions

A deterministic optimization model is used to obtain the maximum capacity of grid-connected multi-wind farms, in which several important limitations are considered including voltage stability limits, active power and reactive power limits of conventional generators, and the thermal limits of transmission lines in the power system. The proposed model is implemented on an IEEE-30 bus test system with two wind farms. The optimal solution is determined by the nonlinear primal-dual interior point method. The results indicate that the proposed model is efficient and the solution is reasonable. In other words, we can get the acceptable maximum capacity of grid-connected wind farms and the optimal allocation of system voltages and generator power outputs while satisfying the static security constraints in electrical grids. Moreover, in order to consider the variable and stochastic characteristics of wind power output, the uncertainty of wind speed will be involved using a probabilistic method in the future works.

Acknowledgments

This project was supported by China Postdoctoral Science Foundation (2012M510805), the Doctoral Start Fund of Liaoning Province (20121023) and the Fundamental Research Funds for the Central Universities (DUT11RC(3)09).

References

[1] Ullah N R, Thiringer T 2007 *IEEE Trans. on Energy Convers.* **22** 52
[2] Wiik J, Gjerde J O, Gjengedal T and Gustafsson M 2002 *IEEE PES winter meeting* p 27
[3] Zhou W, Peng Y, Sun H, Wu J K and Fan F 2008 *Power System Technology* **32** 79
[4] Feijoo A E, Cidras J and Dornelas J L G 1999 *IEEE Trans. on Energy Convers.* **14** 1582
[5] Banakar H, Luo C L and Ooi B T 2008 *IEEE Trans. on Power Syst.* **23** 150
[6] Lei Y Z, Wang W S, Yin Y H and Dai H Z 2002 *Proceedings of the CSEE* **22** 32
[7] Wu J, Li G J and Sun Y Z 2007 *Power System Technology* **31** 15
[8] Zhao M, Chen Z and Blaabjerg F 2006 *Renew. Ener.* **31** 2171
[9] Feijoo A E and Cidras J 2000 *IEEE Trans. on Power Syst.* **15** 110
[10] Torres G L and Quintana V H 1998 *IEEE Trans. on Power Syst.* **13** 1211

Journal of Physics: Conference Series **418** (2013) 012045 doi:10.1088/1742-6596/418/1/012045

Anti-wear additive content in fully synthetic PAO and PAG base oils and its effect on electrostatic and tribological phenomena in a rotating shaft–oil–lip seal system

Juliusz B. Gajewski[*], **Marek J. Głogowski**
Electrostatics and Tribology Research Group, Department of Electrical and Thermal Power Engineering and Technology, Institute of Heat Engineering and Fluid Mechanics, Wrocław University of Technology
Wybrzeże Wyspiańskiego 27, 50-370 Wrocław, Poland

E-mail: juliusz.b.gajewski@pwr.wroc.pl

Abstract. The paper presents the results of experiments on electrostatic and tribological aspects of different anti-wear additive's contents when an additive is blended with different fully synthetic (poly-α-olefin) and PAG (polyalkylene glycol) base oils in a rotating shaft–oil and oil–lip seal interfacial system. The experimental results are the relationships of electric potential induced in a lip seal's stiffening ring to angular velocity of a rotating metal shaft and to temperature of the oils tested. The braking torque of a shaft is measured with a torquemeter sensor connected directly with a microprocessor-based system for controlling the rotational speed and for measuring the shaft's braking torque and oil temperature. The beneficial and promising results are obtained for PAG when an external DC electric field is applied to the system and the braking torque is then reduced for a certain combination of the base oil and additive's contents. On the basis of the former and present research results an analysis is made to permit one to show how the type of the oils and additives tested can affect both interfaces: rotating shaft–oil and oil–lip of the lip seal and especially the braking torque.

1. Introduction

Fully formulated engine oils consist of pure base oils and some part of different additives that improve the tribological properties of a base stock or blend of base stock as used in engine oil. As the base stock, mineral, semisynthetic and synthetic oils, and just recently research is carried out upon the introduction of new engine oils into the market that are based on PAGs (polyalkylene glycols) [1–4]. These base oils have unique properties and among them especially important are low toxicity and improved biodegradability which results in environmental friendliness [5, 6].

Synthetic oils (poly-α-olefins), known as PAOs, are obtained through the polymerization of olefins and next the hydrogenation of products so obtained is carried out to get saturated compounds. These oils are characterized by the high viscosity index, polarity and low toxicity. PAOs are mainly applied to formulate automotive engine oils where the low solidification temperature of oil, low volatility and good thermal-oxidative stability are required.

Polyalkylene glycols (PAGs) are the result of the chemical reaction of substrates that contain hydroxyl groups (alcohols, diols, polyols, carboxylic acids) with olefin oxides. PAGs have high polarity

[*] To whom any correspondence should be addressed.

which results in their surface activity. In addition they have such properties as: good thermal and shear stabilities, god viscosity index (VI), high flash point and auto-ignition temperature, low toxicity and flammability. Lubricants produced on the basis of PAGs have practical applications to braking systems, compressors and hydraulic systems where the low flammability is required.

One of the important additives in engine oils are those of anti-wear (AW) type whose role is the protection of metal surfaces against the effects of boundary friction. The widely used an anti-wear additive is zinc dialkyldithiophosphate (ZDDP). Apart from the anti-wear properties it is also an antioxidant and protects from extreme pressures (EP additive) [7].

Another often used additive to oils is a friction modifier (FM). These agents reduce or increase the coefficient of friction according to a purpose. In engine and gear oils its main task to do is the reduction of friction and at the same time the protection of a metal surface against the friction wear. It use can reduce the fuel consumption and energy losses in friction junctions.

Rizvi in his book [8] says that "Anti-wear and extreme pressure agents have structures that are similar to those of the friction modifiers, with the difference that their hydrocarbon chains are much shorter and are thermally labile. In other words, their polar to non-polar ratio is fairly high, which makes them more surface active. Anti-wear and EP additives both provide protection by a similar mechanism, except that EP additives typically require higher activation temperatures and load than anti-wear additives. Simply stated, anti-wear additives perform under mild conditions and EP additives perform under severe conditions." Also: "In addition to reducing friction, the friction modifiers also reduce wear, especially at low temperatures where the anti-wear agents are inactive, and they improve fuel efficiency." The above statements show the properties of both additives that are interesting and important from the point of view of the research undertaken by the authors.

ZDDP and FM protect metal parts of machinery through the formation of an adherent film on metal surfaces. The film reacts with a ferric oxide layer which adheres to the surface and protects it from wear during metal–metal rubbing. During the relative movement of a metal surface and oil the outermost layer of the film is constantly sheared and replenished after it has been disrupted by the mechanical contact [9]. This layer is a transfer one that reduces friction.

While the metal shaft rotates friction occurs and the outermost ZDDP (AW) or FM layers are disrupted. Each contact and friction generates electric charges during the process of tribocharging. Tribocharging occurs not only at this surface but also in the oil film, as well as at the surface of a lip of a lip seal. It can produce long-range electrostatic effects which affect friction and braking torque in the interfacial system: rotating shaft–oil–lip seal (a friction junction). The charges generated in the system can play an important role in the adhesion at the surfaces in both interfaces: rotating shaft–oil and oil–lip of the lip seal [10]. Tribocharging always takes place in the similar interfacial systems and affects the operation of rotating or moving parts in different types of machinery [11, 12].

Here some experimental results of research upon tribocharging of the synthetic PAO and PAG base oils without and with the AW (ZDDP) and FM agents added to form their blends of different concentrations. The important factors that affect tribocharging are the temperature of pure base oil or the AW/FM–oil blend and the angular velocity of a shaft.

The effect of an external DC electric field applied between the rotating shaft and the stiffening ring of a lip seal on the braking torque of the shaft is also examined. The electric field is used to compensate for the natural electric field generated by net charge of mobile ions and molecules in the zone between both interfaces.

The motivation for doing such research is the introduction of new PAG-based engine oils into the market and the PAG's quite good biodegradability [2, 5, 6, 13] with relatively low kinematic viscosity. Also its relatively low resistivity suggests it be charged during tribocharging less intense than other base oils [14–16]. This in turn should result in the lower friction and reduced braking torque. As a result and as is supposed the effect of a DC electric field on the reduction of the shaft's braking torque should also be more visible, beneficial and even stronger. It can be predicable on the basis of the theory and the authors' many years' experience in the research upon tribocharging of different pure base oils and their blends with different additives under different dynamic and temperature conditions.

2. Test stands, materials tested, and measurements

2.1. Experimental set-ups and measurement procedure

The test stand, as shown in figure 1, permits one to conduct experiments on tribocharging of different fully formulated engine oils, pure engine base oils as well as their blends with additives imparting or improving certain properties of petroleum products and especially engine oils.

The experimental set-up is made up of: the housing of a sump with the seating of a rotary lip seal (1) and base oil or the AW/FM–base oil blend (2); a rotary lip seal (3); a stiffening ring (a metal ring) of a seal (4); a dielectric ring (5); an earthed steel shaft (6) rotating and driven by an electric motor (7) with a regulated angular velocity; an electric oil heater (8) controlled by a microprocessor-based PID controller; a thermocouple (9); a standard capacitor (10); an electrometer Keithley 6517A (11), and a computer (12).

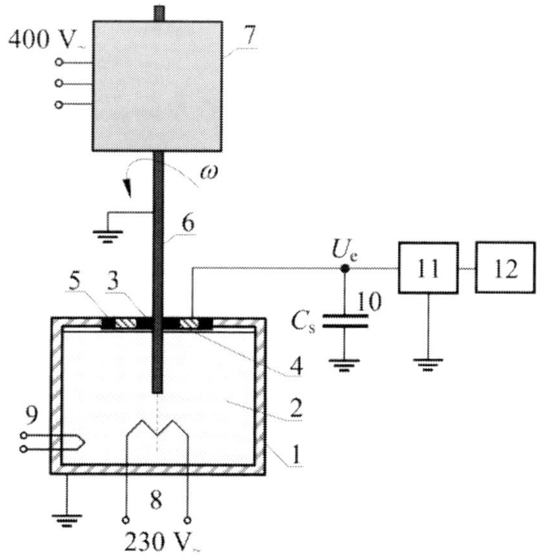

Figure 1. Schematic of the experimental set-up and the computer-based system for measuring the potential.

The potential U_e of a stiffening ring is induced by electric charges in the oil film and is a certain measure of the process of tribocharging that occurs in the gap between both surfaces of the shaft and of the lip seal – its lip. The ring is directly connected with the electrometer the input of which is in parallel to the standard capacitor C_s of relatively high capacitance to attenuate fluctuations of the ring's potential and to assure accurate measurements of the potential with respect to the earth with the use of an electrometer Keithley 6517A.

To determine the effect of a DC electric field on the braking torque of the rotating shaft another test stand is used, which is shown in figure 2. It consists of: the housing of an oil chamber (1); a seal's stiffening metal ring (2); the lip seal tested (3); an insulator (4); the oil tested (5); an air bearing (6); the sensor of a torquemeter (7); an electric engine (8); a steel shaft (9) earthed through carbon brushes; a microprocessor-based system for controlling the angular velocity and for measuring the braking torque of the shaft and the temperature of oil (10); a DC power supply (11); an oil heater (12), and a thermocouple (13). The chamber is filled with the oil tested up to a geometrical axis of the rotating shaft. The whole chamber is a simplified model of the part (oil sump or crankcase) of a real car engine.

The braking torque M is measured with a torquemeter which is connected with its sensor mounted on an earthed rotating metal shaft. The torque sensor is a non-contact and inductive one and is based upon the phenomenon of electromagnetic induction. The DC voltage U_{DC} is applied between the seal's stiffening ring and the earth to produce a DC electric field; here the rotating shaft is precisely earthed as is the housing of an oil chamber. The chamber is filled with the oils and their blends up to the shaft's geometrical axis as are original crankcases of cars.

Both measurement systems, as presented in figures 1 and 2, are microcomputer-based ones to enable the measurement data processing, analysis, acquisition, display, and storage.

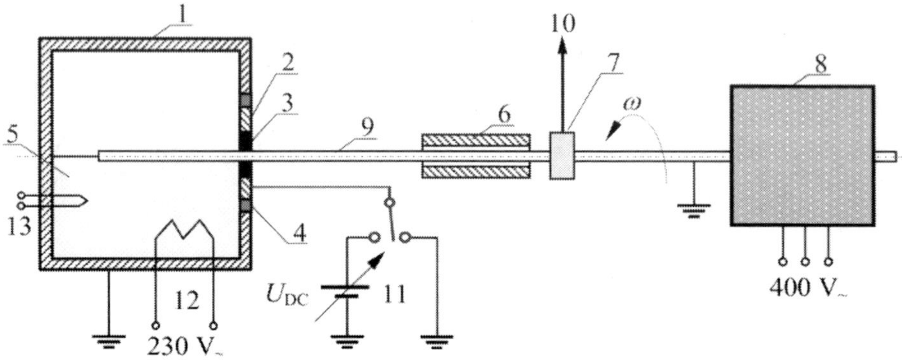

Figure 2. The test stand for measuring the braking torque with and without an external DC electric field.

2.2. Materials tested and experimental procedure

The experiments are carried out for the following pure synthetic base oils: PAO 6 produced by NESTE OIL N.V., Beringen, Belgium and ROKOLUB68® (PAG) produced by PCC ROKITA, Brzeg Dolny, Poland. A specification of the oils tested is shown in table 1. ZDDP Lubrizol 1395 (AW) produced in Austria by LUBRIZOL and IRGALUBE F10A (FM) produced in France by CIBA are additives widely used in engine and gear oils. In the experiments on tribocharging the synthetic PAO and PAG base oils are used without and with the AW and FM agents added to the oils. The agent's contents in the blends are 0.1 and 0.5% by weight.

Table 1. Specification of the base oils tested.

Quantity and unit		ROKOLUB68	PAO 6
Kinematic viscosity			
at 40°C		56.4	30.2
at 100°C,	$[mm^2 \cdot s^{-1}]$	10.5	5.8
Viscosity index,	[–]	176	138
Density at 15°C,	$[kg \cdot m^{-3}]$	990 at 20°C	825
Resistivity			
at 40°C		$1.3 \cdot 10^7$	$2.6 \cdot 10^{11}$
at 100°C	[Ω·m]	$3.5 \cdot 10^6$	$6.4 \cdot 10^{10}$
Relative permittivity			
at 40°C		5.8	1.9
at 100°C	[–]	5.1	1.9

The steel shaft roughness is 0.32 μm. An 88 mm diameter fluorocarbon lip seal is used in all the tests performed.

The pure base oils and their blends with the AW and FM agents tested at the experimental set-up, as shown in figure 1, are heated from 60 to 110°C every 10°C in an oil chamber and are controlled by a microprocessor-based PID controller. For each temperature within the range the measurements of the potential are performed for given angular velocities of 500, 750, and 1000 rpm. It permits one to obtain many various relationships for the stiffening ring's potential as a function of both the temperature and the angular velocity of the shaft.

When the same base oils and the AW/FM–oil blends are tested at the test stand (figure 2), the range of temperatures is 70 to 120°C while the shaft's angular velocity is changed from 500 to 3000 rpm and here the braking torque is a function of DC voltage for the temperature and angular velocity as parameters. All the experiments upon the effect of a DC electric field on the braking torque are carried out according to the design of experiments (DOE).

3. Results and discussion

Some selected experimental results are presented in the form of different characteristics (relationships) which are obtained at both experimental set-ups (figures 1 and 2).

3.1. Measurement of the potential

Figures 3 to 8 show the plots of the stiffening ring's potential U_e as a function of the temperature T for shaft's angular velocities of 500, 750, and 1000 rpm, and for the pure PAO 6 and PAG base oils, and their blends for 0.1 and 0.5 wt% contents of the AW or FM agents, correspondingly.

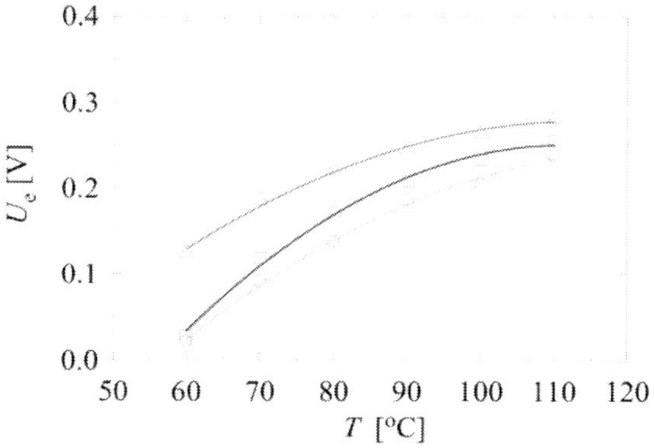

Figure 3. PAO 6. Potential as a function of the temperature for 500 rpm and the different FM contents: ○ – pure, △ – 0.1 wt%, and □ – 0.5 wt%.

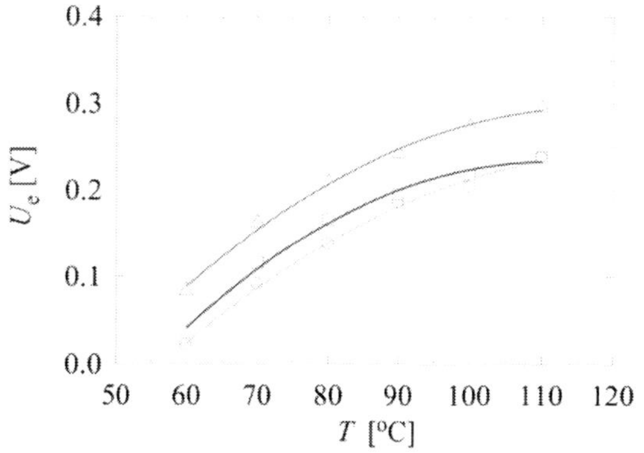

Figure 4. PAO 6. Potential as a function of the temperature for 500 rpm and the different AW contents: ○ – pure, △ – 0.1 wt%, and □ – 0.5 wt%.

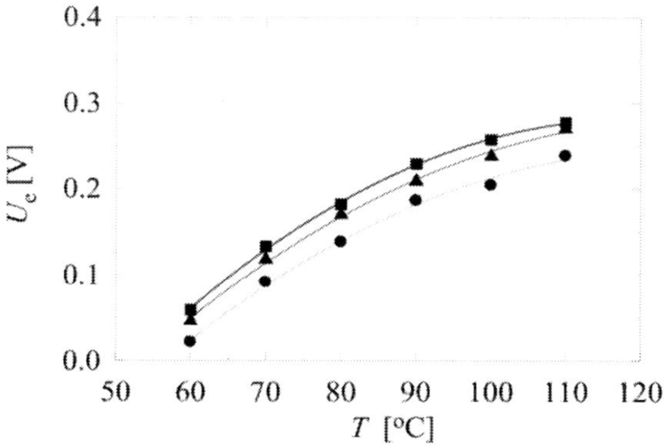

Figure 5. PAO 6 – pure. Potential as a function of the temperature for the different angular velocities: ● – 500 rpm, ▲ – 750 rpm, and ■ – 1000 rpm.

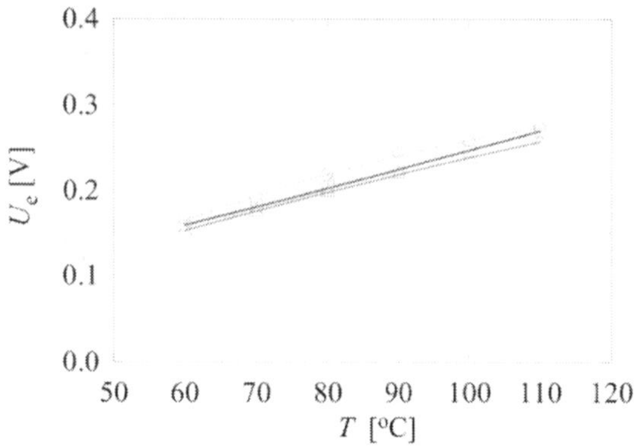

Figure 6. PAG. Potential as a function of the temperature for 500 rpm and the different FM contents: ○ – pure, △ – 0.1 wt%, and □ – 0.5 wt%.

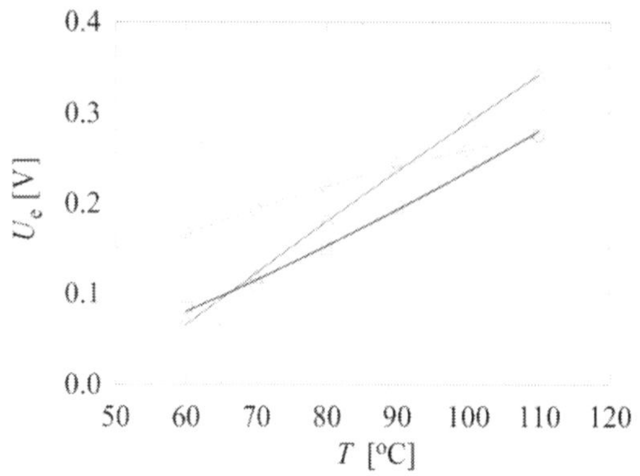

Figure 7. PAG. Potential as a function of the temperature for 500 rpm and the different AW contents: ○ – pure, △ – 0.1 wt%, and □ – 0.5 wt%.

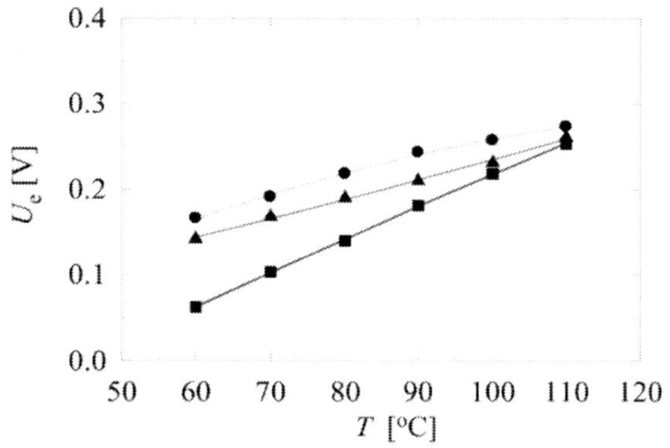

Figure 8. PAG – pure. Potential as a function of the temperature for the different angular velocities: ● – 500 rpm, ▲ – 750 rpm, and ■ – 1000 rpm.

In general the potential of the seal's stiffening ring always increases with increasing temperature and is positive regardless of base oil and its blend with the different additives and their contents in the blends. PAO 6 only exhibits a stronger increase in the potential compared to PAG which results in the broader range of potential changes within the same range of temperatures. It means that in the case of PAO 6 tribocharging is more intense in layers in the gap between the surfaces of the rotating shaft and the lip of the lip seal. It must be attributed to the distinct difference in resistivity of both oils (table 1). It is interesting and strange that for PAO 6 the potential and thus tribocharging is stronger only for 0.1 wt% and moreover the AW agent causes the range of changes in the potential to be greater than for the FM agent. In the case of PAG oil the averaged dependence of the potential on temperature is somewhat stronger for pure base oil than for the AW/FM–oil blends. Here the chemistry of the oil and its blends should play an important role in the process of tribocharging that yields such surprising results.

The effect of the shaft's angular velocity on the potential and tribocharging is also surprising all the more so because one can expect that the process should always be more intense and give the stronger electric field for higher velocities. Here in the case of PAG this prediction confounds and contradicts the generally recognized knowledge and theory of static electrification. Chemistry again?

3.2. Measurement of the braking torque

The braking torque M measured is presented here as a relative one M_r (a dimensionless value) to enable the comparison of the torques for different oils and their blends with additives of different viscosities. The relative torque M_r means the ratio of the braking torque M_{DC} measured under the action of the DC electric field of both polarities (+/–) to the braking torque M_{NDC} without any external DC electric field applied: M_r M_{DC}/M_{NDC}. The trends of the relative dimensionless value of the braking torque M_r as a function of the DC voltage of both polarities are shown in figures 9–12. The graphic symbols used for the data points mean: ● – pure base oil, ▲ – base oil and 0.1 wt% additive, ■ – base oil and 0.5 wt% additive for negative DC voltage; and ○ – pure base oil, △ – base oil and 0.1 wt% additive, □ – base oil and 0.5 wt% additive for positive DC voltage. The research results shown here are selected out of many other obtained during a long series of experiments and are preliminary and most representative ones.

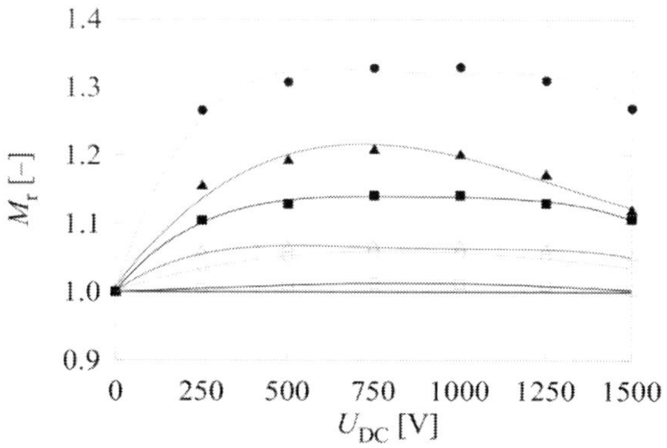

Figure 9. The sample plot of the relative braking torque as a function of DC voltage for PAO 6 and its blend with the FM agent – 70°C and 3000 rpm.

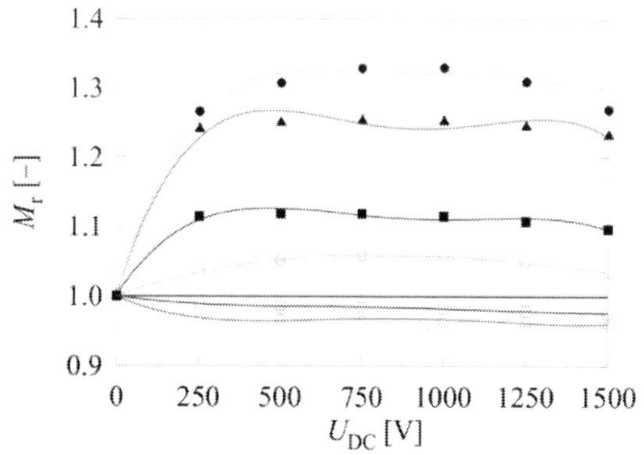

Figure 10. The sample plot of the relative braking torque as a function of DC voltage for PAO 6 and its blend with the AW agent – 70°C and 3000 rpm.

The results obtained for the PAO 6 and PAG base oils as well as for their blends with the FM and AW agents suggest that their chemical structure play a significant role in the process analysed and has an effect on the shaft's braking torque under external DC electric field. The relative braking torque Mr as a function of the DC voltage applied of the negative and positive polarities is shown in figures 9 and 10 for PAO 6 and its blend with the FM and AW agents under steady state conditions for a constant angular shaft velocity of 3000 rpm and a temperature of 70 °C. The positive polarity of the DC voltage applied to the system causes the relative braking torque to decrease only for the AW agent (figure 10). For the FM agent and for the DC voltage of both polarities the braking torque tends to increase in some range of DC voltages (figure 9).

A 33% increase in the relative braking torque of the rotating shaft is observed for PAO 6 blended with the FM and AW (figures 9 and 10) when the DC voltage of both polarities is applied.

The PAG base oil and its blends with the FM and AW agents with the same concentrations are also examined. The results yielded are quite different from those presented for PAO 6. The relative braking torque always decreases both with the negative and positive DC voltage applied.

Maximum reductions of about 7% and 5% in the relative braking torque occur for PAG blended with the same agents, respectively, as shown in figures 11 and 12. These results are obtained under steady state conditions for a constant oil temperature of 70°C and the negative DC voltage.

The chemical constitution of the oils tested results in their polarity, and especially, the polarity of their bonds, as supposed. PAO is non-polar while PAG is polar. It can explain the difference in their behaviour under external DC electric field and effect on the reduction of the braking torque. The external DC electric field affects polar chemical constitutions more strongly than those non-polar, as observed here.

4. Concluding remarks

The results of preliminary experiments presented here demonstrate that the PAG's chemical and physical properties cause the braking torque of the rotating shaft to be reduced under the action of an external DC electric field and especially when the polarity of the DC voltage applied to the interfacial system is negative. It is observed both when it is tested both as pure base oil and in the blend with both additives used. This outcome is similar to the results obtained by the authors earlier [17, 18]. The chemical and physical properties of the oil and additives exert a significant and beneficial effect on the reduction of the braking torque under external DC electric field. This in turn suggests the possibility of such reduction in the braking torque in real engines, in fuel consumption and energy losses.

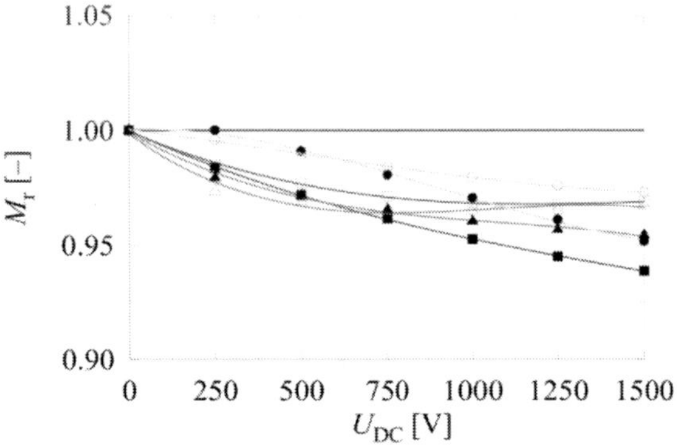

Figure 11. The sample plot of the relative braking torque as a function of DC voltage for PAG and its blend with the FM agent – 70°C and 3000 rpm.

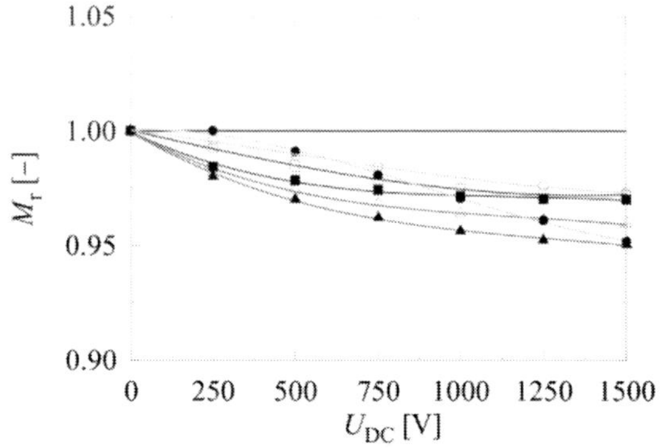

Figure 12. The sample plot of the relative braking torque as a function of DC voltage for PAG and its blend with the AW agent – 70°C and 3000 rpm

References

[1] Woydt M 2007 *J. ASTM Int.* **4** (10) 1
[2] Woydt M 2011 *J. ASTM Int.* **8** (6) 1
[3] Thoen J A, Woydt M and Zweifel D F 2011 US Patent US 2011/0039741 A1
[4] Gili F, Igartua A, Luther R and Woydt M 2011 *Lubrication Sci.* **23** 313
[5] Matlock P L, Brown W L and Clinton N A 1999 *Chapter 6 Polyalkylene Glycols* in: *Synth. Lubr. and High-Perform. Funct. Fluids* 2nd ed. (Marcel Dekker, Inc., NY)
[6] Beran E, Hull S and Steininger M 2012 *J. Polym. Environ.* DOI 10.1007/s10924-012-0445-2
[7] Spikes H 2004 *Tribol. Lett.* **17** 469
[8] Rizvi Syed Q A 2009 *A Compre. Review of Lubr. Chem., Technol., Sel.,and Design* (ASTM International, West Conshohocken, PA, USA)
[9] Hsu S M 2004 *Tribol. Int.* **37** 553
[10] Gajewski J B and Głogowski M J 2009 *J. Electrostat.* **67** 342
[11] Gajewski J B and Głogowski M J 2001 *J. Electrostat.* **51–52** 124
[12] Gajewski J B and Głogowski M J 2005 *J. Electrostat.* **63** 1049
[13] Beran E 2008 *Wpływ budowy chemicznej bazowych olejów smarowych na ich biodegradowalność i wybrane właściwości eksploatacyjne* (*Effect of chem. Struct. on biodegradability and some service prop. of lubricating base oils*) (Wrocław University of Technology Press, Wrocław) in Polish
[14] Gajewski J B, Głogowski M J, Gatner K and Gawliński M J 2010 *Tribol. Int.* **43** 1012
[15] Gajewski J B, Głogowski M J and Paszkowski M 2010 *Proc. 7th Conf. of the French Society of Electrostatics SFE 2010,* (Montpellier, France) p 341
[16] Gajewski J B, Głogowski M J, Paszkowski M and Czarnik-Matusewicz B 2011 *Proc. 13th Int. Conf. Electrostatics 2011,* (Bangor, Wales, UK) *J. Phys. Conf. Ser.* **3-1**

[17] Gajewski J B and Głogowski M J 2007 *Tribol. Int.* **40** 56
[18] Gajewski J B and Głogowski M J 2010 *Tribol. Lett.* **38** 179

7th International Conference on Applied Electrostatics (ICAES-2012) IOP Publishing
Journal of Physics: Conference Series **418** (2013) 012046 doi:10.1088/1742-6596/418/1/012046

The design and simulation test of wireless antenna protection network

Zipeng Chen, Yawen Dai, Peng Li, Zhuoqiu Li*
School of Science, WuHan University of Technology, Wuhan, China

E-mail: tank924@126.com

Abstract. In this paper, a wireless antenna protection program has been designed. In the program, the TVS diode was used as the first lever for protection, and the π-type high pass filtering network as the second lever. As a result, the program not only has the traditional function of ESD protection, which can avoid the high voltage damage to the internal circuit, but also achieves the purpose of load matching, ensuring the signal source not to distort. The ADS simulation software was used to test the ability of this program for filtering and impedance matching, which proved the feasibility of this program. The wireless antenna protection network has been practically used, and its' performance of anti-electromagnetic interference has been validated.

1. Introduction

Wireless Sensor Networks (WSN) are widely used in structural health monitoring, as the WSN device has the characteristics of flexible network topology, and easy for installation and maintenance. But the problem is that the antenna is easy to be damaged or affected by static electricity. The length of the antenna may reach 10 cm which makes the effects of external electromagnetic environment on antenna obvious, especially for the noise, static electricity, or thunder and lightning. This kind of electromagnetic interferences will cause great damage and even destructions once they get into the inside of WSN. Therefore, it is very important to take appropriate protective measures in engineering design, and to improve the anti-interference ability of the antenna, which has become the focus of engineering research [1].

The rapid change in strong magnetic field can be caused by electrostatic discharge and lightning induction, which would produce a high electric potential when the antenna of the WSN is close to the strong magnetic field. Experiment shows that the magnetic field generated by a 10 kV electrostatic spark discharge at 5 cm, is almost the same order of magnitude as the magnetic field generated by lightning induction in tens of meters. But the rising edge is much steeper in electrostatic discharge than that in lightning induction [7-8].

In this paper, a wireless antenna protection system with two levels has been designed. And the ADS simulation software was used to test the ability of this program for filtering and impedance matching, which proved the feasibility of this system.

* To whom any correspondence should be addressed.

Journal of Physics: Conference Series **418** (2013) 012046 doi:10.1088/1742-6596/418/1/012046

2. Protection system design

2.1. Formatting the title Wireless antenna protection system design

In the design of wireless antenna protection, the common measure is to placement electrostatic suppression absorbers, such as discharge tube, varistors or transient voltage suppressor (TVS), which can prevent the destructive current and surge caused by electromagnetic pulse. TVS has the advantages with high corresponding speed, low junction capacitance [2-3].

The ESD0P2RF TVS diode was selected as the first level protection system device, which could adaptation to 2.4 G frequency protection circuit. The maximum compliance voltage of this TVS is 38 V, with maximum current as 40 A, and the inter-electrode capacitance as 200 nF.

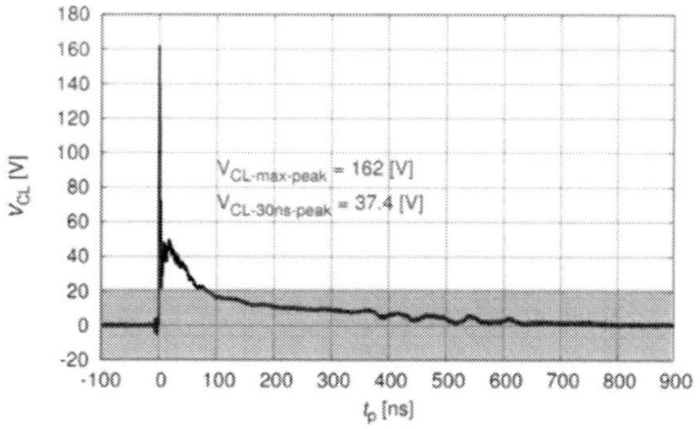

Figure 1. IEC61000-4-2 VCL = f(t), 15 kV positiv pulse from pin 1 to pin 2.

Figure 1 shows the result of IEC61000-4-2 electrostatic discharge immunity test for the TVS diode. The gray part in the figure 1 is standing for the range of withstand voltage of the wireless node. From the discharge waveform, it can be seen that the TVS clamped the voltage between electrodes in 162 V within 3 ns. After 70 ns the voltage was clamped in the safety range (the gray part). As the result, the second level protection system is needed for clamping the impulse voltage in the safe range instantaneously.

The second level protection system has to not only realize the function of instantaneous electromagnetic protection, but also ensure the impedance matching of the load and the source [4]. The common method to ensure the impedance matching is to insert passive matching network between the source and the load. The π-type network is selected in this program, because the bandwidth of π-type is much wider without a blind area. At the same time, the π-type network can also be used as high pass filter.

2.2. Protection circuit design

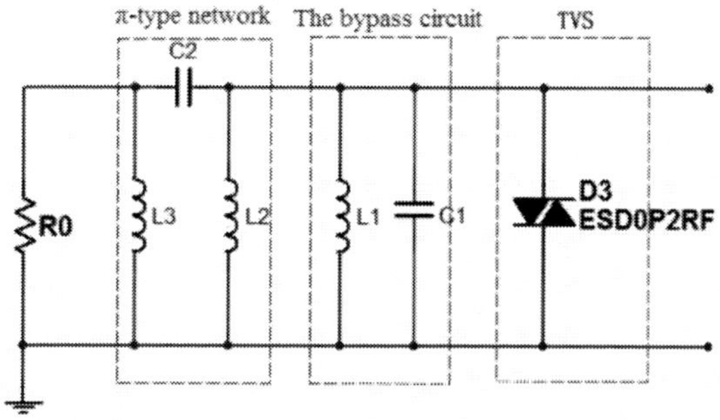

Figure 2. The design of antenna protection system.

Figure 2 shows the design of antenna protection system, in which D3 is the first level protection system device where TVS is used. C1 acts as a bypass capacitor, which can provide a low impedance pathway for 1 GHz electromagnetic pulse. The resonant circuit is composed of L1 and C1, which serves as a parallel frequency selection LC network. The second level is consisted of L2, C2 and L3, which can form the π-type matching network. The main function of this network is to filter out any signals whose frequency is less than 1 GHz, while the frequency of lightning and electrostatic discharge are within the filtered range [5]. Once the frequency of input signal is equal to the resonant frequency of the circuit, the resonance will occur in the parallel frequency selection LC network. The resonant frequency f_0 can be expressed as:

$$f_0 = \frac{1}{2\pi\sqrt{LC}} \tag{1}$$

In this formula, the carrier centre frequency f_0 is 2.4 GHz. The value of the bypass capacitor C1=1/f, in which f means the frequency of interference signal. In this system, f=1 GHz, and C1=1.0 pF. The ESD0P2RF TVS diode can be equivalent to a capacitor of 200 nF in the impedance measurement. Therefore, L1 can be calculated as 3.6 nH.

The π-type network, which acts as the second level protection system, can be calculated by K-type 3 order normalized high-pass filter model. The calculation formulas of capacitance and inductance are as follow:

$$L = \frac{R}{2\pi f_c} \tag{2}$$

$$C = \frac{0.5}{2\pi f_c R} = \frac{1}{4\pi f_c R} \tag{3}$$

$$\Rightarrow LC = \frac{1}{8(\pi f_c)^2} \tag{4}$$

In the formula C is the series capacitor; L is parallel inductance; R is the resistor between both sides; and f_c is cutoff frequency.

The value of resistance capacitance in the π-type matching network can be easily calculated by using the graphic method with Simth chart [9]. First of all, the load impedance Z=R+jX should be normalized with feeder impedance W. Normalized impedance is represented by z=Z/W = r+jx. Secondly, two important circles should be drawn up in the chart. As shown in the figure 3, one is the normalized impedance-admittance circle and the other is the r=1 circle in the circles group.

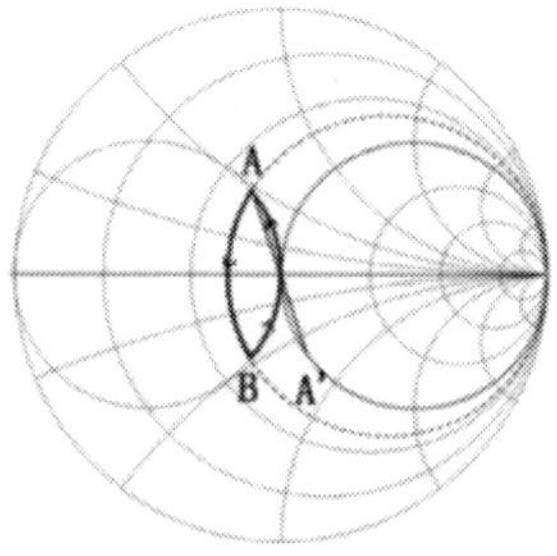

Figure 3. Simth chart.

In this π-type matching network, the impedances of the two sides are 50 Ω, and the normalized impedance z=Z/W=1. The point express for the characteristic impedance in the chart is the center point of z=1. The point of load impedance must be on the admittance circle after parallel connection with the inductance L1, marked as A. The r circle which across point A has the other node with the normalized impedance admittance circle marked as B. Only if the point of load impedance reaches point B after series connection with the capacitor C1, it could return back to the center point after parallel connection with the inductance L2. This allows the point of the load impedance in the same place with the characteristic impedance, which means that they match each other.

The coordinate values of x for A and B point in the chart are $\pm[r(1-r)]^{1/2}$, and the impedance of A is $z = r+[r(1-r)]^{1/2} j$.

The admittance point which is the center of symmetry point for A, is marked as A′. This point is on the circumference of the r=1 circle. The real part of A′ can be calculated as g=1, and the imaginary part as $b = (\frac{1-r}{r})^{1/2}$. Set A′=1-aj, and $a = (\frac{1-r}{r})^{1/2}$. The relationship of a and r can be found: $r = (1+a^2)^{-1}$;

By calculating the reciprocal of A′, the value of point A can be found out: $z = (1+a^2)^{-1} + \frac{a}{1+a^2} j$

Then calculating the values of x for A and B point:

$$x = \pm[r(1-r)]^{1/2} = \pm[\frac{1}{1+a^2}(1-\frac{1}{1+a^2})]^{1/2} = \pm\frac{a}{1+a^2}$$

$$\Delta x = \frac{a}{1+a^2} - (-\frac{a}{1+a^2}) = \frac{2a}{1+a^2}$$

$$X = \Delta x * W = \frac{2a}{1+a^2} \cdot W = \frac{1}{\omega C}$$

$$\Rightarrow C = \frac{1}{X \cdot 2\pi f_0} = \frac{1+a^2}{4aW\pi f_0} \tag{5}$$

Then calculating the values for A′

$$\Delta b = a \qquad B = \frac{a}{W} \qquad B = \frac{1}{\omega L}$$

$$\Rightarrow L = \frac{1}{B \cdot 2\pi f_0} = \frac{W}{2a\pi f_0} \tag{6}$$

According to the formulas of 5 and 6, it can be found that:

$$\Rightarrow LC = \frac{1}{(\pi f_0)^2} \cdot \frac{1+a^2}{8a^2} \tag{7}$$

In order to filter out the disturbance of impulse voltage with frequency less than 2.4 GHz, the value of f_c must in the range of 1 G$<f_c<$2.4 G. According to the formula 4 and 7, the value of a is a$>$0.21. Considering the attenuation characteristics of wave filtering and the specifications of the components, we choose the cutoff frequency f_c as 1.6GHz. The other parameters can be obtained by calculation:

a=0.89 C2=1.3 pF L2=L3=3.7 nH

Figure 3 can be simplified as the figure 4 by combining L2 and L1, which is the final design plan.

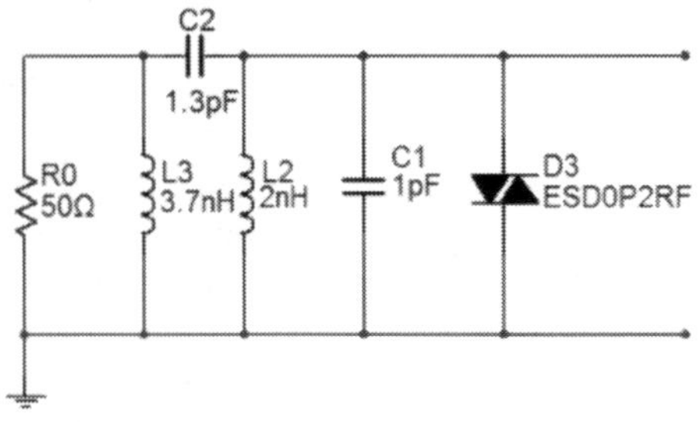

Figure 4. The final design of antenna protection system.

3. ADS software simulation test

The ADS simulation software was used to test the ability of this program for filtering and impedance matching [10-11]. Set the input voltage curve according to the discharge waveform in figure 1, and use the oscilloscope to display the output voltage curve that after the effect of the antenna protection system. The simulation results are shown in figure 5.

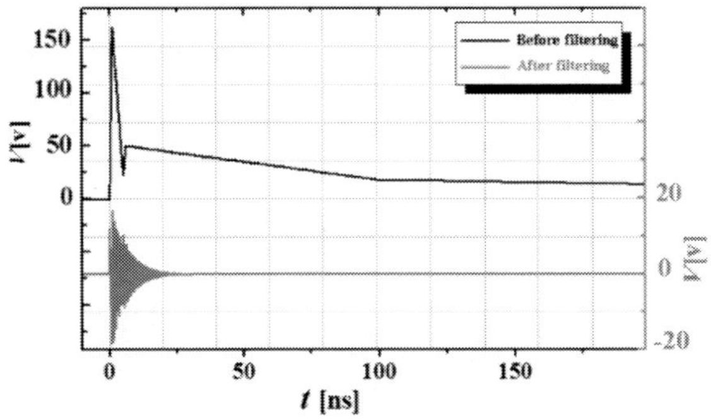

Figure 5. IEC61000-4-2 VCL = f(t), 15 kV positiv pulse.

From the figure 5 we can see that, through the first level protection system, the high voltage of 15 kV as limited and the peak voltage reduced to 162 V. The peak voltage entered the RF antenna was decreased to only 17 V through the second level protection system. This voltage is quite safe for the internal circuit. As the result, this wireless antenna protection system can achieve the purpose of protect the RF antenna circuit against the strong electromagnetic pulse and induction lightning.

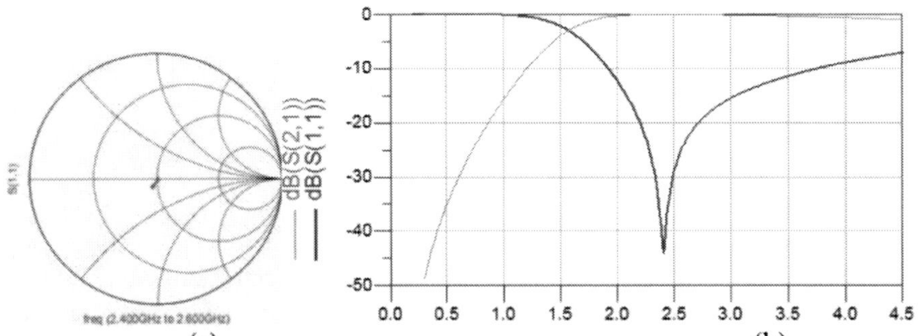

Figure 6. Test of S parameter. (a)Simth chart; (b) S11 S21.

The matching function of the protection system is also be tested by using the ADS simulation software. Set the scan frequency as 0.3 GHz to 4.5 GHz, and the trajectory graph is shown in figure

6(b) with S(1,1) and S(2,1). The frequency bandwidth which meet the requirements with insertion loss S21 > -1dB and echo loss S11 < -10dB is 1.8 GHz~3.7 GHz. This reaches the actual application requirement [6].

The function of the antenna protection network has also been tested in the actual experiment. According to the IEC1000-4-2 standard, the performance test of the protection system was carried out at the experimental environment of temperature of 25. 0 ℃ and humidity of 50%。 The experimental subjects are working wireless nodes, with the antenna of 12 cm in length. The wireless nodes is supplied by battery, and communicated every second.

In the first kind of iterative experiment, the electrostatic discharge simulator was used to indirect discharge on vertical coupling plate, and the wireless node was irradiated by the electromagnetic pulse caused by Electro-Static Discharge (ESD). The voltage in the experiment was increased from low to high, and the maximum is 15 kV. The communication performance of the wireless nodes was tested after each experiment, and the results show that the wireless nodes work normally.

In the second kind of iterative experiment, the electrostatic discharge simulator was used to contact discharge on the antenna of the wireless node. The voltage in the experiment was increased from low to high, and the maximum is 15 KV. The experimental condition was similar to that in indirect discharge, the wireless nodes work normally, too.

4. Conclusion

In this paper, a wireless antenna protection system has been designed. In the system, the TVS diode was used as the first lever for protection, and the π-type high pass filtering network as the second lever. As a result, the system not only has the traditional function of ESD protection, which can avoid the high voltage damage to the internal circuit, but also achieve the purpose of load matching, ensuring the signal source not to distort.

The ADS simulation software was used to test the ability of this system for filtering and impedance matching, which proved the feasibility of this system. The wireless antenna protection system has been practically used, and its' performance of anti-electromagnetic interference has been validated.

References

[1] Siden J, Jonsson P and Olsson T 2001 *IEEE Catalog Number* **01EX484** 371
[2] Albert W 2002 *On-Chip ESD protection for integrated circuits* (Kluwer Boston)
[3] Chen G, Feng H, Xie H, Zhan R, Wu Q, Guan X, Wang A, Takasuka K, Tamura S, Wang Z, and Zhang C 2004 *IEEE RFIC Symp.* 379
[4] Behzad R 2001 *Design of analog CMOS interated circuit* (The McGraw-Hill Companies)
[5] Georgiadis A 2007 *Design of a 2.4 GHz Radio Transceiver for High Speed Wireless Data Networks*
[6] Young J K, Bolme G and Lyles J 2007 *IEEE Particle Accelerator Conference* 2412
[7] Huang J S 1999 *IEEE-IM, AUTOTESTCON'99* Aug.30-Sep.2(San Antonio)
[8] Greason W D 2008 *J. Electrost.* **66** 602
[9] Reinhold L and Bretchko P 2000 *Rf Circuit Design: Theory and Applications* (Prentice Hall)
[10] Finkenzeller K 2004 *RFID Handbook: "Radio-Frequency Identicaion Fundamentals and Applications"* (Wiley)
[11] Li X P, Liu Y and Cao H Y 2006 *J. Beijing University of Posts and Telecommunications* **29** 75

Journal of Physics: Conference Series **418** (2013) 012047 doi:10.1088/1742-6596/418/1/012047

Research on characteristics of electromagnetic radiation of corona discharges from high voltage transmission lines

Lei Wang*, **Shang-he Liu, Ming Wei, Xiao-feng Hu**
Electrostatic and Electromagnetic Protection Research Institute,
Ordnance Engineering College, Shijiazhuang, 050003, China.

E-mail： wl810603@sina.com

Abstract. With the development and application of ultra high voltage electric power transmitting technology, harmful effects of corona discharges to the safe and stable operation of the ultra high voltage (UHV) transmission lines should be considered. In this paper, the radiation law of corona discharges was studied by theoretical analysis and laboratory simulation. Correlated conclusions include that the waveform of corona discharges is in attenuated oscillation mode, the signal of the radiation field increases with increasing charging voltage, whereas the signal amplitude the antenna receives is attenuated with the distance from 3 m to 24 m.

1. Preface

In the field of power systems, the UHV transmission has become the development trend of the power transmission. With the increase in the transmission voltage level, the corona discharge is inevitable, and it has become an important factor which can affect the safe and stable operation of the high-pressure and UHV transmission lines [1-4]. It is a new topic on detecting the fault of high-voltage transmission line through remote detection of the signal radiated from corona discharge, which requires the study of radiation law of corona discharge.

2. Theoretical analysis

The corona discharge radiation field of high voltage transmission line includes two parts: one is the electromagnetic field radiated outward by the rapid charge transfer within the channel of the air corona district, which can be equivalent to the radiation field from a short electric dipole antenna; the other part is radiated by the pulse current generated from the corona discharge excitation conductor, where the high voltage transmission line can be equivalent to pulse transmission line that radiates electromagnetic fields. Both corona discharge radiation fields can be calculated by the electric dipole model and the thin straight antenna model [5].

2.1 Short electric dipole

The dipole with a length of L was set up in the Z-axis, and its center is coincident with the origin (as shown in Figure 1). From figure 1, it can be seen that the current on the dipole is not always acting on the field point P, but there is a delay caused by the difference of the shoot diameter (s) and the distance (r). The current (I) was expressed as a transient form [5] by formula (1).

* To whom any correspondence should be addressed.

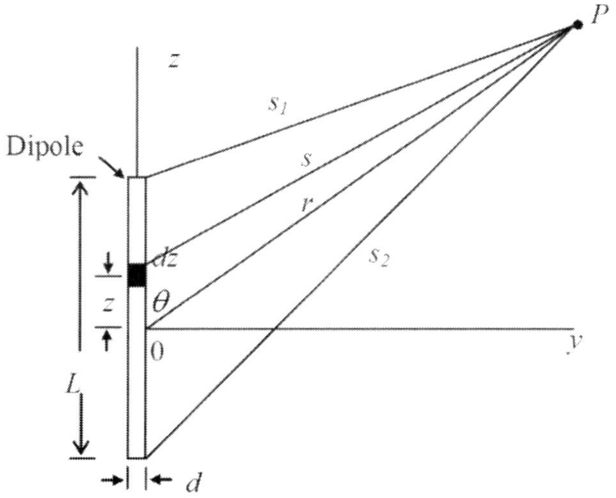

Figure 1. Field strength at any point of short electric dipole.

$$I = I_0 e^{j\omega t} \tag{1}$$

The instantaneous propagation effects of the current can be expressed by using the Lorentz spread or lag time notation, which is written as follows:

$$[I] = I_0 e^{j\omega[t-(r/c)]} \tag{2}$$

In the practical application, we are interested only in the far field, but the distance of the far-field point and the dipole is much greater than the dipole length($r >> L$), and its wavelength is much larger than its length ($\lambda >> L$) . The calculation of the electromagnetic field by the short dipole in the far-field point is as follows:

$$\begin{cases} E = E_\theta = \dfrac{j\omega I_0 L \sin\theta e^{j\omega[t-(r/c)]}}{4\pi\varepsilon_0 c^2 r} = j\dfrac{I_0 \beta L}{4\pi\varepsilon_0 c r}\sin\theta e^{j\omega[t-(r/c)]} \\ H = H_\phi = \dfrac{j\omega I_0 L \sin\theta e^{j\omega[t-(r/c)]}}{4\pi c r} = j\dfrac{I_0 \beta L}{4\pi r}\sin\theta e^{j\omega[t-(r/c)]} \end{cases} \tag{3}$$

The general radiation power of non-uniform current dipole can be calculated using the formula (3) and the Poynting vecor formula

$$P = \sqrt{\frac{\mu}{\varepsilon}} \frac{\beta^2 \left(\frac{I_0}{2}\right)^2 L^2}{12\pi} = \frac{1}{2} I_0^2 R_r \tag{4}$$

For free space ($\mu = \mu_0$, $\varepsilon = \varepsilon_0$) , the availability radiation resistance of electric dipole can be gotten formula (5).

$$R_r = 197(\frac{L}{\lambda})^2 = 197 L_\lambda^2 \ (\Omega) \tag{5}$$

For example, when L_λ is 0.1, R_r will be 1.97 Ω ; when L_λ is 0.01, R_r will be 0.02Ω . At the same time, the wavelength is much larger than the length of the electric dipole under normal circumstances ($\lambda >> L$) , visibility, the radiation resistance of electric dipole is quite small. In certain circumstances, the radiation power radiated by the corona discharge current is also very small.

2.2 The thin straight antenna

The far-field formula of the symmetrical and thin straight antenna with a length of L is derived using figure 2. Taking into account that the antenna is consisted of a string of infinitesimal dipole with length of dz, the field of the antenna can be gotten by integration of all of the dipole field, the result is as follows[5].

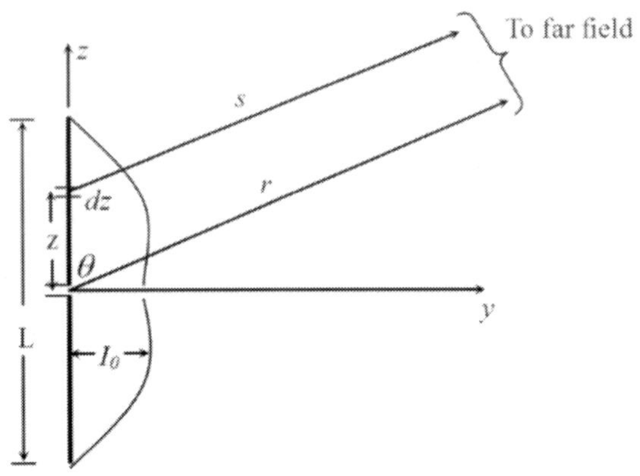

Figure 2. Field strength at any point of thin straight antenna.

$$\begin{cases} E = E_\theta = \dfrac{jI_0 e^{j\omega[t-(r/c)]}}{2\pi r}\left[\dfrac{\cos\left[(\beta L\cos\theta)/2-\cos(\beta L/2)\right]}{\sin\theta}\right] \\ H = H_\phi = \dfrac{j60 I_0 e^{j\omega[t-(r/c)]}}{r}\left[\dfrac{\cos\left[(\beta L\cos\theta)/2-\cos(\beta L/2)\right]}{\sin\theta}\right] \end{cases} \quad (6)$$

Half-wave antenna, for example, the radiation resistance of the long straight antenna was calculated by formula (7).

$$R_r = 30Ci(2\pi) = 30\times 2.44 = 73\Omega \quad (7)$$

These results show that the radiation resistance of thin straight line is much larger than that of the electric dipole. For the same current, the radiation power of thin straight line is much larger than the electric dipole.

By calculation of the equivalent antenna radiated field, it can be obtained that the radiated field of corona discharge is inversely proportional to the distance, and the radiated field strength increases with the discharge current. In addition, the radiation power of space is much smaller than that of the line.

3. Laboratory simulations for radiation law of corona discharge

In order to get a clearer understanding of the radiation field by corona discharge, a test experimental platform was constructed to test the signals radiated from corona discharge and to study its radiation law by testing of signals radiate from corona discharge in different conditions.

3.1 Platform of experiment

Considering that the electromagnetic field produced by corona discharge high voltage transmission line is equivalent to radiation field of dipole and the conductor pulse transmission line, the long straight line is used to simulate the corona discharge, and the experimental layout is shown in figure 3. The corona discharge simulation system in laboratory is divided into two parts.

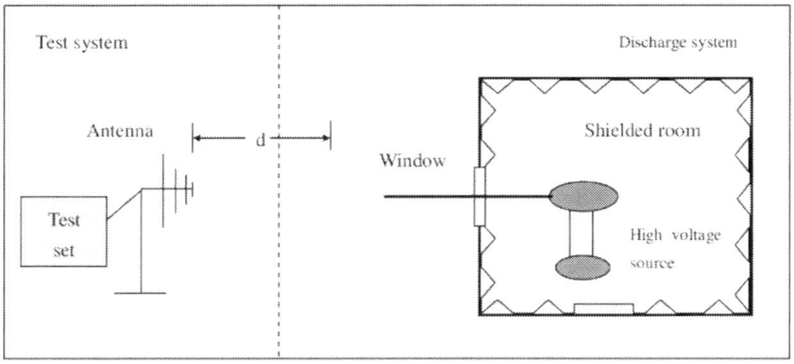

Figure 3. Experimental layout.

(1) Discharge system. The discharge system consists of a long straight line, high voltage static power (300 kV, 3 mA) and high-voltage towers, etc. The main purpose is to simulate the corona discharge of high-voltage line.

(2) Test system. The test system consists of a log-periodic antenna, a corona discharge test system of Tek 7404B high-speed digital storage oscilloscope and a signal pre-amplifier. The signal pre-amplifier is used to test the discharge radiation signal.

The high voltage source with radiation signal seriously interferes with the discharge signal test, so in this experimental system the high-pressure source is set in the shielded room. A high voltage line insulated by high performance material output the high voltage outside the shielded room through a window, and some distance from the walls of the shielded room was left to prevent high-voltage breakdown. The high-voltage line outside shielded room was mounted on the wood bracket and with the distance of 1.5 m to the ground.

One end of the high-voltage line was connected to the high voltage source; the other end was fixed on an insulating stand. The length of high-voltage line outside the shielded room was the effective radiation length of the transmission line. The distance of the antenna and discharge point was d. The high voltage was adjusted by the high voltage pressure control switch.

3.2 The characteristics of corona discharge signal

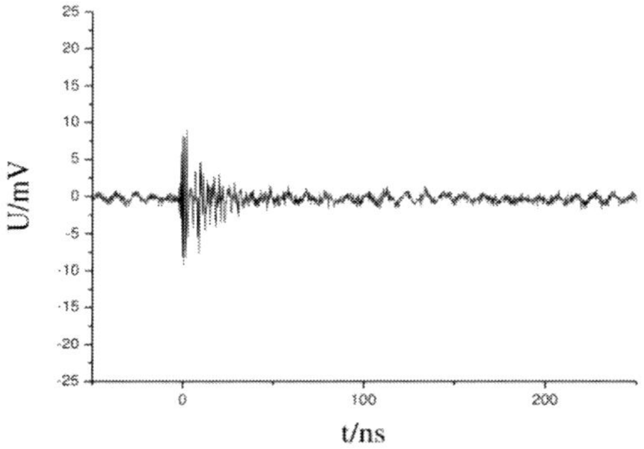

Figure 4. Waveform of 60 kV corona discharging.

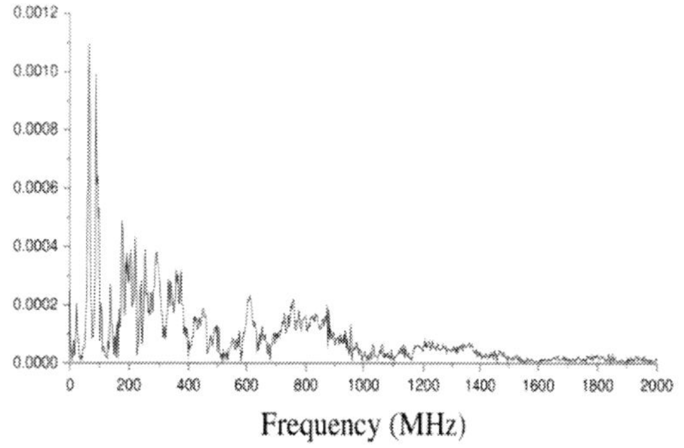

Figure 5. Spectrum of 60 kV corona discharging.

The experiments were conducted under a temperature of 26 °C, a relative humidity of 35%, and a charging voltage of 60 kV. The receiving antenna in horizontal polarization direction was log-periodic antenna, and the digital filters were used to capture the discharge signal.

The length of the high-voltage line outside the shielded room was 150 cm; the test distance was 6 m. When the voltage of corona discharge was 60 kV, the time domain and frequency domain waveforms measured by the receiving system were shown in figure 4 and 5. From figure 4 it can be seen that the rise time of the corona discharge waveform was about 1~ 5 ns, and the duration time of the waveform was about tens of ns. Also, figure 5 pointed that the corona discharge spectrum distribution was extremely wide, the frequency ranged up to 1.5 GHz, and the main energy concentrated below 1 GHz.

3.3 The radiation law of different voltage

Under the same conditions, the discharge voltage was selected from 30 kV to 100 kV by the step of 10 kV, the maximum amplitude of radiation signal was recorded in figure 6. It can be seen from the figure that the signal radiation field increases with increasing the charging voltage. The discharge current would increase with the discharge voltage, which was consistent with the radiation field calculation of the equivalent antenna.

3.4 Discharge radiation law of different distance

Under the same experimental conditions, the distance of the test antenna and high-voltage line was changed from 3 m to 24 m by the step of 3 m. The relation between the maximum amplitude of domain waveform and the test distance was shown in figure 7. From figure 7 it can be seen that the signal amplitude that the antenna received was attenuated with the increase of the distance, which pointed that the received signal power was inversely proportional to the propagation distance, and the result was also consistent with the theoretical analysis.

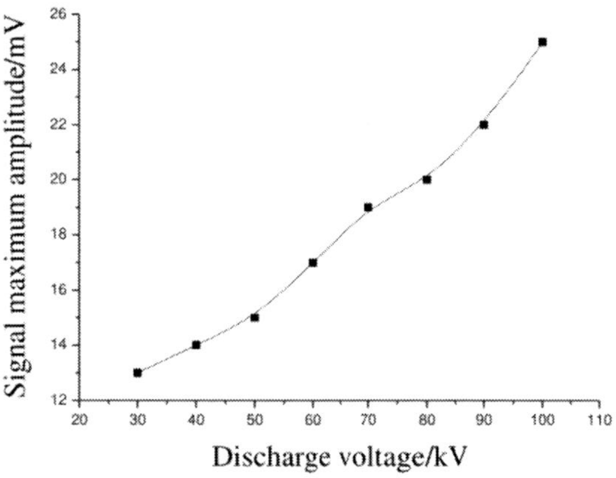

Figure 6. Maximum amplitude varying with discharge voltage.

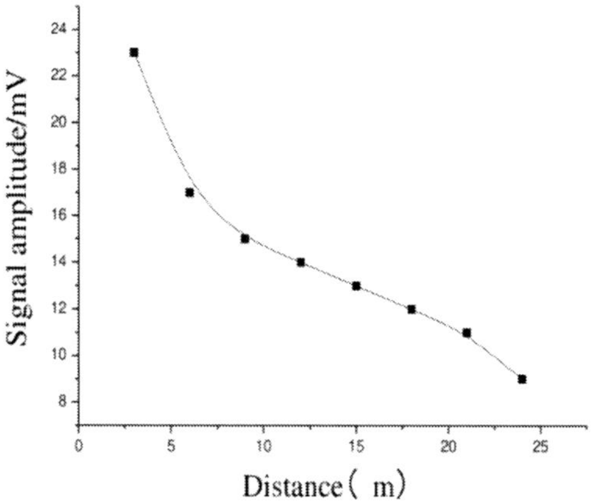

Figure 7. Signal amplitude varying with distance.

4. Conclusion

Through the theoretical analysis and experimental study of radiation law by high voltage transmission line corona discharge, the following conclusions can be drawn: (1) the waveform of corona discharge was attenuated oscillation mode. The waveform rise time was about 1~5 ns, the frequency ranged up to 1.5 GHz, and the main energy concentrated below 1 GHz; (2) with the increase of charging voltage, the signal radiation field increases; (3) the signal amplitude that the antenna received was attenuated with the distance, which showed that the received signal power was inversely proportional to the propagation distance.

Acknowledgments

The paper is supported by the Key Programs of the National Natural Science Foundation of China, No. 61172035

References

[1] Sui X J and Song S X. 2006 *Electric Power Construction (in Chinese)* **27** 37
[2] Xiao D P, He W, Xie P J, Kang P, etc. 2007 *Power System Technology (in Chinese)* **31** 52
[3] Adami J F, Silveira P M, Martinez M L B, Perez R. C and Dallbello A C 2009 *IEEE Trans. Power Delivery* **24** 1515
[4] Jin J D, Sun H H, Cao T and Lin S M 2010 *ICSP 2010 Proceedings* p17
[5] John D K and Ronald J M 2008 *Antennas: For All Applications Third Edition* (Beijing: Publishing House of Electronics Industry).

Journal of Physics: Conference Series **418** (2013) 012048 doi:10.1088/1742-6596/418/1/012048

Research on incentive electrostatic discharge in tankers during oil filling

Gefei Yu *, Xuewen Li
SMU College of Arts and Sciences, Shanghai Maritime University, 1550 Hai Gang Dadao, Shanghai, P.R. China

E-mail: gfyu@shmtu.edu.cn

Abstract. This paper introduces the mechanism of oil charge and discharge and discusses its types and formations during oil loading and transporting. Moreover, the influence factors of the oil discharge were clarified. The charge current in the oil storage tanker was preliminarily calculated and three stages of charge accumulations were discussed and calculated in detail. The staged charge accumulations include charge unfreezing, charge free surface, and charge accumulation.

1. Introduction

More disasters caused by the electrostatic discharge often happened when the oil was filling into the tank. As the flammable oil is ignited by the electrostatic discharge, it relates to two basic physical phenomena. One is the brush discharge on the electrophorus oil surface, and the other is the spark discharge of isolated conductor. The isolated conductor could be sampling pot, tube, metal fittings, human body, and small suspend object, even the tank truck or the tank car, etc. If the two types of discharge could be avoided and the accumulated electrostatic energy could be released safely, the oil would be loaded, unloaded and transported safely. So to prevent electrostatic discharge to the oil is very important.

The circumstances of electrostatic discharge and safety velocity in oil storage tanker are very important. According to the charge conditions of electric field intensity in different dangerous parts and the ignition possibility of oil vapor in the tanker, the whole oil transportation process could be divided into beginning stage, filling stage and finished stage.

2. ESD harm and countermeasure in oil pipe

2.1. ESD basic rule of flowing oil in pipe

Commonly, the pipe transporting liquid petroleum products is made of steel and (or) aluminum, but the joint part and soft connecting part may also be made of rubber, plastic or latex, even sometimes conductive plastic. The characteristic of electrostatic discharge (ESD) of the oil pipe, regardless of being made of any material, is mainly determined by the flow velocity of oil, the diameter and length of the pipe, and also depends on the texture of material of pipe, the degree of roughness inner pipe and the resistivity of liquid.

* To whom any correspondence should be addressed.

If the volume charges density of oil flowing in the pipe is ρ_m, the current of fluid flow would be

$$I = \frac{Q}{t} = \frac{\rho_m \pi D^2 L}{4t} = \frac{\pi}{4} D^2 V \rho_m \tag{1}$$

Considering the saturation density of charge $\rho_{m\infty}$ in pipe and the pipe length long enough the charge density could achieve saturation, equation (1) would be

$$I = \frac{\pi}{4} D^2 V \rho_m \left(1 - e^{-\frac{L}{L_b}}\right) \tag{2}$$

V-Flow velocity of oil in pipe;
D- Diameter of oil pipe;
$\rho_{m\infty}$ -Saturation density of charge of fluid flow;
L-Length of oil pipe;
L_b-Saturation length of oil pipe;
t-Time.

It comes to the conclusion that, while liquid petroleum products flowing in the pipe, the current of fluid flow is proportional to flow velocity and the diameter of the square of pipe, at the same time, also to the length of pipe if it is saturation length of oil pipe. But the current would reach a stable value when the pipe is saturation length.

But the above equation is still different with the actual situation. A more reliable empirical formula is achieved as following:

$$I_\infty = AD^\alpha V^\beta \tag{3}$$

A- Constant of proportionality;
α- Determined by the relationship between the average current and diameter of pipe;
β- Determined by the relationship between average current and flow velocity in pipe.

Data show that it is proportional when A=0.58×10^{-6}~3.75×10^{-6}, α=0.87~2, β=1.6~2.4 and the flow velocity and the diameter of pipe are the main factors to cause ESD in oil pipe.

2.2. Affection of pipe length to ESD

It is commonly considered that the longer the pipe is, the stronger the charge is. Actually, the charge conduction could still exist no matter how much the electrical conductivity of liquid would be. So the longer the pipe is, the greater the conduction effects of charge is, and the more quickly the increase of charge leak and neutralization is. But it is not a linear increase in enlargement of the fluid current, which was caused by oil flowing, with the increase of pipe length. General test has proved that it would be an exponential relationship and reach saturation. In different kinds of pipe length, the fluid current comply with the following rule

$$I_L = I_\infty \left(1 - e^{-\frac{L}{V_\tau}}\right) \tag{4}$$

V-Flow velocity of oil in pipe;
I_∞ -Saturation current of fluid flow;
L-Length of oil pipe;
τ -Time constant of fluid flow ($\tau = \varepsilon\rho$).

If we assume the relative dielectric constant $\varepsilon_r = 2$ (for oil commonly, $\varepsilon_r = 2 \sim 4$), we could calculate the lengths of pipe when fluid flow currents reach saturation as various liquid flowing in the pipe. The average charge accumulated in tanker could be calculated as following equation:

$$Q = I_L \tau \left(1 - e^{-\frac{t}{\tau}}\right) \tag{5}$$

I_L-Saturation current of fluid flow;
t -Time of filling oil;
τ -Time constant of fluid flow ($\tau = \varepsilon\rho$).

3. ESD harm and countermeasure in oil tanker while filling

3.1. Charge diasporas in oil tanker

In order to simplify the problem, we start with the charge diasporas in metal tanker full filled with charged oil. It was assumed that the whole charge in tanker is Q. Under the effect of inner electric field, the electrostatic charge would distribute on the surface of oil and start to loss out through the wall of tanker. According to the law of charge conservation, the reduce volume charge in tanker should equals to the volume charge lost out through the wall of tanker.

$$-\frac{dQ}{dt} = \oint jds \tag{6}$$

$\oint$ - Integral of a closed surface;

j - Charge lost out through the wall of tanker.

$j = KE$ by Ohm's law, K is the electrical conductivity of fluid flow and E is the electrical field strength of the wall. And through Gauss theorem, we could get

$$\oint_S Eds = \frac{Q}{\varepsilon_r \varepsilon_0} \tag{7}$$

$$-\frac{dQ}{dt} = K\frac{Q}{\varepsilon_r \varepsilon_0} \tag{8}$$

The Initial conditions are that when t=0, Q=Q_0, the charge complies with decay law as time goes by.

$$Q = Q_0 e^{-\frac{Kt}{\varepsilon_r \varepsilon_0}} = Q_0 e^{-\frac{t}{\varepsilon_r \varepsilon_0 \rho}} = Q_0 e^{-\frac{t}{\tau}} \tag{9}$$

If we take a tiny volume ΔV , in which the first volume charge density is ρ_0 and at the other time it is ρ , from the charged oil tanker, the electrostatic charge in the volume element at arbitrarily moment is $\Delta Q = \rho \Delta V$ which would be substituted into (8).

$$\frac{d\rho}{dt} = -K\frac{\rho_0}{\varepsilon_r \varepsilon_0} \tag{10}$$

$t = 0, \rho = \rho_0$, after integrating, above equation would be

$$\rho = \rho_0 e^{-\frac{t}{\tau}} \tag{11}$$

It indicates that the volume charge density complies with the exponential decay as time goes by. And the decay is determined by the time constant τ .

3.2. Charge cumulating in oil tanker

3.2.1. Unfree surface of oil It is also named as float top type tank filling. When charged oil was filled into tanker the charge would be conducted out the tanker through the wall, and the charge in tanker would reduce with time. The conduction current density is assumed to be j , the conduction current through the wall should be

$$I = \oint_S jds = \oint KEds = K\frac{Q}{\varepsilon_r \varepsilon_0} = \frac{Q}{\tau} \tag{12}$$

3.2.2. Free surface of oil Taking the circle as a closed surface, the following equation could be gotten by Gauss theorem.

$$Q = \oint_S Dds = \oint_S D_n ds = \int_W D_{nw} ds + \int_A D_{nA} ds \tag{13}$$

$\int_W D_{nw} ds$ is the surface integral of the contact part between oil and wall, D_{nW} is the normal direction component of electric displacement on the interface. $\int_A D_{nA} ds$ is the surface integral of the free surface, D_{nA} is the normal direction component of electric displacement on the free surface. If there is no surface charge on the free surface, we could calculate $D_{nA} = D_{LA} = D_{VA}$, D_{LA} is the normal direction component of electric displacement on the interface between air and oil. D_{VA} is the normal direction component of electric displacement on the air interface.
The current conducted through the wall should be

$$I = \int_W jds = \int_W KEwds = \frac{K}{\varepsilon_r \varepsilon_0}\int D_{nW} ds \tag{14}$$

Substituting equation (13) into (14), the result is

$$Q = \frac{\varepsilon_r \varepsilon_0}{K} I + \int_A D_{nA} ds = \tau I + \int_A D_{nA} ds \tag{15}$$

So

$$I = \frac{1}{\tau}\left(Q - \int_A D_{nA} ds\right) \tag{16}$$

If the inflow current I_s is substituted, we would get following equation by conservation of charge.

$$\frac{dQ}{dt} = I_s - I = I_s - \frac{1}{\tau}\left(Q - \int_A D_{nA} ds\right) \tag{17}$$

Assumed that the current diasporas was stable, then

$$I_s = \frac{1}{\tau}\left(Q - \int_A D_{nA} ds\right)$$
$$Q = \tau I_s + \int_A D_{nA} ds \tag{18}$$

3.2.3. Charge accumulation in tanker while surface charge density is σ Actually, the interface between air and liquid combined with the wall of tanker to form a closed surface. According to Gauss theorem, we would get as following:

$$Q = \oint_S Dds = \int_W D_{nw} ds + \int_A D_{nA} ds = \frac{\varepsilon_r \varepsilon_0}{K}\int_W E_{nw} ds + \frac{\varepsilon_r \varepsilon_0}{K}\int_A E_{nA} ds$$
$$= \tau\left(\int_W j_{nw} ds + \int_A j_{nA} ds\right) \tag{19}$$

D_{nw} is the electric displacement vector on the solid-liquid interface. E_{nw} is electric field strength, j_{nw} is the normal component of current density. The current conducted through closed surface should be

$$\int_W j_{nw}\, ds + \int_A j_{nA}\, ds = \frac{Q}{\tau} \tag{20}$$

Like above calculation, if the inflow current I_s is substituted, then

$$\frac{dQ}{dt} = I_s - \left(\int_W j_{nw}\, ds + \int_A j_{nA}\, ds \right) = I_s - \frac{Q}{\tau} \tag{21}$$

When the current is stable,

$$I_s = \frac{Q}{\tau} \qquad Q = I_s \tau \tag{22}$$

So we get the whole charge in oil except the surface charge in tanker while the situations are stable. If there is surface charge, the result should be the plus of the two parts of charges.

$$Q_{whole} = \tau I_s + \sigma ds \tag{23}$$

This integral includes the whole surface of the liquid. And it is more dangerous if the surface area is bigger because of the increasing of surface charges. In order to reduce the ESD harm in tanker, one method is to reduce the free surface in oil tanker. We could reduce the potential of oil surface by dividing a big tanker into a lot small space using metal material. Another method is to fill with insulating gas into the air space in tanker to reduce the charges of the free surface of liquid.

From above discussing, we know that the electric field strength in air space would be enhanced as the increasing of surface charges. At the same time, the possibility of ESD caused by surface discharge in air space would increase, too. The discharge will happen certainly as the space electric field exceeds the Breakdown electric field.

Acknowledgments
My special thanks and gratitude to the promoter Fund Project (No.20120041) for continues support during my work and while preparing this work.

References

[1] Klinkenberg A, M van der Minne JL 1958 *Electrostat. in the Pet. Ind.* (New York: American Elsevier)
[2] Peng M Wu 1999 *Proc. of the 4th Beijing Int. Conf. on Syst. Simul. and Sci. Coputing* (Beijing: Beijing University Press) p 214
[3] Stokll R L 1967 *IEEE Trans. Magn.* **144** 83
[4] Davies J B, Muilwyk C A 1966 *IEEE Trans. Magn.* **133** 157
[5] Vervalin C H 1963 *Fire Prot. Manual for Hydrocarbon Proc. Plants* (Houseton: Gulf Publishing Co.)
[6] International Oil Tanker Terminal Safety Group 1970 *Int. Oil tanker and terminal Saf. Guide* (Beijing: The Institute of Petroleum Press)

Causes of the ESD immunity testing problems in the IEC 61000-4-2 standard

Qingyun Yuan*, **Shanghe Liu, Yongwei Sun, and Xijun Zhang**
Institute of Electrostatic & Electromagnetic Protection, Mechanical Engineering College, No 97 Heping West Road, 050003 Shijiazhuang, Hebei Province, China

E-mail: qingyuny@163.com

Abstract. To resolve the poor reproducibility of the electrostatic discharge (ESD) immunity test platform which is specified in the standard IEC 61000-4-2, using a newly designed ESD testing system, the effects of the switch action of ESD generator, the discharge current and its derivative, the radiation field and the imperfect specification of the test method on the testing results were studied. It was found that the effect of the switch action of ESD generator on the testing results was very distinct, and the radiation field caused by the switch action was one of the main cause of the different testing results; different discharge current and current derivative were caused by different ESD generators, which would cause different testing results; the imperfect specification of the test method in the standard was the other reason.

1. Introduction

The objective of electrostatic discharge (ESD) immunity testing is two fold [1]: ensuring adequate robustness of the electronic systems against the real world ESD and passing a standardized test as this is often a legal or company's internal requirement for selling a product. When passing a legal requirement an unambiguous pass/fail determination is required. However, it is well known that all EMC tests suffer from reproducibility problems. This is especially true for ESD immunity testing [2-4]. Owing to the large variation nature of the natural ESD phenomena, in the early 1990's testing has been moved from air discharge to contact mode testing to avoid the effect of arc length variations in air discharge [5] and to improve reproducibility. In 2008, a reference ESD event has been introduced in the standard, IEC 61000-4-2 [6]. This document describes the discharge current waveform. However, many users have reported many problems about the IEC 61000-4-2 standard [7-11]:

1) ESD test results depend on the manufacturer of the simulator;
2) Even with the same simulator, the repeatability is also bad;
3) Testing for immunity against ESD of human does not guarantee immunity against furniture ESD;
4) The contact discharge method is specified in detail in the standard, but the air discharge mode is not.

Therefore, many efforts have been done to improve the repeatability. Lutz and Calcavecchio [12]-[13] have focused on defining the right discharge current. Lin [4] concluded that the high frequency components or the current derivatives dominate simulator severity. In [14], it was shown that he transient fields of ESD strongly influence the equipment under test (EUT) response. In order to reduce

* To whom any correspondence should be addressed.

the variation of the test results of the ESD test depending on the ESD generator, in 2006, the additional restriction for ESD current waveform was proposed and considered in IEC SC77B/MT12. The specification states that the width is measured at 60% of the first discharge current peak and should be 1.5 to 3.5 ns. In [15], it was shown that the restriction was effective to reduce the variation; however, the much severe restriction is needed to reduce the dependency of the ESD generator.

In this paper, the switch action of ESD generator, the discharge current and its derivative, the radiation field and the imperfect specification of the test method on the testing results are studied, the reasons which cause the problems about the IEC 61000-4-2 standard are analyzed, and some suggestions are proposed to revise and/or perfect the standard.

2. Measurement Setup

The newly designed ESD testing system includes three parts: the parameter measurement system, the electrode approach speed control system, and the ESD simulator. The measurement setup is depicted in figure 1.

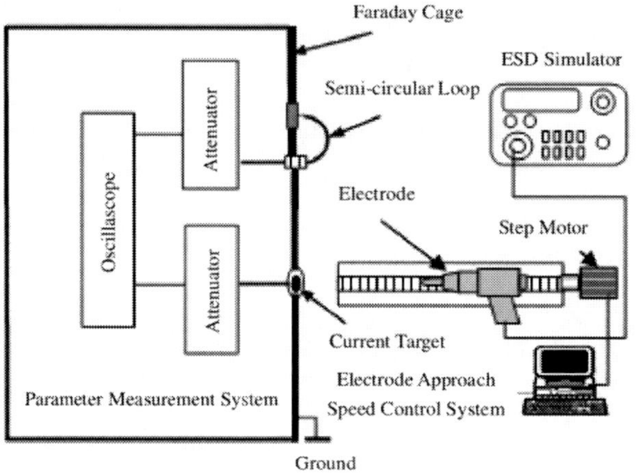

Figure 1. Measurement setup for air discharge.

2.1. Parameter measurement system

The sub-system can capture the discharge current, transient field, and the voltage induced in a semi-circular loop. The wall of a shielded room is used as a ground plane having a current target and field sensors placed upon it while the instrumentation is placed inside. The sub-system includes the current target (Its frequency response is within ±0.3 dB up to 1 GHz and ±1.1 dB up to 6 GHz.), a semi-circular loop (28 mm diameter, 0.7 mm wire diameter) or monopole antenna, a Tektronix TDS 7404B oscilloscope (four channels, 4 GHz, 20 GS s^{-1}, and a Faraday cage.

2.2. Electrode Approach Speed Control System

The electrode approach speed is a key factor affecting the air discharge process. This sub-system is composed of a programmable logic controller (PLC), electrode linking device, step-motor and its driver. The operating principle of the system is: firstly, the input control signals e.g. go forward, fall back, stop, approach speed levels are converted into the output signals such as the direction control level and step pulse through the programming of the PLC. Then, these weak signals are amplified by the step-motor driver into the strong current signals which can drive the step-motor to work. At last, different approach speeds will be achieved through the movement of the electrode linking device which drove by the step-motor. The system can provide 15 speed levels from 0.1 ms^{-1} to 1.5 ms^{-1} and the resolution of speed is better than 0.01 ms^{-1}. During the experiments, the center of the current target and the electrode are in the same axis all the way to ensure that the arc does not initiate at the edge of the target. In addition, when the electrode approaches the target at a high speed of 1 ms^{-1}, the current target or the ESD generator may be damaged, so some measurements are taken to avoid this problem: when the ESD gun reaches the left of the guide track, the position sensor will send PLC a signal to stop the step-motor, the buffer is used for the further protection of the ESD gun or the target.

2.3. ESD simulator

Different ESD simulators can be used in the experiment, e.g., an ESS-200AX simulator (charge voltages range from 0.2 kV to 30 kV) made by Noiseken Company of Japan, a NSG435 simulator (charge voltages range from 0.2 kV to 16.5 kV) made by SCHAFFNER Company of Switzerland, and a NS61000-2A simulator (charge voltages range from 0.2 kV to 28 kV) made by Shanghai Sanji Company of China, these simulators satisfy the qualifications specified in the standard IEC61000-4-2.

3. Analysis of the ESD immunity Testing Problems

3.1. Effect of the switch action of ESD generator on the ESD immunity test

In [16], it was shown that the testing results were varied due to the radiation field in some ESD immunity test. Zhang Xijun [17] also found that the EUT's failure voltage was very low, but became large after the ESD generator was shielded by the grounded wire netting, this change of failure voltage may be caused by the radiation field from the relay switch of ESD generator. Therefore, the effect of the switch action of ESD generator on testing results was studied using the measurement setup depicted in figure1.

3.1.1. Comparisons of the current and the field before and after ESD generator shielded.

In the experiment, the ESS-200AX simulator is used; the testing method is contact mode. Figure 2 shows the tested current waveforms before and after ESD generator shielded at 2 kV. As shown in figure 2, the rise times of the current waveforms well overlaps, but the surging part of the current waveforms after first peak-current slightly change.

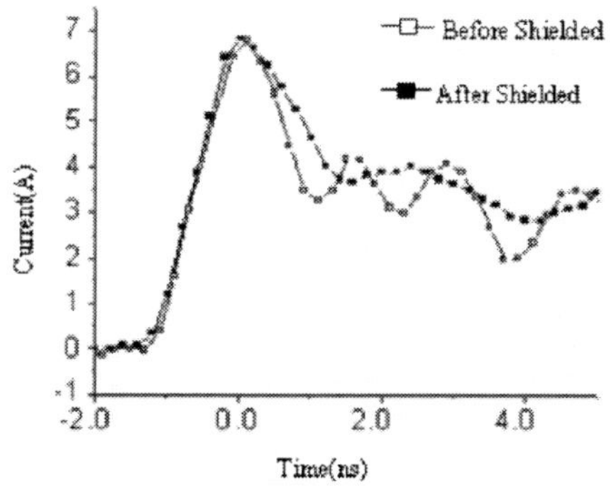

Figure 2. Comparisons of the current waveforms before and after ESD generator shielded.

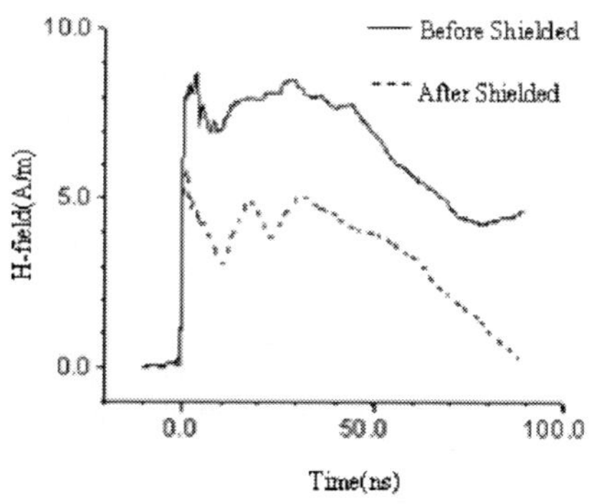

Figure 3. Comparisons of the magnetic field waveforms before and after ESD generator shielded.

Figure 3 shows the H-field waveforms before and after ESD generator shielded using a small loop antenna (the distance from the loop antenna to the discharge point is 10 cm). From the figure 2, it was found that the H-field waveforms largely different before and after ESD generator shielded, the value after shielded is lower than that before shielded. This indicated that a part of the radiation field from switch action was shielded by the wire netting, namely, the switch action of the relay inside the ESD simulator could also generate the radiation field, and moreover, the strength of the field can't be ignored. Therefore, in the ESD immunity test, the electromagnetic interference which EUT suffered was composed of the field radiated from ESD current and the field radiated from the switch action of the relay inside the ESD simulator.

The same results can be gained using the other kinds of ESD simulator, e.g., an NS61000-2A simulator or NSG435 simulator under same testing method. In a word, the electromagnetic field radiated from simulator's switch action could cause interference to a EUT, consequently affect the reproducibility of the testing results. Therefore, the radiation field of the relay inside the ESD simulator was tested in this paper.

3.1.2. Testing of the radiation field of the relay inside the ESD simulator.

In the experiment, the ESS-200AX ESD simulator was adopted, and it was controlled by a remote switch. The process of charging and discharging was divided into three states: charge the capacitance which inside the simulator (thereinafter, this process was called "CHARGE" for short), connect the discharge circuit and the discharge electrode (thereinafter, this process was called "CONNECT" for short), and an air discharge. The discharge current, the radiation field, and the induced voltage corresponding to those three states were tested by the current target, the wide-band pulse field tester, and the monopole antenna, the results were displayed separately in Channels 1, 2 and 3, the waveforms of the discharge current, the radiation field, and the induced voltage under different conditions were shown in figure 4.

As shown in figure 4, when the switch actions of ESD simulator were "CHARGE" and "CONNECT", the ESD radiation field and the induced voltage under those two conditions were very large, though the air discharge was not happened. However, as air discharge happened, the discharge current was very large, the radiation field and the induced voltage were rather lower than that of other conditions. This indicated the electrostatic interference on a EUT was caused by two parts: the fields radiated from ESD current, and the fields radiated from the switch action of the relay inside the ESD simulator.

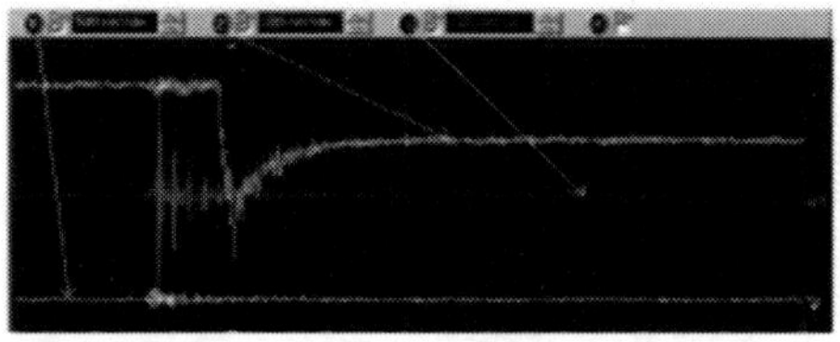

(a) "CHARGE"

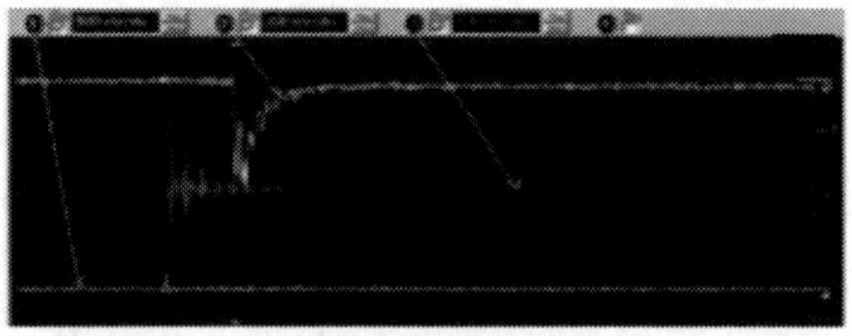

(b) "CONNECT"

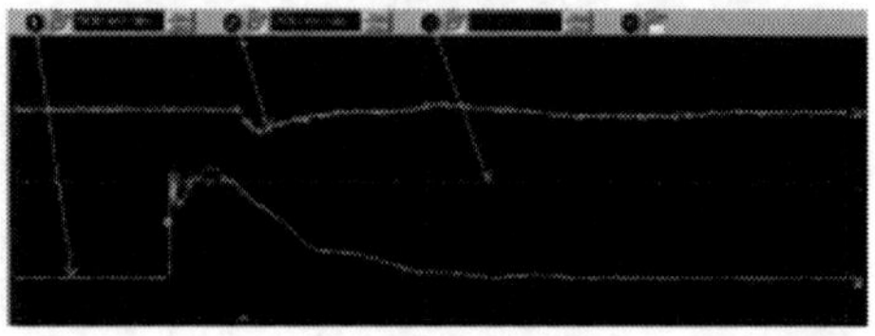

(c) "DISCHARGE"

Figure 4. Comparisons of the discharge current, the radiation field and the induced voltage under different conditions.

3.2. Effect of the discharge currents and its derivatives on the ESD immunity test

Two ESD simulators (NS61000-2A and ESS-200AX) were used in the measurement, Figures 5 and 6 show the waveforms of the measured discharge currents, the current derivatives and its spectrums at 2 kV.

From figure 5, the waveforms and the four parameters of the discharge currents meet the specification in the IEC 61000-4-2 Standard; however, the current waveforms after the first peak deviate significantly. The spectra differ by more than +/- 6 dB below 2.5 GHz, as can be seen in [15]. From figure 6 when different ESD simulators used, the current derivative waveforms and its spectrum also deviate significantly, this phenomena is more distinct in the high frequency range (>500 MHz).

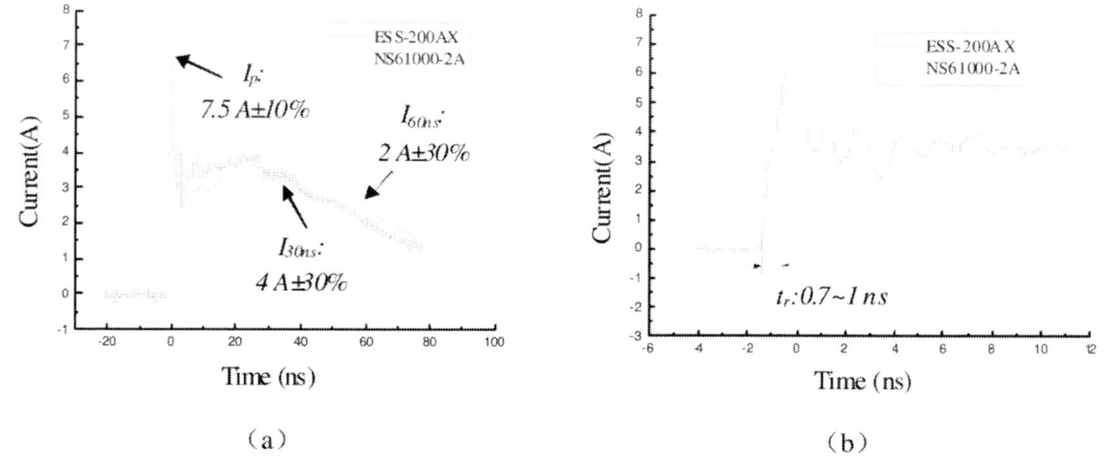

Figure 5. Comparisons of the discharge current waveforms. Two different ESD simulators were used. The four parameters of the discharge currents specification are indicated. (a) the discharge current waveforms. (b) the first 12 ns current waveforms.

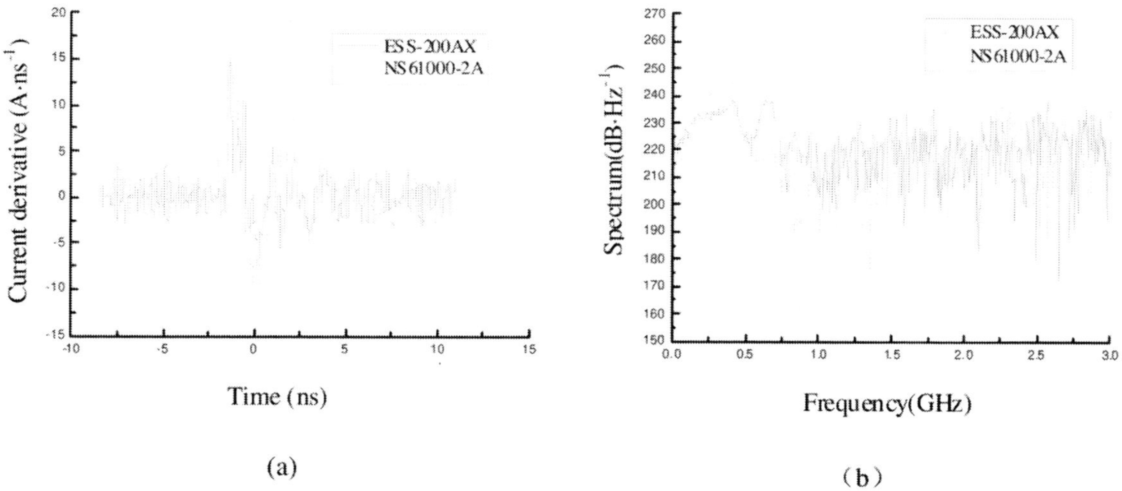

Figure 6. (a) Comparisons of the discharge current derivative waveforms. (b). Comparisons of the current derivative spectrums.

For high-speed, low-voltage digital electronic equipment, which are often designed such that a majority of real world discharges directly go to a grounded part of the system, most coupling is caused by induction (near field) or by radiation (far field). Because the induction field and the radiation field caused by ESD are correlated closely with the discharge currents and the current derivatives, and the high frequency components or the current derivatives dominate the simulator severity. For such EUTs, they will react differently to different ESD simulators because of different discharge currents and different current derivatives. Therefore, if we want to improve the reproducibility of ESD tests, the current derivatives should be also specified in the standard like the discharge current.

3.3. Effect of the radiation fields on the ESD immunity test

Different ESD simulators would cause different current derivatives; therefore, the radiation field caused by different ESD simulators will vary largely. Because of a well correlation between the induced voltages in a semi-circular loop and EUT

ESD failure levels, Romachandran and Pommerenke [18] used the induced voltage to represent the ESD radiation field. However, if the transient field distribution is non-uniform around the ESD simulator, the induced field coupled into the EUTs is correlated with not only the brand of the ESD simulator, but also the angles of the ESD simulator towards the EUTs. Because ESD simulators are not bodies of revolution, if the transient field distribution around the ESD simulators is non-uniform, this will cause a EUT failure level variation. To observe the variation, the induced voltages in a semi-circular loop have been measured using the measurement setup which shown in figure 1. Figure 7 shows the top view of the measurement setup for the induced voltages in a semi-circular loop (28 mm diameter, 0.7 mm wire diameter). The distance between the discharge tip and the center of the semicircular loop is 10 cm. Table 1 shows the variations of the induced voltage at four different revolution-angles of the ESD simulator at 2 kV. From table 1, one can find that different values gained when different ESD simulators used at different revolution-angles even though the same simulator at different revolution-angles. The variation indicates the importance of the transient fields and shows that even when using one simulator there can be repeatability problems.

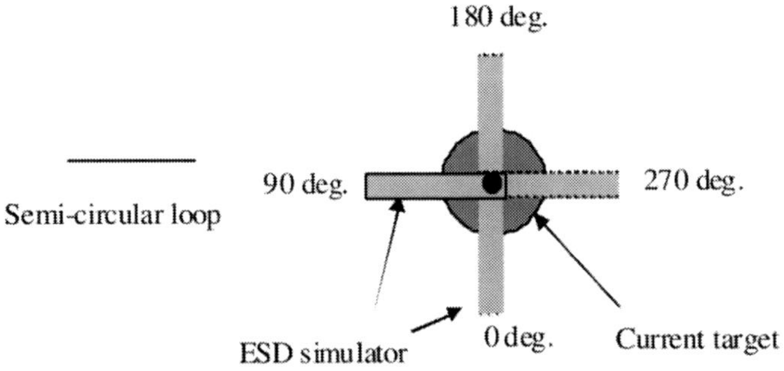

Figure 7. Measurement setup for the induced voltages in a semi-circular loop (top view). The ESD simulators were rotated around the discharge tip. The induced voltage was measured at four angles

Table 1. Values of the induced voltages in a semi-circular loop at different revolution-angles of the ESD simulators.

ESD simulators	Induced voltages at each revolution-angles of ESD simulator (V)			
	0 deg.	90 deg.	180 deg.	270 deg.
ESS-200AX	3.25	3.18	5.69	4.73
NS61000-2A	2.38	2.15	3.54	3.51

Figure 8 shows the spectra of the induced voltages in a semi-circular loop for ESS-200AX simulator at different revolution-angles. Figure 9 shows the spectra of the induced voltages for ESS-200AX simulator and NS610000-2A simulator at 0 degree.

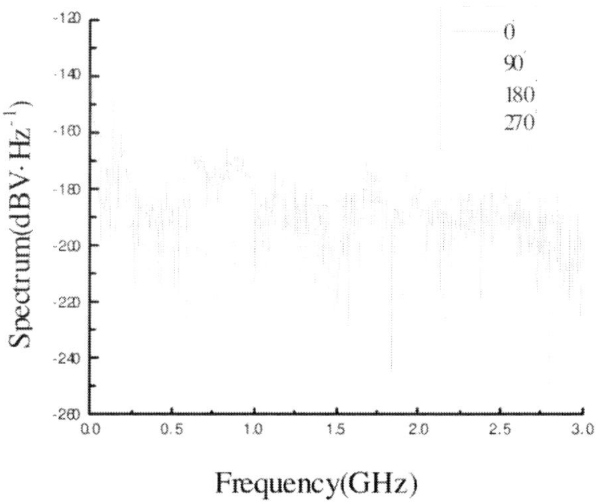

Figure 8. Spectrums of the induced voltage in a semi-circular loop for ESS-200AX simulator and NS61000-2A simulator at different revolution-angles.

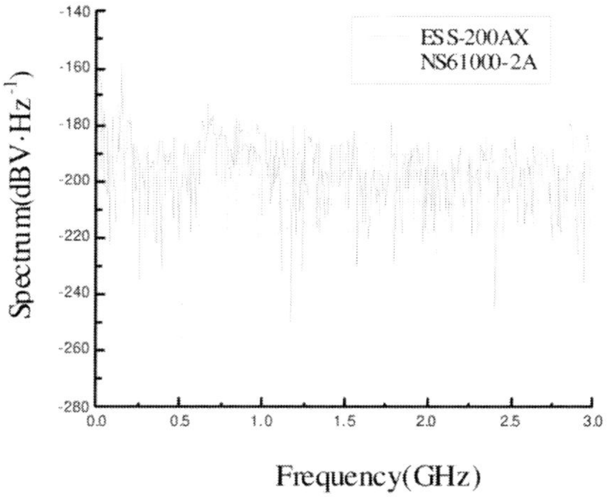

Figure 9. Spectrums of the induced voltage in a semi-circular loop for ESS-200AX simulator and NS61000-2A simulator at 0 degree.

From figure 8, one can distinguish two regions within the spectrum of the induced voltage in a semi-circular loop (the transition occurred at about 500 MHz). In the lower frequency ranges (lower than 500 MHz), the rotation effects are less seen in the spectrum, but strong variations can be observed due to the angle of the rotations in the higher frequency ranges (higher than 500 MHz). This can be explained as follows [14]: in the lower frequency ranges, the induced voltage in a semi-circular loop is dominated by the fields from the discharge current which is not affected by rotating the ESD simulator. The higher frequency components are caused by the relay that initiates the ESD pulse. The voltage collapse time in the relay is less than 100 ps. Thus, a pulse forming network is needed to shape the discharge current into a standard waveform [18]. The currents flowing on this pulse forming network, the relay and the metallic structures in proximity, are not symmetric, therefore, the currents within the ESD simulator will generate non-symmetric transient fields, while the discharge current flowing through the discharge tip generates the symmetric transient field around the ESD simulator. From figure 9, one can see that the variation is larger in the high frequency ranges.

In a word, the variation of the induced voltage in a semi-circular loop is dominated by the brands and the revolution-angles of the ESD simulators, this variation indicates the importance of the transient fields. Thus, better test repeatability will only be achieved by properly controlling the transient field during discharge.

3.4. Effect of imperfect specification of air discharge mode testing method on the ESD immunity test

An air discharge relates to the spark channel forming process, the temperature, the humidity, and the electrode approach speed will cause the discharging process varied obviously, therefore, the repeatability of the air discharge is very bad, and difficult to be controlled. In the early 1990s, testing has been moved from air discharge to contact mode testing to avoid the effect of arc length variations in air discharge and to improve reproducibility [4]. However, the air discharge mode is the most common ESD phenomenon is industry and living environments, and also a main way to damage and/or interfere to electronic equipment. It is one of the most urgent problems needing to be solved for industry. The imperfect specification for the air discharge mode testing method in the standard limits the application of the standard at a certain degree. In addition, the discharge condition for air discharge mode, especially the discharge electrode approach speed is not specified in the standard, which is difficult for the tester to control a relatively consistent testing condition in air mode ESD immunity test, as a result, we can't get a well repeatable testing result. Therefore, it is very significant to study the related factors which affect the air mode ESD, to establish and specify a suitable stern condition for air mode ESD immunity testing method.

In the paper, the repeatability of air discharge was studied using the experimental setup depicted in figure 1. In the experiment, the ESS-200AX simulator is used; the approach speed is about 0.5-1.0 m s^{-1}, the charge voltage from ±2 kV to 28 kV, the temperature is 26 ℃ and the humidity is 33% RH (the fluctuation of temperature and humidity is strictly controlled within ±2 ℃ and ±2% RH). For each combination of approach speed and voltage, 20 successive experiments were performed and the average was calculated (Note: in ESD immunity test, the test result is usually determined by the most serious discharge and not the average value of multiple discharges. Here, the average value is just used to analyze the general rules of the effect of approach speed and voltage on ESD parameters).

The coefficients of variability of discharge current and rise time at different approach speeds and charge voltages are given in tables 2 and 3. Here, the coefficient of variability is defined as the ratio of standard deviation and average of 20 results; it is used to represent the repeatability of test results.

Table 2. The Coefficient of Variation of Discharge Current.

Polarity	Approach Speed ($m\ s^{-1}$)	Charge Voltage (kV) 2	4	8	15	24
Positive	0.5	0.125	0.185	0.249	0.246	0.232
	0.6	0.105	0.196	0.241	0.340	0.223
	0.7	0.108	0.134	0.219	0.314	0.218
	0.8	0.060	0.115	0.142	0.232	0.181
	0.9	0.059	0.132	0.144	0.249	0.183
	1.0	0.076	0.134	0.150	0.230	0.179
Negative	0.5	0.145	0.128	0.128	0.123	0.121
	0.6	0.114	0.116	0.113	0.113	0.119
	0.7	0.166	0.113	0.044	0.036	0.029
	0.8	0.072	0.117	0.025	0.027	0.025
	0.9	0.138	0.115	0.063	0.064	0.058
	1.0	0.139	0.119	0.053	0.049	0.043

From tables 2 and 3, one concludes that, for a constant charge voltage, the coefficients of variability of discharge current and its rise time increase with the growth of approach speed, especially at 0.8 ms^{-1} (this claim is based on many times test results, the same conclusion also get at the humidity of 44% RH and 54% RH). This indicates that the repeatability can be improved by increasing approach speed for the air discharge mode. This important conclusion will provide the foundation of developing a standard method for air ESD immunity tests.

Table 3. The Coefficient of Variation of Rise Time.

Polarity	Approach Speed ($m\ s^{-1}$)	Charge Voltage (kV) 2	4	8	15	24
Positive	0.5	0.304	0.390	0.427	0.377	0.260
	0.6	0.258	0.292	0.393	0.382	0.231
	0.7	0.230	0.300	0.321	0.380	0.204
	0.8	0.181	0.184	0.187	0.237	0.129
	0.9	0.201	0.203	0.178	0.248	0.151
	1.0	0.197	0.201	0.206	0.238	0.133
Negative	0.5	0.402	0.333	0.470	0.430	0.297
	0.6	0.363	0.312	0.395	0.440	0.283
	0.7	0.362	0.245	0.396	0.340	0.255
	0.8	0.216	0.233	0.220	0.292	0.192
	0.9	0.221	0.240	0.232	0.248	0.225
	1.0	0.222	0.228	0.240	0.245	0.222

Except above discussed factors, the non-flat frequency response of current target, cable loss, and the reflections which caused by the scope, the attenuators, and the target are not perfectly matched to the cable impedance, will affect the discharge current measurement, and further affect the ESD immunity testing results.

4. Conclusion

The mostly widely used system level ESD test standard, IEC61000-4-2, causing irreproducibility results, neither ESD standard specifications nor the methodology used to verify the specifications are presently sufficient restrictive to provide acceptable ESD system test repeatability. Therefore, using the newly designed ESD simulation testing system, we find that the switch action of the ESD simulator is one of the main reasons; variation between different brand ESD simulators is also a major reason for the lack of reproducibility in ESD tests. In addition, the brand of ESD simulator and its revolution angles, and the imperfect specification of the test method in the standard can also affect the reproducibility of the test results.

Acknowledgments

This work was supported by National Natural Science Foundation of China under Grant 61172035, Grant 51177173, and Grant 60971042.

References

[1] Joyong K 2008 Ph.D. thesis, the Missouri University of Science and Technology, Missour
[2] Rhoades W and Mass J 1998 *Proc. IEEE Int. Symp. on Electromagnetic Compatibility* p1078
[3] Barth J, Dale D, Hall K, Hyatt H, McCarthy D, Nuebel J and Smith D Proc. 1996 *Int. Symp. on EOS/ESD* p211
[4] Lin D, Pommerenke D, Barth J, Henry L. G, Hyatt H, Hopkins M, Senko G and Smith D 1998 *Proc. Int. Symp. on Electrical/Electrostatic Discharge* (Las Vegas) p29
[5] Pommerenke D 1995 *J. Electrostatics.* **36** 31
[6] EMC—Part 4-2: Testing and Measurement Techniques—Electrostatic Discharge Immunity Test, IEC International Standard 61000-4-2, 2008.
[7] Habiger E 1995 *Proc. IEEE Int. Symp. on Electrical/Electrostatic Discharge* (Zurich) p645
[8] Hall K 1994 *Proc. Int. Symp. on Electrical/Electrostatic Discharge* p161
[9] Pommerenke D 1994 *Proc. Int. Symp. on Electrical/Electrostatic Discharge* p150
[10] Hopkins M 1995 *Test & Measurement world* p31

[11] Smith D 1993 *Proc. Int. Symp. on Electrical/Electrostatic Discharge* p3
[12] Lutz M 1986 *Proc. Int. Symp. on Electromagnetic Compatibility* p461
[13] Calcavecchio R. J Proc. 1986 *Int. Symp. on Electromagnetic Compatibility* p475
[14] Koo J, Cai Q, Wang K, Maas J, Takahashi T, Martwick A and Pommerenke D 2008 *IEEE Trans.Electromang. Compat.* **50** 794
[15] Hirata M, Takahash T and Schibuya N 2007 *Proc. Int. Symp. on Electromagnetic Compatibility* p256
[16] Rowe M 1995 *Test & Measurement World* p15
[17] Zhang X J, Ruan X F and Liu S H 2003 *Asia-Pacific Conf on Environmental Electromagnetics* (Hangzhou) p229
[18] Romachandran C, Pommerenke D, Wang K, Doren T. Van, Centola F.P and Huang J S 2004 *IEEE Trans.Electromang. Compat.* **46** 498

Responses of transmission-line networks to electrostatic discharge electromagnetic pulses

Xing Zhou [1*], Qingguo Wang[1], Yang Li[2]
[1]Electrostatic and Electromagnetic Protection Research Institute, Ordnance Engineering College, Heping West Road, Shijiazhuang 050003,China
[2]Department of Computer Technology, Shijiazhuang Worker University, Shijiazhuang 050054, China

E-mail: zxlwbh@126.com

Abstract. A modelling method of transmission-line networks based on Baum–Liu–Tesche (BLT) equation is investigated, and a use of BLT equation for calculating transmission-line load responses in the time domain is discussed. From the frequency-domain BLT equation, the expressions for the propagation function at each terminal load are derived. Using the propagation function and time-domain convolution calculation, the transient load responses in the time domain are gained. For special cases of a dendriform network and an annular network, the transient load responses excited by electrostatic discharge (ESD) pulse are calculated. Compared with test results, and the agreement of the calculation is excellent. The modelling and calculating method could be used to investigate the load responses of transmission-line networks, and estimate the influence of electromagnetic pulses on networks.

1. Introduction

Electrostatic discharge (ESD) is a natural phenomenon in dry environment. ESD current is a transient pulse with short duration, high peak power and concomitant electromagnetic emission. ESD electromagnetic pulse (EMP) can influence the performance of circuits and devices, and transmission line and its network are the main coupling ways of EMP energy.

In recent years, increasing attention is drawn to the area of ESD effects and ESD protection. It mainly focuses on direct coupling of ESD energy to devices or equipment. This work emphasizes indirect coupling by transmission-line networks, and investigate loads responses of networks when ESD occurred at any place. Different loads, structures and line length of network are also considered.

2. BLT equation

In 1970s, there was an interest in developing transmission-line models that could be used for analyzing large multi-conductor cable harnesses in aircraft. This research resulted in the Baum–Liu–Tesche (BLT) equation, which is a frequency domain matrix equation describing the behavior of voltages and currents at all of the junctions (interconnections) of the multiconductor cables in the network [1–2]. Recently, an extension of the BLT-equation concept was developed to include the effects of electromagnetic (EM) field propagation in space (along trajectories that can be thought of as being an imaginary field propagation "tube") [3-4].

* To whom any correspondence should be addressed.

The multi-conductor transmission-line equations in frequency domain can be expressed as:

$$\begin{cases} \frac{d}{dx}(V_n(x)) = -(Z'_{n,m})\cdot(I_n(x)) + (V_n^{(S)}(x)) \\ \frac{d}{dx}(I_n(x)) = -(Y'_{n,m})\cdot(V_n(x)) + (I_n^{(S)}(x)) \end{cases} \tag{1}$$

where $(Z'_{n,m})$ and $(Y'_{n,m})$ are the per-unit-length impedance matrix and admittance matrix respectively; $(V_n^{(S)}(x))$ and $(I_n^{(S)}(x))$ are the per-unit-length voltage matrix and current matrix respectively. The subscript m and n express the conductor sequences.

The two conductor transmission line models are shown in figure 1. The wave W_1 and W_2 can be expressed in matrix form as:

$$W_1(x) = (V_n(x))_+ = (V_n(x)) + (Z_{C_{n,m}})\cdot(I_n(x)) \tag{2}$$

$$W_2(x) = (V_n(x))_- = (V_n(x)) - (Z_{C_{n,m}})\cdot(I_n(x)) \tag{3}$$

where, $Z_{C_{n,m}}$ is the impedance matrix, $(V_n(x))_+$ is the incidence wave and $(V_n(x))_-$ is the reflection wave. The total voltage and current are as follows:

$$(V_n(x)) = \frac{1}{2}((V_n(x))_+ + (V_n(x))_-) \tag{4}$$

$$(Z_{C_{n,m}})\cdot(I_n(x)) = \frac{1}{2}((V_n(x))_+ - (V_n(x))_-)\cdot \tag{5}$$

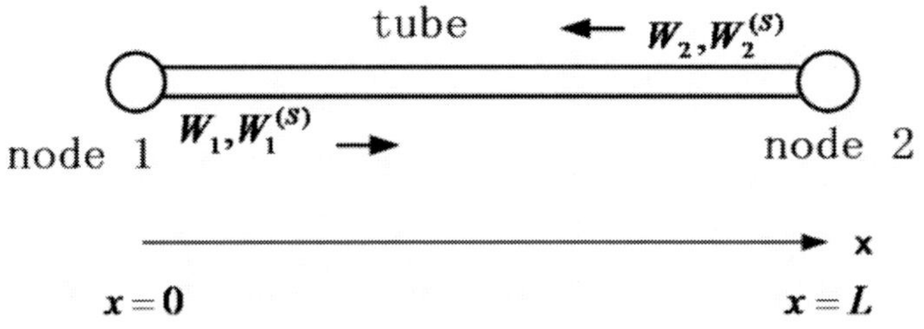

Figure 1. Transmission-line model.

So, the equation (1) can be expressed as:

$$\left(\frac{d}{dx}\cdot(1_{n,m}) + q(\gamma_{C_{n,m}})\right)\cdot(V_n(x))_q = (V_n^{(S)}(x))_q \tag{6}$$

with

$$[\gamma_{C_{n,m}}]^2 = \begin{bmatrix} \gamma_1^2 & 0 & \cdots & 0 \\ 0 & \gamma_2^2 & \cdots & 0 \\ \vdots & \vdots & \ddots & \vdots \\ 0 & 0 & \cdots & \gamma_n^2 \end{bmatrix}, \quad q = \begin{cases} 1 & \text{for incidence wave} \\ -1 & \text{for reflection wave} \end{cases}.$$

According to the foregoing equations, the solution can be obtained as:

$$W_1(x) = e^{-\left(\gamma_{C_{n,m}}\right)\cdot x}\cdot\left(V_n(0)\right)_+ + \int_0^x e^{-\left(\gamma_{C_{n,m}}\right)\left(x-x'\right)}\cdot\left(V_n^{(S)'}(x')\right)_+ dx' \tag{7}$$

$$W_2(x) = e^{\left(\gamma_{C_{n,m}}\right)(x-L)}\cdot\left(V_n(L)\right)_- + \int_L^x e^{\left(\gamma_{C_{n,m}}\right)\left(x-x'\right)}\cdot\left(V_n^{(S)'}(x')\right)_- dx'. \tag{8}$$

Assumed x as u, –x as v, W_u (W_1) and W_v (W_2) could be expressed in new coordinate as:

$$W_u(x_u) = \left(V_n(x_u)\right)_u = e^{-\left(\gamma_{C_{n,m}}\right)\cdot x_u}\cdot\left(V_n(0)\right)_u + \int_0^{x_u} e^{-\left(\gamma_{C_{n,m}}\right)\left(x_u-x_u'\right)}\cdot\left(V_n^{(S)'}(x_u')\right)_u dx_u' \tag{9}$$

$$W_u(L_u) = \left(V_n(L_u)\right)_u = e^{-\left(\gamma_{C_{n,m}}\right)\cdot L_u}\cdot\left(V_n(0)\right)_u + \int_0^{L_u} e^{-\left(\gamma_{C_{n,m}}\right)\left(L_u-x_u'\right)}\cdot\left(V_n^{(S)'}(x_u')\right)_u dx_u' \tag{10}$$

$$W_v(x_v) = \left(V_n(x_v)\right)_v = e^{-\left(\gamma_{C_{n,m}}\right)\cdot x_v}\cdot\left(V_n(0)\right)_v + \int_0^{x_v} e^{-\left(\gamma_{C_{n,m}}\right)\left(x_v-x_v'\right)}\cdot\left(V_n^{(S)'}(x_v')\right)_v dx_v' \tag{11}$$

$$W_v(L_v) = \left(V_n(L_v)\right)_v = e^{-\left(\gamma_{C_{n,m}}\right)\cdot L_v}\cdot\left(V_n(0)\right)_v + \int_0^{L_v} e^{-\left(\gamma_{C_{n,m}}\right)\left(L_v-x_v'\right)}\cdot\left(V_n^{(S)'}(x_v')\right)_v dx_v'. \tag{12}$$

Transmission-line network can be derived by the two conductor transmission-line as:

$$\left(W_u(L_u)\right) = \left(\left(V_n(L_u)\right)_u\right) = \left(\left(\Gamma_{n,m}\right)_{u,v}\right)\cdot\left(\left(V_n(0)\right)_u\right) + \left(\left(V_n^{(S)}\right)_u\right) \tag{13}$$

where

$$\left(\Gamma_{n,m}\right)_{u,v} = \left(1_{u,v}\right)\cdot e^{-\left(\gamma_{C_{n,m}}\right)\cdot L_u} = \begin{cases} e^{-\left(\gamma_{C_{n,m}}\right)\cdot L_u} & u = v \\ 0 & u \neq v \end{cases}$$

and

$$\left(V_n^{(S)}\right)_u = \int_0^{L_u} e^{-\left(\gamma_{C_{n,m}}\right)\left(L_u-x_u'\right)}\cdot\left(V_n^{(S)'}(x_u')\right)_u dx_u'.$$

Scattering matrix equation is defined as:

$$\left(\left(V_n(0)\right)_u\right) = \left(\left(S_{n,m}\right)_{u,v}\right)\cdot\left(\left(V_n(L_u)\right)_u\right). \tag{14}$$

Substituting (14) for (13), we obtain the following equation:

$$\left(\left(V_n(0)\right)_u\right) = \left(\left(S_{n,m}\right)_{u,v}\right)\cdot\left(\left(\Gamma_{n,m}\right)_{u,v}\right)\cdot\left(\left(V_n(0)\right)_u\right) + \left(\left(S_{n,m}\right)_{u,v}\right)\cdot\left(\left(V_n^{(S)}\right)_u\right). \tag{15}$$

Through matrix inverse, equation (15) can be transformed as:

$$\left(\left(V_n(0)\right)_u\right) = \left[\left(\left(1_{n,m}\right)_{u,v}\right) - \left(\left(S_{n,m}\right)_{u,v}\right)\left(\left(\Gamma_{n,m}\right)_{u,v}\right)\right]^{-1}\cdot\left(\left(S_{n,m}\right)_{u,v}\right)\left(\left(V^{(s)}{}_n\right)_u\right). \tag{16}$$

The total voltage is the sum of incidence voltage and reflection voltage:

$$\left(V_n(0)\right) = \frac{1}{2}\left[\left(V_n(0)\right)_u + \left(V_n(L_v)\right)_v\right]. \tag{17}$$

The relation of incidence voltage and reflection voltage is:

$$\left(\left(V_n(L_u)\right)_u\right) = \left(\left(\Gamma_{n,m}\right)_{u,v}\right)\left(\left(V_n(0)\right)_u\right) + \left(\left(V_n^{(s)}\right)_u\right). \tag{18}$$

Combining (16), (17) and (18), we can get the total voltage of node:

$$\left(V_n(0)\right) = \frac{1}{2}\left[\left(\left(S_{n,m}\right)_{u,v}\right) + \left(\left(T_{n,m}\right)_{u,v}\right)\right]\cdot\left[\left(\left(1_{n,m}\right)_{u,v}\right) - \left(\left(S_{n,m}\right)_{u,v}\right)\left(\left(\Gamma_{n,m}\right)_{u,v}\right)\right]^{-1}\left(\left(V_n^{(s)}\right)_u\right). \tag{19}$$

3. Modelling and simulation of networks

The models of dendriform network and annular network are established, and transient responses of loads are analyzed.

3.1 Dendriform network

The dendriform network electromagnetic topology model is shown in figure 2. The lengths of tubes are: T_1=1 m, T_2=3 m, T_3=2 m and T_4=5 m, respectively. The characteristic impedances of all tubes (lines) are 50 Ohm. Using the BLT equation and Fourier transform techniques, transient responses of network to ESD EMP can be analyzed. Figure 3 shows the exciting source, which is the body-metal ESD wave and generated by NS61000-2A ESD simulator. The peak voltage is 18.6 V. The injection place is node 0. The loads of node 2, node 4 and node 5 are all 50 Ohm. The transient responses of node 2, node 4 and node 5 are shown in figure 4.

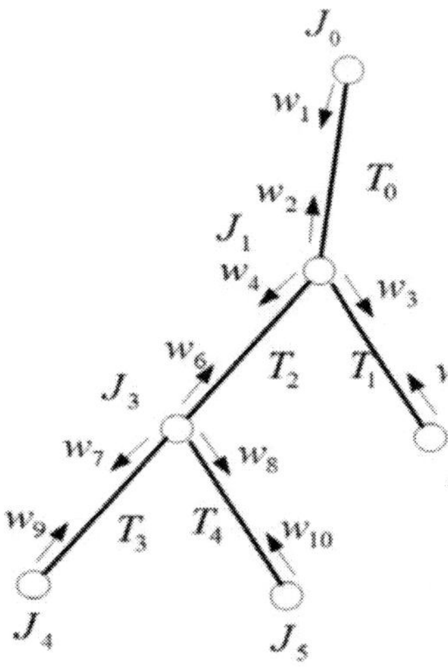

Figure 2. Dendriform network topology model.

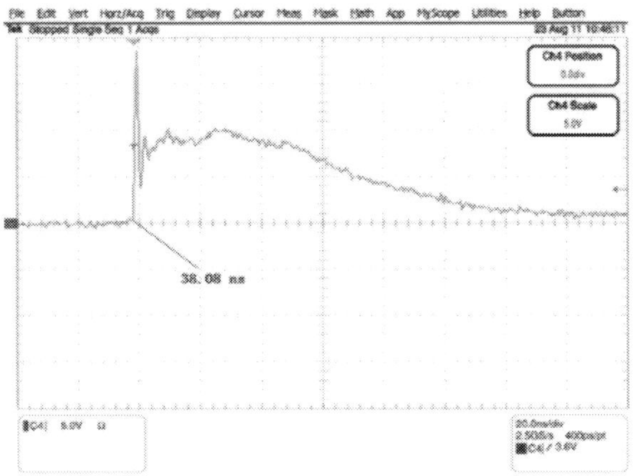

Figure 3. ESD EMP waveform.

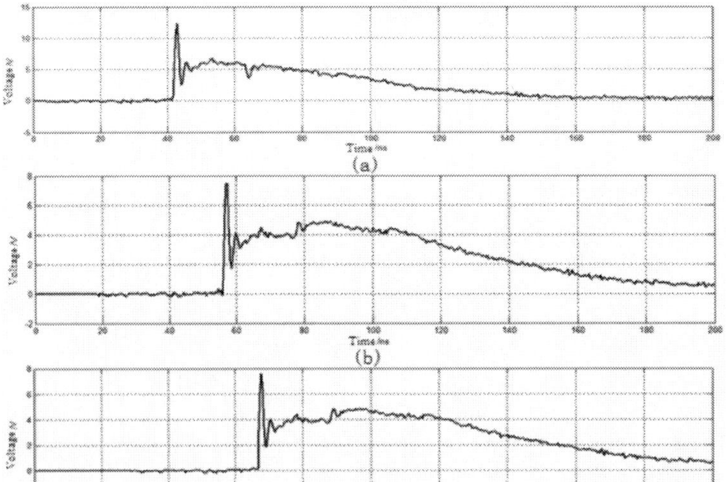

Figure 4. Voltage responses of node 2(a), node 4(b) and node 5(c).

Figure 4 shows the starting times of different loads responses are different. The results imply transmission times are different, which caused by different transmission distances.

Changing the load of node 2, the responses of node 4 are shown in figure 5. When the load doesn't match the transmission line, the mismatching will cause reflection. The second peak of response waves is arisen from the reflection of node 2, when the load of node 2 is 30 Ohm or 1 mega-Ohm. The reflection coefficient of node is defined as

$$\Gamma = \frac{Z_{load} - Z_{line}}{Z_{load} + Z_{line}} \ . \tag{20}$$

The reflection coefficient approximates 1 when the load is 1 mega-Ohm, so the total voltage of node 4 is doubled (blue dot line in figure 5). The reflection coefficient is -0.25 when the load is 30 Ohm. Therefore, the total voltage of node 4 decreased (red dash line in figure 5), due to the negative reflection coefficient.

The responses of node 4 are shown in figure 6 when substitute the length of tube 2 (defined as L). The tube length determines transmission time, so the delays of responses are different at different tube length.

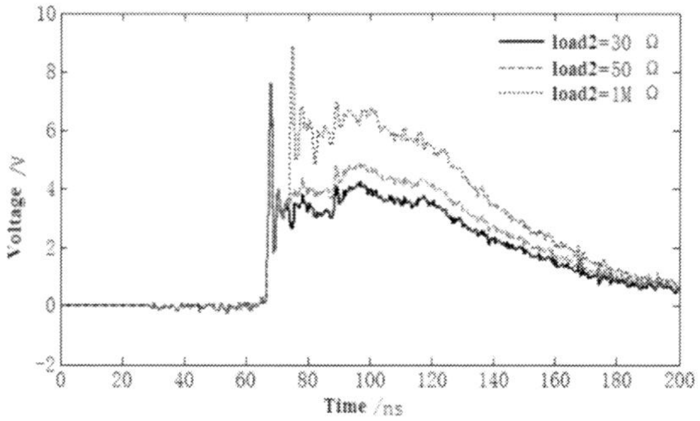

Figure 5. Voltage response at node 4 with different loads of node 2.

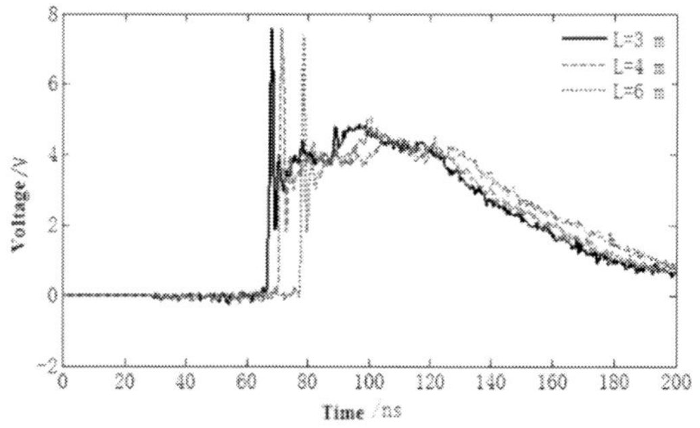

Figure 6. Voltage response at node 4 with different line length of tube 2.

3.2 Annular network

The annular network electromagnetic topology model is shown in figure 7. The lengths of tubes are: T_1=2 m, T_2=1 m, T_3=3 m, T_4=1 m and T_5=1 m, respectively. The characteristic impedances of all tubes (lines) are 50 Ohm. The exciting source is shown in figure 3. The injection place is node 0. The loads of node 4 and node 5 are 50 Ohm. The transient response of node 5 is shown in figure 8.

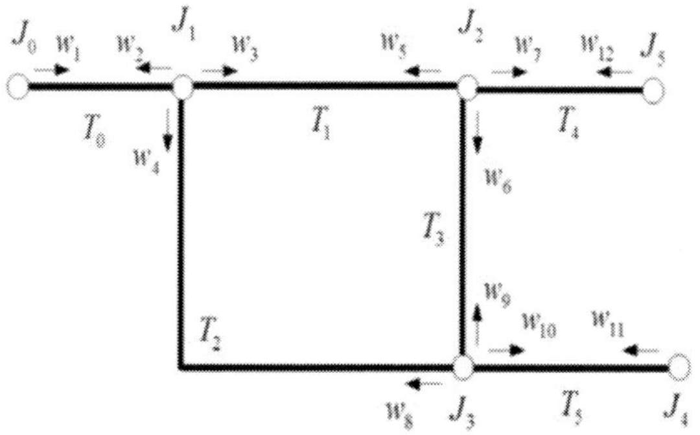

Figure 7. Annular network topology model.

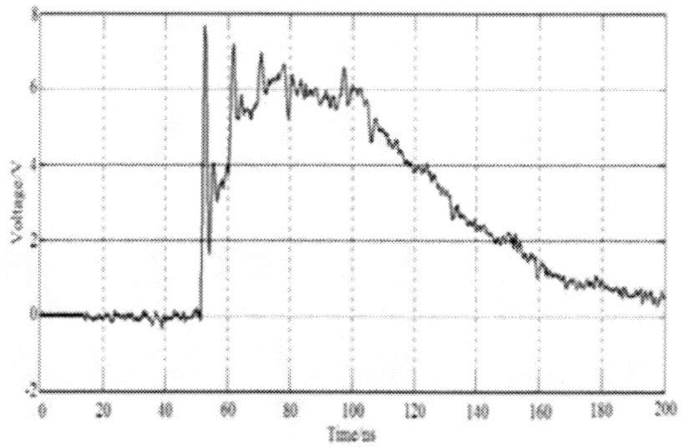

Figure 8. Voltage response at node 5.

The responses of annular network are different from those of dendriform network, by comparing figure 4 with figure 8. In annular network, there are two paths of node 0 (injection node) to node 5 (terminal node), one is $T_0 \rightarrow T_1 \rightarrow T_4$, the other is $T_0 \rightarrow T_2 \rightarrow T_3 \rightarrow T_4$. The lengths of two paths are different, thus transmission times are different. The response of node 5 is the sum of the two paths, so the total response of node 5 has two peaks.

Changing the load of node 4, the responses of node 5 are shown in figure 9. Changing the length of tube 2 (defined as L), the responses of node 4 are shown in figure 10. Different loads could give rise to different reflection (shown in figure 9); different tube length could cause different delay (shown in figure 10), which is similar to the dendriform network.

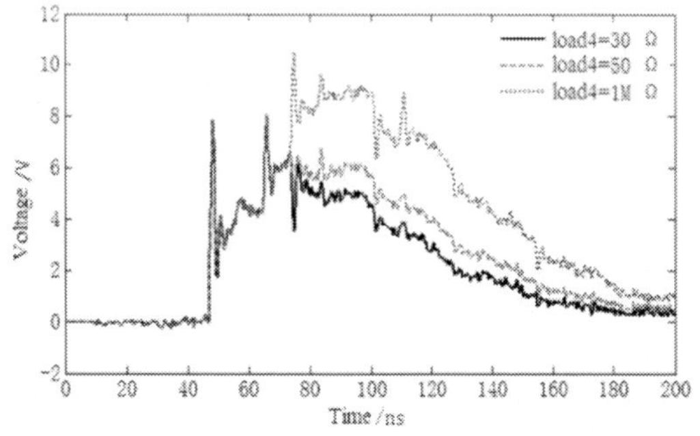

Figure 9. Voltage response at node 5 with different loads of node 4.

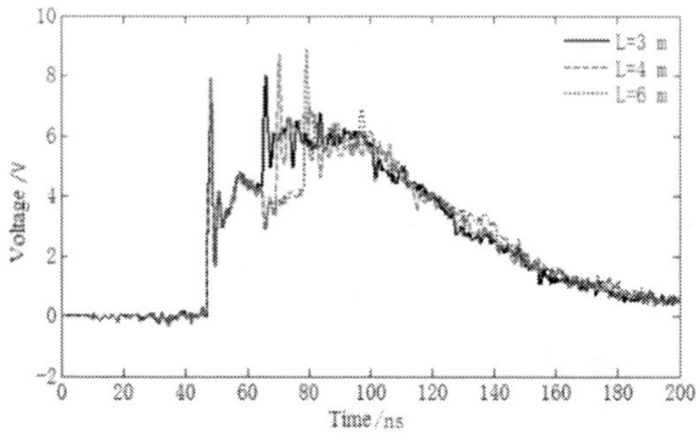

Figure 10. Voltage response at node 5 with different line length of tube 2.

3.3 Experimental validation

To verify the validity of calculating results, the experimental method was designed. The experimental networks are composed of coaxial cables, and the structure and lines length are the same as calculating models (shown in the figure 2 and 7). The parameters of coaxial cable are: $\sigma = 5.8 \times 10^7$ S/m, r=1.955 mm, R=2.375 mm, ε_{rel} =2.1 and $Z_0 = 50\,\Omega$. The comparison of experimental results and computer-simulated results are shown in figure 11.

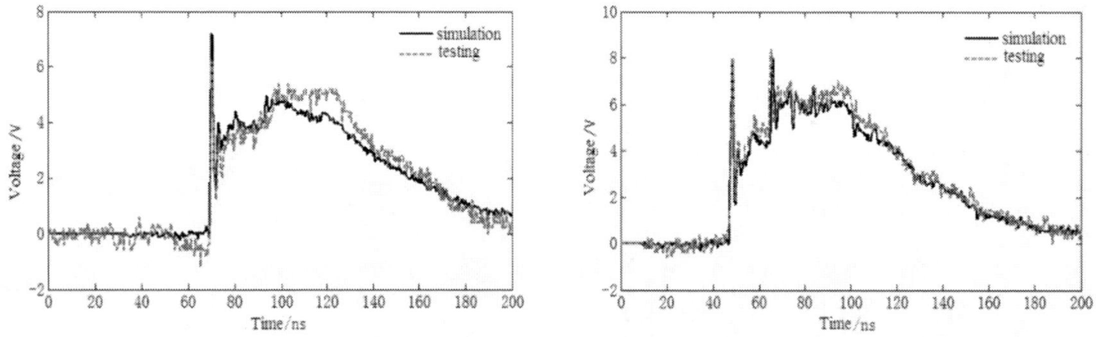

(a) (b)

Figure 11. Comparing of computer-simulated and experimental responses of node 2 (a) dendriform network; (b) annular network.

The agreement of computing results and experimental results are excellent for cable networks. These results show that this modelling method is effective for transmission-line network analysis.

4. Conclusion

This work verified the modelling and calculating method based on BLT equation can be used for analyzing transient responses of transmission-line network, by means of comparing the results of simulation and testing. Simulation results showed that the node responses to ESD EMP are related with network structure, lines length, and nodes load. The method and results of this work are useful for design and protection of networks against EMP.

Acknowledgments
This work was supported by Natural Science Foundation of China (NSFC) under Grant 50977091 and 60971042.

References

[1] Tesche F M 1997 *EMC Analysis Methods and Computational Models* New York: Wiley
[2] Baum C E 1999 *Proc. 13th EMC Symp.* Zurich, p 131
[3] Frederick M T 2007 *IEEE Trans. Electromagn. Compat.* **49** 3
[4] Paul C R 1994 *Analysis of Multiconductor Transmission Lines* NewYork: Wiley
[5] Zhou X, Wang S P, Wei G H 2006 *High Voltage Engineering* **32** 36
[6] Li B Z, He J L, Zhou H, et al 2010 *Information and Electronic Engineering* **8** 36
[7] Frederick M T 2007 *IEEE Trans. Electromagn. Compat.* **49** 427
[8] Zhou X, Li X D, Wang Q G 2011 *Cross Strait Quad-Regional Radio Science and Wireless Technology Conference* Harbin p 97

Journal of Physics: Conference Series **418** (2013) 012051 doi:10.1088/1742-6596/418/1/012051

Effects of electrostatic discharge on the performance of exploding bridgewire detonators

Zhipeng Li*, Ying Liu, Panjun Zhang, Zijian Lv, Yong Tian
Institute of Chemical Materials, China Academy of Engineering Physics, China

E-mail: lzpsunny@sina.com

Abstract. The performance of exploding bridgewire detonators affected by electrostatic discharge (ESD) were studied by microfocus X-ray industrial computed tomography (μCT), and high voltage stimulus to detonators was provided by an ESD simulator. Two stimulation modes were investigated, and the performance indexes of detonators after ESD were tested. The results show that there was no obvious change in bridge resistance when detonators were simulated by the mode of Pin-Pin, while the primary explosive of detonators was partly damaged by the mode of pin-shell, and the performance indexes of 2 detonators were affected by ESD. The reasons for this phenomenon were analyzed. This work contributes to understanding the performance of exploding bridgewire detonators affected by ESD.

1. Introduction

With the development of technology, the application of polymer material, micro-electronic devices and electricity explosion device has been becoming more and more extensive. At the same time, the problems caused by ESD have been paid great attentions [1]. Through the coupling and transfer of energy, ESD could make materials react. The process of heat effect could be considered as adiabatic because ESD can finish in nanoseconds or microseconds. Some serious accidents could happen when detonators were ignited by ESD.

The previous research on safety performance was mainly focused on the effects after ESD of detonators. However, for the weapon systems, the reliability of detonators is also very important. Especially with the reliability of weapon system improved, there are more requirements for both electrostatic safety and reliability. For example, it is clearly required in MIL that the reliability should not be reduced after ESD. The performance of four kinds of high-energy hot bridgewire initiators and three kinds of exploding bridgewire detonators were tested by US researchers, the results show that some problems were caused by ESD stimulus, which included detonators could not be ignited normally, or beyond the range of non-ignition or all-ignition range, the action time and threshold of exploding current were somewhat increased [2]. Qi and co-workers [3] have studied the influences on constant current igniting time of electric igniter caused by multiple ESD stimulation, the results show that constant current igniting time after ESD stimulus tends to shorten. Thus it can be seen that, the influences on performance of detonators after ESD stimulus can not be ignored. So it is very necessary to study and analyze the effects of ESD on the performance of detonators.

An ESD simulator was used to bring high voltage stimulus to exploding bridgewire detonators. The damage effects of bridgewire and primary explosive of detonators caused by ESD were investigated

* To whom any correspondence should be addressed.

by microfocus X-ray industrial volume computed tomography (μCT). The bridge resistance and whether the detonators could ignite after normal ignition pulse were tested, and the test results were analyzed.

2. Method

There are two main modes for ESD stimulus according to structure analysis, as shown in figure 1. The mode of Pin-Pin (A-B in figure) is same as the situation of igniting normally. The breakdown phenomenon could happen while the ESD stimulus was set between pin-shell. For the second mode, the primary explosive may be damaged, and the performance of detonators may be affected.

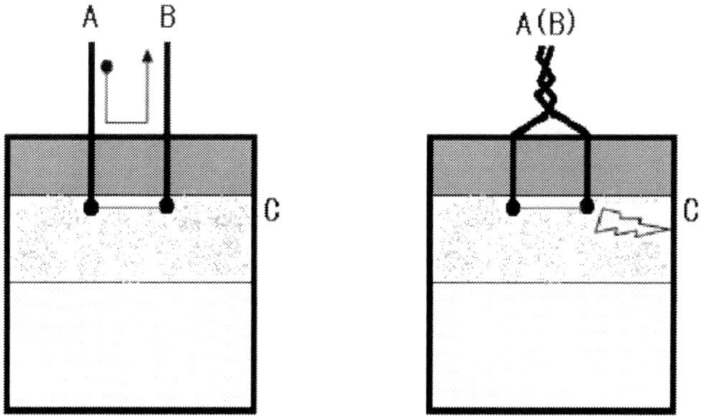

Figure 1. Two modes of ESD stimulation action.

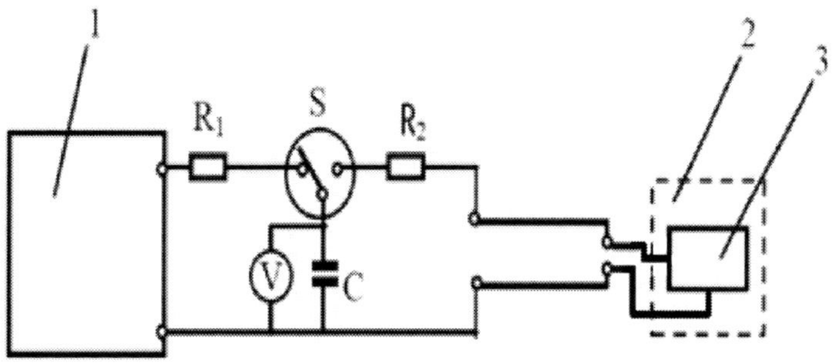

1-Power, 2-defence device, 3-sample,R_1-charging resistor, C-capacitor,

R_2-discharge circuit series resistor, V-voltmeter, S- switch

Figure 2. Schematic diagram of ESD test method.

There are eight kinds of ESD model reported in public literature at present [4], which includes human body electro-static discharge (HESD) model, field enhancement ESD model, machine ESD model, and so on. The electrostatic threat in the process of production, transportation, assembling of detonators is almost related to human action, which makes the electrostatic human body become the main and most common reason for those accidents caused by detonators. Therefore, electrostatic safety evaluation of detonators focuses on the HESD model. According to the reports [5], the maximum electrostatic potential value of the human body is not more than 50 kV. The rigor condition for HESD model is C=500 pF, R=100 Ω, based on these parameters, detonators were stimulated according to the way as shown in figure 2. There were 20 detonators chosen to test for both kinds of modes respectively, and only once stimulus for each sample, and then the performances of detonators were tested.

3. Results and Analysis

Some performances of exploding bridgewire detonators after ESD stimulus were investigated, including the variation of bridge resistance value, whether the detonators can ignite normally after normal ignition pulse, and weather the single function time still meet technical requirements. The results are showed in table 1.

Table 1. Performance test results of detonators after ESD

ESD way	Pin-Pin	Pin-Shell
bridgewire resistance value	no obvious change	no obvious change
ignited	all 20 ignited	18 ignited
single function time	meet technical requirements	meet technical requirements

Test results show that there is no obvious change for bridge resistance when detonators were stimulated by the mode of Pin-Pin, all of 20 samples ignited normally, and the single function time still meet technical requirements. While for the mode of Pin-Shell, although there was no obvious influence on bridge resistance value, two detonators among them were not ignited after normal ignition pulse.

In order to find out the reasons, industrial CT was used to inspect the status of primary explosive. Figure 3 gives the structure diagram of bridgewire and primary explosive in detonators. A-A profile and B-B profile were observed by CT.

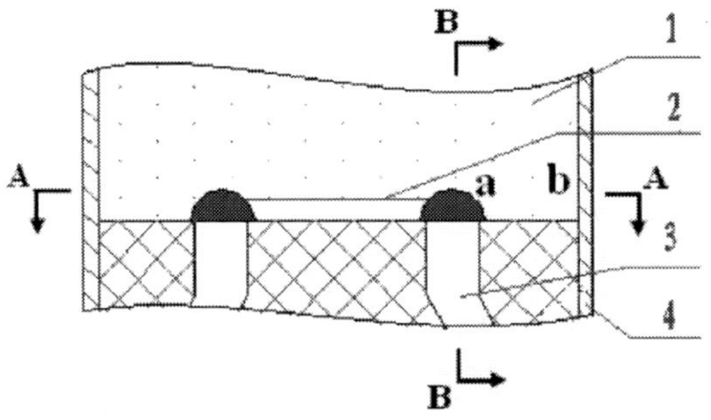

1-primary explosive; 2-bridgewire; 3-pin; 4-shell

Figure 3. Partial sketch of detonator.

3.1. Analysis based on CT results by the mode of Pin-Pin

Figure 4 is the industrial CT picture of primary explosive with ESD acted on detonators by the mode of Pin-Pin. It can be seen that there is no obvious change for grayscale in the area between the pin and shell, indicating that the density distribution of primary explosive still maintains the uniformity, primary explosive is no changed and ESD does not influence on bridgewire.

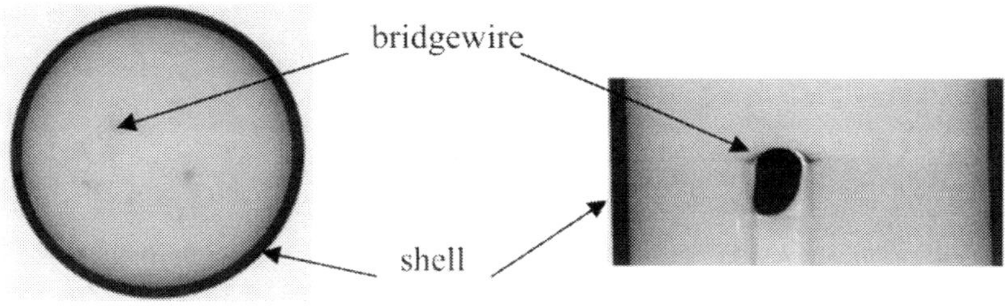

(a) A-A (b) B-B

Figure 4. CT picture of primary explosive after pin-pin ESD.

Stimulus condition with the mode of pin-pin can be simplified as the process detonators ignited by capacitance discharge. However, it distinguishes from the common capacitance discharge. Because discharge time is so short that the heat on bridgewire hardly diffuses outwards, the whole process can be regarded as adiabatic. Capacitance discharge energy is assigned to bridgewire and series resistor according to the ratio of bridgewire resistance and the series resistance. Bridgewire obtains little energy due to less resistance value (only several tens milliohm), so the temperature rise and heat release of bridgewire are so less that it is not enough to damage bridgewire and primary explosive, thus it has no influence on the performance of detonators.

3.2. Analysis based on CT results by the mode of Pin-Shell

CT photos of primary explosive after ESD acted on detonators by the mode of Pin-Shell are showed in figure 5. It can be seen that, grayscale distribution become lighter in some area between pin and shell as a result of the effect of ESD sparks stimulus (part of the arrow in figure 5), indicating primary explosive density decreases and a certain degree damage happens.

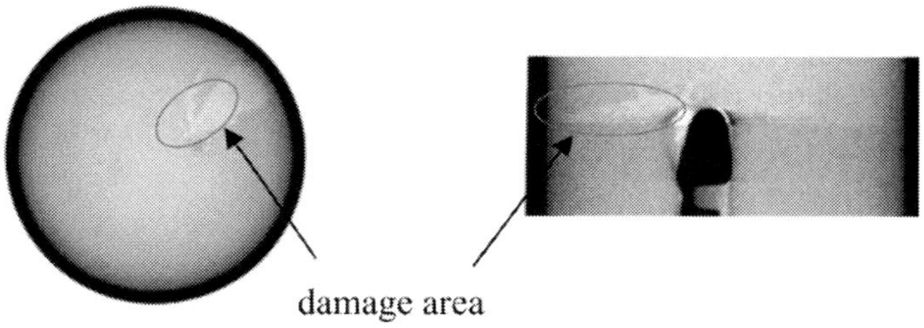

(a) A-A (b) B-B

Figure 5. CT picture of primary explosive after pin-shell ESD.

In order to better understand damage details and distribution of the primary explosive, different parts of the detonators were observed by industrial CT, as shown in figure 6. It can be seen that, the damage modality presents characteristic "branch" shape, which mainly distributed within the areas between bridgewire weld point and shell, and the part closer to the electrode plug is more seriously damaged, it could be related to the more concentrated discharge energy, because static electricity is more likely to release in the area closer to the shell.

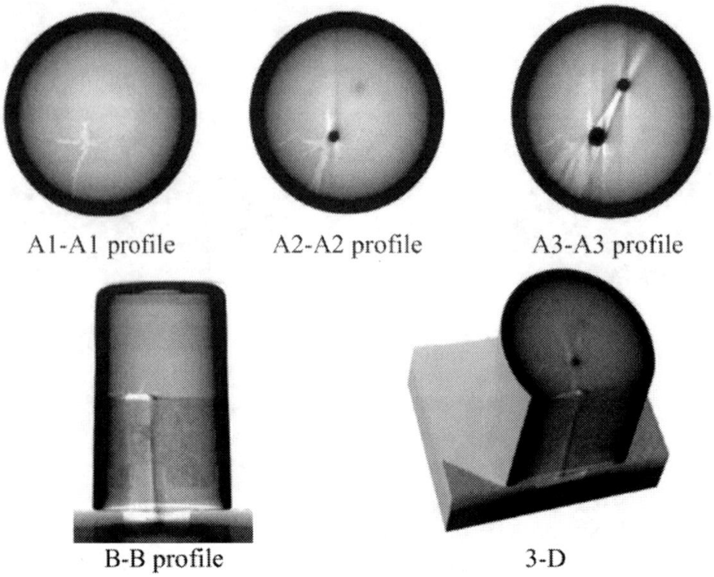

Figure 6. CT photos of different profiles.

When electric spark is forming as ESD acted on pin-shell, the air is ionized instantaneously and heated to a higher temperature, and shock wave is formed after the ionized air expanding rapidly. After that, the discharge breakdown channels and density loose area in the primary explosive are formed, and more rarefaction wave is produced in the transmission process of the area, deflagration reaction can not self-propagate as a result of rarefaction effect. Maybe this is the main reason why several detonators failed to ignite.

4. Conclusion

With the continuous development and application of the insensitive detonators, the electrostatic safety performance of detonators has been improved. However, reliability affected by ESD cannot be ignored. According to ESD test in this paper (50 kV, 500 pF, 100 Ω), some exploding bridge detonators misfired. The analysis indicates that, during the ESD process, the air is ionized instantaneously and heated to a higher temperature, and shock wave is formed after the ionized air expanding rapidly, resulting in density of primary explosive in discharge channels decreased. The detonators fail to ignite because of rarefaction effect. Of course, reliability research on detonators affected by ESD is a complex problem, and it is necessary to do more research work.

Reference

[1] Liu S H and Song X J 2008 *Chin. J. Nature* **29** 63
[2] Wang K M and Wen Y Q 2006 *National Defence Industry Press* **3**
[3] Qi X L and Chen Y Q 1999 *Initiators & Pyrotechnics(in Chinese).* **1**
[4] Wei G H, Guo H G and Sun Y W 2005 *Initiators & Pyrotechnics(in Chinese)* **2** 21
[5] Dan Borovina, Douglas McHugh, Steven Clarke, et al.2006 ESD sensitivity of detonators [A]. The 13th International Detonation Symposium [C]

Research on simulation of human body electrostatic discharge on detonating gas in coal mines

Shengman Liu [*]**, Tingtai Wang, Yuguan Zhang, Jingchang Zhang**
College of science, Zhongyuan University of Technology, Zhengzhou 450007, China

E-mail: liusmlw@163.com

Abstract. Static electricity is one of the reasons for gas explosions in coal mines. When miners leave the narrow tunnel exit of a coal mine, the body voltage may increase by more than ten times with the body capacitance decreasing quickly to 1/10 of its original value, and the body electrostatic discharge (ESD) energy can increase greatly to become a dangerous gas ignition source. In this work, simulation research of human body ESD detonating the gas in a coal mine was conducted. Based on the analysis of the human body electrostatic leakage rules, it is found that electrostatic half-life is mainly relative to the medium resistivity. Therefore it is important to establish a standard for anti-static attire. A series of gas explosion experiments were also conducted by the simulation system, and the results showed that the gas in coal mines might be detonated when the human body capacitance was in the range of 100-1800 pF and the discharge gap between 0.3-1.2 mm. Moreover, gas concentration at approximately 8.5% was most likely to induce an explosion.

1. Introduction

In recent years, many coal mine accidents have happened, especially for gas explosion accidents have happened frequently in the coal mine. Gas explosion occurs for a variety of reasons. As one of the natural hazard sources of society; electrostatic discharge always caused a great of loss and damage [1]. For the gas explosion caused by electrostatic discharge, it is usually think that some insulating materials such as high-polymers produce static spark discharge in the coal, but ignore that the human body is an isolated conductor producing electrostatic discharge [2, 3]. In fact, the human body electrostatic discharge is more danger than that of material in coal mine. Therefore, it is necessary to investigate human body electrostatic discharge [4]. In this work, the relationship between electrostatic discharges of human body and gas explosion under different conditions in coal mine environment has been investigated.

2. Human body electrostatic potential

In this experiment, JDY–3A electrostatic potential dynamic tester and TDS1001B oscilloscope are used to test the electrostatic potential of miners in the coal mine, as shown in figure 1 (a). It is the peak waveform figure of human body electrostatic potential when miners take off rapidly, the peak value of the transient voltage is more than ten times, and the figure 1(b) is the peak waveform of human body electrostatic potential when miners fall suddenly, the peak value of the transient voltage is more than ten times. When the body voltage rise suddenly ten times such as from 50 V to 500 V, the human body

[*] To whom any correspondence should be addressed.

resistance is reduced to 1/10 the original or less, it affords the conditions for electrostatic rapid leakage.

3. Human body capacitance

Miners work is in the narrow tunnel of underground mining. The tunnel around the human body is coal wall, and the coal is conductor. We have designed a stainless steel cylinder of 1 meter in diameter and 3 meters high to simulate the narrow tunnel in mine. Men and women with different height and weight, in miners' clothes and dressed in different resistivity antistatic boots go into the cylinder, and the human body capacitance in different locations are tested simulated in narrow tunnel.

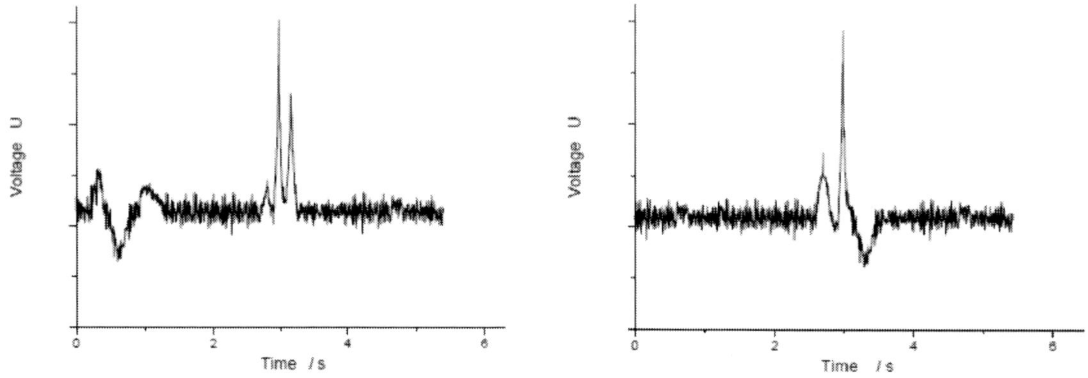

(a) Take off (b) Fall

Figure 1. Peak waveform figure of human electrostatic potential.

Figure 2 shown human's capacitance depend on the surface resistivity of anti-static boots; figure 3 shown human's capacitance depend on the volume resistivity of anti-static boots.

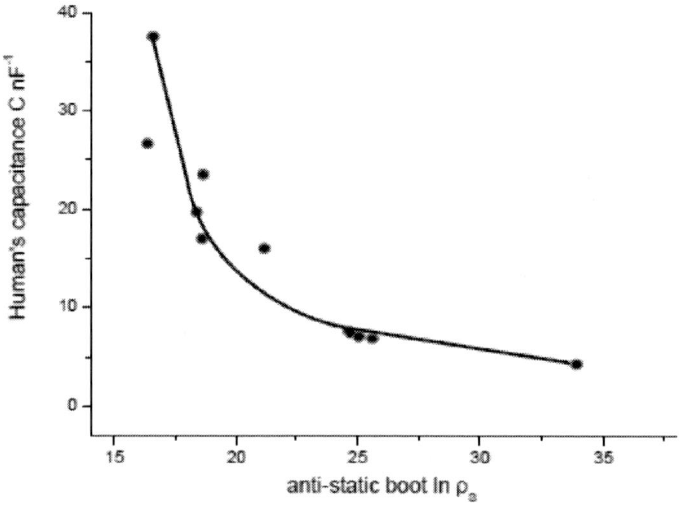

Figure 2. Human's capacitance depend on the surface resistivity of anti-static boots.

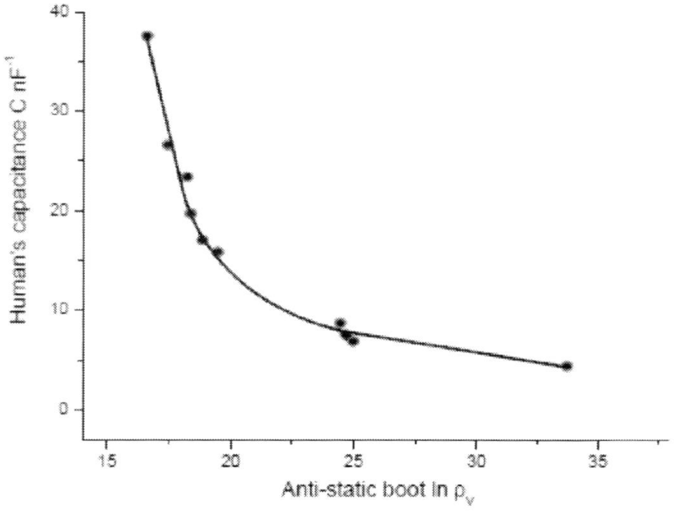

Figure 3. Human's capacitance depending on the volume resistivity of anti-static boots.

Table 1 shows the human body capacitance's results in the different locations of the tunnel. The conditions are as follows: temperature is 19°C, relative humidity 35%, height range 158-181 cm, and weight 50-73 kg. Outside was referred to the capacitance out of the cylinder, which was only 145-175 pF. And inside, the capacitance was not much difference. Obviously, the human body capacitance had no direct relation to human's sex, height and weight. The results shown that the outside capacitance of the cylinder is obviously lower than inside, the human body capacitance increased ten times in tunnel more than it at other place. When the miners are in the narrow tunnel exit, the capacitance reduces to rapidly less than a tenth than it in the tunnel. When the human body static electricity did not change significantly, human body voltage will suddenly increase more than ten times. It's easy to be a source to ignite the gas. Therefore, the narrow tunnel exit is easy to be the source for accidents of human body electrostatic discharge.

Table 1. Results of human body capacitance in different locations of the narrow tunnel. (Temperature: 19°C, Relative humidity: 35%).

Number	Sex	Height (cm)	Weight (kg)	capacitance (pF)				
				Outside	Front	Center	Back	Right
1	male	170	67	155	2610	2820	2620	2730
2	male	173	72	150	2400	2810	2800	2610
3	male	181	73	155	2710	3020	3000	2700
4	male	177	72	160	1700	1550	1520	1300
5	male	170	60	145	2900	3050	2910	2720
6	female	158	54	175	2800	3200	3150	2820
7	female	162	50	150	1800	2180	2100	2200
8	female	167	57	155	1610	1950	1520	1810
Average				156	2316	2573	2453	2361

4. Human body electrostatic leakage rules

Human body static electron leaks in two ways, one is the medium surface, and the other is the medium inner. The former is relatived to surface resistivity, the latter is relatived to volume resistivity. The electrostatic leakage rules can be expressed by discharge half-life.

A gaussian face is taken from inner of the medium, the current through the closed surface between the current density vector $\vec{\delta}$ and medium electrical quantity Q meet:

$$I=\int_S \vec{\delta}\bullet \mathrm{d}\vec{S}=-\frac{\mathrm{d}Q}{\mathrm{d}t}, \quad (1)$$

$$\vec{E}=\rho\vec{\delta} \quad (2)$$

$$\int_S \vec{E}\bullet \mathrm{d}\vec{S}=\frac{Q}{\varepsilon_0\varepsilon_r} \quad (3)$$

$$\int_S \rho\vec{\delta}\bullet \mathrm{d}\vec{S}=-\rho\frac{\mathrm{d}Q}{\mathrm{d}t} \quad (4)$$

or $-\rho\frac{\mathrm{d}Q}{\mathrm{d}t}=\frac{Q}{\varepsilon_0\varepsilon_r}$

$$\frac{\mathrm{d}Q}{Q}=-\frac{\mathrm{d}t}{\varepsilon_0\varepsilon_r\rho} \quad (5)$$

$$Q'=Qe^{-\frac{t}{\varepsilon_0\varepsilon_r\rho}} \quad (6)$$

The time of electrical quantity leakage 1/2 is half-life,

$Q'=\frac{1}{2}Q$, so

$$\frac{1}{2}=e^{-\frac{\tau_{1/2}}{\varepsilon_0\varepsilon_r\rho}} \quad (7)$$

or $\tau_{1/2}=0.69\varepsilon_0\varepsilon_r\rho$

$\tau_{1/2}$ is electrostatic half-life, ε_0 is vacuum dielectric constant, ε_r is relative dielectric constant, the value is from 3 to 9, electrostatic half-life is mainly relative with the medium's resistivity. Therefore the index of the miners' clothes anti-static is vital whether it is qualified.

5. Human body electrostatic discharge model detonated gas experiment

The apparatus is "Human body electrostatic discharge model detonated gas experiment system|". The system is composed with gas mixture cavity, control unit, gas measuring unit, oxygen measuring unit, temperature measuring unit, body ESD model ignition devices, alarm display, PC and gas supply device. The system frame is shown in figure 4. The measurement accuracy of the gas concentration and the oxygen concentration for the system is 0.1%.

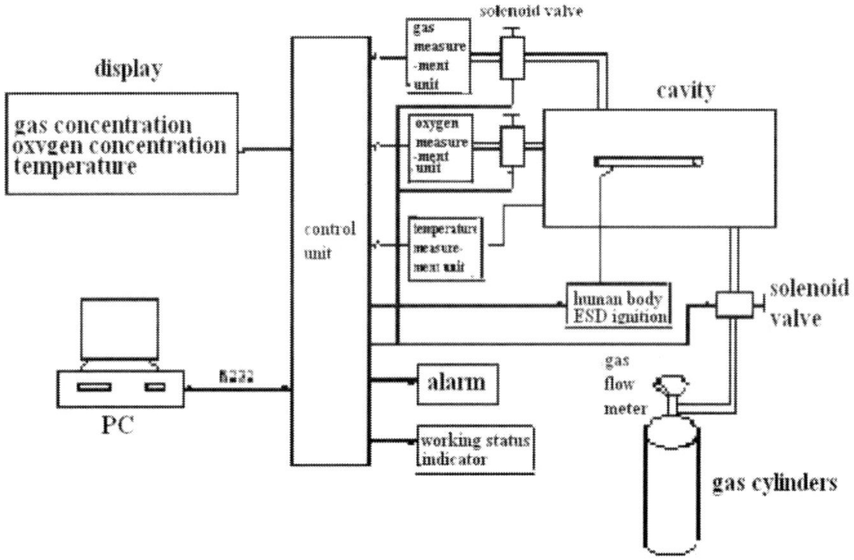

Figure 4. System diagram.

The experiment is carried out in the closed cavity, and the simulation system is controlled by computer, the cavity supply with the gas by gas supply device, and the environmental conditions oxygen content 20.9%, mixed gas temperature 22°C, and humidity 48%. The experiment began from gas concentration 5% slowly increased at a speed of 0.1% s^{-1}, until the gas concentration is up to 16%. In addition, the temperature, the concentration of oxygen and the gas were measured by sensors and infrared analyzer in different time. Different body capacitances are chosen to ignite, simulate different human body capacitance in the coal. Through adjusting the discharge gap, observe the detonation in the tunnel. Detailed data is shown on table 2. It is seen that the gas may be detonated, where the human capacitance range change in 100-1800 pF and the discharge gap is between 0.3-1.2 mm. And the gas concentration at approximately 8.5% is most likely to induce explosion.

Table 2 Results of the experiment system of detonated the gas by human ESD model (Oxygen content: 20.9%; Mixed gas temperature: 22 °C; Humidity: 48%).

Number	Gas concentration (%)	Human body capacitance (pF)	Human body resistance (Ω)	Discharge gap (mm)	Result
1	8.8	1800	250	0.3	explosion
2	8.5	950	350	0.6	explosion
3	8.3	489	350	0.6	explosion
4	8.4	470	350	0.9	explosion
5	8.2	280	350	0.9	explosion
6	8.7	150	350	1.2	explosion
7	8.7	100	350	1.2	explosion

6. Summary

At moving action, the body's electrostatic potential increase instantly more than ten times when people get out of the narrow tunnel exit, the human body's capacitance reduced instantly to only a tenth of the original, so the body's voltage increase instantly more than ten times, discharge energy increase greatly becoming a dangerous source to ignite gas. Based on the analysis of the human body electrostatic leakage rules, electrostatic half-life is mainly relative with the medium resistivity. Therefore the index of the miners' clothes anti-static is very important. The results shown: The human body capacitance range is in 100-1800 pF and the discharge gap is between 0.3-1.2 mm, it easily detonates the gas, and the gas concentration at about 8.5% is most likely to induce explosion.

Acknowledgements

This paper is supported by the National Natural Science Foundation of China (NSFC) with contact No. 50977093 and 51077134.

Reference

[1] Liu S 2004 *Static Discharge and Danger Prot.* (Beijing, Beijing University of Posts and Telecommunications Press)
[2] Nabours R.E. 2003 *Ind. and Commer. Power Syst.* (St. Louis, IEEE Technical Conference) p 66
[3] Mannan M Sam 1999 *Chem. Process Saf.* (AIChE, Chemical Process Safety) p 105
[4] Huang J 2003 *Proc. of ESA* (Arkansas, IEEE Joint Annual Meetings) p 24

Performance improvement of an ESD suppressor by studying its characteristics

H-Y Chen[*], P-K Li
Department of Communications Engineering, Yuan Ze University, 135, Yuan-Tung Road, Nei-Li, Chung-Li, Taoyuan Shian, Taiwan 32003

E-mail: eehychen@saturn.yzu.edu.tw

Abstract. The finite-difference time-domain (FDTD) method was used to study the characteristics of an ESD suppressor. The obtained capacitance of the ESD suppressor filled with air was validated by measurement data and another numerical result obtained by time-domain moment method (TDMM). Under the same conditions, no large differences are found among the obtained capacitances for the ESD suppressor filled with air, neon, argon, and helium. But the ESD suppressor filled with air has a much higher trigger and clamping voltages than the ESD suppressor filled with neon, argon, or helium. Capacitance is found to decrease with the increase of the spark gap distance or the operating frequency. But the capacitance increases with increasing relative dielectric constant of the substrate or the overcoat. t is observed that ESD currents with a transient waveform reach a maximum value and decays sharply as the time increases under the condition of the spark gap distance of 5 μm. Outside the spark gap, the maximum electric field is below the breakdown strength, but the magnitude of the electric field radiated from ESD events is still very large when the distance from the ESD suppressor is less than 2 mm.

1. Introduction

Modern electronic products are becoming smaller, lighter, and simpler while having a higher speed data rate. For these products operating in radio-frequency (RF) and microwave bands, an ESD event can easily damage high sensitivity electronic components which make extensive use of digital technology and fast-speed memory integrated circuit (IC) chips with very high density of multilevel interconnects. Therefore, circuit designers must take ESD problems into account in their quest for a functional and reliable product. For protecting sensitive electronic circuits operating at RF and microwave frequencies, the ESD suppressors are connected in parallel with the signal lines. When an ESD event occurs, the ESD suppressors clamp the ESD voltage to a level where the sensitive electronic circuits can survive and shunt the ESD current away from the data line and the protected sensitive electronic circuits into the chassis ground [1]. The ESD suppressors are designed to have extremely low capacitance which can provide clamping functions for transient suppression and electromagnetic interference (EMI) filtering against unwanted high-frequency signals that couple into the protected sensitive electronic circuits.

In this research work, the FDTD method is used to calculate an ESD suppressor's characteristics under different conditions. Originally, the ESD suppressor is filled with air in the space between two

[*] To whom any correspondence should be addressed.

discharge electrodes. Alternatively, the space between two discharge electrodes is filled with argon, helium, or neon in order to investigate the performance improvement of the ESD suppressor. In FDTD simulations, a double exponential electric field pulse [2, 3] is proposed to model the ESD excitation. The capacitances, ESD currents, maximum electric fields, trigger voltages, and clamping voltages of the ESD suppressor are presented in this report.

2. The FDTD method

In 1966 Yee [4] first proposed the basic FDTD method for electromagnetic analysis. Due to its powerful and versatile features, in recent years the FDTD method has become a very popular technique for solving many electromagnetic problems. In the FDTD solution procedure, the coupled Maxwell's equations in differential form are solved for various points of the scatter as well as its surrounding in a time-stepping manner until convergent solutions are obtained. Following Yee's notation and using centered difference approximation on both the time and space first-order partial differentiations, six finite-difference equations for six unique field components within a unit cell are obtained. In these six finite-difference equations, electric fields are assigned to half-integer (n+1/2) time steps and magnetic fields are assigned to integer (n) time steps for the temporal discretization of fields. To ensure numerical stability, the time step δ_t is set to $\delta/(2C_o)$, where δ and C_o are the cell size and the speed of light, respectively. The center difference approximation ensures that the spatial and temporal discretizations have second-order accuracy, where errors are proportional to the square of the cell size and time increment [4]. Several absorbing boundary conditions (ABC) have been proposed in the FDTD method such as second-order Mur [5], and Liao [6], and perfectly matched layer (PML) [7]. In our formulation, the second-order Mur approximation of absorbing boundary conditions [5] is used for the near-field irradiation problems. We use them because they do not require much memory and have a reasonable accuracy. The details of the FDTD method may be found in many publications and will therefore not be repeated here.

3. Characteristic study on an ESD suppressor

In FDTD simulations, an ESD voltage $V_0(t)$ with a double exponential voltage pulse [2, 3] similar to a standard ESD pulse current waveform, specified in International Standard IEC 801-2, is used to model the ESD excitation in the gap between the two discharge electrodes expressed by

$$V_0(t)= V_{esd}(e^{-at}-e^{-bt}),\ \text{ for } t \geq 0 \tag{1}$$

The double exponential voltage pulse is modelled with the sum of a fast wave and a slow wave, where constants a, and b are given below
For the fast wave:

$$a = 4.55\times10^8,\text{ and } b = 5.00\times10^8 \ \text{ for } 6.81\,ns > t \geq 0 \tag{2}$$

For the slow wave:

$$a = 4.55\times10^7,\text{ and } b = 5.00\times10^7 \ \text{ for } t \geq 6.81\,ns \tag{3}$$

The maximum voltage occurs at time $t_m = ln(b/a)/(b\text{-}a)$, where constants a and b are equal to 4.55×10^8 and 5.00×10^8, respectively. From the calculation, it is found that the maximum voltage occurs at the time t_m =2.09 ns. The waveform has a 1.20 ns rise time (the pulse increases from 10 % to 90 % of its peak value). The ratio of the voltage at 30 ns to the maximum voltage at 2.09 ns is 0.53 and the ratio of the voltage at 60 ns to the voltage at 30 ns is 0.48. It should be noted that the ESD pulse waveform is usually poorly reproducible in practice. Taking the trigger voltage as the maximum voltage, V_{esd} can be calculated for the ESD suppressor filled with air, argon, helium, and neon under the conditions of various spark gap distances. For FDTD simulations, the ESD voltage $V_0(t)$ is transferred into an electric field over the spark gap between the two discharge electrodes by $E(t) = V_0(t)/d$, where d is the distance of the spark gap between the two discharge electrodes. Applying the excited electric fields in FDTD simulations, electric fields at any location inside and outside the ESD suppressor can be calculated. From boundary conditions, electric fields obtained near the two copper electrodes can be used to calculate the surface charge density ρ_s on their surfaces expressed by

$$\nabla \cdot D = \rho_s = \varepsilon_r \varepsilon_0 E_{2n} - \varepsilon_0 E_{1n.} \tag{4}$$

where E_{1n} and E_{2n} are the normal components of electric fields in and outside the discharge electrodes, D is the electric flux density, ε_r and ε_0 are the relative dielectric constant of materials and the dielectric constant of air, respectively. The total charge on any one surface of the two discharge electrodes is determined by integrating the surface charge density over the surface expressed by

$$Q(t) = \int_s \rho_s(t) ds' \tag{5}$$

where s' denotes the surface of the two discharge electrodes. Once the total charges are obtained, the capacitance can also be calculated by

$$C = \frac{Q(t)}{V_0(t)} \tag{6}$$

It should be noted that the capacitance depends only on the geometric structure, material used in the ESD suppressor, and the operating frequency; it does not depend on the applied voltage or current. The ESD current is calculated by the following equation:

$$I(t) = \frac{\partial Q(t)}{\partial t} = \frac{Q(t) - Q(t - \Delta t)}{\Delta t}, \tag{7}$$

where Δt is the time increment and $Q(t)$ is the total charge at time t in one of the two electrodes.

The ESD suppressor is composed of two copper discharge electrodes (suppressor element), an Al_2O_3 substrate, a protective epoxy resin coating (overcoat), two zinc backside electrodes, two nickel connect electrodes, and a solder layer as shown in figure 1. The copper discharge electrodes have a thumbnail structure on the opposite sides of the spark gap and a thickness of 50 μm.

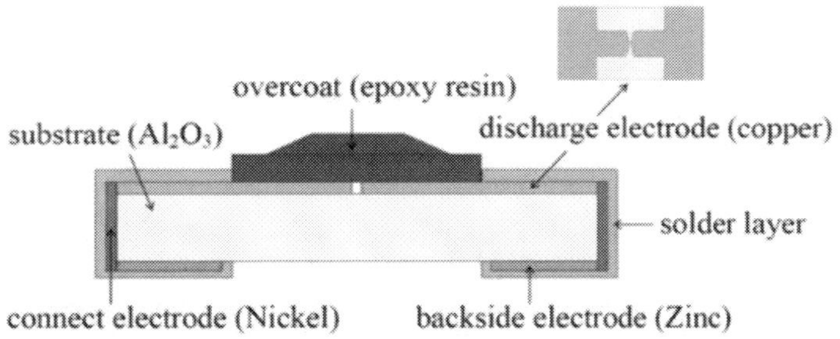

Figure 1. Illustration of the structure of the ESD suppressor.

Originally, the ESD is filled with air in the space between two discharge electrodes. Alternatively, the space between two discharge electrodes is filled with argon, helium, and neon in order to investigate the variation in capacitance, trigger voltage, clamping voltage, and the maximum electric field of the ESD suppressor. The FDTD model including the ESD suppressor, computational region, and the space between the absorbing boundary and the scattering objective for FDTD simulations is constructed with 10434000 cubic cells, where the cell size is taken to be δ=5 μm.

Under the conditions of the ESD suppressor filled with air, a spark gap distance of 5 μm, the ESD suppressor with Al_2O_3 substrate and epoxy resin overcoat, and operating frequency of 1 GHz, the capacitance of 0.0242 pF calculated by the FDTD method as shown in figure 2 is in good agreement with the measurement data of 0.025 pF provided by TA-I Technology Co. [8] and the numerical result of 0.0264 pF obtained by the time-domain moment method (TDMM) [1]. Simulation results of capacitance for the ESD suppressor filled with helium, neon, and argon under the same conditions are also shown in figure 2. The relative dielectric constants only have a small variation among air, helium, neon, and argon. As expected, the obtained capacitance for the ESD suppressor filled with air has a little higher value than that obtained for the ESD suppressor filled with helium, neon, or argon. It is also found that the capacitance increases with the increase of the relative dielectric constants of the substrate and the overcoat. It is also found that the capacitance decreases as the operating frequency increases from 1 to 6 GHz and the spark gap distance increases from 5 to 50 μm.

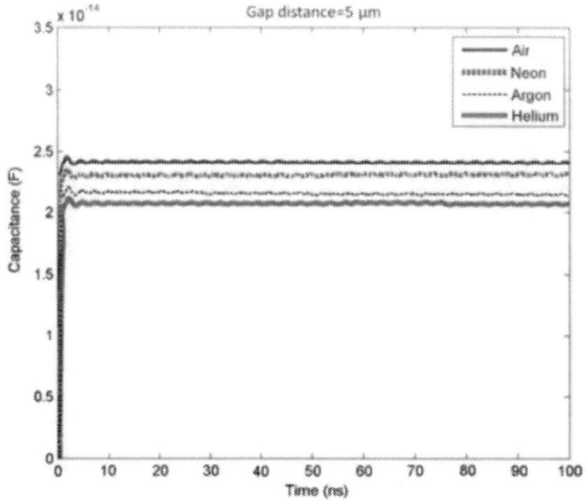

Figure 2. Simulation results of capacitance for the ESD suppressor filled with air, helium, neon, and argon.

All ESD suppressors have a limited life-time span. In order to extend the suppressor's life-time span, the trigger and clamping voltages should be designed as low as possible. Based on breakdown strengths and spark gap distances, the trigger and clamping voltages of the ESD suppressor filled with air, argon, helium, and neon for various spark gap distances are calculated and shown in figures 3 and 4.

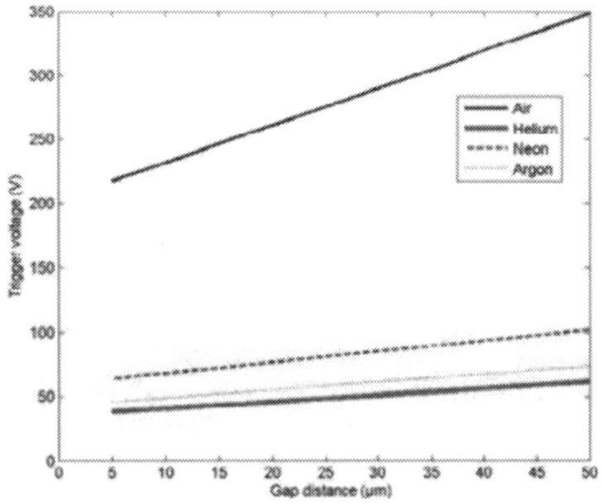

Figure 3. Trigger voltage versus spark gap distance

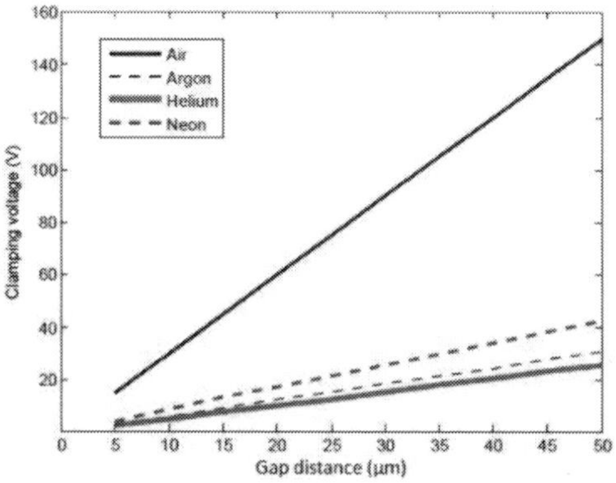

Figure 4. Clamping voltage versus spark gap distance.

Time responses of ESD currents for the ESD suppressor filled with air, helium, neon, and argon under the condition of a spark gap distance of 5 μm are obtained. It is found that ESD currents with a transient waveform reach a maximum value in a variety of 3.2-5.6 mA at 0.9 ns and decays sharply as the time increases. The duration of the transient phenomenon is very insignificant as compared with the operation time of an electronic system. Yet it is very important because the transient ESD current may cause a serious failure in high sensitive information circuits if the ESD current is not removed from the protected high sensitive information circuits. Maximum ESD currents versus spark gap distance for the ESD filled with air, helium, neon, and argon are also obtained. It is found that the maximum ESD current decreases as the spark gap distance increases.

In order to investigate the electric field near the ESD suppressor, electric field distributions on the horizontal plane which consist of the center of the spark gap are calculated. It is found that the maximum electric fields decay from 4.338×10^{7} to 1.436×10^{4}, from 1.266×10^{7} to 8.136×10^{3}, from 9.169×10^{6} to 6.171×10^{3}, and from 7.637×10^{6} to 2.736×10^{3} V m^{-1} for the ESD suppressor filled with air, neon, argon, and helium as the distance from the spark gap increases from 0.0 to 2.125 mm, respectively. Outside the spark gap, the maximum electric field is below the breakdown strength, but the magnitude of the electric field resulting from ESD events is still very large as the distance from the ESD suppressor is less than 2 mm. This observation hints that the high sensitivity electronic circuits should be kept away from the ESD suppressor, because high sensitivity electronic circuits connected to external ports are susceptible to a damaging ESD pulse from the operating environment and from peripheral interference.

4. Conclusions

The obtained capacitance of the ESD suppressor filled with air under the condition of a spark gap distance of 5 m was validated by measurement data provided by TA-I Technology Co. and the numerical result obtained by the TDMM. After checking the validity of the FDTD method, the FDTD method was used to further studies on the capacitance, ESD current, and electric field for the ESD suppressor filled with air, neon, argon, and helium under different conditions. Only a tiny variation of obtained capacitances for the ESD suppressor filled with air, neon, argon, and helium under the same conditions was found. But it should be noted that the ESD suppressor filled with air has a much higher trigger and clamping voltage than the ESD suppressor filled with neon, argon, or helium. From obtained capacitances, it is shown that the substrate or the overcoat with a higher dielectric material used for an ESD suppressor will produce higher capacitances. Higher capacitances have slower response time features so that data waveform distortions and signals without integrity may occur. It is observed that the capacitance decreases as the operating frequency increases from 1 to 6 GHz. It is observed that ESD currents with a transient waveform reach a maximum value in a variety of 3.2-5.6 mA at 0.9 ns and decays sharply as the time increases. The maximum electric fields near the ESD suppressor were also calculated. The maximum electric fields of 4.338×10^{7}, 1.266×10^{7}, 9.169×10^{6}, and 7.637×10^{6} V m^{-1} are obtained for the ESD suppressor filled with air, neon, argon, and helium under the condition of a spark gap of 5 m, respectively. Outside the spark gap, the maximum electric field is below the breakdown strength, but the magnitude of the electric field resulting from ESD events is still very large as the distance from the ESD suppressor is less than 2 mm.

References

[1] Chen H Y, Suo Y, Kuo C T, and Qiu J H 2011 *J. Electrostatics* **69** 604
[2] Camp M and Garbe H2004 *IEEE Trans. Electromagn. Compat.* **46** 675
[3] Prasad Kodali V 1996 *Eng. Electromagn. Compat.: Principle, Meas., and Technol.* (New York: IEEE Press), p 168
[4] Yee K S 1966 *IEEE Trans. Antennas Propag.* **AP-14** 302
[5] Mur G 1981 *IEEE Trans. Electromagn. Compat.* **EMC-23** 377
[6] Liao Z P, Wong H L, Yang B P, and Yuan Y F 1984 *Scientia Sinica (Series A)* **27** 1063
[7] Berenger J P1994 *J. Comp. Phys.* **114** 185
[8] TA-I Technology Co., Ltd. Taoyuan, Taiwan. [Online]. Available: http://www.tai.com.tw.

7th International Conference on Applied Electrostatics (ICAES-2012) IOP Publishing
Journal of Physics: Conference Series **418** (2013) 012054 doi:10.1088/1742-6596/418/1/012054

ESD detection by transient earth voltage

A Ohsawa[1] **and N Ichikawa**[2]

[1] National Institute of Occupational Safety and Health, Japan (JNIOSH), 1-4-6 Umezono, Kiyose, Tokyo 204-0024, Japan
[2] Kogakuin University, 2665-1 Nakano, Hachioji, Tokyo 192-0015, Japan

E-mail: ohsawa@s.jniosh.go.jp

Abstract. We measured transient earth voltages (TEVs) accompanied with electrostatic discharges (ESDs) to investigate the possibility of ESD detection using the TEV. Clear TEV signals could be observed in spark, brush, and propagating brush discharges even when an ESD event detector failed. Therefore, ESD detection using the TEV may be significantly more effective as an auxiliary monitor, while the TEV signals accompanied with cone discharges could not obviously be detected. In addition, the features of the TEV depend on the type of ESDs; therefore, the TEV could also be used to identify the potential of the occurrence of problems and hazards by ESDs.

1. Introduction

Electrostatic discharges (ESDs) sometimes cause problems and hazards in industry, e.g., malfunction of electronic equipment, damage to electronic devices, and occasional fire or explosion. When a gas discharge occurs between an object and the earth, the earth voltage instantly and transiently varies. Monitoring of the transient earth voltage (TEV), therefore, could be effective to detect discharges and thus assess and prevent such problems and hazards. In electric power plants, partial discharge detection using a set of probes including the TEV, UHF, acoustic emission, and current is used to maintain the equipment [1, 2]. In electronics industry, ESD detection using an electromagnetic wave radiated from ESDs is generally used as a monitor in ESD management; thus, ESD event detectors are commercially provided. However, ESD event detectors sometimes fail because the detection depends on the distance from ESDs. In addition, some investigations regarding explosions during the water-jet washing operations of the oil cargo tanks in very large crude carriers showed that ESD detection using electromagnetic radio signals was reasonably and practically inapplicable [3, 4]. In this paper, we report the results of the measurements of TEVs accompanied with spark, brush, propagating brush, and cone discharges and consider the possibilities for detecting ESD using the TEV.

2. Experimental

We used the same apparatus described in [5], in which TEV signals accompanied with corona discharges were previously investigated. ESDs except for cone discharges were produced on an earthed copper plate (1 m disc) placed on an insulator of foamed polystyrene, and their TEV signals on the earthed copper plate were observed with an oscilloscope. In addition, electromagnetic wave accompanied with the discharges were also examined with an ESD event

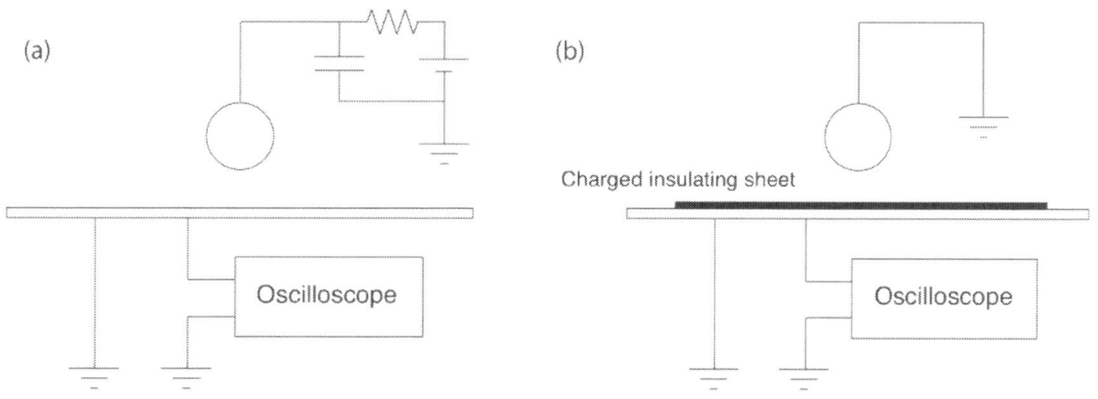

Figure 1. Experimental set-up for measuring TEVs (a) for spark discharges and (b) for brush and propagating brush discharges.

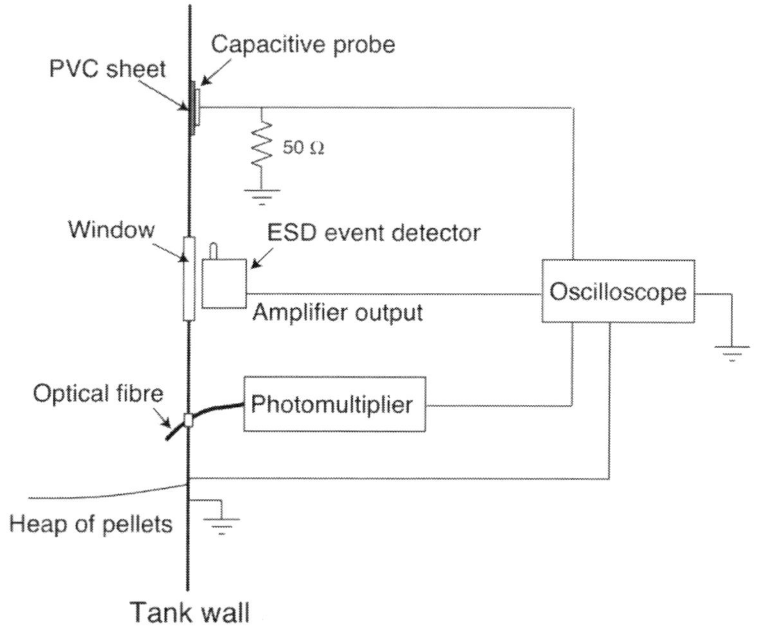

Figure 2. Experimental set-up for measuring signals of the TEV, TEV with a capacitive probe, light emission, and electromagnetic wave accompanied with cone discharges occurred on the heap of PP pellets in a tank during circulated pneumatic pellet transport.

detector. The TEV signals of cone discharges were examined using equipment with a pneumatic transport system [6, 7].

Spark discharges, as shown in figure 1(a), were produced at a gap (with separation of 0.3, 0.6, or 0.9 mm) between the copper plate and a metal sphere (30 mm in diameter) connected to a capacitor (7, 20, or 40 pF) charged by a DC high voltage of a few kV via a resistance of 60 GΩ. Thus, the spark discharges occur repeatedly when the electric field in the gap is over the breakdown field for air. TEV signals accompanied with spark discharges with different capacitors and gap separations were examined.

Brush and propagating brush discharges were produced on a charged insulating sheet (PTFE

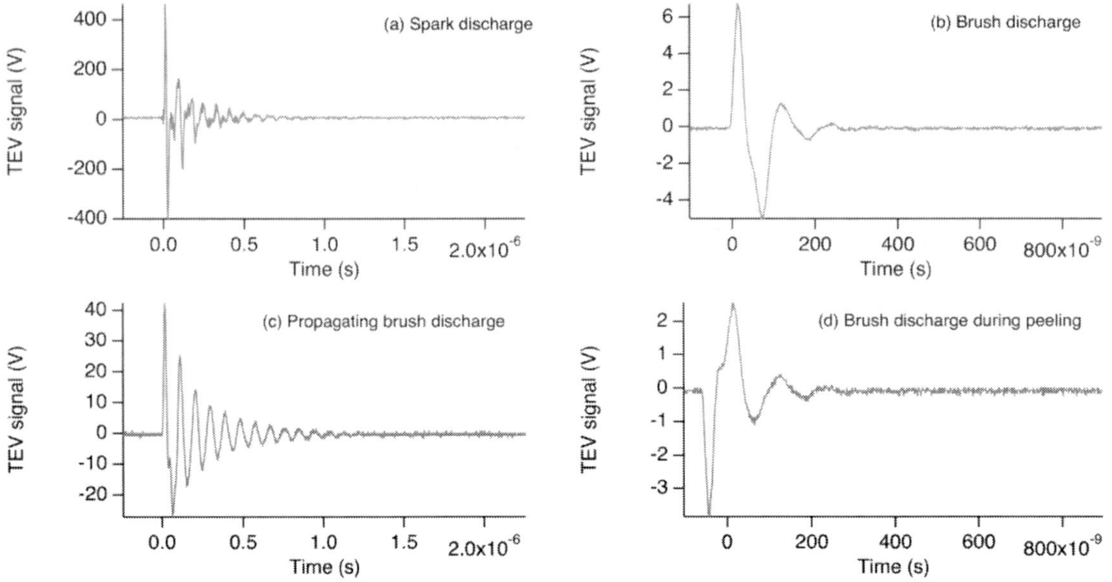

Figure 3. Typical TEV signals accompanied with (a) a spark discharge with a capacitor of 7 pF and a gap separation of 0.3 mm, (b) a brush discharge with a tribo-charged PTEF sheet, (c) a propagating brush discharge with a corona-charged PTFE sheet, and (d) a brush discharge that occurs during peeling of a tribo-charged polyester sheet from the earthed plate.

of 0.2 mm in thickness or polyester of 0.1 mm) placed on the earthed copper plate when an earthed sphere of 30 mm in diameter approaches the sheet, as shown in figure 1(b). Here, the sheets were charged by tribo-charging with an industrial-use tissue paper for brush discharges and by corona charging of -20 kV for propagating brush discharges.

In addition to the ordinary brush discharge above, TEV signals accompanied with brush discharges that took place when a tribo-charged sheet peeled from the earthed copper plate were examined.

TEV signals accompanied with cone discharges that occurred inside an earthed tank during a circulated pneumatic transport of PP pellets (approximately 3 mm in diameter) were examined with the use of the same equipment as in [6, 7], as shown in figure 2. The TEV signals, including those through a capacitive probe using a PVC sheet of 0.3 mm in thickness, were measured at the outside wall of the earthed tank. The light emitted from cone discharges that occurred on the heap of the pellets inside the tank was also observed with a photomultiplier connected to an optical fibre, where the one-side tip of the optical fibre was placed near the heap inside of the tank to introduce the light emitted from the cone discharge. Here, the optical fibre has directivity so that the light emitted from only cone discharges to which the fibre is directed is detectable. The electromagnetic radio signals accompanied with cone discharges were also measured at the outside of a window mounted on the tank by using an output signal of a high-frequency amplifier of the ESD event detector.

3. Results and discussion

Typical TEV signals accompanied with different types of discharges are shown in figure 3. All the signals have an underdamped transient response that is significantly dependent on the type of discharges. Since the TEV is a resultant phenomenon in which the discharge current flows to the earth, representing the discharge as an RLC series circuit, the TEV signal, V_{TEV}, may be expressed by

$$V_{TEV}(t) = Ae^{-\alpha t} \sin \omega_d t, \tag{1}$$

using attenuation α to describe the envelope of the decayed oscillation, and damped resonance frequency ω_d,

$$\alpha = \frac{R}{2L} \text{ and } \omega_d = \omega_0 \sqrt{1 - \zeta^2} \tag{2}$$

where $\omega_0 = \frac{1}{\sqrt{LC}}$ and $\zeta = \frac{\alpha}{\omega_0} = \frac{R}{2}\sqrt{\frac{C}{L}}$. Here, it is likely that R corresponds to the resistance of discharge; L represents the inductance of cables for earthing; and C describes the capacitance of the discharge gap for spark discharges and that of an insulating sheet used for brush and propagating brush discharges. Coefficient A is related to the potential resulting from charge accumulated on an object and the charge itself before ESDs. Furthermore, the time integration of V_{TEV} may be related to charge transferred by a discharge. Thus, we expect that the TEV can provide information of discharges occurred. Such equivalent circuit analyses on the underdamped currents observed in spark, brush and propagating brush discharges were conducted also in [8, 9].

A typical TEV signal of spark discharges is shown in figure 3(a). The signals were underdamped responses having relatively higher peaks and a long envelope. The feature of the TEV signals of spark discharges depends on the gap separation. Their amplitude, which has a maximum peak of several hundreds of volts, increases as the separation increases because the onset voltage of spark discharge increases with increasing the gap separation; an increase from ~400 to ~900 V was observed in the gap separation from 0.3 to 0.9 mm. In addition, the amplitude depends slightly on the capacitor because the charge accumulated in the capacitor is higher with a higher capacitor at the same onset voltage. These results indicate that a TEV signal seems to provide information regarding discharge energy as well as the onset voltage. Here, the onset voltages at different separations are lower than the voltage for charging of 2.0 kV, which was the minimum voltage examined in all separations. Furthermore, the amplitude of the TEV signals of spark discharges was the highest in these measurements in all types of discharges; thus, ESD detection using the TEV appears to be useful because the result implies the fact that spark discharge is most hazardous and incendive in terms of energy density in time and space. However, it should be noted that, if the surface charge on insulating sheets is much higher, higher amplitude of the peaks may be expected for propagating brush discharges. A clear influence of the capacitor and gap separation on the duration of the TEV, however, could not be observed. This denotes that the resistances of spark discharges describing attenuation α are comparable under the conditions investigated.

The TEV accompanied with brush discharges could be detected, as shown in figure 3(b). The duration of the signals was much shorter, resulting in fewer peaks; in addition, the amplitude of the peaks was much lower than that of other types of discharges. The results of these TEV signals are likely to represent lower discharge energies and incendivity characterising brush discharges.

The TEV signals of propagating brush discharges are very different from others, in which a clearly resonant (underdamped) response having relatively higher peaks and a longer envelop was observed, as shown in figure 3(c). The underdamped resonant frequency, ω_d, depends slightly on the sheet materials used. In addition, the amplitude of the maximum peak of the polyester sheet (~100 V) was higher than that of PTFE (~40 V) because the surface potential of a corona-charged polyester sheet (-6 kV) was larger than that of the PTFE sheet (-3.5 kV). These results also suggest that the TEV includes information of propagating brush discharges that are more hazardous and incendive.

TEV signals accompanied with discharges occurred when charged sheets peeling from an earthed plate were similar to those of the brush discharge described above, as shown in figure 3(d). In addition, polarity dependency could be observed, in which the polarity of the first peak

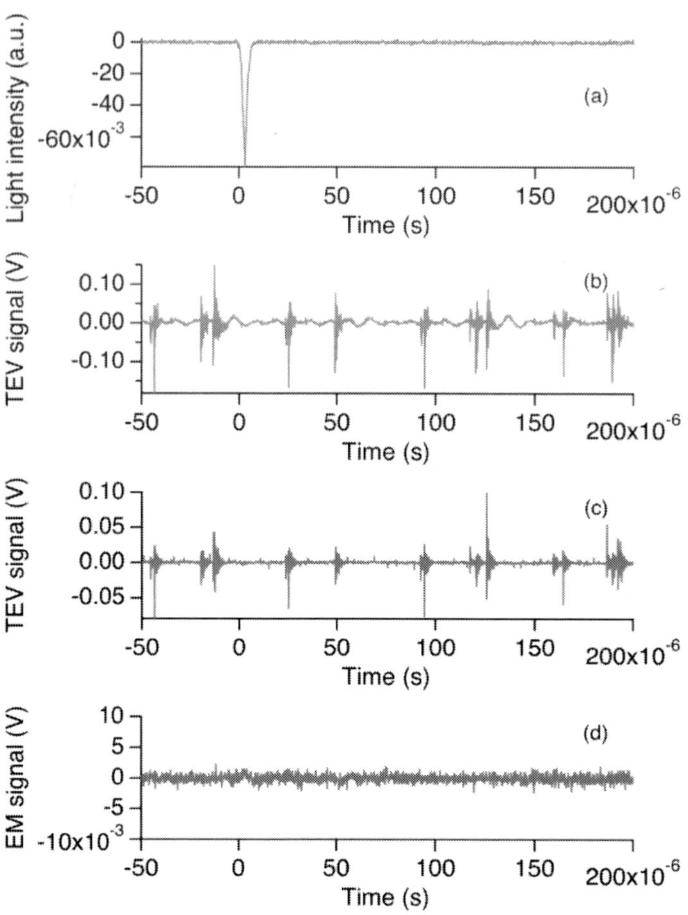

Figure 4. Measured signals with a cone discharge, (a) light emission, (b) TEV: only noises coming from controllers are observed, (c) TEV with a capacitive probe: only noises coming from controllers are observed, and (d) electromagnetic (EM) wave.

depended on the polarity of charged sheets: for example, polyester is positively tribo-charged; then, a negative first peak was observed, and negatively tribo-charged PTFE yielded a positive first peak. This indicates that the TEV could also estimate the polarity of the charged object that is discharged.

Consequently, we found that the TEV signals accompanied with spark, brush, and propagating brush discharges could be clearly observed because the discharge currents flow to the earth in these measurements. Nevertheless, an ESD event detector frequently failed to detect discharges. Therefore, ESD detection using the TEV may be significantly more effective as an auxiliary monitor when discharges occur to earthed objects; in addition, the measurement technique is very easy. Furthermore, since the features of TEV signals, in particular, their amplitude and duration, depend on the properties of discharges, the use of TEV enables not only ESD detection but also identification of the type of discharge as well as the potential for problems and hazards by the ESDs. Such identification is significantly important for reliable assessment because most ESDs whose energies are low, such as corona discharges, are not hazardous and can maintain charges at a safe level in many situations; on the other hand, ESDs are unavoidable phenomena which can lead to problems and hazards.

However, the detection of cone discharges using TEV failed. A set of measured signals of light emission, TEV, TEV with a capacitive probe, and electromagnetic wave accompanied with a cone discharge is shown in figure 4. Only the light emission from cone discharges could be clearly measured; however, the TEVs and electromagnetic radio signals synchronised with the light emission could not be detected. Furthermore, the TEV signals have high-level periodic noises coming from the controllers' clocks used in the pneumatic transport facility, where such noises were observed even without pellet transport. No observation of TEV signals accompanied with cone discharges may occur because most of the current of the cone discharge does not flow to the earth and the discharge is enclosed within the earthed tank as follows: when a cone discharge occurred, the same number of electrons and positive ions are produced; then, most positive ions move toward the heap of negatively charged pellets, and the electrons, including negative ions produced by attachment by these electrons, may move toward the earthed tank wall. During these processes, the total charge in the tank as well as the induced charge on the earthed tank wall does not change, resulting in no response in the TEV signal unless most electrons, including the negative ions, reach the wall because of collisions with a number of pellets existing in space until reaching the wall. This interpretation is similar to the principle of a Faraday cup in that the charges separated or discharged inside the cup cannot be measured. Further investigation, involving noise reduction with a band-path filter and a high-frequency amplifier adapted to TEV signals, will be required to verify the interpretation.

4. Conclusions

In conclusion, we found that the TEV can be used to detect ESDs when ESDs occur to earthed objects; in addition, the TEV could help identify potential problems and hazards on the basis of the detected ESDs. In this regards, ESD detection using TEV is viable and effective as an auxiliary measure when other methods fail. For cone discharge detection using TEV, however, further investigation, including noise reduction, will be required.

References

[1] Renforth L, Seltzer-Grant M and Mackinlay R 2007 *19th Int. Conf. Electricity Distribution* Paper 0675
[2] Makalous S, Tenbohlen S and Feser K A 2008 *IEEE Trans. Diel. El. Insul.* **15** 1576–1583
[3] Chubb J N 1975 *J. Electrostat.* **1** 61–70
[4] Butterworth G J 1979 *Inst. Phys. Conf. Ser. No. 48* 97–105
[5] Ichikawa N 2007 *Plasma Sci. Technol.* **9** 686–689
[6] Kodama T, Suzuki T, Nishimura K, Yagi S and Watano S 2002 *J. Electrostat.* **55** 289–297
[7] Mogami T, Suzuki T, Choi K-S and Ikehata T 2010 *J. Loss Prev. Proc. Ind.* **23** 237–241
[8] Norberg A, Szedenik N and Lundquist S 1989 *J. Electrostat.* **23** 79–88
[9] Greason W D, Kucerovsky Z, Bulach S and Flatley M W 1997 *IEEE Trans. Ind. Applicat.* **33** 1519–1526

Improvements for electrostatic discharge immunity in thermal printing head

Su Ling Liu[1*], Zhi Qiang Liu[1], Wei Guo[1], Ping Guang Duan[1],Ye Qi Sun[1], Geng Zhi[2]

1 Physics Department of Hebei University, No.180 WuSi East Road, Baoding, China
2 College of Mechanics and Electronics, Beijing Institute of Technology, Beijing100081, China

E-mail: lsllgy-20001218@163.com

Abstract. Electrostatic discharge (ESD) is one of the important sources of electromagnetic interference which would cause a reversible or permanent damage in modem electronic equipment. ESD may be introduced by the human body operating the electronic equipment or accumulated by friction, heating or other reasons when the electronic equipment is working. In this paper we use the print head of the thermal printer (TPH) suffered from electrostatic hazard as an example to analyze the reasons of the ESD happening, and some necessary improvements measures for protection of the TPH are also suggested.

1. Introduction

ESD is one important reason that electronic products work disorder or do failure. With the development of the technology and improvement of products' complex degree, people pay more attention to ESD which disturbs electronic circuit and damages components, CMOS circuit and interface circuit. As an important content of electromagnetic compatibility test, ESD of electronic equipment is written into national standards and international standards [1-3].

The resistance to perturbation of electronic products' ESD is directly related to the product, including applicable environment, product performance, product post-sale service cost, product brand image and so on. So manufacturers and designers have made more concentration on ESD immunity and improved the competition ability of production.

Thermal printer is an electronic device using the joule heating for printing. Because of its working environment and the working characteristics, thermal printer is disturbed or damaged by the human body operating or accumulated by friction, heating and other reasons when it is working. This paper explains the interference sources and channels as well as analyses the reasons why ESD products are damaged and then puts forward the specific preventive measures.

2. The analysis of electronic product' static failure

The core part of thermal printer is thermal printing head (shown in figure 1), which includes the PCB, IC and the package, SBS ceramic circuit boards, aluminum heat sink, socket, etc. The main problems of the product confronted in using process includes:

- Heating unit breakdowns due to violation of electrostatic;

[*] To whom any correspondence should be addressed.

- The interior IC of print head is damaged from static intrusion;
- The restarting of CPU and motherboard is caused by static accumulation during printing.

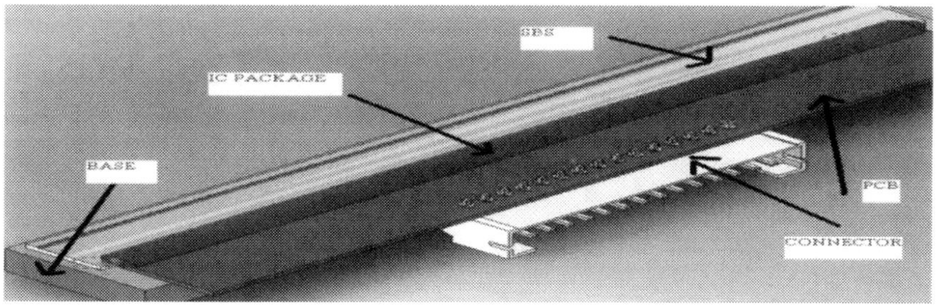

Figure 1. The structure of TPH.

When ESD happens, the mode of energy transmission has two kinds, namely, conduction mode and electromagnetic radiation mode. In the conduction mode, the discharge current spreads along the electronic products without shield conductor and invades circuit, which makes chip produce false action. In the electromagnetic radiation mode, the electrostatic discharge produces large current (human body discharge instant tr = 1 ns). When the current flows through the wire, it also will radiate strong magnetic field (about 300 MHz, upper limit can be more than 1 GHz). The electromagnetic field can directly go through case or the hole, crack, ducts, input/output cables, which couples to the sensitive circuits. When voltage or current is more than the product's resistance to flexibility limit, the circuit performance will decline or lose efficacy [1,2].

According to the result of the electronic product' antistatic failure, we analyze the way of electrostatic invading, which includes three main ways, as shown in figure 2.

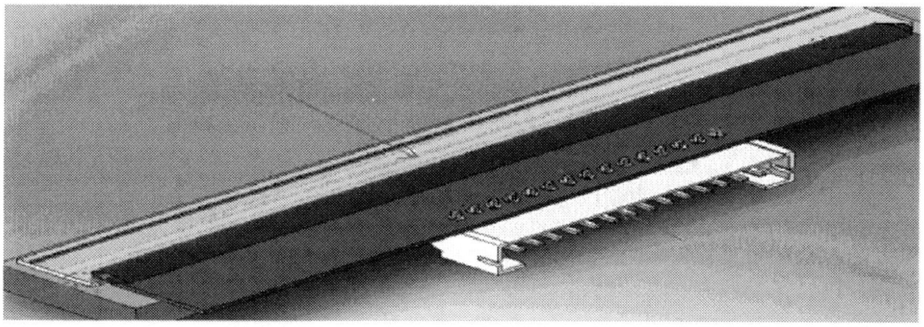

Figure 2. The invasion way of ESD.

- Heating element parts produce static, which resulted from plastic roller, TPH and paper rub during printing;
- Socket parts invade from printer cable;
- Invading through the joint parts between the cooling plate and the printer on the aluminum heat sink.

3. The improvement measures of ESD immunity for electronic product

3.1the general design principle to protect electronic products' ESD [1]

The protection measures of electronic products of ESD are designed according to interference channels, conduction modes and product failure mechanism. Generally speaking, we can improve sensitive equipment from the following several aspects:

The internal sensing device, which may be directly contact to electrostatic, should be buffered with high pressure clamping circuit or filter before we send a signal to the external device or connector. And the sensitive wire of connection parts should be hidden.

Enough insulation or physical spacing should be provided between outside area and any internal conductive parts or in the circuit routing which connect directly or indirectly with ESD sensitivity (ESDS) components. For example, we can add nonconductive films or shell with closed hull or screw hole in the circuit routing. At the same time, the discharge energy should be transferred from ESDS with conductive pathways or a shielded body.

In the process of restraining conduction or radiation noise, the conductive metal shielding body can shield radiation noise. But at the same time conduction coupling will happen because of the second arc from circuit to shielding. So shield should be separated with circuit. Insulation can't prevent radiation ESD noise coupled to the circuit.

The lead of the PCB is the antenna receiving ESD radiation field. So the lead must be as short as possible and the loop area is as small as possible.

ESD in the cable will induce out high voltage and current. Cable is also the biggest antenna receiving ESD radiation field. So cable shielding set should connect to the metal cabinet as a grounding path. And cable should be as short as possible. And each of the electric cable root course should be placed next to the loop line in physics. For example, ribbon cables, each signal line should have a ground wire placed next to it.

3.2 The protection design of ESD for thermal printers

According to the protection design principle of ESD for the electronic products and considering the product structure and site easily damaged by electrostatic, we mainly improve the resistance ESD of the product from the following several aspects.

(1) PCB socket with high pressure clamping circuit can keep invasion of electrostatic from the socket.

(2) The PCB with the ground is short circuit with a large area conductor. Reducing the accumulation of charge and increasing the coupling capacitance can make the static voltage reduce to IC or the range of heating element withstand.

(3) The part which is above SBS to the heating element should be covered with some conductive film. The conductive film connects with power. Electronic charge resulted from rubbing between TPH and paper can be transferred away immediately to avoid the charge accumulation.

(4) The edge of PCB and SBS can not leave a conductive body or isolation. Using multipoint connection fills the ground near the socket.

3.3 The effects after taking antistatic measures

According to the ESD standard IEC61000-4-2 for electronic products, the test for contact discharge is 8kV and for air charge 15kV. Table 1 shows the test standard and test equipment.

Table 1. Test standard and test equipment.

Test items	Test objects	ESD gun and test equipment	Test standard
ESD immunity test for the device (HBM model)	Different position of the thermal printer	TC-815R and ESS-2000	IEC61000-4-2

According to the condition shown in table 1, the good effects of antistatic for the device are obtained.

- Completely using the design of the SBS strengthens anti-static electricity about 20-40%;
- Increasing each signal control area increases anti-static electricity about 40% ~ 60%;
- Exchanging higher anti-static IC enhances anti-static electricity about 20% ~ 40%;
- Adding high pressure clamping circuit between terminal and the ground enhances anti-static above 100%;
- Adding conducting film enhances anti-static electricity above 100%;
- Adding static brushes reduces static voltage about 20% ~ 40%;
- BASE discharge enhances electrostatic anti-static electricity about 20% ~ 40%.

4. Conclusion

ESD is one important reason which can disturb the electronic product and cause the equipment disorder or failure. The static electricity protection is only to reduce the damage, but complete elimination of the ESD is impossible. So enhancement of electronic equipment and protection from system for ESD is our urgent need to address the actual problems. This is the only way to reduce the production cost and improve the quality of product. In addition, how to get the accurate the electromagnetic characteristics of ESD events is the effective method solving the electrostatic interference of electronic product.

Acknowledgments

This work is Supported by National Natural Science Foundation of China (5107735), and supported by the Hebei University Doctor Foundation (2009161), and supported by the NSF of Hebei Province (A2010000182), and supported by program of Scientific Department of Hebei Province (09213519D).

References

[1] Liu S H Wu Z C 2004 *ESD and the defense for Harm* (Beijing)
[2] IEC61000-4-2 1995 Electromagnetic compatibility (EMC) Part 4: *Testing and measurement techniques-Section 2: electrostatic discharge immunity test*
[3] Wan Xiqin Huang Jiusheng Gao Yougang Liu Suling Zhou Feng Wang Langfeng 2006 Pro. Asia-Pacific Conf. on *EMC* vol 1 Dalian China: p71
[4] Wan Ying 2006 *Instrument and Metrological Technology* p 28

Journal of Physics: Conference Series **418** (2013) 012056 doi:10.1088/1742-6596/418/1/012056

Radiation characteristics and electromagnetic compatibility of 5-20 KHz ac plasma antenna

J S Zhao[1], Z T Zhang[1, 2*], Y J Li[2], S J Xu[2], J Yao[2], Y P Tian[2], Z Yu[1]
1 Institute of Marine Engineering, Dalian Maritime University, Dalian, China
2 Department of Physics, Dalian Maritime University, Dalian, China

E-mail: newzhangzhitao@163.com

Abstract. In order to meet the needs of engineering, plasma antenna must meet the requirement in terms of gain, bandwidth, noise, electromagnetic compatibility (EMC) and so on. A plasma antenna prototype was established by 5-20 KHz high voltage ac power, whose radiation characteristics and electromagnetic compatibility was tested and analyzed. Results show that 5-20 KHz ac plasma antenna is of broadband impedance characteristics. Compared with 50 Hz ac plasma antenna, the noise reduces and the gain increases from 150 MHz to 350 MHz. When plasma antenna connected to radio is on, the radio can receive signals. The interference to radio is much lower than RF surface wave plasma antenna and 50 Hz ac plasma antenna. It is of a certain electromagnetic compatibility.

1. Introduction

Plasma antennas are usually radio-frequency (RF) antennas using the plasma column instead of metal conductors or other dielectric materials as elements [1-3]. They are constructed by insulating tubes filled with low pressure gas. Over decades, plasma antenna has attracted much attention since they have numerous advantages over conventional metal antenna [1, 3-6]: its parameters can be reconfigured conveniently by changing discharge power, witch makes it possible for it to be used at wide waveband and to simplify the matching network; plasma can be rapidly created and destroyed by applying an electrical pulse to the discharge tube, hence a plasma antenna can be rapidly switched on and off. When it is on, it exhibits a high conductivity, providing a conducting medium for the applied RF signal. When it is off, it is non-conducting and invisible to electromagnetic (EM) radiations, the radar cross section (RCS) close to 0. So plasma antenna can provide an integral and most important part in both communication and plasma stealth technology.

So far many theoretical and experimental works including impedance, gain and radiation characteristics have been conducted on this subject [1-12]. But few papers concern investigation of their electromagnetic compatibility (EMC). It has been reported that 50 or 60 Hz ac plasma antenna is of very high noise [9], moreover, the antenna efficiency and gain are not satisfactory. In the past decades, the vast majority of plasma antennas have been produced by RF power, which is called surface wave (SAW) plasma antenna and is proved to have a certain gain and low noise [9]. However, the bandwidth of such antenna is limited to the pumping RF frequency, and the cost of pumping RF generator is also high [10, 11]. Electromagnetic interference (EMI) between excitation and signal channels seriously affected the normal work of antenna and other instruments. Hence, the problems of

* To whom any correspondence should be addressed.

EMC, if not addressed, not only can't make the antenna to meet the engineering requirement, but also likely to cause greater losses. Here 5-20 KHz ac power supply is employed to produce plasma antenna, such structure is characterized as square-loop plasma antenna in an experiment and this paper is aimed at presenting the experimental investigation and analysis of the EMC.

2. Antenna structure and experimental setup

2.1. 5-20 KHz ac plasma antenna

Figure 1 shows the home-made 5-20 KHz ac plasma antenna, the main body of which was a commercially available tube with an inner diameter of 10 mm, an outer diameter of 12 mm and a length of 1200 mm. The system was filled with argon and mercury, at a working pressure about 5Torr. The structure of the tube was characterized as square-loop, with two electrodes inserted in an insulating box. A 5-20 KHz ac adjustable power supply was also installed in the insulating box. The capacitive coupling model was applied to the signal coupling system. A capacitive coupler of width 30 mm was placed around the tube and shielded by a well-sealed shielding box (100 mm×60 mm×60 mm) made of cast aluminium. The coupler was connected to transmission line to apply the useful signal. The distance between the center of the coupler and the electrode was longer than 200 mm.

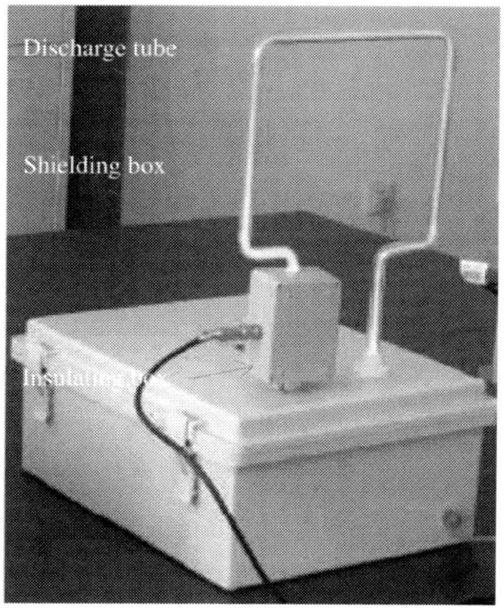

Figure 1. 5-20 KHz ac plasma antenna.

2.2. Experimental setup

To investigate the wireless communication capability and EMC for radio frequency range (100-400 MHz) of 5-20 KHz ac plasma antenna, an experimental setup was constructed and dedicated experiments were carried out to characterize it. A schematic of the experimental setup for 5-20 KHz ac field produced plasma, which acts as a plasma antenna, is shown in figure 2. 5-20 KHz ac power supply could be operated between 5-20 KHz up to 500 W. In the experiment, discharge frequency was 10 KHz. The discharge power of fluorescent was measured with a digital fluorescent oscilloscope. The input impedance was measured with a vector network analyzer which was calibrated. The output impedance of RF port of network analyzer is 50 Ω, which is equal to characteristic impedance of the transmission line. A matching network with variable inductors and capacitors was used to maintain the 50 Ω impedance for transmitting antenna. Then, we collected it to a radio directly. Our aim was now to investigate the radiation characteristics and to analyze the EMC of plasma antenna. In addition, we compared it to 50 Hz ac plasma antenna, 40.68 MHz surface wave plasma antenna and metal antenna in terms of radiation characteristics and EMC.

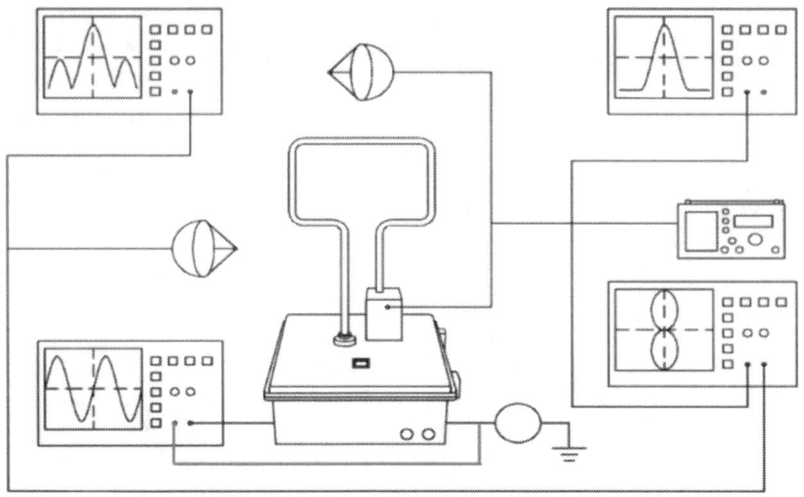

Figure 2. A schematic of the experimental setup of antenna measurement.

3. Results

3.1. Input impedance of plasma antenna

It is measured and calculated the discharge power of 5-20 KHz ac plasma antenna is 6.2 W, which is nearly equal to that of 50 Hz ac plasma antenna (5.8 W) and much lower than that of surface wave plasma antenna (29 W). Figure 3 shows experimental data input impedance $Z_{in}=R+jX$ for a 5-20 KHz ac square-loop plasma antenna driven at 10 KHz. The solid and dashed lines represent resistance and reactance respectively. It is easily found that in a wide frequency range, the resistance components are close to 50 and the reactance ones to 0. It is obvious that 5-20 KHz ac plasma antenna has broadband impedance characteristics.

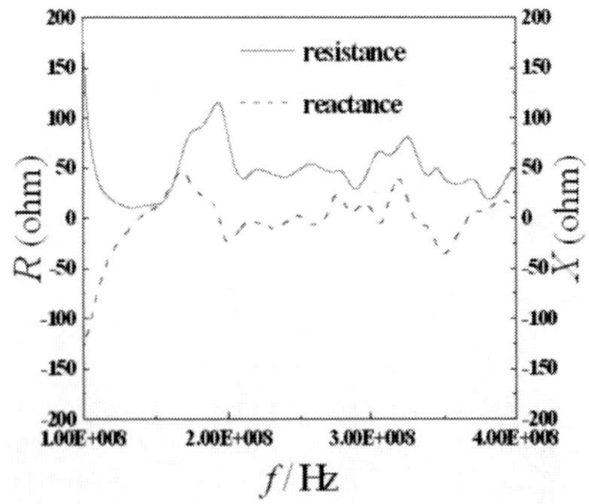

Figure 3. Input impedance of 5-20 KHz ac plasma antenna.

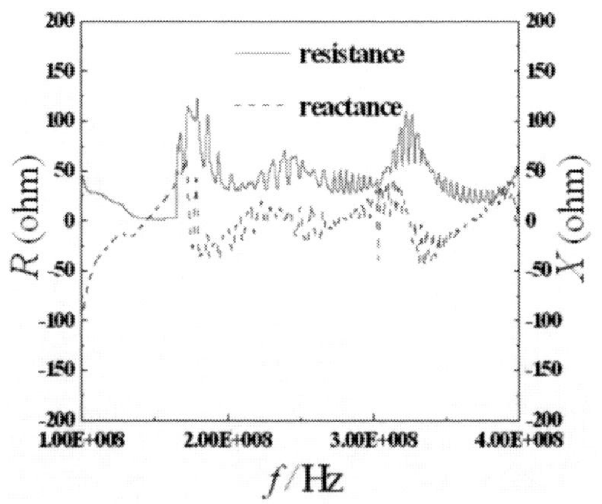

Figure 4. Input impedance of 50 Hz ac plasma antenna.

Figure 4 shows the impedance characteristics of 50 Hz ac plasma antenna. These results display that this antenna also has broadband impedance characteristics, very similar to 5-20 KHz ac plasma antenna. But the difference is that the impedance curves of 50 Hz ac plasma antenna are relatively rough compared with the 5-20 KHz one due to larger disturbance of plasma produced by 50 Hz ac power supply. Because of strong coupling between the two circuits: signal and excitation, we can't measure input impedance of 40.68 MHz surface wave plasma directly.

3.2. Gains and pattern of 5-20 KHz plasma antenna

Figure 5 displays the relative gains of 5-20 KHz ac and 50 Hz ac plasma antennas. In the experiment, the EM waves are imposed with a frequency ranging from 150 to 400 MHz, in this scope the VSWR is lower than 2. It is shown from the results that fluctuation of the value of 50 Hz ac plasma antenna is very large with the frequency changing. In contrast, 5-20 KHz plasma antenna has ideal gains in a wide frequency range. Moreover, varies of value are smaller than 50 Hz plasma antenna. Figure 6 displays radiation pattern in horizontal plane of 5-20 KHz plasma and metal antennas working at 200 MHz, where the VSWR is about 1.2. Because of impacts by shielding layer inside the insulation box, the pattern has directionality. It is obvious that, the two patterns are very similar, and the gain of plasma antenna is a little smaller than that of metal antenna.

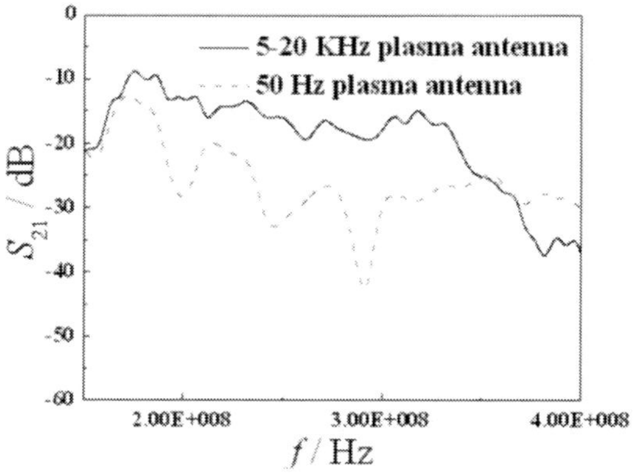

Figure 5. Relative gain of plasma antenna.

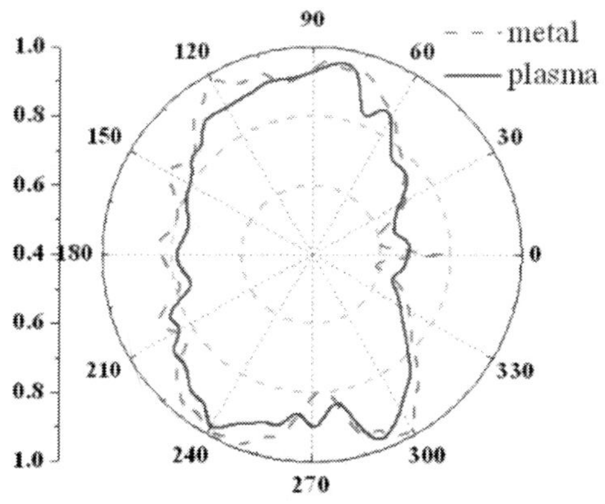

Figure 6. Relative gain of plasma antenna.

3.3. EMC analysis of plasma antenna

Figure 7 is the frequency spectrum of signals measured by spectrum analyzer (N9010, Agilent). The output power of signal source is unchanged -10 dBm (0.1 mW). We choose signal of 100 MHz in this experiment to test disturbance of plasma to signals. 50 Hz ac plasma antenna, 5-20 KHz ac plasma antenna, 40.68 MHz surface wave plasma antenna and metal antenna are used in the experiment. High-pass filter is connected between spectrum analyzer and surface wave plasma antenna to prevent the 40.68 MHz signal from damaging spectrum analyzer. It is so clearly that 50 Hz ac plasma has large disturbance to signal so that the noise level is much higher than other antennas. Besides, the signal level is lower. The gains of 5-20 KHz ac and surface wave plasma antenna are nearly the same, a little lower than conventional metal antenna. And the noise levels of the two antennas are also nearly the same. For a uniform plasma density, the wave vector along plasma column β can be determined from [8]:

$$\varepsilon_r T_0 I_1(T_{pa}) K_0(T_{0a}) + T_p K_1(T_{0a}) I_0(T_{pa}) = 0 \quad (1)$$

Where $T_p^2 = \beta^2 - \varepsilon_r k_0^2$ and $T_0^2 = \beta^2 - k_0^2$, a is plasma column radius, k_0 is free-space wave vector $= \dfrac{\omega}{c}$ and ε_r is plasma dielectric constant. $K's$ and $I's$ are Bessel function. The plasma dielectric constant can be expressed as:

$$\varepsilon_r = 1 - \frac{\omega_{pe}^2}{\omega(\omega + i\nu)} \quad (2)$$

Where ν is electron-neutral collision frequency, electron plasma frequency can be written as

$$\omega_{pe}^2 = \frac{n_e e^2}{m_e \varepsilon_0} \tag{3}$$

When the EM waves propagate in plasma produced by 50 Hz ac power supply, the unstable plasma leads to changes in the wave vector β constantly. In addition, the reflection coefficient is constantly changing. Hence, the measured curves such as impedance and spectrum of 50 Hz ac plasma antenna are unstable. In contrast, there're much smaller changed in parameters of 5-20 KHz ac and surface wave plasma antennas, and the changes of values are relatively small with time.

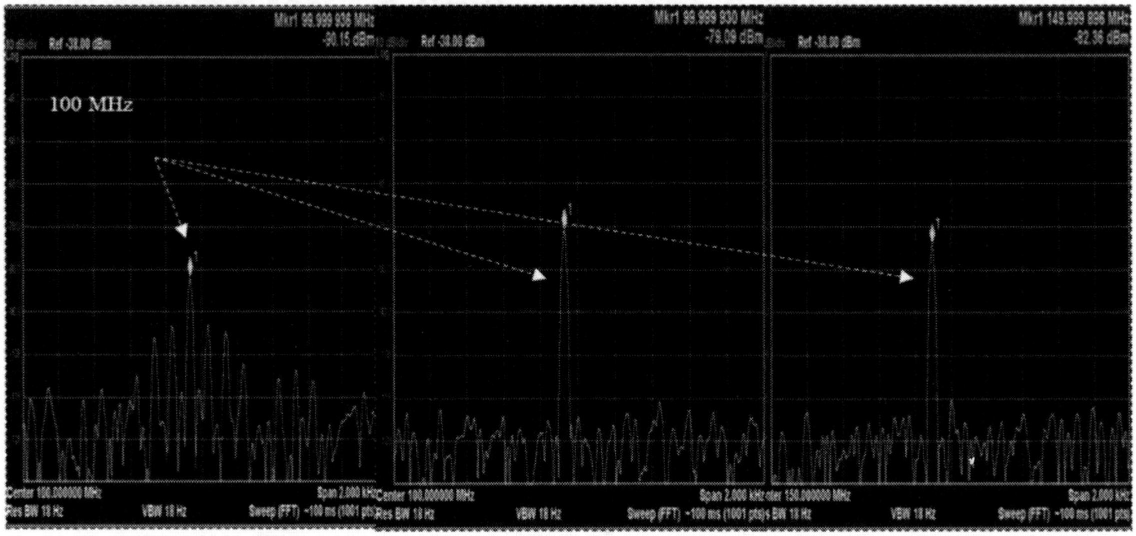

50 Hz ac plasma antenna — 5-20 KHz ac plasma antenna — Surface wave plasma antenna

Figure 7. Signal spectrum of plasma antenna.

Table 1. A table with headings spanning two columns and containing notes.

Radio frequency of Dalian (MHz)	99.1	100.8	105.7	106.7
50 Hz ac plasma antenna	Signal strength : 40%, large noise	Signal strength : 40%, large noise	Signal strength : 40%, large noise	Signal strength : 40%, large noise
5-20 KHz ac plasma antenna	Signal strength : 80%, a little noise	Signal strength : 80%, a little noise	Signal strength : 60%, a little noise	Signal strength : 60%, a little noise
Surface wave plasma antenna	Unable to work	Unable to work	Unable to work	Unable to work
Metal antenna	Signal strength : 100%, small noise	Signal strength : 100%, small noise	Signal strength : 100%, small noise	Signal strength : 100%, small noise

A radio is also used in the experiment to analyze communication properties and EMC of antennas. Table 1 shows the working abilities of radio connected to 50 Hz ac plasma, 5-20 KHz ac plasma antenna, surface wave plasma antenna and metal antenna. Though there is a filter connected between plasma antenna, the EMI is still exist, especially in the system of surface wave plasma antenna. When plasma is on, the radio connected to surface wave plasma antenna cannot work normally. The signal strength is relatively low when the radio connected to 50 Hz ac plasma antenna. And the noise is very large so that we cannot hear the broadcast normally. If the radio is connected to metal antenna, there is very strong signal and low noise. When connected to 5-20 KHz ac plasma, the noise is a litter higher than metal antenna, but much lower than the 50 Hz one. Besides, signal strength is closed to metal in relatively low signal frequency. This antenna maintains a certain EMC.

4. Conclusion

We have presented measured results of input impedance, radiation characteristics and analyzed the EMC of 5-20 KHz plasma antenna. Compared with 50 Hz ac plasma, 40.68 MHz surface wave plasma antenna and metal antenna, it has a certain EMC and ideal gains for relative low signal frequency. Although surface wave plasma can also maintain a certain gain, the strong electromagnetic interference of exciting port lead to bad EMC. Both 50 Hz and 5-20 KHz plasma antennas have broadband impedance characteristics, but the unstable plasma in 50 Hz lead to changed results and larger noise.

Acknowledgments

This work is supported by The National High Technology Research and Development Program of China under Grant No 2012AA062609, The International Science and Technology Cooperation Project of China under Grant No 2010DFA61470, The National Science Foundation for Distinguished Young Scholars of China under Grant No 61025001, The Fundamental Research Funds for the Central Universities under Grant No 2012QN067 and 2012TD061.

References

[1] Rayner J P, Whichello A P and Cheetham A D 2004 *IEEE Trans. Plasma Sci.* **32** 269
[2] Kumar R and Bora D 2010 *J. Appl. Phys.* **107** 053303
[3] Liang Z W, Wang Z J, Zhao G W, Xu J and Xu Y M 2007 *Chin. J.Radio Sci.* **22** 971
[4] Gerri G, Leo D R, Primiani V M and Russo P 2008 *IEEE Trans. Instru. Meas.* **57** 342
[5] Russo P, Gerri G and Vecchioni E 2011 *IEEE Trans. Antenn. Propag.* **59** 1503
[6] Wu Z Y, Yang Y Y and Wang J Y 2010 *Acta Phys. Sinica* **59** 1890
[7] Kumar R and Bora D 2011 *J. Appl. Phys.* **109** 063303
[8] Borg G G, Harris J H, Miljak D G and Matin N M 1999 *Appl. Phys. Lett.* **74** 3272
[9] Borg G G, Harris J H, Matin N M, Thorncraft C, Milliken R, Miljak D G, Kwan B, Ng T and Kircher J 2000 *Phys. Plasmas* **7** 2198
[10] Chung M, Chen W S, Huang B R, Chang C C, Ku K Y, Yu Y H and Suen T W 2007 *IEEE Region 10 Conf. (China)* p 1
[11] Chung M, Chen W S, Yu Y H and Liou Z Y 2008 *Int. Conf. on Microwave and Millimeter Wave (China)* p 1118
[12] Lv J W, Li Y S and Chen Z L 2011 *J. Electromagn. Anal. Appl.* **3** 297

Journal of Physics: Conference Series **418** (2013) 012057 doi:10.1088/1742-6596/418/1/012057

Research on fast rise time EMP radiating-wave simulator

Lisi Fan[1]*, Haitao Liu[2], Yun Wang[1]
1. Electrostatic and Electromagnetic Protection Research Institute, Mechanical Engineering College, 97 Heping West Road, Shijiazhuang, Hebei, 050003, P. R. China.
2. CETC54, Shijiazhuang, Hebei, 050081, P. R. China.

E-mail: fanlisi@sina.com

Abstract. This paper presents an antenna of High altitude electromagnetic pulse (HEMP) radiating-wave simulator which expands the testing zone larger than the traditional transmission line simulator. The numerical results show that traverse electramagnetic (TEM) antenna can be used to radiate HEMP simulation radiating wave, but in low frequency band the emissive capability is poor. The experiment proves the numerical model is valid. The results of this paper show that TEM antenna can be used to HEMP radiating-wave simulator, and can prove the low frequency radiation capability through resistance loaded method.

1. Introduction

High altitude electromagnetic pulse (HEMP) is generated in high altitude nuclear explosive process, with fast rise time and wide frequency band. HEMP can cause severe interference to the electronic equipment even damaged. The HEMP simulator plays an important role for evaluate EUT immunity to high altitude electromagnetic pulse [1]. The traditional transmission HEMP simulator is based on transmission line such as parallel plate [2], GTEM cell [3] and TEM cell [4]; generate HEMP simulation wave similar to the excitation pulse in the transmission line. The rise time of the simulation pulse is restricted by the size of HEMP simulator, so that large type HEMP simulator can't produce fast rise time simulation electromagnetic pulse [5]. The large radiating-wave simulators were built in 1970s, for airplane and missile HEMP testing, such as Germany HPD and VPD [6] simulation system.

The HEMP radiating-wave simulator consists of a biconical-cylindrical cage antenna [7-8], a high voltage pulse source. The biconical-cylindrical cage antenna has some advantages such as constant impedance; wide working frequency band and large uniform test volume; but the disadvantages is obvious: (1) the biconical antenna is no directivity, so the radiation efficiency of antenna is low; (2) The size of antenna is very large, the length is longer than 300 m, so the antenna is very expensive [9].

This paper is to present how a fast rise-time EMP radiating-wave simulator is designed using TEM horn antenna. The experiment results show that can generate EMP as RS105 by TEM horn antenna instead of using transmission line or biconical-cylindrical cage antenna, and the matched load installed at the terminal of the TEM horn antenna can reduce the reflection of the terminal end edge. The merit

* To whom any correspondence should be addressed.

of the fast rise-time EMP radiating-wave simulator using TEM horn antenna is that testing zone is larger and cheaper.

2. TEM antenna
The TEM horn is composed of two triangle plates, as shown in figure 1, l=5.2 m, w=6 m, h=6 m.

The impedance of the TEM antenna is [10]

$$Z_0 = \frac{240\pi}{\frac{2w}{h} + 1.393 + 0.677\ln(\frac{2w}{h} + 1.444)} = 178\Omega \quad (1)$$

Where w is the width of the antenna and h is the height, l is the length.

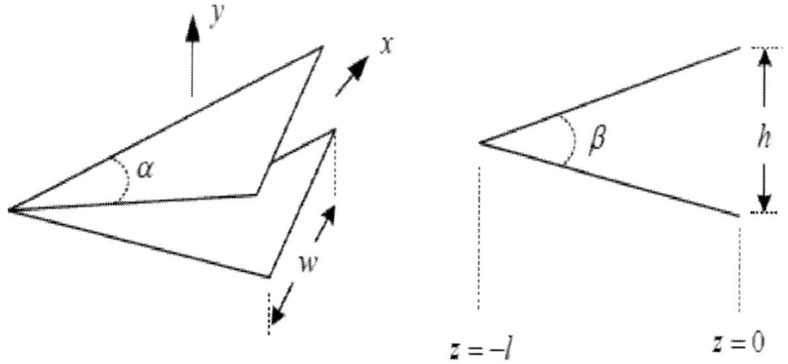

Figure 1. Diagram of the TEM horn antenna.

3. Simulation model and results
The CST simulation model is built as figure 2, the discrete port is applied between two plate apexes. A double exponential excitation pulse is defined as equation 2.

$$V(t) = V_0 \times k(e^{-at} - e^{-bt}) \quad (2)$$

Where V_0=50 kV; a=4.0×10^7 s^{-1}; b=6.0×10^8 s^{-1}; k=1.3.

The radiation pulse waveforms at aperture and 3 m away aperture are shown in figure 3.

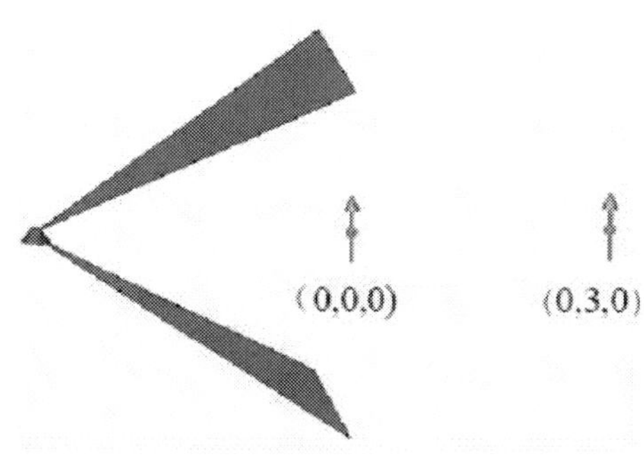

Figure 2. The model of the TEM horn antenna.

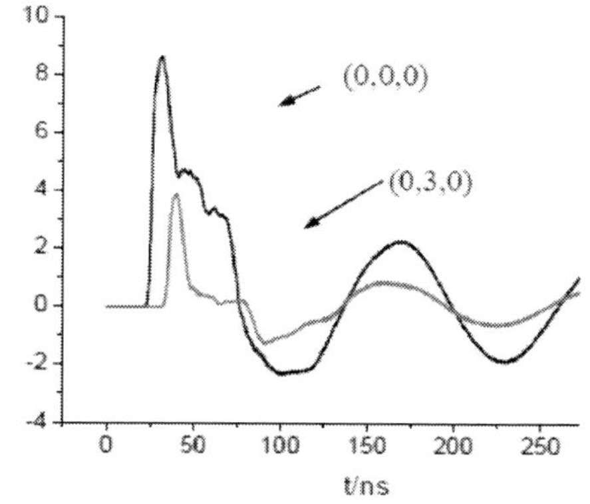

Figure 3. The radiation field waveform without loaded.

From figure 3, we can know that the radiation field strength decreased with the distance away the excitation port. The waveform is distorted and the waveform width becomes narrow. There is zero crossing and damped oscillation at the tail of the waveform, because the antenna end refection and the oscillation period is the pulse current reciprocating time between the excitation port and the end edge. The reverse current from which the pulse source recharge will damage the source, and must avoid in high voltage engineering.

In order to reduce the reflection from TEM antenna end edge, the absorption load should be installed at the TEM antenna apexes. The matched load are composed of two 360 Ω parallel resistors, each resistor is installed at one apex of the TEM horn antenna, as shown in figure 4. The radiation pulse waveforms of aperture and 3 m away the aperture are shown in figure 5.

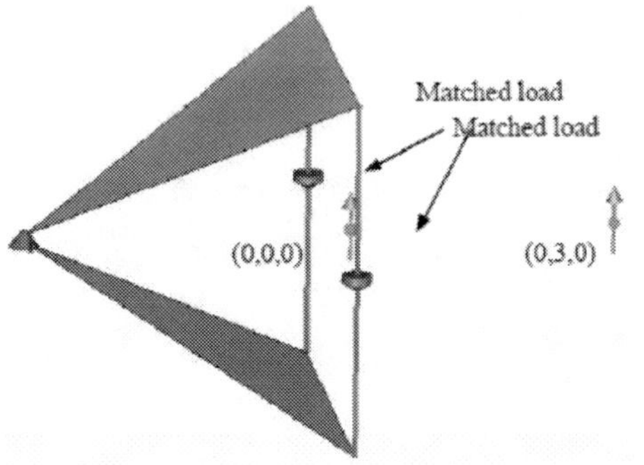

Figure 4. The simulation model TEM horn antenna with matched load.

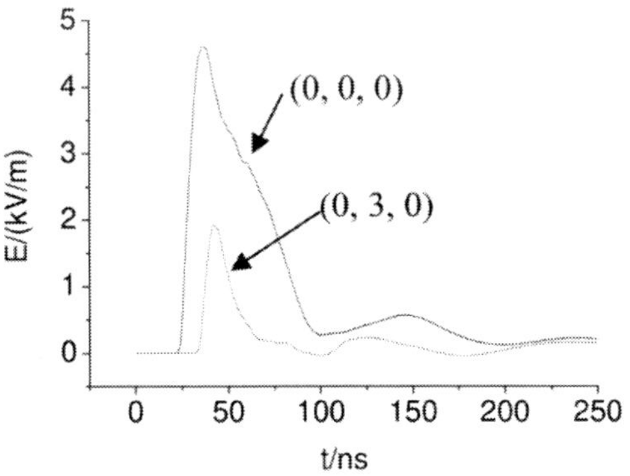

Figure 5. The radiation field waveform with matched loaded.

4. Experiment results

The experimental scheme is shown in figure 6, and the TEM antenna is shown in figure 7. The TEM horn is composed of two triangle plates (L×W×H, 1 m× 2 m×1 m). The high voltage source charges the capacitor through current-limiting resistor. The gas switch breakdown when the voltage is high enough. The high voltage pulse is radiated by TEM horn antenna, the pulse field approximate with HEMP waveform of RS105 is formed. Using optical fibre filed meter test the pulse field, the oscilloscope and optical receiver are placed in EM shielding room.

4.1. Radiation field TEM horn antenna terminal opened

The radiation pulse field of antenna aperture and 3 m away aperture are shown in figure 8 and figure 9. Limit of the optical fibre filed meter only measure the waveform, and the field strength is only a relative value. From reulsts we can know that reflection current from the TEM antenna cause the radiation pulse field oscillation.

4.2. Radiation Field TEM horn antenna terminal match load

In order to minish the terminal reflection, two 360 Ω resistors are installed at the apexes of the TEM horn antenna. The radiation field of aperture and 3 m aperture away are shown in figure 10 and 11.

The results show that the match loads effectively absorb the current, so that the oscillation is less. The radiation field 3 m aperture away is similar with double exponential pulse waveform, and the rise time is about 1-2 ns, the width is about tens of ns. The radiation pulse field waveform is approximate the HEMP RS105 waveform defined by MIL–STD–461F and other international standards.

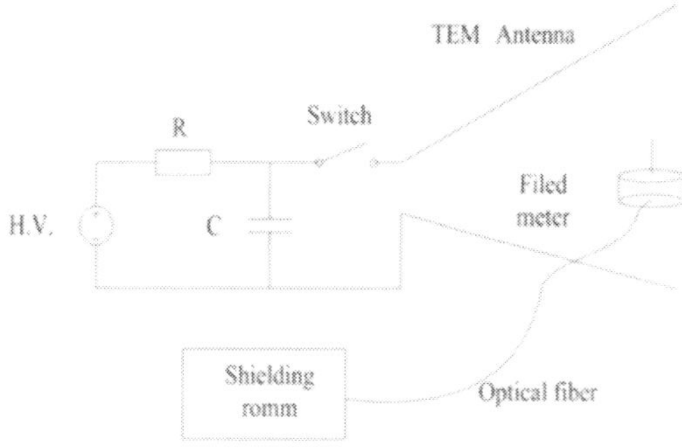

Figure 6. The working principle diagram.

Figure 7. Photo of TEM horn antenna.

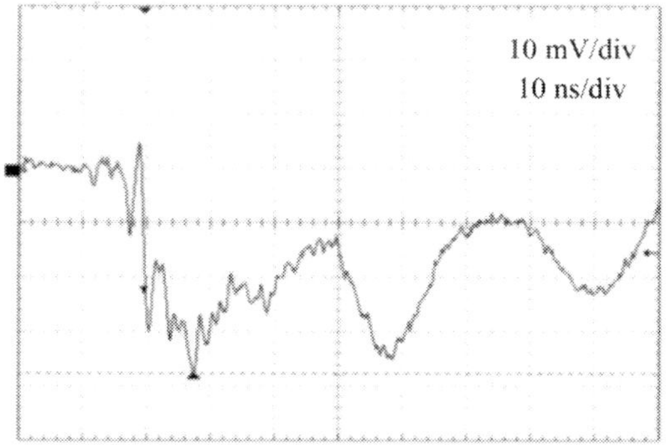

Figure 8. E-field waveform at TEM antenna aperture with terminal opened.

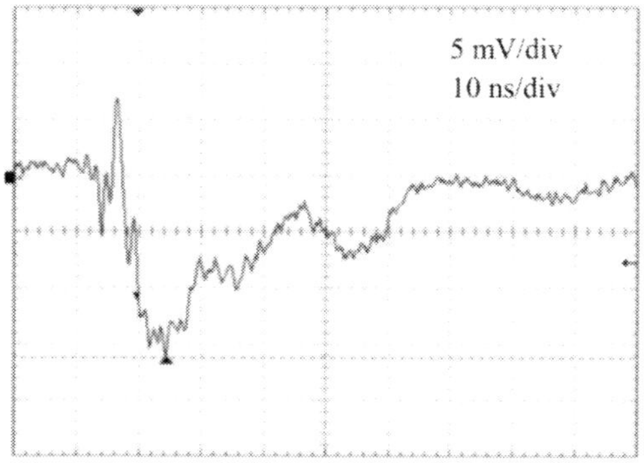

Figure 9. E-field waveform at 3 m away aperture with terminal opened.

If the experiments need higher filed strength, the fast rise time radiating-wave simulator should use Marx generator as excitation pulse source and TEM horn antenna as radiation antenna, and then higher intensity radiation pulse field can be generated by the fast rise time EMP radiating-wave simulator.

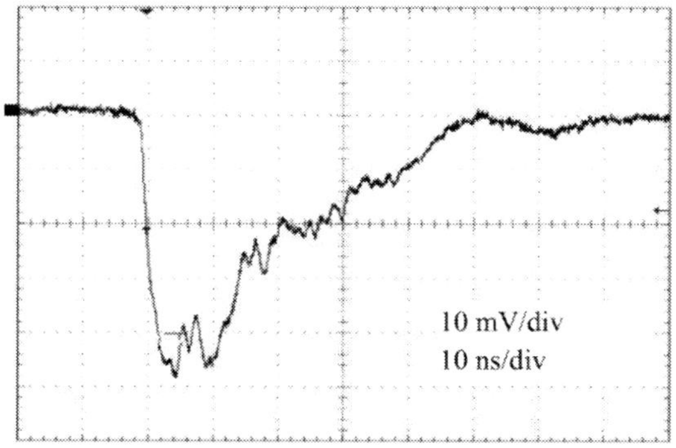

Figure 10. E-field waveform at TEM antenna aperture with matched loaded.

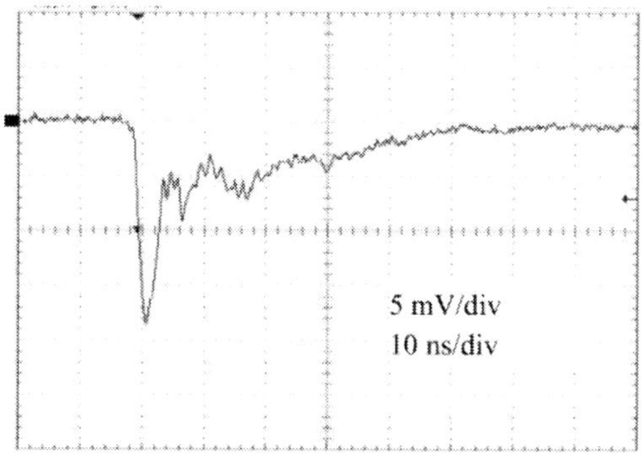

Figure 11. E-field waveform at 3 m away aperture with matched loaded.

5. Conclusions

In this paper, fast rise time EMP radiating-wave simulator is analyzed in numerical simulation and experiment method. The results show that TEM horn antenna can radiate the fast rise time pulse field such as RS105 HEMP. But the TEM antenna radiate ability is relatively poor in low frequency, and low frequency band need install matched loaded to absorb the oscillation current. In this paper, only to introduce the basic principles, the actual fast rise time EMP radiating-wave simulator is designed, and the ultimate goal generates HEMP by TEM horn antenna.

Acknowledgments
This project is sponsored by National Key Library laboratory Foundation (No. 51333030101)

References

[1] Radasky W A 2008 *IEEE International Symposium on Electromagnetic Compatibility* 1
[2] Cai J, Wang B and Hu F 2010 *International Conference on Computer, Mechatronics, Control and Electronic Engineering* p 456
[3] Schilling H, Schluter J and Peters M 1995 *10th IEEE International Pulsed Power Conference* p 1359
[4] Vyvolokin A E and Neskopodov G F 1993 *IEEE International Pulsed Power Conference* (USA: New York) p 163
[5] Zhen J, Hagness S C and Booske J H 2006 *IEEE Transactions on Biomedical Engineering* **53** 780
[6] Baum C E 1978 *IEEE Trans. on EMC.* **26** 35.
[7] Meng F, Xia H and Wang J 2007 *Journal of Microwaves* **23** 6
[8] Tesche F M, Mo C and Shoup R W 1994 *IEEE Transactions on Electromagnetic Compatibility* p 331
[9] Ronald F B and Clayborne D T 1978 *IEEE Transactions on Electromagnetic Compatibility* 240
[10] Meng F 2011 *High power EM pulse technology* (BeiJing: National defense industry press)

Experiments study of coupling parameters characterization of the multiconductor transmission lines

Yingjie Gan, Xiaoying Xu*
School of Sciences, Wuhan University of Technology, Wuhan, China

E-mail: xu_xiao_ying@126.com

Abstract. Although a number of papers have been published on the study of multiconductor transmission lines (MTLs) parameters characterization, little experiment studied on the MTLs, especially in the frequency domain. This paper presented a series of experiments of homogeneous MTLs, got the characteristic parameters such as the capacitance matrix and the inductance matrix, and analyzed the relationship between parameters and frequency. The correctness of the experimental results was verificated by the HFSS software simulation.

1. Introduction

With the requirement of high-speed, broadband, high density and miniaturization in the development of modern electronic technology, the distribution parameters of the transmission lines must be taken into account in power and communication system. In many cases coupling between lines can be attribute to the couple of MTLs [1].

The characteristic parameters determine the propagation of the MTLs and the terminal current and voltage. In most case, when the geometry length of the MTLs is less than 1/10 of the wavelength, the MTLs is considered to be the electrically small, and then the inductance is much lager than the capacitance [2]. Lumped parameters of the resistance and inductance are available to establish the equivalent circuit to analyze and calculate the MTLs coupling response.

One of the major ways, which MTLs electromagnetic interference (EMI) is coupling between lines, is that the degree of coupling can quantitatively characterize the strength of this coupling, using the scattering parameters (S-parameters) to determine the value of it.

This paper made measurements and calculations to the characteristic parameter matrices-scattering matrix ([S]), impedance matrix ([Z]) and the inductance matrix ([L]) of the MTLs. The measurements may generate error due to instrument or other reasons; thus, to verify the validity of the experiment, the measurements are compared with the HFSS simulation results. Mastering the law of the coupling parameters change with the frequency can provide a theoretical basis for the analysis of MTLs coupling and it also can guide engineering application.

2. Experimental tests

2.1 Experimental Principle

It is very convenient to use a network analyzer for measuring the S-parameter of transmission line

* To whom any correspondence should be addressed.

system. When it comes to the analysis and calculation, [S] often needs to be converted to [Z]. At the reference surface of the network port, scattering matrix describes the relationship between the normalized incident wave and reflected wave, and [Z] describes the relationship between the voltage and the current which are not normalized. Since [S] and [Z] describe the same network, there must be a transformational relationship [3].

The relationship between the network parameter Z_{ij} and S_{ij} is shown in Equation 1.

$$\begin{aligned} Z_{11} &= Z_c \frac{(1+S_{11})(1-S_{22})+S_{12}S_{21}}{(1-S_{11})(1-S_{22})-S_{12}S_{21}} \\ Z_{12} &= Z_c \frac{2S_{12}}{(1-S_{11})(1-S_{22})-S_{12}S_{21}} \\ Z_{21} &= Z_c \frac{2S_{21}}{(1-S_{11})(1-S_{22})-S_{12}S_{21}} \\ Z_{22} &= Z_c \frac{(1-S_{11})(1+S_{22})+S_{12}S_{21}}{(1-S_{11})(1-S_{22})-S_{12}S_{21}} \end{aligned} \quad (1)$$

The measurements of the characteristic parameters of MTLs commonly used transmission line terminal short-circuit and open-circuit impedance tests [4]. When the far end shorted and all the near ends but the driven line opened, by measuring current and voltage values of the near end, we can get the transmission lines impedance matrix:

$$[Z_{ij}'(s)] = [R_{ij}'] + s[L_{ij}'] \ (s = j\omega) \quad (2)$$

Then the relationship between the impedance matrix and inductance matrix became as follows:

$$[Z_{INnm}^{sc}(s)] = j\omega[L_{nm}'] \quad (3)$$

Open experimental measurement is the inductance matrix.

2.2 Experimental setup and testing

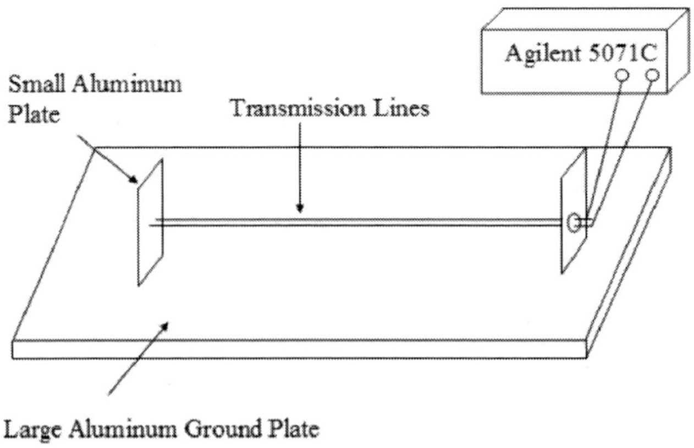

Figure 1. Transmission lines parameter measurement experiment setup.

The experimental device to measure the MTLs S-parameter was shown in figure1. The conductor lines were BV1 type single core copper cable and two conductors arranged in parallel with the same length.

Two small aluminum plate at both ends of the conductor wire screw up the lines, which were welded to the large ground plate to provide a low-impedance current path in the short-circuit measurement. The large ground plate is an aluminum plate, size of 1.5m×2m, which is connected to the ground line of network analyzer by wires [5].

Two transmission lines far ends shorted, after calibrated the network analyzer, probes connected to the near end, sweeping the S-parameter varying with frequency.

The network analyzer is a precise measuring instrument, so it is important to pay particular attention to the impact of its probe on the measurement results. In order to reduce measurement error, it should try to shorten the line length of the probe and ensure the probe fixture contact well with the transmission lines. In the process of measurement, be careful to stay away from the bench to reduce the interference on the measurement of human body electrostatic.

3. The HFSS simulation

Ansoft HFSS software is based on the finite element method. It is a 3D frequency-domain electromagnetic field calculation software, through the calculation of the electromagnetic field components of the grid cells to obtain the various physical quantities and parameters.

HFSS was used to model the experimental setup, which was shown in figure 2. The sizes of all components in the model were the same with the experimental measurement sizes.

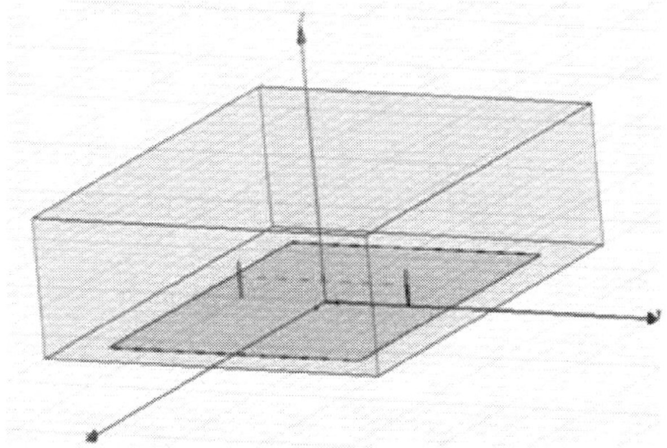

Figure 2. The HFSS model.

To meet the requirements of electrically small, the calculation frequency limits from 1 MHz to 15 MHz, sampling step length is 1 MHz with lumped pot incentives [6]. The software simulation analysis can directly read the S-parameter and Z-parameter matrix data and the sampling frequency points.

4. Data analysis

4.1 S-parameter analysis

Part of the S-parameter data is shown in table 1.

Because the size and placement of the two transmission lines are the same, the scattering matrix is a reciprocal matrix, i.e. $S_{ij}=S_{ji}$. Table 1 gives the values of S_{11} and S_{12} parameters.

In table 1, the measurement and simulation data of S-parameter is slightly different, which is mainly due to the impact of probe fixture on the measure system. The experimental measurement of S_{11} and S_{12} parameters curve are shown in figure 3.

In figure 3, with increasing of frequency, S_{11} continuously reduces, while S_{21} rises, two curves level off after 10 MHz. From 1 MHz to 10 MHz, return loss decrease, while insertion loss increase, the MTLs reflection continuously goes down, meaning the ability of the system to transfer more energy.

Table 1. S-parameters comparison.

f (MHz)	S_{11} (dB) (expt)	S_{12} (dB) (expt)	S_{11} (dB) (sim)	S_{12} (dB) (sim)
1	-0.22	-13.8	-0.22	-13.5
3	-1.36	-5.89	-1.37	-5.84
5	-2.61	-3.56	-2.8	-3.74
7	-3.5	-2.67	-3.43	-2.72
9	-3.98	-2.31	-3.86	-2.4
11	-4.17	-2.19	-3.99	-2.31
13	-4.16	-2.2	-3.94	-2.35
15	-4.02	-2.29	-3.77	-2.47

When the port is matched and the transmission line circuit has no load, the coupling between the transmission lines can be defined as follows:

$$C = 10\lg(|S_{12}|^2) \tag{4}$$

The bigger the value of C, the stronger the coupling between lines, and the greater the mutual interference. The curve that C changes with frequency is shown in figure 4.

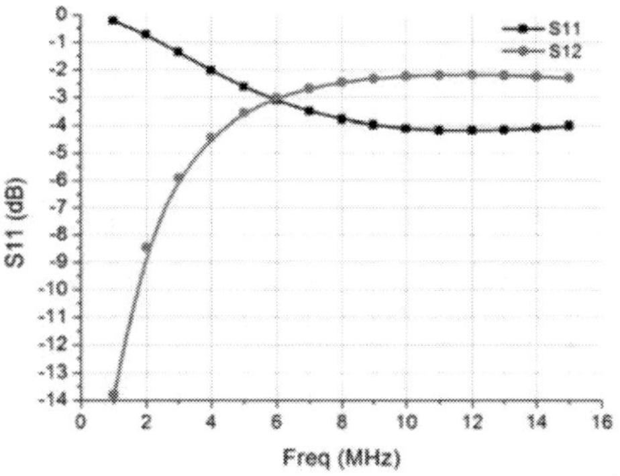

Figure 3. The curve of S-parameters varies with frequency.

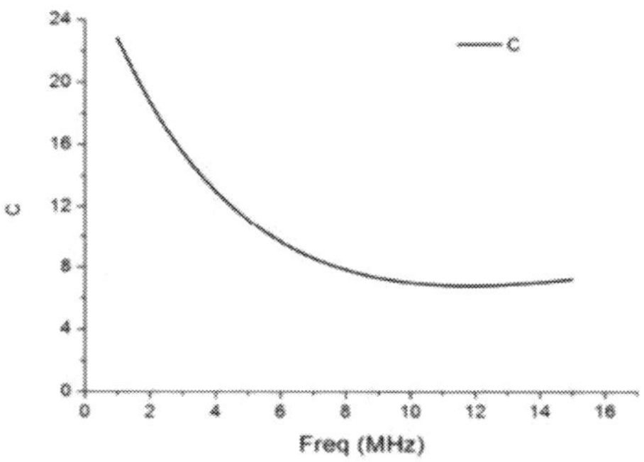

Figure 4. MTLs coupling degree.

From the figure, we can see that with frequency increasing, the degree of coupling reduced, and the transmission performance of the transmission lines became better.

4.2 L-parameter analysis

Table 2. L-parameters comparison.

f (MHz)	L_{11} ($\mu H/m$)(expt)	L_{12} ($\mu H/m$)(expt)	L_{11}($\mu H/m$)(sim)	L_{12}($\mu H/m$)(sim)
1	1.145	0.071	0.131	0.036
3	6.296	0.366	8.228	0.158
5	13.797	0.781	15.677	0.312
7	23.249	1.269	26.096	0.460
9	34.524	1.970	32.411	0.763
11	47.877	3.222	46.892	1.558
13	63.418	4.484	63.554	2.593
15	79.423	4.466	78.071	1.608

The measurement S-parameter matrix and simulation Z-parameter matrix can be converted to L-parameter matrix by (1) and (3). Part of the inductance matrix parameter data was shown in table 2.

The comparison of L_{11} parameter curves was shown in figure 5. L_1 was the simulation curve and L_2 was the experiment data. Figure 6 was about the contrast of L_{12} parameter curves. L_1 was the simulation curve and L_2 was the experiment curve.

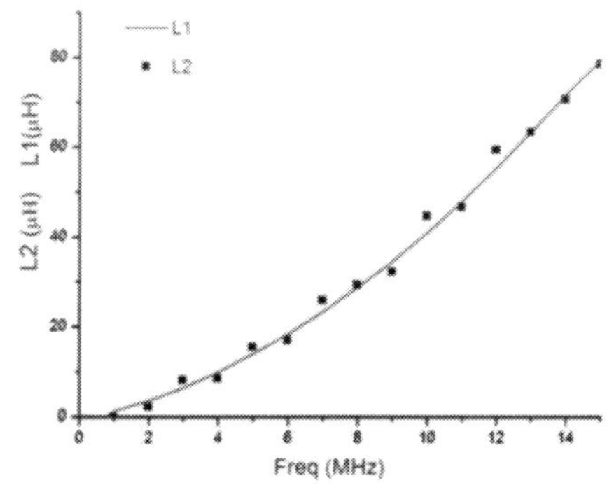

Figure 5. Simulation L_{11} parameters curve in contrasts with experiment data.

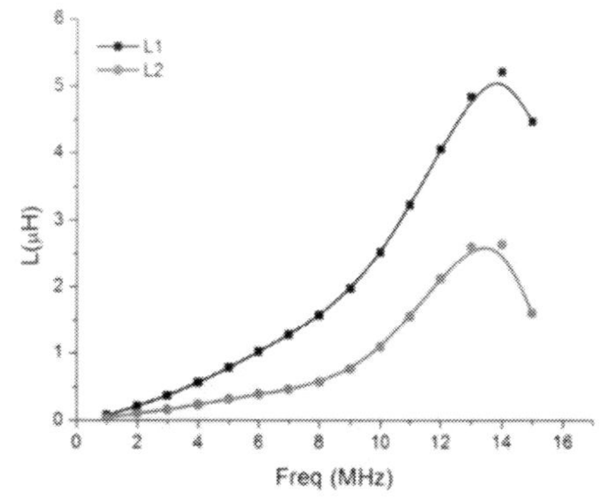

Figure 6. Calculated L_{12} parameter curves contrast.

As it can be seen in figure 5, the experimental and simulation self-inductance of MTLs is basically consistent with the simulation result, so the experimental one is valuable. Using Allometroc1 to fit a curve to the data, the fitted result and the curve in figure 4 mainly overlapped. The fitting formula is:

$$y = 1.129x^{1.564} \tag{5}$$

From (5), we can see that in the measured frequency range, the self inductance of MTLs is exponential growth with the frequency increase.

From the figure 6, we can see that the mutual inductance parameter data of experiments and simulation are quite different, but the trend of the curves are the same, so we can still analyze the relationship between mutual inductance and frequency from the figure. The difference mainly results from the biased error which is caused by the mismatch calibration of network analyzer and load in the measurements. Only if the correction is accurate and the load had total matched, the S-parameter is exact.

The mutual inductance of MTLs increases with frequency in the range of 1 to 13 MHz, thereafter, when the frequency increases, L_{12} decreases. It means that the mutual inductance self-resonant frequency is around 13 MHz. When frequency is below the self-resonant frequency, mutual inductance of MTLs increases with it; at the self-resonant frequency point, inductance reaches its maximum; when frequency over the point, the inductance decreases.

Figure 5 and 6 show that the inductive reactance of transmission lines is not constant while losses exist. Frequency mainly impacts the inductive reactance by the skin effect and proximity effect and this impact grows with the increase of frequency. While the frequency increases, the current through the cross section of transmission lines becomes uneven and moves to the periphery of conductors.

5. Conclusion

This paper introduced a method of measuring and calculating MTLs coupling parameters. When the far ends of MTLs shorted, it is possible to measure S-parameter matrix with analyzer network and then to convert out the L-parameter matrix. This method can help us to analyze the law of the variation of coupling parameters with frequency. What's more, it is convenient and suitable in the laboratory.

When the transmission lines cannot be seen as electrically small, we can still use S parameters to calculate the L-parameter matrices by formulas, so this method is suitable for the whole frequency band.

References

[1] Antonini G and Camillis L D 2009 *Microw. Wirel.* **19** 65
[2] Paul C R 2008 *Analysis of multiconductor transmission lines* (USA: John Wiley & Sons)
[3] Chen X and Kildal P 2011 *Proc. Int. Conf. On AP-S/URSI (Spokane)* (New York: Institute of Electrical and Electronics Engineers) p1885
[4] Paul C R 2007 *IEEE Trans. Eletromagn. Compat.* **49** 237
[5] Rana M and Islam R 2011 *Proc. Int. Conf. On ASEMD (Sydney)* (Chengdu: University of Electronic Science and Technology of China Press) p220
[6] Ryan C G M and Eleftheriades G V 2012 *IET Microwaves Antennas Propag.* **6** 70

A control system based on field programmable gate array for papermaking sewage treatment

Zi Sheng Zhang*, **Chang Xie**, **Yan Qing Xiong**, **Zhi Qiang Liu**, **Qing Li**
Electrostatic Research Institute, Hebei University, No.180 WuSi East Road, Baoding, China

E-mail: zhangzisheng@hbu.edu.cn

Abstract. A sewage treatment control system is designed to improve the efficiency of papermaking wastewater treatment system. The automation control system is based on Field Programmable Gate Array (FPGA), coded with Very-High-Speed Integrate Circuit Hardware Description Language (VHDL), compiled and simulated with Quartus. In order to ensure the stability of the data used in FPGA, the data is collected through temperature sensors, water level sensor and online PH measurement system. The automatic control system is more sensitive, and both the treatment efficiency and processing power are increased. This work provides a new method for sewage treatment control.

1. Introduction

At present, the impact of papermaking wastewater on the environment is more and more serious. In recent years, although sewage treatment has made some achievements, wastewater emission by some enterprise does not satisfy national requirements [1-2]. Wastewater treatment process in paper mills has a very high requirement for analog closed-Loop control system, and the logical relationship of the process is very complex [3-5]. Ensuring the high efficiency is the key of sewage treatment process, so a high level of automation control design is very important.

The principal benefit of the FPGA is unparalleled design flexibility as to where you place the border between hardware and software. The programmability of an FPGA makes it larger (in terms of transistor count) than an equivalent hardwired chip. In this paper, FPGA technology is used to design to improve the efficiency of papermaking wastewater treatment. We have designed a sewage treatment control system based on FPGA. VHDL language is used to design each part of the control program, and Quartus II is used for simulation.

2. The basic process of wastewater treatment

Sewage of paper mill emits from sewer line and flows through the thick grille and fine grille. Large particles in the waste water are filtered effectively by the grille, and the impurities are recycled so as to save raw materials [6]. The process of sewage treatment and reuse is shown in figure 1. Papermaking wastewater treatment process is very complex. In addition to electromechanical equipment, there are also some peripheral equipment, for example digital display devices. The treatment control process designed with FPGA is a very feasible scheme. Since FPGA is used in mechanical equipment control

* To whom any correspondence should be addressed.

system, system operation becomes more convenient, and treatment efficiency and processing power are greatly improved.

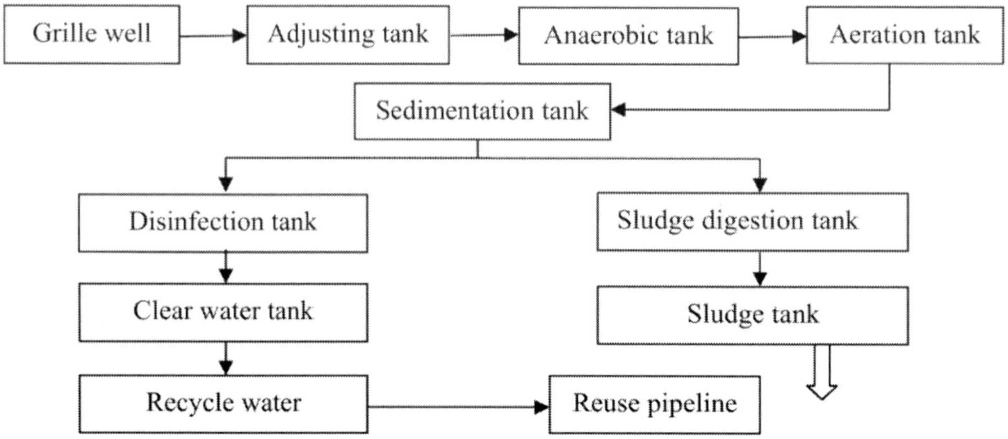

Figure 1. Paper mill sewage treatment and reuse figure.

3. FPGA digital display system

Devices in control room and electrical equipment room are used to monitor and control the whole process. In the control room there is a digital indicator that can display temperature, water depth and pH value. All of the data is collected by sensors and processed by FPGA. The display method of the digital indicator is LED dynamic display, and the digital display system is designed with FPGA, MAX7000AE series model for EPM7064AEFC100-7.

Simulation results of Quartus II software are shown in figure 2. When reset='1', screen displays '0'; when reset='0', scan starts counting. The input signals are din1, din2, din3 and din4, and the output signals are bus4 and shift. When scan_clk= '00', '01', '10' and '11', bus4 outputs the values of din1, din2, din3 and din4, respectively. This figure shows that the simulation results were consistent with the program, and this program met the demand of the design.

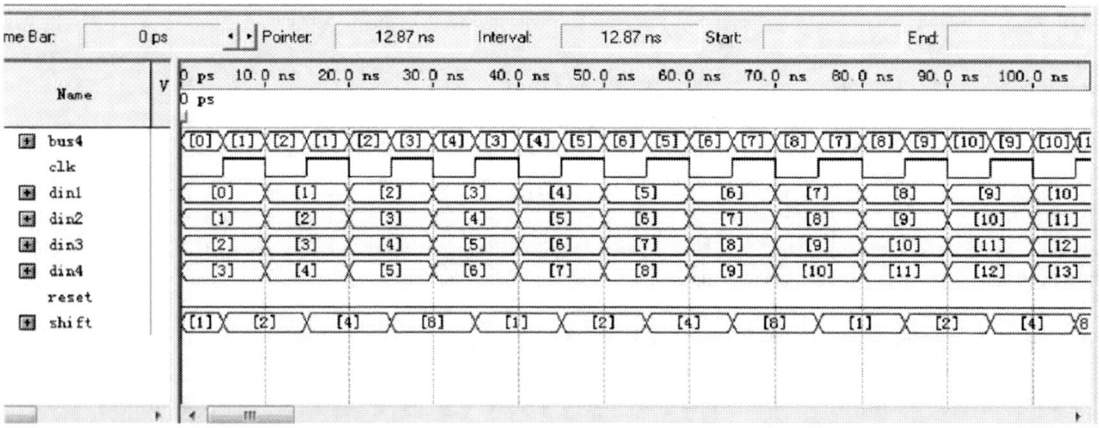

Figure 2. The result of FPGA control LED displays the simulation.

Journal of Physics: Conference Series **418** (2013) 012059 doi:10.1088/1742-6596/418/1/012059

4. Processing system control of FPGA

This process is designed with FPGA, MAX7000AE series model for EPM7032AELC44-7 (Altera Company). This chip can completely meet the needs of the design with low cost. We designed the main part of the whole control process. The main control process is shown in figure 3.

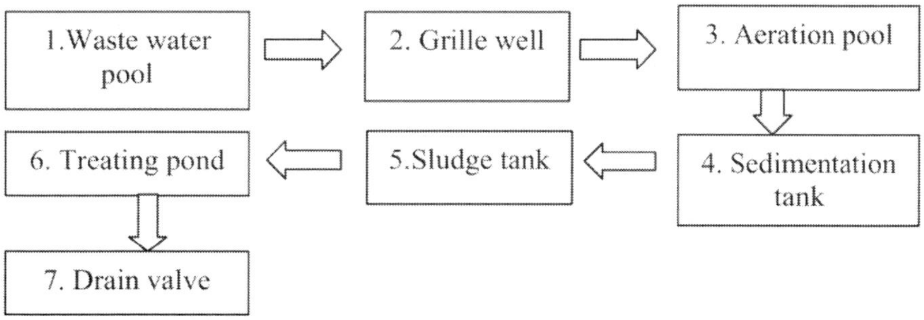

Figure 3. Main control process.

We control start or stop of each step through the switch. Step 1 is waste water pool control process, and step 7 is drain valve control process.

Step 2: This process uses a rake tooth chain rotary grille decontamination machine, which has a strong effect on decontamination. Machine automatically stops after continuing operation one hour and can open again after half an hour.

Step 3: In this design a mechanical brush aerators is used to do aeration work, and an automatic heating device is installed in the pool. Aeration equipment starts heating when the temperature is below 20 ℃, and it stops heating automatically when the temperature is beyond 45 ℃. A temperature sensor which outputs current (simulator AD590) is used to measure the temperature. The current of sensor outputs from 223 μA (50 ℃) to 423 μA (+ 150 ℃), so the sensor can be applied to multi-point temperature measurement and long distance temperature measurement.

Step 4: Considering the difficulty on treating papermaking wastewater, the cycle frequency of mechanical scraping mud can be set to 1 hour. In the settling tank, the inlet valve and the discharge valve can be opened or closed independently. A water level warning device is put in the settling tank. When the level of water in the pool is more than eight meters, the warning device will make an alarm. The water level measurement sensor adopts the model XLH-LHY63 (Beijing), which can work well at the temperature of 40 degrees to + 125 degrees.

Step 5: Mechanical scraping mud device can be open and closed at any time, and the work cycle is 1 hour.

Set 6: Equipments of sewage pool work all day. When the pH value of wastewater is from 6.5 to 7.5, the equipments stop running; when the pH value is out of the range, the equipments open the operating mode automatically. The measurement sensor for pH value uses online pH measurement system (Mettler Toledo), which can satisfy the sewage treatment environment and has certain precision.

Partial program is as follows:

```
P1: PROCESS (CLK,E_IN(1)) -------------------------------------------- ---Waste water pool control
BEGIN
  IF(CLK'EVENT AND CLK='1')THEN
    E_OUT(1)<=E_IN(1);
     IF(E_IN(1)='1')THEN
     SWITCH1_OUT<=SWITCH1;
     ELSE SWITCH1_OUT<='0';
     END IF;
 END IF;
```

Journal of Physics: Conference Series **418** (2013) 012059 doi:10.1088/1742-6596/418/1/012059

```
END PROCESS;

  P2:PROCESS(CLK,E_IN(2))--------------------------------- Grille decontamination machine control
  BEGIN
   IF(E_IN(2)='1')THEN
     IF(CLK'EVENT AND CLK='1')THEN
      PR_STATE2<=NX_STATE2;
      IF (TIMER2=1800) THEN TIMER2<=0;
        ELSE  TIMER2<=TIMER2+1;
       END IF;
     END IF;
    END IF;
  END PROCESS;
  PROCESS(CLK)
  BEGIN
   CASE PR_STATE2 IS
   WHEN STATEA2 =>
     IF(TIMER2=1800)THEN
         NX_STATE2<=STATEB2;
        SWITCH2<='1';
        END IF;
   END CASE;
  END PROCESS;

  P3: PROCESS(CLK,E_IN(3))------------------------------------------------------ -Aeration pool control
  BEGIN
  IF(CLK'EVENT AND CLK='1')THEN
    E_OUT(3)<=E_IN(3);
    IF(E_IN(3)='1')THEN
      IF( TEMPERATURE3>45)THEN SWITCH3<='0';
           ELSIF(TEMPERATURE3<20)THEN SWITCH3<='1';
           END IF;
     ELSE  SWITCH3<='0';
    END IF;
  END IF;
  END PROCESS;

  P7:PROCESS (CLK,E_IN(7)) ----------------------------------------------------- -Waste water discharge
  BEGIN
  IF(CLK'EVENT AND CLK='1')THEN
     E_OUT(7)<=E_IN(7);
      IF(E_IN(7)='1')THEN
          SWITCH7_OUT<=SWITCH7; ---------------------------------------------Open shutoff Settings
              ELSE  SWITCH7_OUT<='0';
      END IF;
  END IF;
  END PROCESS;
  END BEHAV;
```

Simulation results of Quartus II software are shown in figure 4. The working condition of each step is shown by the level of SWITCH1, SWITCH2, SWITCH3, SWITCH6, SWITCH7_OUT, and SWITCH5-OUT. When the signal is at the high level, the equipments stop working; when signal is at the low level, the equipments start working. Take the third part as an example: when TEMPERATURE3>45, SWITCH3<='0', aeration equipment stops heating, but when TEMPERATURE3<20, SWITCH3<='0', aeration equipment starts heating. The figure shows the simulation results were consistent with the program, and this program met the demand of the design.

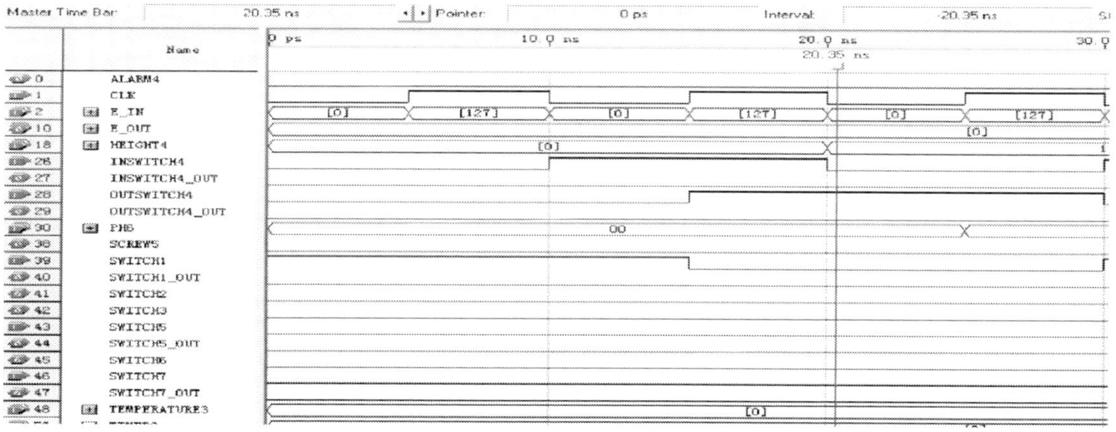

Figure 4. Process simulation results.

5. Conclusions

The principal benefit of the FPGA is unparalleled design flexibility as to where you place the border between hardware and software. In this paper, an automatic control system based on FPGA is designed for sewage treatment. VHDL is used to write the main codes and Quartus II software is used to simulate and compile. Sensors and PH measurement system are used to measure and transfer the current signals to FPGA. The control system operation is convenient, and treatment efficiency and processing power are greatly improved. In a word, this automatic control system based on FPGA for sewage treatment is economical and practical.

Acknowledgments

This paper is supported by the National Science Foundation of China under Grant No.51077035, and the Science Foundation of Hebei province No.A2010000182.

References

[1] Wu S B 2011 *Beijing: CIP (in Chinese).* p12
[2] Li J and Gao P J 2005 Beijing: *CABP (in Chinese).* p22
[3] Jiang H and Li Z Y 2007 *Microcomputer information (in Chinese)* **23** 11
[4] Zhang K, Kang Y and Xiong J 1999 *Power Electronics (in Chinese)* **22** 12
[5] Analog Devices 2004 ADSP-TS101 Tiger SHARCR Embedded Processor, Rev. B.
[6] Huan B G, Lei B and Wang P 2011 *Environ. Sci. Technol.* 9 51

Piezo-electric properties of polypropylene laminates with a non-woven layer

R Kacprzyk[*], **A Kisiel**
Institute of Electrical Engineering Fundamentals, Wroclaw University of Technology, Wybrzeze Wyspianskiego 27, 50-370 Wroclaw, Poland

E-mail: ryszard.kacprzyk@pwr.wroc.pl

Abstract. In the paper piezoelectric properties of non-uniform dielectric structures with elastic layers made from unwoven fabric are presented. Piezoelectric properties of the samples were studied through measurement of piezoelectric d_{33} coefficient, and carried out in dependence on static pressure. The characteristics $d_{33}=f\ (p)$ obtained for the laminates were hyperbolic type. In order to elucidate this kind of relationship, a theoretical model has been proposed. Based on the new model, the calculated d_{33} piezoelectric coefficients showed good agreement with the experimental results. According to the model developed, the $d_{33}\ (p)$ dependence is related to the changes of the unwoven structure elastic modulus Y under the influence of static pressure. The magnitude of this change is strongly influenced by the structure of the non-uniform layer, including the fiber diameter, the number of fibers and their arrangement, and the fiber elasticity modulus.

1. Introduction

During the last years non-uniform materials with foamed or voided structure became the subject of intense research due to their advantages over well-known piezoelectric polymers and ceramics [1]. Flexible cellular polymer films and multi-layer structures may exhibit very high value of a piezoelectric coefficient d_{33} that is one order of magnitude higher than that of reported for traditional ferroelectric polymers like PVDF and comparable to the values obtained for the best ceramic materials [2, 3]. In terms of small loads layered structure can reach a value of d_{33} even above 1000 pCN^{-1}.

Analysis of piezoelectric properties of electromechanical films, as well as of nonuniform layered structures made of dielectrics having very good electret properties, indicates the possibility of using them in the construction of various types of sensors, electro-acoustic and electro-mechanical transducers used for example to convert ambient energy into electrical energy (the harvesters) [4]. Except of very good piezoelectric properties such materials are characterized by the low cost, thus they can be the basis for the development of large-area transducers.

The possible applications of non-uniform structures [5] indicate that they require determining the piezoelectric coefficient dependence on the applied pressure. This realationship was investigated during studies reported in the presented work.

Studies on the nature of the piezoelectric phenomenon occurring in various materials have shown that it may be the result of several mechanisms [6, 7]. Among them it is distinguished piezoelectric effect related to the nonuniformity of the mechanical properties of the dielectric structure, while the

[*] To whom any correspondence should be addressed.

Journal of Physics: Conference Series **418** (2013) 012060 doi:10.1088/1742-6596/418/1/012060

occurrence of space charge. A model describing piezoelectric effect in non-uniform multilayer structures was discussed previously in the literature [8, 9], but to explain the results presented in the paper was insufficient. Therefore a new model of piezo-active laminate with non-woven layer has been developed.

2. Piezoelectric effect in non-uniform dielectric structures

Let's consider three-layer non-uniform "sandwich-like" structure, consisted of two solid films with non-woven (fibers) layer between them. The analysed structure is symmetrical and has an arrangement of layers: hard/ soft/ hard. The charge densities at the interfaces hard/ soft are assumed to be opposite sign and equal in module i.e. $|q_{s1}|=|-q_{s2}|=q_s$.

In the case of such laminate subjected to external stresses applied in the normal direction to its surface the piezoelectric coefficient d_{33} can be expressed by the relation:

$$d_{33} = \frac{2q_s\varepsilon_1\varepsilon_2 xd}{(\varepsilon_1 d + \varepsilon_2 x)^2}\left(\frac{1}{Y_1} - \frac{1}{Y_2}\right) \tag{1}$$

where: d - thickness of "hard" layer, x - thickness of "soft" layer, Y_1 , Y_2 - Young's modules of the "soft" and "hard" layers respectively, ε_1 - dielectric permittivity of non-woven material, ε_2 - dielectric permittivity of polymeric film. Assuming $Y=Y_1<<Y_2$, which means that the external stress p produces strain in the "soft" layer–x (p) only, and validity of the Hook's low, in the form:

$$x(p) = x_0(1-\frac{p}{Y}) \tag{2}$$

where x_0 - the initial thickness of the non-woven layer (without pressure applied), the equation (1) can be rewritten in the form:

$$d_{33}(p) \cong \frac{2q_s\varepsilon_1 d}{Y\varepsilon_2 x_0}\frac{\left(1-\frac{p}{Y}\right)}{\left[\frac{\varepsilon_1 d}{\varepsilon_2 x_0}+\left(1-\frac{p}{Y}\right)\right]^2} = d_{330}F(p) \tag{3}$$

The analysis of relation (3) leads to conclusion that the piezoelectric coefficient $d_{33}(p)$ dependence on static pressure p is determined by the shape of a function $F(p)$. The course of the reduced value of coefficient d_{33} (related to the value of d_{330} when p=0) is determined by the course of the reduced value of the function $F(p)/F(p=0)$. This relationship is shown in figure 1.

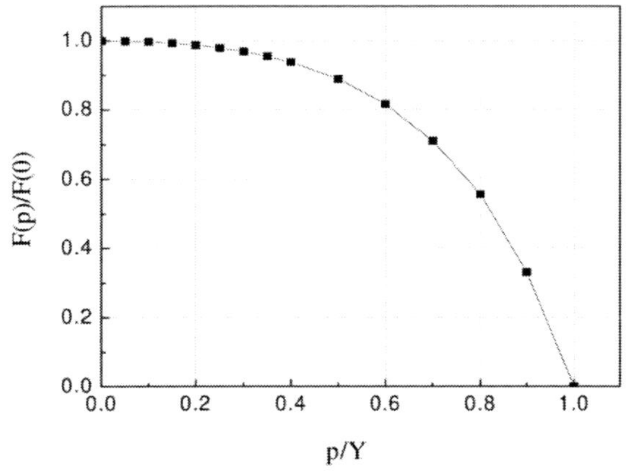

Figure 1. Plot of the F(p)/ F(p=0) function depending on the p/Y reduced pressure value.

3. Experimental

The samples of the three-layer structure were prepared according to the scheme depicted in figure 2.

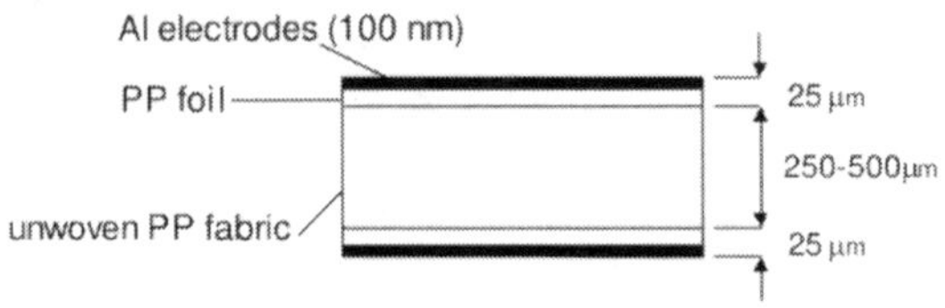

Figure 2. Schematic view of structure with non-woven PP layer.

A non-woven polypropylene (PP) placed between two solid polypropylene films was applied as a "soft "layer. External sides of both foil-solid polymer layers were metalized with a thin layer of aluminium. The model laminate structures were prepared with different thicknesses of the "soft" - non-woven layer, ranged from 0.40 to 3.00 mm. Solid dielectric layers were made of polypropylene 25 μm thick films. The structures were made using thermal or ultrasonic welding. All of laminates exhibited good mechanical properties.

The laminate samples were electrically activated by application of breakdown method. The structures were polarized using different values of polarisation voltage. During polarization process, carried out in high electric fields, partial discharges occurring in air spaces lead to the formation of space charge distributed in the non-woven layer in a required manner, i.e. according to the assumption $|q_{s1}|=|-q_{s2}|=q_s$. Polarization time was kept constant and equal t_p=300±10 s. All samples were charged-polarized at the ambient temperature T=25±2 °C and relative humidity RH=55±10 %. Piezoelectric properties of manufactured structures were investigated by measurements of the piezoelectric coefficient d_{33} with quasi-static method.

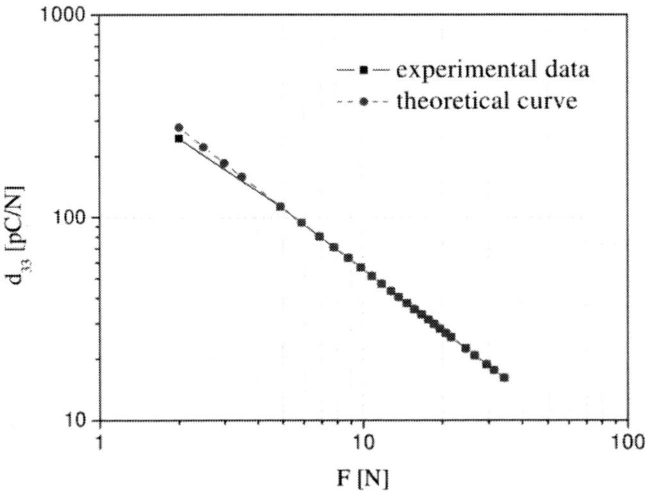

Figure 3. Influence of static load on d_{33} piezoelectric coefficient for tested structure and theoretical calculation.

The dependence of piezoelectric coefficient d_{33} (p) on static pressure p for selected sample of three-layer PP structure is presented in the figure 3. A hyperbolic type of $d_{33}(p)$ dependence was observed for all obtained laminates, which was not consistent with graph of $d_{33}(p)$ predicted by equation (3). For explanation this discrepancy a previous described model has been modified as follows.

4. Model of piezo-active structure with non-woven layer

The basic assumption of the new model is the dependence of elasticity modulus of the "soft"-non woven fabric Y on real contact surface of fibers S, given in the form:

$$Y = \frac{S}{S_0} K_S Y_{pp} \tag{4}$$

where Y - Young's modulus of the "soft" layer, Y_{PP} - Young's modulus of the fibres material (polypropylene), S_0 - the total area of transducer, S - contact area of fibers, K_S - structural factor (dimensionless) depending on the diameter of the fibers, their number and distribution in volume of non-woven material. Additionally it was assumed that:

$$\frac{S}{S_0} = kp \tag{5}$$

where k - constant. Validity of dependence (3) for non-woven layers as well as for fabrics was confirmed experimentally [10]. In the case of three-layer symmetrical structure described above and for the previous made assumptions, the expression (1) can be rewritten in the form:

$$d_{33} = \frac{2q_s\varepsilon_1\varepsilon_2 x(p)d}{Y[\varepsilon_1 d + \varepsilon_2 x(p)]^2} \tag{6}$$

From equations (4) and (5) one can get:

$$Y = kpK_S Y_{PP} \tag{7}$$

Substitution of expression (7) into (2) and next the result into expression (6) leads to the following equation:

$$d_{33} = \frac{2q_s\varepsilon_1\varepsilon_2 x_0 d(kK_S Y_{pp} - 1)}{p[\varepsilon_1 dkK_S Y_{PP} + \varepsilon_2 x_0(kK_S Y_{PP} - 1)]^2} \tag{8}$$

Taking into account that in the above equation, all parameters except p are constant this relation can be written in the form:

$$d_{33} = A\frac{1}{p} \tag{9}$$

where

$$A = \frac{2q_s\varepsilon_1\varepsilon_2 x_0 d(kK_S Y_{pp} - 1)}{[\varepsilon_1 dkK_S Y_{PP} + \varepsilon_2 x_0(kK_S Y_{PP} - 1)]^2} \tag{10}$$

Assuming, that for the applied non-woven structures following relationship is fulfilled [10]:

$$kK_S Y_{PP} >> 1 \tag{11}$$

Equation (8) can be described by:

$$d_{33} \cong \frac{2q_s\varepsilon_1\varepsilon_2 x_0 d}{pkK_S Y_{pp}(\varepsilon_1 d + \varepsilon_2 x_0)^2} \tag{12}$$

As can be seen from the relation (12) for the three-layer structure with non-woven fabric, for which all of the stated assumptions are met, it is expected hyperbolic dependence $d_{33}(p)$. To verify the resulting model, the calculations of piezoelectric coefficients d_{33} were performed for the structure characterized by the following data: PP film thickness d=2.5 ×10^{-5} m; the initial thickness of the non-woven fabric (without static pressure applied) x_0=4.0×10^{-4} m; dielectric permittivity of "soft" layer material (PP non-woven) ε_1=1.5; dielectric permittivity of PP foil ε_2=2.1 the effective surface charge density

q_s=2.7×10^{-5} Cm^{-2}. Calculations carried out for piezoelectric coefficient d_{33}=128±1 pCN^{-1}, measured for static force F=4.9 N, applied to the total sample area S=3.1×10^{-4} m^{-2} allow (see expression (12)) to determine value of the component kK_sY_{PP}. For given data it was equal 1.48. The obtained value of kK_sY_{PP} was constant for a given sample (it was connected with its structure). Subsequently piezoelectric coefficients d_{33} for various static pressure were calculated and plotted as the $d_{33}(p)$ dependence. Resultats obtained from measurements and theoretical calculations are presented in the figure 3. It can be seen that both of plotted curves are almost the same. These results indicate that the developed model of three-layer piezo-active structure with non-woven layer as well as all assumptions are appropriate.

5. Conclusions
Piezoelectric properties of three-layer laminates with non-woven PP layer were investigated. The results of measurements and calculations made on the basis of a new model proposed for the tested structures lead to the following conclusions:

- The tested non-uniform layered structures showed a large piezoelectric coefficient d_{33} (up to several hundred pCN^{-1}), strongly dependent on the applied static pressure.
- For all prepared laminates experimentally obtained $d_{33}(p)$ dependences were of a hyperbolic type.
- The new proposed model of piezo-active structure with a soft layer made from non-woven fabric showed good agreement with the experimental results.
- In accordance with the model developed, the $d_{33}(p)$ dependence is related to the changes of elastic modulus Y of unwoven structure under the influence of static pressure. The magnitude of this change is strongly affected by the structure of the non-uniform layer (the fiber diameter, the number of fibers and their arrangement, fiber elasticity modulus, which are given as kK_sY_{PP} component).

Acknowledgments
This work was funded by the statutory research (No S10026) in 2012.

References
[1] Sessler G M and Hillenbrand J 1999 *Appl. Phys. Lett.* **75** 3405
[2] Hillenbrand J, Xia Z, Zhang X and Sessler G 2003 *Proc. of the 11th Int. Symp. on Electrets* (Melbourne: Australia)
[3] Qiu X, Xia Z and Wang F 2007 P *Front. Mater. Sci.* **1** 72
[4] Sodano R A, Sodano A 2007 *Smart Mater. Struct.* **16** 1
[5] Gerhard-Multhaupt R 2002 *IEEE Proc. ISE* **11** 36
[6] Broadhurst M B, Malmberg C C, Mopsik F I and Harris W P 1973 *Electrochem. Soc. New Jersey* 492
[7] Jiashi Yang 2009 *Specal Topics in the Theory of Piezoelectricity* (Media Springer)
[8] Kacprzyk R, Motyl E, Gajewski J and Pasternak A 1995 *J. Electrostat.* **35** 161
[9] Kacprzyk R 2000 *J. Electrostat.* **56** 111
[10] Kacprzyk R, Lewandowski M 2009 *Fibres Text. East. Eur* **17** 71

7th International Conference on Applied Electrostatics (ICAES-2012) IOP Publishing
Journal of Physics: Conference Series **418** (2013) 012061 doi:10.1088/1742-6596/418/1/012061

Design of high voltage DC power supply based on LCC resonant converter

Z Q Wang*, Z G Liu, J J Wang, G F Li
School of Electrical Engineering, Dalian University of Technology, Dalian, China

E-mail: s200185004@mail.dlut.edu.cn

Abstract. The aim of this paper is to design a small size, light weight high frequency high voltage (HFHV) power supply. It presents a comprehensive procedure for designing a high output voltage power supply based on series-parallel (LCC) resonant converter, aiming to realize the soft-switching. Through mathematical calculation based on an extensive of the first harmonic analysis, the paper derives the approach of determining the resonant parameters of the LCC converter. Then, a 35 kV power supply featuring a series-parallel resonant converter topology to compensate the distributed parameter is built to verify the correctness of the theory.

1. Introduction

Nowadays, high voltage DC power supply is widely used in electrostatic precipitation and medical diagnostic. Most of the high voltage power supplies are made up of power frequency step-up transformer directly. Although it has the advantages of simple fabrication process and low failure rate, it cannot meet the portable requirement.

This paper introduces a high voltage DC power supply featured with higher power and efficiency and less volume. As key parts of high voltage power supply, the size of the transformer should be as small as possible, but the cooling area will be reduced as the volume decreases, thus the losses of the HFHV transformer becomes more significant [1-2]. Moreover, the leakage inductance and the distributed capacitance of the transformer will cause high peak voltage and large charging current, increasing heat, inevitably affecting the normal operation. In this paper, LCC resonant converter is introduced to take advantage of the distributed parameters of the transformer to realize soft-switching within a wide input and load range.

2. Overview of the system

The block diagram of the system is shown in figure 1. It mainly consists of two parts: the power circuit and the control circuit. The power circuit contains a HFHV transformer, a full-wave rectifier, a switching unit, and a low pass filter. The shaded part is the control circuit containing the microcontroller mainly for creating bipolar PWM control signals.

The input for the system is 220 V/50Hz. After rectifier and inverter, a high frequency voltage is increased by HFHV step-up transformer. The high voltage rectifier and the high voltage capacitor are used to reduce ripple or noise of the DC voltage. The specifications of the HFHV power supply are as follows:

Input voltage: 220 V ± 20%.

* To whom any correspondence should be addressed.

Maximum working voltage: 35 kV.
Maximum operating current: 100 mA.

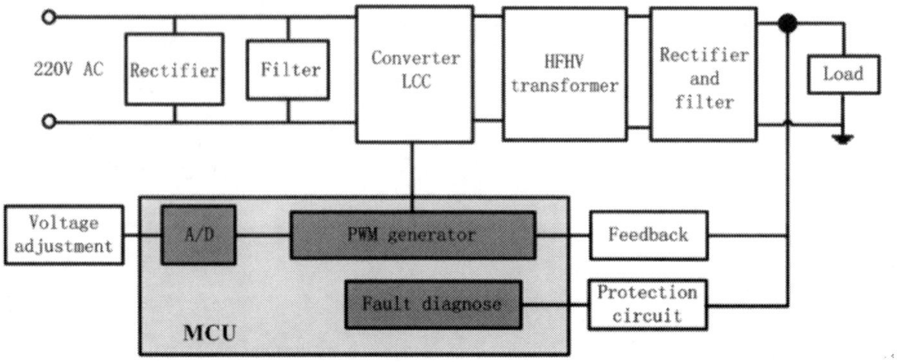

Figure 1. The block diagram of the whole system.

3. Design of the LCC resonant converter

As known to all, the distributed parameter can greatly increase the loss of HFHV transformer, even damaging the switch components [3-4]. Thus how to decrease the adverse impact of the distributed parameters is a key point in designing the power supply. The most widely used and effective method is LCC resonant technique [5-6]. In order to facilitate analysis, the full-bridge LCC resonant converter with capacitance-type filter is given in figure 2.

In figure 2, C_s and L_s are series resonant capacitor and inductor, C_p' is the parallel resonant capacitor; L_t is the leakage inductance while C_t is distribution capacitance including the equivalent distributed capacitance referred to the primary side. In order to achieve zero-voltage switching (ZVS), the circuit is operating above resonance so that the current delivered to the resonant circuit is lagging the voltage applied to the resonant circuit. Each leg has a snubber capacitor directly paralleled across the IGBT [7].

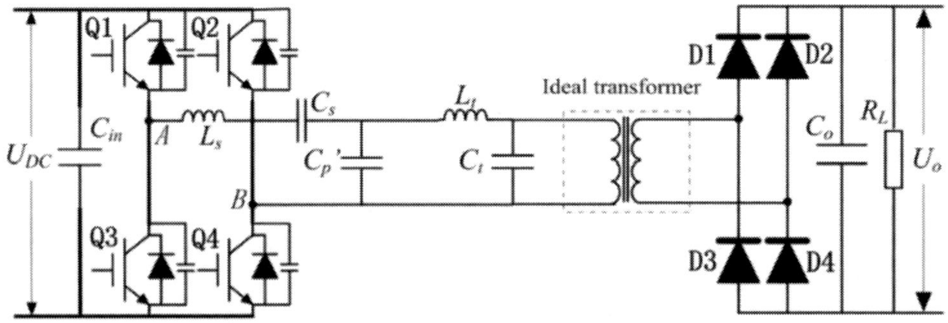

Figure 2. LCC resonant converter with capacitive output filter.

The fundamental voltage of AB in figure 2 is determined by equation (1), where D is the duty cycle:

$$U_{AB1} = \frac{4}{\pi} \cdot U_{DC} \cdot \sin(\frac{D \cdot \pi}{2}) \tag{1}$$

In which, U_{DC} is the DC input voltage of the inverter, D is the duty cycle.

After the Fourier transform of the transformer's primary voltage and current, the transformer, rectifier, output filter and load can be equivalent to a capacitive two-Port network which is represented by a RC parallel circuit as shown in figure 3[8].

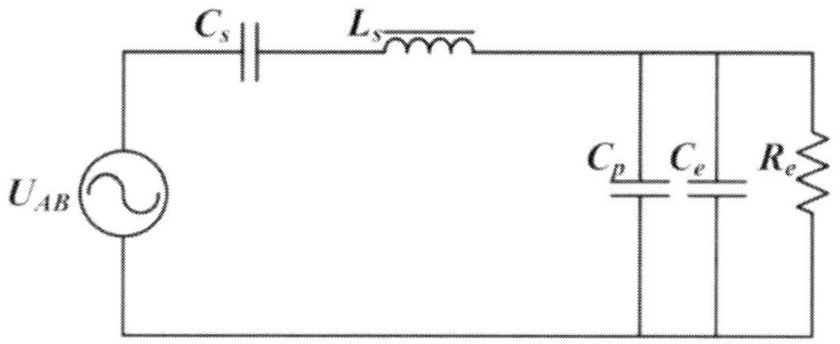

Figure 3. AC equivalent circuit of LCC resonant converter.

Considering the effect of duty cycle, using the extensive of first harmonic analysis, we can get the voltage gain M:

$$M = \frac{U_O}{U_{in}} = \frac{4}{\pi} \cdot \frac{k_{21}}{k_v} \cdot \sin(D \cdot \frac{\pi}{2}) \cdot n \tag{2}$$

In which, n is the turns ratio of the transformer, k_v is a coefficient describing the relationship between output voltage and input voltage, its value is determined as follow:

$$k_v = 1 + 0.27 \cdot \sin\left(\frac{\theta}{2}\right) \tag{3}$$

θ is the normalized value of the conducting angle of the rectifiers, its value is determined by:

$$\begin{cases} \theta = 4\arctan\left(\dfrac{2 \cdot \pi \cdot Q \cdot n^2}{f_{sN} \cdot \alpha}\right)^{1/2} \\ Q = \left(L_s / C_s\right)^{1/2} / R_e \end{cases} \tag{4}$$

$f_{sN} = f_s / f_o$ is the normalized value for the switching frequency, f_o is the resonant frequency of the C_s and L_s, f_s is the switching frequency.

$$k_{21} = \left([1 - \alpha(f_{sN} - 1)(1 + \frac{\tan(|\beta|)}{\omega C_p R_e})]^2 + [\alpha(f_{sN}^2 - 1) \cdot \frac{1}{\omega C_p R_e}]^2 \right)^{-1/2} \tag{5}$$

In which, β=-0.4363sin(θ) is phase difference between fundamental voltage and fundamental current of transformer. α is determined as C_p/C_s and $\omega C_p R_e = \pi k_v^2 / [4\tan(\theta/2)^2]$. Therefore, the relationship between D and other parameters can be obtained:

$$D = 1 - \frac{2}{\pi} \cdot \arctan(\frac{1}{\omega C_p R_e} \cdot \alpha \cdot \left\{ f_{sN}^2 \cdot [1 + (\omega C_p R_e + \tan(|\beta|)^2)] - 1 \right\} - [\omega C_p R_e + \tan\left(|\beta|\right)] \cdot [1 + \alpha \cdot (1 + \frac{\tan(|\beta|)}{\omega C_p R_e})]) \tag{6}$$

According to equation (2)-(6), the relationship between voltage gain and switching frequency under different load (according to equation (4), Q is proportional to R_e) is shown in figure 4 (taking α=2). By adjusting the parameters of the resonant component properly, appropriate gain and the switching frequency range can be obtained. Figure 4 provides the basis curve for the selection of switching frequency under different loads.

Based on the characteristics of figure 4, the adjustment range of f_{sN} is large when the load changes from light-load to full-load. On the other hand, the resonant frequency f_o is fixed with the determined resonant components. Hence, the range of switching frequency f_s is very large which is difficult to achieve due to the design of resonant components.

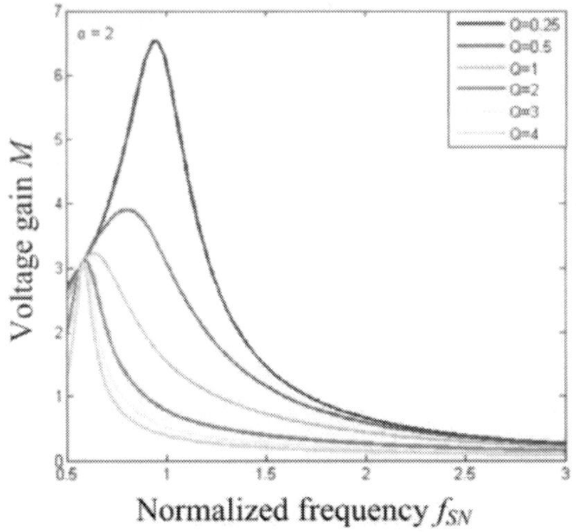

Figure 4. Relationship between voltage gain M and f_{sN} under different load.

Normally, the load changes from full-load to light-load. Therefore, in this paper, phase-shifted control is also adopted to adjust duty cycle at the same time. In this way, the LCC converter can achieve soft-switching over the entire load range with a small adjustment of the switching frequency. Taking the distribution parameters of the transformer into the consideration, an equivalent simplified model of LCC is described in figure 5[9].

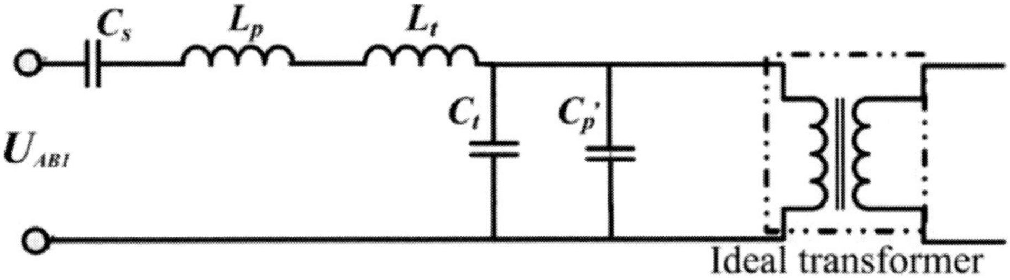

Figure 5. Simplified model of LCC considering the distribution parameters.

Therefore, the actual values of LCC resonant parameters need to be modified as follows:

$$\begin{cases} L_p = L_s - L_t \\ C_p' = C_p - C_t \end{cases} \tag{7}$$

As known to all, transformer design is another key point. The volume of the transformer is small so the effective insulation distance is limited. Therefore, we design a special insulation is adopted in this paper. The distribution parameters of the HFHV transformer are L=75.4 uH and C=44.23 nF.

When we choose Q=3, α=2 and the working frequency is 20 kHz, according to equation (4), the calculated results are L_s=186.7 uH and C_s=307.7 nF. Combining equation (7), we get C_p=153.85 nF, C_p'=19.62 nF and L_p=111.3 uH.

4. Experiment and results

Based on the parameters above, the voltage of the series resonant capacitor and the current of the resonant inductor at full load are illustrated in figure 6. As shown, the series resonant inductor current i_{Ls} and resonant capacitor voltage u_{Cs} are sinusoidal, so the switch can achieve ZVS and the switching stress of the IGBTs is reduced. The resonant inductor current leads the voltage across the resonant capacitor, thus the converter is inductive. The output voltage waveform is shown in figure 7. According to figure 7, we can find that the output DC voltage reaches 35 kV and has low output ripple and noise.

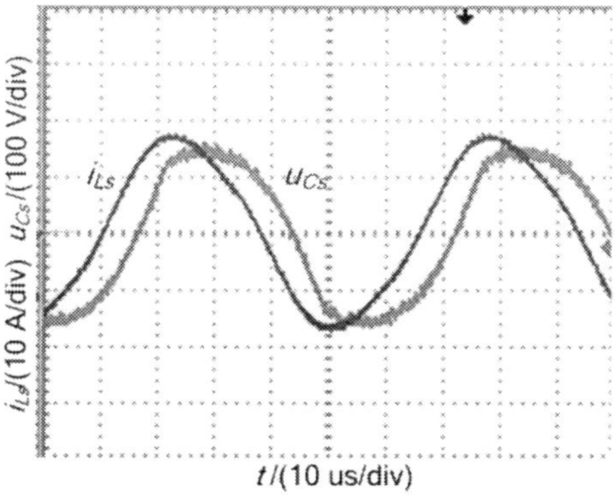

Figure 6. Experimental waveforms at full loads when Uo=35 kV.

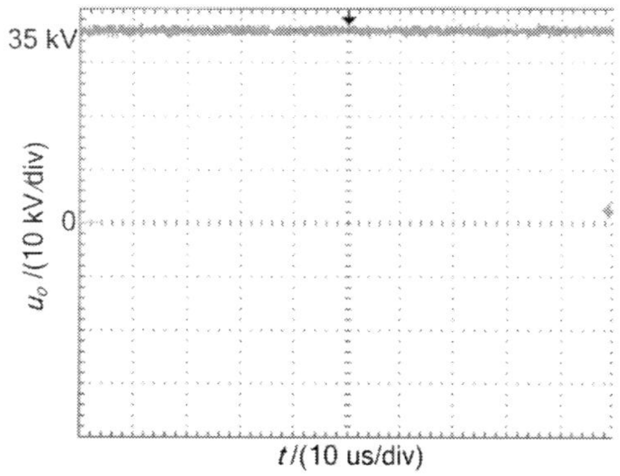

Figure 7. Output DC voltage waveform.

5. Conclusion

This paper gives the specific procedure of determining parameters of high voltage DC power supply based on the LCC resonant converter. Considering the influence of transformer distributed parameters, the extended fundamental approximation is adopted to ensure soft switching. The switching stress of the IGBTs is reduced, improving the efficiency of the converter. Compared with the traditional method, the proposed method is simple and more practical. The paper verified the correctness of the theory and design process through a 35 kV high voltage DC power supply. The efficiency of the DC power supply reached 95%, and the mass of the power supply is about 6 kg.

Acknowledgments

This work is partially supported by the Fundamental Research Funds for the Central Universities [grant NO. DUT11RC(3)08]. The authors also gratefully acknowledge the helpful comments and suggestions of the reviewers, which have improved the presentation.

Reference

[1] F da S Cavalcante and J W Kolar 2003 *Proc. PESC (Acupulco)* vol 4 (New Jersey: IEEE) p 1807
[2] Ivo Barbi and Roger Gules I 2003 *IEEE Trans. Power Electron.* **18** 975
[3] Xia Bing, Ruan Xinbo and Chen Wu 2009 *Trans. China Electrotech. Soc.* **24** 60
[4] Wang Yanan, Xiao Dengming, Li Youhuai, Gu Jin and Chen Guiwen 2009 *High Voltage Engine.* **25** 1049
[5] Mehrdad Jafarboland 2011 *J. Appl. Power Electron.* **11** 622
[6] Bhat A K S 1993 *IEEE Trans. Power Electron.* **8** 1
[7] Johnson S D, Witulski A F and Erickson RW 1988 *IEEE Trans. Aero. Elec. Sys.* **24** 263
[8] Ivensky G, Kats A and Ben Yaakov S 1997 *Proc. PESC (California)* vol 2 (New Jersey: IEEE) p 958
[9] Piazza G L, Alves R L, Font C, and Barbi I 2009 *Power Electron. Con. (Bonito MS)* (New Jersey: IEEE Computer Society) p 326

Journal of Physics: Conference Series **418** (2013) 012062 doi:10.1088/1742-6596/418/1/012062

Trajectory calculation of a trapped particle in electro-dynamic balance for study of chemical reaction of aerosol particles

Miho Okuma[1], Takahiro Itou[1], Azuchi Harano[1*], Takayuki Takarada[1] and Davis E. James[2]
[1] Department of Chemical and Environmental Engineering, Graduate School of Engineering, Gunma University, 1-5-1 Tenjin-cho, Kiryu, 376-8515, Japan
[2] Department of Chemical Engineering, University of Washington, Seattle, WA 98195-1750, USA

E-mail: azuchi@gunma-u.ac.jp

Abstract. Electrodynamic balance (EDB) is a powerful tool for investigating the chemical reactions between a fine particle and gaseous species. But the EDB device alone is inadequate to match the rapid weight change of a fine particle caused by chemical reactions, because it takes a few seconds to set a fine particle at null point. The particle trajectory calculation for the trapped particle added to the EDB is thus a very useful tool for the measurement of the transient response of a particle weight change with no need to adjust the applied DC voltage to set the null point. The purpose of this study is to develop the trajectory calculation method to track the particle oscillation pattern in the EDB and examine the possibility for kinetic studies on the reaction of a single aerosol particle with gaseous species. The results demonstrated the feasibility of applying particle trajectory calculation to realize the research purpose.

1. Introduction

The importance of reactions of a fine particle with gaseous species has been growing in many fields of atmospheric chemistry, combustion, pollution control and material processing during the past decades. The property of the very high surface-to-volume ratios of fine particles is making them readily serve as centres for heterogeneous reactions. Main approaches for heterogeneous reaction of aerosol particles are based on thermo-gravimetric analysis (TGA), drop tube apparatus and fluidized bed reactors. These methods are well developed and provide data over a wide range of experimental conditions. Electrodynamic balance (EDB) is also one of the well established techniques for studying the chemical reactions of a single aerosol particle [1]. EDB is an experimental device for levitating the charged particle in space using electric potential in which DC electric field was used to balance the gravitational force on a charged particle. Hence it is considered to be non-contact type microbalance equipment with great precision for convenience.

The advantages of reaction experiments using EDB device over the conventional approaches of gravimetric measurement using bulk solutions [2] are as follows: 1) the weight change in a particle reactions can be monitored continuously, thus providing unambiguous in situ characterization of them.

* To whom any correspondence should be addressed.

2) The chemical reactions of fine particles can be investigated under well-defined surrounding conditions. 3) A single levitated particle can be investigated over long time periods to have a quantitative understanding of the reaction at an equilibrium state. 4) Supersaturated droplets can be studied since heterogeneous nucleation is suppressed when a droplet is levitated without contacting any foreign surface. 5) The contribution of mass transfer to overall reaction kinetics can be easily estimated and the time scale of mass transfer inside and outside of droplet can be sufficiently short.

However, EDB device is disadvantageous in that it is inadequate for investigation on the relative fast chemical reactions of gaseous species because it takes several seconds to adjust the particle at null point, where the dc field is balanced the gravitational force on a charged particle, by changing the applied DC voltage. Therefore this DC voltage adjustment cannot follow the rapid increase in the particle weight by the chemical reaction (of order few sec).

The AC electric field imposes no net force on a particle when it is stabilized at the null-point, but in the neighbourhood of the null-point, there are the axial components of the AC field which are proportional to the distance from the null-point. Thus, if any particle is positioned away from the null-point, it oscillates by time-dependent forces in the AC field. This oscillation makes it possible to determine the change of the particle weight by following its trajectory.

Maloney et al. [3] generated a step change of DC voltage to obtain the resulting transient response of a particle from an initial steady state using a two dimensional diode array which provides particle position along the EDB centres axis. Then they determined the particle weight by adjusting iteratively to find agreement between measured trajectories and calculated results by a particle dynamic model (PDM), where PDM was a force balance simulation which consisted of electric field forces, gravitational forces, and drag forces acting on a particle. As the gravity is not compensated completely, the resulting oscillatory trajectory is phase lagged with respect to the driving AC field of the EDB, and the phase lag depends only on the drag parameter, particle weight and the AC frequency. Göbel et al. [4] investigated the use of phase lag measurements to size droplets in a four-ring EDB, reporting an uncertainty of less than 6% in the size determination.

Zheng et al. [5] compared four methods for determining the size of particles based on particle oscillation in EDB. The examined methods were: 1) spring point measurements, which corresponds to a sudden large-amplitude oscillation of a levitated particle when the AC electric field is changed to cause the particle to move from a stable to an unstable region based on stability theory for the differential equation of particle motion in EDB. 2) phase lag measurements mentioned above, 3) measurements of the amplitude of the oscillation of an unbalanced particle, and 4) the particle shift from null point as it oscillates due to an unbalance vertical forces. All methods were shown to be in good agreement for spheres (within 3.9%). For non-spherical germanium dioxide particles, the three oscillation methods agreed with the spring point method within 3.4%.

The particle oscillation methods are very useful for a weight measurement of transient response of a particle because adjustment of applied DC voltage for maintaining it at null point in EDB is not necessary. This article describes a development of the gravimetric analysis of the fast heterogeneous reaction of a particle from its oscillation in EDB device. A CCD video camera was used to measure the amplitude of the particle oscillations and numerical calculation of the differential equation of particle motion was performed by adjusting the change of a particle weight to be close agreement with the oscillation patterns of the particle into EDB.

2. Trajectory Calculation Method

The dynamics of the EDB make it possible to determine the particle weight by following particle trajectories. There are two issues that need to be considered when levitated particle is examined. The first is the calculation of an oscillating motion of particle in EDB, and the second concerns the conditions that need to maintain a particle in a stationary state at the null point of the balance [1]. To answer these issues the electric fields inside EDB that affect the particle motions must be determined first and then the differential equation of particle motion must be solved. Assuming that there are no other vertical forces on the particle, a vertical force balance on the levitated particle at null point yields

$$mg = qE_{DC} \tag{1}$$

where m is the particle's weight (kg), g is gravitational acceleration ($m \cdot s^{-2}$), q is charge on the particle (C) and E_{DC} is the DC electric field required to levitate the particle at null point without oscillation ($V \cdot m^{-1}$). For constant charge q, the voltage required to levitate the particle is proportional to its weight. This is the basis for using the device as an analytical balance for micro particles.

When gravity is not compensated completely by the DC electric field due to the particle weight change, a levitated particle starts to oscillate in the vertical direction. As there is no particle motion in the horizontal direction, it is sufficient to know the vertical component of the equation of particle motion. Therefore the dynamics of an oscillating particle trapped in an EDB is governed by the drag force, gravity force, and electrostatic force. The force balance yields

$$m\frac{d^2z}{dt^2} = -mg - 3\pi\eta d_p \frac{dz}{dt} + qE(t,z) \tag{2}$$

where z is the position of the particle with respect to the centre of the balance (m), m is the weight of single aerosol particle (kg), η is gas viscosity ($kg \cdot (m \cdot s)^{-1}$), d_p is diameter of single aerosol particle (m), and $E(t,z)$ is vertical electric field strength at the position of a particle ($V \cdot m^{-1}$) [6]. The aerodynamic drag force was defined for sphere in the Stokes' regime in second term of right side in equation (2). Stokes' law for a sphere particle is valid up to Reynolds number (*Re*) of about 0.1. In our experiments, the Re reached a maximum value of 0.17 for a brief period but was well within the Stokes' regime for most of the particle oscillation. Electric field represents the sum of applied AC and DC electric fields ($V \cdot m^{-1}$) as follows:

$$E(t,z) = E_{AC}(t,z) + E_{DC}(z) \tag{3}$$

in which AC electric field $E_{AC}(t,z)$ is a function of time because of applying the AC voltage.

The main problem is the computation of the electric potential distribution inside EDB, which was necessary for the calculations of electric field (E_{DC}(z)and $E_{AC}(t,z)$) strength alone the central z-axis. For Cartesian coordinate system, the electric potential is determined from the solution of Laplace equation (4) together with appropriate boundary condition.

$$\nabla^2 V = \frac{\partial^2 V}{\partial x^2} + \frac{\partial^2 V}{\partial y^2} + \frac{\partial^2 V}{\partial z^2} = 0 \tag{4}$$

where V is electric potential (V) and x, y, z are Cartesian coordinates (m). The cross section geometry of the double cylindrical tube type EDB used in our experiment was showed in figure 1. This EDB consisted of three stainless cylinders with a coaxial line. An AC high voltage applied to outer cylindrical electrode for stabilization of a particle at centre point (null point) into EDB device, while a DC voltage difference between the inner cylindrical electrodes controls the vertical position of the particle to compensate gravity force on the particle. This vertical z-axis defined the direction of gravity. For cylindrical coordinate with axial symmetry, equation (3) can be easily transform to

$$\frac{\partial}{\partial r}\left(r\frac{\partial V(r,z)}{\partial r}\right) + r\frac{\partial^2 V(r,z)}{\partial z^2} = 0 \tag{5}$$

where r is radial coordinate [m]. Boundary conditions of voltage inside EDB were:

$$\frac{\partial V(0,z)}{\partial r} = 0 \quad \text{(bounded)} \tag{6}$$

$$\frac{\partial V(r,Z_s)}{\partial z} = \frac{\partial V(r,-Z_s)}{\partial z} = 0 \text{ (double cylindrical tubes)} \quad (7)$$

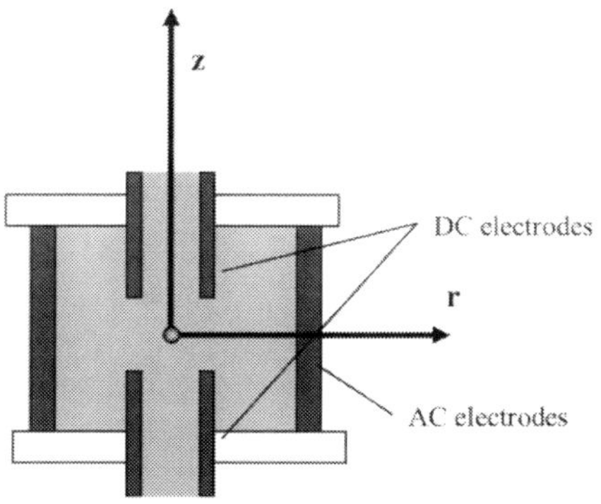

Figure 1. A cross section of the double cylindrical EDB.

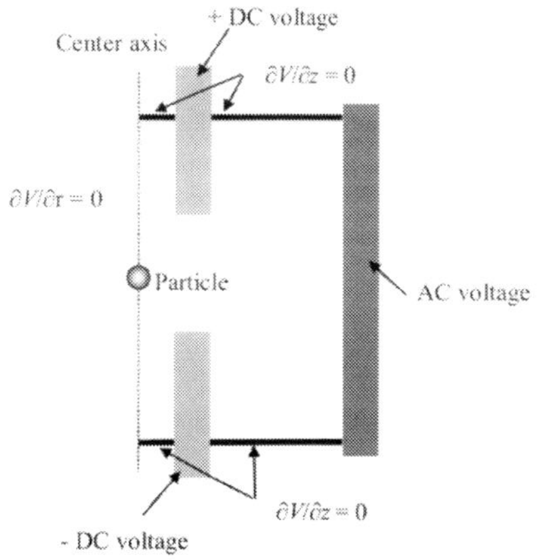

Figure 2. Boundary conditions for electric field calculation inside EDB.

And the potentials are constants over the AC and DC electrode surfaces as shown in figure 2. The boundary value problem can be solved using the Matlab PDE toolbox (The Mathworks Inc.) with finite element method. A contour plot of the equi-potential lines in the EDB as a function of r and z was shown in figure 3, where the origin is at the centre of the balance (null point) and AC peak to peak voltage is 300 V and DC is ±20 V. Grey two rectangles and orange one correspond to the double cylindrical DC electrodes and AC electrode, respectively. It is found that AC potential reached to the centre axis through the gap between the inner cylindrical DC electrodes. In order to calculate the dynamics of a z-axis oscillating particle trapped in an EDB, the only electric potential alone the centre z-axis was necessary.

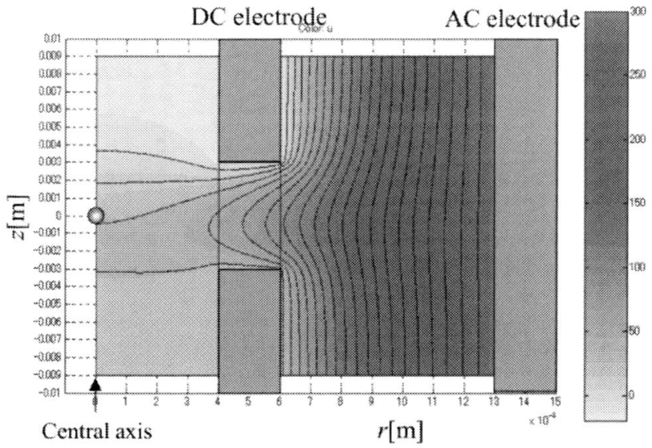

Figure 3. A counter plot of equi-potential lines inside EDB

($-V_{DC}$: −20V, V_{DC}: 20V, $V_{AC\ pk\text{-}pk}$: 300V).

Figure 4 (a) and (b) show the AC and DC electric potential curves alone the centre z-axis respectively, which are extracted from two dimensionally calculated results for potentials of AC peak to peak 300V and DC ±20 V.

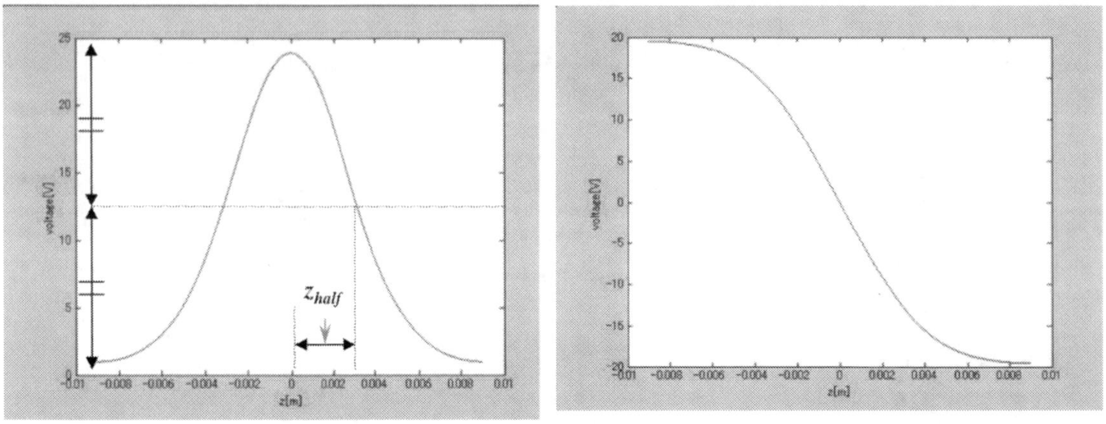

(a) $V_{AC\ \mathrm{peak\ to\ peak}}$ (b) ±DC voltage: 20 V

Figure 4. Axial potential curves calculated by MATLAB PDE Tool.

As particle trajectory calculation requires a huge number of potential calculations, the approximation equations for a time function $V_{AC}(t,z)$ and $V_{DC}(z)$ were derived respectively. A vertical AC electric potential was expressed as Gaussian's equation (8).

$$V_{AC}(t,z) = 0.0797 \times V_{AC,pk-pk} \sin(2\pi f t) \cdot \exp\left(-\frac{z^2}{(1.2012 z_{half})^2} \right) \quad (8)$$

where $V_{AC,pk-pk}$ is the applied AC peak to peak voltage (V), f is the frequency of AC voltage (Hz) and z_{half} is the half value width (m) of the Gaussian distribution curve as shown in figure 4 (a). 0.0797 is a geometric constant which is a function of the specific EDB electrode geometry. The approximation of DC potential curve in figure 4 (b) took the polynomial equation as,

$$V_{DC}(z) = V_{DC} \times (3.0 \times 10^6 z^3 + 4.25 z^2 - 2.4 \times 10^2 z + 1.42 \times 10^{-5}) \quad (9)$$

where V_{DC} is the applied DC voltage to electrodes (V) and z is the distance (m) from null point along with the centre axis of the balance. Therefore the vertical component of electric field is given by the following equation (10).

$$E(t,z) = E_{AC}(t,z) + E_{DC}(z) = -\nabla(V_{AC}(t,z) + V_{DC}(z)) = \frac{dV_{AC}(t,z)}{dz} + \frac{V_{DC}(z)}{dz} \quad (10)$$

Figure 5 shows one cycle of the calculated time variation of electric field along z-axis from equation (10). Electric field changes alternately across the null point with time.

Over a small distance from the null point in EDB in figure 4 (b), the DC electric fields can be approximated by

$$E_{DC} = C_0 \frac{V_{DC}}{Z_0} \quad (11)$$

where C_0 is a geometric constant which is a function of the specific EDB electrode geometry. C_0 for EDB employed in the present study was determined a value of 0.72 by comparing calculated results in figure 4 (b) with equation (11). Numerically, this value is slightly smaller than either that for the flat electrode configuration (C_0=1) or for the bihyperboloidal system ($C_0 \approx 0.8$) [1, 7]. In our EDB a higher

DC voltage must be applied to the electrodes to achieve the same DC field strength as that of bihyperboloidal EDB.

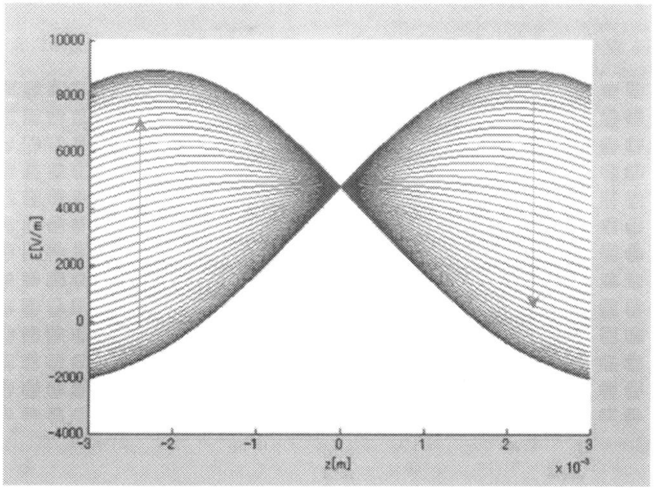

Figure 5. The calculated time variation of electric field along z-axis.

3. Experimental Setup

The coaxial double cylindrical type EDB was used as shown in figure 1. An AC high voltage (Trek: 610C) applied to outer cylindrical electrode for stabilization of a particle at centre point (null point) into EDB device, while a DC voltage difference between the inner cylindrical electrodes controls the vertical position of the particle to compensate gravity force on the particle. The DC voltage was supplied up to 500 V and the typical amplitude of AC field used to drive EDB is about 9-10 kV within the frequency range from 50 Hz to 500 Hz.

The EDB device was placed inside a stainless chamber with equipped with the flexible silicone rubber heaters that can maintain the chamber temperature at about 300 K. A microscope focused on the centre of the EDB was used for visual observation of particle trajectories and their morphologies. Stroboscope was mounted at the opposite side for illuminating the necessary part. A CCD video camera (Sony DCR-TRV70) was equipped with microscope to measure the amplitude of the particle oscillations.

4. Comparison of Particle Trajectory Calculations with Experimental Results

Once the calculation method was determined for the electric field, the trajectory of the particle could be easily obtained by solving the ordinary differential equation (2). The calculation procedure was shown in figure 6. First we calculated the electric field in EDB by the given initial condition $V_{AC,pk\text{-}pk}$, V_{DC} and frequency. Next we informed the particle initial condition such as mass, charge, particle diameter and the initial position and velocity of the particle and ran the program for solving the equation of motion. Finally the calculated trajectories of a particle were displayed on the computer. The Runge-Kutta algorithm for Matlab and Euler algorithm for LabView (National Instruments) were used to calculate numerical solutions. Because the program using Labview is a real time simulator of particle motion, the Euler method was suitable for iteration calculations in order to determination of particle weight by adjusting the fitting parameter to agree between the calculated particle trajectory and the observed amplitude of oscillating particle. These methods enable us to simulate particle trajectories along z-axis for arbitrary initial particle positions and velocities without any approximation.

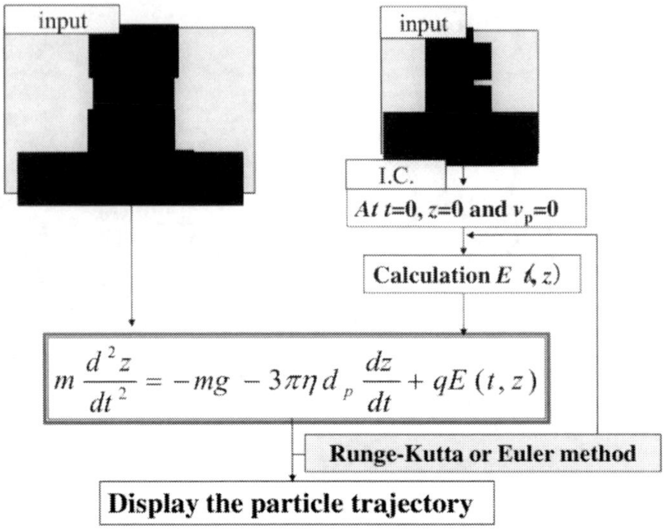

Figure 6. Calculation procedure for solving the equation of particle motion.

The EDB levitation experiments were done for a soda lime-silicate glass particle with a 60 μm diameter (GBL-60, JIS Z 8901) of known density (ρ=2.3 g·cm^{-3}). The particle diameter was measured using a high resolution CCD video camera and the particle charge value was determined from equation (1) and the measured value of DC voltage at null point. In order to check the availability of our particle trajectory calculations, the particle oscillation amplitudes were acquired on the video camera for a step change of DC voltage from that at null point. This caused the particle to move to unbalanced condition where it oscillated with the frequency of the AC electric field. Calculated particle trajectory and oscillating particle photograph, which were obtained by changing from DC 29 V at null point to 19.1 V suddenly, are shown in figure 7. As gravity is not compensated completely by the change of the DC electric field, a levitated particle started to be oscillates in the vertical direction. With increasing distance from the null point, the oscillation amplitude increases linearly due to the z dependence of $E_{AC}(t,z)$. As the Ar ion laser irradiated downward on the transparent particle at null point, two bright glare points can be recognized on top and bottom of the particle. One of them is due to reflected light, the other one is due to refracted light. These points seem to be the bright straight line in the case of oscillating particle as shown on photograph in figure 7 because the particle moves back and forth for several times on one video frame. Therefore the oscillation amplitude could be obtained from distance of this line.

Although the oscillation pattern on photograph agrees with the general trend of the particle trajectory calculation, there are quantitative differences between calculated and experimental results. Oscillation pattern is characterized by two parameter, offset and amplitude. There were some discrepancies of offset and amplitude between calculated results and experimental data. This was considered to be existence of 4 holes drilled through the AC electrodes for optical and observation ports or unreasonable boundary conditions between DC and AC electrodes for the electric potential calculation using Matlab PDE tool.

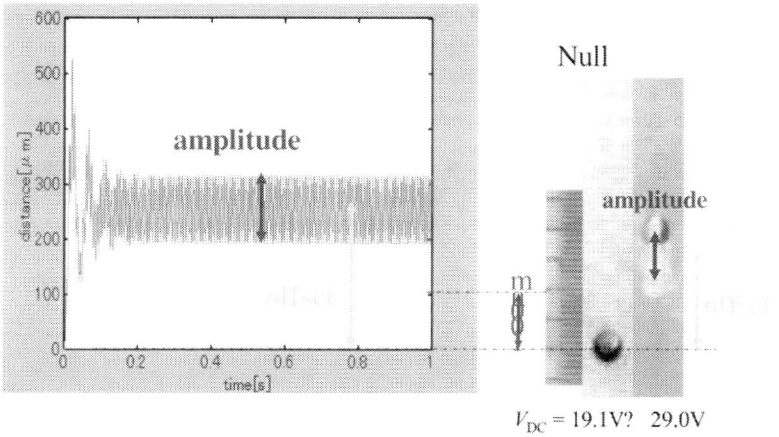

Figure 7. Comparison of the trajectory calculation with the experimental result of the glass particle with 60 μ m when the DC voltage changed from 19.1 V to 29.0 suddenly.

In order to match the particle trajectory with the particle oscillation amplitude, AC electric potential was modified by introducing a parameter C_1 as follows

$$V_{AC}(t,z) = 0.0797C_1 \times V_{AC,pk-pk} \sin(2\pi ft) \cdot \exp\left(-\frac{z^2}{(1.2012z_{half})^2}\right) \tag{12}$$

Determination of correction factor C_1 was performed by the comparison of amplitude on oscillating the glass particles with trajectory calculations by adjusting the AC electric field. The correction factor C_1 was obtained as follows and the modified electric field of DC voltage (blue line) was calculated as shown in figure 8.

$$C_1 = \left|15.28z^{0.5}\right| + 1 \tag{13}$$

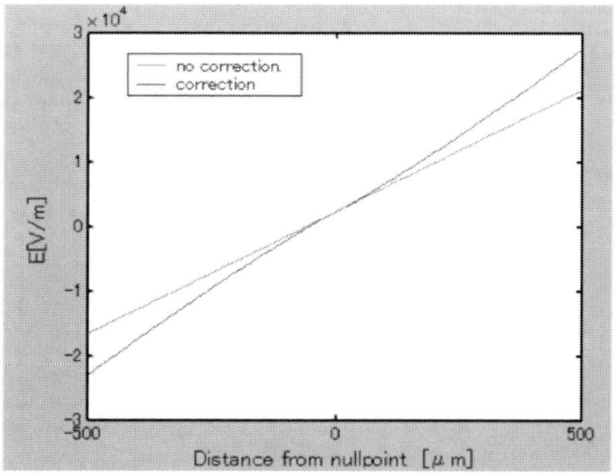

Figure 8. The modified DC electric field alone the centre z-axis.

Journal of Physics: Conference Series **418** (2013) 012062 doi:10.1088/1742-6596/418/1/012062

A constant gas flow of nitrogen was provided vertically upwards through the lower DC electrode tube. Therefore it is necessary to take the existence of drag force against the levitated particle into consideration. A vertical force balance on the levitated particle at null point yields

$$mg = \frac{qC_0V_{DC}}{z_0} + 3\pi\eta d_p u_g \tag{14}$$

The flow velocity u_g (m·s^{-1}) of surrounding gas at the null point in EDB was obtained from the DC voltage difference in the presence or absence of nitrogen gas flow (equations (11) and (14)). However, there is some distribution of flow velocity u_g in space for the oscillating particle and it affects the oscillating particle trajectories. The usual boundary layer theory was adapted for calculation of the velocity distribution equation for circular free jet [8] , and it is defined as

$$u_g(r,z) = \frac{3K'}{8\pi\nu z(1+\xi^2/4)} \tag{15}$$

In this equation, z is a distance from the exit of cylindrical tube which is used as DC electrode (m), ν is kinetic viscosity of gas (m^2·s^{-1}), K' is kinetic momentum (m·s^{-2}) described as

$$K' = \frac{16(Q/d_0)^2}{3\pi} \tag{16}$$

where Q is total flow rate (m^3·s^{-1}) and d_0 is inner diameter of cylindrical tube(m) and ξ(-) is defined as

$$\xi = \frac{r}{4z\nu}\sqrt{\frac{3K'}{\pi}} \tag{17}$$

The velocity distribution along the central axis of EDB device can be calculated from

$$u_g(0,z) = C_2\frac{2(Q/d_0)^2}{\nu z} \tag{18}$$

because of ξ=0 at r=0. C_2 is the adjustment parameter in order to agree with measured velocity at null point. Figure 9 represents the velocity distribution when the flow rate was 50 ml·min^{-1}.

Considering the gas flow, the equation (2) of particle motion was rewritten as follows

$$m\frac{d^2z}{dt^2} = -mg - 3\pi\eta d_p\left(\frac{dz}{dt} - u_g(0,z)\right) + qE(t,z) \tag{19}$$

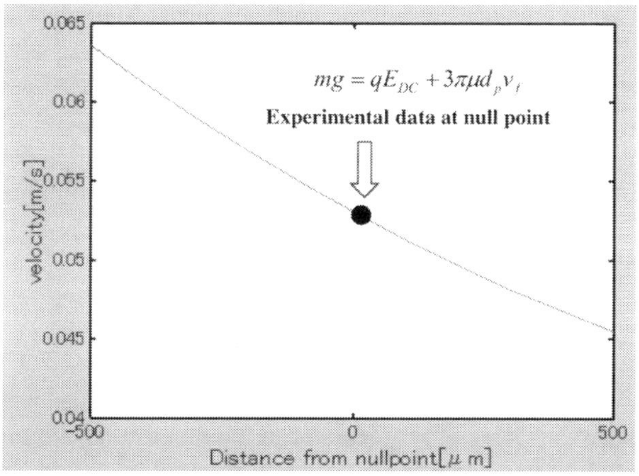

Figure 9. Velocity distribution of flow gas alone the centre z-axis.

Calculation results from equation (19) and photographs of oscillating glass particle are shown in figure 10 in order to verify the validity of trajectory calculation. The calculated result (Green line) with correction for gas flow and electric field was in well agreement with offset and amplitude of the oscillating particle.

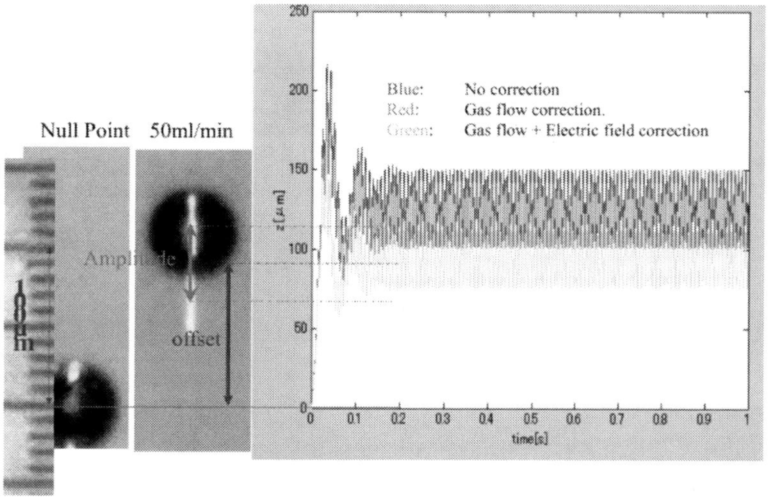

Figure 10. Comparison between the trajectory calculations with correction for gas flow and/or electric field and the photographs of oscillating glass particle.

In this trajectory calculation, it is important that it takes relaxation time about 0.2 sec (about 20 cycles at 100 Hz) to be stable particle oscillation. This fact indicats that the oscillation amplitude measurement based on commercial video camera could not follow the fast weight change which occurred in less than 0.2 sec. However, our particle trajectory calculations could be used in determination of a weight change of the particle in a few seconds. Then we examined the availability of this trajectory calculation for following the evaporation of a water droplet as an example of the quick weight change of a particle.

The diameter change of a water droplet in a steady state vapour field is given by [9]

$$\frac{d(d_p)}{dt} = \frac{2ShD_vM}{R\rho_p d_p}\left(\frac{p_\infty}{T_\infty} - \frac{P_d}{T_d}\right) \tag{20}$$

where d_p is the droplet radius (m), D_v is the gas phase diffusion coefficient ($m^2 \cdot s^{-1}$), M is the molar mass of water ($kg \cdot mol^{-1}$), R is the gas constant ($J \cdot (K \cdot mol)^{-1}$), ρ_p is the density of water ($kg \cdot m^{-3}$), p_∞ is the ambient partial pressure of H_2O (Pa), p_d is the its equilibrium vapour pressure of H_2O (Pa), T_∞ is the ambient temperature (K), T_d is the droplet temperature (K) and Sherwood number (Sh) is the non-dimensional parameters which are expressed as

$$Sh = 2.0 + 0.6Re^{1/2}Sc^{1/3} \tag{21}$$

where Sc is Schmidt number (-) which is defined as the ratio of kinetic viscosity and mass diffusivity, Assuming that the enthalpy of water evaporation is in balanced with the thermal energy conducted to the droplet by the surrounding gas, the droplet temperature may be expressed in the form of

$$T_\infty - T_d = \frac{L_e D_v\left(\frac{P_d}{T_d} - \frac{P_\infty}{T_\infty}\right)}{k_g R} \tag{22}$$

where L_e is the specific enthalpy of water evaporation ($J \cdot mol^{-1}$) and k_g is the thermal conductivity of the surrounding gas ($W \cdot (m \cdot K)^{-1}$). MATLAB program was developed to solve the coupled system of equations (21) and (22). Simulation of the time-dependent droplet weight calculated for specific experimental conditions in humidity of 0 %, temperature of 303 K and initial droplet diameter of 79μm was plotted with the droplet trajectory calculation accompanying simultaneously with its evaporation, and the plot was shown in figure 11. The time changes of amplitude and offset from null point could follow up the weight decrease over the passage of time in a second except for the transient behaviour in less than 0.1 s.

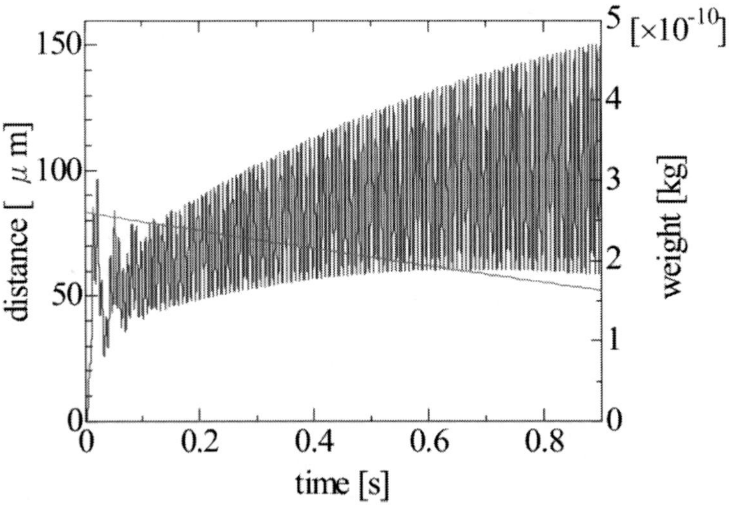

Figure 11. The simulated evaporation curve of the droplet weight (red line) and the calculated result of droplet trajectory (blue line) accompanied with its evaporation.

From the calculation of water droplet evaporation, we found that our trajectory calculation could be useful for following the weight change with the chemical reaction with the order of about 1sec. Furthermore, gravimetric analysis of chemical reactions does not provide direct identification of the chemical species involved in the reactions. In the case of EDB, it is possible to investigate direct measurement of the chemical species of a levitated particle by an EDB device coupled with a Raman spectroscopy [1].

5. Conclusions

This paper describes the preliminary calculation of the particle trajectories for analysis of the rapid weight change of levitated particle in EDB device. The electric fields in EDB were calculated by MATLAB PDE tool, and an approximate equation of electric field along the central axis of cylindrical electrodes was obtained with time variation. By solving the vertical equation of particle motion, the oscillating particle trajectories could be calculated though some modification of electric field and gas flow. It was found that particle trajectory calculation could follow up the weight change within a few oscillations such as the water droplet evaporation. It demonstrates that trajectory calculation of a particle in EDB is very useful for the kinetic studies on the reaction of a single aerosol particle with gaseous species.

Acknowledgments

This research was partially supported by the Ministry of Education, Science, Sports and Culture, Grant-in-Aid for Scientific Research (A), 12558066, 2003 and by CREST of JST (Japan Science and Technology Corporation).

References

[1] Davis E J and Schweiger G 2002 *The Airborne Microparticle - Its Physics, Chemistry, Optics and Transport Phenomena* (Springer)
[2] Peng C, Chow Albert H L and Chan C K 2000 *Pharm. Res.* **17** 1104
[3] Maloney D J, Lawson L O, Fasching E G and Monazam E R 1995 *Rev. Sci. Instrum.* **66** 3615
[4] Göbel G, Wriedt Th and Bauckhage K 1997 *Rev. Sci. Instrum.* **68** 3046
[5] Zheng F, Laucks M L and Davis E J 2000 *J. Aerosol Sci.* **31** 1173
[6] Loyalka S K, Tekasakul P, Tompson R V, Warder R C 1995 *J. Aerosol Sci.* **26** 445
[7] Iwamoto T, Itoh M, Takahashi K. 1991 *Aerosol Sci. Technol.* **15** 127
[8] Schlichting H 1968 *Boundary-layer Theor. 6th Ed.* (McGraw-Hill, New York) p 170
[9] Pruppacher H R and Klett J D 1997 *Microphysics of Clouds and Precipitation* (Kluwer Academic Publishers) p 502

7th International Conference on Applied Electrostatics (ICAES-2012) IOP Publishing
Journal of Physics: Conference Series **418** (2013) 012063 doi:10.1088/1742-6596/418/1/012063

Numerical analysis of electric potential distribution of two dielectric particles in an electrostatic field

J J Wang*, G F Li
School of Electrical Engineering, Dalian University of Technology, Dalian, China

E-mail: 1981wjj@163.com

Abstract. In this paper, a re-expansion method was employed to calculate the electric potential distribution of two Al_2O_3 particles. Then, the distortion of electric field in the vicinity of the particle chain during its formation was researched with theoretical explanation and numerical value calculation. The results show that the electric potential distribution of the polarized particles was similar to that of dipoles. Also, the electric field strength which increased at the particle binding sites promoted the formation of particle chains.

1. Introduction

Minuteness Al_2O_3 particles formed particle chains under the action of electrostatic field [1-6]. Conventional calculation of electric potential distribution between these dielectric particles is based on the equivalent dipole method, in which the contribution of higher-order polarization is difficult to be calculated. The re-expansion technique can be applied for the precise calculation of electric potential distribution in an arbitrary electrostatic field.

In this paper, we set up an interactional model of global dielectric particles under the action of electrostatic field; analyze the electrostatic potential distribution of every two particles by using the re-expansion method, and give theory interpret about the formation of dielectric particle chains, which combine with numerical value calculation by Wolfram Mathematica. It mainly based on the following hypothesis to analyze and use the re-expansion method in theoretical process: (i) Asymmetrical distribution of electric charges on dielectric particles produce multipolar moment; (ii) Interaction of multipolar moment give birth to interactional power between torsion and particles; (iii) There are not net residual electric charges on the particles.

2. Numerical analysis

Two spherical particles, S_1 and S_2, of radius R are located in electrostatic field. On the assumption that the line orientation of the two spheres centre was Z-axis, as shown in figure 1. In the present paper, the surface charge densities on the two spheres are assumed to be uniform.

Two spherical coordinate systems (r,θ) and (r',θ'), are adopted, the origins of which are located, respectively, at the centre of the sphere S_1 and S_2. The respective distance from a random point P to those origins in space were r and r', and the respective angle were θ and θ'. The external electric field E_0 was signed in reference frame which regard center O as origin:

* To whom any correspondence should be addressed.

$$\phi_E = -E_0 z = -E_0 r_0 \cos\theta_0 = -E_0 r_0 P_1(\cos\theta_0) \tag{1}$$

Where P_1 is the Legendre polynomial of order 1, r_0 is distance from point P to centre O.

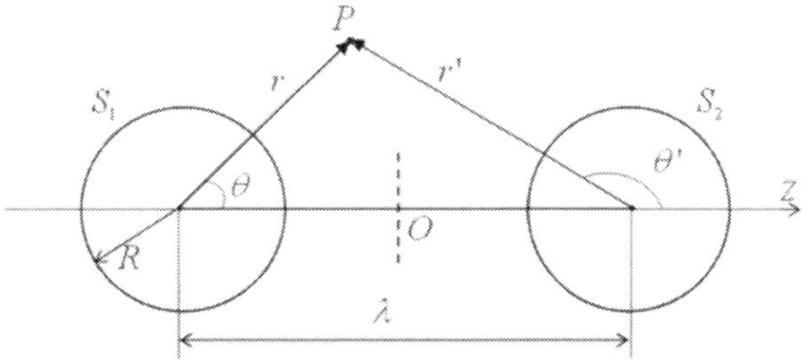

Figure 1. The geometric model of two spherical particles.

According to the re-expansion method, we can transform above equation when the origin moves particle S_1:

$$\phi_E = \frac{\lambda}{2} E_0 P_0(\cos\theta) - E_0 r P_1(\cos\theta) \tag{2}$$

In reference frame which the centre sphere particle S_1 was the origin, electrostatic potential distribution of external particle S1 was that:

$$\phi_1 = \frac{\lambda}{2} E_0 P_0(\cos\theta) - E_0 r P_1(\cos\theta) + \sum_{j=0}^{\infty} \{a_1(j)(\frac{R}{r})^{j+1} + b_1(j)(\frac{r}{R})^j\} P_j(\cos\theta) \tag{3}$$

The term contain $a_1(j)$ can be comprehend electrostatic potential which particle S_1 multipolar moment produce, whereas the term contain $b_1(j)$ can be seen electrostatic potential which particle S_2 multipolar moment produce [7-9].

In reference frame which the centre sphere particle S_2 was the origin, electrostatic potential distribution which particle S_2 multipolar moment produce was that:

$$\phi_2 = \sum_{k=0}^{\infty} a_2(k)(\frac{R}{r'})^{k+1} P_k(\cos\theta') \tag{4}$$

According to the re-expansion method, we can get that by deploying electrostatic potential distribution which particle S_2 multipolar moment produce in reference frame which the centre sphere particle S_1 was the origin.

$$\phi_{21} = \sum_{j=0}^{\infty} (\frac{R}{\lambda})^{j+1} \sum_{k=0}^{\infty} (\frac{-R}{\lambda})^k \frac{(j+k)!}{j!k!} a_2(k)(\frac{r}{R})^j P_j(\cos\theta) \tag{5}$$

According to the symmetry of the two particles [10], we can get that:

$$a_1(k) = (-1)^{k+1} a_2(k) \tag{6}$$

We can obtain that by comparison of the term contain $b_1(j)$ in from equation (3), (5) and (6).

$$b_1(j) = \sum_{k=0}^{\infty} (\frac{R}{\lambda})^{j+1} (\frac{-R}{\lambda})^k \frac{(j+k)!}{j!k!} a_2(k) = -\sum_{k=0}^{\infty} (\frac{R}{\lambda})^{j+k+1} \frac{(j+k)!}{j!k!} a_1(k) \tag{7}$$

In the spherical coordinate systems (r,θ), the potential distributions both inside and outside the spheres are expressed as below,

$$\begin{cases} \phi_{1out}(r,\theta) = \sum_{j=0}^{\infty} \{a_1(j)(\frac{R}{r})^{j+1} + b'_1(j)(\frac{r}{R})^j\} P_j(\cos\theta) \\ \phi_{1in}(r,\theta) = \sum_{j=0}^{\infty} c_1(\frac{r}{R})^j P_j(\cos\theta) \end{cases} \tag{8}$$

Where ϕ_{1in} is the potential distributions inside the spheres, ϕ_{1out} is the potential distributions outside the spheres, the new coefficient $b'_1(j)$ include Influence of external electric fields, different from $b_1(j)$ in equation (3).

By using separation of variables we were able to calculate the boundary conditions as:

$$\begin{cases} b'_1(0) = c_1(0) \\ a_1(0) = 0 \\ b'_1(j) = \{\frac{(2j+1)\varepsilon_m}{j(\varepsilon_m - \varepsilon_p)} - 1\} a_1(j) \\ c_1(j) = \frac{(2j+1)\varepsilon_m}{j(\varepsilon_m - \varepsilon_p)} a_1(j) \qquad (j \ge 1) \end{cases} \tag{9}$$

Where ε_m is the dielectric permittivity of N_2, ε_p is he dielectric permittivity of Al_2O_3.

From equation (3) and equation (9), we can obtain,

$$\begin{cases} b_1(0) = b'_1(0) - \lambda E_0/2 \\ b_1(1) = b'_1(1) + E_0 R \\ b_1(j) = b'_1(j) \qquad (j \ge 2) \end{cases} \tag{10}$$

We can obtain numerical value of $a_1(j)$, $c_1(j)$, $b_1(j)$ and ascertain around particle and its internal electrostatic potential distribution.

Because $a_1(0) = 0$, we calculate the Legendre polynomials of order from 1 to 128. Equations (8) and (10) are combined to obtain a new equation, expressed as follows:

$$-\sum_{k=0}^{\infty} (\frac{R}{\lambda})^{j+k+1} \frac{(j+k)!}{j!k!} a_1(k) = \begin{cases} \{\frac{(2j+1)\varepsilon_m}{j(\varepsilon_m - \varepsilon_p)} - 1\} a_1(j) + E_0 R & (j=1) \\ \{\frac{(2j+1)\varepsilon_m}{j(\varepsilon_m - \varepsilon_p)} - 1\} a_1(j) & (j \ge 2) \end{cases} \tag{11}$$

3. Calculation results

For convenience in numerical calculations, all the actual quantities can be done with per unit values. So E_0 and R are converted to per unit values. In addition, we define the distance of the centre of two spheres is λ. Applied r to above-mentioned research results, we carry on data calculations to international circumstance of two Al_2O_3 particles when λ =2 and λ =3, with the results as illustrated figure 2 to figure 5.

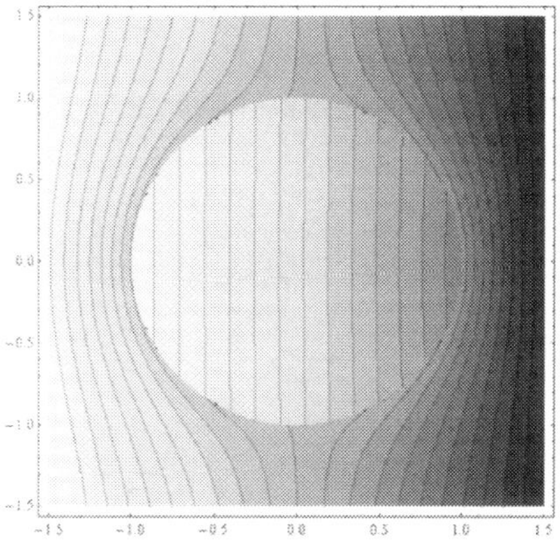

Figure 2. Equipotential line of S_1 with λ=3.

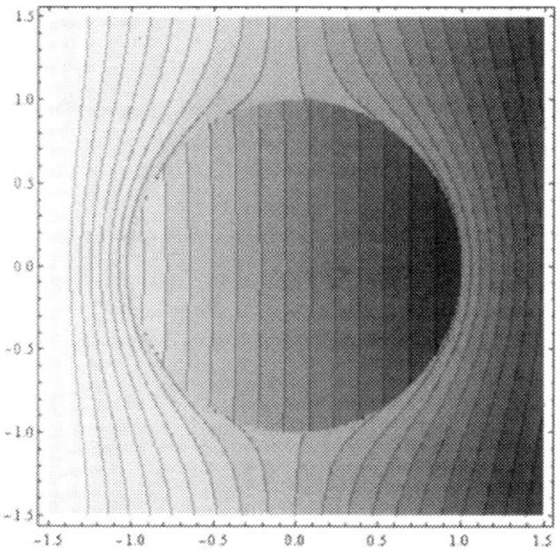

Figure 3. Equipotential line of S_2 with λ=3.

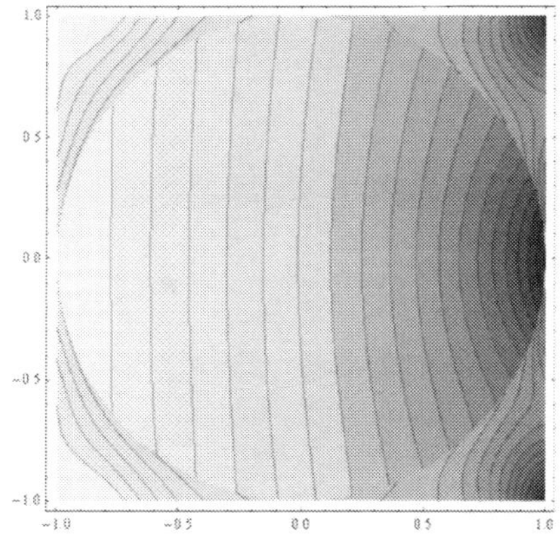

Figure 4. Equipotential line of S_1 with λ=2.

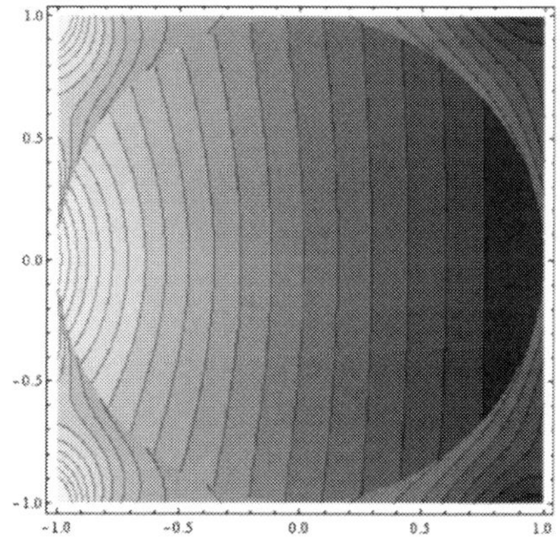

Figure 5. Equipotential line of S_2 with λ=2.

It describes that the electrostatic potential distribution in the vicinity of two Al_2O_3 sphere particles suspending under symmetrical electrostatic field in different condition from figure 2 to figure 5. It is seen that electric field nearby particle chains is strengthen on parallel orientation of external electric field. And on vertical orientation electric field is weaken. Much closer to space in particles, electric field strengthen is much stronger. Asymmetrical distribution of electric field which polarized particles, makes attraction of the particles arraying parallel orientation of electric field, and finally forms particle chains.

4. Discussions

In this paper, dielectric particles are simplified by multipolar particles model. The expression of two electrostatic potential distributions was obtained with the re-expansion method, and the figures of the electrostatic potential distribution of dielectric particle chains was obtained in an electrostatic field by numerical analysis.

Acknowledgments

This work is partially supported by the Fundamental Research Funds for the Central Universities. The authors also gratefully acknowledge the helpful comments and suggestions of the reviewers, which have improved the presentation.

References

[1] LI G, Takashima K, Katsura S and Mizuno A 2003 *J. Inst. Electrostat. Jpn.* **27** 135
[2] LI G, Takashima K, Katsura S and Mizuno A 2004 *Jpn. J. Appl. Phys.* **43** 722

[3] LI G, LI J and WU Y 2004 *Part. Sci. Technol.* **22** 417
[4] LI G, WANG Z Z and WANG N H 2007 *J. Mater. Sci.* **42** 8242
[5] LI G, Takashima K, Katsura S and Mizuno A 2004 *J. Mater. Sci.* **39** 4067
[6] LI G, Takashima K, Katsura S and Mizuno A 2004 *J. Inst. Electrostat. Jpn.* **28** 133
[7] Washizu M, Jones TB and Kaler KV 1993 *Biochim.Biophys.Acta.* **1158** 40
[8] Washizu M and Jones TB 1994 *J. Electrostat.* **33** 187
[9] Matsuyama T and Yamamoto Y 1998 *Kona* **16** 223
[10] Matsuyama T, Yamamoto Y and Washizu M 1995 *J. Electrostat.* **36** 195

Nanoparticles formation and deposition in the trichel pulse corona

R H Amirov [1*], **A A Petrov** [2,3], **I S Samoylov** [1]
1 Department for Gas-Discharge Physics and Plasma Technologies, Joint Institute for High Temperatures of RAS, Izhorskaya 13 Bd 2, Moscow, 125412, Russia
2 Department of Physical Electronics, P.N.Lebedev Physical Institute of RAS, Leninskiy prospekt 53, Moscow, 119991, Russia
3 Moscow Institute of Physics and Technology, Institutskii per. 9, Dolgoprydnyy, 141700, Russia

E-mail: amirovravil@yandex.ru

Abstract. Cathode erosion in the negative corona discharge has been studied in the point-to-plane electrode configuration with Cu cathodes in the Trichel pulse regime. Redeposition of erosion products has been found on the cathode surface in form of agglomerates of 10-nm nanoparticles. Nanocraters and nanoparticles formation in the negative corona discharge has been considered in frames of electro-explosive mechanism of cathode erosion. According to this mechanism the cathode erosion is performed as a consequence of elementary erosion events each of which is caused by a Trichel pulse. A 1-dimentional model of corona-produced nanoparticles dynamics in the gap was elaborated. According to results of the simulation, the redeposition is explained by charging of the nanoparticles due to positive ions adsorption and thermionic emission. The size, temperature and initial velocity of the aerosol nanoparticles have the decisive action on redeposition in the negative corona discharge.

1. Introduction

Negative corona discharge in the Trichel pulse mode [1] is a source of nanometer size aerosol particles which are produced as result of cathode erosion [2]. Nolan and Kuffel [2] proposed that aerosol particles in negative corona were produced due to condensation of cathode material vapors. To the same conclusion Borra [3] came. However, Nolan and O'Toole [4] noted that further work was required to elucidate the nature of the discharge nuclei at the moment of formation.

Cathode erosion in the negative corona discharge was first reported by Bennet [5]. However the mechanism of the cathode erosion in the negative corona is still unclear. The main hypotheses of the cathode erosion mechanism in the negative corona discharge are the next: sputtering by positive ions [6–8], chemical sputtering [7, 9], formation and destruction of dielectric layers [10], ecton and electro-explosive processes [11–13]. Possibility of local heating and evaporation of cathode in corona was discussed by Goldman and Sigmond [7] and Borra [3].

The redeposition of erosion products on the cathode surface may occur [1] in the negative corona discharge. This process was explained by back diffusion [1, 14], divergent electric field action [15], trapping by vortex of electric wind [15, 16]. Redeposition of erosion products in the negative corona

[*] To whom any correspondence should be addressed.

Journal of Physics: Conference Series **418** (2013) 012064 doi:10.1088/1742-6596/418/1/012064

was reported in case of copper cathodes [11, 14], iron and aluminum cathodes [14], platinum cathodes [1]. However redeposition was not found in case aluminum cathodes [17] as well as in case of graphite cathodes [18]. Note that back diffusion and trapping by air vortexes predict the redeposition in case of any cathode material, but the action of the cathode material on the redeposition process seems to occur. So the other mechanisms of redeposition should be considered in the negative corona discharge.

The conclusion is that cathode erosion, nanoparticles formation, and redeposition process in the negative corona discharge should be considered in the context of a single model. The aim of this work is to explain these processes from the point of view of electro-explosive mechanism. For this purpose the cathode erosion in Trichel pulse corona was studied and a 1-dimentional model of erosion products transport in the gap was elaborated.

2. Experimental setup and procedure

Negative corona discharge has been studied in air in the point-to-plane electrode configuration in the Trichel pulse mode. Cathode pins were made of copper with tip diameter 10-100 mkm. Positive dc voltage on the range of 10 kV was supplied to the anode plane, which was an 8-cm diameter silicon disc. Cathode axis was perpendicular to the anode plane, the gap was 10 mm.

Discharge current was registered by S1-75 dial oscillograph with 300 MHz bandwidth which was connected across a 50-Ohm resistor between the cathode and the ground. Average discharge current was registered with use of M-906 dial ammeter; average current was 20-100 mkA. Relative humidity of air was in the range of 60-80 % and did not influence the regimes and typical erosion pictures observed.

Cathode surface during discharge was video-recorded with use of Canon w705i camcorder and MBS-9 binocular microscope with spatial resolution 1 mkm and exposition of a single picture 40 ms. The total time of discharge was 5-90 min so the cathodes became at least 30 mkm shorter due to erosion.

After treatment in discharge the cathodes were analysed with use of Jeol JSM-7001F and FEI Quanta 200 electron microscopes with spatial resolution ~ 1 nm. The X-rays componential analysis of the copper cathode surface with sensitivity 0.01 at.% has also been performed.

3. Results of experiment

Erosion was found in form of reduction of the cathode pin length with typical rate few mkm/min and in formation of micro and nanocraters on the tip. A copper cathode before and after treatment in the Trichel pulse corona is shown in the figure 1 – deleted and redeposited parts are well distinguished with use of optical microscope. Trichel pulse amplitude was 0.3 mA, the frequency of the Trichel pulses was 500 kHz at average current 60 mkA. A typical copper cathode tip after treatment is shown in the figure 2. The tip of the cathode is smooth as result of the discharge erosive action. Redeposited erosion material is observed in form of nanostructures on the side of the tip - figure 2. The edges of the redeposited structures are covered with 10-nm clusters which are forming the crystals probably due to field and temperature assistant surface drift – these clusters are well distinguished in the figure 3. According to results of X-ray analysis the redeposited structures consist of copper and oxygen in the proportion approximately 1:2. No traces of oxygen were detected on the smooth part of the tip.

Redeposition of erosion products has also been found in case of silver and platinum cathodes at atmospheric pressure. Redeposition on graphite cathodes was found at pressure lower than 200 torr.

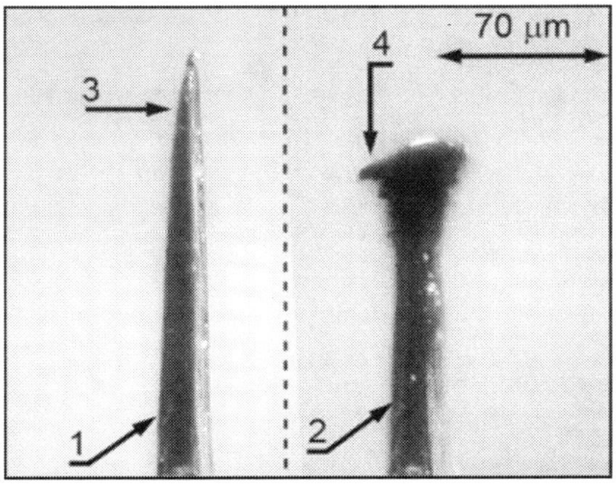

Figure 1. Copper cathode: 1 – before discharge, 2 – after discharge, 3 – deleted part, 4 – redeposited material. Treatment at 50 mkA during 5 min.

Figure 2. A typical copper cathode after treatment in Trichel pulse corona: 1 – smooth tip, 2 – redeposited material.

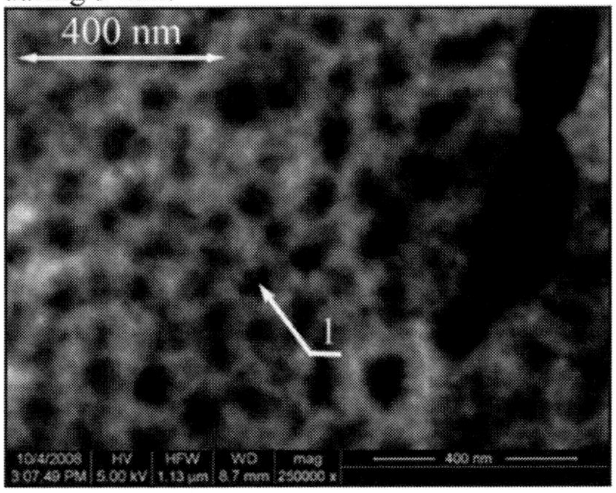

Figure 3. Elementary nanocraters (1) on the smooth tip of the cathode.

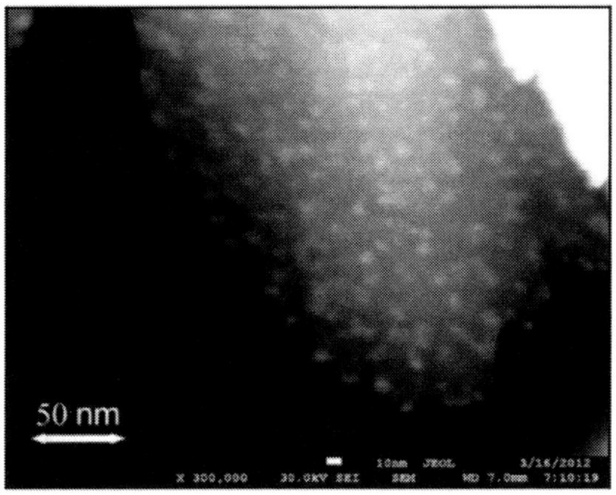

Figure 4. Redeposited material consists of 10-nm nanoparticles.

4. Discussion

4.1 Nanocraters and nanoparticles formation

According to results of Asinovskii et al. [11] and Petrov et al. [18, 19] the cathode erosion mechanism in the negative corona discharge has the electro-explosive nature. This conclusion follows from the estimate of the integral of the specific action of a Trichel pulse for copper and graphite cathodes. According to the electro-explosive theory, the erosion process on the cathode is performed as a consequence of elementary erosion events. Each elementary erosion event is associated with a Trichel pulse and results in the formation of an elementary erosion crater on the cathode surface [19] – these 40-nm craters are shown in the figure 3.

The size of an elementary erosion volume was taken as relation of total erosion volume to total number of Trichel pulses during experiment. This value was calculated as 10^3 nm^3. The current density j on the cathode surface during a Trichel pulse was taken as relation of Trichel pulse amplitude to cross-section of an elementary erosion volume. It was estimated as ~ $5 \cdot 10^8$ A cm^{-2}. Then the value of the integral of specific current action of a Trichel pulse:

$$h = \int_0^{\tau} j^2 dt \sim 10^9 A^2 \cdot s / cm^4 \quad (1)$$

where $\tau = 10$ ns – FWHM of a Trichel pulse.

For graphite cathodes the integral of specific action of a Trichel pulse (1) was measured as 10^8 A^2s cm^{-4} [18]. Values of the integral of Trichel pulse specific action indicate that erosion in the Trichel pulse negative corona discharge is the most probably associated with the electro-explosive mechanism – the same values of the integral of specific current action are known from the experiments on the wire electrical explosion [20].

The mechanism of the cathode erosion in the negative corona discharge is described as follows. Local energy release is associated with the field emission on the micro-asperity on the cathode surface due to positive space charge of the cathode-directed ionization wave. These waves were experimentally registered by Ikuta and Kondo [21] and electric field strength in front of the head of this wave was calculated by Napartovich et al. [22] on the range of 10^6 V cm^{-1}. This value may be increased in 10-1000 times due to micro-asperity geometry [23]. Pulsed electric field with magnitude up to 10^8 V cm^{-1} will result in field emission and local Joule heating of the bulk up to critical temperature 10^4 K [24] in few nanoseconds – this causes the local electrical explosion and dispergation of the bulk. The size of aerosol particles which are presented in the figure 3 coincides with typical electron free path inside the cathode (10 nm) according to the hypothesis discussed by Lebedev and Savvatimskii [25], or critical size of nucleated particles (1-10 nm) according to theory presented by Tkachenko et al. [26].

Lebedev and Savvatimskii [25], and Vorob'ev et al. [27] reported that 10-nm scale particles were produced in the process of wire electrical explosion due to dispergation of the wire. Explosive mechanisms of the cathode erosion in corona [11, 13] admit the aerosol formation due to direct dispergation of cathode bulky media without equilibrium vapor phase. So physics of cathode erosion in the negative corona may be similar to one in the processes of wire electrical explosion.

4.2 Nanoparticles redeposition model

A 1-dimentional model of the cathode erosion products dynamics in the negative corona discharge gap was elaborated. As was discussed in the previous section, the products of the cathode erosion in the negative corona discharge were 10-nm size aerosol particles. According to the electro-explosive hypotheses of the cathode erosion the initial velocity of these particles may be taken from the results of experiments on the wire electrical explosions up to 10^3 m sec^{-1} [25, 28]. According to estimates of Lebedev and Savvatimskii [25], the initial temperature of particles in the process of wire electrical explosion was on the range of critical one (8900 for copper wires). It was proposed that the particles were uncharged in the moment of formation – otherwise the particles would be subjected to redeposition regardless the other initial parameters. The motion of a single aerosol particle in the gap may be described by the next equation:

$$m \frac{dV}{dt} = F_{dr} + F_{el} \quad (2)$$

in which m is the mass of the particle, V – velocity of the particle, F_{dr} – neutral drag force, F_{el} - electrical force. According to estimates the force of gravity and thermophoresis force and the force of electrical polarization are insufficient and may be neglected in the equation of motion of a 10-nm size particle.

The charging of an aerosol particle may be described by the equation:

$$\frac{1}{S} \frac{dq}{dt} = j_T + j_{pl} \quad (3)$$

Here S is the surface of a particle, j_T is the thermionic emission current density, and j_{pl} is the current density of the charged species from the discharge media on the surface of the particle.

The thermionic current density on the aerosol particle surface depends drastically on the particle temperature. So the heat balance of the particle should be considered. Note that the heat redistribution inside a 10-nm size metal particle occurs in ~1 ns. The heat fluxes on the particle surface may be described by the equation:

$$mc\frac{dT}{dt} = P_{th} + P_{drag} + P_{rad} \tag{4}$$

Here c is the heat capacity of the particle material, T is the particle temperature, P_{th}, P_{drag}, P_{rad} is the total heat flux on the particle surface associated with thermionic cooling, interaction with the air molecules and radiation correspondently.

In the free-molecular regime the neutral drag force expression is [29]:

$$F_{dr} = \pi r^2 n m_0 V V_T \tag{5}$$

The Laplace electric field distribution in the point-to-plane gap on the gap axis may be given by the equation [9]:

$$E(x) = \frac{2U}{(r_c + x)\cdot \log(\frac{4h}{r_c})} \tag{6}$$

where U is the gap voltage, r_c is the curvature radius of the cathode, x is the distance from the cathode surface, h is the gap length. According to results of calculations of Napartovich et al. [22], the electric field distribution in the gap is highly disturbed only during the short 20-ns Trichel pulse active phase, and the rest time between pulses it is weakly disturbed. Hence the undisturbed approximation (6) may be used in the aerosol dynamics calculation.

Consider the charging of an aerosol particle by the charged species in the gap. Negative corona discharge flame in average consists of two unipolar zones–positive and negative one. The concentration of the positive ions in the region closer than 100 mkm to the cathode surface is on the range of 10^{12}-10^{13} cm^{-3} [22, 30] during the active phase of a Trichel pulse. The positive ions concentration is negligibly small at the distance further than 100 mkm from the cathode surface. Then the charging of an aerosol particle by the attachment of the positive ions may be expressed by the equation [31]:

$$N = n_i \cdot \pi \cdot r^2 \cdot l \tag{7}$$

where N is the number of the unit charges on the particle, n_i is the concentration of the positive ions in the unipolar positive zone near the cathode surface, l is the path of the particle inside the unipolar charged zone. The approximation has been made that the cross section of the positive ions attachment was equal to the cut of an aerosol nanoparticle.

Charging of the nanoparticles by the negative ions may be described by the same equation (7); the negative ions are distributed in the gap with the maximum concentration approximately 10^{12} cm^{-3} in the region between 100 and 200 mkm from the cathode surface. In the region further than 200 mkm from the cathode surface the concentration of the negative ions is sufficiently lower because of its repulsion spreading.

Electrons are formed in the gap due to secondary processes on the cathode surface and because of the impact ionization of the gas during the Trichel pulse active phase. The duration of the active phase is on the range of 10 ns [22]. The concentration of electrons during the active phase is on the range of 10^{13} cm^{-3} with average electron energy 10 eV; these electrons are mostly localized in the region between 50 and 150 mkm from the cathode surface. After the active phase is gone the electrons concentration diminishes drastically in 10 ns because of the electron attachment to the oxygen molecules [32]. The charge of an aerosol particle received due to attachment of electrons may be found from the equation [33]:

$$N = n_e \cdot \pi \cdot r^2 \cdot V_e \cdot \tau \tag{8}$$

in which n_e is the electrons concentration, V_e is the electrons velocity, τ is the duration of the active phase of a Trichel pulse. The probability to get an elementary negative charge during the active phase of a Trichel pulse is on the range 10^{-1}–10^{-2} for a 10-nm particle. So the aerosol particle should stay in the active region of the discharge during 10-100 mks to get an elementary charge due to electrons attachment.

The ionization threshold of a metal particle is on the range of 5 eV with maximum at 15 eV. According to Smirnov [31] the probability of electrons attachment exceeds the probability of the impact ionization of a particle in the negative glow region of the negative corona discharge flame.

The thermionic current density j_p on the aerosol particle surface may be expressed by equation:

$$j_p = AT^2 e^{-\frac{\varphi}{kT}} \tag{9}$$

where $A=120$ $A \cdot cm^{-2} Coul^{-2}$, φ is the work function of the particle material, $k=1.38 \cdot 10^{-23}$ $J \cdot K^{-1}$ is the Boltzmann constant. The work function depends on the particle radius and charge in the following approximation:

$$\varphi \cong \varphi_0 + \frac{q}{4\pi\varepsilon_0 r} \tag{10}$$

where φ_0 is the work function of the particle with no charge. According to the estimates the external field E does not influence the aerosol material work function in the conditions concerned.

Consider the heat balance of a particle. The thermionic cooling may be specified in the next way:

$$P_{Th} = I \cdot \varphi + 2I\frac{kT}{e} \tag{11}$$

where I is the thermionic current on the particle surface; it may be found from the equation (9).

The heat flux associated with the interaction between the aerosol particle and the gas molecules may be found from the diffuse reflection model of the particle cooling in the free-molecular regime [29]:

$$P_{drag} = \frac{2m_0 n}{\sqrt{\pi}} S \left(\frac{2kT_g}{m_0}\right)^{3/2} \left(\frac{T}{T_g} I_1(\sigma) - I_2(\sigma)\right) \tag{12}$$

Here S is the particle cross section. Parameters $I_1(\sigma)$ and $I_2(\sigma)$ may be found from the next equations:

$$I_1(\sigma) = \exp(-\sigma^2) + \left(\sigma + \frac{1}{2 \cdot \sigma}\right)\sqrt{\pi} \cdot erf\, \sigma \tag{13}$$

$$I_2(\sigma) = \frac{1}{2}\left(\sigma^2 + \frac{5}{2}\right)\exp(-\sigma^2) + \frac{1}{2}\left(\sigma^3 + 3\sigma + \frac{3}{4\sigma}\right)\sqrt{\pi} \cdot erf\, \sigma \tag{14}$$

in which erf is error function,

$$\sigma = \sqrt{\frac{m_0 V^2}{2kT}} \tag{15}$$

According to the equation (12), the heat flux on the particle surface may be either positive or negative depending on the relation between the particle temperature, ambient temperature and particle velocity.

Different approaches to a nm-particle radiation description exist [34]. In any case the radiation power of a particle and evanescent cooling are sufficiently lower than the neutral drag heat flux in the conditions of the task concerned at least at temperatures smaller than 5000 K.

The system of equations (2-4) was numerically solved in a 1-dimentional nonuniform mesh. Two different types of the system behavior have been revealed – the case of the cold clusters with no thermionic emission and the case of the hot clusters when thermionic emission was active.

In the first case the particle motion may be adequately solved analytically. Position of the particle in time may be approached by the equation:

$$x(t) = \frac{V_0}{\pi \cdot r^2 n m_0 V_T}\left[1-\exp(-4\pi \cdot r^2 n m_0 V_T)\right] \qquad (16)$$

in which V_0 is initial velocity of aerosol particle. In the figure 5 the first line is the trajectory of a 10-nm size copper particle with initial velocity 500 m s^{-1}. This particle remains in the region of the positive unipolar region during the Trichel pulses development. Approximately in 0.1 ms it will get a unit positive charge due to ion attachment and will return on the cathode surface. The second line is the path of a 10-nm size copper particle with initial velocity 1100 m s^{-1}. This particle passes the positive space charge zone in few nanoseconds and enters the region of the negative volumetric charge. As result this particle will get a negative charge and will move away from the cathode surface. So the action of the cathode material on redeposition process may be explained by difference of initial velocities of the aerosol particles in case of different cathode materials. Elimination may also be assisted by trapping of a particle by electric wind.

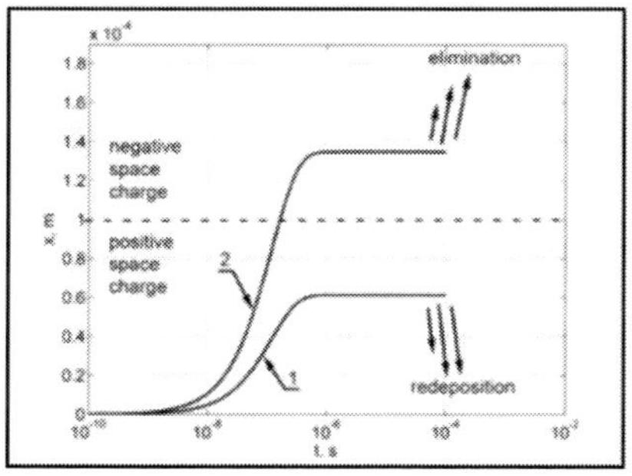

Figure 5. Dynamics of a 10-nm size copper particle in time: 1 – redeposition is observed at initial velocity 500 m s^{-1}; 2 – no redeposition at initial velocity 1100 m s^{-1}.

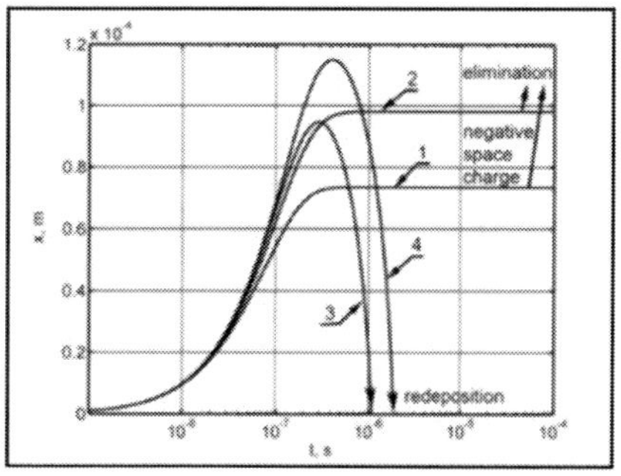

Figure 6. Dynamics of a copper cluster with initial velocity 1000 m s^{-1} with initial temperature 4800 K: 1 – r = 3 nm; 2 – r = 4 nm; 3 – r = 5 nm; 4 – r = 6 nm. In case of 3 and 4 the thermionic charging is responsible for redeposition in 1 mks.

The second type of system behavior takes place when aerosol particles are hot enough initially to get charge due to thermionic emission. In the figure 6 the path of a copper particle with initial velocity 1100 m s^{-1} is shown. Initial temperature of a particle was 5000 K. In this case the particle will get a unit positive charge in the first 10 ns of its way. As result approximately in 1 mks it will be deposited on the cathode surface. So the action of cathode material on the redeposition process may be explained by difference of initial temperatures of the particles in case of different electrode materials. Change of the particle radius may also influence the redeposition effect, as follows from the figure 6.

So the conclusion is that the redeposition is explained by charging of a particle by positive ions and due to thermionic emission. The proposed model of redeposition is in agreement with the electro-explosive mechanism of the aerosol particle formation.

5. Conclusions

Formation and redeposition of nanoparticles has been studied in the Trichel pulse negative corona discharge. According to results of topographical analysis the redeposited structures on the side of a copper cathode consist of 10-nm size nanoparticles So the material transport in the negative corona is associated with the flux of nanoparticles.

Nanoparticles formation and cathode erosion in the negative corona discharge was explained by electrical explosions on the cathode surface. The nanometer-size aerosol particles are formed due to dispergation of the cathode bulky media. Values of the integral of specific action of a Trichel pulse confirm the electro-explosive mechanism of cathode erosion and aerosol formation in negative corona discharge.

A 1-dimensional model of the aerosol particles dynamics was proposed. According to results of the simulation, the redeposition is explained by charging of the aerosol particles due to positive ions adsorption and thermionic emission. Redeposition depends drastically on the size, temperature and initial velocity of the aerosol particles.

Acknowledgements

The work has been made under partial financial support of the Ministry of Education and Science of Russian Federation with use of the equipment of the centre of the collective use of Moscow institute of physics and technology (contract № 16.552.11.7022); the authors would like to thank Evgenii Korostylev, Pavel Starikov and Andrei Baturin for microanalysis of electrodes. The work has been made with partial financial support of the Russian Basic Research Foundation, grant № 12-08-01223 "Research of the cathode erosion mechanism in the negative corona discharge".

References

[1] Loeb L B 1965 Electrical Coronas. Their Basic Physical Mechanisms. Berkeley, CA: Univ. California Press, p 694
[2] Nolan P J and Kuffel E 1957 *Geofisica Pura e Applicata.* **36** 201
[3] Borra J P 2006 *Journal Physics D: Applied Physics.* **39** 19
[4] Nolan P J and O'Toole C P J 1959 *Pure and Applied Geophysics.* **42** 117
[5] Bennet G W H 1940 *Physical Review.* **58** 992
[6] Buchet G and Goldman A 1970 *1st International Conference on Gas Discharges, IEE Conf. Publ.* 459–462
[7] Goldman M and Sigmond R S 1982 *IEEE Transactions on Electrical Insulation.* **EI-17** 90
[8] Weissler G L and Schindler M 1952 *Journal of Applied Physics.* **23** 844
[9] Sigmond R S 1978 Corona Discharges. In: Meek, J.M. and Craggs, J.D. (Ed.) Electrical Breakdown of Gases, Wiley: London, chapt. 4, p. 319–384
[10] Korge H, Laan M and Paris P 1993 *Journal Physics D: Applied Physics.* **26** 231
[11] Asinovskii E I, Petrov A A and Samoylov I S 2008 *Technical Physics.* **53** 279
[12] Laan M, Paris P and Repan V 1997 *J. Phys. France.* **4** 259
[13] Mesyats G A 1995 *Physics-Uspekhi.* 38 567
[14] Giao T N and Jordan J B 1970 *Journal of Applied Physics.* **41** 3991
[15] Laan M and Mirme A (2000). Aerosol and corona discharges. In: Veldhuizen, E.M. (Ed.) Electrical discharges for environmental purposes: fundamental and applications, Nova Science Publisher Inc. Ch. 8, p. 193–220
[16] Sigmond R S, Goldman A and Goldman M 1992 Proceedings 10th International Conference on Gas Discharges and their Applications, Swansea, UK, p. 330–333
[17] Amirov R H, Petrov A A and Samoylov I S 2009 *International Journal of Plasma Environmental Science and Technology.* **3** 35
[18] Petrov A A, Amirov R H, Asinovskii E I and Samoylov I S 2009 *Journal of Plasma Fusion Research Series.* **8** 780

[19] Petrov A A, Amirov R H and Samoylov I S 2009 *IEEE – Transactions on Plasma Sciences*. **37** 1146
[20] Kotov Yu A 2003 *Journal of Nanoparticle Research*. 5 539
[21] Ikuta N and Kondo K 1976 *IEE Conf. Publ.* **143** 227
[22] Napartovich A P, Akishev Y S, Deryugin A A, Kochetov I V, Pan'kin M V and Trushkin N I 1997 *Journal of Physics D: Applied Physics*. **30** 2726
[23] Rees T and Paillol J 1997 *Journal of Physics D: Applied Physics*. **30** 3115
[24] Mesyats G A 2000 Cathode phenomena in a vacuum discharge: the breakdown, the spark and the arc. Moscow: Nauka, 400
[25] Lebedev S V and Savvatimskii A I 1984 *Soviet Physics Uspekhi*. **27** 749
[26] Tkachenko S I, Vorob'ev V S and Malyshenko S P 2004 *Journal of Physics D: Applied Physics* **37** 495
[27] Vorob'ev V S, Malyshenko S P, Tkachenko S I, and Fortov V E 2002 *Journal of Experimental and Theoretical Physics Letters* **75** 373
[28] Nazarenko O 2007 Proc. of European congress of chemical engineering (ECCE-6), Copenhagen, 16
[29] Williams M M R 1973 *Journal of Physics D: Applied Physics*. **6** 744
[30] Chyhin V I and Karpyak S Y 2007 *Condensed Matter Physics*. **10** 209
[31] Smirnov B M 2000 *Physics-Uspekhi*. **43** 453
[32] Nasser E 1971 Fundamentals of Gaseous Ionization and Plasma Electronics. New York: Wiley-Interscience, 442
[33] Reist P C 1984 Introduction to aerosol science. Macmillan, New York
[34] Dedkov G V and Kyasov A A 2007 *Technical Physics Letters*. **33** 305

Effect of the EHD flow on the collection efficiency of fine particles in wire-plate electrostatic precipitators

Qing Li*, Qiaoyan Wang, Wenzhao Li, Jiangshan Meng, Yanqing Xiong
Electrostatic Research Institute, Hebei University, No.180 WuSi East Road, Baoding, China

E-mail: liqing@hbu.cn

Abstract. This paper focuses on the low collection efficiency of fine particles in air pollutant. Based on analyzing the collection rate of fine particles in electrostatic precipitator (ESP), a two-dimensional model of wire-plate discharge was established. The CFD code FLUENT was used to simulate the ionic winds at typical voltages 25, 40, and 45 kV, respectively. In the process of simulating, the ionic wind was replaced by the dynamic wind. According to the state graphs of ionic winds, the abnormal manifestation of the fine particles collection appeared in 40 kV voltage was explained. This work will provide a theoretical support for optimizing electrodes matching.

1. Introduction

The fine particles removal is an emphasis for the study of high voltage electrostatic precipitators in modern dust removal. In the dust source, the smaller particles which occupy a small weight proportion relative to the larger particles have an absolute advantage in the numbers. Especially the fine particles ($PM_{2.5}$) have smaller diameter, bigger specific surface area and they can stay a long time in the atmosphere. They are more likely to be the carriers of other harmful substances leaving harmful substances enter the body. So they have great impact on people's health and the quality of atmospheric environment [1-3]. At present, the ESP has low collection efficiency on $PM_{2.5}$. Therefore, it is an important issue to study the efficient collection methods of fine particles.

The electrohydrodynamic (EHD) field makes the collection process of particulate matter extremely complicate in the discharge space [4-6]. According to the related researches [7-10], the influence that the ionic wind force has on the particulate matter can't be ignored. In this paper, at first we analyzed the collection rate of fine coal particles under different voltages by experiments. Then the distributions of the ion wind were obtained under different voltages by numerical simulation. We explained the abnormal phenomenon of fine particles collection when the voltage was 40 kV. These results finally provide the basis for seeking efficient electrode matching of ESP.

2. Experiment of coal particles collection

The ESP uses wire-plate negative corona discharge device of multi-electric field. The wire-to-plate space was 150 mm. The fine coal particles of different electric fields were collected under different voltages. The distributions of particle diameters were analyzed using BT-9300 laser particle size analyzer.

* To whom any correspondence should be addressed.

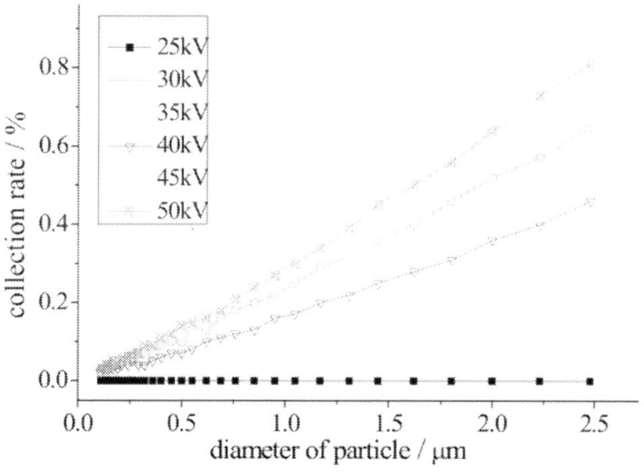

Figure 1. The collection rate of $PM_{2.5}$ under different voltages.

In this experiment, 25 kV was the discharge inception voltage. The collected particles in the first electric field were analyzed. The collection rates (the volume percentage that certain particle diameter takes in all collected particles expresses the collection effect) of different particle diameters were obtained. It is shown in figure 1.

At 25 kV, the fine particles collection was nearly the same as it was in the natural state. This is because the discharge energy, the number of ions, the ion wind strength and electric field strength were weak at 25 kV. Particles were charged incompletely. The coulomb force on the fine particles was weak. The fine particles were deposited due to natural coagulation and electric coagulation with large particles.

As the voltage increased, the electric field strength reinforced and the number of space ions increased. The collection rate of $PM_{2.5}$ increased as well. However, the collection rate was abnormal at 40 kV. The collection rate of $PM_{2.5}$ at 40 kV was only higher than that of 25 kV.

3. Numerical simulation of ionic wind

As discussed above, the collection process of particles is very complex. The particles move under various forces in the electric field. Especially the ionic wind has great influence on the fine particles. Thus the ionic wind in the discharge space is simulated. The simulation also give a reasonable explain for the abnormal phenomenon of collection rate appeared at 40 kV.

3.1. Model building

Several hypotheses:

(1)In the practical application, the upper limit of the inlet concentration should be controlled in the ESP. The concentration of the coal dust was generally set to 20 g m^{-3} [11]. That is 0.01 % if it is converted to volume fraction. As the duty cycle of the particles was very low, the dust stream could be idealized as a single-phase flow (i.e. air). The fluid of the discharge space was regarded as incompressible viscous fluid without regard to the change of the flow density.

(2)In this experiment, the influence among the corona wires was so small that each corona wire could become an independent discharge unit. The wire-plate discharge model was established for one independent unit.

(3)The differences of the flow in different directions of the corona wire were ignored. The flow perpendicular to the plane of the corona wire in the discharge process was considered as isotropic. The three-dimensional discharge model was simplified as two-dimensional. The wire-to-plate space was set to 150 mm. The final model is showed in figure 2.

3.2. The governing equations

According to the established two-dimensional model and the characteristic of the fluid [12, 13], the governing equations of the fluid flow in the discharge space could be described as follows:

The continuity equation: $\operatorname{div}(\boldsymbol{u})=0$ (1)

The momentum equations: $\frac{\partial(\rho u)}{\partial t}+\operatorname{div}(\rho u\boldsymbol{u})=\operatorname{div}(\mu\,\operatorname{grad}u)-\frac{\partial p}{\partial x}+F_x$ (2)

$$\frac{\partial(\rho v)}{\partial t}+\operatorname{div}(\rho v\boldsymbol{u})=\operatorname{div}(\mu\,\operatorname{grad}u)-\frac{\partial p}{\partial y}+F_y$$

k equation:

$$\frac{\partial(\rho k)}{\partial t}+\operatorname{div}(\rho k\boldsymbol{u})=\operatorname{div}\left[\left(\mu+\frac{\mu_t}{\sigma_k}\right)grad\,k-\rho\varepsilon+\mu_t P_G\right] \quad (3)$$

ε equation:

$$\frac{\partial(\rho\varepsilon)}{\partial t}+\operatorname{div}(\rho\varepsilon\boldsymbol{u})=\operatorname{div}\left[\left(\mu+\frac{\mu_t}{\sigma_\varepsilon}\right)grad\,\varepsilon\right]-\rho C_2\frac{\varepsilon^2}{k}+\mu_t C_1\frac{\varepsilon}{k}P_G \quad (4)$$

Where **u** is the velocity vector, u and v the component of the velocity vector in the direction of x and y, ρ the fluid density, μ the kinetic viscosity, p the pressure on the fluid element, F_x and F_y the physical force on the element, k the turbulent kinetic energy, ε the turbulent dissipation rate, μ_t the turbulent viscosity, respectively. $\mu_t=\rho C_\mu\frac{k^2}{\varepsilon}$. In this two-dimensional model, P_G could be written as follows:

$$P_G=2\left[\left(\frac{\partial u}{\partial x}\right)^2+\left(\frac{\partial v}{\partial y}^2\right)\right]+\left(\frac{\partial u}{\partial y}+\frac{\partial v}{\partial x}\right)^2 \quad (5)$$

σ_k and σ_ε are the Prandtl number that the k and ε correspond with. σ_k=1.0, σ_ε=1.3, C_μ=0.09, C1=1.44, C_2=1.92.

3.3. The parameter settings

The boundary conditions were set as follows: the corona wire was set as velocity inlet. AB and CD were set as outflow boundary conditions. The plate electrodes were no-slip side wall boundary conditions.

The governing equations were discrete by the finite volume method (FVM). The discrete scheme for the diffusion term and the convection term were the central difference scheme and the first-order upwind scheme, respectively. The numerical calculation was the SIMPLE algorithm.

3.4. The simulation graphs of the ionic wind

According to our previous research results [14-15] and the analysis above, we simulated the ionic winds of typical voltages (25, 40 and 45 kV) in the discharge space. The results are showed in figure 3, figure 4 and figure 5.

The three figures showed that as the voltage increased, the vortexes of double helix shape neared to the plate electrodes gradually. The scale of helix also increased. The degree of turbulence was intensified in the discharge space. A fully double-helix structure developed until the voltage was 45 kV.

When the voltage was 25 kV, the electric strength was weak and the ionic number was very few. Only near the corona wire formed ionic wind with small spiral scales. The ionic wind made particles rotated around the corona wire, which increased the coagulation probability of fine particulate particles to a certain degree. Because the corona zone was only a few millimetres, and out of the corona zone, the ionic wind velocity reduced quickly. The ionic wind had little influence on the migration velocity of fine particles. All of these evidences supported that the collected fine particles presented a state of natural distribution under the voltage of 25 kV.

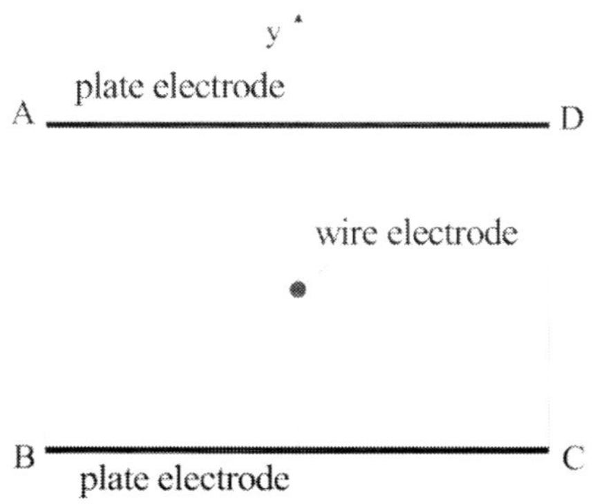

Figure 2. The model of discharge space.

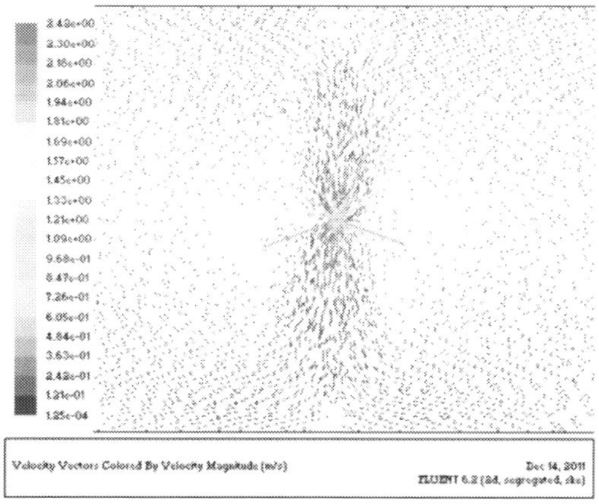

Figure 3. The simulated graph at 25 kV.

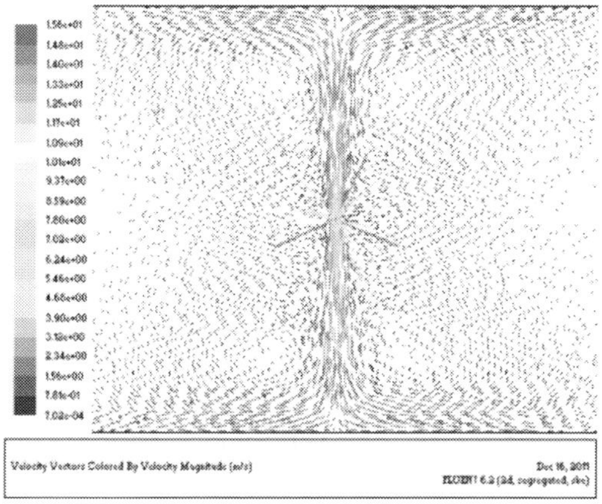

Figure 4. The simulated graph at 40 kV.

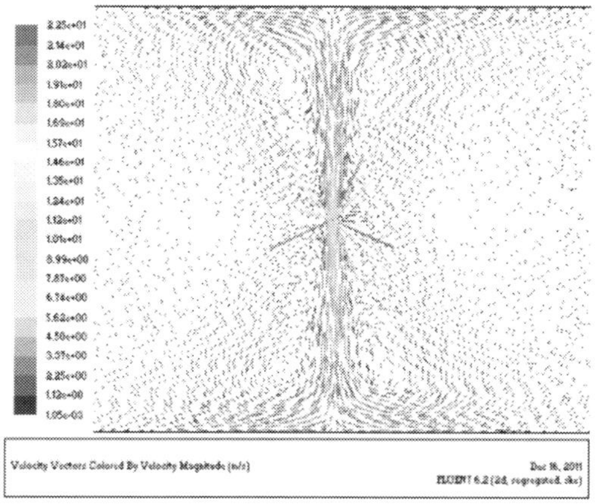

Figure 5. The simulated graph at 45 kV.

When the voltage was 40 kV, large horizontal airflows along the plate electrodes were formed. On the one hand, these airflows just had scouring effect to the collected particles attached to the plate electrodes. Thus the fine particles were easily dropped from the plate electrodes with this airflow force. On the other hand, due to the small scale of vortexes, the dropped fine particles couldn't go into the vortexes to be collected again in good time. Part of the dropped particles only moved along the electrode plates under the ionic wind flow. As we know, the concentration of the ions and the electric strength near the plate were the weakest in the discharge channel. Therefore, these particles didn't have enough electric force. With the ESP operating for a long time, the concentration of gathered particles in the eddy would be more highly. The comprehensive effect led to a low collection rate of fine particles at 40 kV, then an abnormal phenomenon of fine particles collection appeared.

At 45 kV, the scale of vortex was large and the electric field strength was large as well. When the fine particles moved into the vortex, they could be fully charged. Finally, they got rid of the vortex in the strong electric force. They were collected again without washed away by the ionic wind.

4. Numerical simulation of ionic wind

(1)The ionic wind exiting in the discharge space has a great impact on the collection of fine particles. Not only should the influence of electric force be considered, but the role of ionic wind should be considered.

(2)The abnormal phenomenon of fine particles collection at 40 kV reminds us that improving the voltage unilaterally is not the best way to collect the fine particles efficiently.

(3)According to the analysis of the particles collection rate, it showed that the multi-electric field has a great advantage. A single electric field couldn't guarantee the efficient collection of fine particulate matters. Optimizing the electrode matching is the guaranty.

Acknowledgments

This work is supported by the National Science Foundation of China under Grant No.10875036 and No.51077035, Science and Technology Supporting Program of Hebei Provence No.09276712D, Doctor Foundation of Hebei University No.2009-178, and the Science Foundation of Hebei province No.A2010000182.

References

[1] Qiao Y S, Wang J and Wang J Y 2011 Environmental Monitoring in China (in Chinese). **27** 22
[2] Xu Q K and Chen H Y 2010 *Nonferrous Metals Science and Engineering (in Chinese).* **1** 45
[3] Jiang W and Zhao X H 2007 *Chinese Bulletin of Life Sciences (in Chinese).* **19** 78
[4] Neimarlija N, Demirdzic I and Muzaferija S2011 *Numer. Heat Transfer, Part A.* **59** 321
[5] Kim, S.C and Chun, Y.N 2009 *Int. J. Environ . Pollut.* **36** 337
[6] Niewulis A, Podliński J and Mizeraczyk J 2009 *J. Electrostat.* **67** 2
[7] Yuan J X, Yang L J, Zhang Q G, Liu P, Zhao H B, Du X M, Yan J X 2010 *Transactions of China Electrotechnical Society (in Chinese).* **25** 24
[8] Yang Z X, Xiang X D and Chen W S 2008 *High Voltage Engineering (in Chinese).* **34** 1062
[9] Kang Y M and Chi J H 2005 *Journal of Beijing Institute of Technology (in Chinese).* **25** 141
[10] Leonard G,.Mitchner M and Self S. A 1980 *Atmos. Environ.* **14** 1289
[11] Li Q 2008 Study on electrostatic treatment of coal dust pollution based on laser measurement (Baoding: Hebei University)
[12] Wang F J 2004 *Computational Fluid Dynamics Analysis.* (Beijing, Tsinghua University Press) **p** 7-9
[13] Li R X 2008 *The basis of the finite volume method* (Beijing: National Defence Industry Press) **p** 2-3
[14] Li Q, Li H F, Sun X R, Zhang W Y and Wang H 2010 *High Voltage Engineering(in Chinese).* **6** 2729
[15] Li Q, Yang Z Y, Wang Q Y, Liu Z Q and He S J 2011 *Chin. Sci (in Chinese).* **56** 2947

A novel electrostatic precipitator

Minkang Tang[1*], Liqian Wang[2], and Zhigui Lin[2]
1 Jiangxi University of Science and Technology
2 Jingyu Environmental Protection Equipment Manufacturing Ltd. Benxi, Liaoning

Email: tmkjxust@126.com

Abstract. ESP (Electrostatic Precipitation) has been widely used in the mining, building materials, metallurgy and power industries. Dust particles or other harmful particles from the airstream can be precipitated by ESP with great collecting efficiency. Because of its' large size, high cost and energy consumption, the scope of application of ESP has been limited to a certain extent. By means of the theory of electrostatics and fluid dynamics, a corona assembly with a self-cleaning function and a threshold voltage automatic tracking technology has been developed and used in ESP. It is indicated that compared with conventional ESP, the electric field length has been reduced to 1/10 of the original, the current density on the collecting electrode increased 3-5 times at the maximum, the approach speed of dust particles in the electric field towards the collecting electrode is 4 times that in conventional ESP and the electric field wind speed may be enhanced by 2-3 times the original. Under the premise of ESP having a high efficiency of dust removal, equipment volume may be actually reduced to 1/5 to 1/10 of the original volume and energy consumption may be reduced by more than 50%.

1. Introduction

Electrostatic precipitation technology is widely used in industrial production such as mining, the cement manufacturing industry, the production of building materials and boiler dust removal, to separate dust particles in the industrial emissions. Because of its' large size, high cost and energy consumption, the scope of application of ESP has been limited to a certain extent. In this paper, to solve the problem of the deposition of dust on the corona electrode and the shielding of the electric field, the original structure of the ESP is studied and developed according to the theory of ESP and several new technologies. In this way, important technological breakthroughs in improving electric field efficiency and equipment safety, and reducing energy consumption are made which works as a positive guide in expanding the application of ESP technology.

2. The mechanical transformation of ESP and its principle

When the ESP runs, dust particles which are influenced by the electric field will run to the collecting electrode. In practice, however, it is confirmed that the effect on particles of the electric field force near the corona electrode is very small (see figure 1).

According to the analysis of structure of traditional ESP, when dust particles are very close to the

[*] To whom any correspondence should be addressed.

corona electrode, it is very difficult for the electric field force to drive the dust to the collecting electrode. On the contrary, the particles will run to the corona electrode, resulting in abrasion of the corona electrode thus weakening the discharge performance of the corona electrode and reducing the dust removal efficiency [1].

To ensure and improve collection efficiency, in the design of a conventional electrostatic precipitator the electric field wind speed is generally limited to 1.2 m s^{-1}, and the length of the electric field generally to 8-12 m.

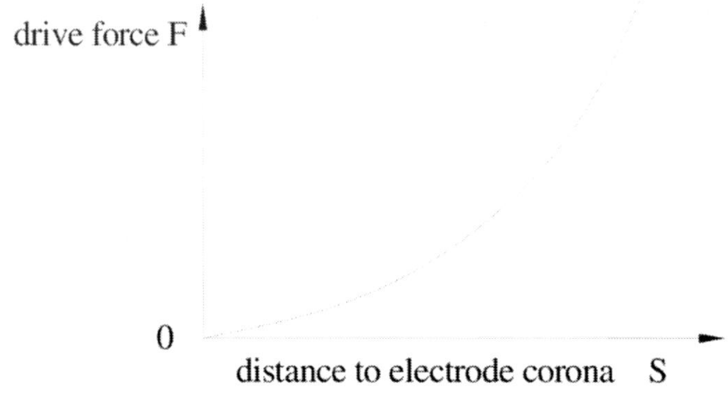

Figure 1. The connection between the position of dust and drive force.

After electric polarization of the dust, the dust and corona electrode have the same polarity. So the corona charge is excluded by the corona electrode. According to Coulomb's law, the force between the point charges is inversely proportional to the square of the distance between them, so the dust particles close to the corona electrode have a greater repulsive force to the charge released by the corona. This is the main reason for the electric field shielding.

According to the electric field distribution of the conventional electrostatic precipitator, the electric field within the electrostatic precipitator is divided into two parts, the one close to the corona electric field is called the inefficient zone, and the one close to the collecting electric field is referred to as the high efficiency zone (see figure 2).

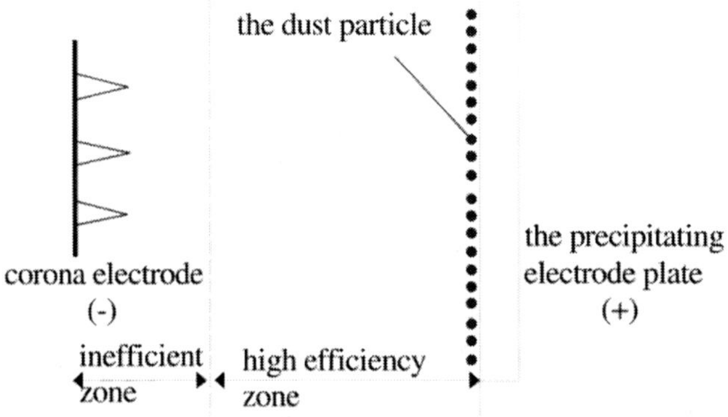

Figure 2. The electric field of disc-shape corona.

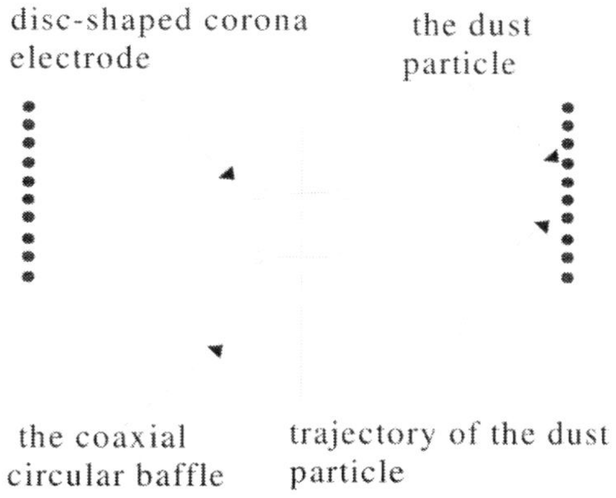

Figure 3. Schematic diagram of electric field parts.

In the cylindrical electrostatic precipitator a disc-shaped corona electrode is installed, and a coaxial circular baffle is inserted into the disc-shaped corona electrode. This may assist the dust stream to avoid the inefficient areas of the electric field at the initial stage as it enters the electric field (shown in figure 3).

The corona is located in the center of the cylinder, and the inner wall of the cylinder is the collecting pole (the ground electrode). The dust stream flows through the electric field in the cylinder from the bottom-up. This device obtained a national invention patent in 2007. Confirmed by several years of practice, while achieving the same level of dust removal, the electric field length of this device can be reduced to about 1/10 of conventional precipitators, and the application of dust concentration can be 80% higher than conventional precipitators [2].

Although the baffle in the center of the cylindrical electrostatic precipitator occupies part of the effective cross-sectional area, according to calculations, when the baffle diameter is half of the cylinder diameter, that is the cross-sectional area is a quarter of the cylinder, dust removal efficacy will increase more than four times the original. Structural analysis shows that this corona electrode can not only be consistent with the technical requirements of producing non-uniform electric field corona current, but also increase the effective area of the corona zone in the electric field (3-5 times larger than traditionally). The driving speed ù (m s^{-1}) of the charged dust particles in the electric field can be four times greater than the empirical value.

According to the patented technology the corona electrode also causes automatic cleaning of the spikes. That is different from the traditional rapping. It can timely and effectively clean out the deposition on the spikes and block board, so that the corona electrode is always in an efficient state. Intense corona discharge causes the electric field current density to improve from 0.35 mA m^{-2} to 1.5-2.5 mA m^{-2}, which greatly improves the collection performance of the electrostatic precipitator [3].

When dust particles with low resistivity are adsorbed to the collecting electrode in the conventional ESP, dust will soon be discharged, so that they lose adsorption ability and return to the electric field. Oscillating repeatedly in such a way, dust particles will eventually escape from the electric field, thereby reducing the efficiency of the Electrostatic Precipitator. Practice shows that because of different characteristics of the dust, even with the same resistivity, various particles will behave differently from each other. The above dust oscillation phenomenon will be evident when the surface of the collecting plate is relatively smooth. To solve the problem of low resistivity dust collecting, the surface of the inner wall of the cylindrical electrostatic precipitator (collecting electrode) was grooved and coated with a layer of polarized material whose resistivity is different from the inner wall of the cylinder, thus improving the dust collection performance. Even if the wind speed of the electric field increased to 2-3 times the original, the dust oscillation phenomenon never occurred.

3. Power transformation principle

The traditional high-voltage power of the corona electrode is DC, rectified and transformed from alternating current. Output voltage waveform is shown in figure 4. The actual corona voltage waveform is shown in figure 5, where "U1" is the compensated voltage of the electric field distribution. Because the initial corona current voltage is higher than U1, the corona current can only generate when the voltage is high voltage waveform. Corona current waveform is shown in figure 6.

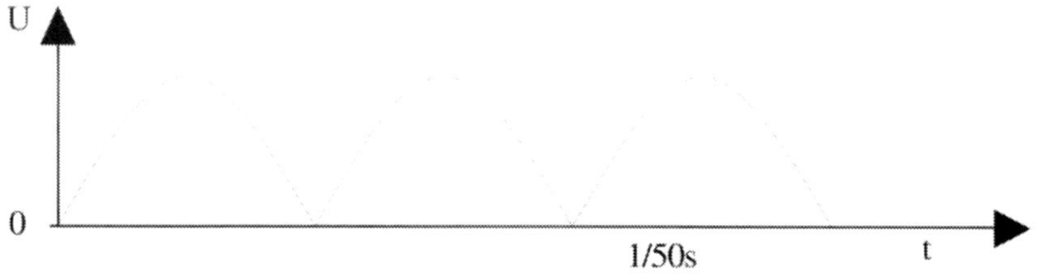

Figure 4. Waveform of voltage output of alternating current 50Hz.

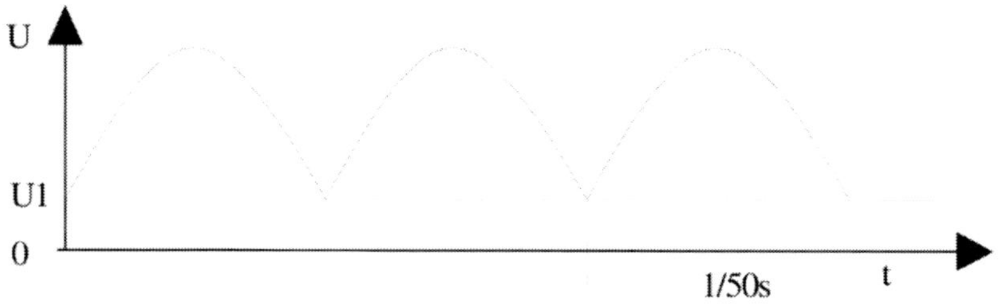

Figure 5. Waveform of corona voltage output of alternating current 50 Hz.

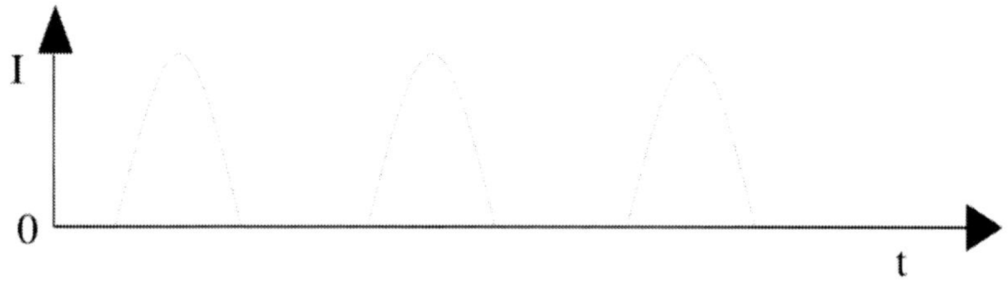

Figure 6. Current waveform of the alternating current 50 Hz.

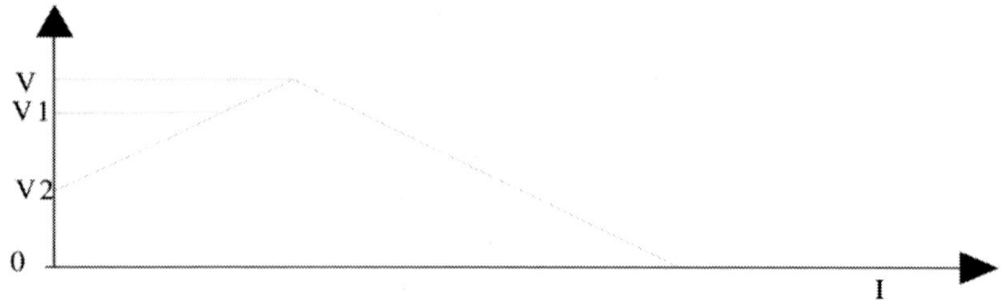

Figure 7. Connection between voltage and current of the electric field.

In the electrostatic precipitator, the motion of dust particles to the corona electrode is mainly affected by the corona current (ion wind) generated by the corona electrode, so the valid value of the electric field is represented by the shaded area of the current waveform (shown in figure 6). It can be deduced that there is the possibility of great variation in the performance of the electric field power supply. In order to improve the discharge effect, the voltage is often increased to the breaking point in a traditional power supply (shown in figure 7, *V* is the breakdown threshold, V_1 is a security value set to prevent the electric field breakdown, V_2 is the initial voltage which can produce corona current), but because the electric field voltage cannot undergo unlimited increase, its effect is limited.

In this study, a high-frequency electronic control boost technology was used, and the electric field voltage is fully compensated at a high frequency state by the internal self-distributed capacitance. So the corona current remains in a steady status and the corona electrode is maintained in a continuous discharge status (shown in figure 9).

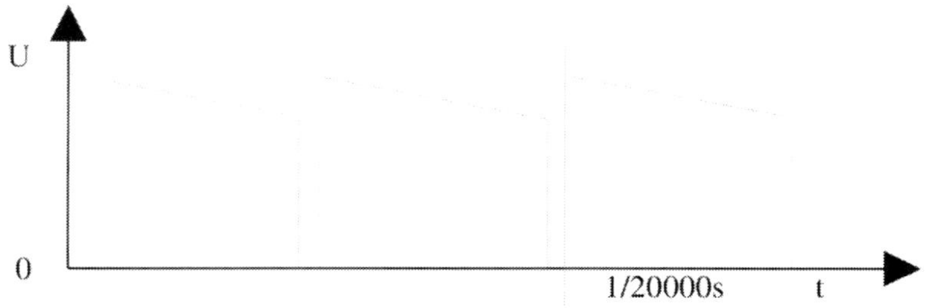

Figure 8. Waveform of 20 kV high frequency voltage.

Figure 9. Current waveform of 20 kV frequency corona electrode.

It is proven that efficiency can increase by 40%, or more, than the frequency power when the high-frequency power is combined with a reliable control system.

The ESP is applied to different working environments. When the operation is complex, it is very easy for damage to be caused by human intervention. But it will be troublesome for users if a hand operation is needed. So, in order to reduce manual operating procedures, dust removal machinery must be equipped with automatic control systems. Maintenance should also be simple.

The threshold voltage automatic tracking technology is one of the most advanced operating modes of the current development of ESP's high voltage electric field power supply system. The electric field voltage will be controlled automatically under the critical point, ensuring the electric field is in a steady and efficient working condition. This technology is superior to the commonly used constant current power supply. It can significantly improve the ESP's work efficiency. Its' high degree of automation can reduce manual procedures. This aspect of performance is particularly important, especially when in constantly changing operating conditions. Components have been developed with the equipment. The control system has auto-complete functions to start the necessary control procedures. The electrical equipment volume is about 1/20 of the same functions of traditional equipment.

 doi:10.1088/1742-6596/418/1/012066

4. Principle of threshold voltage automatic tracking technology

It is easy to cause spark discharge when changes happen in the nature of the dust stream in an actual conventional ESP run-time. That is, the electric field breaks down. If the voltage value remains as originally set, there will be a continuous electric spark discharge phenomenon. This will affect the normal operation of the equipment. If the constant current power was the ESP using, a person will need to adjust and reset the current value.

Threshold voltage automatic tracking technology can not only reset the voltage at a lower value automatically after an electric spark, but also allow the voltage to return to the original value gradually over a certain time, and the equipment will continue running without any effect.

As is shown in figure 10, AB is the initial set voltage value, C is the electric field voltage when at first break down, DE is the set voltage value in the transition time, and EF is the voltage of the slow recovery, which will finally return to the initial voltage value FG. If the electric field varies significantly in a short period of time, there will be multiple breakdowns (see figure 11). Even if the electric field does not return to its original values, the electric field will not breakdown within a short period of time but will continue the normal work status (see figure 11, dashed line). When an accidental short circuit happens in the electric field, power output will cease, and power consumption will be minimized.

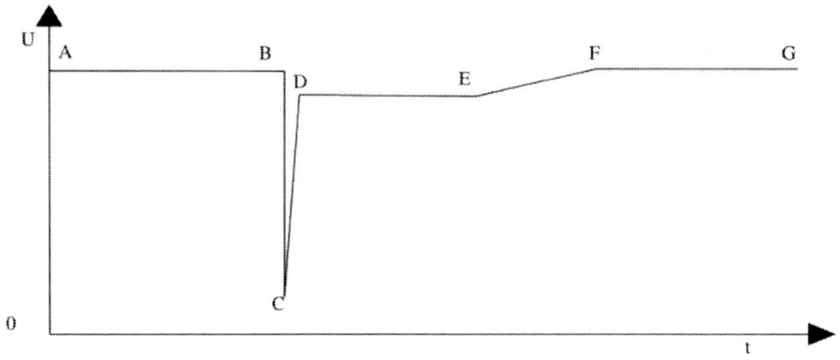

Figure 10. The connection between voltage and time at the first break down.

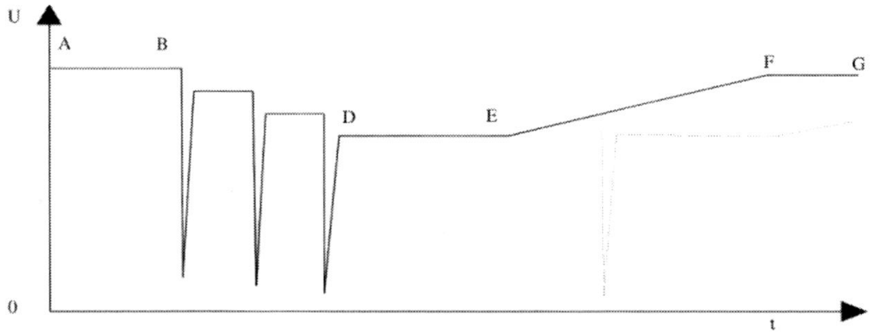

Figure 11. Connection between voltage and time after multiple breakdowns.

At the present time, in order to reduce manual processes, provide intelligent control and achieve power regulation, the voltage is automatically accurately tracked relying on microcomputer intelligent control procedures and also by detecting spark-rate by a complex signal enforcement system.

In order to simplify and miniaturize the complex control facilities of ESP, simulation technology was adopted for all the electronic hardware and the intelligent control system has been designed. The practical application shows that it can meet all design requirements of the intelligent electric field control, and there are no program malfunctions caused by such problems as signal interference. The structure of the facility is greatly simplified.

The new electronic control facilities and equipment are in the same structure, without additional size yet providing good performance. This not only reduces unnecessary operating procedures, but also reduces costs by more than 60% over traditional control devices.

In this structure, there are three different step-up speeds of the electric field voltage as follows:

1. Fast (in approximately 0.2 seconds the voltage returns to 100% rating from zero);
2. Medium speed (in approximately four seconds the voltage returns to 100% rating from zero);
3. Low-speed (in approximately eight minutes the voltage returns to 100% rating from zero). If the power is not turned off after a spark occurs in the electric field, the spark will continue. This will affect the normal operation of the equipment. If the interrupted electric field voltage is in fast recovery, it will enter a transitional period (about 15 seconds) before reaching 90% generated spark voltage, enter the low-speed recovery time.

Aside from the failure or restart of the system voltage caused by a spark, to maintain a normal startup state the electric field voltage will enter a medium-speed state when it reaches 50% of the rated voltage. It can be up to 100% in about two seconds.

Fast, medium-speed, low speed and transition time are a set of values based on experience. They have been identified in the assembly of electronic components, and need no adjustment. If there is indeed a need to change the parameters, they must be adjusted by a professional in the production plant.

5. Technical improvements of a large-scale electrostatic precipitator

Due to large size and high cost, the scope of application of small ESPs has been very limited. ESPs require a large electric field, and the series connection of electric fields. Dust stream has to go through a number of electric fields then can be collected. However, this layout may easily cause a re-entrainment in the dust emission. This may greatly affect the efficiency of the electrostatic precipitator. (Shown in figure 12) [4].

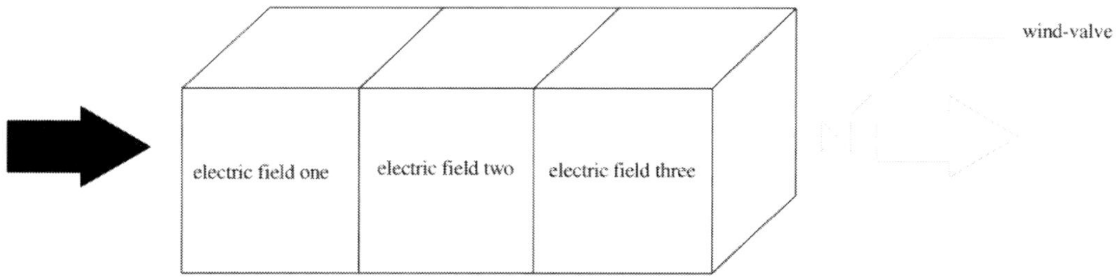

Figure 12. Device will be effected when wind is closed.

When the ESP runs, the main energy consumption is electrical energy for fan. The ventilation resistance of electric field in ESP can be regard as negligible because it is usually less than 15 Pa. However, ESP requires the dust stream wind speed through the electric field to be very small (1 m s^{-1}). Compared with the usual piping wind speed (about 12 m s^{-1}), the uniform airflow facilities are the greatest source of obstacles of ESP. The lower the wind speed, the more demanding the uniform airflow facility is and the higher the resistance facilities (about 200 Pa). Due to the complex system of the present ESP, it can only be applied to large facilities utilizing ventilation ducts. Therefore, it increases the overall resistance of the device.

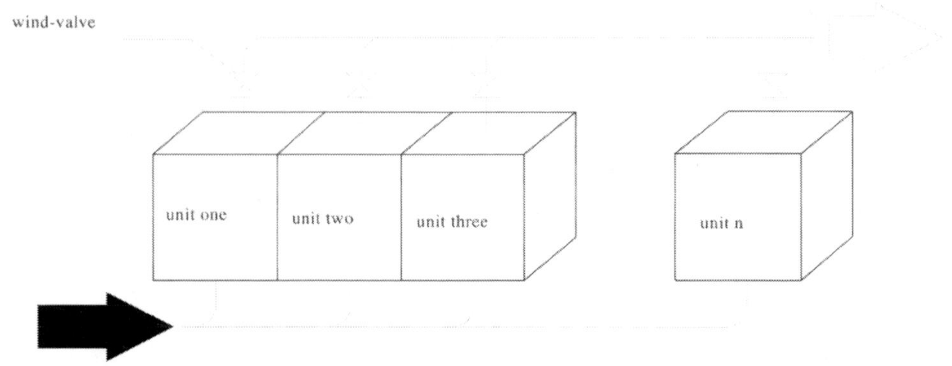

Figure 13. Close a single valve has little effect on the whole run.

An ESP with small size and strong automated-control function is provided in this study. The mechanical and electrical integration structure, which forms an independent dust removal unit, makes the application and the overall layout of the large-scale precipitator is more flexible. Each ESP unit can discharge steam independently. This has little impact on the overall equipment operation. (See figure 13).

Compared with the same functional product, the volume of this ESP can be reduced to 1/5-1/10 of the original, while the wind speed of the electric field can be 2-3 m s^{-1} larger. Its uniform facility of airflow is relatively simple which is suitable for near point installation. Its airflow resistance can be about 50% lower than before and fan power can be more than 50% lower than originally. Coupled with the improvement of the electrical energy efficiency, power consumption can be reduced by 50% for the new product. If installed indoors the purified air can be discharged into the room, the inflow of the outside air will be prevented. A comfortable indoor temperature can be maintained even in winter.

The methods of dust removal by ESP can be either turning or rapping. The practical application shows that turning off the electric field power is a very important measure during rapping. It can significantly improve the cleaning effect. A high-voltage output mode of multiple combinations is more conducive to the control requirements of each unit of the electric field, and the outside line is simple, clear and reliable (see figure 14).

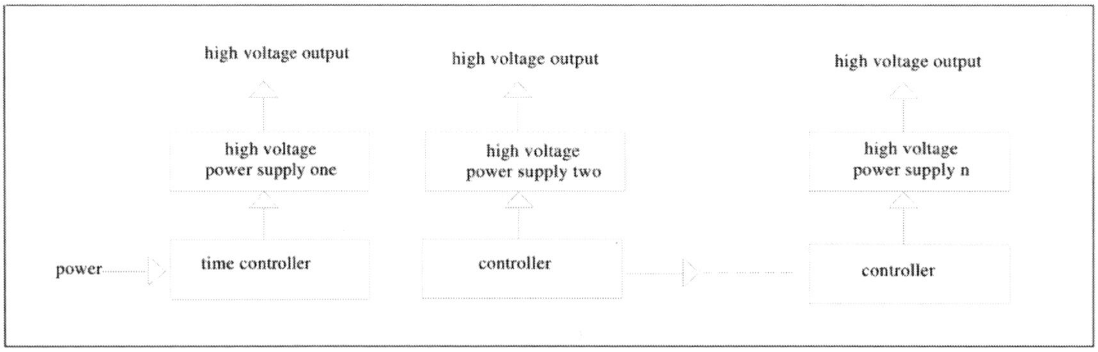

Figure 14. A high voltage output of many rows can manage every electric field neatly.

6. The ESP used in high dew environments

When the ESP is applied in environment of that the air flow will be dewed, the insulators supporting high-voltage corona electrodes will be burned because of the dew, significantly limiting the application of the electrostatic precipitator. The new patented product adopts a new structure, and overcomes this traditional disadvantage (see figure 15).

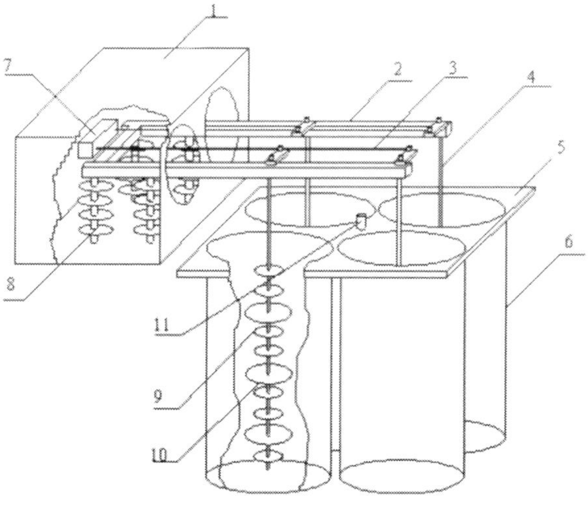

1——box
2——corona electrode stent
3——pull rod of corona electrode
4——component of corona electrode
5——baffle
6——dust collecting box
7——control component of pull rod
8——insulator
9——component of corona disc-shaped corona
10——component of corona disc-shape insulation
11——supply shaft of dust exhausting

Figure 15. Inner structure of the component of an ESP.

The insulator is set on one side of the ventilated area, far away from the air flow area. Therefore, it is also suitable for humid environments.

7. Simplifying the operation and management procedures of the ESP

Due to the complex control procedures, especially the need for a large professional management staff with a traditional ESP, troubleshooting may be very difficult. This ESP, which greatly simplifies operating procedure, needs no professional management staff. The machine is made of modular electrical components, making failure analysis and maintenance easier. Furthermore, the operation is simple.

8. Conclusion

- A component of corona electrode with “self-clean” function has been developed, which improved the corona intensity significantly, increased the density of the corona current and overcame the phenomenon of occlusion of the corona.
- The trench and coating with different resistivity on the inner wall of the ESP cylinder can be adapted to dust collecting with different resistivity. It can avoid re-entrainment of dust particles in ESP.
- The “threshold voltage automatic tracking technology” can control the traditional electrostatic precipitator spark discharge phenomenon and improve the performance of the automatic control of ESP.
- The miniaturized mechatronics integrated into ESP can either be used alone, or in combination for dealing with the big contaminated air stream.

References

[1] Tang M 2010 *Electr. Prop. of The Dust part. in The high voltage Electrost. Field* (Beijing:

Chemical Industry Press)
[2] Lin Z 2010 *The electrode assembly with a self-cleaning function* (invention patent: the People's Republic of China State Intellectual Property Bureau)
[3] Lin Z 2009 *An electrostatic precipitator* (ZL200410021153.0 invention patent certificate the People's Republic of China State Intellectual Property Bureau)
[4] Lin Z 2010 An *uniform electrode assembly* (ZL200710012238.6 invention patent the People's Republic of China State Intellectual Property Bureau)

EHD-assisted plasma electrostatic precipitator for simultaneous PM collection and incineration

T Yamamoto[1*], S Maeda[1], Y Ehara[1], A Zukeran[2], H Kawakami[3]
1 Department of Electronics and Electronics Engineering, Tokyo City University, 1-28-1 Tamatsutsumi Setagaya-ku, Tokyo, 158-8557, Japan,
2 Department of Electronics and Electronics Engineering, Kanagawa Institute of Technology Kanagawa, 243-0292, Japan
3 Road System Engineering Department Fuji Electric Co. Ltd. Tokyo, 191-8502, Japan

E-mail: yama@tapjp.com

Abstract. The electrohydrodynamically-assisted (EHD) plasma electrostatic precipitator (called plasma ESP hereafter) was developed not only to collect low-resistive fine particles generated from diesel engine but also to incinerate captured particulates within the same reactor. The plasma ESP is to utilize EHD to transport the charged particles effectively into the low electrostatic field region where electrostatic repulsion force acting on particles by induction charge is very small, resulting in suppression of PM reentrainment. At the same time, collected particulates are incinerated by ozone and oxidation product of NO_2 in the flue gas by the surface discharge plasma in the same unit so that no particle storage and handling device are required. The particle-size dependent collection efficiency and particle incineration were evaluated by the Scanning Mobility Particle Sizer (SMPS) for particle size of 20-500 nm and the particle counter (PC) for particle size of 300-5,000 nm. The results indicated that the plasma ESP was able to collect more than 90 % collection efficiency and no reentrainment occurred for all particle size measured. The plasma ESP leads to more economical and compact ESP feasible. No PM handling device and storage are required and therefore has high potential for marine engine and automobiles.

1. Introduction

The particulate matters (PMs) emitted from diesel engine exhaust are low resistive and extremely small in the range of 70~100 nanometers (nm). These particles are penetrated into alveolus and extremely harmful to human health. These particles are generated from various emission sources such as diesel automobiles, marine engines, power generation engines and construction machines. The use of diesel particulate filter (DPF) was widely used for the collection of automobile diesel PM but was not cost effective, especially for marine engine emission where PM concentration is often higher than 100 mg m^{-3}. The collection of low resistive PM has been known to be extremely difficult by the conventional electrostatic precipitators (ESPs). The low resistive diesel engine particles are detached from the collection plate where the electrostatic repulsion force due to induction charge exceeds particle adhesion force on the collection

* To whom any correspondence should be addressed.

electrode. These PMs were escaped without collection in the ESP. This phenomenon has been known as particle reentrainment or resuspension, resulting in poor collection efficiency.

There are many references for controlling high resistive particles but few literatures for controlling low resistive particles [1-8]. Two-stage ESP employing charging section by DC field, followed by the collection section by low frequency AC field including the trapezoidal waveforms in the range of 1~20 Hz has been investigated for the collection of diesel particles in tunnel [9-11], while the conventional ESP utilizes DC high voltage. Various collection plate designs such as corrugated collection plates were employed but were exposed to high electric field or not taken into account electrohydrodynamics (EHD) to transport the charged particles into mechanically trapping zone in the ESP. These concepts have limited success for minimizing the particle reentrainment. Recently, electrostatic flocking filter on the collection electrode was developed to capture fine diesel particles [12]. The combination of diesel particulate filter (DPF) combined with sliding discharge was developed to incinerate the captured particles [13]. The wet ESP was another strong candidate for this application but it creates water treatment as opposed to dry process.

Based on fundamentals of reentrainment theory, the new electrohydrodynamically-assisted electrostatic precipitator (EHD ESP) was developed to overcome the reentrainment in the conventional ESP [14, 15]. The EHD ESP, which utilizes the ionic wind to transport the charged particles effectively into the zero electrostatic field zone or pocket zone attached to the collection plate. The captured particles are trapped in the pocket where particles captured in the pocket zone were exposed to zero electric field, so that no electrostatic repulsion force on the particles due to induction charge was taken place. This is the major factor for reduction of particle reentrainment.

As an extension of the EHD ESP, the new EHD-assisted plasma ESP was developed. The low resistive diesel PM is effectively captured in the low electric field region and at the same time, collected particulates are incinerated by ozone and oxidation product of NO_2 in the flue gas by the surface discharge plasma in a single unit so that no particle handling device is required. This novel device is called EHD-assisted plasma electrostatic precipitator (abbreviated as plasma ESP hereafter). Note that the carbon particle incineration by ozone is effective when the flue gas temperature is above room temperature but NO_2 incineration as reaction product becomes effective when the flue gas temperature exceeds 250 °C [16]. When DC and surface discharge for the plasma ESP are operated at the same time within the same reactor, the small carbon particles can be incinerated during the collection process. The particle-size dependent number density and mass collection efficiency, and particle incineration were evaluated by Scanning Mobility Particle Sizer (SMPS) and particle counter (PC). This device is particularity attractive for automobile and marine diesel engine emission control because no additional particle handling device is required.

2. Concept of plasma Esp

The plasma ESP used for this experiment was shown in figure 1 (a)-(b) for cross section views from flue gas inlet and reactor frontal side and 1(c) for three dimensional view. The discharge electrode consists of 6.35 mm in diameter rod and 416 mm long tube where 18 mm diameter star electrode was equally spaced with every 34 mm interval, a total of 12 star electrodes, denoted as the center electrode. The inner electrode was 53 mm in diameter with the thickness of 1 mm, where 30 mm×30 mm square holes were punched every 10 mm spacing. The outer electrode was punched with 30 mm and 13 mm square holes every 4 mm spacing, which was attached to 70 mm OD and 65 mm ID dielectric quartz tube. The dielectric quartz tube was covered by the most outer ground electrode tube. All components were made of stainless steel. Their dimensions were designated in the figures. The minimum distance between the star electrode and the inner electrode was set at 20 mm.

The plasma ESP has two mode operations as mode (A) and (B). As mode (A) operation (DC on and AC off for particle collection), the center electrode was energized as negative DC high voltage, while all inner and outer electrodes as well as the most outer electrodes were grounded. Ionic wind is generated

Journal of Physics: Conference Series **418** (2013) 012067 doi:10.1088/1742-6596/418/1/012067

towards inner electrode but charged particles are escaped from small area of inner electrode and accelerated to move towards outer electrode by EHD action. Majority of particles are collected on the outer electrode and quartz tube surface, and small fraction of particles were collected on the inner electrode on both sides.

The ions for collected charged particles are drained off as soon as particles touch the ground plate for conductive particles or low resistivity particles such as diesel particles. Opposite polarity of charge as positive ions are immediately accumulated on the particles due to the induction charge, which experiences the particle repulsion force, resulting in particle reentrainment. This force is proportional to the square of the electric field and particle diameter [14]. When the repulsion force exceeds the particle adhesion force, particles are detached from the surface and transported into the inter-electrode space, where negative ions exist. Particle are again charged negatively and attracted to the collection plate. These processes are repeated and finally go through the ESP without collection, i.e., particle reentrainment. The detained theoretical investigation including the electrohydrodynamics on particle reentrainment was reported earlier [14].

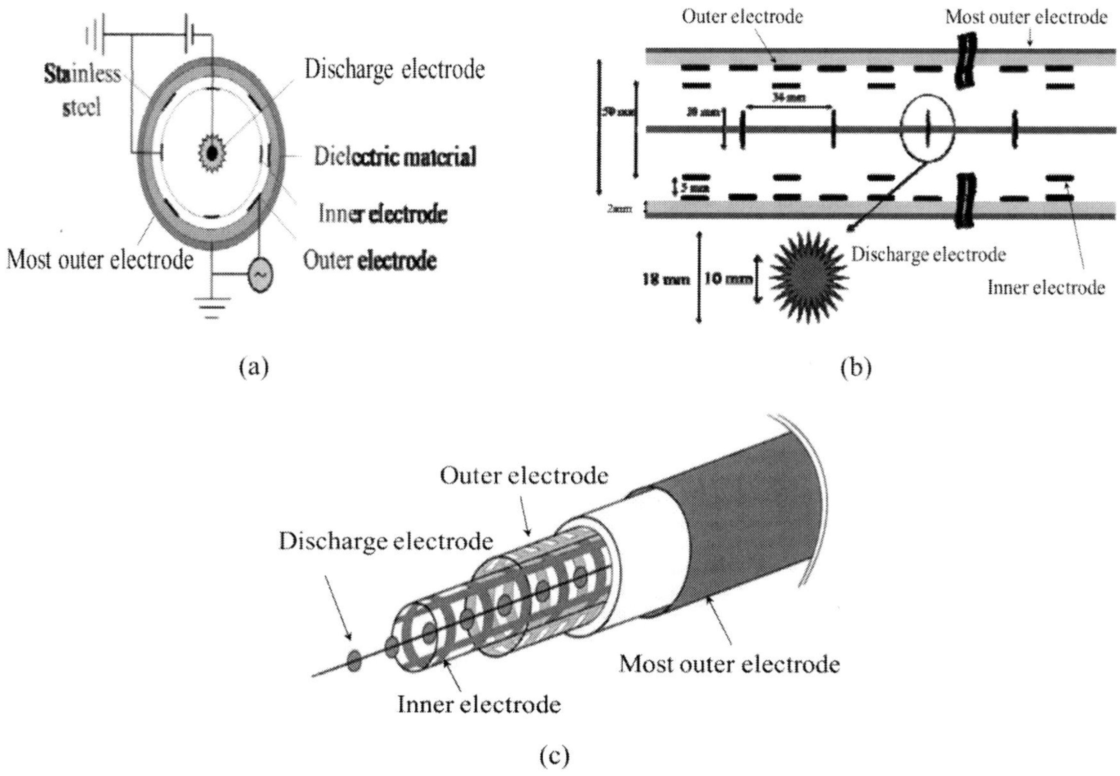

Figure 1. (a) Cross section view for the plasma ESP; (b) Cross section view for the plasma ESP; (c) Three-dimensional view of the plasma ESP.

As for incineration process as mode (A) operation, the surface discharge (6 kV with 10 kHz and 60 mA) was applied between the outer electrode and the most outer electrode, while the inner electrode and the center electrode were grounded for mode (A) operation. Ozone was generated and diffused into the inner and outer electrode surface for particle incineration. When the gas temperature is higher than 250 ^{o}C, NO in the flue gas is oxidized to form NO_2 by ozone and O radical and carbon particles can be incinerated.

Journal of Physics: Conference Series **418** (2013) 012067 doi:10.1088/1742-6596/418/1/012067

The chemical reaction for particle incineration are as follows, where T_g=gas temperature: NO+O or $O_3 \rightarrow NO_2+O_2$ ($T_g > 20$ °C), $C+2NO_2 \rightarrow CO_2+2NO$ ($T_g > 250$ °C , $C+O_3 \rightarrow CO_2+1/2O_2$ ($T_g > 20$ °C). The speed of particle incineration depends on particle size and concentration and ozone concentration. Two stages of plasma ESP are required for continuous diesel emission control for mode (A) operation: one operation is for particle capture and the other is for particle incineration so that two plasma ESPs are required for the continuous emission control.

On the other hand, as for mode (B) operation (DC on and AC on) for the purpose of simultaneous particle collection and incineration, particles collected to the inner and outer electrodes are incinerated by the surface discharge. Small particle may incinerate before reaching to the electrodes, which depends on particle size and ozone concentration. For both mode (A) and (B) operation, particle handing device is totally eliminated. This operation mode may be applicable to light loading at present level. In both cases, particle storage and handling device are totally eliminated so that it is particularly attractive such as automobile and marine engine applications. In the present paper, the results of mode (A) were reported.

3. Experimental setup

A diesel engine generator (Yammer Co., Ltd., YDG200A-5E, direct injection type for a single cylinder, displacement volume of 200 cc, maximum electric power output of 1.7 kW) using light oil was used. The constituents of the diesel PM measured were 99 % of C, 0.1 % of Si, 0.07 % of Fe, 0.1 % of Ca, 0.4 % of S, and 0.03 % of Zn. In order to determine the number particle density in the ESP, the flue gas was diluted approximately 100 times by ambient air and particle size-dependent number densities before and after the ESP was determined by the Scanning Mobility Particle Sizer (SMPS, Model 3034) for the particle size ranged 20-800 nm and the particle counter (PC, Rion KC-01C) for the article size of 300-5,000 nm, respectively. The exhaust gas temperatures were measured at the plasma ESP inlet and outlet by thermocouples. The gas velocity was measured by the hot wire anemometer (Kanomax). Ozone concentration was measured by ozone monitor (Ebara Jitsugyo. EG-2001). Figure 2 shows the experimental setup for evaluation of the plasma ESP.

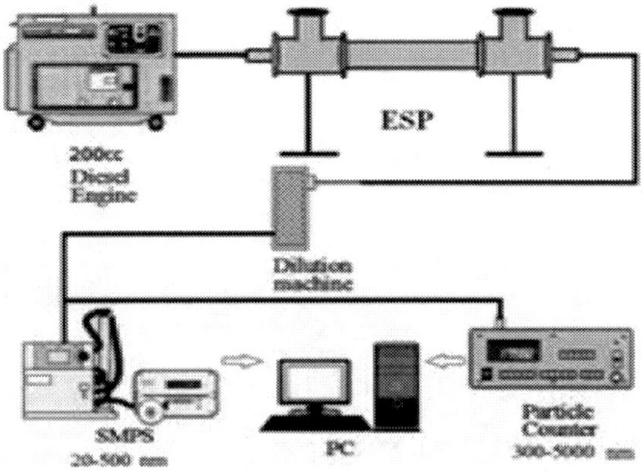

Figure 2. Experimental Setup for the plasma ESP evaluation.

4. Results and discussion

The voltage and current relationship for the plasma ESP was shown in figure 3. The onset voltage was 6.0 kV and the spark-over voltage was 18 kV with 4.2 mA at room temperature before particles were introduced. When the load was 0.5 kW which was 25 % load, the flue gas temperature was 135 °C at the ESP inlet and 65 °C at the plasma ESP outlet so that the gas velocity of the plasma ESP was 1.9 m s^{-1}

based on the average gas temperature.

Figure 4 shows the particle-size dependent number density for the plasma ESP, which was measured by SMPS when the applied voltage was modest V=-10 kV with I=0.5 mA. The maximum number density was 10^8 particles cm^{-3} at particle size of 70 nm. One order of magnitude reduction of number density was achieved for particle size of 20-500 nm.

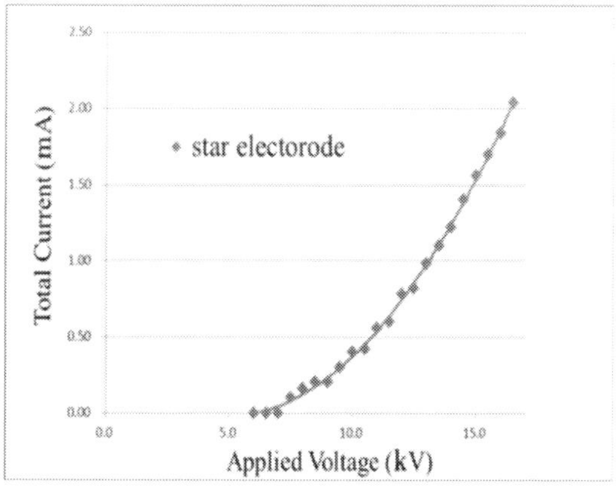

Figure 3. Voltage and current relationship for the plasma ESP.

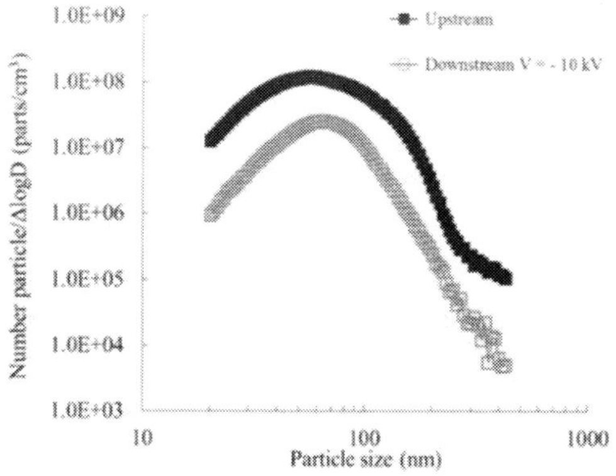

Figure 4. Particle-size dependent number density distributions by the SMPS when the load is 0.5 kW and gas velocity is 1.9 m s^{-1}.

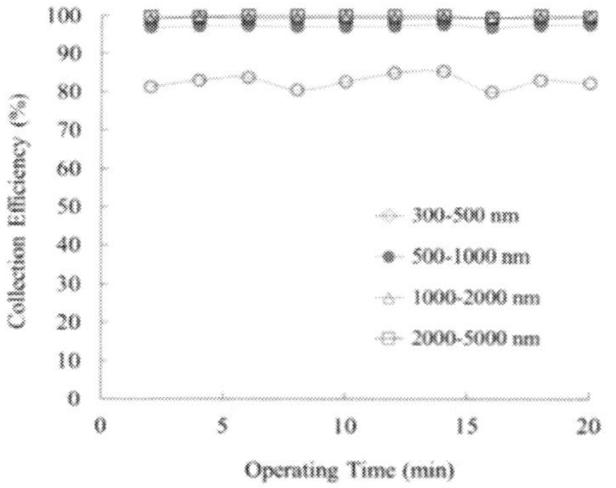

Figure 5. Particle-size dependent number-density distribution for the plasma ESP measured by the PC when the load of 0.5 kW and gas velocity of 1.9 m s^{-1}.

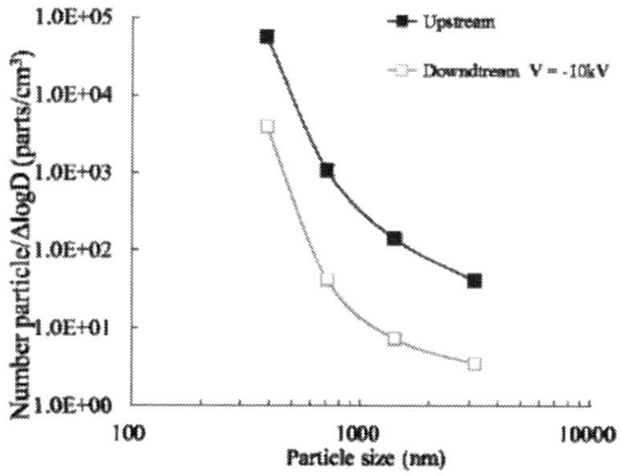

Figure 6. Time-elapsed number density collection efficiency for the plasma ESP measured by PC when the load is 0.5 kW and gas velocity is 1.9 m s^{-1}.

The particle size-dependent collection efficiency for particle size in the range of 300-5,000 nm measured by the PC was shown in figure 5. Again, one order magnitude particle number density reduction was achieved for 300-500 nm and further reduction was observed with increased particle size.

Particle reentrainment occurs for the particle size of greater than 1,000 nm based on previous

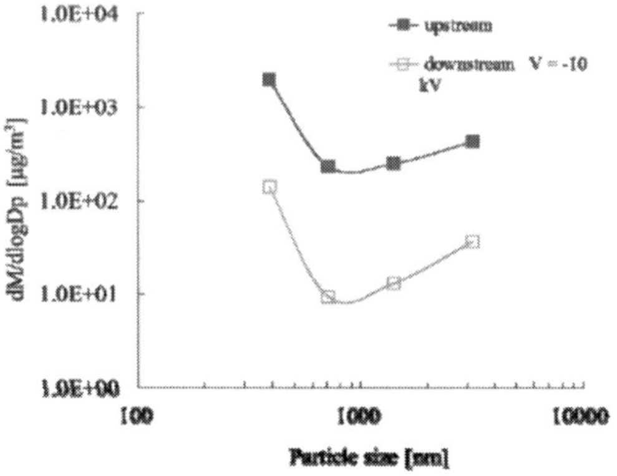

Figure 7. Particle-size dependent mass-base density distribution measured by SMPS for the plasma ESP (η=86.6 %).

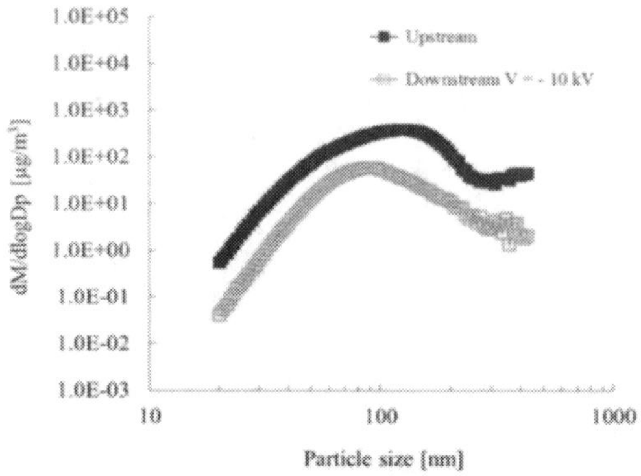

Figure 8. Particle-size dependent mass-base density distribution measured by PC for the plasma ESP (η=93.0 %, η_t=90.0 %).

experimental results [15] so that the time-dependent number density collection efficiency as a function of particle size was evaluated for 300-5,000nm when the load is 25 % as shown in figure 6. The number collection efficiency was greater than 95 % for particle size ranging of 500-5,000 nm and 80~85 % for 300~500 nm during 20 min operation. The collection efficiency was increased with increased particle size and minimum for 300 nm range which implies the classical electrostatic charging theory, indicating no particle reentrainment occurred during 20 min operation. This was attributed to increased adhesion force by moisture content, that is, capillary condensation associated with the contact between particle and the surface, due to low gas temperature when the load was 25 %. Also, the particle surface resistivity becomes more affected over the volume resistivity.

Note that when treating the marine diesel engine emission, A and C grade heavy oil used for diesel emission contains large fraction of soluble organic fractions (SOFs) as well as particulates, When particulates were collected at higher temperature, SOFs and other vapor phase compounds were escaped from PM collection device. SOFs are condensed and converted to solid particulates at lower temperature. For this reason, it is difficult to obtain good collection efficiency when the ESP is operated at higher temperature for treating the marine diesel engine emission. Hence, it is important to remove gas-phase SOFs using nonthermal plasma or other means before the ESP or to operate the ESP at lower temperature (possibly less than 90 °C) to ensure to capture SOFs, bound H_2O, and sulfate as liquid or solids phase because ESP cannot collect gas phase substances. Tsukamoto [17] was also indirectly suggested the similar operation.

Figure 7 and 8 show the corresponding particle-size dependent mass-base collection efficiency for particle size ranging from 20~500 nm measured by SMPS and 300~5,000 nm by PC. The distribution of particle-size dependent mass collection efficiency was significantly changed over particle number density distribution as shown in figure 4 and 5. The collection efficiency (η) was 86.6 % for 20~500 nm and 93.0 % for 300-5,000 nm. The overall mass collection efficiency (η_t) was 90.0 %. Although the collection efficiency was predominantly determined by small particle size range based on previous findings [15], the collection efficiency was not significantly affected by small particle size for this case.

When the load is 1.0 kW (50 % load), the flue gas temperature was increased to 182 °C at the ESP inlet and 70 °C at the ESP outlet so that the average gas velocity was 2.1 m s^{-1} based on the average gas temperature. Figure 9 shows the particle-size dependent number density, measured by SMPS. When the voltage was applied, the peak number density increased to 200 nm, primarily due to particle agglomeration. Two orders of magnitude reduction of collection efficiency were obtained for particle size less than 100 nm. However, the collection efficiency became minimum at 200~300 nm ranges.

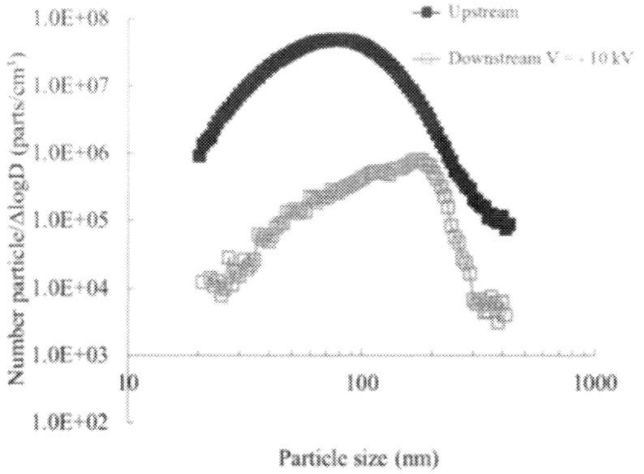

Figure 9. Particle-size dependent number density distribution by the SMPS when the load is 1.0 kW (50 % load) and gas velocity is 2.1 m s^{-1}.

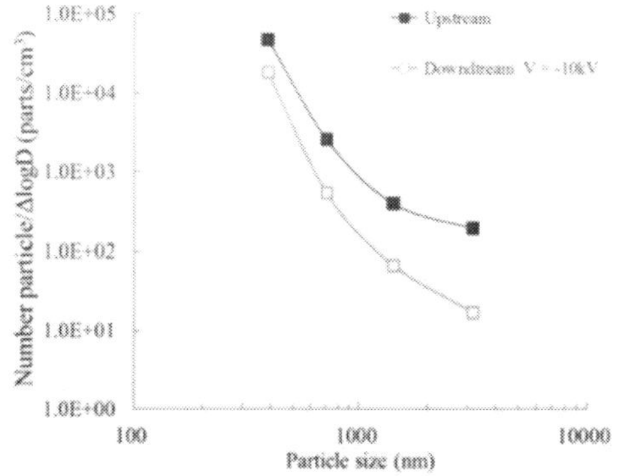

Figure 10. Particle-size dependent number density distributions by the PC when the load is 1.0 kW (50 % load).

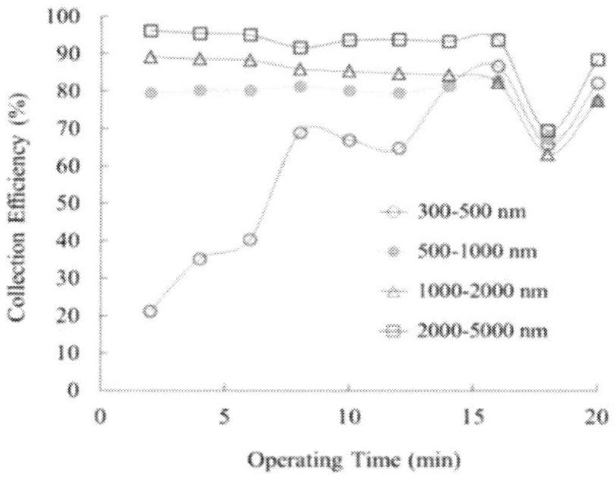

Figure 11. Time-elapsed number density collection efficiency for the plasma ESP when the load is 1.0 kW.

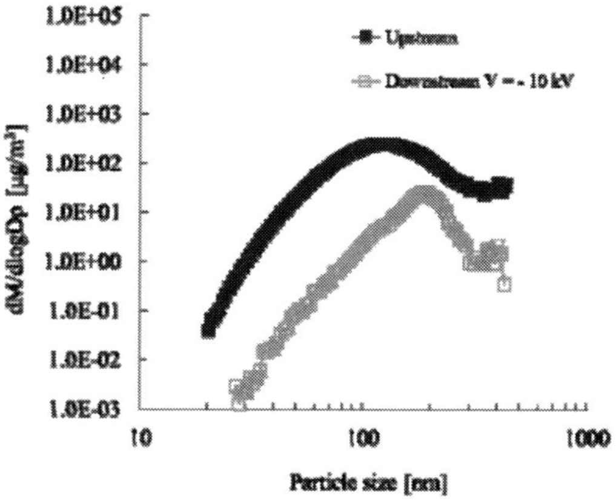

Figure 12. Particle-size dependent mass-base density distribution measured by SMPS for the plasma ESP (η=95.0 %).

The particle size-dependent collection efficiency for particle size in the range of 300~5,000 nm measured by the PC was shown in figure 10. The collection efficiency was lower between 200~500 nm ranges but one order magnitude reduction of particle number density was observed for particle size greater than 500 nm. The increased collection efficiency with increased particle size indicated that the collection efficiency becomes minimum at 200~300 nm by classical charging theory and no reentrainment occurred even for higher load or flue gas temperature.

Figure 11 shows the time-dependent number density collection efficiency as a function of particle size of 300~5,000 nm when the load is 50 %. The number collection efficiency was in the range of 20~85 % for particle size of 300-500 nm 80 % for 500~1,000 nm, 80~90 % for 1,000~2,000 nm, and 90~95 % for 2,000~5,000 nm over 15 min operation. The collection efficiency was increased with increased particle size which again ascribed by the classical electrostatic theory, indicating no reentrainment occurred.

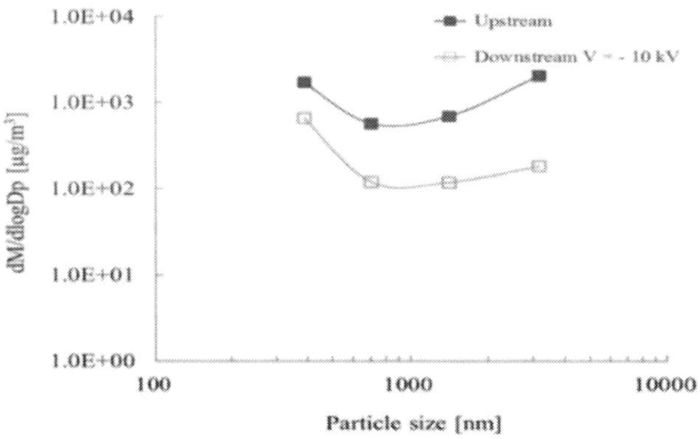

Figure 13. Particle-size dependent mass-base density distribution measured by PC for the plasma ESP (η=81.5 %, η_t =88.3 %).

This was ascribed by classical charging theory for field and diffusion charging regime and increased adhesion force by moisture condensation due to low gas temperature operation or increased particle surface resistivity when the load is even 50 %.

Figure 12 and 13 show the corresponding mass-base collection efficiency for particle size ranging 20~500 nm measured by SMPS and 300~5,000 nm by PC. The collection efficiency was 95.0 % for 20-500 nm and 81.5 % for 300-5,000 nm. The overall collection efficiency was 88.3 %. The collection efficiency was affected by small particle size range owing to high number density. The difference of collection efficiency was due to increased mass loading, gas temperature and primary gas flow velocity.

In order to incinerate the collected diesel particles, surface discharge plasma was induced with the applied AC voltage of 6.0 kV and the total current of 60 mA, whole all other electrodes were grounded for the operation mode (A). The ozone was generated on the surface of the outer electrode and ozone concentration of 47 ppm was obtained during 20 min operation. The amount of ozone generated during 20 min operation was 471 mg, which is able to incinerate 118 mg of carbon particles as described earlier. PM concentration and generation rate during 20 min operation with different load levels are as follows. When the load is 0.5, 1.0 and 1.5 kW, PM concentrations are 1.5, 15.4, and 85.0 mg m^{-3}, and PM generation rate are 8.95, 91.9, and 507 mg during 20 min operation, respectively. Based on the results, the plasma device is able to incinerate PM up to 75 % load with continuous plasma operation. Our plasma device is self-designed reactor and the outer electrode cylinder was not perfectly attached to the dielectric barrier quartz tube, which plasma was generated only at good contact point between the outer electrode and dielectric quartz tube so that strong plasma was not uniformly generated in the plasma reactor. The commercially available surface discharge reactor is able to generate significantly higher ozone generation. However, the fundamental concept of the EHD-assisted plasma ESP was successfully demonstrated. The development of the plasma ESP is particularity important for automobile and marine diesel engine emission control, where no particle storage is required.

Figure 14(a)-(b) show the pictures of the outer electrode before and after incineration and figure 15(a)-(b) also show the pictures of the dielectric quartz tube before 15(a) and after 20 min operation by the surface discharge. It was clear that diesel PM collected on outer and inner electrodes as well as quartz tube surface were perfectly incinerated by ozone during 20 min operation. The concept of plasma ESP was successfully demonstrated for collection of low resistive diesel PM collection as well as incineration, which reentrainment can be minimized.

Figure 14. (a) The picture of the outer electrode after 20 min ESP operation; (b) The picture of the outer electrode after 20 min incineration by the surface discharge operation.

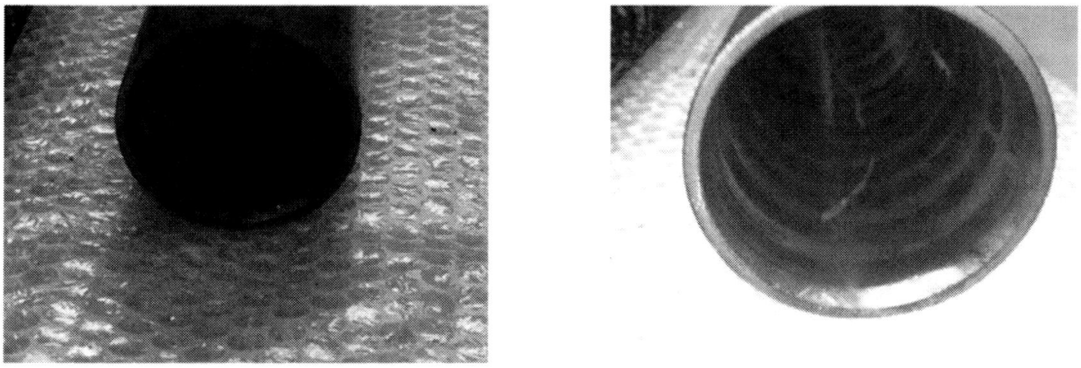

Figure 15. (a) The picture of the dielectric quartz tube after 20 min ESP operation; (b) The picture of the dielectric quartz tube after 20 min incineration by the surface discharge operation.

5. Summary

The novel EHD-assisted plasma ESP was developed to collect low resistive particles generated from diesel engines and simultaneously incinerate collected particles using a single reactor. The plasma ESP showed more than 90 % collection efficiency for low and medium load conditions when the gas velocity was in the range of 1.9-2.1 m s^{-1}. No reentrainment was observed due to low flue gas temperature, where adhesion force was dominated due to high moisture contents over electrostatic repulsion force. The collected particles were incinerated with ozone generated by the surface discharge. The plasma ESP leads to more economical and compact ESP feasible. No PM handling device and storage are required and therefore has a wide range of diesel emission control such as marine engine and automobiles.

Acknowledgments

Authors wish to thank you for support by Grant-in-Aid for Scientific Research (B) of the Japanese Society for the Promotion of Science and the Ocean Policy Research Foundation (OPRF).

References

[1] Bassett J D, Akutsu K and Masuda S 1977 *J. Electrostat.* **3** 311

[2] Felder R M, Medina E A 1985 *J. AIChE* **31** 82
[3] Takahashi T, Kawada Y, Zukeran A, Ehara Y and Ito T 1998 *J. Aerosol Sci.* **29** 1481
[4] Atten P, Pang H L and Reboud J L 2009 *IEEE Trans. IA* **45** 75
[5] Zimon A D 1969 Adhes. of Dust and Powder, Plenum Press (New York, London)
[6] Wen H Y and Gasper G 1989 *J. Aerosol Sci.* **20** 483
[7] Reeks M W, Red J and Hall D 1988 *J. Physics D, Applied Phys.* **21** 574
[8] Williams M M R 1992 *J. Aerosol Sci.* **23** 1
[9] Zukeran A, Ikeda Y, Ehara Y, Matsuyama M, Ito T, Takahashi T, Kawakami H and Takamatsu T 1999 *IEEE Trans. IA* **35** 346
[10] Zukeran A, Ikeda Y, Ehara Y, Ito T, Takahashi T, Kawakami H and Takamatsu T 2000 *EEJ* **130** 30
[11] Yasumoto K, Zukeran A, Takagi Y, Ehara Y, Takahashi T and Ito T 2005 *Int. J Env. Waste Manag.* **2** 399
[12] Sung B J, Aly A, Lee S H, Takashima K, Katsura S and Mizuno A 2006 *Plasma Process. Polymer* **3** 661
[13] Hayashi H, Takashima K and Mizuno A 2011 *Proc. of Int. ESP conf. Nurenburg* (Germany)
[14] Yamamoto T, Abe T, Mimura T, Ito Y, Otsuka N, Ehara Y and Zukeran A 2009 *IEEE Trans. IA*. **45** 2178
[15] Yamamoto T, Mimura T, Otsuka N, Ito Y, Ehara Y and Zukeran A 2010 *IEEE Trans. IA* **46** 1606
[16] Okubo M, Kuroki T, Yoshida K and Yamamoto T 2010 *IEEE Trans. IA* **46** 2143
[17] Tsukamoto T 2006 *J. JIME* **41** 77

Journal of Physics: Conference Series **418** (2013) 012068 doi:10.1088/1742-6596/418/1/012068

Recent progress in experimental studies of electro-hydrodynamic flow in electrostatic precipitators

J Mizeraczyk[1,2*], J Podlinski[1], A Niewulis[1] and A Berendt[1]
1 Centre for Plasma and Laser Engineering, The Szewalski Institute of Fluid Flow Machinery, Polish Academy of Sciences, Fiszera 14, 80-952 Gdansk, Poland
2 Department of Marine Electronics, Gdynia Maritime University, Morska 81-87, 81-225 Gdynia, Poland

E-mail: jmiz@imp.gda.pl

Abstract. Many experimental, theoretical and numerical works were devoted to the electrohydrodynamic (EHD) flow generated in electrostatic precipitators (EPSs). The generally accepted conclusion from these studies is that the EHD flow plays an important role in the particle collection in ESPs. However, despite of the recent fast progress in numerical simulation and experimental techniques used in the ESPs studies many problems related to the influence of the EHD flow on the performance of ESPs are unsolved. In particular, the influence of the turbulent EHD on submicron particles transport and deposition in ESPs is ambiguous. This problem and other related to the particle collection in ESPs are objectives of this paper, basing on the recent progress in experimental investigations of the EHD flow in ESPs.

1. Introduction

1.1. EHD effects in gaseous media - The state of the art

Electrohydrodynamics (EHDs) is devoted to the investigation of motion of electrically charged fluids (gaseous or liquid) subjected to an electric field, usually external. In a single-phase fluid the forces exerted by the external electric field on free or polarization charges present in the fluid are transferred during collisions to the neutral molecules, causing a change of their previous motion (called a primary flow). As a result a secondary flow appears. The secondary flow causes a redistribution of the electrical charges, which in turn modifies the electric field in the fluid. Such a coupling of the electric field, the space charge formed by the electrical charges [1-5], and the fluid flow makes the EHDs a difficult subject [6-8]. The difficulty rises when the fluid consists of two or more matter phases (the case of multi-phase fluid). An example of the two-phase fluid is a flow of flue gas, that can be regarded as a mixture of an after-combustion gas and dust (a particulate matter). In this case, the particulate matter become charged by the gaseous ions and get subjected to the electrical field. Due to it the charged particles form its own secondary flow, usually different from that of the molecular secondary flow. In this case the EHD interaction includes the electric field, the molecular space charge, the particle space charge, and the flow of the carrier-fluid. As a result the EHDs of such a multi-phase

[*] To whom any correspondence should be addressed.

fluid becomes very complex, not only from the phenomenological point of view, but also for theoretical and experimental studies.

EHDs covers also such particle and fluid transport mechanisms as: electrophoresis, electrokinesis, dielectrophoresis, electro-osmosis and electrorotation which occur in gaseous and liquid media.

The subject of this paper is restricted to EHD phenomena in a gaseous medium. Systems with EHD effect in a gaseous medium [9] are used commercially in a wide range of industries. The most important is filtration and separation of dust particles in ESPs, which are by far the most common electrostatic devices used in pollution control.

Another electrostatic system with EHD effect is an electrostatically enhanced cyclone separator [10, 11]. A cyclone separator provides a reasonably effective means of removing dust particles exceeding 5 um in diameter from a gas stream, but its efficiency falls off rapidly as the particle size of the dust decreases. A significant improvement in small-size dust particle removal efficiency has been reported with an electrostatically enhanced cyclone. The particle removal in the electrostatic part of cyclone depends apparently on the flow pattern modified by the EHD forces. However, no in-depth research study are reported on that subject.

EHD effect plays an important role in collection of small pollutant particles in electrostatic scrubbers [12, 13]. In a conventional scrubber, water droplets (collector particles) are sprayed through a polluted gas stream and particles are collected by inertial impaction. The collection efficiency for fine pollutant particles is poor as these tend to follow the gas streamlines round the collector particles, avoiding settling on them. However, the collection of micron and submicron pollutant particles can be significantly improved by charging the pollutant and collector particles to opposite polarities. Some improvement can be still obtained if only the collector particles are charged. Both charging results in the appearance of EHD forces which change the flow conditions inside the scrubber. In particular the EHD forces influence the trajectories of the pollutant particles due to the attraction between them and the collector particles. The improvement of the pollutant particle collection due to the EHD forces concerns mainly the particles of sizes in the range of 0.1 to 1 μm, where the collection efficiency due to inertial impaction is low, but Brownian diffusion does not yet dominate over other collection processes.

Electrostatic atomisation [14-16] is an important example of using EHD forces to disrupt a liquid surface and form a highly charged stream of droplets, which are smaller than those when a liquid is atomised mechanically. EHD atomisation are used in many diverse applications, including paint spraying, spray and powder coating, and electrostatic printing. It has been also proposed as a mechanism for the propulsion of space vehicles.

EHD forces have also been proposed to be used for controlling gaseous (and also liquid) flows in the so called EHD pumps without mobile parts [17-20]. In natural sciences, EHD is indispensable to understand atmospheric electricity.

A newly emerging application of EHD phenomenon is the controlling of air flow [21-24]. For that purpose plasmas of a dielectric-barrier discharge (DBD), corona discharge and spark plasma are employed.

Plasma-based devices for active airflow controlling (called EHD flow actuators) convert electric energy into kinetic energy without involving moving mechanical parts [25-31]. Their response time is very short and they enable a real-time airflow controlling at high frequency [32-39]. Besides, they can reduce the flow-induced noise. EHD flow actuators are very attractive because they do not exhibit many drawbacks typical of mechanical air flow devices. Active air flow control is a hot topic in industrial aeronautics.

Most aspects of EHD phenomenon may be illustrated using an example of electrostatic precipitator [40-42] where the EHD forces are responsible for the collection of dust particles. The majority of EHD effects present in ESPs [43-52] are typical of other gaseous EHD systems. Analyzing the imperfections of the performance of ESPs related directly or indirectly to EHD phenomenon allows us to make a list of the fundamental problems that should be solved to deeply understand this phenomenon, and then implement the attained knowledge in practice.

The ESPs have received the intensive study in their hundred year history. Nowadays they operate with high overall dust particle collection efficiency (99.5 %), however, are not effective in removing the fine particles (below 1 um). This is attributed to poor understanding of the performance of ESPs, operation of which depends on EHD phenomena. Many fine particles of a size of 1 um or less, emitted by ESPs contain toxic trace elements and are dangerous for human beings and animals. Recently, a special environmental concern in the form of tough regulations is directed towards controlling the emission of micron and submicron particles by ESPs. This forces an action towards the improvement of ESPs. However, industrial ways of increasing the fine particle collection are limited, and without studies of the fundamental phenomena occurring in an environment similar to that of ESPs the further progress in this field is doubtful.

A theory for ESPs with turbulent flow was developed by Deutsch in 1922 [53]. In this theory, turbulent mixing is assumed to form a uniform distribution of dust particles, which then have a uniform concentration C throughout a given cross section of the ESP duct. Inside the precipitator the particles experience an electrostatic force which give them a velocity (w) towards the collecting walls. This velocity is called the migration velocity. According to the Deutsch theory the collection efficiency (η) of a precipitator is given by the formula:

$$\eta = (C_{in} - C_{out}) \cdot (C_{in}) = 1 - \exp(-A\, w \cdot (u_g)^{-1}) \qquad (1)$$

where C_{in} and C_{out} are the inlet and outlet concentration of dust particles, A is the total collecting area, and u_g is the volume flow rate of the gas. This formula describes the precipitator collection efficiency in terms of known or measurable parameters.

It has long been known that the Deutsch equation overestimates the efficiency of full-scale precipitators [40]. Not surprisingly many modifications of the Deutsch theory have been made. However, because of a large number of approximations made in the derivation of modified Deutsch's equations, they, as well as the original Deutsch equation, provide only a general approximation for the behaviour of a real precipitator. The most unjustified approximations of the collection efficiency theory which do not hold in the ESPs are as follows.

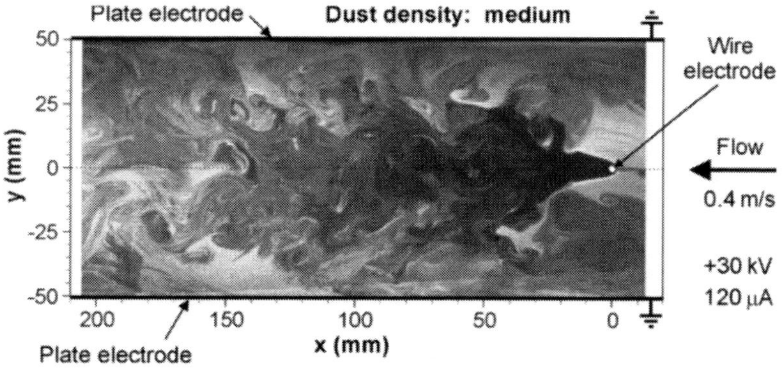

Figure 1. Instantaneous image of the particle flow in the ESP (primary flow velocity: 0.4 m s^{-1}, applied voltage: +30 kV, total discharge current I = 120 μA). The bright places correspond to high density of the dust [54].

It was assumed that the particle concentration across the precipitator cross section is uniform. In reality due to a relatively high charged particle migration velocity brought about by a strong electric field and particle charging near the discharge electrodes, there is a relatively dust-free zone near the discharge electrode, which broadens as the dusty gas moves through the precipitator (figure 1) [54]. Apart from that, the EHD forces produce strong nonuniformity of the particle concentration in the whole volume of precipitator [55-63], forcing the particles to move inside the electrohydrodynamically-formed flow streams and vortices (figure 2) [64, 65]. This fact has been raised by us in several publications and at conferences [66-69];

It was assumed that all particles once deposited on the collecting walls remain there forever and are not re-entrained (i.e. the capture efficiency is one). In practice, significant re-entrainment does occur (e.g. due to the back-corona discharge which tears-off deposited particles). To include this effect into the theory numerous attempts have been made to improve upon the Deutsch equation. Although these modifications accounts for some of the experimentally observed deviations from the original Deutsch equation, they have no theoretical bases;

In the Deutsch theory only the particle size and its dielectric constant are considered to affect the particle collection. However, practice shows that the particle resistivity is of major importance, causing reduction of the particle collection efficiency at a given voltage and decrease in the maximum voltage attainable without an arc. Both these effects are attributed to an electrical breakdown phenomenon within the dust deposited on a collecting electrode, known as back-ionisation. Back-ionisation is more severe the higher the electrical resistivity of the dust. Several attempts to measure an expected correlation between dust resistivity and precipitator performance have not been conclusive.

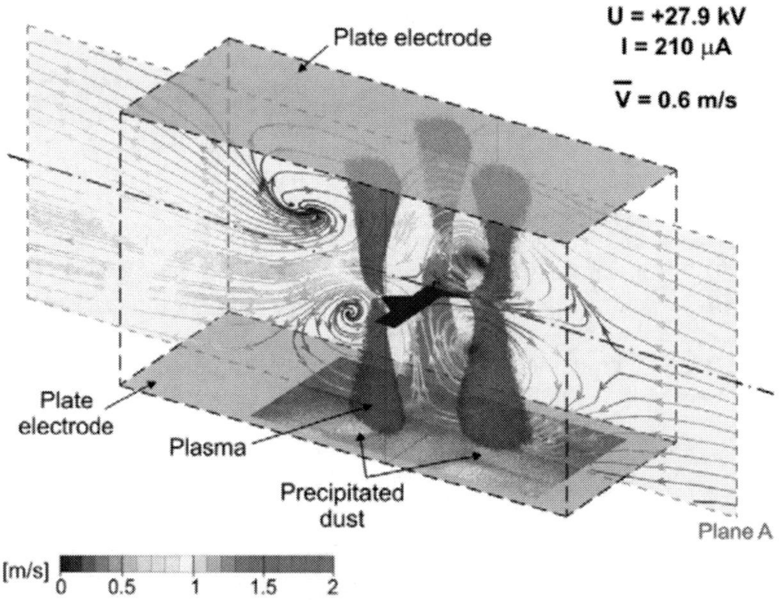

Figure 2. The EHD effects in an ESP with a two-sided spike electrode. Three phenomena are shown: 1) Plasma columns between the spike electrode and collection electrodes (blue–violet colour, 2.5-s photograph exposure), 2) Particle flow streamlines in plane A (coloured continuous lines obtained from PIV measurements, the colour of the streamline in each point corresponds to the magnitude of the velocity vector in this point, see the coloured velocity scale) and 3) Dust deposit on the collecting electrode [65].

Regarding the influence of electrohydrodynamically-induced secondary flow upon the particle collection in a precipitator, the opinions on that are almost equally divided. Some theoretical and experimental papers, rather older, concluded that precipitator efficiency is higher the more laminar the flow, i.e. when the EHD effect is negligible. It means also that precipitators should be more efficient than is predicted by the Deutsch theory, which assumes high turbulence. Those who share this view suggest a careful designing the flow in a precipitator to make it as much laminar as possible. However, the careful design of the laminar flow without taking into account the EHD phenomena cannot be reproduced in a real precipitator, because the EHD forces generate a secondary flow in the discharge electrode zone, which strongly disturb the primary flow (defined as that far before the discharge electrode zone). The EHD secondary flow is the bulk movement of the neutral gas molecules in the direction of the electric field owing to momentum transfer from the molecular ions. This usually results in a bulk velocity of the gas of (1-2) $m \cdot s^{-1}$ which is comparable with the mean velocity of the primary flow. Better or worse performance of aerodynamically well designed precipitators has not experimentally been identified because too many factors, as, for example, re-entrainment, are important. In recent theoretical and experimental papers, which deal with the EHD secondary flow in precipitators there are suggestions that some flow patterns generated by the EHD forces can improve the performance of precipitators. However, still this dilemma is not unambiguously solved. More fundamental research are needed.

Large precipitators are usually designed with negative high-voltage polarized discharge electrodes because this allows a higher voltage to be applied (without an arc). However, small precipitators, such as those designed for the removal of cigarette smoke in rooms, may have positive high-voltage electrodes, to minimise the production of ozone. The behaviour of EHD flows with negatively or positively polarized electrodes was studied by many researchers mainly in terms of practical aspects relevant to the performance of precipitators (breakdown voltage, ozone production, collection efficiency). The investigations of fundamental processes in the EHD flow for both voltage polarities are still unsatisfactory (e.g. the cause of different EHD flow turbulences for negative and positive voltage polarities and their influence on the particle collection efficiency are unclear).

There is a great need of upgrading the existing precipitators to meet new emission regulations (e.g. the regulation regarding particles of diameter lower than 2.5 μm). There have been several proposals of employing a supplying voltage with a pulsed waveform to improve particle collection efficiency, also fine particles, of the existing precipitators. Such an action was called a pulse energisation. If a short duration pulse is superimposed onto the DC voltage it is possible to obtain a higher peak voltage without sparkover. Owing to it the particle charging improves and the migration velocity increases. Therefore the collection efficiency should increase. In practice voltage pulses of different waveforms have been tested. The improved collection efficiency has been reported only for medium- and high-resistivity particles, which are most difficult to collect. In real precipitators it was found that short pulses give the best improvement in collection efficiency and that there should be a sufficiently long time between pulses to stop sparkover and back-ionisation. Typically there have not been much fundamental research done to understand the EHD effect when the flow is energised by voltage pulses.

Precipitator performance undoubtedly depends upon designs of discharge electrodes. Many designs of electrode have practically been used, including plain straight or twisted wires, and pointed electrodes, having barbed or spike-like geometries. No preference for one design rather than another regarding the collection efficiency has been found. The performance differences between the electrode designs are reduced as the electrode becomes coated with dust particles during operation. An interval-regularity of the discharge along the pointed electrode becomes also the case when a plain wire electrode is coated with dust particles. The cause of that has not thoroughly been investigated. Also the way in which the electrodes of different geometries are coated with particle dust in an EHD flow is not clear, although the dust coated electrodes operate differently from those uncoated. However, in many practical cases the design of the electrode is defined by mechanical considerations rather than by electrical performance.

Another parameter related to the precipitator which influences the EHD flow in ESPs is precipitator geometry. It is obvious that the plate-to-plate spacing and discharge electrode spacing influence the precipitator performance, however, there have only been several rather unjustified suggestions related to the effect of precipitator geometry. The geometry effect is met in small precipitators used, for example, for removing cigarette smoke, as dust filters in air conditioning units, and for collecting carbon soot from diesel engines. In such cases several fundamental studies have to be done to clarify the role of small precipitator dimensions on the EHD flow, the coagulation process of small carbon particles, the charging process of carbon particles having a high electrical conductivity, and eventually the collection of the charged carbon particles.

The analysis of the state of the art in electrohydrodynamics in ESPs suggests a list of subjects which still are of interest when the EHD flows in ESPs are concerned:

a. Properties of the corona and back-corona discharges in EHD flows,
b. Fundamental understanding of the physics of the ionic wind produced by the corona and back-corona discharges in EHD flows,
c. EHD flow structures,
d. Blocking the primary flow by secondary EHD flow,
e. Particles migration velocity in EHD flows,
f. Influence of voltage waveform on EHD flows,
g. Influence of the electrode ESP geometries on EHD flows,
h. Turbulences generation, type of turbulences and how they influence the collection of particles and modify the ionic wind produced by the corona discharges.

Although the above list is long, it does not exhaust the subjects of interest. Further progress in the understanding of EHD flows in ESPs depends on employing newly developed or unique diagnostic methods, such as volumetric gas flow velocity measurement method (simultaneous measurement of the velocity in large volumes, the so called Volumetric Velocimetry (VV) method and the time-resolved flow field velocimetry (2-D or 3-D velocity vector measurement in the plane with high time-resolved PIV). Both methods allow investigating the spatial and temporal behaviour of EHD flows in a unique way, with possible observation of the temporal and spatial evolution of the flows on monitors in real time.

Theoretical and numerical investigations play an essential role in elucidating the EHD flows in ESPs. Recently a considerable progress in this field was made by J. S. Chang's [70, 71] K. Adamiak's [72-74] and other groups [75-77]. Comparison of the experimental results with theoretical and numerical findings is an important tool for a better understanding the EHD phenomena in ESPs.

1.2. The paper objectives

In this paper several selected examples of new experimental results on the EHD flow in ESPs are presented. These selected results concern the EHD flow in:

a) narrow circular and rectangular ESPs with longitudinal wire electrode,
b) narrow transverse and longitudinal ESPs,
c) wide multi-spiked electrode ESPs,
d) self-pumped ESP.

These recent results of experimental studies of EHD effects occurring in ESPs enable further progress in a better understanding of the EHD phenomenon and the coupling of the electric field, electric space charge and the flow in ESPs.

2. Diagnostic techniques for EHD flow studies

At present the most popular EHD flow measuring technique, which had been pioneering introduced by us to study the EHD flow in ESPs [66] is Particle Image Velocimetry (PIV). The PIV technique is described in the next chapter.

Over the last decade for measuring the volumetric flow velocity field standard PIV techniques were used. These include scanning a two-dimensional laser sheet through a volumetric region of the flow (which makes the measurement not instantaneous) or using a "thick laser sheet" (which is not truly a volumetric measurement since the thickness of the sheet is limited to less than 1 cm). These techniques were used in many laboratories, but it appeared unsatisfactory and any further progress in deep study and understanding of the EHD flows was difficult, in particular when temporal behaviour of the EHD flows was concerned.

To measure the instantaneous 3-dimensional and 3-velocity components velocity field within a truly volumetric region, in 2009-2010 a VV measuring systems were developed by Dantec Dynamics Inc. and TSI Fluid Mechanics. In their systems the measurement volume is illuminated by a cone of laser light. The laser emits two very short duration light pulses, separated by a known and adjustable amount of time, similarly as in the "conventional" PIV. As the flow with tracer particles (seeding) passes through the measurement volume, the laser light is scattered by the tracer particles during each laser pulse. A system of 3 or 4 high-frame-rate cameras captures separate images of the scattered laser light for each of the respective laser pulses within the entire measurement volume. Then using a sophisticated imaging technique, image transfer system and computer algorithms the complete 3-dimensional 3-velocity component velocity field is determined. For flows with a low particle number density, a volumetric particle tracking velocimetry algorithm is used. Flows with medium seeding density require a tomographic particle tracking velocimetry algorithm, and highly seeded ones use the least-squares matching algorithm. Unfortunately, both the VV measuring systems offered by Dantec Dynamics Inc. and TSI Fluid Mechanics and the measuring procedures are complex (e.g. illumination and observation of the volume from several directions is needed, which in many cases is impractical). Besides the cost of apparatus is rather high. As a consequence the VV measuring systems of this type have not become popular.

Very recently, in November 2011 a novel single camera instantaneous Volumetric Velocimetry was introduced by Dantec Dynamics Inc. and Raytrix. The introduction of this newly developed Volumetric Velocimetry (VV) is essential for instantaneous 3-dimensional (regarding space and velocity) large-volume observation of the EHD flows, which usually change fast in time (EHD flows are turbulent), and which are 3-dimensional over a large volume. In this new VV system an array of thousands of micro-lenses placed on a single imaging sensor observes the same volume from many perspectives. The observed volume can be reconstructed slice-by-slice at a high accuracy. The applied processing algorithm takes full advantage of Graphical Processing Unit (GPU) to produce results at near real-time. This results in a huge speed increase in getting the flow parameters compared to traditional tomographic or other volumetric techniques. Due to all these inventions the newly offered measuring setup has become much simpler (a single camera) and easier. A careful and tedious cameras alignment and calibration as well as adjustment of Scheimpflug adapters, typical of the previous PIV-based techniques with several cameras are not needed anymore. Not only complexity of both the measuring setup and method but also the cost is substantially reduced, compared to the above-mentioned traditional methods.

Using VV enables us getting information about the large-volume EHD flow in a near real-time. With almost on-line volumetric velocity information available, 3-D visualising and viewing the flow dynamics becomes possible. This greatly helps understanding the spatial and temporal flow structures.

Apart from the these sophisticated diagnostics methods a simple laser visualisation method cannot be forgot because it also delivers useful information on EHD flow patterns [78].

3. Particle Image Velocimetry (PIV)

Particle Image Velocimetry [79] is used to determine the flow velocity vectors in a selected cross-section of the flow, called the observation plane The observation plane is set in the flow by introducing into it a laser beam in the form of a laser sheet (figure 3). Then two approaches of PIV are possible, two- or three-dimensional. When using two-dimensional PIV method (2-D PIV), the laser sheet is assumed to be infinitely thin, and only a single CCD camera is used for monitoring the observation plane. This enables measuring two components of the velocity vectors in the observation plane. The velocity vector measurement is based on observation of the movement of seeding particles that cross the observation plane. The flow seeding particle movement is determined by monitoring the laser sheet light scattered by the particles. The scattered laser light forms the flow image that is recorded by the CCD camera. From two successive images the velocity vectors of the seeding particles can be found. When the seeding particles follow the gas flow, the gas flow velocity in the observation plane can be determined. In the other case, PIV method delivers the particle velocity field in the gas.

Three-dimensional PIV method (3-D PIV) is based on the fact that a real laser sheet has a definite thickness. As a result, three velocity vector components in the observation plane can be determined when two CCD cameras observing the same measurement area under different angles are used, as shown in figure 3.

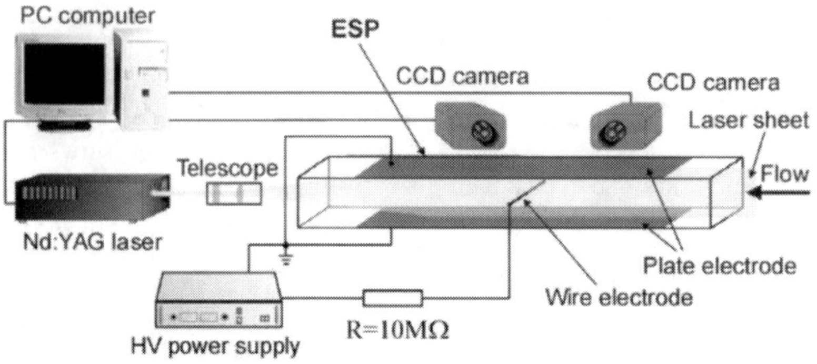

Figure 3. Scheme of set-up for 3-D PIV measurement of flow velocity field in an ESP.

In this paper we present results of new investigations on the EHD flows in ESPs, as mentioned above. In these investigations the PIV equipment consisted of a twin second harmonic Nd-YAG laser system ($\lambda = 532$ nm), imaging optics (cylindrical telescope), one or two CCD cameras and PC computer. The laser sheet of thickness of 1 mm, formed from the Nd-YAG laser beam by the cylindrical telescope was introduced into the ESP to form an observation plane. The images of the particles following the flow in the laser sheet were recorded by two FlowSense M2 cameras. The CCD camera active element size was 1186×1600 pixels. The captured images were transmitted to the PC computer for digital analysis to obtain the velocity map.

The velocity fields presented in this paper resulted from either the instantaneous measurement or the averaging of 100 instantaneous measurements. In the latter case, each velocity map was time-averaged. Based on the measured velocity fields, the flow streamlines were calculated.

4. EHD Flow in Narrow Circular and Rectangular ESPs with Longitudinal Wire Electrode

Recently narrow ESPs have become a subject of interest because of their possible application in diesel engines for the exhausted particulate matter collection [80-82]. Diesel engines emit fine particles, in size range of $7.5 \times 10^{-3} - 1.0$ μm [82], that are harmful for human and animal health. Unfortunately, these submicron particles are not efficiently collected by nowadays ESPs.

The precipitation of particles in the duct of an ESP depends on the dust-particle properties, electric field, space charge, particle physical parameters, electrode geometry and EHD secondary flow. The interaction between the electric field and charge and the flow results in considerable turbulences of the flow which, according to [83, 84], seem to lower the fine particle collection efficiency. They suggested that the flow turbulence should be reduced in order to improve the fine particle collection efficiency. Improving other factors, such as ESP electrode geometry or ESP operating conditions, which influence the flow patterns in ESPs, may also increase the fine particle collection efficiency. In this chapter we present results of 2-dimensional (2-D) PIV measurements of the flow patterns in narrow circular and rectangular ESPs with longitudinal wire-electrode, foreseen for the use in diesel engine exhaust duct to collect fine particles.

4.1. Experimental set-up

The experimental apparatus used in the presented investigation consisted of an ESP, high voltage supply and standard 2-D PIV equipment [85], similar to that shown in figure 3.

The narrow circular ESP consisted of two collecting circular cylinder-electrodes with a longitudinal wire discharge electrode, enveloped by a glass tube (figure 4). The collecting and discharge electrodes were made of stainless steel. Each cylinder-electrode has a length of 100 mm and inner diameter of

25.5 mm. The discharge wire-electrode was 0.23 mm in diameter and 200 mm in length.

The narrow rectangular ESP consisted of four collecting plate-electrodes and a wire discharge electrode, inserted in a glass rectangular cylinder (figure 5). The stainless-steel wire-electrode (diameter of 0.23 mm, length of 100 mm) was mounted in the middle of the ESP, parallel to the main flow direction. The four collecting plate-electrodes (120 mm long and 27 mm wide) were made of aluminium tape of a thickness of 50 μm, glued on dielectric supporting plates.

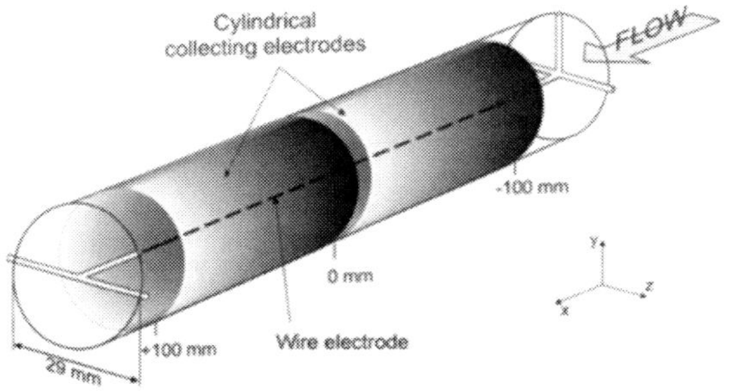

Figure 4. Narrow circular ESP with longitudinal discharge wire-electrode.

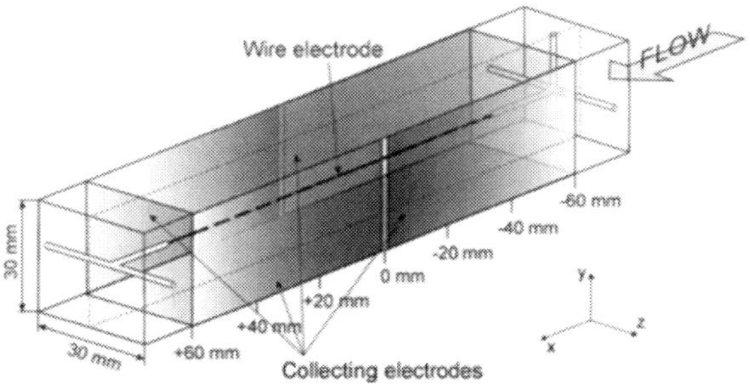

Figure 5. Narrow rectangular ESP with longitudinal discharge wire-electrode.

In both ESPs, positive or negative voltage of 10 kV was supplied to the wire- electrodes through a 10 MΩ resistor. Air flow seeded with cigarette smoke was blown along the ESP duct with an average velocity of about 0.9 $m \cdot s^{-1}$.

The PIV measurements were carried out in the wire electrode mid-plane (i.e. at $x = 0$ mm), perpendicularly to the wire and the collecting electrodes. The velocity fields presented here resulted from the averaging of 100 measurements, which means that each velocity map is time-averaged.

4.2. Results

Instantaneous images of the dust particles and averaged flow patterns (i.e. flow streamlines) in the transverse mid-plane (the y-z plane) of both ESPs are shown in figures 6-7 and 8-9 for circular and rectangular ESPs, respectively.

4.2.1. Circular ESP

The flow patterns in the circular ESP, representing the primary flow motion (i.e. measured without applied voltage not presented here), showed a slight movement of the dust particles in the y-z plane, however, the y- and z- velocity components were very low (below 0.05 $m \cdot s^{-1}$). When a high-voltage was applied to the discharge electrode (figures 6-7), a strong secondary EHD flow was formed in the y-z plane, i.e. transversely to the flow duct. Due to the EHD forces the dust particles moved radially with the migration velocity up to 0.3 $m \cdot s^{-1}$ from the ESP center (i.e. from the discharge wire-electrode) towards the collecting electrodes. The dark circular zones around the wire electrode seen in figures 6a and 7a are the areas from which the particles have already been removed. For the positive voltage polarity, the particle flow towards the collecting electrode was very regular (figure 6b). At the negative polarity, the particle flow pattern (figure 7b) was less regular as for the positive high voltage. The removal of the dust particles from the areas around the wire-electrode may seem faster than in the case of positive polarity (the migration velocity is up to 0.4 $m \cdot s^{-1}$); however, for the negative polarity the discharge current was higher at the same voltage as for the positive polarity.

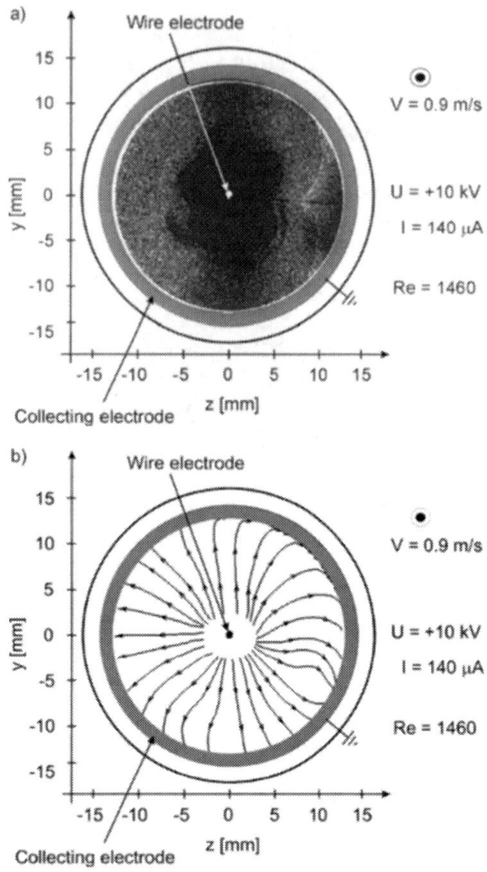

Figure 6. Instantaneous image of the dust particles (a) and averaged flow streamlines (b) in the circular ESP. Average total discharge current 140 μA. Positive voltage 10 kV. ◉ - shows the direction of the primary flow.

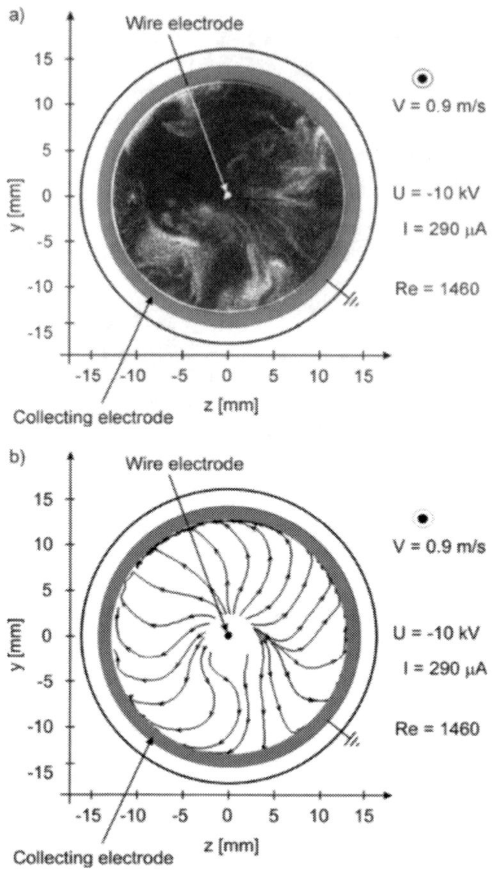

Figure 7. Instantaneous image of the dust particles (a) and averaged flow streamlines (b) in the circular ESP. Average total discharge current 290 μA. Negative voltage 10 kV. ◉ - shows the direction of the primary flow.

4.2.2. Rectangular ESP

Generally, the particle flow were more disturbed by the applied voltage in the case of rectangular ESP than that in the circular one (figures 8-9). For the positive polarity (figure 8), four streams of the dust particles moved towards the plate electrodes. At the plate electrode, each of the four streams met two oncoming neighboring streams, which results in scattering of the streams. As a result, small twin-vortex structures were formed in all 4 corners of the ESP. The particle migration velocity was up to 0.4 $m \cdot s^{-1}$ f, i.e. slightly higher than in the circular ESP. At the negative polarity (figure 9b) the time-averaged flow pattern confirms the features of the particle flow for positive polarity (figures 8b). Several small vortex structures were formed in the ESP cross section. However, the instantaneous particle flow (figure 9a) was less regular and more turbulent than in the case of positive polarity. The more turbulent character of the flow patterns at negative voltage polarity is probably due to nonuniformity of the negative corona discharge, which exhibits the form of tufts irregularly distributed along the wire in time and space.

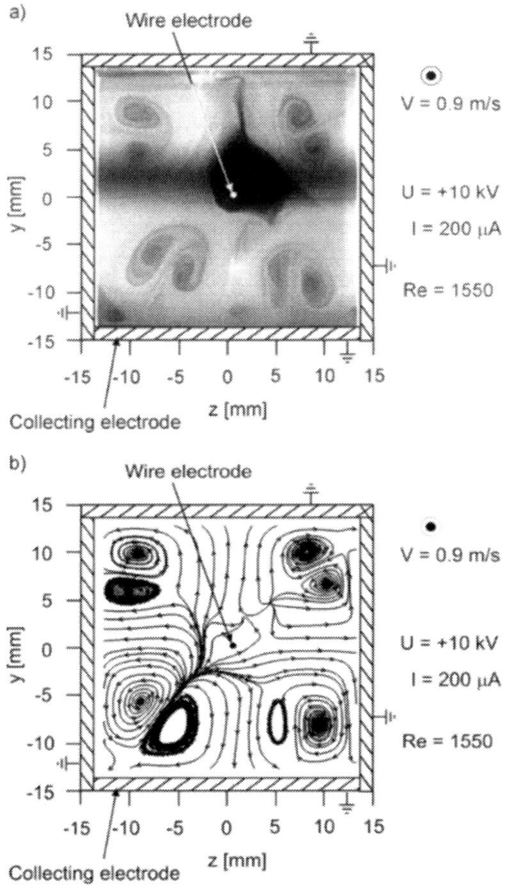

Figure 8. Instantaneous image of the dust particles (a) and averaged flow streamlines (b) in the rectangular ESP. Average total discharge current 200 μA. Positive voltage 10 kV. ◉ - shows the direction of the primary flow.

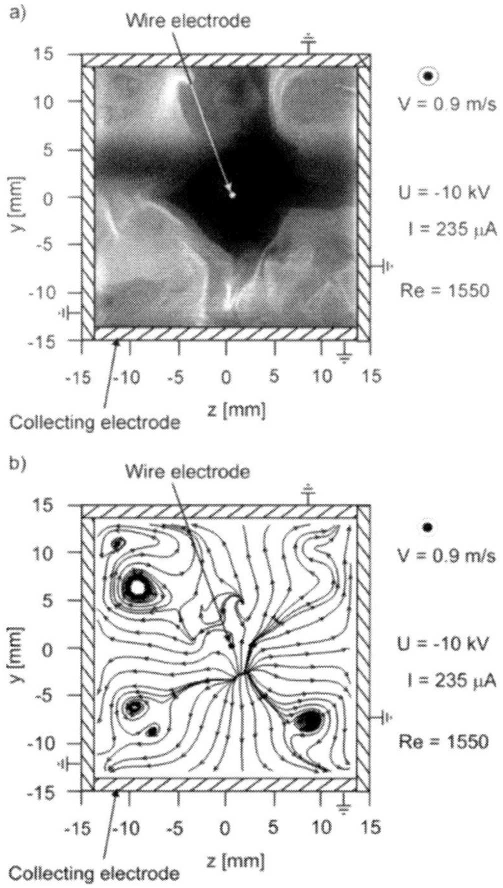

Figure 9. Instantaneous image of the dust particles (a) and averaged flow streamlines (b) in the rectangular ESP. Average total discharge current 235 μA. Negative voltage 10 kV. ◉ - shows the direction of the primary flow.

4.3. Conclusions

The presented results of 2-D PIV measurements of the particle velocity fields in the narrow ESP showed that the electrode geometry significantly changes the particle flow patterns. Comparing the results for both ESPs (circular and rectangular ESPs) one can see, that flow patterns formed by the EHD forces in the plane perpendicular to the main flow are very different. In the rectangular ESP the particle vortex structures are formed mainly due to the presence of four corners in the rectangular electrode geometry, while in the circular ESP the particles move radially towards the collecting electrodes. It would be interesting to measure the flow patterns in both ESPs along their longitudinal cross-sections. However, it is technically difficult for the circular ESP. The longitudinal flow patterns in the rectangular ESPs measured by us are presented in the next chapter (also in [86]).

The different flow patterns in circular and rectangular ESPs may result in different particle collection efficiency. This subject is under investigation.

5. EHD Flow in Narrow Transverse and Longitudinal ESPs

In this chapter we present investigations of the influence of the EHD secondary flow, generated in a narrow rectangular ESP with transversely or longitudinally to the main flow placed discharge electrode, on the submicron dust particle collection efficiency.

5.1. The ESP

The ESP was an acrylic box (300 mm long, 30 mm wide and 30 mm high) equipped with a wire discharge electrode(s) and two plane collecting electrodes. The experiments were carried out for two discharge wire electrode (diameter of 0.23 mm) positions: transversely- or longitudinally-placed in respect to the main flow direction (figure 10 and 11). The collecting electrodes (60 or 120 mm long and 30 mm wide) were placed along the ESP at its bottom and top.

In both ESPs (transverse and longitudinal), positive or negative voltage up to 18 kV was supplied to the wire- electrodes through a 10 MΩ resistor. Air flow seeded with cigarette smoke was blown along the ESP duct with an average velocity of about 0.5 $m \cdot s^{-1}$. The laser visualization and PIV measurements of the EHD flow were performed in the x-y plane along the ESP duct (i.e. along the longitudinal plane) at $z = 0$ mm in the transverse ESP and at $z = 5$ mm in the longitudinal ESP.

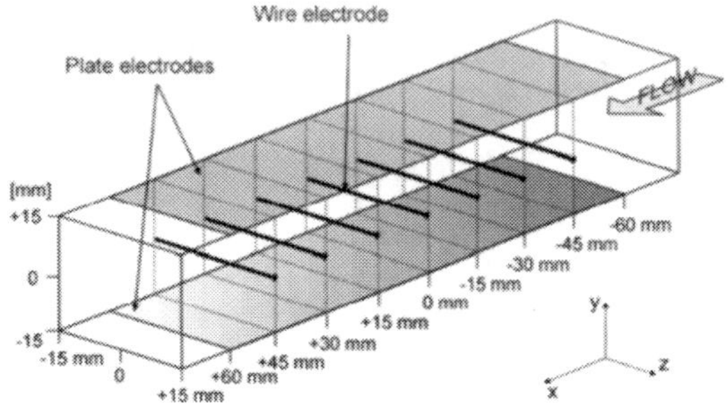

Figure 10. Narrow ESP with transverse wire electrodes.

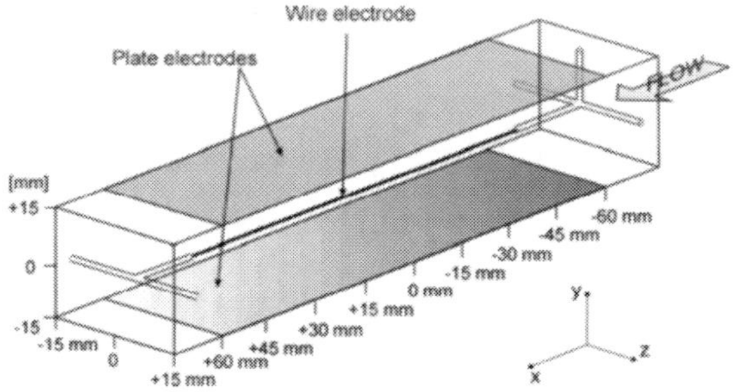

Figure 11. Narrow ESP with longitudinal wire electrode.

5.2. Results

In both narrow ESPs the flow patterns measured without applied voltage (results not presented in this paper) were laminar and the submicron dust particles followed the gas primary flow. After applying a high voltage to the wire electrode/electrodes (figures 12-17), the EHD forces exerted by the electric field on the space charge generated by a corona discharge induced an EHD secondary flow of the gas which altered significantly the primary laminar flow. The seed particles were pushed from the wire electrode(s) outwards by the electric force which was strongest around the wires. The dark areas around the wire electrode/electrodes seen in figures 12-17 are the areas from which the particles have already been removed (the brightness of the image is proportional to the seed particle concentration).

5.2.1. Transverse ESP

In the transverse ESP with one active wire electrode (figure 12), due to the EHD forces the particles moved from the area close to the wire electrode towards the plate electrodes. For the positive voltage polarity the particles flow pattern was regular (figure 12a). The regular dark regions cleared of the dust particles could be easily observed. For the negative voltage polarity the flow structure was rather unstable and turbulent (figure 12b). Moreover, upstream the discharge region near the plate electrodes two vortices were formed (one near the bottom plate electrode and the second one near the top plate electrode).

Increase in the number of the active wire electrodes resulted in a stronger EHD secondary flow (figures 13-15). For the positive voltage polarity the regular structure near the first active wire electrode spread faster towards the collecting electrodes. The flow structures near the subsequent wire electrodes are not so regular and more turbulent. For the negative voltage polarity the flow structures around subsequent wire electrodes are very irregular and turbulent along the channel width. As we show below, the removal of the dust particles in the ESP with negative voltage polarity was faster than in the case of positive polarity.

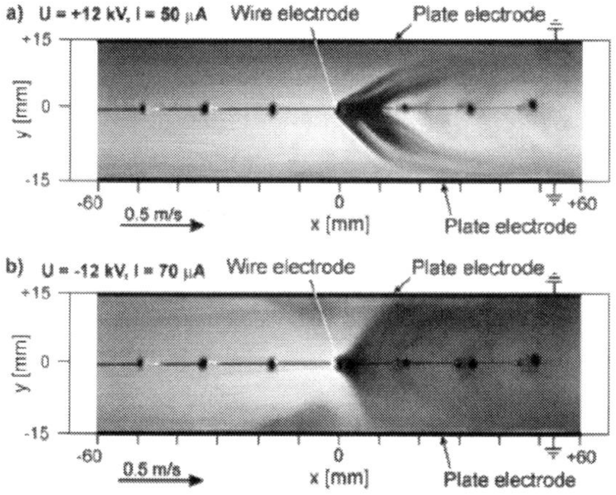

Figure 12. Instantaneous images of the flow structure in the ESP with the one active wire electrode supplied with positive (a) and negative (b) voltage.

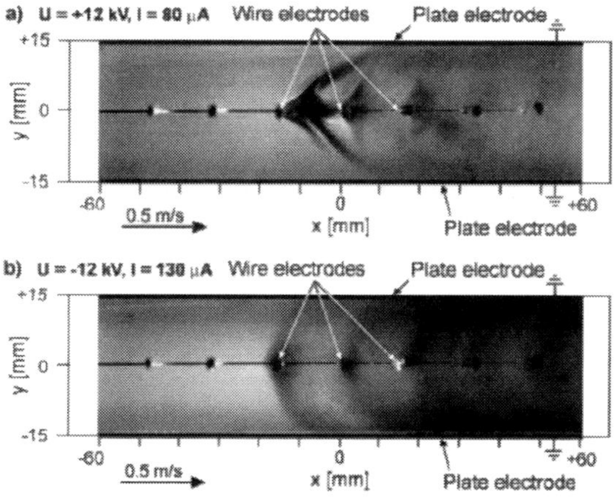

Figure 13. Instantaneous images of the flow structure in the ESP with the three active wire electrodes supplied with positive (a) and negative (b) voltage.

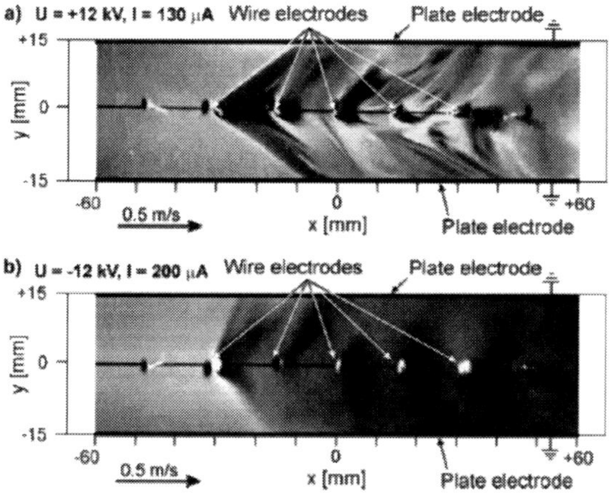

Figure 14. Instantaneous images of the flow structure in the ESP with the five active wire electrode supplied with positive (a) and negative (b) voltage.

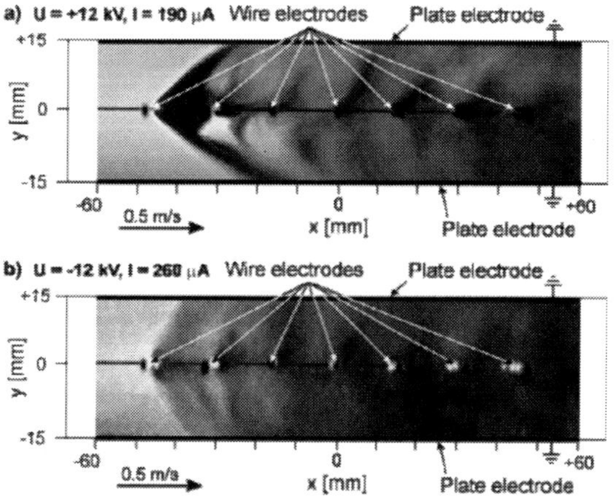

Figure 15. Instantaneous images of the flow structure in the ESP with the seven active wire electrodes supplied with positive (a) and negative (b) voltage.

5.2.2. Longitudinal ESP

Generally, in this case due to the EHD forces the particle flow was formed from the longitudinal wire electrode towards the plate electrodes. The time-averaged flow patterns were very similar for both positive and negative voltage polarities. In both cases, for high voltage of about 9.5 kV, the flow particle trajectories were strongly bent towards the plate electrodes before the discharge region (i.e.at x = -30 mm, figures 16 and 17). However, the most pronounced deflection of the particles towards the plate electrodes occurred between x = -15 mm and x = +15 mm, i.e. in the middle of the discharge region. The particle migration velocity was comparable to the primary flow rate (i.e. $V = 0.5\ \mathrm{m \cdot s^{-1}}$). For positive polarity a vortex was formed near the upper plate electrode (at around x = -30 mm). Despite similar time-averaged particle flow patterns for both positive and negative polarities, we observed that the instantaneous particle flow in the longitudinally ESP was less regular and more turbulent in the case of negative polarity, similarly as in the transverse ESP.

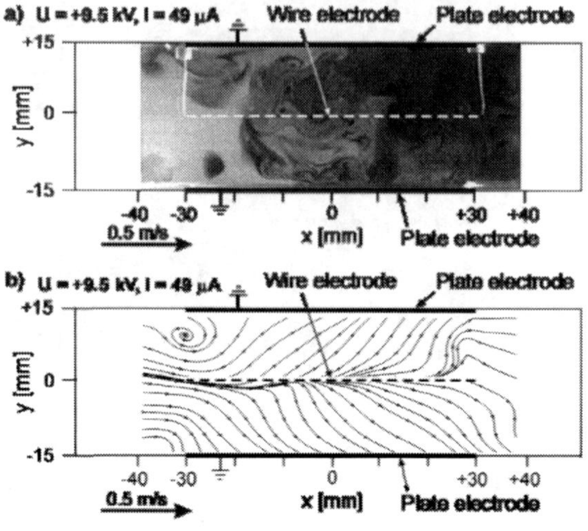

Figure 16. Instantaneous image of the dust particles (a) and averaged flow streamlines (b) in the longitudinal ESP. Average total discharge current 49 μA. Positive voltage 9.5 kV.

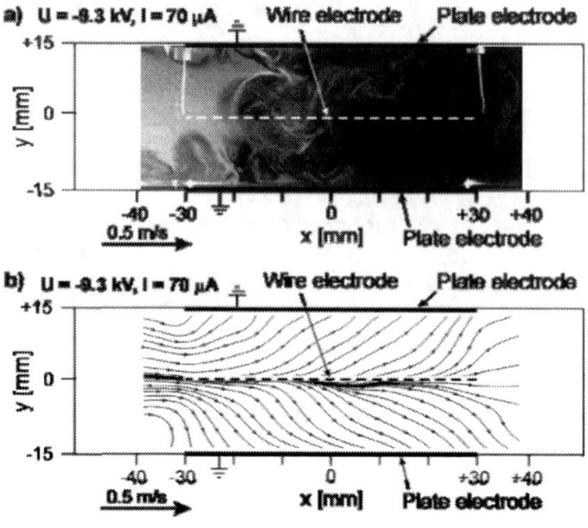

Figure 17. Instantaneous image of the dust particles (a) and averaged flow streamlines (b) in the longitudinal ESP. Average total discharge current 70 μA. Negative voltage 9.3 kV.

Comparison of the collection efficiencies for transverse and longitudinal ESPs for the total discharge current of 140 μA is presented in figure 18. It is seen in figure 18 that the collection efficiency of the transverse ESP with seven active wire electrodes is higher than in the ESP with lower number of the wire electrodes. The higher collection efficiency for arrangement with seven wire electrodes could be explained by the longest active length of ESP (longer discharge region). Comparing the results for the transverse and longitudinal ESPs with the similar active length of the discharge region, the fine-particle collection efficiency was similar for both ESP, transverse and longitudinal.

We found that the fractional collection efficiency in the narrow ESPs is lower for small submicron particles than for larger ones. The particle collection efficiency increases with increasing voltage, and it is higher for negative voltage polarity.

5.3. Conclusions

The experiment showed very complex flow structures in both narrow ESPs, transverse and longitudinal. The EHD secondary flow strongly depended on the applied voltage amplitude and polarity as well as on the arrangement of the discharge electrodes. For the negative voltage polarity the flow structure is more unstable and turbulent. The flow turbulent character at negative voltage polarity is attributed to nonuniformity of the negative corona discharge, which exhibits the form of tufts irregularly in time and space distributed along the discharge wires.

The measurements of the fractional particle collection efficiency confirmed the common view that the particle collection efficiency in the ESP is lower for small submicron particles than for larger ones. The particle collection efficiency increased with increasing voltage, and it was higher for negative voltage polarity. The obtained results showed that the collection efficiency strongly depends on the active length of ESP. It increases with increasing active length of ESP (i.e. with increasing number of wire electrodes in transverse ESP), and is similar for the narrow transverse and longitudinal ESPs of the similar active length.

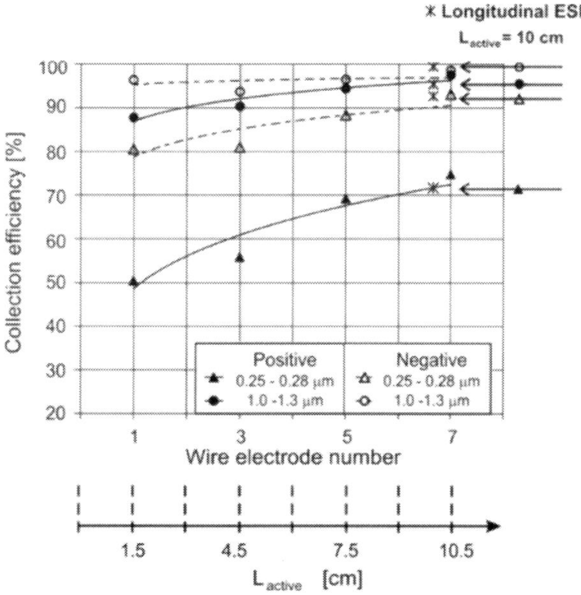

Figure 18. Collection efficiency in the narrow ESPs versus wire electrode number (the total discharge current 140 μA).

6. EHD flow in wide multi-spiked electrode ESPs

In this section investigations of the influence of the EHD secondary flow, generated in an ESP with one-sided and two-sided spike electrodes, on the submicron dust particle collection efficiency are presented.

6.1. The ESP

The duct of the ESP used was made of a transparent acrylic box, 1600 mm long, 200 mm wide and 100 mm high. Such an ESP is classified as a wide ESP. At the top and bottom of the acrylic box two collecting (grounded) stainless-steel plate electrodes (200 mm × 1100 mm) were placed. In the middle of the ESP, between the plate collecting electrodes, a stainless-steel 1 mm thick discharge electrodes were mounted. The distance from the discharge electrodes to the plate electrodes was 50 mm. The distance between neighbouring discharge electrodes was 150 mm. Our experiments were carried out for 2 and 4 discharge electrodes: Two-Sided Spike Electrodes (TSSEs) and a One-Sided Spike Electrodes (OSSEs). The TSSE had 6 spike tips on two sides of the electrode (figure 19a). Two types of the OSSE were used. The first, with 12 spike tips on one electrode side, called an OSSE12 (figure 19b), and the second, with 6 spike tips on one side, called an OSSE6 (figure 19c). In the case of OSSE two electrode arrangements were investigated, i.e. the first, when the spike tips were directed upstream and the second, with the spike tips downstream the primary flow.

The primary flow of air seeded with a cigarette smoke particles was induced by an external fan. The average flow velocity in the ESP duct was controlled using an anemometer. A flow homogenizer was placed before the ESP inlet. During the collection efficiency studies the primary flow average velocity was 0.6 m·s^{-1} ($Re_{0.6}$ = 3800). However, at this primary flow velocity the particles were highly collected in the ESP and the PIV measurements were not possible. Thus, the EHD secondary flow measurements were carried out at the primary flow average velocity of 0.9 m·s^{-1} ($Re_{0.9}$ = 5700).

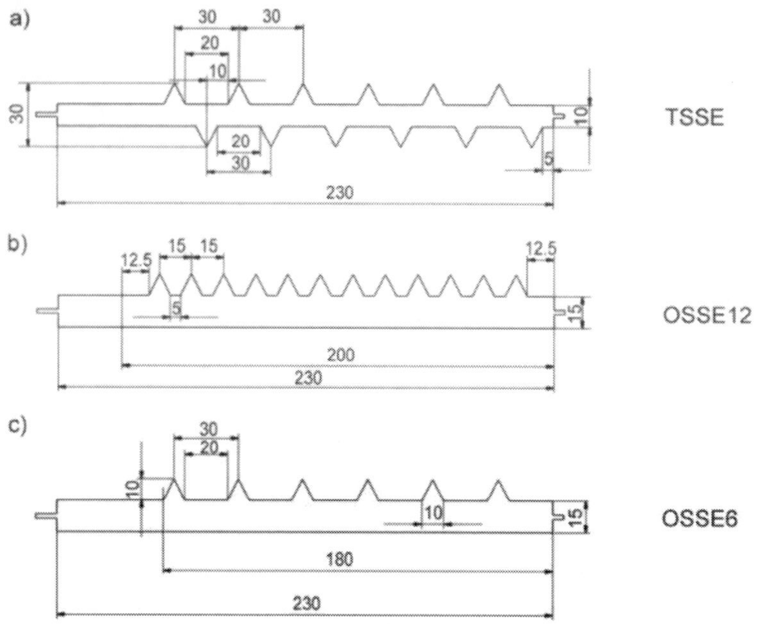

Figure 19. Schemes of the spike discharge electrodes.

The 2-D PIV investigations were carried out in two planes, A and B. In the case of the ESP with TSSEs, plane A passed through the central tip directed upstream; the plane B was passed through the neighbouring tip directed downstream. In the case of the ESP with OSSEs, plane A passed through the central tip; plane B passed in the middle between the central tip and the neighbouring one.

6.2. Results

The EHD secondary flow measurements were carried out for 9 discharge electrode configurations, i.e. with 2 (two) TSSEs, with 2 and 4(four) OSSEs6 directed upstream or downstream, and with 2 and 4 OSSEs12 directed upstream or downstream.

The examples of the flow streamlines in planes A and B obtained for the ESP with 2 TSSEs are presented in figures 20a and 20b, respectively. In this case the negative high voltage of 20 kV was applied and the average total discharge current was 185 μA. As it is seen, in plane A (passing through the spike tip directed upstream) the EHD flow generated near the electrode spike tip (at $z = -130$ mm and $z = 20$ mm) blocked the primary flow in the ESP duct centre and directed it to the plate collecting electrodes. However, about 15 mm downstream the spike tip the flow was again redirected towards the ESP duct centre. This was presumably caused by the sucking effect induced by the EHD flow generated near the spike tip directed downstream (figure 20b). We supposed that this effect could be disadvantageous from the particle collection efficiency point of view. Thus, we proposed to use the discharge electrodes with spike tips only on the one side of the electrode (OSSE12 and OSSE6).

In figures 21a and 21b the flow streamlines measured in the ESP with 2 OSSEs12 directed upstream and downstream are presented. The negative high voltage of 20 kV was applied and the average total discharge current was 130 μA (the average total discharge current was independent of the spike tips direction). When the spike tips were directed upstream (figure 21a) the EHD flow generated from the spike tip of the first OSSE12 blocked the primary flow in the ESP centre and directed it to the plate electrodes. The observed flow structure might suggest that many dust particles were collected on the plate electrodes in the area from $z = -170$ mm to $z = -110$ mm. The particles not collected in this area flowed along the ESP duct close to the plate electrodes. Then, inflowing in the second OSSE12 discharge region they were relatively far from the discharge electrode and the high electric field around it. This might be a cause that majority of these particles passed this region without being collected at the plate electrodes.

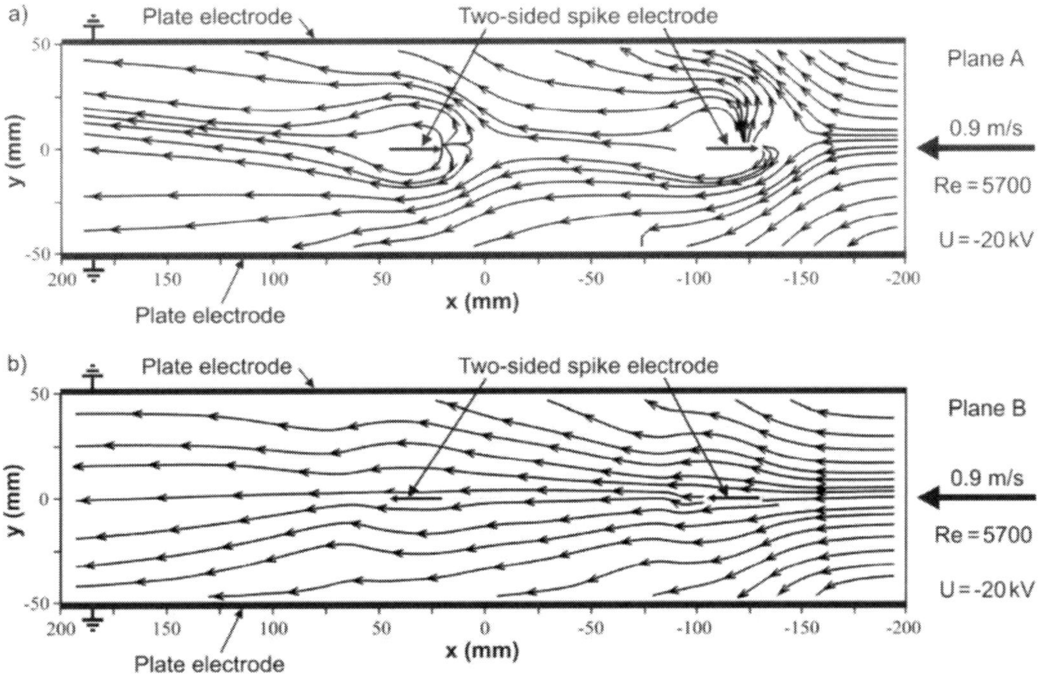

Figure 20. Flow streamlines measured in planes A (a) and B (b) in the ESP with 2 TSSEs. The negative voltage 20 kV, the average total discharge current 185 μA.

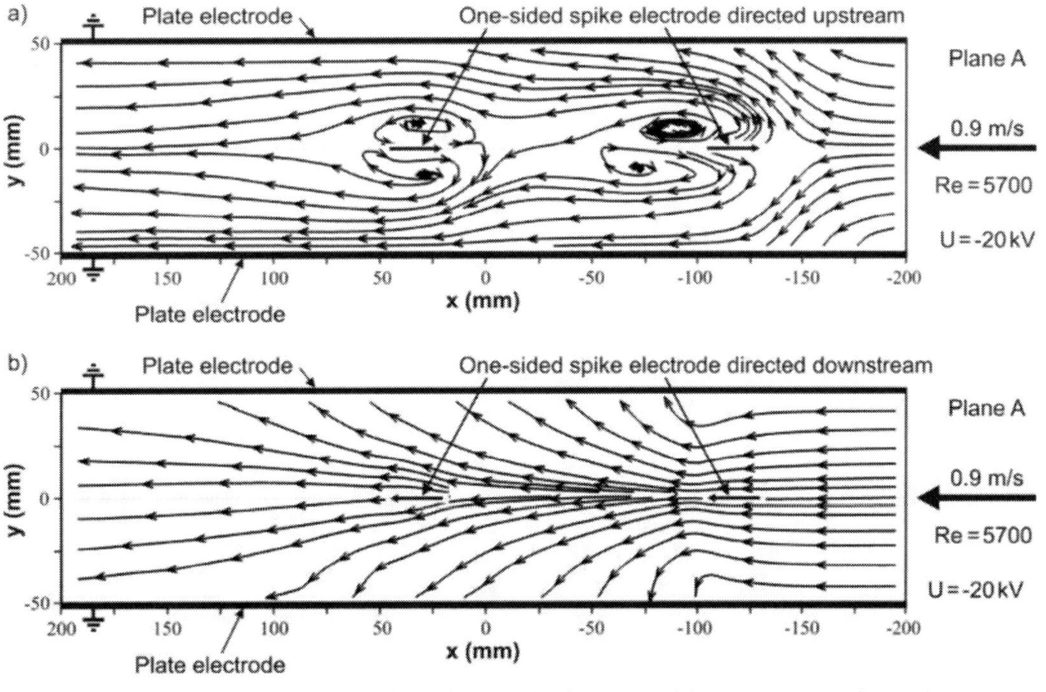

Figure 21. Flow streamlines measured in plane A in the ESP with 2 OSSEs12 directed upstream (a) and dowstream (b). The negative voltage 20 kV, the average total discharge current 130 μA.

The relatively strong EHD flow seen behind the spike tip of the first OSSE12 directed downstream (figure 21b) caused a strong sucking effect. As a result, a flow structure with a very low average velocity near the plate electrodes appeared, and the flow was directed towards the centre of the ESP duct (from z = -130 mm to z = -100 mm). Then, behind the downstream directed spike tip (at z = -105 mm) the dust particles moved through the centre of the ESP duct and quite regularly spread towards the plate electrodes over a large area, eventually reaching the second OSSE12.

The average total discharge current (130 μA) measured when the 2 OSSEs12 were used was much smaller than that of the ESP with the 2 TSSEs (185 μA). Presumably it was due to the short distance between the spike tips of the OSSE12. The discharges generated by each tip interfered with each other which reduced the average discharge current compared to that of the TSSEs. It is known that the lower discharge current results in lower collection efficiency of ESPs. Thus, to avoid this effect we designed another discharge electrode with 6 spikes on one side of the electrode (OSSE6).

Examples of the flow streamlines measured in planes A and B in the ESP with 4 OSSEs6 electrodes directed downstream are presented in figures 22a and 22b, respectively. The applied voltage was – 20 kV and the average total discharge current was 240 μA. The obtained flow streamlines were similar to those obtained for the ESP with OSSEs12. In plane A the flow first moved through the centre of the ESP duct and then scattered towards the plate electrodes. In plane B (in this case placed between the spike tips) only a small flow from the ESP centre to the plate electrodes was observed.

Comparing the flow patterns measured in planes A and B, it can be concluded that for all investigated electrode configurations the EHD secondary flow in the ESP was three-dimensional.

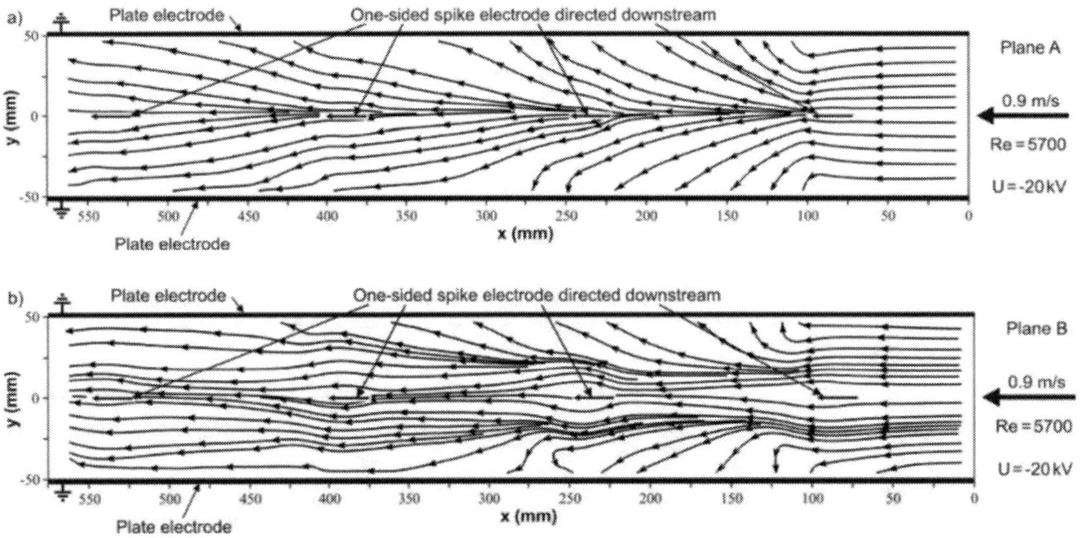

Figure 22. Flow streamlines measured in planes A (a) and B (b) in the ESP with 4 OSSEs6 directed downstream. The negative voltage 20 kV, the average total discharge current 240 μA.

The fractional particle number collection efficiency measurements, in the particle size range from 0.25 μm to 32 μm, were carried out for the ESP with 4 TSSEs, with 2 and 4 OSSEs6 and with 2 and 4 OSSEs12. In the case of OSSEs two electrode arrangements with the spike tips directed either upstream or downstream the primary flow were investigated. Here we present only the results for submicron particles of a diameter from 0.25 μm to 0.28 μm (the average particles diameter 0.265 μm).

The dependence of the submicron particle collection efficiency on the discharge power is shown in figure 23. The results presented in this figure were obtained for the ESP with TSSEs and for the ESP with OSSEs6. For comparison, the collection efficiencies for the ESPs with 1 (single) TSSE and 1 (single) OSSE6 are also presented. These results were obtained in our previous experiment [88].

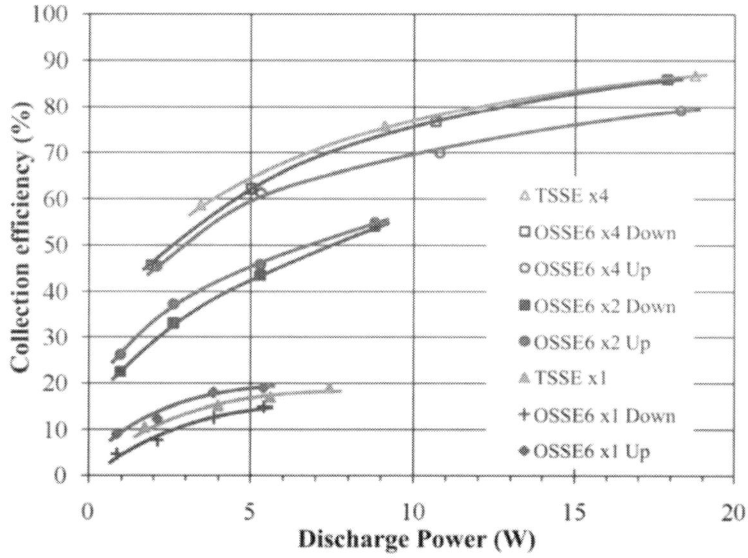

Figure 23. Collection efficiency of submicron particles (a diameter from 0.25 μm to 0.28 μm) in the ESP with TSSEs and OSSEs as a function of discharge power. The negative high voltage. The primary flow average velocity 0.6 $m \cdot s^{-1}$.

As it can be seen in figure 23, in the case of the ESP with a single discharge electrode the highest collection efficiency was obtained when the OSSE6 directed upstream was used. A slightly smaller collection efficiency was obtained when the TSSE was used, and the smallest - in the case of the OSSE6 directed downstream. For the ESP with 2 OSSEs6 higher collection efficiency was again obtained when the spike tips of discharge electrodes were directed upstream (figure 23). However, the difference between collection efficiencies for ESP with 2 OSSEs6 directed upstream and for ESP with 2 OSSEs6 directed downstream was smaller than in the previous case (ESP with a single discharge electrode). The collection efficiencies (figure 23) for the ESP with 4 TSSEs and for the ESP with 4 OSSEs6 directed downstream were almost the same. It is also important to notice that the collection efficiency for the ESP with 4 OSSEs6 directed downstream was higher than for ESP with 4 OSSEs6 directed upstream (unlike the cases of ESPs with 1 OSSE6 and with 2 OSSEs6).

Comparison of the submicron particle collection efficiencies in the ESP with 2 and 4 OSSEs6 and OSSEs12 are presented in figure 24. It can be noticed that a reduction of number of discharge electrode spikes from 12 to 6 almost did not influence the collection efficiency of the ESPs.

6.3. Conclusions

In this chapter the EHD secondary flow and the submicron particles collection efficiency in the spike-plate type ESP with two-sided and one-sided spike electrodes were investigated. The results confirmed that the collection efficiency depends on number, geometry and configuration of the electrodes. The collection efficiency of submicron particles in the ESP with 1 and 2 OSSEs was higher when the spike tips were directed upstream the primary flow. However, when 4 OSSEs were used higher collection efficiency was obtained when the spike tips were directed downstream. Since the ESP working parameters were the same, the obtained results suggest that submicron dust particle collection efficiency in the investigated ESP depended on the induced EHD secondary flow patterns.

From the particle collection efficiency point of view the EHD flow pattern obtained in the ESP with 1 OSSE directed upstream was more beneficial than that formed by 1 OSSE directed downstream. However, in the case of 4 OSSEs directed upstream the EHD flow pattern seemed to hinder the particle collection, and in consequence the collection efficiency of the ESP with 4 OSSEs directed upstream was clearly lower than that with 4 OSSEs directed downstream.

These examples show how the EHD flow is important when the particle collection in ESPs is considered.

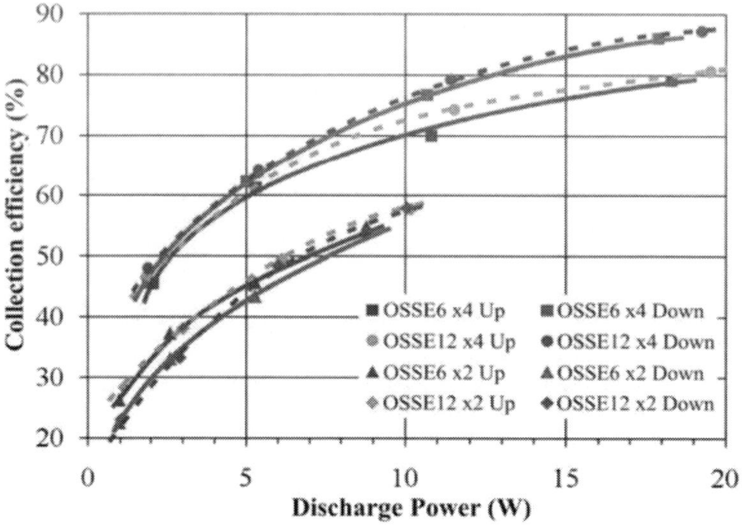

Figure 24. Collection efficiency of submicron particles (a diameter from 0.25 μm to 0.28 μm) in the ESP with OSSEs6 and OSSEs12 as a function of discharge power. The negative high voltage. The primary flow average velocity 0.6 $m \cdot s^{-1}$.

7. EHD flow in self-pumped ESP

In this chapter results of the investigation of EHD flow in a self-pumped ESP e are presented. Our measurements were carried out in an ESP similar to that presented by in [87]. Such an ESP is similar to a typical ESP, however, the gas flow in it is generated by the ESP itself. Therefore, we called this ESP a self-pumped ESP. Such a self-pumped ESP is useful for air cleaning is relatively small rooms, where using a mechanical fan for the ventilation can be bothersome.

7.1. The ESP

The duct of the self-pumped ESP was made of a transparent acrylic box (600 mm long, 150 mm wide and 150 mm high) – figure 25. Inside this box, in the middle of its length, a frame for sustaining spiked high voltage (HV) and grounded electrodes was placed. In the frame two sets of spiked high voltage and grounded electrodes, both made of 1 mm thick aluminium, were mounted. The smooth, grounded electrode and spiked HV electrode were 124 mm wide and 150 mm long. The one of the edges of the HV electrode was ended with 20 spikes. These spikes were 10 mm long. The distance between neighbouring spike tips was 6 mm. The frame sustaining electrodes allowed mounting the HV and the grounded electrodes one above another, thus forming multi-layer electrode arrangement. The distance G between the HV and the grounded electrodes could be changed. The electrodes could also be shifted in the x direction to change the electrode shift X between the edges of the neighbouring electrodes. For the distances G = 15 mm and G = 20 mm the flow patterns near the electrode spikes and the average velocity of the unidirectional flow in the EHD device duct were measured (using 2D PIV method) for the following electrode shift values: X = 8 mm, 15 mm and 50 mm. When the distance G was 20 mm, an electrode set with 3 HV electrodes and 4 grounded electrodes was mounted in the EHD device. In the case of G = 15 mm, an electrode set of 4 HV electrodes and 5 grounded electrodes was formed.

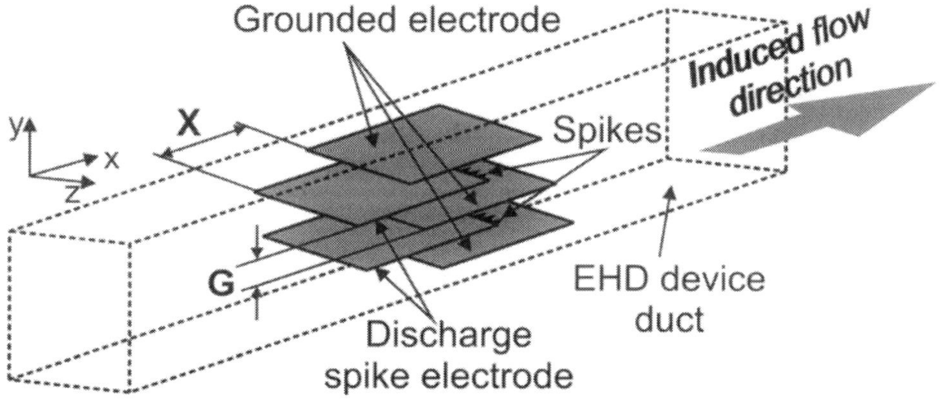

Figure 25. Scheme of the self-pumped ESP for the air cleaning.

7.2. Results

Basing on the PIV measurements, the flow patterns near the spikes of HV electrodes and the average flow velocity induced in the self-pumped ESP duct were determined.

Examples of the obtained flow patterns (streamlines) near HV electrode spikes are presented in figures 26 and 27. The applied DC voltage was 15 kV (positive polarity for G = 15 mm, figure 26; negative for G = 20 mm, figure 27). As it is seen in these figures, a relatively strong EHD flow produced behind the spikes induced significant vortices between the high voltage and the grounded electrodes. For X = 8 mm, the vortices seem to fill almost the whole space between the HV and the grounded electrodes. For X = 15 mm the vortices are slightly smaller than for X = 8 mm, however, the smallest vortices were observed for X = 50 mm. One may expect that strong vortex structures found in the EHD device are capable of hindering the generation of airflow in the EHD device. To check this supposition the average velocity of the unidirectional flow generated in the EHD device for various electrode arrangements was determined.

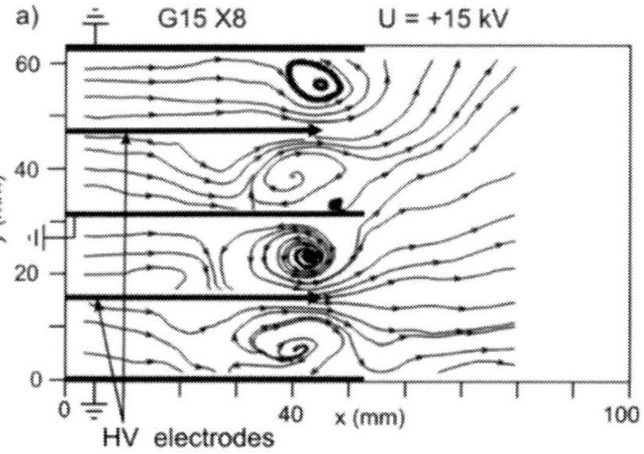

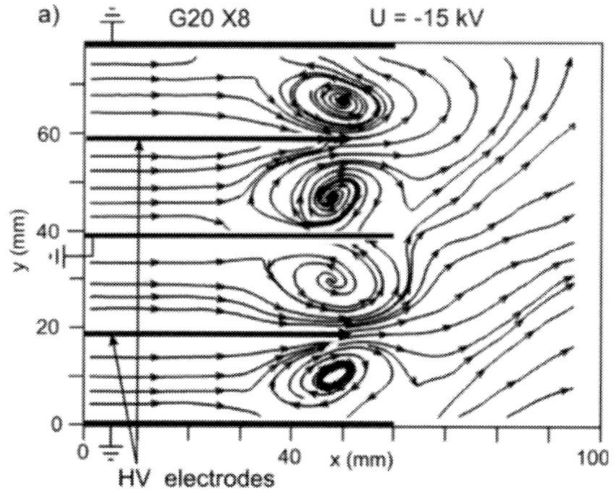

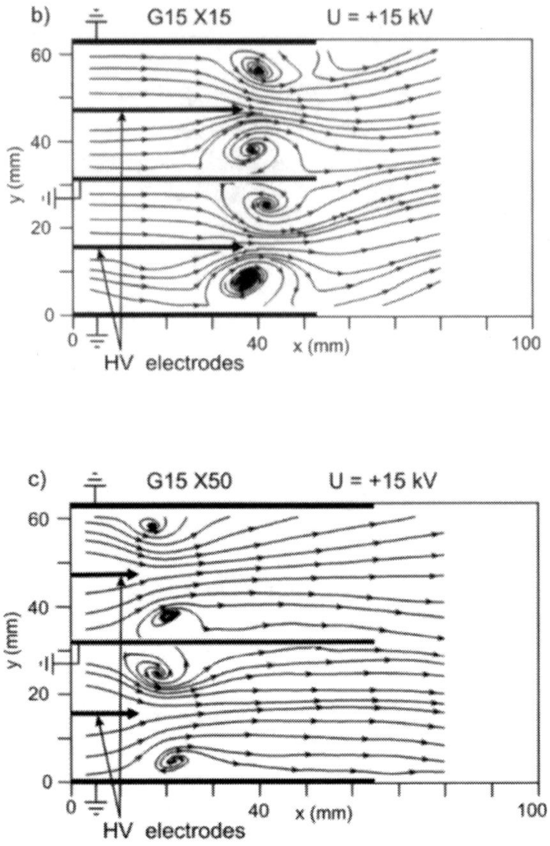

Figure 26. Flow streamlines in the area near the spike tips of the HV electrodes for the electrode arrangements with G = 15 mm, X = 8 mm (a), X = 15 mm (b), X = 50 mm (c). The applied positive high voltage was + 15 kV.

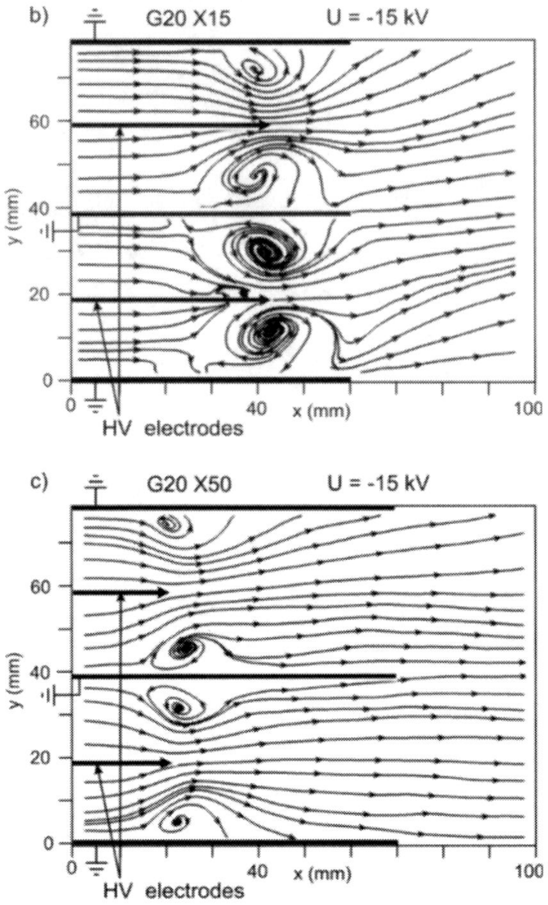

Figure 27. Flow streamlines in the area near the spike tips of the HV electrodes for the electrode arrangements with G = 20 mm, X = 8 mm (a), X = 15 mm (b), X = 50 mm (c). The applied negative high voltage was - 15 kV.

For this purpose, the flow patterns at the EHD device inlet were measured and then the velocity profile was determined at 40 mm before the rear edges of the HV electrodes. Basing on this velocity profile, the average velocity of the induced air flow was calculated. The dependence of the average flow velocity on applied voltage for G = 20 mm and G = 15 mm are shown in figures 28a and 28b. As it can be seen, the average flow velocity increased with increasing applied voltage (with a tendency to saturation). The lowest average flow velocity was found for the X = 8 mm and the highest one for X = 50 mm (as we expected after the streamlines analysis). It seems that in the case of X = 8 mm the majority of applied electrical power was consumed for creating the vortex structures.

The obtained average velocities of the unidirectional air flow (figures 28a and 28b) were slightly higher for negative voltage polarity. The difference in the average flow velocity for the negative and positive polarities was more pronounced for higher voltages. Thus, the electrical efficiency of the unidirectional flow generation the in the EHD device was higher for the negative voltage.

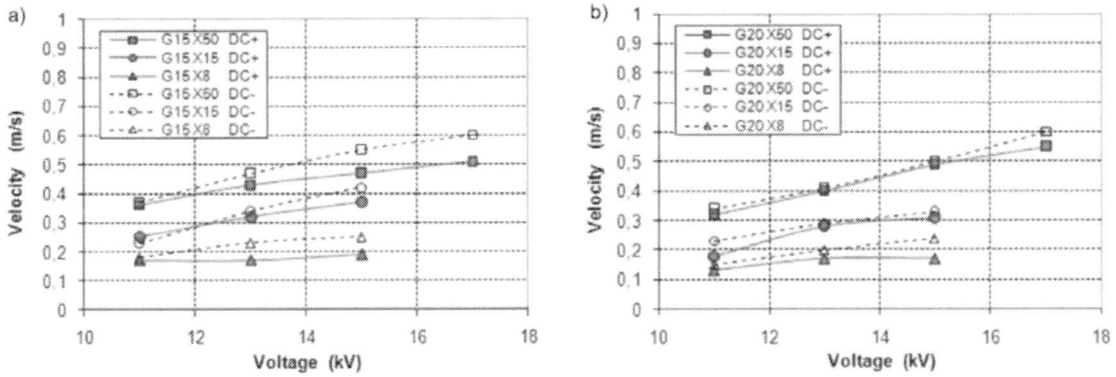

Figure 28. The flow average velocity induced in the self-pumped ESP for (a) G = 15 and (b) G = 20 as a function of the applied voltage

7.3. Conclusions

The results of PIV measurements in the self-pumped spike-electrode ESP showed generation of complex air flow patterns near the spike electrodes and a net unidirectional air flow in the ESP.

The air flow generated by corona discharges behind the spikes induced significant vortices between the spiked electrodes and the grounded electrodes. The diameters of the vortices depended on spiked electrode shift X in relation to the grounded electrode. When the electrode shift X was 8 mm vortices filled the whole space between the HV electrodes and the grounded electrodes. When the electrode shift X was 50 mm, only small vortices near the grounded electrodes were observed. We suggested that the reduction of vortices could be beneficial for inducing strong unidirectional flow in the ESP duct. The results of the measurements of the average flow velocity at the ESP inlet confirmed our suggestions.

The average velocity of the unidirectional flow increased with increasing applied voltage, with a tendency to saturation. The maximum average flow velocity of 0.6 $m \cdot s^{-1}$ was obtained at a voltage of V_{HV} = -17 kV for two electrode arrangements, i.e. with G = 15 mm, X = 50 mm, and with G = 20 mm, X = 50 mm.

Our research showed that the presented ESP can also play a role of a fan, thus, the ESP can be useful not only for air cleaning but also for air pumping.

8. Final summary and conclusions

The presented examples of new experimental results on EHD flow in various ESPs showed complex structures of the flow in ESPs, which depend on the ESP geometry, applied voltage and primary gas velocity. The results confirmed an opinion that the EHD flow plays an important role in the particle collection in ESPs. However, despite of delivering many new results, the recent experimental studies did not answer many questions related the influence of the EHD flow on the performance of ESPs. In particular, the influence of the turbulent EHD on submicron particles transport and deposition in ESPs is still ambiguous. This problem and other related to the submicron particle collection in ESPs are to be objectives of further investigations, also theoretical and numerical.

The presented results can be useful in practical implementation, in particular in these industries that use electrostatic precipitators and are subjected to make changes in their technology due to the new air pollution emission regulations.

Acknowledgments

The research presented in this paper was supported by the Ministry of Science and Higher Education grants PB 281/T02/2010/70, PB 327/T02/2010/70, PB 7475/B/T02/2011/40, and by the R. Szewalski Institute of Fluid Flow Machinery, Polish Academy of Sciences, Gdansk

References

[1] Leny R, Leny A M and Boulloud A 1976 *Proc. 4th Int. Conf. Gas Discharges* (Swansea) p 246
[2] Sigmond R S 1976 *Electr. breakdown of gases* (J. Wiley & Sons)
[3] Yabe A, Mori Y and Nijikate K 1978 *J. AIAA* **16** 340
[4] Crichton G C, McAllister I W and Brengsbo E 1979 *Proc. 14th Int. Conf. Phenom. in Ionized Gases* (Grenoble)
[5] Chang J S 1981 *J. Aerosol Sci.* **12** 19
[6] Chang J S, Lawless P A and Yamamoto T 1991 *IEEE Trans. Plasma Sci.* **19** 1152
[7] Chang J S and Watson A 1994 *IEEE Trans. Diel. Electr. Insul.* **1** 871
[8] Peterson R J and Davidson J H 1994 *Proc. IEEE Ind. App. Soc. Ann. Meeting* (Denver) p 1513
[9] Henry R F, Podolski W F and Saxena S C 1985 *IEEE Trans. Ind. Appl.* **21** 939
[10] Dietz P W 1982 *Powder Technol.* **31** 221
[11] Dodbibat G, Shibayama A, Miyazaki T and Fujita T 2001 *Magn. and Electr. Sep.* **11** 63
[12] Jaworek A, Balachandran W, Krupa A, Kulon J and Lackowski M 2006 *Environ. Sci. Technol.* **40** 6197
[13] Carotenuto C, Nataleb F and Lancia A 2010 *Chem. Eng. J.* **165** 35
[14] Hayati I, Bailey A I and Tadros Th F 1986 *Nature* **319** 41
[15] Okuda H 1996 *Phys. Plasmas* **3** 2191
[16] Abu-Ali J and Barringer S A 2005 *J. Electrostat.* **63** 361
[17] Narayanan N, Akkaraju S, Bloomsburgh J and Zhou L *J. Phy.: Conference Series* **34** 704
[18] Kocik M, Podlinski J, Niewulis A, Mizeraczyk J, Tsubone H and Chang J S 2008 *J. Phy.: Conference Series* **142** 012061
[19] Kocik M, Podlinski J, Mizeraczyk J Urashima K and Chang J S 2009 *IEEE Trans. Diel. Electr. Insul.* **16** 601
[20] Tanski M, Kocik M and Mizeraczyk J 2011 *IEEE Trans. Diel. Electr. Insul.* **18** 1429
[21] Moreau E 2007 *J. Phys. D: Appl. Phys.* **40** 605
[22] Touchard G 2008 *I. J. PEST* **2** 1
[23] Corke T C, Post M and Orlov D 2009 *Exp. Fluids* **46** 1
[24] Corke T C, Enloe C L and Wilkinson S P 2010 *Annu. Rev.Fluid Mech.* **42** 505
[25] Pons J, Moreau E and Touchard G 2007 *Proc. of 28th ICPIG* (Prague)
[26] Boeuf J P, Lagmich Y, Unfer T, Callegari T and Pitchford L C 2007 *J. Phys. D: Appl. Phys.* **40** 652
[27] Takeuchi N, Yasuoka K and Ishii S 2007 *IEEE Trans. Plasma Sci.* **35** 1704
[28] Balcon N, Benard N and Moreau E 2009 *IEEE Trans. Diel. Electr. Insul.* **16** 463
[29] Takeuchi N and Yasuoka K 2009 *IEEE Trans. Diel. Electr. Insul.* **16** 364
[30] Berendt A, Podliński J and Mizeraczyk J 2010 *Proc. 12th Int. Symp. High Pressure Low Temp. Plasma Chem.* (Trencianske Teplice) p 131
[31] Berendt A, Podliński J and Mizeraczyk J 2011 *Eur. Phys. J. Appl. Phys.* **55** 13804
[32] Forte M, Jolibois J, Moreau E and Touchard G 2006 *Proc. 3rd AIAA Flow Control Conf.* p 2006
[33] Benard N, Balcon N and Moreau E 2008 *J. Phys. D: Appl. Phys.* **41** 042002
[34] Benard N, Balcon N and Moreau 2008 *Proc. 39th Plasmadynamics and Lasers Conf.* p 2008
[35] Benard N, Balcon N and Moreau 2009 *Proc. 47th AIAA Aerosp. Sci. Meeting Incl. New Horizons Forum and Aerosp. Exposition* p 2009
[36] Balcon N, Benard N, Lagmich Y, Boeuf J P, Touchard G and Moreau E 2009 *J. Electrostat.* **67** 140
[37] Jolibois J and Moreau E 2009 *IEEE Trans. Diel. Electr. Insul.* **16** 758
[38] Thomas F, Corke T, Iqbal M, Kozlov A and Schatzman D 2009 *AIAA J.* **47** 2169
[39] Benard N and Moreau E 2010 *J. Phys. D: Appl. Phys.* **43** 145201
[40] White H J 1963 *Ind. Electrostat. precipitation* (Addison-Wesley)
[41] Mizuno A 2000 *IEEE Trans. Diel. Electr. Insul.* **7** 615
[42] Jaworek A, Krupa A and Czech T 2007 *J. Electrostat.* **65** 133

[43] Leonard G L, Mitchner M and Self S A 1983 *J. Fluid Mech.* **127** 123
[44] Atten P, McCluskey F M and Lahjomri A C 1987 *IEEE Trans. Ind. Appl.* **23** 705
[45] Chang L and Bai H 1999 *J. Aerosol Sci.* **30** 325
[46] Hoferer B and Schwab A J 2000 *Proc. Conf. Electr. Insul. Diel. Phenom.* p 93
[47] Parasram N T and Taylor A M 2001 *Proc. 8th Int. Conf. Electrostat. Precipitation* (Birmingham)
[48] Blanchard D, Dumitran L M and Atten P 2001 *Proc. 8th Int. Conf. Electrostat. Precipitation* (Birmingham)
[49] Yamamoto T, Okuda M and Ohkubo M 2002 *Proc. Conf. Electr. Insul. Diel. Phenom.* p 228
[50] Fujishima H, Ueda Y, Tomimatsu K and Yamamoto T 2004 *J. Electrostat.* **62** 291
[51] Fujishima H, Morita Y, Okubo M and Yamamoto T 2006 *IEEE Trans. Diel. Electr. Insul.* **13** 160
[52] Adamiak K and Atten P 2009 *IEEE Trans. Diel. Electr. Insul.* **16** 608
[53] Deutsch W 1922 *Ann. Phys.* **68** 335
[54] Podlinski J, Niewulis A, Mizeraczyk J and Atten P 2008 *J. Electrostat.* **66** 246
[55] Yamamoto T and Velkoff H R 1981 *J. Fluid Mech.* **108** 1
[56] Larsen P S and Christensen E M 1986 *Proc. 3rd Int. Symp. Appl. Laser Anemometry Fluid Mech.* (Lisbon)
[57] Self S A, Kihm K D and Mitchner M 1987 *Proc. 3rd Int. Conf. Electrostat. Precipitation* (Padova) p 443
[58] Self S A, Mitchner M and Kihm K D, 1988 *Proc. 7th EPA/EPRI Symp. Part. Contr. Techn.* (Nashville)
[59] Davidson J H and McKinney P J 1989 *Atmosph. Environ.* **23** 2093
[60] Riehle C and Loffler F 1990 *Proc. 4th Int. Conf. Electrostat. Precipitation* (Beijing) p 104
[61] Kanazawa S, Ohkubo T, Nomoto Y and Adachi T 1993 *J. Electrostat.* **29** 193
[62] Hautanen J, Kilpelainen M, Kauppinen E I, Jokiniemi J and Lehtinen K 1995 *Aerosol Sci. Techn.* **22** 181
[63] Soldati A 2000 *J. Aerosol Sci.* **31** 293
[64] Davidson J H and McKinney P J 1991 *IEEE Trans. Ind. Appl.* **27** 154
[65] Podlinski J and Mizeraczyk J 2011 *IEEE Trans. Plasma Sci.* **39** 2260
[66] Mizeraczyk J, Kocik M, Dekowski J, Dors M, Podliński J, Ohkubo T, Kanazawa S and Kawasaki T 2001 *J. Electrostatics* **51** 272
[67] Mizeraczyk J, Dekowski J, Podliński J, Dors M, Kocik M, Mikielewicz J, Ohkubo T and Kanazawa S 2002 *IEEE Trans. Plasma Sci., 3rd Triennial Spec. Issue "Images in Plasma Sci."* **30** 164
[68] Mizeraczyk J, Dekowski J, Podliński J, Kocik M, Ohkubo T and Kanazawa S 2003 *J. Visualization* **6** 125
[69] Mizeraczyk J, Kocik M, Dekowski J, Podliński J, Ohkubo T and Kanazawa S 2004 *Inst. Phys. Conf. Ser.***178** 167
[70] Cotton J S, Brocilo D, Chang J S, Shoukri M and Smith-Pollard T 2003 *IEEE Trans. Diel. Electr. Insul.* **10** 37
[71] Chun Y N, Berezin A A, Chang J S and Mizeraczyk J 2007 *IEEE Trans. Diel. Electr. Insul.* **14** 119
[72] Farnoosh N, Adamiak K and Castle G 2010 *J. Electrostat.* **68** 513
[73] Farnoosh N, Adamiak K and Castle G 2011 *J. Electrostat.***69** 419
[74] Zhao L and Adamiak K 2008 *IEEE Trans. Ind. App.* **44** 683
[75] Schmid H J, Stolz S and Buggisch H 2002 *Flow Turbul. Combust.* **68** 63
[76] Chun Y N and Yeom D S 2006 *Korean J. Chem. Eng.* **23** 560
[77] Kim S C and Chun Y N 2009 *Int. J. Environ. Pollut.* **36** 337
[78] Mizeraczyk J, Dekowski J, Podliński J, Kocik M, Ohkubo T and Kanazawa S 2003 *J. Visualization* **6** 125

[79] Raffel M, Willert Ch and Kompenhans J 2007 *Part. Image Velocimetry: A pract. guide* (Springer-Verlag)
[80] Zukeran A, Ikeda Y, Ehara Y, Matsuyama M, Ito T, Takahashi T, Kawakami H and Takamatsu T 1999 *IEEE Trans. Ind. App.* **35** 346
[81] Saiyasitpanich Ph, Keenera T C, Khangb S J and Lua M 2007 *J. Electrostat.* **65** 618
[82] Kittelson D B, Reinersen J and Michalski J 1991 *SAE Pap.* 910329
[83] Atten P, McCluskey F M J and Lahjomri A C 1987 *IEEE Trans. Ind. Appl.* **23** 705
[84] Yamamoto T 1989 *J. Electrostat.* **22** 11
[85] Westerweel J 1997 *Meas. Sci. Tech.* **8** 1379
[86] Niewulis A, Podliński J and Mizeraczyk J 2009 *J. Electrostat.* **67** 123
[87] Katatani A and Mizuno A 2010 *J. Inst. Electrostat. Jpn.* **34** 187
[88] Podlinski J, Niewulis A, Shapoval V and Mizeraczyk J 2011 *IEEE Trans. Dielectr. Electr. Insul.* **18** 1401

Journal of Physics: Conference Series **418** (2013) 012069 doi:10.1088/1742-6596/418/1/012069

Collection of low resistivity fly ash in an electrostatic precipitator

M Jędrusik*, A Świerczok[1] and A Jaworek[2]
[1] Wroclaw University of Technology, Institute of Heat Engineering and Fluid Mechanics, Wyb. Wyspianskiego 27, 50-370 Wroclaw, Poland
[2] Polish Academy of Science Poland, Institute of Fluid Flow Machinery, Fiszera 14, 80-231 Gdansk, Poland

E-mail: maria.jedrusik@pwr.wroc.pl

Abstract. Due to increasing restrictions on dust emission limits (IED directive), particularly in the fine particle size range, wider application of electrostatic precipitators (ESPs) in cleaning the combustion gases from stoker boilers can be anticipated. The objective of the model studies in this paper was to select the optimal construction of the discharge electrode in ESP for obtaining high collection efficiency of fly ash leaving stoker boilers. In these studies a test bench was constructed, which comprised one-stage model ESPs with a set of discharge and collecting electrodes. The main dimensions of the precipitator chamber were as follows: length of electric field 2.0 m; active height 0.45 m and spacing between the collecting electrodes 0.4 m. Four constructions of discharge electrode were tested for fly ash of different fractional sizes and chemical compositions. The aim of the tests was to determine the current-voltage characteristics and the discharge current distribution on the collection electrode so as to find out the optimal construction and ensure the maximal collection efficiency of ESP. The results of the collection efficiency measurements in these tests were compared with those obtained from an ordinary industrial ESP. The comparison shows that it is necessary to optimise the discharge electrode construction for a specific physico-chemical property of fly ash so as to obtain the highest collection efficiency.

1. Introduction

Stoker boilers are usually equipped with mechanical dust collectors or bag filters for fly ash removal, and electrostatic precipitators (ESP) are scarcely used. More restrictive dust emission limits (under IED Directive) as well as expected requirements regarding fine particles emission (PM10 and PM2.5) will require wider application of ESPs of improved parameters for the cleaning of combustion gases after stoker boilers.

Improvement of the collection efficiency of mechanical collectors (cyclones and multi-cyclones) requires higher gas flow velocities in the collector that increases the pressure drop and shortens its lifetime. In the last years, bag filters or a combination of bag filters with cyclone were introduced after stoker boiler for exhaust gas cleaning. However, the application of such equipment for exhaust gas cleaning after stoker boiler appeared to be economically ineffective because of high cost of bag filters. For these reasons, electrostatic precipitators, which are competitive with respect to the investment and

* To whom any correspondence should be addressed.

operational costs and can better comply with environment protection regulations, seems to be an alternative to other devices.

Fly ash leaving stoker boiler has high contents of unburned coal (8-25%). From technical literature, it is known that with an increase of unburned coal contents in fly ash, the collection efficiency of an ESP decreases. The first information about an effect of unburned coal in fly ash on ESP collection efficiency appeared in the seventieth of XX century. Hagemann and Ahland [9] have presented the results of experimental studies showing that a content of unburned coal in fly ash up to 15 % causes an increase of ESP collection efficiency, but the collection efficiency started to decrease when the coal content was higher than 20 %. Many recent studies [10, 11] also pointed out a significant influence of unburned coal on the collection efficiency of ESP. The unburned coal strongly influences the electric resistivity of fly ash, decreasing its value even below 10^9 Ωcm that has positive effect on the collection efficiency, and suppresses the back discharge. However, for too low resistivity ($<10^6$ Ωcm) the force of cohesion radically drops that results in particles re-entrainment and decreases the collection efficiency. It is particularly apparent for particles of diameter larger than 100 μm, for which the unburned coal content is ten times higher than in smaller particles, having diameter below 38 μm [10].

Many years of experience with industrial ESPs and proprietary studies [1-4] indicate that a proper selection of the construction of discharge electrode for a specific technological process, taking into account the physical and chemical properties of fly ash, has a significant influence on the collection efficiency of ESP.

In late ninetieth of XX century, many laboratory studies on discharge electrode constructions as well as numerical modelling of the phenomena occurring in corona discharge were carried out for different forms of the electrode spike [5-8]. The main purpose of those studies was to optimize the inter-electrode spacing, discharge electrode shape and electric supply parameters.

The aim of these studies was to determine the current-voltage characteristics and the discharge current distribution on the collection electrode so as to find the optimal discharge electrode construction, for which the collection efficiency of ESP is the highest.

2. Laboratory tests

In this section, the results of an effect of discharge electrode construction on the collection efficiency of ESP, and used for the removal of fly ash of low resistivity (due to high content of unburned coal), are presented. These studies extend and refine our earlier results presented in [1].

2.1. Current-voltage characteristics

Four discharge electrode constructions, three of which where of the RDE type (figure 1), based on results of our former studies [1-4], were proposed for the investigation.

It is shown that the construction of discharge electrode influences the collection efficiency of ESP because of different values of the discharge current specific to the electrode type. Depending on the discharge electrode construction, for the same ESP geometry and volume, by the same supply voltage it is possible to obtain different discharge currents, which maximal magnitude is limited only by the breakdown voltage. The current-voltage characteristic $I=f(U)$ shows the relationship between the discharge (corona) current and high voltage between the discharge and collecting electrodes.

Depending on fly ash resistivity, the optimal current-voltage characteristics, particularly the required discharge current, can vary that makes it practically almost impossible to design a 'universal' discharge electrode for ESP. For this reason, in order to improve the charging process and increase the collection efficiency of fly ash particles, higher values of supply voltage are, therefore, advised [12-14].

In figure 2 are presented the current-voltage characteristics of various types of discharge electrodes for the spacing between collection electrodes 2h=400 mm.

Based on the current-voltage characteristics, with respect to the discharge current, the electrodes can be characterised as follows:

- high current at high voltages ('current aggressive' characteristics): RDE-2 and RDE-3

- high current at medium voltage: barbed tape
- medium current at high voltage ('soft' characteristics): RDE-1

High content of unburned coal in fly ash from stoker boilers (from several up to several dozen percents) results in low electric resistivity of fly ash layer (the resistivity was in the range of 10^7-10^8 Ωcm, measured in laboratory conditions). By this resistivity, the electric charge on fly ash particles is easily conducted to ground after the deposition of these particles onto the collecting electrodes that leads to particles reentrainment. In this case, to keep the particles on the electrode, a high discharge current and uniform current density distribution on the collecting electrode are required. The primary task is, therefore, to design a discharge electrode, which generates high discharge current and ensures high magnitude of electric field over the collection electrode. With respect to discharge current, the 'current aggressive' discharge electrodes (RDE-2 and RDE-3) are recommended. However, these electrodes not always can ensure the desired uniform current distribution.

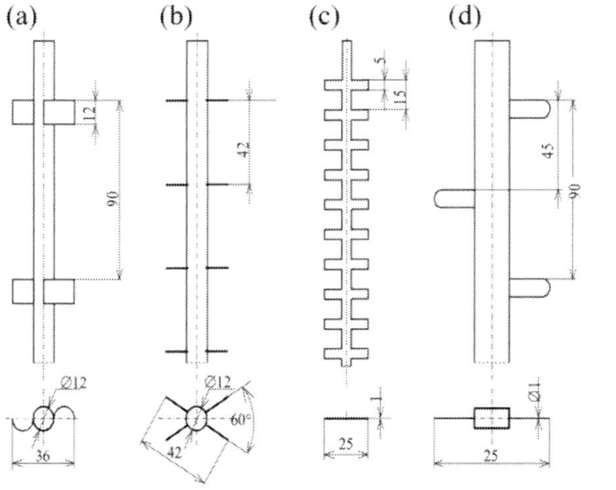

Figure 1. Tested discharge electrodes: a) RDE-1; b) RDE-2; c) barbed tape; d) RDE-3 (dimensions in mm).

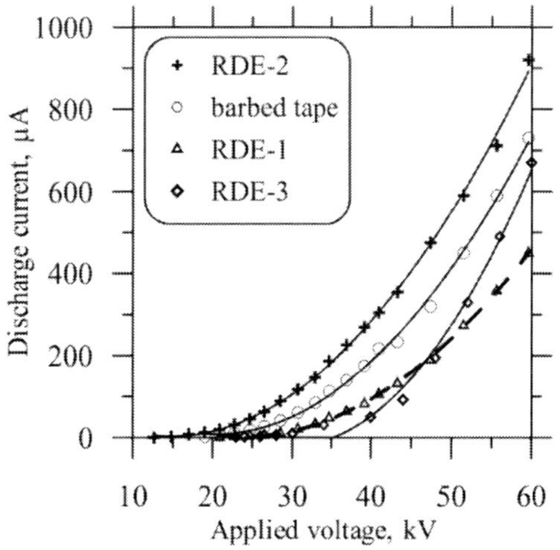

Figure 2. Current-voltage characteristics of the tested discharge electrodes.

2.2. *Current distribution on collecting electrode*

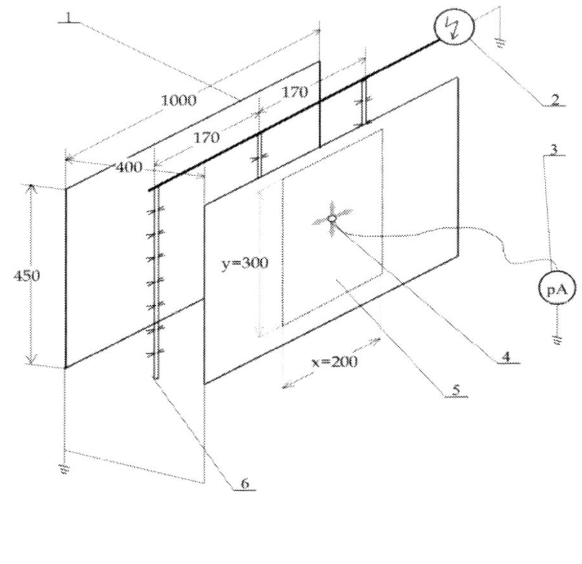

Figure 3. Scheme of laboratory bench for measuring discharge current distribution on the collecting electrode surface: 1-collecting electrodes, 2-high voltage supply unit, 3-pico-ammeter, 4-moving measuring probe, 5-measuring area, 6-discharge electrodes (dimensions in mm).

Measurements of discharge current distribution on collecting electrode have been carried out on laboratory bench with a moving current-measuring probe (figure 3). Examples of the measurements are shown in figure 4 for RDE-2 and RDE-3 electrodes.

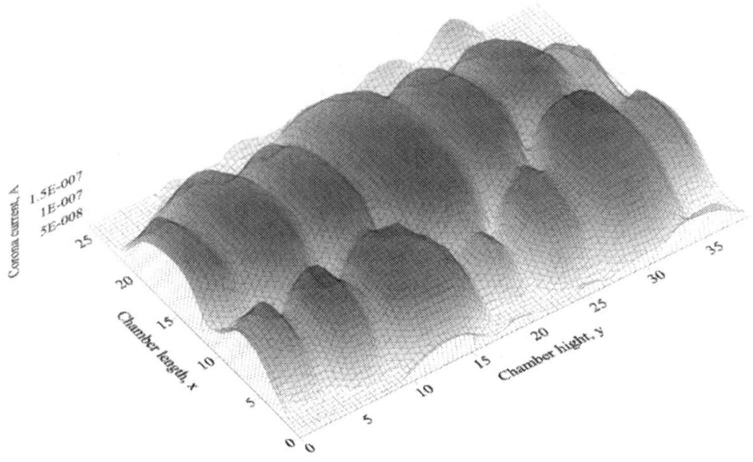

(a)

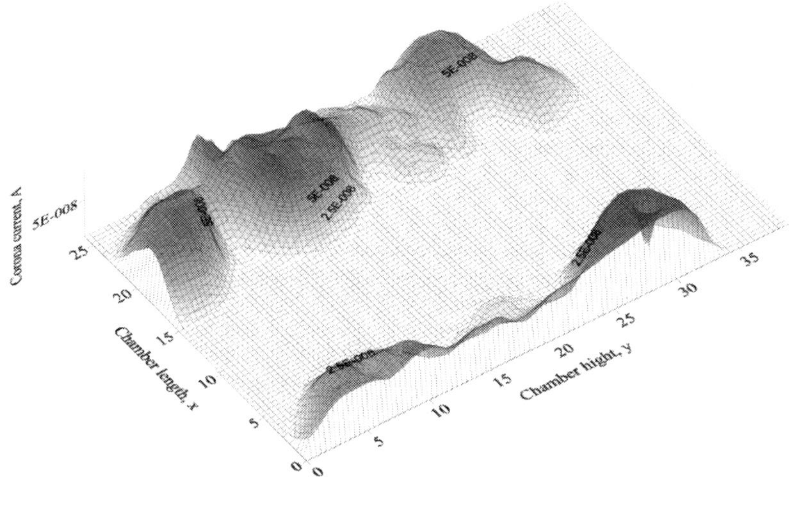

(b)

Figure 4. Discharge current distribution on the collecting electrode surface for: (a) RDE-2 electrode, (b) RDE-3 electrode, by a supply voltage of U=50 kV.

From these measurements results that the RDE-2 electrode is characterized by relatively uniform coverage of the collecting electrode by discharge current, by low relative standard deviation (RSD) of its values, and also ensures the highest average value of the discharge current as compared to the others three constructions by the same supply voltage. With this respect, the worst construction seems to be the RDE-3 electrode, providing poor uniformity of coverage of the collecting electrode with discharge current, by high RSD, and low average discharge current.

2.3. Physical and chemical properties of fly ash from stoker coal-fired boilers

Various fly ashes from different stoker boilers have been collected for the measurements of collection efficiency of ESP. The coal burned in these boilers had the granulation of larger than 10 mm that significantly influenced the structure of fly ash and particle size distribution. In table 1 are given the physical and chemical properties of the tested fly ashes, in particular, unburned coal content, and content of alumino-silicates, iron compounds, and SO_3. Figure 5 presents cumulative size distribution curves for fly ash particles entering the ESP, determined by *Mastersizer S* (*Malvern Instruments Ltd.*), operating on a principle of laser light scattering on the measured dust particles. The presented characteristics show considerable differences in fly ash particle size distribution, depending on particle source. Usually when multi-cyclone is used as the first stage of dust control, nearly all larger fly ash fractions are precipitated in it, and only small particles enter the ESP. For example, the fly ash sample P-3 (after a multi-cyclone) has the median diameter d_{50}=20 μm, and fly ash sample P-2 (without first-stage cleaner) d_{50}=200 μm. Fly ashes P-1 and P-17 contain mainly larger particles, and characteristic diameters are d_{50}=250 μm and d_{50}=560 μm, respectively. These fly ashes contain less unburned carbon and more silica than ash P-2 and P-3 and, furthermore, fly ash P-1 contains a minimum of unburned carbon (ca 13.8 %) and is characterized by higher resistivity (10^8 Ωcm) than the others.

The studies had also proved that the content of unburned coal in fly ash changes with the particle diameter and gets the highest value for fractions above 126 μm. For example, for fly ash sample P-2 the content of unburned coal was: d<32 μm–23%, d=32–80 μm–22%, d=80–125 μm–18%, d>125 μm–37%, respectively.

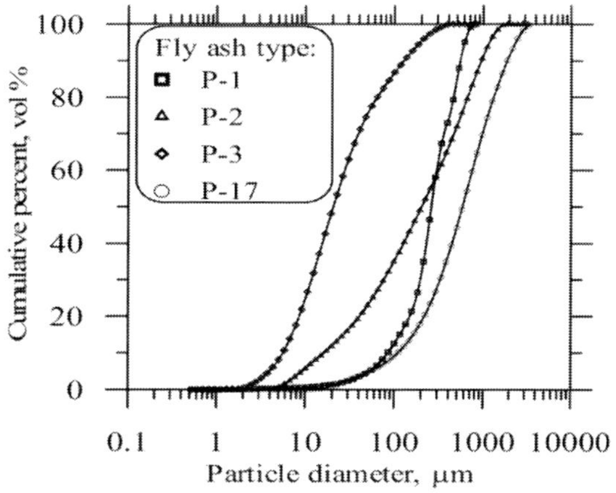

Figure 5. Cumulative size distribution of fly ash particles at ESP inlet.

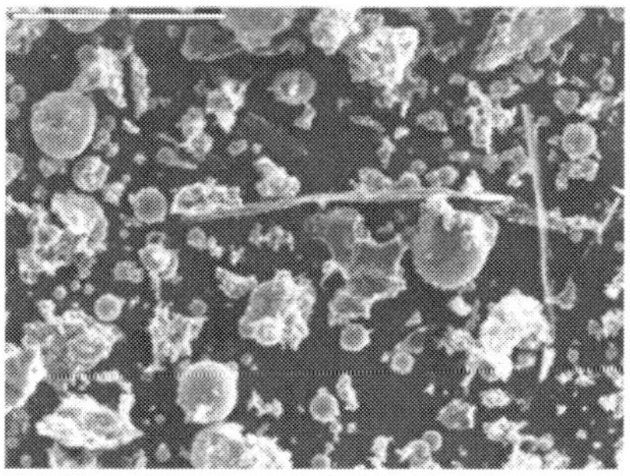

(a)

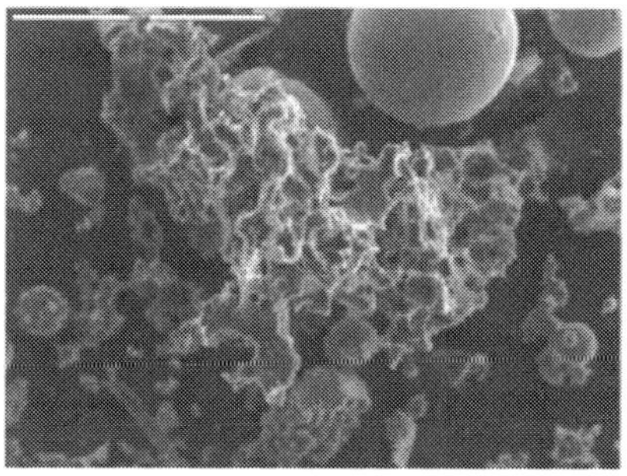

(b)

Figure 6. Scanning electron microscope micrographs of fly ash particles: (a) sample P-2: particles of irregular shapes and dimensions. (b) sample P-1: in the central area is visible irregular particle of high coal contents (magnification 230x).

The fractional particle size distribution analysis determines only one geometric parameter of fly ash it's the equivalent diameter, regardless of the method of measurement used. The actual shape of fly ash particles is rare spherical (figure 6), that also affects the process of their separation in ESP. For the analysis of collection efficiency of ESP it is essential to take into account the content of fly ash particles of irregular shape and porous large unburned coal particles, which favor the particle re-entrainment from the collecting electrode [15].

Table 1. Analysis of physical and chemical properties of fly ashes from selected stoker boilers.

Parameter	Unit	Fly ash type			
		P-1	P-2	P-3	P-17
SiO_2	%	47,44	28,99	28,64	40,39
Fe_2O_3	%	6,91	3,67	8,62	10,01
Al_2O_3	%	19,65	17,14	18,06	18,09
TiO_2	%	0,99	0,86	1,09	0,96
CaO	%	3,98	2,82	2,84	5,44
MgO	%	1,41	1,01	0,98	1,47
SO_3	%	0,73	2,26	6,48	1,27
K_2O	%	3,03	2,68	2,48	2,08
P_2O_5	%	0,01	0,01	0,01	-
Na_2O	%	1,33	1,14	1,17	1,47
Unburned carbon	%	13,77	28,60	27,72	17,92
Density	kg m^{-3}	1850	2090	2350	1760
Resistivity	Ω·cm	$5,1\times10^8$	$5,0\times10^7$	$6,4\times10^7$	$5,0\times10^7$

2.4. Collection efficiency measured in laboratory conditions
The test bench (figure 7) comprises of model ESP with discharge and collecting electrode system, fly ash feeding system, exhaust gas fan, supplying and drain-off gas ducts, high voltage supply unit and outlet filter. The main dimensions of the model ESP are: length of electric field L=2000 mm, height of the field H=450 mm, collecting electrode spacing 2h=400 mm and distance between the discharge electrodes s=170 mm.

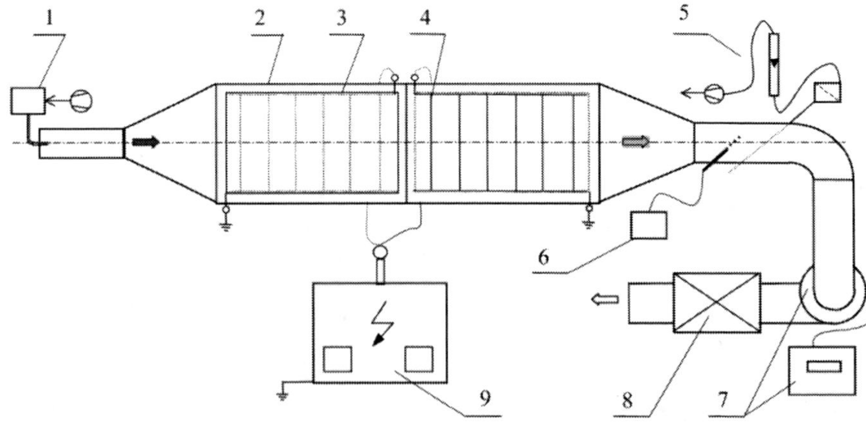

Figure 7. Scheme of laboratory ESP test bench: 1–fly ash feeder, 2–model chamber, 3–collecting electrodes, 4–discharge electrodes, 5–gravimetric dust-meter, 6–thermo anemometer, 7–exhaust gas fan with controlled rotational speed, 8–oulet filter, 9–high voltage supply (of negative polarity).

A fly ash collection efficiency tests were carried out at the stand shown in figure 7 in order to determine the effect of physico-chemical properties of fly ash, in particular, its resistivity, and fractional size distribution on the ESP collection efficiency. The collection efficiency has been measured for four different fly ash samples, collected from stoker boilers, and also for four constructions of discharge electrode. For each magnitude of supply voltage, three test of the collection efficiency have been carried out.

The total collection efficiency was estimated from the measured fly ash mass at the inlet and the outlet of ESP and calculated from the following formula:

$$\eta_c = 1 - \frac{\dot{m}_{out}}{\dot{m}_{in}} \tag{1}$$

where: $\dot{m}_{in}$ –mass flow rate of fly ash at the inlet of ESP, g s^{-1}; $\dot{m}_{out}$ –mass flow rate of fly ash at the outlet of ESP, g s^{-1}.

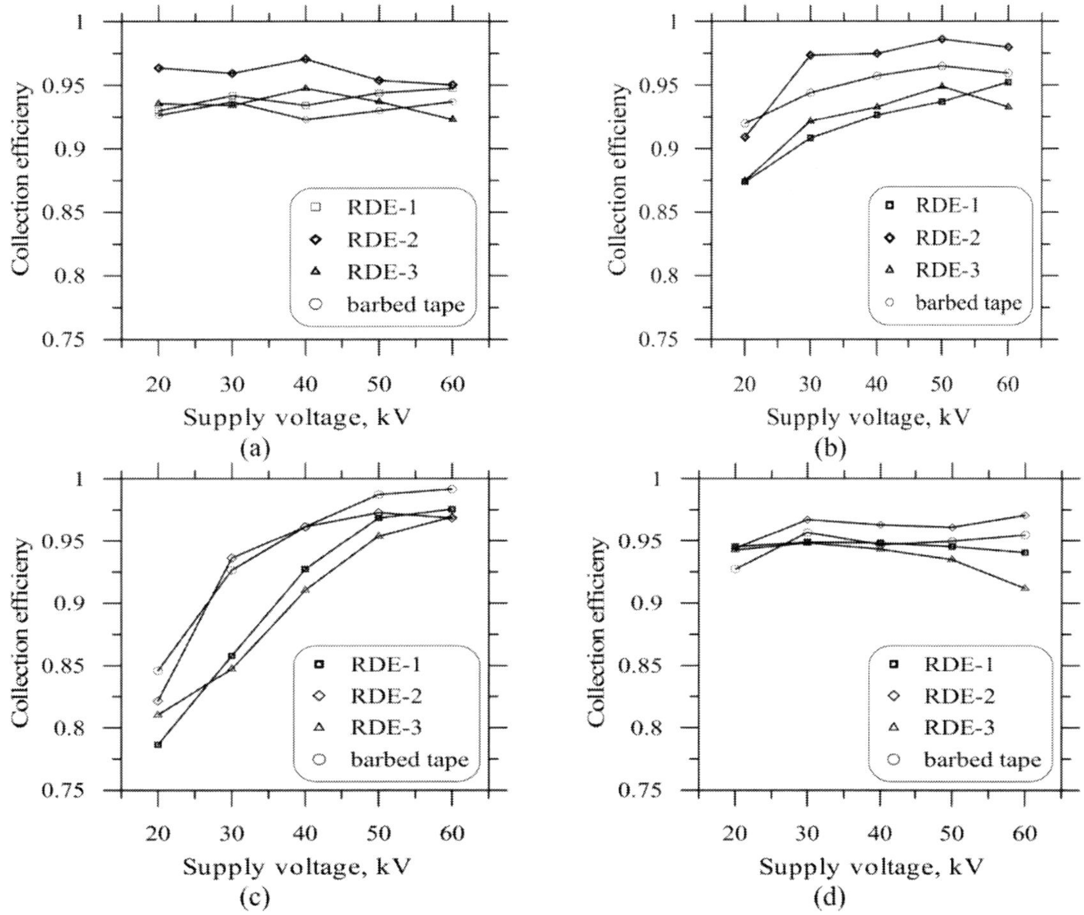

Figure 8. The effect of supply voltage on average collection efficiency of ESP for four constructions of discharge electrodes, and four samples of fly ash: (a) P-1, (b) P-2, (c) P-3, (d) P-17 (air flow velocity in the ESP v=0.8 m s^{-1}).

The fly ash particles concentration at model ESP outlet was measured by means of gravimetric dust meter, after isokinetic sampling with a multi-hole probe. The sampled dust was collected on a highly efficient absorbent paper of a measuring filter. The ash was injected by the vibration feeder with concentration of 0.2 g m^{-3}. All tests were carried out in laboratory in ambient conditions: air temperature of 293 K and relative humidity of φ=60%. The gas flow velocity inside model ESP was measured with thermo-anemometer. Power supply *Sörensen HV Supply 1121* (0-100 kV) was used for the excitation of discharge electrodes with rectified and smoothed DC current.

Test results have shown that fly ashes with high content of large particles (about 90 % above 100 μm; samples P-1 and P-17) and having resistivity in a range of 10^7-10^8 Ωcm are difficult to collect by ESP as compared to fine fly ash (sample P-3) and fly ash with low silica content (sample P-2), regardless of low content of unburned coal. Furthermore, in the case of fly ash with high content of large particles, the collection efficiency is less sensitive to supply voltage than in a case of fine fly ash (P-3). Most probably it is because of high content of unburned coal (more than 20 %) in particles having diameter above d>100 μm [10].

For fly ash samples taken from ESP without a cyclone as the first stage, the highest collection efficiency was obtained when RDE-2 type of discharge electrodes was used, however, for fly ash sample P-3 (from a boiler with an initial mechanical collector stage) the best collection efficiency was obtained for barbed-tape discharge electrode.

On the other hand, an optimal value of high voltage supply was about 50 kV. Further increase in supply voltage does not improve the collection efficiency, and even sometimes makes it worse. This result indicates that the optimal supply voltage has to be determined for specific fly ash parameters, which are related to a boiler type and the fired coal parameters.

3. Industrial tests

The primary aim of the experiments was to estimate the influence of operational parameters of boiler (boiler capacity) and ESP (average values of discharge current, secondary voltage, method of ESP energizing: conventional DC, semi-pulsed) on the ESP collection efficiency.

For these experiments, an industrial ESP with two sections in series and having inter-electrode spacing of 400 mm was selected. Each section has been energized with H.V. supply unit with nominal secondary parameters of 250 mA/ 106 kV. The ESP has been equipped with the collecting electrodes of SIGMA VI type, and discharge electrodes made of spiked pipes. The source of raw gas has been stoker boiler with capacity of 32 MW. The fly ash fractional size analysis of sample P-2, as well as its chemical analysis are presented in Table 1.

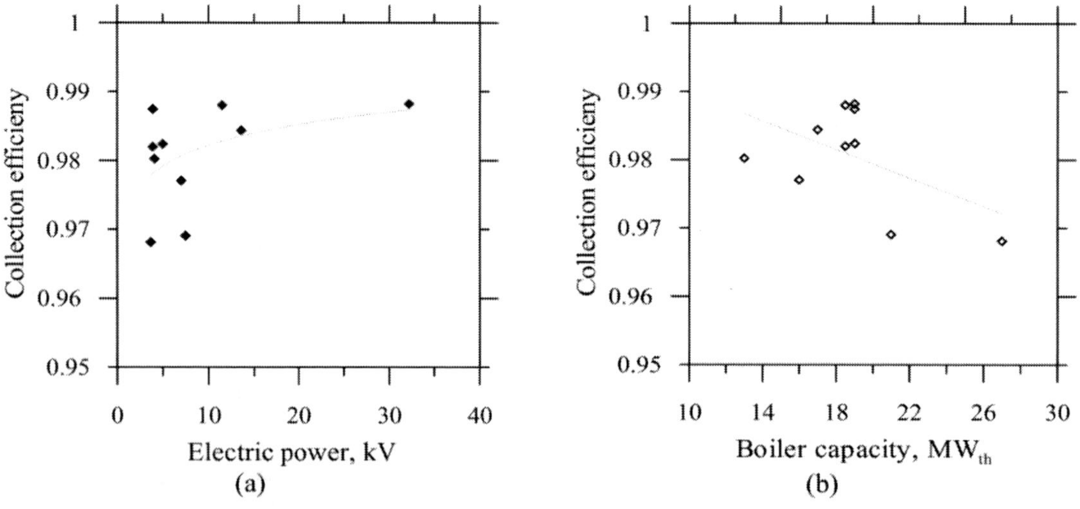

Figure 9. Collection efficiency characteristics of ESP after stoker boiler as a function of: (a) average electric power supplied to the ESP, (b) operational capacity of the boiler.

In order to estimate the ESP collection efficiency a gravimetric measurements of dust concentration were carried out at the ESP inlet and outlet [16]. It is worth noting that experimental tests carried out on industrial ESPs have many limitations, and the measurement results cannot be statistically processed because of uniqueness of operational conditions. Nevertheless, from the collected measuring results some trends, graphically presented in figure 9, can be noticed.

Results presented in figure 9(a) show, that there exists a limit of power supplied to ESP, above which the collection efficiency no further increases that was also shown by the measurements carried out during laboratory tests. However, a decrease in the collection efficiency with an increase of boiler load, shown in figure 9(b), results from an increase of dust concentration in flue gas and an increase of gas flow velocity in ESP.

In order to determine the influence of the method of ESP energization on the collection efficiency, two energizing methods have been tested: semi-pulse energization (with periodically switched-off supply voltage) and conventional DC supply. Characteristic supply voltage waveforms are shown in figure 10.

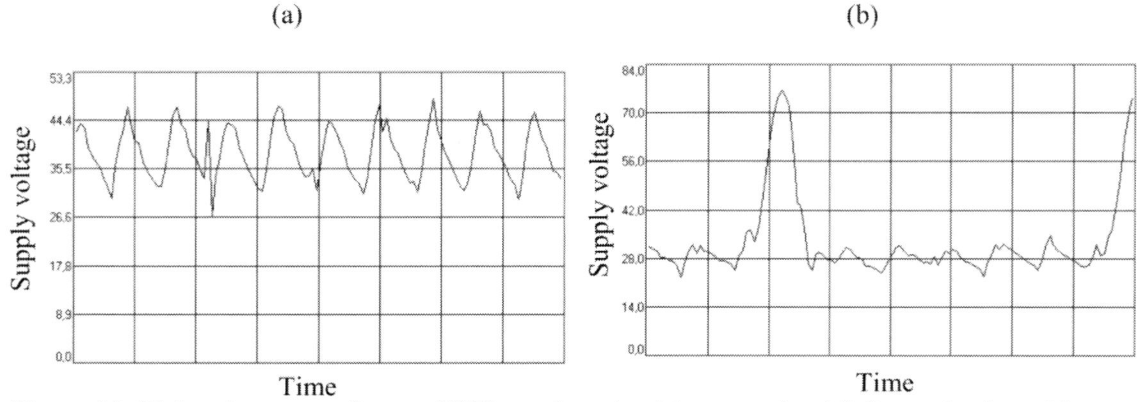

Figure 10. High voltage waveforms of ESP supply units: (a) conventional DC energization with ripples, (b) semi-pulse energizing (with each seventh supply voltage half-cycle ON, superimposed on DC voltage, and six half-cycles OFF).

The results shown that in order to get the same collection efficiency of 98.2 % for both energizing methods it is necessary to deliver 5 kW of discharge power to ESP with semi-pulsed method of energization, and 3.9 kW for conventional method. It seems that the semi-pulse energizing option is not justifiable from the point of view of energy efficiency. However, this method is successfully applied in large industrial ESPs as a means of energy-efficient energizing or even as a method of increasing collection efficiency of ESP for collecting high-resistivity dust.

4. Discussion and conclusion

Fly ash leaving stoker boiler has its characteristic particle morphology (chemical composition, unburned coal contents, particle size distribution and particle shape) which can cause significant differences in electrostatic precipitation process in comparison with operation of ESP after PC or fluidal boilers.

The results obtained from the tests carried out in laboratory-scale ESP and in industrial conditions have shown that special construction of ESP discharge electrode has to be used as well as the optimization of electric supply parameters and energizing method (taking into consideration the physical characteristics and chemical composition of fly ash) has to be carried out for the efficient collection of low-resistivity fly ash coming from coal fired stoker boilers.

Analysis of the collection efficiency measured in laboratory scale ESP for various fly ashes has shown that:

- An increase of the number of particles larger than 100 μm (samples P-1 and P-17 in this specific case) in fly ash decreases the maximal obtainable level of collection efficiency. Particles of this size are easy to charge and its collection is possible even at low supply voltage levels. An increase of supply voltage does not improve significantly the collection efficiency, and sometimes, at certain conditions can even decrease the efficiency. The most probably it is caused by a large percentage of larger porous particles having well developed surface, light weight, and great content of coal (figure 6), which in contact with collection electrode can easily lost its charge and are re-entrained to the flow.
- For fine particles, the collection efficiency can be increased via increasing supply voltage, however, some measurements indicate that the optimal voltage (for example about 50 kV for the sample P-2) can exist, for which the collection efficiency is the maximal. There was also noticed that a relatively high content of unburned coal (comparing to fly ash with large particles) does not worsen the collection process. This effect may suggest that more important in the precipitation process is the size of individual particles than their chemical composition, mainly because fine particles are not easily re-entrained from the collecting electrode.
- The optimal construction of discharge electrode for the tested fly ash samples was that of type RDE-2, for which the current-voltage characteristics can be approximated by the following equation: $I=101-13.2\cdot U+0.44\cdot U^2$. This type of electrode generates high discharge current, which is uniformly distributed on the collection electrode. It may be concluded that this type of discharge electrode, of current-voltage characteristics described with this equation, can be applied in ESPs for the collection of fly ash particles of similar physical and chemical properties to those given in table 1.

Finally, it may be said that it is possible to electrostatically precipitate fly ash from exhaust gases after stoker boilers with high collection efficiency. But in order to obtain the best results for actual physical and chemical properties of fly ash, a careful selection of discharge electrode construction and electric energization parameters will be required. The proposed discharge electrode of type RDE-2 may be an alternative solution to typical electrodes with barbed tape or similar constructions generating non-uniform current distribution on the collection electrode. This alternative should be considered during ESP modernization or retrofitting via replacement of existing electrodes without a need of changing HV supply unit.

Acknowledgments
These studies were carried out as a part of project no. N R06 0014 06/2009 funded by the Polish National Centre for Research and Development (NCBiR)

References
[1] Jędrusik M and Świerczok A 2011 *J. Phys. Conf. Ser.* **301** 1742
[2] Jędrusik M, Świerczok A and Teisseyre R 2003 *Powder Technol.* **135** 295
[3] Jędrusik M and Świerczok A 2006 *Int. Conf. on Air Pollut. Abatement Technol.–Future Challenges. ICESP X* (Cairns: Queensland/ Australia)
[4] Jędrusik M and Świerczok A 2009 *J. Electrostat.* **67** 105
[5] Brocilo C, Chang J S and Findlay R D 2001 *Procee. 8th ICESP* Vol. **1** (Southern Comp. Services Inc., Birmingham, Alabama, USA)
[6] Caron A and Dascalescu L 2004 *J. Electrostat.* **61** 43
[7] Bai H, Lu C and Chang C L 1995 *J. of Air and Waste Manage. Assoc.* **45** 908
[8] Mc Cain J D 2001 *Procee. 8th ICESP* Vol. **1** (Birmingham, Alabama, USA)
[9] Hagemann H and Ahland E 1973 *Staub-Reinhalt. Luft* **33** 367
[10] Clocke M, Hanson S, Lester E and Thompson A 2007 *IC of Coal Science, San Francisco* (USA)
[11] Barranco R, Gong M, Tompson A, Cloke M, Hanson S, Gibb W and Lester E 2007 *Fuel* **86** 2521
[12] Crynack R R 1991 *Proc. of the Ninth Part. Control Symp.* (Wiliamsburg)

[13] Grieco G J 1994 *Power Gen'94 Conf.* (Orlando: Florida)
[14] Jędrusik M 2008 *Dev. of sel. tech. to increase collect. effic.*
[15] Jaworek A, Jędrusik M, Świerczok A, Czech T, Sobczyk A T and Lackowski M 2011 Int. J. Plasma Environ. Sci. and Technol. **5** 161
[16] PN-Z-04030-7, Badania zawartości pyłu, pomiar stężenia i strumienia masy pyłu w gazach odlotowych metodą grawimetryczną – The study of dust content, measurement of concentration and mass flow by gravimetric method, grudzień 1994

A new power regulator control system based on verilog for electrostatic precipitators

Zisheng Zhang*, **Guan Li, Taotao Liu, Pengbo Ge, Zhiqiang Liu**
Electrostatic Research Institute, Hebei University, No.180 WuSi East Road, Baoding, China

E-mail: zhangzisheng@hbu.edu.cn

Abstract. In order to improve the inefficient response of the power system in traditional electrostatic precipitators, such as long design cycles and low safety, a new power regulator control system is designed to tackle the deficiencies. The working voltage system of an electrostatic precipitator consists of an L-C component, a rectifier bridge group and a step-up transformer. The Verilog hardware description language is used to complete the design of the feedback systems. Continuous steady current can be obtained automatically through changing the number of steady voltage control units. The results show that control systems can accurately feed back the changes of the voltage signal of the electrostatic precipitator. Comparing with other control systems, it has the advantages of faster response, higher accuracy, better monitoring performance and superior anti-interference capacity.

1. Introduction

Comparing with traditional wet and sack-style dust-cleaning apparatus, the electrostatic precipitator has the advantages of high efficiency and cheap operation. The control system of electrostatic precipitator mainly uses integrated circuit and relays as the control unit of electrostatic precipitator, making the line voltage signal could feed back and send an alarm in real time [1]. But because of the integrated circuit's limitation, especially the interference of the strong feedback signal, has already blocked the development of the control system. We designed a voltage control system with the electronic design automation (EDA) to solve this problem. This method changed the traditional electrostatic precipitator control unit [2]. EDA technology has the incomparable advantage since it uses the modular language, independent description of the structure and the top of the design concept. It uses the Verilog HDL hardware description language to complete the feedback control system in the EDA software platform, and proceeds the logic compile, logic division, and optimization, the logical layout wiring, logic simulation, the logical mapping and programming download, making the system control unit can apply in different type of coal dust environments. The maximum corona voltage and constant current can be obtained through adjusting the L-C steady voltage control unit and silicon rectifier bridge, which ensures the electrostatic precipitator work effectively. At the same time, the feedback signal of high pressure electrostatic precipitator can control the number of L-C steady voltage control unit through the program written by Verilog HDL language so as to achieve the real-time detection and early warning signal voltage [3].

* To whom any correspondence should be addressed.

2. Main circuit design of the ESP

The main functions include two requirements: (1) the design of the working voltage and current of electrostatic precipitator (2) real-time monitoring of voltage and current under the environment and adjustment of operating voltage adaptively. Figure 1 shows the main circuit of the electrostatic precipitator. According to the design standard, the power uses the voltage of 380 V and the working current with a frequency of 50 Hz, and the working current is in the range of 15-20 A. We can get the constant current with L-C steady voltage control unit. Work voltage and current are boosted through the high impedance booster transformer and silicon rectifier bridge group, and applied on the corona electrodes in the electrostatic precipitator. The corona electrodes make the dust in flue gas be electrified, and produce the negative and positive ions. When the dust particles was charged by these ions, dust particles with the positive and negative ions is absorbed by the corona electrode and the collection plates [4]. Analog voltage and current signals are tested by voltage and current detection sensor in this electrostatic precipitator. The Verilog HDL language design of feedback control chip receives the digital switching device into digital signals of the voltage and the current. And through the optimal comparative method, the feedback control chip changes the number of the L-C steady current units in order to achieve safe power supply. If the voltage signal is unusual, this feedback system will alarm and cut off the power automatically, which prevents electrostatic precipitator from damage.

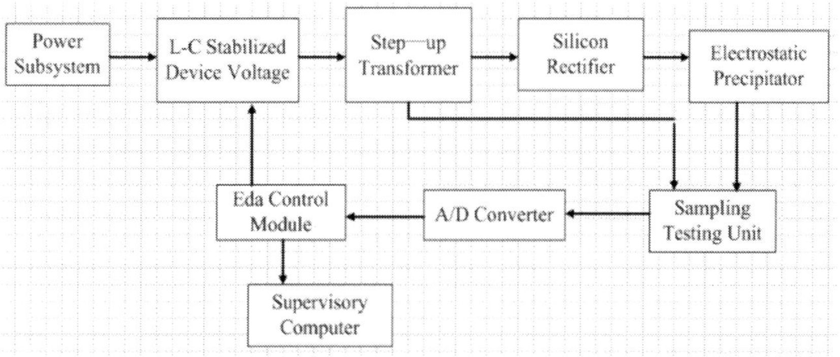

Figure 1. High pressure constant current ESP main circuit.

3. L-C steady voltage control unit

The L-C steady voltage control unit is composed of inductance and capacitance. Capacitance and inductance present a pure resistance because of their storage characteristics. It makes the reactive power of control system reduce to the minimum so as to achieve the stable current and voltage. When the electrostatic precipitator system develops into the short circuit, the L-C steady voltage control unit can absorb the power energy. And it protects the electrostatic precipitator system itself. At the same time, the feedback control unit of the electrostatic precipitator increases or decreases the number of the L-C steady voltage control unit for realizing the regulation of the voltage and the current.

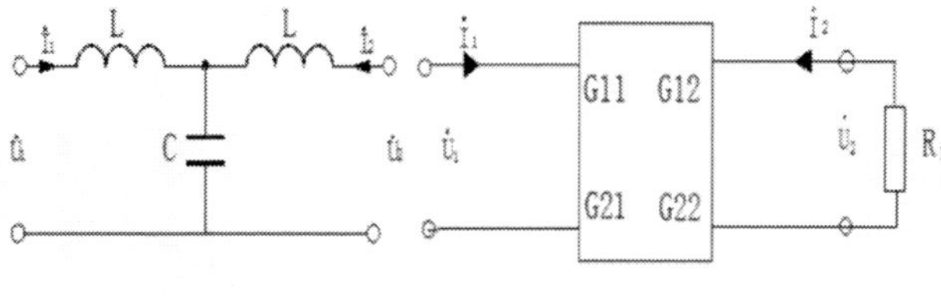

Figure 2. LC parallel resonant circuit module and two port model.

According to two port network theory and two port model (G11, G12, G21, G22 are plural coefficient, RL is the equivalent load, RL impedance) we can put forward:

$$\begin{cases} \dot{U}_1 = Z_{11}\dot{I}_1 + Z_{12}\dot{I}_2 \\ \dot{U}_2 = Z_{21}\dot{I}_1 + Z_{22}\dot{I}_2 \end{cases} \tag{1}$$

According to the two port models, we can get the formula of the input/ output voltage and current:

$$\dot{U}_1 = G_{11}\dot{U}_2 + G_{12}\dot{I}_2 \;,\; \dot{I}_1 = G_{21}\dot{U}_2 + G_{22}\dot{I}_2 \tag{2}$$

According to the superposition theorem, U1, U2 should be equal to the sum of their voltage:

$$\dot{I}_2 = \frac{\dot{U}_1}{G_{12}+G_{11}R_L} \;,\; \dot{U}_2 = \frac{\dot{U}_1 R_L}{G_{12}+G_{11}R_L} \tag{3}$$

If LC chooses the appropriate value and connecting method, G11 will be 0. The load current is determined by the voltage and G12, which is irrelevant to the impedance.

If the input voltage is Usinωt, we can get the load current.

$$\dot{I}_2 = -j\dot{U}_1 / 2\omega L + R_L(\omega^2 LC - 1) \tag{4}$$

When the circuit produces resonance, according to $\omega^2 LC - 1 = 0$,we can get:

$$\dot{I}_2 = -j\dot{U}_1 / 2\omega L \tag{5}$$

We can see that the load current is unrelated to the load size, and it is only related to the resonance frequency and the internal matching in the LC module. According to the resonance theory, the size of the resonant frequency only depends on the value of capacitor and reactor. Therefore, when the number of the LC module is determined, the circuit's output current will be stable. Similarly, U2 is related to RL.

4. Feedback control unit

The voltage and current signals from the sensor are received by the feedback control unit, in order to achieve the effect of tracking the feedback. The feedback control unit regulates the number of the L-C steady voltage control unit in an optimal way to make the electrostatic precipitator operate safely. It uses the Verilog HDL hardware description language to complete the feedback control system in the EDA software platform, and proceeds the logic compiled, logic division, and optimization, the logical layout wiring, logic simulation, the logical mapping and programming download, achieving the simulation. Because of the characteristics of simulation and flexibility, the feedback control system can keep efficient removal efficiency of dust, and adapt to different dust environments [5].

Figure 3 shows the feedback control unit circuit. The feedback control circuit mainly includes the frequency divider, the comparator and the counter. Its macroscopic principle is that in the unit time, the control unit realizes the feedback control by comparing the output work voltage and the rated voltage. If the input voltage is greater than the rated voltage, the comparator will count. When the voltage signal is greater than the rated voltage, the number of the counter will be more than 4, the feedback control unit alarms and cuts off the power, in order to protect the electrostatic precipitator. The microscopic description is: The power signal produces the clock signal through the frequency divider. By comparing the size of the rated voltage U1 and the work voltage U2, the feedback control unit realizes the real-time control of the voltage signal. When U2>=U1, the comparator outputs the high electrical level, and makes the counter start count. If the counter counts 4 times in 20s, the counter will produce the alarm signals. On the contrary, when U2<U1, the comparator outputs the low electrical level, the counter resets, and the electrostatic precipitator will work normally [6].

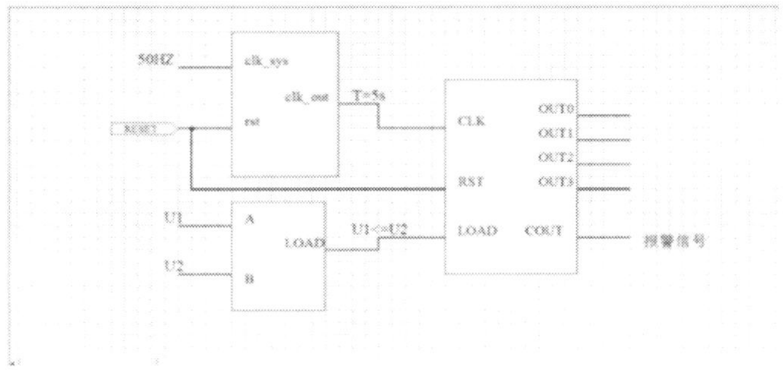

Figure 3. Voltage control circuit.

Because the detection unit sampling and the protection potential signal are all below 12v, this paper is based on the 4-bit comparator to compare the rated voltage and work voltage in order to realize the signal input of the counter.

The reset signal makes the "out" of the counter be "1100", and the "cout" of the counter be "0". When the clock signal of the counter comes, the counter adds 1, and by parity of reasoning. When the counter counts 4 times in 20s, the counter outputs the alarm signal ("cout"=1) .We use the signal of 50Hz as the divider's input signal. By 250 frequency divider's frequency division, the cycle of the clock signal will be 5s.

5. The simulation and download of the control unit

The design adopts Altera Company's software (Quartus) for program's compiles and simulation, and then we download the program to the SOPC chip for the experiment simulation. When the rated voltage a > the work voltage b, the "Load" of the comparator will be 0. On the contrary, when the rated voltage a <= the work voltage b, the "Load" of the comparator will be 1. When the clock signal comes, we compare the signal a and b. If the work voltage is greater than the rated voltage, the comparator outputs the high electrical level, and the counter starts to count. If the counter counts 4 times continuously, the electrostatic precipitator alarms and cuts off the power. The results of the simulation are shown in Figure 4.

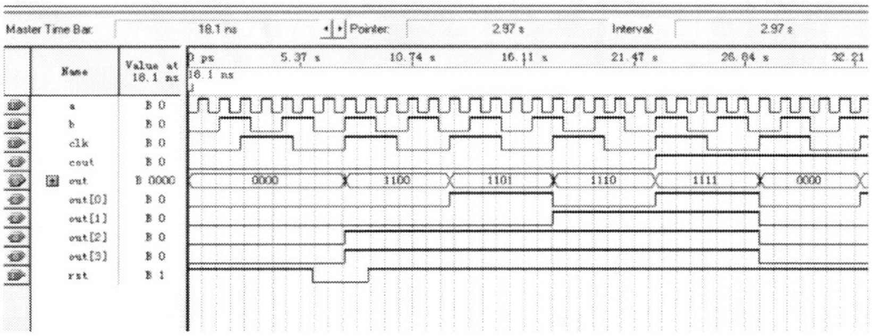

Figure 4. The result of the filter's power alarm and simulation.

The function of the divider is to get the required clock cycle. On the basis of our field experimentation, the frequency of 50Hz is the best clock signal. Choosing this clock signal, we can not only keep the efficiency of the dust removal, but also protect the electrostatic precipitator [7]. As the Verilog HDL language changes the traditional control method, and makes the experiment accomplish successfully. The divider's results of simulation are shown in Figure 5.

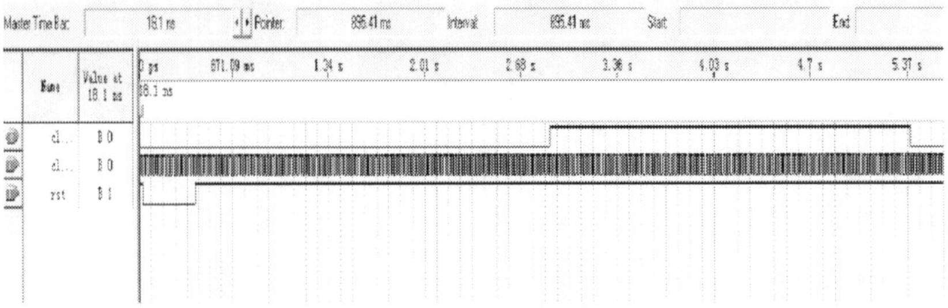

Figure 5. The simulation results of the divider.

Using the SOPC chips for the analog control, we can realize the real-time monitor. If the electrostatic precipitator has an accident, the feedback control unit can detect and alarm. The experiment results are shown in Figure 6. The electrostatic precipitator reaches the work voltage in 10s. If the environment has the first breakdown, the voltage appears arc, and the feedback control system controls the numbers of the L-C steady voltage control unit. In this way, the work voltage of the electrostatic precipitator declines to 45kv. When the fourth breakdown happens in the electrostatic precipitator, the feedback control unit cuts off the power supply and alarms [8].

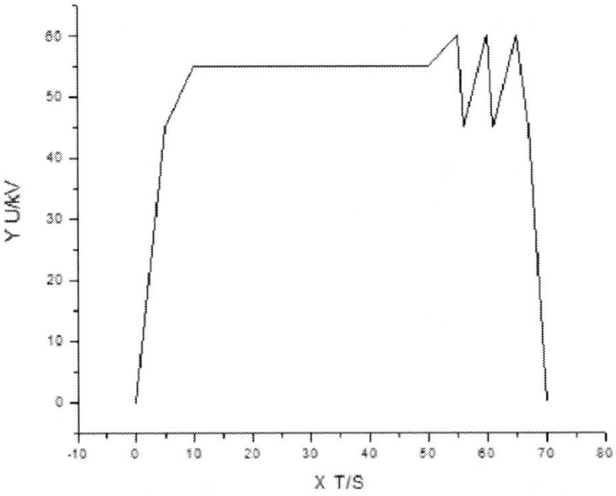

Figure 6. The experiment result.

6. Conclusion

The design of the electrostatic precipitator's intelligent control system can complete requirements of experiments. A complete electrostatic precipitator power feedback system integrated in a chip (SOPC) has more advantages than the classical electronic design method. This design uses the computer software to improve the design efficiency. Through the test, the feedback control unit can track the changes of the electric field and output the best voltage. This feedback control unit improves the alarm ability and dust removal efficiency of the electrostatic precipitator, and it is faster than the electric precipitator feedback unit which is controlled by PLC. It's especially suitable for the dust removal equipment in thermal power plant, so it has a broad application prospect in the market.

Acknowledgments

This paper is supported by the National Science Foundation of China under Grant No.51077035, and the Science Foundation of Hebei province No.A2010000182.

References

[1] Li Q 2010 CHIN. POWER **43** 60
[2] Cao X 2008 *High voltage technol. (in Chinese)* **34** 89
[3] Mao T 2009 *Control & Autom. (in Chinese)* **25** 74
[4] Xie W 2010 *Mod. to gold.* **38** 58
[5] Cao X 2008 *High-Voltage Technol. (in Chinese)* **34** 525
[6] Wu B 2009 *Inst. of Technol. Press (*Beijing*)* **20** 09
[7] Patterson D, Hennessy J 2007 *The hardware software interface* **23** 87
[8] Tian L 2008 *Electr. Eng. (*Jiangsu*)* **27** 70

Particle deposition in a full scale hybrid electrostatic-filtration collector

Zhengwei Long[1], Qiang Yao[2*]
[1] School of Environmental Science and Technology, 92 Weijing Road, Tianjin University, Tianjin 300072, China
[2] Department of Thermal Engineering, Tsinghua University, Beijing 100084, China

E-mail: longzw@tju.edu.cn

Abstract. This paper presents a numerical model for a full scale hybrid particulate collector (HPC), which combines the ESP technology and the filtration technology together. The corona discharge is solved by using a finite volume method. The turbulent flow equations are solved by using the Fluent package. The effect of the electric field on the fluid field named electro-hydrodynamic is considered. The particle motion is calculated by using the Lagrangian method. The particle charging rate is calculated by using a filed-diffusing combined model. The bag is modelled as a porous media but the pressure drop across the particle cake is neglected. Eight particle sizes are considered: 0.01, 0.1, 0.3, 0.5, 1, 2, 5 and 10 μm. The applied voltage is fixed at 60 kV. The results show that the collection efficiency of the electrostatic zone changes a little when the particle size is below 0.3 μm but increases clearly when the particle size exceeds 0.3 μm. Totally, the bag collection efficiency decreases from the first one to the last one in the same row.

1. Introduction

China has the biggest thermal power installed capacity, which reached up to 76.5 million kW by the end of 2011. It is a big challenge for the control of the particulate emissions from those power plants, especially for the $PM_{2.5}$. Both the conventional electrostatic precipitator and the bag filter have some disadvantages for the $PM_{2.5}$ control. A new type precipitator that combines the ESP and the baghouse together has been proposed for many years, named the advanced hybrid particulate collector (HPC) [1], as shown in figure 1. The HPC can provide ultrahigh collection efficiency and solve the problem of reentrainment. The flue gas first flows into the ESP zone to remove the most particles (about 90 % by mass), and then goes through the holes in the perforated plates into the fabric filtration zone to remove the left particles. Hence, the pressure drop across the collector increases far more slowly than the conventional bag filter.

Zhuang et al. have done lots of experimental research on the HPC including the voltage-current characteristics, collection efficiency of the electrostatic zone, different perforated-plate design, different types of bag and the pulse cleaning performances [1]. Long et al. have developed a finite volume method [2] to analyze the three dimensional electric field and the space charge density distributions in the HPC [3]. Though so many studies have been done, the particle deposition characters in the full scale HPC is still not clear. Experimental study can provide reliable data but

[*] To whom any correspondence should be addressed.

usually lack of some details. Numerical method has been a powerful tool for the study of the ESP [4-7] and the filtration [8-9]. Numerical results could show the particle trajectories in the ESP and the structures of the particle layer in the filtration. It is hard to obtain that information from the experiments.

In this paper, a numerical model of the HPC is presented to study the particle deposition characters inside the full scale HPC, as shown in figure 1. The model is two dimensional and its geometrical parameters are close to an industrial one. The model HPC includes twenty-five discharge wires, eight perforated-plates with many staggered holes and thirty-two bags. Eight particle sizes are considered: 0.01, 0.1, 0.3, 0.5, 1, 2, 5 and 10 μm. For simplicity, the pressure drop across the particle cake is not considered in this paper.

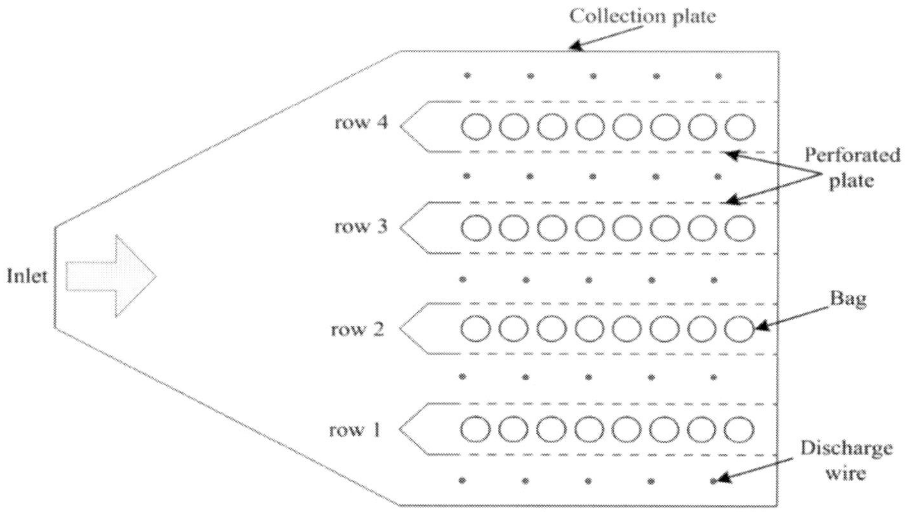

Figure 1. Demonstration of the hybrid particulate collector.

2. Model description

In the HPC, the physical processes include the fluid flow, the corona discharge, the particle charging, the particle dynamics and the filtration. The coupled effects between the fields should be considered.

The governing equations describing the corona discharge include the Poisson's equation and the current continuity equation [10]:

$$\nabla^2 V' = -\left(\frac{\rho_w}{\varepsilon_0 V_w}\right)\rho_s' \tag{1}$$

$$\nabla \bullet \boldsymbol{J} = 0 \tag{2}$$

$$\boldsymbol{J} = b_i \boldsymbol{E} \rho_s' \tag{3}$$

$$\boldsymbol{E} = -\nabla V' \tag{4}$$

where V is the electric potential (V), V_w the electric potential at the corona wire, $V' = V / V_w$, ρ_w is the space charge density at the corona wire (C/m^3), ρ_s is the space charge density and $\rho_s' = \rho_s / \rho_w$, ε_0 is the air permittivity ($8.85\times10^{-12}\ C^2/N\cdot m^2$), $\boldsymbol{J}$ is the current density (A/m^2), b_i is the ion mobility ($m^2/V\cdot s$), $\boldsymbol{E}$ is the electric field (V/m).

This paper adopts the steady flow model. The steady, incompressible Navier-Stokes equations can be written as:

$$\frac{\partial \boldsymbol{U}_i}{\partial \boldsymbol{x}_i}=0 \tag{5}$$

$$\rho_{\mathrm{f}}\frac{\partial \boldsymbol{U}_i}{\partial t}+\rho_{\mathrm{f}}\frac{\partial \boldsymbol{U}_i\boldsymbol{U}_j}{\partial \boldsymbol{x}_j}-(\mu+\mu_{\mathrm{t}})\frac{\partial}{\partial \boldsymbol{x}_j}\left(\frac{\partial \boldsymbol{U}_i}{\partial \boldsymbol{x}_j}+\frac{\partial \boldsymbol{U}_j}{\partial \boldsymbol{x}_i}\right)=-\frac{\partial P}{\partial \boldsymbol{x}_i}+\boldsymbol{F}_{ci}+\rho_{\mathrm{f}}g \tag{6}$$

$$\boldsymbol{F}_{ci}=\boldsymbol{E}_i\rho_s \tag{7}$$

where $\boldsymbol{U}_i$ (i=1, 2, 3) is the fluid velocity (m/ s), ρ_{f} is the fluid density (kg/ m^3), t is the time (s), μ is the laminar viscosity (kg/ m·s), μ_{t} is the turbulent viscosity (kg/ m·s), P is the fluid pressure (Pa), $\boldsymbol{F}_{ci}$ is the electric body force (N/ m^3), g is the acceleration of gravity (m/ s^2).

For the turbulence, the RNG $\kappa-\varepsilon$ model is employed.

The particle deposition adopts the Lagrangian method. The electric field force and the gravity force are the primary forces. According to the Newton's second law, the equation that describes the particle motion is written as:

$$\frac{d\boldsymbol{U}_{\mathrm{p}}}{dt}=C_{\mathrm{D}}\frac{3\rho_f\left|\boldsymbol{U}-\boldsymbol{U}_{\mathrm{p}}\right|(\boldsymbol{U}-\boldsymbol{U}_{\mathrm{p}})}{4\rho_{\mathrm{p}}d_{\mathrm{p}}}+g+\frac{3\boldsymbol{E}q_{\mathrm{p}}}{4\rho_{\mathrm{p}}d_{\mathrm{p}}^3} \tag{8}$$

where $\boldsymbol{U}_{\mathrm{p}}$ is the local particle velocity, $\boldsymbol{U}$ is the local fluid velocity, ρ_{p} is the particle density, d_{p} is the particle diameter, q_{p} is the particle charge (C), C_{D} is the drag force coefficient, g is the gravity gradient.

To solve the Eq. (8), the particle charge q_{p} must be known. In this paper, the FMD model developed by Lawless [11] is employed to calculate the particle charge number along the trajectories in the HPC. The charging rate is expressed in the following dimensionless form.

$$\frac{\mathrm{d}v}{\mathrm{d}t_q}=\begin{cases} f(w)\dfrac{v-3w}{\exp(v-3w)-1}, & v>w \\ \dfrac{3w}{4}\left(1-\dfrac{v}{3w}\right)^2+f(w), & -3w\le v\le 3w \\ -v+f(w)\dfrac{-v-3w}{\exp(-v-3w)-1}, & v<-3w \end{cases} \tag{9}$$

where $f(w)$ is the fraction of the surface covered by the diffusive band:

$$f(w)=\begin{cases} \dfrac{1}{(w+0.475)^{0.575}}, & w\ge 0.525 \\ 1, & w<0.525 \end{cases} \tag{10}$$

where v ($=\dfrac{q_{\mathrm{p}}e}{2\pi\varepsilon_0 d_{\mathrm{p}}kT}$) is the dimensionless particle charge, w ($=\dfrac{K_{\mathrm{p}}}{K_{\mathrm{p}}+2}\dfrac{Ed_{\mathrm{p}}e}{2kT}$) is the dimensionless electric field, and $t_q(=\dfrac{\rho_s b_i t}{\varepsilon_0})$ is the dimensionless particle charging time, k is Boltzmann's constant ($1.38\times10^{-23}J/K$), T is the temperature (K), e is the charge on an electron ($1.6\times10^{-19}C$), K_{p} is the particle dielectric constant, E is the local electric field (V/m), ρ_s is the space charge density (C/m^3).

The discharge equations (1)-(4) are solved by a finite volume method [2]. The fluid equations (5)-(6) and the particle motion equation (8) are solved by using the CFD package Fluent. The particle charge equations (9)-(10) are solved by using the user defined program. Figure 2 is the comparison between the numerical predictions and the experimental data in a laboratory electrostatic precipitator [12]. The model predictions agree well with the measurements of the hasten speed along a line near the collection plate.

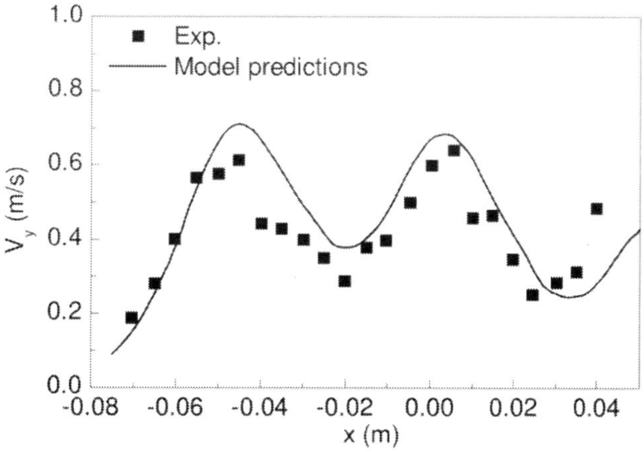

Figure 2. Comparison of the model predictions with the experimental data [12].

3. Results and discussion

The inlet boundary condition is the velocity-inlet and the outlet one is the outflow. The boundary condition of the walls is the wall function. The mean inlet velocity is 1.5 m/ s. The particle density is 2610 kg/ m^3. The particle size is monodisperse. The bag is modeled as a porous media, whose permeability is 3.6e-11 and thickness is 1.5 mm. The applied voltage and the corona current are 60 kV and 68.4 μA/ m, respectively.

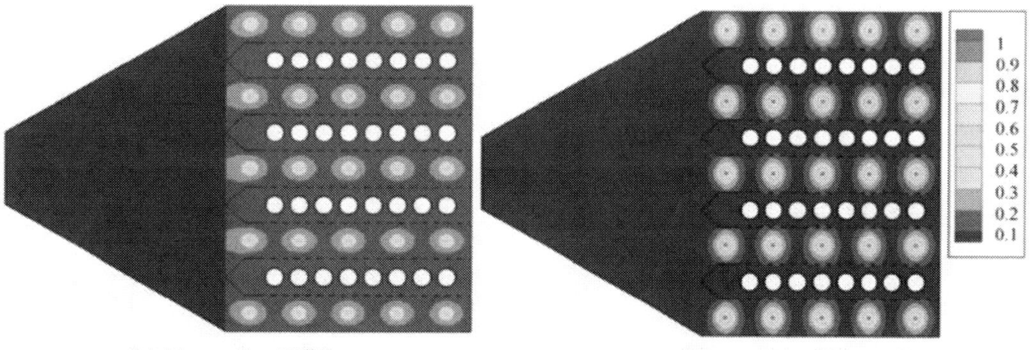

(a) Normalized Voltage (b) Normalized Space Charge

Figure 3. The voltage and the charge density distribution in the model HPC.

Figure 3 shows the voltage and the space charge density distribution of the model HPC. It clearly shows that the wire around area has the biggest voltage and space charge density. Besides, many ions go through the holes of the perforated plates. This indicates that the electric field intensity in the bags in not zero. Both the sides of the perforated plate can collect the particles. Figure 4 presents the streamlines in the model HPC. The electrohydrodynamic exists in the corona discharge field. But the electrohydrodyanmic in the model HPC is not very obvious. This may be due to the low electric field intensity. There are some vortexes near the end of the model HPC due to the end wall [13].

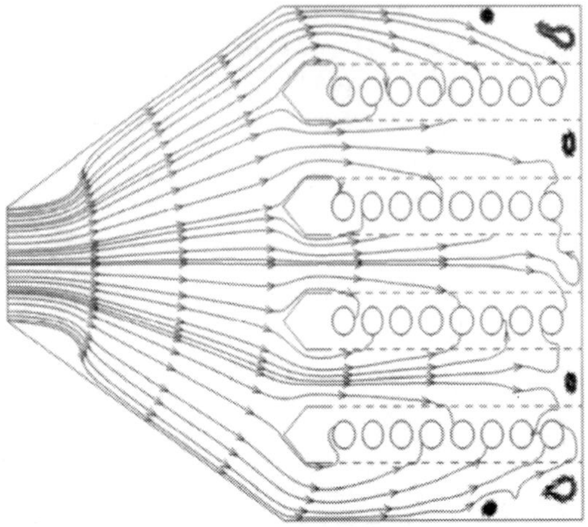

Figure 4. The streamlines in the model HPC.

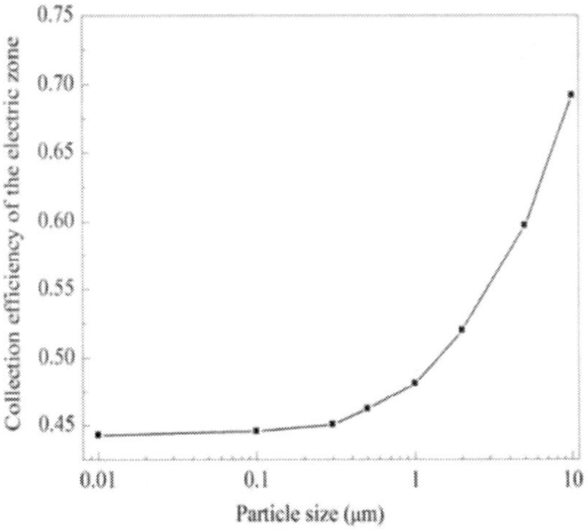

Figure 5. The relation of the electric zone collection efficiency with the particle size.

Figure 5 presents the collection efficiency of the electric zone of the model HPC. The electrostatic field zone includes the two collection plates and the eight perforated plates. The results show that the electrostatic zone collection efficiency changes a little when the particle size is below the 0.3 μm but increases clearly when the particle size exceeds 0.3 μm. The particle dynamics in the model HPC is decided by the drag force and the electric force. For the particles lower than 0.3 μm, the differences between two forces are small. Thus, the collection efficiency is similar. But the drag force and the electric force differ a lot for the larger particles. Thus, the collection efficiency increases obviously. Therefore, it is still a big challenge to increase the collection efficiency of the electric zone of the model HPC for the fine particles. Almost half fine particles are deposited onto the bag surfaces. The porosity of the cake on the bag surface will be smaller.

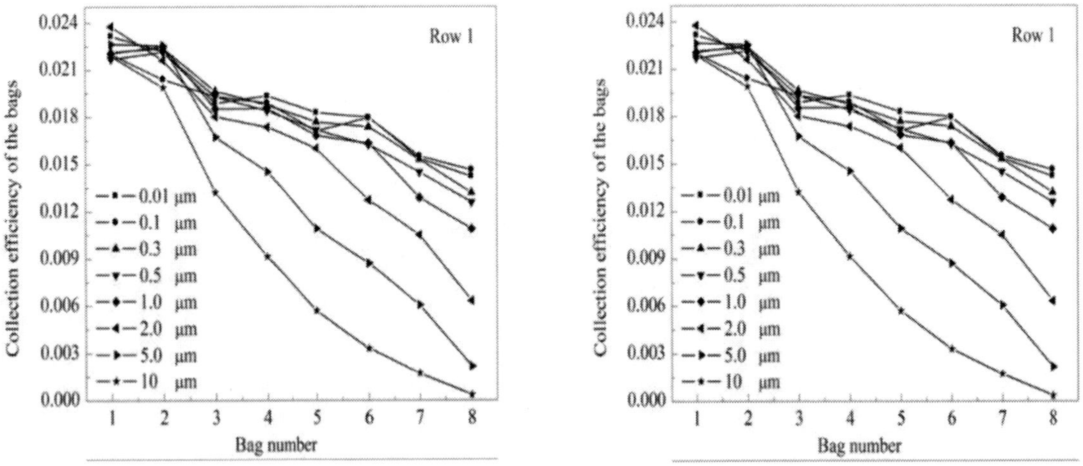

Figure 6. The comparison of the bag collection efficiency for different particle sizes.

Figure 6 shows the collection efficiency of each bag in rows 1 and 2. The bag is numbered 1-8 from left to right. The rows 3 and 4 are symmetrical with the rows 1 and 2. In row 1, the bag collection efficiency decreases from the bag 2 to the bag 8 for all the particle sizes. The collection efficiencies of the bags 1 and 2 are nearly the same for particle size 0.3, 0.5, 1.0 and 5.0 μm. Besides, the collection efficiencies of the bag 1 for the eight particle size are at the range of 0.021 to 0.024, which is very narrow. The other seven bags also have the similar characters but for different particle size range. The eight bags have similar collection efficiencies for the particle size 0.01, 0.1 and 0.3 μm but significant difference for the other particle sizes. In row 2, the collection character is different with the row 1. The collection efficiencies decrease from the bag 2 to the bag 7 for the eight particle sizes. But the collection efficiencies increase slightly from the bag 1 to bag 2 for the particle size 0.1 and 10 μm. And the collection efficiencies increase obviously from the bag 7 to bag 8 for the particle size 0.01, 0.1 and 0.3 μm. Each bag has also similar collection efficiency for different particle size range. For example, the collection efficiencies of the bag 1 have a range of 0.02 to 0.021 for the particle size 0.01, 0.1, 0.3, 0.5, 1 and 2 μm.

4. Conclusion

This paper presents a numerical model for a full scale hybrid particulate collector. The corona discharge field is solved by using a finite volume method. The turbulent fluid flow equations are solved by using the Fluent package. The effect of the electric field on the fluid field named electro-hydrodynamic is considered. The particle motion is modelled using the Lagrangian method. The particle charging rate is calculated by using a filed-diffusing combined model. The bag is modelled as a porous media but the pressure drop across the particle cake is neglected. The results show that the collection efficiency of the electrostatic zone changes a little when the particle size is below the 0.3 μm but increases clearly when the particle size exceeds 0.3 μm. Totally, the bag collection efficiency decreases from the left one to the right one in the same row except the first bag and the last bag. But each bag has similar collection efficiency for a special particle size range.

Acknowledgments

This project was supported by the National Natural Science Foundation of China (Grant NO. 51106105) for funding this project.

References

[1] Zhuang Y, Stanley J M 2001 *Advance Hybrid Particulate Collector final topic report for phase III* p 20
[2] Long Z W, Yao Q, Song Q, Li S Q 2009 *J. Electrostat.* **67** 597
[3] Long Z W, Yao Q, Song Q, Li S Q 2009 *J. Electrostat.* **67** 835
[4] Oglesby S, Nichols G B 1978 *Electrostatic Precipitation* (M. Dekker, New York)
[5] Choi B S, Fletcher C A J 1997 *J. Electrostat.* **40** 413
[6] Gallimberti I 1998 *J. Electrostat.* **43** 219
[7] Nikas K S P, Varonos A A, et al. 2005 *J. Electrostat.* **63** 423
[8] Schmidt E 1996 *Powder technol.* **86** 113
[9] Thomas D, Contal P, et al. 2001 *Chem. Eng. Sci* **56** 3549
[10] McDonald J R, Smith W B, Spencer III H W, Sparks L E 1977 *J. Appl. Phys* **48** 2231
[11] Lawless P A 1996 *J. Aerosol Sci.* **27** 196
[12] Parasram N 2001 *Particle Motion in Electrostatic Precipitators* (Ph.D. Thesis, Thermofluids section, Mechanical Engineering Department, Imperial College of Science, Technology and Medicine)
[13] Kallio G A, Stock D E 1992 *J. Fluid. Mech.* **240** 133

Computational modeling and simulation of electro-hydrodynamic (EHD) ion-drag micropump with planar emitter and micropillar collector electrodes

Shakeel Ahmed Kamboh[1*], Jane Labadin[1*] and Andrew Ragai Henry Rigit[2*]
1,2 Faculty of Computer Science & Information Technology,
2 Faculty of Engineering Universiti Malaysia Sarawak, 94300 Kota Samarahan, Sarawak, Malaysia

E-mail: shakeel.maths@yahoo.com, ljane@fit.unimas.my, arigit@feng.unimas.my

Abstract: Computational models can be used to simulate a prototype of electrohydrodynamic (EHD) ion-drag micropump with planar emitter and micropillar collector electrodes. In this study, a simple and inexpensive design of an ion-drag micropump was modeled and numerically simulated. A three-dimensional segment of the microchannel was simulated by using periodic boundary conditions at the inlet and outlet. The pressure and velocity distribution at the outlet and in the entire domain of the micropump was obtained numerically. The effect of the gap between the emitter and the collector electrode, width and the height of micropillar and flow channel height was analyzed for optimum pressure and output flow rate. The enhanced performance of micropump was compared with existing designs. It was found that the performance of micropump could be improved by decreasing the height of micropillar and the gap between both electrodes. The numerical results also show that a maximum pressure head of about 2350 Pa and maximum mass flow rate 0.4 g min^{-1} at an applied voltage 1000 V is achievable with the proposed design of micropump. These values of pressure and flow rate can meet the cryogenic cooling requirements for some specific electronic devices.

1. Introduction

The prominent features of micropumps inspired their use in a wide variety of applications ranging from biotechnology and chemical analysis to space exploration and cooling of micro electro mechanical systems (MEMS) such as sensors and detectors [1-2]. EHD ion-drag micropump is a non-mechanical type of micropump that uses the electric forces (mainly the Coulomb force) in the liquid bulk. The fluid flow is produced due to the friction between moving ions and the working fluids drag [3-4]. Ion-drag micropumps are preferred for pumping single phase incompressible flows because of their attractive features. These pumps are extensively utilized for the applications of fuel injection loops and cryogenic cooling of microdevices [5]. In the last decade, different designs of ion-drag micropumps have been proposed and developed to enhance the output flow rate and pressure. Unfortunately, with the shrinking size of many electronic products still the high pressure output is required in certain applications. On the other hand, the development of new compatible designs of micropumps becomes more challenging because of microfabrication complexity and instruments cost. In such circumstances, the numerical simulation of micropumps plays important role not only to

[*] To whom any correspondence should be addressed.

understand the different working principles but also enables to model the designs with better performance.

Numerical models for ion-drag pumping have been developed extensively to simulate and predict the fluid behavior. A number of numerical analyses and simulation of the EHD ion-drag mechanism inside liquids have been reported in literatures. Many [3, 6-11] were used to investigate the performance of different designs of ion-drag micropumps. To understand the effect of the electrode geometry on the electric field and fluid pumping effect, Chaudhary et al. [12] performed numerical simulations using FemLab 2D software. Darabi and Ekula [13] performed a three dimensional electric field analysis using ANSYS/Multiphysics to determine electric field distributions. Benetis et al. [14] did numerical modeling using fluent software to create a hydrodynamic model for the micropump. They found relation between internal pressure drop of micropump and mass flow rate, and investigated the relation between pump pressure head and delivered mass flow rate. Darabi and Rhodes [15] performed a two-dimensional numerical model using commercial CFD software FIDAP to model EHD ion-drag pumping between an array of emitter and collector electrodes. The focus of their simulations was to study the effects of electrode gap, stage gap, channel height, and applied voltage. Lee et al. [16] presented parametric numerical studies on the 3D micropillar electrodes that were performed using commercial Finite Element package COMSOL Multiphysics to model the EHD pumping between one emitter and collector stage. A numerical study using COMSOL Multiphysics was conducted by Kazemi et al. [17-18] and demonstrated for the first time, that an asymmetry in the electrode geometry (both 2D and 3D) will result in significantly higher pressure generation with lower power consumption than conventional symmetric electrode designs. A numerical model was developed by Hasnain et al. [19] for an EHD ion-drag micropump that takes into account the effect of the electric field on the charge distribution at the electrodes. The simulations of model were performed on COMSOL and validated using the experimental data of [18]. Rigit et al. [20] used COMSOL Multiphysics and obtained initial 2D simulation for the planar ion-drag micropump.

Most of the numerical models used to simulate and investigate the ion-drag pumping are two-dimensional which restrict to the designs of micropump with 3D geometric features. A 3D modeling and simulation of ion-drag micropump with three different designs of collector electrodes was presented by the same authors of this paper [21]. The simulation was carried out on COMSOL Multiphysics 4.2 and they concluded that a design with planar emitter and 3D micropillar collector may produce better output pressure and flow rate. This paper further analyses and optimizes the works of [21, 17] for enhanced performance and also validates the simulation results by comparing with the recent experimental data of [17].

The paper is arranged by first presenting the motivation of this research and its overview. In the next section, the three dimensional computational model with its boundary conditions is clearly discussed and the values for simulation are also presented. Finally, the numerical results are deliberated where the overall aim is to investigate the effect of various geometric parameters on the performance of a prototype design of ion-drag micropump. Particularly, the dimensions of micropillar collector electrode and the flow channel height were investigated for optimum output pressure and flow rate.

2. Computational Modeling and Simulation

The EHD phenomenon is a complicated process, since various electrostatics and hydrodynamics factors are involved. These factors include geometrical parameters, working fluid and actuator electrodes material properties, electro-chemical reaction factors, electric and fluid dynamics variables and many others. Therefore, the modeling of the EHD phenomenon is quite difficult unless several reasonable assumptions and simplifications are made. The mathematical models that govern the EHD phenomenon are derived from the Maxwell's general quasi-static electromagnetism equations and fluid flow conservative equations [7,10] and then simplified by using certain realistic assumptions [21]. The complete set of governing equations for incompressible steady state flow in 3D geometry is listed as follows:

$$\vec{F}_e = -q_e(x,y,z)\left(\frac{\partial V}{\partial x}+\frac{\partial V}{\partial y}+\frac{\partial V}{\partial z}\right) \quad (1)$$

$$\varepsilon\left(\frac{\partial^2 V}{\partial x^2}+\frac{\partial^2 V}{\partial y^2}+\frac{\partial^2 V}{\partial z^2}\right) = -q_e(x,y,z) \quad (2)$$

$$\vec{E} = -\left(\frac{\partial V}{\partial x}+\frac{\partial V}{\partial y}+\frac{\partial V}{\partial z}\right) \quad (3)$$

$$\left(\frac{\partial J}{\partial x}+\frac{\partial J}{\partial y}+\frac{\partial J}{\partial z}\right) = 0 \quad (4)$$

$$J = \mu_e\, q_e(x,y,z)\left(\frac{\partial V}{\partial x}+\frac{\partial V}{\partial y}+\frac{\partial V}{\partial z}\right) \quad (5)$$

$$\left(\frac{\partial u}{\partial x}+\frac{\partial v}{\partial y}+\frac{\partial w}{\partial z}\right) = 0 \quad (6)$$

$$\rho\left(u\frac{\partial u}{\partial x}+v\frac{\partial u}{\partial y}+w\frac{\partial u}{\partial z}\right) = -\frac{\partial p}{\partial x}+\eta\left(\frac{\partial^2 u}{\partial^2 x}+\frac{\partial^2 u}{\partial^2 y}+\frac{\partial^2 u}{\partial^2 z}\right)-q_e\left(\frac{\partial V}{\partial x}\right) \quad (7)$$

$$\rho\left(u\frac{\partial v}{\partial x}+v\frac{\partial v}{\partial y}+w\frac{\partial v}{\partial z}\right) = -\frac{\partial p}{\partial y}+\eta\left(\frac{\partial^2 v}{\partial^2 x}+\frac{\partial^2 v}{\partial^2 y}+\frac{\partial^2 v}{\partial^2 z}\right)-q_e\left(\frac{\partial V}{\partial y}\right) \quad (8)$$

$$\rho\left(u\frac{\partial w}{\partial x}+v\frac{\partial w}{\partial y}+w\frac{\partial w}{\partial z}\right) = -\frac{\partial p}{\partial z}+\eta\left(\frac{\partial^2 w}{\partial^2 x}+\frac{\partial^2 w}{\partial^2 y}+\frac{\partial^2 w}{\partial^2 z}\right)-q_e\left(\frac{\partial V}{\partial z}\right) \quad (9)$$

Eq. (1) represents the electric body force $\vec{F}_e$ (Coulomb force) which is the product of space charge density q_e and the gradient of electric potential V. Eq. (2) states the Gauss's law that combines the space charge density with electric potential field. The gradient of electric potential results in electric field $\vec{E}$ as given in Eq. (3). The conservation of charges is described by Eq. (4) where J is the current density dominated by the migration current as defined in Eq (5). Eq. (6) is the well-known flow continuity equation with mass conservation and the Eq. (7)-Eq. (9) are the Navier-Stokes momentum equations which describe the fluid flow. Equations (1-9) have to be solved iteratively for potential field V, electric field $\vec{E}$, current density J, charge density q_e, fluid velocity components (u,v,w) and pressure field p.

The initial values for the entire domain of micropump are set to zero charge, zero current flux, no flow and atmospheric pressure. The boundary conditions are of Dirichlet type and are periodic because only one segment of the whole micropump is simulated. Figure 1 shows the dimensions of a representative segment with emitter and collector shapes. The emitter surface is set to a constant applied voltage and collector is set to ground i.e., zero voltage. The boundary conditions for charge density are not known because the charge injection process is not well understood. Therefore, approximate value for charge boundary condition at emitter surface is calculated from certain experimental value of current [15] that will also provide the boundary conditions for current density between emitter and collector [21-22]. The boundary conditions for fluid flow at both inlet and outlet are assumed equal to zero velocity and pressure. The walls of flow channel are set as no-slip boundary.

After setting up the problem the three dimensional numerical model was performed using commercial finite element based simulation program COMSOL Multiphysics 4.2. The working fluid is selected as HFE-7100 (Hydrofluoroether, $C_4H_9OCH_3$) into the flow channel with density ρ, 1480 kg m^{-3}, viscosity η, 6.51×10^{-4} Pa.s, ion mobility μ_e, 1.31×10^{-8} m^2 V^{-1}.s and dielectric constant k, 7.39; the ambient conditions are set to room temperature 25°C and the atmospheric pressure 101.325 kPa.

Electrode material is set to gold (Au) with predefined default properties. The mesh structure of entire segment of micropump consists of 33194 free tetrahedral mesh elements. The simulation results presented here were performed with extremely coarser grid. First the Electrostatics module was used to compute electric body force and then laminar fluid flow module was used to couple the effect of Coulomb force. The electric current flowing between electrodes is obtained by the relation $I=\int_s J\,dA$ and the power consumption is computed by $P=I.V$ [23]. Pressure generated within single segment was obtained from the difference between the average pressure at the inlet and outlet [15]. The proposed design of micropump consists of 100 pairs of electrodes, consequently the total generated pressure was computed by multiplying the number of segments by the pressure generated from single segment; the flow rate was calculated by $\dot{m}=\rho\bar{u}\,A$. In order to optimize the design, extensive simulation trials were performed by varying the different controlling parameters. The effect of the gap between the emitter and the collector electrode, width and the height of micropillar and flow channel height was analyzed for optimum pressure and output mass flow rate. The simulation results were compared with experimental data of [17].

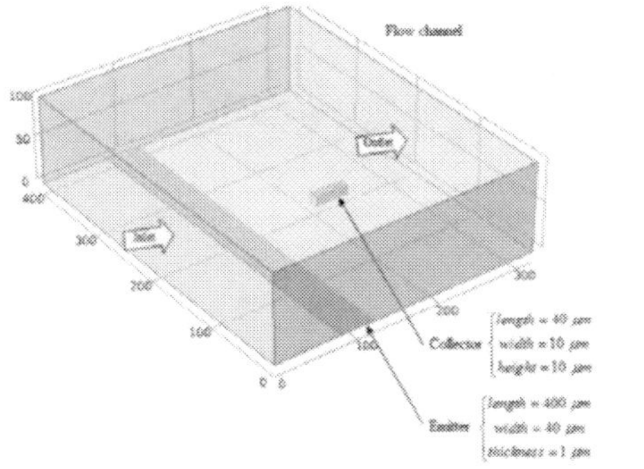

Figure 1. Schematic of a segment of EHD ion drag micropump showing electrode geometry.

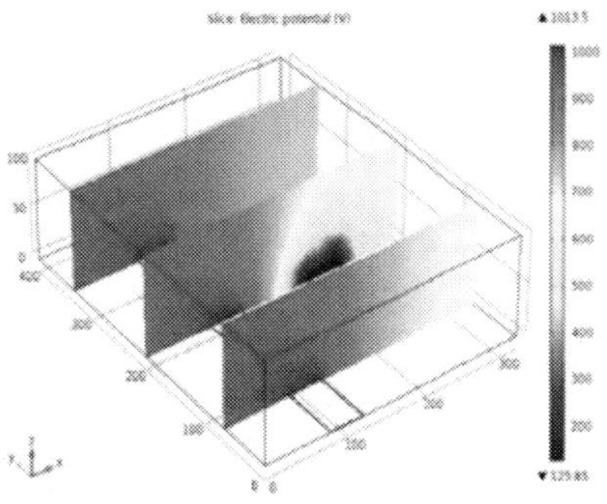

Figure 2. Local electric potential distribution along zx planes.

3. Results and Discussion

The electric potential profiles are obtained from the numerical solution of Poisson's equation (2) and are useful to understand the behavior of electric field, electric current flux and charge density within the entire domain. It is observed that there is a continuous transition of electric potential from emitter to collector (figure 2). The values of electric potential remain higher in zx planes and near the surface of emitter but lower near the collector. It demonstrates that the charge concentration will be higher at the emitter surface and lower at the collector surface. Electric field is the most important factor in the ion-drag pumping because it is responsible for the drag force (Coulomb force). Therefore, in order to understand the electric field behavior, its simulation profiles are obtained. Figure 3 exhibits the electric field direction and contour lines. It can be seen that the direction of electric field lines is from emitter to collector and the field lines converging to the collector surface. It is also shown that the high magnitude of electric field occurs at the edges of emitter and collector near the bottom of flow channel. The behavior of fluid velocity can be understood by the simulated local velocity field profiles (figures 4-6).

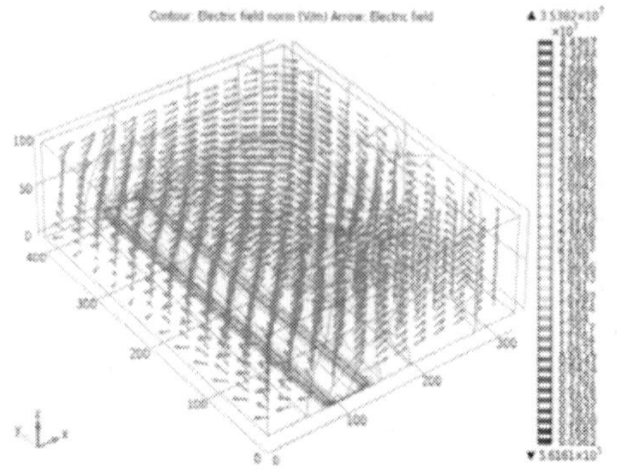

Figure 3. Electric field profiles within the entire domain of micropump.

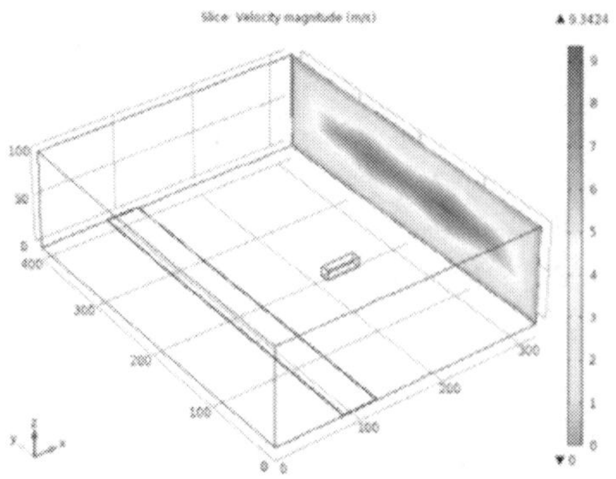

Figure 4. Local velocity field distribution at the outlet.

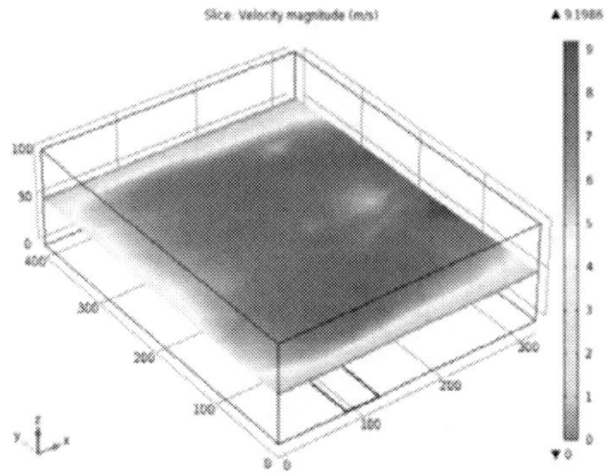

Figure 5. Local velocity field distribution along xy plane.

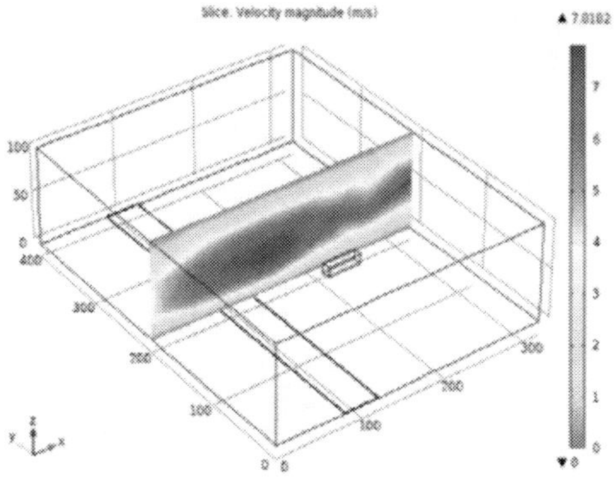

Figure 6. Local velocity field distribution along zx plane.

Both the fluid velocity and pressure depend upon the volume forces that create the fluid flow in the direction of electric field. It is observed that the magnitude of fluid velocity varies in different regions of the entire domain. The distribution of fluid velocity changes from inlet towards outlet of flow channel and the high magnitude of velocity is noted at the central regions of the entire domain. The local pressure distribution patterns are shown in figures 7-9. Due to the high electric field near the surface of micropillar collector, the local pressure distribution is higher around the collector and at the centre of outlet. Both the current and power consumption in the micropump increase exponentially as the applied voltage increases linearly as shown in figure 10. Power consumption is of the order of mW (milliwatt) as the current is of the order μA (microampere). The average electric field between emitter and collector is approximately linear with respect to the increase in applied voltage (figure 11). The pressure and mass flow rate can be controlled by applied voltage; the generated pressure and mass flow rate rise with the increase in voltage (figure 12).

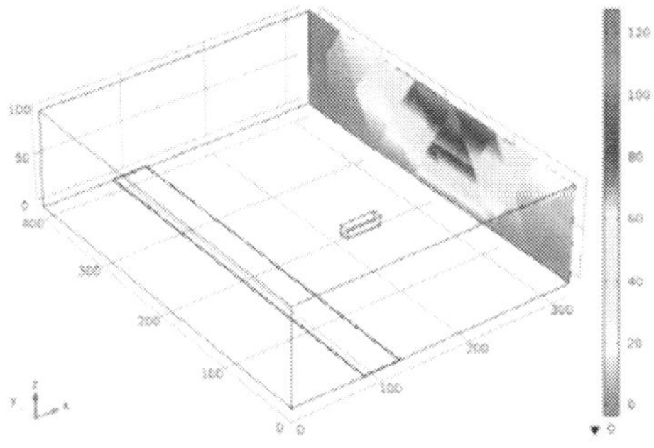

Figure 7. Local pressure field distribution at the outlet.

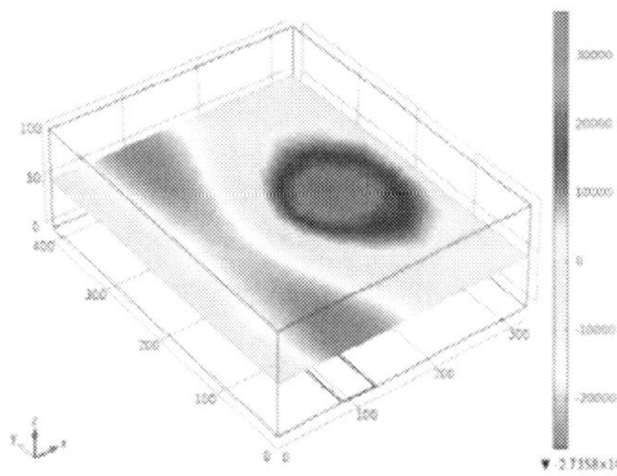

Figure 8. Local pressure field distribution along xy plane.

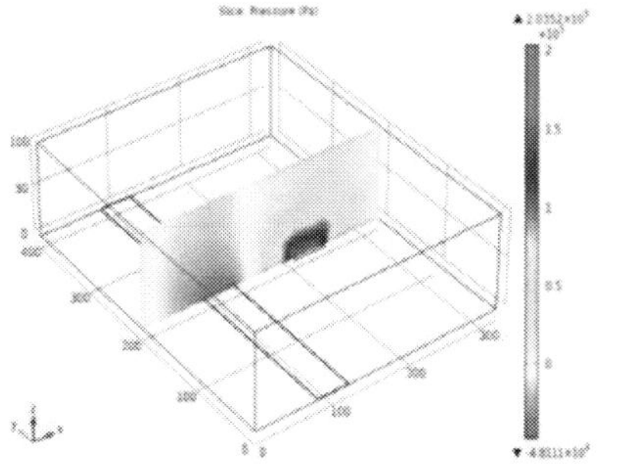

Figure 9. Local pressure field distribution along zx plane.

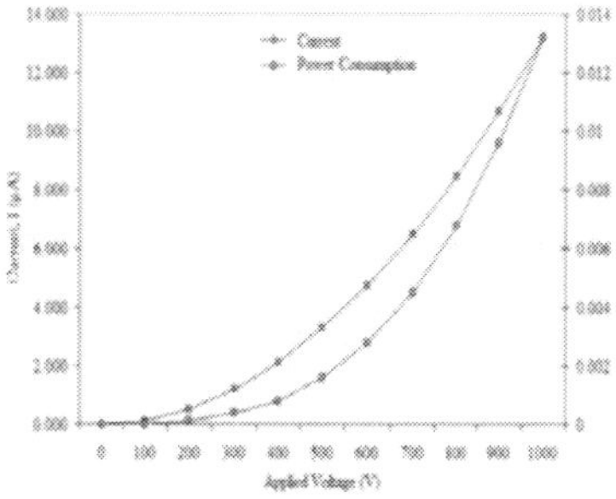

Figure 10. Variation of current and power consumption at applied voltage.

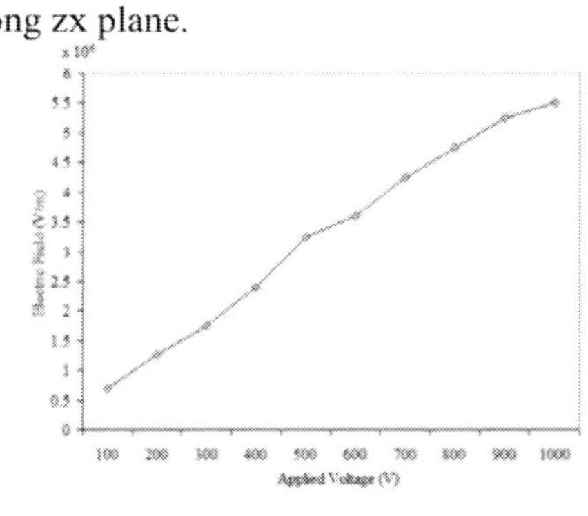

Figure 11. Average electric field between emitter and collector at applied voltage.

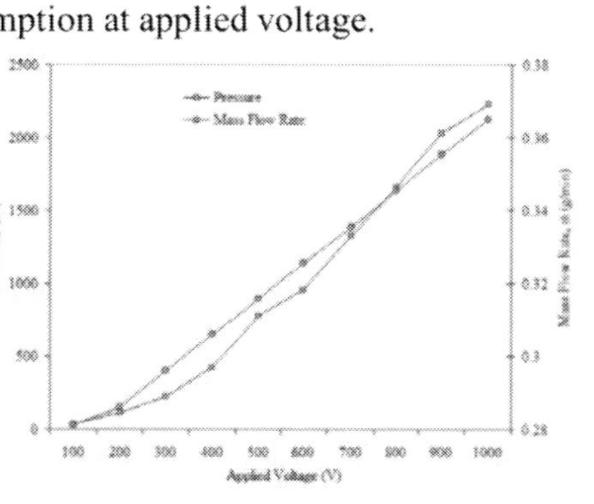

Figure 12. Variation of pressure and mass flow rate at applied voltage.

In order to optimize the design, a parametric study was conducted by analyzing the effect of different controlling parameters on the performance of micropump. Both the mass flow rate and pressure increase with the decreasing of the electrode gap up to a limited extent. In some earlier experimental studies [5], it was noted that the liquid breakdown may occur at very small gap of 5-10 μm that may affect the life time of the micropump. In this study the electrodes gap is varied from 20 μm to 120 μm (figure 13) and as a result the pressure and mass flow rate vary from 2650 to 1840 Pa and 0.45 to 0.32 g min^{-1}, respectively. The width and height of micropillar are also reduced to reduce the possible resistance to flow as shown in figures 14-15. When the width of the micropillar is decreased from 40 μm to 10μm the pressure and flow rate have ranges of about 1990-2520 Pa and 0.1-0.4 g min^{-1}, i.e., more thinner micropillar will provide better pressure and mass flow rate will be obtained. Similarly, with the decrease in the height of micropillar from 100 μm to 10 μm the pressure and flow rate increase 1490 to 2250 Pa and 0.158 to 0.35 g min^{-1} respectively. The effect of microchannel variation is also analyzed as shown in figure 16. There is an inverse relation between pressure variation and mass flow rate variation as the flow channel height is reduced from 100 μm to 50 μm. The pressure increases however the flow rate decreases. This is due to the fact that the mass flow rate depends upon the area of the outlet. Hence, by decreasing the flow channel height the area of outlet also decreases. On the other hand, the magnitude of the driving force increases with the decrease in the flow channel height that results in high pressure output. A number of simulation trials were done and the simultaneous effects of all four parameters viz. electrode gap, micropillar width, micropillar height and microchannel height were analyzed. It was found that the maximum pressure head ~2350 Pa and flow rate ~0.4 g min^{-1} is achievable at the electrode gap of 40 μm, micropillar width and gap at 10μm and flow channel height of 100 μm.

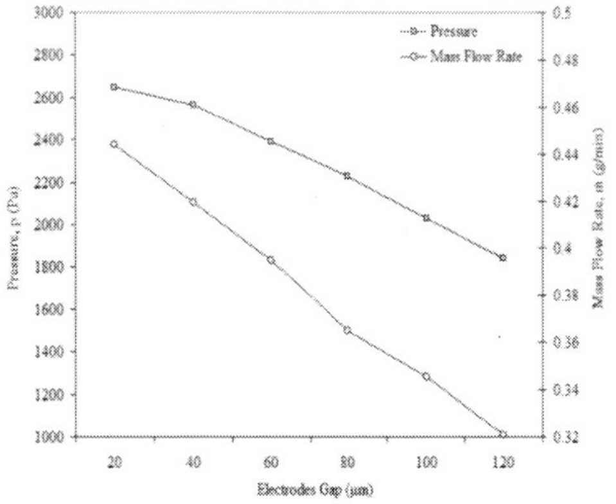

Figure 13. Variation of pressure and mass flow rate at gap between electrodes.

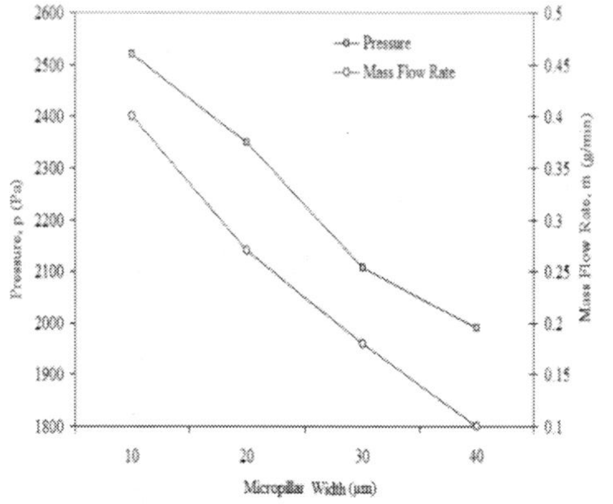

Figure 14. Variation of pressure and mass flow rate at micropillar width.

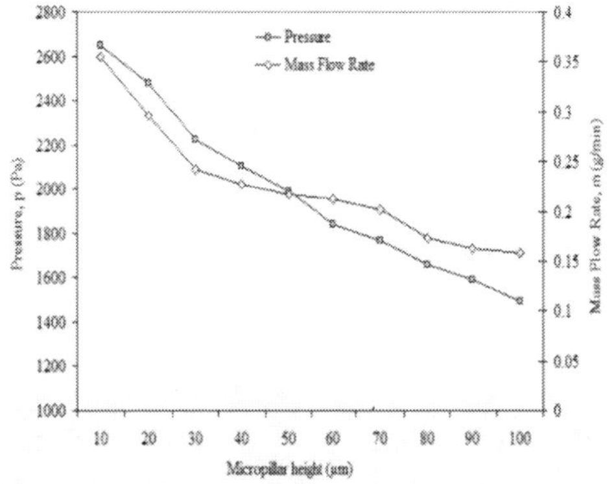

Figure 15. Variation of pressure and mass flow rate at micropillar height.

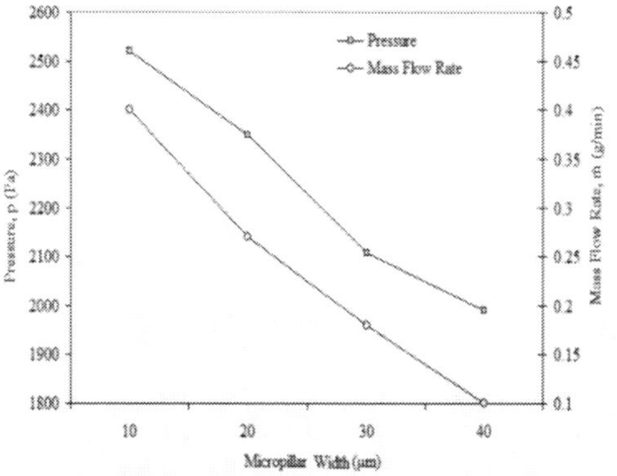

Figure 16. Variation of pressure and mass flow rate at micro channel height.

To validate the model the simulation results were compared with the experimental data of [17] obtained from four different configurations of electrodes. The output pressure achieved from the simulation of proposed design was compared with the planar symmetric, planar asymmetric, 3D symmetric and 3D asymmetric designs of [17] as shown in figure 17. The simulation results are significantly better than the first three designs and clearly, the results are almost equivalently good as the 3D asymmetric design. The proposed design is simple and inexpensive because it uses only one micropillar collector whereas [17] used two wider and thicker micropillars. It can also be deduced that by increasing the number of micropillars as optimized in the proposed design will increase the pressure and mass flow rate appreciably, unfortunately the fabrication and computational cost may also increase significantly.

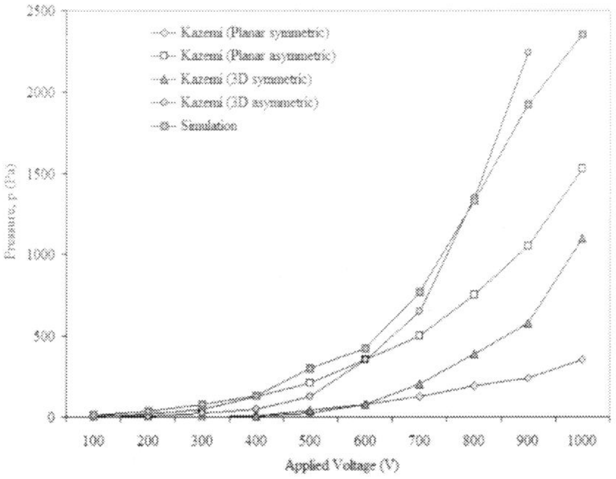

Figure 17. Comparison of simulation results with experimental results of Kazemi et al. [17]

4. Conclusions

A modeling and simulation study was performed for a simple prototype design of EHD ion-drag micropump with planar emitter and 3D micropillar collector electrode. The simulation methodology was applied to obtain and understand the behavior of electric potential, electric field, velocity field and pressure field within a representative segment of the micropump. Simulation results were used to optimize the design by analyzing the effect of different geometric parameters. The numerical results reveal that a maximum pressure of about 2350 Pa and a mass flow rate 0.4 g min^{-1} at an applied voltage 1000 V is achievable with the proposed design of micropump. These values of pressure and flow rate can meet the cryogenic cooling requirements for some specific electronic devices. Finally, the optimized design was compared with experimental results and found that output pressure showed fairly good agreement with experimental data.

Nomenclature:

A	Area of outlet (m^2)	x	Length of flow channel (m)
$\bar{u}$	Average velocity at outlet (m/s)	y	Width of flow channel (m)
q_e	Charge density (C/m^3)	z	Height of flow channel (m)
J	Current density (A/m^2)	μ_e	Ion-mobility $(m^2/V\,s)$
I	Current (A)	ρ	Mass density (kg/m^3)
k	Dielectric constant	$\dot{m}$	Mass flow rate (kg/s)
η	Dynamic viscosity (Ns/m^2)	p	Pressure (Pa)
F_e	EHD body force density (N/m^3)	V	Voltage (V)
ε	Electric permittivity (F/m)	S	Surface area b/w electrodes (m^2)
$\vec{E}$	Electric field intensity (V/m)	(u, v, w)	Velocity components (m/s)

Acknowledgment

This work is supported by the Ministry of Science, Technology and Innovation (MOSTI), Malaysia, under the Sciences project 06-01-09-SF0066. The research is carried out at the Faculty of Computer Science and Information Technology, University Malaysia Sarawak.

References

[1] Laser D and Santiago J 2004 *J Micromech. Microengineering* **14** 35
[2] Tsai N C and C Y Sue 2007 *Sens. Actuators A Phys.* **134** 555
[3] Stuetzer O M 1959 J. Appl. Phys. **30** 984
[4] Antonio R 2007 Microfluidic Technologies for Miniaturized Analysis Systems. New York: Springer
[5] Darabi J and Wang H 2005 *J. Microelectromech Syst.* **14** 747
[6] Pickard W H 1963 J. Appl. Phys. **34** 246
[7] Melcher J R 1981 "*Continuum Electromechanics*," Cambridge: MIT Press
[8] Crowley J M, Wright G S and Chato J C 1990 *IEEE Trans. Ind. Appl.* **26** 42
[9] Bryan J E and Yagoobi J S 1992 *IEEE Trans. Ind. Appl.* **28** 310
[10] Castellanos A 1998 International Centre for Mechanical Sciences, p 363
[11] Darabi J, Rada M, Ohadi M and Lawler J 2002 *J. Microelectromech Sits.* **11** 684
[12] Chowhury S, J. Darabil M, Ohadi and Lawer J 2002 *IEEE Trans. Compon. Packag. Technol.* **31** 1
[13] Darabi J and Ekula K 2003 *Microelectron J.* **34** 106
[14] Benetis V, Shoushtari A, Foroughi P and Ohadi M 2003 *Annu. IEEE Semicond Therm. Meas. Manage Symp.* 236
[15] Darabi J and Rhodes C 2006 *Sens. Actuators A Phys.* **127** 94
[16] Lee C K, Robinson A J and Ching C Y 2007 *13th International Workshop on THERMal Investigation of ICs and Systems* p 48
[17] Kazemi P Z, Selvaganapathy P R and Ching C Y 2009 *IEEE Transactions on Dielectrics and Electrical Insulation* **16** 483
[18] Kazemi P Z, Selvaganapathy P R and Ching C Y 2009 *J. Microelectromech. Syst.* **18** 547
[19] Hasnain S M, Bakshi A, Selvaganapathy P R and Ching CY 2011 *J Fluids Eng. Trans. ASME* **133**
[20] Rigit A R H and Chiong M C 2010 *Sens Actuators A Phys* **134** 650
[21] Kamboh S A, Labadin J and Rigit A R H 2012 *Intelligent Systems, Modeling and Simulation (ISMS)* p 417
[22] Rada M 2004 *Mechanical Engineering* University of Maryland
[23] Grant IS, Phillips WR 2008 *Manchester Physics*. John Wiley& Sons, Second edition

Journal of Physics: Conference Series **418** (2013) 012073 doi:10.1088/1742-6596/418/1/012073

Physical model of granule adhesion to the belt-electrodes of a tribo-aero-electrostatic separator

Jia Li[1*], Lucian Dascalescu[2], Mohamed Miloudi[2], Mihai Bilici[2], Zhenming Xu[1]
[1]School of Environmental Science & Engineering, Shanghai Jiao Tong University
800 Dongchuan Road, Shanghai 200240, China
[2]P' Institute, UPR 3346, CNRS – University of Poitiers – ENSMA
IUT, 4 avenue de Varsovie, 16021 Angoulême Cedex, France

E-mail: weee@sjtu.edu.cn

Abstract. Recent studies have demonstrated the effectiveness of tribo-aero-electrostatic separation technologies, which consist in the selective sorting of mixed granular insulating materials in a fluidized bed affected by an electric field orthogonally oriented to the direction of the fluidization air. The aim of the present paper is to put the theoretical bases for the optimization of this process, i. e. maximize the total mass of the granules collected at the two electrodes that generate the electric field. The various forces that drive a granule of given mass and electric charge through the electric field and make it stick to an electrode are expressed as functions of the several input variables and parameters of the process, such as the applied high-voltage or the surface roughness, the size and the position of the electrodes. The concepts of "critical electrostatic field" and "virtual climbing distance" are introduced. The prediction of the theoretical model are confirmed by the results of three sets of experiments, carried out on samples of a granular mixture consisting of 50% Acrylonitrile Butadiene Styrene (ABS) and 50% High Impact Polystyrene (HIPS), originating from the recycling of waste electric and electronic equipment. Higher separation efficiency was obtained when the electric field in the active zone was intensified by the use of an additional electrode connected to the ground and when the collecting electrodes were covered by a thin insulating layer.

1. Introduction

The tribo-electrostatic separation has been widely studied in view of many different industrial applications [1-3], including the recycling of granular plastic wastes [4-6]. A novel tribo-aero-electrostatic separator was designed [7] to avoid the drawbacks of classic tribo-charging separation by the simultaneous usage of triboelectricity, Coulomb force and electric image force [8]. In such a device, the granule charging and separation process takes place in a fluidized bed affected by an electric field orthogonal to the direction of the fluidization air.

The main parts of a tribo-aero-electrostatic separator are (figure 1): an air-flow chamber, a separation chamber and two collecting systems. The walls of the air-flow and separation chambers are made of transparent insulating boards, to facilitate the observation of the phenomena. A fine insulating mesh (air

---

[*] To whom any correspondence should be addressed.

diffuser) separates the two chambers and prevents the particles from falling into the air ducts.

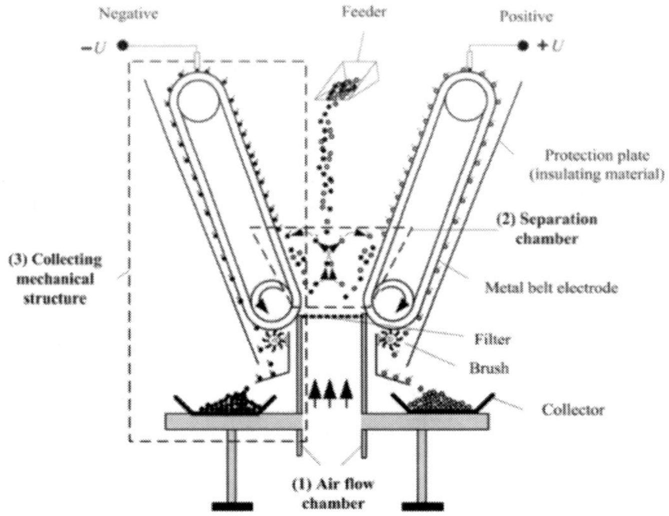

Figure 1. The standard structure of tribo-aero-electrostatic separator.

Each of the two collecting mechanical structures is composed a metal-belt conveyor, a protection plate, a brush and a collector. The metal belts are also the electrodes of the installation, as they are connected to two DC high-voltage power supplies of opposite polarities, and generate an electric field across the separation chamber. The protection plates are made of insulating materials to secure the operators and guide the particles fell into the collectors. The brushes are set close to the bottom edge of two electrodes in order to remove the charged granules that would otherwise remain stuck to the metal belts. A vibratory feeder introduces the granules in the separation chamber from the top of it. The free-falling granules are affected by the upwards-directed fluidization air flow. The collisions that occur between the granules in the mixture represent the main tribo-charging mechanism [9-11].

The electrodes generate electric field that is quasi-orthogonally-oriented to the direction of the fluidization air. Under the action of the Coulomb force, the charged granules are driven to the surface of the belt electrodes. The well-charged ones will stay pinned onto the surface of the electrodes by the electric field and image forces, and move along with the metal conveyors; the poorly-charged granules would rebound at the impact with the electrodes and fell back into the fluidized bed. They stay there until they get enough charged to be collected at the electrodes.

The factors that might influence the efficiency of tribo-electrostatic separation process have been estimated using the experimental design methodology [12, 13]. The experiments and the numerical simulations pointed out that the separation can be optimized by adjusting the control variables of the process, such as the high voltage or the speed of the fluidization air [14-16].

The aim of the present paper is to establish a physical model that would take into account the various forces that act on the charged granules, drive them through the electric field and make them stick to the electrodes. This model can be used to optimize the design of a novel tribo-aero-electrostatic separator.

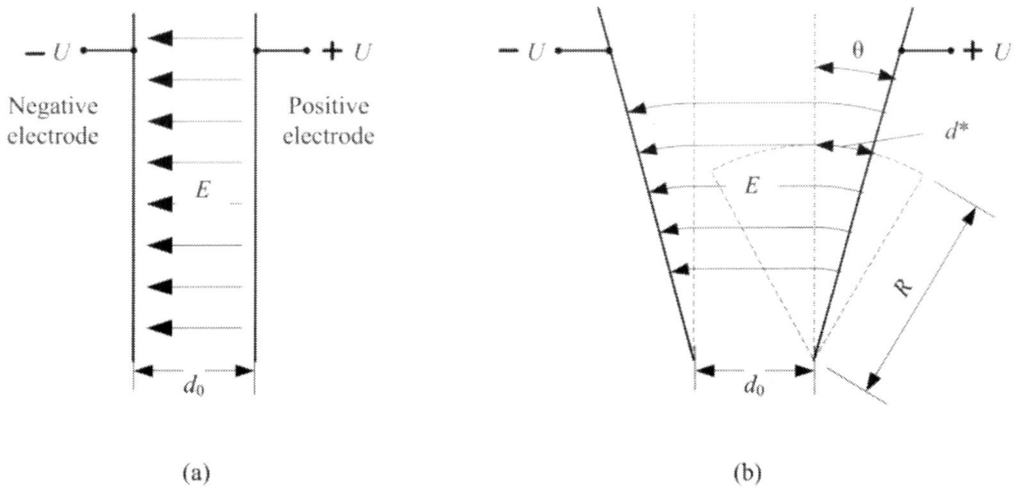

Figure 2. Electric field between vertical (a) and inclined (b) plate electrodes.

2. Physical model

The electrostatic separation is produced by the simultaneous action of various electrical and mechanical forces. The former depend on the local intensity of the electric field. Therefore the first step in the elaboration of a physical model of granule behavior in an electrostatic separator consists in the computation of the electric field.

2.1. Electric field computation

Figure 2 (a) shows the distribution of electrostatic field between two vertical parallel plate electrodes, distanced at d_0 and energized from two high voltage supplies of opposite polarities ± U. The electrostatic field lines are perpendicular to the parallel plates and the strength E of the uniform electric field at the surface of plates can be computed as:

$$E = 2U / d_0 \tag{1}$$

In case that the electrodes make an angle θ with the vertical plane as shown in figure 1, the electric field at the surface of electrodes cannot be computed with (1). Each field line can be approximated by a segment of length d_0 and two arcs of length d^*, as shown in figure 2 (b). The distance d^* is computed as:

$$d^* = (\theta / 360) \times 2\pi R \tag{2}$$

where R is the radius of the arc.

As shown in figure 3, the length d_L of the electrostatic field line corresponding to the point (x_L,y_L) can be computed from (2), with $R = L$:

$$d_L = d_0 + 2d^* = d_0 + (\theta / 90) \times \pi L \tag{3}$$

The electrodes being supplied with symmetrical high-voltages of opposite polarities ± U, the electrostatic field E_L at the position of (x_L,y_L) can be computed approximately as:

$$E_L = 2U / d_L = 180U / (90d_0 + \pi\theta L) \tag{4}$$

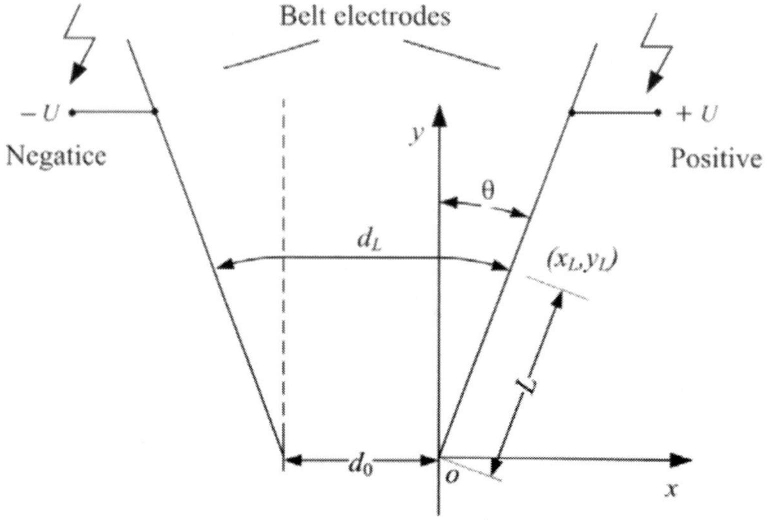

Figure 3. Geometry of non-parallel electrodes system.

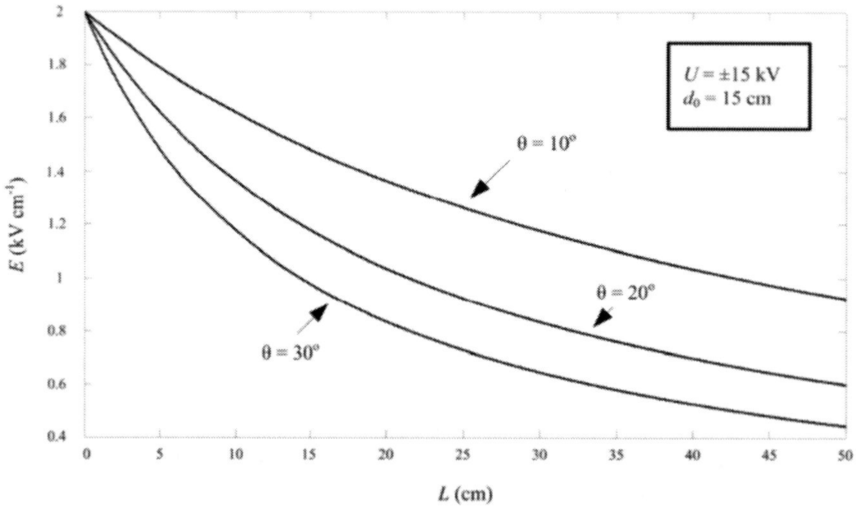

Figure 4. The repartition of the electric field strength E as function of the distance L at the surface of the metal belt electrodes, for three values of the angle θ.

Figure 4 shows the repartition of the electric filed strength E at the surface of electrode. Because the electrodes are inclined at an angle θ, the electric field at their top edge L_{top} is weaker than at $L = 0$ (i.e., at the bottom edge of the electrodes). The decay of the electric field strength E along the electrode length is steeper for θ = 30° than for θ = 20° or θ = 10°.

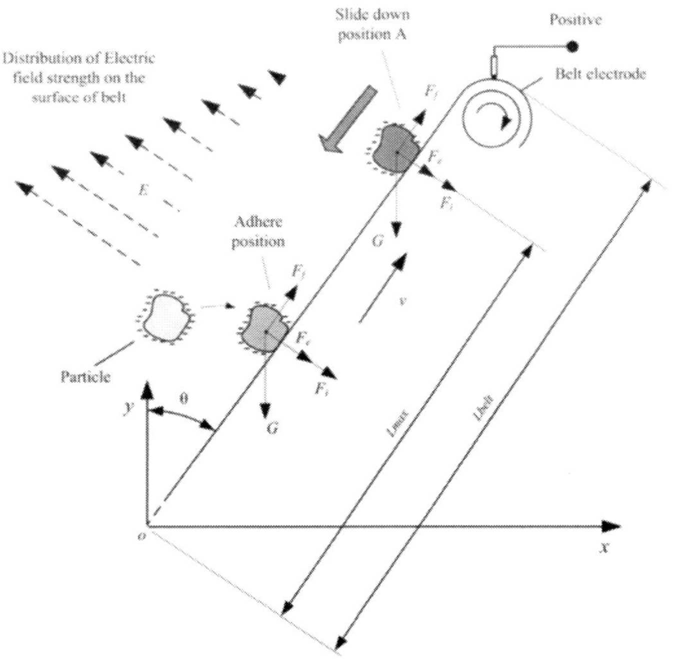

Figure 5. Schematic representation of particle adhesion to the surface of the electrode.

2.2. *Forces analysis*
The physical model is established for granules that are assumed to be perfect insulating spheres of radius r and specific mass ρ, carrying a charge Q uniformly distributed on their surface. This means that Q can be considered as being concentrated in the center of the sphere. When compared with the distance d_L (figure 3), the radius of sphere can be neglected and the granules can be represented as mechanical points of constant charge Q. The space between the electrodes is homogeneous and isotropic, of permittivity ε_0.

As shown in figure 5, the belt electrode moves at a constant speed v. A charged granule is attracted to the electrode and got in contact with it. The collision is assumed to be perfectly plastic (i.e., there is no granule rebound after the impact). As a consequence, the charged granules stick to the electrodes and move with them as long as the mechanical and electrical forces acting on it satisfy the condition:

$$G \times \cos(\theta) \le [G \times \sin(\theta) + F_i + F_e] \times \mu_{\max} \qquad (5)$$

where G is the weight of particle:

$$G = (4\pi r^3 / 3)\rho g \qquad (6)$$

F_i is the electric image force:

$$F_i = Q^2 / [4\pi\varepsilon(2r)^2] \qquad (7)$$

F_e is the electric field (Coulomb) force:

$$F_e = QE \qquad (8)$$

With $g = 9.81$ m s^{-2}, $\varepsilon = 1.5/(4\pi 9\times10^9)$ and $\mu_{\max}$ is the maximum coefficient of static friction between the granule and the electrode.

The above condition can be rewritten as:

$$(4\pi r^3/3)\rho g \times \cos(\theta) \le \{(4\pi r^3/3)\rho g \times \sin(\theta) + Q^2/[4\pi\varepsilon(2r)^2] + QE\} \times \mu_{max} \quad (9)$$

Hereafter, the left and right members of (9) will be designated as:

$$Y_1 = (4\pi r^3/3)\rho g \times \cos(\theta) \quad (10)$$

$$Y_2 = \{(4\pi r^3/3)\rho g \times \sin(\theta) + Q^2/[4\pi\varepsilon(2r)^2] + QE\} \times \mu_{max} \quad (11)$$

The function Y_1 is constant for granules of given size r and mass density ρ, attached to an electrode inclined at a well-defined angle θ. As shown in figure 4, for any angle $\theta > 0^o$, the electric field strength E decreases while the granules get closer to the upper edge of the belt electrode. To ensure the balance between Y_1 and Y_2 in a given point A (figure 5) and for a well-defined value of the coefficient of static friction μ_{max}, the local electric field should have the minimum value E_{min} computed from (4) and (9):

$$E_{min} = \{[G \times \cos(\theta)/\mu_{max}] - G \times \sin(\theta) - F_i\}/Q \quad (12)$$

If the electric field strength E in any point at the surface of the electrode is higher than the critical electrostatic field E_{min}, the charged granule will pass over the top of the belt and will be collected in an appropriate compartment. Otherwise (i.e. $Y_2 < Y_1$), the granule will slide down and fall back into the fluidization bed. For a given electrode configuration (i.e., electrode length L, electrode angle θ and electrode distance d_0), the corresponding value of the critical high-voltage U_{min} can be computed from (12):

$$U_{min} = (90d_0 + \pi\theta L)\{[G \times \cos(\theta)/\mu_{max}] - G \times \sin(\theta) - F_i\}/(180Q) \quad (13)$$

For example, in the case of a granule characterized by r = 1 mm, ρ = 1050 kg m^{-3} and Q = 330 pC, if L = 0.5 m, θ = 10° and d_0 = 0.15 m, then U_{min} = 20.535 kV. This means that if the voltage applied to the electrode system were lower than ±20.535 kV, the respective granule would not be collected.

The position L_{max} of the highest point on the electrode where the granule still adheres to it can be computed from (4), (5), (10) and (11):

$$L_{max} = \{180UQ/[G \times \cos(\theta)/\mu_{max} - G \times \sin(\theta) - F_i] - 90d_0\}/(\pi\theta) \quad (14)$$

If the critical climbing distance $L_{max} > L_{belt}$, the granule passes the top of the belt electrode and the separation is successful. With Q, G and F_i constant in (14), the adhesion of the granules to the electrode can be improved by adjusting the high-voltage U, the inter-electrode distance d_0, the angle θ. and the coefficient of static friction μ_{max}.

The charge/mass ratio Q/m that ensures a virtual climbing distance L_{max} for a granule r = 1 mm, ρ = 1050 kg m^{-3} has been calculated for different values of the angle θ (figure 6). The longer is the distance L_{max}, the higher should be the charge/mass ratio Q/m. The granules climb a longer distance L_{max}, if θ is larger. For a belt of given length L_{top} = 0.4 m, the charge/mass ration Q/m of particles should be higher at small angle setting θ =10°, than at θ =20°. However, this does not mean that the larger is the angle θ, the better it is. The experiments show that at large angle θ, the operating conditions of the fluidized bed are severely altered and the attainable charge/mass ratio significantly decreases. As shown in figure 6, the L_{max} is very sensitive to the value Q/m. For example, in the case of the large angle θ =20°, the L_{max} reduces by 50%, from 0.4 m to 0.2 m, at a tiny diminution of Q/m from 25 nC g^{-1} to 23 nC g^{-1}.

The figure 7 shows the relationship between the climbing distance L_{max} and the charge/mass ratio Q/m for a granule r = 1 mm, ρ = 1050 kg m^{-3}, at the different applied high voltages U. The granule requires lower charge (26.2 nC g^{-1}) at higher voltage ($U = \pm 19$ kV) for climbing to the same distance $L_{max} = 0.4$ m. Setting the voltage at higher values is better. However, it should be noted that the effect of a change in the charge/mass ratio is more important than the applied voltage. The climbing distance L_{max} is more sensitive to the changes in Q/m than in U. In the development of a new industrial application, the choice to make is between a more expensive high-voltage generator, capable to provide higher U, or a longer residence time of the granules in the fluidized bed, so that they acquire a larger charge Q, to compensate for the lower E in the formula of the Coulomb force (8).

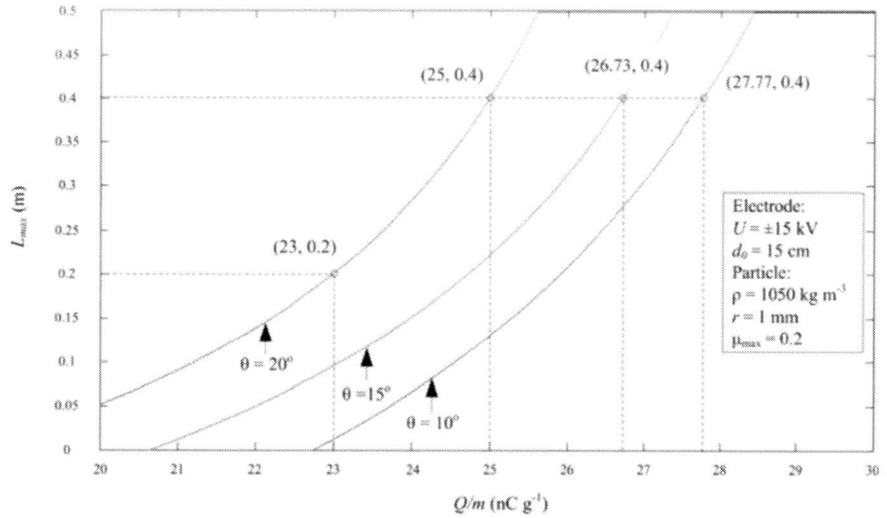

Figure 6. Computed critical climbing distance Lmax as the function of the charge/mass ratio Q/m, for a granule r = 1 mm and ρ = 1050 kg m^{-3} , at U = ±15 kV, d_0 = 15 cm, μ_{max}= 0.2, and different electrode angles θ.

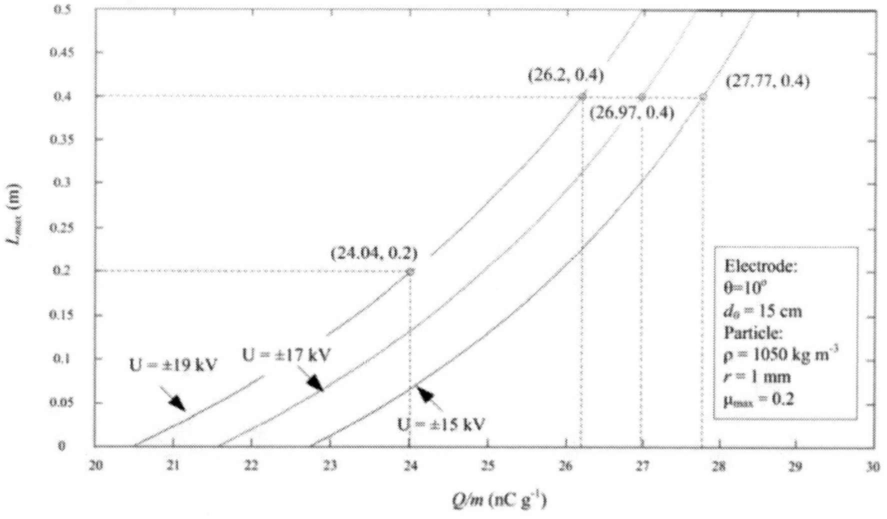

Figure 7. Computed critical climbing distance Lmax as the function of the charge/mass ratio Q/m, for a granule r = 1 mm and ρ = 1050 kg m^{-3} , at θ = 10°; d_0 = 15 cm, μ_{max}= 0.2, and

different high-voltages U.

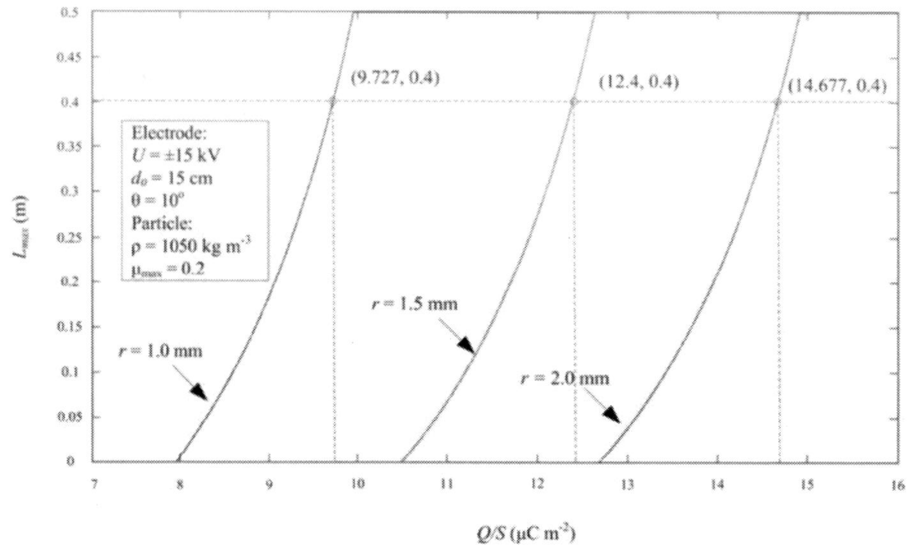

Figure 8. Computed critical climbing distance Lmax as function of the surface charge density Q/S, for granules of ρ = 1050 kg m^{-3} and various radius r, at U = ±15 kV; θ = 10°; d_0 = 15 cm, μ_{max}= 0.2.

At the end of the tribo-charging process, the charge Q is uniformly distributed on the surface S of the granules. The relationship between the critical surface charge density (Q/S) and climbing distance L_{max} for granules of different sizes r was computed from (14) and represented in figure 8. The larger granules (r = 2.0 mm) need significantly higher surface charges densities than the smaller ones (r = 1.0 mm), for the same climbing distance L_{max}. That means that coarser particles are much more difficult to be separated.

3. Experimental study

3.1. Materials and method

In order to validate the conclusions of the physical modeling reported in the previous section of the paper, several experiments were carried out on a granular mixture consisting of 50% Acrylonitrile Butadiene Styrene (ABS) and 50% High Impact Polystyrene (HIPS) originating from the recycling of waste electric and electronic equipment (figure 9). The particle size ranged between 2.5 and 4 mm.

The three electrode settings employed for this experimental study are described in table 1 and displayed in figures 1, 10 and 11.

Table 1 Explanation for four different electrodes settings.

Setting	Designation	Descriptions
I	Standard	Standard setting (figure 1).
II	Additional grounded electrode	A "V" type ground electrode was placed parallel to the surface of metal belt electrodes to increase the electrostatic field strength E (figure 10).
III	Insulator	The surface of the metal belt is covered by an insulating layer to limit

layer	charge leakage to the ground and change the coefficient of static friction μ (figure 11).

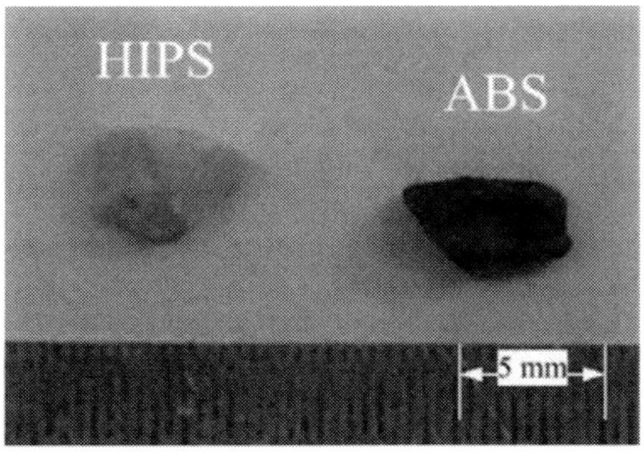

Figure 9. Aspect of HIPS and ABS granules originating from WEEE.

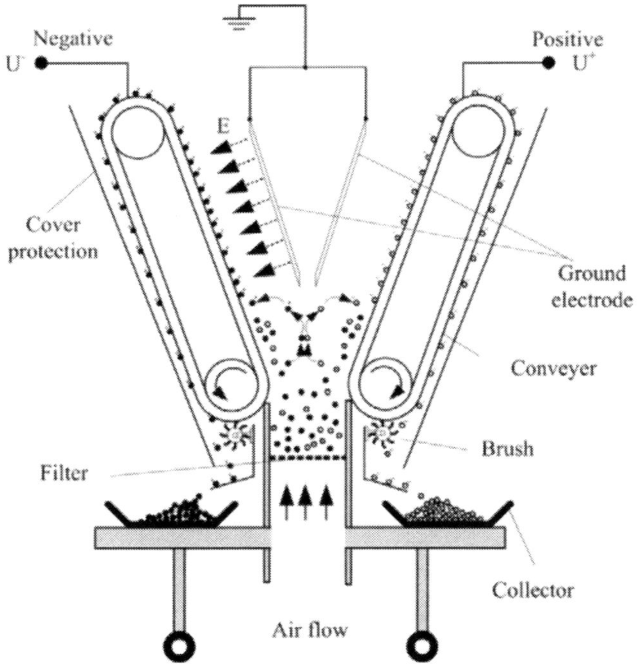

Figure 10. Electrode setting II, V-shaped grounded electrode placed at equal distances from the two metal belt electrodes.

At first, 2 kg of a (50% ABS +50% HIPS) granular mixture were fed into the separation chamber to be charged for 4 min at the maximum air flow. Then the high voltage was applied to the electrodes and the belt speed was set at 16 cm/sec. to collect the products. Finally, the mass of products in the collector were measured at 4 min, 10 min and 15 min after high-voltage switch-on. After 15 minutes, there were no more products to be collected.

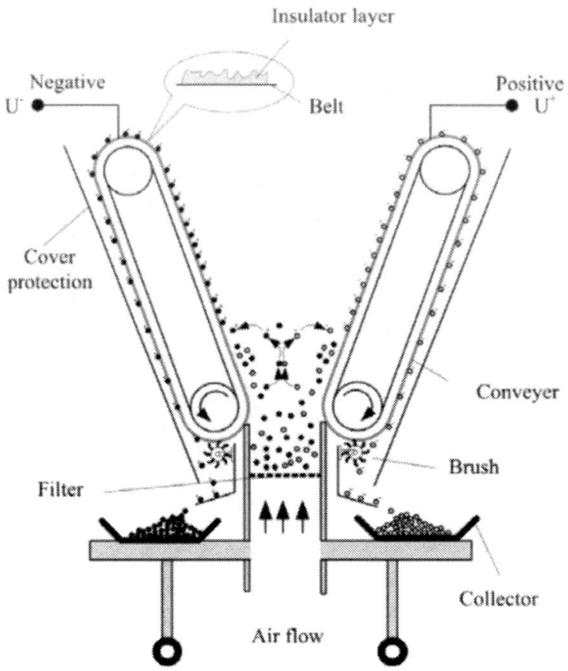

Figure 11. Electrode setting III, metal belt electrodes covered with an insulator layer.

3.2. Results and discussion

The recovery R_A of product A was computed with:

$$R_A = (m_A / m_{Atotal}) \times 100\% \qquad (15)$$

m_A is the mass of the A granules accumulated in the collector, and m_{Atotal} is the total mass of A in the feed. The separation results are given in table 2, figure 12 and figure 13. In all cases, ABS granules separate better than HIPS, as the granule-to-wall collisions increase the total charge of the former, but have an adverse effect on the latter. Both the ground electrode (II) and the insulator layer (III) settings increased the recovery of ABS granules (figure 12). For this kind of granules, the setting III leads to better results than the setting II. In the case of HIPS granules (figure 13) the best results were obtained for the setting II. Surprisingly, the results obtained with the setting III were worse than with the standard setting (I).

Table 2. Cumulated recoveries of ABS and HIPS at 4, 10, and 15 min after high-voltage switch-on, using different electrode settings.

Settings	I Standard		II Ground electrode		III Insulator layer	
Time points	ABS(%)	HIPS(%)	ABS(%)	HIPS(%)	ABS(%)	HIPS(%)
4 min	29.26	12.84	32.65	30.42	30.3	10.95
10 min	39.76	16.14	43.65	38.92	49.95	14.42
15 min	46.76	18.14	48.15	42.4	57.83	15.87

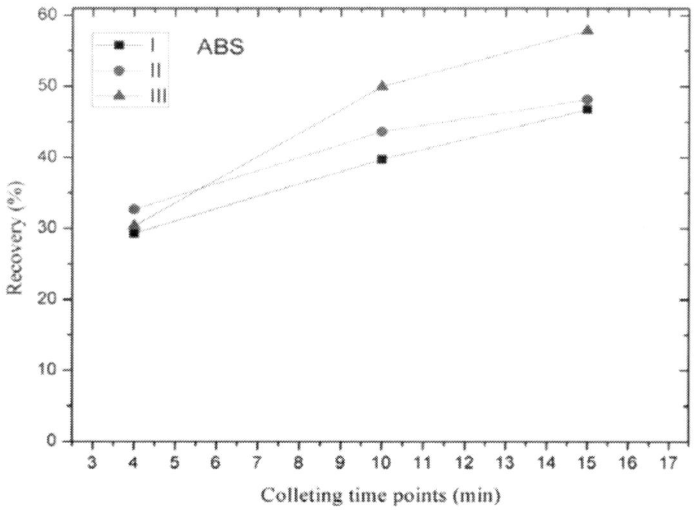

Figure 12. ABS recovery as function of time, for three different electrode settings (I: standard setting; II: additional grounded electrode setting; III: insulator layer setting).

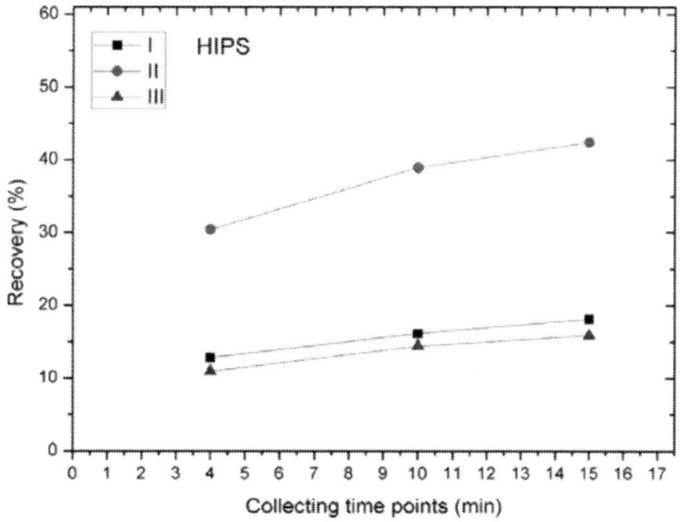

Figure 13. HIPS recovery as function of time, for three different electrode settings (I: standard setting; II: additional grounded electrode setting; III: insulator layer setting).

These experimental findings can be discussed in the light of the physical model established in the previous section. Thus, the setting II increased the intensity E of the electric field at the surface of the electrodes. Granules with lower charge Q satisfy (9), adhere to the electrodes, and are collected as distinct products. The recoveries are improved. However, the effect of setting II on ABS is not as spectacular as on HIPS (figures 12 and 13). The ABS granules being better charged than HIPS, they will benefit less of the increase in the intensity of the electric field: in their case, the virtual climbing distance already satisfies the condition $L_{max} > L_{top}$.

The insulating layer in setting III increased the static friction coefficient μ_{max} of both types of granules and was expected to improve the separation, due to the increase of L_{max} given by (14). No such effect was obtained in the case of HIPS granules, as in their case it was surpassed by the decrease of the electric image force F_i. Indeed, the expression of F_i is no longer the one given by (7), as it should take into account the thickness a of the insulating layer:

$$F_i = Q^2 / \left\{4\pi\varepsilon[2(r+a)]^2\right\} \quad (16)$$

According to (16), the presence of the layer diminishes F_i, and hence the virtual climbing distance L_{max}.

A third physical mechanism should be taken into account in order to explain the experimental observations in figures 12 and 13: the decay of the charge carried by the granules. In the standard setting, where the granules are in direct contact to the electrode, the charge of ABS granules decays faster than that of HIPS. The insulating layer in setting III limits the charge leakage. In this way, the charge of ABS remains practically constant and equal to the initial value. The increase of Q as compared to its value in setting I is accompanied by an increase of L_{max} in (14), which explains the better separation results. HIPS particle benefit less of this slow down of the charge decay and their behavior is more affected by the decrease of the electric image force F_i, as discussed above. The insulating layer has an opposite effects on the collection of ABS and HIPS granules, positive for the former and negative for the latter.

4. Conclusions

The physical model of granule adhesion on the belt-electrodes of a tribo-aero-electrostatic separator enables the definition of two critical parameters for the design of such equipment: the minimum electric field strength and the maximum electrode length for which a charged granule can be safely collected. Both parameters depend on the charge of the granules and the geometry of the electrode system. By increasing the angle between the belt electrodes and the vertical plane, the minimum electric field strength diminishes.

By placing a V-shaped grounded electrode equally-distanced from the two high-voltage electrodes, it is possible to improve the outcome of the electrostatic separation, as it increases the electric field strength and hence the electric forces that make the particles adhere to the metallic belts and be transferred to the collecting boxes. Particle adhesion can also be enhanced by covering the belt electrodes with an insulating layer that would reduce the charge leakage and increase the friction coefficient. The effect of the insulating layer is more important in the case of the insulating materials that are characterized by relatively-fast charge decay, such as ABS.

Acknowledgements

This project was supported in part by the National Natural Science Foundation of China (51008192), Doctoral Fund of Ministry of Education of China (20090073120041), Shanghai Natural Science Foundation (10ZR1415900) and Open Project of Key Laboratory of Solid Waste Treatment and Resource Recycle (09gk01).

References

[1] Boland D and Geldart D 1971/1972 *Powder Technol.* **5** 289
[2] Dwari R K and Rao K H 2009 *Minerals Eng.* **22** 119
[3] Rokkam R G, Fox R O and Muhle M E 2010 *Powder Technol.* **203** 109
[4] Calin L, Mihalcioiu A, Iuga A and Dascalescu L 2007 *Part. Sci .& Technol.* **25** 205
[5] Calin L, Caliap L, Neamtu V, Morar R, Iuga A, Samuila A and Dascalescu L 2008 *IEEE Trans. Ind. Appl.* **44** 1045

[6] Miloudi M, Medles K, Tilmatine A, Brahami M and Dascalescu L 2011 *J. Electrostat.* **69** 631
[7] Calin L and Dascalescu L 2010 *Patent FR2943561, WO2010109096*
[8] Miloudi M, Remadnia M, Dragan C, Medles K, Tilmatine A and Dascalescu L 2011 *Conf. Rec. IEEE IAS Ann. Meet. (Orlando, FL, USA)* DOI: 10.1109/IAS.2011.6074274
[9] Ciborowski J S and Wlodarski A 1962 *Chem. Eng. Sci.* **17** 23–32
[10] Guardiola J, Rojo V and Ramos G 1996 *J. Electrostat.* **37** 1
[11] Ali F S, Inculet I I and Tedoldi A 1999 *J. Electrostat.* **45** 199
[12] Frigon N L and Mathews D 1996 *Practical Guide to Experimental Design* (New York: Wiley)
[13] Eriksson L, Johansson E, Kettaneh-Wold N, Wikström C and Wold S 2000 *Design of Experiments. Principles and Applications* (Stockholm: Learnways AB)
[14] Dragan C, Fati O, Radu M, Calin L, Samuila A and Dascalescu L 2011 *IEEE Trans. Ind. Appl.* **47** 1922
[15] Bilici M, Dascalescu L, Dragan C, Fati O, Iuga A and Samuila A 2011 *IEEE Trans. DEI* **18** 1476
[16] Rahou F Z, Tilmatine A, Bilici M and Dascalescu L 2012 *IEEE Trans. Ind. Appl.* **48** 816

New design of high performance ionizing bar

Ronggang Wang *, Yurong Sun
Suzhou TA&A Ultra Clean Technology Co. Ltd., Suzhou, Jiangsu 215121, China

E-mail :wangronggang@canmax.com.cn

Abstract. This paper introduces a new design of DC-pulse ionizing bar to solve the problem of imbalance offset voltage for the AC ionizing bar, which is easily affected by the environment, as well as indicate the final tests. The new design mainly includes five parts: power supply circuit, main control unit, logic circuit, high frequency transformer unit, and feedback unit. The ionizing bar can automatically adjust the discharge voltage, pulse frequency and pulse width to balance the positive and negative ions. The final test results indicate that the DC ionizing bar owns good effect in electrostatic elimination.

1. Introduction

In the electronic industries, the products always carry on the positive and negative charges due to the relative motion such as friction, contact and separation. The static charge can cause problems in production, for example, the surfaces of the products adsorb tiny dust particles, plastic film products bind hard to separation in the process, accumulated charges cause electronic components breakdown for electrostatic discharge [1]. In order to minimize or eliminate static electricity damage, the best way is to build up electrostatic protected areas (short for EPA) as a systematic project [2]. In this project, the effective protection requires the trinity measures of environment, human body and manufacturing process control. For the performance of electrostatic eliminator, offset voltage, decay time, stability and durability is considered the important parameters [3-4]. As the advantage of installation convenience, size extensibility, environment adaptability and good performance, the ionizing bars is applied in industrial production widely. This paper introduces a new design of ionizing bar which can achieve the desired effect to remove static electricity, and makes analysis of the performance combining with measured data. It employs DC pulse discharge and closed-loop feedback system to adjust high voltage output to guarantee stability of the offset voltage.

2. New design of ionizing bar

2.1 Principle of DC pulse discharge

The DC pulse-discharge is different from the other AC-discharge products in the market. The discharge mode is showed in figure 1. The core is control circuit of high-voltage generator which employs closed-loop feedback system. Figure 2 (A) shows the positive and negative pulse voltage is same value in one period, and figure 2 (B) shows the adjustment of pulse frequency, amplitude and width. This technique can guarantee the ion balance in multiple environment conditions.

---

* To whom any correspondence should be addressed.

Journal of Physics: Conference Series **418** (2013) 012074 doi:10.1088/1742-6596/418/1/012074

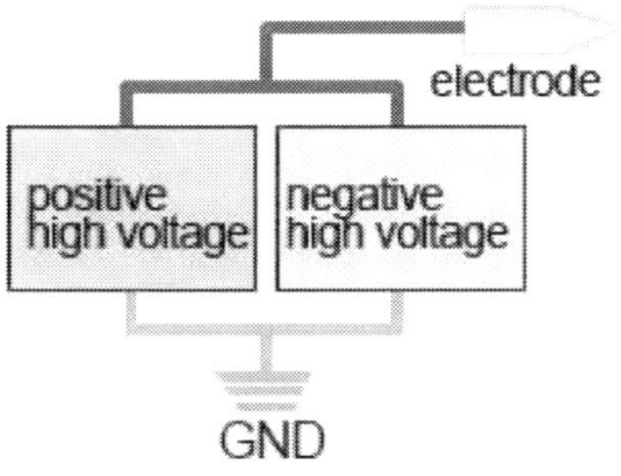

Figure 1. DC ionizing bar principle.

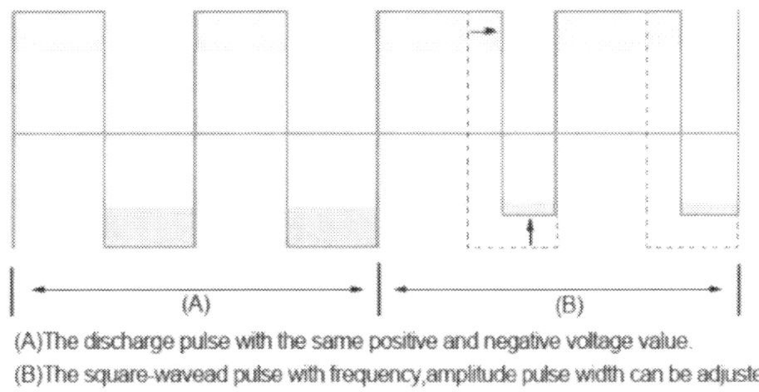

Figure 2. Power supply principle.

2.2 Design of Control Circuit

The circuit mainly includes five parts: power supply circuit, main control unit, logic circuit, high frequency transformer unit, and feedback unit. The specific design principle is shown in figure 3.

Power supply circuit includes HV+ & HV- amplitude adjustment circuit. It cooperates with feedback unit to supply the power of high frequency transformer unit.

With the help of good performance MCU, the main control circuit realizes the feedback signal collection and related control. The A/D converter samples the human-setting and environment information and sends them to MCU. MCU analyzes and contrasts with the inner database to generate the appropriate pulse wave. This wave includes two frequencies: 1-60 Hz (adjustable) and 50 kHz signal. 50 kHz signal is the work signal of the high-frequency transformer unit. 1-60 Hz (adjustable) signal is the discharge wave frequency. Logic circuit synthesizes the 1-60 Hz (adjustable) signal and 50 kHz signal and convert them to one signal as seen in figure 4. High frequency transformer unit receive the signal which generated by logic circuit and power supply which generate by power supply unit, and then transform to corresponding high voltage pulse output. The high voltage pulse output transmits to the electrode for corona discharge to release positive and negative ions.

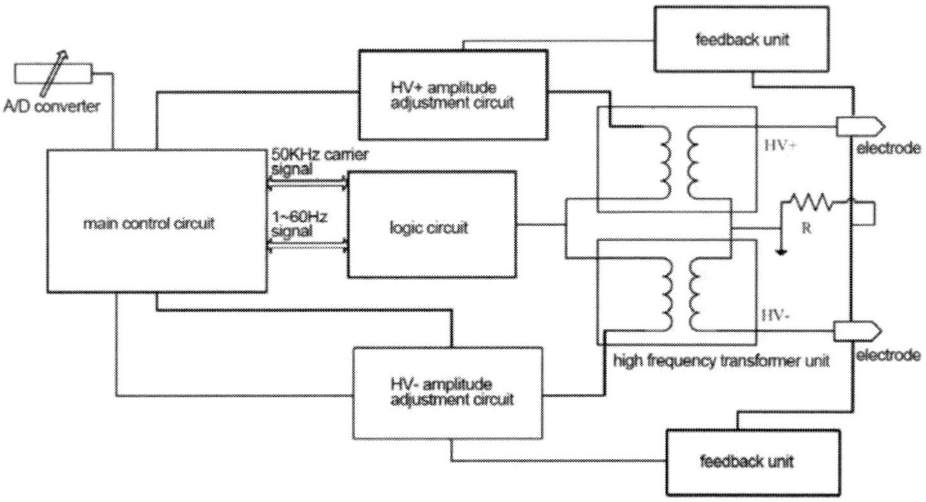

Figure 3. Design of Control Circuit.

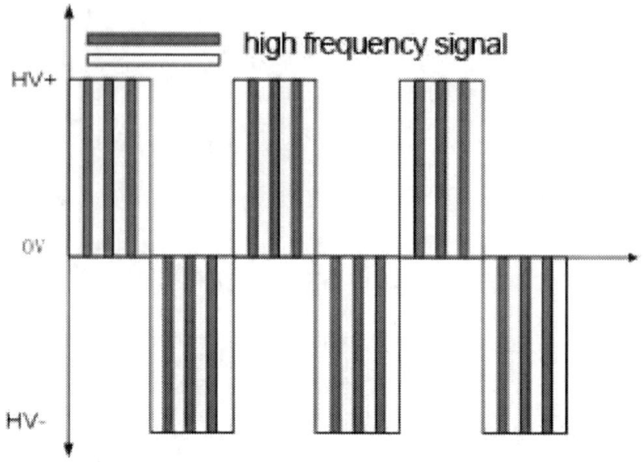

Figure 4. High voltage square-wave pulse.

The feedback unit installed near the electrode can detect the balance of positive and negative ions, and sends the detected signal to the main control unit as the conference of next adjustment. The main control unit bases on the detected signal to make appropriate adjustment of frequency, amplitude and pulse width, thus ensuring the offset voltage is near 0 V. The whole process performs the function of the closed-loop feedback and control.

The new design makes the EPA voltage control within the secure range effectively, and it greatly improves the stability, anti-interference and dust-resistant performance of the ionizing bar.

3. Performance test

The performance of the ionizing bar consists of two aspects: one is Electrical performance, including offset voltage and decay time; the other is stability and anti-interference.The standards IEC61340 and ANSI/ESD STM 3.1 provide electrical performance test method [6-7]. The test equipment is the charged plate monitor (CPM), which has installed continuous sample and analysis software, and it can show and remember the real-time offset voltage in the computer.

The test operates in a class 10,000 cleanroom which control temperature & humidity. The temperature is 23±2℃ and the humidity is 40%~60%. The stability and anti-interference is the capabilities that the ionizing bars worked in a period time within the voltage range. The test time can be set as necessary, for example, 12 hours. Test diagram as shown in figure 5, the ionizing bar hangs in 60 cm above the work surface. The CPM is placed just below the middle of the ionizing bars and the electrode is perpendicular to the work surface.

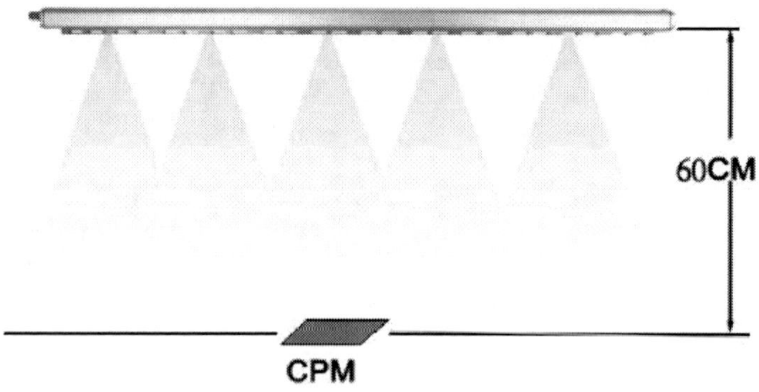

Figure 5. Test diagram.

4. Test data analysis

The test included the decay time test, offset voltage test, stability and anti-interference test. The result of the ionizing bars decay time was less than 3s. The offset voltage of the continuous work in 15 hours was showed in figure 6. The offset voltage was controlled within 10 V. It validated the ionizing bar owned high stability and anti-interference performance. At the same time, it shows the continuous test method is better than the points measure. Setting voltage range ±30 V (the high-performance ionizing bars requirements value), the process capability index was 6.6. It showed the work operation was very stable, and the result was showed in figure 7.

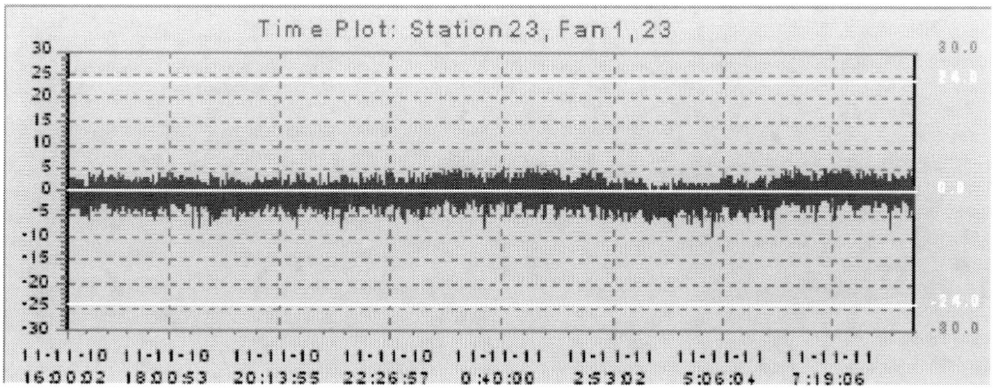

Figure 6. Offset voltage test.

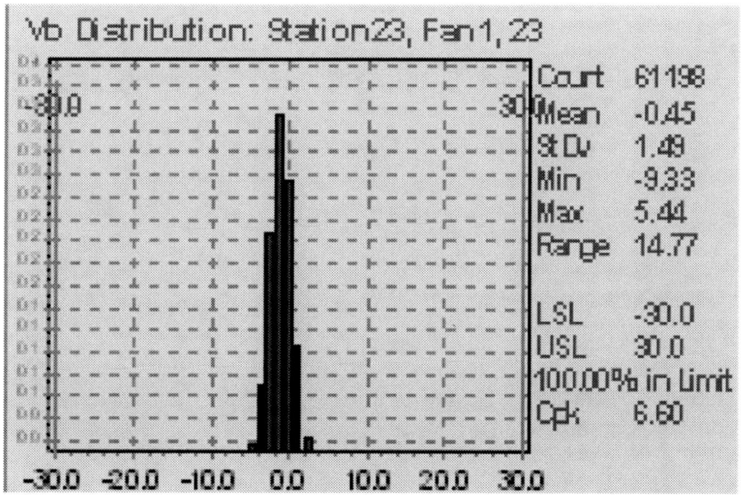

Figure 7. CPK result.

5. Conclusion

This paper introduces the design of new type ionizing bar using DC pulsed high-voltage discharge.The control circuit can adjust frequency, amplitude and pulse width according to the environment changing. Compare with existing AC transformer discharge, the new design has many advantages, such as, high discharge efficiency, good electrical performance, excellent stability and anti-interference under various working condition.

References

[1] Liu S 1999 *Electrostat. Theor. & its Prot.* vol 5 (Beijing: Weapons Industry Press) p 135
[2] GJB/Z105 *Electrostatics discharge control handbook for protection of electronic products* 1998 (Beijing: COSTIND)
[3] Sun Y, Cheng Y and Zhao L 2009 *Trans. of Beijing Inst. of Technol.* **74** 178
[4] Guo S,Wu J 2005 *Meas. & Test.* **22** 120
[5] Yang Jinji 1983 *Gas discharge* vol 3 (Beijing: Science Press) p 110
[6] IEC 61340-5-1 *Protection of electronic devices from electrostatic phenomena-General requirements* 2007 (Geneva, Switzerland: IEC Central Office)
[7] ANSI/ESD STM3.1 *ESD association standard test method for the Protection of Electrostatic Discharge Susceptible Items-Ionization* 2006 (New York: Electrostatic Discharge Association)

Research of vibration controlling based on programmable logic controller for electrostatic precipitator

Zisheng Zhang*, **Yanhu Li, Jiaojiao Li, Zhiqiang Liu, Qing Li**
Electrostatic Research Institute, Hebei University, No.180 Wusidong Road, Baoding, China

E-mail: zhangzisheng@hbu.edu.cn

Abstract. In order to improve the reliability, stability and automation of electrostatic precipitator, circuits of vibration motor for ESP and vibration control ladder diagram program are investigated using Schneider PLC with high performance and programming software of Twidosoft. Operational results show that after adopting PLC, vibration motor can run automatically; compared with traditional control system of vibration based on single-chip microcomputer, it has higher reliability, better stability and higher dust removal rate, when dust emission concentrations ≦ 50 mg m^{-3}, providing a new method for vibration controlling of ESP .

1. Introduction

At present, the dust pollution in China is mainly produced from cement, coal-fired power and metallurgy, and the emission of coal-fired power plant is increasing day after day. The process of coal-fired power will produce a large number of coal-dust. What's more, the coal-dust is harmful to health of human [1]. Every year, coal-dust not only causes air pollution, but also leads to huge waste of resources. Therefore, we must take control of coal-dust pollution.

Electrostatic precipitator (ESP) is a kind of popular device which can capture particles; which has obvious advantage [2]. The motor's vibration control system is directly related to efficiency of ESP. The traditional vibration is based on single-chip microcomputer, but it has some disadvantages. For example, its circuit is complex, consumption of power is high, and program of single-chip microcomputer is easy to break down when electromagnetic interference (EMI) occur. The reliability and stability of traditional vibration control system is lower, and the result is to reduce efficiency of removing dust. In recent year, programmable logic controller (PLC) is widely used in industrial environment; PLC is substituting for single-chip microcomputer as control core of ESP, which can improve the efficiency of removing dust, and is a kind of ideal design scheme.

2. Improved plan of vibration control

Schneider PLC has powerful functions, which adopts I/ O photoelectric isolation and filter on the hardware; and adopts fault detection and alarm to insure reliability and stability on the software. Besides, it has many communication modules, just like RS232/485. Therefore, we choose Schneider PLC as cybernetics core [3].

Diagram of vibration controlling is shown in figure 1 [4]. The main control process includes four parts: vibration of gravity separator chamber, vibration of the 1st, 2nd and 3rd electric field. PLC can

* To whom any correspondence should be addressed.

communicate with computer through RS485 port, and then sends the collecting data to the computer to monitor the site. Meanwhile, computer also appropriately modifies internal timer and counter of PLC, to make ESP work at the best state.

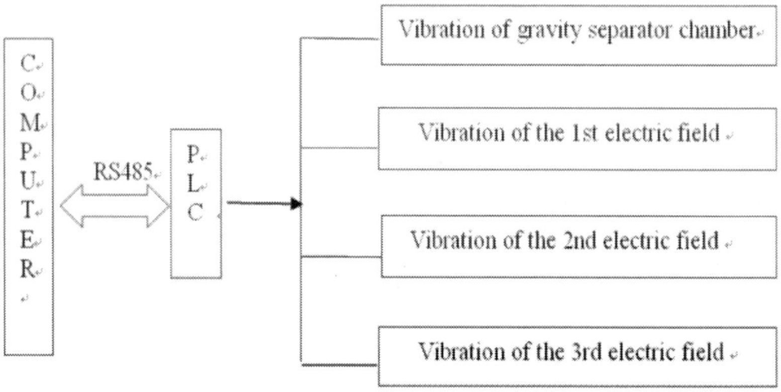

Figure 1. Diagram of vibration controlling.

2.1. Experiment device

Figure 2 is the schematic diagram of vibration device. The ESP has the strict request for resistance of dust, the dust resistance is in the range of 105-1011 Ω· cm. In the case of the resistance of dust is above 1011 (high resistance type), the electric field will against electric corona, leading to reducing the efficiency of removing dust. The coal-dust belongs to the type of high resistance, and the temperature is about 350 ℃. Before coal-dust goes into the draught fan, gas goes through the humidifying tower to reduce the resistance of dust first, and then passes through gravity separator chamber and three electric fields; finally, the gas is expelled by the flue.

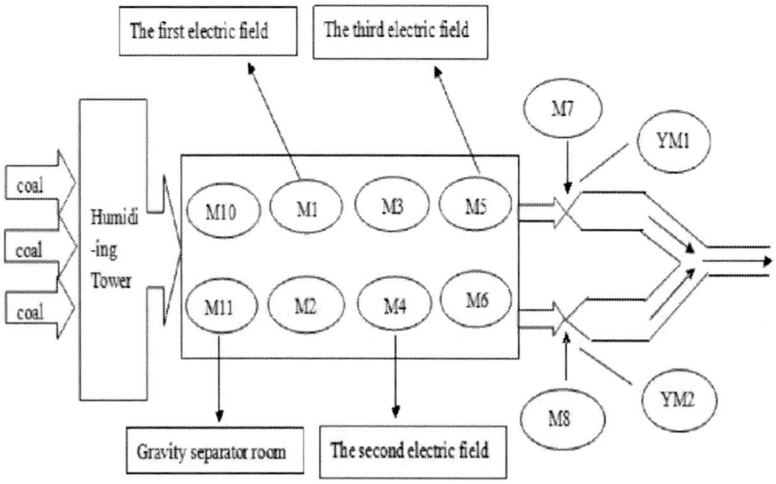

Figure 2. Schematic diagram of vibration device.

2.1.1. Control Circuit

The equipment has a gravity separator room and 3 electric fields. The applied voltage is fixed at 75 kV. As shown in figure 2, M1, M3 and M5 are motors which are at the anode, while M2, M4 and M6 are motors which are at the cathode. M10 and M11 are motors which are on both sides of gravity separator room. M7 and M8 are motors which control YM1 and YM2 that are electromagnetic valves [5]. Through motor's forward and inversion, M7 and M8 enable YM1 and YM2 to be opened and closed.

In order to prevent dust flying again, both of two motors in the same electrical field and the adjacent motors located at the adjacent electric fields can not work at the same time.

The timing design of gravity separator chamber: When the signal of starting arrives, M10 works immediately. It works for 3 minutes, and stops for 7 minutes; next, M11 stops for 7 minutes, and then works for 3 minutes. The first cycle is over. The after cycle is triggered by timer [6].

The timing design of the 1st electric field: When the signal of starting arrives, M1 also works immediately. It works for 2 minutes, and then stops for 8 minutes. M2 vibros cathode, stops for 8 minutes, then works for 2 minutes. The first cycle is over. The after cycle is triggered by timer.

The timing design of the 2nd electric field: When the signal of timer triggers, the first vibration of M3 starts. It stops for 18 minutes, then works for 2 minutes. M4 vibros cathode, works for 2 minutes, and then stops for 18 minutes. The first cycle is over. The after cycle is triggered by timer.

The timing design of the 3rd electric field: When the signal of timer triggers, the first vibration of M5 starts. It works for 2 minutes, and then stops for 28 minutes. M6 vibros cathode, stops for 28 minutes, then works for 2 minutes. The first cycle is over. The after cycle is triggered by timer [7].

When M5 and M6 work, the dust may be taken away by the airflow. To get a better efficiency of dedusting, when M5 works, the vavle of YM1 shuts off; YM1 opens after M5 stops for 2 minutes. In the case of M6 works, the vavle of YM2 shuts off; then YM2 opens after M6 stops for 2 minutes.

2.1.2. The design of wiring

According to the control requirements, this system chooses TWDLCAE40DRF type of PLC. The wiring diagram is shown in figure 3.

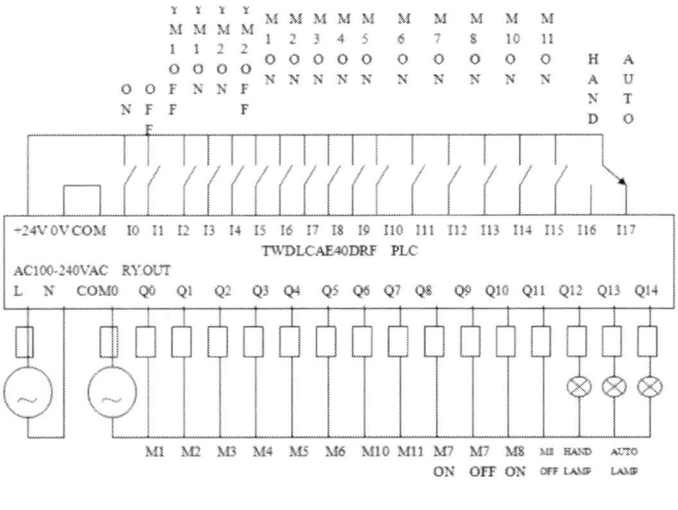

Figure 3. The wiring diagram of I/ O for PLC.

2.2. The design of program

When the switch of system is turned to automatic switch (I17 is on), system will work according to the settled time. However, when the switch of system is turned to hand switch (I16 is on), the system is controlled by hand switch. At the same time, PLC don not work.

As shown in figure 4, if the button of start (I0) is on, M0 switches to open. System carries out 5 subroutines. They are L0, L1, L2, L3 and L4.

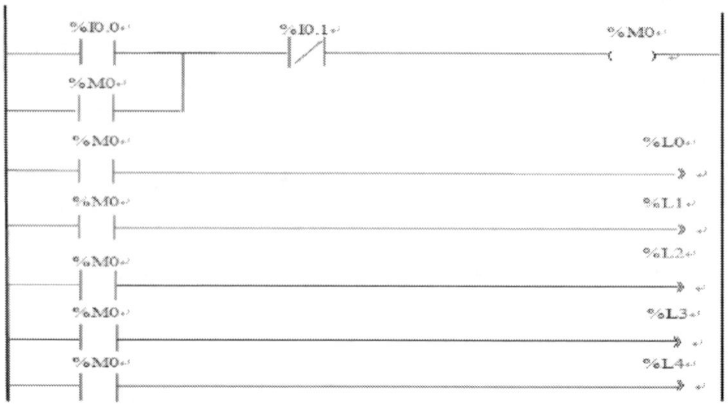

Figure 4. The main program of Vibration.

As shown in figure 5, the state of M0 switches to open when the start button is pressed. At the same time, Q0.6 is flicked to "on", M10 starts the first vibration. The timer (TM0) starts to woks for 2 minutes. When the time is up, Q0.6 is reset [8], and then M1 stops working. Meanwhile, Q0.7 switches to "on", M11 starts to work, and the timer (TM1) works for 2 minutes; when the time is up, the timer (TM2) starts to work for 6 minutes, M1 woks again. The first cycle is over. The programs of three electric fields are similar with the gravity separator chamber, so we do not repeat any more [9].

As shown in figure 6, when M5 woks, M7 turns forward and YM1 is closed; after M5 works for 2 minutes and delay for 2 minutes, M7 starts to inverse, then open YM1 to discharge the gas; when M6 works, M8 turns forward and YM2 is closed. After M6 works for 2 minutes and delays for 2 minutes, M8 starts to inverse then open YM2 to discharge the gas [10].

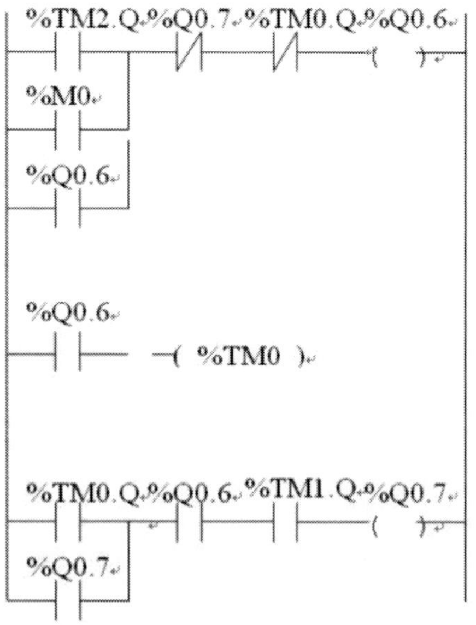

Figure 5. Program of settlement room wall.

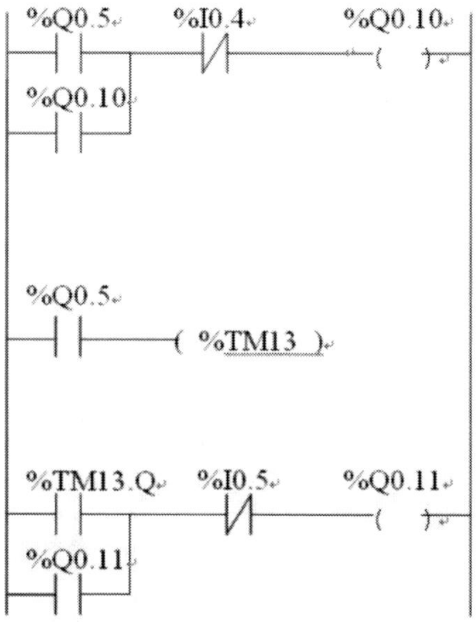

Figure 6. Program of YM1 and YM2.

3. Conclusion

The innovation point of the paper is substituting PLC for single-chip microcomputer control vibration of motor in the electric field of ESP, to improve stability and reliability. One the one hand, this design optimizes original vibration control system of electrostatic precipitator (ESP), makes the gas go through the humidifying tower to reduce resistance, and then go into the ESP. On the other hand, adding vibration of gravity separator room to improve dust removal efficiency. The operation's result shows that the design is reasonable and effective, has strong anti-interference and reliability. The design can meet the National effluent standard which is no more than 50 mg m^{-3}

Acknowledgments
This paper is supported by the National Science Foundation of China (51077035), the science and technology research projects of Colleges and Universities in Hebei province (2011166).

References
[1] Bai M Y, Yi C W and Yang B 2007 *Environ. Eng. (in Chinese)* **8** 15
[2] Mo X H and Guo H Q 2006 *Power Autom. Equip. (in Chinese)* **10** 81
[3] Zhu C J, Ou M S and Liang Z 2009 *Environ. Eng. (in Chinese)* **11** 2095
[4] Liu X Q 2002 *North Chin. Electr. Power (in Chinese)* **10** 46
[5] Zhang Z S 2002 *Hebei Univ. (in Chinese)* **15** 70
[6] Xiong T Z and Sun C Z 2010 *Micro Comput. Inf. (in Chinese)* **1** 61
[7] Giere R, Carleton L E and Lumpkin G R 2003 *Am. Mineral.* **88** 1853
[8] Ye Z,Yang J K,Tai G L and Pratim B 2000 *J. Electrostat. (in Chinese)* **4** 48
[9] Gallimberti I 1998 *J. Electrostat. (in Chinese)* **9** 43
[10] Pavageau M P, Pecheyran C, Krupp E M, Morin A and Donard O F X 2002 *Environ.Sci Technol.* **36** 1561

Separation of photosensitive substances in humic acids using molecular imprinting method based on electrostatic interactions and hydrogen bond

Xiaoxia Ou[1*], Chunyan Yu[2], Chong Wang[1], Fengjie Zhang[1]
[1]College of Environment and Resource, Dalian Nationalities University, Dalian, China
[2] National Marine Environmental Monitoring Center, Dalian, China

E-mail: ouxiaoxia@dlnu.edu.cn

Abstract. The goal of this research was to provide an improved understanding of the existence of photosensitive structure in humic acids (HAs) that are extracted from Changbai Mountain soils. Molecular imprinting technique was used to separate phthalocyanine-like substances from HAs with the mechanisms of electrostatic interactions and intermolecular hydrogen bond. Copper phthalocyanine (CuPc) was used as template molecule and the fraction bound by CuPc-imprinted polymers (MIP) named F (bind) presented higher spectroscopic activity than that of effluent fraction named F (eff). The fluorescence intensity of F(bind) at emission wavelength of 462 nm was 5.5 times as high as that of F(eff) at 458 nm, and the UV-vis absorbance at 254 nm of F(bind) had been increased to 2.5 times as compared with F(eff). The results of this work show the key role of humic substances with special structures in the light or photo involved process.

1. Introduction

Humic acids (HAs) are one of the important components in natural environment. It has generally been accepted that the photodegradation of organic pollutants were affected significantly by the presence of HAs since these processes such as absorbing light by HAs were always accompanied by the formation and consumption of reactive oxygen species (ROS, e.g. •OH, $^1O_2$, and ROO•).

However, previous studies have shown that HAs often control many photochemical reactions of pollutants in two ways. In brief, one theory assumed the presence of HAs could facilitate the photochemical degradation of organic compounds, which can be attributed to the excited triplet states ($^3HA^*$) and ROS produced through irradiation of HAs [1, 2]. The second theory assumed HAs could inhibit the photolysis of organic pollutants due to the competitive absorption for available photons and the hydrophobic binding interaction between pollutants and humic materials [3, 4]. These results reveal that different compositions of HAs display different photochemical behavior and photochemistry of HAs is closely related to their structures [5]. Therefore, to evaluate the contribution of different structures in HAs to the sensitized photodegradation of pollutants, recognition and separation of the substances with specific structure from HAs that are related to certain photochemical behavior become crucial [5, 6].

Molecular imprinting technique, widely applied in separation, synthesis, and catalysis, is usually created by imprinting template molecules in a rigid porous polymers matrix through the non-

[*] To whom any correspondence should be addressed.

covalent/covalent interactions between template molecule and functional monomer [7, 8]. Molecular imprinting technique is beneficial for the separation of molecules with the photosensitive structure from HAs. Chlorophyll, the key component during photosynthesis of plants, might absorb the sunlight and result in a series of electron transfer due to the presence of photosensitive structure, and would remain in humic substances through humification after plants' life cycle finishes [9]. In this paper, copper phthalocyanine (CuPc) is selected as the analogue of chlorophyll due to their similar photochemical property and the chemical labilization of chlorophyll, and then CuPc is served as the template molecule during molecular imprinting.

The study aimed to contribute to a better understanding of photosensitive structure in humic substances by separating HAs components using the molecular imprinting technique. The spectroscopic techniques was used to characterize the different components that constitute the major fraction of HAs extracted from Changbai Mountain soils, making it possible to elucidate the relationship between photochemical properties and the structure of humic substances.

2. Materials and methods

2.1. Soil samples and humic acids analysis

HAs were extracted from broadleaf forest soil samples by the International Humic Substances Society (IHSS) methodology [10]. The sample sites were situated in Changbai Mountain, Jilin province, China. The HAs powder was dissolved in Milli-Q water that was adjusted by NaOH to pH 7.5–8.0, and the HAs concentrations were measured by a total organic carbon analyzer (TOC-V_{CPH}, Shimadzu, Japan).

2.2. Standards and reagents

All chemicals were analytical reagent grade. α-Methacrylic acid (MAA) and 2,2′-azobis (isobutyronitrile) (AIBN) were purchased from Fuchen Chemistry Reagent Co. (Tianjin, China). Ethylene glycol dimethacrylate (EGDMA) was supplied by Alfa Aesar, A Johnson Matthey Co. (U.S.). Copper phthalocyanine (CuPc, purity⩾98%) was obtained from Tokyo Chemical Industry Co., Ltd. (Japan). Oxalic acid, citric acid, glucose, and sucrose were obtained from Kemiou Chemistry Reagent Co. (Tianjin, China). Milli-Q water (resistivity ≥ 18 MΩ•cm) was used in all experiments.

2.3. Characterization

UV-vis spectra were measured using a spectrophotometer (Jasco UV-550, Japan). The fluorescence was measured using a fluorescence spectrophotometer (RF5301-PC, Shimadzu,, Japan) with the excitation wavelength at 350 nm and the emission wavelength ranged from 360 nm to 600 nm (10 nm bandwidth). The morphology of samples was observed using scanning electron microscope (SEM, Hitachi S-4800). FTIR spectra were recorded on an FTIR spectrometer (Prestige-21, Shimadzu, Japan).

2.4. Preparation of MIP

CuPc was used as the template molecule to prepare molecularly imprinted polymers (MIP). The functional monomer, cross-linking agent, and initiator are MAA, EGDMA, and AIBN, respectively. The preparation process of MIP was described in reference 11. Simply, 1 mmol CuPc was mixed with 4 mmol MAA, 20 mmol EGDMA, and 40 mg AIBN in a glass tube containing 12 mL chloroform. The mixture solution was sonicated for 5 min, flushed with nitrogen gas for 15 min, and then polymerized by heating at 60 °C in thermostatic water bath for 24 h. The obtained MIP was removed the template molecules by washing with methanol/acetic acid (9:1, v/v) solution. Nonimprinted polymers (NIP) were prepared in the same manner only without CuPc template. MIP and NIP after elution were dried and used for further experiments.

2.5. MIP-SPE cartridges preparation and fractionating procedure of HAs by MIP

CuPc-MIP (0.25g) was placed into empty polytetrafluoroethylene (PTFE) solid phase extraction (SPE) cartridges (6 mL) to prepare MIP-SPE cartridges. The fractionating procedure is that HAs aqueous solution (20 mL, 5 mg L^{-1}) passed through CuPc-MIP cartridges at a flow rate of 0.5 mL min^{-1}, effluent fractions were collected and named F(eff), and the bound fractions were washed using 60 mL of fresh Milli-Q water, and then collected and named F(bind). All fractions were concentrated by evaporation at 60 °C in a thermostatic water bath, and used for further experiments.

3. Results and Discussion

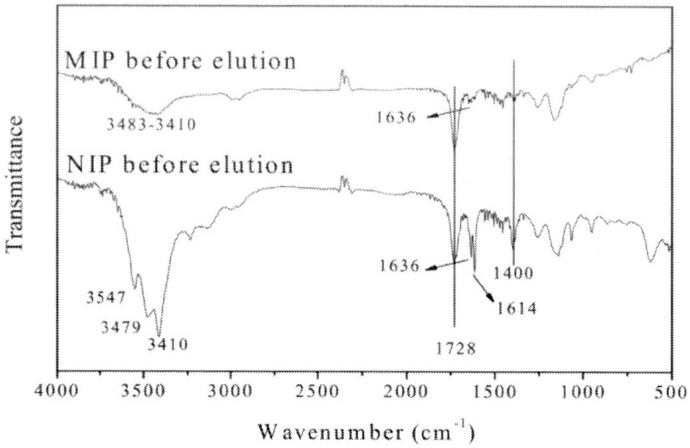

Figure 1. FTIR spectra of MIP and NIP before elution.

3.1. FTIR characterization of polymers

FTIR spectra of polymers (MIP and NIP) were recorded before template molecule CuPc removal to investigate intermolecular interactions between template molecule CuPc and functional monomer MAA. Figure 1 gave the FTIR spectra of MIP and NIP before elution. The changes of MIP compared with NIP were as follows: (i) decreasing in the intensity of ν(–OH) at 3400–3600 cm^{-1}; (ii) decreasing the bands assigned to the asymmetric and symmetric –COOH stretching at 1636 cm^{-1}, 1614 cm^{-1} and 1400 cm^{-1} [6]. Furthermore, the band at 1728 cm^{-1} assigned to C=O stretching of carbonyl and carboxyl had the same intensity for MIP and NIP, which may be due to the rich carbonyl in EGDMA. These observations indicated that the interaction between CuPc and MAA was possible contributed by –OH···N hydrogen bonds and electrostatic attraction resulted from the interaction between protonated nitrogen and ionized carboxyl [7, 11].

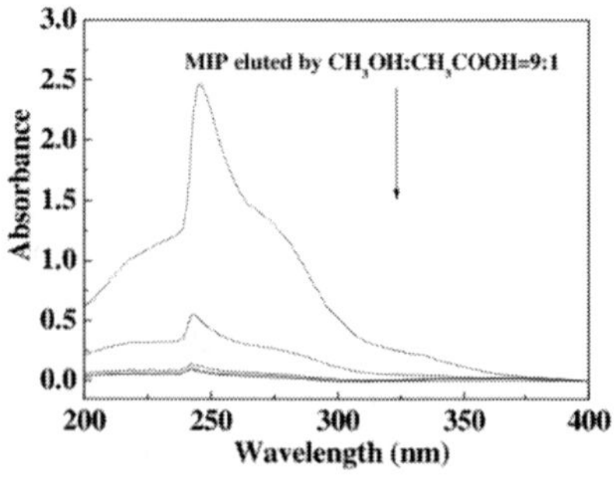

Figure 2. UV-vis spectra of MIP eluted by $CH_3OH:CH_3COOH$ (9:1).

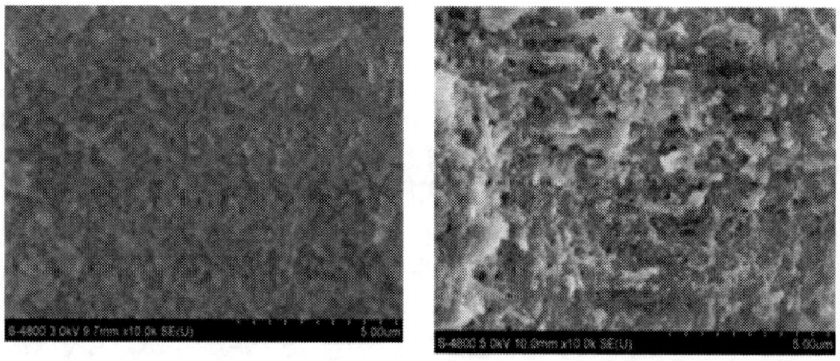

(a) (b)

Figure 3. SEM image of MIP (a) before and (b) after elution.

3.2. Elution of MIP and SEM characterization of polymers

The obtained MIP was removed the CuPc molecule by washing with methanol/acetic acid (9:1, v/v) solution until no template molecules were detected in the effluents with the UV-vis measurement. It can be seen from figure 2 that the mixture of methanol and acetic acid (9:1) was effective to remove the template molecule from MIP. The surface topography of MIP before and after elution was observed by SEM, which was shown in figure 3a and b. It is clear that both MIP before and after elution are obviously network materials with microporous structures. The surfaces of the MIP before elution appear to be smooth and porous, whereas the SEM micrographs of MIP after removal of template produced rough surfaces and more pores of different sizes. The result provided additional evidence for the efficient removal of template molecules with organic solvent.

3.2. Selectivity of MIP

To investigate the selective binding of Pc-like substances from HAs for the eluted MIP, some elementary structures in HAs, such as oxalic acid, citric acid, glucose, and sucrose, were used for binding tests onto MIP (NIP)-SPE. The chemical structures of these substances may be form hydrogen bonds and have electrostatic interactions with MIP because of the presence of carboxyl or hydroxyl groups in these substances. The concentrations of original solutions and effluents from MIP (NIP)-SPE cartridges were determined by UV-vis spectrophotometer. Figure 4 showed that differences of absorbance from 200 nm to 400 nm between original solutions and effluents for four organic matters were negligible, indicating that these substances did not interact with the eluted MIP or NIP according to hydrogen bonds or electrostatic interactions. Thus the four substances did not effectively bind onto MIP and the prepared MIP possessed high selectivity.

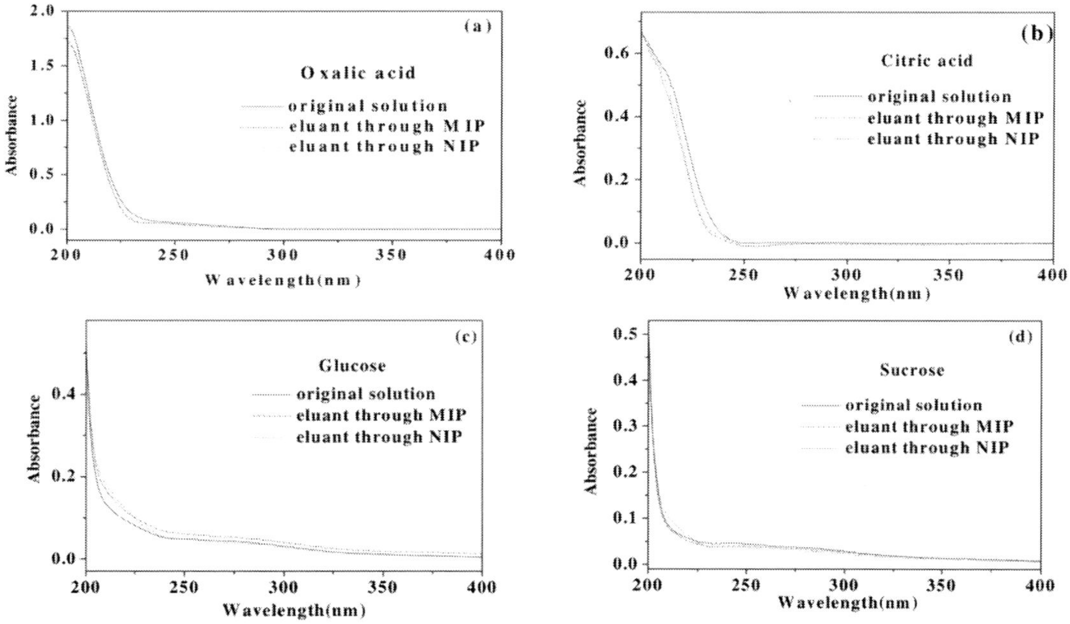

Figure 4. UV-vis spectra of original solutions and effluents through MIP and NIP for oxalic acid, citric acid, glucose, and sucrose.

3.3. Fluorescence and UV-vis spectra of HAs fractions

Figure 5 showed the synchronous fluorescence spectra of HAs fractions recorded upon excitation at 350 nm. Compared with F(eff), the spectrum of F(bind) presented some differences as follows: (i) fluorescence intensities of F(bind) were obviously more intense than that of F(eff) and maximum intensity of F(bind) was 5.5 times as high as that of F(eff); (ii) the emission bands of F(bind) was broad and that of F(bind) was narrow; (iii) the band with maximum intensity shifted from 458 nm for F(eff) to 462 nm for F(bind). These phenomena showed that the structures of F(eff) and F(bind) were indeed different and the fraction with similar structure of Pc such as extended π-conjugated ring structures had more fluorophores. This result is of interest for better understanding the relationship between photochemical activities and the structure of HAs in respect that the substances with more fluorophores could display better photoinductive activity [12].

The UV-vis absorption spectra of HAs fractions were presented in figure 6. The spectra of F(eff) and F(bind) exhibited a featureless increase in absorbance during the wavelength decreasing and were dominated by broad humps at 254 nm. The shoulder at 254 nm is attributed to aromaticity structures or $\pi-\pi^*$ electron transitions of unsaturated bonds [13]. The UV-vis absorbance at 254 nm of F(bind) had been increased to 2.5 times as compared with F(eff), indicating that F(bind) possessed the better electron transfer capacity. It was clear that humic substances with special structures (e.g. Pc) would present the key role in the light or photo involved process.

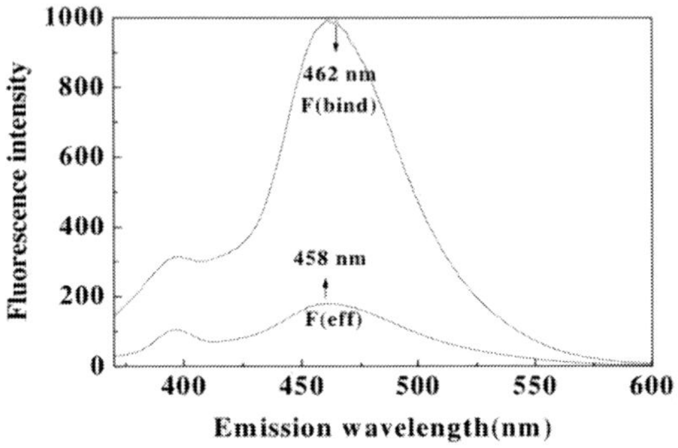

Figure 5. Fluorescence emission spectra of HAs fractions (5 mg L^{-1}).

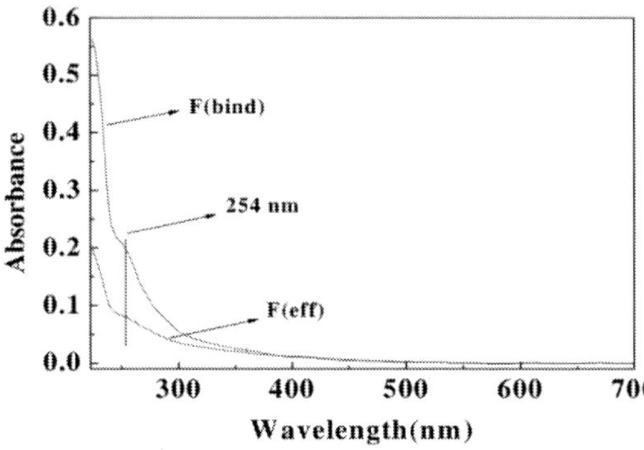

Figure 6. UV-vis absorption spectra of HAs fractions (5 mg L^{-1}).

4. Conclusion

The separation of HAs extracted from Changbai Mountain soils can be achieved well with molecular imprinting method. The functional monomer MAA can interact with N atoms of template molecules CuPc according to hydrogen bonds and electrostatic interactions. Based on the spectroscopic characterization of different HAs fractions by synchronous fluorescence and UV–vis, the fluorescence intensity and UV-vis absorbance of HAs frations binding on CuPc-MIP were both much higher as compared with effluent fractions when HAs solutions passed through CuPc-MIP SPE cartridges. The results provide an improved understanding of the existence of photosensitive structure with better spectroscopic properties in humic material.

Acknowledgments

The study was supported by the Key Laboratory of Industrial Ecology and Environmental Engineering, China Ministry of Education (China, KLIEEE-10-04) and the National Natural Science Foundation (China, No. 20907006).

References

[1] Canonica S, Jans U, Stemmler K and Hoigné J 1995 *Environ. Sci. Technol* **29** 1822
[2] Gerecke A C, Canonica S, Muller S R, Scharer M and Schwarzenbach R P 2001 *Environ. Sci. Technol* **35** 3915
[3] Torrents A, Anderson B G, Bilboulian S, Johnson W E and Hapeman C J 1997 *Environ. Sci. Technol* **31** 1476
[4] Bachman J and Patterson H H 1999 *Environ. Sci. Technol* **33** 874
[5] Yu C, Zhang Y, Quan X, Chen S, Han J, Ou X and Zhao J 2010 *Environ. Sci. Technol* **44** 5812

[6] Ou X, Chen S, Quan X and Zhao H 2008 *Chemosphere* **72** 925
[7] Gao D M, Zhang Z P, Wu M H, Xie C G and Guan G J 2007 *J. Am. Chem. Soc* **129** 7859
[8] Alexander C, Davidson L and Hayes W 2003 *Tetrahedron* **59** 2025
[9] Chen J, Gu B H, LeBoeuf E J, Pan H J and Dai S 2002 *Chemosphere* **48** 59
[10] Swift R S 1996 Organic matter characterization *Methods of Soil Analysis* (Madison: Soil Science Society of America) pp 1018–1020
[11] Yu C, Chen S, Quan X, Ou X and Zhang Y 2009 *J. Agr. Food Chem* **57** 6927
[12] Richard C, Trubetskaya O and Trubetskoj O 2004 *Environ. Sci. Technol* **38** 2052
[13] Zhang H, Peng Q and Li Y 2005 *Modern Spectral Analysis of Organic Matters* (Beijing: Chemical Industry Press) pp 228

The numerical simulation of air distribution in the hybrid electrostatic-fabric precipitator

Minkang Tang*, Ping Yang, Haiping Guo
School of resources and environmental engineering of Jiangxi University of Science and Technology

Email: tmkjxust@126.com

Abstract: In this paper, the computational fluid dynamics software is adopted to simulate the fluid field of the cylindrical hybrid electrostatic-fabric precipitator, by doing that distributions of static pressure and flow velocity are gained. By increasing the opening rate of the internal circular collecting plate, the uniformity of the air can be improved, and then RMS identifying method is used to judge the air homogeneity. Finally, the basic characteristics of the toroidal electric field are gained, all of these provide a reference to design and improve the function of hybrid electrostatic-fabric precipitator.

1. Introduction

With the development of society, the environmental issue which has a great importance of mankind survival attracts an improving attention. The hybrid electrostatic fabric filter, an effective control of fine dust emission, has a promising future [1, 2]. The dust stream in electrostatic fabric filter is a complex three-dimensional flow. It is very difficult to make out the internal flow field with the conventional test precipitator. The distribution of internal airflow in the dust removal device can be analysed by computer simulation, and numerical simulation methods have the advantage of reproducibility when compared with the experimental research methods. Numerical simulation technology attracts more and more attention, while it is applied widely [3]. The purpose of this thesis is to deepen people's understanding of the electric bag filter performance in theory, intensify the effect of electric bag filter dust through the guidance of the CFD simulation and optimize the structure of the electrostatic fabric filter.

2. Model of the hybrid Electrostatic-Bag precipitator

The laboratory homemade hybrid electrostatic-bag precipitator by Jiangxi University of Science and Technology is used as the physical prototype, and the whole calculation model of hybrid electrostatic-bag precipitator is built by the Gambit software.

2.1. The physical prototype

The Laboratory homemade hybrid electrostatic-bag precipitator includes two parts, the electricity dust removal and bag dust removal, and both parameters of the parts are shown in table1.

* To whom any correspondence should be addressed.

2.2 Calculation Model

2.2.1 Simplify the Physical prototype

The main function of the diversion holes on the circular plate is to channel the air of the annular channels to bag dust removal area even, the arrangement of diversion hole is shown in figure 1. As there are many diversion holes, fine grid is needed in the precise computation. However, because the computer capacity is limited and direct calculation is very difficult, the holes can be equivalent to the gap strip in the case of the opening rate unchanged.

Table 1. Design value of dust removal and the bag.

Design value of dust removal		Design value of the bag	
Radii of external circular plate	560 mm	Bag number	7
Radii of the inner circular plate	280 mm	Length of the bag	800 mm
Corona electrode number	12	Center distance of the bag	50 mm
Rows of the diversion holes	4 rows	Distance of the outlaying bags and inner circular plate	50 mm
Radii of the big diversion holes	40 mm		
Radii of the little diversion holes	25 mm		
Distance of the inner and external circular plates	280 mm		

Table 2. Operating parameters of the dust removal.

Name	Data
Air volume need handling	$Q = 2600\ m^3\ s^{-1}$
Cross-sectional area of the electric field	$S = 280\ mm \times 800\ mm = 0.224\ m^2$
Velocity of the electric field	$v = \frac{Q}{S} \times \frac{1}{2} = 1.6\ m\ s^{-1}$

The 3D model is built according to the structure characteristics of the hybrid electrostatic-bag precipitator, the geometry model can be divided into several parts: inlet, the electric field, bag dusting space, bag, internal circular collecting plate, and pure air box and ash hopper [5]. Local mesh division method is used to deal with the model, the bags and the bag space grids can be structural grids, the rest can be the unstructured grids (shown in figure 2 and figure 3).

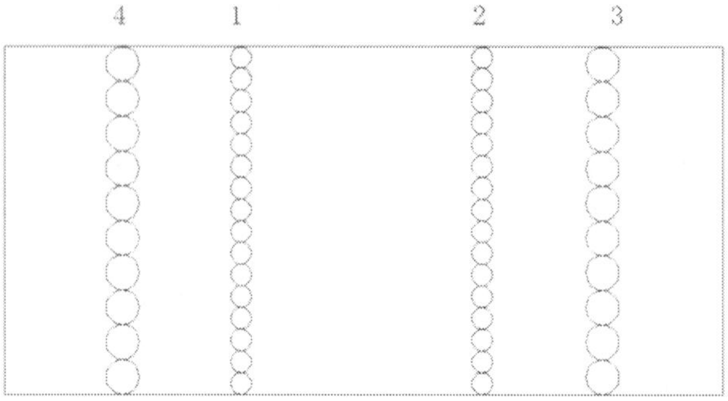

Figure 1. Pareto diagram of the diversion holes on the inner circular plate.

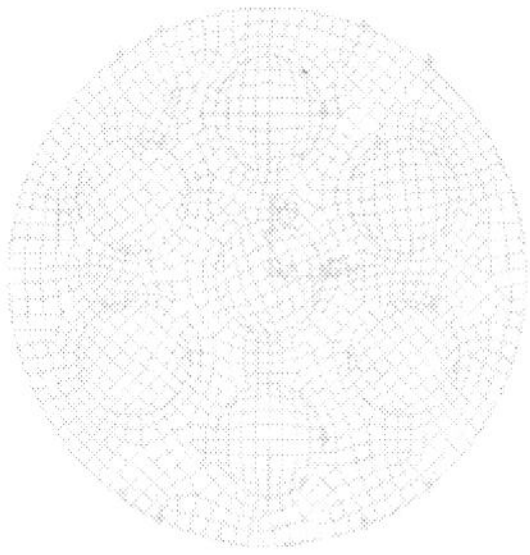

Figure 2. Bags and the grids.

Figure 3. The whole grids.

2.2.2 Definite the boundary conditions

- Boundary conditions of air inlet;

Velocity is the condition of inlet boundary. Define velocity and scalar of the in the flow inlet boundary. Suppose the inlet gas of the model pipeline is a uniform fully developed turbulence and the inlet velocity is constant. Then the inlet turbulent kinetic energy and turbulent dissipation conditions are respectively:

$$k = \frac{3}{2}(\bar{u}I)^2 \tag{1}$$

$$\varepsilon = C_\mu^{3/4} \frac{k^{3/2}}{l} \tag{2}$$

$$I = 0.16(\mathrm{Re}_{D_H})^{-1/8} \tag{3}$$

Where: I is the Turbulence intensity; $\bar{u}$ is average velocity of the entrance (m / s); l is the turbulence (Turbulence length scale), l=0.07 L (m); C_u is Constant and $C_u = 0.09$. Re_{DH} is the Reynolds number calculated by hydraulic radius (D_H); L is the associated scale; for the fully developed turbulent flow, L is generally the hydraulic diameter.

- Boundary conditions of outlet;

Pressure condition of the pipe outlet is the boundary condition of the outlet.

- Filter bag.

Boundary condition of the filter bag is porous jumping. In this paper, the hybrid filter bag is at the beginning, the coated filter bag is clean and air permeability is good, too.

- Boundary conditions of wall

The wall is stationary, and is at the zero-gradient condition, the standard function of wall is adopted by the wall.

- Gas phase parameters

No compressed air, at normal temperature.

3. Results and analysis of the simulation

3.1. Simulation of the convergence condition

Two-equation model RNG and SIMPLE algorithm is used for the simulation. Second-order upwind is used for the flow differential.

Usually, there are two meanings for the convergence calculation. One is the approaching trend of the exact solution of the solutions of differential equation; the other is the approximation of the solution of algebraic equations to the true solution of the set of coefficients. Ether in Iterative solution of nonlinear problems or algebraic equations, iterative calculation is considered convergent if the deviation of the consecutive solutions or the relative deviation of each node is less than a set allowed values.

In the calculation process of this paper, the calculation convergence is monitored by the checking residual changes of each variable dynamic. There are 500 iterations, iteration will stop at about 10-4 middleweight convergence, after 500 iterations, variables will meet the convergence conditions at the 851st iteration. Stop calculating, the Fluent console prompt solution is converged [6].

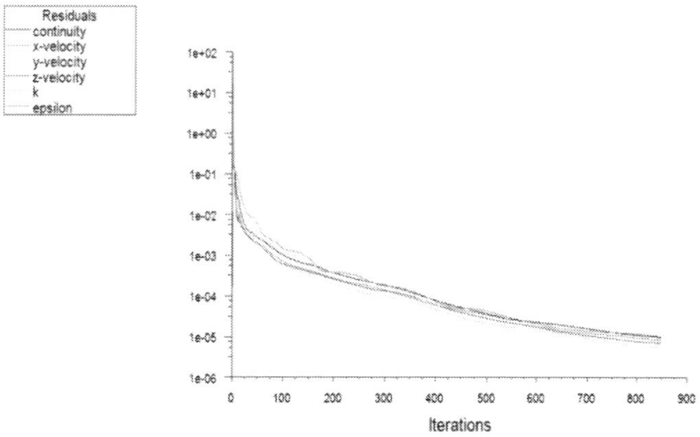

Figure 4. The residual curve.

3.2. Results of the model simulation

To analyze the internal flow of the hybrid electric bag, first of all, key section of the specific location in the dust space should be selected, and the measurement points of the critical section should be layout too. Analyze the airflow distribution of the whole equipment by observing the velocity distribution of the critical section shown in figure 5, the values of the key section are shown in table 4.

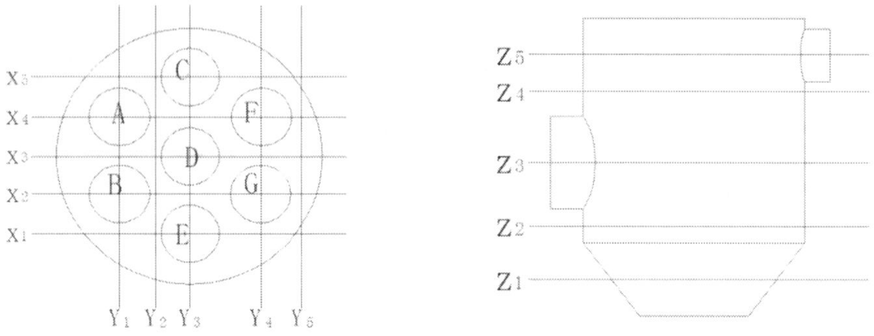

Figure 5. Schematic diagram of the calculation section and the bags.

Table 3. Value of the key section of X, Y, Z direction.

X direction	Y direction	Z direction
X_1=-170	Y_1=-147.22	Z_1=-1000
X_2=-85	Y_2=-73.6	Z_2=-700
X_3=0	Y_3=0	Z_3=-400
X_4=85	Y_4=73.6	Z_4=0
X_5=170	Y_5=147.22	Z_5=150

4. Analysis of the model simulation

4.1. Electric field

The analysis of speed distribution of electric field, bag dust removal area and hopper area were shown in figure 6-8. The electric field is unevenly distributed: the dust stream will move forward at a speed of 1.6 m s^{-1} after it passed the entry. The stream will divided into two streams quickly when it hit the inner circle collecting plate. One will spread quickly to both sides of inner circle plate. Due to a fast speed, the flow will enter the bag at a speed of 2 m s^{-1}, in the case of it passes the first and second hole. It can be seen from figure 8 that only little air can reach the back-end of the electric field where the field strength is weak. Another part of the airflow diffuses into the dust hopper from the posterior of the electric field. The air velocity of the posterior of the dust hopper is fast, it can be in the range of 1-1.5 m s^{-1}. The airflow will form a notable vortex. The vortex of the dust hopper seriously impacted the whole equipment. Deposited dust will be brought to the bag by the vortex. This will not only increase the dust concentration, but will also increase the filter load. What's more, it will cause erosion to the bag, and is not conducive to the operation. From figure 8 we know that the air enters the above box evenly and discharged from the outlet.

The flue gas velocity will reduce significantly after it enters the bag (from 1.6 m s^{-1} to about 0.4 ~ 0.6 m s^{-1}), however, airflow distribution is more uniform. Cross section Y1=-147.22, Y3=0 is the velocity distribution map of plane X-Z. Air velocity increases gradually along Z-axis of the bag. The middle map of figure 6 is the velocity distribution map of the bag clearance, the airflow rising along the bag clearance, the velocity is between 1and 1.5 m s^{-1}, and what's more, the velocity is relatively uniform. Such airflow can reduce secondary blowing dust in the bag room.

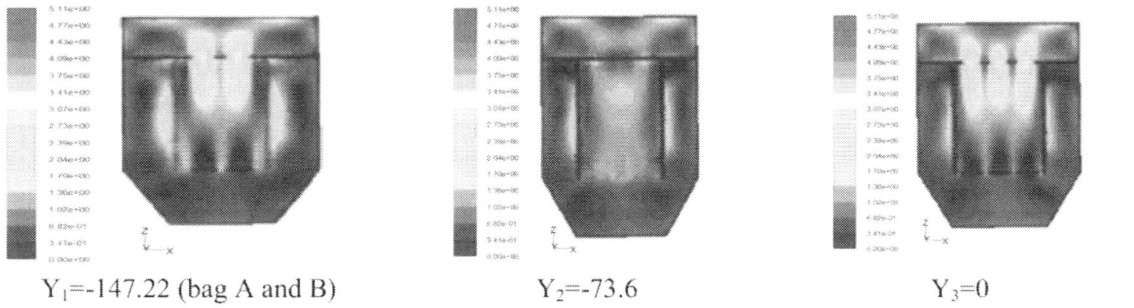

Figure 6. Velocity cloud of Y direction (section X-Z).

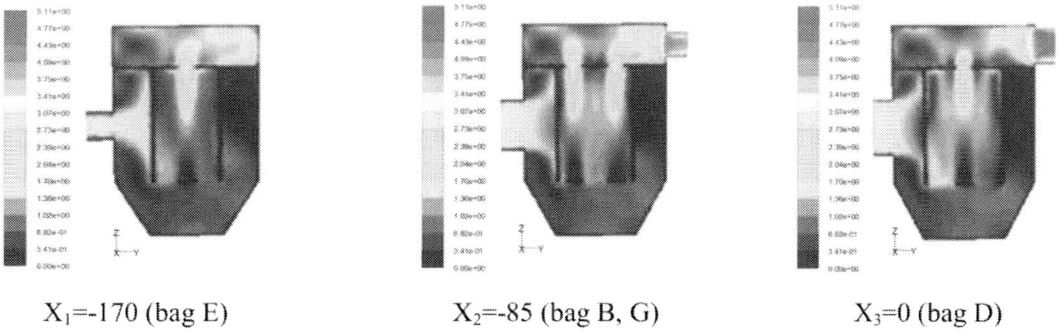

Figure 7. Velocity cloud of X direction (section Y-Z).

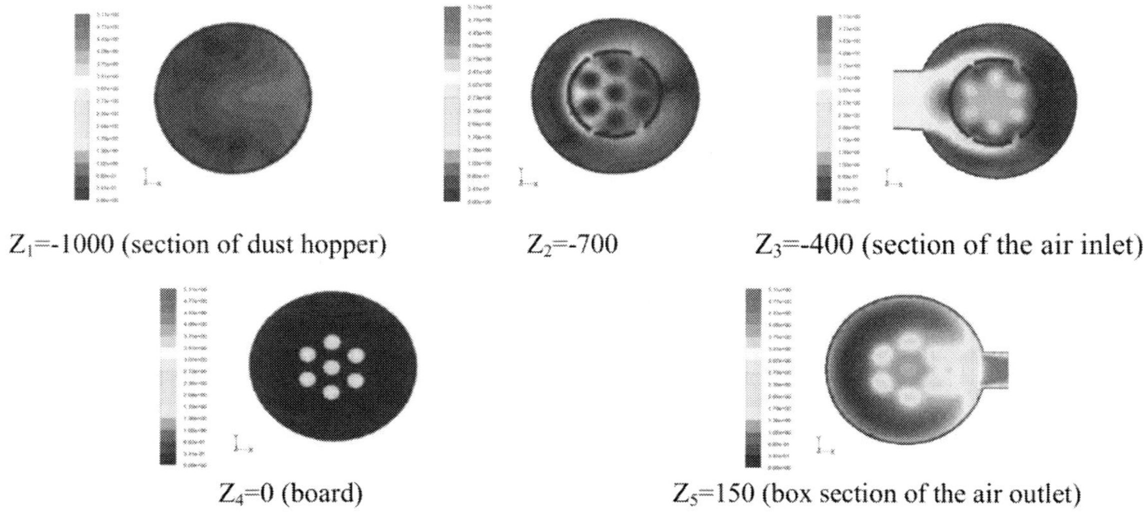

Figure 8. Velocity cloud of Z direction (section X-Y).

Nowadays, the most commonly used evaluation method is the RMS standards in the domestic. Generally, the air homogeneity of the electric field cut is judged, and the RMS standards are shown in table 4 [7].

Table 4. Criteria.

σ	σ≤ 0.10	σ≤ 0.15	σ≤ 0.25	else
Criteria	excellent	fine	qualified	unqualified

The relative RMS formula:

$$\sigma = \left(\frac{\sum_{i=1}^{n} \left(v_i - \bar{v}\right)^2}{n \times \bar{v}^{-2}} \right)^{\frac{1}{2}} \tag{4}$$

Here: n is the total number of the point; v_i is the flow velocity of the i point, m s^{-1}; $\bar{v}$ is the mean flow velocity of the point, m s^{-1}.

As shown in figure 9, at X=-150, that is, diversion holes 1 and 2 close to the air inlet, the maximum speed of airflow is about 2.5 m s^{-1}, which can be accounted for two aspects: firstly, the distance of the guiding holes are too small, thus there is jet effect when air enters the bag; secondly, there is energy loss due to the friction effect when air flows along the circle, which will make the air velocity near the diversion holes 3 and 4 be less than the two front rows.

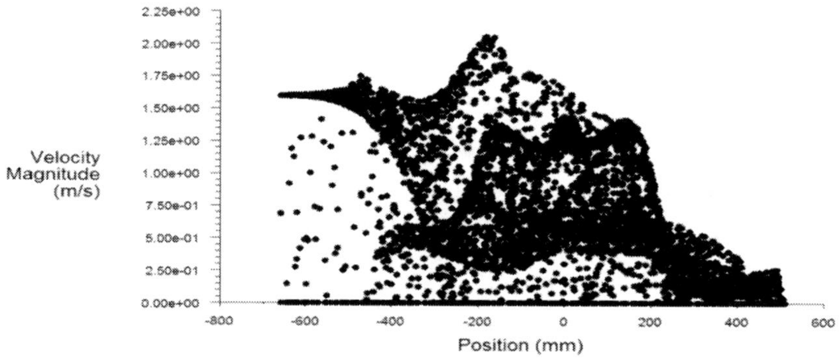

Figure 9. Velocity of the section where Z=-400.

4.2. The velocity distribution of the filter bag

As shown in figure 10 and 11, although the velocity of the filter bag concentrated between 0.5 and 1.7 m s^{-1}, the velocity distribution of the bags are not uniform. The velocity at the bottom of the bags is slow, and which is much faster at the top, particularly near the outlet. Local area is with a high speed, which has seriously impact on the service life of the filter bag.

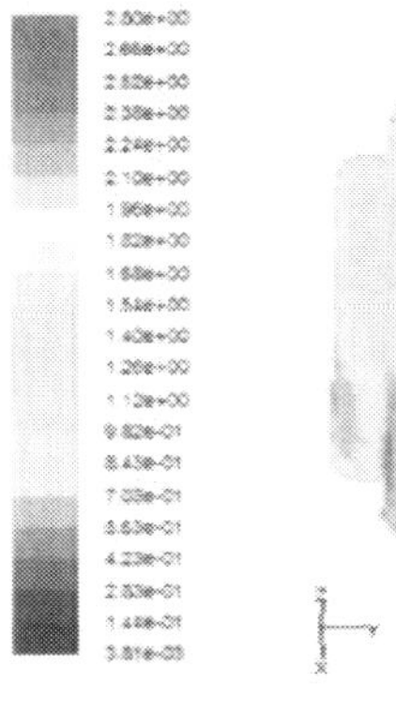

Figure 10. Velocity cloud chart.

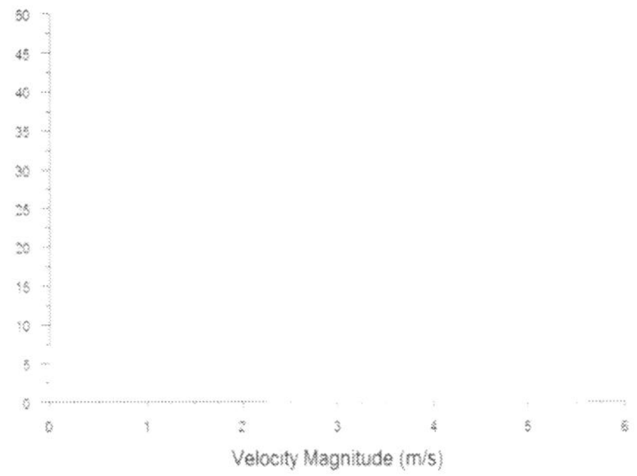

Figure11. Velocity distribution of the bag.

4.3. The velocity distribution of the ash hopper
It can be seen from figure 12 and 13 that airflow velocity is mainly distributed between 0.2 and 0.9 m s^{-1}, this is less than the design requirement. Even though scouring wear is small due to the slow flow speed, air speed along the ash hopper wall far away from the entrance is much greater than the one in the center of the dust hopper; the maximum speed is around 0.9 m s^{-1}.

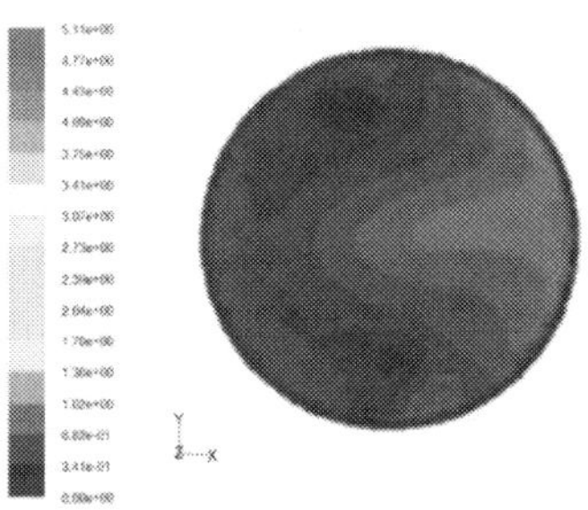

Figure 12. Velocity chart of the section of dust hopper where Z=-1000.

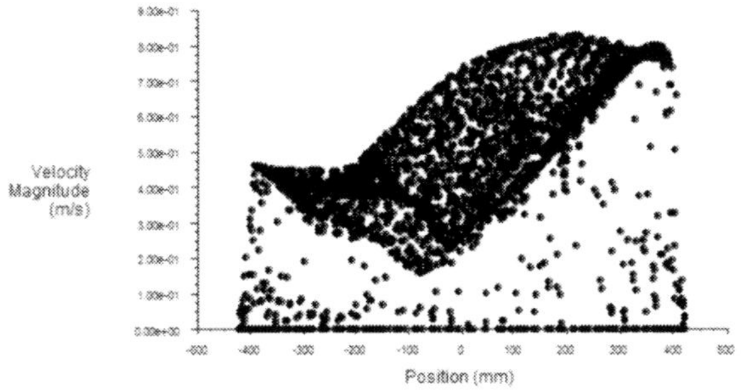

Figure 13. Velocity of the section where Z=-1000.

4.4. Optimize the model
There are 4 rows of diversion holes on the inner circle dust collection plate of the physical prototype, the collecting plate opening rate is relatively small, that is, area ratio of the effective collecting area of the inner circle collecting plate and the area of the diversion holes is large. And the larger the ratio is, the more effective the collecting area is, so is the ability of dust collecting of the electric field area. In addition, the speed of the electric field is 1.6 m s^{-1}, the opening rate is small, and the collecting plate can help to protect the bag from the air scour and the electric field of corona discharge.

However, analysis of the simulation results shows that due to the uneven distribution of the airflow which is at the back-end of the precipitator field area, a re-allocation of the airflow in this part is needed [8]. Uneven distribution of airflow is caused by the speed of the inlet and the unreasonable set up of the circular collecting plate deflector. Therefore, to distribute the airflow of the electric field uniformly, the airflow of the backend of the electric field can run into the dust bag evenly, improvement in two aspects is needed: (1) increasing the opening rate; (2) reducing the flow velocity.

4.4.1 Increase opening rate
After the equivalent processing of the internal circular plate, the diversion hole and dust collection

plate in of the inner circle are in the same height, so the opening rate equals to the ratio of the openings length and the circumference of the inner circle dust collection plate.

The opening rate of prototype is:

$$\zeta = \frac{80\text{mm} \times 2 + 50\text{mm} \times 2}{2\pi \times 280\text{mm}} \approx 15\% \tag{2}$$

Supposing the rate is 20 % after increasing the diversion hole. By the calculation, the required length of the openings is 91.68 mm. In order to simplify the design, length of the openings is 100 mm.

4.4.2 Reduce the flow velocity

Airflow in the bag is mainly composed of two parts, one entered the bag from the deflector holes on the inner collecting plate, and the other entered the bag from the bottom of the bag. These two parts of air develops the windward side speed and the updraft speed at the end of the bag. If the speed of the windward side is too fast, there will be serious wear and tear to outer part of the bag. The ESP is a small one, the original wind speed of the electric field (1.6 m s^{-1}) is a little fast. The definite velocity should be decided by the air that needs to deal with.

4.5. Optimizations

Optimization program one: improve precipitator internal structure on the basis of the original model, distribute the airflow by opening hole on the inner circle plate (see figure 14), and the effect is notable as shown in Figure 15. The improved equipment re reduced the load caused by run of air flow from the bottom of the bag to the dust removal area. It can be seen from Figure 16 that stream the bag mouth ran out upward vertically, the direction of the electric field point to the inner circle speed uniformly by the outer circle collecting plate.

Optimization program two: air diffusion area increases, the distribution around the inner circle board is uniform than the calculation of the original model. The filtration rate decreases along the Z axis. The turbulent flow decrease markedly. And the airflow of the bottom is gentle (see figure17).

Optimization program three: the velocity of inlet airflow is 0.4 m s^{-1}. This is not only being good to the airflow distribution of the whole equipment, the flow field is also smoother when compared with Option II (shown in figure 21).

After the simulation of re-meshing and iteration convergence, contrast of the flow field velocity distribution of the internal part of dust collector is shown in figure 15-18.

4.6. Comparison of optimization

Export the airflow velocity data of the key points of the electric field section where Z=-400 in Option II and Option III.

Export the key airflow data of section Z=-400 in Option II and Option III, the calculation of the data showed that the max air velocity of optimization two is 1.65 m s^{-1}, the min is 0.10 m s^{-1}, mean air flow rate is 0.472; relative RMS is 0.184656. In optimization three, the maximum flow velocity is 1.34 m s^{-1}, the minimum flow velocity is 0.06 m s^{-1}; mean air flow rate is 0.352; relative RMS is 0.094235. According to common standards of the current domestic electric field airflow uniformity judgment (see Table 4). Optimization three was adopted. That is: adjust the inlet flow rate is 0.4 m s^{-1}, increase the entrance width of electrostatic fabric filter circular plate to 100 mm.

4.7. Velocity distribution of the improved bag gap

Rising velocity of bag clearance mainly refers to the flow velocity paralleled to the bag direction (Z axis) of the dust stream. This velocity decided whether the bag can get a normal and efficient operation by pulse gray or not. The faster the rising velocity is, the less possible the dust will deposit on the surface of the filter. Though the dust deposit on the filter surface, they will move again when pulse jet clean the bag, then captured by the filter and generate a secondary adsorption, resulting in a difficult deashing [9, 10]. In order to study the variation of the equipment of the bag flow field, due to the special symmetry of the device, select the speed on Y_2=-73.6 section as the study object.

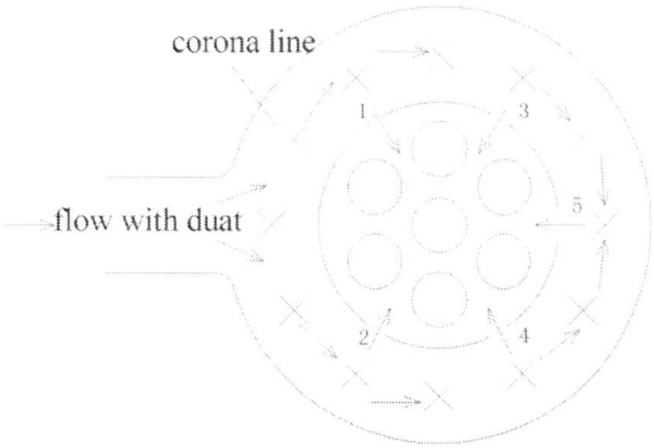

Figure 14. Distribution of the diversion holes.

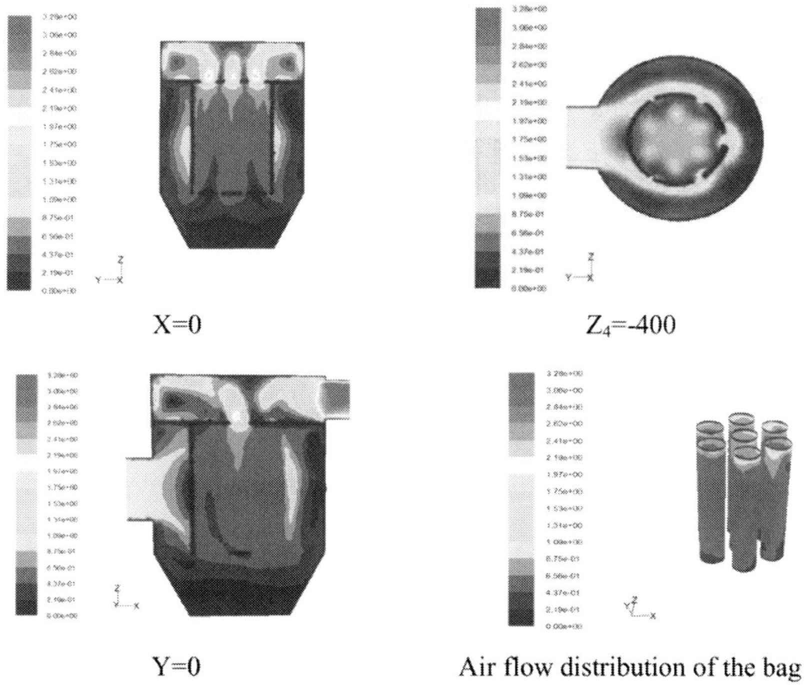

Figure 15. Section velocity of program one (v=1.0 m s^{-1}).

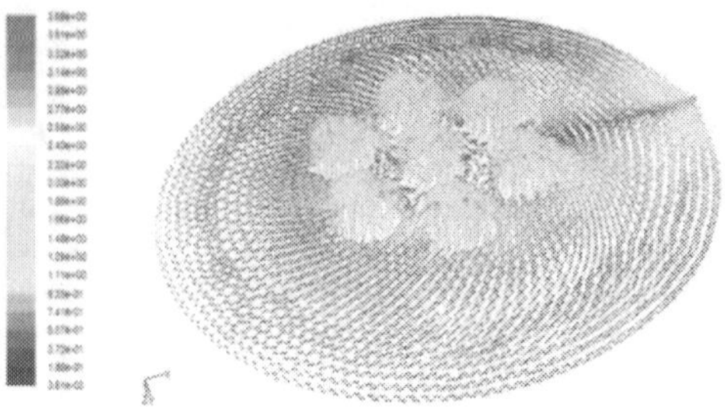

Figure 16. Vectogram of velocity where Z=0 of program one (v=1.0 m s^{-1}).

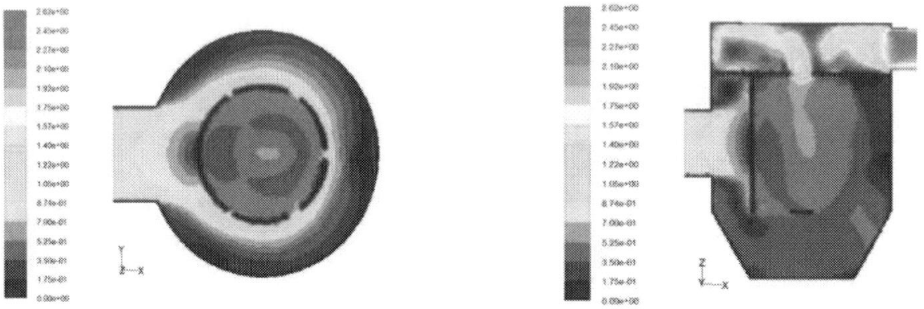

Figure 17. Section velocity of program two (v=0.8 m s^{-1}).

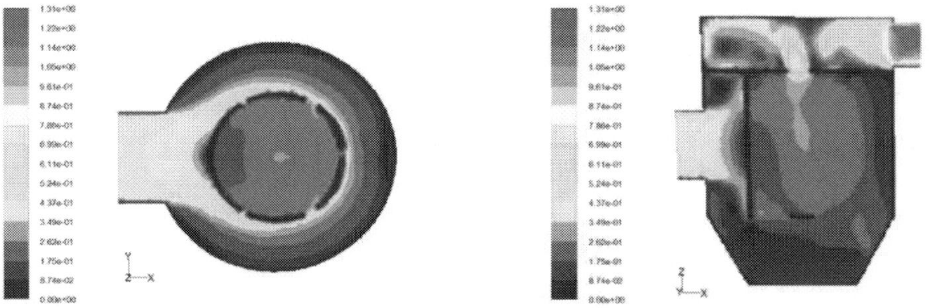

Figure 18. Section velocity of program three (v=0.4 m s^{-1}).

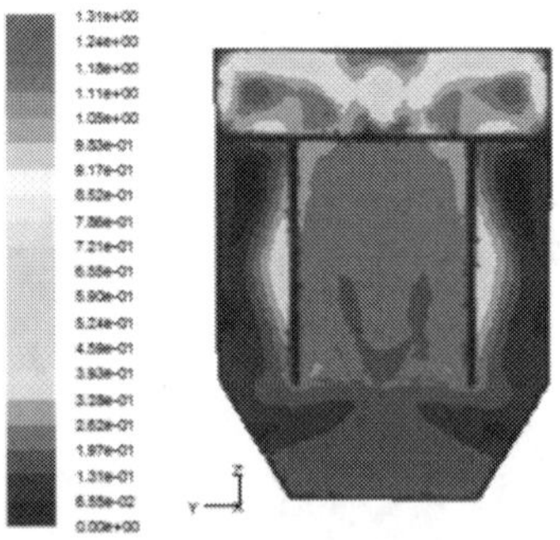

Figure 19. Velocity cloud of the section where Y_2=-73.6 (v=0.4 m s^{-1}).

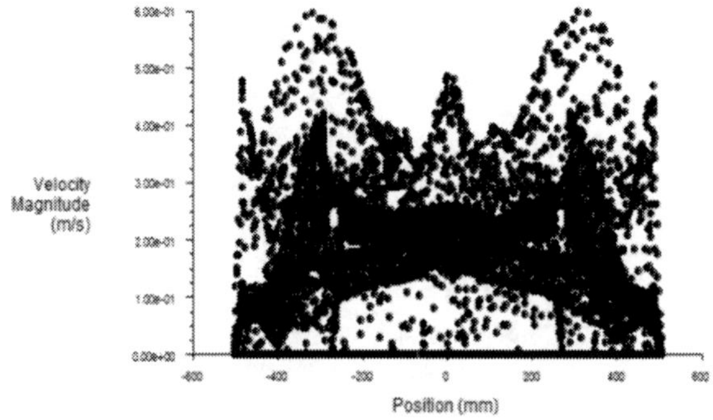

Figure 20. Section velocity where Y_2=-73.6 (v=0.4 m s^{-1}).

As can be seen from the figure 19 and 20, the maximum speed along the Y axis appears twice, the maximum 0.6 m s^{-1} appears at the two rows of symmetrical diversion holes of the collecting board, respectively. Airflow velocity distribution is more uniform after the air flow enters the filter space from the diversion holes on both sides, there is a maximum value at Y=0. The majority of the flow velocity distribute between 0.1 and 0.3 m s^{-1}. According to the design, the air flow velocity should be less than 1 m s^{-1}, which is conducive to dust settlement and form a loose layer of dust.

5. Main conclusion

5.1. After the analysis of the electric current whose field channels is toroidal, using CFD software, basic characteristics of the toroidal electric field are got:

- The flow velocity between internal and external circular collecting plate shows a gradient distribution, that is, the flow velocity gradually decreases along the center of the circle outward. Airflow velocity near the cylindrical dust collection plate is relatively small, so dust particles can stay in the electric field for a relatively long time, thus dust particles have enough time to charge in the electric field, which is conducive to the settlement of dust particles of the internal and external cylindrical dust collection plate.
- The electric field flow velocity near the entrance is relatively large, while the one away from the entrance is relatively small. The entrance aperture of deflector hole on the inner circle of the collecting plate is small, while the one away from the imported aperture is big, what's more, the circumference symmetric structure is good to the gas flow evenly distributed when the airflow from the farm area transition to the filter bag dust area.
- The airflow along the inner circle collecting plates will affect the dust layer formation of the inner circle collecting board, thus reducing the collection efficiency of dust collection plate in the inner circle.

5.2. Bag area

In the bag area, the air flow rate introduced by the collecting plate deflector is bigger than the one enter from the bottom of bag, this is unfavorable to the bag close to the collecting plate. The longer the bag, the more obvious the trend which filtration velocity increasing along longitudinal axis is, resulting in the local area near the outlet side of the bag stay a long period of high filtration velocity state.

During engineering design, the above two parts seriously affect the service life of the bag; it is recommended that thicken these areas in the engineering design.

Acknowledgments
The project is supported by National Natural Science Foundation of China (51064008).

References

[1] Guo X L 2008 *North Chin. Electr. Power Technol.* **9** 5
[2] Tang M K 2010 *Chem. Ind. Press* (Beijing)
[3] Bai H, Lu C, Chang C L 1995 *J. Air Waste Manage. Assoc.* **45** 900
[4] Di J H 2005 *Hebei Univ. Sci. Technol.* (in Chinese) **2** 160
[5] Meng G Z, Guo L J 2002 *Chem. Mach.* **29** 197
[6] Wang Y F 2010 *Donghua Univ.*(Shanghai)
[7] Xiang X D 2002 *Metall. Ind. Press* (Beijing)
[8] Wei J P, Zhang Y H 2009 *Shanxi Electr. Power* **4** 32
[9] Marzio P, Enrico N, Thomas J 2002 *J. Fluid Mech.* **458** 419
[10] Wissink J G 2003 *Int. J. Heat Fluid Flow* **24** 626

Measurements of global current density and electric field in urban environment

R Kacprzyk*, **A Pelesz**
Institute of Electrical Engineering Fundamentals, Wroclaw University of Technology. Wybrzeże Wyspiańskiego 27, 50-370 Wrocław, Poland

E-mail: ryszard.kacprzyk@pwr.wroc.pl

Abstract. Results of continuous and simultaneous measurements of global current density (flowing between ionosphere and the earth) and electric field intensity (in the vicinity of the earth surface) are presented. The results were obtained in a measuring system placed 20 m above ground level. Using the continuity equation, the conductive and displacement components of current density (as well as conductivity of the atmosphere) were determined. It was found that the displacement part of the total current is much smaller in comparison to the conduction one. Analysis of measurement showed a high coincidence of the electric field derivative with the time course of the global current. Temporal courses of conductivity indicate the presence of space charge and suggests the continuity equation is supplemented with a space charge component. The calculated average value of the short-period (10 min long) current density was found to be on the level of 2-3 pA m^{-2} (average conductivity of the order of 15-20 fS m^{-1}). At the same time period the average value of field intensity was on the level of 150-300 V m^{-1}.

1. Introduction

In fair weather conditions, through the atmosphere flows so called global electric current, which density is estimated to be on a level of 1-2 pA m^{-2} [1, 2]. The global current flows between ionosphere and the ground due to charge carriers produced during atmospheric electrical discharges (in thunderstorm regions), as well as the effect of their generation by cosmic rays and others sources of high-energy radiation and by radon. In the first approximation ionosphere may be considered as a specific, weak-conducting "electrode" with potential on the level of (+) 250-300 kV [3]. Values of electric field created in the ionosphere-earth space as well as global current density, measured locally, are (or can be) "modulated" by the temporary presence of clouds in the space. Additional component is the displacement current, flowing due to natural fluctuations of the electrical field intensity in the ionosphere – earth space as well as in the nearest vicinity of the measuring electrode employed for charge carriers collecting.

2. Global current density

The measurement of global current density j_T is carried out by measurements of total current I_T flowing between an isolated (from a ground) collecting electrode and the earth. The electrode is

---

* To whom any correspondence should be addressed.

exposed to natural (global) electric field E and charge carriers stream moving in the electric field. Basic measurement system is shown in figure 1.

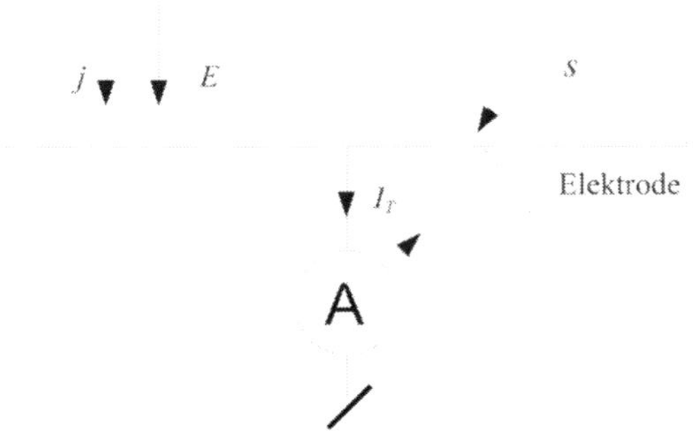

Figure 1. Basic set-up of current density measurement system.

The average current density j is associated with the measured total current value I_T by a relationship:

$$I_T = j \cdot s \tag{1}$$

where s is a surface of the measuring electrode.

The current density j measured in the plane of the earth is determined by two components: drift (conductivity) current, resulting from the motion of electrons and atmospheric ions in electrical field $E(t)$ and a displacement current resulting from $E(t)$ changes with time. For the total current density j one can write:

$$j = \sigma_T \cdot E(t) + \varepsilon_0 \varepsilon \frac{dE(t)}{dt} = \frac{I_T}{s} \tag{2}$$

where σ_T – conductivity of the atmosphere (air) surrounding the measured electrode, ε – relative permittivity of the atmosphere; $\varepsilon_0 = 8.85\times10^{-12}$ Fm^{-1} - permittivity of vacuum. Assuming for the air $\varepsilon=1$, the expression (2) can be transformed into the form allowing determine the conductivity σ_T:

$$\sigma_T = \left[\frac{I_T}{s} - \varepsilon_0 \frac{dE(t)}{dt}\right] \cdot \frac{1}{E(t)} \tag{3}$$

3. Measurements system

The prototype E-U transducer FM-09 field-mill type [4], was used for the continuous measurements of the electric field $E(t)$ intensity. The transducer offers sensitivity on the level of 3-4 V m^{-1} and provides linearity of the conversion characteristic for the field intensity measured in a range of 0 ± 10 kV m-1. With conversion characteristic slope equal to $1.0\pm0.05\times10^{-3}$ [m] (1[V]/1[kV m^{-1}]). The transducer (FM) was mounted at the edge of a building, at the height of about 20 m above the ground. Details of the whole installation (geometry) are given in figure 2.

The measuring electrode (ME), collecting charge creating the total current $I_T(t)$ was mounted at a distance of 0.85±0.01 and parallel to the conductive surface of the terrace – floor. Details of construction of the collecting electrode and its installation are shown in figure 3. The electrode is made in the form of an aluminum plate with dimensions 1.00 x 2.00 m and 3.0 mm thick.

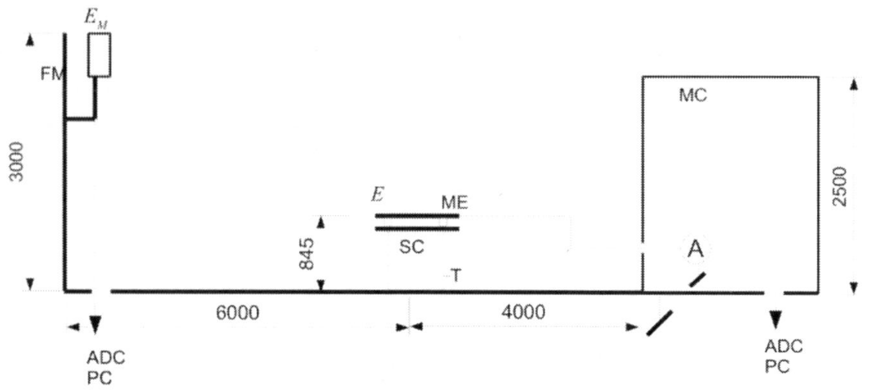

Figure 2. Distribution of the whole devices applied for simultaneous measurements of the global current and field strength. Symbols: FM - field meter, ME - measuring electrode, A - current meter, ADC - analog-digital converter and multiplexer, PC - a computer. MC - measuring chamber, SC – screen, T – table.

The plate is supported in six places by polymeric insulators (I) with a diameter of 50 mm and a height of 80 mm. The insulators are mounted on a grounded steel bars. Surfaces of insulators submitted to the direct contact with the measuring electrode were covered with a graphite layer, in order to avoid the "microphone" effect due to a charge deposited on insulators and natural vibrations of the whole electrode. Steel bars were earthed and placed on the screen (SC) created by a second aluminum plate. The screen dimensions were identical to the dimensions of the measuring electrode. The whole system is mounted on the surface of week-conducting tables (T), to ensure the lack of its own electric field. The analog current-meter (A) (Pico-ammeter, type PA-100) placed in the earthed

.

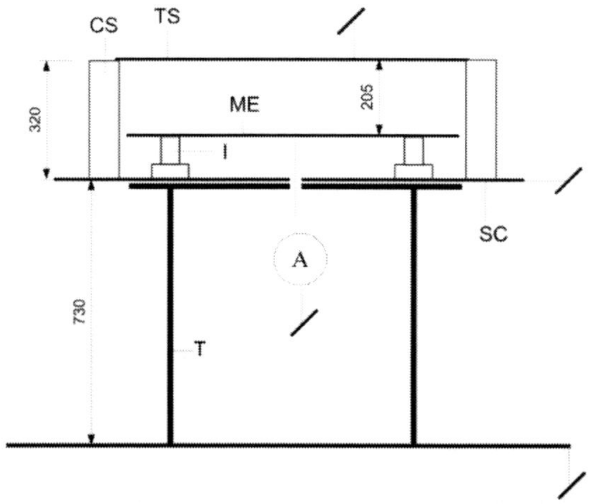

Figure 3. Sketch of the electrode system for measuring the global current – side view. Symbols: ME – measuring electrode, T – table (conducting), SC – screen, A – current meter tip I - support insulators, TS – temporary shielding, CS – conducting shielding support.

Measuring chamber (MC) was employed to measure the total current $I_T(t)$. Registration of the field strength $E_M(t)$ and the current $I_T(t)$ was performed using a PC and Agilent 34970 (used as an A/D converter and multiplexer).

Before current measurements the system was examined for the influence of parasitic signals. Current components due to electrochemical and Volta potential, piezo- or pseudo-piezoelectric effects (related to the charge that may be accumulated during assembly and manipulation of the measuring electrode on support insulators) were considered. In order to verify the mentioned disturbances an additional screen (TS) was temporarily mounted on conductive supports (CS) at a distance of 0.20 ± 0.01 m above the surface of the measuring electrode (ME). This procedure allowed to determine the real value of the parasitic part of the total current $I_T(t)$.

Due to a field distribution in space of the whole measuring system (including the field-meter (FM) and the electrode (ME) for measuring the global current as well as due to a field distribution on the surface of the electrode (ME), the average value of the field strength E at the surface of electrode (ME) was determined from relationship:

$$E = E_M K_E K_D \tag{4}$$

where: E_M - field strength measured by the transducer-meter placed on the edge of the building (see figure 2), K_E - constant, determined from the formula:

$$K_E = \frac{E_E}{E_M} \tag{6}$$

K_D – constant, determined from the formula:

$$K_D = \frac{E}{E_E} \tag{7}$$

where: E_E - the value of field strength measured in the center of the measuring electrode (ME) and in its plane. Value of K_E constant was determined from equation (6) using results of simultaneous measurements of fields E_M (measured as described above) and E_E measured in the middle of the measuring electrode (ME). The JCI-141F field-meter, with measuring aperture co-planar with the surface of the measuring electrode (ME) was applied for the temporal measurement of E_E value. The value of K_D (the ratio of the average field strength over the whole surface of the electrode E to the intensity E_E in its central part) was determined from relation (7) using simulation of the field distribution along a flat electrode, assuming the geometry as shown in figure 3. Received values for $K_E = 0.075 \pm 0.001$ (field strength at the edge of the building is much higher than at the plane electrode) and $K_D = 1.4 \pm 0.1$ (mean value of the field strength is higher than the values measured in the central part of the electrode) allow to determine the real average value of the field $E(t)$. The global current density was calculated from the relation (2) assuming area of the surface of measuring electrode (ME) $s = 2.00\ m^2$.

4. Measurement results and discussion

An example of measurements results of the average value of the field strength $E(t)$ in the plane of measuring electrode (ME), obtained for the case of "fair weather", is shown in figure 4. Before measurements of global current density, the influence of parasitic signals was examined. Current value measurements, in situation of shielding the measuring electrode (ME) with applied temporary screen (TS) showed, that the average current density $j_{TS}(t)$ after shielding does not exceed 10% of the value after removal of the screen (TS). Measurements confirmed that the electrochemical and piezoelectric components of the total current were negligible.

Comparison of measurement results of the global current density $j_T(t)$ and field strength, given in the form of derivative $dE(t)/dt$, for the case of "fair weather", is shown in figure 5. Similar nature of the temporal course of the total value of global current density $j_T(t)$ and the derivative of the field strength $(dE(t)/dt)=f(t)$ indicates the dominant role of the displacement component $(dE(t)/dt)$. Similar nature of the obtained time characteristics seems to confirm (in this particular case – time course and low conductivity current component) the internal consistency of results of measurements as well as measurement technique applied.

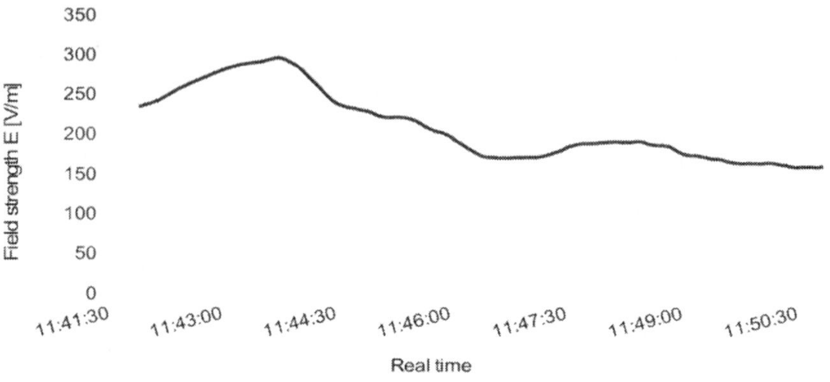

Figure 4. The intensity of the normal component of electric field $E(t)$ at the measuring electrode-collector, determined from the relation (5). The average value of the field strength during the registration (approx 10 min), E = 209 V m^{-1}. Results obtained for a sunny day with transient small clouds (05.10.2011).

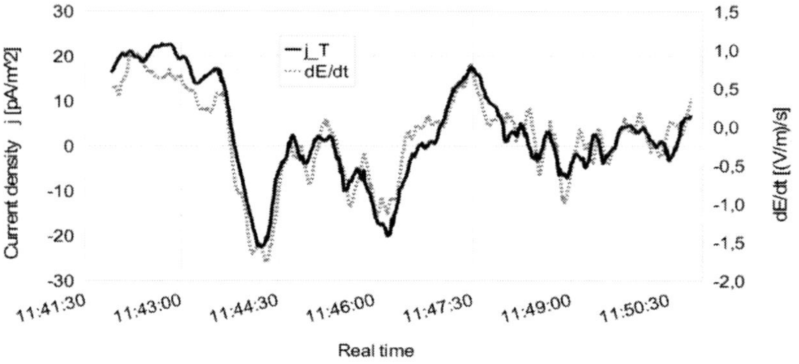

Figure 5. Time variation of the average electrode current density $j_T(t)$ and the time derivative of the average intensity of electric field $dE(t)/dt$ at the measuring electrode. Results obtained for a sunny day with transient small clouds (05.10.2011). Current density average value j_T=2,7 pA m^{-2}.

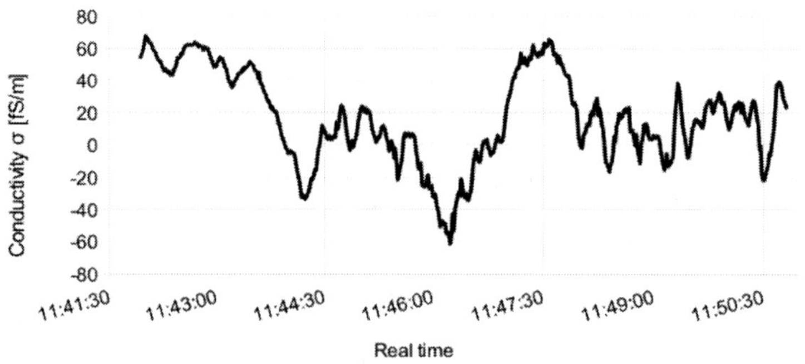

Figure 6. Time variation of air conductivity $\sigma(t)$ in the region of measuring electrode. The conductivity average value (for the period of 600 s) σ = 16.6 fS m^{-1}. Results obtained for a sunny day with transient small clouds (05.10.2011).

The air conductivity $\sigma(t)$ time variation (which can be considered as a parameter of air quality), received from the formula (3) is shown in figure 6. The result indicates that the conductivity $\sigma(t)$ may vary in relatively wide limits. Negative values of $\sigma(t)$, obtained for certain time periods, may result from different reasons. The reasons include: 1) errors resulting from the small difference of large numbers (a small component of the conductivity current in the total current value), 2) transient space charge ("clouds of ions"), 3) emission of electrons from the electrode-collector during it exposure to direct sunlight (UV) (when not screened by clouds).

The obtained results (for global current density and electric field strength) indicate the need to correlate them with the local (in the vicinity of the collector) measurements of ion density and light intensity, especially in the UV region.

Acknowledgments

This paper has been written as a result of realization of the project entitled: "Detectors and sensors for measuring factors hazardous to environment – modeling and monitoring of threats". The project financed by the European Union via the European Regional Development Fund and the Polish state budget, within the framework of the Operational Program Innovative Economy 2007÷2013. The contract for refinancing No. POIG.01.03.01-02-002/08-00.

References

[1] Harrison R G 2011 *J. Phys. Conf. Ser.* **301** 012001
[2] Siingh D, Gopalakrishan V, Singh R P, Kamra A K , Singh S , Pant V, Singh R and Singh A K 2006 *Atmos. Res.* **84** 91
[3] Markson R, Ruhnke L H and Williams E R 1999 *Atmos. Res.* **51** 315
[4] Czapka T, Dymczyk J, Kacprzyk R and Seredyniecki W 2011 *Transducer for the measurements of dc and low frequency electrical fields*. Raport nr I-7/SPR-5/2011, Institute of Electrical Eng. Fundamentals, Wroclaw Univ. of Technology, Wroclaw 2011 (in Polish)

Measurement of dust concentration based on VBAI

You Fu[*], **Ninghui Wang**
School of EEE, Dalian University of Technology, Linggong Street, Dalian, China.

E-mail:benjamin85100@126.com

Abstract. There are almost no economic and management solutions in automated measurement of dust concentration of magnesium oxide smelting electric arc furnace mouth. Nearly a thousand degrees of heat and the strong magnetic fields created by high-current Constitutes a serious interference with the sensor. This makes close-contact measurement method becomes almost impossible. To solve the above problem, this paper proposes a measurement scheme based on VBAI (Vision Builder for Automated Inspection, a software developed by National Instruments) to calculate the dust concentration. It uses visual analysis of the means to calculate the number of dust particles per unit volume of. And then it calculates the dust concentration. Though the analysis of the actual photo of the smelting site, the relative dust concentration in the mouth of the furnace is calculate. This method will be applied to the precipitator 70 kW motor inverter control and has a high recognition rate and significant application prospects.

1. Introduction

A lot of researches on dust concentration measurement technology have been done around the world. And a series of monitor machine for dust concentration have been made, such as dust sampling machine, direct reading dust measurement device, dust concentration sensor and so on [1]. However, the main principle of these sensors is light scattering theory. The dust concentration is changed into 4 to 20 mA current signal. And then these signals are transmitted to the computer terminal to deal with. These are all belonging to the contact measuring method. So they can not be applied to the measurement of dust concentration of electric arc furnace mouth, where the temperature is quite high (1800 degree centigrade). In response to these issues, we propose to directly collect of the images of electric arc furnace mouth. And then we can use the image processing ability of VBAI to calculate the number of dust particles accurately on PC. In this way, the on line detection of dust concentration in the high temperature can be achieved.

2. Software and hardware platform for the dust particles calculate based on VBAI

VBAI (Vision Builder for Automated Inspection) is a fully interactive configuration of machine vision developing software launched by National Instruments. It has the image acquisition, processing algorithms, the full set of visual output control module. VBAI can be used to configure and modify the real time parameter and build machine vision applications.

In recent years, more and more machine vision applications in large-scale has automated production line as well as higher manufacturing equipment in order to achieve the reliability, accuracy,

[*] To whom any correspondence should be addressed.

Journal of Physics: Conference Series **418** (2013) 012079 doi:10.1088/1742-6596/418/1/012079

and degree of automation, which can not be achieved in the past. Machine vision applications often involve special algorithm development and system environment, however, this development is relatively complex. Therefore, in this paper we chose Vision Builder AI NI interactive visual development software to quickly build a machine vision system to calculate the dust concentration. In fact VBAI is quite simple to use, its configuration interface just contain 4 main parts, main window, overview window, inspection steps palette and state configuration window. In fact the users do not have to write any program, they just need to configure the parameters and drag each block to the state configuration window.

According to actual needs, we chose the latest NI smart camera as the target hardware object to collect the real-time conditions in the mouth of the electric arc furnace.

The NI 177x smart cameras offer a range of sensors, a powerful Intel Atom 1.6 GHz processor, IP67 housing, M12 connectors, a lens cover, and multiple I/O options to perform in the most demanding applications. This high-performance hardware is paired with a real-time operating system to create a high-performance, deterministic machine vision system. In addition to high-performance image acquisition and processing, you can use built-in digital I/O and industrial communication options for dynamic, real-time communication and integration with industrial automation devices including programmable logic controller (PLC), human machine interface (HMI), robotics, sensors, and industrial machinery. NI vision software, including vision builder for automated inspection and the vision development module, provides a menu-driven or graphical programming option to develop application for the NI 177x smart cameras [2].

In order to analysis the experiment result as well as to monitor and control the dust concentration from the real Arc furnace, a NI PXI embedded controller instead of PC was selected to build the experiment platform together with the smart camera. PXI-8106 controller with PXI-1042Q Chassis with universal AC source has internet Ethernet interface, therefore it can communicate with the smart camera in a long distance (about 100 m) very conveniently. What is more, with the help of network routers, the PXI-8106 controller even can communicate with several smart cameras at the same time. And a PXI-6259 multifunction data acquisition board is also added to the platform so the arch furnace's working state can be controlled according to the dust concentration if it is needed, just shown as figure 1.

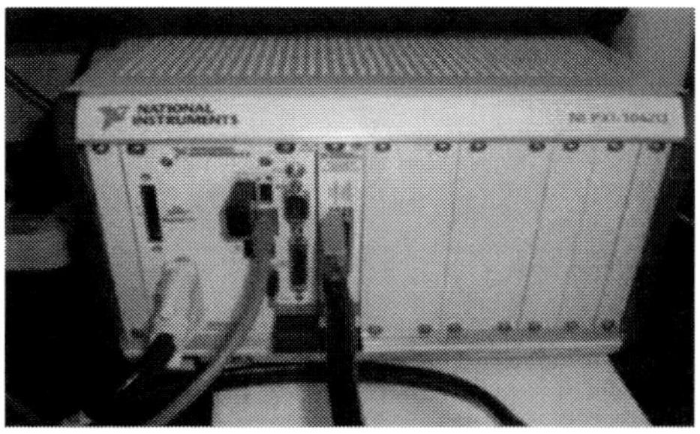

Figure 1. NI embedded controller.

Finally, we have the basic structure of the dust concentration measurement platform. It is shown as figure 2.

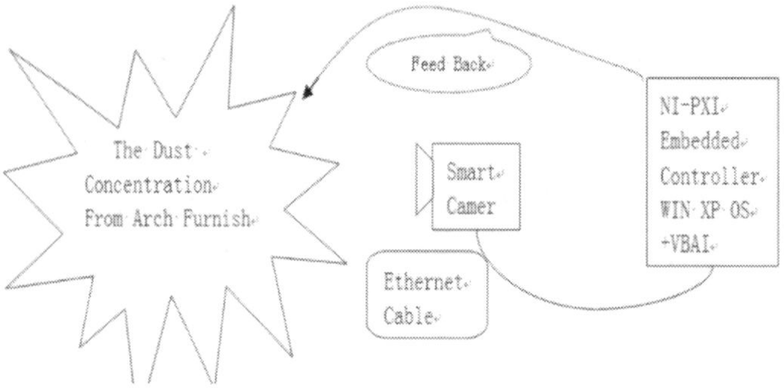

Figure 2. The experimental platform block diagram.

As figure 2 shows, firstly, the smart camera takes a picture of the dust from the mouth of the arch furnace every 10 minutes (the interval can be longer or shorter). And then the PXI embedded controller (in fact, it can be used as a normal PC) deals with the picture with the help of VBAI, in addition, the signal of the picture is transmitted through the Ethernet cable in a distance which can be as long as 100 meters. Last but not least, the result is shown on the screen, and we can also add a feed back to the arch furnace system to control the dust concentration if it is required.

3. Algorithm flow of dust measuring based on VBAI

There are generally three steps to complete the calculation of the dust concentration. They are image acquisition, binarization processing and object detection.

Image acquisition, in VBAI, the acquire images tab contains several acquisition steps you can use to acquire images from many different types of cameras. The tab also contains a simulate acquisition step, which simulates image acquisition by loading images from file. The select image step enables you to switch to a previously acquired image that you need to process later in the algorithm [3]. In this application, a group of real industrial site photos of the actual dust image in the smelting site is imported in the form of simulation to the main window of VBAI. It is shown as figure 3.

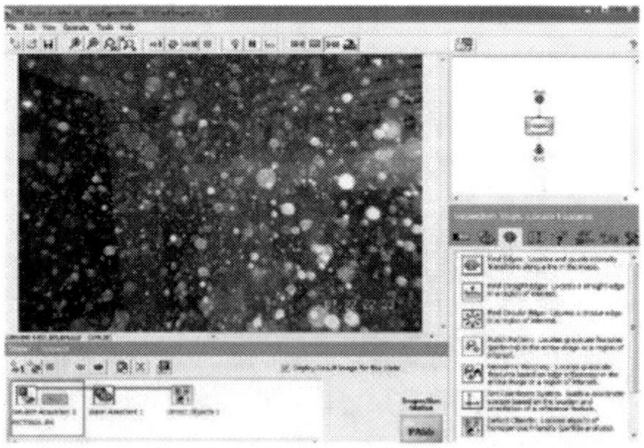

Figure 3. Actual dust image in the smelting site.

Binarization processing, it is the technique of converting a gray scale image to a binary image by using threshold selection techniques to categorize the pixels of an image into either one of the two classes [4]. In this application, binarization processing is used to change the dust picture of the industrial site into image that only contains black and white pixels. This is a very critical step in the whole algorithm, which is laying the groundwork for the following object detection. However it is not complicated in VBAI. It is so simple that it is just need to drag a small black called vision assistant and configure the parameters as binarization processing. VBAI will automatically deal with the complex calculation process. In figure 5 it is obvious that the colour of the dust showing bright white. So it is appropriate to add filter function to filter out the dim background colour. And then the result after binarization processing is as fallows, white dots represent the dust and black parts represent the background. It is just like figure 4 shows in the VBAI.

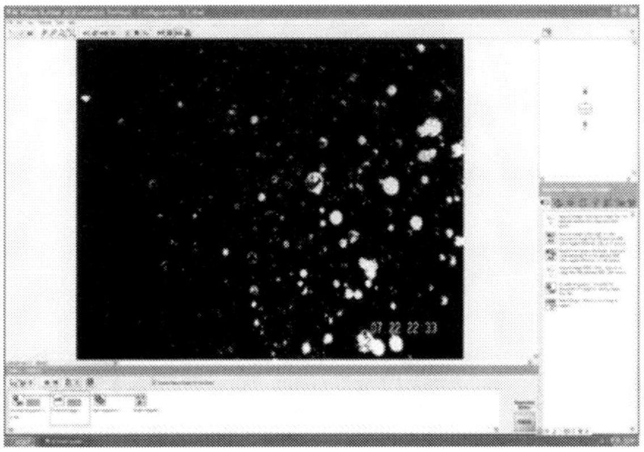

Figure 4. The result of binarization processing.

Object detection, if some kind of object is wanted to be detected from a binary image, the simplest and the most effective mean maybe setting the minimum and the maximum pixel value of the objects. Actually this simple theory is tried in this application and seems quite effective. That is to say:

If $N\ min<N<N\ max$, the object is a dust (1)

Else $N> N\ max$, or $N< N\ min$, the object is not a dust (2)

N is the size of the object the unit of which is pixel;

N min is the minimum pixel value;

N max is the maximum pixel value;

The main question is how to find minimum and the maximum pixel values. And maybe they are not unique. In our application several groups of values have been tried. In return, the group with the best results is selected as the minimum and the maximum pixel values. The minimum value for the dust is 20 pixels while the maximum is 500 pixels. In this way, the number of the dust is successfully detected. As figure 5 shows.

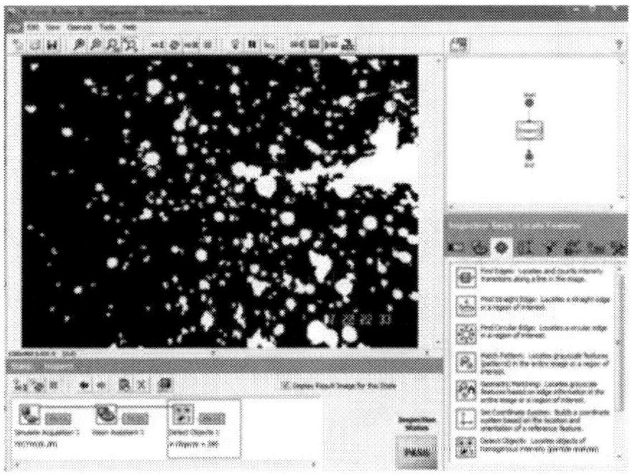

Figure 5. The result of object detected.

And the total algorithm flowchart is as shown below.

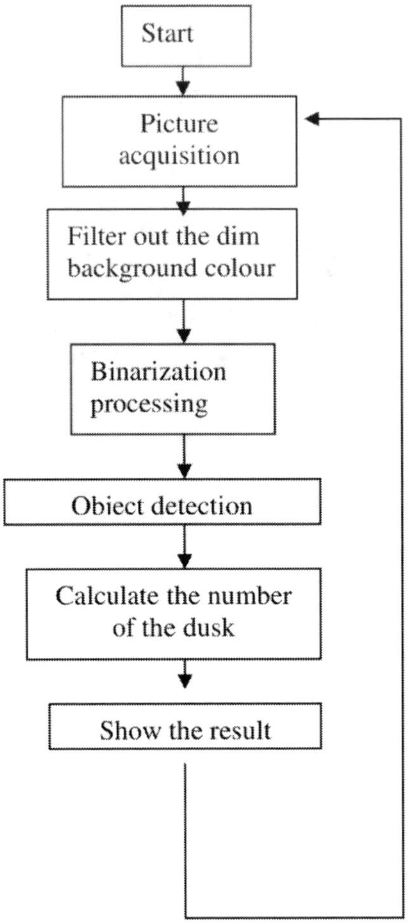

Figure 6. The total algorithm flowchart.

4. The result and the application of the measurement of dust concentration based on VBAI

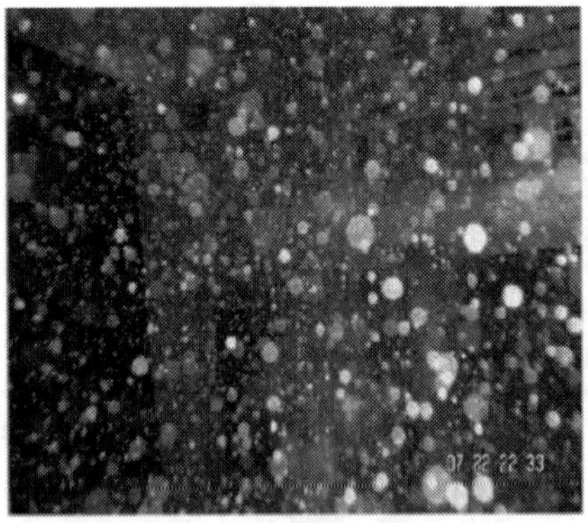

Figure 7. Melting site with higher density.

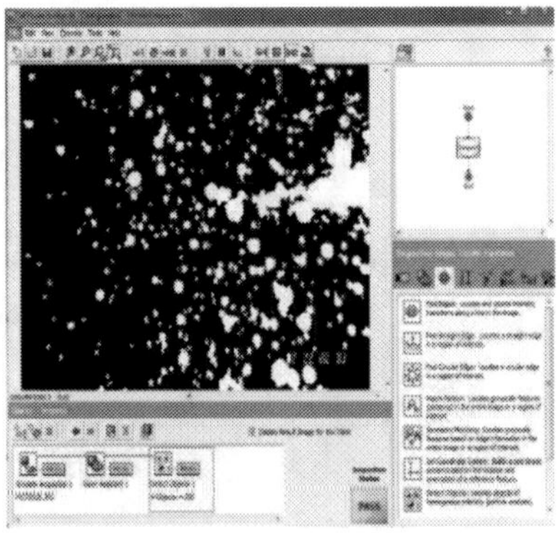

Figure 8. 290 dusts detected.

If the dust number is divided by the volume of the smelting site, the dust concentration will be got by the controller. And then it is very easy for the controller to control the dust concentration through the PXI bus. At least it is possible to distinguish whether the dust concentration has changed, that is very important to control the dust concentration automatically, just as shown below.

Figure 9. Melting site with lower density.

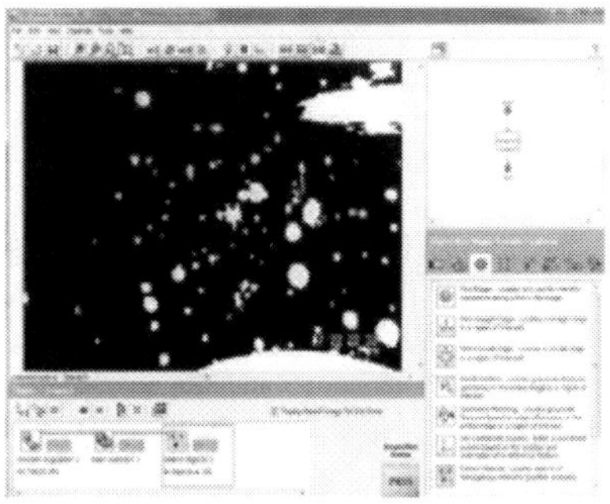

Figure 10. 100 dusts detected when density decrease.

From figure 7 to figure 10, it is obvious that the detected dust changes when the density changes. In figure 8, 290 dusts are detected from figure 7, but in figure 10 only 100 dusts are detected when the density decreases. So it is easy to use this result to monitor the change of the dust density.

5. Conclusion

This paper gives a new method of the real time measurement of the dust concentration. And through calculating the dust concentration of the photos from the real smelting site, it is proved that this new method is feasible and effective. Measurement of dust concentration based on VBAI is also a new non-contact measurement method, which is going to be used in the control system of high-temperature plasma arc furnace control system to control a 75 kW main motor for the bag filter.

Acknowledgments

This work is supported by The National High Technology Research and Development Program of China ("863") under Grant No 2011AA060103.

References

[1] Tang J 2009 *Mining Safety & Environmental Protection (in Chinese)* . **36** 69
[2] National Instruments Corporation 2012 High-Performance NI Smart Cameras (National Instruments Official Corporation).
[3] National Instruments Corporation 2011 NI Vision Builder for Automated Inspection Tutorial (National Instruments Corporation).
[4] Rapeeporn Chamchong and Chun Che Fung 2010 *Th. Int. Con on Knowledge Discovery and Data Mining* p 236

Shielding effectiveness testing method of pulse electric field based on GTEM cell

Long Zhang*, **Guanghui Wei, Xiaofeng Hu, Xinfeng Li**
Institute of electrostatic and electromagnetic protection, Mechanical Engineering College, Heping west Road, Shijiazhuang, China

E-mail: zhang80415@tom.com

Abstract. In order to evaluate the shielding ability of materials under the radiation of electromagnetic pulse (EMP), a test method based on GTEM cell is put forward. And the test theory and the components of test system are introduced. The shielding box method is verified to be reasonable with simulation. The shielding effectiveness (SE) of one material is tested with this method, and the test results are compared with the results based on the coaxial method. The results show this test method can be used to test the effect of SE on EMP, and the test results can reflect the true shielding ability of the test material.

1. Introduction

With wide application of the microelectronic device, electric device and electric system become sensitive and vulnerable under the radiation of EMP. The hazards of EMP on the microelectronic device are realized by people. Shielding technique can effectively improve the vulnerable parts in the bad electromagnetic environment. The test methods of SE are vital to evaluate the SE of materials. There are different results depending on different test methods [1].

The test methods of SE are introduced in GJB6190-2008 in detail, in which input wave is continuous wave. At present, the SE of materials mainly formulates with the changing frequency. With the research of high power EMP jamming effect, frequency domain SE of materials can not substitute the time domain SE [2, 3]. On the shielding problem of EMP, due to the "dispersion" effect, the SE of EMP could not apply the single frequency SE formula to calculate. More appropriate define is worthy of study [4].

Through a large number of experiments, Bihua Zhou analyzed the EMP SE and the relation of electromagnetic pulse rise time, pulse width and peak value, and put forward an SE calculation formula including pulse rise time [5]. L. Klinkenbusch got the EMP SE with the calculation of the energy with and without test materials according to the same position [6]. Waveform reconstruction method based on the minimum phase system is introduced [7-9], using Hilbert transform to evaluate system impulse response characteristics according to frequency domain measurement results. This method provides an approach for using the system transfer function of the amplitude-frequency characteristic to evaluate system time response characteristics, especially suitable for solving electromagnetic pulse measurements. A "specimen" method is proposed to test the EMP SE [10]. This method overcomes the shortcoming of "windows" method, such as bad low frequency response and measurement error. And it solves the problem of engineering material EMP SE. Pulse electric field SE

* To whom any correspondence should be addressed.

measurement system is reported [11] and it is mainly for the metal bridge, control cabinets, shielding tents, conductive cement concrete. The SE of material is got according to the pulsed electric field peak attenuation [12].

Above-mentioned test methods are mainly for larger buildings or material products, and this paper presents a pulse electric field SE measurement method aiming at a smaller plane material, namely shielding box method. The shielding box method is proved to be reasonable through simulation analysis and one material was tested with this method. Comparing test results of this method with flange coaxial method, it proved the test method feasible.

2. Test theory and component of test system

In order to get the SE under the radiation of EMP, the shielding box is placed in the GTEM cell, using pulse source and GTEM cell to produce a high power EMP radiation. The SE is got by testing the waveform in the centre of shielding box with and without material on the test window.

2.1. *Test principle*

Generally, SE refers to the ratio on the same incentive level, the received electric field or power without shielding material and the electric field or power with shielding material. And the computation formula is:

$$S = 20\lg\frac{E_0}{E_1} = 10\lg\frac{P_0}{P_1} \tag{1}$$

In the formula, S refers to SE; E_0 and P_0 refer to the electric field strength and the received power at a point without shielding material respectively; E_1 and P_1 refers to the electric field strength and the received power at the same point with shielding material respectively.

As for shielding room method, signal transmitting equipments are standard signal source and the antenna, and the signal receiving equipments are antenna and spectral analyzer. Signal generating devices of the shielding box method are mainly the standard pulse source and GTEM cell, and signal receiving device is a pulse sensor. A test box with window (equivalent to a small shielding room) is placed in the GTEM interior. Through testing the waveform of the same point loading and unloading material on the test window, the SE of material is got. Test schematic diagram is shown in figure 1.

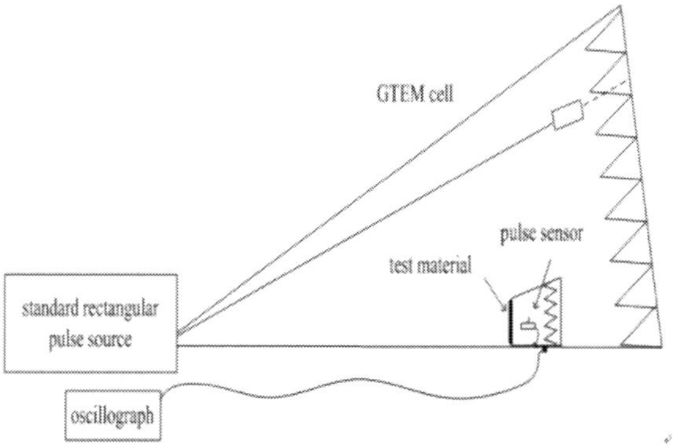

Figure 1. Sketch of test system.

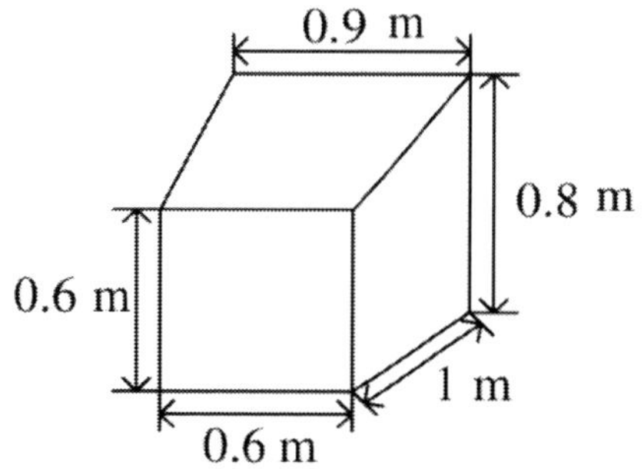

Figure 2. Model of shielding box.

Shielding box is placed in the GTEM cell, and the peak value and pulse width of the pulse are set up. Firstly, the test waveform is derived from pulse sensor when there is no test material. Then test material is loaded and the waveform is got, keeping the pulse sensor position unchanged position. According to the formula (1), the SE of the material is calculated from the pulse peak value of waveform.

2.2. Test equipments

- GTEM cell. GTEM cell is Gigahertz transverse electromagnetic cell, and the working frequency ranges from DC to gigahertz. It has the trait of internal large free field area and several advantages of other test method, such as open field shielding room and TEM cell. And it overcomes the limitations of other methods, with the advantages of high quality and low price.
- High voltage pulse source. Pulse source used in high-frequency noise generator which made in Japan NOISENKEN Company. Model: INS-4040. It can generate square wave pulse, the peak value is less than 4 kV, the rise time is less than 1 ns, and the pulse width is optional.
- Pulse sensor. The bandwidth of the sensor is from DC to 1 GHz, dynamic range is 44 dB, and it has good linearity.

2.3. Shielding box development

The shielding box is placed in the GTEM cell. And in order to minimize the electromagnetic reflection in GTEM cell caused by shielding box, the box designs into a certain wedge-shaped cavity. The angle of tilt of the box surface is the same as the surface of GTEM cell. The box model is shown in figure 2. Top opening of shielding box is 0.6 m×0.6 m, bottom size is 0.8 m×0.9 m, and height is 1 m. In order to suppress the electromagnetic reflection and resonance in the cavity, absorbing materials are placed in the bottom of shielding box.

Only high SE of the shielding box itself, the SE of test material should not be affected by the test system. Generally, the SE of the shielding box is 6 dB more than the test material.

3. Simulation analysis

In order to analyze the wave of the pulse sensor in the centre of shielding box, the pulse waveform through simulation analysis is illustrated as follows. It is helpful to analyze the SE of EMP.

CST-MWS software is based on the time domain finite integral. The simulation model of shielding box corresponds to the actual model. The thickness of test material is 0.001 m, relative dielectric constant is 1, the relative permeability is 1, and the conductivity is 1000 S m^{-1}.

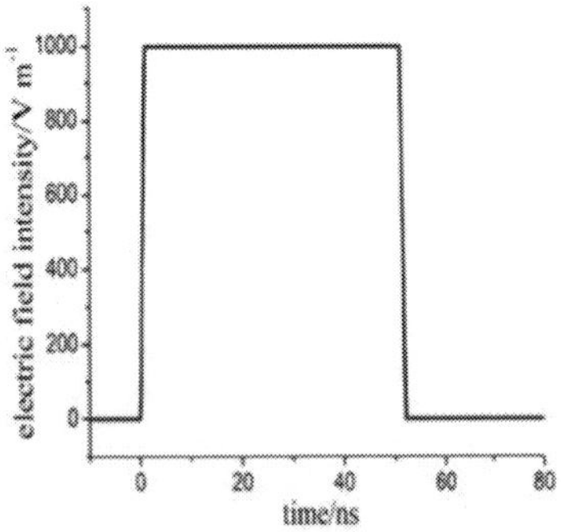

Figure 3. Input waveform.

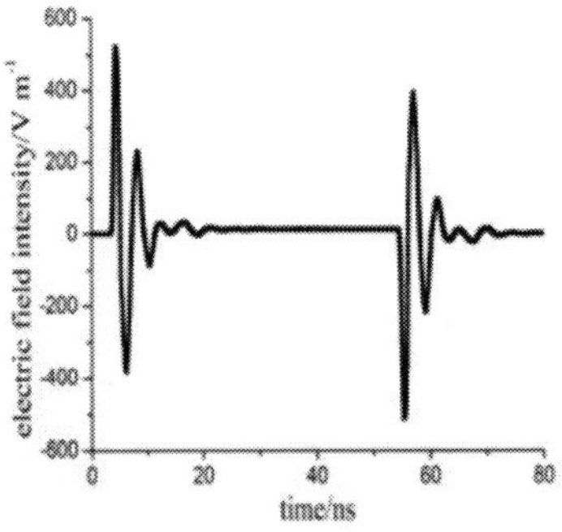

Figure 4. Waveform unloading test material.

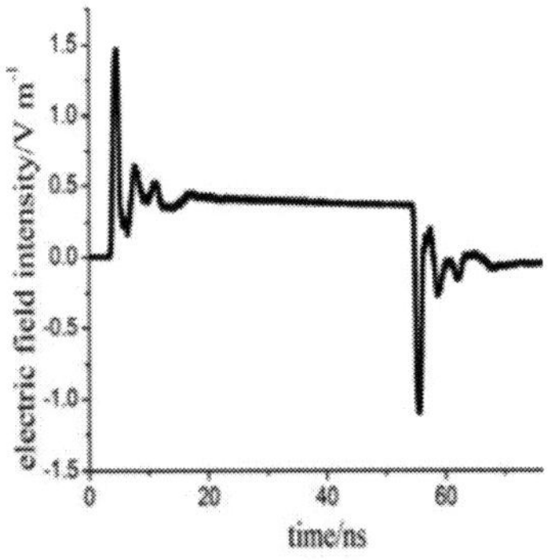

Figure 5. Waveform loading test material.

Figure 3 shows that the input pulse is ideal square wave pulse, the peak value is 1 kV m^{-1}, and the pulse width is 50 ns. When the input pulse transmits inside the box, the waveform tested by the pulse sensor is shown in figure 4. The pulse waveform changes greatly mainly due to the shielding box. The reason is that the low frequency signals are difficult to access. Figure 5 shows that, because of the test material, the amplitude of wave becomes lower, and the shape of wave changes little. So the signal inside the shielding box is mainly high frequency signal corresponding to rising and falling edges of the square wave pulse. With and without test material, the amplitude of pulse changes only, and the waveform changes little. According to the amplitude changes of waveform with and without test material, the SE for the pulse of the material can be calculated.

4. Test experiment

Last section shows that the shielding box method is validated to test the SE of the material with simulation. Generally, the SE could be got to evaluate the shielding ability through the experiment.

4.1. Waveform analysis

The test system is build up according to figure 1. The waveforms tested by the pulse sensor with and without test material in the shielding box are shown in figure 6 and figure 7.

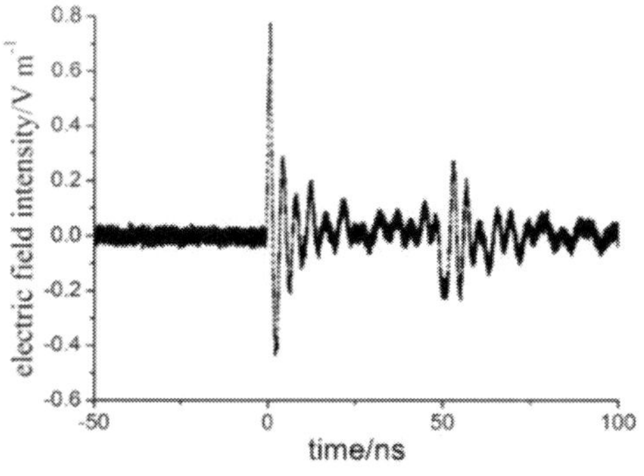

Figure 6. Waveform unloading test material.

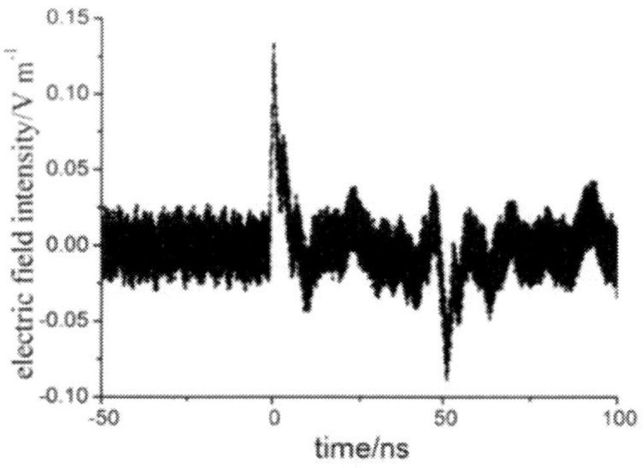

Figure 7. Waveform loading test material.

Figure 6 and figure 7 show that the waveform has little change with and without test material, and the amplitude of the pulse changes only. Based on the definition of SE, the SE can be calculated for 15.8 dB according to the maximum peak value of the pulse.

The peak voltage of input square wave is 3 kV when the SE of the material is tested, and the electric field strength at the test point in the GTEM cell is about 1.5 kV m^{-1}. From the figures, we know that the material is under the radiation of high electromagnetic pulse. The shielding box method can realistically simulate the bad electromagnetic environment. So the SE of the material is tested in close proximity to the actual environment of strong electromagnetic field, and it is helpful to accurately grasp the SE of material.

4.2. Time domain, frequency domain comparison

In order to further analyzing test results, the SE of the material is tested again with flange coaxial method. The test result is shown in figure 8.

Comparing the coaxial test curve with peak value SE, it is found that the peak value SE and the SE of frequency 700 MHz are consistent, and it shows that the pulse peak value SE is corresponding to a frequency point in the frequency range of the SE. According to the Schelkunoff shielding theory, the SE is related to the test frequency. The pulse peak value SE is corresponding to the SE of single frequency in frequency domain test results, but the pulse SE is tested under the radiation of strong electric field. It has the unique characteristics of strong electromagnetic field test.

For linear material, the SE curve tested through a weak field can reflect the SE of the material; as for the nonlinear material, the SE is related to the input pulse peak value, and the SE which is measured by inputting weak electromagnetic field was not a true SE that can not represent the shielding ability for strong electromagnetic field. So, it is very effective to input strong electromagnetic pulse to simulate actual environmental when testing SE, and it can make the SE of the materials to be more accurate and reliable for electromagnetic field shielding capability.

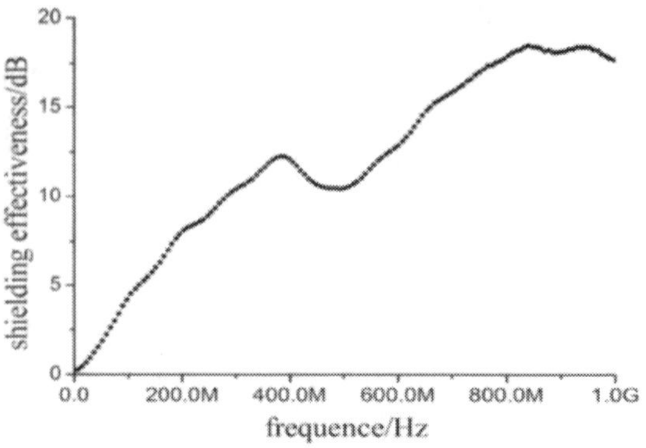

Figure 8. Frequency domain SE.

5. Conclusion

This paper presents pulse electric field SE test method. A shielding box with absorbing material inside the box was made and placed in the GTEM cell, and the pulse sensor is placed in the centre of the shielding box. The SE is got through testing the waveform with pulse sensor with and without test material. And the shielding box method is verified by simulation and experiments to illustrate the application. In this method, the SE is got under the radiation of strong electromagnetic pulse. And using this method, SE will be more accurate with simulating the bad electromagnetic environments, and it is the special characteristic of this method.

Acknowledgments

The authors are grateful to an anonymous reviewer; and to Xiaodong Pan and Xiang Chen for comments that helped improve the paper.

References

[1] Chen S F, Ma W Y and Ma X Q 2001 *Electromagn. Compat. Exp. Technol.* (Beijing: Beijing University of Posts and Telecommunications Press)
[2] M P Robinson, T M Benson and C Christopoulos 1998 *IEEE Trans. on EMC.* **40** 240
[3] Andrew C Marvin, John F Dawson and Simon Ward 2004 *IEEE Trans. on EMC.* **46** 459
[4] Zhou B H, Chen B and Shi L H 2003 *Electromagn. pulse and Eng. Prot.* (Beijing: National Defense Industry Press)
[5] Dai C Y, Chen C X, and Xie Z F 1997 *High Voltage Eng.* **23** 3
[6] Gao C, Liu X and Shi L H 2010 *High Voltage Eng..* **36** 2272
[7] Zhou B H, Gao C and Ren H M 2003 *Asia-Pacific Conf. on EMC* p 562
[8] L Klinkenbusch 2005 *IEEE Trans. on EMC.* **47** 589
[9] Xie Y Z, Wang Z J, and Wang Q S 2004 *High Power Laser & Part. Beams.* **16** 320
[10] Guo J, Zou J and He J L 2004 *J. Tsinghua. Univ.* **44** 1302
[11] Shi L H, Zhou B H and Chen B 2000 *Chin. J. Radio Sci.* **15** 467
[12] Liu S K, Han J and Chen X Y 2009 *Nucl. Electron. & Detect. Technol.* **29** 1456

Time-frequency analysis of nonlinear and non-stationary weak signals of corona discharge

Lei Wang*, Shanghe Liu, Ming Wei, XiaoFeng Hu
Electrostatic and Electromagnetic Protection Research Institute, Ordnance Engineering College, Shijiazhuang, China, 050003.

Email: wl810603@sina.com

Abstract. It is very useful to study the signals radiated from corona discharges for the purposes of high-voltage line monitoring. Time-frequency analysis can clearly reveal the time-varying spectrum characteristics of such signals, which is very useful for analyzing and processing the non-linear and non-stationary weak signals, such as the signals radiated from corona discharges. Several time-frequency analysis methods, such as the Short-Time Fourier Transform (STFT), Wigner-Ville distribution and the Hilbert-Huang Transform (HHT) and so on, are used in this paper. The simulation data with the same and different amplitudes are comparatively analyzed by these time-frequency distribution methods. It can be concluded that the time-frequency analysis method based on HHT is more efficient to identify and suitable for the non-linear and non-stationary weak signals.

1. Introduction

In the field of power systems, the ultra-UHV transmission has become the development trend of the power transmission. With the improvement of the transmission voltage level, the corona discharge is inevitable, and has become an important factor to affect the safe and stable operation of the high-pressure and UHV transmission lines. It is a new topic for the fault detection of high-voltage transmission line by remotely detecting the signal radiated from corona discharge. However a lot of noise appears in the testing process. So we need to process these signals by some method and detect the weak signals. The time-frequency analysis can clearly reveal the time-varying spectrum characteristics of signals, which is very useful for analyzing and processing the non-linear and non-stationary weak signals. At present, the time-frequency analysis method mainly contains the linear time-frequency analysis method and the bilinear time-frequency distribution based on Fourier transform, and the other method based on Hilbert-Huang transform [1]. Among them, the linear time-frequency analysis method includes the time-frequency analysis of short-time Fourier transform [2], the wavelet transform time-frequency analysis [3] and time-frequency distribution of Gabor [4]; the typical representation of the bilinear time-frequency distribution is the Wigner-Ville time-frequency distribution [5]. The Hilbert-Huang Transform is different from the above two time-frequency analysis methods, which does the frequency decomposition based on

* To whom any correspondence should be addressed.

Journal of Physics: Conference Series **418** (2013) 012081 doi:10.1088/1742-6596/418/1/012081

the local time-varying characteristics of the signal. This paper selects three kinds of time-frequency analysis methods: the short-time Fourier transform time-frequency analysis method, the Wigner-Ville time-frequency distribution and the Hilbert-Huang transform time-frequency analysis method. By analyzing the simulation data and comparing the advantages and disadvantages of this time-frequency analysis methods, the method which is more suitable for time-frequency analysis of weak signals radiated from corona discharge can be obtained.

2. The time-frequency analysis

2.1 The short-time Fourier transform [2, 6]

This is a basic time-frequency analysis. The short-time Fourier transform is the Fourier spectral analysis with a time window. The time-frequency analysis can be obtained by continuous sliding window on the timeline. The defined formula is

$$STFT(t,\omega) = \int_{-\infty}^{+\infty} X(\tau) w^{*}(\tau - t) e^{-j\omega\tau} d\tau \tag{1}$$

Where, $X(\tau)$ is the time series signals to be analyzed, and $w(\tau - t)$ is the sliding time window.

2.2 Wigner-Ville distribution [5, 6]

The time and bandwidth product of the Wigner-ville distribution reached the lower limit of the Heisenbergde uncertainty principle. So the Wigner-ville distribution has good time-frequency resolution rate. It is defined as the instantaneous autocorrelation function of the Fourier transform, which is bilinear distribution proposed by Ville in the study of quantum mechanics. The expression is as follows:

$$WVD(t,\omega) = \int_{-\infty}^{+\infty} X(t + \frac{\tau}{2}) X^{*}(t - \frac{\tau}{2}) e^{-j\omega\tau} d\tau \tag{2}$$

In the formula, the signal x appears twice, and without any window function, which avoids the mutual restraint of the time resolution and frequency resolution in the time-frequency linear representation.

2.3 Hilbert-Huang Transform [1]

For any time series, X(t), in order to determine the phase and eliminate the negative frequency effect in the signal analysis, the Hilbert transform should be done

$$H(t) = \frac{1}{\pi} \int_{-\infty}^{+\infty} \frac{X(\tau)}{t - \tau} d\tau \tag{3}$$

The analytic signal $Z(t)$ corresponding to the original signal $X(t)$ is:

$$Z(t) = X(t) + jH(t) = A(t) e^{j\theta(t)} \tag{4}$$

Where, $A(t)$ is the amplitude function, $\theta(t)$ is the phase function. Their expressions are

$$A(t) = \sqrt{X(t)^2 + H(t)^2} \tag{5}$$

$$\theta(t) = \arctan \frac{H(t)}{X(t)} \tag{6}$$

So the instantaneous frequency ω can be defined as follow:

$$\omega = \frac{d\theta(t)}{dt} \tag{7}$$

From the definition of Hilbert transform, it can be seen that the Hilbert transform is the convolution of the signal on $\frac{1}{t}$. So it stressed the local characteristics of the signal, the instantaneous frequency can be given in some extent. However, many signals may contain multiple oscillation modes, which makes the simple Hilbert transform unable to give the full frequency content of these signals.

For this phenomenon, Huang [1] proposed a new time-frequency analysis method. By using it to analyze the data, two basic steps are required: First, the signal should be decomposed into the Intrinsic Mode Function (IMF) by using some decomposition method. Then, the energy distribution spectra in the time-frequency plane can be obtained by doing the Hilbert transform to the IMF.

The IMF is a function that satisfies two conditions: (1) in the whole data set, the number of extremum and the number of zero crossings must either equal or differ at most by one; and (2) at any point, the mean value of the envelope defined by the local maxima and the envelope defined by the local minima is zero.

The signal is decomposed into several IMF, and a residue, r_n which can be either the mean trend or a constant. We finally obtain,

$$X(t) = \sum_{i=1}^{n} C_i + r_n \tag{8}$$

In the formula, C_i is the IMF component. The IMF reflects the inherent the internal volatility of signals, at any time it only has a single frequency component. So it can be carried out the Hilbert transform to calculate the instantaneous frequency.

3. The analysis of simulation data

The simulation data should be analyzed by different time-frequency analysis methods. So the advantages and disadvantages of these time-frequency analysis methods can be compared.

3.1 The analysis of simulation data with the same amplitude

The simulation signal can be obtained by the equation (9), which contains the frequency modulation signal and the sine signal.

$$x(t) = \cos(2\pi 30t + 0.8\sin(2\pi 15t)) + \sin(2\pi 150t) \tag{9}$$

The frequency modulation (FM) signal and the sine signal have the same amplitude. The frequency of the sine signal is 150Hz and the frequency of the FM signal is:

$$f(t) = 30 + 12\cos(2\pi 15t) \tag{10}$$

From the above equation, the frequency of the FM signal is from 18 Hz to 42 Hz. The time domain waveform of the simulation signal is shown as figure 1.

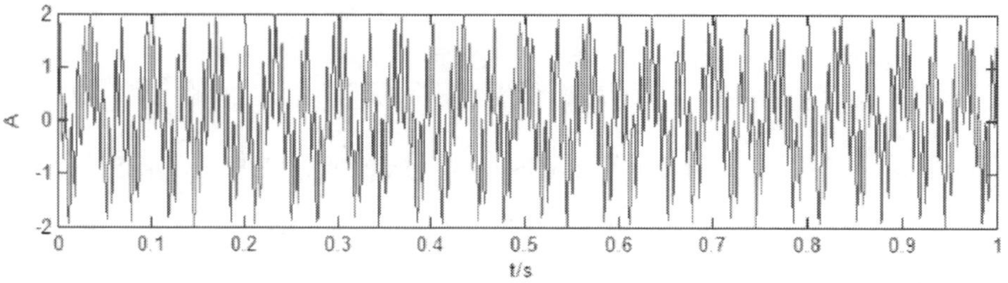

Figure1. The simulation signal with the same amplitude.

The above simulated signal should be analyzed respectively by short-time Fourier transform time-frequency analysis method, Wigner-Ville time-frequency distribution, and Hilbert-Huang transform time-frequency analysis method. The results are illustrated in figure 2.

It can be seen from figure 2, these three time-frequency analysis methods can reflect intact for the sine signal in frequency of 150Hz, and basically reflect the frequency of this signal.

As for the FM signal in frequency of $30+12\cos(2\pi 15t)$, the results are different. The short-time Fourier transform time-frequency analysis methods can only reflect the carrier wave in frequency of 30Hz, the upper frequency 42Hz and the lower frequency 18Hz; the sine characteristics in this frequency can not be reflected. The Wigner-Ville time-frequency analysis can basically reflect the sine characteristics in the frequency of the FM signal, but it has a serious interference near the frequency of 90 Hz. The HHT time-frequency analysis integrated shows the sine characteristics in the frequency of the FM signal.

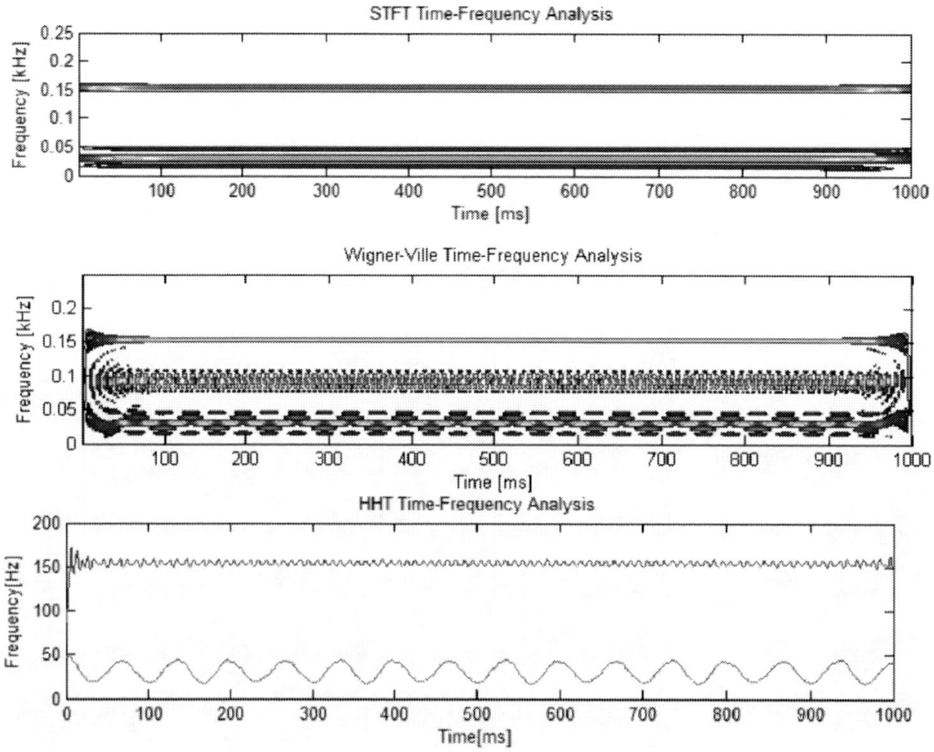

Figure 2. Time-frequency analysis of signal with the same amplitude.

As the short-time Fourier transform depending on the traditional Fourier spectrum analysis, it must assume the data to be piecewise smooth, that is, the data is stationary at least in the analysis time window. By the constraint of the Heisenberg uncertainty principle, it can not be minimized simultaneously in time and frequency resolution, which is only suitable for approximating describing to the data that is stationary in subsection. Essentially, short-time Fourier transform is a kind of fixed bandwidth band-pass filter. For the short-time Fourier transform is divided into Grid in time-frequency plane, and its reliability relies on people's subjective choices, so it is difficult to analyze the complex non-stationary signals. The Wigner-ville distribution is not linear, which contains serious cross-term. If the characteristics of complex signals are not known before analysis, the Wigner-ville distribution will be difficult to distinguish the true nature of the signal. The Hilbert-Huang transform time-frequency analysis does the frequency decomposition based on the local time-varying characteristics of the signal. The definition of the frequency is based on the local features and transient features of the signal waveform, which can be given, from point to point variation between the instantaneous frequency values at each time point of the signal data. It dose not need several oscillation periods of the waveform to give a frequency value. The shortcomings of expressing the nonlinear and the non-stationary signals by using the meaningless harmonic components in traditional methods are overcome. By using this method, we can obtain the high frequency resolution and good time-frequency concentration, and achieve the complete, accurate analysis of the nonlinear and non-stationary signals.

Therefore, the Hilbert-Huang transform time-frequency analysis method is more suitable for the nonlinear and non-stationary signals.

3.2 The analysis of simulation data with the different amplitudes

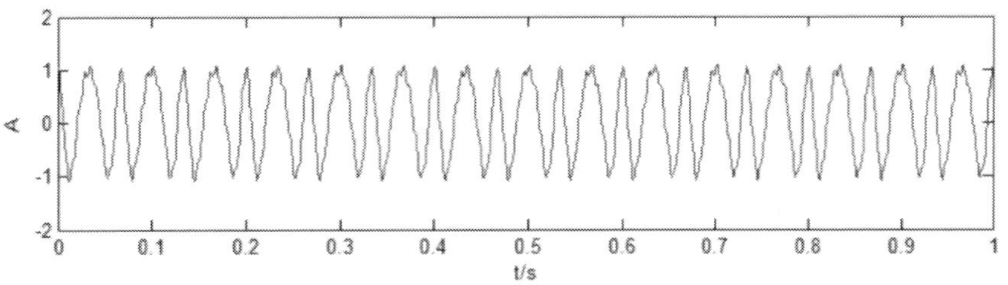

Figure 3. The simulation signal with the different amplitude.

The amplitude of the sine signal in formula (9) is reduced to 10%, which is given by formula (11). The frequency of this simulation signal doesn't chang, but the intensity the sine signal becomes one-tenth of the original. The time domain waveform of the simulation signal is shown in figure 3.

$$x(t) = \cos(2\pi 30t + 0.8\sin(2\pi 15t)) + 0.1\sin(2\pi 150t) \tag{11}$$

It can be seen from the figure, the sine signal completely submerges in the FM signal. The time-frequency analysis is shown in figure 4.

In figure 4, the sine signal of 150 Hz can not be identified by the short-time Fourier transform time-frequency analysis and Wigner-Ville time-frequency analysis, while it can be identified by the Hilbert-Huang transform time-frequency analysis. Compared with other methods, the Hilbert-Huang transform method has the higher resolution, which is more effective to detect and analyze the weak signal.

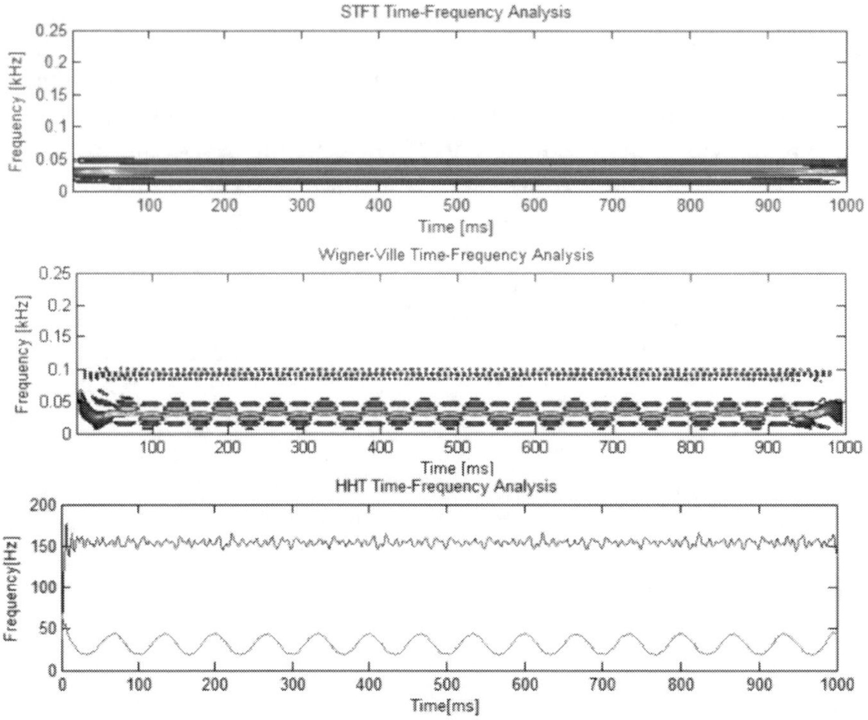

Figure 4. Time-frequency analysis of signal with the different amplitude.

4. Conclusions

From the analysis of the simulation data, it can be concluded: compared with other time-frequency analysis methods, the Hilbert-Huang transform method can be more comprehensive and accurate to analyze the time-frequency characteristics of the nonlinear and non-stationary signals; and the Hilbert-Huang transform method has the higher resolution, which is able to detect the weak signal component from the strong background, so this method is more significant in the detection of weak signals radiated from corona discharge.

Acknowledgment

The paper is supported by the Key Programs of the National Natural Science Foundation of China, No. 61172035.

References

[1] Huang N E, Shen Z and Long S R 1998 *Proc. R. Soc. Lond*, p 903
[2] Kramer M L and Jones D L 1994 *Proc. the IEEE-SP Int. Symp. on Time-Freq. and Time-Scale Anal.* p 264
[3] Kawada M and Isaka K 2008 *Time-Freq. Conf. Rec. of the 2008 IEEE Int. Symp. on Electr. Insul.* p 381
[4] Subbanna N K and Eldar Y C 2004 *Proc. the 11th IEEE Int. Conf. on Electron.,Circuits and Syst.*, p 650
[5] Tan J L and Sha'ameri bin Z 2008 *6th National Conf. on Telecommun. Technol. and 2008 2nd Malaysia Conf. on Photonics*, p 260
[6] Shie Q and Chen D P 1996 *Prentice Hall PTR.*

Electromagnetic interference analysis of magnetic resistance sensors inside a projectile under complex electromagnetic environments

Qingwei Guo[1*], Min Gao[1], Zhicai Lu[1], Peijie Yang[2]
1 The 4rd Department of Shijiazhuang Mechanical Engineering College Shijiazhuang, China
2 Institute of Electrostatic and Electromagnetic Protection Mechanical Engineering College Shijiazhuang, China

E-mail: 718335001@qq.com

Abstract. Accurate measurement of angular motion has long been recognized as a daunting task. In recent years the measurement of projectiles utilizing magnetic resistance sensors has become a hot research field. Electromagnetic interference on attitude measurement cannot be ignored in complex electromagnetic environments such as battlefield conditions. In this paper, the influence and function pattern of electromagnetic interference on the measuring performance are theoretically analyzed, and the shielding effectiveness (SE) simulation of projectile is conducted via software Computer Simulation Technology (CST). Considering the specific tests, the intensity of the influence is judged. The simulation indicates that the battlefield's complex electromagnetic environment influences the environment inside the projectile, especially its electronic components and capability. The research results can provide important theoretical support on the errors compensation and precision improvement of the projectile attitude measurement with Magnetic Resistance sensor.

1. Introduction

With the improvement of new concept weapons, especially the high power microwave bombs, electromagnetic bombs, electromagnetic missiles and nuclear electromagnetic weapons, they make the battlefield environments more complex. In the recent battlefield environments, it is necessary to improve the combat power of weapon system, so it is important to analyze more precisely the battlefield electromagnetic environments and take action to protect from the harm. With the advance of ammunition, more intelligent, smaller, and more accurate electronic products are used more and more popularly in the projectile, and these products are becoming complicated and composite. Nowadays the magnetic resistance sensors broadly used in the ammunition are one kind of important sensor to measure the magnetic [1-3]. And researches pay more attention to the electromagnetic environment change in the projectile and its influence on the sensors, which provides technological support for the sensors applications.

* To whom any correspondence should be addressed.

2. Action Principle of Magnetic Resistance Sensor

Magneto resistance effect (MRE) is a phenomenon that the resistance of material will change in the magnetic field. It can be divided into normal MRE based on Hall Effect and anisotropy MRE. The latter one aimed at strong ferromagnetic metal (iron, cobalt, nickel and the alloy of them and so on), the resistance remains when the outer magnetic field is in parallel with the magnetization direction of the metal, but it diminishes when the magnetic field direction departures. If the metal is made into a zonal lead and given the current, its resistance will change following the relationship between the magnetization direction of inside and outside magnetic field and the current direction, the same will be the bigger, and the different will be the smaller. As the resistance changing rate of perm-alloy is biggish, it is appropriate to low magnetic field. As shown in the figure 1 and figure 2.

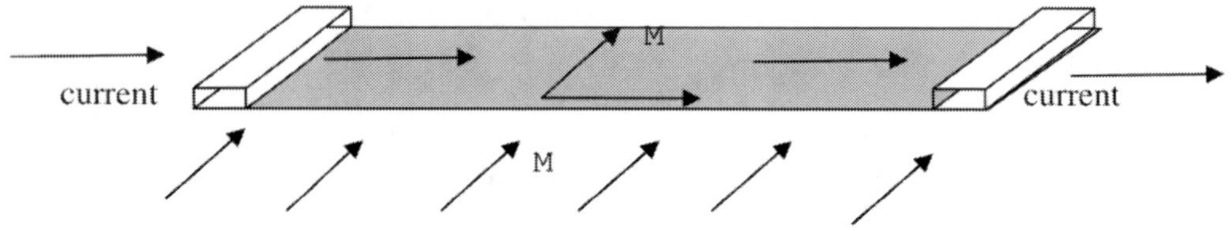

Figure 1. The component of MRE.

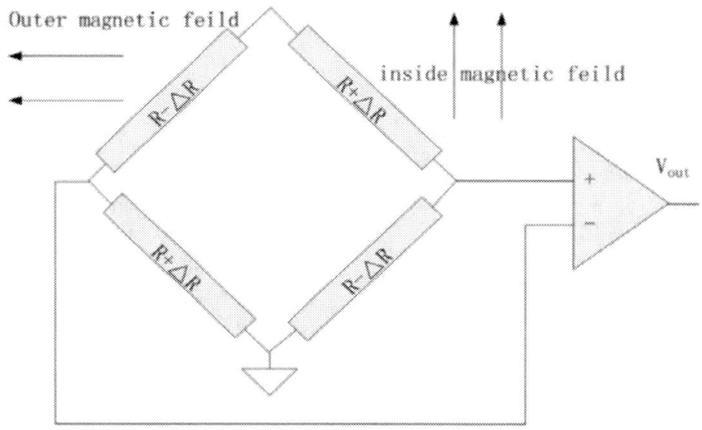

Figure 2. The Wheatstone bridge and the sensor action.

From figure 2, the Wheatstone bridge is the foundation of the sensor and it gives the action principle. Supply power voltage V_b provides the resistance with the current. If there is no magnetic field, the resistances of the four brachium points are the same, they are in balance and the output is zero. If there is magnetic field, the resistances change and the balance will be broken up, the resistances of brachium point 1 and 3 decrease and the brachium point 2 and 4 amplify, the output voltage is V_{out}. The output voltage will be $-V_{out}$ when the magnetic field reverses.

From the variation of the resistance, the relation function can be build:

$$\Delta V_{out} = \left(\frac{\Delta R}{R}\right) V_b \tag{1}$$

$$\Delta V_{out} = SMV_b \tag{2}$$

R is film resistance;
ΔR is relative variation of the resistance;
S is sensitivity of MRE;
M is the outer magnetic field value.

3. Electromagnetic Interference Models Analysis

3.1. Electromagnetic SE Analysis of Projectile Cavity

Based on the Electromagnetic SE theory [2, 5], the projectile can be classified as circular cylinders cavity. Supposing that the inside and outside radius of projectile are r_1 and r_2, and dielectric constant is ε_r, the electromagnetic E_i (the direction supposed to be perpendicular to projectile axle, u_t) is uniform inside and outside of the projectile, as well as the circular cylinder is perfect, as shown in the figure 3.

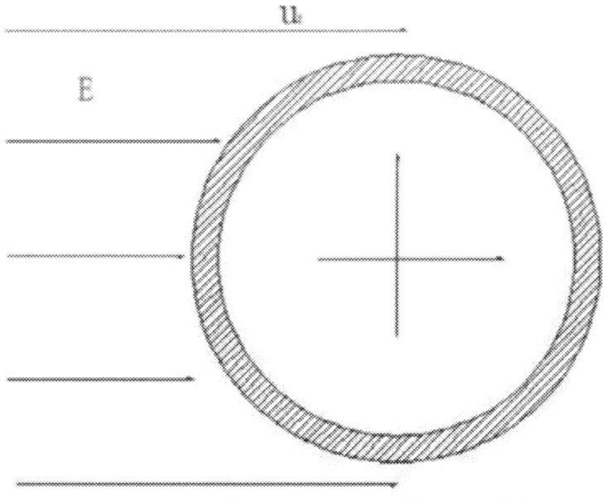

Figure 3. Equality electromagnetic outside the projectile.

The similar method is employed with the roundness cavity to solve the calculation, ignoring the tiny parts, and the electric field is given by

$$E(r) = -\nabla V\ (r) = \alpha E_i u_t \tag{3}$$

Where, α and D_c is given by:

$$\partial = 4\frac{\varepsilon_r}{D_c} \tag{4}$$

$$D_c = (\varepsilon_r + 1)^2 - \left(\frac{r_1}{r_2}\right)^2 (\varepsilon_r - 1)^2 \tag{5}$$

In that case, the thickness of projectile is $h = r_2 - r_1$; and $\varepsilon_r >> 1$. Analyzing the function (3), (4), (5), electromagnetic SE is gotten by

$$SE = 20\lg\left|\frac{1}{\alpha}\right| \approx 20\lg\left(\frac{1}{2}\frac{\varepsilon_r h}{r_1}\right) \tag{6}$$

From the equation, the conclusion can be gotten: the bigger dielectric constants ε_r, the better shielding effectiveness.

3.2. Electromagnetic Leak Analysis of Projectile Aperture

The projectile is comprised of many components; every component joins each other with the apertures. Especially the screw thread, it is one of the main reasons which result in the weakening of SE [4]. The small aperture will play a role as the antenna, which will receive and launch the electromagnetic signal and bring bad influence to the SE when it is in the field, as shown in figure 4.

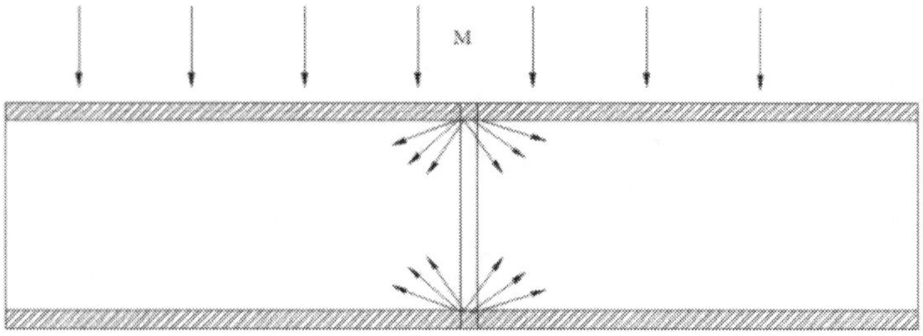

Figure 4. The Wheatstone bridge and the sensor action

Where the apertures can be looked as a linear interstice, the thickness is described as b. So the electromagnetic of the point that leaves the aperture about l in distance are then given by

$$E_g = \left(\frac{2g}{\pi e}\right)^2 e^{-\pi l/b} H_0 \frac{1}{r^2} = \left(\frac{2}{\pi e}\right)^2 \left(\frac{b}{r}\right)^2 e^{-\pi l/b} H_0 \tag{7}$$

4. Electromagnetic SE Simulation Analysis of Projectile

In the recent years, the normal SE simulation methods are the Electromagnetic module of ANSYS and CST EM STUDIO. CST EM STUDIO is the software that simulates the static electricity, static magnetic, constant current and low frequencies electromagnetic [6]. Also it can be used in the DC-100MHz EMI, sensors, induction heating, non-destructive flaw detection and electromagnetic SE. The CST EM module can be used to simulate the electromagnetic SE of projectile and study the influence on the projectile of the outside complex electromagnetic.

4.1. Outside Electromagnetic Environment

In the battlefield environment, complex electromagnetic environment has become a significant component, especially after the improvement of operational forms and the renewal of the weapons, the study on the influence of the complex electromagnetic environment becomes a new topic. With analysis of the battlefield environment, the electromagnetic environments are composed of natural electromagnetic (such as thunderbolt, geomagnetism, and static electricity) and artificial electromagnetic (such as radar, broadcast television, and radio stations). As we know, the nature of electromagnetic are difficult to predict and control, then the attention is paid on the artificial factor. If the EMI is not deliberate, the radar signal will be the main object, so it will be used in the simulation.

4.2. Simulation Model

The simulation model is based on the projectile of US army Hydra-70, the particular parameters are shown in the table 1.

Table 1. The parameters of Hydra-70

parameters	rocket	Warhead	caliber	length	HARO	Fin blades
value	MK66	Javelin	70mm	1060mm	yes	three fin blades landscape orientation roll

Base on this projectile and the built model, the math shape model is built using the parameters in table 1 and the figure building capability of CST.

4.3. The Influence on Projectile of Different Electromagnetic Direction

The EMI and electromagnetic SE will make a difference with different action direction. So we make the projectile axle as the datum line and detect the influence, and there are three different directions to be taken, namely, perpendicular to projectile axle, in parallel with projectile axle and with an intersection angle 45°. In the actual condition, the frequency cannot be a steady value; it is a variation which always ranges from 0 to 200 MHz.

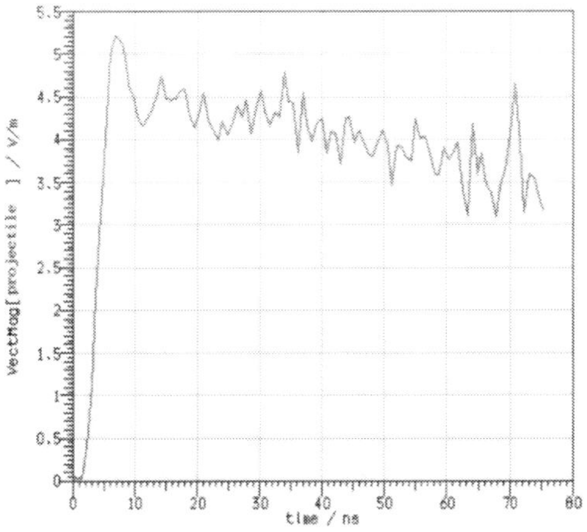

(a) perpendicular to projectile axle;

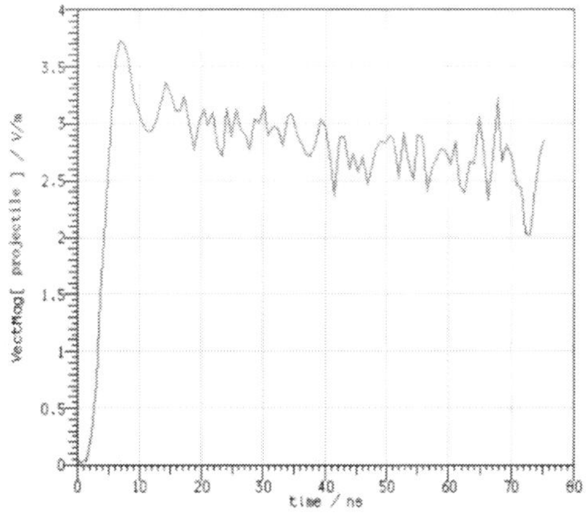

(b) in parallel with projectile axle;

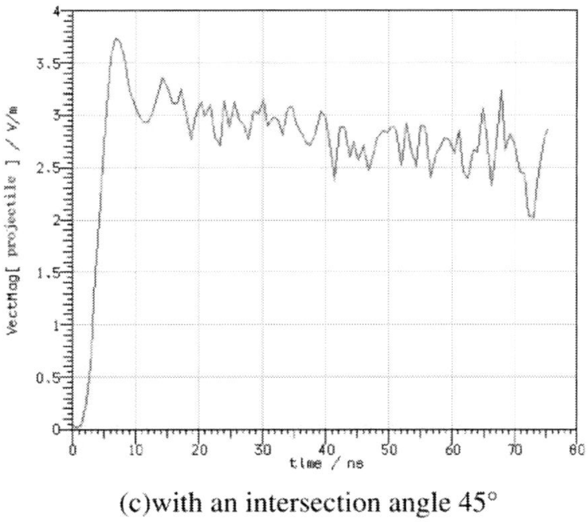

(c)with an intersection angle 45°

Figure 5. Electromagnetic SE with different action direction

From the simulation (see Figure 5), the electromagnetic in different directions has an influence on the environment inside the projectile. In the actual condition, the force value is around 4 v m^{-1} and the rule of value change is similar. They all increase at first and surge in the following time. The intension of the electromagnetic which is perpendicular to projectile axle can reach 5 v m^{-1} and larger than the other two.

4.4. The Influence on Projectile of Different Frequencies

With the different outer environment, the influence will be changed in different areas or different conditions, so the range of signal frequency must be taken into consideration and the simulation can be closer to the actual battle environment. The radar signals are based on the US Army standard with different frequencies 0~200 MHz. Because the operational forms change, the both sides will not contact with each other. So the radar signal can be regarded as the planar waves.

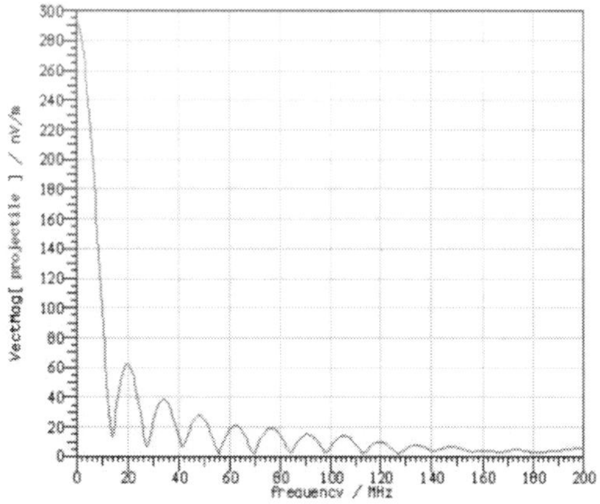

(a) perpendicular to projectile axle;

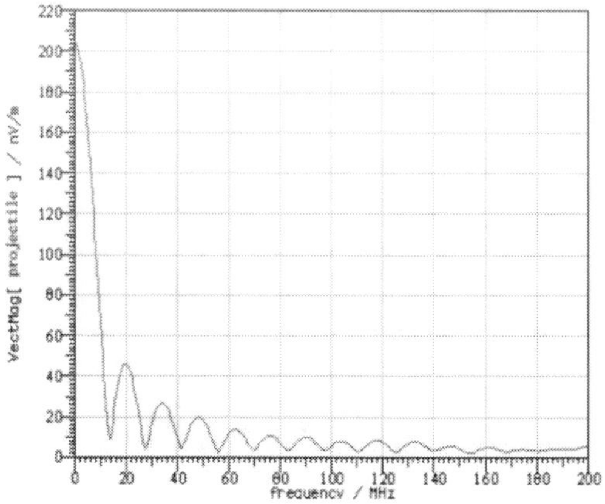

(b) in parallel with projectile axle;

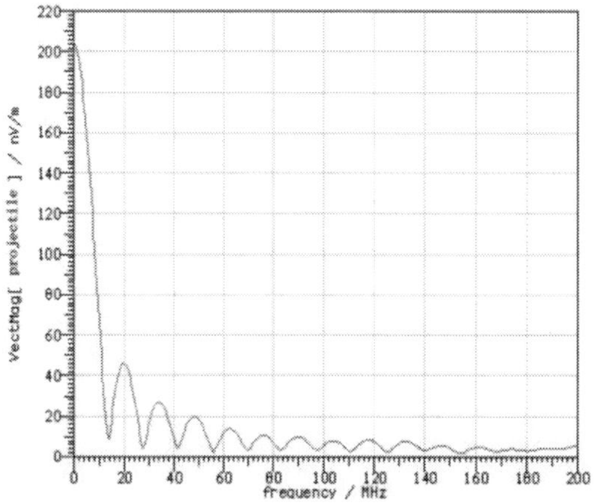

(c) with an intersection angle 45°

Figure 6. Electromagnetic SE with different frequency

From the figure 6, the influence is not intense that it just drops down so fast in the frequency range, especially in 0~15 MHz the value changes 95%. And the value can be ignored when the frequencies reach 160MHz. The rules are similar, which is changing from the beginning to the end, and the value has a little difference because the directions are different.

5. Conclusion

CST simulation analysis indicates that the battlefield complex electromagnetic environment influence the environment inside the projectile, especially for the its electronic components and capability, but the influence depends on different condition. When the frequency or the direction changes, the influence varies small because the SE is so well, and the changing rules are similar. If the value is small, the influence to magnetic resistance sensor will be small too. With different conditions, it makes a small difference and the rules are similar. From the simulation results, the values are small. The battlefield complex electromagnetic environment can change the environment inside the projectile, but the changes are not obvious for the SE is working. As the situation is clear, it is useful for the researchers to make an improvement and perfect the design of future ammunition.

References

[1] Lenz J E 1990 *Solid-State Sensor and Actuator Workshop, 4th Technical Digest IEEE(Hilton Head Island, SC)* vol 78 p 973
[2] Salvatore C, Rodolfo A and Giampiero L 2010 *Electromagnetic Shielding* (China: China Machine Press)
[3] Shi L Y, Song W Y and Yao Z M 2009*International Conference on Measuring Technology and Mechatronics Automation ICMA 2009 (Changchun)* p 4105
[4] Wang Z, Mi D and Lu Z C 2008 *Electrical Measurement&Instrumentation(in chinese)* **2** 20
[5] Agrawal A K, Price H I and Gurbaxani S H 1980 *IEEE Trans. Electromagn Compat EMC* **2** 119
[6] Zhang M 2004 *CST Microwave Studio* (Chengdu: Press of Electronic Science and Technology University)

Anti-electromagnetic interference analysis of equivalent circuit of ion channel based on the Hodgkin-Huxley model

J Chu[1*], X L Chang[1], M Zhao[1], M H Man[1], L Yuan[2], M Wei[1]
1 Institute of Electrostatic & Electromagnetic Protection, Mechanical Engineering College, Number 97 of Peace west road, Shijiazhuang, China
2 Department of Computer Engineering, Mechanical Engineering College, Number 97 of Peace west road, Shijiazhuang, China

E-mail: chujie2000@163.com

Abstract. With the continuous improvement of circuit integration and working clock frequency in the electronic system, it is increasingly easy for the system to be affected by electromagnetic waves, and electromagnetic susceptibility and vulnerability become more severe. However, living beings in nature have shown extraordinary compatibility, immunity and adaptability to the electromagnetism at the same time. In addition, the ion channel on the neuron cytomembrane is a typical representation of "bioelectrical immunity". So the Hodgkin-Huxley circuit model with one capacitor in parallel with some power supplies and resistors was adopted to simulate the ion channel on the neuron cytomembrane. Through analysis, the circuit model can be used to simulate some electrical characteristics of biological neuron cells, and then acquire a certain level of anti-electromagnetic interference ability. This method will be useful for improving the reliability, compatibility and anti-interference capability of the electronic system in the complicated electromagnetic environment.

1. Introduction

With the development of electronic technology, integrate circuits are widely used in the field of military and aerospace. At the same time, there are always all kinds of electromagnetic interference (EMI) in the application environments [1-2], such as lightning electromagnetic pulse (LEMP) and nuclear electromagnetic pulse (NEMP). Lightning electromagnetic effect consists of electrostatic induction and electromagnetic induction. Integrate circuits are especially susceptible to some short pulse disturbances aroused from electromagnetic induction of LEMP. Lightning is the result of sparks between electrical clouds and the ground. Its current reaches tens to thousands amperes. Lightning current has very high steepness of pulse edge and peak. There will be a transient electromagnetic field around it. The transient electromagnetic field will induce a big induced electromotive force to interfere with nearby electronic equipments. Furthermore, LEMP can also disturb electronic equipments in a short pulse disturbances way. Especially, circuits are especially susceptible to these short pulse disturbances whose rising edges and pulse lengths are in the range of microseconds. In the same way, NEMP will also disturb integrated circuits. Traditional electromagnetic protection measures consist of grounding, shielding and filtering etc. There are many unexpected kinds of short pulse disturbances in

* To whom any correspondence should be addressed.

the war field and the space environment. Thus relying solely on traditional protective effect is not good.

Consequently, bionic defensive way for EMI was brought forward [3-4]. The way is to imitate related biological structure and characterises to improve the anti-electromagnetic interference, fault tolerance and self-repairing ability of electronic systems further [5-6]. This paper analyses the ion channel equivalent circuit of classic Hodgkin-Huxley model to improve the anti-interference ability of circuits for short pulse disturbances.

2. Hodgkin-Huxley mathematical and circuit model

As shown in figure 1, this creates a simple circuit for measuring the potential difference between the intracellular space and the extracellular fluid. This potential difference is called the membrane voltage. A stimulus current (the 'input' to the cell) can be delivered through the electrode and the resulting changes in membrane voltage (the 'output' from the cell) recorded for later study.

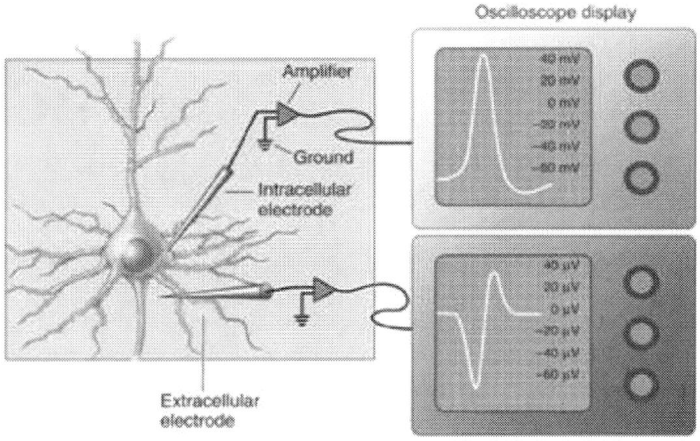

Figure 1. Simple electrophysiology setup for testing membrane voltage.

On the basis of such observations, we can find that the membrane voltage of cell rises and falls rapidly, as shown in figure 2. The short-lasting event is called an "action potential". It looks as if a digital pulse. The action potential can propagate over long distance without weakening, and make the neuron realize the long distance transmission of neural information. Therefore, cell membrane's transferring signal performance is better than that of the circuit.

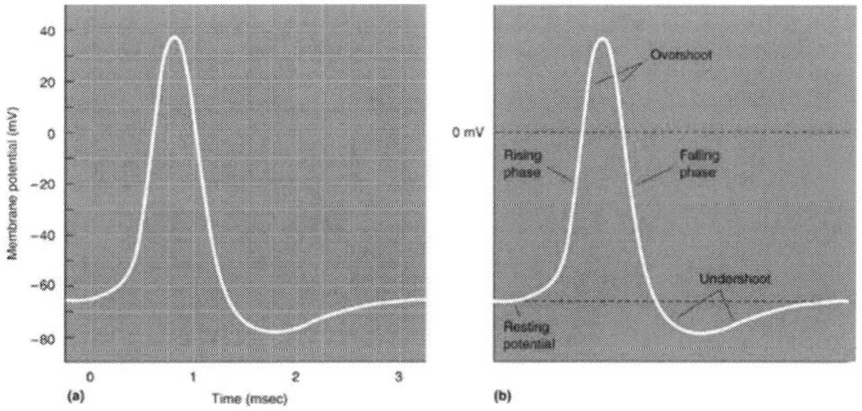

Figure 2. The ups and downs of an action potential of some cell membrane.

Journal of Physics: Conference Series **418** (2013) 012083 doi:10.1088/1742-6596/418/1/012083

The Hodgkin-Huxley model (H-H model) is a mathematical model (a type of scientific model) that describes how action potentials in neurons are initiated and propagated. It is a set of nonlinear ordinary differential equations that approximates the electrical characteristics of excitable cells such as neurons and cardiac myocytes. Alan Lloyd Hodgkin and Andrew Huxley described the model in 1952 to explain the ionic mechanisms underlying the initiation and propagation of action potentials in the squid giant axon [7]. They received the 1963 Nobel Prize in Physiology or Medicine for this work.

The components of typical Hodgkin-Huxley model are shown in the figure 3 [8]. Each component of an excitable cell has a biophysical analogy. As shown in Figure 3, we can find that the lipid bilayer membrane separating the intracellular space from the extracellular fluid is capable of storing charge and so acts as a capacitor (C_M). Voltage-gated ion channels are represented by nonlinear electrical sodium (R_{Na}) and potassium (R_K) resistance. This was later shown to be mediated by voltage-gated cation channel proteins, each of which has an open probability that is voltage-dependent. Leak channels are represented by linear resistance (R_L). The electrochemical gradients driving the flow of ions are represented by batteries (E_L, E_{Na} and E_K), the values of which are determined from the Nernst potential of the ionic species of interest. Finally, ion pumps are represented by stimulus current sources.

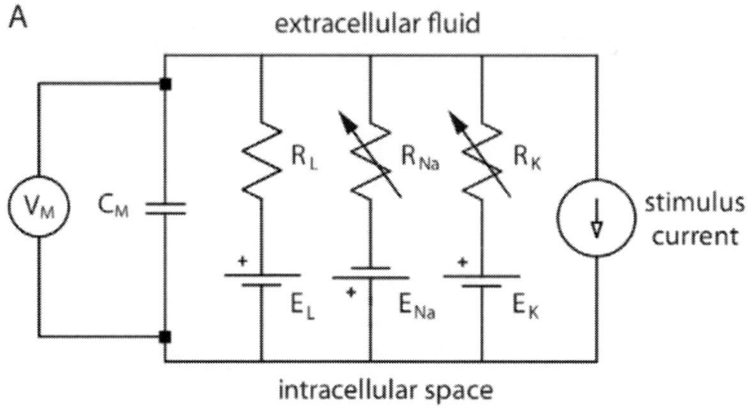

Figure 3. The Hodgkin-Huxley equivalent circuit.

In this case, the behaviour of the nonlinear resistors is described by a complex system of coupled differential equations (1), (2), (3) and (4) [8].

$$\frac{dv}{dt} = f(v,h,m,n,i) = -120m^3h(v-45) - 36n^4(v+82) - 0.5(v+60) \tag{1}$$

$$\frac{dh}{dt} = f(v,h) = 0.07\mathrm{e}^{-\frac{1}{20}v-\frac{7}{2}}(1-h) - \frac{1.0h}{1+\mathrm{e}^{-\frac{1}{10}v-4}} \tag{2}$$

$$\frac{dm}{dt} = f(v,m) = \frac{1}{10}\frac{(v+45)(1-m)}{1-\mathrm{e}^{-\frac{1}{10}v-\frac{9}{2}}} - 4.0\mathrm{e}^{-\frac{1}{18}v-\frac{35}{9}}m \tag{3}$$

$$\frac{dn}{dt} = f(v,n) = \frac{0.1\left(\frac{1}{10}v+6\right)(1-n)}{1-\mathrm{e}^{-\frac{1}{10}v-6}} - 0.125\mathrm{e}^{-\frac{1}{80}v-\frac{7}{8}}n \tag{4}$$

The variable v corresponds to the observed membrane voltage, the variable i corresponds to the stimulus current, and the hidden variables h, m, n correspond to gating parameters of the sodium and potassium conductance. The equivalent circuit with H-H model can reproduce the behaviour of the modelled systems to a large degree, such as the ups and downs of action potential of the cell membrane, as shown in the figure 2.

3. The analysis of anti-electromagnetic interference ion channel circuit with Hodgkin-Huxley model

It is found that the equivalent circuits with H-H model are not sensitive to short pulse disturbances, which rising edges and pulse lengths are in the range of microseconds (10^{-6} s). This is because the capacitor C_M in equivalent circuits plays a stabilizing role. It is well known that the voltage of an ideal capacitance is continuous. So capacitor's voltage can't spring. When a short pulse disturbance comes, the capacitor will charge and delay the cell membrane voltage V_M up. If the short pulse is enough short, V_M will not spring rapidly.

The LEMP and NEMP can generate many kinds of short electromagnetic pulse. The average wave first time of surge aroused from LEMP is $25\pm2.5\mu s$ [9]. The average half peak time of LEMP surge is $55\pm15\mu s$. NEMP electric field change more quickly. It can reach a peak value in several nanoseconds (10^{-9} s) [10]. Pulse width of NEMP continues between dozens of nanoseconds and microseconds. The time of action potential of the cell membrane is milliseconds (10^{-3} s). We can see that the pulse width of LEMP and NEMP is so short that the action potential of equivalent circuits will not change sharply in a short time. Accordingly, the equivalent circuits of H-H model have some degree anti-electromagnetic interference ability for short pulse disturbances like LEMP and NEMP.

4. Conclusions

According to the conclusion of "capacitor voltage can't spring", the equivalent circuits of H-H model can resist the short pulse disturbances aroused from LEMP and NEMP. If the design is applied to the electronic system in the war field and the space environment, it will improve the anti-electromagnetic interference ability of the circuit system.

Acknowledgments

This project has been supported by the National Key Lab Fund (9140C8702020803) and the National Natural Science Fund(51207167).

References

[1] Liu S H 2005 *Journal of Institute of Command and Technology (in Chinese)* **16** 1
[2] Liu S H and SUN G Z 2008 *Journal of the Academy of Equipment Command & Technology (in Chinese)* **19** 1
[3] Liu S H, Chu J and Yuan L 2009 *Journal of Academy of Armored Force Engineering (in Chinese)* **23** 1
[4] Liu S H, Yuan L and Chu J 2009 *Chinese Journal of Nature (in Chinese)* **31** 1
[5] Man M H, Ju Z Q, Yuan Q Y, Zhou Y X, Yuan L and Liu S H 2011 *High Voltage Engineering (in Chinese)* **37** 375
[6] Chu J, Man M H, Liu S H, Wei M, Yuan L, Ju Z Q and Chang X L 2011 *Proc. Int. Conf. on Network Engineering and Computer Science.* (Xi'an China) vol 11 (library of congress) p 2670
[7] Hodgkin A L and Huxley A F 1952 *J. Physiol.* **117** 500
[8] Cornforth T W, Kim K J and Lipson H 2010 *Proc. Int. Conf. on Evolvable Systems:From Biology to Hardware* (York UK) vol LNCS 6274 (Springer-Verlag Berlin Heidelberg) p 157
[9] Wei M 2009 *Lightning Electromagnetic Pulse and Defence* (Beijing: National Defence Industry Press)
[10] Fan L S and Cui Y Z 2009 *Simulation Technology of Electromagnetic Environment* (Beijing: National Defence Industry Press)

Experimental investigation on time-domain testing of shielding effectiveness of composite materials against EMP

QingguoWang*, Zhaoming Qu, SiliangQin
Research Institute of Electrostatic and Electromagnetic Protection, Mechanical Engineering College, Heping West Road, Shijiazhuang 050003, China

Email: qwang_99@hotmail.com

Abstract: The shielding effectiveness (SE) of materials with different thickness is experimentally investigated using an improved time-domain testing system of SE and a window on a shielding enclosure in a GTEM cell. Experimental results show that the proper test position of transient electric probe (E-probe) within the enclosure is determined to make the system capable of finding out the maximum peak value of pulsed electric field inside enclosure. The SE test experiments for several shielding composite materials fabricated using nickel-plated carbon fiber proved that the SE of materials for electromagnetic pulse (EMP) is highly dependent on the thickness of materials and the proposed testing method can obtain high repeatability, and thus it can be used as a standard test system to assess the shielding effectiveness of materials for pulsed electric field.

1. Introduction

Electromagnetic pulses (EMP), as a kind of wideband and high intensity transient signal, can strongly interference electronics or cause damages, so shielding materials were normally applied to protect sensitive electronic system against electromagnetic environment. However, how to characterize the shielding effectiveness for EMP of materials is still needing further research and attracting more and more attention. In tradition, shielding effectiveness (SE) is defined based on frequency domain. But this definition is not able to provide the amplitude of field that the electronics is exposed to and so cannot assess the possible interference or damage to electronics. So the shielding effectiveness based on frequency domain is failed in dealing with protection ability of electronics against EMP. Currently, time-domain testing methods are mainly dependent on the improved technique based on traditional standard frequency domain methods. Koc and Camp from Hannover University investigated the shielding properties of texture-based composite and carried out time-domain testing of SE using open TEM waveguide [1-2]. Herlemann and Koch measured the SE of conductive texture bags using open TEM waveguide and transient field testing system and, according to the formula proposed by Klinkenbush [3], calculated the transient SE [4]. Afterward, they carried out SE tests of a metal enclosure (size 3 m × 3 m × 2.5 m, with 1.9 m × 1.9 m window) using frequency domain in anechoic chamber and using time domain in GTEM cell for different shielding materials [5].

EMP simulator is normally big and testing working volume is small, so it is difficult to be applied for SE test directly using windows method. For this reason, an improved method using an enclosure with window for time domain SE test for materials is proposed. The test installation is setup and the

* To whom any correspondence should be addressed.

probe position is determined. Furthermore, the SE of several composite materials (mixed with nickel-plated carbon fiber) with different thickness are measured. Test results show that the developed time-domain SE system reached high repeatability and could be applied for the SE assessment of materials.

2. The definition of SE for EMP

EMP, especially for short time rise-time EMP, has abundant frequency components. Because of the dispersion effect for different frequency electromagnetic wave of materials, EMP field will change not only in amplitude, but also in EMP waveform after transmitting through shielding materials. In ideal cases, the definition of SE should consider both the change of amplitude and waveform (frequency components) of EMP. However, because the interferences or damage effects for electronics by EMP may vary with frequency, it is hardly to define the SE for EMP using frequency changes. At this moment, the authors of the present paper will simply use the peak radiation field of EMP to define the SE of materials. Therefore, SE of materials is defined as the ratio of peak values of field (E or H field) or peak values of power of EMP with and without shielding materials [6], which is indicated by $SE_{E(H)tp}$, although some other researchers advised the definition to use rise-time [7]:

$$E_{E(H)tp} = 20\log\frac{\left|E(H)_{tp-unshielded}\right|}{\left|E(H)_{tp-shielded}\right|} \quad (1)$$

$$\text{or } SE_{P_{tp}} = 10\log\frac{\left|P_{tp-unshielded}\right|}{\left|P_{tp-shielded}\right|} \quad (2)$$

where $E(H)_{tpunshielded}$ is the peak value of electric field or magnetic field without shielding materials, and $E(H)_{tpshielded}$ is the peak value of the electric or magnetic field when shielding materials applied, SE_{PtP} is shielding effectiveness if using power of measured radiation field, $P_{tp\text{-}unshielded}$ and $P_{tp\text{-}shielded}$ is the peak power measured without and with shielding materials applied.

3. Development of testing system

To measure the shielding effectiveness of materials for EMP, a testing system is developed based on a window method with a shielding enclosure.

3.1 Testing setup

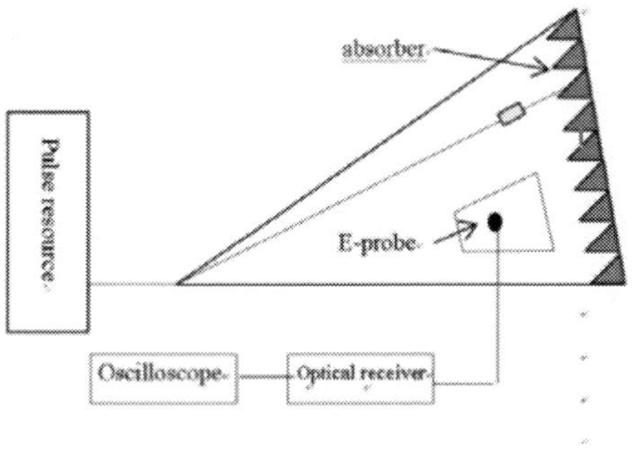

Figure 1. Time-domain testing system for shielding effectiveness.

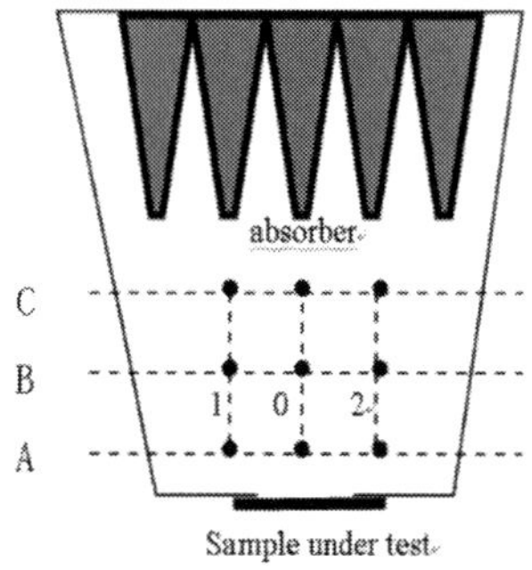

Figure 2. Different test points in the enclosure.

The developed time-domain SE testing system mainly includes GTEM cell, square wave generator, E-probe for transient EMP electric field, optical receiver and oscilloscope etc, as shown in figure 1. The GTEM cell is feed by coaxial cable using square wave pulse generator and produces pulse radiated field. The applied transient electrical field probe (E-probe) as electrical-optical converting unit is composed of sensor, transmitting fiber and optical receiver. The optical fiber is applied in transmitting field signals to reduce the field interferences by field-cable coupling. To avoid possible resonance within the enclosure, a trapezium shaped enclosure is applied to provide unparallel side walls of the enclosure. For the parallel two walls of front panel and bottom of the enclosure, absorbers are mounted on the opposite bottom of the front panel to reduce the resonance effects between the two walls. (figure 2). During the installation of absorbers, conductive glue is applied to keep electrical connection well with the enclosure.

3.2 The selection of field sampling positions within enclosure

When a shielding box is applied to measure SE, a key problem is to find out the correct sampling position within the box which is capable for probe to detect the maximum peak value of field. For this purpose, nine locations within the box were tested. The nine locations are chosen according to the symmetry property of box at three lines (A, B, C) at the distance of 10 cm, 15 cm and 20 cm from window (figure 2). The central line vertical to the front panel (with window) is labeled "0" and the measured field is labeled A_0, B_0 and C_0. The other two lines (10 cm away from central line "0") are labeled "1", "2" and the measured field is labeled A_1, B_1, C_1 and A_2, B_2, C_2, respectively. Figure 3 gives the waveform of input waveform in the GTEM cell and figure 4 gives the measured pulsed field (described by voltage reading from oscilloscope) after shielding materials installed. Table 1 presents the measured values of field (voltage from oscilloscope, mv) at nine locations. By comparing with figure 3 and 4, it can be found that the square wave within GTEM cell is changed to decayed oscillating wave at points A_0, B_0 and C_0 inside enclosure because of coupling of electromagnetic wave with window and shielding materials. It is also clearly indicated that, alone the central line and at the nearest location from test window (A_0), the measured field is higher than all other locations. So, in the following measurements, the field will be measured at the point A_0, which is at central line and close to the window and material under-test at 10 cm distance from the window.

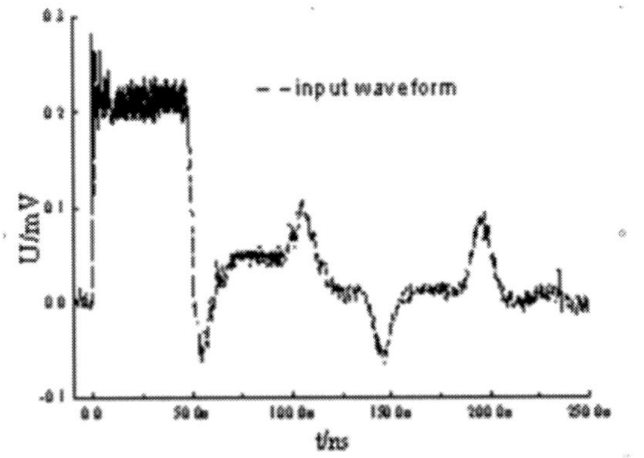

Figure 3. Input field waveform in the GTEM cell.

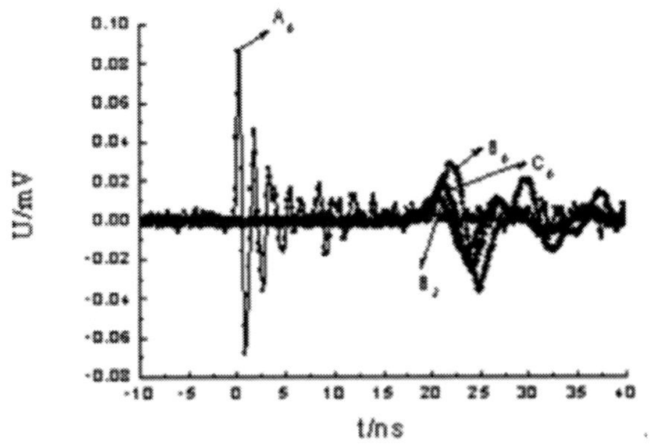

Figure 4. Waveform measured within enclosure at different test locations.

Table1. Measured field value (indicated by voltage from oscilloscope) at different test points in box.

Test points	A_0	B_0	C_0	A_1	B_1	C_1	A_2	B_2	C_2
voltage mv^{-1}	0.093	0.086	0.065	0.075	0.058	0.043	0.074	0.056	0.044

3.3 Verification examination

To examine the effectiveness of the developed system for measuring SE of materials for EMP, the experimental system is setup (figure 5) and three shielding materials of nickel-plated carbon fiber mixed composites with thickness 1 mm, 2 mm and 3 mm are chosen to do the SE tests. During the test, the pulse-width of square wave resource is adjustable, the output of resource has good repeatability and the rise-time of square wave is less than 1 ns. The testing box with window is placed in the rare position and in the central line of GTEM cell and the E-probe is place in the central coaxial ling (vertical to front panel with window) of the box and 10 cm away from the window and sample under test.

a) Test system b) Unshielded c) Shielded

Figure 5. The testing system for transient pulsed field SE measurement.

Figure 6 presents the time domain SE test results for three samples with different thickness. The three shielding materials are made of short nickel-plated carbon fiber and rubber materials at a known volume ratio. From the results, it can be found that the SE increases with the increase of thickness of materials. But with the increase of applied input voltage of square wave, i.e. with the increase of electric field E, the SE of materials decreases. This is reasonable because the materials have nonlinear response to applied electric field. It can be also concluded from test results that, in the design of electromagnetic shielding materials, the thickness of materials should be greater as possible and the content of inclusion particles should be higher to obtain higher conductivity and shielding ability.

For a standard test method, the test repeatability is a key requirement. This is because, for a stand test method, the test results should be the same when different researchers do the same particular test. For this purpose, the repeatability test is carried out using a 3 mm thickness shielding rubber material under the condition of 4 kV input square wave.

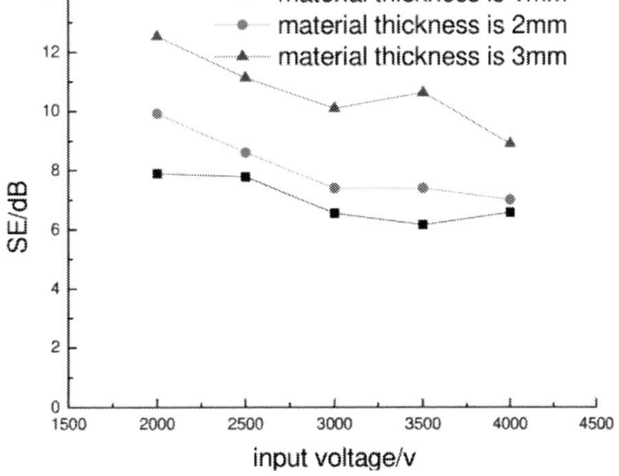

Figure 6. Dependence of Shielding effectiveness on input voltage of square wave

for different thickness of materials.

Table 2 shows the test results. It can be concluded that the developed testing system achieves highly repeatability and it is suitable to be applied as a standard testing system for pulsed SE test of materials.

Table 2. Test results of repeatability.

test	1	2	3	4	5
SE/dB	8.88	8.86	8.87	8.85	8.85

4. Conclusions

An improved time domain SE testing system is proposed for materials under the condition of pulsed electric field, high repeatability has achieved and can be applied as a kind of standard test system. The EMP shielding effectiveness has also been investigated for materials with different thickness. It is found that the SE of materials for pulsed field depends not only on the thickness of materials, but also on the peak value of input pulse voltage, because of dispersion effect of materials to different frequency wave, it can be predicted that the SE of materials also depends on the rise-time or waveform of input EMP. These problems will be further investigated in the future research.

References

[1] Koch M, Camp M, Kebel R, Garbe H, Sabath F and Nitsch D 2003 *Proc. 15th International Zurich Symposium and Technical Exhibition on Electromagnetic Compatibility* (Zurich, Swiss) p 379
[2] Herlermann H, Koch M, Camp M, Bausen A and Sabath F 2004 *Proc. International Conference and Exhibition on Electromagnetic Compatibility* (York, England)
[3] Klinkenbusch L 2005 *IEEE Trans. On Electromagnetic Compatibility* **47** 589
[4] Herlermann H and Koch M 2007 *Advances in Radio Science* **5** 75
[5] Herlermann H and Koch M 2008 *Advances in Radio Science* **6** 293
[6] Zhou B, Chen B, Gao C and Chen Z 2000 *Chines Journal of Radio Science (in Chinese)* **15** 251
[7] Zhou B, Gao C and Ren H 2003 *Proc. Asia-Pacific Conf. on Environmental Electromagnetics* (Hangzhou), p 562
[8] Celozzi S, Araneo R and Lovat G 2007 *Electromagnetic Shielding* (John Wiley Ltd)

The overview of measuring the shielding effectiveness of the materials in complex electromagnetic environment

E W Cheng*, **Q G Wang, Y F Liu**
Institute of electrostatic and electromagnetic protection, Mechanical Engineering College, Heping west road, Shijiazhuang, China

E-mail: ew_cheng@163.com

Abstract. The wide-spreading use of innovative lightweight materials like advanced composites and metal coated plastics in several fields of applications has pushed toward an increasing interest in the characterization of the shielding performances of test materials. In the paper, we revisit the conventional methods of measuring shielding effectiveness of materials, and introduce the nested reverberation chamber methods (two reverberation chambers). The deficiencies in the existing approaches were illustrated. We advised the new approach should account for aperture, cavity size, and chamber loading effects, for which some current reverberation chamber approaches do not correctly account. This work is useful in estimating the shielding ability of the materials in complex electromagnetic environment.

1. Introduction

Materials used for the shielding of electromagnetic fields range from very simple metallic wire meshes to more sophisticated advanced composite materials (ACMs). ACMs are typically made by pressing and binding together several matrix-reinforced laminates with different fiber orientations. ACMs are becoming a very popular material because the main advantages of these materials with respect to metals are superior mechanical and chemical properties, such as lower weight, higher stiffness and strength, lower corrosion, lower tooling costs, and greater ease of fabrication. ACMs have been used for some time within the spacecraft and aircraft structure industry. In recent years, composite materials have begun to spill over into other applications, such as cases and housings used in electronics, computer and business equipment, and in naval and industrial environments, to mention only a few applications [1-3].

Shielding effectiveness (SE) is typically used to assess the shielding property of a material [4, 5], and is defined as the ratio of the transmitted power through the material to the incident power [6-10]. The SE can be obtained through numerical modelling or through measurements.

2. Shielding effectiveness test methods

2.1. Conventional methods of measuring shielding

The coaxial fixture is a commonly used measurement technique for determining far-field equivalent SE [11]. Dual TEM cells have been used to measure near-field SE [12]. However, those approaches determine SE for only a very limited set of incident wave conditions. Figure 1 illustrates such

[*] To whom any correspondence should be addressed.

conditions. But in most applications, ACMs are exposed to complex electromagnetic environments where fields are incident on the material with various polarizations and angles of incidence [13, 14]. This is illustrated in figure 2, where the materials are incident on multiple-incidence.

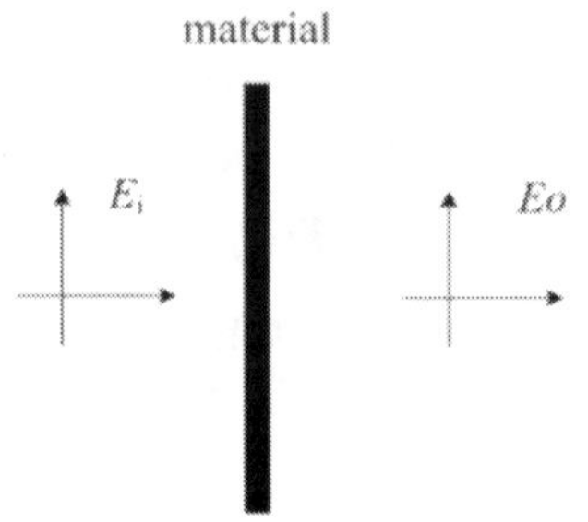

Figure 1. Conventional methods of measuring the SE of the materials.

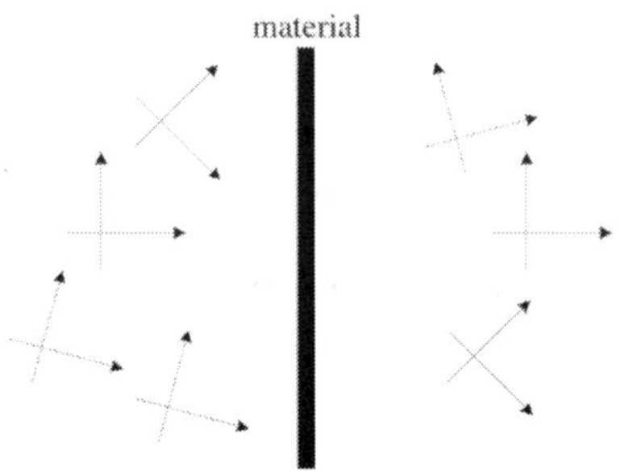

Figure 2. The materials of exposure in the realistic electromagnetic environment.

The electromagnetic field and the electromagnetic radiation exists in lots of instruments such as inside the boxes for avionics and electronic equipment, aircraft cavity, and instrument enclosures, the electromagnetic field coupling from exterior and the electromagnetic radiation of the interior electronic device, which will reflect and refract for times with inner surface, so the electromagnetic environment is very intricate. In those applications, the materials are exposed to complex electromagnetic environments where fields are incident on the material with various polarizations and angles of incidence [15-17]. A reverberation chamber SE test which differs from traditional SE measurement technique offers such an environment [18, 19].

2.2. Nested reverberation chamber method

A reverberation chamber is an electrically large, highly conductive enclosed cavity or chamber used to perform electromagnetic measurements (both emissions and immunity) on electronic equipment. When the chamber is excited with RF energy, the resulting multi-mode electromagnetic environment can be “stirred” by the mechanical tuner/stirrer. The resulting environment is statistically uniform and statistically isotropic (i.e., the energy was arrived from all aspect angles and at all polarizations) when the field is averaged over a sufficient number of positions of the mechanical tuner/ stirrer [20].

There are many test techniques which are available to evaluate the shielding performance of gaskets, window materials, and other system configurations designed to provide shielding. Typically, a SE measurement compares the electromagnetic environment with and without the shielding material in place. SE test techniques are described in many sources [6-10]. Unfortunately, many SE measurements are noted for their lack of repeatability and facility-to-facility comparability. Some of these discrepancies can be attributed to the test technique while the test article itself causes others. Many factors such as the condition of mating surfaces and torquing of fasteners can markedly affect the repeatability of SE measurements.

For many shielding designs (e.g., gaskets, windows, etc.), reverberation testing uses a “nested chamber” approach (e.g., a reverberation chamber within a reverberation chamber). A receiving antenna and paddle wheel tuner are installed in a test fixture to detect any RF energy that "leaks" into the fixture. Figure 3 illustrates such an experimental setup.

Nested reverberation chambers have been used in the past to determine the SE of materials [21, 22]. However, after we revisited the methods, we found that all the existing standards for the SE measurements leave some open questions.

Firstly, the SE definition gives incorrect results.

The SE is commonly defined as:

$$SE = -10\log_{10}(P_{ic,s} / P_{oc,s}) \tag{1}$$

Where $P_{ic,s}$ is the power received inside the inner chamber with a sample, and $P_{oc,s}$ is the power received in the outer chamber with a sample in the aperture and source in the outer chamber. Throughout this paper, received powers and power densities are actually ensemble averages that are obtained from a large number of stirrer (tuner) positions, where stirrers are located in both the inner and outer chamber. Theoretically, if no sample is in the aperture, the SE should go to zero [18]. The problem with this approach is that it does not accurately account for the amount of power that can actually couple into the inner chamber's aperture; as a result, the measurement SE based on equation (1) will not in general go to zero with no sample in the aperture for all possible chamber and aperture sizes. This is illustrated in figure 4, where the SE is plotted as a function of frequency.

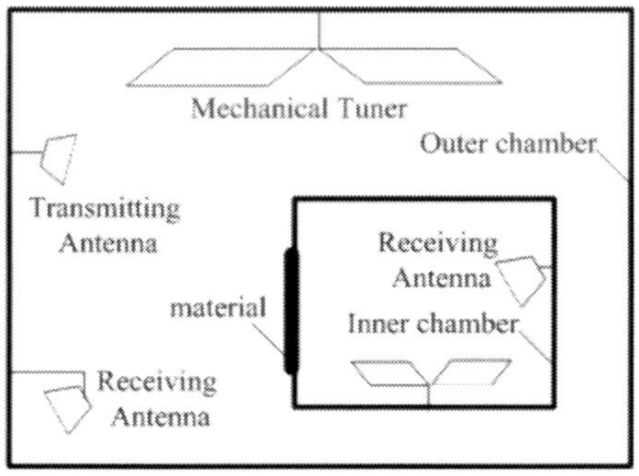

Figure 3. Typical test fixture installation for material testing.

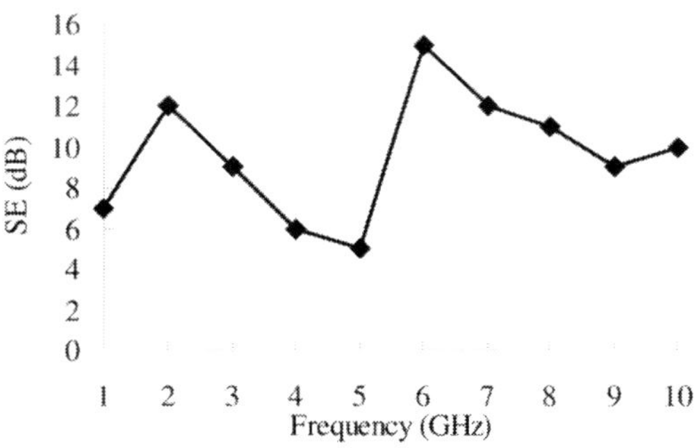

Figure 4. SE obtained with no sample in the aperture.

These measurements were performed in a nested chamber with no sample in the aperture. The outer chamber has dimensions of 2.8 m × 3.2 m × 4 m, and the inner chamber has dimensions of 1.6 m × 1.2 m × 1.2 m, with an aperture size of 0.25 m × 0.25 m. Ridged horn antennas were used as the receiving and transmitting antennas and the inner chamber was placed on the center of the floor of the outer chamber. This figure shows that SE does not yield zero in the no-sample case.

Secondly, when the inner chamber is placed into the outer chamber, there is the possibility that the small chamber will "load" the chamber. If the small chamber loads the big chamber, the energy absorbed by the inner chamber is no longer available to generate the desired environment, and the EM environment may be not a statistically uniform environment [23]. For this reason, the inner chamber is always small with linear dimensions less than 2 m. The difficulty here is that in practice, tested chambers are usually too small to have a complex mechanical stirring system and a receiving antenna. On the other hand, it is still problematic even if the tested enclosure is large enough to be installed with a mechanical stirrer since this adds internal structure to the enclosure in an uncontrolled way.

The principle of nested reverberation chamber is the creation of a statistically uniform field distribution in the inner chamber. To achieve a statistically uniform and statistically isotropic environment, means other than mechanical tuners may be acceptable. Frequency stirring is an alternative way to mechanical stirring to achieve statistical field uniformity in an over-moded chamber or enclosure, and has the potential to be used in the shielding effectiveness measurements (SE) of materials [24]. Holloway et al. have validated the frequency stirring approach by comparing it with the results obtained from mechanical stirring [25, 26], and the testing approach was illustrated in figure 5.

Finally, the use of reverberation chambers for the SE measurement of materials should fulfill the following requirements: to be account for aperture, cavity size, and chamber loading effects, to be easy to be implemented by technical personnel, to be not too expensive in terms of testing time and equipment, to be reliable and meaningful of the shielding performance of the materials in case of real electromagnetic environment.

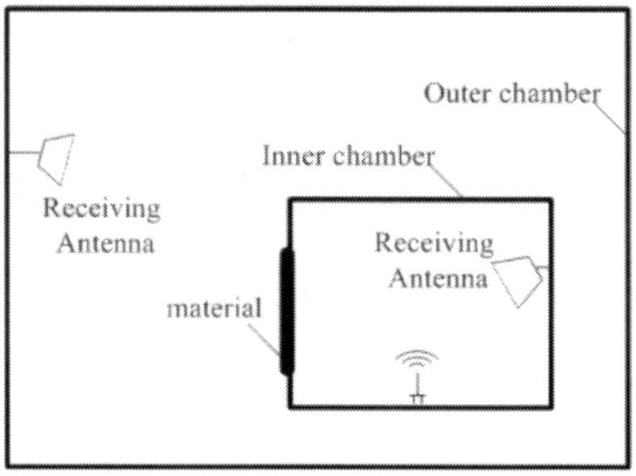

Figure 5. Frequency stirring approach of material testing.

3. Conclusions

In this paper, we have overviewed some techniques for measuring the SE of materials, and also made comparisons to existing reverberation techniques. The nested reverberation chamber approach can provide the realistic environment, but there are some open questions, such as loading effects, normative aperture and cavity size, shielding effectiveness definition, the correlation with conventional method results. In these comparisons, deficiencies in the existing approaches were illustrated.

Acknowledgments

This work is supported by National Science Foundation of China under the grant number of No. 51107147.

References

[1] Pilato L A and Michno M J 1994 *Advanced Composite Materials Berlin* (Springer-Verlag)
[2] Flower H M 1995 *High Performance Material in Aerospace London* (Chapman & Hall)
[3] Evans R W 1997 *Washington DC NASA Contractor* **p** 4784
[4] Casey K F 1978 Proc *IEEE Int. Symp. Electromagnetic Compatibility (Atlanta)* **p** 228
[5] Wilson P F and Ma M T 1986 *U.S. Department of Commerce National Bureau of Standards Boulder, CO, NBS Tech. Note* **p** 1095
[6] *ASTM-ES7 and ASTM-D4935 Standard for Measuring the Shielding Effectiveness in the Far Field* vol. 10.02, ASTM, Philadelphia, PA, 1995
[7] *IEC/TS 61587-3, Mechanical structures for electronic equipment - Tests for IEC 60917 and IEC 60927 - Part 3: Electromagnetic shielding.*
[8] *GAM T20, Method of testing shielding effectiveness and shielded rooms.*
[9] *VG 95373-15, Elektromagnetische Vertraglichkeit - Elektromagnetische Vertraglichkeit von Geraten - Teil 15: Messverfahren für Kopplungen and Schirmungen.*
[10] *IEEE STD 299-2006, IEEE standard for Measuring the Effectiveness for Electromagnetic Shielding Enclosures* 2007
[11] Wilson P F And Ma M T 1988 *IEEE Trans.Electromagn. Compat.* **30** 239
[12] *U.S. Department of Commerce National Bureau of Standards Boulder, CO, NBS Tech. Note* **p** 1095
[13] Wilson P F And Ma M T 1988 *IEEE Trans. Electromagn. Compat.* **30** 251
[14] Loughry T A and Gurbazani S H 1995 *IEEE Trans. Electromagn. Compat.* **30** 449
[15] Holloway C L, Garzia R, Ladbury J, Hill D A and Koepke G 2002 *Int. Symp. Electromagnetic Compatibility (Sorrento) (Proc. of EMC EUROPE)* **p** 365

[16] Urban L 2004 *Lulea University of Technology Sweden*
[17] Robin S L 2010 *IEEE Trans. Electromagn. Compat.* **52** 682
[18] Greco S and Sarto M S 2006 *IEEE Trans .Electromagn. Compat.* **28** 502
[19] Holloway C L, Hill D A, Ladbury J and Garzia R 2003 IEEE *Trans.Electromagn. Compat.* **45** 350
[20] Angelo G and Maurizio M 2008 IEEE *Trans.Electromagn. Compat.* **36** 783
[21] IEC61000-4-21 *International Electrotechnical Commission Std.* 2003
[22] Frank B J, Leferink C and Hans B 2006 *IEEE Trans. Electromagn. Compat.* **26** 505
[23] Coder J Ladbury J and Holloway C L 2007 *IEEE Trans. Electromagn. Compat.* **33** 453
[24] Cancela M A, Romero S F and Morales A L 2009 *IEEE Trans.Electromagn. Compat.* **16** 1
[25] Andrew C M and He Y H 2008 *IEEE Trans.Electromagn. Compat.* **38** 1
[26] Holloway C L 2005 IEEE Trans. *Electromagn. Compat.* **33** 128
[27] Sandra G and Maria S 2007 *IEEE Trans. Electromagn. Compat.* **38** 16

Journal of Physics: Conference Series **418** (2013) 012086 doi:10.1088/1742-6596/418/1/012086

Statistical investigation and experimental verification of frequency stirring reverberation chamber

Yifei Liu*, Yongguang Chen, Qingguo Wang, Erwei Cheng

Key laboratory of Electromagnetic Environmental Effects, Shijiazhuang Mechanical Engineering College, Shijiazhuang 050003, P. R. China

E-mail：liuyifei5693305@163.com

Abstract. A statistically uniform E-field can be created in a reverberation chamber by frequency stirring. The statistical properties of frequency stirring reverberation chamber in over-moded are analyzed, and the ideal probability density functions (PDF) of E-field and field power are presented. In order to verify the statistical model, the S21 parameter which is the normalized power is measured between transmitting and receiving antenna. The result of the examination and processed data shows that the histogram of S21 can be good agreement with the theoretical model when the number of sample is enough.

1. Introduction

Reverberation chambers are becoming more and more popular for electromagnetic compatibility and immunity testing [1]. Such chambers can generate high field strengths using modest power sources. The goal of reverberation chambers is to create a statistically uniform field that eliminates the need to rotate the test object [2]. A mechanical mode stirrer (paddle wheel) is usually used to vary the chamber boundary conditions in an attempt to obtain statistical field uniformity. Mechanical stirring can be quite effective, but it is fairly slow [3]. It has been pointed out that the continuous changing of the mechanical tuner position has some equivalent effect with frequency modulation of the source and experimental work has been done by Loughry in 1991 to show the possibility of using frequency stirring in a reverberation chamber [4]. A computational investigation on frequency stirring in an ideal two-dimensional rectangular cavity has been done by Hill [5] and the good field uniformity results imply that frequency stirring could be an alternative to mechanical stirring in reverberation chambers. In this paper we are going to investigate the statistics of the field in the frequency stirring reverberation chamber and the organization of this paper is as follows: In section 2, the statistics of frequency stirring reverberation chamber is analyzed, and the theoretical E-field PDF model is presented. In Section 3 the experimental setup used in this paper is discussed. Section 4 presents experimental and theoretical results to illustrate the validity of the E-field PDF model. Section 5 thanks the funded projects and Section 6 summarizes the research presented here briefly and discusses future work.

---

* To whom any correspondence should be addressed.

2. Statistical characteristics of frequency stirring reverberation chamber

The theory of reverberation chamber derived from wave-guide resonance cavity. In reverberation chamber, a mode stirrer is commonly used to change the boundary conditions for the electromagnetic field. Each location of the stirrer creates an independent field distribution. Similarly, independent electromagnetic environment can be provided by the frequency stirring process, because mechanical stirring leads to the physical dimension alter, meanwhile, the frequency-stirred mode makes the electrical dimension change.

The E-field at a location in the frequency stirring reverberation chamber varies greatly depending on the electrical dimension alter by changing the operating frequency. When the dimensions of the chamber are large compared to the wavelength at the operating frequency, that is, when a sufficient number of resonant modes in the chamber exists, the E-field vector at a location in the chamber is the sum of multipath plane waves with random phases. The number of resonant modes (N) is given approximately by using Weyl's formula [6].

$$N \cong \frac{8\pi abc}{3\lambda^3} \tag{1}$$

where a, b, and c are the dimensions of the chamber and λ is the wavelength.

The E-field E as the vector sum of three rectangular components, each of them consists of a real and imaginary part, that is, as six parameters in total, are represented by

$$E = E_x + E_y + E_z \tag{2}$$

$$E_{x,y,z} = \mathrm{Re}(E_{x,y,z}) + i\,\mathrm{Im}(E_{x,y,z}) \tag{3}$$

As the large number of multipath waves arrives at a location in the chamber by moving the stirrers, each real and imaginary part of three rectangular components is normally distributed by the central limit theorem. Since these real and imaginary parts are not correlated, they are independent random variables. Therefore, six parameters of the E-field are independent random variables and each parameter is normally distributed with zero mean and variance σ^2

$$f\left(E_{xR}\right) = \frac{1}{\sqrt{2\pi}} e^{-\frac{E_{xR}^2}{2\sigma^2}} \tag{4}$$

The other parameters ExI, EyR, EyI, EzR, EzI also obey the law of normal distribution.

$E_x^2 = E_{xR}^2 + E_{xI}^2$, E_x^2 is the chi-square distributed with two degrees. The PDF of E_x^2 is

$$f\left(E_x^2\right) = \frac{1}{2\sigma^2} e^{-\frac{E_x^2}{2\sigma^2}} \tag{5}$$

The power is in proportion to the square values of E-field, so each rectangular components of the power also obey the chi distributed with two degrees. The PDF of P_x is

$$f\left(P_x\right) = \frac{1}{2\sigma^2} e^{-\frac{P_x}{2\sigma^2}} \tag{6}$$

The square valuesof E-field is

$$E^2 = E_x^2 + E_y^2 + E_z^2 = E_{xR}^2 + E_{xI}^2 + E_{yR}^2 + E_{yI}^2 + E_{zR}^2 + E_{zI}^2 \tag{7}$$

The sum or the square values of the six parameters, i.e. the squared magnitude of the resultant field, is the chi-square distributed with six degrees of freedom. Similarly, the root value of the chi-square distribution, i.e. the magnitude of the resultant field, is chi distributed with six degrees of freedom. The probability density functions of these distributions are as follows:

$$f\left(E^2\right)=\frac{E^4}{16\sigma^2}e^{-\frac{E^2}{2\sigma^2}} \tag{8}$$

$$f\left(E\right)=\frac{E^5}{8\sigma^2}e^{-\frac{E^2}{2\sigma^2}} \tag{9}$$

The PDF of the power is

$$f\left(P\right)=\frac{P^2}{16\sigma^2}e^{-\frac{P}{2\sigma^2}} \tag{10}$$

3. Experiment method and setup

In order to verify the statistical properties of frequency stirring reverberation chamber, We test the power at a location by sweeping the S21 measurement between the transmitting antenna and the receiving antenna in the test region.

The measurement setup is shown in figure1. A two-port vector network analyzer (VNA) is used to measure the S21 parameters. Port 1 connected to a transmitting antenna, port 2 connected to a receiving antenna and both of the two horns are Log-periodic antenna.

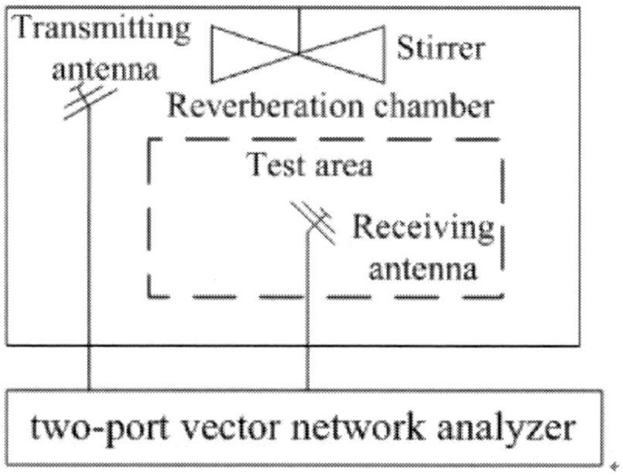

Figure 1. Diagram of the testing system.

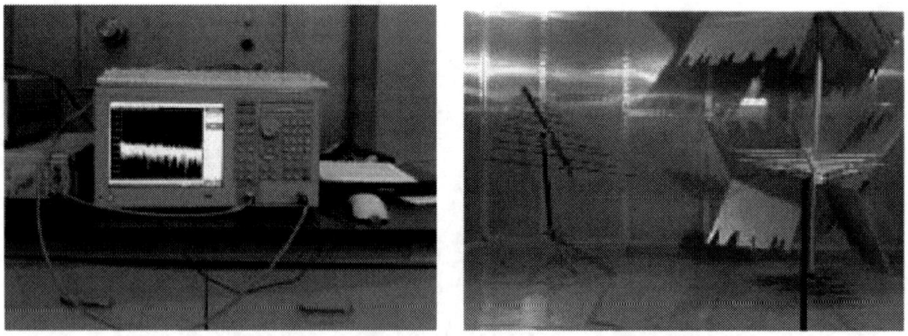

Figure 2. Configuration of the experiment.

The configuration of the experiment is shown in figure 2. In the test, the dimension of the reverberation chamber is 10.5 m by 8 m by 4.3 m and the lowest useable frequency is 80 MHz. The operating frequency of VNA is from 300 KHz to 1.5 GHz and the number of frequency points set on the NVA is always chosen to be the maximum 1601. We take the example of the swept frequency region from 500 MHz to 1 GHz, so the gap between each frequency point is about 0.625 MHz.

4. Experimental Results and analysis

Figure 3 shows the measured S21 parameter fluctuates with the operating frequency changes, and the result approves that frequency stirring is an alternative way to mechanical stirring to achieve statistical field uniformity in an over-moded chamber.

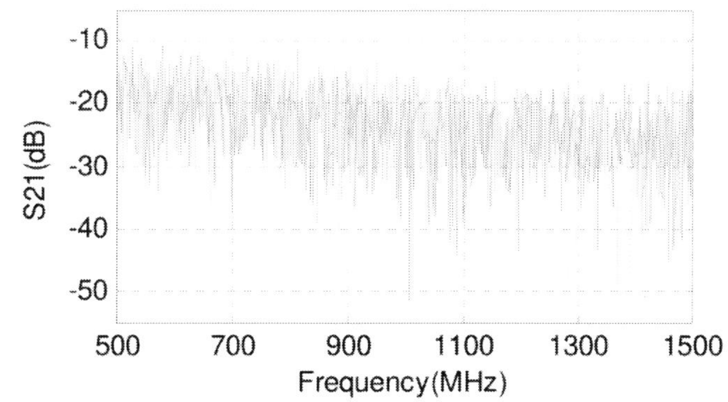

Figure 3. The S21 curve at a location.

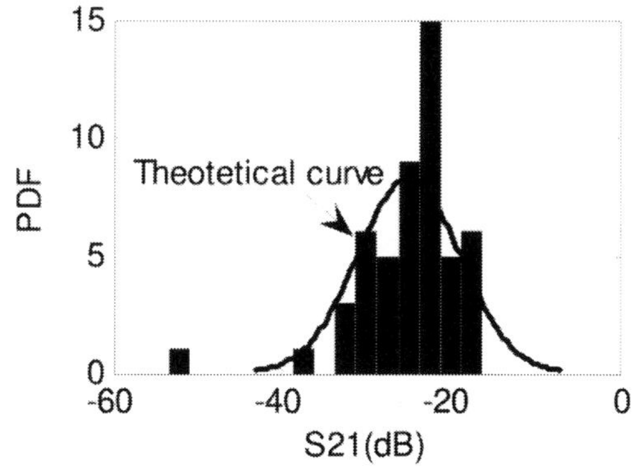

Figure 4. Comparison of the histogram and PDF when the number of sample is 50.

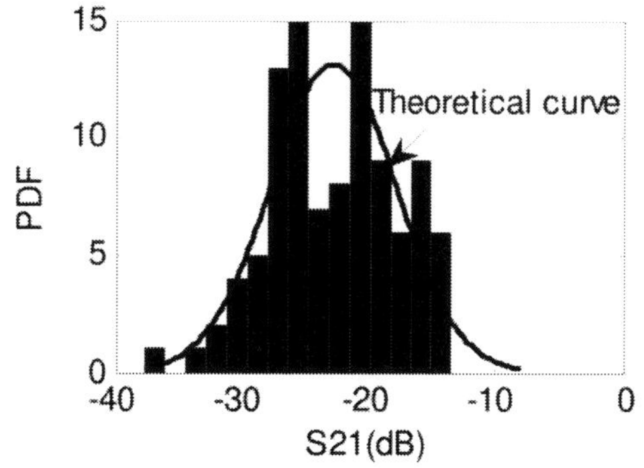

Figure 5. Comparison of the histogram and PDF when the number of sample is 100.

When the operating frequency is 1 GHz, the reverberation chamber is in over-moded status. Figure 4~7 shows the measured S21 histogram of 50, 100, 200 and 300 samples and is compared with the theoretical PDF which is presented in formula 10. When the number of samples is 50, the characteristics of the measured distribution is not agreement with the theoretical value, because the independent sample is not enough, and when the number of samples increases to 200, the PDF of the S21 distribution consists with the theoretical curve well, even more better when the number reach to 300. The processed data proves that the statistical characteristics of frequency stirring reverberation chamber can be satisfied with the theoretical model, therefore statistically uniform spatial E-field can be created efficiently by frequency-stirred mode.

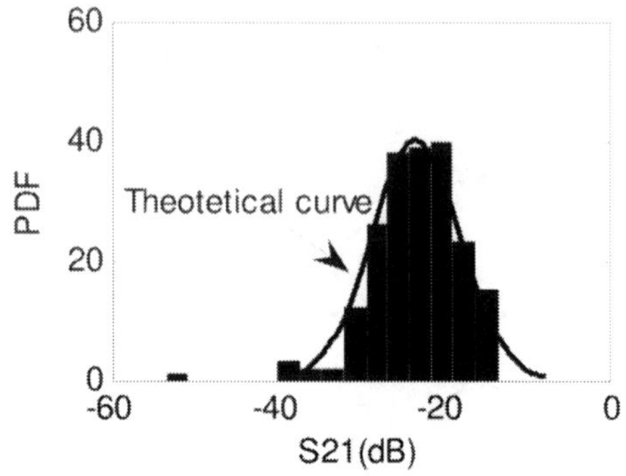

Figure 6. Comparison of the histogram and PDF when the number of sample is 200.

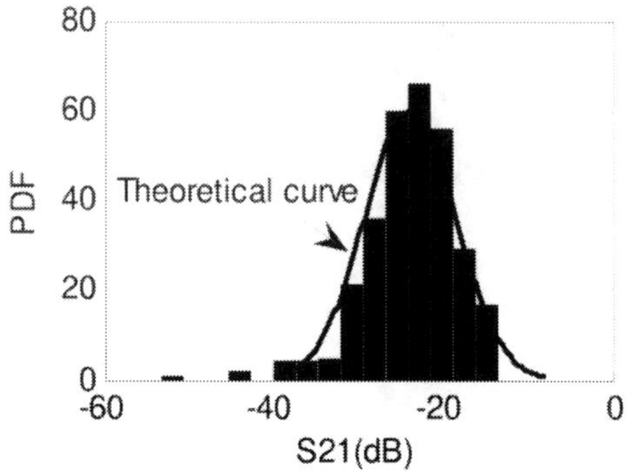

Figure 7. Comparison of the histogram and PDF when the number of sample is 300.

5. Conclusion

In the paper, the statistical property of a frequency stirring reverberation chamber is analyzed, and in order to verify the theoretical PDF model, the S21 parameter (normalized power) is measured by a two-port vector network analyzer in the paper. The processed data shows that the histogram of S21 can be good agree with theoretical model when the number of sample is more than 200, and this proves that frequency stirring can achieve spatial uniformity efficiently. Meanwhile we should see that the effect of the nonzero bandwidth of frequency stirring in performing immunity measurements, particularly for test objects with sharp resonances and the frequency stirring method used in practical application needs further theoretical and experimental studies.

Acknowledgements

This work is supported by National Science Foundation of China under the grant number of No. 51177174 and 51107147.

References

[1] Hill D A 1994 *IEEE Trans. Electromagn. Compat.* **36** 294.
[2] Marvin A C and Cui Y 2005 *IEEE Trans. Electromagn. Compat.* **1** 200
[3] Hallbjörner P 2002 *Microwave Opt. Technol. Lett.* **33** 25
[4] Lemoine C, Besnier P and Drissi M 2007 *IEEE Trans. Eectromagn. Compat.* **49** 745
[5] Yuan Z Y, Li T and Chen S M. 2007 *Chinese Journal of Radio Science* **22** 571
[6] Arnaut L R 2003 *IEEE Trans. Electromagn. Compat.* **45** 61

Signal integrity analysis on discontinuous microstrip line

Qingyang Qiao[1*], Yawen Dai[2], Zipeng Chen[2]
1 School of Automation, WuHan University of Technology, Wuhan, China
2 School of Science, WuHan University of Technology, Wuhan, China

E-mail: qiaoqingyang@yahoo.com.cn

Abstract. In high speed PCB design, microstirp lines were used to control the impedance, however, the discontinuous microstrip line can cause signal integrity problems. In this paper, we use the transmission line theory to study the characteristics of microstrip lines. Research results indicate that the discontinuity such as truncation, gap and size change result in the problems such as radiation, reflection, delay and ground bounce. We change the discontinuities to distributed parameter circuits, analysed the steady-state response and transient response and the phase delay. The transient response cause radiation and voltage jump.

1. Introduction

In PCB design, transmission lines are used to transmit the signal from one side to the other side. But it has become more difficult to keep the signal integrity (SI) during transmission as the clock frequency get faster and signal rising time became shorter. The good SI requires the signal to reach the correct voltage within required time. When the signal in transmission line encounter a change of transient impedance, part of the signal will be reflected and distorted. This is the common reason of bad signal integrity.

The transient impedance of the signal propagated along the transmission line is named as characteristic impedance. The characteristic impedance is the inherent properties of transmission line which is only related to the material properties, dielectric constant, the unit length capacitance, and the length of the transmission line, and it is a key factor for SI in transmission line circuit. Therefore, it is very important to study the relationship between the characteristic impedance and the SI.

Microstrip line is consisted of the dielectric substrate, the conductor and the return path. The two most important parameters of microstrip lines are characteristic impedance and phase velocity. The former one related to the impedance matching, while the latter determines its electrical length and geometric length. Characteristic impedance means the ratio of the voltage travelling wave to the current travelling wave transmitted on the transmission line, or the ratio of the incident voltage wave to the current [1]. And the phase velocity represents the moving speed of electromagnetic waves in transmission lines. The characteristic impedance and phase velocity are just effective for a specific wave. The microstrip lines are developed from the parallel two-lane with the dielectric substrate was added between the conductors, which make the analysis more complex. In the dielectric region, the phase velocity of the TEM is $C/\varepsilon_r^{-0.5}$ and ε_r is the relative permittivity. While in the air, the phase velocity of the TEM is c. Thus, between the interface of the air and the dielectric, the TEM wave can not reach a phase matching [2]. In most practical applications, dielectric substrate is so thin that the

[*] To whom any correspondence should be addressed.

field can be regarded as quasi TEM field. The equation of the characteristic impedance and the phase velocity are as follows:

$$\begin{cases} Z_0 = 60\ln\left(\dfrac{8H}{W}+\dfrac{W}{H}\right)\Omega\cdots\cdots\cdots\cdots\cdots\cdots W/H \le 1 \\ Z_0 = \dfrac{120\pi}{W/H + 2.42 - 0.44H/W + (1-H/W)\Omega}\cdots W/H \ge 1 \end{cases} \quad (1)$$

When W/H is really large, all power lines can be considered as within the media region.

The current of the return path concentrated bellow the signal path. The current become more centralized with the frequency become higher [3].The current distribution always tends to reduce the circuit impedance. The return current must flow through this impedance. A voltage drop on the return path has been generated [4].

When there is a continuous microstrip line, a large plane capacitance was generated between the signal path and return path. So the return path is a low inductor signal path. It is known that the signal is propagated along the lowest inductor path, and a complete return path can effectively reduce the signal propagation delay.

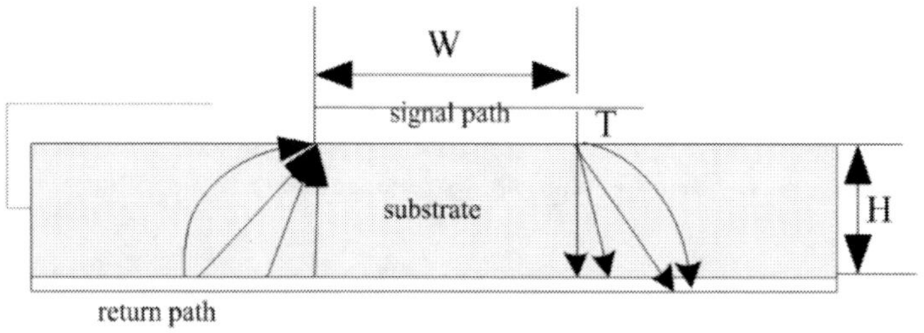

Figure 1. The model of a microstrip line.

2. The influence of discontinuity

In the process of signal transmission, when the signal has an electric transition, will cause a path discontinuity inevitably. The discontinuity can be equalled to a circuit in some way.

Micorstrip circuits invariably incorporate transmission line discontinuities of one type or another. Some of the most common forms of microstrip discontinuities are open ends, gaps, steps in width, right-angled bends, T junctions, and cross junctions. Since discontinuity dimensions are usually much smaller than the wavelength in microstrip, they may be modeled by lumped element equivalent circuits.

Signal distortion caused by discontinuities was affect by two parameters: the rise time and the impedance mutation. The transient impedance on capacitor and inductor is determined by the change rate of voltage and current.

A discontinuity in a microstrip is caused by an abrupt change in the geometry of the strip conductor. Therefore, electric and magnetic field distributions are modified near the discontinuity. The altered electric field distribution gives rise to a change in a capacitance, and the changed magnetic field distribution can be expressed in terms of an equivalent inductance. The analysis of a microstrip discontinuity involves the evaluation of these capacitances and inductances and can either be based on quasi-static considerations or carried out more rigorously by full wave analysis [5].

2.1. Open end

Near the open end, the distribution of the electric field changed, the power line extended out of the open end. So the energy is deposited in this region. In figure 2, microstrip open end is not just an open end, so we can assume that there is a capacitor. For a very short ideal open line is equivalent to a capacitor. As the result of the spread of the electric field to the free space outside the microstrip open end, a surface wave was generated on the surface of the dielectric plane, and a wave radiated to the free space. If the dielectric thickness of the plate reaches a certain level, a high mode wave has been generated in the microstrip line transmitted on the reverse direction. Consider the characteristic impedance, couple with the c_{oc}, they form an RC circuit. In the rising, there is a damped oscillation to form the signal ringing [4, 6].

$$u(t) = \frac{1}{c}\int_{0_-}^{t} i(x)dx + u_{0_-} \tag{2}$$

$$Z_{oc} = \int_0^t u(t)^2 dt \Big/ c_{oc} \tag{3}$$

In equation 2, we know that *u(t)* is relation to the previous state, and it have an integration time t. For the voltage between capacitance can't jump, the transient response was disturbed. For example, if there is no capacitance, $u(0)_-$=1 v, at $u(0)_+$, the voltage should be 0 v, if we have a excitation δ, at 0_+, the voltage $u(t)=u(0)_+ +\delta=0$ v. While if there this a c_{oc}, $u(0)_+$=1 v, we make a integration with δ, $u(\delta)$=1 v, so at 0_+, $u(t)$=1+1=2 v, the response has been seriously damaged. In a 2.5 v LVTTL circuit, $V_{OH}\geq2.0$ v, $V_{IH}\geq1.7$ v. In the circuit, thus a logic error was generated. The circuit also has a buildup time, during this time, the circuit has an oscillation state, sometimes it can cause a logic error either, thus in the building time, a phase delay was generated. So, signal integrity problems were generated.

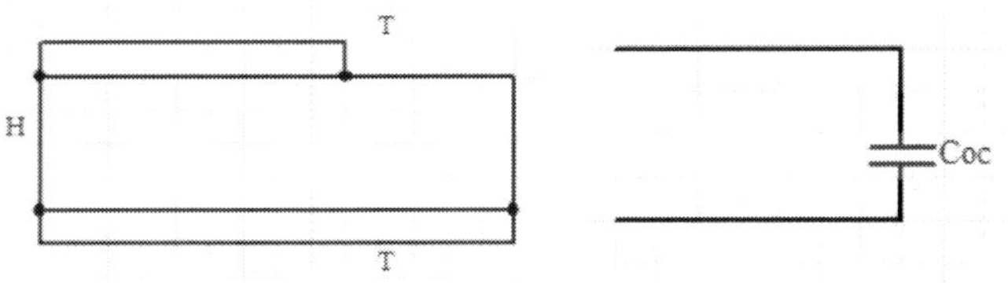

Figure 2. Open end and equivalent circuit.

2.2. Gaps in a microstrip line

When there is a gap in a microstirp line, we can get the equivalent circuit as figure 3, in the circuit, two lines were linked by a series capacitor. At the open end of the two lines, two capacitors in a parallel connection [6]. So, the gap can equivalent to a ∏ Capacitor network. The equation circuit consists of parallel capacitor C_{p1}, C_{p2}, and a series capacitor C_g.

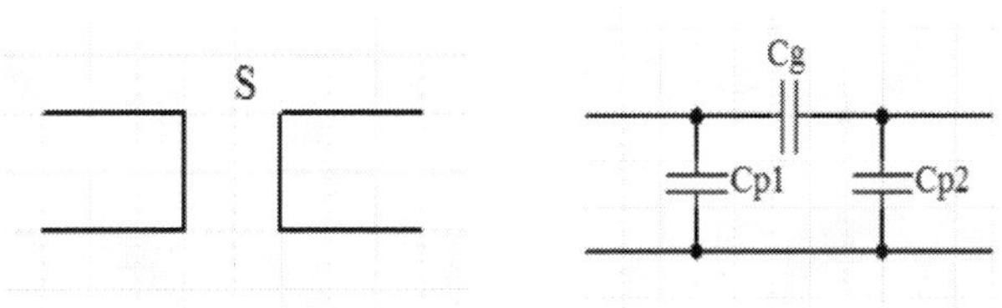

Figure 3. Gap and equivalent circuit.

We can't calculate the circuit with Kirchhoff theorem because this circuit is consists of capacitors without other elements. If this circuit works at a high frequency, we assume all the capacitors were short-circuit with a building time. During this time, the capacitors involve in a charging and discharging process, and the circuit works in an unknown state. In some cases, the circuit would works in a self-excited oscillation state. When C_g is so tiny that we can ignore it, the electric field E_{p1} energy storage in C_{p1}, on C_{p2}, an induced electric field E_{p2} was formed. For $E_{p1} \neq E_{p2}$, between E_{p1} and E_{p2}, there is a voltage jump. So, near the gap, there is radiation, voltage jump and unknown states.

2.3. Step in width

Steps in width exist a junction of two microstrip lines that have different impedances. In the cut off area, the electric charge is decreased. The configurations of a step discontinuity and its equivalent circuit are shown in figure 4. In terms of distributed elements, the discontinuity capacitance C_s has the effect of an increase in the wide line's length and an equal decrease in the narrow line's length. The series inductor shows that the magnetic energy increase considerable while the electricity decreased [4, 6]. With the energy exchange from the inductor to the capacitor, the energy is exchanged between the electric field and the magnetic field, electromagnetic waves were generated and radiate to the open area, the EMI emerged.

When the bend has a load Z, the

$$VSWR = C + \left(C^2 - 1\right)^{1/2} \tag{4}$$

The reflection coefficient

$$\Gamma = \frac{VSWR\ \ 1}{VSWR + 1} = \frac{C - 1 + \left(C^2 - 1\right)^{1/2}}{C + 1 + \left(C^2 - 1\right)^{1/2}}. \tag{5}$$

Γ is depending on C, when C changes, Γ is changed.
We assume there is a voltage source U_s in this circuit, we can get a 3 order equation:

$$U_s - L_1\left\{dC_s d\left[(L_2 \frac{di_{L2}}{dt} + R_L i_{L2})/dt\right] + i_{L2}\right\}/dt - L_2 \frac{di_{L2}}{dt} - R_L i_{L2} = 0 \tag{6}$$

$$U_s - L_1 C_s L_2 i'''_{L2} - L_1 C_s R_L i''_{L2} - L_2 i'_{L2} - R_L i_{L2} = 0 \tag{7}$$

When U_s=0, the response:

$$i_{L2} = 1 + \left(A_1 + A_2 t + A_3 t^2\right) e^{-t} \tag{8}$$

I_{l2} have a constant 1, an index, thus this will works in an oscillation state. In some cases, this circuit works in an over damping state, and the steady state is damaged.

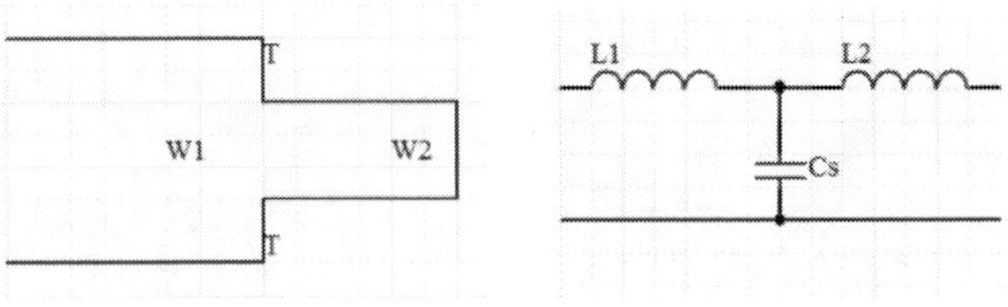

Figure 4. Step in width and equivalent circuit.

2.4. Bends

The current line of right angle bend shown in figure 5: Around the corner there is a parallel capacitor, the extend length just as two serious inductors [4, 6]. We can convert the microstrip bend to a uniform bend. Then use the duality theorem transform to a duality waveguide, we got the equivalent circuit as figure 5. For the circuit is similar with figure 4, the analysis is similar with the former situation.

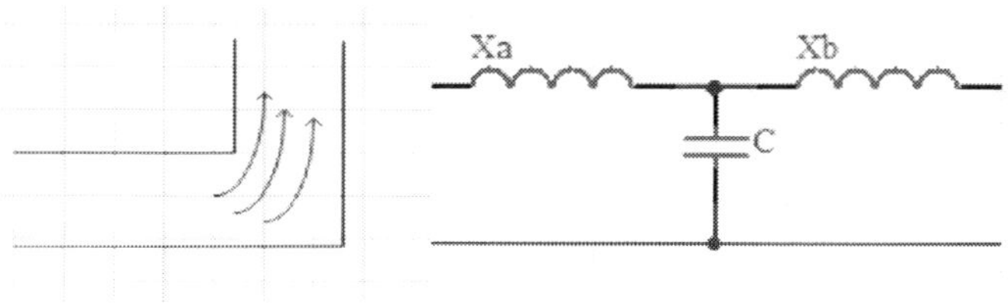

Figure 5. Bend and equivalent circuit.

3. Conclusion

The microstrip discontinuity generates lots of electromagnetic effects, and cause signal integrity problems. Near the open end, the impedance has changed, and a reflection wave was generated, as well as time delay. Near the gap, there is a time delay and a voltage transition, the signal integrity problem comes. Step in width in microstrip lines cause signal integrity problems such as radiation, delay, ringing and reflection.

References

[1] Eric B 2005 *Signal Integrity* (Simplified Publishing Beijing: House of Electronics Industry)
[2] David M P 2005 *Microwave Eng.* (Beijing: House of Electronics Industry)
[3] Stenphen H H and Howard L H 2011 *Adv. Signal Integrity for High-Speed Digital Des.* (Beijing: House of Electronics Industry)
[4] Howard J and Martin G 2010 *High-Speed Digital Des.: A Handb. of Black Magic* (Beijing: House of Electronics Industry)
[5] Gupta K C, Ramesh G, Inder B and Prakash B 1996 *Microstrip Lines and Slot Lines* (London: Aerech House)
[6] Tsinghua University 1996 *Microstrip Circuits* (Beijing: Posts&Telecom Press)

Novel injection methods for substitution of irradiation under different coupling routes of electromagnetic interference

Xinfu Lu*, Guanghui Wei, Xiaodong Pan
Institute of Electrostatic and Electromagnetic Protection, Ordnance Engineering College, Shijiazhuang, 050003, China

E-mail: luxinfu123@126.com

Abstract. At present, the maximum electromagnetic radiated field strength which can be simulated is not able to satisfy the requirements of electromagnetic environment effects tests on electronic systems. To broaden the effective frequency range of the current injection technique, new current injection methods were put forward for substitution of the radiation. A wide-band current injection device was designed based on the symmetric directional coupler principle. With a typical antenna receiving system under test, the validity of the methods were analyzed by theory and verified by experiments when the Equipment Under Test (EUT) had different coupling routes of electromagnetic interference (EMI). The results indicate that: firstly, the relationship between radiated fields and injected voltage is linear, even if the responses of the system are non-linear; secondly, the method is valid even if the test frequency is up to gigahertz; thirdly, the novel methods of current injection are able to substitute for irradiation effectively when the antenna and the interconnected cable are the main coupling routes of EMI respectively.

1. Introduction

With the appearance of electromagnetic pulse (EMP) weapons, electronic systems suffer from a quite worse electromagnetic (EM) environment. So there is a corresponding increase in electromagnetic compatibility (EMC) clearance levels. The need to test these systems at higher levels is posing great problems for test community, because the levels of radiated strength simulated by laboratories are limited. As an alternative to the standard radiated field susceptibility test, current injection has been researched for many years. Bulk current injection (BCI) technique is widely used for interconnected systems. However, BCI test is proved to be effective for frequencies below 400MHz [1-4]. Above 400 MHz, errors are significantly increased and test results become sensitive to the position of the probe [5-7]. At present, the operating frequency of many electronic systems such as some radar and communication systems is up to gigahertz. Hence, in order to broaden the current injection technique's effective frequency range, new current injection methods should be put forward.

2. Relationship between radiated field strength and injected voltage

In general, radiation and injection cannot be made equivalent in the strict sense for interconnected systems. This stems from the fact that radiated fields give rise to distributed coupling along the length of the cable, whereas injection devices act as lumped sources of voltage. However, as the equipment attached to the cable ends is the real victims of electromagnetic (EM) disturbance, it is an

*To whom any correspondence should be addressed.

appropriately equivalent method to ensure that the injection-induced noise currents at the ends of cable are equal to the radiation-induced noise currents. Considering the limitations of the BCI technique, new current injection method is put forward based on the differential mode coupling principle. Symmetric directional couplers are used as injection device. The test system employing this device is sketched in figure 1. Equipment A and B is auxiliary equipment (AE) and EUT respectively, and they are connected by coaxial cable. Ports from 1 to 4 of directional coupler are the input port, transmitted port, coupled port and isolated port, respectively. The current source attached to port 4 is employed to inject noise current into equipment B.

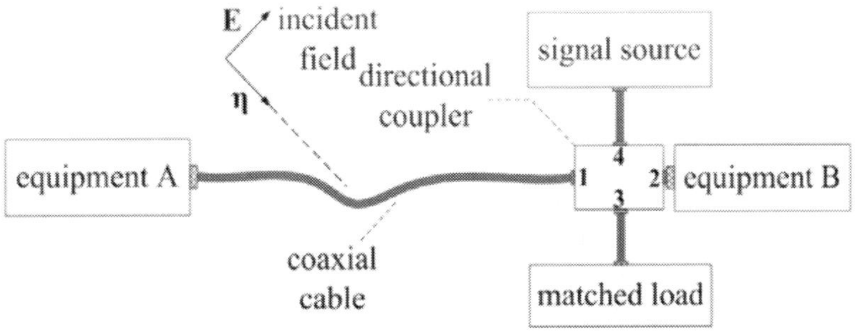

Figure 1. Block diagram of system for current injection.

If the response of equipment B is concerned only, the Thevenin equivalent circuits of the system for radiation and injection can be seen in figure 2 and 3 respectively. U_{SR} and U_{SI} represent the equivalent voltage source for radiation and injection. Z_{SR} and Z_{SI} represent the corresponding equivalent source impedance. Z_{SR} equals to Z_{SI} since the entire system for radiation and injection are the same. Namely,

$$Z_{SR} = Z_{SI} = Z_0 \frac{1 + S_{12}{}^2 \Gamma_A e^{-2\gamma L}}{1 - S_{12}{}^2 \Gamma_A e^{-2\gamma L}} \tag{1}$$

where S_{12} is the scatter parameter of directional coupler, Z_0 and L denote the characteristic impedance and the length of the coaxial cable respectively, and Γ_A is the reflection coefficient of equipment A. According to the Thevenin theorem, the responses of equipment B are equal for radiation and injection if U_{SR} is equal to U_{SI}, no matter whether the response of EUT is linear or non-linear. The way to ensure the equivalence between U_{SR} and U_{SI} is to make the responses of EUT equal by adjusting the current source. The process of electromagnetic field coupling to antenna and cable is linear, as well as the coupling process of directional coupler. Due to this fact the relationship between the electric field strength E and the injected voltage U_C is also linear. Namely, the proportional coefficient $k = U_C E^{-1}$ is constant. On account that the levels of electric field strength can be simulated is limited, the value of injected voltage equivalent to high level electric field strength should be linearly extrapolated. By this means the current injection method can substitute for high level field illumination equivalently. So the entire test procedures are as follows: (1) obtain the output voltage of equipment B U_B through low level field radiation test at electric field strength E_1; (2) adjust the output voltage of current source U_1 to impose equivalence between U_B and the output voltage of equipment B in the injection test; (3) calculate the proportional coefficient $k = U_1 E_1^{-1}$, repeat aforementioned procedures at other levels electric field strength, and obtain the average value k_{av} of the proportional coefficient k; (4) if the radiation experiment is carried out at higher levels of electric field strength E_2, the corresponding injected voltage U_2 yields $U_2 = k_{av} E_2$, and the output of the system in the injection experiment is equivalent to that in the radiation experiment with electric field strength E_2.

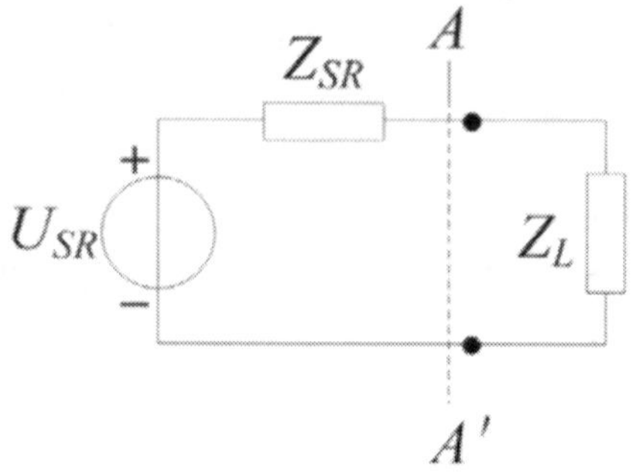

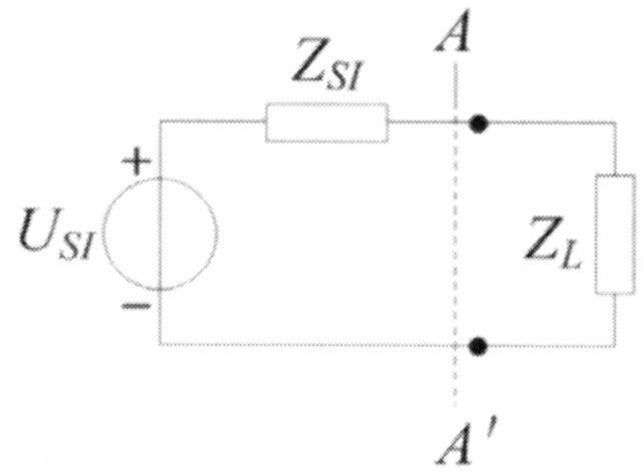

Figure 2. Equivalent circuits of radiation. **Figure 3.** Equivalent circuits of current injection.

In order to demonstrate the validity of the above-mentioned theoretical analysis, a receiving antenna and a low-noise amplifier (LNA) were employed as the equipment A and B respectively. They were connected with a 9-m-long RG-58 coaxial cable. The antenna and LNA are typically linear and non-linear equipment respectively. A symmetric directional coupler was used as the injection device. Connect the whole system according to figure 1, the output of equipment B was connected to spectrum analyzer. Radiation experiment was finished in an open area test site. The antenna and coaxial cable are in the radiation area except the equipment B. The aim is to ensure the input port of equipment B is the only way for the interference current to enter into the equipment. The working frequency band of the whole system is from 2 to 8 GHz. 4.6 and 7.2 GHz were chosen as the test frequency points. A horn antenna was chosen as the transmitting antenna which was vertically polarized. The experiment was carried out according to the aforementioned procedures.

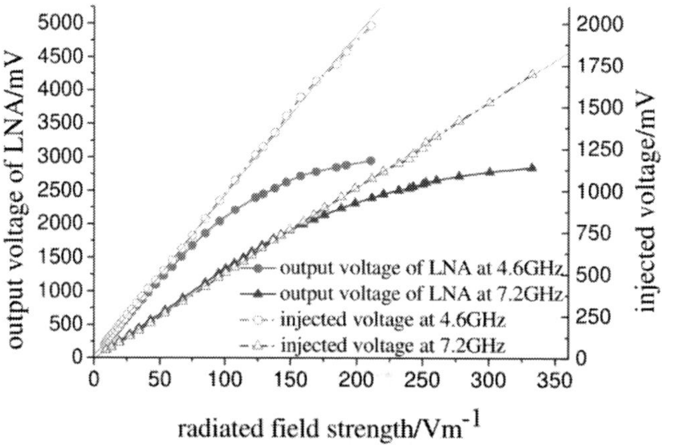

Figure 4. Output voltage of LNA and injected voltage with different radiated fields strength.

The relationship between the output voltage of LNA and the electric field strength is illustrated in figure 4. The figure also plots the injected voltage versus the electric field strength when the output voltages of LNA are equal for radiation and injection. As shown in figure 4, the relationship between the output of LNA and the electric field strength is typically non-linear. However, at the same time, the curve of the injected voltage versus the electric field strength is linear. Perform straight-line fitting for the latter curve. The results show that the coefficient of determination is 0.999 for both 4.6 and 7.2 GHz. Accordingly, the proposed equivalent methods are able to reproduce radiation-induced effects even if the EUT is nonlinear. In addition the BCI test is based on the injection of common-mode noise currents in the cable. Hence, it cannot substitute for radiation if the interference is differential-mode noise currents just like this experiment. However, this injection method which is able to inject differential-mode currents can solve this problem well, as demonstrated by the experimental results.

3. Equivalence with different coupling routes of EMI

Aforementioned analysis has demonstrated the equivalence between radiation and injection when the antenna is the main coupling route of EMI. Additionally, the cable is another main coupling route. Hence, in this situation, the question whether this injection method can substitute for radiation equivalently or not should be studied, especially for coaxial cable. When considering shielded cables, the noise currents induced at the terminal units by distributed field coupling can be directly related to the current and voltage distribution along the external surface of the cable. It seems to be different from the new injection methods which inject noise currents into the terminal units directly.

For the interconnected system as seen in figure 1, the left side of equipment B can be modeled as nonideal voltage generators (i.e., series connection of ideal voltage generators U_S and internal source impedance Z_S) if the response of equipment B is concerned only. As shown in equation (1) Z_S equals to Z_{SR}. According to the BLT equations and Thevenin theorem [8], the equation (2) and (3) can be obtained.

$$U_{1S} = \frac{2\left(e^{-\gamma L}S_1 + \Gamma_A e^{-2\gamma L}S_2\right)}{1-\Gamma_A e^{-2\gamma L}} \tag{2}$$

$$\Gamma_{1S} = \Gamma_A e^{-2\gamma L} \tag{3}$$

where U_{1S} and Γ_{1S} are the equivalent ideal voltage generators and source impedance at the left side of port 1. Hence, U_S yields

$$U_S = \frac{2S_{12}\left(e^{-\gamma L}S_1 + \Gamma_A e^{-2\gamma L}S_2\right)}{1-S_{12}^2\Gamma_A e^{-2\gamma L}} \tag{4}$$

Consequently, it is feasible to substitute injection for radiation in this situation since the response of equipment B is able to be seen as the effect of a lumped voltage generator.

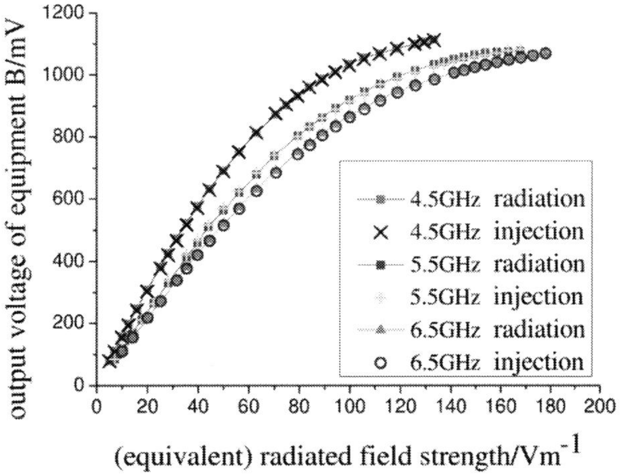

Figure 5. Output voltage of equipment B for radiation and injection with cable as the main coupling route.

In order to assess the aforementioned analysis, an experiment was carried out. The test setup is as follows. The system under test was the same as that shown in figure 1. The coaxial cable was placed in the shielded room to ensure the cable was the only way to couple EM interference. The remaining parts were placed outside of the shielded room and connected to the coaxial cable through connectors on the wall of the shielded room. Equipment A and B were a receiving antenna and radio frequency front-end respectively. The cable was illuminated by a horn antenna in the shielded room. The experiment was performed according to the aforementioned procedures. 4.5 GHz, 5.5 GHz and 6.5 GHz were chosen as the test frequency points. The results are shown in figure 5. For simplicity, the abscissa of the injection experiment's data curve is the electric field strengths which are equivalent to the corresponding injected voltages. The mean errors of equipment B's output voltage for radiation and injection are 0.8%, 0.71% and 0.49% for 4.5, 5.5 and 6.5 GHz respectively. The results indicate that the injection method in this paper can substitute for radiation even if the cable is the main coupling route of EMI. It also shows that the noise currents injected into the terminal units are differential-mode interference. This fact makes it possible for the proposed equivalent scheme to substitute for radiation.

4. Conclusion

In this paper, a new current injection method substituting for radiation has been developed and discussed. The proposed methods are aimed at broadening the current injection technique's effective frequency range. The results indicate that the equivalent relationship between radiated field strength and injected voltage is linear for interconnected systems even if the responses of the system are non-linear. Secondly, the method can substitute for radiation equivalently even when the frequency is up to gigahertz. Thirdly, the method is valid when the antenna and coaxial cable are the main coupling route of EMI respectively. It has no problems in representing the stresses that will be seen on the internal cores of the cable. The mean errors of the EUT's outputs for radiation and injection are less than 2%. It proves the validity of the proposed methods.

References

[1] Grassi F, Spadacini G and Marliani F 2008 *IEEE Trans. Electromag. Compat.* **50** 524
[2] Orlandi A, Antonini G and Rizzi R. 2006 *IEEE Trans. Electromag. Compat.* **48** 701
[3] Pignari S and Canavero F 1996 *IEEE Trans. Electromag. Compat.* **38** 469
[4] Rasek, Guido A and Loos S, 2008 *IEEE Trans. Electromag. Compat.* **50** 499
[5] Grassi F, Marliani F and Sergio A. P. 2007 *IEEE Trans. Electromag. Compat.* **49** 563
[6] Spadacini G and Pignari S 2004 *IEEE Trans. Electromag. Compat.* **46** 446
[7] Hill D 1992 *IEEE Trans. Electromag. Compat.* **34** 445
[8] Frederick M and Michel V 1997 *EMC analysis methods and computational models* (New York: Wiley)

Research on the test method of using injection as an equivalent substitute for electromagnetic radiation

X D Pan*, **G H Wei, X F Lu, K Li**
Institute of electrostatic and electromagnetic protection, Mechanical Engineering College, heping west road, shijiazhuang, China

E-mail: xiaodong-pan@hotmail.com

Abstract. This paper presents a method to carry out high intensity radiated field (HIRF) effect experiments by using injection as an equivalent substitute for electromagnetic radiation. In allusion to typical interconnected system, the equal response voltage on the equipment cable port is regarded as an equivalent basis of injection and radiation methods. The equivalent relation formula between injected voltage and radiated field is derived theoretically. The conditions needed for extrapolating injected voltage in HIRF are confirmed, and the extrapolation method is proposed. On the basis of the above research, the electromagnetic environment effect test new method combined injection with radiation for interconnected system is summarized. The typical nonlinear interconnected system is selected as equipment under test, and the single frequency continuous wave radiation and injection effect experiments are carried out separately. The test results indicate that the relation between radiated field and injected voltage is linear, and the equivalent injected voltage used to substitute HIRF can be obtained by linear extrapolation.

1. Introduction

With the development of high power radio equipment, electronic warfare system, electromagnetic pulse bomb, high power microwave weapon and so on, the electromagnetic environment in limited space is on the slide [1]. High intensity radiated field (HIRF) has become a new challenge faced by weaponry and civil electronic system [2]. "MIL-STD-464A electromagnetic environmental effects requirements for systems" illuminate external electromagnetic environment (EME) for system. The electric field intensity is higher than 200 V m^{-1} in vast frequency band. In the frequency band of 2.7~3.6 GHz, the peak external EME for shipboard operations in the main beam of transmitters is 27460 V m^{-1}, and the average value is 2630 V/m. If carrying out electromagnetic interference safety margin test, the much higher electric field intensity should be provided. Yet the needed electric field intensity cannot be simulated under the condition of present laboratory. Thus, it is hard to carry out electromagnetic radiated susceptibility and safety margin test for system with single electromagnetic radiation method [3].

Therefore, it is necessary to explore a new injection method to substitute for radiation test. Under this condition, the theory and method of using coupling module injection as an equivalent substitute for radiation is proposed in this paper.

* To whom any correspondence should be addressed.

2. Theoretical research

2.1. Equivalence analysis between injection and radiation

Strictly speaking, injection cannot be equivalent to radiation completely. Because radiated response is the result of a good many distributed source, but injected response is the result of lumped source. However, as for the system under test with definite coupling channel, such as interconnected system and antenna feeding system, the electromagnetic energy acts on the inner circuit of the system mainly through cable or antenna port with conducted interference mode [4]. Hence, if examining the electromagnetic susceptibility of the equipment connected by the cable, the injection test method can be adopted to substitute for radiation effect test equivalently [5].

The strict equivalent base in theory between injection and radiation method is equal response on the equipment under test, and the equivalent base in engineering is the same effects caused by the two test methods. Based on the above-mentioned viewpoint, if the response voltage or induced current acted on the cable port of the equipment, it could be ensured equal, the equivalence of the two test methods can be guaranteed [6]. So the equivalent base of the two test methods in this paper is identified as the equal response voltage on the cable port of the equipment under test [7].

The purpose of presenting a new injection test method is to substitute for HIRF radiated effect experiments. How to acquire the corresponding relation between injected voltage and HIRF intensity is the crucial problem because HIRF is difficult to simulate under the condition of laboratory presently, the equivalent injected voltage could not be obtained by practical experiment. Therefore, the equivalent injected voltage used to substitute HIRF test can only be obtained by linearly extrapolating with the correspondent relation in lower radiated field.

2.2. Equivalent model of using injection as a substitute for radiation

The structure of typical interconnected system is shown in figure 1. Assuming B is the equipment under test, and A is either interconnected equipment or receiving antenna. Under the condition of electromagnetic radiation, the interconnected system can be reduced to a transmission network shown in figure 2. Where Z_A is the equivalent impedance of equipment A, Z_L is the equivalent impedance of equipment B, and Z_C is the characteristic impedance of transmission line. If the transmission line is coaxial cable, the whole system can be regarded as two independent transmission line models. The one is external transmission line model composed of shielding layer and ground, and the other is inside transmission line model composed of shielding layer and inner conductor. When the interconnected system is radiated by electromagnetic field, the inside transmission line is excited by external transmission line through transfer impedance and transfer admittance. Accordingly, the interconnected transmission line is coaxial cable, and the circuit model shown in figure 2 also could be adopted to analyze the radiated response of the terminal load.

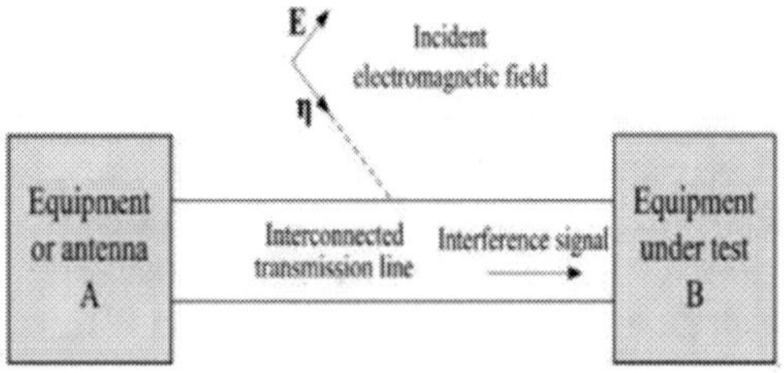

Figure 1. Scheme of typical interconnected system structure.

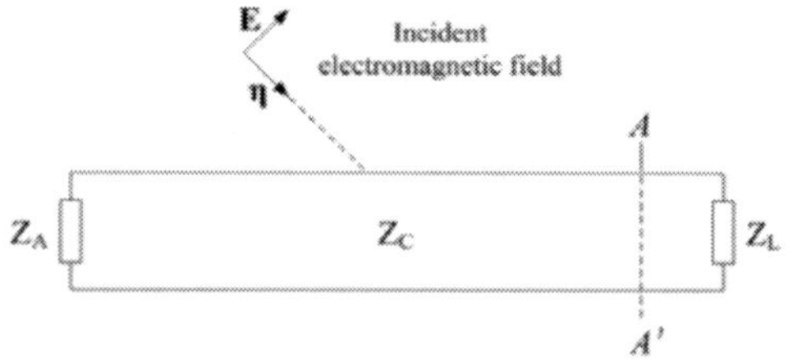

Figure 2. Reduced transmission network of interconnected system.

In order to calculate the response voltage on the equivalent impedance Z_L of equipment B, the interconnected system is divided into two parts at the position of A-A' in circuit analysis. The left branch of A-A' can be equivalent to Thevenin equivalent circuit, as shown in figure 3. In this way, the interference on the external port of equipment B can be equivalent to impedance element Z_{SR} and voltage source U_{SR}. Where Z_{SR} is the output impedance of A-A' left branch, and U_{SR} is the open-circuit voltage. Therefore, the entire radiated response equivalent circuit combined the left branch with equipment B is shown in figure 4.

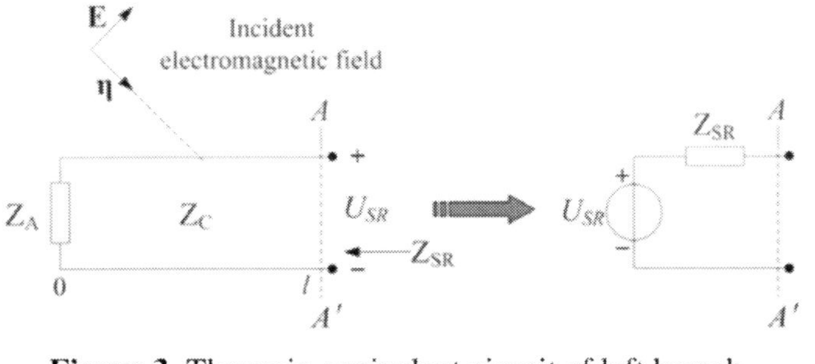

Figure 3. Thevenin equivalent circuit of left branch.

Defining the transmission line length between A-A' port and impedance Z_A is l, and the reflection coefficient is ρ_A. According to transmission line theory, the internal impedance Z_{SR} of equivalent voltage source is as follows:

$$Z_{SR} = Z_C(1+\rho_{A-A'})(1-\rho_{A-A'})^{-1} = Z_C(1+\rho_A e^{-j2\gamma l})(1-\rho_A e^{-j2\gamma l})^{-1} \quad (1)$$

Under the condition of electromagnetic radiation, the open-circuit voltage U_{SR} can be calculated with BLT equation. Set A-A' port is open, i.e., $\rho_L = 1$. The open-circuit voltage at the position $x=l$ is as follows:

$$U_{SR} = 2(1-\rho_A e^{-2\gamma l})^{-1}(e^{-\gamma l}S_1 + \rho_A e^{-2\gamma l}S_2) \quad (2)$$

where S_1, S_2 are the source vector in BLT equation. If adopting Agrawal model [8], there is a linear relation between the source vector and incident electric field intensity. In order to simplify the expression, defining the transfer function between source vector U_{SR} and incident electric field E is f. Hence, the equation (2) can be simplified as follows:

$$U_{SR} = f(E) \quad (3)$$

The transfer function f satisfies the equation $f(k\cdot E) = k\cdot f(E)$. By solving the impedance element Z_{SR} and voltage source U_{SR}, according to equivalent circuit in figure 4, the radiated response on equivalent impedance Z_L can be derived as follows:

$$u_{LR} = Z_L(Z_{SR}+Z_L)^{-1}U_{SR} = Z_L(Z_{SR}+Z_L)^{-1}f(E) \quad (4)$$

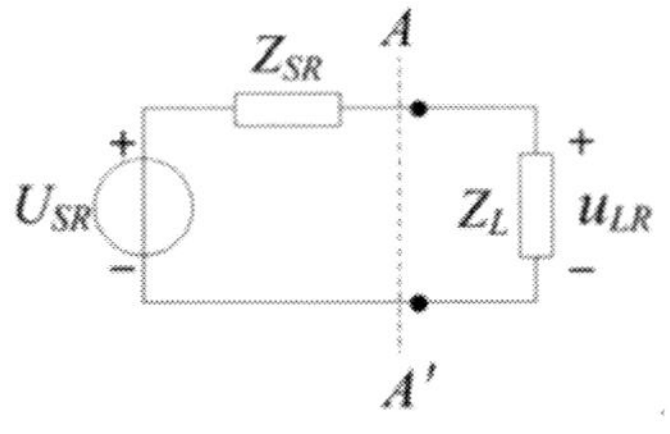

Figure 4. Radiated response equivalent circuit of system.

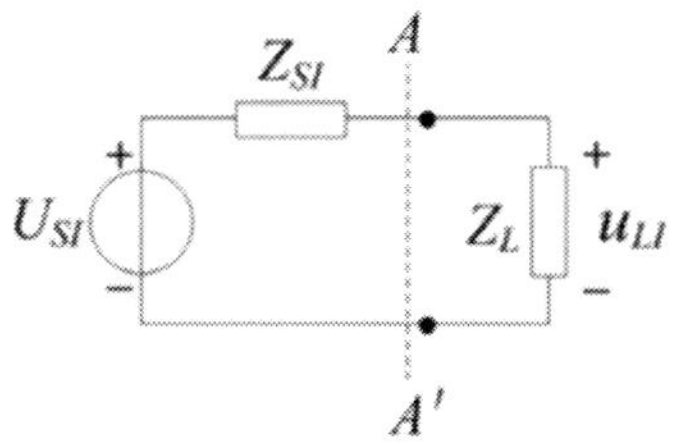

Figure 5. Injected response equivalent circuit of system.

Journal of Physics: Conference Series **418** (2013) 012089 doi:10.1088/1742-6596/418/1/012089

Consulting the above analysis process, the injected response equivalent circuit can be easily obtained, which is shown in figure 5. Where U_{SI} is injected voltage source, Z_{SI} is the inner impedance of injected voltage source, and u_{LI} is the injected response of equipment B.

According to the equivalent circuit in figure 5, the injected response on equivalent impedance Z_L can be expressed as:

$$u_{LI} = Z_L(Z_{SI} + Z_L)^{-1}U_{SI} \tag{5}$$

The equivalent basis for injection and radiation test methods presented in section 2.1 is the equal response on equipment under test, i.e., $u_{LR} = u_{LI}$. The injected voltage source can be derived as follows:

$$U_{SI} = (Z_{SI} + Z_L)Z_L^{-1}u_{LR} \tag{6}$$

$$U_{SI} = (Z_{SI} + Z_L)(Z_{SR} + Z_L)^{-1}U_{SR} = (Z_{SI} + Z_L)(Z_{SR} + Z_L)^{-1}f(E) \tag{7}$$

Here, Z_{SR} and Z_{SI} can be analyzed according to equation (1), and Z_L can be obtained by practical testing using radio frequency impedance analyzer. The lumped voltage source U_{SR} can be calculated according to equation (2). Finally, according to equation (7), the correspondent relation between theoretical equivalent injected voltage U_{SI} and radiated field E is confirmed, which ensures the equivalence between injection and radiation.

2.3. Extrapolation model for injected voltage source in high intensity field

The test method using injection as a substitute for radiation mainly solves electromagnetic radiation effect test for interconnected system in HIRF. According to the above equivalent principle, the low power equivalent injected voltage U_{SI} obtained through pretest need to be extrapolated in high intensity field, and the high power equivalent injected voltage V_{SI} can be obtained in this way. So the high intensity radiated effect experiment which is difficult to accomplish in present laboratory condition can be substituted by high power injection experiment.

The majority of systems under test are nonlinear response system. It means the relation between voltage response and the input signal is not linear because of the working state, material performance and parasitic parameter change. As to nonlinear system, how to extrapolate the injected voltage source is the crucial problem. Therefore, on the basis of theoretical analysis and experiment research to the typical nonlinear system, the interference (degradation, failure and damage) caused by electromagnetic radiation is divided into two process. The one is field coupling process to the cable, and the other is the response process on the module and device. As can be known from electromagnetic field theory, the field coupling process is linear, and the device response process is nonlinear. As long as the exciting effect on the device input port excited by injected voltage source and the radiated equivalent voltage source can ensure equal, the response voltage in injection experiment will be the same as it in radiation experiment, and the equivalent injected voltage source used to substitute high intensity field radiation test can be extrapolate with linear relation. In order to ensure the exciting effect on the input port is the same under the condition of high intensity field, two conditions must be satisfied. The first one is the same open circuit voltage of injected voltage source and radiated equivalent lumped voltage source in high intensity field, viz. $U_{SI}=U_{SR}$, the other is the same voltage dividing rate for device input port response in the radiation and injection equivalent circuit, viz. $Z_L/(Z_{SI} + Z_L) = Z_L/(Z_{SR} + Z_L)$.

The equation (2) gives the theoretical calculation method for radiated equivalent lumped voltage source. However, for practical engineering test, it is inconvenient to obtain U_{SR} through complicated calculation process and carry out equivalent substitute injection test. Moreover, because of the effect of surrounding environment and the existence of error, it may result in a certain gap between theoretical calculation result and practical situation. Therefore, it is very important to get injected voltage source U_{SI} by using practical test method. In order to solve the above problem, the forward voltage $u^+(0)$ on the interconnected transmission line is regarded as equivalent parameter in low intensity field pretest. The equivalent injected voltage U_{SI} in low intensity field pretest can be obtained when the forward voltage u^+ (0) in injection pretest is equal to the one in radiation pretest. Finally, the equivalent injected voltage V_{SI} in high intensity field can be obtained by linear extrapolate from U_{SI} ($V_{SI}=kU_{SI}$), and then the equivalent injection test substituting for high intensity field radiation test can be accomplished in this way. It has more universal meaning and engineering application value for interconnected system to carry out radiated effects test.

3. Experiment verification

By carrying out radiation and injection effect test to antenna receiving system, the linear relation between radiated electric field and equivalent injected voltage source is verified. So the injected voltage source substituting for high intensity field can be obtained by linear extrapolation from the corresponding relation in low intensity field.

3.1. Interconnected system under test

The interconnected system under test is typical nonlinear system. It is composed of receiving antenna, coaxial cable and RF front end assembly. The connected mode is shown in figure 6. The RF front end assembly is integrated in one box including clipping filter module, directional coupler, low-noise amplifier (LNA), sensitivity control module, clipping amplifier and so on.

3.2. Verification experiment method

The interconnected system under test is carried out radiation and injection effect experiments separately. The equipment B in figure 6 appears nonlinear response in radiation and injection test. The corresponding relation between radiated field, injected voltage and the response on equipment B is recorded. Because the response voltage value on equipment B can be monitored, the response voltage on equipment B is regarded as equivalent principle directly. The corresponding relation between radiated field and equivalent injected voltage can be obtained when the equipment B appears linear and nonlinear response. If the corresponding relation keeps linear, the injected voltage source substituting for high intensity field can be obtained by linear extrapolation from the corresponding relation in low intensity field.

3.3. Experiment results and analysis

Three frequency points (3.3 GHz, 4.6 GHz and 7.2 GHz) in interconnected system working frequency band are selected to carry out verification experiments.

According to the radiation test method talked above, the relation curve between equipment B output response and radiated electric field intensity is obtained with different frequency points. It is shown in figure 7.

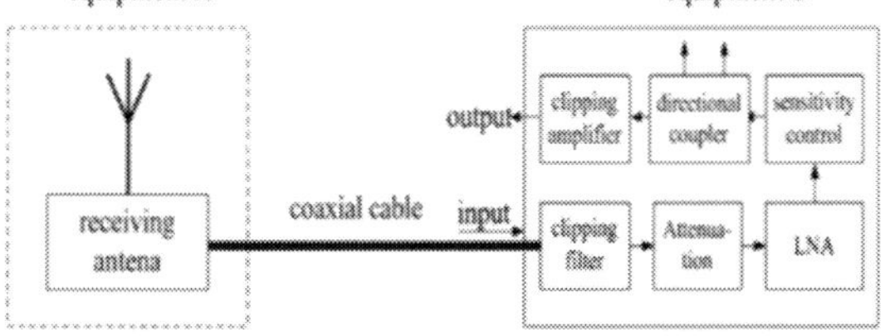

Figure 6. Scheme of connection mode for antenna receiving system.

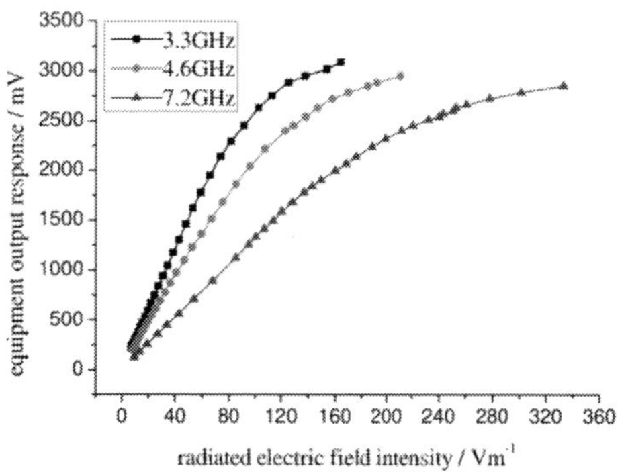

Figure 7. Relation curve between output response and radiated electric field intensity.

The injection effect test is carried out according to the above test method. The relation curve between equipment B output response and injected voltage value is obtained with different frequency points. It is shown in figure 8.

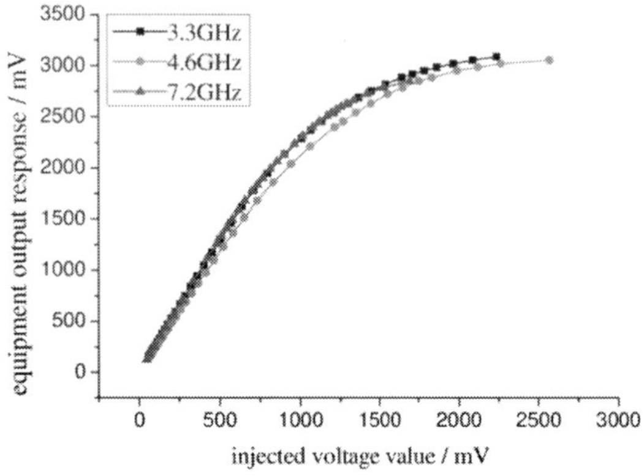

Figure 8. Relation curve between output response and injected voltage value.

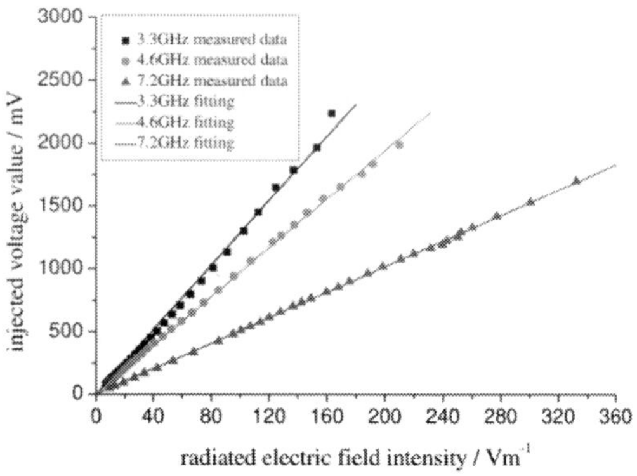

Figure 9. Relation curve between injected voltage and radiated field intensity.

In order to quantitatively analyze the equivalent correspondent relation between injected voltage and radiated field intensity, on the premise of equal equipment B output response, the correspondent relation between injected voltage and radiated field intensity is obtained with different frequency points. It is shown in figure 9.

It can be seen from the variation trend of the measured data in figure 8. The correlation between radiated field intensity and equivalent injected voltage value is linear. The measured data is processed by straight-line fitting ($y=mx$). The gradient m and correlation coefficient R under different frequency test condition is shown in table 1.

Table 1. The gradient and correlation coefficient of straight-line fitting with measured data.

system B	f / GHz	straight-line fitting gradient	correlation coefficient
B1	3.3	12.8281	0.99886
B1	4.6	9.7117	0.99972
B1	7.2	5.08199	0.99985

In sum, even the response of the equipment under test is nonlinear, on the premise of equal equipment output response, the correspondent relation between equivalent injected voltage and radiated field intensity is linear. Therefore, the equivalent injection voltage used to substitute HIRF test can be obtained by linearly extrapolating the correspondent relation in lower radiated field condition.

4. Conclusions

The equal response voltage on the equipment cable port is regarded as equivalent principle. The equivalent substitute model and injection source extrapolation model in HIRF condition are established. The equivalent correspondent relation between injection voltage and radiated field is put forward theoretically. The couple conditions of using injection as a substitute for radiation are determined. One is the same equivalent source, and the other is the same output impedance.

The forward voltage on interconnected transmission line is regarded as indirect equivalent parameter, and then the equivalent injected voltage source U_{SI} in low intensity field is obtained. The equivalent injected voltage source V_{SI} in high intensity field can be obtained by extrapolating U_{SI} linearly（$V_{SI}=kU_{SI}$）.

The typical nonlinear interconnected system is selected as equipment under test, and the single frequency continuous wave radiation and injection effect experiments are carried out separately. The test results indicate that the relation between radiated field and injected voltage is linear, and the equivalent injected voltage used to substitute HIRF can be obtained by linear extrapolation.

Acknowledgments

This paper is supported by Arm Pre-Research Program (51333040101), National Defence 973 Program (6131380301) and national natural science foundation of China (61040003).

References

[1] Tao J Y and Chen Y 2011 *Journal of China Academy of Electronics and Information (in Chinese)* **6** 111
[2] Sun G Z, Liu S H and Chen J P 2006 *Military Operations Research and Systems Engineering (in Chinese)* **20** 43
[3] Flavia G, Giordano S and Filippo M 2008 *IEEE Trans. on EMC* **50** 524
[4] Li B Z, He J L and Zhou H 2009 *High Voltage Engineering (in Chinese)***35** 2753
[5] Rasek G A and Loos S E 2008 *IEEE Trans. on EMC* **50** 499
[6] Spadacini G and Pignari S A 2004 *IEEE Trans. on EMC* **46** 446
[7] Xu J Z, Jiang Q X and Wang W 2006 *Chinese Journal of Radio Science(in Chinese)* **21** 488
[8] Ni G Y, Luo J S and Li C L 2007 *High Power Laser and Particle Beams(in Chinese)* **19** 1521

Design of an insulator leakage current measurement system based on PLC

Changhai Sun*, **Yan Wu, Wenqi Han**
College of Electrical Engineering, Dalian University of Technology, Dalian, China

E-mail: duttd@dlut.edu.cn

Abstract. It is usually difficult to detect a small current signal in a high-pressure environment with strong electromagnetic interference. The paper introduces a high-voltage electrical equipment that is used to measure the small current. The system consists of three parts including the DC high voltage generator, data acquisition modules and PC data display section. The experimental results show that the device can acquire weak current signal effectively. Data acquisition module can communicate with the PC software with Ethernet, and the users can store, query the data through a database easily.

1. Introduction

In recent years, with China's sustaining and rapid economic growth, electricity demand is increasing and the voltage level of transmission line is grading, which has put forward higher performance requirements for the insulator used in transmission line. Under the promotion of national policy, Dalian Insulator Group focuses on the manufacturing industry of insulators, and the annual output has reached 150,000 pieces. Especially the suspension porcelain insulator used in the DC transmission line plays a key role in several 500 kV EHV transmission engineering projects. There are more than 1,000,000 pieces of XWP-70, XWP-160 and XWP-210 with different contaminated levels running on these 500 kV transmission lines in China, which keep good running records and guarantee the safe and reliable operation of transmission lines [1].

At the same time, Dalian Insulator Group and Dalian University of Technology carried out the research on the technology of detecting the insulator. Based on the enterprise's requirement for the measurement technique, it's necessary to conduct typical tests and sampling tests to ensure the products' eligibility when the products are dispatched from factory. But currently when measuring the insulation resistance and volume current, enterprises are facing some problems including poor and obsolete equipments, too large leakage current when the element is exposed in the air, and very backward and poor accuracy of the artificial test method.

This paper designs an insulator leakage current measurement system based on PLC in order to realize the intelligent, accurate and efficient inspection work and solve problems for the enterprises.

2. The overall design of the system

The system mainly consists of three parts including a DC high voltage generator, a data acquisition module and a control PC. The DC high voltage generator is responsible for providing DC high voltage;

---

* To whom any correspondence should be addressed.

data acquisition module is responsible for collecting and processing data and communicating with the host computer; and control PC is responsible for data display and saving the queries.

This system uses PEC8000 data acquisition module to realize the measurement of weak leakage current of insulator. The overall system structure is shown in figure 1.

Taking the insulation resistance of XP-70 insulator as an example, the experimental process consist three steps. In the first step, the test system was built as shown in figure1 and the insulator was put in the artificial high voltage chamber (constant temperature). In the second step, the temperature of incubator was maintained at 60 °C for 2 h in order to avoid the influence of humidity. In the third step, controlling the DC high voltage generator to output negative high voltage to insulator, the voltage was boosted to 15 kV and kept at 15 min to charge the capacitor stably. Then, data acquisition module will process signals, establish communication with host computer and measure the weak current using the software. The software can realize the intelligent management for the measured data.

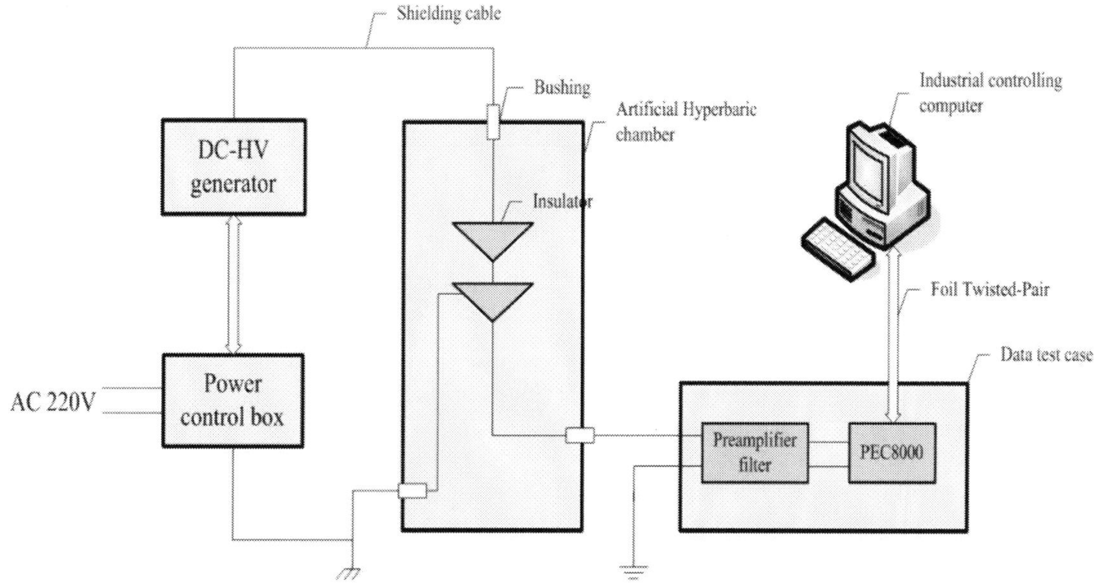

Figure 1. System architecture diagram.

Figure 2. The insulator of XP-70.

Figure 3. Data testing box.

3. The design of the hardware for the system

3.1. High voltage DC generator

ZGF-200 high voltage DC generator, which is developed by the Special Power Plant in Dalian University of technology, is used to provide the high voltage. This power supply uses containerized integration technology for sealing well and avoiding of dust and moist. Voltage is output from the embedded type electrode and it can be adjusted continuously with small pulsating factor, small leakage current and high testing precision. In addition, the power supply has a protection measure for overload voltage and current, which is suitable for enterprise field test.

The main principle diagram is shown as in figure 4.

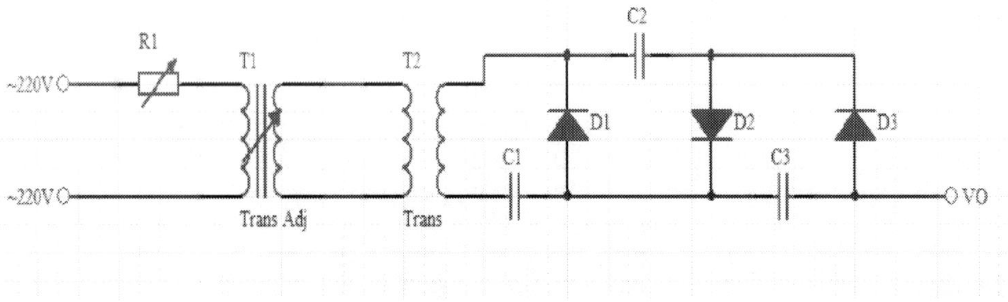

Figure 4. Schematic diagram of the DC supply.

As shown in figure 4, the input is 220 V AC, and after the conversion of the self coupling voltage regulator T1, 0~70 kV voltage is output. Then, the AC voltage is converted to DC voltage and is amplified three times after three times rectifier circuit. The V_0 point outputs negative polarity DC high voltage ranging 0 ~ 200 kV. It can greatly reduce the volume of the power using this kind of rectifier circuit. And output voltage pulsation can be improved by the different size of the capacitance group.

3.2. The design of data acquisition module

The system mainly uses the PLC to collect data for processing. Weak signal will be preprocessed by prefilter amplifier. Generally, the most important thing for the amplification is to select an appropriate amplifier which has a slight bias current. If amplification is for nA level of current signal, it must choose pA level bias current of the op-amp. This system mainly aims at the insulator leakage current measurements. Under a few kilo-volts high voltage, insulator volume current changes from dozens to hundreds of nA with the change of temperature and humidity.

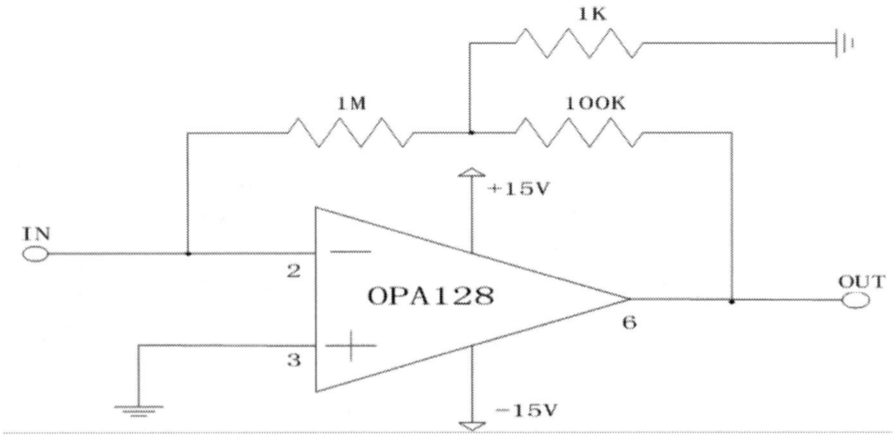

Figure 5. I-V Conversion circuit.

In the design of sampling amplifier, I-V conversion circuit with good linearity performance is selected. Burr Brown Company OPA128 is chosen for the op-amp, the bias current is not more than 75 fA, and the offset voltage is 500 μV. To reduce the noise interference and op-amp load, usually output voltage should be two times or more compared with the input voltage, but if the output voltage is too high, feedback resistance (R_f) must be increased and op-amp performance requirements need to be increased at the same time. So it will be suitable for the design of the input voltage range form 50~100 mV. From the formula $V_O = - I_S R_f$, R_f should be 100 MΩ when input current is nA range. But it can easily produce resistance thermal noise current and increase the distributed capacitance once the R_f is large enough. This system design uses the T resistance feedback network to improve resistance noise, as shown in figure 5.

There is some burr due to the noise interference after amplification. In order to filter out the interference, the system used a 20 Hz second order passive low pass filter. The effective signal is provided for the PLC after being filtered.

PLC module chose PEC8000 industrial Ethernet programmable controller, with 32-bit LPC2378 microprocessor as the core, with one 10 m/100 m industrial Ethernet communication interface, two RS485 communication interface, support MB + high-speed bus, eight digital input (support 4 top 50 kHz high-speed input), eight digital output (support 2 highest frequency 50 kHz high speed output), six analogue input, and two analogue output [2].

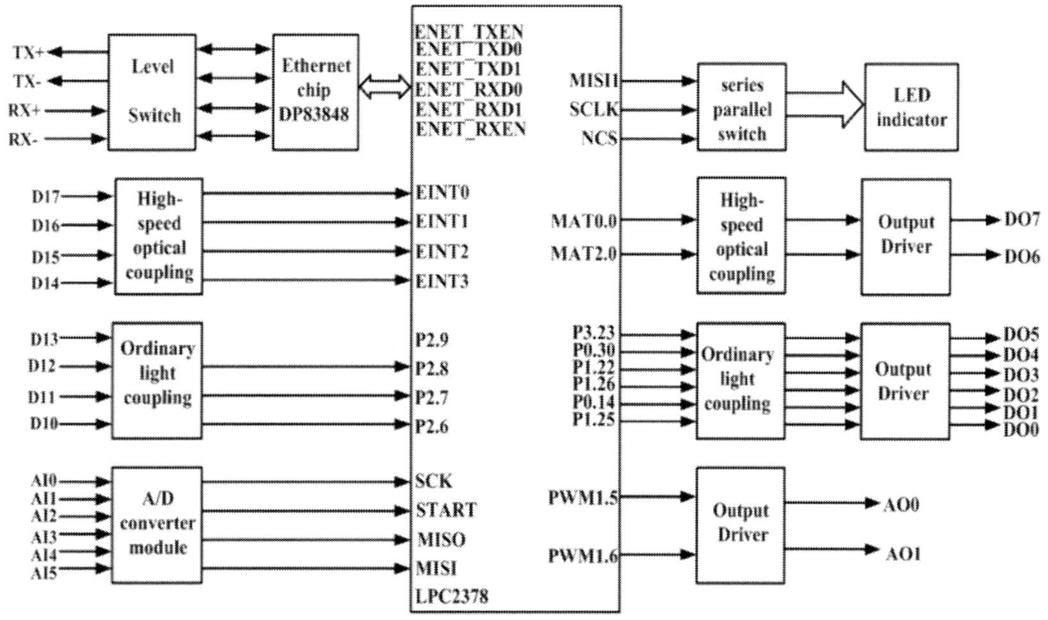

Figure 6. Architecture flow of the PEC8000.

As shown in figure 6, in the PEC8000, DI0 - DI3 for ordinary switch input, DI4 - DI7 for four high speed switch input, AI0 - AI5 for six analog input signals, can input thermocouple/thermal resistance/standard current/standard voltage signal. There is a light coupling isolation between physical input and output between LPC2378. And LPC2378 built-in the Ethernet MAC layer, by connecting the physical DP83848 chip, after level transformation, can realize Ethernet communication function. DQ0-DQ7 is for switch output, AQ0 and AQ1 are for analogue output [3].

In this system, the analog current signal after the preamplifier is sent into the interface circuit, then after AD converter ICL7135 realized the analog to digital conversion, photoelectric isolation, and then LPC2378 input port is connected. That means the input signal needs sampling, amplification, filtering, modulus conversion and photoelectric isolation before sending into the CPU input port. PEC8000's input code value range is 0 ~ 10000 after analogue-to-digital converting. This paper's design according to different size of the current signal use different analogue input terminals, corresponding to different code value range. When the input current range is 0 ~ 1000 nA, code value accuracy is 10 yards/nA; when the input current range is 0 ~ 200 uA, code value accuracy is 50 yards/uA. Current signal after internal processing is sent to DP83848 Ethernet chip, and then can be connected to computer through the twisted pair.

4. The design of system software

4.1. PLC module programming

PEC8000 programming based on the IEC61131 standard, the matched programming software provides a functional block diagram language edit mode. The controller will contact up program and physical input/output together, through the logic procedures to complete the corresponding control. The PLC config software have given many algorithms of function block, such as logic operation, compare instruction, timer, counter, etc., using the function block, the user can composite module freely and realize complex algorithm.

The main program block diagram of the system is shown in figure 7. It should obtain input operator EN fristly, if EN is 0, don't allow the function block execution, then will assign ENO value 0, if not, allow the function block executive, then read input parameter, according to the function block algorithm to carry on the calculation, and the calculated results will be written to output parameters, and will assign ENO value 1, meaning that it can allow the output.

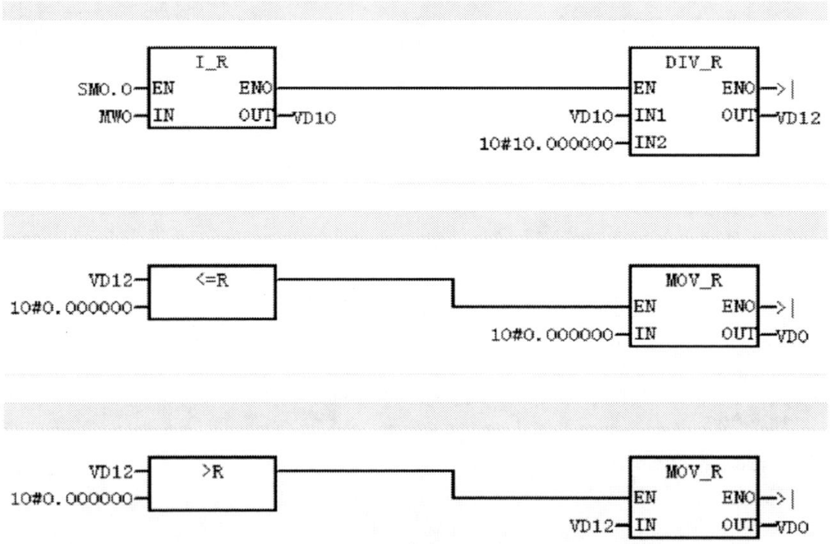

Figure 7. Main function chart of the system.

4.2. The design of interactive interface of the host computer

This system adopts Visual Basic6.0 to design the user measurement interface of the host computer. VB is a kind of object-oriented, visual and event-driven programming language, which could write out a user-friendly operator interface. Before measurement, user can set external environment variables, including the type of electrical equipment and temperature. User can also set the parameters in measurement, such as the sampling time, number of samples, accuracy and channel selection etc. Measurement interface is shown in figure 8.

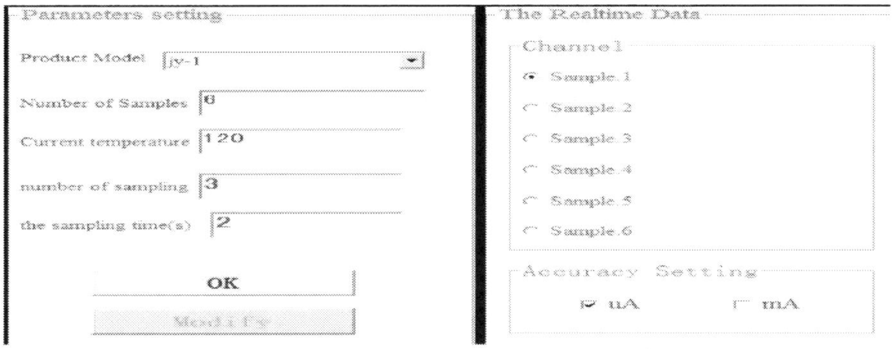

Figure 8. Interactive interface map.

4.3. The design of database.

In order to be convenient for the operator to record, analyze and query the measurement data, this system adds on the function of database management, which can use SQL Server to build relational data table and link the characteristics amount of the sample models, temperature, voltage, current and test time etc. Then the application of host computer can use ADODB control in VB to operate the database [4].

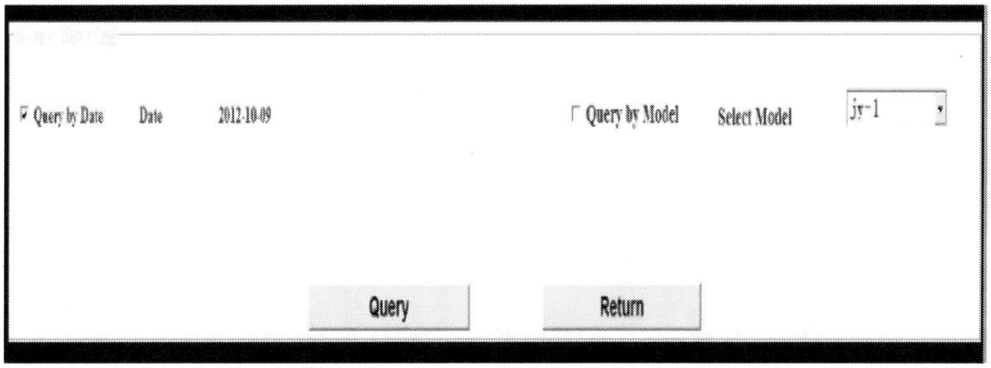

Figure 9. Historical data query page.

4.4. Communication between PLC and host computer

The communication protocol used in this system is Ethernet MODBUS protocol. Compared with traditional Ethernet communication protocol, MODBUS protocol extends new packet header in the data link layer, which can identify the data unit using this protocol and is very suitable for industrial control. PEC8000 module connects to the computer via twisted pair, the communication port is 11000 by default and the transport layer uses the UDP mode to transmission packets. Host computer configures the properties of the relevant agreements and triggers the events via the Winsock control of VB.

5. Conclusion

The measurement system using Ethernet can solve the difficulties of intelligence and accuracy during the test of insulator leakage current in this paper. This system has been in use since 2010 and keeps stable and efficient until now. In addition, this system also plays an important role during the measurement of actual voltage and current in high-voltage electrical equipment.

Acknowledgments

This work is supported by the CNPC Petroleum Science &Technology Innovation Foundation (Grant No. 2011D-5006-0402) and Construction Science & Technology Project of Dalian city.

References

[1] Kang Y 2007 *Electric appliance industry* **5** 45
[2] Zhong H, Wang Z and Wen R 2005 *J. Polytechnic. Control.* **237-239** 1032
[3] Xue X 2010 *J. Automation Control. Electron.* **37-42** 234
[4] Wang H, Han W Q, You L and Yang J H 2011 *J. High Voltage Tech. Measure* **79-84** 196

Trichel pulses in a negative corona discharge in air at low pressure

Shou Jie He [1*], Jing Ha [2], Yu Hong Han [1], Zhi Qiang Liu [1], Shu Qing Guo [1]
1 College of Physics Science and Technology, Hebei University, Baoding , China; 2 Institute of Science, Hebei Agriculture University, Baoding, China

E-mail: heshouj@hbu.edu.cn

Abstract. Most works focused on the Trichel pulses at an atmosphere pressure in the past. In this paper, the characteristics of Trichel pulses at low pressure (several Torr) in air used by a needle-to-plane discharge structure are investigated. The current-voltage (*I-V*) characteristics, the images of discharge and the time evolution of the current and voltage of Trichel pulses are measured under different conditions. Results show that Trichel pulses exist in the negative differential resistance of *I-V* curves, and disappear when the discharge transits to the abnormal glow discharge mode with an abrupt contraction of discharge area around the needle tip. The pulses are consisted of a very short rising time (about 10 μs) and a short durations time (tens of μs) separated by longer inter-pulse periods (more than 100 μs). The frequency of pulses can be tuned continuously from a few kilohertz to more than ten kilohertz by adjusting the averaged current and the pressure. It increases with the increase of averaged current at the same pressure, and rises as the pressure increasing at the same averaged current. These results will be helpful to understand the mechanism of Trichel pulses.

1. Introduction

Trichel pulses are an interesting phenomenon with regular voltage and current pulses, existing in the direct negative corona discharge [1-2]. Trichel reported this phenomenon firstly in 1938[1]. He argued that the pulses of discharge seemed to derive logically from space-charge formation and subsequent clearing. He predicted the build up of a strong negative space charge away from the point and positive space charge near the point. The positive space charge enhanced the field near the point. While the field beyond the space charge was decreased, so that almost no ionization could take place. The positive ions moved toward the point, and the region of enhanced field near the cathode would become too narrow to sustain the discharge and, finally, the ionization process stopped.

In the following work, Trichel pulses were investigated extensively both from experimentation and theories [3-5]. Loeb et al [6] stated that the Trichel pulses existed only in the electronegative gases. He also found the time of negative ion drift to the anode was much longer than the pulse period. Lama and Gallo [5] investigated the dependence of pulse frequency, charge per pulse, and time averaged corona current with the applied voltage, needle tip radius, and needle to plane spacing. Cross et al [7] pointed out that the corona current, point material, and pressure also had important influence on the pulses. He stressed that the first pulse was distinctly different from the regular pulse. The Trichel pulses also have been investigated by simulation. In 1985, Morrow [8-9] proposed a one-dimension model for the

[*] To whom any correspondence should be addressed.

development of Trichel pulses. Continuity equations for electrons and positive and negative ions in a one-dimensional form were numerically solved together with Possion's equation. Recently, some two-dimensional models have developed to simulate the Trichel pulses [10-11].

In the last, most work focused on the Trichel pulses at the high pressure (more than 100 Torr). Few works have been done at the low pressure (below 20 Torr). In this work, we report the Trichel pulses at low pressure in air used by a needle-to-plane discharge structure.

2. Experimental set up

The experimental setup is schematically shown in figure1. A needle electrode of 0.12 mm radius of curvature is utilized and connected to a negative dc high voltage supply. The needle is placed 10.0 mm away from the plane electrode. The plane electrode is grounded. The discharge system is located in a sealed vacuum chamber, with the pressure ranging from 1 to 50 Torr. A ballast resistor R_1=4 MΩ and a measurement resistor R_2 = 50 kΩ were connected in series at the cathode side and the anode side, respectively. An oscilloscope (Tektronik TDS3054B) is used to read out temporally the resolved potential, V_1 and V_2, and the discharge current I can be deduced through V_2/R_2. An Ampere-meter is connected in series with the anode to measure the averaged discharge current. A digital CCD camera (Canon A950) is used to record the plasma images.

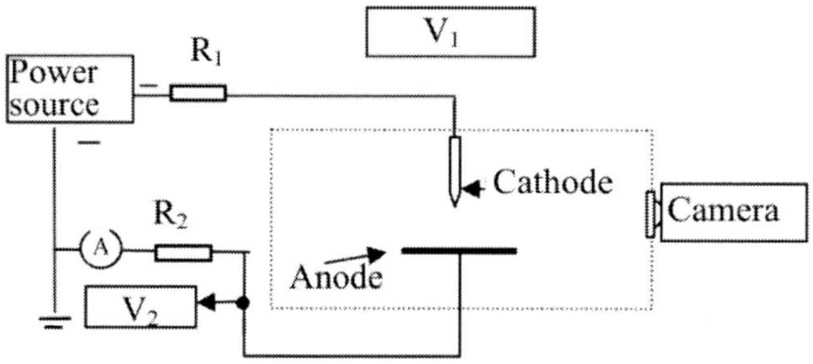

Figure 1. Schematic of experimental setup.

3. Results and discussion

3.1 I-V curves and discharge images

Figure 2 shows the I-V characteristic at three gas pressures. For all pressures, the I-V curves show three stages as the current increasing, where V=V1-V2.

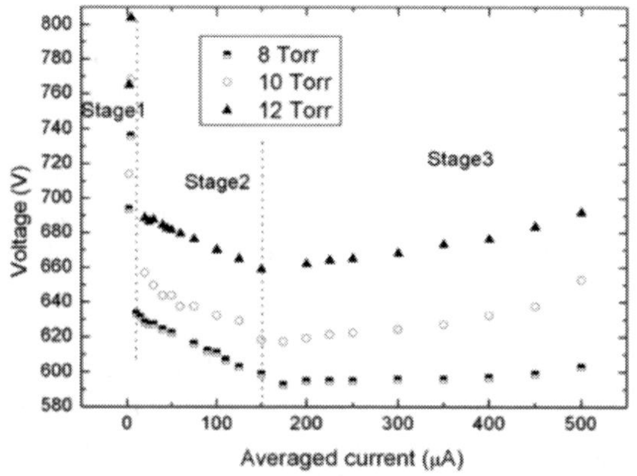

Figure 2. The *I-V* characteristics of negative corona discharge at different pressure.

In stage1, the voltage quickly increases with the averaged current, and the I-V curve shows a high positive slope. The highest averaged current is no more than 15 μA. This stage should correspond to the pre-discharge. When we increase the voltage continuously, the averaged current increases quickly to about 15 μA and the discharge transits to the second stage. In stage2, both the current and the voltage are unstable and self-pulsed, the Trichel pulses appear. The pulses are very regular in behavior, which will be discussed below. The averaged voltage decreases when the averaged current increases. When the averaged current reaches about 140 to 160 μA depending on the gas pressure, the pulse disappears abruptly. Then, the voltage has a weak increase with the averaged current rising. This stage corresponds to the abnormal glow discharge.

The images of discharge at different averaged current have been obtained for the pulsing stage and stage 3 at 8 Torr. It can be seen that the images show a round structure around the tip of needle with a higher brightness, and a conical structure from the tip of needle to the anode plane with a very lower brightness. In the pulsing stage, as shown in figure 3 (a), (b), (c) and (d), the brightness of discharge images falls gradually far away from the center of images, and has no abrupt change. Moreover, the brightness and the areas of discharge rise continuously with the increases of averaged current. When the averaged current rises to about 175 μA, the brightness of discharge has an abrupt increase compared with that in stage2, which correspond to the location in which the pulsing stage transits to the abnormal glow stage. While the diameter of discharge falls obviously, from 0.6 cm at 150 μA to 0.4 cm at 175 μA. Furthermore, we can see that there is an apparent abruption of brightness from the centre to the border of discharge. The dot line shows the boundary between the two parts. This great increase of brightness means that the ionization and excitation rates are significantly enhanced. Subsequently, the brightness of discharge increase with the current again.

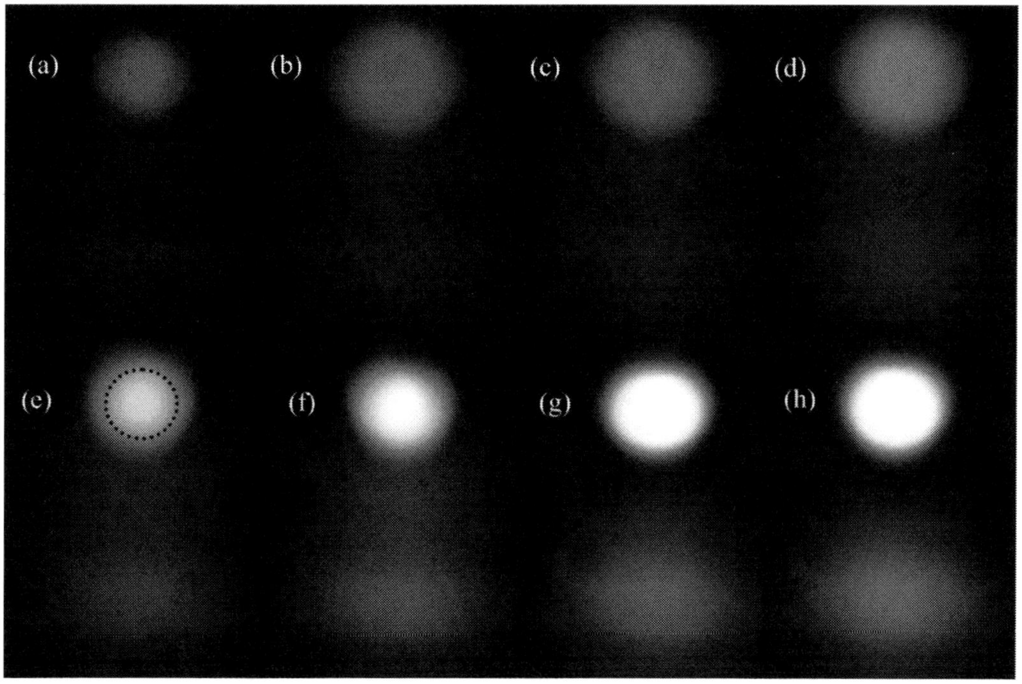

Figure 3. The discharge images at different averaged current. (a) 50 μA, (b) 100 μA, (c) 125 μA, (d) 150 μA, (e) 175 μA, (f) 200 μA, (g) 250 μA and (h) 300 μA.

3.2 Typical current and voltage curves

As mentioned above, the Trichel pulses appear in stage 2. Both the discharge current and discharge voltage oscillate. These oscillations are highly reproducible and can be tuned continuously from a few kilohertz to more than ten kilohertz by adjusting the averaged current.

Figure 4(a) and (b) show the typical time evolution of the voltage and the current of the Trichel pulse for pressure p=8 Torr at averaged current 30 μA. The pulse is very stable with frequency of 5.0 kHz in figure 4. The regular current pulse has a shorter rising time of t_r = 11 μs, a longer decay time of t_d =35 μs, and a great long waiting time t_w=154 μs, respectively. The peak current is about 212 μA and the lowest current is nearly zero, corresponding to a lowest voltage of 565 V and a peak voltage of 295 V. The shape of pulses in the present is similar to that obtained at high pressure, but has a smoother structure. This shows that it is easy to get stable pulses at lower pressure. In addition, the peak value of current in the present is much lower than that of pulses produced at high pressure. The peak current is on the order of several milliamperes at high pressure [7]. This difference originates from the difference of particle density. Higher pressure leads to higher particles density, and more electrons can be produced. Therefore, higher peak current appears at high pressure.

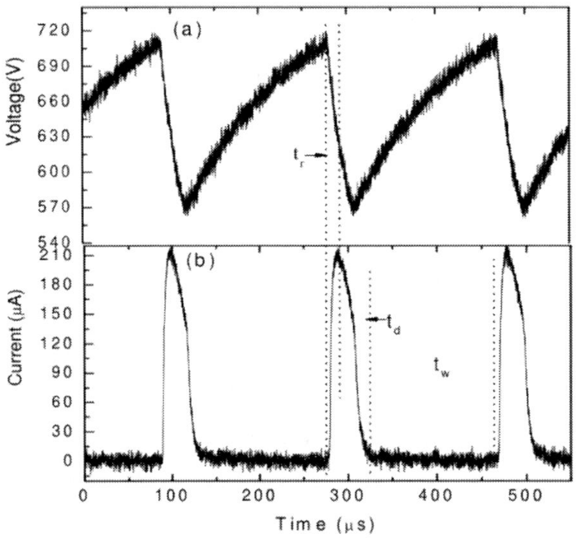

Figure 4. Typical waveforms of (a)current and (b) voltage of Trichle pulse at averaged current of 30 μA for p=8 Torr.

3.3 The influence of averaged current on the Trichel pulses.

Figure 5 shows the Trichel pulsing frequency as a function of the averaged discharge current at 8 Torr, 10 Torr, and 12 Torr. For all the investigated pressures, the self-pulsing frequency is nearly proportional to the averaged current and increases with the rising in pressure.

Figure 6 shows the current waveforms of Trichel at different averaged discharge current. All of the current pulses have the same shape for different averaged current. It is seen both the peak and the lowest current increase with the rising of averaged current. The increase of peak current is because that the electric field increases with the averaged current. The lowest current should correspond to the direct component of the pulses. It means that the direct component of current become stronger with the increase of averaged current. It is noticed that all of the current pulses have the same rising time. The decay time has a weak increase, while the waiting time has an obvious decrease with the increase of averaged current. This indicates that the formation mechanism of the discharge should be the same for different conditions. The result in figure 6 also indicates that the change of frequency is caused by the change of waiting time of current pulses.

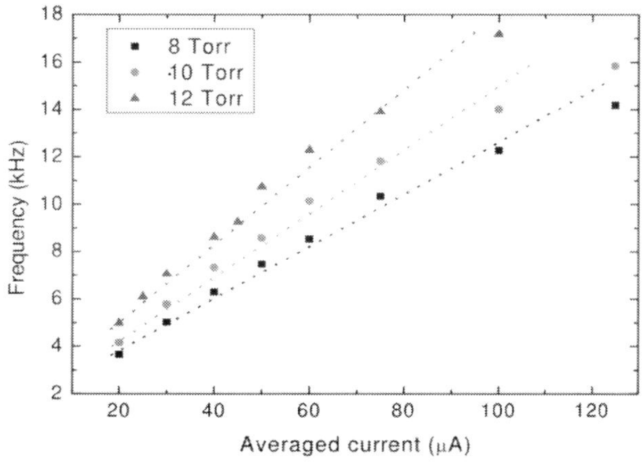

Figure 5. The self-pulsing frequency as a function of the averaged discharge current at different pressures.

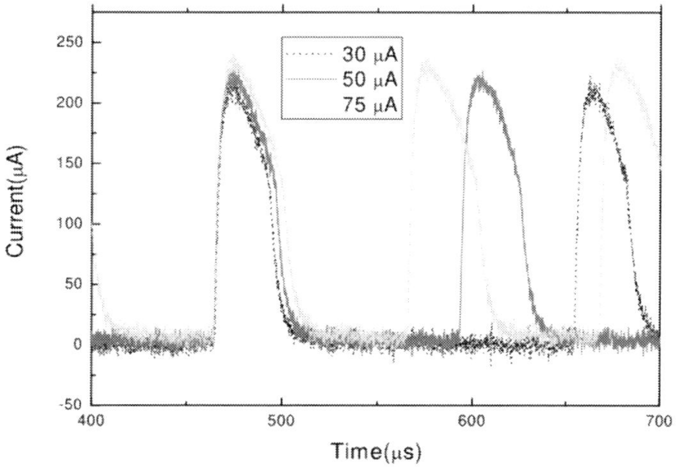

Figure 6. The current waves of Trichel at different averaged discharge current for p=8 Torr.

3.4 Discussion

As discussed above, the averaged current has important influence on the Trichel pulses. This can be explained as the following. The rising of current originated from the positive space charge layer to enhance the electric field. And the quench of current is caused by the build up of the negative space charge. In the other words, the decay time of current depends on the time used to concentrate enough negative ions. The high current means the stronger positive space charge layer. Therefore, this requires that more negative ions be created to yield field that quench the discharge, a longer time for the formation of the negative space charge is necessary. In another, the waiting time between two pulses depends on the time used to clear the negative ions. As the current increasing, the field rises, which enhance the velocity of ions to the anode. So the negative ions can reach the anode in a shorter time, the waiting time decreases. Because the frequency of pulses mainly depends on the waiting time, the frequency of pulses increases as the averaged current rising.

4. Conclusion

Trichel pulses at low pressure in air used by a needle-to-plane discharge structure are investigated firstly.

(1) Regular voltage and current waveforms can be obtained at the negative differential resistance of I-V curves.

(2) Both the brightness and areas of discharge have an obvious change when the discharge transits from the Trichel pulsing stage to the abnormal glow stage.

(3) The pulses are consisted of a very short rising time, a short duration time, and a longer inter-pulse periods.

(4) The frequency of Trichel pulses increases with the increase of averaged current at the same pressure. As the current increasing, the rising time is invariable nearly, the decay time increase weakly, and the waiting time decrease greatly.

Acknowledges
The authors would like to acknowledge the helpful discussion and advices of Professor Ou Yang. This work is supported by the National Science Foundation of China under Grant No. (11005009; 51077035) and the science and technology research projects of colleges and universities in Hebei province No.2011166, and the Science Foundation of Hebei province No.A2012201037.

References
[1] Trichel G W 1938 *Phys. Rev.* **54** 1078
[2] Van Brunt R J and Leep D 1981 *Phys. Rev.* **55** 6588
[3] Soria C, Pontiga F and Castellanous A 2007 *J. Phys. D.* **40** 4552
[4] Sattari P Castle G S P and Adamiak K 2011 *J. Phys: Conference Series* **301** 012058
[5] Lama W L and Gallo C F 1974 *J. Appl. Phys.* **45** 103
[6] Loeb L B, Kip A F and Hudson G G 1941 *Phys. Rev.* **60** 714
[7] Cross J A, Morrow R and Haddad G N 1986 *J. Phys. D.* **19** 1007
[8] Morrow R 1985 *Phys. Rev .A, Gen .Phys.* **32** 1799
[9] Morrow R 1985 *Phys. Rev. A, Gen. Phys.* **32** 3821
[10] Tran T N, Golosnoy I O, Lewin P L and Georghiou G E 2011 *J. Phys. D.* **44** 015203
[11] Sattari P, Gallo C F, Castle G S P and Adamiak K 2011 *J. Phys. D.* **44** 155502

Journal of Physics: Conference Series **418** (2013) 012092 doi:10.1088/1742-6596/418/1/012092

Comparison of point-to-plane corona in different gases

Lijuan Liu, Ziliang Zhang, Zulin Peng, Jiting Ouyang
School of Physics, Beijing Institute of Technology
5 South Zhongguancun Avenue, Haidian District, Beijing 100081, China

E-mail: jtouyang@bit.edu.cn

Abstract. The point-to-plane DC corona discharges at atmosphere pressure have been investigated experimentally in different gases including argon, helium, N_2, O_2 and air in a needle-to-plate configuration. The breakdown voltage (the onset voltage of corona and the inception voltage to spark), current-voltage characteristics and the corona images have been measured in the gases under positive and negative polarity. Generally, the breakdown voltage of the positive corona is higher than the negative in the gases except O_2 in which the results is reversed. The current of the negative corona discharge is much higher than the positive in all gases. There exists a stable negative corona glow in the gases except air or O_2 before the spark occurs, while the pulsed positive corona will transit into the spark directly when the voltage increases, without the stable glow stage.

1. Introduction

Corona discharge is a partial discharge where the geometry confines the gas ionizing processes to high-field ionization region around the active electrode [1]. The field near the discharge electrode is much stronger than the rest of the gap. This situation typically arises when the characteristic size of the electrode radius is much smaller than the inter-electrode distance [1, 2]. Corona discharge is also one of the common low-temperature, non-equilibrium plasmas. Because it is easy to generate, stable at atmospheric pressure, and operate at low currents (on the order of several micro-amperes) and power consumption [3], and also rich in the active radicals, excited and ionized species, corona discharge has been widely used for many traditional applications such as spray coating copy machines, drying separation systems and radiation detectors, surface treatment, electrostatic precipitators [4-6], as well as in some new fields such as probe of microwave (MW) plasma jet, treatment of skins diseases and airflow control [6-8], etc.

The electrode configuration has great influence on the characteristics of the corona discharge. The typical configurations include point-to-plane or point-to-point, wire-to-wire, wire-to-plane or wire-to-cylinder, etc. Among them, the point-to-plane (or needle-to-plate) is the most typical and popular configuration. The corona discharge with the point-to-plane configuration has been investigated widely in air under various conditions. In this paper we investigate the DC corona discharges in different gases at atmospheric pressure. The aim is to make a comparison of characteristics of the point-to-plane corona in air and other gases.

To whom any correspondence should be addressed.

2. Experimental set-up

The experimental setup of the discharge system is shown in figure 1.

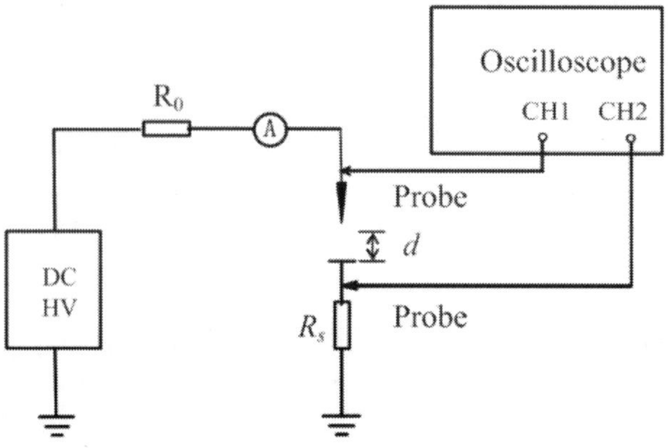

Figure 1. Schematic of experimental setup.

The discharge electrode is made of stainless steel with a diameter about 0.05 mm in tip. A copper plane is used as the grounded electrode through a resistor R_S (1 MΩ) which is also the current sampling one. The gap distance between the needle and plane is 5 mm. A high voltage (HV) DC power supply was used as the power source and connected to the needle through a protection resistor R_0 (2 kΩ). The output of the power source may be positive or negative. The applied voltage was measured by using a digital oscilloscope (Tektronix TDS3054B, 500 MHz) with a voltage probe P6015A. The discharge current was measured by a micro-Ampere-meter. It was also monitored by the voltage drop V_R across sampling resistor R_S which was measured with the same oscilloscope. A digital camera (Canon EOS 550D) was used to record the corona images.

3. Results and discussion

There are two breakdown voltages for the corona discharge, i.e., the onset voltage of the corona, named as V_C in this paper and the inception voltage of spark, named as V_{spark} in this paper. We tested the corona discharge under different gases at atmospheric pressure and different voltage polarity.

3.1. The breakdown voltage in different gases

Table 1 gives out the measured onset voltage V_C, the inception voltage of spark V_{spark} and the corresponding transition current I_{spark}.

Table 1. The breakdown voltage and current in different gases and voltage polarity.

	Ar		He		Air		N_2		O_2	
	+	-	+	-	+	-	+	-	+	-
V_C (kV)	1.91	1.23	2.37	1.02	3.10	2.05	2.94	1.68	3.08	2.50
V_{spark} (kV)	3.24	2.15	5.07	2.24	5.42	5.08	5.10	4.32	5.82	6.69
I_{spark} (μA)	6.00	32.0	50.0	101.0	20.0	145.0	15.0	365.0	33.0	94.0

Generally, the onset voltage is much higher for the positive corona than the negative in all the investigated gases. For instant, the onset of positive corona is 1.91 and 3.10 kV in Ar and air, larger than that of negative corona (1.23 kV and 2.05 kV, respectively). This is also true for the breakdown voltage to spark except in O_2 in which the result is reversed. This unexpected result might relate to the existence of the electronegative oxygen molecules. The attachment of electrons on O_2 might cause a decrease of the avalanche under negative polarity and cause increase in breakdown voltage. This is also evidenced by the result in air that the breakdown voltage under negative polarity is close to the positive. The corresponding transition current to spark is also different for the positive or negative corona. The transition current of the negative corona is much larger than the positive in all the investigated gases.

3.2. Current-voltage characteristics

The current-voltage (*I-V*) characteristics were measured in all the tested gases under applied voltage between the corona onset V_C and the spark breakdown V_{spark}.

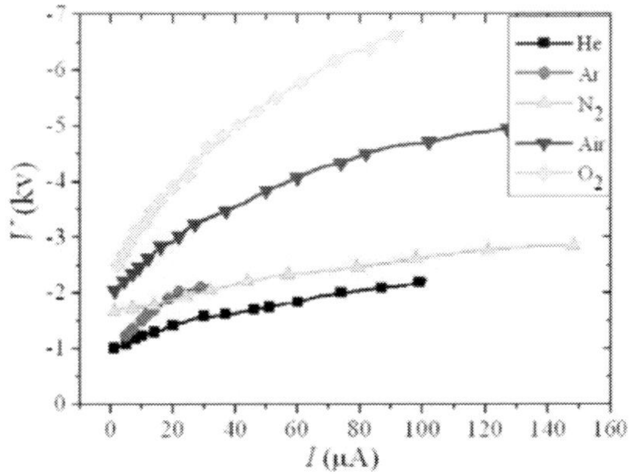

Figure 2. *I-V* curves of negative corona in various gases.

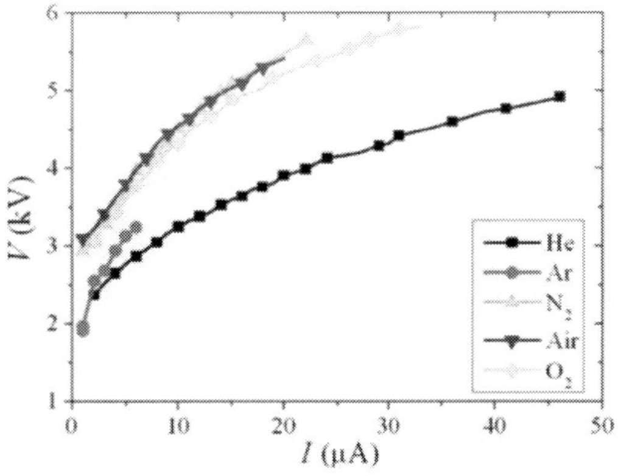

Figure 3. *I-V* curves of positive corona in various gases.

It is seen that the *I-V* is similar to the previous results [10, 11]. The corona current increases with the applied voltage in case of negative (see figure 2) or positive polarity (see figure 3). The *I-V* dependence of negative corona in all gases is nearly according with the Townsend's relation [2] for a stable corona, or $I = kV\ (V - V_C)$ where k is constant that depends on the geometric parameters. The plot of *I/V* as a function of *V* is almost a linear. For the positive corona, the *I/V* curve is nearly linear in large range of current but not true in the whole measured range.

The discharge current of negative corona is much larger than the positive, showing a significant polarity effect. For example, the current under negative polarity can reach as large as 100 A before spark in gases except Ar, while the current under positive polarity is generally less than 30 A except in He. This strong polarity effect would come from the asymmetry of the electrode geometry.

As expected, the voltage should be higher in N_2, air and O_2 to create the same discharge current voltage than in the noble gases. But the operation voltage in N_2 is lower under negative polarity, close to that of the noble gases (see figure 2). This might relate to the electronegativity of oxygen, which has a large electron-attachment cross section to make large probabilities of loss of electrons [1, 2].

Moreover, the current pulse (if occurs) is generally random under positive polarity. But the negative corona shows a regular repeat of the current pulse in air and O_2, known as Trichel pulse [1, 2]. The electronegative oxygen molecules are suggested to play an indispensable role for the Trichel pulse [1, 2]. But the Trichel pulses could also be observed in nitrogen as well as the noble gases in our experiments. This would be confirmed in the future.

3.3. Images of corona

The images of corona discharge in different gases at various voltages are shown in figures 4-8, where frames (a-c) are the negative corona, while frames (d-f) are the positive. All the images were recorded from side-view, with exposure time of 0.8 s. The electrodes are also shown in the figures.

It is seen that the negative corona occurs firstly around the tip of needle at lower voltage. As the voltage increases, the corona area increases, until a stable glow forms between the electrodes before the spark bridges the gap. Only in O_2, the glow discharge has not been observed in experiments. This would also relate to the role of electronegative oxygen. The stable glow has 3 characterized regions of DC normal glow, i.e., the light negative glow near cathode, the Faraday dark space and the followed positive column, as shown in figures 4-7. The glow discharge may diffuse outside toward the anode, forming a bell-shaped plasma region.

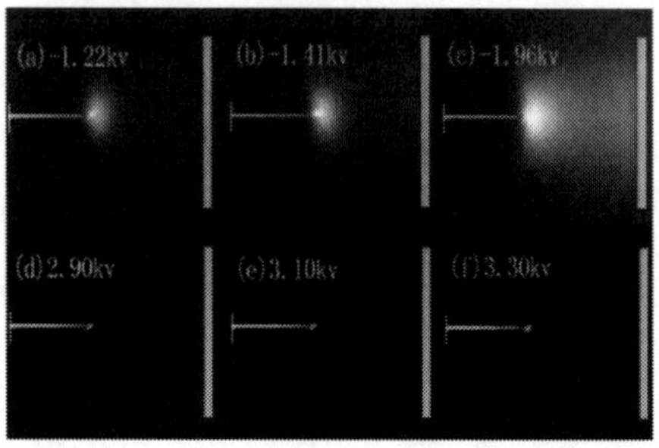

Figure 4. Images of (a-c) negative corona and (d-f) positive corona in Ar.

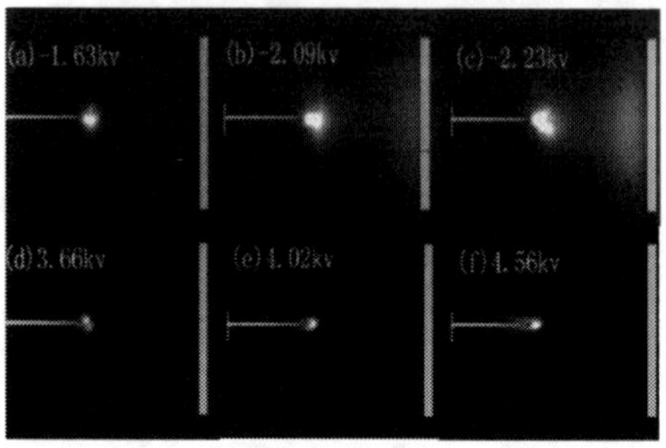

Figure 5. Images of (a-c) negative corona and (d-f) positive corona in He.

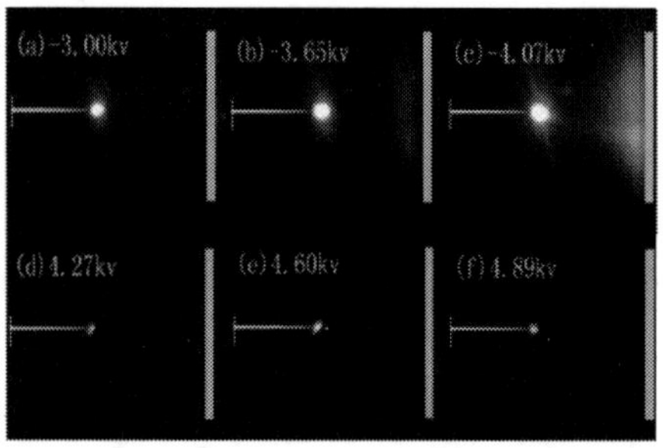

Figure 6. Images of (a-c) negative corona and (d-f) positive corona in N_2.

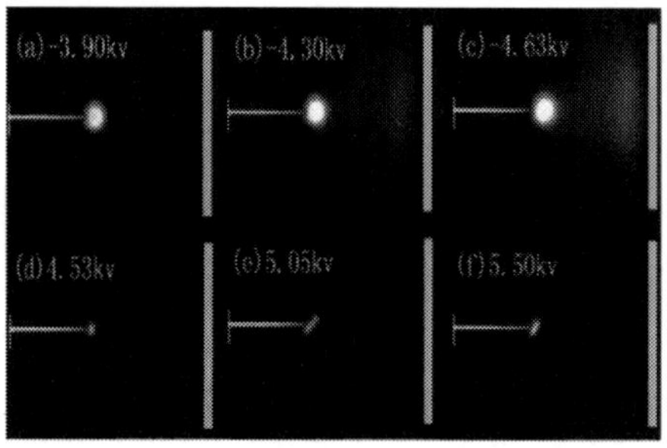

Figure 7. Images of (a-c) negative corona and (d-f) positive corona in air.

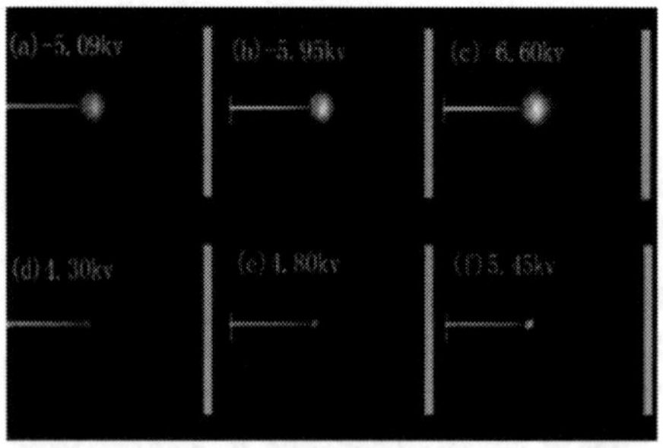

Figure 8. Images of (a-c) negative corona and (d-f) positive corona in O_2.

Under positive polarity, the corona generally occurs in small region around the tip of the needle. The bright area is 0.5-1 mm in diameter, and it is very similar in all the gases. In this case, the positive corona is generally pulsed, and will directly transit into the spark as the voltage increasing without stable glow. However, a positive corona glow may also be seen in air or O_2 sometimes, but the current is smaller than the negative.

4. Conclusion

We have investigated the DC corona discharge with different gases at atmospheric pressure in a needle-to-plane configuration. The results show a significant polarity-effect. In all gases the onset voltage of positive corona is much higher than the negative corona. The breakdown voltage of positive corona to spark is also higher than the negative except in O_2 that the result is inversed. The current of the negative corona is much larger than the positive in all gases. The current-voltage dependence of negative or positive corona shows the Townsend's relation. The negative corona has a large luminous area than the positive in all gases and shows a stable bell-shaped glow before spark, except in case of O_2 in which the negative corona exists near the tip of the cathode. The positive corona in all gases occurs only in a small region around the anode needle. The electronegative oxygen is suggested to play an important role in the characteristics of negative corona discharge.

Acknowledgments

This work was partly supported by the State Key Laboratory of Mechatronics Engineering and Control, China, under Grant No. 2011C3606.

References

[1] Golden M and Golden A. *Corona Discharge*. In: 1978 Hirsh M and Oskam H J ed. *Gaseous Electronics* (New York: Academic)
[2] Raizer Yu P 1991 *Gas Discharge Physics* (Berlin: Springer-Verlag)
[3] Tirumala R, Li Y, Pohlman D A and Go D B 2011 *J Electrostatics* **69** 36
[4] Lestinska L, Martisovits V and Machala Z 2011 *J Quant Spectrosc Radiat Transfer* **112** 2779
[5] Chang J S, Lawless P A and Yamamoto T 1991 *IEEE Trans Plasma Sci.* **19** 6
[6] El-Koramy R A, Yehia A and Omer M 2011 *Phys. Plasmas* **18** 023502
[7] Bussiahn R, Brandenburg R, Gerling T, Kindel E and Lange H 2010 *Appl Phys Lett.* **96** 143701
[8] Magnier Pierre, Hong Dunpin, Leroy-Chesneau Annie, Bauchire Jean-Marc and Hureau Jacques 2007 *Exp Fluids* **42** 815
[9] Zhang Q and Wu Z 2006 *Journal of Ordnance Engineering College* **18** 260
[10] Goossens O, Callebaut T, Akishev Yu, Napartovich A, Trushkin N and Leys C 2002 *IEEE Trans Plasma Sci.* **30** 176
[11] Antao D S, Staack D A, Fridman A and Bakhtier F 2009 *Plasma Sources Sci Technol.* **18** 035016
[12] Akishev Yu, Grushin M, Kochetov I, Karal'nik V, Napartovich A and Trushkin N 2005 *Plasma Sources Sci Technol.* **14** 18
[13] Belevtsev A A, 1991 *Proc. Int. Conf. on Properties and Applications of Dielectric Materials, (Tokyo, Japan, 8-12 July 1991)* p 517
[14] Tendero Claire, Tixier Christelle, Tristant Pascal, Desmaison Jean and Leprince Philippe 2006 *Spectrochim. Acta. Part B* **61** 2

Characterization of millimetre magnitude atmospheric pressure streamer discharge in pin-to-plane dielectric barrier discharge

S J Xu [1], Y H Zhang [1], Z Yu [2], J Yao[1], Z T Zhang [1, 2*]

1 Department of Physics, Dalian Maritime University, Dalian 116026, China
2 Institute of Marine Engineering, Dalian Maritime University, Dalian 116026, China

E-mail: newzhangzhitao@163.com

Abstract. The streamer regime of pin-to-plane dielectric barrier discharge in air was studied by means of fast photography, electrical measurement and photoelectricity. The fast photographs of positive streamer were obtained by CCD camera with micro lens. The exposure time is one microseconds. The images illustrate that the streamer is non-axisymmetric because of some random factors, such as surface charge position, space charge distribution, gas liquidity and so on. In fact, the streamer propagates along bend discharge channel. The bending degree increases with the electric field strengthen. By surveying a mass of images, the diameter of streamer, height of surface charge effect and scope of surface charge was estimate used to describe the shape of streamer.

1. Introduction

Dielectric Barrier Discharge is an economy and executable way of non-equilibrium plasma [1], and this way has been used in many fields like preparation of highly reactive oxygen, surface treatment, UV light source, carbon dioxide laser, plasma chemistry, environmental protection and so on [2-11]. It always has lots of different firing patterns [12], because there are many factors affect the discharge in the system, such as electrode structure, shape of the reactor, dielectric materials and structures, and medium gas. At atmospheric pressure, as high collision frequency [13] and electron attachment, electron avalanche spread some distance, then it turn into the streamers. Streamer's discharge life is just only 10~100 ns, the front electron's spread is 107~108 cm s^{-1}, and it has higher electric field strength [14], so it is widely used in industrial production

Streamers are developed by Raether [14, 15] and Loeb [16]. At present photo-ionization is considered as the best theoretical [17-19]. Balsur Eliasson and Ulrich Kogelschatz had established the numerical model of streamers and explain the principles of streamers in 1991 [20]. The model is axisymmetric [21] and people see with the naked eye, the streamers develop from one pole to another, at last the front electron spread evenly around the streamer in the dielectric layer surface. In fact, the streamer is non-axisymmetric, because the electric field which caused by the deposition charge will affect the streamers' electron. The development processes of the streamer of pin-to-plane dielectric barrier discharge under different pressure are shooting by T. M. P. Briels [22], and some of the

[*] To whom any correspondence should be addressed.

photographs show the existence of the streamer. Unfortunately, this term didn't take enough attention, and explain the reasons for this phenomenon. In this study, the experimental device is pin-to-plane dielectric barrier discharge in air, and its discharge gap is 0.9 mm. Morphology changes under the effect of different factors were collected and analyzed by CCD camera with a macro lens.

2. Experiment setup and method

Figure 1 gives an overview of the experimental set-up. This experimental set-up used here consists of the needle high voltage electrode, the plate electrode and the dielectric layer. The needle high voltage electrode is made by stainless steel, and its diameter is 0.75 mm, its curvature radius is 0.05 mm. The plate electrode is a size of 50 mm×50 mm×15 mm aluminium, the dielectric layer is made of purity of 96%α-Al_2O_3. The AC voltage is applied using 10 kHz power, for voltage between 0 and 10 kV. CCD camera recording units is made of CCD camera and a computer, a zoom multiples of 2×SE-16SM macro lens is mounted on the CCD camera, and it can observe the image detail of the micro-discharge. The streamer is random generation. Therefore, we use burst mode, its exposure time is 1 μs, and we can receive the integral image, we can use these images the structure and development process of the streamer.

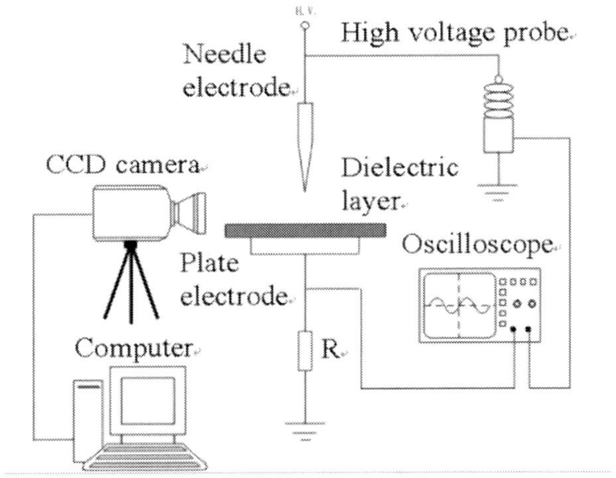

Figure 1. Experiment setup.

3. Experiment results and discussions

Figure 2 gives a numerical simulation of the streamer's formation process [23], due to the electric field of the high voltage electrode tip strong, the streamer start from the tip of the needle electrode to the plate electrode, at last run though the entire channel. In the whole process, the streamer has been developed along the axial. The shape of the discharge channel approximate as a cylinder.

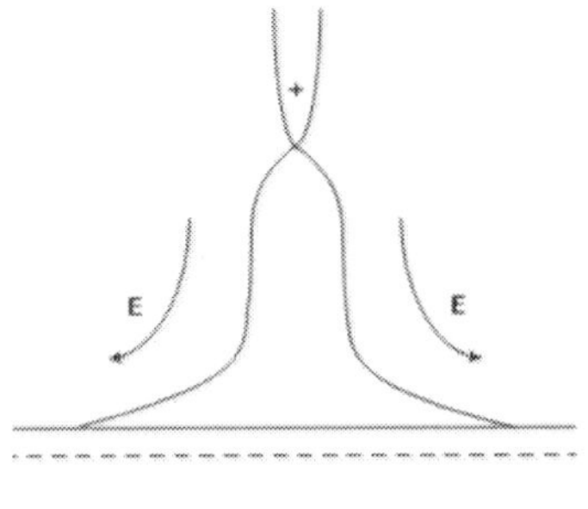

Figure 2. The mode of streamer.

Figure 3 gives the images of the streamer under different applied voltage. Under lowerapplied voltage the streamer almost show the shape of the axisymmetric. With the increasing of the applied voltage, non-axisymmetric is more and more obvious. In the higher applied voltage the non-axisymmetric is mainly manifested in the following aspects. First of all, though discharge gap is very small, bending and offset phenomenon is obvious. With increasing applied voltage and the deposition charge, reverse electric field is gradually increasing. Reverse electric field affect the charge. So the higher the voltage is, the more obvious the offset and bending phenomenon is. In addition, the streamer walks around the deposited charge layer edge at high-speed. The cumulative effects are what we observed in the macro cone discharge form. At last, as a whole, discharge on the surface of the dielectric layer shows a bright light-emitting state, the most region of the charge deposition luminous dark, only the streamers mainly shines brightly. In charge deposition region, the charge energy is not high, and most are the slow electrons. Their mobility in the surface of the dielectric layer is poor, this is more conducive to form local reverse electric field, and further affects the development of the streamer.

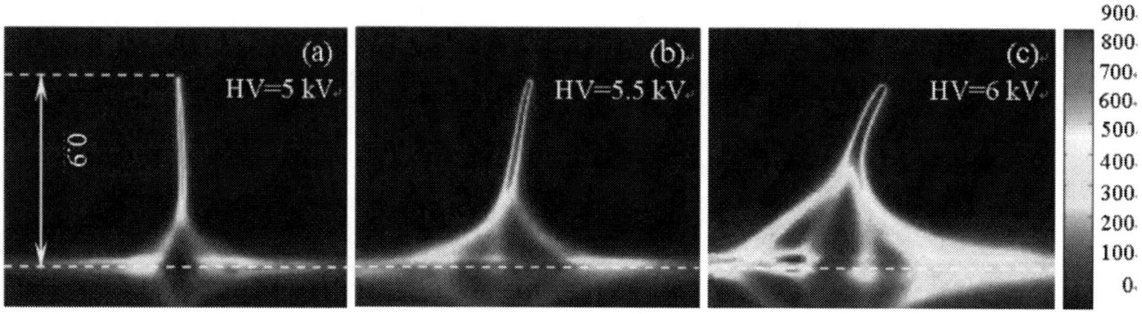

Figure 3. The images of streamers with different applied voltages.

In addition to the applied voltage, patterns of development of the streamer will be affected by the dielectric barrier discharge structure. Figure 4 gives the images of different discharge gap. The applied voltage is 6 kV, and the frequency is 10 kHz, the thickness of the dielectric layer is 0.64 mm, exposure time is 1 μs, the discharge gaps are 0.3, 0.6 and 1.2 mm. Under the different discharge gaps, the larger the discharge gap is, the longer straight part of the discharge is. From this, the discharge gap is relatively small impact on the form of the streamer.

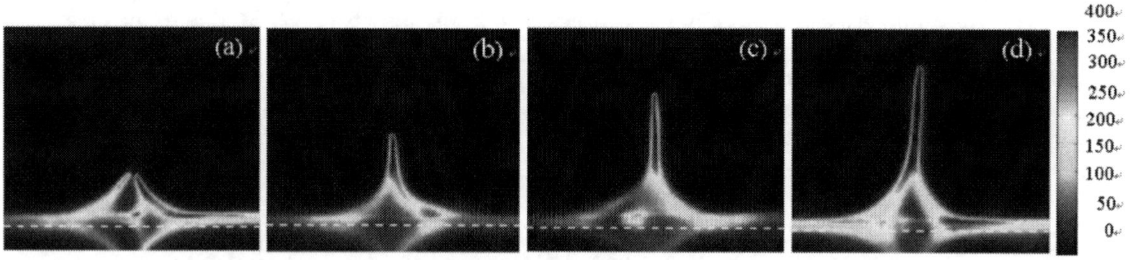

Figure 4. The images of streamers in different discharge gaps.

Figure 5 gives the images of different thicknesses of dielectric layer. From these images we can find the streamer's form isn't changed. When the dielectric layer is different, it only affects the optical radiation intensity does not affect its shape.

From this experiment, we find that the development of the streamer with the simulation is not the same. To be a theoretical analysis, three parameters are used as characteristic quantities. They are half-height diameter of the streamer $d_{1/2}$, the diameter of the deposited charge region D_{max}, and the height of the bending.

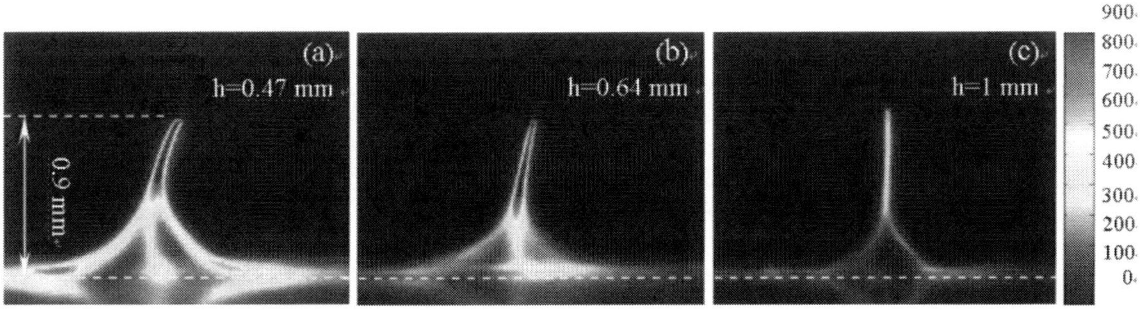

Figure 5. DBD images at ambient condition. (a) 0.47 mm, (b) 0.64 mm, and (c) 1 mm

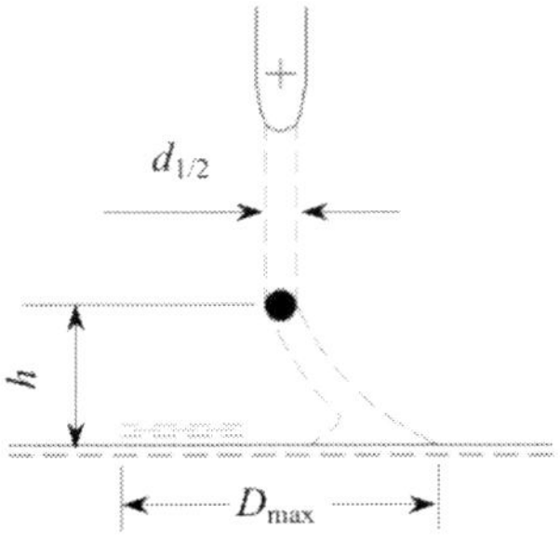

Figure 6. Schematic diagrams of the streamer.

Figure 6 gives the physical model of the streamer, and figure 7 gives the diagram of the applied voltage and the three parameters. With the increase of the applied voltage, the half-height diameter $d_{1/2}$ increase significantly. The applied voltage is increased from 4 kV to 6 kV, although the voltage only improve by 50%, $d_{1/2}$ has increase by over 70%, the diameter of the deposition zone increased more than twice, the height of the bending also increased.

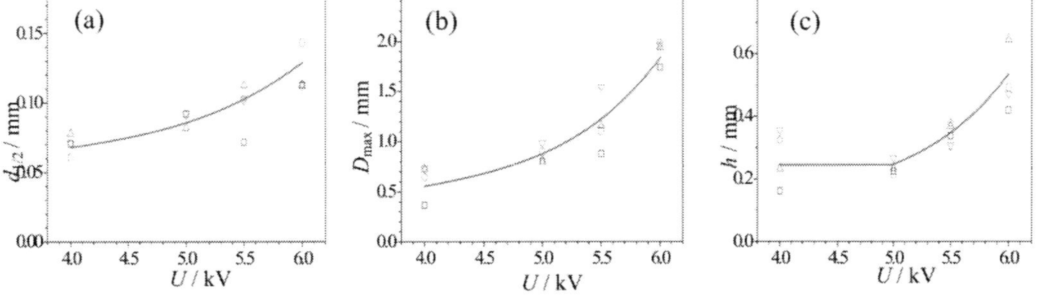

Figure 7. Plots of (a) the half-height diameter of the streamer $d_{1/2}$, (b) the diameter of the deposited charge region D_{max}, and (c) the height of the bending h vs the applied voltage

The increase of the half-height diameter of the streamer $d_{1/2}$ is increased, because the increase of the applied voltage to enhance the electric field strength of the needle electrode, meanwhile, electron temperature and density are all increased. Radial scale of the streamer increase with increasing applied voltage

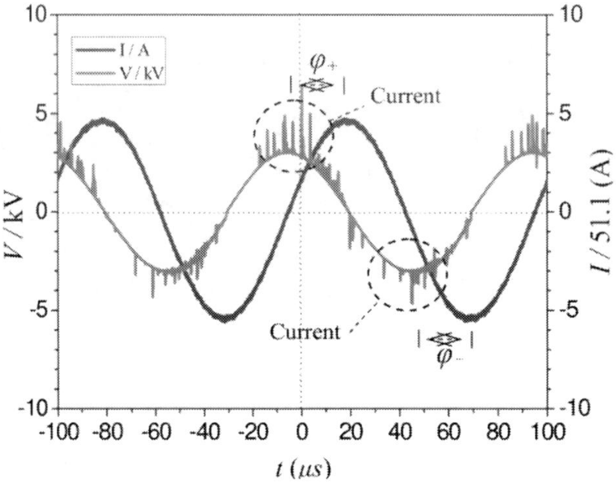

Figure 8. Plots of the applied voltage and current vs the time.

Figure 8 shows the voltage and current waveforms of the pin-to-plane DBD when the applied voltage is less than 4 kV. At lower voltage, the pin-to-plane DBD shows typical characteristics of the streamer, a large number of current pulses superimposed on the current curve. The amplitude and quantity of the current pulse in the positive half cycle are larger than in the negative half cycle mainly due to structural asymmetry. When the streamer complete, electron charge accumulate in the dielectric surface, it will produce the opposite charge field.

4. Conclusions

Under the streamer regime of pin-to-plane dielectric barrier discharge in air, space charge field, the charge field of electrolyte deposited on the surface, and the incentives electric field make the streamer asymmetry. With the rise of the applied voltage, the development of the streamer will deflect and bend. Charges on the surface of the dielectric layer are randomly formed along the surface of the dielectric layer by the streamer. The conical shape of the streamer is only the integral result of the above process.

Acknowledgments
This work is supported by The National High Technology Research and Development Program of China under Grant No 2012AA062609, The International Science and Technology Cooperation Project of China under Grant No 2010DFA61470, The National Science Foundation for Distinguished Young Scholars of China under Grant No 61025001, The Fundamental Research Funds for the Central Universities under Grant No 2012QN067 and 2012TD061.

References
[1] Wang X X 2009 *High Voltage Engineering* (in chinese) **35** 1
[2] Bai M D, Zhang Z T, Xue X H, Yang X L, Hua L S and Fan D 2010 *Plasma Chem. Plasma Process.* **30** 831
[3] Roth J R, Rahel J, Dai X and Sherman D M 2005 *J. Phys. D: Appl. Phys.* **38** 1

[4] Kogelschatz U, Eliasson B and Egli W 1999 *Pure Appl. Chem.* **71** 1819
[5] Hu J H, Fang Z and Zhang C 2008 *High Voltage Engineering* **34** 883
[6] Ma H, and Qiu Y 2003 *Ozone Science and Engineering* **25** 127
[7] Fan W L, Dong L F and Li X CH 2007 *Acta Phys. Sin.***56** 1467
[8] Zhang H Y, Wang D Z and Wang X G 2007 *Chin.Phys.***16** 1089
[9] Ma Y, Zhang G J and Shi X M 2008 *High Voltage Engineering* **34** 363
[10] Huang J Q, Xue L and Lv T 2010 *High Voltage Engineering* **2** 444
[11] Tang J, Duan Y X and Zhao W 2010 *High Voltage Engineering* **3** 733
[12] Kogelschatz U 2002 *IEEE Trans. Plasma Sci.* **30** 1400
[13] Kogelschatz U August 2002 *IEEE Trans. Plasma Sci.***30** 4
[14] Raether H 1939 *Z Phys.***112** 464
[15] Raether H 1940 *Arch. Elektrotechno.* **34** 49
[16] Loeb L B and Meek J M 1940 *J. Phys. D: Appl. Phys.* **11** 438
[17] Morrow R and Lowke J J 1997 *J. Phys. D: Appl. Phys.* **30** 614
[18] Kulikovsky A A 2000 *J. Phys. D: Appl. Phys.* **33** 1514
[19] Naidis G V 2006 *Plasma Sources Science and Technology* **15** 253
[20] XU X J and Zhu D CH 1996 *Shanghai: Fudan University Press*
[21] Eliasson B and Kogelschatz U 1991 *IEEE Trans. Plasma Sci.* **19** 309
[22] Georghiou G E and Papadakis A P 2005 *J. Phys. D: Appl. Phys.* **38** 303
[23] Briels T.M.P. van Veldhuizen E.M. and Ebert U 2008 *IEEE Trans. Plasma Sci.***36** 908

Characterization of millimeter magnitude atmospheric pressure glow discharge in pin-to-plane dielectric barrier discharge

Z Yu[1], Z T Zhang[1,2*], S J Xu[2], J Yao[2], Q X Yu[2], Y J Li[2], J S Zhao[1]

1 Institute of Marine Engineering, Dalian Maritime University, Dalian 116026, China
2 Department of Physics, Dalian Maritime University, Dalian 116026, China

E-mail: newzhangzhitao@163.com

Abstract. Discharge mode in pin-to-plane dielectric barrier discharge at atmospheric pressure was investigated by means of the electrical measurement, the photo-electricity and the fast macro photography. The discharge was operated in atmospheric air. Streamer and glow discharge were generated in the positive half-period and negative half-period of discharge respectively with a 10 kHz AC power supply when the applied voltage reached a high value. The subject of this paper is the study of the characterization of the glow discharge. It was found that the glow discharge have the hierarchical structure, although the discharge gap is small (only 0.9 mm). The positive column is shortened as the distance of discharge gap decreases, but the dimensions of other discharge area do not change. The discharge current waveform illustrates that the positive half-period discharge current is very short, which is streamer current waveform, and the negative half-period one is longer, which is glow discharge current waveform. The photo-electricity signal waveform corresponds with discharge current waveform. Two main glow discharge generating reasons are proposed in this paper. One is a sufficient number of space charge produced in the very uneven electric field around needle electrode; the other is the effective secondary electron emission process on the exposed needle electrode.

1. Introduction

Dielectric barrier discharge (DBD) is widely used in ozone generation, material surface treatment, pollution abatement, UV light source, plasma display technology, etc [1, 2]. Generally, there are two main discharge modes in the DBD system, streamer and homogeneous mode. In atmospheric air, streamer mode is common mode, due to the higher density of gas and electronegative gas such as oxygen, water vapour, etc. Since the 1980s, homogeneous mode became a new research hotspot because of the industrial application needs. In 1988 [3], the homogeneous mode in atmospheric pressure helium was first found by Okazaki in Japan. The homogeneous mode includes Townsend discharge and glow discharge. In the Townsend mode, the strongest luminous layer is observed close to the anode, because the electric field is almost uniformly distributed along the gap. In glow discharge,

[*] To whom any correspondence should be addressed.

on the contrary, the strongest luminous layer is closed to the cathode, because the space charge around the cathode is distorted the electric field [4]. In 2003 [5-7], Radu in Canada realized the glow discharge with the pin-to-plane and cylinder-to-plane DBD system, but the discharge gas only focused on noble gas. And then, some researchers also realized the homogeneous discharge with some special methods and electrode form, such as pre-ionization of spark, Mesh electrode, etc [8-11]. But the discharge in the most of the research results does not have the hierarchical structure that like typical low pressure glow discharge, therefore some researchers only regard it as Townsend discharge [12]. Streamer is often adopted in the industrial application. Research shows that the productivity of active particles in homogeneous discharge is higher than that in streamer [13]. So, the efficiency of chemical reaction will be improved if the homogeneous mode introduces into discharge system.

In 1938 [14], G W Trichel reported a negative corona discharge (named Trichel pulses) with only few micro amps of current, occurred in the extremely uneven electric field around the cathode. When the electric field rises, Trichel pulses discharge transforms to glow discharge. This glow discharge has hierarchical structure like glow discharge in low pressure, albeit on a very small scale. In the pin-to-plane electrode without dielectric layer, because the discharge current increases rapidly, the glow discharge is not stable and transforms to arc easily. But if there is a dielectric layer between pin and plane electrode (i.e. pin-to-plane DBD), on the one hand, extremely uneven electric field forms around the cathode, on the other hand, discharge current does not increase too fast, due to the existence of the dielectric layers. However, needle electrode won't always keep negative potential because of the AC power source, the streamer will happen when the needle electrode is positive potential. So, two kinds of modes (streamer and glow) happen alternately in pin-to-plane DBD system.

In this paper, the glow discharge in the two modes is the main research object. The detail of glow discharge morphology is acquired by the CCD camera with macro lens. The mechanism is expounded qualitatively at last.

2. Experiment setup and method

The experiment setup is schematically shown in figure 1. The discharges were generated between exposed needle electrode and plane electrode covered by rectangular Al_2O_3 dielectric layer with thickness 0.47 mm. An adjustable high voltage power source with amplitude from 0 to 10 kV and frequency 10 kHz is connected to the needle electrode, the plane electrode is grounded. The pin-to-plane DBD reactor was placed in a chamber which is connected to a rotary pump. In our experiments, only air without additional gas is used.

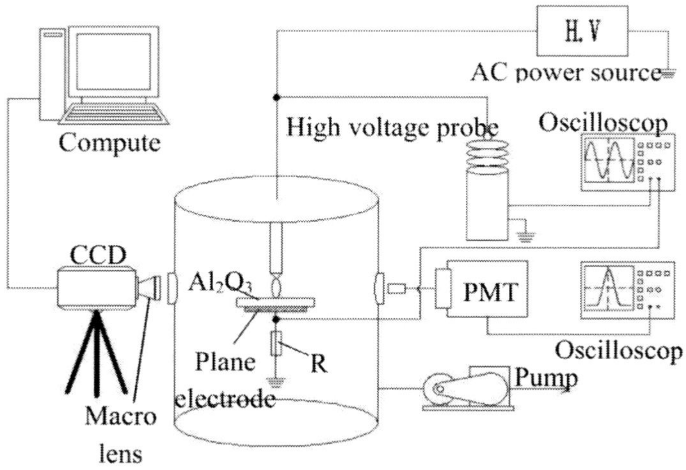

Figure 1. Experiment setup.

The light emission images from the discharge are recorded by a CCD camera (CoolSNAP EZ) which is mounted macro lens (2×, SE-16SM), with the exposure time 1 μs and spectrum response from 300 to 1000 nm. The voltage applied to the needle electrode is measured by a high voltage probe (HV-P60, Japan). The discharge current is measured by a 51.1 Ω non-inductance resistor (R) placed between plane electrode and ground. Photoelectric signal acquired via a PMT (SENS-TECH DM0047C, English). The voltage, current waveform and photoelectric signal are recorded by a digital oscilloscope (Tektronix TDS3032B, America).

3. Experiment results and discussions

Figure 2 shows the discharge character of pin-to-plane DBD system in which the applied voltage is 6 kV, the discharge gap is 0.9 mm, and the thickness of dielectric layer is 0.47 mm. The applied voltage waveform and the discharge current waveform without displacement current are shown in figure 2(a) and figure 2(b) respectively. It can be seen that there are numerous short current pulses in the positive half-period of the discharge current waveform and only one wide current pulse in the negative half-period. Figure 2(c) shows the photoelectric signal waveform which has the characteristics like the discharge current waveform. So, it can be sure that there are two kinds of discharge modes generated alternately. Figure 2(e) and figure 2(f) are the discharge morphology acquired by CCD camera in the moment of figure 2(b)I and figure 2(b)II respectively. Two kinds of discharge mode is clearly distinguished in this two pictures. Figure 2(e) shows a typical streamer mode, but in the figure 2(f) there is a distinct different discharge mode which can be characterized by hierarchical structure orderly consisted of negative glow, Faraday dark space, positive column, Anode glow, and cathode dark space. We call it micro-glow discharge momentarily, because it is only less than a millimeter.

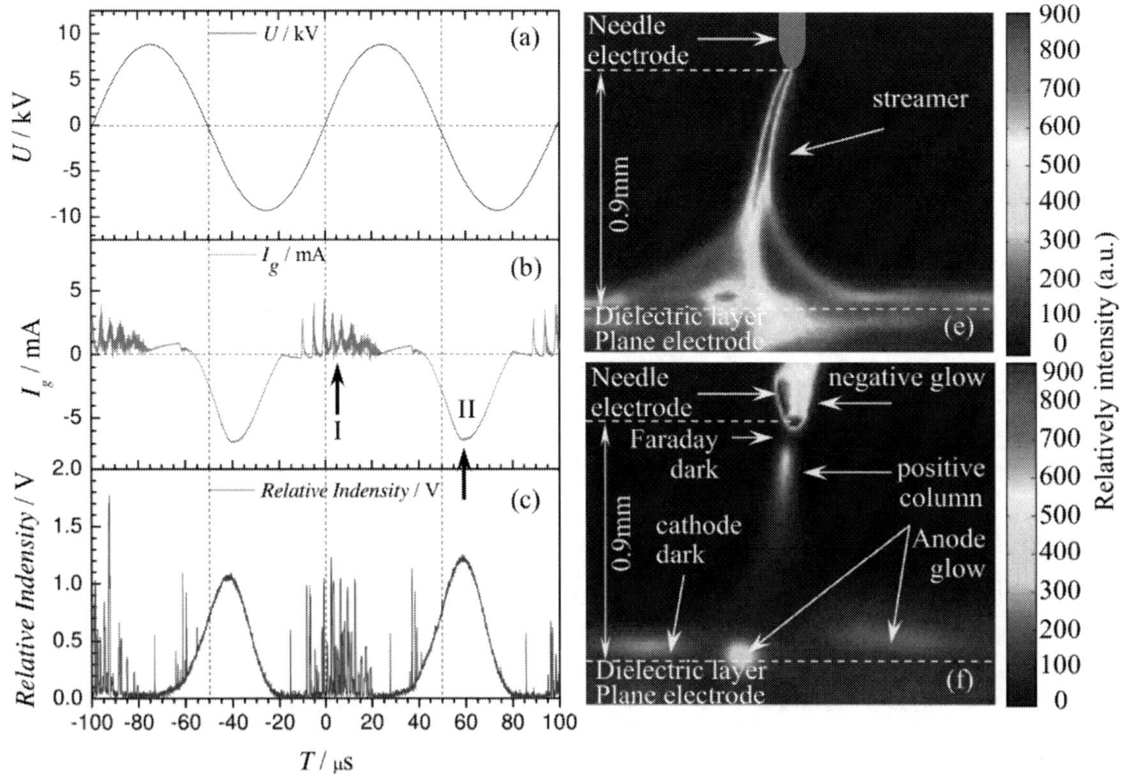

Figure 2. The waveform of applied voltage (a), discharge current (b) and photoelectric signal (c), the images of streamer (e) and micro-glow discharge (f).

In the experiment, only when the applied voltage reach to a certain value U_c, the micro-glow discharge is generated in the negative half-period of discharge current waveform. If the applied voltage is less than U_c, the discharge is only corona or Trichel pulse in the negative half-period. But in this paper, we focus on micro-glow discharge, rather than discuss what discharge mode it is, corona or Trichel pulse. The waveform of photoelectric signal with increase of applied voltage is shown in figure 3. The photoelectric signal is all short pulses in positive and negative periods, as shown in figure 3(a). When the applied voltage reach to 4 kV (figure 3(b)), a wide and week photoelectric signal appears in the negative half-period, that is the photoelectric signal of micro-glow discharge. And then, the photoelectric signal strengthens gradually with increase of applied voltage. In figure 3(b), (c), (d), moreover, there are some short pulses generated before the wide pulse of micro-glow discharge. These short pulses are corona or Trichel pulse. Owing to the AC power source, corona or Trichel pulse are generated before the applied voltage reach to U_c. Then, the transition from corona or Trichel pulse to micro-glow discharge is happened when the applied voltage is higher than U_c.

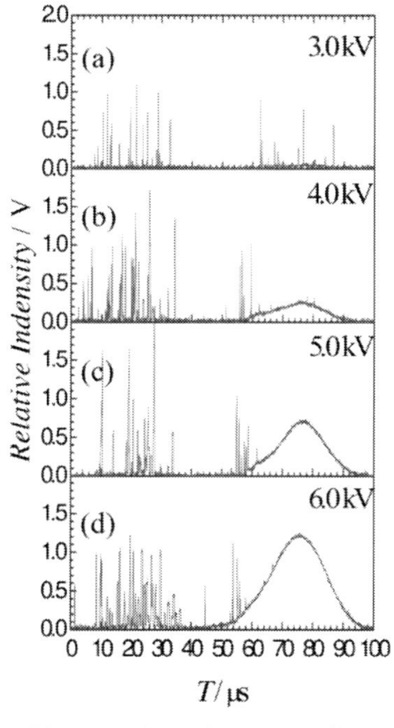

Figure 3. The waveform of photoelectric signal when the applied voltage is 3kV, 4kV, 5kV and 6kV.

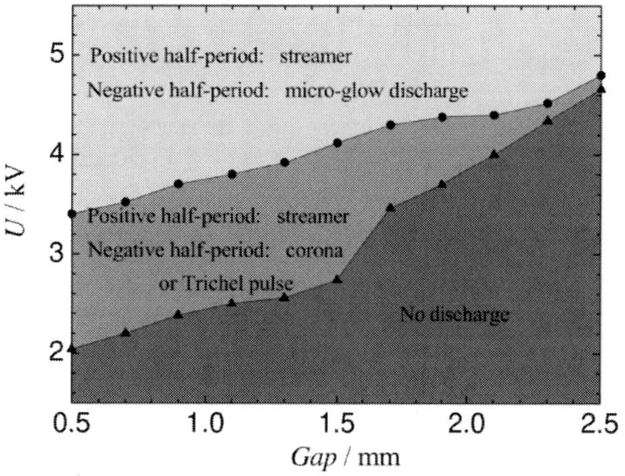

Figure 4. The transition applied voltage of different discharge modes in different discharge gap.

So, the discharge modes change is observed clearly through the waveform of photoelectric signal. With the increase of the applied voltage, there are only streamers with different discharge strength in the positive half-period. But in negative half-period, the discharge mode is more complex. First, when the applied voltage is less than discharge ignition voltage U_d, there is no discharge in the discharge gap. Secondly, when the applied voltage is between U_d and U_c, corona or Trichel pulse is generated. And then, the applied voltage is more than U_c, the transition from corona or Trichel pulse to micro-glow discharge is observed in the negative half-period. U_d and U_c are higher with the increase of discharge gap, as shown in the figure 4.

Figure 5 shows the images of micro-glow discharge with different gap of 1.2 mm, 0.9 mm, 0.6 mm and 0.3mm, applied voltage 6 kV. It is noted that the size and strength of micro-glow discharge features is unchanged except positive column. Like the glow discharge in the low pressure, the positive column of the micro-glow discharge is shortening with the decrease of discharge gap, and disappearing when the discharge gap is reduced to 0.3 mm.

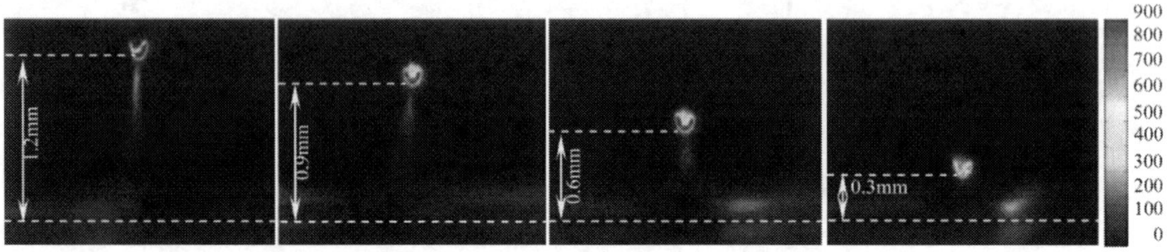

Figure 5. Images of micro-glow discharge with the gap of 1.2 mm, 0.9 mm, 0.6 mm and 0.3 mm.

The mechanism leading to micro-glow discharge is speculative at two points. First, there are enough space charges in the discharge gap. A lot of space charge, generated in the positive half-period, is not disappeared in the negative half-period. So, the cathode potential drop is formed between the space charge and needle cathode. Bright negative glow appear following the area of cathode potential drop. Secondly, secondary electron emission is efficient on the needle cathode. Strong electric field strength, formed around the needle cathode, is conducive to secondary electron emission.

4. Conclusions

In this paper, we have presented the micro-glow discharge in the pin-to-plane DBD device. The hierarchical structure of micro-glow discharge is clearly revealed by the CCD camera with the macro lens. Enough space charge and efficient secondary electron emission is two critical factors for generating of micro-glow discharge.

Acknowledgments

This work is supported by The National High Technology Research and Development Program of China under Grant No 2012AA062609, The International Science and Technology Cooperation Project of China under Grant No 2010DFA61470, The National Science Foundation for Distinguished Young Scholars of China under Grant No 61025001, The Fundamental Research Funds for the Central Universities under Grant No 2012QN067 and 2012TD061.

References

[1] Kogelschatz U 2003 *Plasma Chem. and Plasma P.* **32** 1
[2] Wagner H E, Brandenburg R, Kozlov K V, Sonnenfeld A, Michel P and Behnke J F 2003 *Vacuum* **71** 417
[3] Kanazawa S, Kogoma M, Moriwaki T and Okazaki S 1988 *J Phys. D: Appl. Phys.* **21** 838
[4] Luo H Y, Liang Z, Lv B, Wang X X, Guan Z C, and Wang L M 2007 *Appl. Phys. Lett.* **91** 231504
[5] Radu I, Bartnikas R and Wertheimer M R 2003 *J. Phys. D: Appl. Phys.* **36** 1284
[6] Radu I, Bartnikas R and Wertheimer M R 2004 *J. Phys. D: Appl. Phys.* **37** 449
[7] Radu I, Bartnikas R, Czeremuszkin G and Michael R *IEEE Trans. Plasma Sci.* **31** 411
[8] Golubovskii Y B, Maiorov V A, Behnke J F, Tepper J and Lindmayer M 2004 *J. Phys. D: Appl. Phys.* **37** 1346
[9] Luo H Y, Liang Z, Wang X X, Guan Z C and Wang L M 2010 *J. Phys. D: Appl. Phys.* **43** 155201
[10] Qi B, Ren C S, Wang D Z, Li S Z, Wang K and Zhang Y T 2006 *Applied Physics Letters* **89**

131503
[11] Wang X X, Luo H Y, Liang Z, Mao T and Ma R 2006 *Plasma Sources Sci. Technol.* **15** 845
[12] Luo H Y, Liang Z, Wang X X, Guan Z C and Wang L M 2010 *J. Phys. D: Appl. Phys.* **43** 155201
[13] Kogoma M and Okazaki S1985 *J. Phys. D :Appl. Phys.* **27** 1985
[14] Trichel G W 1938 *Physical Review* **54** 1078

Characteristics of gas phase discharge above water surface

Xiaoping Wang, Zhongjian Li, Lecheng Lei*
Key Laboratory of Biomass Chemical Engineering of Ministry of Education, Department of Chemical and Biological Engineering, Zhejiang University, Hangzhou 310027, China

E-mail: lablei@zju.edu.cn

Abstract. Characteristics of gas phase discharge with both electrodes suspended above the water surface were investigated. Discharge channels propagated from electrode to electrode or from electrode to water surface were determined by the distance between electrodes, the distance from the high voltage electrode to water surface and the water conductivity. The voltage-current relation was almost not affected by water conductivity, the distance between electrodes and the distance from electrode to water surface unless spark occurred. However, the voltage-pulsed energy relation was affected by these factors. The larger water conductivity and smaller distance between electrodes and distance from electrode to water surface, the larger pulsed energy.

1. Introduction
A typical water surface discharge reactor suspended the high voltage (HV) electrode over the water surface and put the ground electrode in water [1, 2]. The discharge occurred in the air and on the air-water interface. Because of no directive contact between the HV electrode and the water, the leakage current was cut off by the air gap. The problem caused by large leakage current and joule heating could be alleviated in water surface discharge. So for high conductivity water treatment, water surface discharge was much more suitable. Numerous articles reported the discharge characteristics and applications of water surface discharge [3-8] .However, this arrangement of electrodes could not avoid the corrosion of the ground electrode by high conductivity water. A solution for this problem was suspending both the HV electrode and the ground electrode above the water surface. Depending on the distance between electrodes, the distance between the high voltage electrode and water surface, and the water conductivity, two kinds of discharge paths existed. Discharge channels from electrode to electrode or from electrode to water surface would afford different water treatment potentials. Therefore, transition criterions and electric characteristics of these two kinds of discharge channels must be considered before the utilization.

2. Experimental

2.1. Power supply
Schematic diagram of the main circuit of the power supply was shown in figure 1. The main components including a low-voltage rectifier (D_1), a filter capacitor (C_1), a DC /AC inverter ($IGBT_{1\text{-}4}$, L_1 and C_2), a primary transformer (T_1), a high-voltage rectifier (D_2), energy storage capacitors (C) , a pulse transformer (T_2) and a hydrogen thyratron switch (Th). This system could output electrical pulses with peak voltage from 0 to 70 kV, repetition frequency from 0 to 200 Hz.

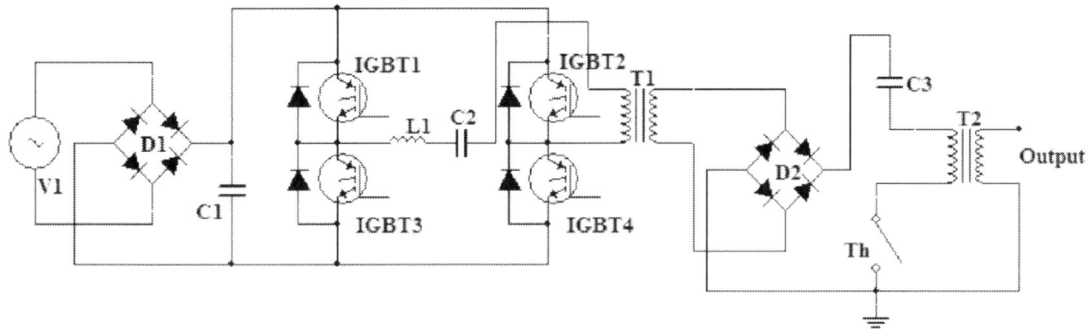

Figure 1. Main circuit of the power supply.

2.2. Reactor

The discharge reactor composed of a pair of graphite needle electrodes with tips of about 0.2 mm and the water surface. The distance between electrodes (L), the distance between the high voltage electrode and water surface (d_1) and the distance between the ground electrode and water surface (d_2) were labelled in figure 2 and could be changed. The area of the water surface was a constant (with a length of 20 cm and a width of 10 cm).

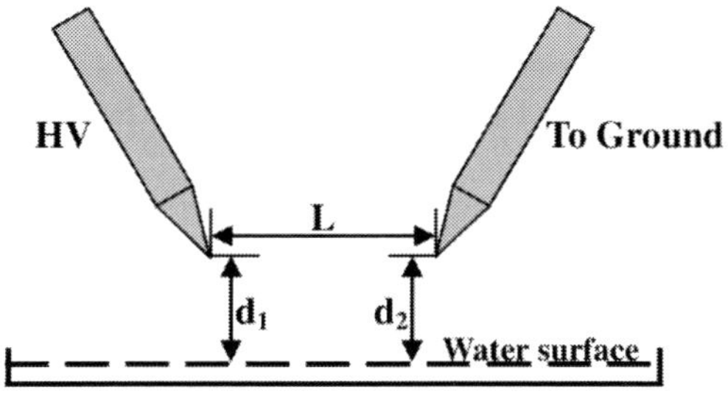

Figure 2. Arrangement of the discharge system.

2.3. Analytical procedures

The waveforms of pulsed voltage and current applied to reactors were monitored with a digital oscilloscope (Tektronix DPO 3000, USA), a 1:1000 high voltage probe (Tektronix P6015A, USA) and a current probe (Tektronix TCP 0150, USA).

3. Results and discussion

3.1. Discharge channel transition criterion

For spark discharge, along with the increase of electrode distance, discharge channels existed along L transitioned to along d_1 and d_2. Transition criterions were shown in figure 3 (a) as curves, above which spark channels propagated along d_1 and d_2, but below which spark channels propagated along L. The transition criterion curve moved downward when increasing the water conductivity.

For streamer discharge, discharge channels propagated along d_1 and d_2 existed invariably when decreasing the electrode distance. However, if the electrode distance was large enough, streamer channels along L disappeared. Curves in figure 3 (b) indicated the streamer channel transition criterions above which streamer channels propagated only along d_1 and d_2, but below which streamer channels propagated along d_1 and d_2 and along L coexisted.

The electrode distance, the distance from the electrode to water surface and the water conductivity influenced the direction of discharge channel. Increase of water conductivity, increase of electrode distance, and decrease of the distance from the electrode to water surface would be beneficial to the propagation of discharge channels from the electrode to the water surface. Contrarily, propagation of discharge channels from the HV electrode to the ground electrode was more likely. That is to say, influence of the water surface could be enhanced by increasing the electrode distance and the water conductivity, and decreasing the distances from the electrode to the water surface.

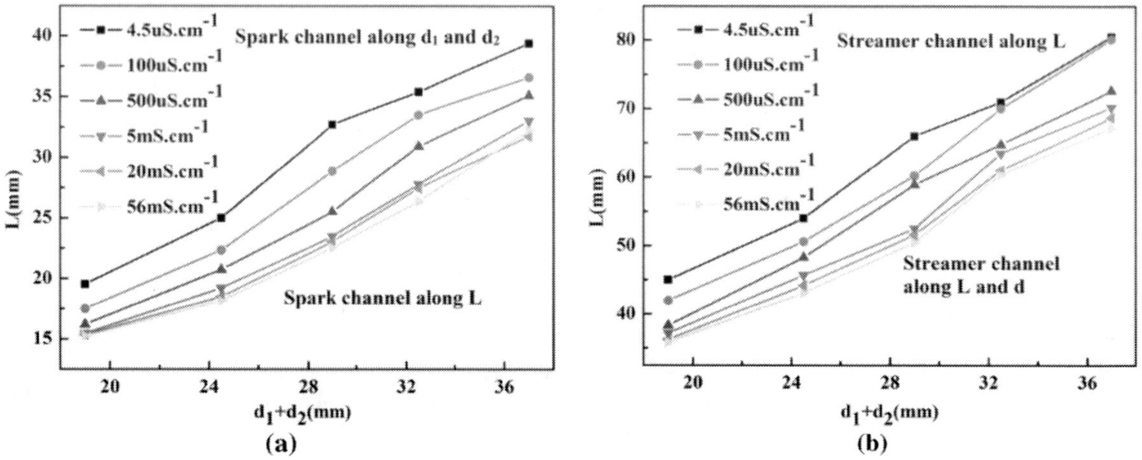

Figure 3. Discharge channel transition criterion (a) spark channel, (b) streamer channel.

3.2. Effect of electrode distance on electrical characteristics

Experiments were conducted at three different electrode distances (15 mm, 22.5 mm and 50 mm), other conditions were kept identical. These different electrode distances would result in different directions of discharge channel which could be seen from the transition criterions indicated in figure 3. For streamer discharge, channels along L and d were coexist for electrode distances of 15 mm and 22.5 mm but propagated only along d_1 and d_2 for the electrode distance of 50 mm. For spark discharge, channels propagated along L for 15 mm but along d_1 and d_2 for 22.5 mm and 50 mm.

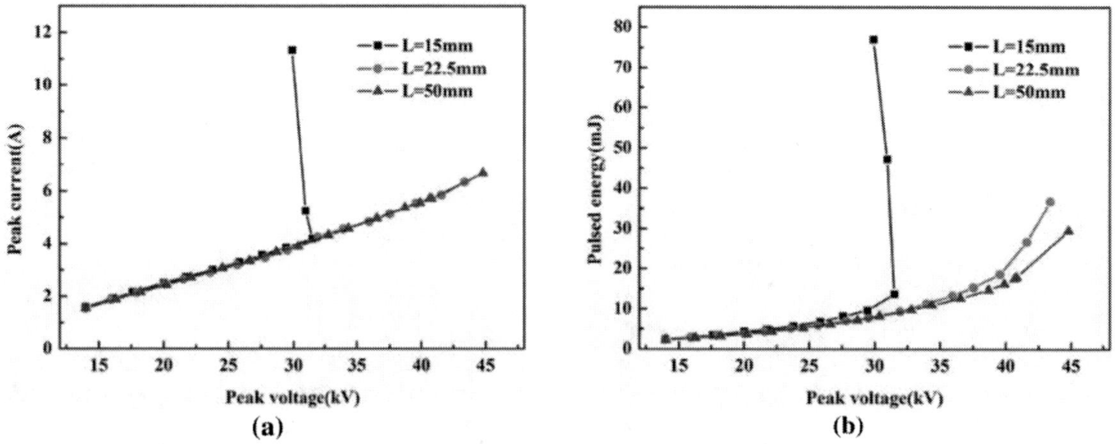

Figure 4. Effect of electrode distance on electrical characteristics, (a) voltage-current, (b) voltage-pulsed energy. (d_1+d_2=24.5mm, water conductivity=20mS·cm^{-1})

Peak current was not influenced by the electrode distance unless spark discharge occurred which was shown in figure 4 (a). The peak current of spark channels propagated along L was much larger than that along d_1 and d_2. If the spark channels propagated only along d_1 and d_2, the peak current was not affect by electrode distance. Effect of electrode distance on pulsed energy was similar to that on peak current. But the pulsed energy decreased with electrode distance even though the spark channels propagated only along d_1 and d_2.

3.3. Effect of the electrode-water surface distance and water conductivity on electrical characteristics

Effect of the distance from electrode to water surface was shown in figure 5. Experiments were conducted at five different electrode-water surface distances as shown in figure 5 (a), other conditions were kept identical. Spark channels propagated along L and streamer channels coexisted along L and d. Effect of electrode-water surface distance on peak current could be ignored as shown in figure 5 (a). But pulsed energy decreased with the increase of the distance between the electrode and the water surface. For spark channels propagated along d_1 and d_2, the result was the same as figure 5 indicated.

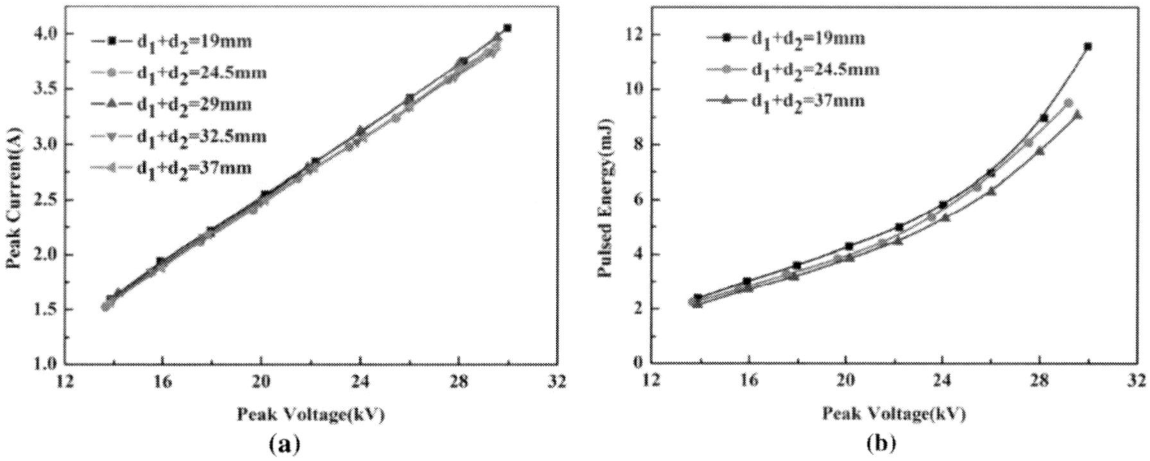

Figure 5. Effect of electrode-water surface distance on electrical characteristics, (a) voltage-current, (b) voltage-pulsed energy. (L=15 mm, water conductivity=20 $mS \cdot cm^{-1}$)

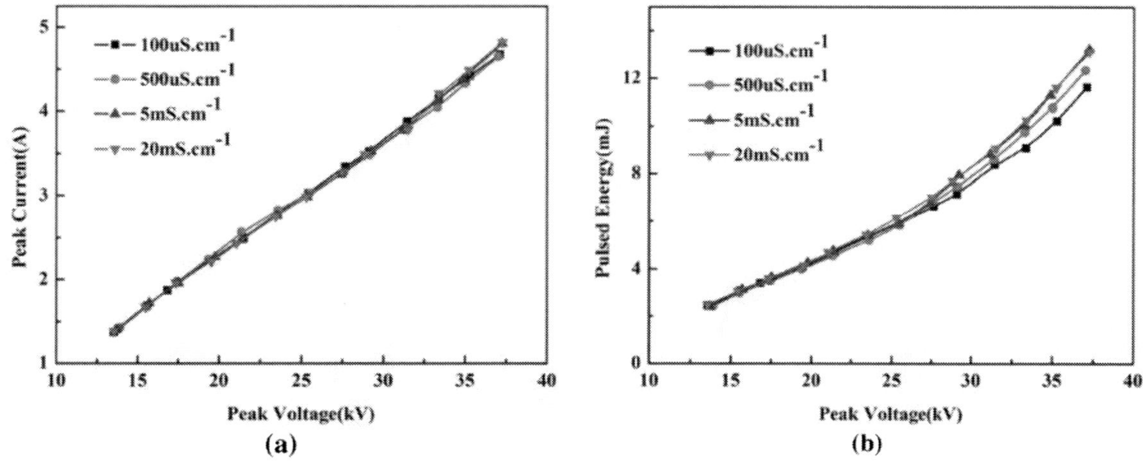

Figure 6. Effect of water conductivity on electrical characteristics, (a) voltage-current, (b) voltage-pulsed energy. (d_1+d_2=24.5 mm, L=55 mm)

The influences of water conductivity on electrical characteristics were shown in figure 6. Spark channels propagated along d_1 and d_2 but streamer channels coexisted along L and along d1 and d2. Influence of water conductivity on peak current was little. But the pulsed energy increased with water conductivity for high voltages. When discharge channels only existed between electrodes, the effect of water conductivity was much smaller.

4. Conclusions

The existence of the water surface influenced the gas phase discharge. For gas phase discharge without water surface, discharge channels propagated only between the electrodes. However, the existence of water surface would attract the discharge channels to propagate from electrode to the water surface. Increase of water conductivity, increase of electrode distance, and decrease of the distance from the electrode to water surface were beneficial to the propagation of discharge channels from the electrode to the water surface. Influence of these factors on streamer current and pulsed energy was little. However, occurrence of spark discharge would amplify the influence of these factors on electrical characteristics.

Acknowledgments

We appreciated the financial support to this work by NSFC of China (No. 20836008 and 21076189).

References

[1] Grymonpre D R, Finney W C, Clark R J and Locke B R 2004 *J. Ind. Eng. Chem. Res.* **43** 1975
[2] Bruggeman P and Leys C. 2009 *J. Phys. D: Appl. Phys.* **42** 053001
[3] Verreycken T, Bruggeman P and Leys C. 2009 *J. Appl. Phys.* **105** 083312
[4] Sato M, Tokutake T, Ohshima T and Sugiarto A T 2008 *IEEE Trans. Ind. Appli.* **44** 1397
[5] Bruggeman P, Ribezl E, Maslani A, Degroote J, Malesevic A, Rego R, Vierendeels J, and Leys C 2008 *Plasma Sources Sci. Technol.* **17** 025012
[6] Bruggeman P, Liu J J, Degroote J, Kong M G, Vierendeels J and Leys C 2008 *J. Phys. D: Appl. Phys.* **41** 215201
[7] Bruggeman P, Guns P, Degroote J, Vierendeels J and Leys C 2008 *Plasma Sources Sci. Technol.* **17** 045014
[8] Titov V A, Rybkin V V, Smirnov S A, Kulentsan A L, Choi H S 2006 *Plasma Chem. Plasma Process.* **26** 543

Experimental research on V-I characteristic of a novel cavity atomizing corona discharge device

Qing Li*, Jiangshan Meng, Qiaoyan Wang, Jiaojiao Li, Zhiqiang Liu, Shoujie He
Electrostatic Research Institute, Hebei University, No.180 WuSi East Road, Baoding, China

E-mail: liqing@hbu.cn

Abstract. In order to seek a new control method which can control the dust, SO_2, NO_x and other pollutants comprehensively, a novel atomizing corona discharge device without external dynamic force was invented. The voltage-current (V-I) characteristics in different electrode structures were studied. In the cavity-wire type discharge structure, the voltage-current characteristics of de-ionized water, tap water and saturated solution of $Ca(OH)_2$ were compared and analyzed. The results showed that: the atomizing effect of cavity-wire type discharge structure was better than that of cavity-plate under the same electrode diameter and voltage. A better atomizing effect could be obtained when the curvature radius of the cavity atomizing electrode was reduced. Active ions in solution had damaging effect on hydrogen bond among water molecules, which is in favour of solution atomization. This work will provide theoretical basis and experimental supports for a new field that the electrostatic and the wet flue gas desulphurization (WFGD) are combined together at the same time.

1. Introduction

Atmosphere is our essential survival elements. In recent years, with the development of the living and producing standard of the people, the carrying capacity of environment is increasing. China is a coal production and coal-fired country. Coal accounts for 70% of the total energy consumption [1]. This situation will not improve in a short time. Therefore, a series of environmental problems caused by coal-fired can not be ignored. In the process of coal combustion, not only is the fly ash produced, but the harmful gases polluting the atmosphere environment are produced such as SO_x and H_2S. SO_2 is the main material for the formation of acid rain. Acid rain affects human life and cause serious damage on the Earth's environment. Therefore, coal-fired desulphurization is essential.

At present, as one of the desulphurization areas, atomized discharge technology has already made some progress. Binlin Dou and Y M Kang [2-3] studied SO_2 from the perspective of mass transfer and catalytic, respectively. J L Wen [4] used pressure nozzles to compare the desulphurization efficiency between water and lime slurry. It showed that the static gain is obvious. Z T Wang [5] analyzed the influence that the atomizing charged characteristic of two-fluid and various factors had on the charge - mass ratio of droplet. X L Zou [6] studied the velocity and particle size distribution of lime slurry droplet using pressure nozzles. However, there is rarely research on the atomized nozzles without external dynamic force at home and abroad.

* To whom any correspondence should be addressed.

In the present, a new atomized discharge device without external dynamic force was used. The influence that the electrode matching and solution property have on the corona discharge was studied experimentally. This method will provide theoretical and experimental basis for the new area that the electrostatic and wet flue gas desulphurization (WFGD) are combined together at the same time. After technological design is mature, this technology can join to the ESP to achieve dust removal and desulphurization process simultaneously. This also makes the flue gas purification process simplified, practical and effective.

2. Experimental devices, materials and methods

The experimental device was based on the atomized discharge device [7-8]. After being improved, the device is shown in figure 1. The grounding electrodes are plate structure and circular wire, respectively. They are shown in figure 3.

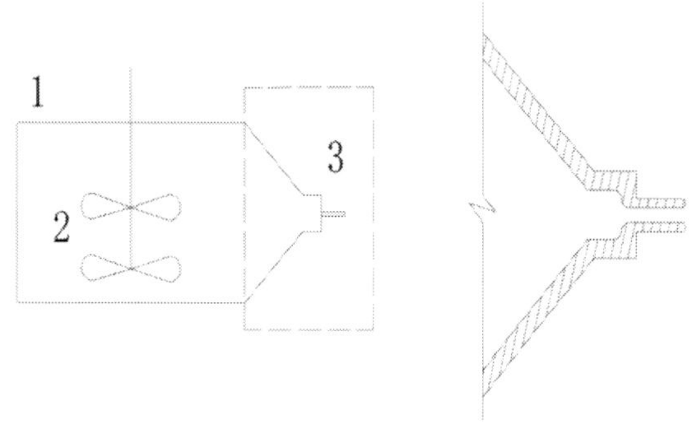

1-liquid storage pot, 2- blender, 3-atomizing nozzle

Figure 1. Cavity atomizing device.

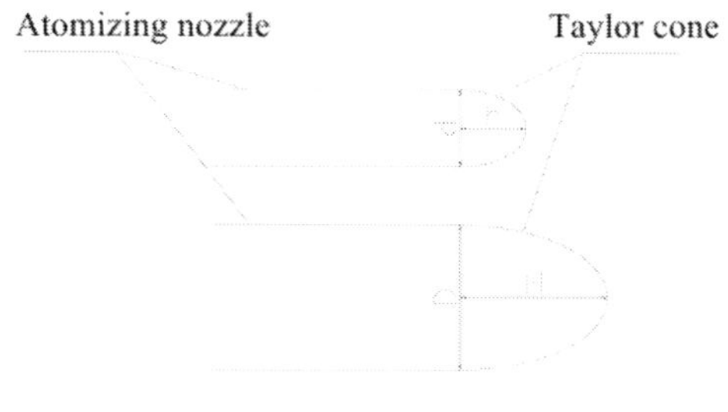

Figure 2. Atomizing nozzle and Taylor cone.

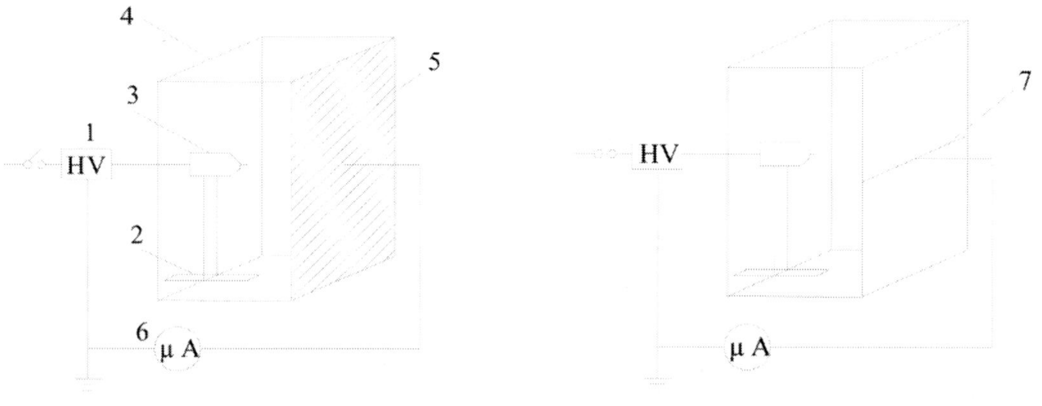

Cavity-plate discharge system. Cavity-line discharge system.

1- High-voltage supply device, 2- Adjustable rail, 3-Cavity atomizing device, 4-Insulator bracket, 5-Grounding plate, 6- Microampere ammeter, 7- Grounding wire

Figure 3. The experimental device of atomizing corona discharge system.

The grounding wire with 1.9 mm diameter was set in horizontal. The size of grounding plate electrode placed perpendicular to the ground was 1.0×0.6 m^2. The space between cavity atomizing electrode and ground electrode was set to 10 cm. 1# and 2# micro-cavity atomizing electrode of different diameters were used. The diameter of 1# is 1.26 mm and 2# is 0.5 mm. The length of cavity is 4.5 mm. Experiments were performed at room temperature. Atomized liquid are de-ionized water, tap water and saturated solution of $Ca(OH)_2$. The solution of $Ca(OH)_2$ consisted of ionized water and analytically pure $Ca(OH)_2$ powder. The devices above were used to measure the V-I characteristics under different voltages. The influences that electrode structures and liquid property had on the atomizing discharge were studied.

3. Experimental results and analysis

3.1. The influences that the cavity diameter and electrode structure had on the atomizing discharge

In this experiment, the systems of cavity-line and cavity-plate and the cavity electrode of 1# and 2# were used. The atomizing liquid was de-ionized water. The discharge characteristics under different discharge conditions were investigated. The result is shown in figure 4.

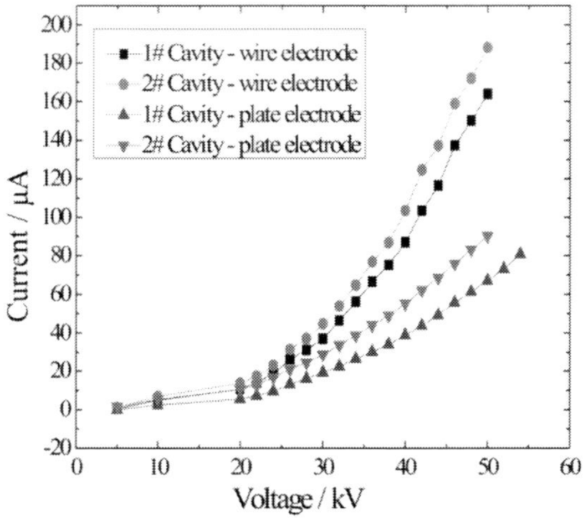

Figure 4. The V-I characteristics of different electrode matching and cavity diameter.

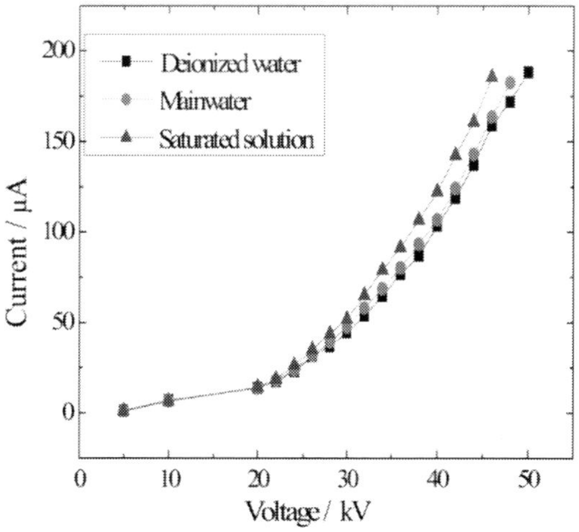

Figure 5. The V-I characteristics of solutions with different ion concentration.

For two different discharge structures, if the wire electrode diameter and voltage were all the same, the cavity-line structure could obtain higher corona current. With the voltage increasing, the current difference between two types of discharge structure rise. The relation between the electric field strength and the electric fluxline can be written as follows:

$$E = \frac{dN}{dS} \tag{1}$$

Where dS is cell area, dN is the number that the electric fluxline through the cell area.

In the same voltage, the electric field strength between the electrodes of cavity-line structure is stronger. Correspondingly, the number of space charges increases. The space charges have higher velocity. Then the rate of droplet charged increases. The formula of critical droplet breakage charge is as follows:

$$q_0 = 8\pi\left(\varepsilon\sigma r^{1.5}\right)^{1/2} \tag{2}$$

Where q_0 is the droplet charge, ε the permittivity of liquid, σ the surface tension of droplet, r droplet diameter.

When the droplet charge is greater than q_0, the binding effect of surface tension will be less than the expansion of the electrostatic force [9], which makes the droplet break. Therefore, in the strong discharge conditions, the droplet can quickly reach the Rayleigh limit [10]. In the cavity-line structure, the spraying droplets can form an unstable jet even under a lower voltage. Besides, the raise of space charges not only increases the possibility of the droplet charge, but increases the chance that the space charges strike the ground electrode due to air entrainment. These factors make the discharge current of cavity-line structure is much larger than that of cavity-plate structure in the same voltage.

For the same electrode structure, the cavity having small curvature radius (2#) could obtain larger current. There are two reasons for this phenomenon. The first reason is according to the electric field strength formula of the needle electrode [11]:

$$E = 2U\left[(r+2x)\ln(\frac{r+2d}{r})\right]^{-1} \tag{3}$$

Where E is the electric field strength in the vicinity of the corona electrode, U the applied voltage, r the curvature radius of needle, d the space between the electrodes, x the distance of a point to the needle tip in the electric field.

With the curvature radius of the tip decreasing, the electric field strength in the vicinity of the corona electrode increase under the same voltage. Then the cavity electrode having smaller diameter could generate stronger electric field strength. The number of space charges in this structure was larger. These increase the probability of droplets charged.

The other reason is when the Taylor cone forms, the ration of long axis and short axis of Taylor cone belonging the same liquid is constant, namely, D/H=d/h (it is shown in figure 2). When the long axis of the Taylor cone is 1.9 times the short axis [12], the shape of droplet tends to unstable and breaks. Being restricted by the cavity electrode wall, different cavity electrodes form different short axis diameters of Taylor cone. This make the droplet volume formed in the tip of the cavity difference. After the Taylor cone breaks, the liquid forms initial spherical droplet due to the effect of the surface tension. Because of the electric field, droplets have collision charged with the space charges and the charges attach to the outer surface of the droplets. With the electric field strength increasing, the charges accumulating in the surface of the droplet are more and more and gradually reach saturation. Due to the electrostatic repulsion, the surface tension of the droplet can't maintain its inherent form and the droplet breaks. When a single large droplet is broken into several small droplets, the specific surface of the droplet increases resulting in a single droplet charged unsaturated. Broken droplets will continue to be charged to the saturation and then broken again under the electric field. This process will be repeated until the electrostatic repulsion and surface tension reach to balance. When the droplets are no longer broken, the entire process of breaking charge is completed. If it is the same liquid to reach to the balance of breaking charge, the droplets size and charge basically the same. In the same voltage, the droplet having small initial diameter can quickly reach to the balance of breaking charge. The number of droplet and the total surface area increase, so that the amount of charge reaching to the ground plate increase rapidly.

These two aspects above (the electric field strength and the rate of droplets breaking) make the smaller diameter cavity generate larger current in the discharge process.

Basing on the analysis above, we can imagine that different properties of discharge liquid and different ion concentrations make the time reaching to the balance of breaking charge different. This will have influence on the atomizing corona discharge.

3.2. The influences that the ion concentration had on the atomizing discharge

Calcium-based desulphurization technology is one of the world's most widely used and most mature desulphurization processes [13-14]. Then $Ca(OH)_2$ was selected as the experimental material.

According to the experiments above, the cavity-line atomizing system was selected and the 2# atomizing electrode was used. The solutions were de-ionized water, tap water and saturated solution of $Ca(OH)_2$. The V-I characteristics of these solutions are shown in figure 5.

The figure shows that in the same voltage, the order of the three solutions discharge current was saturated solution of $Ca(OH)_2$> tap water> de-ionized water. With the voltage increasing, the intensity of corona discharge increased. Therefore, the discharge current rose. Increasing the ion concentration in liquid (ie, the ionic number in the per unit volume of aqueous solution) had a positive impact on the liquid charge. Then the discharge growth trends of these three solutions were different.

There are two reasons for the difference of discharge current. The first one is water molecules have strong polarity. There is a strong hydrogen bond among molecules [15]. The ions added to the de-ionized water will damage the original hydrogen bond of water molecules. Different ion concentration has different damaging effect for hydrogen bond. This causes the reduced degree of the liquid viscosity and the surface tension different. Increasing the ion concentration will enhance the damaging rate of the hydrogen bond. Thus increases the possibility of droplets breaking.

The second reason is from the formula which is as follows:

$$E^2 = 77 \times \frac{\sigma}{2\pi\varepsilon_0 d} \tag{4}$$

Where E is the electric field strength, σ surface tension of the droplet, d droplet diameter.

If the ion concentration increases, the droplets break into small droplets more easily. This will increase the total surface area of the droplets and the space electric field strength. Then the total droplets charges increase. Finally the current is larger in the same voltage. In the de-ionized water, tap water and saturated solution of $Ca(OH)_2$, the ion concentration is from small to big, so the discharge current will change with the change of ion concentration.

4. Conclusion

(a) In the same electrode diameter and same voltage, cavity-line electrode can get better atomizing effect than cavity-plate electrode. The cavity-line electrode can be selected as the electrode matching of flue gas desulphurization.

(b) The influence that solution properties and electrode shape of the cavity having on the particle size and surface tension of the atomizing droplets can make discharge current change. With the electrode radius of curvature decreasing and the number of ions in the solution increasing, the discharge current increases.

(c) The active ions in the solution can damage the hydrogen bonds among water molecules. Improving the concentration of active ions can enhance the ability of droplets breaking.

Acknowledgments

This work is supported by the National Science Foundation of China under Grant No.10875036 and No.51077035, Science and Technology Supporting Program of Hebei Provence No.09276712D, Doctor Foundation of Hebei University No.2009-178, the science and technology research projects of Colleges and universities in Hebei province No.2011166, and the Science Foundation of Hebei province No.A2010000182.

References

[1] Gao L F, Liu G J, Lin C C, Zheng L G and Wang A H 2005 *Bull. Mineral., Petrol. Geochem. (in Chinese)* **24** 79
[2] Binlin D, Young C B and Wang J H 2008 *Energy Fuels* **22** 1041
[3] Kang Y M and Li S L 2001 *Acta Scientiae Circumstantiae (in Chinese).* **21** 538
[4] Wen J L, Wang J, Wang Z Q and Sun X N 2008 *High Voltage Eng. (in Chinese)* **34** 2167
[5] Wang Z T, Wen J L, Song X N and Luo T Q 2008 *High Voltage Eng. (in Chinese)* **34** 2172
[6] Zou X L, Chen H L, Jin C H, Liu X A, Cen Q G 2007 *Drain. Irrig. Mach. (in Chinese)* **25** 57
[7] Li Q, Sun X R, Li H F, Zhang W Y and Wang J G 2010 *High Voltage Eng. (in Chinese)* **36** 1520
[8] Li Q, Liu Z Q and Zhang X J *ZL 200920350237.7*
[9] Xu Y 2004 *Electromagnetism* (Beijing, Science Press) p 34
[10] Wang Z T, Jia W D and Huang J W 2010 *High Voltage Eng. (in Chinese)*.**36** 1525
[11] Durham K G 1985 *Trans. the ASAE* **28** 658

[12] Xu X, Na R, Yang J and Liu J J 2009 *J. Inner Mongolia Univ. (in Chinese)* **40** 600
[13] Li X Y and Shen M L 2005 *Non-ferrous Mining and Metallurgy (in Chinese)* **21** 46
[14] Deng X Y, Xiao H Q and Yu L Y 2008 *Guangzhou Chem. Ind. (in Chinese)* **36** 24
[15] Zhou G D and Duan L Y 1995 *The basis of struct. Chem.* (Beijing Peking University press) p 478

Journal of Physics: Conference Series **418** (2013) 012097 doi:10.1088/1742-6596/418/1/012097

Voltage-current characteristics of a pin-plate system with different plate configurations

Zhuangbo Feng, Zhengwei Long*

School of Environmental Science and Technology, 92 Weijin Road, Tianjin University, Tianjin 300072, China

E-mail: longzw@tju.edu.cn

Abstract. In this paper, the voltage-current (V-I) characteristics of a pin-plate system with four types of collection plate configurations are studied experimentally. The collection plates consider a single metal plate, a metal plate with a fly ash cake layer, a metal plate with a clean filter media and a metal plate with a dirty filter media. The results show that the clean filter media has no obvious effect on the V-I characteristics. But the dirty filter media reduces the current density because of its high resistance. The thick fly ash cake layer increase current density because of the anti-corona effect but the increment is not very obvious.

1. Introduction

The electrostatic precipitator and filtration are widely used in industrial project and civil construction. However, both methods have their own defects. The efficiency of electrostatic precipitator for submicron particulate, especially ultra-fine particles sizes from 0.1 to 1 mm is below 90%. Although some types of filter media, such as HEPA, have high removal efficiency for submicron particulate, the pressure drop across media filter increase fastly with the cake formation. Clausen [1] found that the dust particulates in filter media and cake layer on the surface of filter were also a harmful pollution source because that the air went through the particulate layer. Therefore, the hybrid filtration system combining electrostatic precipitator and filtration become more promising and are used more widely. Huang [2] studied the removal efficiency of viable bioaerosol particles with a low-efficiency filter enhanced by continuous emission of unipolar air ions. Park [3-4] investigated the characteristics of carbon fiber ionizer-assisted medium air filter removing the submicron particles. Zuraimi [5] suggested that electrostatic precipitation filters enhanced with a media pre-filter has better effect on the particulate removal in tropical buildings.

The electrostatic precipitator and media filter in hybrid system have interaction with each other. Electrostatic field may change the flow field in and around filter media; the media filter and the cake on it affect the electric filed and voltage-current characteristics. Stanley [6] revealed that incorrect design of hybrid electrostatic and bag filter damaged the filter material electrically. Long [7-8] stated the advantages of industrial hybrid electrostatic and bag filter particulate collector and investigated the

* To whom any correspondence should be addressed.

parameters of perforated-plate and filter material on the voltage-current characteristics with experimental methods. Adamiaka [9-10] investigated the numerical methods on the corona discharge process and studied electrical field strength and electrical charge distribution in point–plane configurations.

In this study, we design a simple experimental system and investigate the effect of media filter material and the fly ash cake on the voltage-current characteristics of hybrid system. Because of various patterns of electrostatic and filter media zone, such as advanced hybrid electrostatic and bag filter particulate collector (AHPC) or electrostatic-HEPA (E-HEPA) system, this study just consider special and simple condition and put the filter or fly ash cake layer on the whole plate surface. The study investigated four dust collection plate conditions: iron plate with nothing on it; iron plate with clean filter media; ion plate with used dirty filter media where most of the dust particle is in the filter media rather than on the media surface; iron plate with 1 mm fly ash cake layer. This paper also studied effect of different thickness of cake layer on the system performance. The result can give basis for industrial engineering design and boundary conditions for computional fluid dynamics.

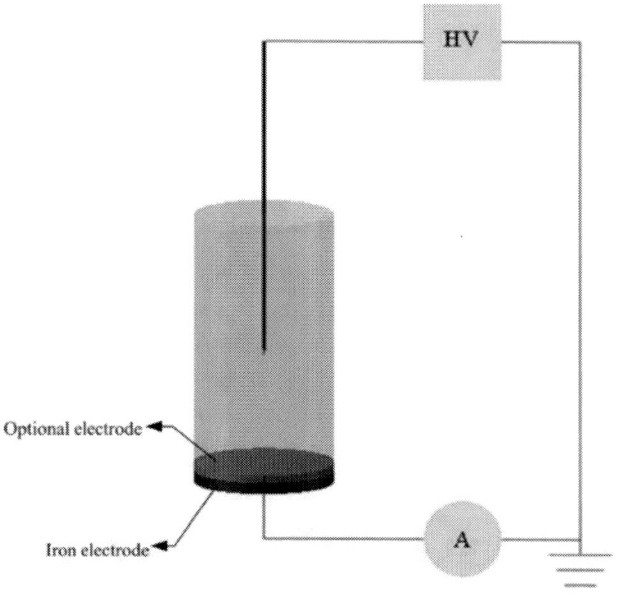

Figure 1. Setup of the experimental system.

2. Experimental system

The system consists of high voltage equipment, electrode, dust collection plate, insulation cylindrical tank and ammeter. Figure1 shows the structure of experimental system.

The high voltage equipment is made up of control box, high voltage generator and grounding system. The high voltage range is from 0 to 50 kV. Length of Electrode is 224 mm and it has a discharge peak. The insulation cylindrical tank providing discharge space has radius of 100 mm and height of 245 mm. The dust collection plate on the bottom of insulation tank is made of iron, its radius is 100 mm and thickness is 3 mm. Indoor environment of the laboratory is controlled by HVAC system, the air temperature is 22 ℃ and relative humidity is 45%.

There are four dust collection plate conditions: iron plate without anything on it; iron plate with plat clean filter media; iron plate with plat dirty filter media; iron plate with cake layer of 1 mm thickness. The thickness of plat filter media is 0.3 mm, permeability is $3.565\times10^{-9}(m^{-1})$ and porous ratio is 0.8-0.85. Weight of clean filter media with diameter of 200 mm is 1.8530 g. The dust cake is made up of fly ash, the grain diameter obey Rosin-Rammeler distribution and the dust density is 2460 kg m^{-3}. Weight of cake with 1mm thickness is 43.2177 g and the cake porosity is 0.64. Total fly ash weight in the used dirty filter media is 1.2252 g and most of it is in the filter media rather than on the filter surface.

This paper also investigate the effect of different thickness of cake on the system's V-I characteristics. The cake thickness includes 0.2 mm, 0.4 mm, 0.6 mm, 0.8 mm, 1 mm. Figure 2 describes how to change the thickness of cake layer on the dust collection plate. The accuracy of thickness is controlled by the fine finishing. The cake porosity of all the cases is 0.6. The cake zone is divided into six subzones equally and each subzone is filled up with 1/6 of total cake weight in order to ensure the relative uniform distribution of porosity.

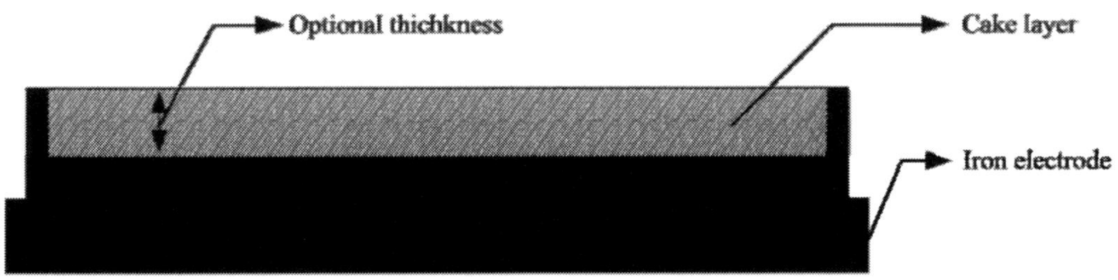

Figure 2. Different thickness of cake layer on iron plate.

3. Results and discussions

3.1 The effect of distance between electrode peak and plate

Figure 3-6 show the V-I characteristics of different distances between the electrode and the dust collection plate with or without material on it. The distance D varies from 20 mm to 60 mm and the material include clean filter media, dirty filter media, and dust cake. For the different material condition, V-I curve patterns are similar. In hybird electrostatic filtration system, the operating voltage should below 15 kV in order to be widely and safely in civil and industrial HVAC system; the breakdown voltage my bring great damage to the high voltage power supply whose accuracy is 0.00001, so the highest corona current in this experiment is about 5 μA.

For all the cases showed in figure 3-6, the current decreases when the electrode and plate distance increases at the same voltage, this phenomenon can be attributed to the high resistance caused by thicker air layer. Figure 7 reveal linear relationship between the current and voltage when the distance is 60 mm and the plate condition is iron with nothing on it. Linearity of V-I characteristics is well above the corona onset voltage and the V-I relationship can provide basis for industrial design.

The onset voltage increases while the D increases from 20 mm to 60 mm and value of 60 mm is 2 kV higher than 60 mm. For the system that the electrostatic part's main role is releasing ion or plasma, reducing electrode and plate distance may not improve the ion or plasma generation potential significantly because the breakdown voltage may be lower than bigger distances, the much more efficient method is to change the geometry and material characteristics of electrode and plate.

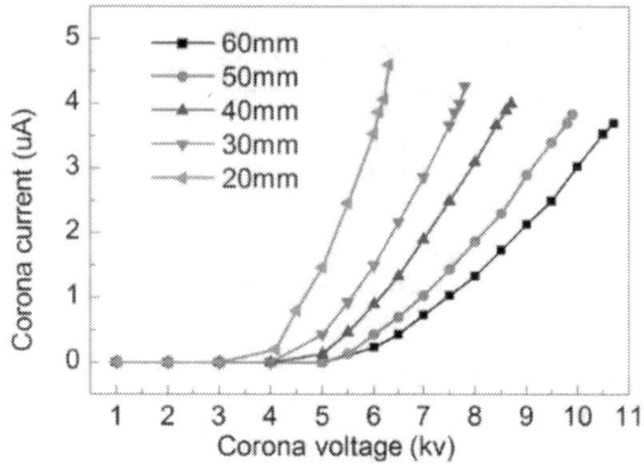

Figure 3. The V-I curves for iron plate condition.

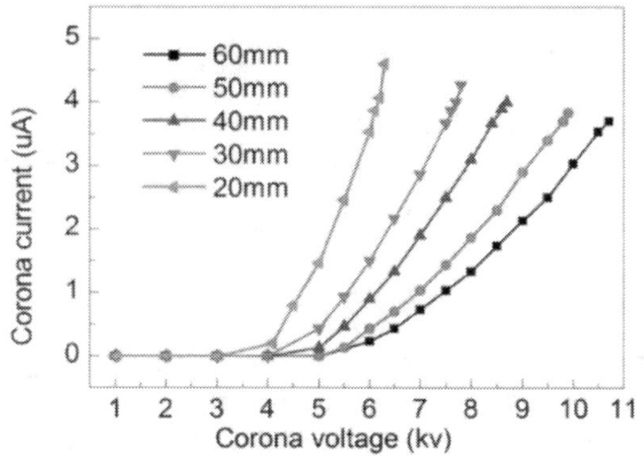

Figure 4. The V-I curves for clean filter condition.

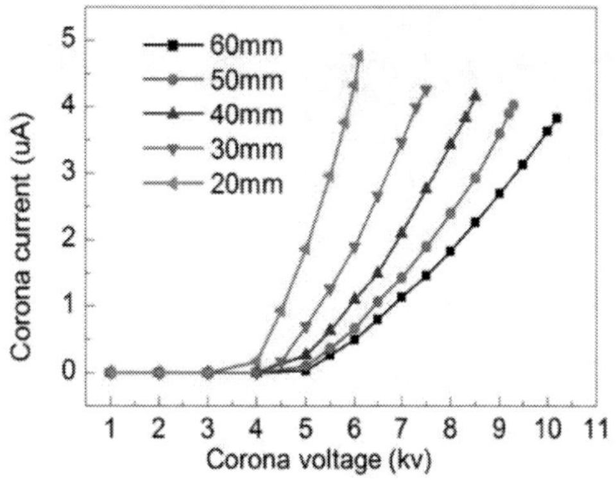

Figure 5. The V-I curves for cake condition.

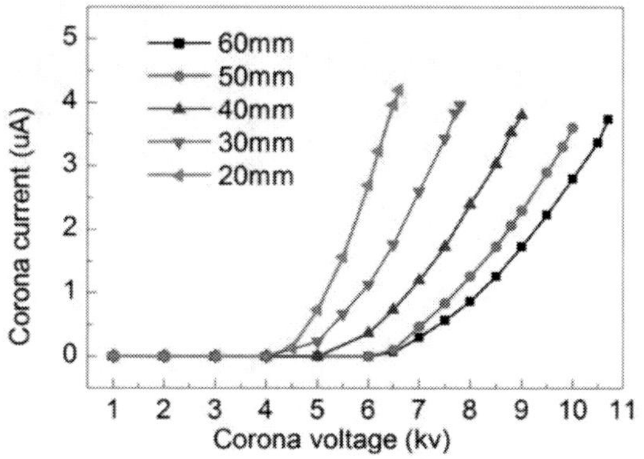

Figure 6. The V-I curves for dirty filter condition.

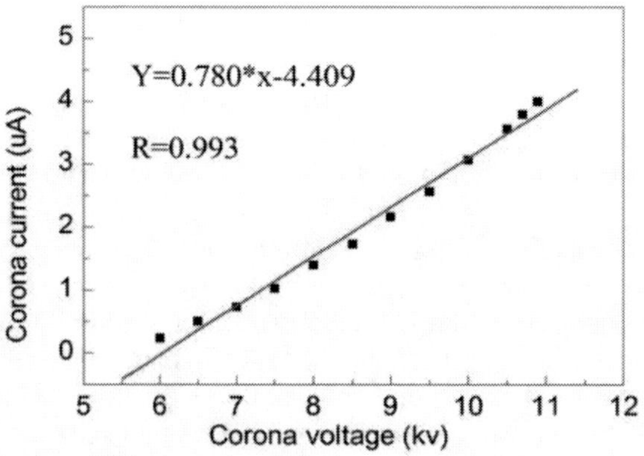

Figure 7. The V-I curve for iron condition and D=60 mm.

3.2 The effect of material on dust collection plate

Figure 8-12 show the V-I characteristics of different materials on the plate while distance of electrode peak and plate vary from 20 mm to 60 mm.

For the iron plate and clean filter condition, the V-I characteristics are same while distance D between electrode and plate are relative longer (D=40, 50, 60 mm). When D is 20 mm or 30 mm, the current of clean filter condition is less than iron plate under the same voltage but the difference is under 10%; The corona onset voltage are approximately same. The effect of filter on plate on V-I characteristics can be ignored when the distance between electrode and plate is longer and the resistant effect of air is strong. So the clean filter media on the dust collection plate has slight effect on performance of electrostatic zone in hybrid filtration system.

For the iron plate and cake condition, the current of cake condition is higher than the iron plate and the onset voltage is same. But the current difference is not very obvious and the current increment is about 0.3-1 μA. Although the dust cake add resistance, the current still increase because of the anti-corona effect. Process of anti-corona effect is described as follows: the high resistivity make the particles in dust cake layer remain charged and added electric field strength is created; the air in cake will break down while added electric field strength in cake layer is high enough; Positive ions transport from cake layer to electrode so the total current increase. Equation (1) describes the anti-corona occurance condition. Ep is electrical field strength in fly ash cake layer; Ecr is breakdown electrical field strength for cake; Ea is breakdown electrical field strength for atmospheric air. The effect of Anti-corona increasing current density and decreasing breakdown voltage should be avoided in design and operation of hyhird electrostatic filtration system.

$$E_p \geq E_c \approx E_a \quad (1)$$

For the dirty filter condition, the current is lower than iron plate and clean filter conditions. While D=40-60 mm, the difference of current is more obvious. The reason is that the dust in filter media increase resistance but the cake layer is not thick enough to generate anti-corona phenomenon. The corona onset voltage is approximately same between dirty filter and iron plate condition. Contrary to the clean filter media, used dirty filter has negative effect on the performance of electrostatic zone.

For the different cake thickness, the V-I characteristics remain constant. When the dust collection plate condition is cake with different thickness, the current value under the same voltage is little bigger than the condition with nothing on the plate. But the current increment is about 0.3-1 μA and the change is not very obvious, so the current difference can be ignored when operation voltage is big. Figure 13 describes the V-I characteristics of iron plate, 0.2 mm and 1 mm cake thickness. Two curves of iron plate and 0.2 mm cake coincide approximately, the V-I characteristics of 0.2 mm and 1 mm cake thickness look same.

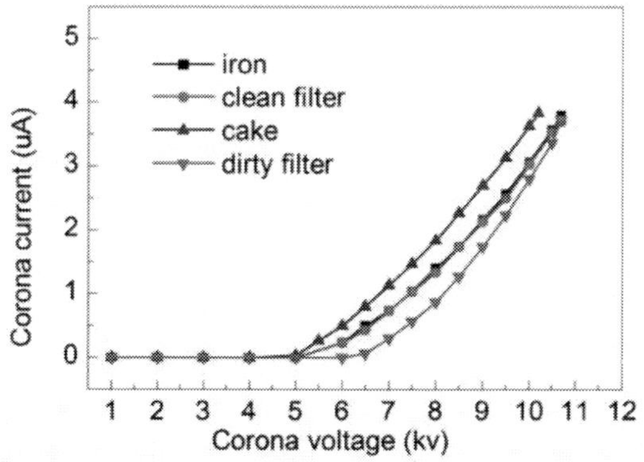

Figure 9. The V-I curve for D=60 mm.

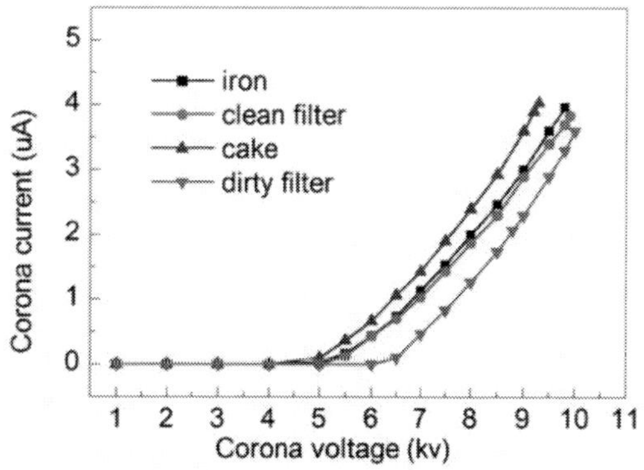

Figure 10. The V-I curve for D=50 mm.

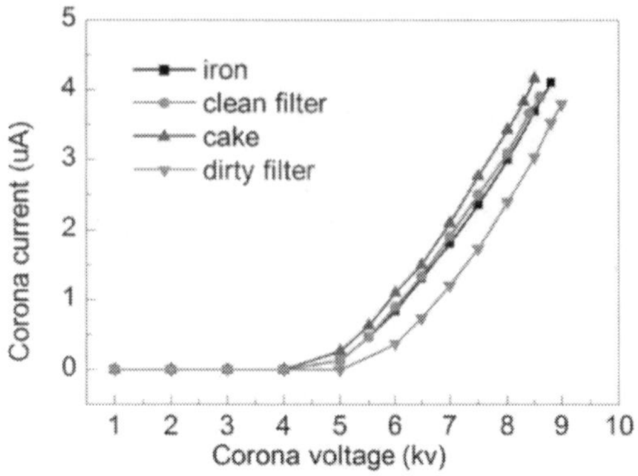

Figure 11. The V-I curve for D=40 mm.

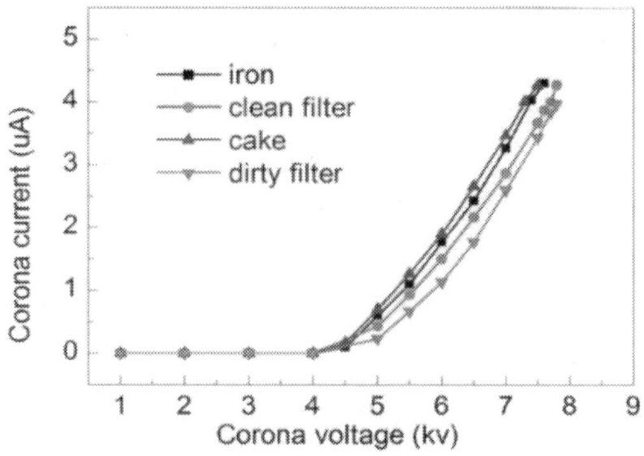

Figure 12. The V-I curve for D=30 mm.

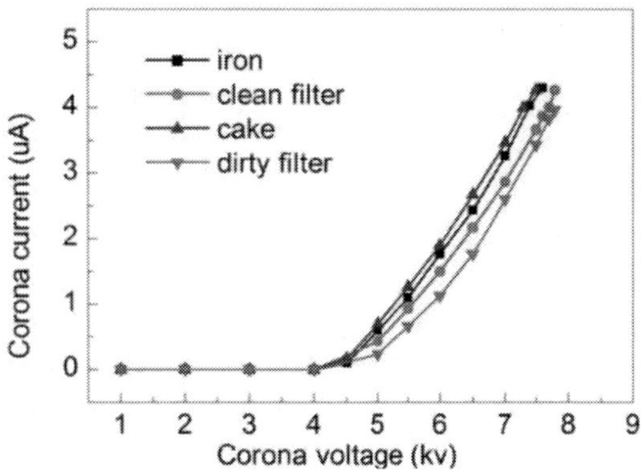

Figure 13. The V-I curve for D=20 mm.

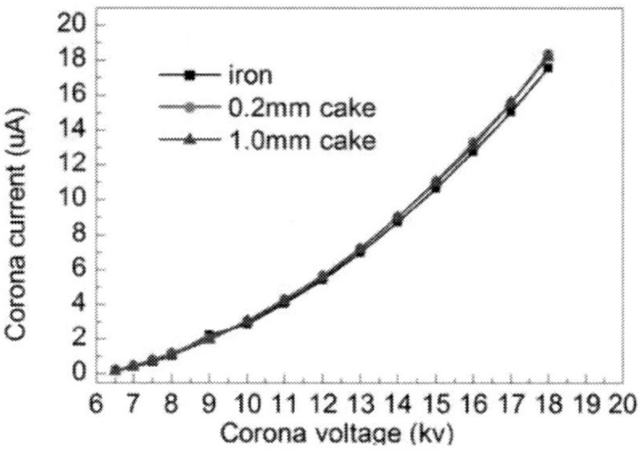

Figure 14. The V-I curves for iron plate, 0.2 mm and 1 mm cake condition.

4. Conclusions

The experimental results show that clean filter media on dust collection plate does not have obvious effect on the V-I characteristics of electrostatic zone. However, the dirty filter media reduce current density because of the high resistance. Although resistance of cake layer should not be ignored, the thick fly ash cake layer increase current density because of the anti-corona effect but the increment of current is so small. For different cake thickness, the V-I characteristics of system vary little from 0.2 mm to 1 mm. All the results can support engineering design and CFD simulation.

However, it is necessary to investigate the effect of different thickness, permeability and porosity of cake layer and filter media on the performance of electrostatic filtration zone, due to the anti-corona phenomenon is determined by the material and thickness of the media or cake.

Acknowledgement

The research presented in this paper was financially supported by the National Key Basic Research and Development Program of China (the 973 Program) through grant No. 2012CB720100.

Reference

[1] Clausen G 2004 *Indoor Air* **14** 202

[2] Huang R, Agranovski I, Pyankov O and Grinshpun S 2008 *Indoor Air* **18** 106

[3] Park J, Yoon K, and Hwang J 2011 *Build. Environ.* **46** 1699

[4] Park J, Yoon K, Kim Y, Byeon J and Hwang.J 2009 *Mech. Sci. and Technol.* **23** 184

[5] Zuraimi M and Tham K 2009 *Build. Environ.* **44** 2475

[6] Miller J, Grant L, and Grant E 2000 *J.Electrostat.* **58** 197

[7] Long Z W and Yao Q 2012 *Powder Technol.* **26** 215

[8] Long Z W, Song Q, Li S Q, and Yao Q 2010 *Proc. of the CSEE* **30** 13

[9] Adamiaka K and Atten P 2004 *J. Electrostat.* **61** 85

[10] Adamiaka K 1994 *IEEE Trans Ind Appl* **30** 381

Optimization of discharge types and electrode structure in a cylinder discharge reactor with saw-wheel array electrodes

Chunyang Zhang, Kefeng Shang, Na Lu, Jie Li*, Yan Wu
Institute of Electrostatics and Special Power, Dalian University of Technology, Dalian, China

E-mail: lijie@dlut.edu.cn

Abstract. The application of corona discharge technology in gas purification and wastewater treatment has been received great attention in recent years. The configuration of discharge electrode and the discharge types directly affect the discharge power and the power density, and then affect the generation of active species as well as the removal efficiency of pollutants. A novel cylinder-type discharge reactor with saw-wheel-array electrodes was developed for removal of SO_2/NO_x from flue gas, and influence factors such as electrode structure (ratio of spacing of saw-wheel slices and discharge distance, herein defined as R) and power supply types (positive DC, negative DC, and pulse power) on discharge characteristics and the output power was discussed. The experimental results show that the current and output power of three types of discharges firstly increased with R increasing from 0.3-1.7, and then tended to a stability from 1.7-2.5; while the power density reached a maximum at the ratio of 1.7.

1. Introduction

Non-thermal plasma technologies (NTP) including corona discharge, dielectric barrier discharge etc were widely used in electrostatic precipitators, electrostatic separation, modification of materials and other fields. In recent years, extensive attention has been paid to NTP for gas purification, soil purify and wastewater treatment [1-3].

The corona discharge types and electrode structure directly affect the generation of free radicals and active species and then affect the treatment efficiency of pollutants. Therefore, it is very important to study and optimise the discharge types and electrode structure. Many scholars [4-7] have investigated the optimal configuration of wire-plate and needle-plate reactor, but little has been reported to investigate optimal configuration of the wire-cylinder reactor about corona discharge especially the cylinder-type discharge reactor with saw-wheel-array electrodes.

In the present study, the effect of three kinds of discharge types such as positive DC, negative DC and positive pulse on discharge current, discharge power and power density are investigated using the cylinder-type discharge reactor with saw-wheel-array electrodes, and optimization of discharge types, and electrode configuration in a cylinder reactor with saw-wheel type array electrodes was determined by the discharge characteristics.

* To whom any correspondence should be addressed.

2. Experimental

The corona discharge system was illustrated in figure 1. A stainless steel cylinder with the inner diameter of 56 mm and height of 800 mm was used as ground electrode, and a stainless steel rod, equidistantly linked with several saw-wheel slices, was used as the discharge electrode. The outer diameter of the saw-wheel slices was 14 mm and the thickness of each saw-wheel slice was 1 mm. The schematic of saw-wheel slice was illustrated in figure 2, the tooth number of every saw-wheel slice was 20. Discharge distance between tooth point of saw-wheel slice to cylinder inside was 14 mm.

The pulsed high-voltage supplied by spark gap type pulse power and the circuit principle was shown in figure 3. The voltage of positive and negative DC ranged from 0 kV- 60 kV. The voltage were measured with oscilloscope (Tektronix TDS2014) equipped with high voltage probe (Tektronix P6015A), the positive and negative DC current measured with DC microammeter, pulse discharge current measured with the current probe (Tektronix P6021). The pulse frequency and pulse-forming capacitance Cp were 50 Hz and 500 pF, respectively. The air flow rate was 0.5 $m^3 h^{-1}$, relative humidity (RH) in air was 50%-60%. The power density was defined as:

$$Power\ density = \frac{P}{Nl} \tag{1}$$

Where P is discharge power (W) , N is saw-wheel slice number, l is spacing of adjacent saw-wheel slices (mm).

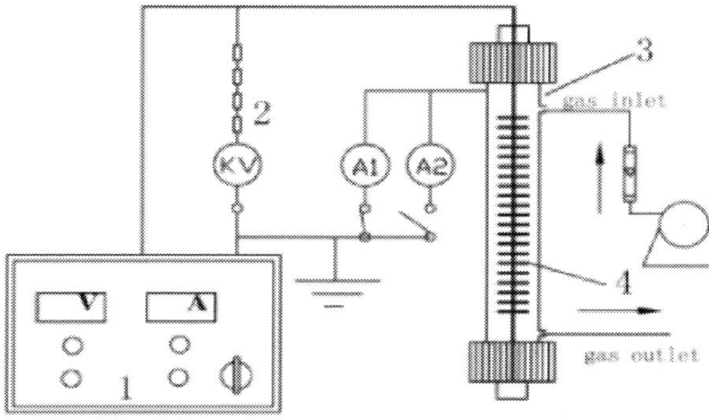

1. Power supply; 2. High voltage probe; 3. The reactor; 4. High voltage electrode; 5. Gas pump.

Figure 1. Schematic diagram of the experimental set up.

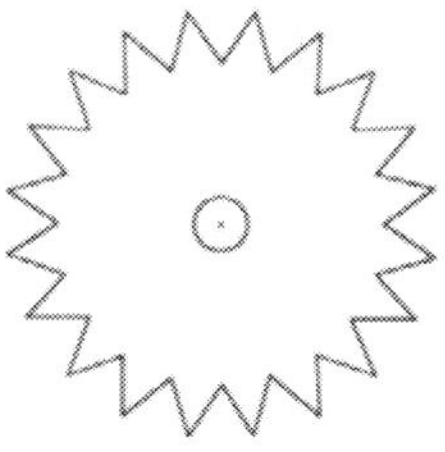

Figure 2. Tooth stainless steel pieces.

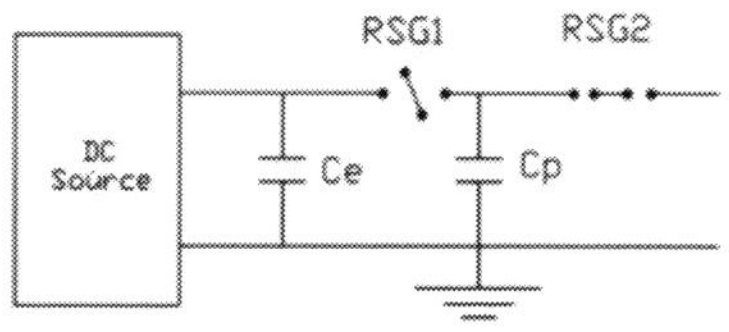

Figure 3. The circuit principle of pulse power.

3. Result and discussion

3.1. Current-voltage characteristics of three discharge types

Different discharge phenomenon were produced by the different discharge types, the experimental results were shown in figure 4, 15 discharge saw-wheel slices linked at a spacing of 15 mm were used in the experiment. In fact, the images of positive pulse, positive DC and negative DC discharges (figure 4) showed brilliant streamer corona across the entire inter-electrode space with the same voltage (13 kV). In addition, the glow corona can be observed only in the vicinity of discharge tooth tine for negative DC discharge with the same applied voltage. The voltage-current waveforms of the corona discharge for positive DC, negative DC and positive pulse were show in figure 5, figure 6 and figure 7, respectively.

As showed in figures 5-7, the discharge current increased with the applied voltage, but it firstly increased slowly, and then increased dramatically. When the spacing between saw-wheel slices increased from 5 mm to 35 mm, the breakdown voltage decreased. Moreover, at the same voltage, with the increase of tooth slices spacing from 5 mm to 35 mm, the discharge current firstly increased gradually, and then tended stability (figure 5 and figure 6) or even decreased (figure 7). The phenomenon indicated that with the increase of the spacing between saw-wheel slices, the inhibitory effect between tooth slices gradually declined, and the instability of electric field rose, thus, the discharge current increased. Compared with the positive and negative DC discharges, the discharge current of negative DC was larger than that of positive DC under the condition of the same applied voltage, with lower initial voltage, higher break-up voltage, and larger glow discharge area.

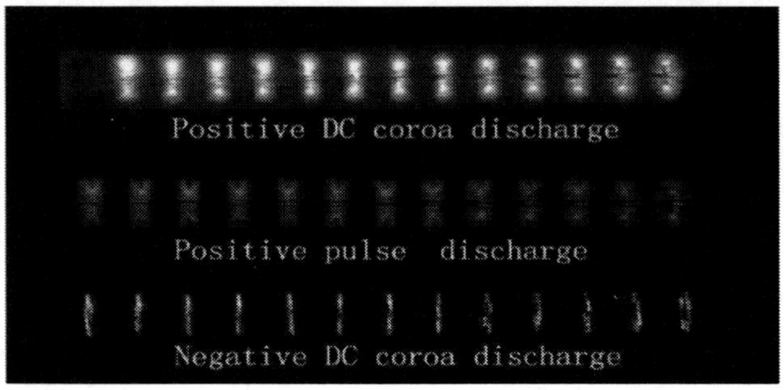

Figure 4. The images of three types of discharge with the same voltage (13 kV).

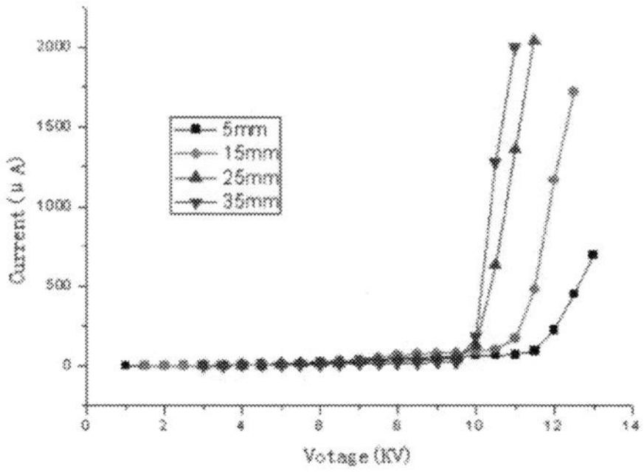

Figure 5. V-I waveforms of positive DC.

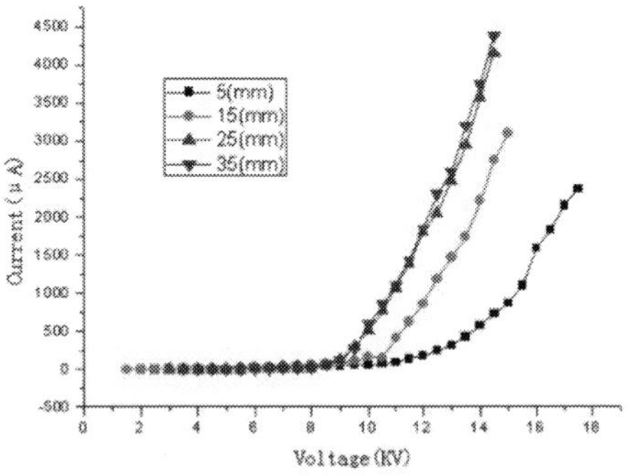

Figure 6. V-I waveforms of negative DC.

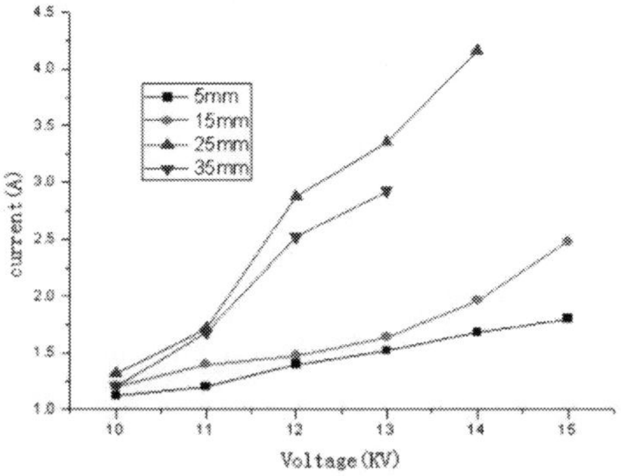

Figure 7. V-I waveforms of positive pulse discharge.

3.2. The influence of saw-wheel slice spacing on the output power and power density

In order to investigate the influence of *saw-wheel slice* spacing on output power and power density of three discharge types, 15 saw-wheel slices linked at spacing of 5, 10, and 20 mm were used. A high voltage of 11.5 kV was used in this study, and the pulse frequency and pulse-forming capacitance C_p were 50 Hz and 500 pF, respectively. The effects of saw-wheel slice spacing on the output power and power density were showed in figure 8 and figure 9, respectively.

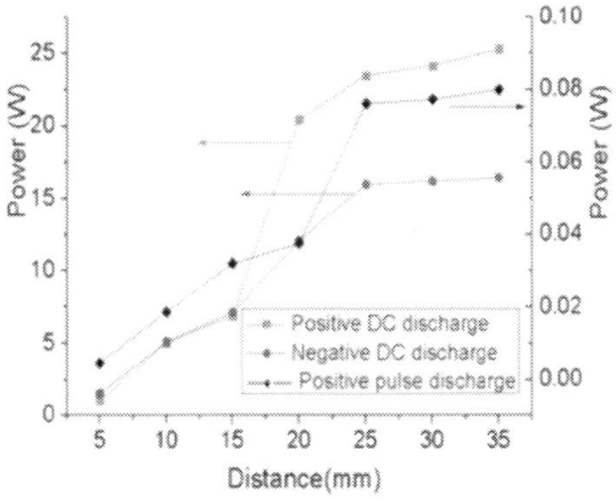

Figure 8. Changes of discharge power with saw-wheel slice spacing.

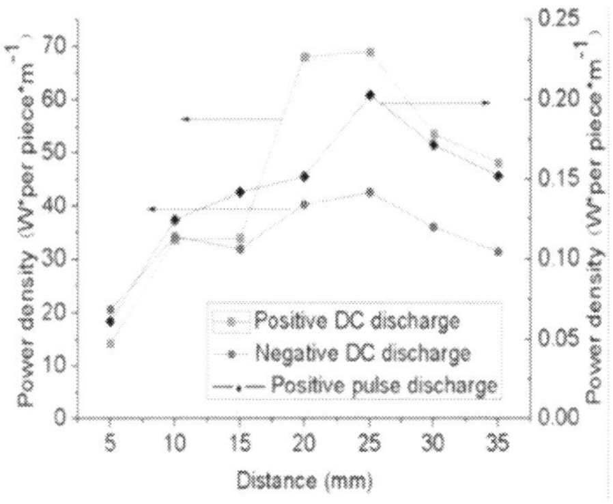

Figure 9. Changes of power density with saw-wheel slice spacing.

As showed in figure 8, at the same voltage of 11.5 kV, the discharge output power of three discharge types increased with the saw-wheel slice spacing, and it was more remarkable with the spacing increasing from 5 to 25 mm, correspondingly, the ratio (r) of saw-wheel slices spacing and discharge distance changed from 0.3 to 1.7. Then it tended stability when the saw-wheel slice spacing increased from 25 to 35 mm, with the ratio (r) range from 1.7 to 2.5. When the saw-wheel slice spacing was less than 25 mm, the electric field between saw-wheel slices could be inhibited, and it was also reported that too many discharge points might lead to corona shield phenomenon. When the spacing was larger than 25 mm, the shield phenomenon did not occur in this study. Additionally, it was found that discharge output power of positive and negative DC was significantly larger than that of positive pulse discharge.

As showed in figure 9, the power density firstly increased with the saw-wheel slice spacing increased from 5 -25 mm, and then it decreased with the further increase of the saw-wheel slice spacing. Therefore, the power density of three types of discharge reached a maximum at the distance of 25 mm, and the corresponding ratio of the saw-wheel slice spacing and discharge distance was 1.7.

Furthermore, as could be seen in figure 8 and figure 9, the change of the saw-wheel slice spacing had the most remarkable impact on the positive DC discharge, and followed by pulse discharge, and the weakest occurred in negative DC discharge. For positive DC discharge, low mobility of free electrons from the head of electron avalanche generated a larger high electric area, but for negative DC discharge, electron avalanche only produced a small area near the tooth point of the saw-wheel slice, however, the high electric area of positive pulse discharge (as a special form of corona discharge) was bigger than that of negative DC discharge, but it was smaller than that of positive DC discharge.

4. Conclusions

1) The current and output power of three types of discharges increased with the ratio (r) of saw-wheel slices spacing and discharge distance increasing from 0.3 to 1.7, then reached stability or even decreased when the ratio (r) changed from 1.7 to 2.5. The initial voltage increased but the breakdown voltage decreased with an increase in the spacing (l) between the neighbouring saw-wheel slices. In addition, the change of l had the greatest impact on the positive DC discharge, and had the least impact on the negative DC discharge.

2) The power density firstly increased with l and then decreased, and the optimum power density was obtained at the ratio (r) of saw-wheel slices spacing and discharge distance of 1.7, and it did not affect by the discharge types.

Acknowledgments

This work is supported by the Fundamental Research Funds for the Central Universities [Grant NO. DUT12RC(3)12], and partially supported by Natural Science Foundation of Hebei Province (Grant No. E2012501026). The authors thank for the financial supports.

References

[1] Wu F 2009 *J. Environ. Sci.* **21** 328
[2] Hartmann W, Roemheld M, Rohde K D and Spiess F J 2009 *IEEE Dielectrics and Electrical Insulation Society* **16** 1061
[3] Li Z 2006 *J.Hazard. Mater.* **135** 218
[4] Yang J Y 2008 *High voltage engineering* **34** 95
[5] Wang X C 2003 *High voltage engineering* **29** 40
[6] Li J 2003 *High voltage engineering* **29** 37
[7] Zhu Y M 2005 *Transactions of Beijing institute of technology* 25 41

Effect of reactor parameters on matching properties of microwave discharge in liquid

Bo Wang, Bing Sun*, Xiaomei Zhu, Zhiyu Yan, Yongjun Liu, Hui Liu
Environmental Science and Engineering College, Dalian Maritime University, Dalian, China. 116026

E-mail: sunb88@dlmu.edu.cn

Abstract. In the paper, the effects of physical and chemical liquid properties in the reactor on matching properties of microwave discharge are studied. The liquid in the reactor is water or water solution; the effects of pH, pressure, conductivity, size of bubble over the inner electrode, water temperature on the standing wave ratio (SWR) are analysed using vector network analyzer. The results show that the SWR increased from 1.512 to 2.623 as the water temperature rose, and the SWR increased as the bubble around the inner electrode grew, while the pH, pressure and conductivity had no effects on the SWR. The main cause for the change of the SWR is the change of the dielectric constant of liquid.

1. Introduction

Plasma is the fourth existence state of materials, and it is widely used in the field of environmental protection and material synthesis, as it contains high-energy particles. At present, the main methods of generating plasma include glow discharge, arc discharge, microwave discharge in gas, and so on. Microwave plasma in liquid is a new plasma technology and becomes more and more concerned due to its large spatial distribution and high density of plasma. At present, little international research work of microwave plasma in liquid has been done except Ehime University of Japan, and study on characteristics of plasma and material synthesis is the main research direction [1-6]. The process of generating plasma is affected greatly by matching properties, and physical and chemical properties of the liquid in reactor have great effects on the matching properties of microwave discharge in liquid, but less research has been done to study that at home and abroad.

The Standing wave ratio (SWR) used to indicate that whether the impedance of the antenna and feeder are matched, it is also an important parameter of the efficiency of injected energy and has great significance to study microwave discharge in liquid. The relationship between SWR and reflection coefficient K is as follows:

$$\mathrm{SWR} = \frac{1+K}{1-K} \tag{1}$$

Therefore, when the SWR is too large, it indicates that matching properties are unsatisfactory, so the SWR has great significance to study microwave discharge in liquid. The height and diameter of the inner electrode and outer electrode, dielectric constant of the insulating material between the inner electrode and outer electrode play decisive roles in the matching process of the electrode and feeder. In

* To whom any correspondence should be addressed.

this paper, the effects of the water temperature, pH, conductivity, size of bubble over the inner electrode and pressure on the SWR of microwave are discussed to lay the foundation for the later study of characteristics of microwave discharge in liquid.

2. Experimental methods

Figure 1 is a schematic representation of the plasma reactor. The height and diameter of the reactor are 16 cm, 15 cm respectively. The height and diameter of the inner electrode are 63 mm, 2 mm respectively, and the height and diameter of outer electrode are 56 mm, 8 mm respectively. The insulating material between the inner electrode and outer electrode are the ceramic tube and silicone. The liquid in the reactor is pure water. The SWR is measured by agilent network analyzer .External pressure is controlled in the range of 0-100 kPa by the water circulation vacuum pump. The pH of water is controlled in the range of 2-12 by HCl and NaOH. The water temperature which ranges from 20 °C to 70 °C is measured by the thermocouple. The conductivity is controlled in the range of 0-0.7 ms/cm using anhydrous Na_2SO_4.

In the process of making electrode, the top of the ceramic tube flushes with the inner electrode as much as possible, because it is conducive to the adhesion of bubbles, even to study the effect of bubbles on the SWR. When the water temperature is 20 °C, the pressure is 100 kPa, the SWR of pure water equals 1.512.

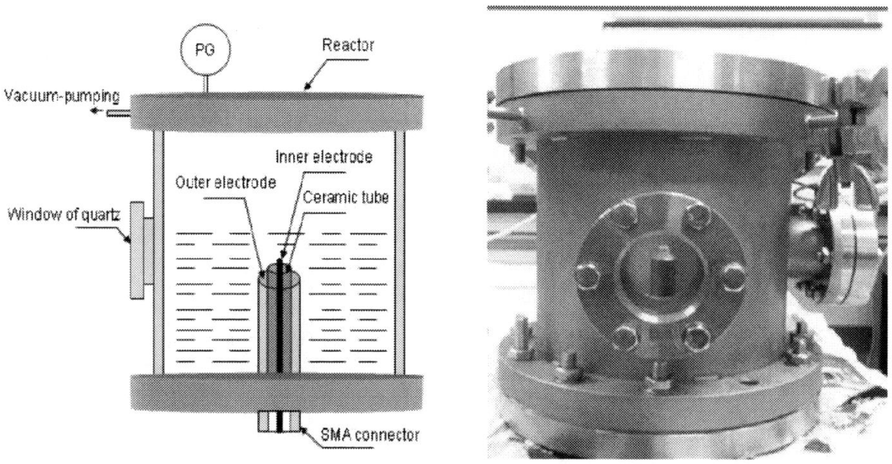

Figure 1. Schematic representation of the plasma reactor.

3. Results and discussion

In the process of matching, the entire reactor can be regarded as a complex capacitor. When the proportion of the reactor structure is determined, the change part is equivalent to the dielectric materials in capacitor. The dielectric constant is not only related with materials, but also with temperature and frequency of electric field.

As shown in figure 2, 3, 4, the effects of conductivity, pH, pressure on the SWR rise and fall around the original SWR which is 1.512, and they are entirely within the fluctuation range of the network analyzer, so solution conductivity, pH value and pressure have no effects on the SWR of microwave discharge in liquid. From figure 5, it can be seen that the SWR increases from 1.512 to 2.011 with the increase of the temperature. It indicates that the temperature leads to the change of the dielectric constant which makes the match of the electrode and the feeder get worse.

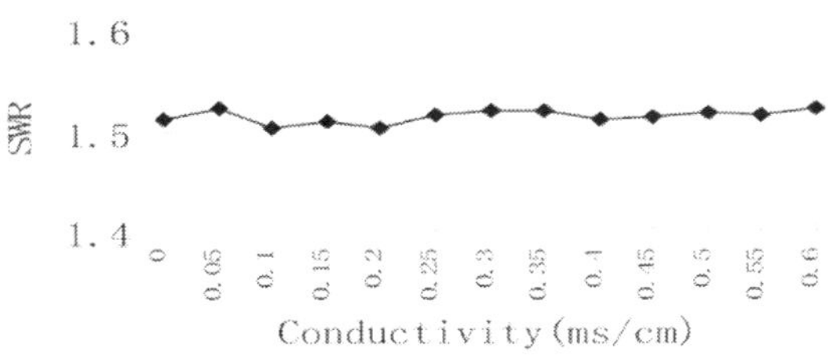

Figure 2. Effect of conductivity on the SWR.

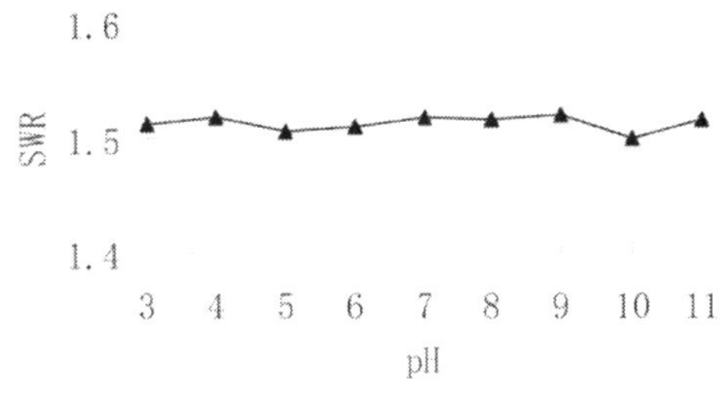

Figure 3. Effect of pH on the SWR.

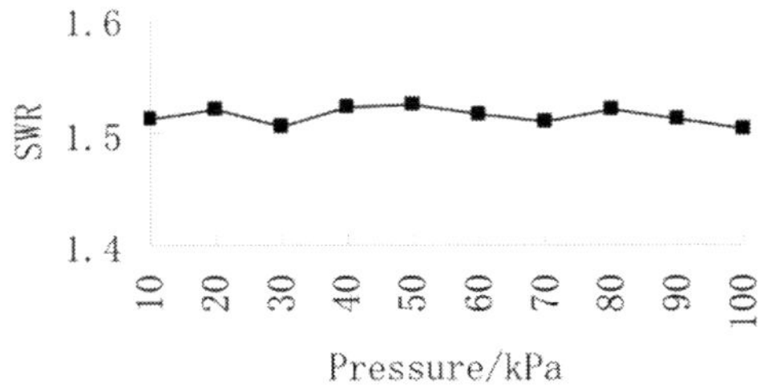

Figure 4. Effect of pressure on the SWR.

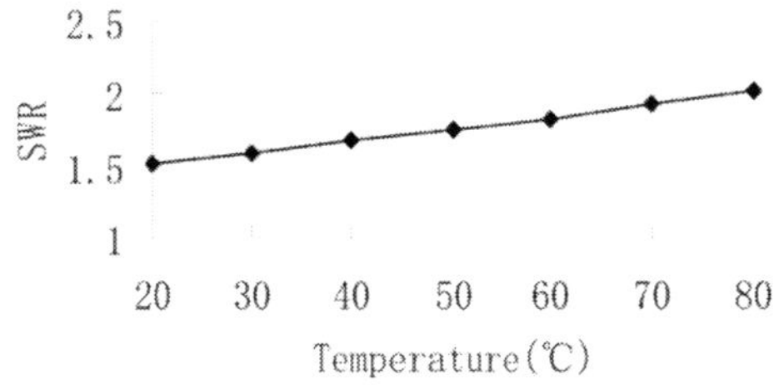

Figure 5. Effect of temperature on the SWR.

It can be seen from table 1 and figure 6 that the SWR increases with the increase of the bubble size, and when the tip of the inner electrode is not completely shrouded by bubble, the SWR has small change, and the SWR ranges from 1.512 to 2.623, only when the inner electrode is completely shrouded by the bubble, the SWR has tremendous change. So it indicates that the significant effect on the SWR is not the size of the bubble, but whether the inner electrode contacts with water. From the viewpoint of the dielectric constant, the introduction of bubble is equivalent to that the new insulating material fills in capacitor, and that makes the matching of the electrode and the feeder get worse.

Table 1. Effect of size of bubble over the inner electrode on the SWR.

Size of bubble over the inner electrode	SWR
No bubble	1.512
Half a bubble	2.012
A bubble	9.032
In the air	12.165

4. Conclusions

Based on the study of the effects of temperature, pH, conductivity, size of bubble over the inner electrode and pressure on the SWR, it is found that conductivity, pH and pressure have no effects on the SWR, while the SWR increases with the increase of temperature. The SWR increases as the bubble size increases, and the significant effect on the SWR is not the size of the bubble, but whether the inner electrode contacts with water.

Therefore, in the process of matching the electrode and the feeder in the future, the main parameters that should be considered include the relative dielectric constant of the liquid in reactor and design parameters of the electrode structure, not the pH, conductivity, pressure and temperature.

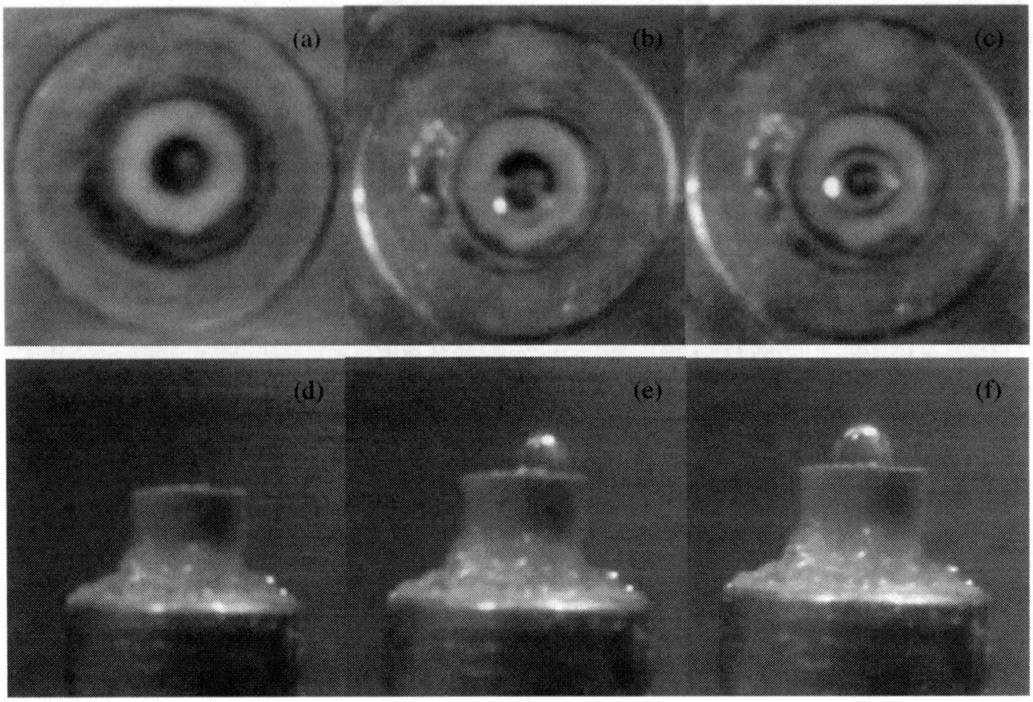

Figure 6. Photos of size of bubble over inner electrode.
(a): top view of no bubble (b): top view of half a bubble (c):top view of one bubble
(d):side view of no bubble (e):side view of half a bubble (f):side view of one bubble

Acknowledgements
We thank the support of the National Natural Science Foundation (NSFC-10875019, 41005079). It is supported by "The Fundamental Research Funds for the Central Universities". The authors thank Professor M. Sato for his valuable suggestion.

References

[1] Sato S, Mori K and Ariyada O 2011 *Surf. Coat. Technol.* **206** 955
[2] Toyota H, Nomura S and Takahashi Y 2008 *Diamond Relat. Mater.* **17** 1902
[3] Hattori Y, Mukasa S and Toyota H 2011 *Mater. Chem. Phys.* **131** 425
[4] Hattori Y, Mukasa S and Nomura S 2010 *J. Appl. Phys.* **107** 063305
[5] Nomura S, Toyota H and Mukasa S 2006 *Appl. Phys.Lett.* **88** 211503
[6] Finke B, Schroder K and Ohl A 2009 *Plasma Processes Polym.* **6** S70

Journal of Physics: Conference Series **418** (2013) 012100 doi:10.1088/1742-6596/418/1/012100

Experimental investigation of spark discharge energy

Zulin Peng, Yu Zhang, Dongjie Chen, Jinsong Miao, Jiting Ouyang*
School of Physics, Beijing Institute of Technology
5 South Zhongguancun Street, Haidian District, Beijing 100081, China

E-mail: jtouyang@bit.edu.cn

Abstract. Minimum ignition energy is one of the important characteristics for electrostatic sensitivity of explosives and determination of the spark discharge energy is necessary for decision making of explosion prevention. In this paper, the real spark discharge energy was determined by the integration of the measured voltage and discharge current in a needle-to-plate configuration. The influence of the charging voltage and polarity, the charging capacitance and the electrode gap on the spark discharge energy was investigated. It is shown that the energy deposited into the discharge system is always smaller than that stored in the charging capacitor, especially at lower voltages that the discharge shows one current pulse and the energy efficiency is less than 30%. The discharge becomes complete at higher voltages when the discharge shows double pulses and the energy efficiency is close to 100%. For given charging capacitance and voltage, the real discharge energy is determined by the discharge mode rather than the electrode gap or the voltage polarity.

1. Introduction

Minimum ignition energy (MIE) is one of the important characteristics for the electrostatic sensitivity of explosives [1]. It is important for the evaluation of the sensitivity of fireworks, firecrackers, and explosive substances to the electrostatic spark, the MIE of combustible gases, flammable liquid vapours and dust/air mixtures, etc. Accurate determination of the minimum ignition energy is necessary for decision making of explosion prevention and design of intrinsically safe electrical apparatus. The determination of the MIE using a capacitive electrostatic discharge (ESD) is well established in the literatures as well as in standards [2-9]. Thereby, a capacitor with a defined capacitance is supplied by a defined voltage. Then the capacitor is discharged in a test vessel filled with a certain flammable gas/air mixture via an electrode configuration with a defined electrode diameter, electrode material and electrode gap width. The energy stored in the capacitor is calculated using the equation $E_0 = CU_0^2/2$, where E_0 is the energy, C is the capacitance and U_0 is the voltage supplied to the capacitance, and the transferred charge is calculated as $Q_0 = CU_0$. The MIE is determined by varying the capacitance, the charging voltage, the electrode diameter, the electrode gap width and the mixture composition in order to find the lowest energy necessary to ignite the given mixture. In many cases, the ESD energy is assumed to have the amount of energy or charge stored in the capacitance before the discharge.

However, the electrostatic spark discharge energy is normally not equal that stored in the capacitor (or the nominal energy). In fact the stored charge might not discharge completely after the spark, with a small amount remaining in the capacitor. Moreover, the loss of the charge stored in the capacitance

* To whom any correspondence should be addressed.

might also be taken by some means (e.g., corona) before or after the spark discharge happening. Hence the discharge energy is generally smaller than stored energy. For this reason, the stored charge was sometime not used for the definition of the transferred charge thresholds or MIE (e.g., Ref. [7, 8]). Practically, the spark discharge energy hence the ratio of the measured discharge energy to the nominal energy depends on the used electrode radius, the gap distance between the electrodes, the charging voltage, the charging capacitance, and usually the property of the tested explosive.

The aim of this work is to examine the correlation between the energy stored in the capacitance before the discharge and the real spark ESD energy. The dependence of the spark energy on the charging voltage, the capacitance, the electrode gap is studied.

2. Experimental setup

The schematic of the ESD system used in experiments is shown in figure 1.

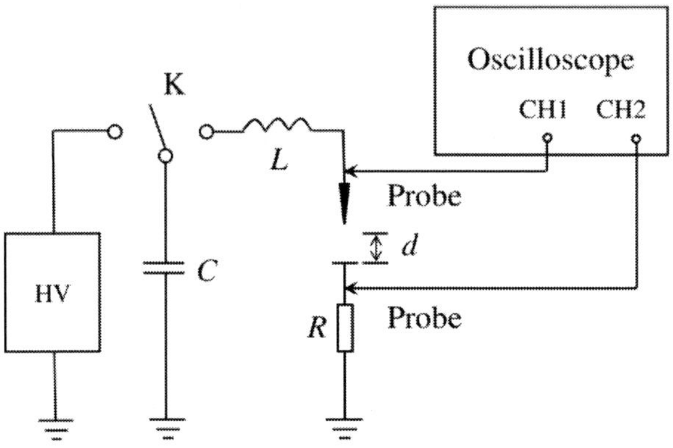

Figure 1. Schematic of experimental setup.

The system consists of a needle-to-plate electrode, a discharge capacitor *C*, a high voltage (HV) power source and a HV relay switch K. The needle is made of Tungsten and the radius is about 0.05 mm. The electrode gap *d* is from 1 to 4 mm. The capacitor C is firstly connected to the HV source through switch K, to be charged to a voltage level of U_0, then connected to the needle electrode through an inductor L to trigger the ESD spark. The inductance *L* is 2.3 mH in our experiments. The plate electrode is made of copper and grounded. Both the negative and positive high-voltage DC power sources are tested in experiments. The voltage is measured by a digital Oscilloscope Tektronix TDS-3054B and voltage probe, with hardware bandwidth of 500 MHz and sample rate of 5 G/s. The voltage *U* on the needle is measured by a Tektronix voltage probe P6015A, while the discharge current $I(t)$ is sampled by the measured voltage on a non-inductance resistor R, or $I = U_R/R$ (where U_R is measured by Tektronix voltage probe P6139A). The sampling resistance *R* is 1.5 Ω. The charging capacitance C is variable in experiments, ranging from 250 pF to 5000 pF. The real discharge energy is determined by the integration of the voltage $U(t)$ and current $I(t)$ with time, or

$$E = \int_0^{tf} U(t)I(t)dt = \int_0^{tf} U(t)\frac{U_R(t)}{R}dt \tag{1}$$

The each discharge is repeated 5 times and the total energy is averaged. The discharge energy *E* is compared with the initial stored energy in the capacitor $E_0 = CU_0^2/2$.

3. Results and discussion

3.1. Waveform of discharge current and voltage

We tested the ESD sparks in different conditions. The temperature and relative humidity is 25℃ and 50% respectively. The spark occurs when the applied voltage is greater than a critical value. The typical waveforms of voltage and current are shown in figure 2 for gap of 1 mm and capacitance of 250 pF. The voltage is of 7 kV and the polarity is positive.

It is seen that the discharge current shows a typical ESD current, with a stronger pulse firstly and following inversed small pulse. The rising time of the current pulse is about 10 ns and the duration is about 50 ns. The calculated discharge energy according to equation (1) is $E = 0.61$ mJ which is much lower than the nominal one in the capacitor, with a discharge efficiency of $E/E_0 = 10\%$.

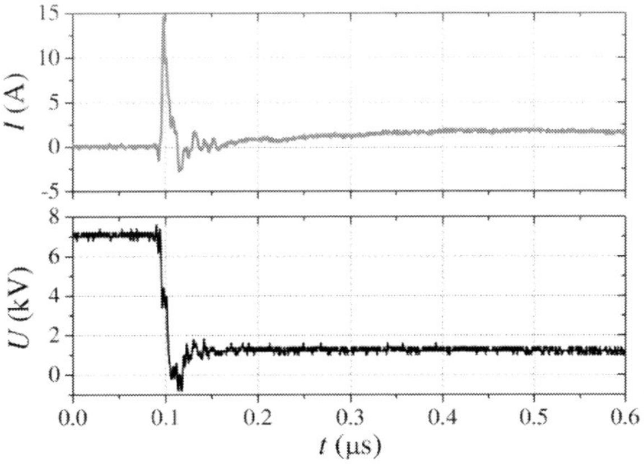

Figure 2. Waveforms of spark discharge current and voltage at $U_0 = 7$ kV, $d = 1$ mm.

Note that there is a remaining voltage about 1200 volts on the capacitor after the spark and there is a low current of 1.5 A continues through the circuit, resulting a transient power consumption of 1.7 kW in the circuit after the spark. This current lasts for several microseconds and would release the major part of the energy stored in the capacitor.

When the voltage increases, the discharge current may show double peaks. Typical waveforms are shown in figure 3 for voltage of 9 kV.

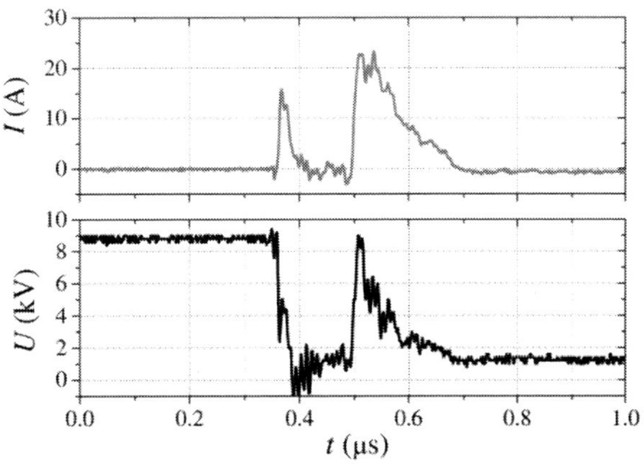

Figure 3. Waveforms of spark discharge current and voltage at $U_0 = 9$ kV, $d = 1$ mm.

It is seen that the spark is firstly started by a small current, followed by a stronger second pulse. The interval between the two pulses is from several to tens of nanoseconds. The first pulse has a rising time of about 10-20 ns and lasts for about 50-70 ns. The energy in the first pulse is small, with value of 0.9 mJ or 9% of the stored energy. The second has a rising time of 10-30 ns and lasts for as long as 200 ns. The discharge energy in the second pulse is about 8.7 mJ or 87% of the stored one. In this case, there is also a remaining voltage (about 1200 V) on the charged capacitor, but the remaining current is zero after spark. The total discharge energy is E = 9.6 mJ which is close to the nominal one in the capacitor, with a ratio of E/E_0 = 96%. This indicates that the discharge is almost complete.

3.2. Effect of polarity

For given conditions, the discharge energy is not much different for positive or negative voltage. Table 1 shows the comparison at gap spacing of 1 mm and capacitance of 250 pF. The discharge energy is averaged in several times. Generally the energy of negative voltage is slightly larger than the positive. When the discharge is complete at higher voltages, the energy is almost the same for both polarities.

Table 1. Positive or negative discharge energy.

U_0 (kV)	Positive		Negative	
	E (mJ)	E/E_0 (%)	E (mJ)	E/E_0 (%)
5	0.36	11.4	0.47	15.0
6	0.48	10.6	0.71	15.7
7	0.66	10.8	0.77	12.6
8	0.82	10.3	5.49	68.6
9	9.32	92.0	9.31	95

3.3. Effect of discharge capacitance and voltage

The discharge energy and the energy efficiency at different voltages and capacitances are shown in figure 4 and 5, respectively.

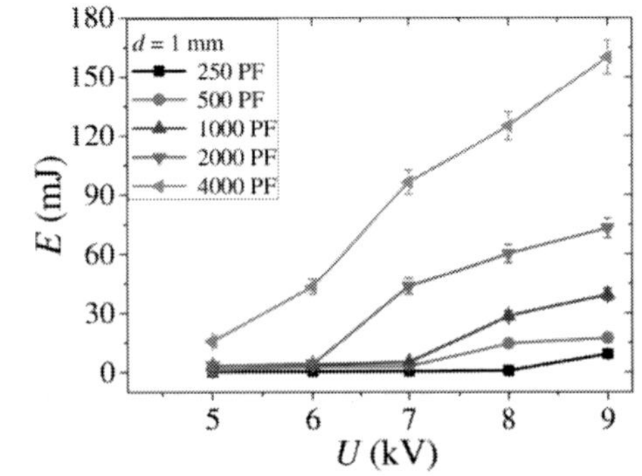

Figure 4. Discharge energy changing with voltage at different capacitances.

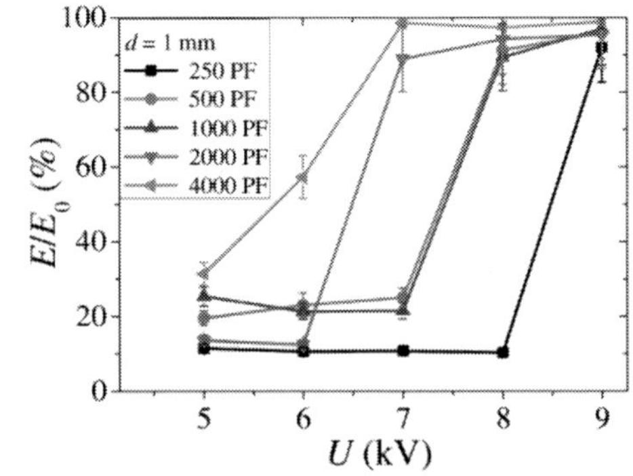

Figure 5. Energy efficiency changing with voltage at different capacitances.

As expected, the discharge energy generally increases with the charging capacitance and the voltage. However, the discharge is not complete at lower voltages, or only a part of the charge on the capacitor is consumed. Consequently, the energy efficiency is low, typically less than 30%. In this case, the discharge energy does not depend strongly on the capacitance or the voltage (see the case of 250-1000 pF/5-7 kV in figure 4). Only when the voltage increases above a critical value so that the discharge tends to be complete (showing a character of two discharge pulse), the discharge energy will be close to the stored energy in the capacitor and the efficiency becomes nearly 100%.

3.4. Effect of discharge gap

The spark discharge energy at different gap is nearly the same if the charging capacitance and the voltage is the same. Table 2 shows the examples of discharge energy at capacitance of 250 pF and two voltages of 7 kV and 10 kV. It is seen that the discharge energy at 7 kV is smaller compared with the energy in the capacitor and the energy efficiency is small with value around 11%. At 10 kV, the discharge energy is close to the stored energy in capacitor and the efficiency is nearly 100%. This result indicates that the discharge energy relates to the discharge mode rather than the electrode gap. For this reason, the measurement of the ESD energy might be taken under different configurations.

Table 2. Discharge energy at various gap.

d (mm)	7 kV		10 kV	
	E (mJ)	E/E_0 (%)	E (mJ)	E/E_0 (%)
1	0.66	10.8	12.0	96.0
2	0.75	12.2	11.7	93.6
3	-	-	11.8	94.4

4. Conclusion

We have investigated the ESD spark discharge in air with a needle-to-plane configuration. It is shown that there are two discharge modes at low or high voltages, characterized by single or double current pulses respectively. The discharge energy at lower voltages is generally small and the efficiency is less than 30%. The discharge at higher voltages is complete and the efficiency is nearly 100%. The spark discharge energy increases with the changing capacitance as well as the voltage, but it does not depend on the voltage polarity or the electrode gap.

Acknowledgments

This work was partly supported by the State Key Laboratory of Mechatronics Engineering and Control, China, under Grant No. 2011C3606.

References

[1] Eckhoff R. 2002, *J. Loss Prevention Proc. Indu.* **15** 305
[2] Lewis B, von Elbe G. 1987 *Combustion, Flames and Explosions of Gases* (3rd ed. London: Academic Press INC)
[3] IEC 61241-2-3. 1994 *Method for determining minimum ignition energy of dust/air mixtures*
[4] IEC 60079-0. 2007 *Explosive Atmospheres - Part 0: Equipment - General Requirements (Ed 5.0)*
[5] AQ/T 4120-2011. 2011 *Pyrotechnic composition used for fireworks and firecrackers test method for sensitivity to electrostatic spark* (in Chinese)
[6] European Standard EN 13463-1. 2007 *Non-electrical Equipment for Use in Potentially Explosive Atmospheres e Part 1: Basic Method and Requirements*
[7] MIL-STD-1751(USAF). 1982 *Safety and performance tests for qualification of explosives*
[8] Von Pidoll U, Brzostek E, Froechtenigt H R. 2004 *IEEE Trans. Ind. Appl.* **40** 1467
[9] Langer T, Gramse G, Möckel D, von Pidoll U, Beyer M. 2012, *J. Electrostatics* **70** 97

7th International Conference on Applied Electrostatics (ICAES-2012) IOP Publishing
Journal of Physics: Conference Series **418** (2013) 012101 doi:10.1088/1742-6596/418/1/012101

Optical diagnosis of high pressure non-thermal plasma

T Oda[*], **Y Nakagawa, R Ono**
Department of Electrical Engineering and Information Systems, School of Engineering, the University of Tokyo, 7-3-1 Hongo, Bunkyo-ku, Tokyo, 113-8656, Japan

E-mail: oda@ee.t.u-tokyo.ac.jp

Abstract. Optical diagnosis of high pressure plasma is one of the most powerful tools for investigating chemical reaction mechanisms in the high pressure plasma region. The authors developed various optical measurement systems by using tunable lasers, such as (two-photon-)laser-induced-fluorescence, coherent-anti-Stokes Raman scattering, time evolution of optical emission imaging and others for detecting O_3, O, OH, N, NO* and other radicals in atmospheric pressure plasma. Outline of basic measuring techniques developed by the authors for high pressure plasma diagnosis are explained and real examples of plasma diagnosis are demonstrated in this paper. For example, density distributions of single nitrogen (N) and excited nitrogen molecule ($N_2(A)$) below the discharge needle generated by the pulse plasma suggest that single N might be generated in the secondary streamer, while $N_2(A)$ might be generated in the primary streamer, and single N decomposes NO more than $N_2(A)$.

1. Introduction

The high pressure plasma has been developed for more than 40 years by late Prof. Masuda, at the University of Tokyo and many researchers have learned by him, including Prof. Wu and all Japanese invited speakers. Our common target is the environmental protection and providing a new tool for biological handling. Related with the ESP (Electrostatic precipitation) technology [1], DeNOx [2, 3] and $DeSO_x$ processes for combustion flue gas cleaning by using the non-thermal plasma were developed [4-7]. Moreover, combined with the ozone-hole protection and sick-house syndrome, VOC (volatile organic compounds) removal from the air by the plasma has been also developed by many researchers [8-14]. Recently high pressure glow-like plasma is also investigated for cleaning, mechanical plasma processing (cleaning, cutting, polish), and biological or medical processing tools. However, still now, the high pressure plasma process has been never understood. What kinds of radicals play any role in the process? It is not so easy to identify that radical behaviour generated by the plasma.

The authors' group has developed various kinds of optical radical measurement systems by using UV tunable (variable wavelength) laser systems. UV absorption measurement with a nanosecond UV laser can detect the ozone generation with time resolution of nanoseconds and spacial resolution of mili-meter in the streamer. Laser-Induced Fluo-rescence (LIF) technique can detect various kinds of radicals with the time resolution of nanoseconds which cannot be observed as the direct optical images without the discharge excitation. Several radicals, such as O, N, OH and excited N_2 were observed by LIF techniques in the non-thermal plasma. Those results are listed in table 1.

[*] To whom any correspondence should be addressed.

2. Streamer image and ozone generation for pulse discharge

2.1. Streamer of pulse discharge

The needle to plate streamer discharge image was observed by a gated nanosecond ICCD camera or streak camera where submicrosecond high voltage pulse was applied. Some examples are shown in figure 1 in the nitrogen and oxygen mixture condition. The streamer is usually divided into two parts. The sphere-like bright point runs from the needle point to the plate for the positive pulse application. The streamer speed is pretty high, (10 – 20 nanoseconds for 13 mm gap which means that the velocity of the primary streamer is about 1 mm ns^{-1}). Then the bright secondary streamer progresses from the needle point to the flat plate. A streak camera picture is also shown in figure 2 where the brightness is shown at the same sensitivity indicating the secondary streamer is long and much bright.

Table 1 Measured active species or parameters observed by the authors.

Active species	Measuring method	Ref.
O_3	Photo absorption	15
OH	LIF	16,17
O	LIF	18
N	LIF	19
$N_2(A)$	LIF	20
NO	LIF	21
$O_2(V)$	LIF	22
$N_2(V)$	CARS	23
Temperature	LIF	17

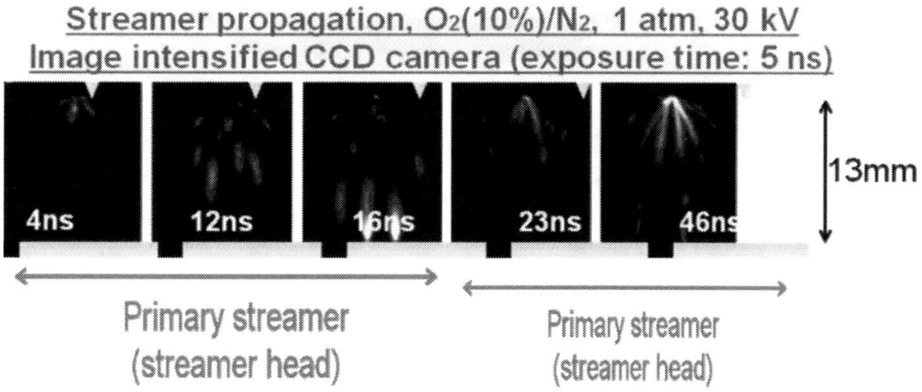

Figure 1. Streamer development after the pulse energization.

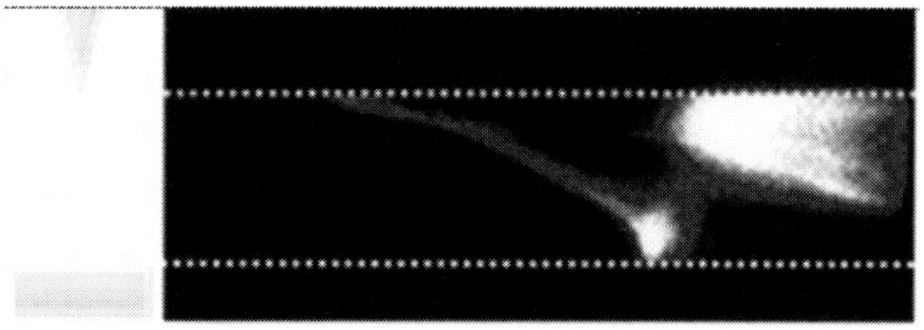

Figure 2. Streak photo of streamer.

2.2. Ozone generation

UV light is easily absorbed by the ozone gas and we can know the ozone generation in the streamer as shown in figure 3 where a strong excimer laser of 248 nm was used for observation of the ozone. It is obvious that the ozone is surely generated in the streamer. If we estimate the ozone region of about 1 mm in diameter, the ozone concentration in the streamer region is about 3,000 ppm at maximum. If we check the ozone generation area exactly, most ozone is generated in the secondary streamer region which is shown in figure 4. The ozone generation is mostly near the needle electrode and that pattern is very similar to the secondary streamer photo emission as shown in figure 4. In previous simulation and streak camera measurements reported by others suggested that the most ozone should be generated in the primary streamer because the electron energy in the primary streamer is much larger (about 10 eV) than that in the secondary streamer (about 1 eV = electron Volt). Our results are far from that estimation but might be because the electron density is much large in the secondary streamer region and the electron energy is also more than a few eV estimated our new simulation.

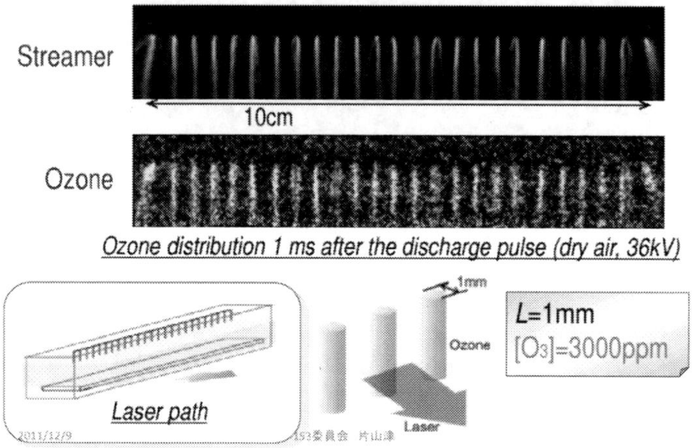

Figure 3. Ozone generation in the streamer by UV absorption.

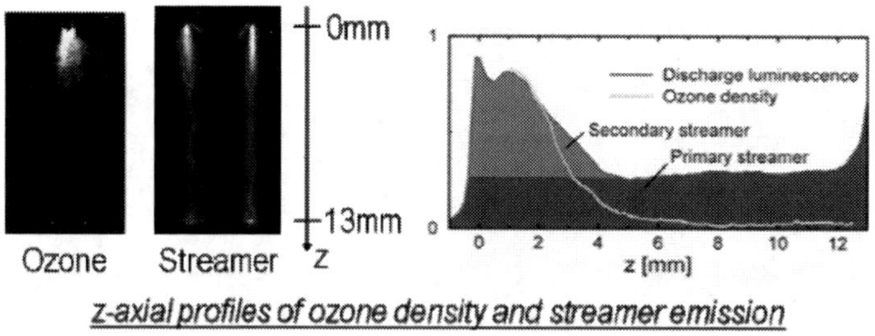

Figure 4. Ozone distribution compared with the streamer illumination.

2.3. Optical emission spectroscopy (OES)

Optical emission spectroscopy is also very effective in observing various radicals generated by the pulse discharge but the illumination is limited by the discharge excitation only. Some data for the Corona discharge, barrier discharge and plasma jet will be demonstrated at the conference. One example of the OES of the plasma jet (gas flow is 40 l min^{-1} and the discharge power is roughly 500 W) where upper position is just below the outlet of the plasma jet and lower level is about 40 mm below the jet is shown in figure 5. The centre is about 309 nm and spectra width is about 20 nm.

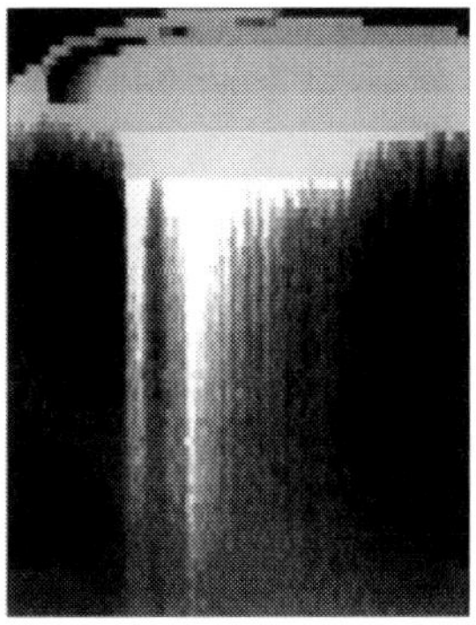

Figure 5. Optical emission spectra of the plasma touch near 310 nm.

2.4 Others
Schlieren image can also show the molecular density perturbation by the plasma-discharge, especially shock wave can be displayed well.

3. LIF Observation

3.1. What is LIF
One example of the electron energy level of OH radical related with LIF (Laser-Induced Fluorescence) is shown in figure 6. In the figure, the horizontal position means distance from the other atom (between O and H). v means vibration mode (electron energy level is separated by the quantized vibration mode of two atoms) and each level is also separated by the rotational mode. Some of excited electros from the ground state ($X^2\Pi$) to upper state ($A^2\Sigma$) lose their energy by predissociation and quenching. So $A^2\Sigma$ state electrons are easily lose energy by predissociation, and the quenching effect is rather negligible which means LIF of OH is not affected by the composition of surrounding gas but the sensitivity of OH is rather small. On the other hand, LIF signal of OH excited by 281 nm laser is very large (sometimes, more than 1,000) but the absolute value comparison among OH radical densities in different mixture of oxygen and nitrogen, for example. In the case of rare gas, OH density of ppb can be detected.

Table 2 kinds of LIF.

laser wavelength (nm)	detected radical
205	H
207	N
211	N
226	NO, O
243	H
248	$NO_{(v=2)}$, OH, O_2^*
254	Hg
281	OH
292	H
308	OH
336	NH
388	CN
473,512,563	C2
607	N_{2vib}

At present, various wavelength lasers are used as LIF exciting lasers. Typical laser wave lengths and detected radicals are shown in table 2 where gothic means that we tested.

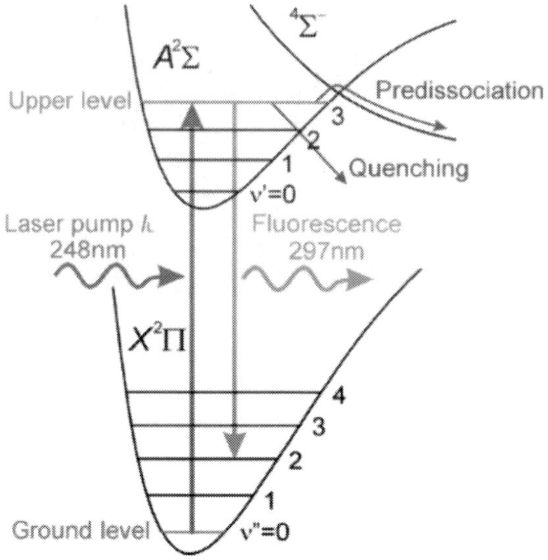

Figure 6. Electron energy level for OH radicals related with LIF.

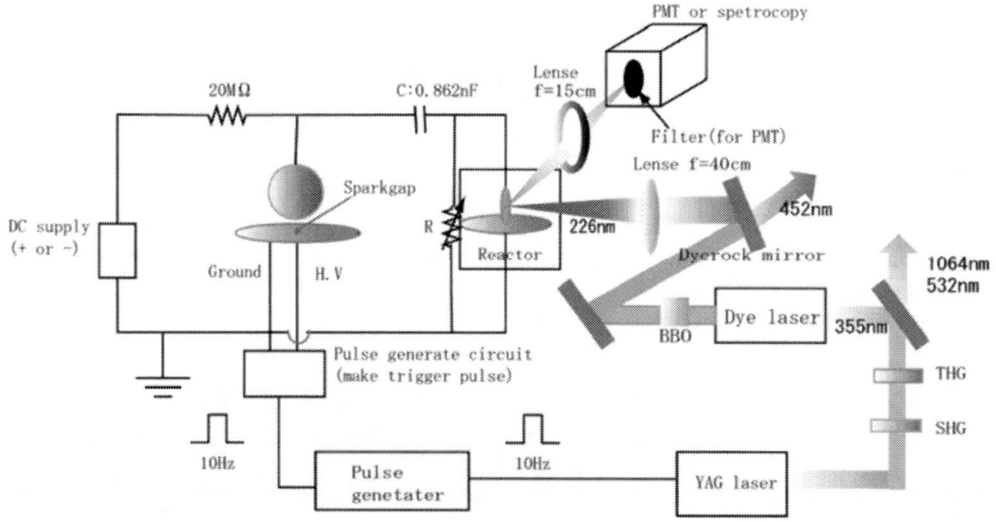

Figure 7. Example of Laser-Induced Fluorescence measuring system.

One example of our experimental setup is shown in figure 7 for NO and O measurement where a dye tunable laser system is used as a laser source in this figure. For LIF signal imaging, a high sensitive gated Image-Intensified CCD Camera (IICCD) with narrow band pass filter is used.

3.2. LIF examples

3.2.1. NO image Ground state NO is sensitive for 226 nm laser irradiation where the experimental system is in figure 8. Sample gas is nitrogen with 50 ppm NO which flows slowly from left to right in sight. Because of optical reason, the viewing area is circle like and a sheet pattern laser beam is only at the centre position (bright area of original gas: 50 ppm). Left isolated picture is a streamer illumination photo just at the discharge (during pulse). Pulse width is about 500 ns but nothing occurs just after the pulse discharge. About 30 microsesonds later, a slight decay of NO (black area) can be detected as a string which becomes large with time and right side becomes dark which means most NO might be decomposed. At 5 ms after the discharge, right side is mostly dark (left side is new NO supplied from left side. At 20 ms later, all gas is exchanged and fresh NO is detected. What (radicals) decompose NO after the discharge? The life time of that radical should be much longer than milliseconds. For analyze that, some new experiments have been done.

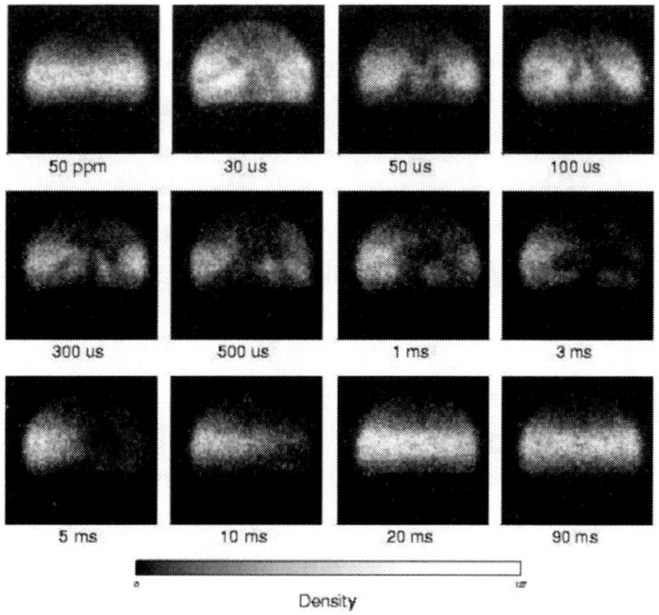

Figure 8. LIF image of NO after the pulse Corona discharge.

3.2.2 OH radicals We tested two different OH LIFs. One is used tunable excimer laser of 248 nm which is very strong of 450 mJ/pulse and it is possible to take LIF picture. Figure 9 shows OH LIF image after the pulse corona discharge. Each picture is taken at different discharge but it is apparent that OH is generated in the streamer. From the LIF signal intensity by calculation with photomultiplier sensitivity, detecting solid angle and transition ratio, we can estimate the OH density, if OH density distribution is assumed. We can also observe the decay time of OH radicals and simulation including 14 reactions can suggest the original OH density. Time decay of OH signal is shown in figure 10 with simulation curves [24]. Simulation curve (theory 100 ppm) fits experimental data mostly and we can estimate the concentration of OH at this data should be about 100 ppm. From the estimation calculation by the transition coefficient is also about 100 ppm and both data are in good agreement. This error range is sometimes 50 % but we can say the order is the same. Those LIF signals are separately observed with each oscillation mode. If at that timing, OH energy might be equilibrium, we can estimate the gas temperature from the Boltzmann plot as shown in figure 11. Log plot of LIF signal intensity gradient versus electron energy difference shows the temperature. After some time delay from the discharge, OH radical temperature change with time can be observed. OH temperature is very different for the gas composition. From figure 11 and other data, we can know the gas temperature change with time after the discharge as shown in figure 12. OH gas temperature increases with time after the discharge and be maximum at about 40 or 50 microseconds and decreases after that time. In this experiment, back ground (balance) gas is nitrogen and humidity is roughly close to 100 %. Just after the discharge, highly excited states of higher vibration mode might be produced by the high energy electron impaction and high electric field. After several tens microseconds, many collisions transfer the high energy to oscillation mode and ground state molecular temperature might be increase. Without the oxygen, that energy transfer is not so large but a little bit oxygen molecules enhance that transfer so much. More oxygen cannot increase the temperature so much but that reason is not yet known still now. Anyway, peak gas temperature might be close to 1,000 K which may enhance oxidation process so much.

The authors constructed a practical VOC decomposition plasma reactor and the relation between the real TCE (trichloro-ethylene) decomposition efficiency and OH temperature. Some data will be presented at the Conference.

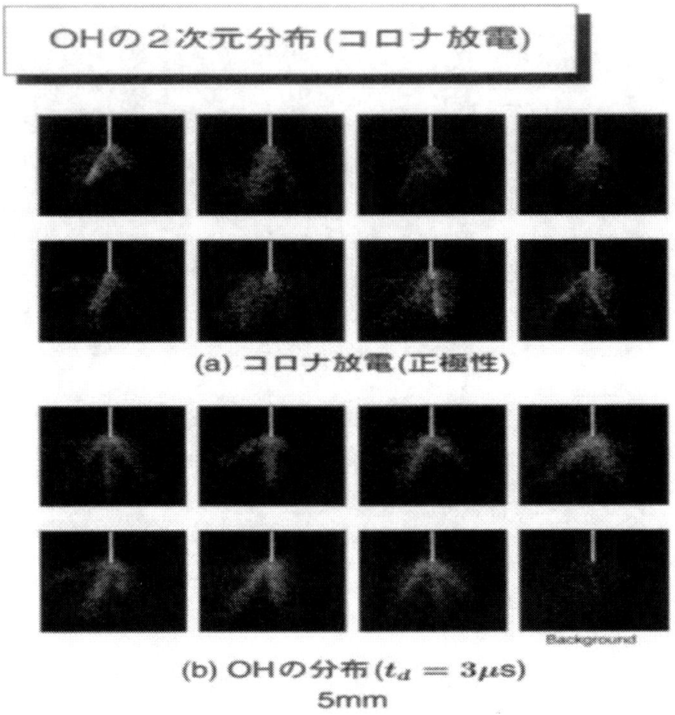

Figure 9. (a) Corona streamer and (b) OH LIF images excited 248 nm.

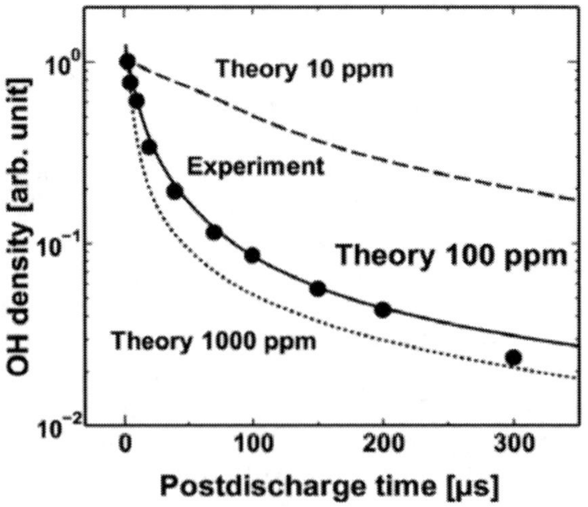

Figure 10. OH LIF signal decay after the pulse discharge (281 nm laser).

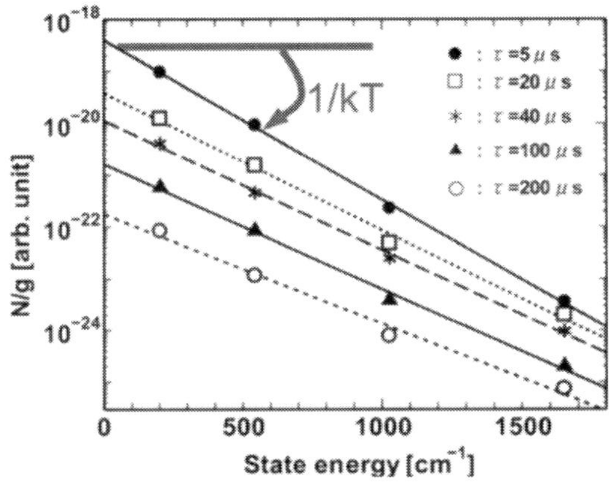

Figure 11. Log plot of OH LIF signals versus electron energy level difference wher humidity is roughly 50 % and charging voltage is 28 kV observed at 3.6 mm below the discharge needle.

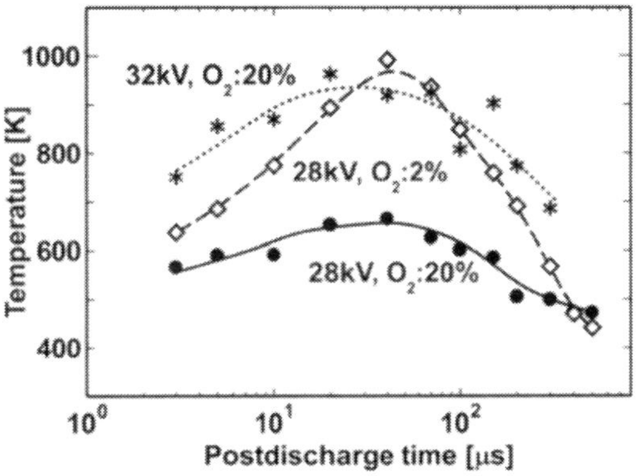

Figure 12. OH gas temperature change with time after the discharge for different gas and discharge conditions of 28 kV or 32 kV wher oxygen concentration is 20 % or 2 %.

3.2.3. Atomic oxygen The atomic oxygen shows two-photon-assisted laser-induced fluorescence (TALIF) which energy level is shown in figure 13. As shown in figure 13, two photons were absorbed at the same time meaning strong photo density is essential (TALIF signal proportional to the square of photon density. If photon density is too strong, photo-desorption (laser beam decomposes O_2 molecule and generates atomic oxygen) occurs and real atomic oxygen signal cannot be separated from that signal. We must be care for that error. Atomic oxygen distribution from the corona needle point to the flat plate is shown in figure 14 where most atomic oxygen is generated in the secondary streamer which is in good agreement with the ozone generation profile. Uniform atomic oxygen generated by the primary streamer is also recognized from the zero position (just below the needle). Former simulation suggested that the decomposition of oxygen molecule, O_2, needs about 5 eV energy and that electron energy can be generated only in the primary streamer. However, these results and our recent simulation suggested that a several eV electrons are produced also in the secondary streamer and they can also contribute to decompose oxygen molecule to atomic oxygen.

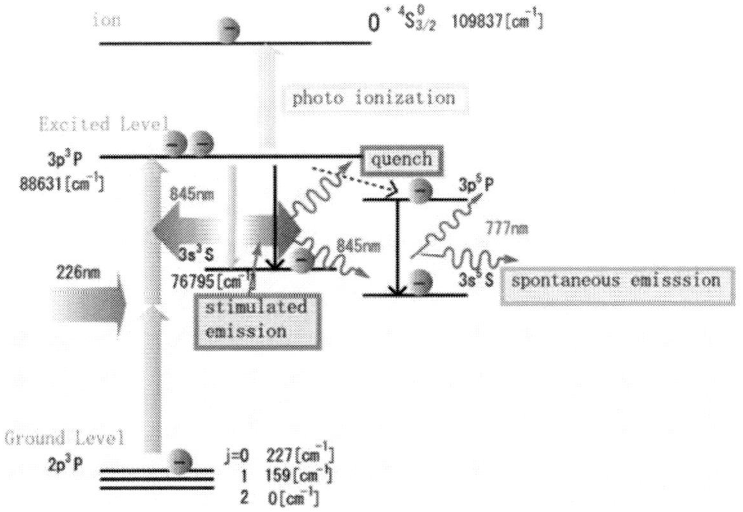

Figure 13. Energy level of atomic oxygen related to TALIF.

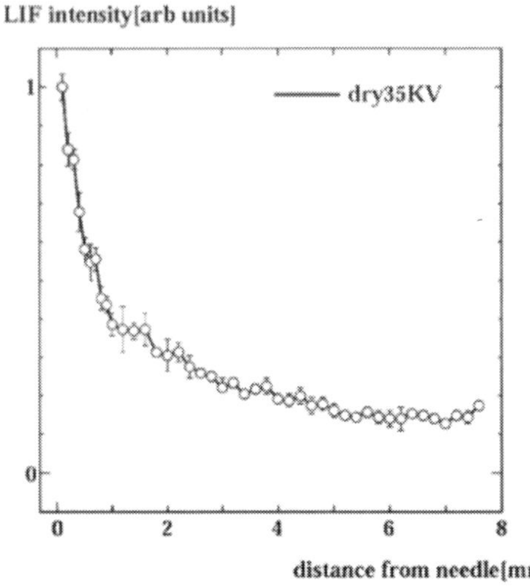

Figure 14. Atomic oxygen profile from the needle point (observed by TALIF).

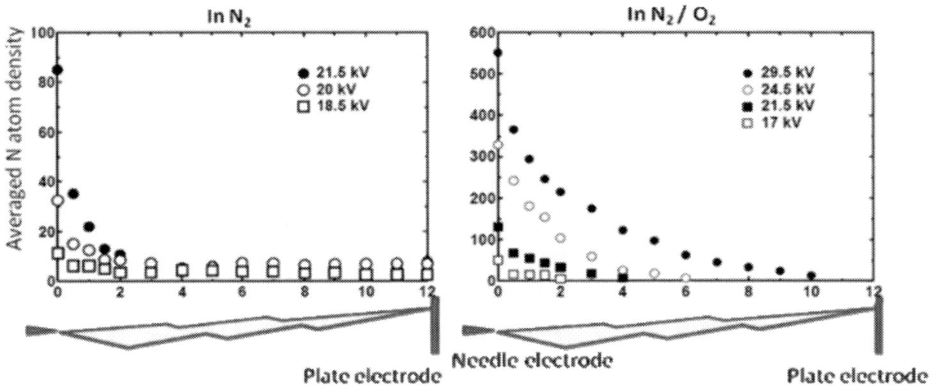

Figure 15. Atomic nitrogen distribution observed by TALIF (207 nm) for pulse corona.

3.2.4. Atomic nitrogen LIF signal of atomic nitrogen is also detected by TALIF where the excitating laser wavelength is 211 or 207 nm which is a little bit smaller than that for atomic oxygen but the energy level is quite similar to figure 14 and an experimental system is also roughly the same as that for an atomic oxygen TALIF measurement. One example of N atom distribution by TALIF is shown in figure 15 for in pure nitrogen and in air. That N distribution is also similar to the atomic oxygen distribution and the atomic nitrogen is also generated mostly in the secondary streamer where all signal is the integration of all TALIF except the distance from the needle point because the laser beam is a sheet pattern of 1 mm thickness. We simulate the time-decay by the parameters of the atomic nitrogen concentration and the streamer diameter, the atomic nitrogen concentration is 300 ppm and a streamer width is 240 micrometer in nitrogen with 200 ppm NO. If in the pure nitrogen, the nitrogen atom density is 300 ppm and the streamer diameter is 200 micrometers. That diameter determined by TALIF time decay is in good agreement with the light illumination measurement of the streamer diameter. That pretty high concentration of the atomic nitrogen is rather too large compared with the former reported simulation results, but our recent simulation results supports rather high atomic nitrogen and the decomposition of NO by the plasma might be due to that atomic nitrogen effect, although many papers suggested the NO decomposition by excited nitrogen molecule, maybe N_2 (A), is dominant, which will be discussed in next section.

3.2.5. N_2 (A) We use N2 ($A^3\Sigma^+_u$)-LIF, that is, the transition to N_2 ($B^3\Pi_g$). Indeed, about 618nm laser beam is selected for many different wave lengths (transitions). That energy diagram is shown in figure 16.

LIF signal (N_2 (A)) distribution below the needle point is shown in figure 17 with discharge photoemission intensity. As shown in the figure, the photo-illumination is very strong just under the needle point and might be due to the secondary streamer propagation. On the other hand, N_2 (A) signal is roughly constant from the needle to the ground flat plane suggesting that most N_2 (A) is generated in the primary streamer region which is rather constant from the needle to the plane. If we estimate several equations for N_2 (A) decay with time for different gas composition, we can estimate the absolute density of that excited metastable N_2 (A). Example of that log-decay plot for N_2/O_2 mixture gas is shown in figure 18.

In pure nitrogen, typical concentration of N_2 (A) by about 20 kV pulse for 13 mm gap discharge is about $1\text{-}5\times10^{13}$ cm^{-3} which is rather very small compared with previously estimated density of 10^{15} cm^{-3} of N_2 (A) which decompose NO in N_2 plasma.

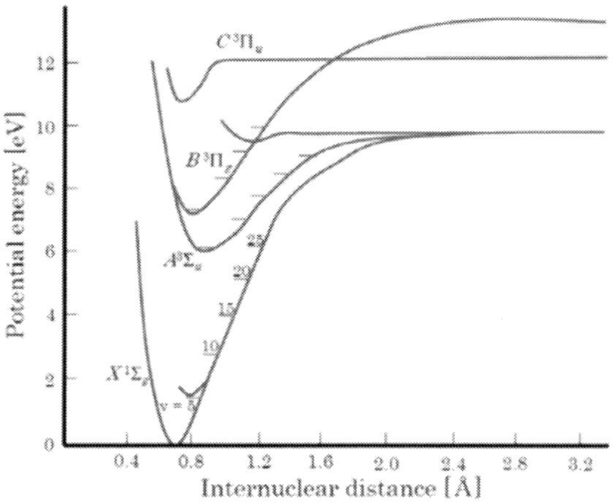

Figure 16. Energy diagram of nitrogen molecule [25].

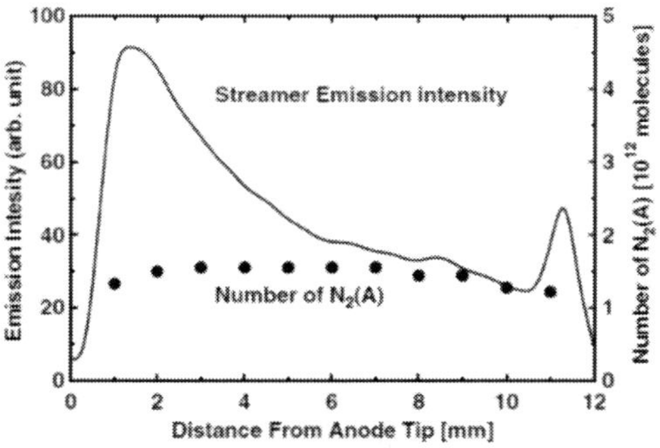

Figure 17. Axial distribution of N_2(A) metastable observed by LIF from the needle point to the plate with the light illumination intensity.

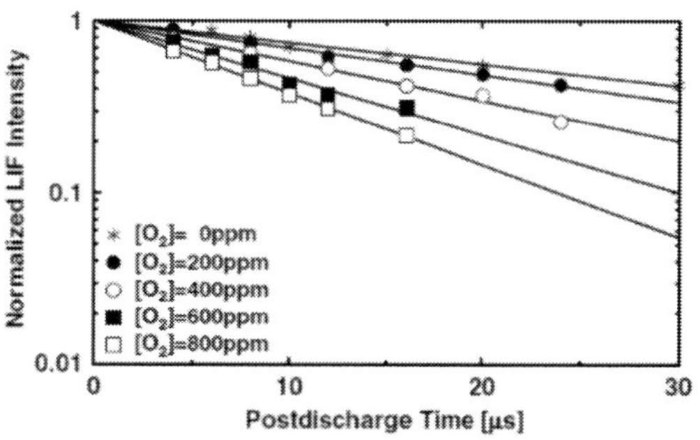

Figure 18. Deca plot of N2(A) at 6 mm below the needle point.

4. Raman Scattering

If the photon collides the molecule, most photon is scattered elastically but a little bit photon is scattered inelastically (energy loss or gain). Raman scattering is one of that. Photon energy loss or gain is due to the molecular quantum mode (vibration of plural atoms). The merit of Raman scattering is that does not need narrow special photon energy but short wavelength is preferable because the collision cross section is proportional to 4 powers of frequency. Sensitivity is not so large and detection is not so easy. Moreover, Raman scattering does not occur for single atom molecules (no atom vibration mode). Raman shift (shift of inelastic energy loss) is dependent not only the molecule but also their mode which means exciting state can also measurable. The authors observed Raman for hydrogen molecules and water molecules after the discharge explosion with time.

Recently the authors also succeed to observe CARS (coherent-anti-stokes Raman scattering) which needs at least two laser beams as shown in figure 19. Higher vibrating mode density of nitrogen was observed by this measurement after the pulse plasma.

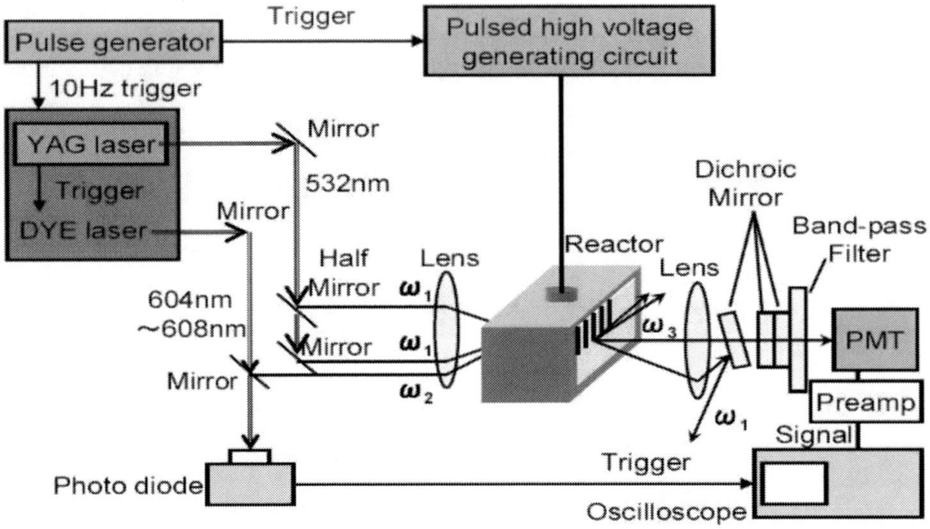

Figure 19. CARS experimental setup.

5. Conclusions

Some of our recent activities on high pressure plasma by using laser technology are demonstrated in this paper and recent other results will be discussed at the conference.

In air plasma condition, atomic nitrogen, oxygen and OH radical behaviours are a little bit understood by using the laser diagnosis; especially laser-induced-fluorescence technique is very effective to observe many radicals at any timing and temperature or density can be estimated by LIF only. It was surprising that

(a) Ozone is mostly generated not in the primary streamer but in the secondary streamer;

(b) Decomposition of NO and other reactions, atomic nitrogen works so much and density of metastable N_2 (A) is rather small in our experiment.

Acknowledgments

We appreciate Dr. Y.Teramoto and other students for their large contribution of N measurement.

References

[1] Ohtsuka K, Yukitake T, and Shimoda M 1985 *J.Inst.Electrost.Jpn.* **9** 346 (in Japanese).

[2] Masuda S and Nakao H 1986 *Conf. Records of the IEEE/IAS Annual Meeting,* (Denver, CO) p 1173

[3] Mizuno A and Clements Davis R H 1984 *Conf. Records of the IEEE/IAS Annual Meeting*, p 1015

[4] Chang J S, Lawless P A and Yamamoto T 1991 *IEEE Trans. Plasma Sci.* **19** 1152
[5] Ohkubo T, Kanazzawa S, Nomoto N, TAdachi Yan T K, and Chang J S, 1997 *Proc. of the second Non-Thermal Plasma Technol. and Gaseous Pollut. of Control* (Salvador) p 70
[6] Shimizu K, and Oda T 2002 *J.Adv.Oxid.Technol.* **5** 217
[7] Yoshioka Y, Sano K and Teshima K 2003 *J.Adv.Oxid.Technol.* **6** 143
[8] Yamamoto T, Ramanathan K, Lawless P A, Ensor D S and Newsome J R 1992 *IEEE Trans. Ind. Appl.* **28** 528
[9] Oda T, Takahashi T, Nakano H, Masuda S 1993 *IEEE Trans. Ind. Appl.* **29** 787
[10] Ogata A, Shintani N, Mizuno K and Kushiyama S 1998 *Proc. of the Asia-Pacific Workshop on Adv. in Oxid. Technol.* (Tsukuba) p 91
[11] Einagar H, Ibusuki T and Futamura S 2000 *IEEE Trans. Ind. Appl.* **37** 1476
[12] Oda T, Takahashi T and Yamaji K 2004 *IEEE Trans. Ind. Appl.* **40** 1249
[13] Oda T, Kuramochi H, and Ono R, 2008 *Proc.11th. ICESP* p 638
[14] Oda T, 2003 *J.Electrost.* **57** 293
[15] Ono R and Oda T 2004 *J. Phys. D,* **37** 730
[16] Ono R and Oda T; 2003 *J. Appl. Phys.* **93** 5876
[17] Ono R and Oda T 2008 *J. Phys. D* **41** 5204
[18] Ono R, Takezawa K and Oda T, 2009 *J. Appl. Phys.* **106** 3302
[19] Ono R, Teramoto Y, and Oda T 2009 *Jpn. J. Appl. Phys.* **48** 122302
[20] Teramoto Y, Ono R and Oda T 2009 *J. Phys. D* **42** 235205
[21] Ono R and Oda T 2007 *Int. J. Plasma Environ. Sci. Technol.***1** 123
[22] Ono R, Teramoto Y, and Oda T 2010 *Plasma Sources Sci. Technol.* **19** 015009
[23] Tanaka K, Teramoto Y, Ono R and Oda T 2011 *Annual Spring Meeting IESJ*, p 41 (in Japanese)
[24] Nakagawa Y, Ono R, and Oda T 2011 *J .Appl. Phys.* **110** 073304
[25] Makabe T ed. 1999 *Plasma Electronics* (Baihukan, Tokyo) (in Japanese)

Journal of Physics: Conference Series **418** (2013) 012102 doi:10.1088/1742-6596/418/1/012102

Application of chemical dosimetry to hydroxyl radical measurement during underwater discharge

S Kanazawa*, T Furuki, T Nakaji, S Akamine and R Ichiki
Department of Electrical and Electronic Engineering, Oita University, 700 Dannoharu, Oita, 870-1192, Japan

E-mail: skana@oita-u.ac.jp

Abstract. The pulsed streamer discharge was generated in the liquid by using point-to-sphere electrode geometry. In this study, the OH radicals were measured indirectly by a chemical dosimetry. Terephthalic acid (TA) was used as an OH radical scavenger and fluorescence of the resulted 2-hydroxyterephthalic acid (HTA) was measured. Instead of the use of conventional fluorescence spectrophotometer, a novel fluorescence observing system by using a light-emitting diode (LED) as a light source and a simplified spectrometer as a detector was developed. As a result, the fluorescence intensity increased with time elapsed during the discharge process. The production rate of OH radicals in the liquid was estimated to be of the order of 10^{-10} M s^{-1} to 10^{-9} M s^{-1} under the present experimental conditions. As an advantage of terephthalate dosimetry, it was found that the discharge system in the small reactor used in our experiment was a useful means for *in-situ* evaluating formation of OH radicals in an underwater discharge.

1. Introduction

The discharges in liquids or on the liquid surfaces are of great interest both from the point of view of discharge phenomena and their various applications [1-4]. A dynamics of streamers in liquid has been investigated by many researchers [5-9]. It was found that the streamers in liquid was composed of primary streamer with the velocity in the range of 100 m s^{-1}- a few km s^{-1}, followed by secondary streamer with the velocity greater than 10 km s^{-1} when positive pulsed high voltage was applied to the needle electrode. Streamers in liquid generate ultraviolet rays, shock waves, microbubbles and reactive oxygen species (ROS) such as the hydroxyl radical (OH) and free radical oxygen. Especially, the OH radical plays an important role in plasma chemistry and plasma medicine due to a higher oxidation potential and stronger disinfection power compared to other oxidative species. The identification of OH radicals in water has been observed mainly by optical emission spectroscopy (OES) [10, 11]. Recently, the optical absorption technique has been employed to investigate the density of the OH radicals in liquid solutions [12]. However, a precise measurement of OH radical density in liquid is still in a line of research.

In this study, the OH radicals were measured indirectly by a chemical dosimetry. Terephthalic acid (TA) was used as an OH radical scavenger and fluorescence of the resulted 2-hydroxyterephthalic acid (HTA) was measured. Instead of the use of conventional fluorescence measurement equipment, a novel fluorescence observing system by using a light-emitting diode (LED) as a light source and a

* To whom any correspondence should be addressed.

simplified spectrometer as a detector was developed. This paper also describes the merit of this method compared to other detection methods.

2. Experimental

2.1. Underwater discharge system

Figure 1 shows schematic diagram of the experimental setup. A pulsed high-voltage circuit with a self-trigger spark gap switch was used to generate streamers in liquid. A needle-to-sphere discharge electrode system was inserted into terephthalic acid (TA) solution, filling a cylindrical silica-glass tube reactor (20 mm in inner diameter and 23.8 mm in outer diameter). A stainless-steel needle (0.14 mm in inner diameter and 0.32 mm in outer diameter) with an insulating cover was used as the stressed electrode, while a stainless-steel sphere (15 mm in diameter) was used as the grounded electrode. The gap distance between the needle and sphere electrodes was set to be 20 mm.

The applied voltage and the current were measured by a high voltage probe (Iwatsu, HV-P30) and a current probe (Pearson Electronics, 2877), respectively. A spectrometer (Ocean Optics, USB2000) was used to measure the emission spectra of the streamers in liquid.

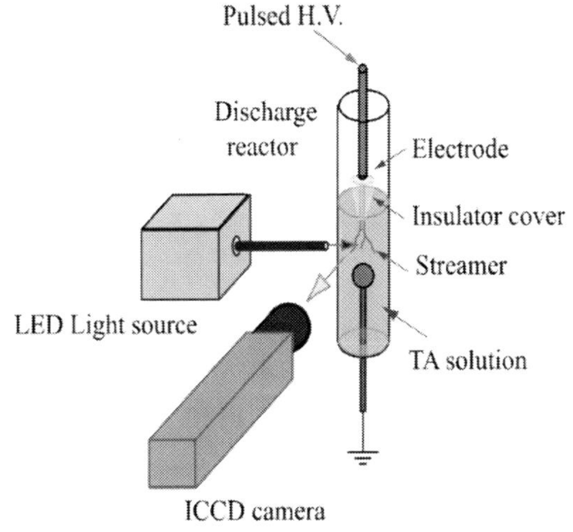

Figure 1. Schematic of experimental setup.

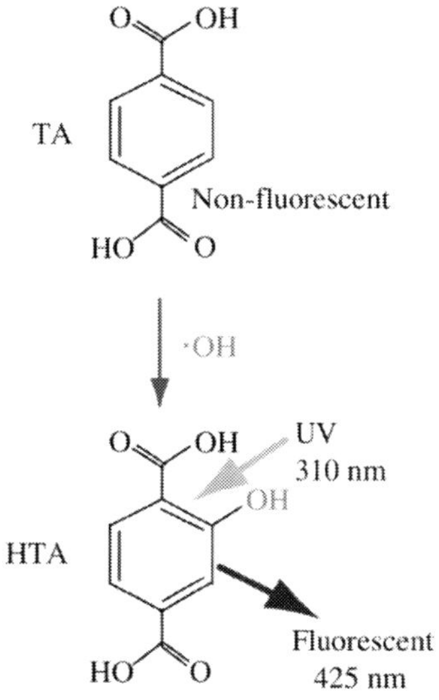

Figure 2. Formation of HTA through the reaction of TA and OH radical.

2.2. Chemical dosimetry

In order to measure the OH radicals dissolved in the liquid, we used chemical dosimetry [13] based on terephthalic acid (TA). Terephthalic acid is a well known OH scavenger which does not react with other radicals, such as O_2^-, HO_2 and H_2O_2. As shown in figure 2, the OH radical can convert terephthalic acid to 2-hydroxyterephthalic acid (HTA), which can be detected by fluorescence measurement. When the solution containing TA and HTA molecules is irradiated by UV light (λ=310 nm), HTA molecules emit light at λ=425 nm, while TA molecules do not. The fluorescence intensity of HTA is independent of pH in the range of 6-11. Since TA (Aldrich) does not dissolve in acidic/neutral liquid, aqueous solution of TA was prepared by dissolving TA in the distilled water containing NaOH (Wako). The initial concentrations of TA and NaOH were 2 and 5 mM except for the case of fluorescence imaging, respectively. The initial values of pH and conductivity of the solution were 10 and 368 μS cm^{-1}, respectively. The solution volume in the cylindrical silica-glass tube reactor was ca. 33 mL. The LED (Sandhouse Design, λ=310 nm, FWHM 10 nm) was used as a light source to excite HTA. During the discharge, a collimated beam from the LED output is passed through the liquid in the reactor, just below the needle electrode. The fluorescence (λ=425 nm, center wavelength) image was captured by an ICCD camera (Andor, i-Star) at various time intervals during the process, liquid was sampled into a cuvette and fluorescence spectrum around λ=425 nm was recorded through an optical fiber by the spectrometer (Figure.1). In order to quantify the OH radical concentration in the liquid, a calibration curve for known OH radicals concentrations was prepared using the standard HTA (Atlantic Research) solution. Typical calibration curve is shown in figure 3. The linear relationship between the peak area of fluorescence spectrum and the concentration of HTA was obtained.

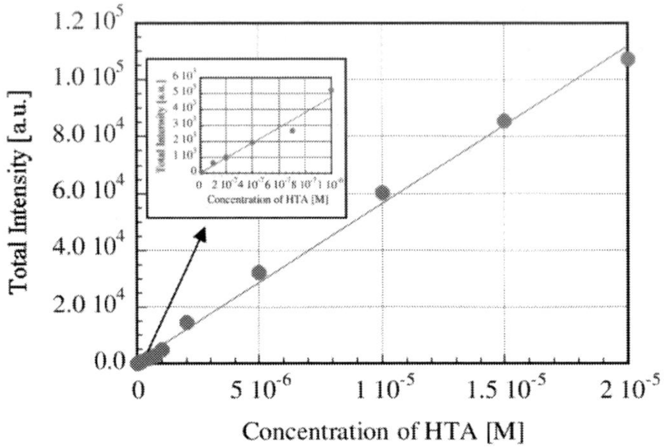

Figure 3. Calibration curve for hydroxyterephthalic acid (HTA).

3. Results and discussion

Typical voltage and current waveforms for the discharge in TA solution are shown in figure 4. The rise time of the pulsed high voltage used in this study is about 70 ns and pulsewidth is about 10 μs. After a displacement current appears at the rising edge of the pulsed voltage, the discharge current with a 6 A peak and 5 μs pulsewidth flows. The average of pulse repetition rate is about 10 pps. Since the solution conductivity is relatively high, a streamer discharge consists of a few filaments with the length less than 5 mm as shown in figure 5.

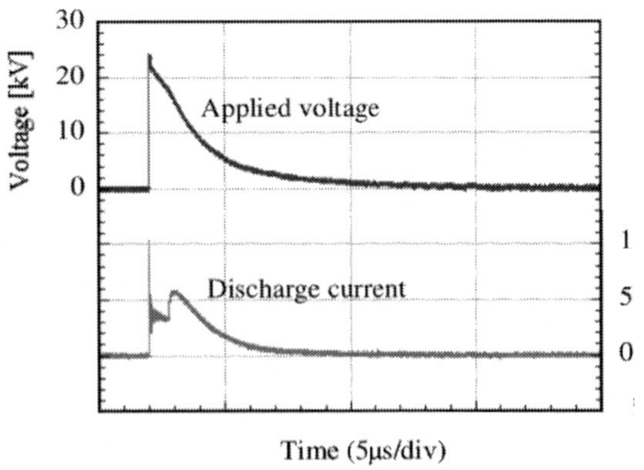

Figure 4. Applied voltage and current waveforms for the discharge in TA solution.

Figure 5. Streamer in TA solution. (Applied voltage: 24.5 kV)

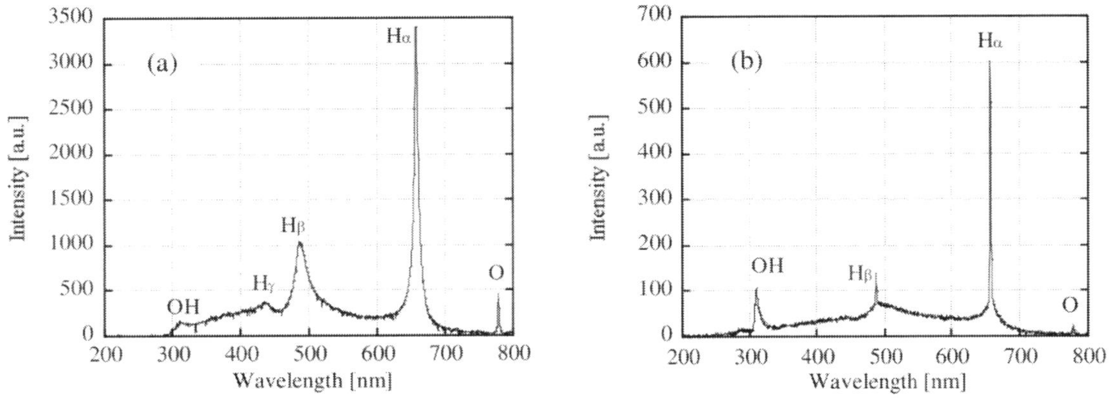

Figure 6. Optical emission spectrum of underwater discharge: (a) discharge in TA solution and (b) discharge in distilled water. (Applied voltage: 24.5 kV)

Figure 6 shows the typical emission spectra of the streamers in TA solution and in distilled water. In both spectra, we found the emissions of hydrogen atomic lines (Hα, Hβ, Hγ) and OH radicals from the excited state ($A^2\Sigma^+$). Also the atomic line of oxygen (OI) was found. A strong emission was observed for the discharge in TA solution. This is due to the higher conductivity of TA solution. The line broadening of hydrogen atomic lines such as Hα, Hβ indicates higher electron density plasma for underwater discharge, especially in the case of TA solution, compared to other discharge-induced plasmas in contact with liquids.

Figure 7 shows time-dependent fluorescence images under the illumination of the LED light source during the underwater discharge. Each image is an accumulation of 20 captured images. OH radicals are formed along the streamer propagation and trapped by TA molecules. The intensity of the fluorescence due to the HTA formation increased with time elapsed and its part is almost uniform even though no stirring of the liquid was performed, indicating homogeneous HTA diffusion in the solution. The light scatterings by the needle electrode and reactor wall are also observed. For the case of the discharge operating longer than 5 min the fluorescence can be observed by our naked eye. Furthermore, the possibility of the decomposition of HTA by the underwater discharge was investigated. Using a HTA solution we confirmed that HTA decomposition is negligible under the present experimental condition.

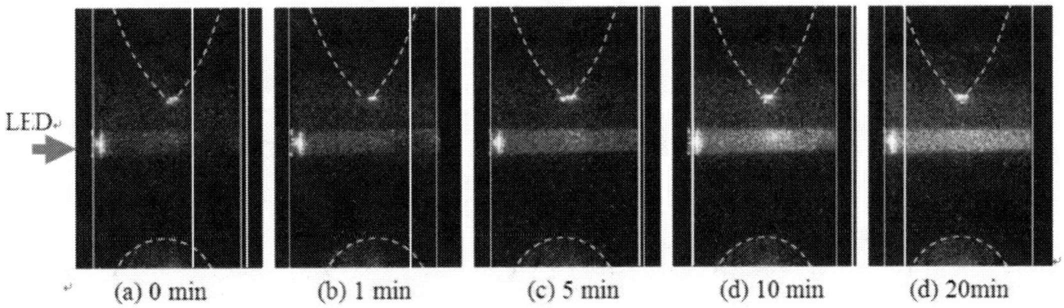

Figure 7. Time evolution of the fluorescence from TA solution during underwater discharge.(Applied voltage: 24.5 kV, TA solution: 0.2 mM, pH 9.6)

Figure 8 shows the fluorescence spectra as a function of treatment time. The fluorescence intensity corresponds to a time integrated OH radical concentration in the liquid. As the time elapsed, the fluorescence intensity increased, indicating the increase of the total amount of OH radicals trapped by TA. In order to evaluate the density of OH radicals in the solution, using the fluorescence intensity integrated over the wavelength (shown in figure 8) and the calibration curve (figure 3) for known concentrations of OH radicals, we calculated the OH radical density in the solution as a function of treatment time as shown in figure 9. The concentration of OH radicals in the reactor almost linearly increases with increasing the time. The tendency of saturation is probably related to the decreasing TA in the liquid. Besides trapping of OH by TA, however, other reactions that consume OH radicals also occurred. In the present case, the HTA yield was assumed to be 35%, according to [14]. As a result, the production rate of OH radicals in the liquid was estimated to be of the order of 10^{-10} M s^{-1} to 10^{-9} M s^{-1} under the present experimental condition.

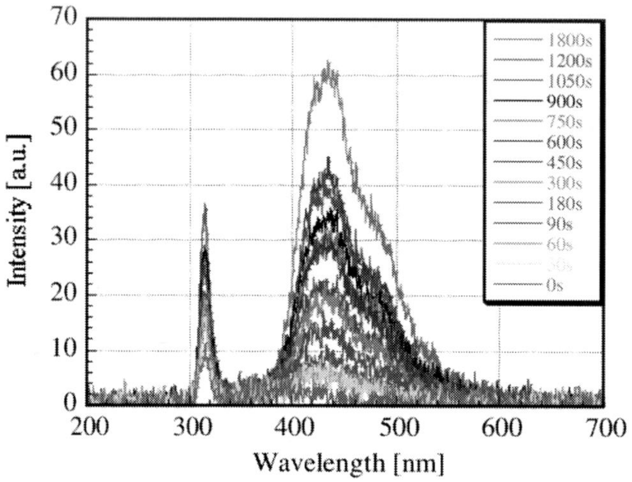

Figure 8. Fluorescence spectra of aqueous TA solutions irradiated by underwater discharge.(Applied voltage: 24.5 kV, TA solution: 2 mM, pH 10.16, volume 32.5 mL)

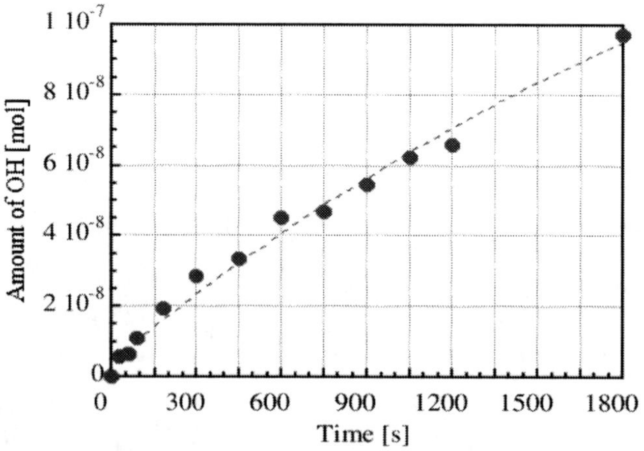

Figure 9. Formation of OH radicals in aqueous TA solution as a function of treatment time.(Applied voltage: 24.5 kV, TA solution: 2 mM, pH 10.16, volume 32.5 mL)

Under our another experiment, the rates of production of OH radicals are of the order of 10^{-9} M s^{-1} for the surface pulsed streamer discharge on the liquid [15] and of the order of 10^{-8} M s^{-1} for the plasma jet on the liquid [16], respectively. Sahni et al. [17] have reported that the production rate of

OH radicals was 1.67×10^{-8} M s^{-1} for the direct discharge in water at an applied voltage of 45 kV and input power delivered to the water of 64 W. Joshi et al. [18] have determined the rate of formation of OH radicals using the free radical scavenging property of carbonate ions. They reported the value of 9.25×10^{-10} Ms^{-1} for the pulsed streamer corona discharge in an aqueous solution. Recently, Li et al. [19] have estimated OH production rate of the order of 10^{-10} Ms^{-1} for pulsed discharge in water by using a salicylic acid dosimetry. At the present stage, we attribute these differences to various factors, such as discharge types, reactor size, operating conditions, and different measuring methods.

Finally, other chemical dosimetries are reviewed in connection with TA dosimetry. There are various dosimetries which might be used to monitor the OH radicals. The Fricke dosimetry and iodide dosimetry are widely used in sonochemistry and radiation chemistry. However, Fricke or iodide dosimeters are non-specific since iodide and Fe (II) ions are oxidized not only by OH radicals but also by the H atom, HO_2 radical, and H_2O_2. The salicylic dosimetry based on salicylic acid is similar method to TA dosimetry. In recent years, this method is used to measure the OH radicals generated by the discharges in water [19] or in air [20]. In this case, liquid chromatography was used to detect fluorescent substances. Moreover, the electron spin resonance (ESR) spin-trapping has been also recognized as the radical monitoring technique with high sensitivity and lower detection limits [21]. The ESR requires an expensive instrument and a skillful operator. Taking into account the sensitivity and simplicity in data collection, TA dosimetry seems to be a suitable method for characterizing the discharge-induced OH radicals.

4. Conclusions

A novel method of measuring OH radicals in aqueous solution was developed and applied to the measurement for underwater discharges. In the case of terephthalate dosimetry, the apparatus is simple and cheap, especially chemical reagent of terephthalic acid, compared to other dosimetries. The time-integrated amount of OH radicals in the reactor was visualized by using fluorescence of TA solution irradiated by the UV LED. The production rate of OH radicals in the liquid was estimated to be of the order of 10^{-10} M s^{-1} to 10^{-9}M s^{-1} under the present experimental condition.

Acknowledgments

This study was partly supported by the Japan Society for the Promotion of Science, Grant-in-Aid for Scientific Research (No.23360127).

References

[1] Akiyama H 2000 *IEEE Trans. Dielectr. Electr. Insul.* **7** 646
[2] Locke B.R, Sato M, Sunka P, Hoffmann M.R and Chang J.-S. 2006 *Ind. Eng. Chem. Res.* **45** 882
[3] Kolb J.F, Joshi R.P, Xiao S and Schoenbach K H 2008 *J. Phys. D: Appl. Phys.* **41** 234007
[4] Bruggeman P and Leys C 2009 *J. Phys. D: Appl. Phys.* **42** 053001
[5] An W, Baumung K and Bluhm H 2007 *J. Appl. Phys.* **101** 053302
[6] Kanazawa S, Abe Y, Kihara Y, Kocik M and Mizeraczyk J 2008 *IEEE Trans. Plasma Sci.* **36** 922
[7] Ceccato P. H, Guaitella O, Gloahec M. R. L and Rousseau A 2010 *J. Phys. D: Appl. Phys.* **43** 175202
[8] Kanazawa S, Ichihashi Y, Watanabe S, Akamine S, Ichiki R, Ohkubo T, Sato T, Kocik M and Mizeraczyk J 2012 *Intl. J. Plasma Environ. Sci. Technol.* **6** 49
[9] Wen X. Q, Liu G. S and Ding Z. F 2012 *IEEE Trans. Plasma Sci.* **40** 438
[10] Sun B, Sato M, Clements J.S 1997 *J. Electrostat.* **39** 189
[11] Bruggeman P, Verreycken T, Gonzalez M.A, Walsh J L, Kong M G, Ley C, Schram D C 2010 *J. Phys. D: Appl. Phys.* **43** 124005
[12] Miron C, Bratescu M.A, Saito N and Takai O 2011 *J. Appl. Phys.* **109** 123301
[13] Mason T J, Lorimer J P, Bates D M, Zhao Y 1994 *Ultrason. Sonochem.* **1** S91
[14] Fang X, Mark G, and Sonntag C. V 1996 *Ultrason. Sonochem.* **3** 57

[15] Kanazawa S, Kawano H, Watanabe S, Furuki T, Akamine S, Ichiki R, Ohkubo T., Kocik M. and Mizeraczyk J 2011 *Plasma Sources Sci. Technol.* **20** 034010
[16] Kanazawa S, Furuki T, Nakaji T, Akamine S and Ichiki R 2012 *Electrostat. Joint Conf.* (Cambridge, Canada)
[17] Sahni M and Locke B. R 2006 *Ind. Eng. Chem. Res.* **45** 5819
[18] Joshi A. A, Locke B. R, Arce P and Finney W. C 1995 *J. Hazard. Mater.* **41** 3
[19] Li S, Hu S and Zhang H 2012 *IEEE Trans. Plasma Sci.* **40** 63
[20] Liao X, Guo Y, He J, Ou W and Ye D 2010 *Plasma Chem. Plasma Process* **30** 841
[21] Tahara M and Okubo M 2012 *Electrostat. Joint Conf.* (Cambridge, Canada)

Diagnosis of hydroxyl radicals and the treatment of Acid Orange Π in negative pulsed discharge

M Sun*, **Y Zhang, L J Cai, H Y Wang**
Institute of Electrostatics, Shanghai Maritime University, Shanghai, China

E-mail: mingsun@shmtu.edu.cn

Abstract. A kind of device supplied with negative pulsed high voltage was designed for water treatment. The nozzle discharge electrode was supplied with high voltage and the cylinder wall of the treatment chamber was grounded. Since the non-thermal plasma, especially, hydroxyl radicals played the dominant role in removing pollutants in the water by discharge plasma, hydroxyl radicals were measured in the negative pulsed discharge process by the optical emission spectroscopy (OES) effects of the repetition rate and the peak value of the pulsed discharge on the hydroxyl radicals' production and the decomposition rate of the Acid Orange Π solution were studied.

1. Introduction

The application of pulsed discharge technology in the removal of pollutant in the air or in the water has been widely studied [1, 2]. It is well known active species, such as OH, HO_2, O and O_2* play a dominant role to remove the contamination during the discharge plasma technology. Especially, hydroxyl radicals are the most significant active species on decoloration or different kinds of pollutants' removal for its very strong oxidation. So it is very useful for optimizing the device to measure the hydroxyl radicals' production in different experiment conditions. In this paper, the structure of nozzle-cylinder electrode supplied with negative pulsed high voltage was adopted. The effects of the repetition rate of the pulsed discharge, the peak value of pulsed discharge voltage on the hydroxyl radicals' production and the decomposition rate of the Acid Orange Π solution were studied.

2. Experimental set-up

The experiment system consists of power supplied system, treatment chamber, discharge voltage and discharge current monitor system etc. The structure of treatment chamber is shown in figure 1. Discharge electrode is a single nozzle supplied with negative pulsed high voltage. A steel cylinder treatment chamber is grounded. The space between the nozzle and the wall of the chamber grounded is 35 mm. Air is injected into chamber though the nozzle discharge electrode. Observation of the discharge phenomenon and the measurement of the hydroxyl radicals were completed through the window for observing in the wall of the chamber.

3. Results and discussion

In air discharge, high energy electrons are produced when free electrons gain energy from the strong electric field. A great number of collisions happen between different kinds of particles for particles

* To whom any correspondence should be addressed.

thermal movement. During the collisions between high-energy electrons and the air molecules such as O_2, N_2 and H_2O, high-energy electrons transfer the energy to the molecules. Chemical bonds of these molecules are broken. Ions, substance in metastable state, atoms and radicals with strong oxidation such as OH, HO_2, O and O_2^* were produced. Some instable substance in excitated states, for an example, O (^1D), continues the reactions with other species and transfers the energy [3-6]:

$$e+O_2 \rightarrow O + O + e \quad (1)$$

$$e+O_2 \rightarrow O+O\ (^1D) + e \quad (2)$$

$$e+H_2O \rightarrow OH + H + e \quad (3)$$

$$e+N_2 \rightarrow N + N + e \quad (4)$$

$$O\ (^1D) + H_2O \rightarrow OH+OH \quad (5)$$

$$O\ (^1D) + H_2O \rightarrow O+H_2O \quad (6)$$

$$O\ (^1D) + O_2 \rightarrow O+O_2 \quad (7)$$

$$O\ (^1D) + N_2 \rightarrow O+N_2 \quad (8)$$

……

During these reactions, reactions (3) and (5) describe the formation mechanism of hydroxyl radicals. In Reaction (3), hydroxyl radicals were produced by the collision between the electrons and water molecules. To the reaction (5), hydroxyl radials were produced by the propagation reaction of O (^1D) radicals.

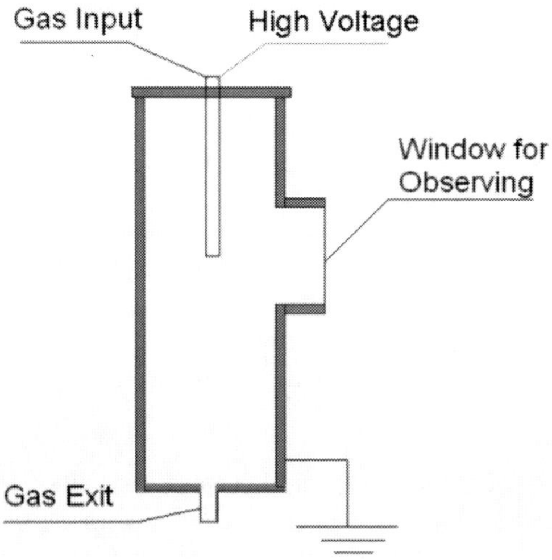

Figure 1. Schematic diagram of the treatment chamber.

In the paper, we want to learn about the effect of the different experiment conditions on the hydroxyl radicals' production in order to optimize our experiment setup in the future. The point cared is when more hydroxyl radicals produced, not the acute quantity. So, in this work, we paid attention to the relative intensity of hydroxyl radicals' OES, which is effected directly by the quantity of hydroxyl radicals.

3.1. The effect of peak value of the negative pulsed high voltage on the hydroxyl radicals' production

Hydroxyl radicals' OES was measured with the absolute peak value of the negative pulsed high voltage being from 18 kV to 34 kV, the outer diameter of nozzle electrode being 3 mm and the repetition rate of the pulsed discharge being 80 Hz. The OES of the hydroxyl radicals with different peak value of the pulsed high voltage is shown in figure 2.

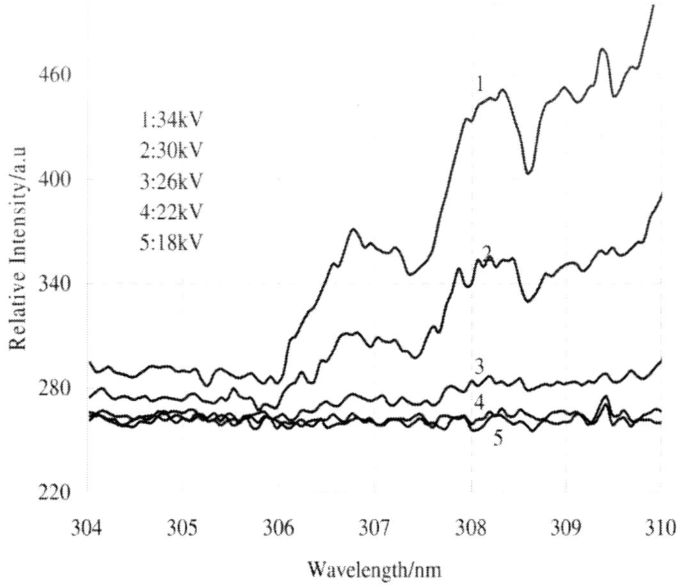

Figure 2. OES of hydroxyl radicals with different absolute peak value of pulsed discharge high voltage.

From figure 2, we can see the relative intensity of hydroxyl radicals' OES rises with the absolute peak value of the pulsed high voltage. The onset corona discharge should be near negative 26 kV, for the hydroxyl radicals can be detected at this peak value.

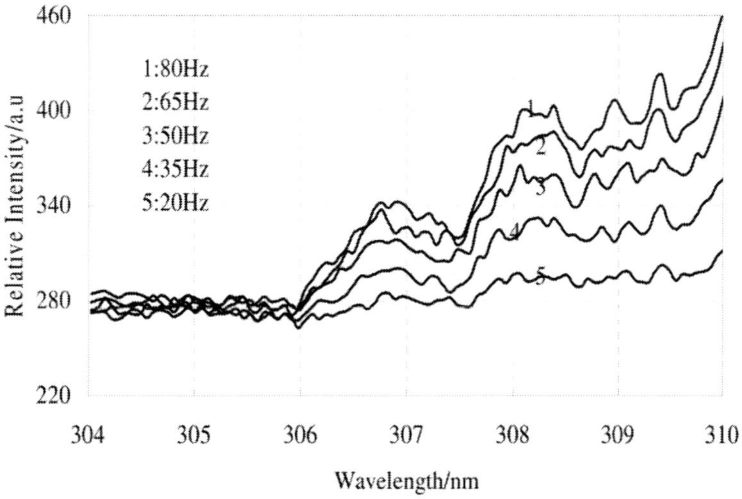

Figure 3. OES of hydroxyl radicals with different repetition rate of pulsed discharge.

3.2. The effect of the repetition rate of the pulsed discharge on the hydroxyl radicals' production

OES of hydroxyl radicals with the repetition rate of pulsed discharge at the range of 20 Hz to 80 Hz is shown in figure 3. In the experiments, the outer diameter of nozzle electrode was 3 mm. And the absolute peak value of pulsed high voltage was -28 kV.

From the figure 3, the conclusion can be drawn that the relative intensity of hydroxyl radicals' OES increases with the repetition rate of the pulsed high voltage. High energy electrons increase with the repetition rate of the pulsed high voltage supplied to the nozzle discharge electrode because the input energy increases.

3.3. The effect of the peak value of the pulsed discharge voltage on the decomposition rate of the Acid Orange Π solution

The effect of the peak value of the pulsed discharge voltage on the decomposition rate of the Acid Orange Π solution is shown in 'figure 4'. The experimental conditions were the discharge time, initial concentration of the solution, the pulsed repetition rate, solution volume and the electrode diameter being 20 min, 10 mg L^{-1}, 50 Hz, 460 mL and 3.5 mm, respectively.

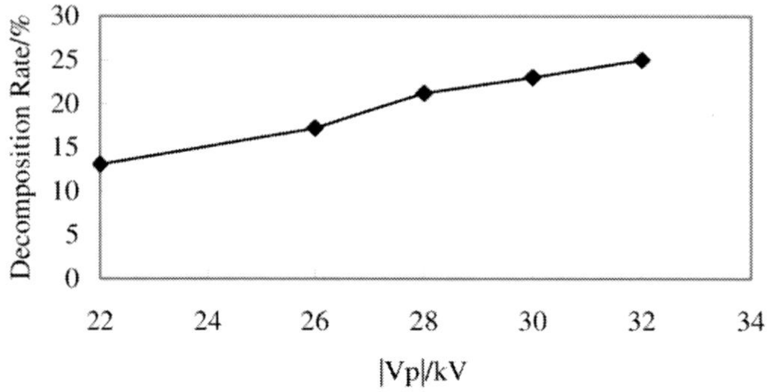

Figure 4. Decomposition rate of Acid Orange Π as a function of the peak value of the pulsed discharge voltage.

The result indicated that the decomposition rate of Acid Orange Π solution increased with the peak value of the pulsed discharge voltage. When the discharge voltage was - 32 kV, the decomposition rate was about 25%. The input energy increases with the discharge voltage, the increased input energy improving the number of active species which reacted with Acid Orange Π.

3.4. The effect of the pulsed repetition rate on the decomposition rate of Acid Orange Π solution

The decomposition rate of the Acid Orange Π solution as a function of the pulsed repetition rate (20 ~ 60 Hz) is shown in 'figure 5'. The experimental conditions were the discharge time, the peak value of the pulsed discharge voltage, initial concentration of the solution, solution volume and the electrode diameter being 20 min, -32 kV, 10 mg L^{-1}, 460 mL and 1.2 mm, respectively.

The result shows that the decomposition rate of Acid Orange II solution increased with the pulsed repetition rate. The decomposition rate improved from 12% to 28% with the pulsed repetition rate being 20 Hz and 60 Hz, respectively. The input energy increases with the pulsed repetition rate when the discharge voltage being constant. The increased input energy improves the number of active species which reacted with Acid Orange Π.

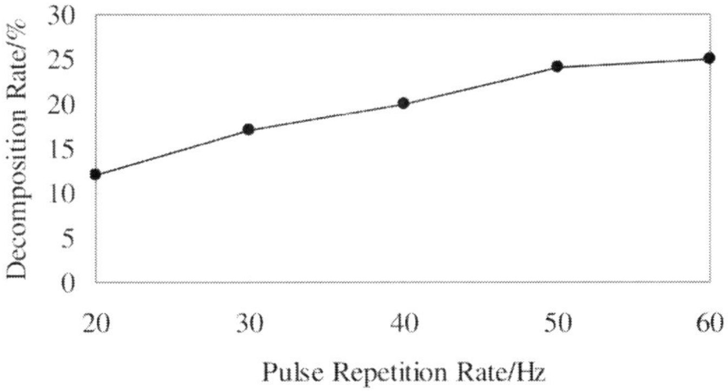

Figure 5. Decomposition rate of Acid Orange Π as a function of pulsed repetition rate.

4. Conclusions

Hydroxyl radicals were measured in air negative pulsed discharge with nozzle-cylinder electrodes by the OES. The effects of the repetition rate and the peak value of the pulsed discharge high voltage on the relative intensity of the hydroxyl radicals' OES and the decomposition rate of the Acid Orange Π solution were studied. Conclusions drawn from the result of the experiments are hydroxyl radicals' production and the decomposition rate of the Acid Orange Π solution increase with the absolute peak value and the repetition rate of the pulsed. The conclusions of this work will be helpful for the optimizing the experimental setup in the next step work of this project.

Acknowledgments

Authors thank Shanghai Maritime University (No. 20120097) for financial support.

References

[1] Chen C W, Lee H M and Chang M B 2009 *J. Electrostat.* **67** 703
[2] Zhang L, Sun B and Zhu X M 2009 *J. Electrostat.* **67** 62
[3] He Z G, Liu J S and Cai W M 2005 *J. Electrostat.* **63** 371
[4] Sun M, Wu Y, Li J, Wang N H, Wu J, Shang K F and Zhang J L 2005 *Plasma Chem. Plasma Proc.* **25** 31
[5] Mok Y S and Ham S W 1998 *Chem. Eng. Sci.* **53** 1667
[6] Lowke J J and Morrow R 1994 *Pure Appl. Chem.* **66** 1287

Detection of hydroxyl radicals during regeneration of granular activated carbon in dielectric barrier discharge plasma system

Shoufeng Tang, Na Lu, Kefeng Shang, Jie Li*, Yan Wu
Key Laboratory of Industrial Ecology and Environmental Engineering (MOE), Institute of Electrostatics and Special Power, Dalian University of Technology, Dalian 116024, PR China

E-mail: lijie@dlut.edu.cn

Abstract. To understand the reactions taking place in the dielectric barrier discharge (DBD) plasma system of activated carbon regeneration, the determination of active species is necessary. A method based on High Performance Liquid Chromatography with radical trapping by salicylic acid, has been developed to measure hydroxyl radical (•OH) in the DBD plasma reactor. The effects of applied voltage, treatment time, and gas flow rate and atmosphere were investigated. Experimental results indicated that increasing voltage, treatment time and air flow rate could enhance the formation of •OH. Oxygen atmosphere and a suitable GAC water content were contributed to •OH generation. The results give an insight into plasma chemical processes, and can be helpful to optimize the design and application for the plasma system.

1. Introduction

To understand the reaction mechanisms taking place in the dielectric barrier discharge (DBD) plasma system of activated carbon regeneration, the determination of active species is necessary [1]. It gives a perception into plasma chemical processes, and can be helpful to optimize the design and application for the plasma system.

The hydroxyl radical (•OH) is an important reactive agent in various plasma applications. It has extremely high oxidation potential (2.8 V), which is only lower than that of F_2 (3.0 V) [2]. The existence of •OH in the aquatic systems and the atmospheric phase has been universally believed [3]. •OH is also believed to be the key reactive species for the oxidation of environmental contaminants in the plasma system [4].

---

* To whom any correspondence should be addressed.

To analyze the mechanism of pollutants removal in plasma reactions, a diagnosis technique of the atmospheric pressure non-thermal plasma process is very important to develop in practice [1]. One of the plasma diagnoses is the detection of radicals and active species produced in the atmospherical pressure plasma. Though the importance of •OH in the pollutants' decomposition by plasma have been affirmed, •OH measurement techniques have long been a difficulty due to its short lifetime. In recent studies, several methods have been developed to detect •OH in non-thermal plasma systems, including electron spin resonance (ESR) [5], laser-induced fluorescence (LIF) [6], colorimetric spectrophotometry [7], a radiocarbon method [8], and quantitative analysis of •OH by oxidating CO to CO_2 [9]. However, these reported methods have lots of limitations for actives species determination.

Because •OH has an extremely short lifetime ($<10^{-3}$ s), it is difficult to determine it directly. In this paper, salicylic acid (SA, o-hydroxybenzoic acid) was chosen as the molecular probe to capture •OH. When •OH exists in the solution, it will rapidly react with SA and produce 2, 3-dihydroxybenzoic acid (2, 3-DHBA) or 2, 5-DHBA, which can be determined by High Performance Liquid Chromatography (HPLC). Jen [10] first applied the HPLC method to determine •OH concentration in an advanced oxidation process, which proves to be a simple and sensitive way as the measurement of •OH in a biological system [11]. Guo and Liao [12, 13] employed the HPLC method to detect •OH in non thermal plasma system.

Our previous bench-scale work reported that using the DBD reactor to regenerate granular activated carbon (GAC) exhausted with organic pollutants, and the experiments yielded desirable results. However, there is little information available in literature about generation of active species in the DBD regeneration reactor, and the pollutants' degradation mechanisms have not been fully understood. The present work aimed to detect the generation of •OH by HPLC method in DBD regeneration system, and to study the effect of applied voltage, treatment time, gas flow rate, gas type and water content of GAC on •OH production.

2. Experimental

2.1 DBD plasma system

The schematic diagram of the DBD plasma regeneration system was shown in figure 1. It mainly consisted of alternating current high-voltage power supply and the DBD reactor. The frequency of high-voltage power of this experiment was 1000 Hz. The voltage was 0-50 kV adjustable. The DBD reactor (as shown in figure 1) was comprised of two parallel-plate stainless steel electrodes ($\varphi 45\times 2$ mm) as the high-voltage and ground electrode respectively, and each electrode was covered by one quartz glass dielectric ($150\times 150\times 2$ mm). The distance between two dielectrics was 6 mm. One Plexiglas chamber ($\varphi 40\times 6$ mm), between the two dielectrics, was used as the GAC packed bed. A Plexiglas TM cylinder ($\varphi 4$ mm) was embedded in GAC to provide air as reaction gas source for the DBD reactor. In each experiment, adsorbed GAC (approximately 2.8 g, moisture content 10 %) was introduced into the reactor and evenly filled up the packed bed, and then the DBD process began when the AC power supply was turned on.

2.2 Materials

The GAC (approximately 2.0 mm diameter and 1.0-5.0 mm length) used in this study was columned coal-based carbon and was manufactured by Shenyang Chemical Reagent Factory, China. Prior to use, the GAC was immersed in deionized water for 24 h, and then liberally rinsed with deionized water to remove fine particles and impurities, at last dried in an oven at 378 K overnight and then kept in a desiccator. All chemical reagents used in this study were analytical grade reagents supplied by Tianjing Beilian Fine Chemical Product Co. Ltd, China.

Salicylic acid (SA, 99.5 % purity) was purchased from Shanghai Chemical Reagent Company, China. 2, 3-DHBA (Sigma) and 2, 5-DHBA (Sigma) were prepared as standard solution according to desired concentration.

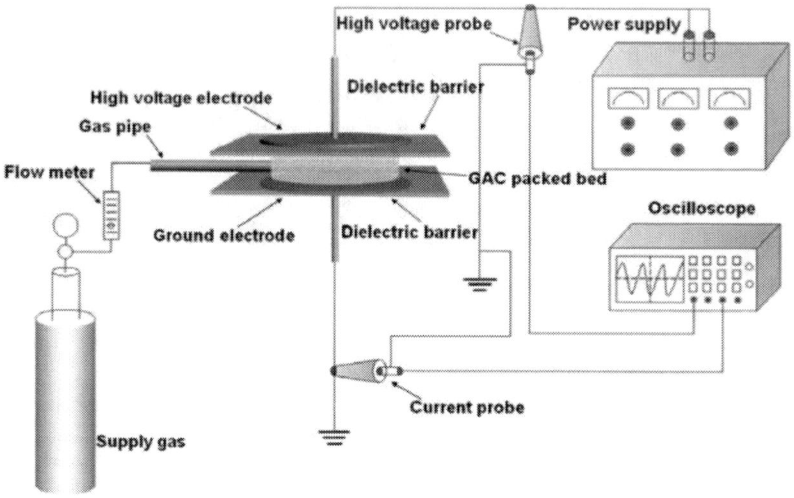

Figure 1. Schematics of experimental setup.

2.3 Experimental procedures and analysis

For •OH determination, the aqueous solution of SA with an initial concentration of 5000 mg L^{-1} for experiments was prepared from deionized water and SA reagent. Then, 20 g of GAC were added into the sealed flasks with SA solution and placed in the water-bath shaker with an agitation speed in the range of 150 rpm to adsorb SA on GAC. The adsorption capacity (q, mg g^{-1}) was calculated using the following equation:

$$q = (C_o - C_t)\, V/m \quad (1)$$

Where q (mg g^{-1}) is the adsorption capacity of GAC, C_o (mg L^{-1}) is the initial concentration of SA in aqueous solution, C_t (mg L^{-1}) is SA concentration in aqueous phase at time t, V (L) is the solution volume and m (g) is the mass of the adsorbent. After that, the adsorbed SA GAC were dried in an oven under very mild condition (313 K) until a constant certain mass was achieved. After that, a definite mass of prepared GAC (approximately 2.8 g) was introduced into the DBD reactor and was evenly spread over the packed bed. The DBD process began when the AC power supply was turned on, and the regeneration would last for different experimental conditions.

During the DBD process, the adsorbed SA on GAC would be reacted with •OH generated in the discharge space. After each discharge experiment, the SA residues on GAC samples were extracted in a conical flask using a 50 mL ethanol solution. The conical flask was kept shaking at 150 rpm at 298

K for 60 min, and then ultrasonic extracting for 30 min. The extraction procedures were repeated for three times, and the extract was collected finally. The recoveries were about 85 % for SA on GAC samples in this study. The determination of SA concentrations in solution were performed by a High Performance Liquid Chromatography (LC-10Avp Shimadzu) using a Hypersil ODS (25 μm, 4.6×250 mm) reverse phase column. The DHBA yields represented hydroxyl radicals' concentration.

3. Results and discussion

3.1 Effect of applied voltage on •OH production

Discharge voltage has a significant effect on the generation of plasma (such as active radicals, high-energy electrons and ultraviolet light, etc.) [14]. Therefore, the discharge voltage was regarded as one of the most important parameters for the degradation of pollutants loaded on GAC and regeneration of GAC. The effect of applied voltage on •OH production was determined, and the results were revealed in figure 2. The other operation conditions were as follows: air flow rate 0.8 L min^{-1} and water content of GAC 10 %. Three levels, 18, 23, 28 kV, were investigated. It was found that •OH production increased with the applied voltage, At pulse voltage of 28 kV, about 47.1 μmol DHBA was generated within 100 min of treatment, while approximately 46.6 and 44.8 μmol DHBA could be measured at voltages of 23 and 18 kV within the same time, respectively. It is generally believed that more energetic electrons are produced at higher voltage, leading to the accelerated formation of active species. Furthermore, •OH radicals and hydrogen peroxide can be generated by DBD due to the humidity on GAC samples [15]. On the other hand, ozone reacts with hydrogen peroxide and finally readily dissociates into •OH radical [16]. These active radicals are strong oxidants, which play a significant role in contaminates' degradation [17]. Hence, the increase in applied voltage contributes to the increase in active species, which in turn enhances the degradation of pollutants adsorbed on GAC.

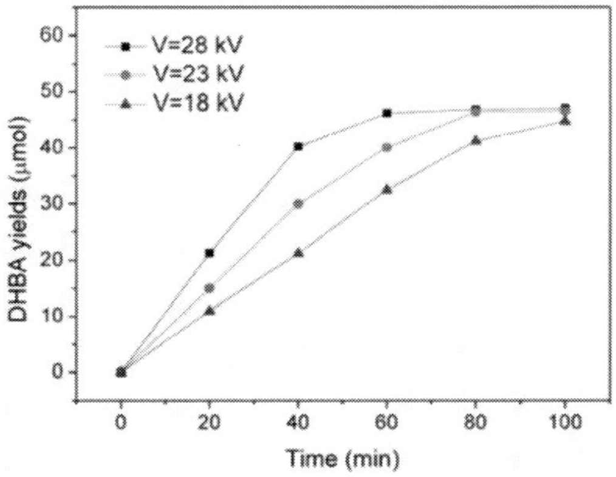

Figure 2. Effect of applied voltage on •OH production.

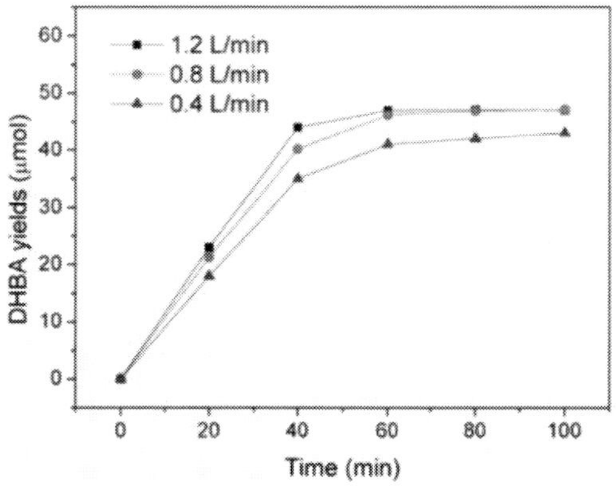

Figure 3. Effect of air flow rate on •OH production.

3.2 Effect of gas flow rate on •OH production

The effect of air flow rate on DHBA yields was studied in this part, and the experimental result was illustrated in figure 3. The other operation conditions were as follows: applied voltage 28 kV and water content of GAC 10 %. When air flow rate increased from 0.4 to 1.2 L min^{-1}, DHBA yields increased from 43 to 47 μmol after 100 min of discharge time. However, the maximal DHBA yield at 1.2 L min^{-1} has little differences than 0.8 L min^{-1}.

It could be suggested that with the increasing of air flow rate, the generation of •OH radicals are correspondingly increased. In contrast, the residence time of active species is shortened and their utilization efficiencies on GAC are correspondingly decreased. So, the combined effects of above two aspects resulted in that there is a preferable compromise flow rate 0.8 L min^{-1} for DBD plasma reactor.

3.3 Effect of gas type on •OH production

The varieties and numbers of oxidative species are considerably different during discharge processes under different gas atmospheres [18]. In order to investigate the roles of different gas atmospheres in •OH production, experiments were conducted separately under oxygen, nitrogen, and air atmospheres. The experimental results for •OH production under different gas atmospheres were shown in figure 4. The other operation conditions were as follows: applied voltage 28 kV, gas flow rate 0.8 L min^{-1} and water content of GAC 10 %. The DHBA yields obtained under oxygen and air atmospheres were much higher than those under nitrogen atmosphere. DHBA yields of 57.0 and 47.1 μmol were achieved in oxygen and air atmospheres after 100 min of discharge treatment, respectively; while only 21.3 μmol DHBA was obtained in nitrogen atmosphere at the same treatment time.

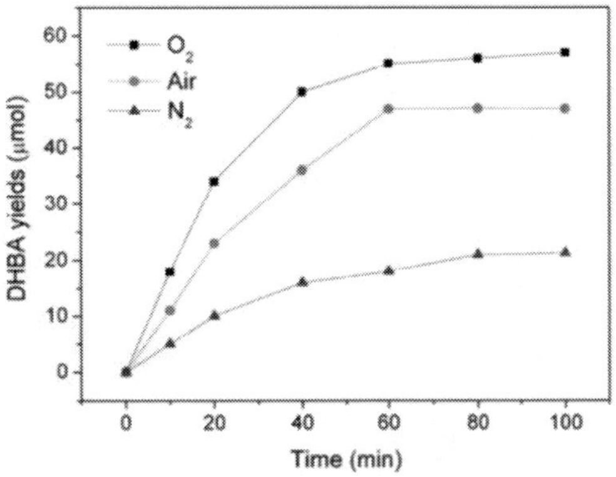

Figure 4. Effect of gas type on •OH production.

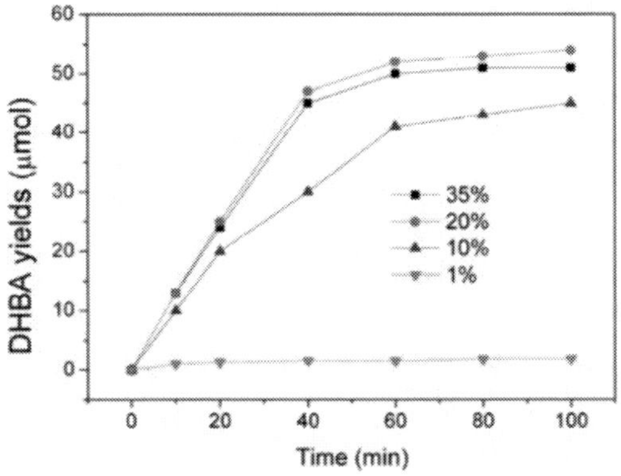

Figure 5. Effect of GAC water content on •OH production.

Under oxygen atmosphere, the input energy was mainly consumed to produce the O-reagents, whose concentrations were larger than those under air atmosphere, resulting in the greater generation of •OH radicals under oxygen atmosphere than those under air atmosphere. In general, N_2 is inactive and cannot participate in many chemical reactions. However, when discharge occurred in gas atmosphere containing N_2, it can be dissociated to the atomic N, N_2^+ and N^+ [19]. The bond dissociation energy of N_2 (9.82 eV) is higher than that of O_2 (5.12 eV), and thus it is much easier for O_2 to dissociate than N_2 when discharge occurs under air atmosphere. Therefore, more •O and •OH radicals can be generated under air atmosphere than N_2. Furthermore, the oxidative potentials of •O and •OH radicals are much stronger than those of atomic N, N_2^+ and N^+, so the DHBA yield was higher under air atmosphere that under N_2 atmosphere.

3.4 Effect of GAC water content on •OH production

Under different water content of GAC, the concentration of •OH radical was detected, and the results are shown in figure 5. The other operation conditions were as follows: applied voltage 28 kV and air flow rate 0.8 L min^{-1}. The humidity of GAC increased from 1 % to 35 %. It shows that •OH radical increased with the increase of GAC water content within 100 min discharge treatment.

Water plays a very important role in the reaction since it decomposes into •OH and H free radicals in the plasma system. The initiating reactions involving the H_2O excitation are given in reactions (2)–(5) [20]. When the humidity is high, more H_2O molecules collide with high-energy electrons, and •OH radicals are formed.

$$H_2O \rightarrow H, \quad e_{aq}^-, OH, H_2, H_2O_2, H_3O^+, OH^- \tag{2}$$
$$H + O_2 \rightarrow HO_2 \quad K = 1.2 \times 10^{10} \tag{3}$$
$$e_{aq}^- + O_2 \rightarrow O^-_2 \quad K = 1.9 \times 10^{10} \tag{4}$$
$$HO_2 + O^-_2 + H^+ \rightarrow H_2O_2 + O_2 \quad K = 9.7 \times 10^7 \tag{5}$$

However, water also has an adverse effect on pollutants' removal due to its electronegative characteristics. Increasing the humidity limits the electron density in the system and quenches some activated chemical species. Therefore, the optimal water content of GAC for •OH generation is 20 % rather than 35 % in this experiment.

4. Conclusions

A HPLC method with radical trapping salicylic acid, was used to detect indirectly the concentration of •OH radical on granular activated carbon in the DBD regeneration system. Under the studied conditions, increasing applied voltage and treatment time resulted in higher •OH generation, due to the enhancement of energy input. The quantity of •OH increased with an increase of gas flow rate. The role of gas atmosphere for •OH formation was identified. The highest •OH yield was achieved with a GAC water content of 20 %. At a higher humidity, the concentration of •OH declined because of a decrease in electron mean energy by water molecules. This work supplies a simple and rapid determination method for the generation of •OH in the DBD regeneration system. As •OH is an active and environmental friendly oxidant, it will have great significance in the fields of green chemistry, environmental protection, and other related fields.

Acknowledgments

The authors thank the Ministry of Science and Technology, P.R. China (Project No. 2008AA06Z308) and the National Natural Science Foundation, P.R. China (Project Nos. 51177007 and 40901150) for their financial supports to this research.

References:

[1] Oda T *J. Electrostat.* 2003 **57** 293.

[2] Bach R D and Dmitrenko O *J Am Chem Soc* 2006 **128** 1474.
[3] Brezonik P L and Fulkerson-Brekken J *Environ. Sci. Technol.* 1998 **32** 3004.
[4] Chen Z and Mathur V K *Ind. Eng. Chem. Res.* 2002 **41** 2082.
[5] Miura Y, Ueda J and Ozawa T *Inorg. Chim. Acta.* 1995 **234** 169.
[6] Dilecce G, Ambrico P F, Simek M and De Benedictis S *Chem. Phys.* 2012 **398** 142.
[7] Bektaşoğlu B, Özyürek M, Güçlü K and Apak R *Talanta* 2008 **77** 90.
[8] Campbell M J, Hall B D, Sheppard J C, Utley P L, Obrien R J and Hard T M et al. *J. Atmos. Sci.* 1995 **52** 3421.
[9] Su Z and Ito K *J. Phys. D: Appl. Phys.* 2002 **35** 3192.
[10] Jen J, Leu M and Yang T C *J. Chromatogr. A* 1998 **796** 283.
[11] Freinbichler W, Bianchi L, Colivicchi M A, Ballini C, Tipton K F and Linert W, et al. *J. Inorg. Biochem.* 2008 **102** 1329.
[12] Guo Y, Liao X and Ye D *J. Environ. Sci.* 2008 **20** 1429.
[13] Liao X, Guo Y, He J, Ou W, Ye D *Plasma Chem. Plasma Process* 2010 **30** 841.
[14] Mok Y S, Jo J O, and Lee H J *Plasma Sci. Technol.* 2008 **10** 100.
[15] Klamt A *Chemosphere* 1996 **32** 717.
[16] Hoigne J and Bader H *Water Res.* 1976 **10** 377.
[17] Hernandez-Leal L, Temmink H, Zeeman G and Buisman C *Water Res.* 2011 **45** 2887.
[18] B. Sun, M. Sato, J. Sid Clements, J Electrostat 39 (1997) 189-202.
[19] Fancey K S *Vacuum* 1995 **46** 695.
[20] Drzewicz P, Trojanowicz M, Zona R, Solar S and Gehringer P *Radiat Phys Chem* 2004 **69** 281.

Experimental research on spectral intensity of needle-plate corona discharge

Zhiqiang Liu*, **Tie Liu, Xinyu Miao, Wei Guo, Shoujie He**
Electrostatic Research Institute, Hebei University, No.180 Wusidong Road, Baoding, China

E-mail: lzq@hbu.cn

Abstract. A needle-plate corona discharge is investigated at atmospheric pressure in air. The current-voltage characteristics are measured in three kinds of needle-plate gaps. The corona discharge process is analyzed to obtain the relationship between the needle-plate gap and discharge current. The corona discharge emission spectrum is given through collecting emission light signal from the discharge area of needle-plate corona by emission spectroscopy. The spectral intensity of N_2 ($C^3\Pi_u \rightarrow B^3\Pi_g$) of 337.1 nm spectrum line is studied with varying the discharging voltage. By measuring the spot size of light emission, the relationship is discussed between the thickness of corona layer and voltage. The spectral intensity dependence on the voltage approximately shows a linear relationship. Spectral intensities in different positions are compared by analyzing the grey value of the discharge image. It shows that spectral intensity turns stronger and the stability of the grey value is greater under the same voltage with decreasing the distance from the needle tip.

1. Introduction

In recent years, the atmospheric pollution caused by energy utilization attracted much concerns. At the same time, particulate matter and gaseous pollutants (Sulfur dioxide and nitrogen oxides) have become the most important atmospheric environment pollution sources in China. The high energy electron and many active particles can be produced by the process of needle-plate corona discharge, which can oxidize the gaseous pollutants molecules and convert them into acid mist or aerosol particles. Then, the atmospheric pollution was controlled by collecting the poisonous gas. The needle-plate corona discharge has attracted considerable attention because of its low dynamic and high removal rate. It is widely used for elimination of the atmospheric pollutants. Although a lot of research has been studied on the needle-plate corona discharge, these works are mainly focused on the mechanism analysis, numerical simulation and influencing factors [1-3]. Up to now, few works has been done on the research of the thickness of corona layer and spectral intensity in different positions. Spectral intensity is related to electronic density. The thickness of corona layer and electronic density produced by corona discharge are the main influence on pollutant removal efficiency. In this paper, a self-made needle-plate discharge device is used to generate corona discharge in the atmosphere. The needle-plate corona discharge process and current-voltage characters curve are analyzed. Through the method of emission spectrum and reading the discharge image grey value data, the study is performed, which

* To whom any correspondence should be addressed.

contains the relations among the voltage, the electron density and thickness of corona in different position of the corona.

2. Experimental set-up

The experimental set-up is shown schematically in figure 1. It is done at atmosphere pressure. A needle with 0.1 mm curvature radius of needle tip is used as the cathode electrode, which connects to the negative high voltage DC power with 0~60 kV continuous adjustable. A rotundity copper plate with 65 mm diameter connecting to ground is used as anode. The microamps current value is measured in different voltages and needle-plate gaps. A camera is used to record the discharge images. Photoshop is used to measure dimension and gray level of spot of light-emitting photos, and get the thickness of corona layer and spectral intensity of corona luminous area. Grating spectrometer is used to get spectrum of the corona.

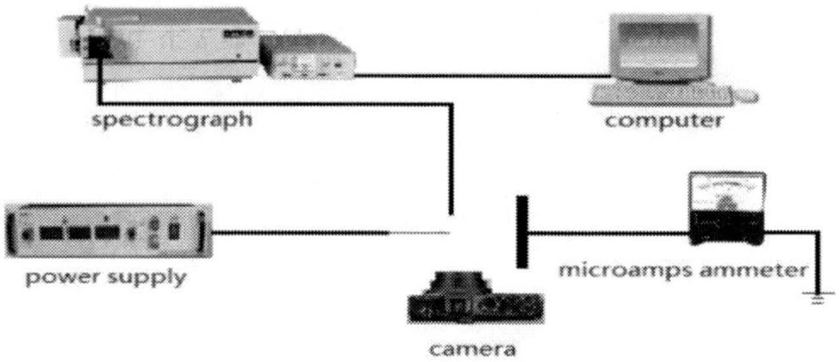

Figure 1. Experimental set-up.

3. Results and discussions

In the experiment, the current-voltage characters curve is measured at 2cm, 3cm, 4cm needle-plate gap, respectively, as shown in figure 2.

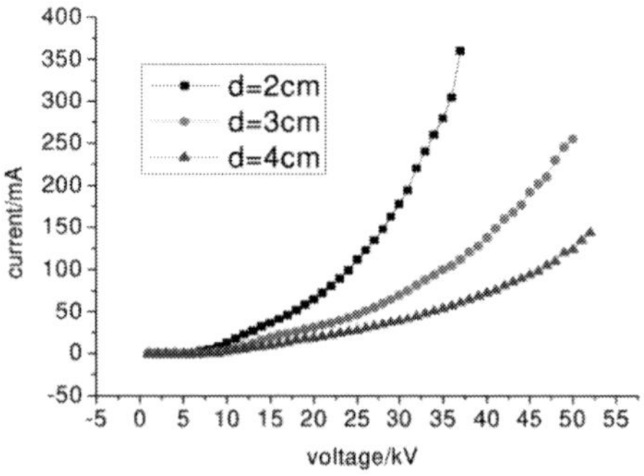

Figure 2. Current-voltage curves at different needle-plate gap.

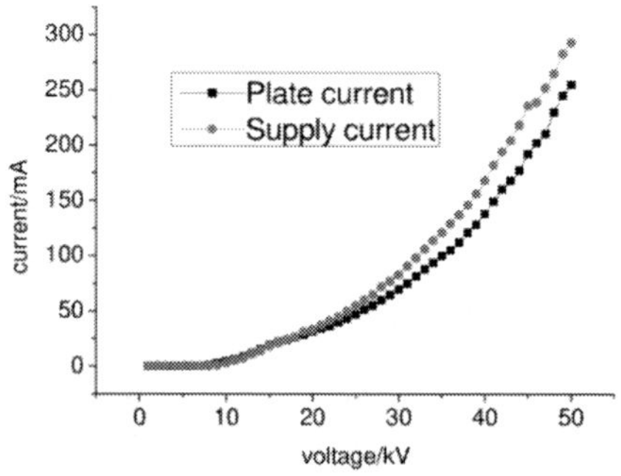

Figure 3. Supply current and plate current contrast.

The current-voltage curve of negative corona discharge is analyzed which is under the condition of 3 cm needle-plate gap. When the voltage increases from zero gradually, the silent non-self-sustaining dark discharge will firstly occur. The current is very weak under this voltage, and its intensity depends on the space remaining amount of ionization. As shown in figure 3. The plate current appears at 7 kV firstly. When the voltage is added to 8 kV, the supply current will appear. The plate current is greater than the supply current at the same voltage. Townsend electron avalanche theory can give some explanations. When an electrons run from cathode to the anode, they will ordinal collide with gas atomic. Then it leads to electronics grow with avalanche, and much more electrons reach the plate. When the voltage increases to 7.4 kV, corona discharge began. That voltage is called onset voltage, accompany with a sudden increase of current between the electrodes (from 10^{-14} to 10^{-6}A) and in a smaller radius of curvature of the electrode appeared the hazy glow. Adamiak [4] studies current-voltage characteristics of the needle-plate corona discharge, getting a needle-plate corona discharge current-voltage relation for, $I=AK\varepsilon U(U-U_0)\ d^{-1}$, K for ion migration rate, ε for gas dielectric constant; U_0 for onset voltage; d for electrode gap. According to the literature [5], onset voltage can be estimated through the I-U curve-fitting. This paper adopts straight-line fitting onset voltage of needle-plate corona discharge, and get theoretical onset voltage is 5.04 kV. After onset voltage, when the voltage is 8 kV, the supply current greater than the plate current 1.7 μA, then the two currents approach. When the voltage is 17 kV, the two currents are equal. This is because space charge effect is around the needle electrode, which hindered the electronic transfer to the anode, leading to two current gradually closed. With the increase of voltage, the supply current is greater than the plate current gradually. There are three main reasons. First, with the increase of the voltage, a large number of neutral air molecules to accumulate before the board, high-speed movement of electrons can not directly reach the plate, due to the blocking of the plate electrode. Second, with the increase of electronic near the plate, electronic exclusive each other, so part of the electrons can not reach the plate. Third, the movement of ions caused a strong ion wind in the discharge space [6], so the movement of electrons by the ion wind generated disturbance. When the applied voltage is much higher than the onset voltage, corona discharge will be transformed into spark discharge.

The relations between corona voltage and needle-plate gap is also investigated. The corona onset voltage increases with increasing the gap of the needle-plate. When the needle-plate gap is 2 cm, the onset voltage is 6.2 kV. When the needle-plate gap is 4 cm, the onset voltage is 8.7 kV. It is because the needle-plate gap increases, the electric field between the electrodes become weak, which weaken the positive and negative ions or electrons, positive ions compound and energy level transition caused by electronic collision of air molecules, and electronic avalanche growth. Figure 2 shows that the discharge current increases with the increase of voltage, and the current difference also increased gradually with different needle-plate gap.

Figure 4 shows a spectrum of the needle-plate corona discharge. The spectral analysis proves that there is a strong ionization and inspire in corona luminous layer. Cathode surface exist the positive ion secondary electron emission, optical emission and the secondary emission caused by metastable atoms. Around high-voltage needle electrode, the main spectral intensity is second positive system of N_2. According to the literature [7], the second positive system of $N_2(C^3\Pi_u \rightarrow B^3\Pi_g)$ spectral distribution is: $\triangle$v=+2,297.7 nm; $\triangle$v=+1,315.9 nm; $\triangle$v=0,337.1 nm; $\triangle$v=-1,357.7 nm; $\triangle$v=-2,380.5 nm; $\triangle$v=-3,405.9 nm; $\triangle$v=-4,434.4 nm. The experiment results are consistent with it.

Microscopically speaking, filamentous streamer is not continuous on the time distribution in DC corona discharge. There is a time interval between two adjacent streamers. Streamer frontier is the enrichment of the high energy electron. However, from the macroscopic viewpoint, DC corona discharge is a continuous discharge. DC corona discharge is a continuous process, formation and demise of $N_2\,(C^3\Pi_u)$ will reach a balance if the corona is stable. The experiment found that the spectral intensity of 337.1 nm have a certain fluctuation in fixed voltage and needle-plate gap. In this paper, we scanned wavelength with 337.1 nm in the time scale of 20 s. Through calculation the average of data, the spectrum line strength is obtained. The relationship between electronic density and voltage is shown in figure 5. Results show that the spectral intensity increases linearly with the rising of voltage approximately.

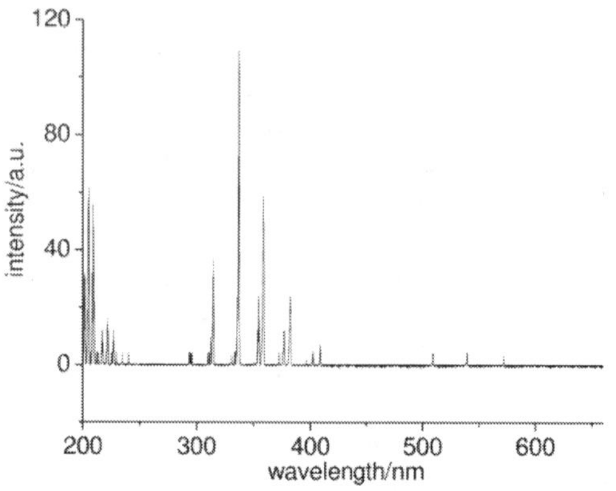

Figure 4. Spectrum of needle-plate corona discharge.

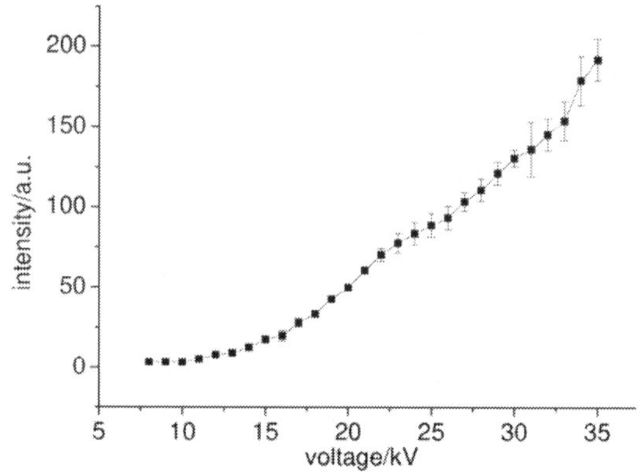

Figure 5. The relationship between spectral intensity and voltage.

Figure 6.light-emitting photo of needle-plate corona discharge.

Transition of strong electric field in corona layer to weak in periphery is not a mutation, but a gradual change. So space ionization and excitation change from strong to weak is a gradual transition [8]. Figure 6 shows light-emitting photo of needle-plate corona discharge. After onset of the discharge, the images of discharge are recorded by taking photos of the corona discharge in needle electrode. We select luminous images of grey value 5 as the corona area and peripheral area boundaries. The length of needle direction light is the standard of the thickness of corona layer. The experiment measured the thickness of corona layer diagram at different needle-plate gaps and voltages. Figure 7 shows the relationship between thickness of corona layer and voltage.

When the voltage between the poles gradually increasing from zero, electronic move slowly and direction relatively scattered, which caused it can not inspire air particles and produce radiative transition in the process of electron moving to the grounding plate .So we don't observe corona light around the needle electrode. After onset voltage, light of corona discharge appears around the needle electrode, with the rapid thickness of corona layer growth. The spontaneous radiation transition of stimulate particles is the cause of corona layer light-emitting. If continue to increase the voltage, the current and electric field strength will increase, with increase of the charged particle velocity. It enhances collision particle energy, the impact ionization rapid development. At the same time, the process of electronic and positive ions, positive ions and anion composition reinforced. And more non-steady-state particle transit to the ground state or lower energy state, so that radiation photon number was increase significantly, with the increase of the thickness and brightness of corona layer. It is shown that in a stable of corona discharge. When thicknesses of corona layer attain and maintain the flat part with little change. With the increase of needle-plate gap, the stable area is obviously, the corresponding to external voltage range get big. In stable area, the thickness of corona layer is close to each other in different needle-plate gap. With the voltage rising continuously, the thickness of corona discharge development from flat rise, which is shown that corona discharge transit to the pre-breakdown streamer. From figure we can observe that, along with the increase of voltage and reduce of needle-plate gap, the thickness of corona layer is increased obviously.

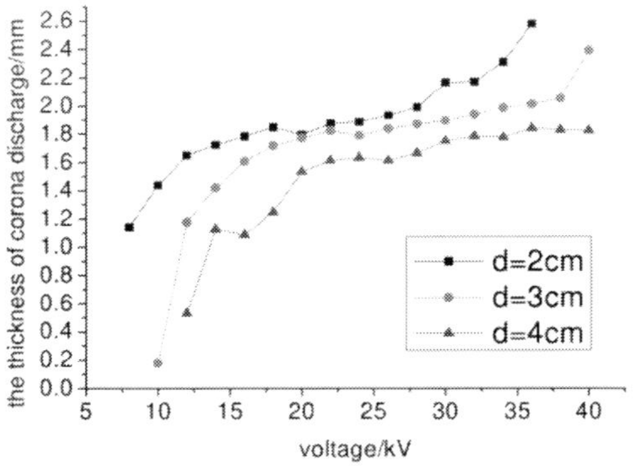

Figure 7. The relationship between thickness of corona layer and voltage at different needle-plate gap.

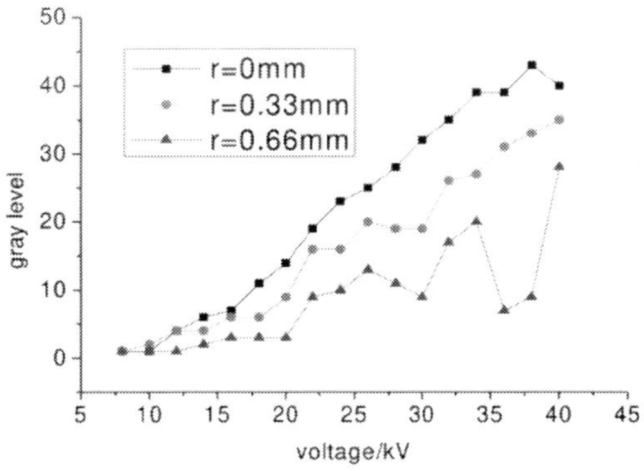

Figure 8. The relationship between gray level and voltage in different positions.

The change of space ionization and excitation from strong to weak is a gradual transition. So we can get the spectral intensity in different positions by reading the corona luminous gray value of the photos in different locations at 3 cm needle-plate gap. Figure 8 shows the relationship between gray level and voltage in different position. The light-emitting photo data is consistent with measurement of spectral intensity. So reading light-emitting photo data instead of spectral intensity in different corona area position is feasible. The figure shows that in the same voltage, the closer from the needle tip, the bigger of spectral intensity and the greater stability of the value. The farther away from the tip, the smaller of spectral intensity, and the higher of fluctuating value. Results show that spectral intensity with the growth of the voltage is close to a relation with linear strengthen.

4. Conclusion

a) The corona onset voltage increases with increasing the gap of the needle-plate. With the increase of voltage, discharge current increases, and with different needle-plate gap the current difference also increased gradually.

b) Around High-voltage needle electrode, the main spectral intensity is second positive system of N_2. The thickness of corona layer grows rapidly after onset voltage. And then there will be a stable region. When the voltage continues to increase, the thickness of corona layer rises rapidly. At the same voltage, the thickness of corona layer is bigger with decreasing of the needle-plate gap.

c) In the same voltage the closer from the needle tip, the bigger of spectral intensity and the greater stability of the value. The farther away from the tip, the smaller of spectral intensity, and the higher of fluctuating value. Results shown that spectral intensity with the growth of the voltage is close to a relation with linear strengthen.

Acknowledgments

This paper is supported by the National Science Foundation of China (51077035). The science and technology research projects of Colleges and Universities in Hebei province (2011166).

References

[1] Su P H, Zhu Y M and Chen H F 2008 *Spectrosc. Spect. Anal.* **28** 1998
[2] Koh W H and Park I H 2010 *Vacuum* **84** 550
[3] Gao X, Yu Q, Wu Z L and Tan Y J 2007 *High Power Laser & Particle Beams (in Chinese).* **19** 700
[4] Adamiak K and Atten P 2004 *J. Electrostat.* **61** 85
[5] Zhao Y M, Ma M H, Cchu J W, Guan Z C and Wang L M 2008 *High Voltage Eng. (in Chinese)..* **33** 1567
[6] Yuan J X, Yang L J, Zhang Q G, Liu P, Zhao H B, Du X M, Yan J X. 2010 *Trans. Chi. Electrotechnical Soc. (in Chinese).* **25** 24
[7] Herzberg G 1983 *Mol. Spectra and Mol. Struct.* (Beijing: Science Press)
[8] Hu Z Q, Zhen H S and Shi Y N 1985 *Gaseous Electron.* (Electronic Industry Press)

Journal of Physics: Conference Series **418** (2013) 012106 doi:10.1088/1742-6596/418/1/012106

Oxygen atomic density of atmospheric Ar plasma jet generated with syringe needle-ring electrodes

Y Hong[1,2], J Li[1*], J Pan[2], N Lu[1], K F Shang[1], and Y Wu[1]
1 Institute of Electrostatics and Special Power, Dalian University of Technology, Dalian 116024, People's Republic of China
2 Departments of Physics and Electrical Engineering, Weinan Teachers University, Weinan 71400, People's Republic of China

E-mail: lijie@dlut.edu.cn

Abstract: Atmospheric-pressure argon plasma jet is generated with syringe needle–ring electrodes in an 8 kHz sinusoidal excitation voltage. It is found that the rotational temperature of nitrogen is in the range of 333 - 373 K obtained by comparing the simulated spectrum with the measured spectrum at the $C^3\Pi_u \rightarrow B^3\Pi_g(\Delta v=-2)$ band transition, the electronic excitation temperature is in the range of 3187 - 3243 K determined by the Boltzmann's plot method, and the oxygen atomic density is in the order of magnitude of 10^{16} cm^{-3} estimated by the actinometry method, respectively.

1. Introduction

Recently, cold atmospheric-pressure plasma jets (CAPPJs) have received much attention owing to their attractive features that make them indispensable in a large number of applications, such as surface modification [1], thin film deposition [2-4], etching [5], sterilization [6-9], and biomedical and chemical decontamination [10-11]. One of the most prominent features of the CAPPJs is their enhanced plasma chemistry at low gas temperatures [12]. The CAPPJs can be generated using the different electrode configurations and the generators [13]. However, the discharge of CAPPJs is ignited in the inside of dielectric (such as quartz tube and Pyrex tube) rather than ambient air [5, 12-15]. It is due to the fact that the most of electrons are attracted by the oxygen in ambient air, which impede the electrons impacting the working gas to generate the plasma [16]. Wang et al. [17] observed that as the fraction of O_2 in radio-frequency (RF) driven He/O_2 plasma jet up to 3.2%, which causes the amplitude of the RF current to drop rapidly to its original value before the gas is ignited. In

* To whom any correspondence should be addressed.

addition, the quantification of O in discharge center of CAPPJs has been widely reported. But, few studies have reported the density of O radicals in downstream of a CAPPJ [18].

For these reasons, a specially designed Ar plasma jet is generated with syringe needle-ring electrodes by a sinusoidal excitation voltage at frequency of 8 kHz, and the oxygen atomic density in downstream of plasma plume is calculated by the actinometry method. The discharge of Ar plasma jet is ignited in ambient air, which reduces the dissipated power of plasma bullet through the inside of dielectric and the time of plasma bullet reached to the object to be treated.

2. Experimental setup

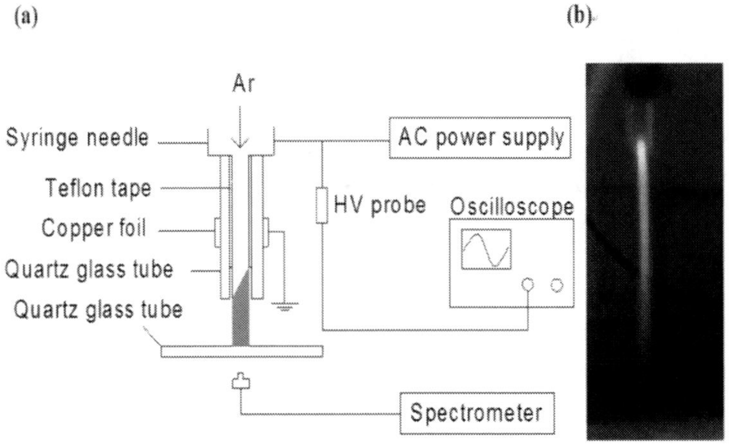

Figure 1. Experimental setup of device (a) and discharge photograph (b) in peak applied voltage of 6.2 kV and Ar flow rate of 0.8 lpm.

A schematic of experimental apparatus and discharge photograph are shown in figure 1. The high voltage electrode is a syringe needle with an inner diameter of 1.2 mm, an outer diameter of 1.6 mm and a length of 51 mm, which is powered by a sinusoidal excitation voltage at 8 kHz. The syringe needle is inserted to a quartz glass tube and sealed tightly by wrapping Teflon tape on it. The quartz glass tube is an outer diameter of 3 mm, an inner diameter of 2 mm and a length of 38 mm. The bottom of quartz glass tube and the tip of syringe needle are in the same height. A copper foil of 7 mm wide is wrapped on the outside of quartz glass tube used as a ground electrode, which is in the position of 12 mm away from the tip of syringe needle. A quartz glass plate (thickness of 1 mm) is placed in the position of 5 mm away from the tip of syringe needle.

The working gas of Ar (99.999%) is injected through the syringe needle controlled by mass flow controller at flow rate of 0.8 lpm. The applied voltages are measured by using a high voltage probe (Tektronix P6015A), and its signals are recorded via a digital oscilloscope (Tektronix TDS 2012B). An optical fiber located in the position of 2 mm away from the quartz glass plate is used to collect the optical emission of the plasma plume and the signals are recorded by a spectrometer (Acton INS-300-122B) with a grating of 1200 grooves per millimeter and slit width of 20 μm. The discharge image is obtained by a Nikon digital camera COOLPIX S600.

3. Experiment results and discussions

3.1 Optical emission characteristics

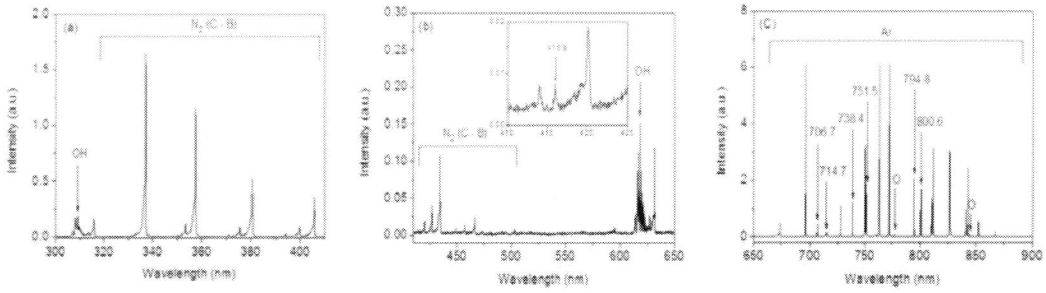

Figure 2. Optical emission spectrum 300 – 410 (a), 410 – 650 (b), and 650 – 900 nm (b) in peak applied voltage 6.2 kV and Ar flow rate of 0.8 lpm.

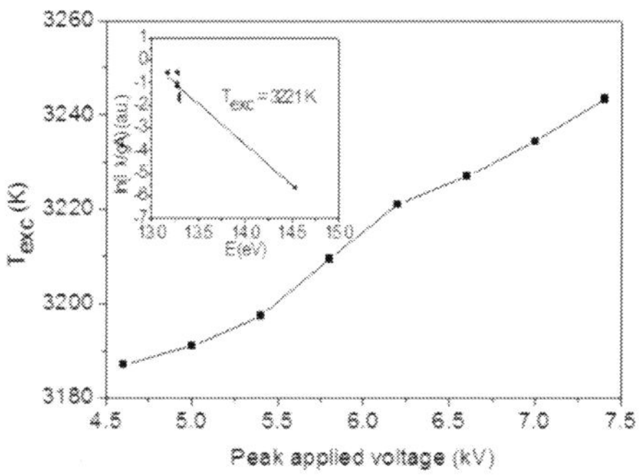

Figure 3. Excitation temperature as a function of peak applied voltage at Ar flow rate of 0.8 lpm and the inset graph is Boltzmann plot of excitation temperature at peak applied voltage of 6.2 kV.

The measured optical emission spectrum from 300 nm to 900 nm is shown in figure 2. As the results shown in figure 2, the plasma plume is dominated by the excited OH, N_2, Ar, and O, in which the excited OH and O are active species. It is indicating that the Ar plasma jet can be used to sterilization [6-9], surface modification [1], etching [5], etc. The characteristic spectral lines 415.9 nm for the transition $(3s^23p^5)5p \rightarrow (3s^23p^5)4s$ and 706.7, 714.7, 738.4, 751.5, 794.8, and 800.6 nm for the transition $(3s^23p^5)4p \rightarrow (3s^23p^5)4s$ are chosen to determine the electronic excitation temperature T_{exc} (K) under a Boltzmann approximation [19-22]. Figure 3 shows the variation in excitation temperature with peak applied voltage, and the inset graph is the Boltzmann plot of excitation temperature at peak applied voltage of 6.2 kV. Regarding the results in figure 3, as the peak applied voltage increases from 4.6 to 7.4 kV, the excitation temperature increases from 3187 to 3243 K. It is in well agreement with the conclusion reported by Wei et al. [22].

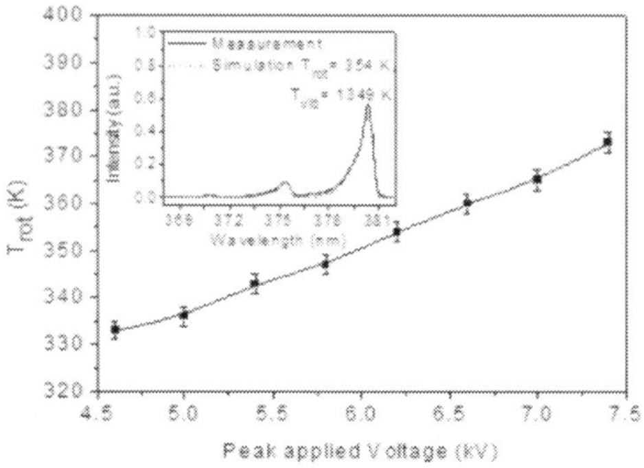

Figure 4. Rotational temperature of N_2 vs peak applied voltage at Ar flow rate of 0.8 lpm and the inset graph is measured and simulated spectrum of the $C^3\Pi_u \rightarrow B^3\Pi_g(\Delta v = -2)$ band transition of N_2 from 368 nm to 382 nm at peak applied voltage of 6.2 kV.

Figure 4 shows the variation in rotational temperature versus peak applied voltage and the inset graph is measured and simulated spectrum of N_2 second positive system 0-2 transition at peak applied voltage of 6.2 kV. The rotational temperature T_{rot} (K) is determined by comparing the simulated spectrum with the measured spectrum at the $C^3\Pi_u \rightarrow B^3\Pi_g(\Delta v = -2)$ band transition, which is much close to the gas temperature [12, 23]. As the results presented in figure 4, as the peak applied voltage increases from 4.6 to 7.4 kV, the rotational temperature increases form the 333 to 373 K. In peak applied voltage of 6.2 kV, the vibrational temperature T_{vib} (K) is 1349 K, which is much higher than the rotational temperature. It is shown that the plasma plume is under extremely nonequilibrium condition.

3.2 Oxygen atomic density and nitrogen molecular density

The density of atomic oxygen can be determined by choosing the atomic oxygen line λ = 844.6 nm and the argon line λ = 750.4 nm [16, 24-25] and the applicable conditions of this actinometry method are as follows [24]:

(1) The excitation cross sections have the same shape, particularly close to the threshold.
(2) The population of excited levels from higher levels is negligible.
(3) Two-step excitation, e.g., via metastables, is negligible.
(4) There is no population of atomic levels via dissociation.
(5) Radiationless de-excitation (quenching) of excited levels is negligible.

To improve the accuracy of calculation, the collisional induced quenching of the upper Ar ($2p_1$) and O ($3p^3P$) states in the argon gas is considered in this study. The intensity of argon line λ = 750.4 nm can be expressed as

$$I(Ar) = C_{Ar} h\gamma_{Ar} A_{ij}^{(Ar)} n_e k_e^{(Ar)} n_{Ar} (A_i^{(Ar)} + k_{Arq} n_{Ar})^{-1}. \tag{1}$$

The intensity of atomic oxygen line λ = 844.6 nm is given by

$$I(\mathrm{O}) = C_{\mathrm{O}} h\gamma_{\mathrm{O}} A_{ij}^{(\mathrm{O})} n_e k_e^{(\mathrm{O})} n_{\mathrm{O}} (A_i^{(\mathrm{O})} + k_{\mathrm{O}q} n_{Ar})^{-1}. \tag{2}$$

Where the quantities C contain all optical and geometrical Parameters; $h\gamma$ is the photon energy of respective transition; A_i denotes the sum over all optical transition probabilities A_{ij} for the excited state, and is equal to the reciprocal of the natural lifetime τ_0; A_{ij} is equal to $A_i = 1/\tau_0$, since the only optically allowed transition from each of the excited states is observed, respectively [25]. n_e, n_{O}, and n_{Ar} represent the electron density, oxygen atomic density, and argon density, respectively, in which the argon density n_{Ar} is determined by the ideal gas equation $n = p/kT_g$, where the gas pressure p (Pa) takes the value of one atmospheric-pressure and the gas temperature T_g (K) is obtained from the figure 4. $k_{\mathrm{O}q}$ and k_{Arq} represent the quenching rate coefficients of the upper Ar ($2p_1$) and O ($3p^3P$) states in argon gas, respectively. Table 1 summarizes the data of radiative lifetimes and quenching rate coefficients for O ($3p^3P$) and Ar ($2p_1$) atoms taken from the literature [26-27].

Table 1. Radiative lifetimes and quenching coefficients of O ($3p^3P$) and Ar ($2p_1$) atoms.

	Radiative lifetime (ns)		
	O ($3p^3P$)	Ar ($2p^1$)	
	34.7[a]	24[b]	[a] [26]
Reagent	Quenching coefficients (10^{-10} cm^3 s^{-1})		
	O($3p^3P$)	Ar($2p^1$)	
Ar	0.14[a]	0.16[b]	[b][27]

The electron excitation rate coefficient k_e is determined by integrating the cross sections over an assumed Maxwellian distribution

$$k_e = \int_0^\infty \sigma(\varepsilon) f(\varepsilon) \sqrt{\frac{2e}{m_e}} \varepsilon d\varepsilon, \tag{3}$$

where $\sigma(\varepsilon)$ is the electron-impact cross section for excitation [28-29], and the variations in it versus electron energy are shown in figure 5. $f(\varepsilon)$ is the electron energy distribution function, and m_e is the electron mass. Taking the intensity ratio of Eq. 2 to Eq. 1, we obtain

$$n_{\mathrm{O}} = \frac{C_{Ar} h\gamma_{Ar} I_{\mathrm{O}} k_e^{(Ar)} a_{ij}^{(Ar)}}{C_{\mathrm{O}} h\gamma_{\mathrm{O}} I_{Ar} k_e^{(\mathrm{O})} a_{ij}^{(\mathrm{O})}} n_{Ar} \tag{4}$$

with

$$a_{ij}^{(Ar)} = \frac{A_{ij}}{A_i + k_{Arq} n_{Ar}} \quad \text{and} \quad a_{ij}^{(\mathrm{O})} = \frac{A_{ij}}{A_i + k_{\mathrm{O}q} n_{Ar}}. \tag{5}$$

Eq. 4 is used to determine the density of atomic oxygen. Figure 6 shows the variation in oxygen atomic density with peak applied voltage. Regarding the result presented in figure 6, as the peak applied voltage increases from 5 to 7.4 kV, the oxygen atomic density increases form the 1.26×10^{16} to 1.48×10^{16} cm^{-3}. Besides, the curves of oxygen atomic density and electronic excitation temperature versus peak applied voltage (in figure 3) present a similar tendency, which is indicating that the oxygen atomic density increases with the electronic excitation temperature. It is in well agreement with conclusions reported by Qian et al. [16].

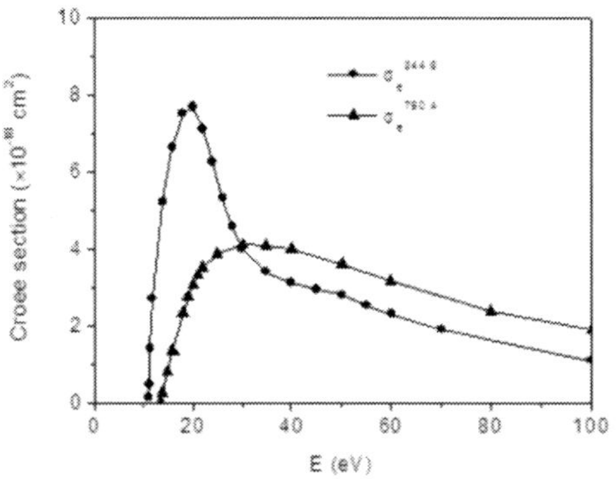

Figure 5. Electron-impact excitation cross section of atomic oxygen $\sigma_e^{844.6}$ and argon $\sigma_e^{750.4}$.

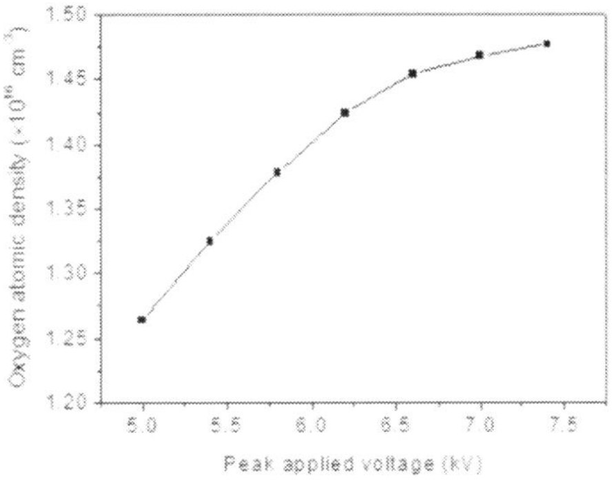

Figure 6. Oxygen atomic density as a function of peak applied voltage at Ar flow rate of 0.8 lpm.

4. Conclusions

A cold argon plasma jet generated with syringe needle–ring electrodes in atmospheric-pressure has been studied. It is found that the excited Ar, N_2, OH, and O are existed in the plasma plume, and the rotational temperature, electronic excitation temperature, and oxygen atomic density increase linearly with peak applied voltage. In addition, the oxygen atomic density in the position of 5 mm away from the tip of syringe needle is in the order of magnitude of 10^{16} cm^{-3}.

Acknowledgments

The authors thank the National Natural Science Foundation, P.R. China (Project No. 51177007) and the Ministry of Science and Technology, P.R. China (Project No. 2009AA064101-4).

References

[1] Dorai R and Kushner M J 2003 *J. Phys. D: Appl. Phys.* **36** 666
[2] Benedikt J, Focke K, Yanguas-Gil A and Keudella von A 2006 *Appl. Phys. Lett.* **89** 251504
[3] Benedikt J, Raballand V, Yanguas-Gil A, Focke K and Keudell von A 2007 *Plasma Phys. Control. Fusion.* **49** B419–B427
[4] Leduc M, Coulombe S and Leask R L 2009 *IEEE Trans. Plasma Sci.* **37** 927-933
[5] Walsh J L and Kong M G 2008 *Appl. Phys. Lett.* **93** 111501
[6] Lu X P, Ye T, Cao Y G, Sun Z Y, Xiong Q, Tang Z Y, Xiong Z L, Hu J, Jiang Z H and Pan Y 2008 *J. Appl. Phys.* **104** 053309
[7] Lu X, Xiong Z, Zhao F, Xian Y, Xiong Q, Gong W, Zou C, Jiang Z and Pan Y 2009 *Appl. Phys. Lett.* **95** 181501
[8] Huang J, Chen W, Li H, Wang X Q, Lv G H, Latif Khosa M, Guo M, Feng K C, Wang P Y and Yang S Z 2011 *J. Appl. Phys.* **109** 053305
[9] Fridman G, Brooks A D, Balasubramanian M, Fridman A, Gutsol A, Vasilets V N, Ayan H and Friedman G 2007 *Plasma Process. Polym.* **4** 370-375
[10] Zirnheld J L, Zucker S N, DiSanto T M, Berezney R and Etemadi K 2010 *IEEE Trans. Plasma Sci.* **38** 1317
[11] Kolb J F, Mohamed A, Price R O, Swanson R J, Bowman A, Chiavarini R L, Stacey M and

Schoenbach K H 2008 *Appl. Phys. Lett.* **92** 241501

[12] Xiong Q, Lu X P, Ostrikov K, Xian Y, Zou C, Xiong Z and Pan Y 2010 *Phys. Plasmas.* **17** 043506

[13] Jarrige J, Laroussi M and Karakas E 2010 *Plasma Sources Sci. Technol.* **19** 065005

[14] Park H S, Kim S J, Joh H M, Chung T H, Bae S H and Leem S H 2010 *Phys. Plasmas.* **17** 033502

[15] Kim S J, Chung T H and Bae S H 2010 *Phys. Plasmas.* **17** 053504

[16] Qian M Y, Ren C S, Wang D Z, Feng Y and Zhang J L 2010 *Plasma Sci. Technol.* **12** 561

[17] Wang S G and Wan J 2009 *IEEE Trans. Plasma Sci.* **37** 551-554

[18] Waskoenig J, Niemi K, Knake N, Graham L M, Reuter S, Gathen V Schulz-von der and Gans T 2010 *Pure Appl. Chem.* **82** 1209-1222

[19] Qian M Y, Ren C S, Wang D Z, Zhang J L and Wei G D 2010 *J. Appl. Phys.* **107** 063303

[20] Li S Z, Huang W T, Zhang J L and Wang D Z 2009 *Phys. Plasmas.* **16** 073503

[21] Nie Q Y, Ren C S, Wang D Z and Zhang J L 2008 *Appl. Phys. Lett.* **93** 011503

[22] Wei G D, Ren C S, Qian M Y and Nie Q Y 2011 *IEEE Trans. Plasma Sci.* **39** 1842-1848

[23] Laroussia M and X Lu 2005 *Appl. Phys. Lett.* **87** 113902

[24] Katsch H M, Tewes A, Quandt E, Goehlich A, Kawetzki T and Döbele H F 2000 *J. Appl. Phys.* **88** 6232-6238

[25] Niemi K, Reuter S, Graham L M, Waskoenig J, Knake N, Gathen V Schulz-von der and Gans T 2010 *J. Phys. D: Appl. Phys.* **43** 124006.

[26] Niemi K, Gathen V Schulz-von der and Döbele H F 2005 *Plasma Sources Sci. Technol.* **14** 375

[27] Sadeghi N, Setser D W, Francis A, Czarnetzki U and Döbele H F 2001 *J. Chem. Phys.* **115** 3144

[28] Hayashi M 2003 online available (http://jila.colorado.edu/˜avp/collision_ data/electronneutral/ hayashi. txt).

[29] Laher R R and Gilmore F R 1990 *J. Phys. Chem. Ref. Data.* **19** 277

Research on active particle distribution in the crop stalk plasma reaction under the condition of streaming and discharging micro electric current

C L Song[1]*, S G Pi[2], X Y Liu[1], H B Xu[1]
1.The Mine Security & Environmental Engineering Department, Jixi University, He Ping South Street, Jixi, China;
2. Jixi No. 1 Senior Middle School, Xing Guo Street, Jixi, China

E-mail: songchunlian@163.com

Abstract. In order to further study sugar-production mechanism in the crop stalk plasma reaction under the condition of streaming and discharging micro electric current with N_2 and H_2O as the electro-discharging gases, the active particles distribution is researched and analyzed. Through the study of methods and theories of statistical mechanics and statistical thermodynamics, the ionization process of the mixed discharging gas and the distribution features of the active particles changing with temperature in the plasma reactor are discussed. The constant pressure form of Saha equation that applies to this experiment is deduced. And the active particles generated in the dense ionization discharging process and the crop stalk conversion processes are also analyzed. The experimental results show that N_2 and H_2 can ionize synchronously. And it is found that during the crop stalk conversion process, while one gas is ionised, the other gas will not take the energy away, which will not affect the conversion. Among the discharging gases consisting of water vapour and nitrogen, the main components are ions of H_2O^+, H^+, N_2^+ and molecules of N_2, H_2O, which are all beneficial to a high sugar yield.

1. Introduction

Presently, development and application of the crop stalk biomass in the chemical industry has become a hotspot and a leading-edge research subject in the field of biomass resources study [1]. In the preliminary stage of this experiment, sugar is produced in the plasma conversion process of crop stalks. [2]. It is found that the active particles generated during the discharging process and the energy distribution is often related to atmospheric gas, which will directly affected the sugar yield in crop stalks. To study the distribution features of the active particles, this paper analyzes the discharging process of dense ionization in the plasma reactor and the active particles generated during the stalk conversion process, which has a significant meaning for improving the sugar yield in the crop stalk plasma reaction.

* To whom any correspondence should be addressed.

2. The Experiment

2.1. The Experimental Materials

The crop stalks, the main experimental material used in this experiment were produced in the countryside of Heilongjiang province in 2010, which were naturally preserved before being used and not moldy. For the experiment, the stalks were cut into segments of an inch or so and then were crushed and grounded into particles of an average size of less than 0.2 mm.

2.2. The Experimental Operation

The devices and equipments in this experiment was described in detailed previously [2]. The N_2/ H_2O mixture is prepared in the range: $N_2/H_2O>5$.

3. The experimental result

3.1. The Particle Types in the Crop Stalk Plasma Reactor

To analyze the complex physical and chemical processes occurred in the plasmas, the reactants, intermediates and products in the excited state must be taken into account [3]. Under the condition of N_2/ H_2O mixture, the following types of active particles may exist: N_2^+ atoms, N atoms or ionized atoms, ie, N atoms in the ground state, N atoms in the excitation state, N ions and H_2O^*, H_2O^+, H_3O^+ in the stimulated state, etc.. The reacting processes in the plasma reactor all obeyed the fundamental laws of energy conservation and mass conservation. In the stalk plasma reactor, on the basis of the above two conservation laws, the thermodynamics laws and characteristics of this system can be explored through theoretically analyzing.

3.2. Analyzing the Ionization Degree of the Particles by Statistical Thermodynamics Method

When the system is in or approaches the equilibrium stage, the chemical, thermal and ionization equilibriums will reach, at least in one tiny element volume [4, 5]. When the stable plasma reaches ionization equilibrium, it obeys the Saha equation. But generally, the Saha equation is introduced in the constant-volume process. The stalk plasma reactor used in this study operated under the constant voltage condition, thus it is necessary to derive the Saha equation in the form of constant voltage.

It can be assumed that the plasma is generated by A, that is, the following reaction occurred: $A=A^+ + e$. In this formula, A^+ and e denote to positive ion and electron, respectively. The above reaction reaches equilibrium at a certain temperature. It should be assumed that in the state of equilibrium, with the total pressure of P, molecule A will be ionized, which is caused by the external electric field's effect or the electron collision. The ionization degree is α. Thus, when ionization equilibrium is reached, the partial pressures of all the gases will show the following relationship:

$$\frac{P_{A^+}P_e}{P_A P^0} = K^0(T) \tag{1}$$

In formula (1), P_A^+, P_e and P_A represent the partial pressures of each component, respectively. P^0 is the standard pressure with the value of 101325 Pa. And K^0 (T) is the thermodynamic equilibrium constant of the system. When the total pressure reaches P, in accordance with the Dalton law, the partial pressures of each component are as follows:

$$P_{A^+} = P_e = \frac{\alpha}{1+\alpha} \tag{2}$$

$$P_A = \frac{1-\alpha}{1+\alpha} \tag{3}$$

If we substitute formula (1) and (2) into (3), the following expression can be obtained:

$$\frac{P\alpha^2}{1-\alpha^2} = K(T) \qquad (4)$$

In accordance with the basic thermodynamic equation, the following formula can be obtained:

$$\ln K(T) = -\frac{\Delta_r G^0}{RT} = \frac{\Delta_r S^0}{R} - \frac{\Delta_r H^0}{RT} \qquad (5)$$

In the above formula, $\Delta_r G^0$, $\Delta_r H^0$ and $\Delta_r S^0$ denote to the standard Gibbs free energy variable, the standard enthalpy (heat) variable and the standard entropy variable of the reaction system, respectively. Under the condition of gas ionization, the standard enthalpy (heat) variable $\Delta_r H^0$ can be represented by I_1, which stands for the first ionic energy of the gases. See the material I_1 in the following table [6]:

Table 1. The first ionic energy of some gases I_1.

Units	N_2	H_2	H_2O	Ar
eV	15.59	15.42	12.6	15.75
kJ mol^{-1}	1504.1	1489.5	1215.1	1519.9

From this table, we can see that I_1 changes little with temperature. Then we can decide that it is independent on temperature. The relationship between the standard entropy variable $\Delta_r S^0$ and temperature can be calculated as following when the temperature of the ideal gases change merely:

$$\Delta_r S^0 = \int \frac{\Delta_r C_p{}^0}{T} dT = \int \frac{\lambda kR}{T} dT = \lambda kR \ln T \qquad (6)$$

In formula (6), $\Delta_r C_p{}^0$ stands for the reaction's thermal capacity variation in constant voltage and can be calculate as follows:

$$\Delta_r C_p{}^0 = \sum_i v_i C_p{}^0(i) \qquad (7)$$

For ionization, when regarding any matter in it as the ideal gas, C_p of A and A^+ is equivalent. And for linear molecule it will always be 7/2R. Electron e can be regarded as monatomic molecule, whose heat capacity in constant voltage is 5/2R. The above relationship can be expressed as following:

$$\ln K(T) = \frac{5\lambda \ln T}{2} - \frac{I_1}{RT} \qquad (8)$$

$$K(T) = \kappa(T)^{\frac{5}{2}} \exp(-\frac{I_1}{RT}) \qquad (9)$$

If we substitute the expressions of ionization constant and degree of ionization α into the above formula, the Saha equation in constant voltage can be obtained:

$$\frac{P\alpha^2}{1-\alpha^2} = \kappa(T)^{\frac{5}{2}} \exp(-\frac{I_1}{RT}) \qquad (10)$$

In the above expression, m_e, h and k represent electron mass, Plank constant and Boltzmann constant respectively.

For the molecule system, the chemical bonds will crack and become dissociations, which form molecules in high temperature. We can take the reacting gas N_2 in plasma reactor as the sample. It can be assumed that the molecules and atoms of N_2 have reached equilibrium as follows:

$$N_2(g) = 2N(g) \tag{11}$$

The dissociation degree x can be expressed as follows:

$$x = \frac{n}{n_0} \tag{12}$$

In expression (12), n and n_0 stand for the dissociated molecules and the mol number of the total molecules, respectively.

The molecule dissociation degree x, the total pressure P and temperature T all obey the following equation which is similar to Saha equation:

$$\frac{Px^2}{1-x^2} = \exp\left(\int \frac{[\Delta H^0(T_1) + \int_{T_1}^{T} \Delta C_p dT]}{RT^2} dT + I \right) \tag{13}$$

In expression (13), ΔH^0 (T_1) is the enthalpy variable of the dissociation process in temperature T_1, and ΔC_P is the heat capacity variable in constant voltage, which can be expressed by relevant temperature polynomial. Here the linear part of the temperature is: $\Delta C_P \approx \alpha + \omega T$. I stands for the determined integral constant.

The dissociation process demands large number of energy consumption. In the general form (13), ΔH^0 (T_1) in the exponential molecules is much larger than the latter. But in the case of the less accurate, ΔH^0 (T_1) can be replaced by molecule dissociation energy $E_{A\text{-}A}$ (or bond energy $D_{A\text{-}A}$). Then (14) can be:

$$\frac{Px^2}{1-x^2} = T^{\omega} \exp(-\frac{D_{A-A}}{RT}) \tag{14}$$

In expression (14), ω stands for an empirical parameter whose approximation is 2.

3.3. Ionization of the Mixed Discharging Gas

In this experiment, the plasma reaction is a process of dielectric barrier discharge under 80 kPa. In the high frequency alternating electric field, N_2 partially discharges, generating a small number of electrons and positive ions, which results in the ionization and dissociation of N_2 and H_2O, ionization equilibrium is reached subsequently. Thus, the ionization degree of the reacting gas at different temperatures α can be obtained by the Saha equation as following:

$$\alpha = \left(\frac{\kappa(T)^{\frac{5}{2}} \exp(-\frac{I_1}{RT})}{p + \kappa(T)^{\frac{5}{2}} \exp(-\frac{I_1}{RT})} \right)^{\frac{1}{2}} \tag{15}$$

The obtained relationship between the temperature and the ionization degrees of the gases is shown in figure 1. From experimental results, the ionization degrees of all the gases with temperature show the same trend with two transition temperatures. The first situation appears in a certain low temperature range, when the temperature exceeds this range, ionization degree can be significantly increased. The second situation appears in a high temperature range. Under the operating pressure condition of the calculation for N_2 and H_2O, these two group values are: 2800 K, 2500 K and 3200 K, 2800 K. The ionization energy of N_2, H_2, Ar are all reached to 250 kJ mol^{-1}, which is significantly higher than H_2O. This result indicates that N_2, H_2 and Ar can be ionized synchronously, which can be confirmed by Saha equation.

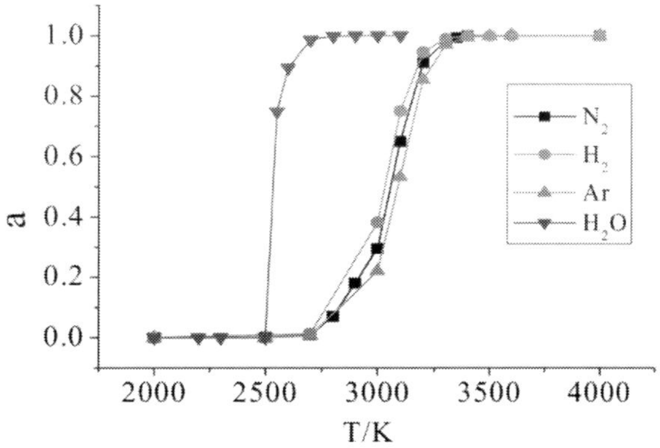

Figure 1. The relationship between the ionization degrees of the gases and the temperature under the plasma hydrolyzation condition.

It is assumed that when there is a less current conveyance through the system, namely, the ionization degree α_i a small value can meet the demand of $1+\alpha_i^2\approx 1$. Under this situation the ratio between α_i/α_j, the two ionization degrees, can be expressed by the following equation:

$$\frac{\alpha_i}{\alpha_j} = \exp\left(\frac{I_{1j} - I_{1i}}{2RT}\right) \quad (16)$$

The ionization ratio between the two different gases at a certain temperature can be calculated by the formula (16). In the discharging gas mixed by nitrogen and water vapour, the current rate of N_2 is five times larger than H_2O in the same temperature and pressure. So we can conclude that N_2 is the main discharge gas. The ionization of H_2O is 289 kJ mol^{-1}, which is lesser than that of N_2. So when H_2O is ionized, the main discharge gas is ionized too.

The mixed discharging gas composed of nitrogen, hydrogen and water vapour. The first ionization energy of N_2 and H_2 are 1504.1 kJ mol^{-1} and 1489.5 kJ mol^{-1}. The difference between them is 14.6 kJ mol^{-1}. So at a certain temperature, such as 4000 K, the ionization degree ratio between N_2 and H_2 can be expressed as following:

$$\frac{\alpha_{H_2}}{\alpha_{N_2}} = \exp\left(\frac{14600}{2\times 8.314\times 4000}\right) = 0.2195 \quad (17)$$

This number increases with increasing temperature, so when H_2 ionizes N_2 also ionizes. In the discharging process, N_2 and H_2 are the main gases, the summed current rate of them are five times larger than that of H_2O under the same conditions. Even when very small part of water vapour ionizes, H_2 and N_2 will also be ionized.

4. Conclusion

In this paper, through the studying methods and theories of statistical mechanics and thermodynamic, the molecules, atoms, ions and mixture of N_2/H_2O and $N_2/H_2/H_2O$ in plasma reactor were analyzed.

The experimental results show that, when H_2O was ionized, N_2 will be ionized too. And N_2 and H_2 can ionize synchronously. What's more, it is found that the mixed gases are very favourable for producing H^+ sheath layer, which can enhance the acidity and the sugar yield in the crop stalk.

Because of the large ionization energy difference between these two gases, when a gas ionizes the other one can not follow. Under this condition, a large number of heat energy will be released. The gas combination in this experiment can avoid the loss of the heat energy and increase the stalk conversion rate in a large extent.

Acknowledgements
This paper was supported and sponsored by The Support Plans of Natural Science Foundation of Heilongjiang Province (B201118).

References

[1] Yaman S 2004 *Energy Convers. Manage.* **45** 651
[2] Song C L, Zhang Z T, Chen W Y and Liu C 2009 *IEEE Trans. Plasma Sci.* **37** 1817
[3] Bai M D, Zhang Z T and Bai X Y 1999 *Journal of Dalian Maritime University (in Chinese).* **25** 84
[4] Miller T A 1981 *Plasma Chem. Plasma Process.* **3** 20
[5] Gernogora G, Hochard L and Touzeau M 1981 *Plasma Chem. Plasma Process.* **13** 15
[6] Bai X Y, Li S and Bai M D 2000 *Plasma Phys.* **239** 615

Inactivation of *Escherichia coli* phage by pulsed electric field treatment and analysis of inactivation mechanism

Takanori Tanino, Tomoki Yoshida, Kazuki Sakai and Takayuki Ohshima[*]
Department of Chemical and Environmental Engineering, Gunma University, 1-5-1 Tenjincho, Kiryu, Gunma 376-8515, Japan

E-mail: tohshima@cee.guunma-u.ac.jp

Abstract. Inactivation of bacteriophage by pulsed electric field (PEF) treatment, one of the effective procedures for bacteria nonthermal inactivation, was studied. Model phage particles *Escherichia coli* bacteriophages M13mp18 and λ phage, were successfully inactivated by PEF treatment. The survival ratios of both bacteriophages decreased depending on the PEF treatment time when applied peak voltage was 5 or 7 kV, and the survival ratios after 12 min PEF treatment were 10^{-4}- 10^{-5}. Electrophoresis analyses of biological molecules of inactivated λ phage detected no degradation of total protein and genomic DNA. These results suggested that the factor of phage inactivation by PEF treatment was not based on the degradation of protein or DNA, but on the destruction of phage particle structure. Sensitivity of *E. coli* phage to PEF treatment was compared with that of *E. coli* cell. Phage and MV1184 cell were treated with same condition PEF at 5 kV, respectively. After 12 min treatment, the survival ration of λ phage and MV1184 were 4.0×10^{-5} and 1.7×10^{-3}, respectively. The survival ratio of phage was lower than that of MV1184. *E. coli* cell is more tolerant to inactivation with PEF treatment than *coli* phage.

1. Introduction

Fermentation by bacteria is one of the most important unit operations in the industry. In addition to the production of traditional foods, recently, enzymes, pharmaceuticals and chemicals production by fermentation using bacteria including genetic modified one has been applied to practical use. In these industrial scenes, infection with a bacteriophage to the fermentation bacteria would result in the decrease of the product amount and quality. Inactivation procedure of bacteriophages has been becoming increasingly important to prevent the contamination of the bacteriophage into the fermentation culture. Bacteria inactivation procedures, especially heat treatment, are also applied for inactivation of bacteriophage. The inactivation mechanism of heat treatment resulting from the biomolecules denaturation of bacteria and bacteriophage is secure. However, heat treatment also denatures and degrades other heat sensitive content such as vitamin, sugar, protein, flavour and color components of fermentation feedstock. To avoid these problems, application of nonthermal bacteria inactivation procedures for inactivation of bacteriophage has also been investigated [1-4].

As the one of nonthermal inactivation procedure of bacteria, pulsed electric treatment (PEF) has been attracting attention. PEF characterized by nanoseconds rising time and microseconds of half

[*] To whom any correspondence should be addressed.

bandwidth of applied voltage dose not generate large amount of ohmic heat, and this characteristic phenomenon was applied for the liquid sterilization technology. The inactivation mechanism of microorganism in the liquid by PEF is the destruction of biological outer membrane structure of microorganism resulting from the electrical compression and fenestration [5, 6]. PEF dose not produce active species that react to components of fermentation feedstock.

In this study, inactivation of bacteriophage by PEF treatment was investigated as the one of nonthermal inactivation procedure of bacteriophage. *Escherichia coli* and its bacteriophage were selected as the model host and bacteriophage. Inactivation profile of the *E. coli* bacteriophage by PEF was compared with inactivation profile of heat treatment. And damage of bacteriophage DNA and protein was analyzed by electrophoreses. Sensitivities for PEF of *E. coli* cell and bacteriophage were also compared using mixture solution of cell and bacteriophage.

2. Materials and methods

2.1. Strains and media

E. coli phage M13mp18 phage (Takara Bio Inc., Ohtsu, Japan) [7] and λ phage (NBRC 20016) were used as the inactivation target by the PEF treatment. *E. coli* MV1184 (Takara Bio) was used as the host strain of the *E. coli* phage infection, and cultivated in Luria-Bertani (LB) medium (1 % (w/v) bacto peptone, 0.5% yeast extract and 0.5% sodium chloride) until OD_{660} = 0.5-0.8 at 35 °C. Then the phages were inoculated into *E. coli* broth, respectively, and cultivated 4 h. The supernatants centrifuged for 10 min at 18,000 × g and 4 °C were used as the phage stock solutions.

2.2. Phage titer measurement

A phage titer (plaque formation unit: PFU) was measured by the plaque counting method. In Brief, the *E. coli* cells cultivated for 12 h were collected by centrifugation for 5 min at 1,000 × g and suspended into sterilized distilled water. An appropriate diluted phage stock solution, *E. coli* solution and LB medium containing 0.6% agar, which was autoclaved and cooled down to 45 °C, were mixed well and overlaid onto LB medium plate containing 1.5% agar. The phage titer was calculated from the number of the plaques formatted on plate after 12 h cultivation at 37 °C.

2.3. PEF generation and PEF treatment apparatus

A high-voltage pulse generator with a rotary spark gap described in our previous report was used [8]. The capacitor capacity, pulse frequency and peak applied voltage were 8 nF, 50 Hz and 5 and 7 kV, respectively. The schematic diagram of the treatment chamber used in this study is shown in figure 1. Treatment chamber consisted of two parallel acrylic plates that have plate stainless steel electrodes facing each other with a gap of 31 mm and an acrylic cylinder. The phage solution diluted to 10^7 PFU/ ml was poured into the treatment chamber, and pulsed voltage was applied for 0-12 min. The phage titer of the treated phage solution was measured as described above. *E. coli* cell suspension was also treated by PEF, and viable cell number (colony formation unit: CFU) was measured by colony counting. The mixture of *E. coli* cell (10^7 CFU/ mL) and phage (10^7 PFU/ mL) was treated by PEF at 5 kV, and then treated sample was inoculated into LB medium. The mixture was cultivated at 35 °C and the time course of culture turbidity (optical density at 660 nm: OD_{660}) was measured.

The capsid proteins of phage were concentrated by ultrafiltration (30 kDa cut-off) and applied to SDS-polyacrylamide gel electrophoresis (SDS-PAGE) using 12% gel. The proteins separated on the gel were stained with Silver stain II kit Wako (Wako Pure Chemical Industries, Ltd., Osaka, Japan).

The phage DNA was purified form the proteins by phenol-chloroform extraction and concentrated by ethanol precipitation. The purified and concentrated DNA was applied to electrophoresis using 0.7% agarose gel and stained by ethidium bromide.

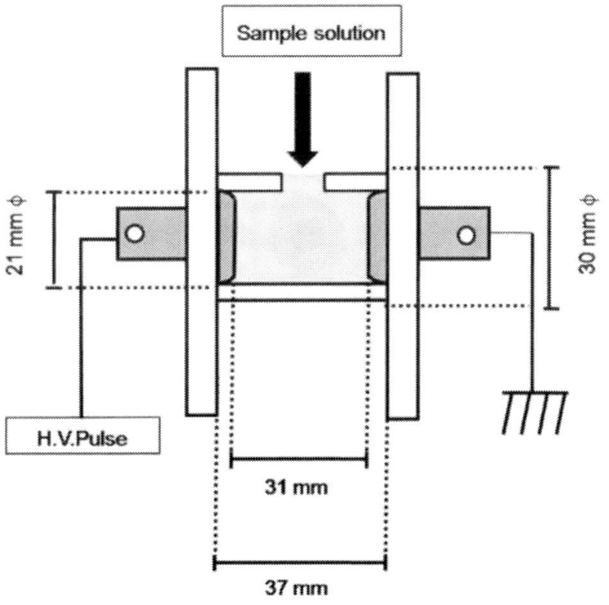

Figure 1. Schematic illustration of the PEF reactor used in this study.

3. Results

3.1. Inactivation of bacteriophage by PEF treatment

The effect of PEF treatment on inactivation of *E. coli* bacteriophage was confirmed using M13mp18 phage at room temperature. Figure 2 shows the time course of phage survival ratio in the PEF treatment at 5 and 7 kV. The survival ratio of M13mp18 phage was decreased with the increase of treatment time. The survival ratio of M13mp18 phage after 12 min PEF treatment at 5 and 7 kV were decreased to 9.0×10^{-5} and 2.0×10^{-6}, respectively. Increase of PEF applied voltage improved the inactivation efficiency of M13mp18 phage. The temperature of phage solution after 12 min PEF treatment at 5 and 7 kV reached to 48 and 71 °C, respectively. It was also confirmed that the survival ratio of m13mp18 phage did not decreased by heat treatment at 80 °C for 12 min (data not shown). From these results, it was confirmed that *E. coli* bacteriophage could be inactivated by PEF treatment.

Inactivation of λ phage, which has different structure of the phage particle from M13mp18 phage, by PEF treatment, was also studied. The decrease of λ phage survival ratio with the increase of the treatment time was also confirmed (figure 3). In addition to M13mp18 phage, PEF treatment could inactivate λ phage. The survival ratio of λ phage in the PEF treatment at 5 and 7 kV were almost same until 9 min. The survival ratio of λ phage after 12 min PEF treatment at 5 kV was decreased to 4.0×10^{-5}. Temperatures of phage solution after 12 min PEF treatment at 5 and 7 kV were 51 and 68 °C, respectively, and no inactivation by heat treatment at 70 °C for 12 min were also confirmed. Combination of PEF applied voltage and solution temperature above some threshold might increase efficiency of λ phage inactivation.

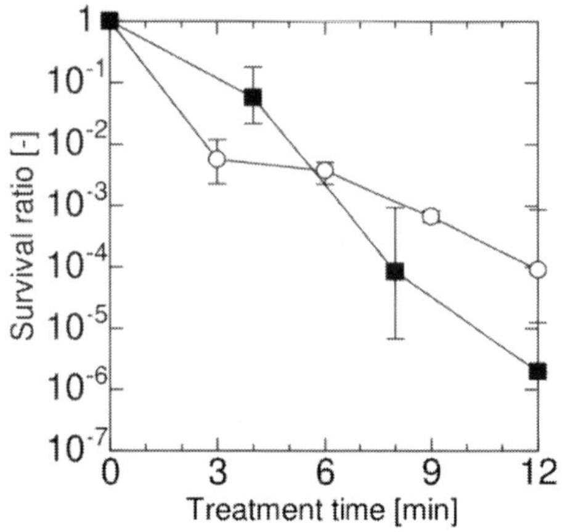

Figure 2. Time courses of M13mp18 phage survival ratio in the PEF treatment at 5 (open cyrcles) and 7 (closed squares) kV.

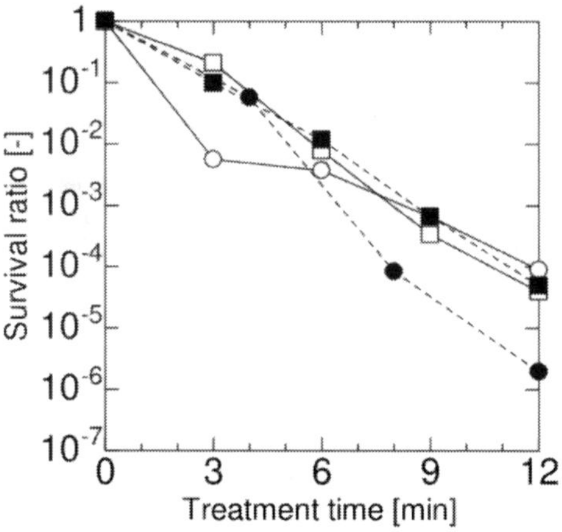

Figure 3. Time courses of phage survival ratio in the PEF treatment. Symbols show M13mp18 phage treated at 5 kV (open circles), M13mp18 phage treated at 7 kV (closed circles), λ phage treated at 5 kV (open squares) and λ phage treated at 7 kV (closed aquares).

3.2. Analysis of biological molecules of inactivated phage

Total proteins and DNA extracted from λ phage inactivated by PEF treatment were analyzed by the SDS-PAGE. Figure 4 shows the result of SDS-PAGE of PEF inactivated λ phage total protein. Some bands indicating capsid proteins were detected in the extracted samples from all PEF treated phage. Comparison between the band patterns of PEF untreated (lane 1) and PEF treated λ phages (lanes 2-5) showed no significant change and increase of small protein band generated by capsid protein digestion is not confirmed though survival ratios of λ phage were quite different each other (figure 3). From these results, it was confirmed that PEF treatment did not cause the protein degradation of λ phage during the inactivation process. In the agarose gel electrophoretic analysis of total DNA extracted from PEF treated λ phage, only single DNA band indicating genomic DNA of λ phage and no smear band were detected (figure 5). Meanwhile, purified bare λ phage genomic DNA was degraded by PEF treatment [9]. From these results, it was also confirmed that genomic DNA was not released from the phage particle and PEF had not degraded the genome DNA in the phage particle in the inactivation process.

From our electrophoresis analysis, degradation of biological molecules, which were capsid protein and phage DNA, of PEF inactivated λ phage was not detected. Inactivation of λ phage by PEF treatment might result from the denaturation of the protein or phage particle structure change.

3.3. Inactivation of E. coli phage in E. coli cell suspension

Sensitivity of *E. coli* phage to PEF treatment was compared with that of *E. coli* MV1184 cell. λ phage and MV1184 cell were treated by same PEF treatment at 5 kV, respectively (figure 6). The survival ratio of λ phage was lower than that of MV1184 at any treatment time, the survival ration of λ phage and MV1184 were 4.0×10^{-5} and 1.7×10^{-3}, respectively, after 12 min treatment. This result shows that λ phage is more sensitive to the PEF treatment than MV1184, i.e. *E. coli* cell is more tolerant to inactivation of PEF treatment than *E. coli* phage. This difference of sensitivity (tolerance) to PEF treatment might allow preferential inactivation of *E. coli* phage in the mixture of *E. coli* phage and *E. coli* cell.

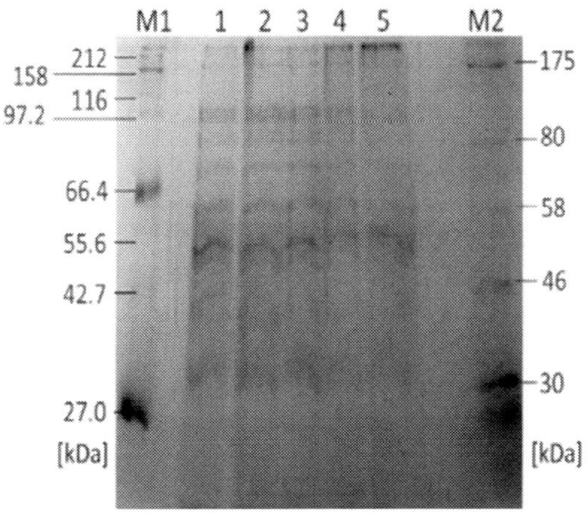

Figure 4. Picture of SDS-PAFE electrophoresis with total protein of native λ phage (lane 1) and PEF treated λ phage at 7 kV for 3 (lane 2), 6 (lane 3), 9 (lane 4) and 12 (lane 5) min.

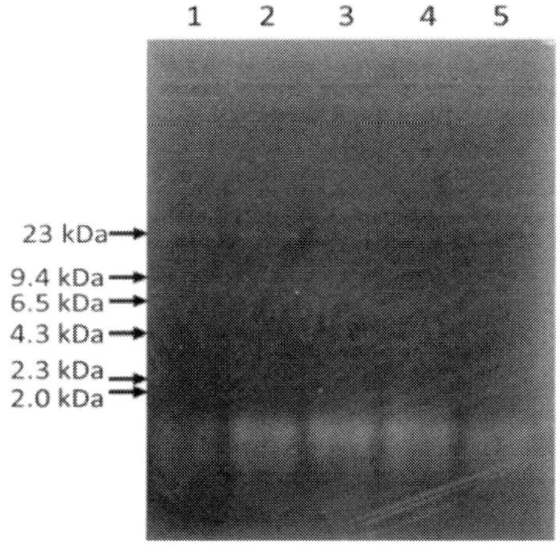

Figure 5. Picture of agarose gel electrophoresis with λ phage genomic DNA extracted from native λ phage (lane 1) and PEF treated λ phage at 7 kV for 3 (lane 2), 6 (lane 3), 9 (lane 4) and 12 (lane 5) min.

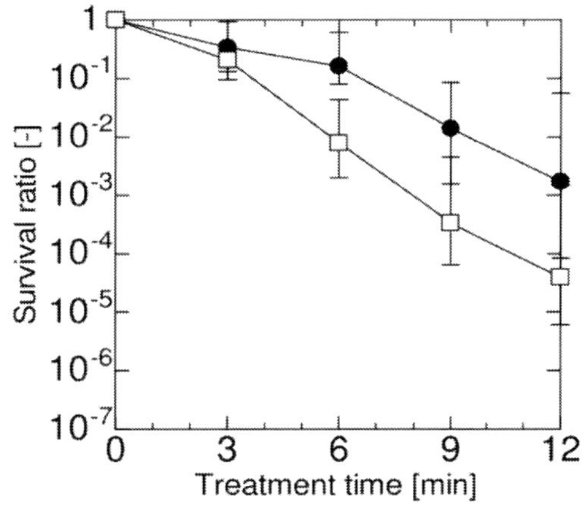

Figure 6. Time courses of *E. coli* MV1184 (closed circles) and λ phage (open squares) survival ratio in PEF treatment.

The mixture of λ phage and MV1184 was treated by PEF at 5 kV. And then, treated suspensions were inoculated into LB medium and cultivated at 35 °C. Figure 7 shows the time course of culture turbidity (OD_{660}) of each treated and cultivated suspension. In the cultivation of the phage and cell mixture without PEE treatment, the turbidity of culture reached peak value (0.36) at 2 h, and then decreased due to the cell lysis result from the infection of phage. In the cultivation of the mixture suspension treated by PEF for 3, 6, and 9 min, the time required to reach peak value of culture turbidity increased with the PEF treatment time. The culture turbidity of mixture suspension treated by PEF for 3, 6 and 9 min reached at 3, 4 and 6 h cultivation, respectively. This elongation of time resulted from the decrease of viable cell number existed in the mixture suspension used for the inoculum by the PEF treatment. However, the peak value of culture turbidity increased with the PEF treatment time. The peak values of culture turbidity of mixture suspension treated by PEF for 3, 6 and 9 min were 0.55, 0.68 and 0.74, respectively. These increases of attainable cell growth indicated the inactivation of phage in the phage and cell mixture suspension by PEF treatment. In the cultivation of mixture suspension treated by PEF for 12 min, significant cell growth did not detected until 7 h. However, cell growth reached to the high value as same as that of pure *E. coli* cell cultivation and no decrease by the cell lysis was determined. This result shows that phage in the mixture suspension was successfully inactivated by PEF treatment without complete inactivation of *E. coli* cell.

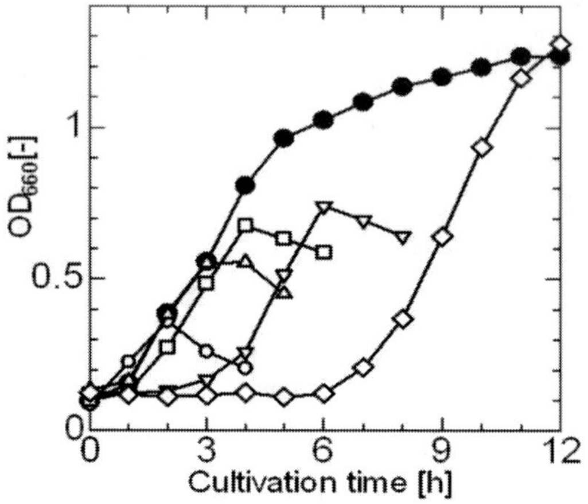

Figure 7. Time courses of turbidity of cell and phage mixture culture. Symbols show pure cell (closed circles), cell and phage mixture (open circles) and mixture treated by PEF at 5 kV for 3 (open triangles), 6 (open squares), 9 (open reverse-triangles) and 12 (open diamonds) min.

4. Discussion

In this study, the inactivation of the bacteriophage by PEF treatment was investigated using *E. coli* bacteriophage M13mp18 and λ as the model bacteriophage. The survival ratio of M13mp18 phage treated by PEF decreased with the increase of treatment time and applied voltage (figure 2). The heat treatment at 80 °C for 12 min did not decrease the survival ratio of M13mp18 phage (data not shown). Phage inactivation ability of PEF treatment was confirmed from these results. λ phage was also successfully inactivated by PEF treatment (Figure 3). M13mp18 phage particle is simple liner form (filamentous) bacteriophage harboring cyclic single strand genomic DNA surrounded by capsid protein [10-12]. λ phage particle is complicated form bacteriophage harboring double strand genomic DNA in the polyhedral head structure [12, 13]. The successful inactivation of both *E. coli* bacteriophages by PEF treatment demonstrated that PEF treatment could inactivate bacteriophage without dependence of bacteriophage particle form. However, inactivation profile of λ phage was a little different from that of M13mp18 phage. Increase of applied voltage of PEF did not affect the decrease of survival ratio of λ phage though M13mp18 phage inactivation was depended on the applied voltage. The difference of the dependence on the applied voltage for λ phage and M13mp18 inactivation was not clear from our results. The structural difference of phage particle seems to affect the dependence on the applied voltage for PEF inactivation. Further study should be necessary to discuss the dependence on the applied voltage.

Electrophoretic analysis of biological molecules, total protein and genomic DNA, of inactivated λ phage were carried out to investigate inactivation mechanism. Significant degradations of protein (figure 4) and DNA (figure 5) were not detected. In our previous study, it was reported that PEF treatment degrades the genomic DNA [9]. These results indicated that phage particle was not broken down by PEF treatment and genomic DNA was not release to the solution. Therefore, inactivation of phage might result form the disruption of phage function. Because the PEF treatment could denature several proteins [14, 15], part of capsid protein might be denatured, then phage particle lost infection ability to the cell.

The comparison of survival ratio of *E. coli* cell and phage in the PEF treatment showed that *E. coli* phage exhibited more sensitive behavior to the PEF treatment (figure 6). Meanwhile, in the heat treatment at the temperature below 70 °C, decrease of survival ratio of *E. coli* cell was larger than that of phage (data not shown). As the application of this difference, inactivation of *E. coli* phage in the mixture suspension of *E. coli* cell and phage without complete inactivation of *E. coli* cell was investigated (figure 7). The time required to detect significant cell growth was increased due to the cell death by the PEF treatment, however, the peak value of the culture turbidity before the cell lysis start was increased with the PEF treatment time. In the cultivation of the 12 min PEF treated suspension; no cell lysis by phage infection was detected. This result demonstrated that the inactivation of *E. coli* phage below the level to exhibit an influence for cell growth by PEF treatment without complete inactivation of cell. This phenomenon with PEF treatment could be applicable for the phage-free fermentation process because the phage could be inactivated completely in the suspension of bacteria and phage.

In conclusion, we demonstrated that PEF treatment could inactivate *E. coli* phage without dependence on the phage form. Unlike the heat treatment, *E. coli* phage was inactivated easier than *E. coli* cell during the PEF treatment. These finding could be available for inactivation of contaminated *E. coli* phage in the culture of gene modified *E. coli* to produce useful protein. Furthermore, it could be useful for other bacteria used in the food and chemical industry if this phenomenon used by PEF treatment could be confirmed in other phage-host system.

Acknowledgments
This work was supported partly by Grants-in-Aid for Scientific Research, Japan (No.22510082).

References

[1] Chen H, Joerger R D, Kingsley D H and Hoover D G 2004 *J. Food Protect.* **67** 505
[2] Capra M L, Patrignani F, Quiberoni A L, Reinheimer J A, Lanciotti R and Guerzoni M E 2009 *Int. Diary J.* **19** 336
[3] Müller M M, Rauscher T and Hinrichs J 2005 *Int. Dairy J.* **15** 777
[4] Lee C, Kim J and Yoon J 2011 *Chemosphere* **82** 1135
[5] Weaver J C, Chizmadzhev Y A 1996 *Bioelectrochem. Bioenerg.* **41** 135
[6] Locke B R, Sato M, Sunka P, Hoffmann M R and Chang J S 2006 *Ind. Eng. Chem. Res.* **45** 882
[7] Perron C Y, Vieira J and Messing J 1985 *Gene* **33** 103
[8] Ohshima T, Sato K, Terauchi H and Sato M 1997 *J. electrostat.* **42** 159
[9] Ohshima T, Ono T, Sato M, 1999 *J. Electrostat.* **46** 163
[10] Marvin D A, Hale R D, Nave C and Helmer C M 1994 *J. Mol. Biol.* **235** 260
[11] Marvin D A 1998 *Curr. Opin. Struct. Biol.* **8** 150
[12] Aksyuk A A, Rossmann M G 2011 *Viruses* **3** 172
[13] Hohn T, Katsura I 1977 *Curr. Top Miicrobiol. Immunol.* **78** 69
[14] Vega-Mercado H M, Powers J R, G. V. Barbosa-Ca ́novas G V and Swanson B G 2001 *Technomic Publishing* (Lancaster) p 121
[15] Castro A J, Swanson B G, Barbosa-Ca ́novas G V, Dunker A K 2011 *Technomic Publishing,* (Lancaster) p 83

The *Chlorella* killed by pulsed electrical discharge in liquid with two different reactors

Z Y Gao, B Sun*, Z Y Yan, X M Zhu, H Liu, Y J Song, M Sato
Environmental Science and Engineering College, Dalian Maritime University, Dalian, 116026, China

E-mail: sunb88@dlmu.edu.cn

Abstract. The application of pulsed high-voltage discharge in liquid has attracted wide attention as an effective water treatment. In this paper, two different liquid high-voltage discharge systems were constructed with plate-hole-plate and needle-plate electrode structures, and the inactivation behaviors of *Chlorella* were studied in the two reactors. The results show that the killing rates of algae in both reactors all increased significantly with increasing discharge voltage and the killing rates were intensely related to discharge power, instantaneous power and single pulse input energy. Furthermore, the inactivation effect in needle-plate reactor was superior to that in plate-hole-plate reactor under the same experimental conditions.

1. Introduction

The indiscriminate discharge of ship ballast water and sediments is a serious threat to the marine environment, resulting in the spread of harmful aquatic organisms and pathogens [1, 2]. The high voltage pulse discharge has attracted wide attention as an effective treatment of ship ballast water.

Bing Sun *et al.* [3] studied the removal of organic compound (phenol) using pulsed-streamer corona discharge and the effects of various parameters on phenol removal efficiency. The results showed that the organic contaminants in aqueous solution could be destroyed effectively by the pulsed-streamer corona discharge and the removal of the organic compound (phenol) was greatly influenced by gas injection (bubbling). M. Sato *et al.* [4] investigated the decoloration of some dyes in aqueous solution by pulsed discharge plasma in water through the pinhole. They concluded that the decoloration occurs as a result of the attack of high-speed electron and radical reaction in the plasma. Bing Sun *et al.* [5] reported the inactivation effect of algae with pinhole pulsed discharge reactor. The results showed that the chlorophyll a in the treated algae solution decreased with the increase of peak voltage. Bing Sun *et al.* [6] also investigated the decoloration of Methyl Orange and the degradation mechanism with multi-needle reactor. G. M. El-Aragi *et al* [7] proposed the effect of hybrid gas-liquid electrical discharge on liquid foods (milk). Their results showed that the electric field, ultra-violet radiation, shock wave, ozone and free radicals destructed the microorganisms, and the hybrid discharge treatment was effective in the destruction of different types of bacteria. M. Sato *et al.* [8] studied the sterilization of microorganisms by high-voltage pulsed discharge in water. They found that the great advantage of sterilization by pulsed discharge in water is that feed liquid can be selectively

* To whom any correspondence should be addressed.

sterilized, with little damage to any immobilized catalytic organism. Chih-Wei Chen [9] attempted the influence of pH on the inactivation of aquatic microorganism with gas-liquid pulsed electrical discharge using brush-plate type discharge reactor.

This paper investigated the inactivation of *Chlorella* with two different liquid discharge systems, and the effects of discharge power, instantaneous power and single pulse input energy on algae inactivation rates were compared with these two reactors.

2. Materials and methods

A plate-hole-plate reactor and a needle-plate reactor were used in this experiment. The schematic diagrams of the experimental apparatus were shown in figure 1 and figure 2. The plate-hole-plate reactor was made of two Plexiglas cylinders containing two 28 mm diameter stainless steel plate electrodes with thickness 3 mm and one 1 mm thick insulating board with a 1 mm hole in the center. The distance between the high voltage electrode and the insulation board was 7 mm, and the distance between the ground electrode and the insulating plate was 6 mm. The needle-plate reactor was made of a Plexiglas cylinder containing a hollow needle electrode with a diameter of 1 mm and a 28 mm diameter stainless steel plate electrode with thickness 3 mm. The volume and inner diameter of both two reactors were 40 mL and 32 mm, respectively. The electrode distances of the two reactors were both 14 mm. A pulsed power supply with a rotating spark-gap switch was used to generate high voltage pulses (0–60 kV, 0–300 Hz, 3 nF). In this experiment, the frequency of pulsed power supply was kept at 30 Hz.

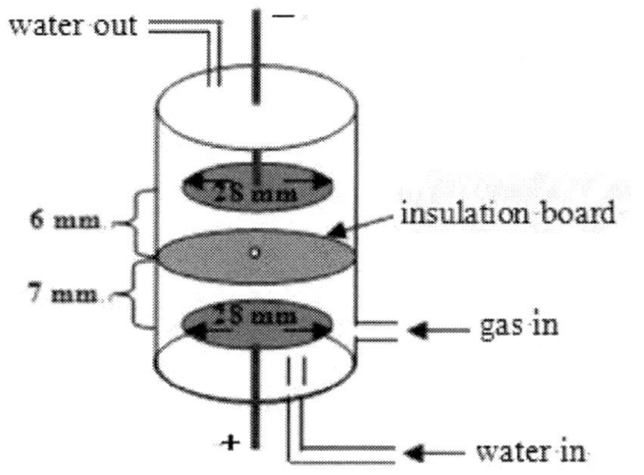

Figure 1. The plate-hole-plate reactor.

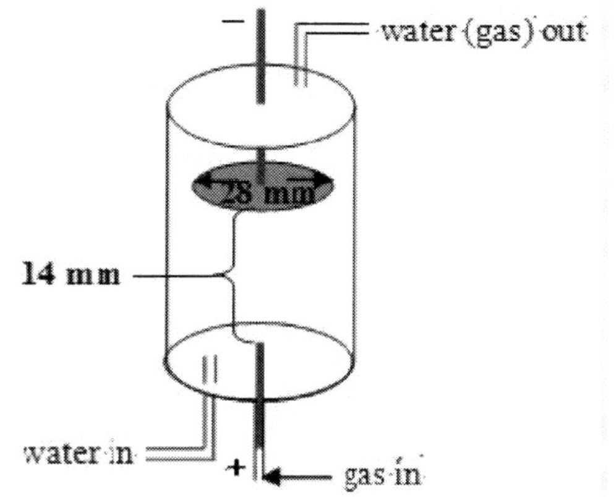

Figure 2. The needle-plate reactor.

In each experiment, 200 mL of *Chlorella* solution with an initial concentration of 8×10^5 per L^{-1} was circulated by a peristaltic pump at a flow rate of 53 mL min^{-1}. The electrical conductivity of the algae solution was 500 μS cm^{-1}. The viability of the algae remaining after discharge treatment was determined by re-culturing for five days. The number of *Chlorella* was counted by the optical microscope (OLYMPUS MA5201). The individual experiments were carried out by introducing air into the liquid in the reactor with the gas flow rate of 0.04 L min^{-1}.

In order to calculate the single pulse input energy of the reactor and discharge power, discharge voltage and current were measured by Digital Storage Oscilloscope (TDS 2024 B). The current and voltage waveforms of the discharge system were recorded by the current probe (Tektronix A 22) and voltage probe (Tektronix P6015), respectively.

3. Results and discussion

3.1. Comparison of discharge characters of plate-hole-plate reactor and needle-plate reactor

The discharge characteristic parameters of these two discharge reactors in table 1 and table 2. The values of peak current, maximum instantaneous power and single pulse input energy of two reactors both increased with increasing discharge voltage. The energy efficiency of needle-plate reactor was superior to that of plate-to-plate reactor because of higher peak current, maximum instantaneous power and single pulse input energy under the same voltage.

Table 1. The input voltage, peak instantaneous power and total input energy of a single pulse into the different pulse peak voltages (plate-hole-plate reactor).

Voltage (kV)	Peak current (A)	Maximum instantaneous power (MW)	Single pulse input energy (J)
18	122	2.02	0.54
22	138	2.71	0.82
26	164	3.51	1.08
28	164	3.74	1.29
30	182	2.48	1.43

Table 2. The voltage, peak instantaneous power and total input energy of a single pulse into the different pulse peak voltages (needle-plate reactor).

Peak voltage (kV)	Peak current (A)	Maximum instantaneous power (MW)	Single pulse input energy (J)
18	148	2.45	0.59
22	228	3.65	1.01
26	252	4.54	1.27
28	278	6.13	1.40
30	284	6.72	1.50

3.2. Effects of pulse peak voltage upon the inactivation of Chlorella sp.

During the experiment, the both reactors pulse peak voltage was set at 18, 22, 26, 28 and 30 kV, the pulse frequency was 30 Hz, the treatment time was 10 min, and the introducing air was 0.04 L min^{-1}.

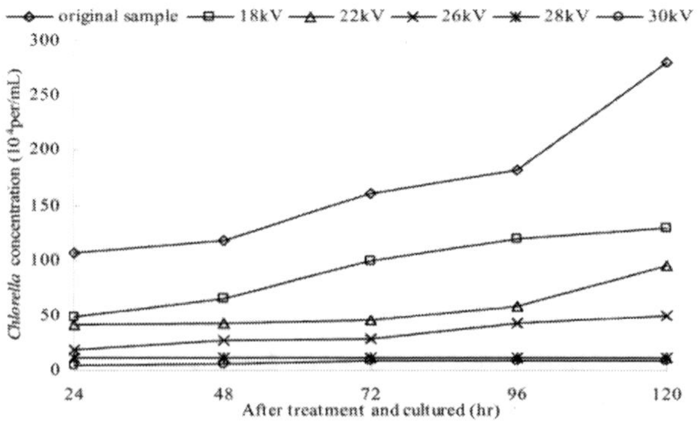

Figure 3. The effects of plate-hole-plate reactor pulse peak voltage changes upon the inactivation of the *Chlorella sp.* after re-culturing for 120 hr (treatment time of 10 min, frequency of 30 Hz, introducing air).

As shown in the figure 3, the viability of algae gradually decreased with increased pulse peak voltage by plate-hole-plate reactor. The maximum inactivation rate of algae nearly reached to 100% at pulsed voltage of 28 kV. The reason was probably that the pulse input power and the amount of high-energy electron generation increased in a single pulse with increased pulse peak voltage, which leading to more production of high-energy electrons, hydroxyl radicals, active oxygen, ozone and other strong oxidants [10]. In addition, the intensity of ultraviolet radiation also increased when the pulse discharge voltage, further contributing to the activation of algae.

From figure 4, the variation of viability of algae affected by pulse discharge voltage in needle-plate reactor was similar to that in plate-hole-plate reactor. There was no re-growth of algae when the pulse peak voltages were higher than 26 kV in the needle-plate reactor, even if the treated algae were re-cultured for five days. That is to say, the inactivation rate of algae nearly reached to 100% at pulsed voltage of 26 kV. The inactivation rate of the *Chlorella sp.* in the needle-plate reactor is higher than that in plate-hole-plate reactor under the same voltage. This result also confirmed that energy efficiency of discharge reactor played a key role in inactivation of algae.

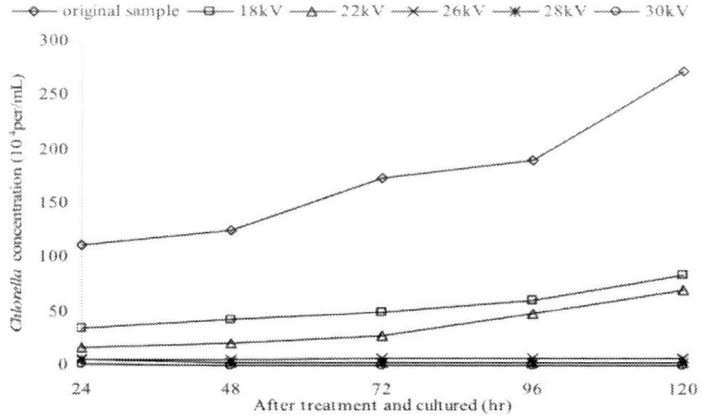

Figure 4. The effects of needle-plate reactor pulse peak voltage changes upon the inactivation of the *Chlorella sp.* after re-culturing for 120 hr (treatment time of 10 min, frequency of 30 Hz, introducing air).

3.3. Color changes of Chlorella sp. solution by pulse peak voltage

After discharge treatment, the algae solution was re-cultured for 120 hours. The color changes of *Chlorella sp.* solution with different pulse peak voltage in two reactors were shown in figure 5 and figure 6. The leftmost conical flask labeled "Sample" contains the solution without discharge treatment and other flasks labeled with various pulse peak voltages contain the solutions that have been discharged. The results showed that the color of the treated algae solution gradually became colorless with increased pulse discharge voltage, and the solution became nearly colorless at 28 kV in plate-hole-plate reactor and at 26 kV in needle-plate reactor, even if the treated algae solutions were re-cultured again for 5 days. It again proved that the maximum inactivation rates reached almost 100% in both reactors with discharge voltage higher than 28 kV, and the energy efficiency of needle-plate reactor was remarkable compared with plate-hole-plate reactor under the same experimental condition.

Figure 5. The photo of *Chlorella sp.* solutions re-cultured for 120 hours after plate-hole-plate reactor discharge treatment with the pulse frequency of 30 Hz and the treatment time of 10 min.

Figure 6. The photo of *Chlorella spp.* solutions re-cultured for 120 hours after needle-plate reactor discharge treatment with the pulse frequency of 30 Hz and the treatment time of 10 min.

4. Conclusions

The conclusions can be obtained as follows.

(1) Compared with plate-hole-plate reactor, needle-plate reactor had higher current intensity and instantaneous power in the discharge process, and total single pulse input energy was also higher under the same discharge condition.

(2) The killing rate significantly increased with increasing voltage in both the discharge reactors.

(3)The killing rate of algae reached to 100 % at voltage of 28 kV in plate-hole-plate reactor and at voltage of 26 kV in needle-plate reactor, and the color of the treated algae solution was nearly colorless. Generally, the inactivation rates of *Chlorella sp.* by needle-plate reactor are higher than that by plate-hole-plate reactor.

Acknowledgments

We would like to express our gratitude for the support of the National Natural Science Foundation (NSFC-10875019, 41005079). It is supported by "The Fundamental Research Funds for the Central Universities".

References

[1] Tamelander J, Riddering L, Haag F and Matheickal J 2010 (GloBallast Partnerships Project Coordination Unit /International Maritime Organization)
[2] Lin W, Zhou G F, Cheng X Y and Xu R M 2007 *Fast Economic Development Accelerates Biological Invasions in China PLoS* 2 1208
[3] Sun B, Sato M and Clements J S 1999 *J. Phys. D: Apply. Phys.* **32** 1908
[4] Sato M, Yamada Y and Sugiarto A T 2000 *Trans. Inst. Fluid-Flow. Mach.* **107** 95
[5] Sun B, Aye N N, Wang X M, Zhu X M, and Sato M 2011 *IEEE Trans. Ind. Appl.* **47** 1079
[6] Sun B, Aye N N, Gao Z Y, Lv D, Zhu X M and Sato M 2012 *J. Environ. Sci.* **24** 840
[7] El-Aragi G M and Abedel Rahman Y M 2008 *African J. Food. Sci.* **2** 11
[8] Sato M, Tokita K and Sadakata M 1990 *Int. Chem. Eng.* **30** 695
[9] Chen C W 2009 *J. Electrostatics* **67** 703
[10] Sun B, Sato M and. Clements J S 1997 *J. Electrostatics* **39** 189

The inactivation of Chlorella spp. with dielectric barrier discharge in gas-liquid mixture

Dan Song, Bing Sun*, Xiaomei Zhu, Zhiyu Yan, Hui Liu, Yongjun Liu
Environmental Science and Engineering College, Dalian Maritime University, Dalian, China. 116026

E-mail: sunb88@dlmu.edu.cn

Abstract. The inactivation of *Chlorella spp.* with high voltage and frequency pulsed dielectric barrier discharge in hybrid gas-liquid reactor with a suspension electrode was studied experimentally. In the hybrid gas-liquid reactor, a steel plate was used as high voltage electrode while a quartz plate as a dielectric layer, another steel plate placing in the aqueous solution worked as a whole ground electrode. A suspension electrode is installed near the surface of solution between high voltage and ground electrode to make the dielectric barrier discharge uniform and stable, the discharge gap was between the quartz plate and the surface of the water. The effect of peak voltage, treatment time, the initial concentration of *Chlorella spp.* and conductivity of solution on the inactivation rate of *Chlorella spp.* was investigated, and the inactivation mechanism of *Chlorella spp.* preliminarily was studied. Utilizing this system inactivation of *Chlorella spp.*, the inactivation rate increased with increasing of peak voltage, treatment time and electric conductivity. It was found that the inactivation rate of *Chlorella spp.* arrived at 100% when the initial concentration was 4×10^6 cells mL^{-1}, and the optimum operation condition required a peak voltage of 20 kV, a treatment time of 10 min and a frequency of 7 kHz. Though the increasing of initial concentration of the *Chlorella spp.* contributed to the addition of interaction probability between the *Chlorella spp.* and O_3, H_2O_2, high-energy electrons, UV radiation and other active substances, the total inactivation number raise, but the inactivation rate of the *Chlorella spp.* decreased.

1. Introduction

The gas-liquid mixture dielectric barrier discharge (DBD) is a leading advanced oxidation water treatment technology that can produce large amounts of active substances which contain high energy electron, ion and plasma, and generate O_3, H_2O_2. This technique have the advantages of high treatment efficiency, time-energy saving and without secondary pollution [1-4]. Currently, there is none research of using gas-liquid mixture DBD to kill *Chlorella spp.*.

This paper utilized non-thermal plasma generated from gas-liquid mixture DBD to kill the *Chlorella spp.*, investigated the impact of parameters on the inactivation of *Chlorella spp.*, concluding discharge peak voltage, the treatment time, solution conductivity and *Chlorella spp.* initial concentration, and preliminarily investigated the killing mechanism.

2. Experiment

* To whom any correspondence should be addressed.

2.1. Instrument

The experimental setup is shown in figure 1. The reactor made of Plexiglas was 160×160×150 mm, the quartz plate was as a dielectric layer and the electrodes plate was stainless. Suspended electrode was immersed in the surface of solution with the mesh size of 5×5 mm. Discharge gap covered between the quartz plate and the solution surface. The voltage of high voltage pulse power was 0-20 kV, the frequency was 0-10 kHz, voltage probe model was Tektronix P6015A, the current probe model was Pearson 2878, and the oscilloscope model was Tektronix TDS2024B which detected the output voltage and current waveforms.

The concentration of *Chlorella spp.* in laboratory was 4×10^{7} cells mL^{-1}, and the conductivity was 12 $mS\cdot cm^{-1}$. In this paper, d stands for the discharge gap, *f* stands for the frequency, U is the peak voltage, σ is conductivity of the solution, c represents the initial concentration of *Chlorella spp.*, t represents the discharge time.

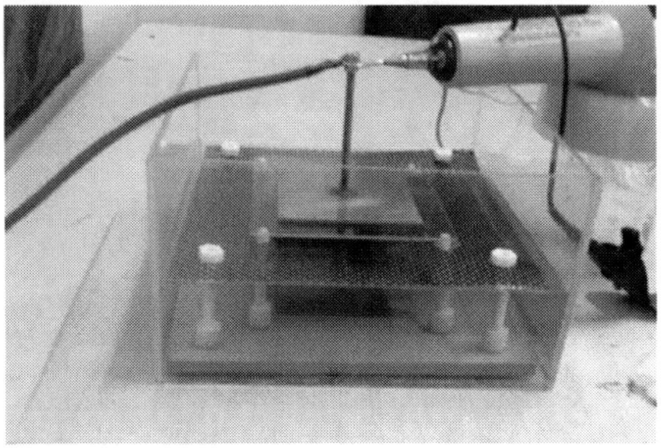

Figure 1. Scheme of experimental device.

2.2. Methods of test

Each experiment recycling treated 800 mL *Chlorella spp.* solution in the reactor, then took 100 mL solution after discharge treatment and 100 mL without discharging to be cultured for 5 days under the same conditions. The microscopy was used to count the number of *Chlorella spp.* directly, and calculated the killing rate of the *Chlorella spp.* as shown in equation 1.

$$\text{The inactivation rate}(\%) = \frac{C_0 - C_i}{C_0} \times 100\% \qquad (1)$$

C_0 is the concentration of *Chlorella spp.* without discharging treatment after 5 days, C_i is the concentration of *Chlorella spp.* under discharge treatment after 5 days.

3. Results and discussion

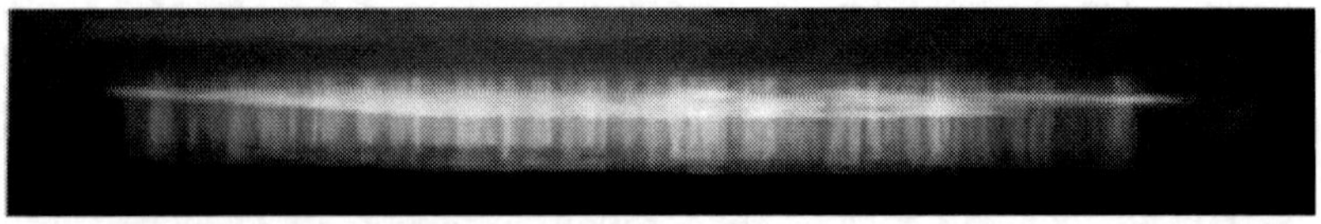

Figure 2. The gas-liquid mixture DBD topography (d=5 mm, *f*=7 kHz, U=20 kV, pH=6.85, σ=1.0 μS cm^{-1}).

Figure 2 is the topography of gas-liquid DBD while the ultra-pure water was injected into the reactor, it was clear that there were numbers of discharge plasma channels, and the discharge was intense, uniform and stable.

Figure 3 shows the voltage and current waveforms under the same conditions as in figure 2, there were two discharges in a pulse period; the power was applied to start the first discharge, while the surface charge accrued. When the electric field direction changed, and the surface charge accumulated to a certain extent to lead the second discharge.

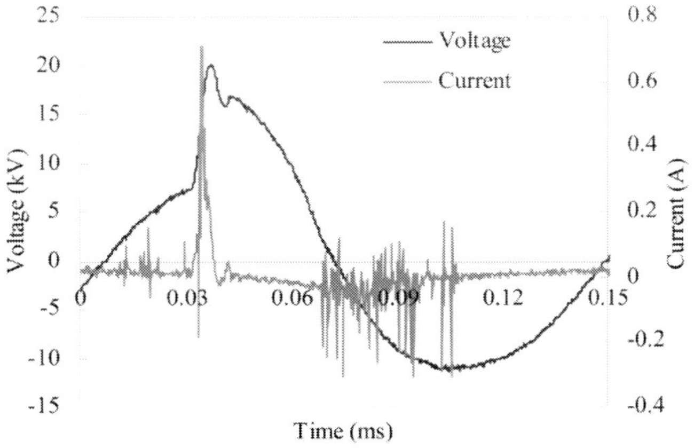

Figure 3. The gas-liquid mixture DBD voltage-current waveforms (d=5 mm, f=7 kHz, U=20 kV, pH=6.85, σ=1.0 μS cm^{-1}).

3.1. *Effect of discharge peak voltage*

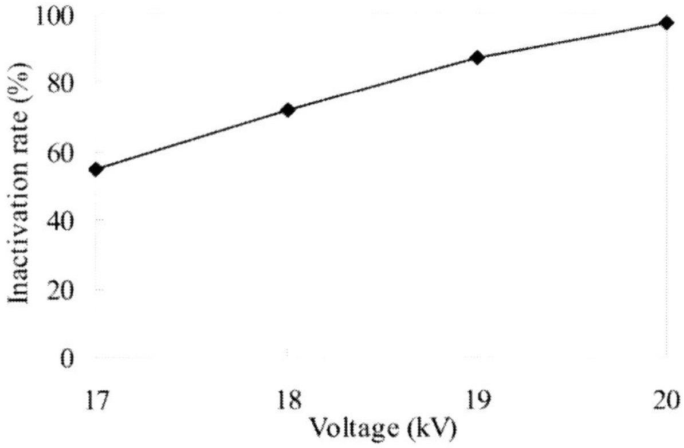

Figure 4. Effect of peak voltage on the inactivation rates of *Chlorella spp.* (t=10 min, c=4×10^6 cells mL^{-1}, σ=3 mS cm^{-1}).

The higher the discharge peak voltage was, the lower survival number of *Chlorella spp.* was, and the higher the kill rate was. When the discharge intensity increased, the number of oxidizing groups concluding the hydrogen peroxide, hydroxyl and ozone increased, the effect of membrane permeability increased, resulting to cell broken [5, 6]. UV light and the shock wave enhanced killing effects of *Chlorella spp.*

In figure 4, the kill rate of *Chlorella spp.* was 97.58% with the peak voltage of 21 kV. The cells of *Chlorella spp.* cultured for 5 days were observed by microscopic; the numbers were essentially the same in each day. The cells may had died, but not broken and suspended in solution, so the microscope cannot judge that they were dead. Therefore, we identified the inactivation rate was 100%.

3.2. Effect of treatment time

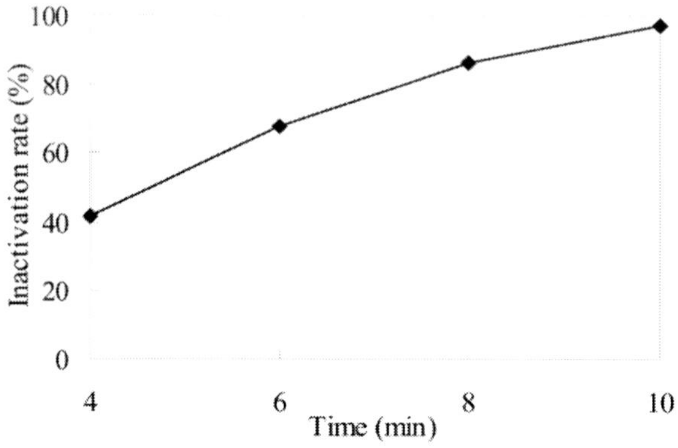

Figure 5. Effect of treatment time on the inactivation rates of *Chlorella spp.* (U=20 kV, c=4×10^6 cells mL^{-1}, σ=3 mS cm^{-1}).

Figure 5 shows that the longer the discharge treatment time was, the higher the inactivation rate was. The various factors of UV, reactive species, electric field strength and the shock wave had accumulated superimposed effect with the time continuing [7], which affected the activity of *Chlorella spp.* and resulted in an increase of the kill rate.

3.3. Effect of solution conductivity

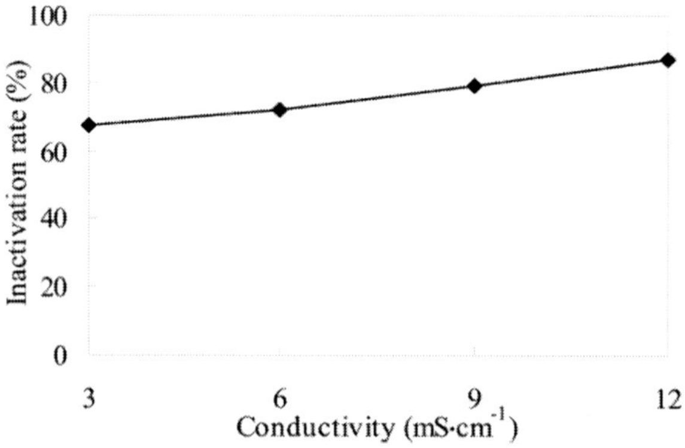

Figure 6. Effect of conductivity on the inactivation rates of *Chlorella spp.* (U=20 kV, c=4×10^6 cells mL^{-1}, t=6 min).

In figure 6, after discharging 6 min, the inactivation rate of *Chlorella spp.* with the conductivity of 12 mS cm^{-1} was 13% more than the conductivity of 3 mS cm^{-1}. The conductivity increased and the number of survival cells decreased, the kill rate raised up. The addition of conductivity led to the electrical resistance decreasing, the total current in the circuit rising, the voltage applied to discharge gap increasing, so the discharge was more intense and resulted a higher inactivation rate.

3.4. Effect of initial concentration

In Figure 7, the inactivation rates of *Chlorella spp.* with the initial concentration of 4×10^5 cells mL^{-1}, 4×10^6 cells mL^{-1}, 4×10^7 cells mL^{-1} were 100%, 97.58% and 94.38%. The inactivation rate of *Chlorella spp.* decreased with the addition of the initial concentration. Though the cells had more probability to contact with hydrogen peroxide, ozone and other active substances in the solution, the number of death cells increased, the initial concentration was too large to decrease the killing rate.

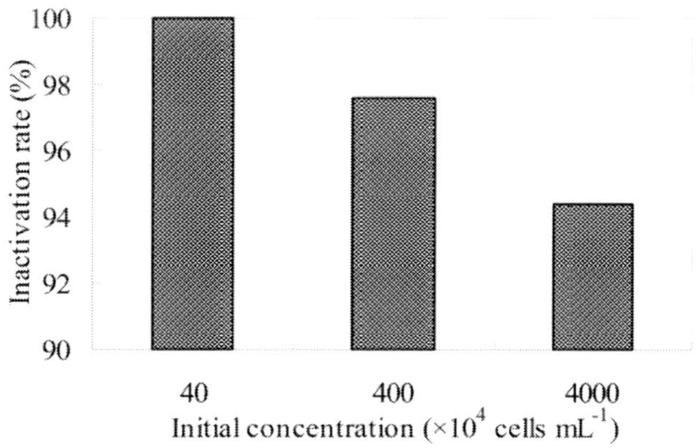

Figure 7. Effect of initial concentration on the inactivation rates of *Chlorella spp.* (U=20 kV, t=10 min, σ=12 mS cm^{-1}).

4. Conclusion

(1) The concentrations of hydrogen peroxide and ozone in solution by gas-liquid mixture DBD increased with the increasing of the discharge peak voltage and treatment time.
(2) Utilizing this system inactivation of *Chlorella spp.*, the inactivation rate increased with increasing of peak voltage, treatment time and electric conductivity, and decreased with the addition of initial content.
(3) The inactivation rate of *Chlorella spp.* arrived at 100% when the initial content was 4×10^6 cells mL^{-1}, and the optimum operation condition required a peak voltage of 20 kV, a treatment time of 10 min.

Acknowledgement

We thank the support of the National Natural Science Foundation (NSFC-10875019, 41005079). It is supported by "The Fundamental Research Funds for the Central Universities". The authors appreciate to Professor M. Sato for his valuable suggestion.

References

[1] Clments J S and Dubbelman T M 1987 *IEEE Trans. Ind. Appl.* **23** 225
[2] Sun B, Sato M and Harano A 1998 *Journal of Electrostatics.* **43** 115
[3] Sato M and Ohgiyama T 1996 *IEEE Trans. on Industry Appl.* **32** 106

[4] Laroussi M and Dobbs F C 2002 *IEEE Trans. on Plasma Science.* **30** 1501
[5] Zuckerman H, Krasik Y E and Felsteiner J 2002 *Innovative Food Science and Emerging Technologies.* **3** 329
[6] Sun B, N.N.Aye, Wang X, Zhu X and Sato M 2011 *IEEE Trans. Ind. Appl.* **47** 1079.
[7] Sun B and Sato M 1999 *Phys. D: Appl. Phys.* **32** 1908

Study on the killing of oceanic harmful micro-organisms in ship's ballast water using oxygen active particles

C Chen[1], X Y Meng[1], M D Bai[1, 2*], Y P Tian[1], Y Jing[1]

1 Institute of Marine Engineering, Department of Physics, Dalian Maritime University, China

2 College of Environment and Ecology, Xiamen University, China

E-mail: Mindong-bai@163.com

Abstract. Global Environment Facility has identified that the spread of marine invasive alien species is one of the four major risk factors threatening the safety of global marine environments. Ballast water discharge is the main cause of biological invasion. With physical methods of strong electric field ionization discharge at atmospheric pressure, O_2 and sea water (gaseous) were ionized, and then dissociated to a number of oxygen active particles (ROS) such as ·OH, O_2^+, H_2O^+, etc. ROS was injected into 0.6 t h^{-1} ballast water treatment system to form high concentration ROS solution in order to kill the harmful micro-organisms in ballast water. According to the land-based test standard of International Maritime Organization (IMO) Guidelines for Approval of Ballast Water Management Systems (G8), this paper concludes that single-cell algae of 3.0×10^4 cell ml^{-1} and bacteria of 2.0×10^4 cfu ml^{-1} were killed by ROS solution of 2.0 ppm. Death rate could reach almost 100%. The results meet the requirements of Regulation D-2 of International Convention for the Control and Management of Ships' Ballast Water and Sediments completely.

1. Introduction

Invasion of native biotas by non-indigenous species is a threat to the integrity of biotic communities, the economy and even human health. The introduction of invasive marine species into new environments by ship's ballast water has been identified as one of the four greatest threats to the world's oceans [1]. With the development of economic globalization, marine shipping industry has occupied an important part of world trade, and its cargo has carried more than 90 percent of all goods in world trade [2]. Ships' ballast water as a major bio-invasion pathway is increasing environmental threats [3]. Ballast water containing marine plankton is carried by ship. When the ship reaches other harbor, the ballast water with the plankton including bacterial virus, algae, sea fish, small invertebrates, spores, etc is discharged into local water [4]. According to report, approximately 80000 ships all over the world carry more than 10 billion tons of ballast water every year [5]. Migratory species in the ballast is about 10000 every day in the global scope. So far, the number is still increasing. Every year, biological invasion of ballast water of the global scope causes the economic losses about billions of dollars, and the number is increasing year after year.

* To whom any correspondence should be addressed.

In order to reduce the effects of discharge of ballast water on ecological environment, International Maritime Organization (IMO) issued "Ships ballast water and sediment control and management of International Convention" in February 2004. In the Convention, from 2009 onwards, all new ships should be installed the ballast water treatment equipments. By 2016, all ships must be installed the ballast water treatment equipments. At present, there are several methods for treating ballast water. The most common techniques are as follows: ballast water exchange, ultraviolet, electrolytic and ozone [6-9]. However, these methods can't kill the microorganism in ballast water completely. The treated ballast water can't reach D2 emissions standards formulated by IMO.

Advanced oxidation technology (AOT) based on the classical chemical oxidation method [10], which can produce a large number of •OH and induce a series of •OH chain reactions, resulting in removing the organisms and other pollutants and eventually become H_2O and CO_2 [11, 12]. The technique provides a green pollution-free new approach for the treatment of ballast water.

2. Materials and methods

2.1 Discharge process of strong electric field at atmospheric pressure

In our research, the plasma discharge of strong electric-filed was consist of large numbers of micro-streamers. This process involved a well-distributed, diffuse and stable discharge phenomenon akin to glow discharge. In micro-streamer channels, lots of active particles produced giving rise to a series of plasma chemistry reactions.

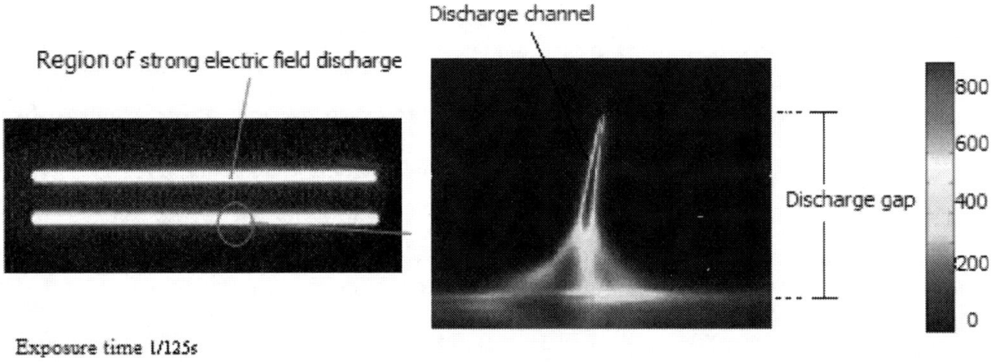

Figure 1. Schematic diagram for the formation of micro-streamer in strong-field ionization discharge.

Under the conditions of the 0.47 mm discharge gap, a power supplying with the frequency in the range of 5-20 kHz and the peak voltage in the range of 8-20 kV, the strong electric field could be obtained in the discharge gap. When electric field strength (E) was higher than 100 $kV \cdot cm^{-1}$, the average electron energy was higher than 10 eV, electron density was higher than 10^{14} cm^{-3} and the ionization duty cycle was higher than 2%. As shown in figure 1, the average electron energy of the electrons in the electric field was greater than 10 eV. Particularly in the discharge channel, the electron temperature was more than 12 eV. Electron energy was greater than the ionization energy of N_2 and H_2O. Figure 2 was a structure diagram of plasma source.

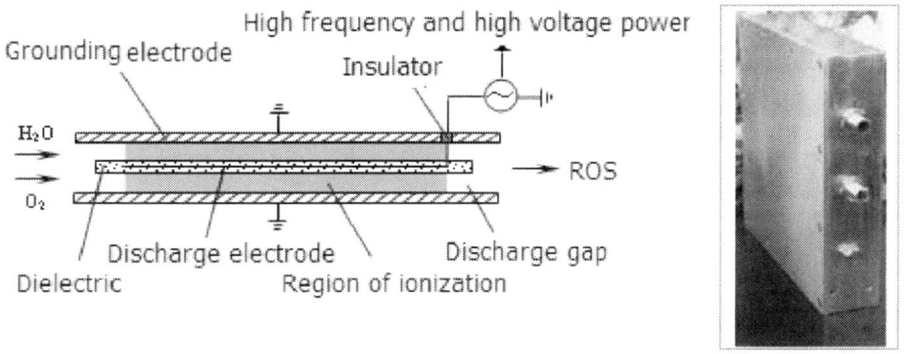

Figure 2. Schematic diagram for the formation of micro-streamer in strong-field ionization discharge.

2.2 Formation of oxygen active particles

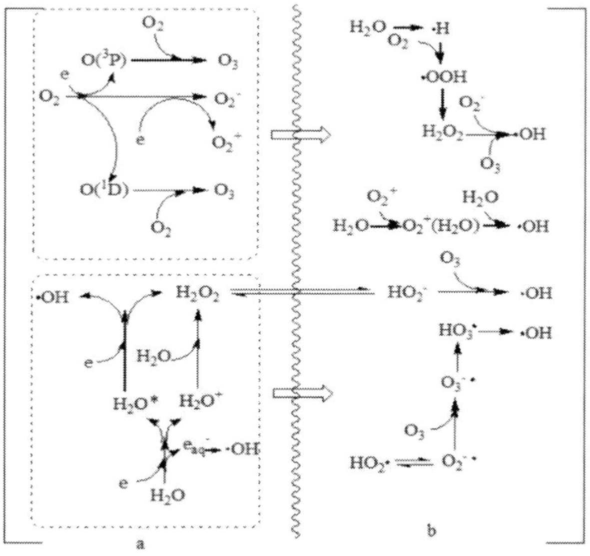

Figure 3. Chemistry mechanism of hydroxyl radical formation.

In the atmospheric electric field discharge, the active particles ROS contained •OH, O_2^+, $O_2^{-\cdot}$, O_3, H_2O^+, H_2O_2, $HO_2\cdot$, etc, and •OH was the primary active particle. •OH was generated by three reaction pathways, as shown in figure 3 [9, 13]. Firstly, H_2O_2 can be decomposed inside the plasma to give HO_2^- radical. HO_2^- was an important initiator of generating of •OH, which serves as the main reaction. The reaction rate constant of producing •OH was 2.2×10^6 L (mol s) $^{-1}$ which was three times higher than the HO^- average reaction rate constant of 70 L (mol s) $^{-1}$. Under the action of HO_2^-, active particles such as O_3, O_2^-, etc, were efficient to generate •OH.

2.3 Experimental systems and parameters

In our research, we developed 0.6 t h^{-1} ballast water treatment equipment and the core technology was plasma source (figure 4). This equipment was used to deal with harmful micro-organisms in ballast water. Original sea water was obtained from Guanglu Island, Dalian. The salinity was 28.1 with practical salinity unit (PSU) and the temperature was 18.7 °C. As the detection index for concentration of ROS, TRO was defined as all the oxidants which could kill harmful micro-organisms. TRO was mainly composed of ·OH and other active oxygen particles such as HO_2^-, $O_2\cdot^-$, $HO_3\cdot$, O_2^+, H_2O, etc, which were generated during the ROS chain reactions in ballast water. Experiments lasted for 5 days, including control tank and treated tank tests.

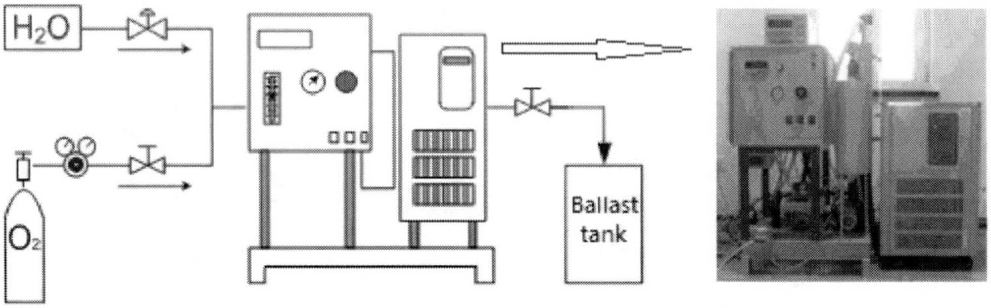

Figure 4. Schematic diagram for 0.6 t h^{-1} ballast water treatment equipment.

According to G8, phytoplankton as the research objects should be selected in less than 50 microns in size and five species in three phyla. Therefore, phytoplankton were selected as follows: *Tetraselmis*, *porphyridiun*, *Cylindrotheca*, *Alexandrium* and *Karenia mikimotoi*, whose concentration were 0.70×10^4 cell ml^{-1}, 0.72×10^4 cell ml^{-1}, 0.68×10^4 cell·ml^{-1}, 0.70×10^4 cell·ml^{-1} and 0.67×10^4 cell·ml^{-1}, respectively. Total concentration of algae was 3.47×10^4 cell·ml^{-1}. Heterotrophic bacteria should be present in a density of at least 10^4 living bacteria per millilitre, so three kinds of bacteria (*Heterotrophic bacteria*, *Escherichia coli*, *Enterococcus*) were selected as the experimental bacteria, concentration were 2.03×10^4 cfu ml^{-1}, 0.96×10^4 cfu ml^{-1} and 0.72×10^4 cful ml^{-1}, respectively. Concentrations of ROS solution were ranged from 0-2.0 ppm.

3. Results and Discussion

The experimental results for the ballast water treatment were shown in table 1. The result presented that single-cell algae of 3.0×10^4 cell·ml^{-1} and bacteria of 2.0×10^4 cfu·ml^{-1} were killed by ROS solution of 2.0 ppm. After detecting 150 mL treated water sample, the results showed that the concentration of algae and bacteria were 7 cell·ml^{-1} and 3 cfu ml^{-1}, respectively. Death rate reached almost 100 %. It reached the requirement of Regulation D-2 of International Convention for the Control and Management of Ships' Ballast Water and Sediments completely. The ROS concentration in the treated tank declined over time during storage for 5 days. The TRO in the treated tank was advantageous to inhibit the re-growth of algae and bacteria. Until the fifth day, the micro-organisms did not reactivate, indicating that harmful micro-organisms in ballast water were rapidly killed by ROS solution.

Different algae and bacteria had different morphology, so the treatment effects of ROS were also different. As shown in figure 5, with the increase of ROS solution concentration, death rate of *Alexandrium*, *porphyridiun* and *Cylindrotheca* increased gradually, which indicated the survival rate of algae declined gradually. But the same kind of algae with different original concentration had different lethal concentration of ROS solution. For the concentration of 0.7×10^4 cell mL^{-1} of *Alexandrium*, when ROS solution concentration increased to 1.2 ppm, its lethal effect reached to inflection point and a large number of deaths appeared. Equally, for the concentration of 0.72×10^4 cell ml^{-1} of *porphyridiun*, its death inflection point also appeared in 1.2 ppm of ROS solution. However, for *Cylindrotheca*, whose concentration was 0.68×10^4 cell ml^{-1} and less than concentrations of *Alexandrium* and *porphyridiun*, the death of inflection point extended to 1.25 ppm. In comparison to three kinds of algae with the same concentration, *Cylindrotheca* died with higher concentration of ROS solution due to the morphology structure. *Cylindrotheca* belonged to the *Bacillariophyta*, and its surface had chitin carapace that could resist ROS, therefore, greater dose of ROS solution was needed

Table 1. Five day ballast water treatment experiment.

Storage time	0 h		48 h		120 h	
item	control	treated	control	treated	control	treated
TRO (mg L^{-1})	0	2.0	0	0.21	0	0
Tetraselmis (cell ml^{-1})	0.70×10^4	1	0.40×10^4	0	0.21×10^4	0
Porphyridiun (cell ml^{-1})	0.72×10^4	0	0.38×10^4	0	0.20×10^4	0
Cylindrotheca (cell ml^{-1})	0.68×10^4	5	0.42×10^4	0	0.25×10^4	0
Alexandrium (cell ml^{-1})	0.70×10^4	1	0.37×10^4	0	0.19×10^4	0
Mikimoto (cell ml^{-1})	0.67×10^4	0	0.41×10^4	0	0.21×10^4	0
Total algae(cell ml^{-1})	3.47×10^4	7	1.98×10^4	0	1.06×10^4	0
Heterotrophic.bacteria (cfu ml^{-1})	2.03×10^4	3	1.96×10^4	1	1.92×10^4	1
Escherichiacoli(cfu ml^{-1})	0.96×10^4	1	0.91×10^4	0	0.87×10^4	0
Enterococcus (cfu ml^{-1})	0.72×10^4	0	0.7×10^4	0	0.66×10^4	0

In figure 6, two kinds of bacteria *Escherichia coli* and *Enterococcus* had no significant difference in the lethal dose due to the simple structure of bacteria and ROS was effective for them. *Escherichia coli* and *Enterococcus* began to show a large number of deaths in 0.8 ppm of ROS solution. The death rate reached approximately 100 % when the concentration increased to 1.5 ppm.

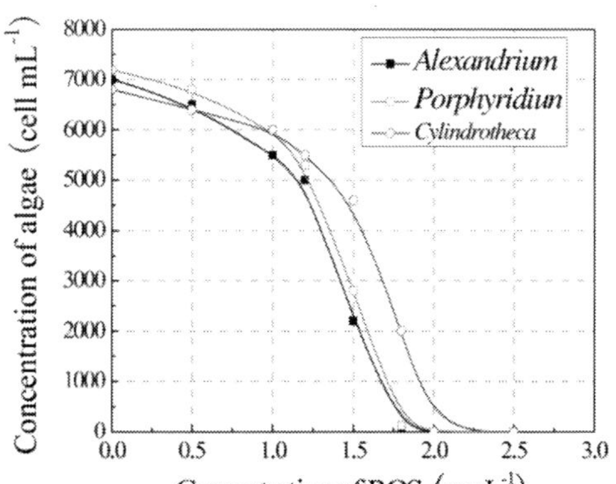

Figure 5. Relationship between the concentration of algae and the lethal concentration of ROS.

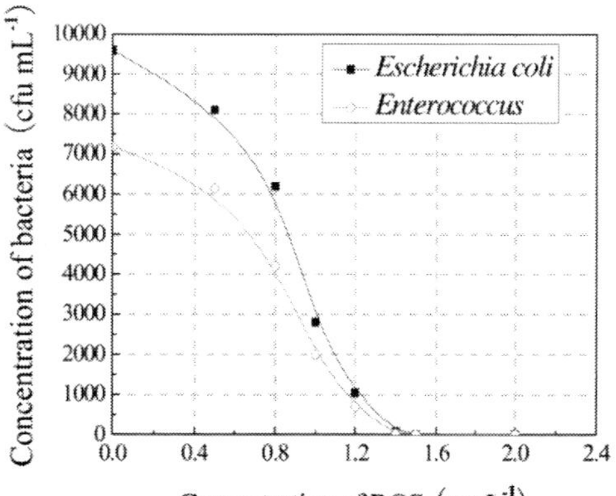

Figure 6. Relationship between the concentration of bacteria and the lethal concentration of ROS.

4. Conclusions

With physical methods of strong electric field ionization discharge at atmospheric pressure, O_2 and sea water (gaseous) are ionized, and then dissociate to a number of oxygen active particles (ROS) such as ·OH, O_2^+, H_2O^+, etc. In order to kill the harmful micro-organisms in ballast water, ROS is injected into 0.6 t h^{-1} ballast water treatment system to form a high concentration ROS solution. In transport process of the ship ballast water, harmful micro-organisms are killed rapidly and efficiently.

Algae of 3.47×10^4cell ml^{-1} and bacteria of 2.03×10^4 cfu ml^{-1} are killed by the 2.0 ppm of ROS solution. The experimental results show that the concentrations of micro-organisms are all less than 10 unit ml^{-1}. Death rate reaches approximately 100%. The micro-organisms do not reactivate within 120 h. The treated water reaches the requirements of regulation D-2 of International Convention for the Controlling and Management of Ships' Ballast Water and Sediments completely.

Acknowledgments

This work is supported by The National High Technology Research and Development Program of China under Grant No 2012AA062609, The International Science and Technology Cooperation Project of China under Grant No 2010DFA61470, The National Science Foundation for Distinguished Young Scholars of China under Grant No 61025001, The Fundamental Research Funds for the Central Universities under Grant No 2012QN067 and 2012TD061.

References

[1] Pughuic D, Raaymakers S and Maker M 2001 *1st Int. Ballast Water Treatment Standards Workshop, Workshop Report* p 1
[2] Chen L Q, Li Y K and Hou J L 2005 *J. East China Normal University (Natural Science)* **12** 40
[3] Ruiz G M, Rawlings T K, Dobbs F C, Drake L A, Mullady T, Huq A and Colwell R R 2000 *Nature* **408** 49
[4] Therriault T W, Orlova M I, Docker M F, MacIsaac H J and Heath D D 2005 *Heredity* **95** 16
[5] David M and Perkovic M 2004 *Mar. Pollut. Bull.* **49** 313
[6] Rigby G R and Hallegraeff G M 1997 *water exchange trials and marine plankton distribution on the MV* (Australian Government Publishing Service)
[7] Rigby G R and Hallegraeff G M 2001 *J. Marine Environ. Eng* **6** 211
[8] Zhang S, Chen X, Yang D, Gong W and Wang Q 2003 *2nd Int. Ballast Water Treatment R&D Symposium* p 26
[9] Oemcke D J and Van L J 2005 *Water Res.* **39** 5119
[10] Hoigné J and Bader H 1976 *Water Res.* **10** 377
[11] Buxton G V, Greenstock C L, Helman W P and Ross A B 1988 *J. of Phys. and Chem. Ref.* **17** 513
[12] Spanggord R J, Yao D and Mill T 2000 *Environ. Sci. and Tech.* **34** 450
[13] Bai X Y, Zhang Z T, Yang B and Bai M D 2005 *Plasma Chem. Plasma P.* **25** 41

Research on killing *Escherichia Coli* by reactive oxygen species based on strong ionization discharging plasma

Y J Li [1], Y P Tian [1*], R H Li [2], J Y Gao [3], L J Cai [2], Z T Zhang [1, 2, 3]

1 Department of Physics, Dalian Maritime University, Dalian 116026, China
2 Marine Engineering College, Dalian Maritime University, Dalian 116026, China
3 Environmental Engineering Institute, Dalian Maritime University, Dalian 116026, China

E-mail: yiping_tian@163.com

Abstract. Reactive oxygen species solution produced by strong ionization discharging plasma was used to kill *Escherichia coli* by spraying. Several effect factors such as pH value, solution temperature, spraying time and exposure time were observed in this study, and their effects on killing rate of *Escherichia coli* were discussed and analysed. Results show that the treating efficiency of ROS solution for *Escherichia coli* is higher in alkaline solution than that in acid solution. The killing rate of *Escherichia coli* increases while the spraying time and exposure time are longer and the temperature is lower. The effects of different factors on killing rate of *Escherichia coli* are as follows: spraying time > pH value > exposure time > solution temperature.

1. Introduction

Reactive oxygen species (ROS), also known as oxygen radicals, refer to ordinary molecules and oxygen radicals with stronger oxidizing capability, which mainly include: superoxide anion free radical ($\cdot O^{2-}$), hydroperoxyl radical ($\cdot HO_2$), hydroxyl radical ($\cdot OH$) and ozone (O_3). ROS are widely applied as strong oxidant and broad-spectrum bactericide [1, 2]. In this study, highly effective and non-selective ROS solution was produced by ionizing oxygen and water through a strong electric field discharge plasma generator with special dielectric layer and processing technology under the atmospheric pressure and temperature [3-5]. The disinfection effect of ROS solution was also investigated through the killing experiment of bacteria.

Escherichia coli (*E. coli*) as the simulated warfare agent were used in the killing experiment according to the Technical Standard for Disinfection. To well study the disinfection effect of ROS solution on *E. coli*, the impact factors like the solution pH value, temperature, spray time, exposure time and their interactions were observed and analyzed in the experiment. Therefore, the optimized technical parameters of ROS solution could be obtained and provided for practical use.

2. Materials and methods

2.1. Reagents

Sodium thiosulfate solution (1 %), 0.03 mol L^{-1} phosphate buffer solution, pH=7.2.

* To whom any correspondence should be addressed.

2.2. *Strains*
E. coli (ATCC8099) using in the killing experiment is preserved in the laboratory strains, purchased from the Management Center of the China Industrial Microorganisms.

2.3. *Experimental system*
Figure 1 shows the ROS solution process system. According to Reference [6], the homemade plasma device was used to send O_2 an H_2O into the plasma reaction chamber after pre-treatment, where ROS is produced under the strong electric field ionization discharge condition. Then ROS solution went through a mixer with injector. The effective mass transfer efficiency could be obtained by the high-pressure injecting. After a series of complex plasma chemical reactions, the ROS solution was formed.

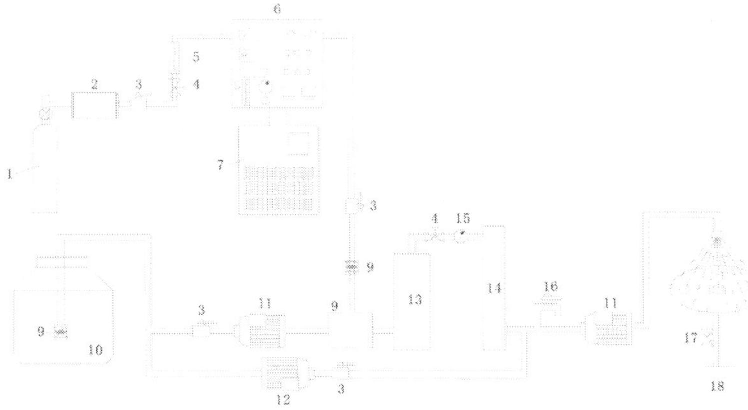

Figure 1. Experimental flow chart.

1. Oxygen 2. Filter 3. Ball Valve 4. Solenoid Valve 5. Flow Meter 6. Control Module 7. Cooling System 8. Sprinkler 9. Check Valve 10. Water 11.Pump 12. Booster Pump 13. Buffer 14. Mixer 15. Pressure Gage 16. Sample Connection 17. Altitude Controller 18. Sampling Unit

2.4. *Experimental methods*
The ROS solution preparation device was turned on first. When the system was stable (about 24min later), the quantitative bactericidal spray experiment was conducted in laboratory.

2.4.1. *Neutralizer experiment*
1.0 % sodium thiosulfate was chose as the neutralizer to carry out the neutralizing experiment according to the Technical Standard for Disinfection with *E. coli* ATC8099 as the test organism [7].

2.4.2. *Carriers quantitative spray sterilizing*
The concentration of bacterial suspension was diluted to 5×10^7 cfu mL^{-1}. Bacterial infectant disc was prepared in accordance with the Technical Standard for Disinfection. Then put them in sterile glass petri dishes with the diameter of 55 mm. Each petri dish contained three pieces of disc arranging in a triangle. The petri dishes were placed in the sample positions of disinfection and then sprayed with ROS solution. After certain minutes, dose of neutralizer was dropped onto the disc immediately. 10 minutes later, bacterium suspension was diluted to proper concentration and the numbers of bacteria were counted. The average value of three paralleled samples in each group was adopted. The control group replaced the disinfectant with aqueous solution without oxygen active particles. The other operations were the same.

2.4.3. Effect of Bactericidal

$$killing\ rate = \frac{bacterium\ amount\ in\ the\ positive\ control\ group}{bacterium\ amount\ in\ the\ negative\ control\ group} \times 100\ \%$$

Positive control group: results of the experiment when replacing disinfection agent solution with hard water; negative control group: the experimental group. Other conditions were equal.

3. Results and Discussion

3.1. The effect of pH value on killing E. coli by ROS solution

Experimental condition: ambient temperature T_1=23 °C ambient relative humidity D=42.6 %; solution temperature T_2=20 °C; spray time t=2 s; the oxygen ventilation volume of preparation device maintains 4 L min^{-1}.

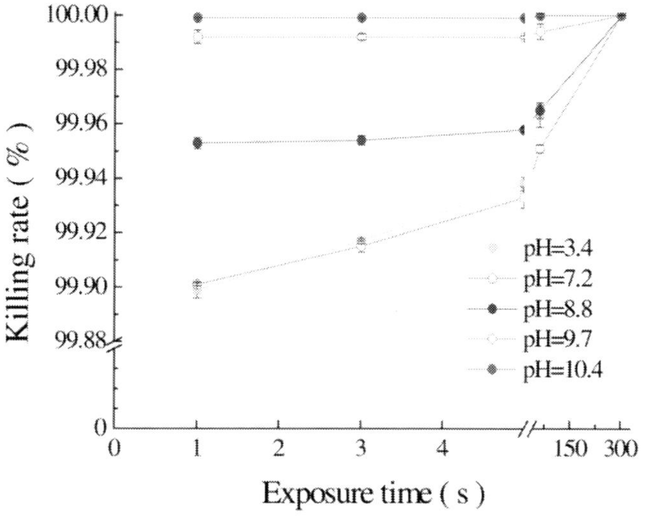

Figure 2. The effect of pH on *E. coli* killing rate.

Figure 2 shows the effects of ROS solution at different pH (from 3.4~10.4) on the killing rate of *E. coli* under the same spray time of 2 s. The results indicate that the killing rate keeps rising while the pH value of ROS solution increases, which mainly results from the rising content of ·OH in ROS solution. The reaction rate constant of ·OH is 10^9 L (mol s)$^{-1}$. It could rapidly react with lipid, protein and nucleic acid of bacterial cells, which only last from 1~10 s [8]. The pH value falls while the increasing rate of sterilizing goes up significantly with the reaction continuing. That is mainly because that the ROS solution contains not only strong oxidizing and unstable chemicals such as ·OH, but also stable active substance, such as ozone being the major one. The concentration of ozone water increases while the pH value of the solution decreases. To sum up, the disinfectant effect of ROS solution on *E. coli* strengthens with the solution pH value rising. The ROS solution is more effective in alkaline condition than in acid condition.

3.2. The effect of temperature on killing E. coli by ROS solution

Experimental condition: ambient relative humidity D=42.6 %; pH=7.2; spray time t=2 s; the oxygen ventilation volume of preparation device maintains 4 L min^{-1}.

The effects of ROS solution at different temperature (from 8~23 °C) on killing rate of *E. coli* are shown in figure 3, under the same spray time of 2 s and pH of 7.2. Results show that the killing rate of *E. coli* keeps rising while the water temperature falling. The killing rate maintains 100 % regardless of the temperature with the reaction continuing. There are two possible reasons for these results: the first one is that lower temperature helps to increase the mass transfer efficiency, therefore the initial concentration of ROS solution is relatively high; the second one is that the decay rate of active substance in ROS solution decreases with the temperature dropping, therefore when the ROS solution is sprayed out after efficient atomization, the lower temperature leads to higher killing rate. In addition, lower temperature may lead the bacteria at dormant state, which results in suspended animation. To sum up, lower temperature is advantageous to kill *E. coli*.

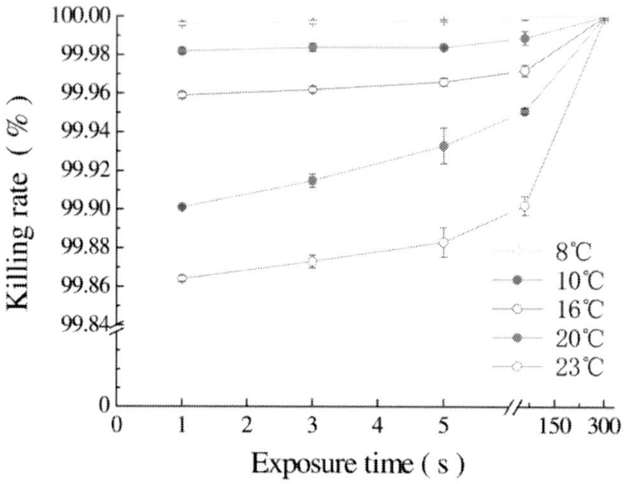

Figure 3. The effect of temperature on *E. coli* killing rate.

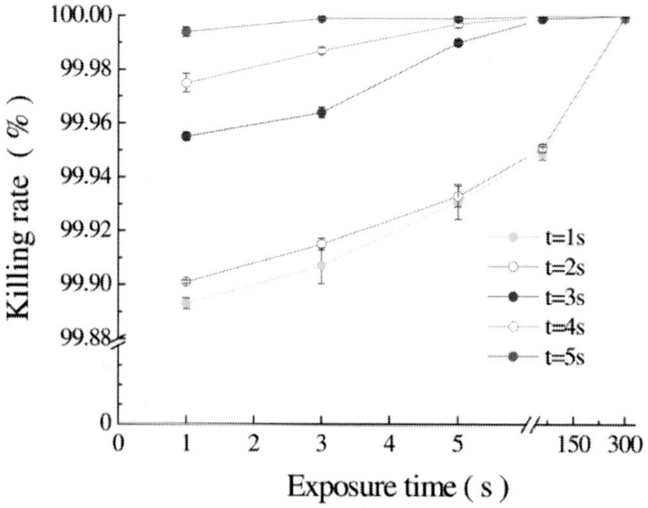

Figure 4. The effect of spray time on *E. coli* killing rate.

3.3. The effect of spray time on killing E. coli by ROS solution
Experimental condition: ambient temperature T_1=23 °C; pH=7.2; ambient relative humidity D=42.6 %; solution temperature T_2=20 °C; the oxygen ventilation volume of preparation device maintains 4 L min^{-1}.

Figure 4 shows the effects of spray time on killing *E. coli* by ROS solution with pH of 7.2. The killing rate keeps rising with the spray time extending (regardless of water loss). In addition, the longer spray time leads to shorter exposure duration of ROS solution. This is probably caused by higher density of ROS solution on bacterial contamination disc with longer spray time, which promotes biochemical reaction between ROS solution and bacteria, and results in higher killing rate. However, considering energy consumption, the dose and requirements in Technical Standard for Disinfection, it is economical and effective to maintain spray time for 2 s and allow the reaction lasting for over 2 s.

3.4. Orthogonal experiment
In order to find out the main factor influencing the killing rate of *E. coli*, and the optimum reaction condition, the following orthogonal experiments were conducted in the hope of providing technical parameter for massive biological contamination. The results are shown in table 1 and 2

Table 1. League table of experimental factors.

Influencing factors	1	2	3	4
Contents	pH	solution temperature (°C)	spray time (s)	exposure time (s)
Level	1, 2, 3	1, 2, 3	1, 2, 3	1, 2, 3
Value	3.4, 7.2, 10.4	8, 16, 23	1, 2, 3	5, 60, 300

Table 2. *E. coli* orthogonal experiment table.

Number	A	B	C	D	Killing rate (%)
1	1	1	1	1	99.931
2	1	2	2	2	99.971
3	1	3	3	3	100.000
4	2	1	2	3	100.000
5	2	2	3	1	100.000
6	2	3	1	2	99.938
7	3	1	3	2	100.000
8	3	2	1	3	100.000
9	3	3	2	3	99.996
k1	99.967	99.977	99.956	99.976	K stands for the average value of parallel samples
k2	99.979	99.990	99.989	99.970	
k3	99.999	99.978	100.000	100.000	
R	0.032	0.013	0.044	0.030	R stands for range
Optimum choice	A_3	B_2	C_3	D_3	A_3 B_2 C_3 D_3

The range in experimental result shows that the effects of different factors on killing rate of *E. coli* are as follows: spraying time > pH value > exposure time > solution temperature. The optimum reaction conditions are: ambient temperature=19 °C, ROS solution temperature=16 °C, solution pH=10.4, spray time=3 s, reaction time=5 min. In this case, the bacteria solution with original concentration of 1.0×10^8 cfu mL^{-1} could be disinfected by 100 %.

3.5. Resurrection experiment

were selected according to Technical Standard for Disinfection, i.e. In order to further test the effectiveness and feasibility of ROS solution in killing *E. coli*, the resurrection experiments were carried out in this study under the following conditions: Ambient temperature=23 °C; solution pH value=7.2; ambient relative humidity D=42.6 %; solution temperature T_2=20 °C; the oxygen ventilation volume of preparation device maintains 4 L min^{-1}. After spraying for 2 s, we observed whether there was any survival *E. coli* in 24 h, 36 h, 48 h, 72 h, and 96 h. The observation result shows that no *E. coli* resurrects after disinfection by ROS solution, which confirms the effectiveness and feasibility of ROS solution in killing *E. coli*.

4. Conclusion

ROS solution prepared through strong electric field ionization and efficient immiscible can kill *E. coli* by spraying. This is a disinfection and epidemic prevention technology which is efficient and environmental friendly. The experiment results show that: ① Alkaline condition is advantageous to the killing of *E. coli* by ROS solution. ② The killing rate of *E. coli* by ROS solution increases with the solution temperature decreasing. ③ The killing rate of *E. coli* keeps increasing with the spray time extending and meets the requirements of *Technical Standard for Disinfection* when the spray time lasts for 2 s. ④ The effects of different factors on killing rate of *E. coli* are as follows: spraying time > pH value > exposure time > solution temperature.

Acknowledgments

This work is supported by The National High Technology Research and Development Program of China under Grant No 2012AA062609, The International Science and Technology Cooperation Project of China under Grant No 2010DFA61470, The National Science Foundation for Distinguished Young Scholars of China under Grant No 61025001, The Fundamental Research Funds for the Central Universities under Grant No 2012QN067 and 2012TD061.

References

[1] Zuo C H, Du J Q and Pu Y Q 2009 *J. Beijing Univ. Chem. Technol. (Nat. Sci. Ed.)* **37** 104
[2] Sun H F and Du J Q 2010 *Mod. Chem. Indus.* **30** 102
[3] Buxton G V and Clive L G 1988 *J. Phys.Chem.Ref.Data.* **17** 513
[4] Zhang Z T, Xian Y Z, Bai M D and Xu B 2002 *J. Northeastern Univ.(Nat. Sci.)* **23** 507
[5] Bai X Y, Bai M D, Zhou X J and Yang B 2002 *Chem. Indus. Times.* **16** 1
[6] Alexander H R, Ronald E E, Claude J A and Philip S b 1960 *J. Am. Chem. Soc.* **82** 1801
[7] Health Law and Oversight Division of the Ministry of Health 2002 Technical Standard for Disinfection
[8] William H G, Joon W K and Douglas H C 1987 Ozone Sci. Eng. : *J. Int. Ozone. Assoc.* **9** 335

Fresh water disinfection by pulsed low electric field

C Zheng, Y Xu, Z Liu and K Yan *
Key Laboratory of Biomass Chemical Engineering of Ministry of Education, Industrial Ecology and Environment Research Institute, Zhejiang University, Tianmushan Road 148, Hangzhou 310007, China

E-mail: kyan@zju.edu.cn

Abstract. In this paper, we describe a pulsed low electric field process for water disinfection. Electric intensity of 0.6-1.7 kV cm^{-1} is applied. Experiments are performed with a 1.2 L axis-cylinder reactor. A bipolar pulsed power source with pulsed width of 25 μs and frequency of 100-3000 Hz is used. Water conductivity of 3-200 μs cm^{-1} is investigated, which can significantly affect pulsed voltage-current waveforms and injected energy. Energy per pulse rises with increased water conductivity. The initial *E. Coli* density and water conductivity are two major factors influencing the disinfection. No disinfection effect is performed with deionized water of 3 μs cm^{-1}. When water conductivity is 25 μs cm^{-1} and bacteria density is 10^4-10^6 cfu ml^{-1}, significant disinfection effect is observed. More than 99% of the cells can be disinfected with an energy density of less than 70 J ml^{-1}, while water temperature is below 30 ℃.

1. Introduction

Pulsed electric filed (PEF) is an emerging non-thermal technology in the field of liquid disinfection and food preservation [1]. Fruit juices, liquid egg and derivatives, milk, grape wines and water are widely tested [2-6]. Comparing to traditional thermal pasteurization for liquid food disinfection, use of this technology can satisfy consumer demand for fresh looking products with minimal losses of nutrients and changes in flavour [7-8]. The effectiveness of PEF technology in reducing microorganisms has been widely demonstrated [9-10]. Disinfection usually depends on a number of factors, such as PEF intensity, pulse shape and duration, PEF energy density, products parameters and microbial characteristics [11]. A peak electric field of around 20-90 kV cm^{-1} is usually applied, while the pulse width is 1-10 μs. The volume of PEF reactor is limited to few millilitres to realize its high intensity electric field, which is a drawback for industrial scale up. Wan recently present a review for liquid food disinfection, the energy density is in the range of 30-7541 J ml^{-1} for achieving 1-6 orders of microbial reduction [12].

This paper presents our recent work on PEF liquid disinfection. Fresh water and *E. Coli* ATCC25922 are selected for fundamental research. We found that electric field as low as 0.5-2 kV cm^{-1} can also effectively inactivate the bacteria.

2. Experimental system

Figure 1 shows the schematic diagram of the PEF reactor. It consists of a grounded stainless steel cylinder and a HV electrode. Its inner diameter and volume are 100 mm and 1.2 L. The diameter of

* To whom any correspondence should be addressed.

HV electrode is 30 mm. It is separated by a ceramic insulator to the grounded cylinder. The distance between the HV electrode and ground is 35 mm. A 10 kW bipolar pulsed power source is used. It can provide a pulsed voltage of 2-6 kV and a pulsed frequency of 100-3000 Hz (200-6000 pps). The pulse width is 25 μs, which is controlled by IGBT. All electrical measurements are recorded with Tektronix oscilloscope (TDS 2014B, 100 MHz). Typical bipolar pulsed voltage waveform is shown as figure 2.

The used *E. Coli* is firstly cultivated for 8-10 hours at 37 ℃ in a prepared LB medium (peptone of 10 g L^{-1}, yeast extract of 5 g L^{-1} and sodium chloride of 10 g L^{-1}). By this cultivation, the cell density usually rises to about 10^9-10^{10} cfu ml^{-1}. The cell number density is obtained by means of heterotrophic plate count (HPC) method. Then, *E. Coli* is harvested by centrifugation and suspended in deionized water for PEF treatment. Water conductivity is adjusted by adding potassium chloride.

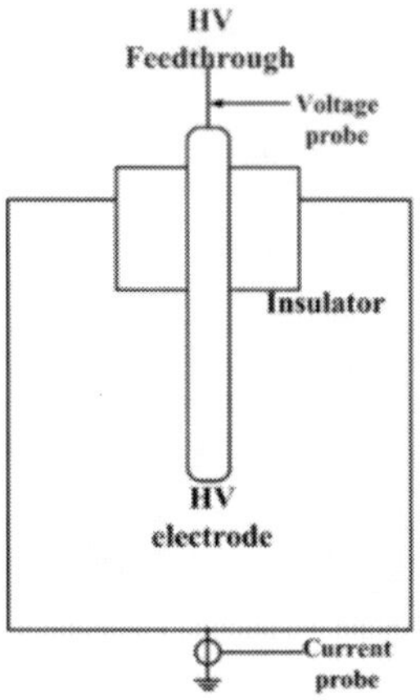

Figure 1. Schematic diagram of PEF reactor.

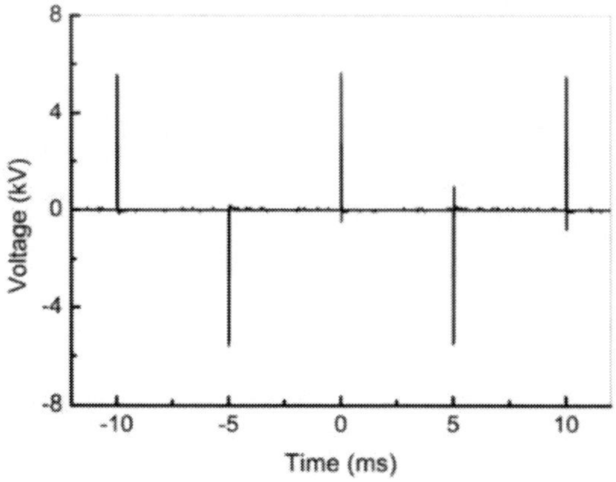

Figure 2. Typical bipolar pulsed voltage waveform when frequency is 100 Hz (200 pps).

3. Results and discussion

3.1. Effect of water conductivity on pulsed waveforms

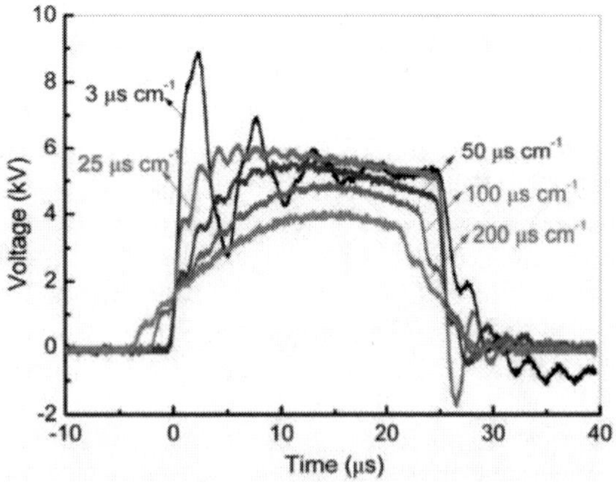

Figure 3. Pulsed voltage waveforms under water conductivity of 3-200 μs cm^{-1}.

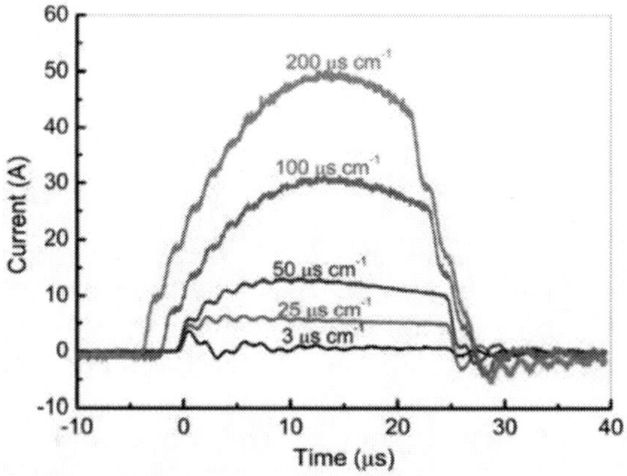

Figure 4. Pulsed current waveforms under water conductivity of 3-200 μs cm^{-1}.

Water conductivity can significantly affect pulsed waveforms. When the pulsed power source has the same output, figure 3 and figure 4 show the variation of voltage-current waveforms under different water conductivities within 3-200 μs cm^{-1}. As water conductivity rises, the pulsed voltage decreases slowly and corresponding current increases sharply. When water conductivity is 25 μs cm^{-1}, the pulsed peak voltage and current are 6 kV and 6 A, respectively. At the same time, both its voltage and current waveforms are fine square waves. When conductivity rises to 200 μs cm^{-1}, its peak voltage is 4 kV, but the peak current is as high as 50 A. Meanwhile, the waveforms become smooth.

From voltage-current waveforms in figure 3 and figure 4, we can obtain the single pulse energy under different water conductivities, as results shown in figure 5. As water conductivity rises from 3 μs cm^{-1} to 200 μs cm^{-1}, injected energy per pulse increases from 0.09 to 3.5 J.

3.2. Variation of water temperature during PEF treatment

Water temperature rises during PEF processing. When initial water temperature is 20 ℃, water volume is 1.2 L, pulsed waveforms are as figure 3 and figure 4, pulsed frequency is 200 pps and treatment time is 10 min, the increment of temperature is detected as shown in figure 6. If water conductivity is 3 μs cm^{-1}, it is 1 ℃. From figure 5, we can calculate the corresponding injected energy density is 9 J ml^{-1}. As water conductivity rises, both the injected energy density and temperature persistently increase. When water conductivity is 200 μs cm^{-1}, the energy density and temperature increment are 350 J ml^{-1} and 18 ℃, respectively.

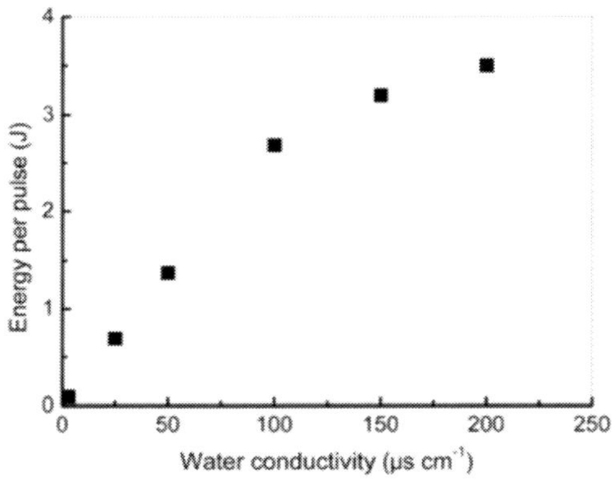

Figure 5. Injected energy of single pulse under water conductivity of 3-200 μs cm^{-1}.

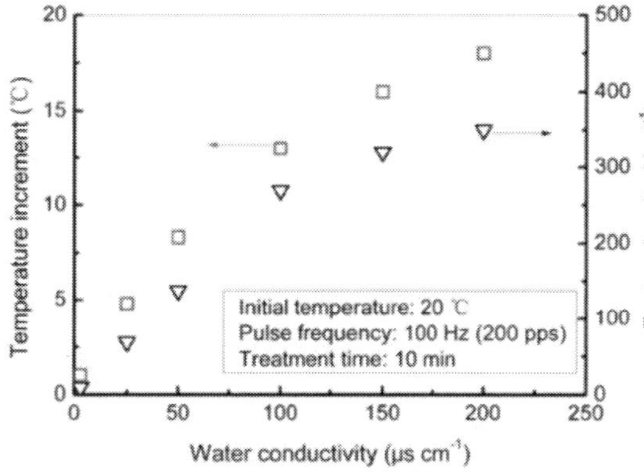

Figure 6. Increment of water temperature and injected energy density under water conductivity of 3-200 μs cm^{-1}.

3.3. Effect of water conductivity on PEF disinfection

Water conductivity can affect the PEF disinfection. When *E. Coli* is suspended in deionized water of 3 μs cm^{-1}, no disinfection effect is observed. If water conductivity is 25-200 μs cm^{-1}, the cell survival curves are shown in figure 7. The initial cell density, pulsed frequency and treatment time are 2×10^5 cfu ml^{-1}, 200 pps and 10 min, respectively. It is found that best disinfection effect is performed when water conductivity is 25 μs cm^{-1}. More than 99% of the cells can be disinfected, while the temperature is around 25 ℃. Disinfection efficiency decreases with raised water conductivity. Meanwhile, the corresponding PEF intensity is decreased from 1.7 kV cm^{-1} to 1.1 kV cm^{-1}, which can be seen from figure 3 and figure 4. Figure 8 shows the relationship between injected energy density and order reduction of cell density, which is defined as the logarithm ratio of the initial cell density over its survival value. When water conductivity is raised, the injected energy density increases obviously but without improved disinfection efficiency. The higher the water conductivity, the lower the energy efficiency of disinfection. Major of the energy is dissipated for water heating. We think that PEF disinfection efficiency is depend more on electric intensity and pulsed waveforms than injected energy density in present experiments. Square wave is supposed better than smooth one for disinfection.

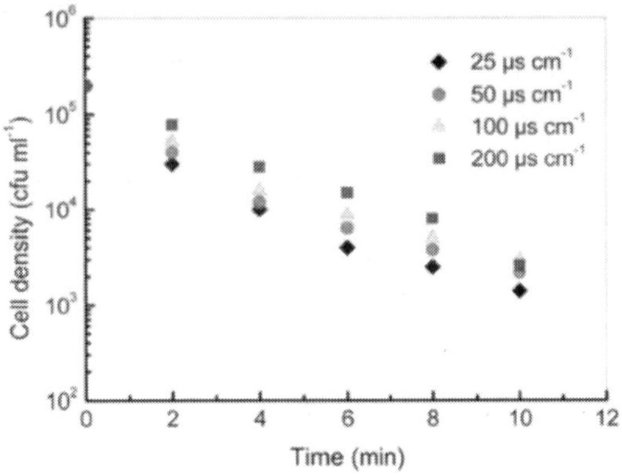

Figure 7. Effect of water conductivity on PEF disinfection.

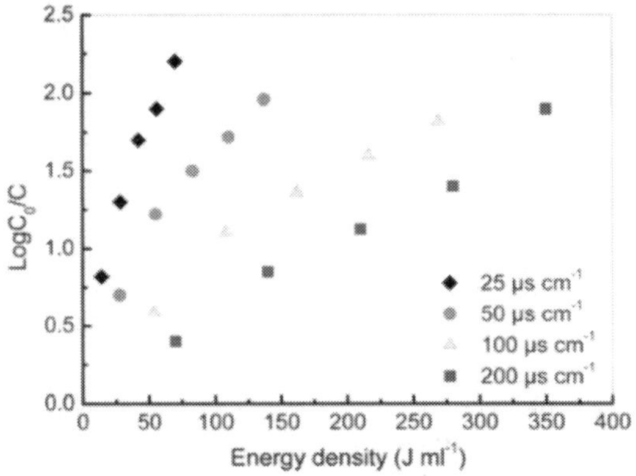

Figure 8. Effect of injected energy density on PEF disinfection under different water conductivities.

3.4. Effect of initial cell density on PEF disinfection

Figure 9 shows cell survival curves in terms of the treatment duration and/or the PEF energy density. Experiments are performed in water conductivity of 25 μs cm^{-1}. Its PEF intensity is 1.7 kV cm^{-1} and pulsed frequency is 200 pps. All the *E. Coli* cells can be inactivated in 10 min or under 70 J ml^{-1} when initial density is below 1.3×10^4 cfu ml^{-1}. Disinfection becomes less efficient when initial density rises. The orders reduction is 2.3 after 10 min when initial cell density is 4.5×10^4 cfu ml^{-1}. For an initial density of 1.6×10^6 cfu ml^{-1}, only 0.9 order reduction can be achieved.

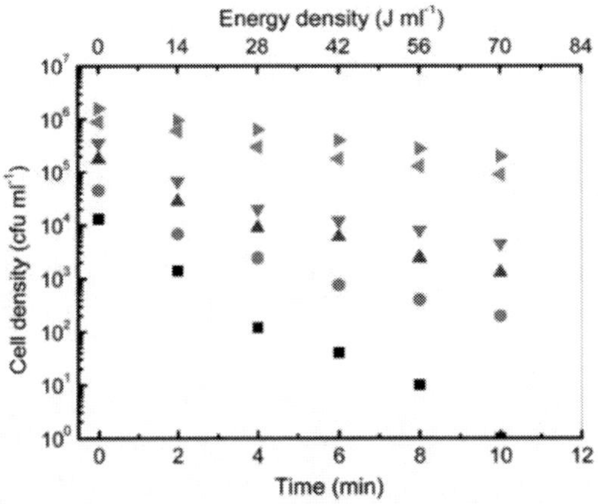

Figure 9. Effect of initial cell density on PEF disinfection.

3.5. Effect of PEF intensity on disinfection

Figure 10 shows the effect of PEF intensity on disinfection. The water conductivity, initial cell density, pulsed frequency and treatment time are 25 μs cm^{-1}, 1.3×10^4 cfu ml^{-1}, 200 pps and 10 min, respectively. PEF intensity is controlled by adjusting the output of the power source. The waveforms are all fine square waves as figure 3 and figure 4 but with different amplitudes. If PEF intensity is 0.6 kV cm^{-1}, only 0.5 order reduction can be achieved after 10 min. Disinfection efficiency rises with increased PEF intensity. The order reductions are 1.0 and 2.2 when PEF intensity are 0.9 kV cm^{-1} and 1.3 kV cm^{-1}, respectively. All the cells can be disinfected under a PEF intensity of 1.7 kV cm^{-1}. The energy per pulse for PEF intensity of 0.6 kV cm^{-1}, 0.9 kV cm^{-1}, 1.3 kV cm^{-1} and 1.7 kV cm^{-1} are 0.10 J, 0.22 J, 0.45 J and 0.70 J, respectively. When re-plotting the curves shown in figure 10 in terms of the PEF energy density and the order reduction as shown in figure 11, the disinfection efficiency is mainly dependent on the energy density.

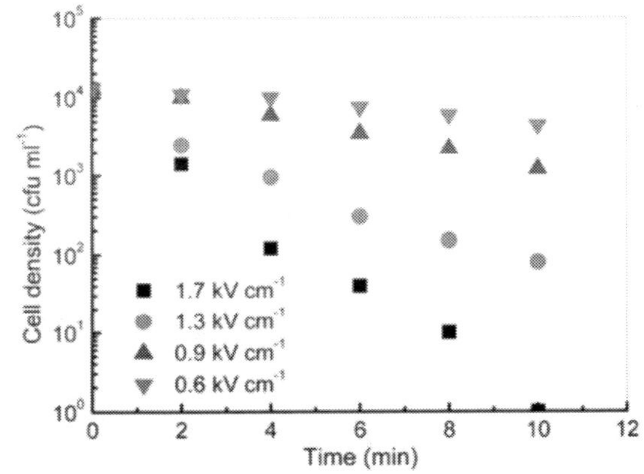

Figure 10. Effect of PEF intensity on disinfection.

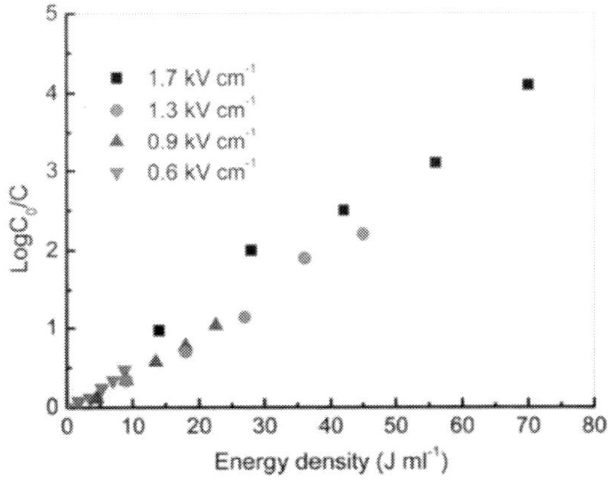

Figure 11. Effect of injected energy density on PEF disinfection under different electric intensity.

3.6. Effect of pulsed frequency on PEF disinfection

Figure 12 shows the effect of PEF frequency on disinfection. Water conductivity, PEF intensity and initial cell density are 25 μs cm^{-1}, 1.7 kV cm^{-1} and 5.0×10^4 cfu ml^{-1}, respectively. All experiments are performed with 120000 pulses. The treatment duration drops to 1 min when pulsed frequency is 2000 pps. Results show that PEF disinfection is not depends on pulsed frequency but depend on the total pulse number. 2.1-2.2 orders reduction is achieved after 120000 pulses. If we raise the pulsed frequency to the maximum of 6000 pps, few seconds is enough for the process.

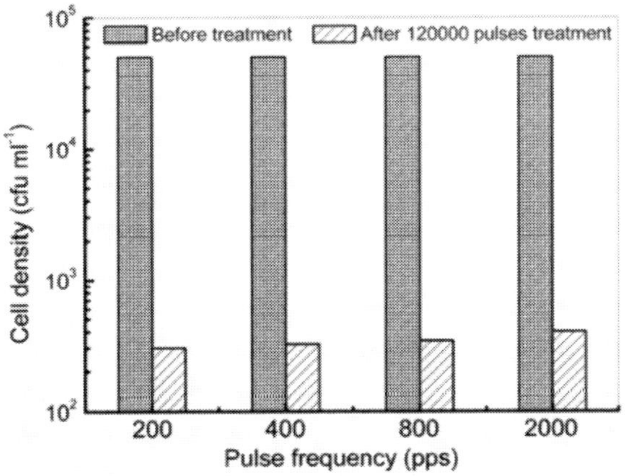

Figure 12. Effect of pulse frequency on PEF disinfection.

4. Conclusions

This paper gives a preliminary investigation on pulsed low electric field process for liquid disinfection. Based on present experiment on the *E. Coli* disinfection, we can conclude as follows:

1. PEF intensity of 0.6-1.7 kV cm^{-1} has disinfection effect. Disinfection efficiency rises with increased PEF intensity and energy density.

2. Water conductivity significantly affects the PEF waveforms and disinfection efficiency. When conductivity is within 3-200 μs cm^{-1}, best disinfection efficiency is performed under the water conductivity of 25 μs cm^{-1} and its fine square wave.

3. All the cells can be disinfected under an energy density of 70 J ml^{-1} after 120000 pulses when initial bacteria density is below 1.3×10^4 cfu ml^{-1}. Disinfection efficiency drops as initial cell density rises.

References

[1] Zhang Q H, Barbosa-Canovas G V and Swanson B G 1995 *J. Food Eng.* **25** 261
[2] Mosqueda-Melgar J, Raybaudi-Massilia R M and Martin-Belloso O 2007 *Int. J. Food Microbial.* **117**192
[3] Sampedro F, Rodrigo D, MartInez A, Barbosa-Canovas1 G V and Rodrigo M 2006 *Food Sci. Tech. Int.* **12** 397
[4] Jaeger H, Schulz A, Karapetkov N and Knorr D 2009 *Int. J. Food Microbial.* **134** 154
[5] Bermudez-Aguirre D, Yanez J A, Dunne C P, Davies N M and Barbosa-Canovas G V 2011*Food Res. Int.* **43** 2201
[6] Puertolas E, Lopez N, Condon S, Raso J and Alvarez I 2009 *Int. J. Food Microbial.* **130** 49
[7] Aguilo-Aguayo I, Oms-Oliu G, Soliva-Fortuny R and Martin-Belloso O 2009 *Food Chemistry* **116** 59
[8] Jia M, Zhang Q H and Min D B 1999 *Food Chemistry* **4** 445
[9] Heinz V, Toepfl S and Knorr D 2003 *Innov. Food Sci. Emerg.* **4** 167
[10] Huang K, Tian H, Gai L and Wang J 2012 *J. Food Eng.* **111** 191
[11] Wouters P C, Alvarezb I and Rasob J 2001 *Trands Food Sci Tech.* **12** 112
[12] Wan J, Coventry J, Swiergon P, Sanguansri P and Versteeg C 2009 *Trands Food Sci. Tech.* **20** 414

Effect of rising time of rectangular pulse on inactivation of staphylococcus aureus by pulsed electric field

Ruobing Zhang[1 *], **Dapeng Liang**[1], **Nanchen Zheng**, **Jianfu Xiao**[1], **Mengbin Mo**[1], **Jing Li**[2]
1. Graduate School at Shenzhen, Tsinghua University, Shenzhen 518055, China;
2. College of Food Science & Nutritional Engineering, China Agriculture University, Beijing 100083, China

E-mail: zhangrb@sz.tsinghua.edu.cn

Abstract. Pulsed electric field (PEF) is a novel non-thermal food processing technology that involves the electric discharge of high voltage short pulses through the food product. In PEF study, rectangular pulses are most commonly used for inactivating microorganisms. However, little information is available on the inactivation effect of rising time of rectangular pulse. In this paper, inactivation effects, electric field strength, treatment time and conductivity on staphylococcus aureus inactivation were investigated when the pulse rising time is reduced from 2.5 μs to 200 ns. Experimental results showed that inactivation effect of PEF increased with electric field strength, solution conductivity and treatment time. Rising time of the rectangular pulse had a significant effect on the inactivation of staphylococcus aureus. Rectangular pulses with a rising time of 200 ns had a better inactivation effect than that with 2 μs. In addition, temperature increase of the solution treated by pulses with 200 ns rising time was lower than that with 2 μs. In order to obtain a given inactivation effect, treatment time required for the rectangular pulse with 200 ns rise time was shorter than that with 2 μs.

1. Introduction

Pulsed Electric Field (PEF) is a newly developed technology in non-thermal food processing, and is based on a pulsing power delivered to the product placed between a set of electrodes confining the treatment gap of the PEF chamber. The basic definition of PEF relies on the use of high intensity pulsed electric field (10-50 kV cm^{-1}) for cell membrane disruption where induced electric fields perforate microbial membranes by electroporation. Conventional pulse width applied in PEF technology is in microsecond level, thus very little energy will be consumed during the process and the temperature increase of food can be kept at very low. Therefore, PEF is a non-thermal processing technology with low energy consumption and is capable of keeping food's flavors and nutrition, particularly suitable for foods processing which can be easily affected by temperature.

Because PEF technology has the advantages of protecting physical & chemical characteristics of foods with low energy consumption and temperature rise, it has been considered as a non-thermal processing technology and has a good industrial application prospect. The degree of inactivation

[*] To whom any correspondence should be addressed.

strongly depends on several constraints. The lethality factors contributing to the effectiveness of PEF technology include technical (electric field strength, shape and width of the pulse, frequency, and treatment time), media (pH, electrical conductivity and temperature) and biological factors. Typical voltage waveforms of PEF are exponential or rectangular pulse. The rectangular pulse is now a focus in researches because of its higher energy efficiency and better inactivation effect compared with the exponential voltage.

Effects of the pulse width, amplitude and frequency of rectangular pulses on the inactivation have been detailed studied in previous studies. However, little information is available on the effect of pulse rising time of the rectangular pulse. In this paper, we have studied the inactivation effect of the rising time of rectangular pulse on staphylococcus aureus inactivation.

2. Instruments and experimental equipments

2.1 Experimental Equipment

In this study, a high voltage PEF equipment PEF-3 (shown in figure 1) recently developed by Graduate School at Shenzhen, Tsinghua University was used. The pulse generator could output a bipolar rectangular pulse in the range of 2-10 μs in width and frequency in the range of 0-1 kHz. The electric field strength between the two electrode of the treatment chamber is in the range of 0~45 kV cm^{-1}. Normally, the rising time of rectangular pulse is in the range of 2.0-2.5 μs, and is reduced to about 200 ns by adding the optimization unit.

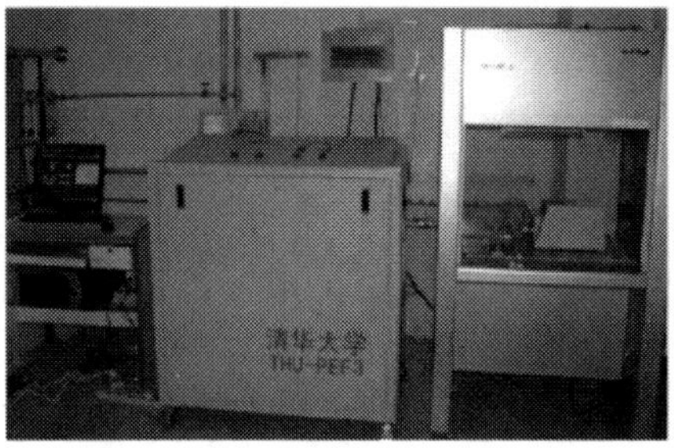

Figure 1. Photograph of THU-PEF3.

Typical characteristic of the equipment is that pulse parameters including frequency, width, number and polarity can be precisely adjusted through the upper-control personal computer (PC). The general structure of the pulse generator is shown in figure 2.

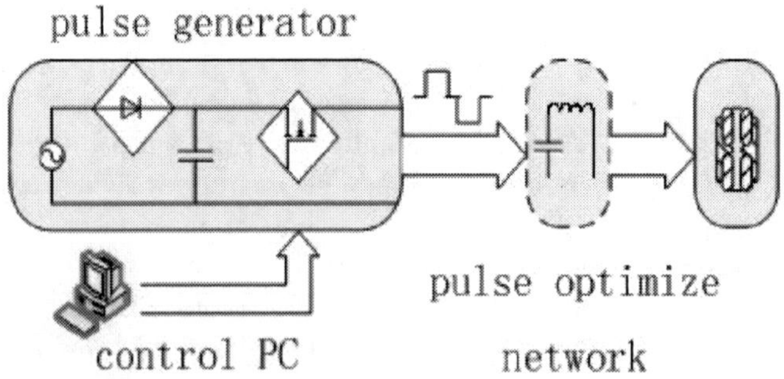

Figure 2. Diagram of THU-PEF3 system.

Co-field treatment chambers, which have a lower temperature increase and a lower probability of electric breakdown used in this study. The structure of the treatment chamber is shown in figure 3.The inner diameter of the electrode is 4 mm and the distance between electrodes is 4 mm. Equivalent resistance of the chamber depends on the electrical conductivity of the solution treated.

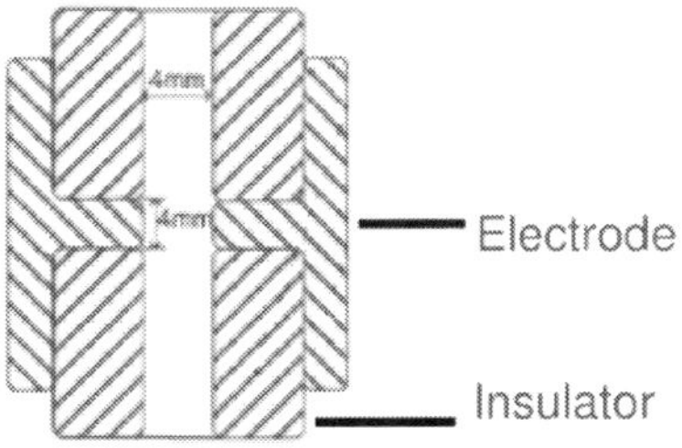

Figure 3. Treatment chamber.

2.2 Material and method

2.2.1 Preparation of apple juice

Huiyuan 100% pure apple juice (i.e. the volume fraction of this apple juice is 100%) was used and kept at 4 ℃ in refrigerator. Unfreeze the juice at room temperature for two days before the experiment, and then adjust the pH value and the conductivity using NaOH and NaCl of 1 mol L^{-1} respectively. Finally, inactive the juices at 120 ℃ for 20 min, and then cool it to prepare for inoculating microorganism.

2.2.2 The culture of microorganism

Staphylococcus aureus (CGMCC1. 1861, bought from the China General Microbiological Culture Collection Center) was used as target microorganism

Activate the freeze-dried powder of staphylococcus aureus and then preserve it on inclined plate. Put the strains into an appropriate amount of nutritional broth medium. Centrifuge the bacteria liquid after culturing at 37 ℃ for 18 h and pour away the clear-water. Add some sterile water and then take out the bacteria after centrifuging the bacteria liquid again. Finally, dilute by equivalent inactivation apple juice to make the number of microorganisms fall in 10^6~10^8 cfu mL^{-1}.

2.2.3 PEF treatment method

In this paper, we have individual studied the effect of electrical conductivity and treatment time on inactivation. Hence, the specific experiment methods were slightly different.

For studying the effect of electrical conductivity, a constant equivalent treatment time of 75 μs was used. The juice solution was pumped into the treatment chamber by gear pump at a constant flow speed of 6.4 mL min^{-1}. The electrical conductivities of the solution were 1.5 mS cm^{-1}, 2.0 mS cm^{-1} and 2.5 mS cm^{-1}, respectively. Pulse voltage was kept at 10 kV, 12 kV and 14 kV, corresponding electric field strength is 25 kV cm^{-1}, 30 kV cm^{-1} and 35 kV.cm^{-1}, respectively. The rising time of the rectangular pulse was reduced from 2 μs to 200 ns after optimization, as shown in figure 4. Solution temperature was measured at the inlet and outlet of the treatment chamber.

For studying the effect of equivalent treatment time, juice solution with an electrical conductivity of 2.0 mS cm^{-1} was selected. The equivalent treatment time was changed by adjusting pulse frequency at a constant pump flow rate of 6.4 mL min^{-1}. The pulse width was kept at 5 μs, while the pulse frequency was set at 8 Hz, 10 Hz, 12 Hz, 14 Hz and 16 Hz, respectively. For each treatment time, juice solution was treated at a pulse voltage of 12 kV and 14 kV respectively, and the corresponding electric field strength is 30 kV cm^{-1} and 35 kV cm^{-1}.

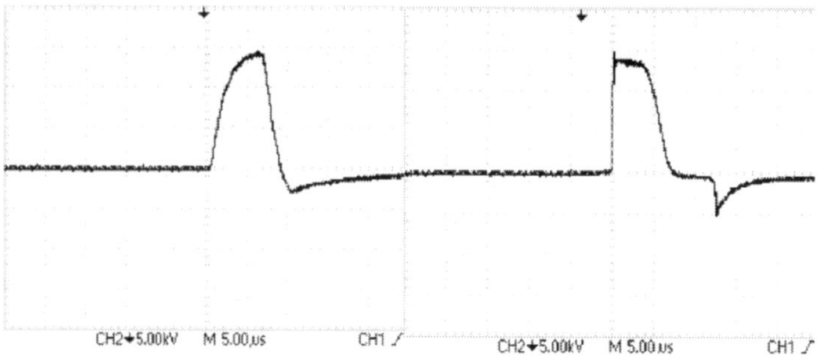

Figure 4. Waveform of different rise time.

2.2.4 Detection of microorganism

The plate count method was adopted to detect the microorganisms. Dilute the apple juice by 10 time's gradient dilution method with the NaCl solution (0.85% of volume fraction). Pour 1 mL of the diluted solution and 15 mL of the nutrient agar medium onto each plate, and then put the plate into the incubator at 37 ℃ for 24 h. Count the microorganisms after culturing and take an average from three Parallel specimens. The inactivation effect can be expressed by residual rate S, calculated by following formula:

$$\log S = -\log \frac{N}{N_0}$$

Where N is the number of residual microorganisms after treating by pulse electric field, and its unit is cfu mL^{-1}; N_0 is the number of microorganisms before processed by PEF, cfu mL^{-1}; the residual rate S represents the decline of the microorganisms calculated by logs.

3. Results and discussion

3.1 Influence of rising time

The influence of rising time on the inactivation effect of the solution was shown in figure 5. As shown in figure 5, for all the solution conductivities, inactivation effect of PEF increased as the electric field increased. In addition, for a constant applied electric field, higher inactivation effect was achieved for 200 ns rising time rectangular pulse than 2 μs. Generally, survival ratio of the microorganisms could be improved by 0.5~1 logs in all the experiments, indicating that short rising time was favorable to improve the inactivation effect in PEF process.

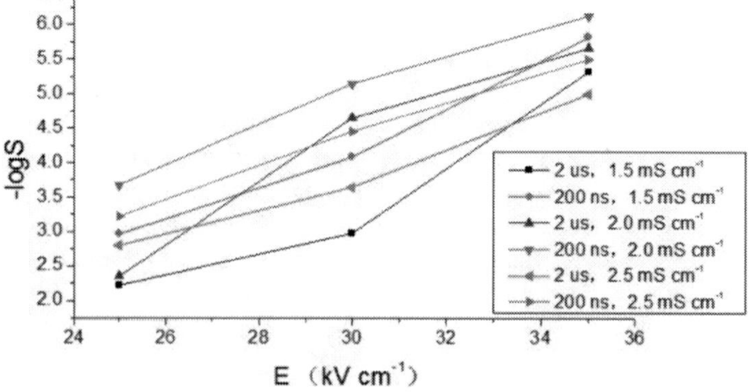

Figure 5. The relationship between electric field and survival ratio under the conductivity of 1.5 mS cm^{-1}, 2.0 mS cm^{-1} and 2.5 mS cm^{-1}.

Meanwhile, we had compared the temperature increase of treated solutions with three different conductivities before and after the PEF treatment, as shown in figure 6. With the increase of the electric field strength, temperatures of the treated solutions increased under all conductivities. However, the temperature increase of the sample treated by the short rising time pulse of 200 ns was lower than that of 2 μs. For example, temperature increase of juice solutions (2.5 mS cm^{-1}) treated by the short rising time pulse of 200 ns was 14.2 ℃, 18.2 ℃ and 25.5 ℃, correspondingly the electric field strength was 25 kV cm^{-1}, 30 kV cm^{-1}, 35 kV cm^{-1} respectively. However the temperature increase was 16.7, 19.8 and 31.5 ℃ when the 2 μs rising time pulse was used. The maximum temperature difference of the solution even reached 6 ℃ at the 35 kV cm^{-1}(solution conductivity is 2.5 mS cm^{-1}). It should be noted that the lower temperature increase has a significant benefit for ensuring the non-thermal characteristic of PEF technology.

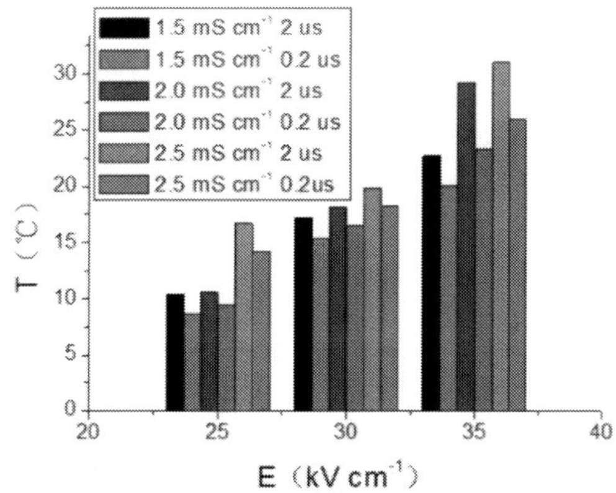

Figure 6. Temperature increase before and after treatment by PEF.

3.2 Influence of treatment time

Figure 7 showed the effect of equivalent treatment time on survival ratio under the electric field strength of 30 kV cm-1 and 35 kV cm-1. It could be concluded that survival ratio of the microorganisms reduces with the increase of the treatment time. No obvious difference can be found in the survival ratio treated by the two kinds of waveforms when the treatment time is less than 55 μs. However, when the equivalent treatment time was longer than 55 μs, the advantage of inactivation by the shorter rise time rectangular pulse becomes obvious, when the maximum treatment time 75 μs was reached, the survival difference can even be enlarged by more than 1 log.

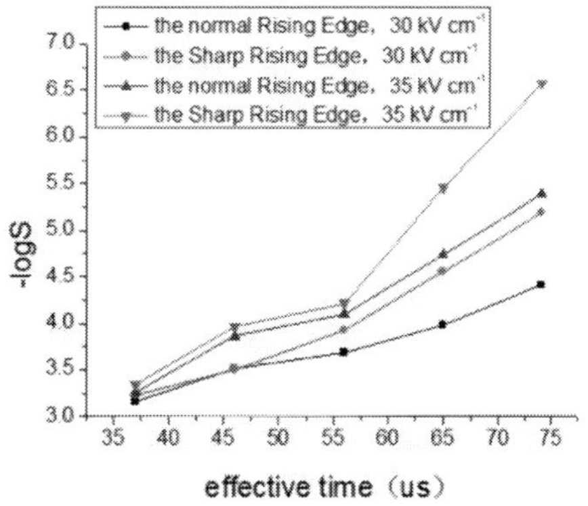

Figure 7. The relationship between equivalent treatment time and survival ratio under the electric field strength of 30 kV cm^{-1} and 35 kV cm^{-1}.

4. Conclusion

a) For the present PEF treatment with rectangular pulse, better inactivation effect can be achieved at larger electric field strength, longer treatment time and higher solution conductivity.

b) The rising time of the rectangular pulse has significant effect on the inactivation of staphylococcus aureus. Rectangular pulse with a rising time of 200 ns has better inactivation effect than that with 2 μs.

c) Temperature increase of the solution treated by pulse with 200ns rising time is lower than that with 2μs.

d) In order to achieve a given inactivation effect, lower treatment time was required when the rising time of the rectangular pulse decreases, indicating that less energy consumption are needed to obtain the same inactivation effect.

Acknowledgement

This work was partly supported by NSFC, Grant No. 51177082, by the Research Fund for the Doctoral Program of Higher Education of China, Grant No.20100002110032, and by the special fund for the emerging strategic industries of Shenzhen Grant No.JCYJ20120616220833092.

References

[1] Beveridge J R, MacGregor S J, Marsili L, et al. 2002 *IEEE Trans. Plasma Sci.* **30** 1525
[2] Castro A J, Barbosa-Canovas G V and Swanson B G 1993 *Microbial in activation of foods by pulsed electric fields* (Food Proc Preserve)
[3] Chen J, Zhang R B, Xiao J F, Li J, Wang L M and Guan Z C 2010 *IEEE Trans. Plasma Sci.* **38** 1935
[4] Redondo L M, Margato E and Silva J F 2002 *IEEE Trans. Power Electr.* **17** 196
[5] Redondo L M, Silva J F and Margato E 2007 *IEEE Trans. Magn.* **43** 1973
[6] Chen J, Zhang R B, Xiao J F, Wang L M and Guan Z C 2011 *IEEE Trans. Dielect. El. In.* **18** 1163
[7] Zhang R B, Chen J, Xiao J F, Liao X J, Wang L M and Guan Z C *High Voltage Engineering* **03** 2011
[8] Wang L M, Mo M B, Zhang R B, Chen J, Li J, Xiao J F, et al. *High Voltage Engineering* **04** 2010
[9] Chen J, Zhang R B, Wang X Q, Luo W, Mo M B, Wang L M, Guan Z C *Spectroscopy and Spectral Analysis* **01** 2010

Towards Ideal NO_x and CO_2 Emission Control Technology for Bio-Oils Combustion Energy System Using a Plasma-Chemical Hybrid Process

M Okubo [1*], H Fujishima [1], Y Yamato [1], T Kuroki [1], A Tanaka [2], and K Otsuka [2]
[1]Department of Mechanical Engineering, Osaka Prefecture University, 1-1 Gakuen-cho, Naka-ku, Sakai 599-8531, Japan
[2]Engineering Department, Takao Iron Works Co., Ltd., 1-3-29 Shimae-cho, Toyonaka-city, Osaka 561-0826, Japan

E-mail: mokubo@me.osakafu-u.ac.jp

Abstract. A pilot-scale low-emission boiler system consisting of a bio-fuel boiler and plasma-chemical hybrid NO_x removal system is investigated. This system can achieve carbon neutrality because the bio-fuel boiler uses waste vegetable oil as one of the fuels. The plasma-chemical hybrid NO_x removal system has two processes: NO oxidation by ozone produced from plasma ozonizers and NO_2 removal using a Na_2SO_3 chemical scrubber. Test demonstrations of the system are carried out for mixed oils (mixture of A-heavy oil and waste vegetable oil). Stable combustion is achieved for the mixed oil (20 – 50% waste vegetable oil). Properties of flue gas—e.g., O_2, CO_2 and NO_x—when firing mixed oils are nearly the same as those when firing heavy oil for an average flue gas flow rate of 1000 Nm^3/h. NO_x concentrations at the boiler outlet are 90 – 95 ppm. Furthermore, during a 300-min continuous operation when firing 20% mixed oil, NO_x removal efficiency of more than 90% (less than 10 ppm NO_x emission) is confirmed. In addition, the CO_2 reduction when heavy oil is replaced with waste vegetable oil is estimated. The system comparison is described between the plasma-chemical hybrid NO_x removal and the conventional technology.

1. Introduction

A series of studies have been conducted on laboratory-scale nonthermal plasma-chemical hybrid processes for the removal of NO_x from gases emitted from various stationary sources such as boilers, diesel power generators and refuse incinerators. These processes can now successfully remove almost all traces of NO_x with negligible amounts of byproducts [1]–[7]. However, because the flue gases must be treated directly, treatment of large volumes requires both a large plasma reactor and a large amount of power. An alternative process has been developed that involves the injection of ozone (O_3) or radicals generated from oxygen (O_2), ammonia (NH_3), nitrogen (N_2) and methane (CH_4) by using plasma (indirect plasma method). This process has been studied in laboratory-scale and pilot-scale experiments [8]–[18], where it has been found to be extremely effective for NO_x removal, because only the necessary amounts of gas—e.g., ambient air, O_2, NH_3, CH_4 and N_2—are treated externally at ambient temperature and pressure. Therefore, the indirect plasma method is more economical than the direct plasma method in terms of construction and operating costs for plasma generation; it also

*To whom any correspondence should be addressed.

requires less installation space. NO_x removal using the ozone injection method for diesel engine exhaust gas and the simultaneous removal of NO_x and SO_2 from the flue gas of a thermal power station have been reported [14]–[18].

On the basis of the laboratory-scale experimental studies, tests on the removal of both NO_x and SO_x from the gas emitted from a boiler were carried out using our first pilot-scale apparatus employing the indirect plasma-chemical hybrid process [19]. This apparatus consisted of a pulsed corona plasma reactor that produced mainly O_3 from ambient air and a chemical scrubber (diameter: 0.9 m) that required a manual supply of chemical solutions. The NO_x removal efficiency during the combustion of city natural gas exceeded 90%. However, during the combustion of heavy oil, the efficiency was only about 60% because sufficient O_3 could not be generated with the available pulse power supply; in addition, the supply of chemical solutions was incomplete. The SO_2 removal efficiency was 85 – 90%. The waste water from the scrubber could be drained because it met the waste water regulation.

Our second pilot-scale apparatus [20]–[22] consisted of an ozonizer that was more compact and economical than the original pulsed corona reactor and a chemical scrubber that had approximately one-third the volume of the original. The scrubber was made smaller on the basis of the test results of the NO_x removal process and included an automatic injection control for the chemical solutions. Continuous operation for 3 – 5 h was successful, and the NO_x removal performance was confirmed during a preliminary test [20]. The performance characteristics were then investigated at flue gas flow rates of 410 – 1480 Nm^3/h (N: Standard state) [21]. The amount of NO_x removed was clearly dependent on the oxidation reduction potential (ORP) of the scrubbing solution, and it increased with a decrease in the ORP. To remove the required amount of NO_x from the flue gas, the ORP should be maintained below 0 mV or as low as possible. However, because the ORP fluctuates over time, to maintain a stable NO_x removal performance, the injection rate of fresh Na_2SO_3 solution should be sufficient to keep the rate of change in ORP at less than zero (0 mV/min). Furthermore, the influence of the packing material height in the scrubber on NO_x removal performance characteristics was investigated [22]. The correlation between NO_x removal efficiency and gas retention time was obtained; this provided an important index for scrubber design.

Our third generation demonstration plant was constructed from scratch. It consisted of a boiler that can use multi-fuels—not only compressed natural gas or heavy oil but also waste vegetable oil (WVO), etc.—and an improved chemical scrubber. WVO, which is mainly collected from snack food factories and fast food restaurants, differs from heavy oil in properties such as dynamic viscosity, flash point and fuel components. The scrubber size is optimized in terms of the flue gas condition, and an automatic injection control for both ozone and the chemical solution is included.

The objective of this study is to investigate the combustion characteristics of the third demonstration multi-fuel boiler system, confirm the performance characteristics of NO_x removal when using plasma-chemical hybrid process and obtain practical operation details for the purpose of industrial application. In the first stage of the study, operation tests of the system are carried out for heavy oil, mixed oil (heavy oil and WVO) and city natural gas fuel.

2. Experimental apparatus and methods

Figure 1 shows an overview of the third demonstration plant. **Figure 2** shows a schematic diagram of the plant. The boiler (Takao Iron Co., Ltd.) had a new type of a rotary burner for gas or oil fuels and is operated by using city natural gas (13 A type, Osaka Gas Co., LTD., Japan) at 157 Nm^3/h, heavy oil (A type) at 147 kg/h and WVO as a fuel; this is the rated value of the boiler and is equivalent to a thermal output value of 1.57 MW. The boiler has a steam generation rate of 2.5 ton/h. An economizer (exhaust gas heat exchanger) is used to recover the heat energy from the flue gas and sufficiently reduce the temperature of the flue gas to less than 150°C at the scrubber inlet to keep the plasma-chemical reaction effective.

(a) Frontal view (b) Details

Figure 1. Frontal view and details of the third generation pilot-scale low-emission boiler system

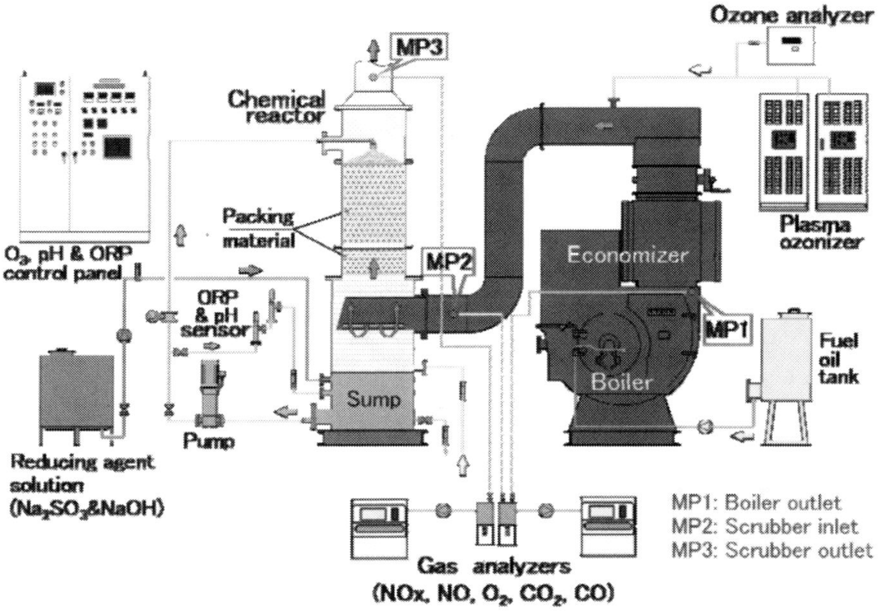

Figure 2. Schematic diagram of the low-emission boiler system.

One or two sets of commercial ozonizers (Ebara Jitsugyo Co., Ltd., EW-90Z) with a pressure swing adsorption (PSA) oxygen generator (power = 1.6 kW) are employed for ozone generation as described in **Table 1**; when the O_3 gas flow rate is 0.9 Nm^3/h and the discharge power is 1.5 kW, 90 g/h of ozone is generated, and its concentration is approximately 4.7%. The energy efficiency for ozone generation is 60 g/kWh.

The ozone is injected into a flue gas duct for NO oxidation ($NO + O_3 \rightarrow NO_2 + O_2$). The flue gas is then introduced into the scrubber with a height of 3.7 m to reduce NO_2 ($2NO_2 + 4Na_2SO_3 \rightarrow N_2 + 4Na_2SO_4$). The diameters are 0.9 m at the sump part of the scrubber and 0.7 m at the packing material layer part. The scrubbing solution is pumped from the sump to the top of the scrubber and sprayed through a nozzle over the packing material. Two types of packing materials (Packing A and Packing B) are used. The total height is set to 1.2 m, which includes Packing A (0.3 m) and Packing B (0.9 m). Packing A is a No. 1-type Raschig Super-Ring (RSR) made of SUS304 (specific surface: 163 m^2/m^3; void fraction: 98%) piled beneath Packing B in the scrubber. Packing B is an S-II-type Tellerette packing made of polypropylene (specific surface: 150 m^2/m^3; void fraction: 92%). After the NO_x in the flue gas is removed in the scrubber, the cleaned flue gas is discharged in the air through the smokestack. On the other hand, the scrubbing solution passes through the packing layer into the sump is circulated by a multistage centrifugal pump operating at 0.5 kW. A small amount of scrubbing solution (50 – 100 L/h) is continuously drained from the scrubber to keep the activity of the solution, i.e., remove the reaction product from the scrubber solution. The liquid flow rate is set to 3.0 m^3/h by controlling the valve at the outlet of the circulating pump. Meters installed on a tributary line connected to the sump monitored the ORP and pH of the scrubbing solution. To effectively remove NO_x, maintaining a pH of more than 7 and an ORP of less than 0 mV for the scrubbing solution is essential. An aqueous solution of Na_2SO_3 and NaOH (concentrations: 200 g/L and 10 g/L) obtained from a chemical factory as a byproduct is continuously added into the sump. The flow rate of the fresh aqueous solution is carefully controlled according to the pH and ORP of the scrubbing solution. The initial Na_2SO_3 concentration of the scrubbing solution in the sump is set to more than 16 g/L. It is necessary to drain the waste aqueous solution from the scrubber. The waste fluid rate of aqueous solution depends on the flow rate of the flue gas. The waste liquid rate is approximately 100 L/h in this pilot plant. It is important to maintain the proper Na_2SO_3 concentration of the solution in order to continue stable plant operation.

Table 1. Specifications of plasma ozonizer

Type	EW-90Z
O_2 generation	PSA method (O_2 more than 90%)
O_3 generation	Air-cooled tubular Silent discharge
O_3 mass rate	0 – 90 g/h
O_3 concentration	0 – 100 g/Nm^3 = 4.7%
O_3 flow rate	0.9 Nm^3 / h
Power consumption	1.6 kW (PSA), 1.5 kW (Discharge)
Dimension	W740 × D925 × H1840 mm

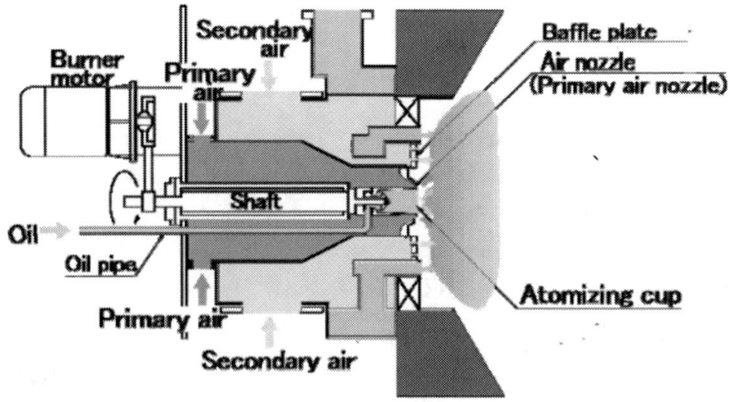

Figure 3. Structure of rotary-type boiler burner.

Figure 3 shows the rotary-type burner. This burner provides highly atomized characteristics for many kinds of oils such as light and heavy ones through centrifugal force generated by a high-speed rotating cup and primary combustion air injection. Secondary combustion air from the baffle plate helps enhance the flame-stabilizing and diffusion effects of the combustion air. Even fuel oils with unstable combustibility can be almost completely burned.

Experiments are performed after the temperature of the flue gas reached a steady state. The boiler is operated at 30 – 100% of the boiler rated load. The flow rate of the flue gas is not measured but determined by a practical formula based on the fuel consumption rate and O_2 concentration of the flue gas at the boiler outlet. The flue gas temperature is measured using thermocouples. The concentrations of the gaseous components (O_2, CO_2, CO, NO_x and NO) are measured by using a gas analyzer (Horiba PG-240). The temperature of the flue gas and concentrations of its component are measured at the following three sampling points, as shown in **Fig. 2**: MP1 (boiler outlet), MP2 (scrubber inlet) and MP3 (scrubber outlet). The sampling points MP1 and MP2 are switched over after every measurement. The O_3 concentration generated by the ozonizer is measured by using an O_3 monitor (Ebara Jitsugyo Co., Ltd., EG-550) in the pipeline between the ozonizer and O_3 injection point of the flue gas duct.

Table 2. Oil Fuel properties.

Fuel	Unit	Heavy oil (A type)	Waste vegetable oil (WVO)
Low calorific value	J/kg	44940	38880
Kinetic viscosity at 40°C	mm^2/s	5~10	39
Flash point	°C	< 60	296
Residual carbon	%	0.06	0.15
Nitrogen	%	0.01	< 0.01
Sulfur	%	0.059	< 0.005

Table 2 shows the fuel properties of both heavy oil and WVO. Because of relatively high viscosity, bio-fuel must be often preheated in the fuel tank when the temperature is low and/or filtered in the pipeline.

The experiments are carried out by firing heavy oil as the first step and then firing mixed oil as the second step. The mixture ratios of WVO to heavy oil are increased to 20%, 40% and 50% (vol.) in a stepwise fashion. The mixed oils are made up of 20%, 40% and 50% WVO corresponding to 80%, 60% and 50% (vol.) heavy oil, respectively (hereafter called 20%, 40% and 50% mixed oil). The experiment is carried out for 180 – 240 min on a single day.

3. Experimental results and discussion

Figure 4 shows the temperature of the flue gas at MP1, MP2 and MP3 when flow rates of the flue gas while firing the heavy oil are 930, 1560 and 2250 Nm^3/h, respectively. The temperature of the flue gas at MP1 reached 261°C when the flow rate of the flue gas is 2250 Nm^3/h. However, the temperate of the flue gas decreased to 126°C at MP2 when the economizer is in operation. The temperature of the flue gas at MP3 is below 56°C under all conditions.

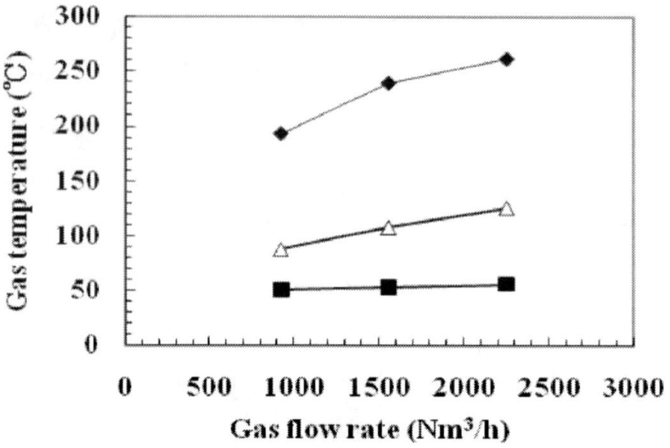

Figure 4. Flue gas temperature vs. gas flow rate.

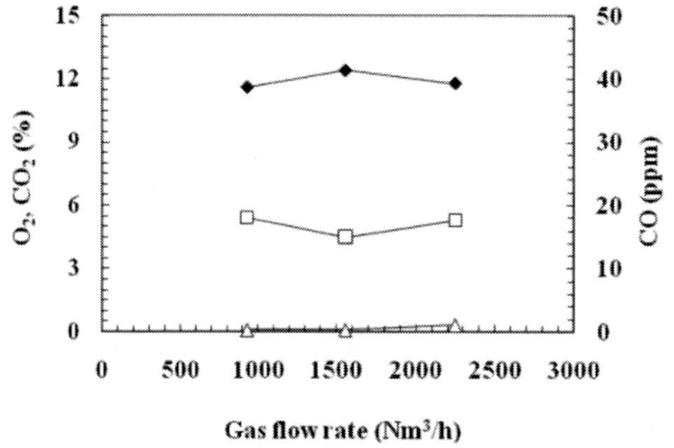

(a) O_2, CO_2 and CO

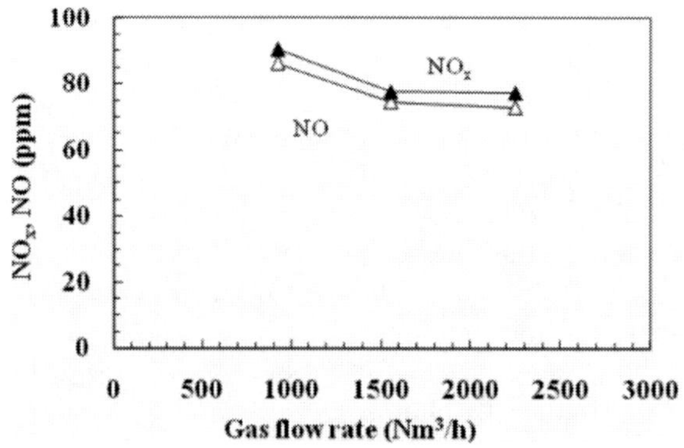

(b) NO_x and NO

Figure. 5 Gas components vs. gas flow rate when firing heavy oil. (Without plasma-chemical hybrid aftertreatment)

Figures 5(a) and **(b)** show the concentrations of O_2, CO_2, CO, NO and NO_x at MP1 at flue gas flow rates of 930, 1560 and 2250 Nm^3/h, respectively. The concentration of O_2 is maintained in the range of 4.5%–5.4% in order to prevent the generation of both CO and particulate matter (PM). The concentration of CO is less than 1 ppm, and the concentration of CO_2 is 12.0% on average. On the other hand, the concentration of NO_x decreased from 90 ppm to less than 80 ppm with increasing flow rate of the flue gas. The NO concentrations are 3 – 4 ppm less than those of NO_x.

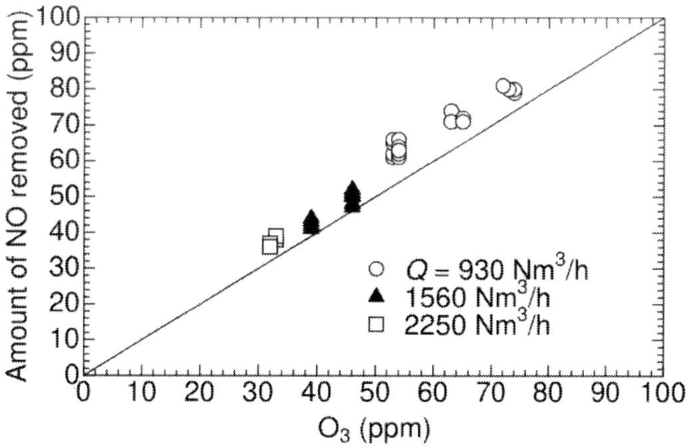

Figure 6. Relation between the amount of NO removed and the ozone concentration estimated in the gas duct.

Figure 6 shows the relationship between the amount of NO removed from the flue gas and the ozone concentration estimated in the flue gas duct where O_3 is injected. Flow rates of the flue gas are set to 930, 1560 and 2250 Nm^3/h, respectively, when firing heavy oil. The amount of NO removed is almost the same as the amount of the corresponding ozone concentration to oxidize NO to NO_2 (1:1 stoichiometric ratio) regardless of the flue gas flow rate. This proved that the distribution of ozone injected in the flue gas duct is good enough because that the flue gas duct is relatively small. However, in the case of scale-up of the system, a proper distribution of ozone is to be considered for larger gas duct.

After the heavy oil combustion test, mixed oil combustion tests are carried out to measure the flue gas components. **Figures 7 (a)–(c)** show the time-dependent flue gas components at MP1 when firing the mixed oils. The flow rates of the flue gas are 1090, 960 and 970 Nm^3/h when firing 20%, 40% and 50% mixed oils, respectively. For comparison between the mixed oils and heavy oil, the flue gas components when firing heavy oil are shown in these figures for the flow rate of 930 Nm^3/h of the flue gas. The average O_2 concentrations are 5.7%, 5.5% and 5.6% for 20%, 40% and 50% mixed oils, respectively. On the other hand, the average is 5.4% for heavy oil. The CO_2 concentrations are 11.5%, 11.8% and 11.7% for the mixed oils and 11.6% for heavy oil. Moreover, the CO concentrations are less than 1 ppm for all cases. Therefore, the mixed oils are fired under nearly the same conditions as heavy oil. Both the NO_x and the NO concentrations plotted in **Figs. 7(b) and (c)** are the converted data at 4% O_2 in the flue gas; the measured concentrations of flue gas components at different O_2 concentrations are converted into the equivalent values at 4% O_2 to compare flue gas emissions. The NO_x concentrations are almost the same at 97, 98 and 99 ppm on average for 20%, 40% and 50% mixed oils, respectively, and 98 ppm for heavy oil. The NO concentrations are also similar to NO_x. **Figures 7 (a)–(c)** indicate that the concentrations of flue gas components remained almost unchanged and stable throughout the experiment for all cases, which is quite satisfactory for boiler combustion.

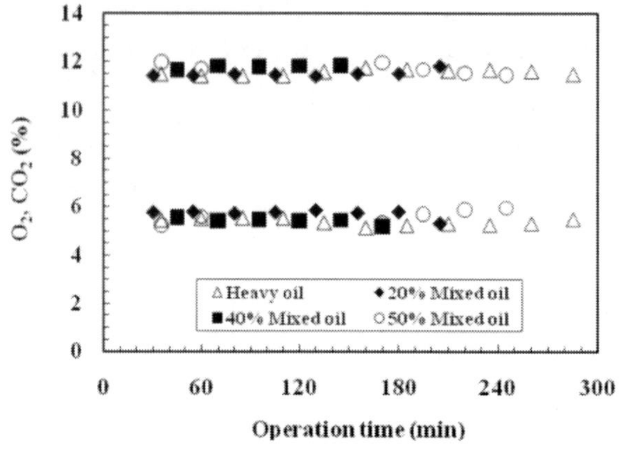

(a) O_2 and CO_2

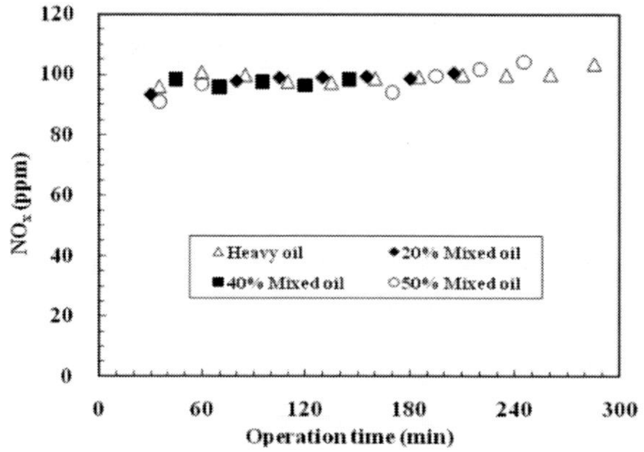

(b) NO_x

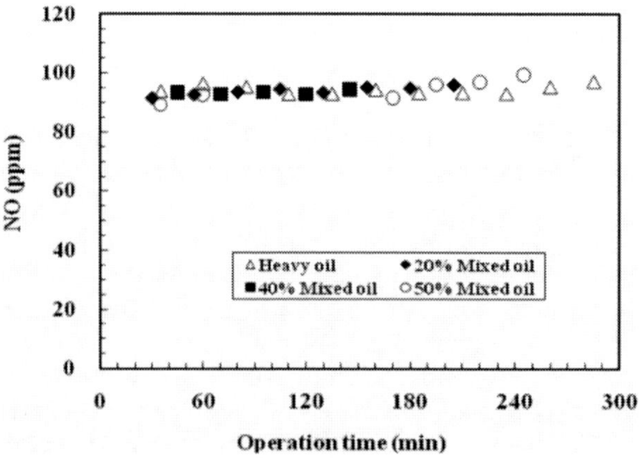

(c) NO

Figure 7. Gas components vs. gas flow rate when firing 20%, 40% and 50% mixed oils (without plasma-chemical hybrid aftertreatment)

Furthermore, boiler combustion and flue gas components need to be confirmed when firing other kinds of bio-fuels.

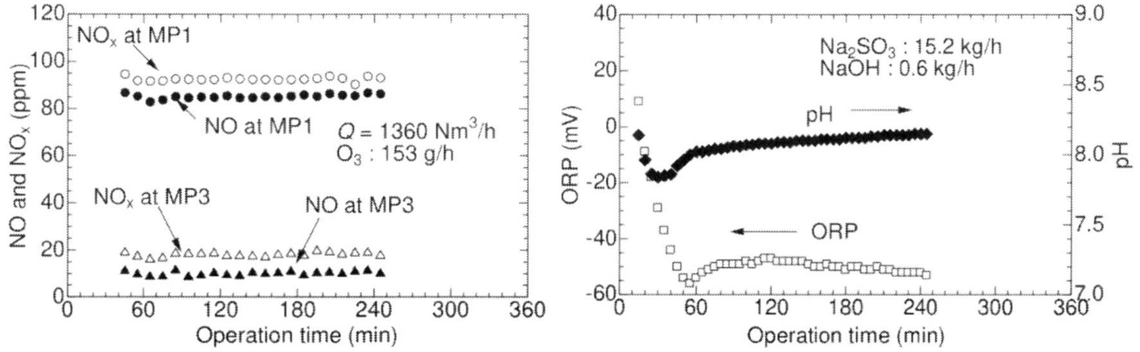

(a) (b)

Figure 8. Result of the operation test of the system when the system is operated with heavy oil and plasma-chemical hybrid aftertreatment is carried out: (a) NO_x and NO concentrations at MP1 (boiler exit) and MP3 (scrubber exit), (b) ORP and pH of the aqueous solution in the scrubber

Figure 8(a) shows the time-dependent NO_x and NO concentrations at MP1 and MP3, respectively, during a 200-min operation period when firing heavy oil. For the first 20 min, the boiler is warmed up. Then after the boiler is operating at the corresponding fuel flow rate, measurement begins. The flow rate of the flue gas is set to 1360 Nm^3/h. O_3 is injected at a rate of 153 g/h; this corresponded to an ozone gas flow rate of 1.8 Nm^3/h. The scrubber solution flow rate is kept at 3.0 m^3/h by maintaining the average injection rate of the Na_2SO_3 solution at 15.2 kg/h and that of the NaOH solution at 0.6 kg/h throughout the experiment. The concentrations of NO and NO_x at MP1 are nearly constant until the end of the experiment. NO removal is on average from 85 ppm at MP1 to 10 ppm at MP3, and NO_x removal is on average from 92 ppm at MP1 to 18 ppm at MP3; this is equivalent to a NO_x removal efficiency of more than 80%. **Figure 8(b)** shows the time-dependent ORP and pH of the scrubbing solution when firing the heavy oil. The values of ORP and pH varied widely for 30 min after the plant started but then became stable owing to proper Na_2SO_3 injection control. The ORP and pH are maintained at approximately –50 mV and 8.2, respectively, which are satisfactory for scrubber operation. Stable performance of NO_x removal is confirmed to have been achieved for a 200-min operation when firing heavy oil.

Figure 9(a) shows the time-dependent NO_x and NO concentrations at MP1 and MP3, respectively, during a 300-min operation period when firing 20% mixed oil. The flow rate of the flue gas is set to 870 Nm^3/h on average. O_3 is injected at a rate of 122 g/h. The scrubbing solution flow rate is kept at 3.0 m^3/h by maintaining the average injection rate of the Na_2SO_3 solution at 11.7 kg/h and that of the NaOH solution at 0.5 kg/h throughout the experiment. Owing to a few changes in the O_2 concentration of 4.7 – 5.1%, NO and NO_x at MP1 varied in the ranges of 80 – 93 and 90 – 105 ppm, respectively. However, NO is removed to less than 2 ppm at MP3. NO_x is removed to less than 10 ppm at MP3; this is equivalent to a NO_x removal efficiency of more than 90%. **Figure 9(b)** shows the time-dependent ORP and pH of the scrubbing solution when firing the 20% mixed oil. The values of ORP and pH varied widely for 30 min after the plant started but then became stable. The values of ORP and pH are maintained at approximately –30 mV and 8.1, respectively. A stable performance for NO_x removal is confirmed to have been achieved for a 300-min operation when firing 20% mixed oil.

Journal of Physics: Conference Series **418** (2013) 012115 doi:10.1088/1742-6596/418/1/012115

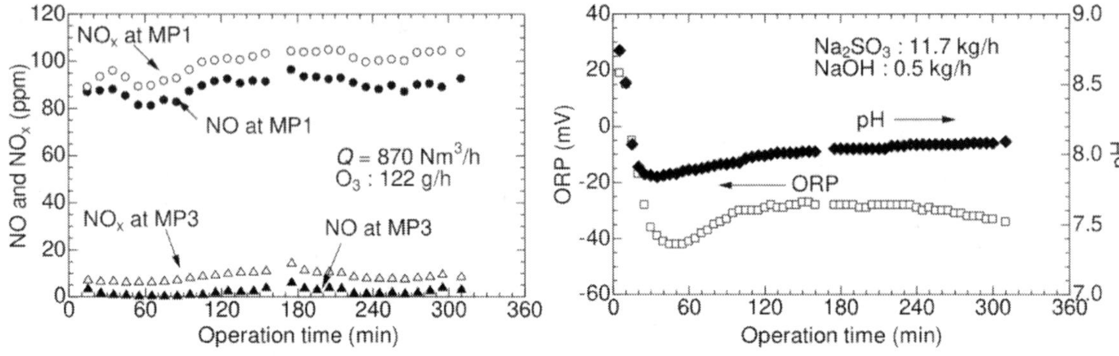

(a) (b)

Figure 9. Result of the operation test of the system when the system is operated with 20% mixed WVO fuel and plasma-chemical hybrid aftertreatment is carried out: (a) NO_x and NO concentrations at MP1 (boiler exit) and MP3 (scrubber exit), (b) ORP and pH of the aqueous solution in the scrubber.

Table 3 lists estimation results of the CO_2 reduction when firing bio-fuel. Waste vegetable oil is carbon neutral and is acknowledged to have less CO_2 emissions than other fuel types. When heavy oil (fuel rate: 147 kg/h) is burned at the rated value of the boiler, the concentration of CO_2 is calculated to be 11.6% in the flue gas, and the total amount of CO_2 emitted from the boiler is estimated to be 1,390 ton/year for 3,000 h of operation, as indicated by case 2 in the table. Therefore, if 50% mixed oil (50% heavy oil) is burned, CO_2 emission is reduced by 50% to 695 ton.

.

Table 3. Estimation of CO_2 reduction for bio-fuel firing.

Items	Unit	Case 1	Case 2
Boiler load	%	40	100
Fuel flow rate	L/h	68	171
Total CO_2 emission (100% heavy oil)	t/y	555	1,390
Total CO_2 reduction (20% WVO + 80% heavy oil)	t/y	444	1,110
Total CO_2 reduction (50% WVO + 50% heavy oil) as carbon neutral	t/y	278	695

Remarks: Annual operation = 3,000 h

Figure 10 shows an example of measurement result on time-dependent NO_x at MP1 and MP3 when the system is operated with city natural gas fuel. Flow rate of exhaust gas is Q = 690 Nm^3/h with CO_2 = 9.5%, O_2 = 4.5%. NO_x concentrations are based on the equivalent value with O_2 = 5%. It is known from the figure that ultra-low NO_x emission of less than 5 ppm is realized for 5 h.

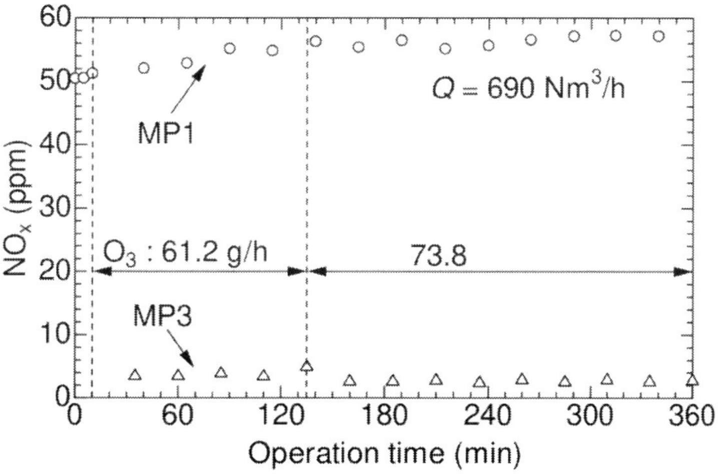

Figure 10. Time-dependent NOx at MP1 (boiler exit) and MP3 (scrubber exit) (city gas fuel, Q = 690 Nm^3/h, CO_2 = 9.5%, O_2 = 4.5%, NO_x concentrations are based on the value with O_2 = 5%, with plasma-chemical hybrid aftertreatment).

4. Conclusions

The third pilot-scale low-emission boiler system for firing heavy oil and mixed oil (heavy oil and waste vegetable oil) is examined. Stable combustion is achieved for the mixed fuel (20 – 50% waste vegetable oil). Properties of flue gas—e.g., O_2, CO_2 and NO_x—when firing mixed fuel are nearly the same as those in the case of firing heavy oil when the flue gas flow rate is 1,000 Nm^3/h. NO_x emission concentrations are 90 – 95 ppm. The effect of ozone injected into the flue gas is confirmed. The amount of NO removed is almost the same as the amount of the corresponding ozone concentration for oxidizing NO to NO_2.

Continuous operation tests of the system are successfully carried out for both heavy and mixed oil. When firing heavy oil, more than 80% of the NO_x is demonstrated and confirmed to have been removed from the flue gas (less than 18 ppm at MP3) during a 200-min continuous operation for which the flue gas flow rate is 1360 Nm^3/h and the ORP and pH are maintained at approximately 50 mV and 8.1, respectively. Furthermore, when firing 20% mixed oil, more than 90% of the NO_x is removed from the flue gas (less than 10 ppm at MP3) during a 300-min continuous operation for which the flue gas flow rate is 870 Nm^3/h, and the ORP and pH are maintained at approximately –30 mV and 8.0, respectively.

Mixed oil is successfully fired. The effect of replacing heavy oil with waste vegetable oil can reduce CO_2 emissions from the boiler. If 50% mixed oil (50% heavy oil) is burned, the amount of CO_2 discharged into the air is reduced by 50% to 695 ton for an 3000-h operation in 1 year.

The present De-NO_x system is developed and effective for small size boiler application as compared to the conventional selective catalytic reduction (SCR) De-NO_x system. For scale-up of the system there are necessary tunings to be confirmed; proper distribution of ozone injected in the flue gas duct and of scrubbing solution by spraying nozzle in the scrubber. For commercializing the system, the long-term operation tests are to be carried out to confirm the performance reliability of NO_x removal and operational characteristics of the system.

Acknowledgments

The authors thank Mr. Y. Yoshioka, Mr. K. Takekoshi, Mr. Y. Nakagishi, Mr. Y. Fukumoto, and Mr. H. Mochizuki, all students of Osaka Prefecture University, for assistance with carrying out the experiments. This study is supported by the Regional Research and Development Resources Utilization Program conducted in the Japan Society Technology Agency (JST).

References

[1] Yamamoto T, Yang C L, Kravets Z, and Beltran M 2000 *IEEE Trans. Ind. Appl.* **36**(3) 923–927
[2] Yang C L, Yamamoto T, Beltran M, and Kravets Z 1998 *Env. Prog.* **17**(3) 183–189
[3] Yamamoto T, Okubo M, Hayakawa K, and Kitaura K 2001 *IEEE Trans. Ind. Appl.* **37**(5) 1492–1498
[4] Kuroki T, Takahashi M, Okubo M, and Yamamoto T 2002 *IEEE Trans. Ind. Appl.* **38**(5) 1204–1209
[5] Yoshida K, Yamamoto T, Kuroki T, and Okubo M 2009 *Plasma Chem. Plasma Process.* **29**(5) 373–386
[6] Chang M B, Lee H M, Wu F and Lai C 2004 *J. of Air & Waste Manage. Assoc.* **54** 941–949
[7] Okubo M, Kuroki T, Kitaura K, and Yamamoto T 2006 *JSME, J. of Environment and Engineering* **1**(1) 29–38
[8] Yamamoto T, Kajimoto A, Okubo M, Kuroki T, and Yoshida K 2008 *IEEE Trans. Ind. Appl.* **44**(5) 1431–1435
[9] Zhou Q, Yao S C, Russell A, and Boyle J 1992 *J. of Air & Waste Manage. Assoc.* **42**(9) 1193–1197
[10] Nishida M, Yukimura K, Kambara S, and Maruyama T 2001 *Jpn J. Appl. Phys.* **40** 1114–1117
[11] Nishida M, Yukimura K, Kambara S, and Maruyama T 2001 J. Appl. Phys. **90**(6) 2672–2677
[12] Dinelli G, Civitano L, and Rea M 1990 *IEEE Trans. Ind. Appl.* **26**(3) 535–541
[13] Chang J S, Urashima K, Arquillq M, and Ito T 1998 *Combust. Sci. and Tech.* **133** 31–47
[14] Sano K and Yoshioka Y 2003 *IEEJ Trans. FM* **123**(10) 1030–1035
[15] Yoshioka Y 2007 *International J. Plasma Environmental Science and Technology* **1**(2) 110-122
[16] Simachev V Y, Novoselov S S, Svetlichnyi V A, Gavrilov A F, Gorokhov M V, Semenov V I, Ryzhikov V A, and Demchuk V V 1988 *Thermal Engineering* **35**(3) 67–70
[17] Novoselov S S, Gavrilov A F, Simachev V Y, and Svetlichnyi V A 1986 *Thermal Engineering* **33** 496–499
[18] Slater S M and Rizzone M S 1980 *Fuel* **59** 897–899
[19] Yamamoto T, Fujishima H, Okubo M, and Kuroki T 2010 *IEEE Trans. Ind. Appl.* **46**(1) 29–37
[20] Kuroki T, Fujishima H, Otsuka K, Ito T, Okubo M, Yamamoto T, and Yoshida K 2008 *Thin Solid Films* **516** 6704–6709
[21] Fujishima H, Kuroki T, Ito T, Otsuka K, Okubo M, Yamamoto T, and Yoshida K 2010 *IEEE Trans. Ind. Appl.* **46**(5) 1707–1714
[22] Fujishima H, Kuroki T, Tatsumi A, Tanaka A, Otsuka K, Yamamoto T, and Okubo M 2010 *IEEE Trans. Ind. Appl.* **46**(5) 1722–1729

Plasma induced toluene decomposition on alumina-supported Mn-based composite oxides catalysts

Lingling Ye, Fada Feng, Ji Liu, Zhen Liu, Keping Yan*

Key Laboratory of Biomass Chemical Engineering of Ministry of Education, Industrial Ecology and Environment Research Institute, Zhejiang University, Tianmushan Road 148, Hangzhou 310007, China

E-mail: kyan@zju.edu.cn

Abstract. This paper presents a two-stage hybrid plasma catalyst system for toluene decomposition in air. It consists of a dielectric barrier discharge reactor and Mn-based composite oxides catalysts. Experimental results show that removal efficiency of toluene was 10 % by only plasma, but a 100 % destruction efficiency as well as 100% ozone removal was achieved by plasma catalysis at a specific energy density of 28.8 J L^{-1}. The catalytic decomposition efficiency of toluene and ozone produced in the first stage is in the following order: NTP+ Mn-Co/Al > NTP + Mn-Ce/Al > NTP+ Mn-Ni/Al > NTP. Compared to other catalysts coupled with non-thermal plasma in a two-stage plasma catalytic reactor, the alumina-supported Mn-based composite oxides catalysts showed a better performance for toluene and ozone removals, especially the Mn-Co/Al catalyst.

1. Introduction

Air pollution with volatile organic compounds (VOCs) has been one of the serious environmental problems because of its toxicity, destruction of the ozone layer and through photochemical smog. As an alternative to low concentration VOCs abatement techniques, non-thermal plasma has been widely investigated for its unique properties such as moderate operating conditions and high-energy electrons with short residence times[1-3]. The energetic electrons excite, dissociate and ionize the gas molecules producing chemically active species such as O and OH radicals, which will result in VOC conversion to CO_2, H_2O and other products [4]. However, plasma alone has many disadvantages such as the plasma energy consumption, CO_2 selectivity, and toxic by-products [5, 6]. A new attempt to overcome these limitations is the combination of the NTP method with catalysis. The hybrid reactor can be a single-stage (SPC) or a two-stage (TPC) system. For many cases, a better performance could simply be achieved by a catalytic post-treatment of the NTP effluent in a two-stage system. The main role of the catalyst bed is to recycle O radicals via the slipped O_3 [7-9]. Alumina-supported transition metals have been widely investigated based on the high reactivity for ozone decomposition [10, 11].

In this work, various catalytic materials based on Mn supported on Al_2O_3 were combined with a dielectric barrier discharge (DBD) for toluene abatement at room temperature. Detailed comparison experiments of toluene decomposition are performed in order to evaluate any synergistic effects between the plasma and the catalyst.

* To whom any correspondence should be addressed.

2. Experimental system

2.1. Plasma/catalyst system

The reaction system is as reported previously [12]. The non-thermal plasma is produced by using a dielectric barrier discharge (DBD) reactor. A stainless steel high-voltage electrode is placed inside a silica tube with a copper mesh grounded electrode. The active length and the discharge gap are 10 cm and 2 mm, respectively. A 50 Hz AC high-voltage power source is used for energizing the reactor. All experiments are performed under atmospheric pressure and room temperature.

Toluene is flushed out from a tubular saturated at zero centigrade degree with dry air under a flow rate of 6.0 mL min^{-1}. After mixing with air, the total gas flow rate is 500 mL min^{-1} for present work, which corresponds to the gas hourly space velocity (GHSV) of the catalysis bed of 150,000 h^{-1}, and initial toluene concentration of 107 parts per million (ppm). Gaseous analyses are carried out with an on-line gas chromatograph (Fuli GC-9750) and iodometric titration.

2.2. Materials

Various catalytic materials based on Mn supported on Al2O3 in powder form have been tested to select the highest active catalyst. The Co-Mn/Al, Ce-Mn/Al and Ni-Mn/Al catalysts were all prepared by the deposition precipitation method, and Co-Mn, Ce-Mn, Ni-Mn mol ratio were 1:1. The catalysts' BET surface areas are 181.53, 183.23 and 166.02 m2 g-1, respectively.

3. Results and discussion

3.1. Effect of catalysts on toluene conversion and ozone decomposition

Figure 1 shows the conversion of toluene using Mn-Co/Al catalysts of different Mn-Co loading amounts. At SIE of 19.9 J L^{-1}, an increase in the Co-Mn amount increases the decomposition efficiency of toluene. For example, toluene conversion of 10wt% Co-Mn loading amount is 70%. When increasing the loading amount to 30 wt%, toluene conversion increases to 88%. On the other hand, when SIE>28 J L^{-1}, toluene can be almost removed irrespective of the Mn-Co loading amounts. This means that a good synergistic effect between the plasma and the Mn-Co/Al catalyst can also be achieved with a low metal loading amount.

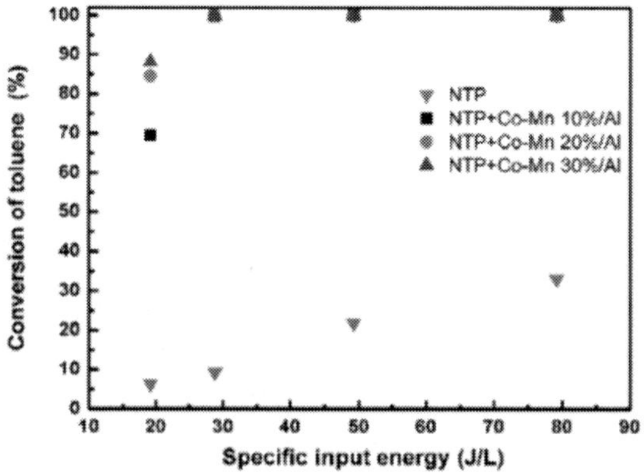

Figure 1. Effect of Mn-Co loading amount on the conversion of toluene.

Figure 2 and figure 3 show conversion of toluene via the SIE and residual ozone concentration using γ-Al_2O_3 catalysts supported with Co-Mn, Ce-Mn and Ni-Mn. Hereafter, the catalysts with 30 wt% are used for the catalytic tests. For both toluene and ozone decomposition, the catalyst reactivity is in the order of Co-Mn/Al >Ce-Mn/Al > Ni-Mn/Al. For example, at the SIE value of 28.8 J L^{-1}, the plasma reactor induced decomposition rate is about 10%, but increases to 100 % by integrating the

Co-Mn/Al catalyst. At the same time, 100% of ozone is destroyed. Ozone is the only oxidant in the post-plasma which can be utilised to oxidise toluene via its decomposition to produce oxygen atom [13]. These findings indicate that the γ-Al_2O_3 plays an important role in the decomposition of both toluene and ozone. This might be due to the larger BET specific surface area with Al_2O_3 supported.

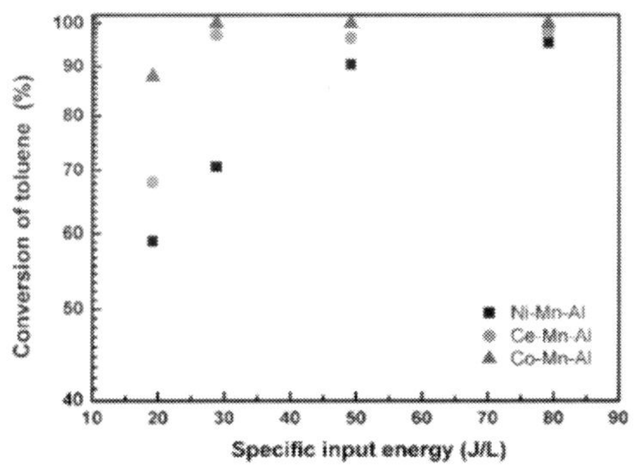

Figure 3. Conversion of toluene via the SIE over using γ-Al_2O_3 supported with Co-Mn, Ce-Mn and Ni-Mn catalysts.

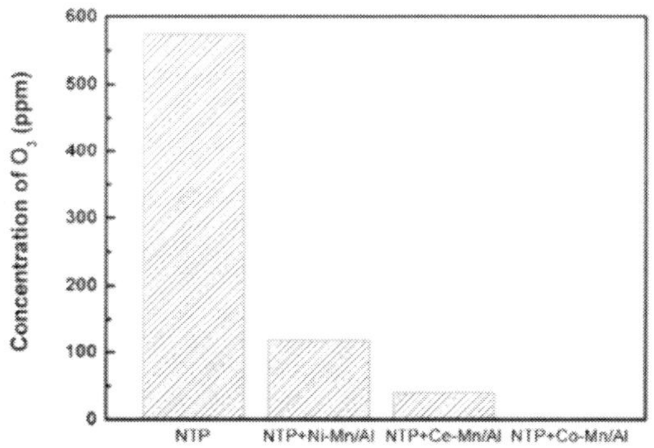

Figure 4. Residual ozone concentration using γ-Al_2O_3 catalysts supported with Co-Mn, Ce-Mn and Ni-Mn.

4. Conclusions

Based on present experimental investigations, we can give the following remarks:

(1) The alumina-supported Mn-based composite oxides catalyst is found to be very effective for the post-plasma system due to its larger specific surface area.

(2) The Mn-Co/Al catalyst possesses both high activity toward toluene removal and ozone decomposition. Toluene can be almost removed at SIE of 28.8 J L^{-1} with no ozone residue.

References

[1] Chang J S 2008 *Plasma Sources Sci.Technol.* **17** 2
[2] Mista W and Kacprzyk R 2008 *Catal.Today* **137** 345
[3] Van Durme J, Dewulf J, Sysmans W, Leys C and Van Langenhove H 2007 *Chemosphere* **68** 1822
[4] Than Quoc An H, Pham Huu T, Le Van T, Cormier J M, and Khacef A 2011 *Catal.Today* **176** 475
[5] Mok Y S and Kim D H 2011 *Curr.Appl.Phys.* **11** S58
[6] Subrahmanyam Ch, Renken A and Kiwi-Minsker L 2007 *Chem.Eng.J.* **134** 80
[7] Guo Y F, Ye D Q, Chen K F and He J C 2007 *Cata. Today* **126** 332
[8] Roland U, Holzer F, Kopinke F D 2005 *Appl. Catal. B: Environ* **58** 225
[9] Van Durme J, Dewulf J, Demeestere K, Leys C and Van Langenhove H 2009 *Appl. Catal. B: Environ* **87** 81
[10] Einaga H and Futamura S 2004 *J.Catal* **227** 306
[11] Einaga H and Futamura S 2006 *J.Catal* **243** 450
[12] Tang X J, Wang M, Feng W Q, Feng F D and Yan K P ISPC **20** 2011 USA
[13] Harling A M, Glover D J, Whitehead J C and Zhang K 2009 *Appl. Catal. B: Environ* **90** 158

Journal of Physics: Conference Series **418** (2013) 012117 doi:10.1088/1742-6596/418/1/012117

NO removal by nonthermal plasma with modified sepiolite catalyst

M G Chen[1*], D X Yu[1], J F Rong[1], Y L Wan[1], G C Li[1], Y M Ni[1], X Fan[1], G H Hou[2], N Xu[2]

1. School of Chemical Engineering of Anhui University of Science and Technology, Huainan Anhui 232001, China;

2. Key Laboratory for Advanced Technology in Environmental Protection of Jiangsu Province of Yancheng Institute of Technology, Yancheng Jiangsu 224051, China

E-mail: mgchen@aust.edu.cn

Abstract. Non-Thermal Plasma (NTP) combined with a catalyst is one of the effective ways to remove NO from auto exhaust gas. Sepiolite Ore Powder (SOP), which was modified by acid washing, copper nitrate soaking, drying and calcinations, served as the Modified Sepiolite Catalyst (MSC) for NO removal in a rod-cylinder Dielectric Barrier Discharge (DBD) reactor. The characteristic of the MSC was characterized by X-ray powder diffraction (XRD) and scanning electron microscopy (SEM). The experiment showed that the acid concentration, washing time, the packed site of MSC and input voltage of the NTP impacted the NO removal rate effectively. The NO removal rate increased and then decreased with an increase in the acid concentration and the washing time, and the NO removal rate increased monotonously with the increased input voltage. The NO removal rate was higher at the beginning, decreased gradually then maintained stability after 10 min. Thus, the result indicated that MSC has a good ability for adsorption and storage of NO.

1. Introduction

Non-Thermal Plasma (NTP) is an effective way to produce a large amount of free electrons, ions, active free radicals and various excited free particles, which have high active energy to improve the speed of chemical reaction. Dielectric Barrier Discharge (DBD) is one of the methods to generate NTP. The structures of DBD reactor are usually line-tube, line-plate, and plate-plate [1-3]. It has been researched that the DBD combined with catalyst can remove NO_x from the exhaust gas [4, 5], but the step of oxidizing reaction from NO to NO_2 is very difficult and slow [6]. The concentration of NO_x in auto exhaust is common $(200\text{-}300)\times10^{-6}$. In order to remove NO_x effectively, the catalyst not only has a good adsorption and storage capability, but also has a good activity and high selectivity [7, 8]. The catalyst of γ -Al_2O_3, TiO_2 and molecular sieve as a carrier have been widely researched, however, these kinds of

[*] To whom any correspondence should be addressed.

catalyst are not only expensive but also easy to lose its activity when at higher temperature [9, 10].

Sepiolite Ore Powder (SOP) is a one kind of natural clay which contains the element of magnesium and silicate. Its micro structure is a large number of fiber-like gathered together to form irregular network, so it has a large of specific surface area and good capability of adsorption [11-13]. The SOP has a good thermal stability and low price [14-16]. The SOP had been modified by acid washing, copper nitrate soaking, drying and calcination to prepare Modified Sepiolite Catalyst (MSC) to instead of traditional catalyst in this paper. The characteristic of the MSC was detected by XRD and SEM. The plasma-catalysis process for NO removal took place in rod-cylinder DBD reactor in this study [17].

2. Experiment

2.1. Catalyst preparation

2.1.1 Reagents and experimental equipments

Reagents: SOP (Henan LeiBao Sepiolite Ore Company, 280 mesh, industrial grade); Nitric acid (the AR), Copper Nitrate, Air Cylinder, NO/N_2 Cylinder (the content of NO 2011×10^{-6}, the balance is N_2).

Main equipments: Automobile exhaust analyzer (Guangdong Fushan analysis instrument factory, FGA-4100; NO measurement range:$0\sim4000\times10^{-6}$; Constant Temperature Oven (Model DHG-9240); Scanning Electron Microscope (SEM, SHIMADZU, KYKY-EM3200), X-ray Diffraction (XRD, XD-3 type crystal X-ray powder diffraction was produced by Beijing General Company); Muffle Furnace, Gas Mass Flow-meter; Transformer; Power Meter; High Voltage Power Supply (JNB-208B); High-voltage Probe; Digital Oscilloscope.

Rod-cylinder DBD reactor: The reactor is made of Borosilicate Pyrex Glass Tube of inner diameter 27 mm, wall thickness 2 mm and length 120 mm. The tube outer wall was pasted aluminum film of 60 mm length and 0.6 mm thick, served as cathode; the aluminum rod with diameter of 8 mm arranged in the center of the tube as anode. The MSC was loaded inside the tube; both end of the reactor was sealed by silicone rubber plug [18]

2.1.2. Catalyst preparation

Purification of SOP: 200 g SOP was mixed with 120 mL distilled water and stirred for 10 min. After a stillness moment, the SOP was deposited from the suspension and washed for four times with distilled water in the same way. The suspension was filtrated by vacuum filter, and then, the filter cake was dried in the oven at 105 °C and grinded and washed by nitric acid for 24 h, 48 h, and 72 h, respectively. The ratio of the SOP quality to the nitric acid volume is 1:10 (g mL^{-1}). The acidified SOP was washed for four times with de-ionized water and filtered. The filter cake was dried in Muffle Furnace at temperature of 120 °C for 4 h, then, grinded after natural cooling.

Modification and Activation of SOP: Every sample of acidic and original SOP 50 g was immersed in $Cu(NO_3)_2$ solution for 24 h at room temperature (ensuring the CuO content of 8%), evaporated and dried at 60 °C in water bath and grinded to 240 mesh, and then the SOP was dried at the room temperature for 24 h and calcined in the Muffle Furnace at 500 °C for 5 h, respectively.

2.2. Experimental methods

The 30 g MSC was loaded into the same fixed bed reactor with NTP. The reactor was installed in a thermo-tank at the 180 °C. The simulation gas of the NO concentration 297×10^{-6} (297 ppm) went into the reactor from the bottom side. The gas flow was kept at 7.0 L min^{-1}, the input voltage ($V_{p\text{-}p}$ peak-peak) and the frequency were fixed at 35 kV and 50 Hz, respectively. The NO removal rate =100× (inlet NO concentration-outlet NO concentration) /inlet NO concentration. The schematic experimental process was shown in figure 1.

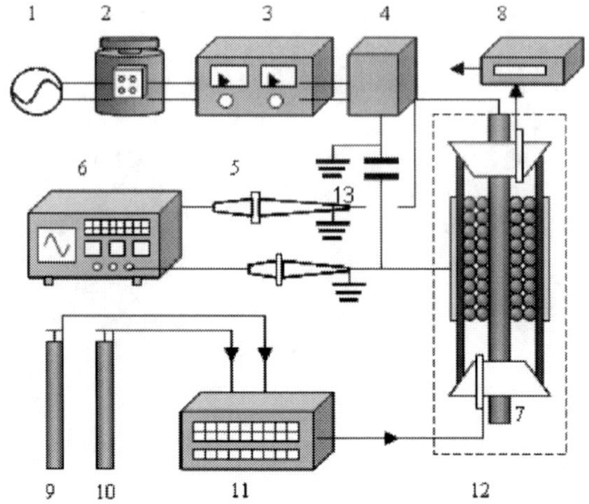

Figure 1. Schematic experimental process
1 Power, 2 Transformer, 3 Power meter, 4 High voltage power generator, 5 High-voltage probe, 6 Digital oscilloscope, 7 Non-thermal plasma catalytic reactor ,8 NO concentration online detector, 9 Air cylinder, 10 NO gas cylinders, 11 Gas quality controller, 12 thermo-tank, 13 capacity

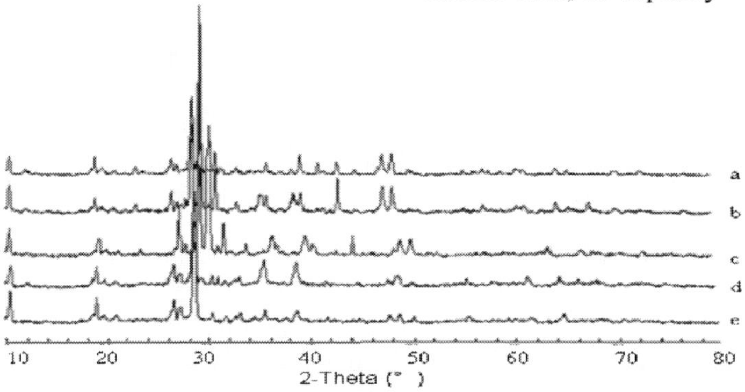

Figure 2. XRD of the SOP and the MSC.
a. the original Ore; b. without acid washing, only soaking copper nitrate for 24 h and calcination for 5 h at 500 °C; c. with HNO_3 of concentration 1.0 mol L^{-1} washing for 2 d and soaking copper nitrate for 24 h, calcination for 5 h at 500 °C; d. with HNO_3 of concentration 1.5 mol L^{-1} washing for 2 d and soaking copper nitrate for 24 h, calcination for 5 h at 500 °C; e. with HNO_3 of concentration 3.0 mol L^{-1} washing for 3 d and soaking copper nitrate for 24 h, calcination for 5 h at 500 °C

3. Results and discussions

3.1. Catalyst characterization

3.1.1 XRD analysis

Results of X-ray diffractometry for the SOP and the MSC are shown in figure 2. Diffraction conditions are target of Cu, tube voltage of 36 kV, tube current of 40 mA.

Figure 2 shows that the SOP contains a large amount of Si, Mg, Al and other elements, which exist in the different compounds. The peak values of 35.7° (2θ) and 39.1° (2θ) are corresponding of CuO, after soaking and calcination, a new material of CuO is formed on the surface of MSC, the Sepiolite keeps its basic structure. Compared with c, d, e, it shows that 2 day is the optimization acid washing time.

3.1.2. SEM analysis

The micro structure of the MSC is more improved than the SOP from the figure 3. The cross section and the number of micro-pore have been enlarged in the MSC surface, so the MSC has a good ability of adsorption and is benefit to carry the active component of CuO. Meanwhile, the acid washing can remove impurities from the SOP surface effectively. With the electronic probe, the active component of CuO has been detected in many bright points and distributed evenly in the SOP surface from the SEM in figure 3.

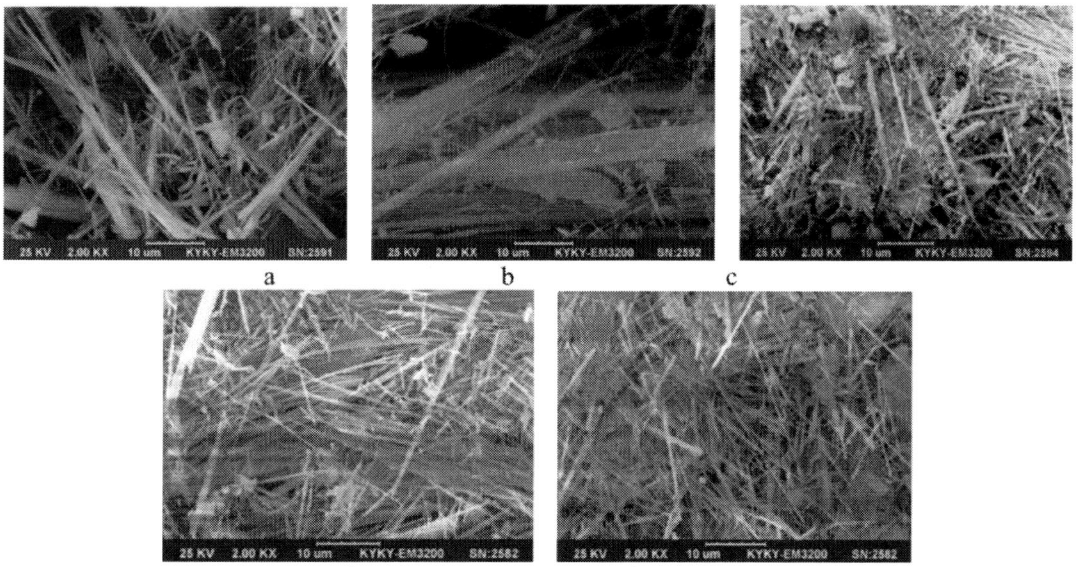

a b c

c d

Figure 3. SEM photos.

a. the original Ore; b. without acid washing, only soaking copper nitrate for 24 h and calcination for 5 h at 500 °C; c. with HNO_3 of concentration 1.0 mol L^{-1} washing for 2 d and soaking copper nitrate for 24 h, calcination for 5 h at 500 °C; d. with HNO_3 of concentration 1.5 mol L^{-1} washing for 2 d and soaking copper nitrate for 24 h, calcination for 5 h at 500 °C; e. with HNO_3 of concentration 3.0 mol L^{-1} washing for 3 d and soaking copper nitrate for 24 h, calcination for 5 h at 500 °C

3.2. Effect of nitric acid concentration on NO removal rate

The concentrations of nitric acid solution are 1.0, 1.5 and 2.0 mol L^{-1}, respectively, the SOP is soaked for 2 days, the weight ratio of acid to SOP is 1:10 and to prepare the MSC. The relationship between nitric acid concentration and NO removal rate is shown in figure 4.

When the HNO_3 concentration is above 1.5 mol · L^{-1}, the NO removal rate reaches the maximum value stably. When the SOP has no processes of acid washing, only soaking and calcination, the NO removal rate is lower than others condition. Firstly, it is because that micro-pore (<1.5 nm) and part of medium-pore (1.5-5.0 nm) on the surface of SOP is developed and enlarged by acid washing. Next, magnesium is removed from the SOP surface by acid washing and leaved behind many pores.

When the HNO_3 concentration is smaller than the optimum value (1.5 mol · L^{-1}), some of magnesium are apart from the SOP and reproduced many new micro-pores which connected each other forms many pore-nets in the SOP surface. The new micro-pore and pore-net is benefit to improve the capability of adsorption and the activity of the MSC. With the acid concentration increasing, the surface area of SOP is been increased.

When the HNO_3 concentration is bigger than the optimum value (1.5 mol · L^{-1}), the most of the micro-pore and the medium pore have been developed and become the big-pore with the diameter bigger than 10 nm. So the surface area of the MSC and the NO removal rate is decreased.

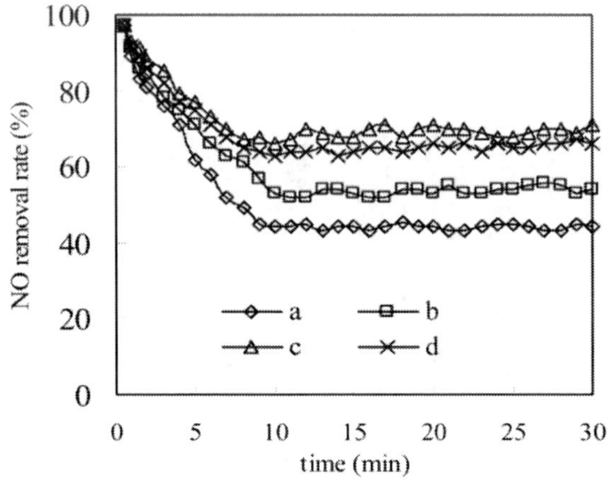

Figure 4. The relationship between HNO_3 concentration and NO removal rate.
a: plasma+ SOP without acid washing, only soaked $Cu(NO_3)_2$ for 24 h and calcination for 24 h at 500 °C; b: plasma+ SOP with HNO_3 of concentration 1.0 mol L^{-1} washing for 2 d and soaked $Cu(NO_3)_2$ for 24 h and calcination; c: plasma+ SOP with HNO_3 of concentration 1.5 mol L^{-1} washing for 2 d and soaked $Cu(NO_3)_2$ for 24 h and calcination; d: plasma+ SOP with HNO_3 of concentration 2.0 mol L^{-1} washing for 2 d and soaked $Cu(NO_3)_2$ for 24 h and calcination.

3.3. Effect of washing time on the NO removal rate

The concentrations of nitric acid solution 1.5 mol L^{-1}, the SOP is soaked for 1 day, 2 days and 3 days, respectively. The weight ratio of acid to SOP was 1:10. The relationship between the acid washing time and NO removal rate is shown in figure 5.

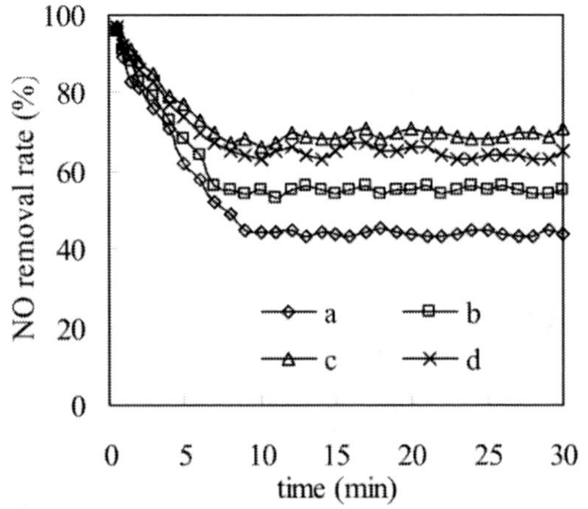

Figure 5. The relationship between the wishing time and NO removal rate.
a: plasma+ SOP without acid washing, only soaking $Cu(NO_3)_2$ for 24 h and calcination for 5 h at 500 °C; b: plasma+ SOP with HNO_3 of concentration 1.5 mol L^{-1} washing for 1 d and soaking $Cu(NO_3)_2$ for 24 h and calcination; c: plasma+ SOP with HNO_3 of concentration 1.5 mol L^{-1} washing for 2 d and soaking $Cu(NO_3)_2$ for 24 h and calcination; d: plasma+ SOP with HNO_3 of concentration 1.5 mol L^{-1} washing for 3 d and soaking $Cu(NO_3)_2$ for 24 h and calcinations.

When the SOP without acid washing used as catalyst, the NO removal rate is the lowest. When the MSC of washing is for 2 days, the NO removal rate reaches the highest value. The reason is that the acid washing of the SOP is a fast chemical reaction [19]; the magnesium ions are apart easily from the SOP to form the new micro-pore. As the washing time prolong, the ion of H+ has diffused to the deep channels of micro-pore. When the time is smaller than 2 days, as the effect of diffusing resistance, the acid concentration decreased gradually in the deep hole, so it is difficult to remove magnesium from the deep channel of the SOP. When the washing time is bigger than 2 days, the impurities which washed away may be adsorbed on the hole well again.

3.4. The relationship between NO removal rate and the MSC packed condition

At the same experimental condition: input voltage is 35 kV, NO initial concentration 297×10^{-6}, inlet gas volume flow 7 L min^{-1} (space-velocity 10.2 s^{-1}). The MSC packed condition is: (a) only used plasma, without any stuff, (b) plasma combined with the SOP, (c) plasma combined with the MSC of without acid washing, only soaking $Cu(NO_3)_2$ and calcination, (d) plasma with MSC of washing by HNO_3 of concentration 1.5 mol L^{-1} for 1 day and soaking $Cu(NO_3)_2$ for 24 h and calcination, (e) without plasma, only used MSC of washing by HNO_3 of concentration 1.5 mol L^{-1} for 1 day and soaking $Cu(NO_3)_2$ for 24 h and calcination. The relationship is shown in figure 6.

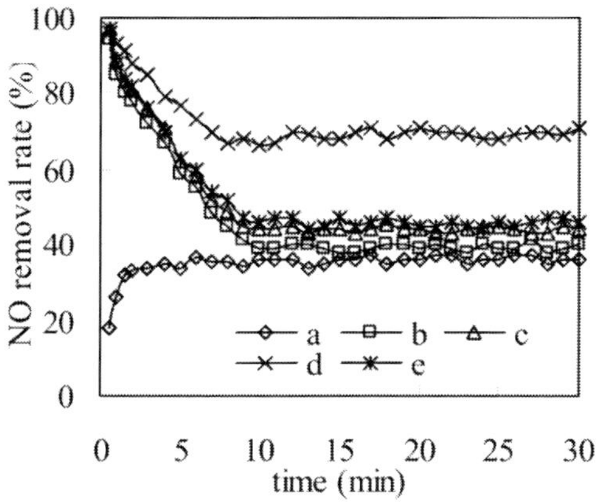

Figure 6. The relationship between NO removal rate and time in different of stuff condition.
a: only plasma, without any stuff; b: plasma+ packed SOP; c: plasma+ MSC made of without acid washing, only soaking $Cu(NO_3)_2$ for 24 h and calcination for 5 h at 500 °C; d: plasma+ MSC made of washing by HNO_3 of concentration 1.5 mol L^{-1} for 1 d and soaking $Cu(NO_3)_2$ for 24 h and calcination for 5 h at 500 °C; e: MSC made of washing by HNO_3 of concentration 1.5 mol L^{-1} for 1 d and soaking $Cu(NO_3)_2$ for 24 h and calcination for 5 h at 500 °C

Figure 6 shows: (1) Either the SOP or the MSC is packed in the NTP reactor, NO removal rate is higher at the beginning and then reducing gradually with time going on, until a plateau is reached after 10 minutes. It attributes that the SOP and the MSC can adsorb and store the NO, when the quantity of adsorption and storage reaches its equilibrium value; the NO removal rate decreases and reaches stability state. (2) After 15 minutes, when the chemical reaction reaches stable situation, the NO removal rate order is: (d) > (e) > (c) > (b) > (a). The information shows that the NTP combined with MSC (washing by HNO_3 of 1.5 mol L^{-1}, soaking Cu $(NO_3)_2$ and calcination) has a good ability to remove NO. (3) The NO removal rate in the condition of the NTP combined with the MSC is bigger than only NTP. This informs that NTP with the MSC is an effect way to remove NO.

4. Conclusions

Relating with the XRD and SEM results, it is found that the Mg and other impurities have been removed effectively from the SOP and the number of the micro-pore and its cross-section area in the MSC surface has been increased by acid washing. The new active CuO has been formed stably and distributed evenly in the MSC surface and micro-pores by the process of acid wishing, copper nitrate solution soaking, drying, calcination and activation. The MSC combined with NTP has a good ability to absorb, store and remove NO from exhaust gas. The NO removal rate increases with the increasing of input voltage. The NO removal rate order is: (plasma with MSC made of washing by HNO_3 of concentration 1.5 mol L^{-1} and soaking Cu $(NO_3)_2$ and calcination) > (MSC made of washing by HNO_3 of concentration 1.5 mol L^{-1} for 1 d and soaking Cu $(NO_3)_2$ and calcination) > (plasma with MSC made of without acid washing, only soaking Cu $(NO_3)_2$ and calcination) > (plasma with SOP) > (only plasma, without any stuff).

Acknowledgements
This research has been supported by the National Natural Science Foundation of China (No.21076002), research fund of Key Laboratory for Advanced Technology in Environmental Protection of Jiangsu Province (AE201002) and International Collaboration Project of Anhui Province (12030603003)

References

[1] Kwak J H, Szanyi J, and Peden C H 2004 *Catal. Today* **89** 135
[2] Niu J H, Yang X F, and Shi C 2006 *Catal. Commun.* **7** 297
[3] He F, Liu C J, Eliasson B and Xue B 2001 *Sur. Interface Anal.* **32** 198
[4] Santillan M M, Vincent A, and Santirso E 2008 *J. Cleaner Prod.* **16** 198
[5] Vinogradov J, Rivin B, and Sher E 2007 *Energy* **32** 174
[6] Yukimura K, Kawamura K, and Hiramatsu T 2007 *Thin Solid Films* **515** 4278
[7] Matsi Y, Takasima K, and Mizuno A 2005 *J. of Advanced oxidation* **8** 255
[8] Matsur Y, Takashima K, and Mizuno A 2004 *J. Institute of electrostatics Japan* **28** 35
[9] Kundakovie and Fiytzani S M 1998 *Appl. Catal. B* **171** 13
[10] Fernandez G M, Rodriguez R I and Ferreira A P.1998 *J. Catal* **178** 253
[11] Aznar A J, Gutierrez E, and Diaz P 1996 *Microp Mater* **6** 105
[12] Li S J, Luo L T 2001 *Modern Chemical Industry (China)* **21** 21
[13] Li L, Li Q Y, and Wang D 2005 *J. Chem. Indust. Engin. (China)* **56** 64
[14] Suarez S, Yates M, and Petre A L 2006 *Appl. Catal. B: Environ.* **64** 302
[15] Corma A, Perez P J, and Soria J 1985 *Clay Miner* **20** 467
[16] Yebra R A, Martin R J D, and Del R F 2003 *Clay Minerals* **38** 353
[17] Mok Y, Koh D J, and Shin D 2004 *Fuel Process. Technol.* **86** 303
[18] Chen M G, Takashima K, and Mizuno A 2008 *Int. Conf. on Electrostatic Precipitation* (Hangzhou) vol 2 p 681
[19] Jin S M, Yand W J, Tang M T 2001 *Modern Chemical Industry (China),* **21** 26

7th International Conference on Applied Electrostatics (ICAES-2012) IOP Publishing
Journal of Physics: Conference Series **418** (2013) 012118 doi:10.1088/1742-6596/418/1/012118

Oxidation efficiency of elemental mercury in two DBD plasma reactors

Yuze Jiang[1*], Jiutao An[2], Kefeng Shang[2], Diwen Jiang[3], Jie Li[2], Na Lu[2], Yan Wu[2]
1Shandong Electric Power Research Institute, Jinan, P.R. China
2 Institute of Electrostatics and Special Power, Dalian University of Technology, Dalian, P.R. China
3 Jinan Jiyang Power Supply Company, Jinan, P.R. China

E-mail: jiangyuze@163.com

Abstract. Configuration of plasma reactors influences the generation of active species including the energized electrons, active radicals and the distribution of active species in reactor, and thus influences the removal efficiency of pollutants. Oxidation efficiency of elemental mercury (Hg^0) in two different DBD plasma reactors was studied in this paper. One plasma reactor is a surface discharge reactor (SDR) with a spiral stainless steel thread as the high voltage electrode, and the other plasma reactor is a concentric cylinder type DBD reactor (CCDR) with a copper screw rod as the high voltage electrode. The oxidation efficiencies of Hg^0 under different specific energy density (SED), oxygen content, flue gas residence time and the temperature of flue gas indicate that SDR had a better performance than CCDR in oxidation of Hg^0, which can be attributed to the higher generation efficiency of ozone in SDR than in CCDR.

1. Introduction

The combustion of fossil fuels, especially coal, often emits pollutants such as SO_2, NO_x, heavy metals including Hg and As, and so on. Since fossil fuels contain many natural elements including sulfur (S), mercury (Hg) and arsenic (As). Hg emission can also bring serious harms to organisms and humans during its transfer and conversion in nature. So control of mercury emission is very important and necessary.

In recent years, active carbon adsorption [1], photocatalysis and adsorption [2], catalytic oxidation [3] etc have been studied to purify the flue gas containing Hg. Adsorption of Hg by active carbon is effective on flue gas purification. However, the expensive absorbents limit the application of adsorption process. Moreover, the particle matters, SO_2, etc in flue gas might poison the expensive catalysts to the extent to lose its catalytic activity, which results in the difficulty in application of catalysis process to purify flue gas containing Hg.

Non-thermal plasma technologies (NTP) have been studied to abate the pollutants in flue gas [4-10]. During the process of plasma generation, strong oxidative species such as O, ·OH, O_3, etc are produced by electrical discharge, which are responsible for oxidation of $SO_2/NO_x/Hg^0$. The present

[*] To whom any correspondence should be addressed.

studies indicate that non-thermal plasma technologies hold the potential of simultaneously purifying the flue gas containing gaseous pollutants and heavy metal elements.

Dielectric barrier discharge (DBD) is a kind of effective method for production of O_3, one of the important species for oxidation of Hg^0. In this paper, the effect of electrode configuration on the purifying efficiency of flue gas containing Hg^0 was studied. One of the electrode configurations was a typical concentric cylinder DBD type reactor (CCDR), and the other was a surface discharge reactor (SDR). The Hg^0 oxidation efficiency in the two different DBD plasma reactors was compared, and the reason of different Hg^0 purification efficiency of two DBD reactors was analyzed.

2. Experimental

2.1. Experimental method

Removal of Hg^0 from simulated flue gas was conducted in two DBD plasma reactors, and the schematic of Hg^0 removal process was shown in figure 1. The simulated flue gas was synthesized by air and Hg^0 vapor. The Hg^0 vapor was generated by liquid Hg^0 bubbling method with N_2 as carrier gas, in which a bottle containing liquid Hg^0 was put into water bath (306 K) to keep a stable evaporation rate of Hg^0. The Hg^0 concentration was regulated by controlling the flow rates of air and N_2 carrier gas. Flue gas sampled at the outlet of plasma reactor was analyzed by Ontario Hydro Method (OHM) [11], and then analyzed by Lumex RA-915+ and RP91. All the chemicals used for analysis of samples were analytically pure.

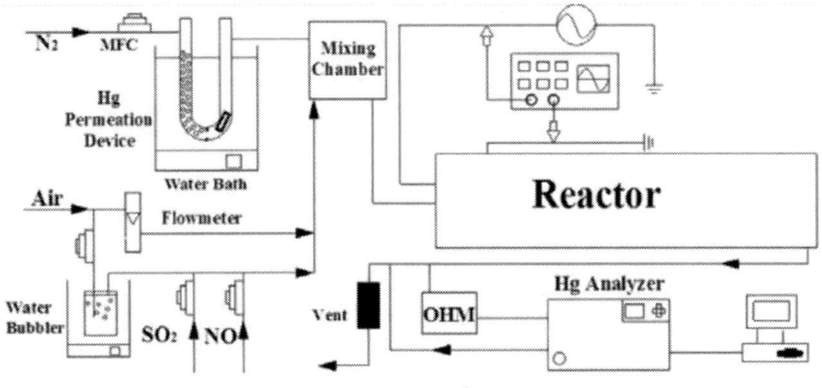

Figure 1. Schematic of Hg^0 removal process

2.2. DBD plasma reactors

1. grounded electrode; 2. quartz dielectric; 3. high-voltage electrode.

Figure 2 (a). Schematic of concentric cylinder type DBD plasma reactor.

Figure 2 (b). Schematic of surface discharge plasma reactor.

One of the plasma reactors, illustrated in figure 2 (a), was a typical concentric cylinder type DBD reactor, in which a copper screw rod with a diameter of 5.5 mm was fixed at the centre of a quartz glasses tube. The inner diameter and the wall thickness of quartz glasses tube was 12 mm and 3 mm, respectively. The copper rod was used as high voltage electrode and the quartz tube wrapped in foil was grounded. The effective zone of plasma generation was about 200 mm in length.

The other plasma reactor, illustrated in figure 2 (b), was a surface discharge reactor, in which a spiral electrode made of stainless steel of 1 mm diameter was used as the high-voltage electrode. The high-voltage electrode was attached tightly to the inside wall of the cylindrical quartz tube with an outer diameter 15 mm and an inner diameter 12 mm and a length of 260 mm. A silver paper with a length of 200 mm wrapping outside the quartz tube was used as the ground electrode. In the two plasma reactors, the high voltage electrodes were supplied by an AC power supply of 50 Hz.

2.3. Measurement method

The discharge voltage and discharge current waveforms were measured by a digital oscilloscope (TDS2014, Tektronix) equipped with a high voltage probe (P6015A, Tektronix) and a current probe (A6021, Tektronix). The consumed discharge energy of DBD reactor was obtained by voltage-charge curve (Lissajous figure). The charge was measured using a capacitor (2 μF) inserted between the reactor and the ground.

3. Results and Discussion

Configuration of discharge reactors has an important influence on the generation of active species and their distribution in reactor, and thus impact the removal of pollutants. The removal efficiency of Hg^0 in two different DBD reactors under different conditions including specific input energy, residence time, O_2 content and the temperature of flue gas was studied.

The oxidation removal rate of Hg^0 under different specific energy density (SED) was shown in figure 3, and SED was calculated by equation (1).

$$\mathrm{SED}\,(\mathrm{J\,L^{-1}}) = \frac{P}{Q_s} \tag{1}$$

Where P (W) and Q_s (L s^{-1}) was the input power and the flow rate of simulated flue gas, respectively.

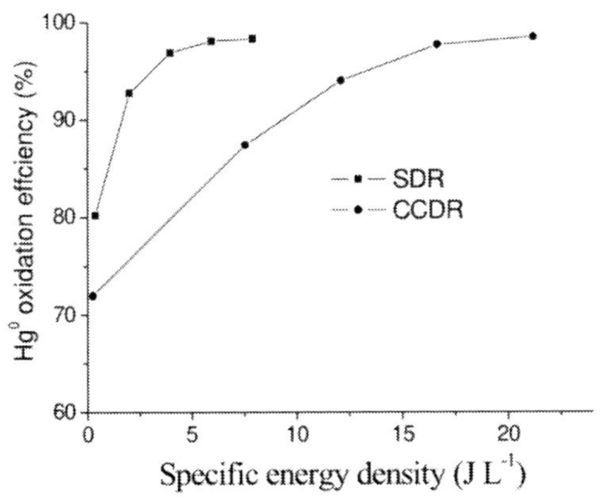

Figure 3. Hg^0 oxidation efficiency in two DBD reactors.

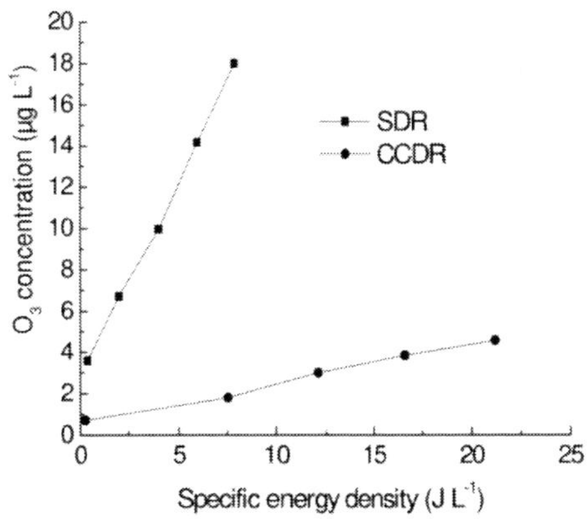

Figure 4. O_3 generated in two DBD reactors.

In this experiment the flow rate of flue gas was 4.5 L min^{-1}, and the initial concentration of Hg^0 was 110 μg m^{-3}. The oxidation efficiency (η) of Hg^0 was calculated according to the equation (2).

$$\eta = (1 - Ct / Co) \times 100\% \tag{2}$$

Where C_t and C_0 was the concentration of Hg^0 (μg m^{-3}) at any time and the initial concentration, respectively.

Figure 3 indicates that removal of Hg^0 increased with the SED both in SDR and CCDR. Figure 3 also indicates that higher removal rate of Hg^0 was obtained with the surface discharge reactor under the same SED, and the Hg^0 oxidation efficiency in SDR was about 8%~20% higher than that in CCDR.

When the voltage exerted on discharge electrode is high enough, electrical discharge is initiated between the electrodes and leads to the production of energized electrons. During the collision of energized electrons and background gas (e.g. air), active species including O, ·OH and O_3 will be generated by the dissociation and excitation of air. Since O_3 is the important species for Hg^0 oxidation, the higher generation quantity of O_3 by surface discharge reactor can explain the phenomenon that higher Hg^0 removal was gained by SDR, and the measurement result of ozone concentration testified our speculation. The data shown in figure 4 indicates that the ozone concentration produced by SDR was 3 ~ 9 times higher than by CCDR.

Figure 5 shows the removal rate of Hg^0 under different oxygen content. The flow rate of flue gas and the initial concentration of Hg^0 were similar to that in figure 3. Figure 5 presents a similar phenomenon like in figure 3, namely, higher Hg^0 oxidation removal rate was obtained by SDR even under lower SED. Under a SED of 5.93 J L^{-1}, the Hg^0 removal rate in SDR could be improved by 4%~10% than that in CCDR under a SED of 12.1 J L^{-1}. The experiment in figure 5 also demonstrates the higher removal efficiency of Hg^0 could be obtained by surface discharge reactor.

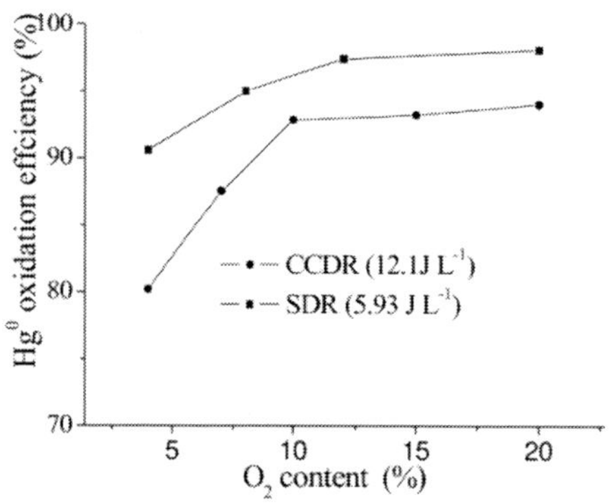

Figure 5. Hg^0 oxidation efficiency under different oxygen contents.

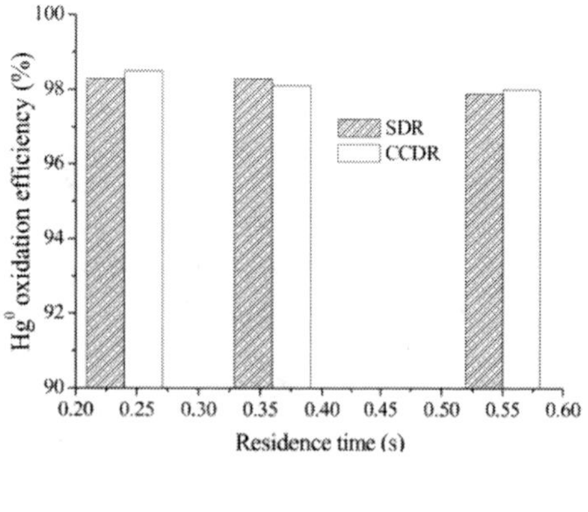

Figure 6. Hg^0 oxidation efficiency under different residence time of flue gas.

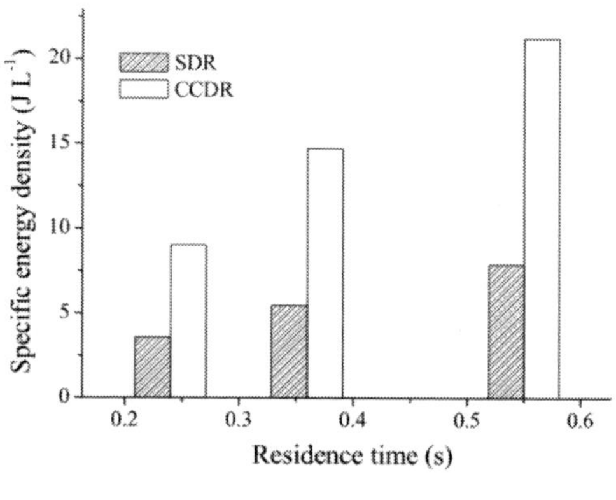

Figure 7. SED corresponding to the experiment in fig.6.

Figure 6 and figure 7 show the removal rate of Hg^0 and SED values under different residence time, respectively. Figure 6 shows that the removal rate of Hg^0 in SDR and CCDR differed very little, but figure 7 indicates that the SED value in SDR was far lower than that in CCDR. The results in figure 6 and 7 show that almost the same removal rate can be obtained under lower SED in SDR than in CCDR, which indicates the energy efficiency of Hg^0 removal in SDR was higher than CCDR.

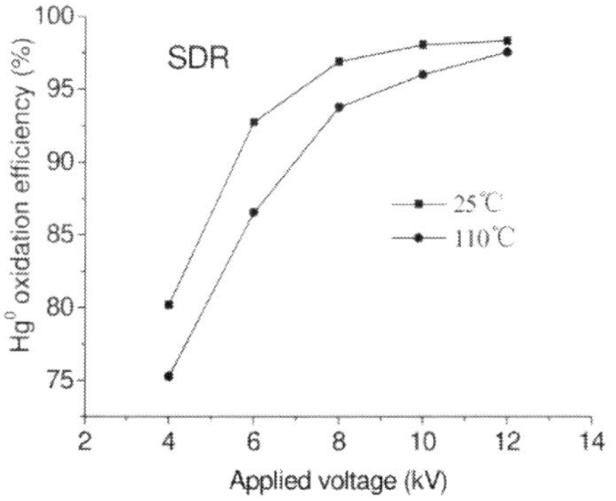

Figure 8. Effect of temperature of flue gas on Hg^0 oxidation efficiency in SDR.

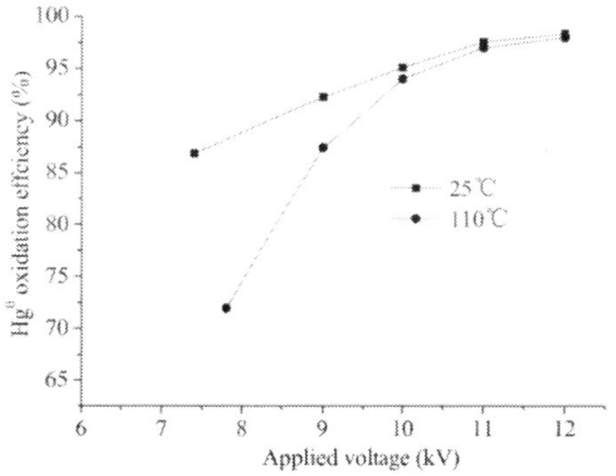

Figure 9. Effect of temperature of flue gas on Hg^0 oxidation efficiency in CCDR.

Temperature of flue gas has important influence on the discharge characteristics and ozone production. Since the decomposition of ozone speeds up with the increasing temperature, the high temperature should be disadvantageous for oxidation of Hg^0 in flue gas. Figure 8 and figure 9 demonstrate the disadvantageous influence of high temperature on Hg^0 oxidation. In addition, higher Hg^0 oxidation efficiency was obtained by SDR under the same applied voltage when comparing with CCDR.

4. Conclusions

Oxidative removal of Hg^0 from simulated flue gas was conducted in two DBD plasma reactors. The experimental results indicate that higher oxidation efficiency of Hg^0 could be obtained in a surface discharge reactor than in a concentric cylinder DBD plasma reactor, which mainly contributed to the higher ozone concentration produced by surface discharge reactor. Decomposition of ozone speeds up with the increasing temperature, so the high temperature is disadvantageous for oxidation of Hg^0 in flue gas.

References

[1] Serre S D, Gullett B K and Ghorishi S B 2001 *J. Air Waste Manage Assoc.* **51** 733
[2] Wang H, Zhou S, Xiao L, Wang Y, Liu Y and Wu Z 2011 *Catal. Today* **175** 202
[3] Stergarek A, Horvat M, Frka P and Stergarek J 2010 *Fuel* **89** 3167
[4] Penetrante B M and Schultheis S E 1993 *Non-thermal Plasma Techniques for Pollution Control* (Springer-Verlag, Berlin)
[5] Lee Y H, Jung W S, Choi Y R, Oh J S, Jang S D, Son Y G, Cho M H, Namkung W, Koh D J, Mok Y S and Chung J W 2003 *Environ. Sci. Technol.* **37** 2563
[6] Shang K F, Wu Y, Li J, Li G F, Li D and Wang N H 2006 *Plasma Chem. Plasma Process.* **26** 443
[7] Jeong J and Jurng J 2007 *Chemosphere* **68** 2007
[8] Byun Y and Shin D N 2011 *Chemosphere* **83** 69
[9] Ko K B, Byun Y, Cho M, Namkung W, Shin D N, Koh D J and Kim K T 2008 *Chemosphere* **71** 1674
[10] Wang Z H, Jiang S D, Zhu Y Q, Zhou J S, Zhou J H, Li Z S and Cen K F 2010 *Fuel Process. Technol.* **91** 1395

[11] Standard Test Method for Elemental, Oxidized, Particle-Bound and Total Mercury in Flue Gas Generated from Coal-Fired Stationary Sources (Ontario Hydro Method), 2001, ASTM Standard

Experimental study of removing benzene from indoor air by needle-matrix to plate streamer discharge

H Ge [1], R Yu [1], D Mi [2], Y M Zhu [1*]
1. Institute of Environment of Remediation, Dalian Maritime University., Lingnan Road Dalian, China
2. Department of Physics, Dalian Maritime University., Lingnan Road Dalian, China

E-mail: ntp@dlmu.edu.cn

Abstract. The degradation of benzene by needle-matrix to plate streamer discharge was investigated at normal temperature and pressure in indoor air. The effects of benzene initial concentration, air speed, discharge power and relative humidity (*RH*) on benzene removal rate were systematically studied. Meanwhile, the benzene removal efficiencies by adding MnO_2/SiO_2-active carbon catalyst to the system were also studied. The results showed that the benzene removal rate increased with the rise of the air speed and discharge power, decreased with the rise of the benzene initial concentration, and firstly increased and then decreased with the rise of the of *RH*. Under the same experimental conditions, adding MnO_2 catalyst to the system did not significantly improve the removal efficiency of benzene.

1. Introduction

With the continuous improvement of living standards, more and more people are realizing the importance of indoor air quality. As one of the common volatile organic compounds(VOC_s) pollutants in indoor, benzene has caused wide concern on its damage for human health [1,2]. Non-thermal Plasma has attracted a lot of attentions in the field of VOC_s treatment by its high efficiency, simple process and low energy consumption [3-5]. Until now, many researchers have focused on dielectric barrier discharge (DBD) and pulsed corona discharge, but they are not too concerned about direct current (DC) corona discharge. In this study, the degradation of benzene by needle-matrix to plate DC streamer discharge was investigated in a closed chamber of 0.4 m^3, and the effects of discharge parameter, air parameter and catalyst on benzene removal efficiency had also been discussed.

2. Experimental Setup

The experimental system is shown in figure 1. The needle-to-plate distance is 25 mm, and the needle-to-needle distance is 20 mm. There are 97 needles in the matrix with needle radius 1 mm and length 5 mm. Discharge power is supplied by ZGF-DC high-voltage generator and the applied voltage is measured by a Tek P6015A divider connected with an HP 54503 digital oscilloscope. The flow rate in electrical field is measured by ZRQF anemometer. The manganese oxides catalyst is 10 mm thickness per layer with hexagon holes of 2 mm side length. Benzene solution is placed into a water bath and benzene vapor is fed into closed chamber by air pump. The concentration of benzene in closed chamber is controlled by air flow rate as well as water temperature, and then is measured by

* To whom any correspondence should be addressed.

PID 312 gas chromatograph.

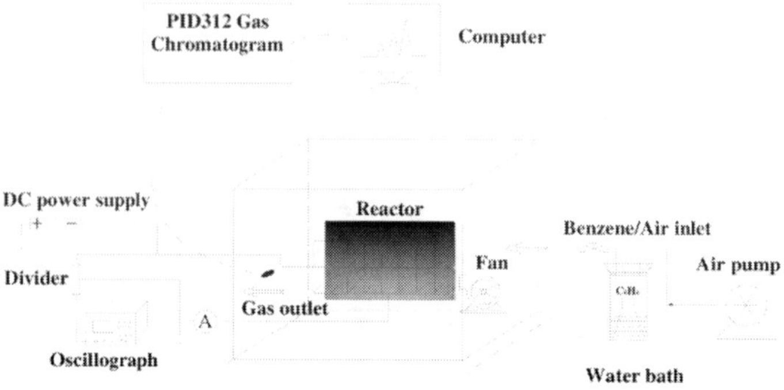

Figure 1. Experimental setup.

3. Results and Discussion

3.1. Benzene natural attenuation

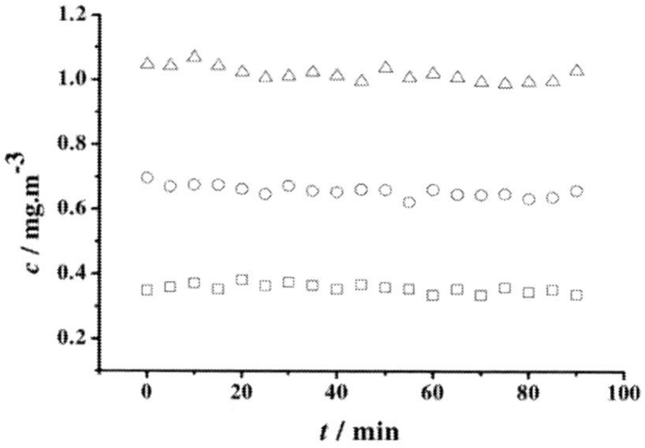

Figure 2. Blank experimental results of the natural reduction of benzene concentration. (□ 0.35 mg m^{-3}; ○ 0.70 mg m^{-3}; △ 1.05 mg m^{-3})

The concentrations of Benzene in closed chamber were kept at 0.35, 0.70 and 1.05 mg m^{-3}. The flow rate in the electric field was kept at 2.8 m s^{-1}, and the discharge power and relative humidity (*RH*) were 9 W and 20 %, respectively. The variation trend of benzene degradation rate was studied under three concentrations.

As shown in figure 2, each of three benzene concentrations has a slight fluctuation during discharge 90 min. Benzene concentration slowly decreased with discharge time (t), but was not exceeding 9 %. Considering that some benzene could collide with aerosol and then adsorbs at the chamber surface in the discharge process, the theoretical benzene value should be lower than the obtained results. Hence, the natural reduction does not significantly affect benzene removal efficiency by discharge, and its value at the same time will be deducted in the following experiments.

3.2. The effect of air speed on benzene removal rate

The concentration of benzene in the closed chamber was kept at 0.70 mg m^{-3}. The flow rate of air in the electric field was adjusted by auto-transformers and its values were set as: 0.5 m s^{-1}, 1.0 m s^{-1}, 2.0 m s^{-1}, 2.4 m s^{-1} and 2.8 m s^{-1}, respectively. The variation trend of benzene degradation rate under different flow rates were studied with discharge power 9 W and *RH* 20 %.

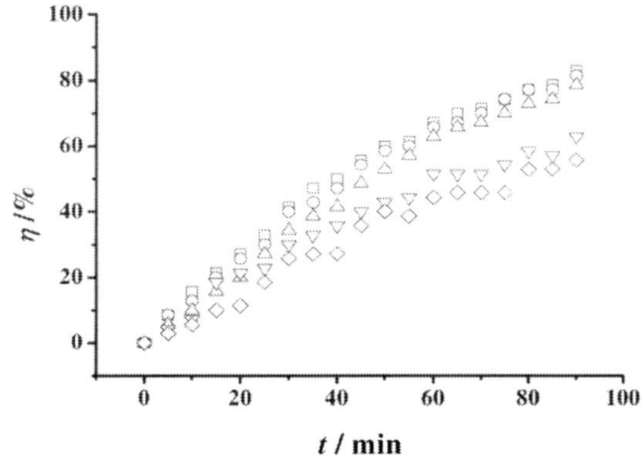

Figure 3. Relationship between benzene removal efficiency and discharge time. (◇ 0.5 m s^{-1}; ▽ 1.0 m s^{-1}; △ 2.0 m s^{-1}; ○ 2.4 m s^{-1}; □ 2.8 m s^{-1})

As shown in figure 3, the benzene removal efficiency increased with improved flow rate, but this trend was not obvious when its value exceeded 2.4 m s^{-1}. The reason is probably that the benzene removal efficiency by discharge under unit time reached dynamic balance with increased air speed, because the energy input in the closed chamber was fixed. The removal efficiency under five flow rates were 55.7 %, 62.9 %, 78.6 %, 81.4 % and 82.9 %, respectively. It was thus clear that the benzene removal efficiency did not significantly change when air speed reached a certain value. Therefore, the air speed should be selected at 2.4 m s^{-1} if needle matrix to plate corona discharge reactor is applied in indoor air purification.

3.3. The effect of initial concentration on benzene removal efficiency

Within discharge 90 min, the variation trends of benzene removal rate under initial concentration 0.35, 0.70 and 1.05 mg m^{-3} were shown in figure 4. The relationship between discharge time and removal rate was discussed by trend line equations of 0-30 (stage1) min and 30-90 (stage2) min obtained from experimental data. The relevant experimental parameters were set as: power 9 W, air speed 2.4 m s^{-1} and *RH* 20 %.

As shown in figure 4, the benzene removal efficiency under three initial concentrations increased with discharge time. The removal speed rate of stage 1 was faster than that of stage 2, and the corresponding first order coefficient on a low to high initial concentration were 1.49, 1.32 and 1.26 respectively. The benzene removal speed almost kept unchanged in stage 2, the corresponding coefficients were 0.68, 0.66 and 0.69, respectively. In addition, the benzene removal efficiency decreased with the rise of the initial concentration and it reached maximum value at 0.35 mg m^{-3}. The removal efficiencies on a low to high initial concentration were 88.6 %, 82.9 % and 81.9 %, respectively. This is because that the collision probability between each benzene molecular and high-energy electrons improved with the reduction of benzene concentration.

3.4. The effect of discharge power on benzene removal efficiency

The variation trend of benzene removal efficiency under different discharge power was shown in figure 5. The powers were kept at 9, 6.5, 4.5 and 2 W, respectively, and the other experimental parameters were same to section 3.3. As shown in figure 5, the benzene removal efficiency increased with the rise of the discharge power under different powers, but the magnification range gradually decreased. At 90 min, the efficiencies from 9 W to 2 W were 82.9 %, 77.1 %, 67.1 % and 44.3 %, respectively. Moreover, the fluctuation of removal efficiency did not significantly change with the rise of discharge power.

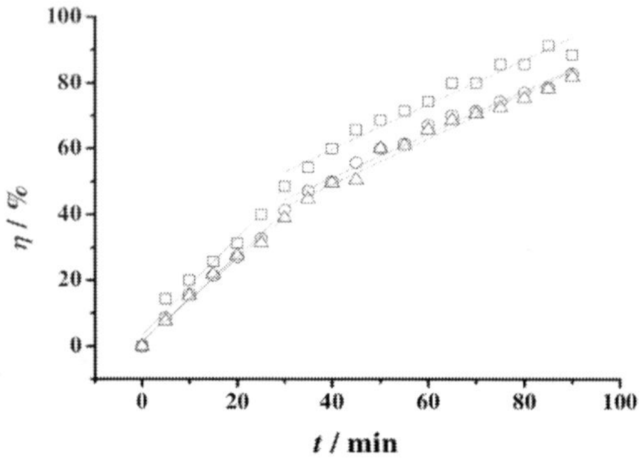

Figure 4. Relationship between benzene removal efficiency and discharge time. (□ 0.35 mg m^{-3}; ○ 0.70 mg m^{-}; △ 1.05 mg m^{-3})

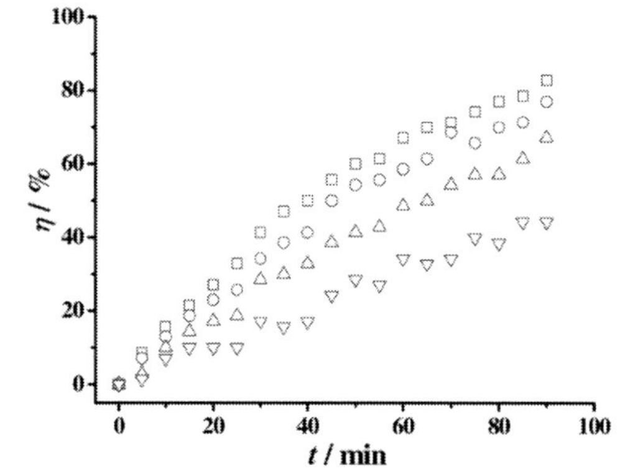

Figure 5. Relationship between benzene removal efficiency and discharge time. (□ P=9 W; ○ P=6.5 W; △ P=4.5 W; ▽ P=2 W)

The discharge process can be divided into three parts: initial streamer discharge, glow discharge and spark discharge [6]. The characteristics of initial streamer discharge are low current and discharge instability, which leads to the fluctuation of removal efficiency in this stage. Then, more and more high energy electrons, active species and free radicals are produced with the rise of discharge power, leading to the improvement of benzene removal efficiency.

3.5. The effect of RH on benzene removal efficiency

In this part, the parameter *RH* was changed from 20 % to 60 % in order to investigate its effect to benzene removal efficiency. Other experimental parameters were the same with those in section 3.3. As known from figure 6, along with the rise of *RH*, the removal efficiency firstly increased and then decreased. The efficiencies of three *RH* were 82.9 %, 85.7 % and 80 %, respectively. The best *RH* value under our experimental condition should be 40 %.

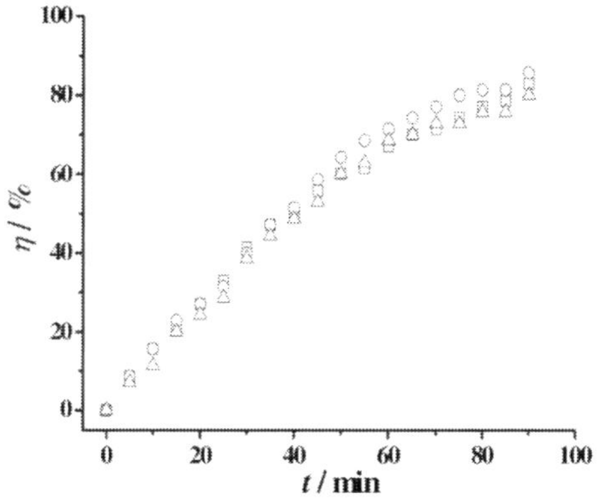

Figure 6. Relationship between benzene removal efficiency and discharge time. (□ RH= 20 %; ○ RH= 40 %; △ RH= 60 %)

This observation can be explained as follows: It is known that OH and O could effectively dispose benzene molecular. The amount of OH and O radicals increases with increased collision frequency between energetic electrons and water molecules [7, 8]. When the relative humidity is low, the collision frequency will be also less; on the contrary, the collision frequency will be high. On the other hand, with the rise of *RH*, more electrons will be absorbed by the increased H_2O molecular due to its electronegativity. This will lead to the reduction of density of energetic electrons, and the reduction of the collision frequency between energetic electrons and water molecules. At the same time, streamer discharge will be instable and the discharge current will sharply decrease when *RH* is relatively high. These will make the reduction of benzene removal efficiency.

In addition, the synergistic effect of plasma and MnO_2 catalyst on benzene degradation has also been investigated in this study. The comparison of the removal effects by plasma and plasma-MnO_2 had been obtained. However, comparing with plasma, the addition of MnO_2 catalyst did not significantly improve the benzene removal efficiency.

4. Conclusions

(1) There is a linear relationship between the benzene removal efficiency and discharge time, and the removal efficiency decreased with the rise of initial benzene concentration. Benzene removal rate slightly increased with the deduction of initial concentration in discharge stage 0-30 min, but did not significantly change in discharge stage 30-90 min.

(2) The benzene removal efficiency gradually increased with the rise of air speed. This trend was not obvious when air speed reached a certain value. The air speed should be selected as 2.4 m s^{-1} under this experimental condition.

(3) The benzene removal efficiency gradually increased with the rise of discharge power, but magnification range gradually decreased. The fluctuation of removal efficiency weakened with the rise of power.

(4) Along with the rise of RH, the benzene removal efficiency firstly increased and then decreased. The best benzene removal efficiency appeared when RH is 40 %.

(5) Comparing with only plasma, the MnO_2-Plasma system probably could not significantly improve benzene removal efficiency.

Acknowledgments

This work was supported by NSFC 51078048, the Fundamental Research Funds for the Central Universities and Dalian Maritime University culture plan Funds for excellent technological innovation group.

References

[1] Yu B F, Hu Z B, Liu M, Yang H L, Kong Q X and Liu Y H 2009 *Int. J. Refrig.* **32** 3
[2] Mo J H, Zhang Y P, Xu Q J, Jennifer J L and Zhao R 2009 *Atmos. Environ.* **43** 2229
[3] Lu B, Zhang X, Yu X, Feng T and Yao S 2006 *J. Hazard. Mater.* **137** 633
[4] Ye Z L, Zhang Y N, Li P, Yang L Y, Zhang R X and Hou H Q 2008 *J. Hazard. Mater.* **156** 356
[5] Wang H C, Li D, Wu Y, Li J and Li G F 2009 *J. Electrostat.* **67** 547
[6] Ge H, Zhang L L, Yan L, Mi D and Zhu Y M 2011 *J. Electrostat.* **69** 529
[7] Sun M, Wu Y, Zhang J L, Lie J, Wang N H, Wu J and Shang K F 2005 *Spectrosc. Spect. Anal.* **25** 108

Removal of NO_2 and O_3 generated from corona discharge in indoor air cleaning with MnO_2 catalyst

H Ge [1], R Yu [1], D Mi [2], Y M Zhu [1*]
1. Institute of Environment of Remediation, Dalian Maritime University., Lingnan Road Dalian, China
2. Department of Physics, Dalian Maritime University., Lingnan Road Dalian, China

E-mail: ntp@dlmu.edu.cn

Abstract. The production rules and removal efficiency of harmful byproducts such as NO_2 and O_3 generated from DC corona discharge in indoor air cleaning were investigated. The production behaviours of NO_2 and O_3 and the relationship between the amount of catalyst (MnO_2) and the removal rate of harmful byproducts were experimentally studied. Further, indoor application tests were carried out in a closed room with 90 m^3. The results showed that the concentrations of NO_2 and O_3 produced by corona discharge linearly increased with discharge time. The NO_2 yield is larger than O_3 by almost one order of magnitude under the same discharge power. To satisfy the demand of Standard of Indoor Air Quality (GB/T18883-2002), the power consumption of unit volume should be less than 1 W m^{-3} and the catalyst MnO_2 consumptions in positive-negative corona discharge were 200 cm^3 W^{-1} and 100 cm^3 W^{-1}, respectively.

1. Introduction

Corona discharge in air could produce many active species, and then could be widely applied in environmental pollution control [1-5]. Corona discharge has multiple purification effects in indoor air cleaning process, including particulate matter trapping, sterilization as well as VOCs removal. It also has many advantages comparing with other plasma technologies, such as low energy consumption, discharge stability and easy to use etc. However, the harmful by-products O_3 and NO_2 are also produced simultaneously by corona discharge during purification process. If these by-products can not be effectively eliminated, the practical application of corona discharge in air pollution control will be subject to considerable restrictions.

Some studies show that MnO_2 could accelerate the conversion of O_3 to O_2, and then can be used as

* To whom any correspondence should be addressed.

plasma air purification aftertreatment to improve discharge quality [6]. Mix oxides of MnO_2 and Fe_2O_3 attached to active carbon (AC) can effectively decompose O_3 [7]. As far as O_3 catalytic decomposition ability is concerned, as active component, MnO_2 is better than CuO and Fe_2O_3. Meanwhile, AC is a kind of superior carrier over γ-Al_2O_3 and molecular sieve [8]. In addition, MnO_2 can also be used to effectively remove NO_2 produced by discharge [9].

In this study, the production behavior of O_3 and NO_2 generated by corona discharge in a closed chamber was investigated. On this base, a manganese catalyst was used to decompose the harmful by-products O_3 and NO_2. Finally, the applied research was carried out in a closed 84 m^3 room to study the decomposition law and removal method of by-products.

2. Experiment Method

Figure 1 shows the experimental setup of multi-needle-to-plate corona discharge in a closed chamber of 0.4 m^3. The needle-to-plate distance is 25 mm, and the needle-to-needle distance is 20 mm. There are 97 needles in the matrix with needle radius of 1 mm and length of 5 mm. Discharge power is supplied by ZGF-DC high-voltage generator and the applied voltage is measured by a Tek P6015A divider connected with an HP 54503 digital oscilloscope. The flow rate in electrical field is kept at 2.8 $m{\cdot}s^{-1}$, which is measured by ZRQF anemometer. The manganese oxides catalyst is of 10 mm thickness per layer with hexagon holes of 2 mm side length. The accumulated concentrations of NO_2 and O_3 in the closed chamber are measured by PGM-7840 Five-Gas Monitors (RAE Co.) with resolution of 0.1 ppm and GASTEC Detector tube(0.05～0.6 ppm, 0.01 ppm), respectively.

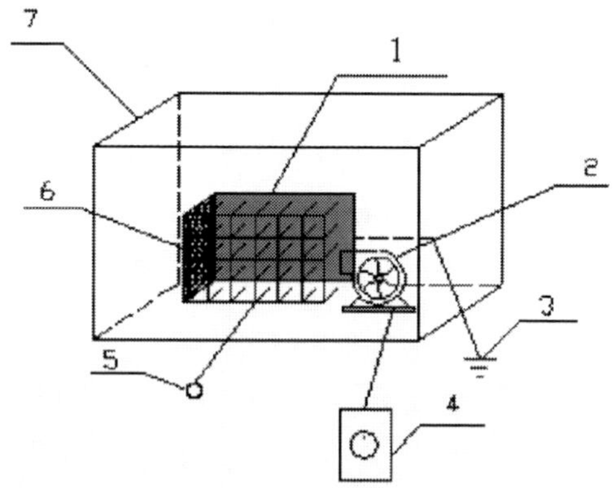

Figure 1. Experimental setup.
1. needle-matrix electrode 2. blower fan 3. ground electrode 4. transformer 5. HV source 6. catalyst grid 7. closed chamber.

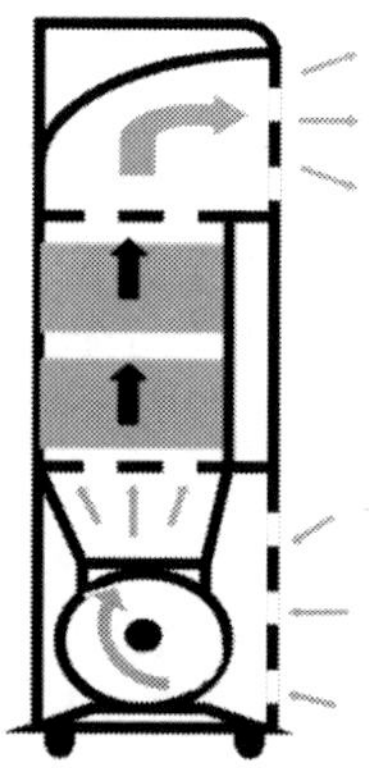

Figure 2. Diagram of indoor air cleaning.

As shown in figure 2, the indoor air is cycled and purified in the following way by the air cleaner: firstly, air is taken into the air cleaner by blower fan, then it is processed in the discharge area and catalyst layers, and finally, it returns to the room with dimension of $5\times6\times3$ m^3. The discharge reactor possesses 10 channels with discharge distance of 25 mm, and each channel has 194 needles in the matrix. The air circulation can be calculated and it is about 400 $m^3\,h^{-1}$.

3. Results and Discussion

3.1. Production of NO_2 and O_3 generated by corona discharge

Figure 3 shows the effect of discharge time on the productions of NO_2 and O_3 under different electrode polarity. The temperature and relative humidity in closed chamber are approximately kept at 20 ℃ and 83 %. The applied voltage in positive corona discharge is 15.6 kV and the corresponding current and power are 0.04 mA and 0.624 W, respectively. Two sets of experiments were made in the case of negative corona discharge, and the measured voltage, current and power are 17 kV, 0.045 mA, 0.765 W and 17.4 kV, 0.05 mA, 0.87 W, respectively.

As shown in figure 3, the concentrations of both NO_2 and O_3 produced by positive-negative corona discharge increase linearly with discharge time at initial stage, then slightly increase after 1 h, and finally achieve a dynamic balance at 1.5 h. The productions of NO_2 and O_3 generated by negative corona discharge are a little more than that by positive corona discharge. This means that the concentrations of NO_2 and O_3 at dynamic balance are related to discharge power but not electrode polarity. The production of NO_2 is one order of magnitude higher than O_3.

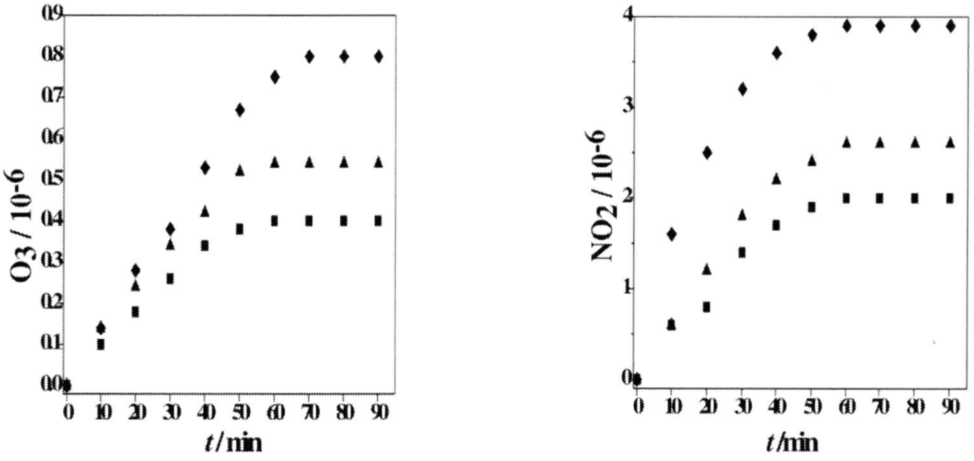

(a) O_3 (b) NO_2

Figure 3. Concentrations of NO_2 and O_3 produced as a function of time in corona discharge.
■ +0.624 W; ▲-0.765 W; ◆-0.87 W

The standard values of NO_2 and O_3 for 1 h in indoor air quality standard (GB-T18883-2002) are 0.24 mg m^{-3}(0.12 ppm) and 0.16 mg m^{-3} (0.07 ppm), respectively. In indoor air purification research, it is thought that generating amount of gas pollutant should be one order of magnitude less than those in GB-T18883-2002, because only the results obtained from experiment are more meaningful and representative for practical application. Thus, under present experimental condition, the discharge power in positive-negative corona discharge should be a little higher than 0.5 W when the closed chamber is 0.4 m^3 and the estimated power consumption of unit volume should be less than 1 W m^{-3}.

3.2. Catalytic removal of NO_2 and O_3

Figure 4 shows the effect of two layer manganese catalyst on decomposition of NO_2 and O_3 when discharge power is lower than 1 W in the closed chamber with 0.4 m^3. As shown in figure 4, the concentration of O_3 generated by positive-negative corona discharge reach dynamic balance and kept at 0.05 ppm after discharge 10 min, satisfying the national standard of indoor air quality (O_3, 0.07 ppm, 1 h). The concentration of NO_2 simultaneously produced with O_3 also reach dynamic balance after discharge for a period of time, and kept at 0.2 ppm for positive discharge as well as 0.3 ppm for negative discharge, slightly exceeding the national standard of indoor air quality (NO_2, 0.12 ppm, 1 h).

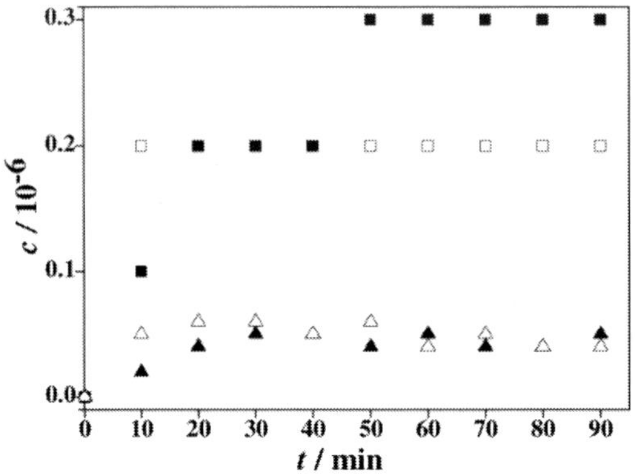

Figure 4. Concentrations of NO_2 and O_3 produced in corona discharge with 2-layer catalyst.
■NO_2 in negative corona (0.765 W); ▲O_3 in negative corona (0.765 W);
□NO_2 in positive corona (0.624 W); △O_3 in positive corona (0.624 W).

3.3. Practical Application of decomposing NO_2 and O_3

In practical application of catalytic decomposition of NO_2 and O_3, using two layer catalyst, the production behavior of by-products during discharge under different electrode polarity was studied.

As shown in figure 5(a), the concentrations of NO_2 and O_3 produced by positive corona discharge under different discharge power reach a dynamic balance after 1 h. Under the conditions of both 4.8 W and 8.58 W, the concentrations of O_3 all satisfy GB-T18883-2002. However, the concentrations of NO_2 slightly exceed the standard for 4.8 W. Especially, the concentrations of NO_2 are two or three times higher than the standard value for 8.58 W. Therefore, the discharge power of positive corona discharge should be less than 4 W under this experiment condition.

The variation trend of concentrations of NO_2 and O_3 produced by negative corona discharge is shown in figure 5(b). The concentrations of both by-products satisfy the standard at 8.25 W when their values reach a dynamic balance. The concentration of O_3 is less than the standard value, while the concentration of NO_2 slightly exceeds the standard at 12 W. Thus, the discharge power of negative corona discharge should be less than 12 W under this experiment condition.

In order to satisfy the standard GB-T18883-2002, the estimated values of the dosage of manganese for decomposing NO_2 and O_3 produced by positive or negative corona discharge are as follows: 200 $cm^3\,W^{-1}$ for positive corona discharge and 100 $cm^3\,W^{-1}$ for negative corona discharge.

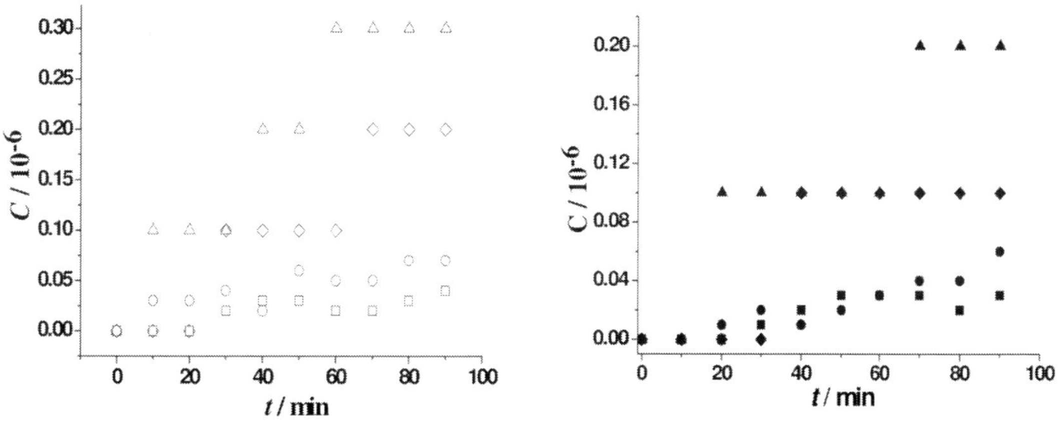

(a) Positive discharge (b) Negative discharge

Figure 5. NO_2 and O_3 reduced by catalyst in indoor air cleaning.
1. +4.8 W NO_2(◇); 2. +4.8 W O_3(□); 3. +8.58 W NO_2 (△); 4. +8.58 W O_3(○)
5. -8.58 W NO_2 (◆); 6. -8.58 W O_3(■); 7. -12.0 W NO_2(▲); 8.-12.0 W O_3(●)

4. Conclusion

(1) The concentration of O_3 produced by needle matrix to plate corona discharge linearly increases in 1.5 h and its production is directly related to discharge power. The production of NO_2 is one order of magnitude higher than O_3 under the same discharge power.

(2) The power consumption of unit volume in our experimental condition is less than 1 W m^{-3}.

(3) The O_3 and NO_2 produced by corona discharge can be rapidly and easily decomposed by MnO_2 catalyst. The dosages of catalyst in positive and negative corona discharge in our experimental condition are approximately 200 cm^3 W^{-1} and 100 cm^3 W^{-1}, respectively.

Acknowledgments
This work was supported by NSFC 51078048, the Fundamental Research Funds for the Central Universities and Dalian Maritime University culture plan Funds for excellent technological innovation group.

References

[1] Schmid S, Jecklin M C and Jecklin Z R 2010 *Chemosphere* **79** 124
[2] Van D J, Dewulf J, Sysmans W, Leys C and Van L H 2007 *Chemosphere* **68** 1821
[3] Chen H F, Su P H, Yang S and Zhu Y M 2008 *J. Adv. Oxid. Technol.* **11** 116
[4] Sari A H and Fadaee F 2010 *Surf. Coat. Tech.* **205** 385
[5] Zhu Y M, Kong X P, Zhang Z R, Zhang M X and Zheng C L 2005 *Journal of Beijing Institute of Technology (in Chinese)***128** 165
[6] Li W, Gibbs G V and Oyama S T 1998 *J. Am .Chem. Soc.* **120** 9041
[7] Liu R Q, Zhao M, Wang R F 2008 *Jiangsu Environmental Science and Technology (in Chinese)* **21** 12
[8] Li W and Oyama S T 1998 *J. Am. Chem. Soc.* **120** 9047
[9] Huang L P, Yang S, Liu S J, Wang Q H and Zhu Y M 2009 *J. Adv. Oxid. Technol.* **12** 238

Journal of Physics: Conference Series **418** (2013) 012121 doi:10.1088/1742-6596/418/1/012121

Experimental research of pulsed discharge plasma and TiO_2/Zeolite coupling technology for formaldehyde removal

Bingyan Dong[*], **Shuirong Lan**
College of Resources & Environmental Engineering, Jiangxi University of Science & Technology, Ganzhou 341000, Jiangxi, China

E-mail: dongbingyan1@sina.com

Abstract. The pulsed discharge plasma combining with catalyst to remove formaldehyde is a novel type of advanced oxidation technology. In the present work, taking wire-tube pulsed discharge plasma and TiO_2/Zeolite coupling technology for formaldehyde removal. The studies have investigated the wire-tube reactor with zeolite, TiO_2, TiO_2/Zeolite for formaldehyde removal respectively. Results show that in the optimal experimental conditions and the baking time is 120 min, the baking temperature is 450℃, that TiO_2/Zeolite catalyst which made by sol-gel shows higher photocatalytic activity and efficiency. The pulsed discharge with TiO_2/Zeolite catalyst for formaldehyde removal has higher removal efficiency than pulsed discharge with zeolite or TiO_2. Therefore, pulsed discharge plasma with TiO_2/Zeolite for the removal of formaldehyde can greatly increase the removal efficiency.

1. Introduction

Formaldehyde, a colourless volatile liquid with a pungent odour, is the main poisonous and harmful substance in the room. Formaldehyde is of great harm to human health, it ranks second in toxic chemicals list of China and it is identified as suspicious teratogenic carcinogen by the World Health Organization.

Long-term release of organic pollutants which volatilize from decorating materials and furniture can endanger people's health seriously, and formaldehyde is one of such pollutants. Nowadays, photocatalysis and low-temperature plasma are the most promising air purification techniques, their functions and advantages on degradation of formaldehyde have been gradually recognized by people with further development of their application and studies, therefore, they have become a hot spot in the field of environmental protection [1-7].

[*] To whom any correspondence should be addressed.

At present, compared with other traditional methods, pulsed discharge plasma and catalyst coupling technology for indoor toxic gas removal is promising [8-9], its advantage includes simple devices, oxidation to almost all pollutants, high efficiency, fast oxidation rate, less secondary pollution and catalyst recycling.

Taking TiO_2/Zeolite catalyst which made by sol-gel and wire-tube pulsed discharge plasma coupling technology for formaldehyde removal, the influence of various factors on the characters of such photocatalyst in the sol-gel preparation process and the influence of electrical parameters on the removal efficiency of formaldehyde were studied. In the optimal experimental conditions, further research of pulsed discharge plasma and TiO_2/Zeolite catalyst coupling technology for formaldehyde removal was carried on, that shows the synergy of ultraviolet light and pulsed discharge non-thermal plasma can improve formaldehyde removal efficiency.

2. Experimental Facility and Method

2.1. Experimental Facility

High-voltage pulsed power supply was provided by the Institute of Electrostatic and Special Power, Dalian University of Technology. The electrical parameters of high-voltage pulsed power supply are mainly involved with positive pulse, pulse width less than 500 ns, pulse rise lower than 100 ns, and pulse repetition rate of 0-200 Hz.

The wire-tube reactor is adopted in the experiment, tube is made of stainless steel cylinder, inner diameter is 58 mm, outer diameter is 60 mm and the length is 310 mm. The tube is used as ground electrode, and star wire which made of stainless steel is used as discharge electrode and its diameter of the wire is 4 mm. Wire electrode is fixed in the cylinder circle and connected with high-voltage pulsed power supply.

The whole experimental system diagram as shown in figure 1, the system is composed of four parts including high-voltage pulsed power supply, gas supply system, reactor and detecting instrument.

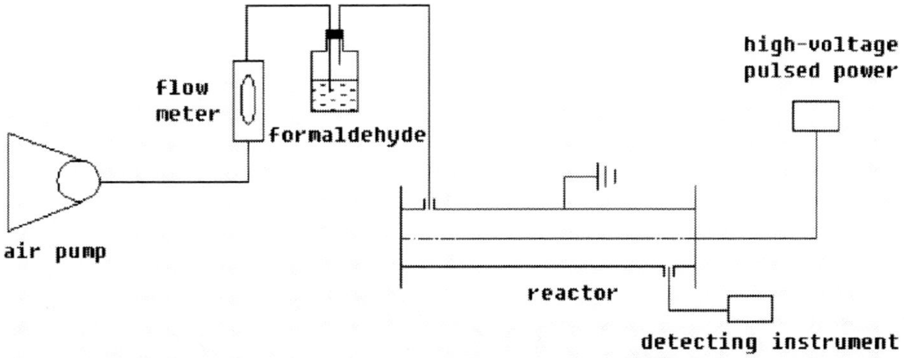

Figure 1. The Experimental System Diagram.

2.2 .Experimental Instruments and Reagents

(1) Instruments

XL30 Scanning Electron Microscope is supplied by Philips; Miniflex X-ray Diffractometer is supplied by Rigaku; AR2130 Electronic Balance (0.001 g) is supplied by Ohaus Corp. Pine. Brook, NJ.USA; Muffle with the type of HZC4-LX-1; LZB-10 Glass Rotameter is supplied by Zhejiang Yuyao Zhenxing Flow Meter Factory; ACO-003 Electromagnetic Air Pump is supplied by Guangdong Risheng Group Company; SM95 PPM HTV, America.

(2) Reagents

Formaldehyde Solution (analytical reagent) is supplied by Beijing Wuxing Chemical Plant; Titanium Dioxide (anatase TiO_2): Particle size is about 10 nm, which BET specific surface area is about 130 m^2 g^{-1}, and supplied by Ganzhou Chengxin Direct Marketing Factory; Tetrabutyl Titanate ($C_{16}H_{36}O_4Ti$): Chemically pure, it is supplied by Shanghai Kefeng Chemical Reagent Co., Ltd; Anhydrous Ethanol (CH_3CH_2OH): Analytical reagent, it is supplied by Tianjin Damao Chemical Reagent Plant; Concentrated Nitric Acid (HNO_3), analytical reagent, it is supplied by Jiangxi Hongdu Biochemical Co., Ltd; Glacial Acetic Acid ($C_2H_4O_2$): Analytical reagent, it is supplied by Tianjin Chemical Reagent Plant; Zeolite is supplied by Sinopharm Chemical Reagent Co., Ltd; Deionized water which made in lab has the effect of Hydrolysis.

2.3. Experimental Method

(1) Preparation of Load-Type Titanium Dioxide

Zeolite was rinsed by distilled water repeatedly till upper solution became clear, then sediment will be heated drying at 105 ℃ for 5 hours.

First, 30 ml $C_{16}H_{36}O_4Ti$ was added into 115 ml CH_3CH_2OH, then 1 ml $C_2H_4O_2$ and zeolite were put into them. A solution was made up with them by stirring until smooth. 7.5 ml $C_2H_4O_2$ and 2.5 ml HNO_3 inhibitors were added into 52.5 ml distilled water under shaking, then pH was adjusted to 2, B solution was formed. A solution was dropwise put into B solution about 30 minutes, transparent sol with light yellow was made after stirring by magnetic stirrer for one day and hydrolyzing for three days. Then sol should be heated drying at 130 ℃ for 2 hours. After it was crushed down, it was calcined in muffle at 450 ℃, at last, TiO_2/Zeolite catalyst was made.

(2) Formaldehyde Removal Efficiency

Experiment is focus on volatile formaldehyde, formaldehyde concentration within the reactor is controlled through intake flow which is regulated by precision flow meter. The removal efficiency could be obtained by this formula.

$$\eta = \frac{C_0 - C}{C_0} \times 100\% \tag{1}$$

Where η is removal efficiency, C_0 is formaldehyde concentration of reactor inlet, and C is formaldehyde concentration of reactor outlet.

3. Experimental Results and Analysis

3.1. Structural Characteristics of TiO_2/Zeolite

When bath temperature is 30 ℃, volume ratio of $C_2H_4O_2$ and $C_{16}H_{36}O_4Ti$ was 0.1, water to $C_{16}H_{36}O_4Ti$ was 0.15, HNO_3 to $C_{16}H_{36}O_4Ti$ is 0.1, CH_3CH_2OH to $C_{16}H_{36}O_4Ti$ was 4.0, TiO_2/Zeolite photocatalyst was made under such conditions, and then TiO_2/Zeolite photocatalyst was calcinated in the muffle for 2 hours at 375 ℃, 450 ℃ and 500 ℃ respectively. After that, photocatalyst was analyzed through electron microscope scanning and X-ray Diffraction.

Structural characteristics of catalyst were observed by using XL30 scanning electron microscope at different calcination temperature. Figure 2 shows that there is white material on the surface of carrier. It means that TiO_2 has been loaded on carrier successfully. TiO_2 loaded on zeolite has larger surface area, so surface area and activity of TiO_2 catalyst increase. X-ray Diffraction shows that preparation temperature has great effect on crystal form and phase composition of TiO_2. When solvent is the same, pure anatase TiO_2 could be generated at calcination temperature of 450 ℃ for 2 hours. Figure 3 shows that crystalline state of anatase TiO_2 is in good order and intensity reaches the maximum at 25°.

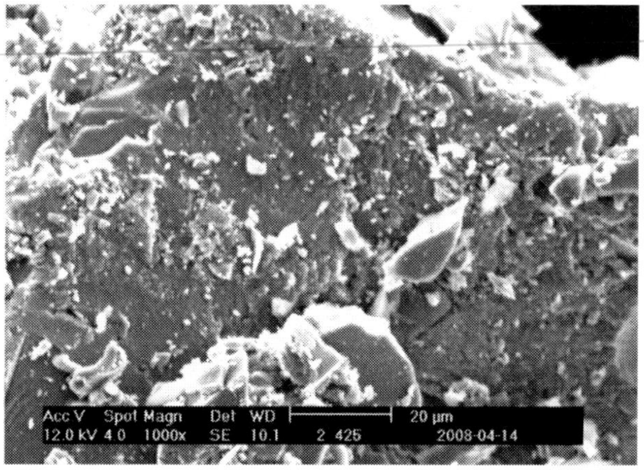

Figure 2. SEM pictures of TiO_2/Zeolite at 450℃.

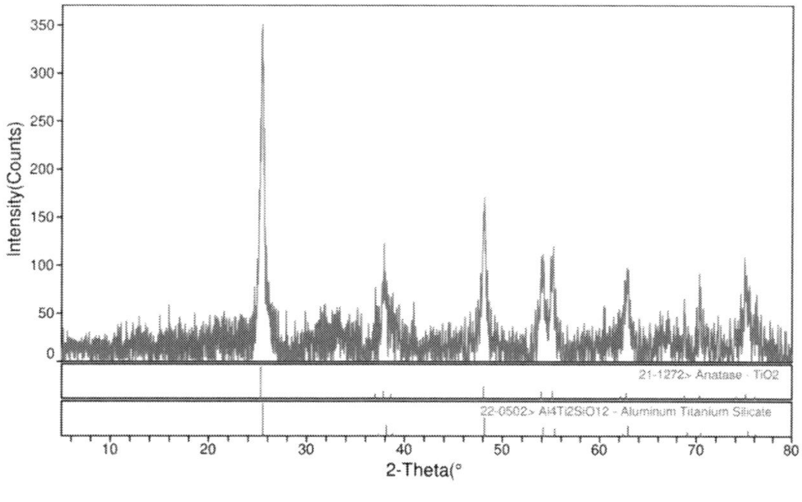

Figure 3. TiO_2/Zeolite diffraction pattern at 450℃.

3.2 Pulsed Discharge Plasma and TiO_2/Zeolite Coupling Technology for Formaldehyde Removal

3.2.1 .The Effect of Pulse Voltage on Formaldehyde Removal Efficiency Figure 4 shows that formaldehyde removal efficiency increases with the increase of voltage, when voltage is 30 kV, removal efficiency is 43.32%, after voltage reaches 35 kV, efficiency arrives to 46.93%, while increase rate gradually slows down when voltage increases from 35 kV to 37.5 kV. That is because when the pulse voltage increases, the average electric field strength in the discharge gap increases markedly, and led to discharge become stronger, which can lead to generate radicals and reactive substances by the increasing of pulsed energy in reactor. But when voltage reaches a certain value, increases the voltage will decreases energy utilization efficiency, and removal efficiency enhance is not obvious. So the optimal discharge voltage of the reactor is 35 kV.

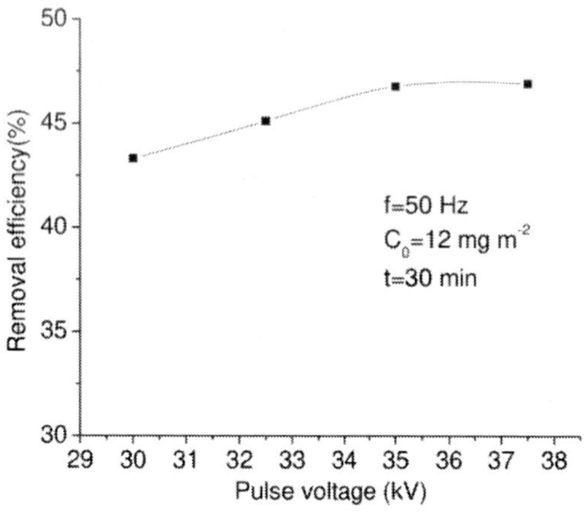

Figure 4. The effect of pulse voltage on formaldehyde removal efficiency.

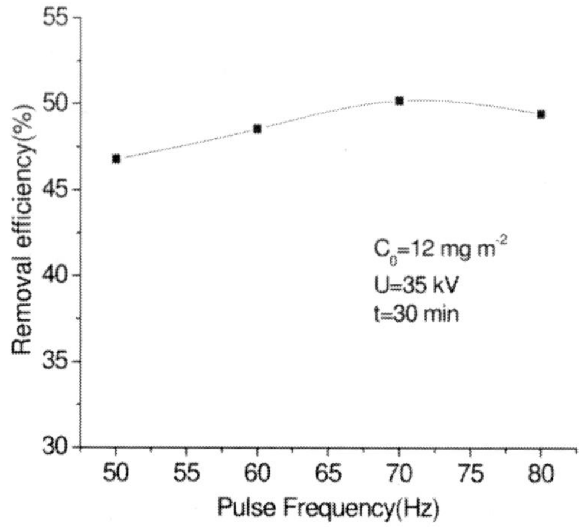

Figure 5. The effects of pulse frequency on formaldehyde removal efficiency.

3.2.2 .The Effects of Pulse Frequency on Formaldehyde Removal Efficiency As figure 5 shows, formaldehyde removal efficiency increases first and then decreases as the increase of pulse frequency, when pulse frequency is 50 Hz, after 30 minutes of pulsed discharge treatment, the removal efficiency is 46.78%; when pulse frequency increases to 70 Hz, the efficiency arrives to 50.21% under the same condition. Increased pulse frequency means the increase of discharge rate and discharge number, the increase of pulse frequency will always increase high-energy electron per unit time, and also increase actives concentration, all of these will enhance removal efficiency. While pulse frequency is too high, discharge time shortens in unit pulse, which hinder energy from discharging rapidly, so energy efficiency and removal efficiency decreases. Under such experimental conditions, formaldehyde removal efficiency is best when pulse frequency is 70 Hz.

3.2.3. Pulsed Discharge Plasma Coupling with Zeolite and TiO_2 for Formaldehyde Removal A certain amount of anatase TiO_2 or zeolite were put into discharge reactor respectively, then formaldehyde removal efficiency of pulsed discharge plasma coupling with zeolite and TiO_2 were tested at different time and analyzed.

As figure 6 shows, in the condition of different additives, formaldehyde removal efficiency improves along with the increase of treatment time, when time lasts 40 minutes, formaldehyde removal efficiency of pulsed discharge plasma coupling with zeolite catalyst is 57.87%, efficiency of pulsed discharge plasma coupling with TiO_2 catalyst is 65.34%, the latter is higher than the former obviously. Because photocatalyst is motivated by discharge UV-light, in the process, photogenerated electrons and holes are produced, then they form a redox system, and a mass of highly reactive radicals are generated after a series of reactions. In the meantime, catalyst activity would also be motivated by pulsed discharge plasma. Zeolite additions only play a part of adsorption, which will gather formaldehyde on the surface of zeolite and strengthen oxidation of pulsed discharge plasma, but pulsed discharge plasma coupling with zeolite do not have real synergy, so pulsed discharge plasma coupling with zeolite has less function than pulsed discharge plasma with TiO_2 photocatalyst. At last, through the experimental research shows that pulsed discharge plasma coupling with TiO_2 photocatalyst do have synergy, discharge plasma and UV-light will motivate photocatalyst activity to improve formaldehyde removal efficiency.

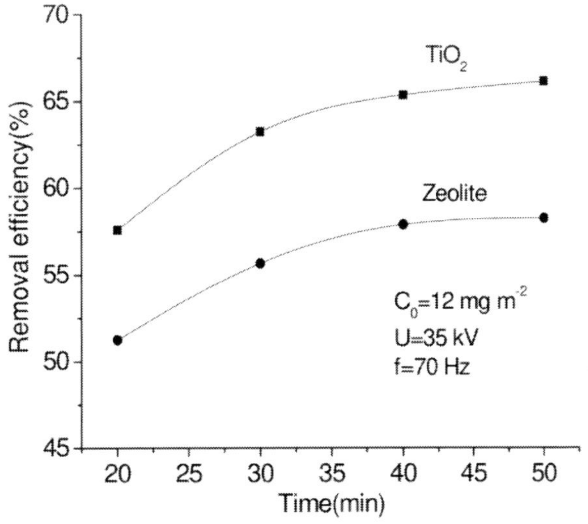

Figure 6. Pulsed discharge plasma coupling with Zeolite or TiO_2 for formaldehyde removal.

3.2.4. Pulsed Discharge Plasma Coupling with TiO_2 and TiO_2/Zeolite for Formaldehyde Removal

A certain amount of anatase TiO_2 or TiO_2/zeolite were put into discharge reactor respectively, then formaldehyde removal efficiency of pulsed discharge plasma coupling with TiO_2 and TiO_2/zeolite were tested at different time and analyzed.

Figure 7 shows that the pulsed discharge with TiO_2/Zeolite catalyst for formaldehyde removal has higher efficiency than pulsed discharge with TiO_2. Because discharged UV-light contacts with TiO_2/Zeolite catalyst directly, which has photocatalytic function. Moreover, active radicals which produced from pulsed discharge plasma have synergistic effect with TiO_2/Zeolite catalyst, thus increasing formaldehyde removal efficiency. Furthermore, pulsed discharge could prevent electron-hole from recombining and enhance catalyst efficiency. Electron-hole pairs will be motivated more easily in light for changing catalyst bandwidth and shape and accelerating electron speed on the band by pulsed discharge. Besides, using zeolite as carrier, carrier could provide photocatalytic reaction environment for high concentration formaldehyde and accelerate the rate of photocatalytic degradation on formaldehyde by adsorption and surface gathering. Formaldehyde adsorbed by carrier will move to TiO_2 surface through diffusion, then TiO_2 can resolve formaldehyde on the carrier, carrier will be regenerated in situ. So pulsed discharge plasma coupling with TiO_2/Zeolite catalyst for formaldehyde removal can bring the coupled effect of discharge plasma, UV-light and photocatalyst to improve formaldehyde removal efficiency.

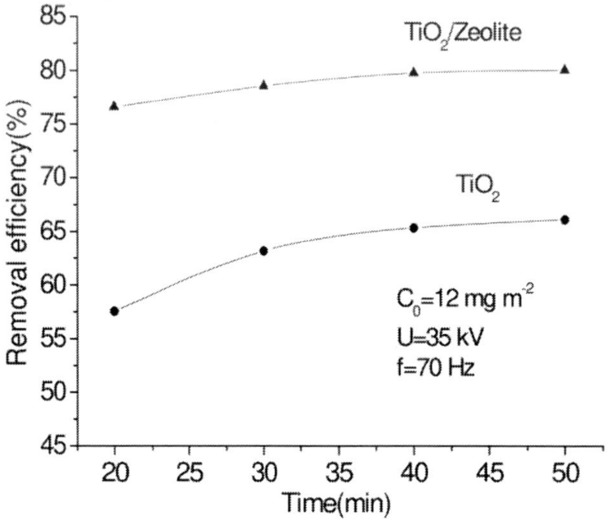

Figure 7. Pulsed discharge plasma coupling with TiO_2 or TiO_2/Zeolite for formaldehyde removal

4. Conclusions

(1) TiO_2/Zeolite catalyst was made by sol-gel method. The volume ratio of CH_3CH_2OH to $C_{16}H_{36}O_4Ti$ is 4.0, $C_2H_4O_2$ to $C_{16}H_{36}O_4Ti$ is 0.1, water to $C_{16}H_{36}O_4Ti$ is 0.15, HNO_3 to $C_{16}H_{36}O_4Ti$ is 0.1, when the baking temperature is 450 ℃ and the baking time is 120min, TiO_2/Zeolite catalyst shows higher photocatalytic activity.

(2) Pulse voltage and frequency do have great effect on formaldehyde removal efficiency, within a certain scope, removal efficiency increases with the increase of pulse voltage and frequency, when pulse voltage and frequency exceed a certain value, increase rate of the efficiency will slow down.

(3) The pulsed discharge plasma coupling with TiO_2 photocatalyst do have synergy, discharged plasma and UV-light will motivate catalyst activity, which could improve formaldehyde removal efficiency.

(4) The pulsed discharge with TiO_2/Zeolite catalyst for formaldehyde removal has higher photocatalytic activity and efficiency than pulsed discharge with TiO_2.

Acknowledgement

The authors would like to thank the National Natural Science Foundation Committee (PR China, No. 50867003), the Education Department of Jiang Xi (No. GJJ11455) and Jiangxi University of Science & Technology (No. jxxjzd10004) for their financial support of this research.

References

[1] Lakshmi S, Renganathan R and Fujita S 1995 *J. Photo-Chem. and Photobio. A:Chem.* **88** 163

[2] Amirov R H 1993 *The second in. conf. on appl. Electrost.* p 139

[3] Mizuno A, Kisanuki Y and Noguchi M 1999 *IEEE Trans. on Ind. Appl.* **35** 1284

[4] Oda T, Takahashi T and Yamaji K 2002 *IEEE Trans. on Ind. Appl.* **38** 873

[5] Yang Q and Guo F 2005 *J. Environ. Sci.* **26** 35

[6] He Z H, Ding L H and Deng D 2005 *High Voltage Eng.* **31** 37

[7] Yang C H, Zhou F and Gao D L 2007 *High Voltage Eng.* **33** 30

[8] Chavadej S and Kiatubolpaiboon W 2007 *J. Molec. Cataly. A: Chem.* **263** 128

[9] Huang B, Yang Y and Zhang X 2008 *J. South Chin. Univ. of Technol.: Nat. Sci.* **36** 56

Treatment of formaldehyde by corona plasma cooperated with photo-catalysis

X G Yue[1], Z H He[1*], P Y Ma[2], L Zhu[1], J Ma[1], H Zhao[1]
[1]State Key Laboratory of Advanced Electromagnetic Engineering and Technology, Huazhong University of Science and Technology, Wuhan 430074, China
[2]School of Science, Wuhan University of Technology, Wuhan 430070, China

E-mail: hzh@mail.hust.edu.cn

Abstract. Formaldehyde, as a main indoor pollutant, is a serious hazard for human health. Here the experimental results of dielectric discharge, together with photo-catalysis to treat formaldehyde, were presented. The modified photocatalyst was loaded onto the electrode surface for degradation of formaldehyde in an artificially enclosed room. The effect of catalyst quantity, inlet rate, temperature, humidity and other process variables were examined in order to explore the relative purification factors of plasma cooperated with photo-catalysis. It is found that formaldehyde degradation efficiency is proportional to photo-catalyst quantity, and reciprocal to gas inlet rate and humidity. Experimental results confirm that plasma discharge combined with photo-catalysis can significantly improve the treatment efficiency of formaldehyde compared to a single plasma discharge. This research may form a basis for practical development of an efficient, low-power-cost, and relatively comprehensive indoor pollutant purification system.

1. Introduction

Indoor air pollution problems gain an increasing concern. Formaldehyde, as a main indoor pollutant, possesses serious hazards for human health [1]. However, most previous studies focus on benzene and its derivatives such as total volatile organic compounds (TVOC), comparing with less researches on formaldehyde removal. Previous studies have confirmed that the corona discharge plasma accelerating with photo-catalytic effect could degrade TVOC [2]. In order to provide the reference for understanding the mechanism of coordination, this study attempts to combine gas discharge with photo-catalysis, namely, using low temperature plasma, instead of simply ultra-violet light, during gas discharge process as the driving force for photo-catalytic materials, and further explores the influencing factors of synergistic purification and the relative impact mechanism.

Existing mainstream toward plasma discharge cooperating with photo-catalytic purification has accepted that the co-purification process mainly relied on the following three ways [2, 3]. Firstly, organic gases are to be excited, ionized, dissociated, etc. after the bombardment treatment by high energy electron, which leads to the chemical bond rupture and undermines the organic gases; Secondly, a large number of free radicals and ozone with strong oxidization abilities were produced by gas discharge to oxidize organic gases into CO_2, CO, and H_2O; Thirdly, when ultraviolet light produced by gas discharge irradiates nano photo-catalyst, the surface of the photo-catalyst will produce electron-

[*] To whom any correspondence should be addressed.

hole pairs, which react with O_2 or H_2O absorbed on the catalyst surface. The reaction will generate OH radicals which can oxidize the organic gases.

2. Experiment setup

The reactor used in the experiment for plasma corona discharge cooperating with photo-catalytic reaction is shown in figure 1. The reactor consists of one stainless steel rod (diameter of 22.5 mm) and one ceramic dielectric (inner diameter of 24.5 mm, outer diameter 31.5 mm). The ceramic tube wrap around with an aluminium foil of 83 mm length. Stainless steel aluminium foil as the induced pole, were energized with 5 ~ 10 kV AC voltage, and photo-catalyst TiO_2 thin film were deposited on the outer surface of the rod electrode and inner surface of the ceramic tube.

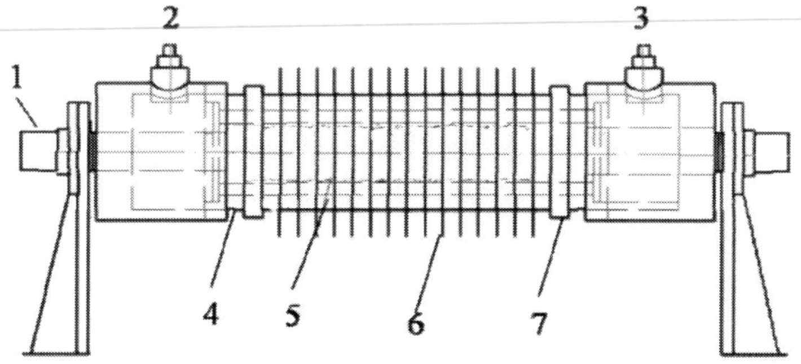

Figure 1. Reactor structure (1-discharge electrode high voltage input, 2-air inlet, 3-air gate, 4-induced electrode aluminium foil, 5-TiO_2 thin film, 6-thermal slug,7-ground connection).

The photo-catalyst TiO_2 powder, which is characterized by high urging efficiency and stable non-toxic, is a well-recognized and efficient environment-friendly material [4, 5]. In order to improve its catalytic efficiency, a small amount of transition metal ions is doped in the mixture. Firstly, the precursor of iron and other transition metal are dissolved in isopropanol, and we mix tetrabutyl titanate and isopropanol with concentrated nitric acid by stirring; then mixture forms a white precipitated after aging 12 h at room temperature. By electrophoretic deposition technique we load the catalyst on the electrode, which is heated at 450 °C for 2 h. After detected by X-ray diffraction analysis equipment, the XRD spectrum of catalyst samples is shown in figure 2. We also obtain the spectra of catalyst samples by energy dispersive spectrometer, which is shown in figure 3.

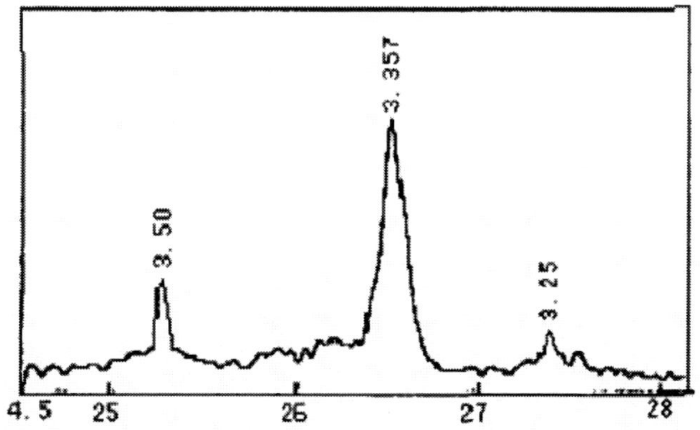

Figure 2. The spectra of catalyst by X-ray. The two main peaks could be introduced catalyst samples' particle size of 9.39 nm and 10.04 nm.

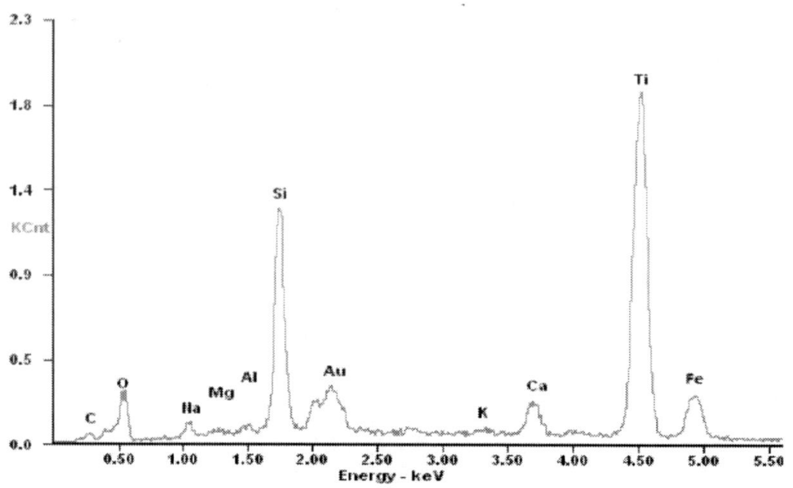

Figure 3. The spectra of catalyst by energy dispersive spectrometer. Due to the glass vector of the electron microscope, Si element also could be detected.

3. Result and discussion

Experimental results confirm that the plasma discharge combining with photo-catalytic treatment has a degradation effect (figure 4) on formaldehyde, and which can significantly improve the treatment efficiency compared to a single plasma discharge. The degradation eventually tends to saturation since the oxidation reactions is high-energy process and the formaldehyde molecule is reversible, plus radicals' transient existence and the ozone saturated concentration, which resulted in an oxidation equilibrium; Ultimate product H_2O would increase the humidity of air, so that the corona onset voltage increases, and the final concentration of formaldehyde gas is no longer reducing.

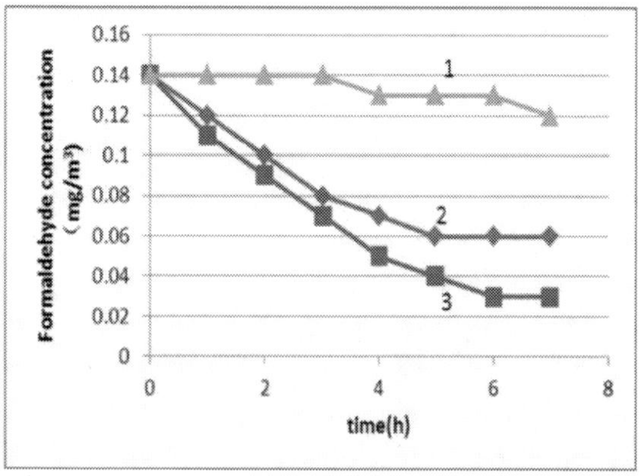

Figure 4. Comparison of formaldehyde concentration after different treatment.
(Curve 1, 2, 3 reflect conditions of un-treatment, simply-discharge treatment and cooperate treatment, respectively).

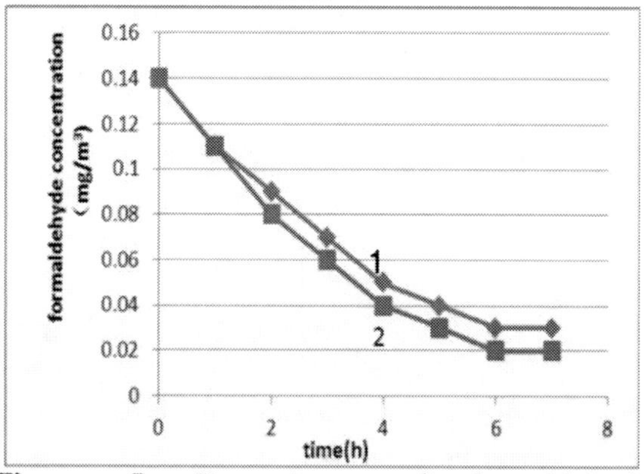

Figure 5. Dependence of formaldehyde degradation efficiency on catalyst quantity.
(Curve 1 and 2 respectively reflect TiO_2 thin film with single layer and bi-layer).

The detailed influence of loading quantity of catalyst on purifying effect is shown in figure 5. We can see from the figure 5, at the very beginning of the experiment, purifying efficiencies with different catalyst quantity almost show no difference; however, with the ongoing reaction, the purification efficiency of the bi-layer membrane is greater than the monolayer. It may be because electric field energy levels is relatively low on the initial discharge stage, and the TiO_2 catalyst has not all fully activated, which leads to little impact on the treatment of formaldehyde. When the electric field energy level rises enough to fully activate the photo-catalyst and prompts the organic degradation process, the electron-hole pairs possess a great reactivity. The oxidative ability of hydroxyl radical is higher than that of ozone, so the photo-catalyst will play a dominant role during latter reaction. Therefore, more amount of catalyst generates more electron-hole pairs, and thereby the degradation efficiency of the bi-layer membrane is greater than the monolayer.

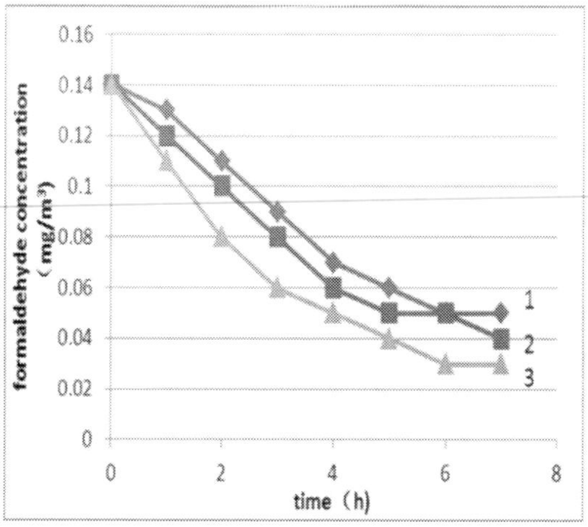

Figure 6. Dependence of formaldehyde degradation efficiency on gas inlet rate.

(Curve 1, 2, 3 reflect respectively inlet rate at 1000 mL min^{-1}, 800 mL min^{-1} and 600 mL min^{-1}).

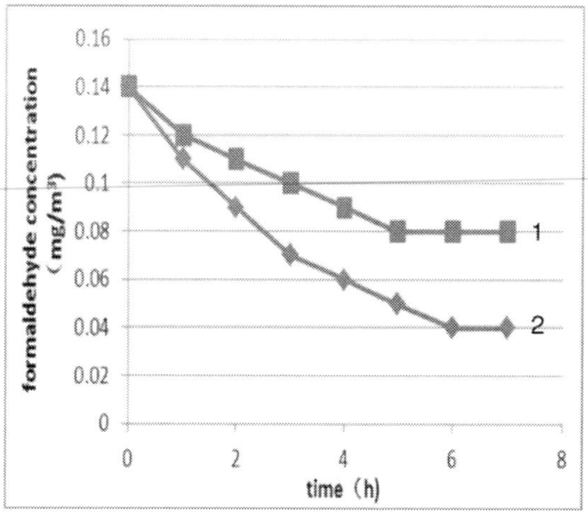

Figure 7. Dependence of formaldehyde degradation efficiency on ambient humidity.

(Curve 1 and 2 reflect relative humidity of about 30% and 72% (sunny and rainy), respectively).

The intake rate reflects the length of gas molecules' residence time in the reactor, under the condition that electric field length is constant. When changing the rate of intake during experiment, we can see the purifying effect of intake rate in figure 6. Greater inlet rate, the more transient airflow stays in the reactor, and lower probability of collision between the gas molecules and free electrons are, which eventually results in the decrease of the probability of gas molecules' dissociation and ionization, and thus the removal rate will drop.

The relationship between humidity and co-purifying efficiency is showed in figure 7. From one aspect, during the degradation process water vapour and reactants compete for absorption on the catalysts' water active centre, thus the reaction rate is influenced by absorption competition. Once the humidity increases, the amount of pollutants adsorbed is reduced, so is reaction rate. From another aspect, electrons would be changed into anion when meeting water vapour. Due to anion's low moving rate and small ionization ability, the effective number of electrons involved in the avalanche is thus reduced. Reduction of effective electron indirectly leads to plasma energy falling, hence results in the drop of purifying rate.

Previous experiments demonstrate almost no relationship between ambient temperature and formaldehyde purifying effect. On the one hand, the final decision on the collaborative purification efficiency is the direct reaction of the hole and pollutant gases. The reaction is a key step in the photo-catalytic reaction, which requires low apparent activation energy. When the temperature varies, the apparent activation change is smaller, and the reaction rate varies smaller responding to temperature, so is the temperature dependence of nano-TiO_2. On the other hand, the average field strength in the discharge space exceed 50 kV cm^{-1}, and the temperature in the reactor is proportional to electric field strength and reaction time length. The surface temperature of the reactor rose to 60 °C, which is 2~4 times than that of the external environment, so the influence of ambient temperature compared to the temperature of the reactor can be negligible. Considering two factors above, there is almost no ambient temperature influence during the experiment for co-purification efficiency.

4. Conclusion

The project loads modified photo-catalyst on the electrode surface, and builds an artificial enclosed room to treat formaldehyde, we concentrate that the plasma discharge cooperating with photo catalytic truly has degradation effect on formaldehyde. It is found that the formaldehyde degradation efficiency is proportional to photo-catalyst quantity, and reciprocal to gas inlet rate and humidity. Experimental results confirm that the plasma discharge combining with photo-catalysis can significantly improve treatment efficiency compared to a single plasma discharge. We hope our research will provide the basis for understanding the mechanism of coordination and a viable direction for a multi-purpose, high efficiency, low consuming dealing system for indoor pollutants.

Acknowledgments

The work is supported by 'the Fundamental Research Funds for the Central Universities', HUST: 2010JC018.

References

[1] Bigda R J 1995 *Chem. Eng. Prog.* **91** 62
[2] Hammer T 2002 *Plasma Sources Sci. Technol.* **11** A196
[3] Sullivan P D, Rood M J and Dombrowski K D 2004 *J. Environ. Eng.* **130** 258
[4] Holzer F, Roland U and Kopinke F D 2002 *Appl. Catal. B: Environ.* **38** 161
[5] Sano T, Negishi N, Sakai E and Matsuzawa S 2006 *J. Mol. catal. A, Chem.* **245** 235

Regeneration of nitrobenzene-exhausted granular activated carbon by dielectric barrier discharge method

Tian Lan, Wenli Gao, Zhongjian Li, Lecheng Lei*
Key Laboratory of Biomass Chemical Engineering of Ministry of Education, Department of Chemical and Biological Engineering, Zhejiang University, Hangzhou 310027, China

E-mail: lablei@zju.edu.cn

Abstract. A novel method for the regeneration of nitrobenzene-exhausted granular activated carbon using dielectric barrier discharge (DBD) was proposed in this study. The influence of several parameters including voltage, frequency, and plasma medium on the regeneration efficiency were studied. Under optimum conditions, regeneration efficiency can reach over 80% and remain nearly stable after 5 times of regeneration cycle. The texture characteristic and surface chemistry of Granular Activated Carbon (GAC) samples were also investigated. Analysis shows that the pore volume and specific surface area of regenerated GAC is strongly recovered compared to the exhausted GAC, but the discharge can cause some pores to diminish. Acidic functional groups on GAC's surface especially carboxylic groups had a growing tendency after DBD process. Experimental results show that the regeneration of GAC by DBD method mainly attributes to high active species and thermal effect, while O_3 has minor effect.

1. Introduction

Activated carbon (AC) is a widely used adsorbent material due to its extended surface area, tailored pore distribution, and high degree of surface reactivity [1]. It has been proved as an efficient and versatile adsorbent used in environmental pollution control fields such as water treatment and removal of gaseous pollutants.

In many pollution treatment processes, when AC absorbent reach to exhaustive limit, the spent AC was usually discarded or dumped into landfill. But this is an uneconomic way; moreover, the spent AC usually contains high content of pollutant [2], so discard way could cause secondary pollution. Regeneration or reuse of AC is an environmentally sound option to ensure its economic and environmental acceptability [3].

In the past two decades, many techniques for regeneration of AC have been established, such as thermal volatilization, chemical extraction, microwave irradiation, electrochemical methods, solar regeneration and bio-regeneration [2-12].

Thermal regeneration is the most widespread method [8]. This method needs to heat the spent carbon to the temperature up to 800 °C under mildly oxidizing or inert gas medium. So the thermal regeneration method is practically energy intensive and time consuming, resulting in economically unfavorable. According to a previous study, the regeneration efficiency is low and the adsorption

* To whom any correspondence should be addressed.

capacity drop very furiously in merely 6 regeneration cycles [4]. Moreover, quite a few of AC might be burn-off during the regeneration process [13].

Chemical regeneration involves chemical extraction or advanced oxidation process. Desorption of adsorbents with an extracting agent or decomposition of adsorbed species using oxidizing chemical agents usually need to conduct under subcritical or supercritical conditions [14]. Extra investment cost is acquired due to the extraction agent itself need to be regenerated after the regeneration [12].

Some absorbed organics such as phenol or 2-Chlorophenol [15] can be served as substrates for microorganism growth. Although bio-regeneration has been proposed as an economical regeneration process, it usually requires long reaction time and can only be applied to the biodegradable substances [16].

In recent years, microwave irradiation and ultrasonic regeneration have been widely investigated [2, 3, 17, 18]. But in most of the cases, complicated processes were required [19] and the economic cost is high.

Recently a novel AC regeneration method: dielectric barrier discharge (DBD) plasma regeneration has been presented [19-21]. This method utilizes the DBD process to create non-thermal plasma under atmospheric pressure. The non-thermal plasma can provide active species such as O_3, high-energy electron and free radicals, which might be effective for the oxidation of pollutants adsorbed on AC. Published researches, have proved that the DBD method seems to be a promising process [19-21].

Nitrobenzene and its degradation intermediate are seemed as hazardous pollutants. AC has already widely used in water treatment field to remove nitrobenzene [22-24]. In this paper, a study of regeneration of nitrobenzene-exhausted granular activated carbon by DBD method was presented. Since there were very few reports concentrated on DBD regeneration method, we studied the influence of several parameters on DBD discharge, including voltage, frequency and the plasma medium to provide useful information for further research. And the texture characteristic, surface chemistry of AC samples were also investigated.

2. Experiment and materials

2.1. Materials

Commercial nutshell-based granular activated carbon purchased from a local filtrate material factory was used for the experiments. Granular AC (GAC) samples were washed with distilled water to remove adhering impurities from its surface, and then boiled it in distilled water for 2 h, then washed again till its pH reached around 7. The cleaned samples were screened to the desired mesh size of 2-3 mm and dried at 105 °C for 2 h. Table 1 shows the characteristics of virgin GAC.

Liquid nitrobenzene concentration was detected by UV–vis spectrophotometer (cintra 303, GBC Scientific Equipment Pty Ltd., Australia) at the characteristic absorption peak of 268 nm. Gasous O_3 concentration was tested by iodimetry method.

Nitrobenzene was selected for this study as a typical pollutant. 1000 mg L^{-1} working solution of nitrobenzene was prepared by dissolving appropriate quantity in distilled water. The chemicals used for acid–base properties estimation and other tests were all reagent grade.

Table 1. Physical and chemical characteristics of virgin GAC

Packing density (g L^{-1})	$S_{Langmuir}$ (m^2 g^{-1})	Average pore diameter (Å)	$V_{Micropore}$ (cm^3 g^{-1})	$V_{Mesopore}$ (cm^3 g^{-1})	Acidic groups (mmol g^{-1})	Basic groups (mmol g^{-1})	pH_{PZC}
570	654	2.33	0.3051	0.0923	0.703	0.825	6.72

2.2. DBD experiment system

The schematic of experiment device is given in figure 1. The experimental system consists two parts: AC power supply and DBD reactor. The CTP-2000K HVAC power supply is produced by Suman electricity Ltd. (Nanjing, China) with the voltage 0-50 kV adjustable and the frequency 0-10 kHz adjustable. The reactor uses quartz glass tube as dielectric barrier (10 mm inner diameter, 350 mm long). A copper net is spiral wounded outside of the tube as HV electrode, a stainless steel threaded rod equipped in the middle of the inside tube as ground. The other space of the tube can loosely filled with GAC, so that different gas can pass through the tube to generate non-thermal plasma.

Typical voltage and current waveforms data of DBD reactor are recorded by a digital oscilloscope (TDS 1012B-SC, Tektronix, USA). An AvaSpace-2048 spectrometer (Avantes, Netherlands) is used to detect the radical emission spectrum during experiments.

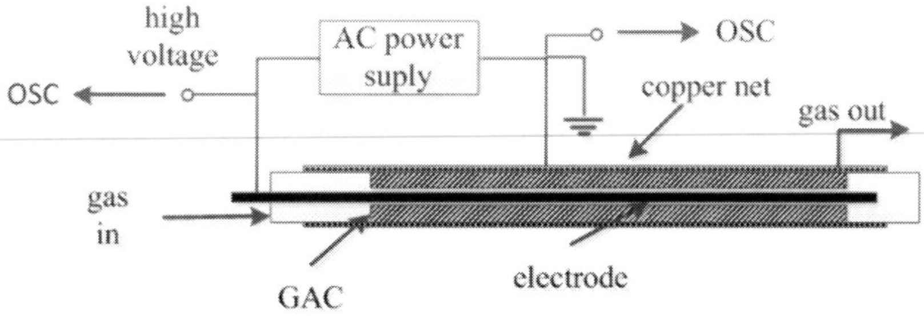

Figure 1. Scheme of DBD regeneration experiment system.

2.3. The adsorption equilibrium and adsorption isotherm

The kinetics adsorption of nitrobenzene on virgin carbon was studied in this section. Virgin GAC samples were placed into a continuous cycle flow adsorption column. Different concentration of nitrobenzene solution was circulated through the adsorption column by a peristaltic pump at a stable flow rate. Samples were withdrawn to measure the amounts of nitrobenzene adsorbed on GAC. From figure 2 one can draw a conclusion that the adsorption equilibrium state reached in about 1h. According this result we set 10 h equilibrium time for later adsorption isotherm study.

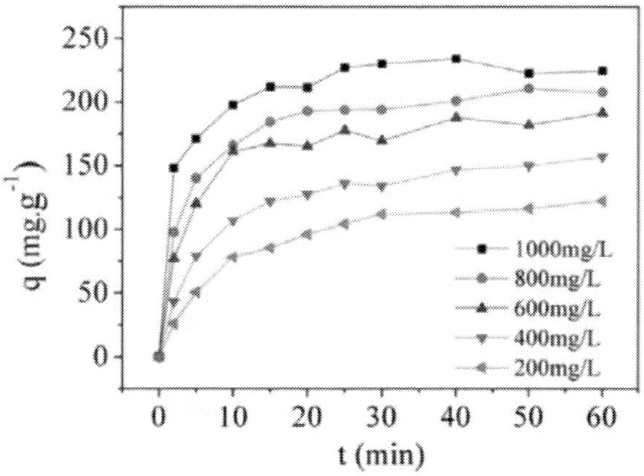

Figure 2. Adsorption kinetics at different initial concentrations.

Adsorption isotherm of nitrobenzene on GAC was studied by batch method in oven oscillator. Different amounts of GAC samples were added into conical flasks containing 150 mL of nitrobenzene solution of 1000 mg L^{-1}. The flasks were sealed and shaken at a constant speed of 120 rpm at 25 °C for about 10 hours to reach adsorption equilibrium. 3 mL samples of suspensions were withdrawn after experiment to analysis the nitrobenzene remaining in the fluid phase. The amounts of nitrobenzene adsorbed on GAC samples were inferred from the mass balance as follows:

$$q_e = \frac{V(C_0 - C_e)}{m} \quad (1)$$

Where q_e is the amount of nitrobenzene per gram of adsorbent, V the volume of liquid phase, C_0 is the initial concentration of nitrobenzene in the bulk phase, C_e is the concentration of the solute in the bulk phase at equilibrium status, and m is the amount of GAC.

2.4. Experiments
For each single experiment, 1.5 g exhausted GAC was filled into the reactor. Different kind of gas passed through a washing bottle in advance to increase its humidity and then the damped gas was impregnated to the reactor with the flow rate controlled in the range of 0.06-0.1 L min^{-1}. Each regeneration experiment lasted for 30 min. Other experimental parameters were sternly under controlled in desired level.

O_3 and several kind of free radicals such as •OH, HO_2• , RO• , O_2• generated by DBD process might be important oxidants for GAC regeneration, Moreover, the thermal effect caused by plasma might also plays an important role on the regeneration process. In order to ascertain contributions of different kinds of factors on the GAC regeneration process separately, parallel experiments without discharge and with the same concentration of O_3, or the same temperature were carried out.

2.5. Characterization of activated carbons
The structural properties of virgin and regenerated carbon were obtained from the physical adsorption of N_2 at 77 K determined by a Quanta chrome autosorb-1 adsorption apparatus (Autosorb-1, Quanta chrome Com. USA). The surface area and pore volume distribution calculated using the Langmuir equations and the BJH model, respectively. The micro pore volume and micro pore area were obtained from the de Boer's t-plot method. The surface acidity was estimated by Boehm titration method.

3.Results and discussion

3.1 GAC regeneration by DBD process
Non-thermal plasma conditions can be found in DBD process. DBD demonstrates flexibility with respect to the geometric shape, working gas shape or mixture composition, and operational parameters [25]. In this section, we studied the GAC regeneration efficiency at different operational parameters and different gas ambience.

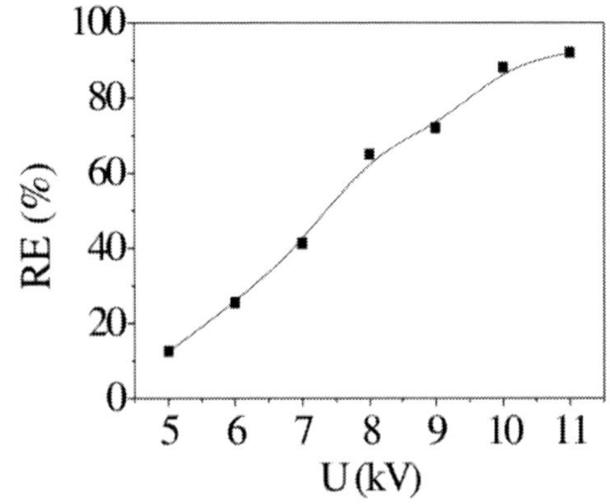

Figure 3. Effect of voltage on regeneration efficiency (damped O_2 ambience).

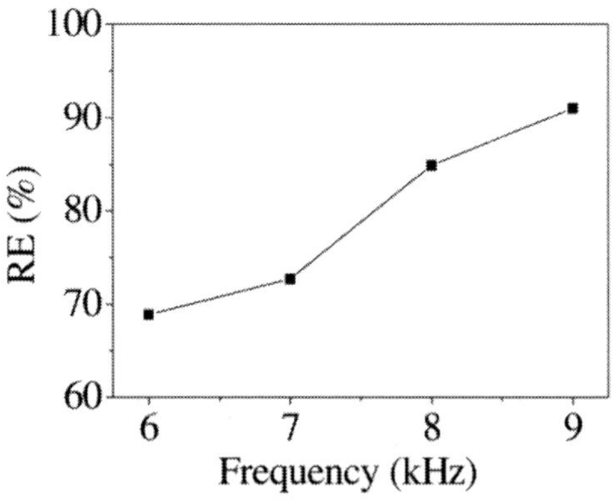

Figure 4. Effect of frequency on regeneration efficiency (damped O_2 ambience).

Figure 3 illustrates the influence of driving voltage on regeneration efficiency in damped O_2 ambience. In our experimental system, when the driving voltage was lower than 5 kV, discharge can hardly take place. When the driving voltage was above 11 kV, the silica barrier would easily be breakdown, for this reason, the experiment was carried out with the voltage ranged from 5 kV to 11 kV. Experiment result turns out that the RE% increase almost linearly with the driving voltage under 10 kV, but the increase tendency seems slowdown from 10 kV to 11 kV, owing to discharge become unstable in high voltage.

Figure 4 shows the effect of power frequency on regeneration efficiency. It can be concluded that the higher frequency, the better RE% can achieved.

It is important to study the GAC regeneration efficiency in different gas ambience. It is generally believed that discharge in the gas which contains high percentage of oxygen can completely oxidize pollutants to CO_2 [26]. Gaseous pollutants can aggregate to micrometric particles and be removed by filtration using N_2-DBD [27]. But discharge in pure nitrogen ambience cannot create ozone; it might result in less reactive in GAC regeneration. Besides, N_2's bonding energy (15.63 eV) is much higher than O_2 (5.12 eV), N• radicals is harder to be generated than O• radicals in DBD process. N_2-DBD is usually used for surface treatment or surface modification [25]. N_2-DBD might not suitable for GAC regeneration. Because the moisture in gas is a source of •OH radicals, we also compared the RE% of dry O_2 to the damped O_2, experiment results sees in figure 5 below:

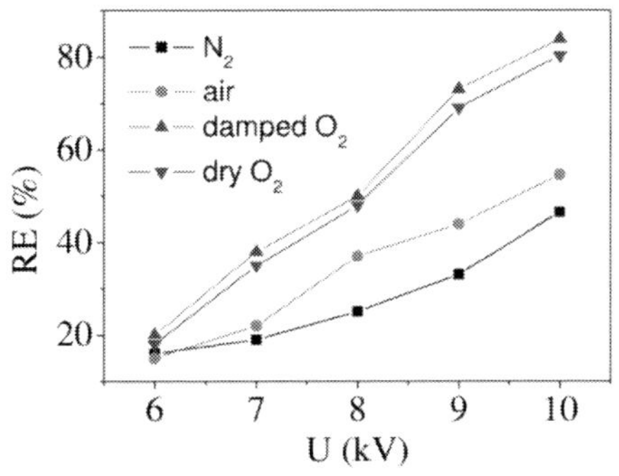

Figure 5. Regeneration efficiency in different gas ambience.

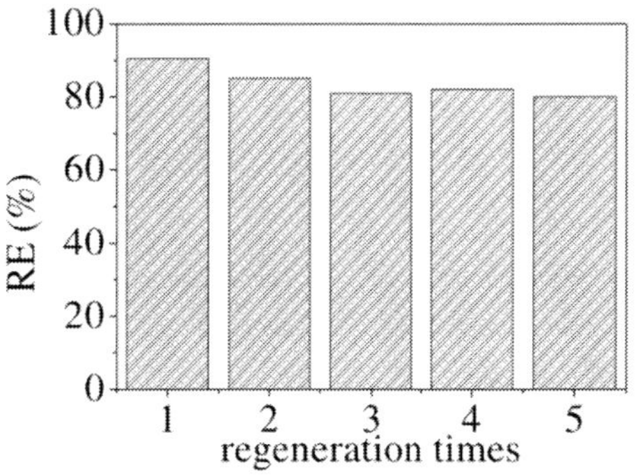

Figure 6. Regeneration efficiency at 1-5 regeneration cycles.

As figure 5 illustrations, DBD regeneration in wet O_2 atmosphere get highest RE% as expected, which can be up to 80%.

According to former researches, discharge could cause some physical change to the GAC [19], the GAC reuse time after DBD regeneration process is an essential issue. Regeneration yields in different regeneration cycles are presented in figure 6.

From figure 6, it can be concluded that despite a little decrease of RE% after the first regeneration cycle, the recovery of regenerated GAC's capacity of adsorption is basically achieved. In 2-5 cycles, the RE% is almost stable. This result indicates that the GAC regeneration by DBD method have industrial potential.

3.2. Characterization of activated carbons

Table 2 summarizes the relevant micropore structural properties of virgin, exhausted, and regenerated GAC after the second regeneration cycle. It could be concluded that the S_{BET}, V_{micro}, V_p and S_{micro} of exhausted GAC all greatly diminished compared to virgin GAC, but after DBD regeneration, all of the structural index were well recovered.

The PSD curves of GAC samples are shown in figure 7. It could be concluded that the exhausted GAC has significantly porous channel blockage, especially at the radius around 20 Å. In the regenerated GAC, 20 Å micro pores were mostly recovered, but the pores size ranged from 25 to 100 Å did not turn out to be significantly recovered. The small quantity of pore diminish might be attribute to channel collapse. Parts of the micropore have been destroyed cause by DBD process. The pore diminish result in somewhat decrease of equilibrium adsorption capacity of regenerated GAC compared to virgin GAC.

Table 2. Comparison of structural properties.

	S_{BET}	R_p	V_{micro}	V_p	S_{micro}
virgin	538	1.36	0.219	0.366	252
exhausted	160	1.74	0.042	0.139	41
regenerated	438	1.35	0.176	0.296	203

Table 3 presents the acid–base titrations of the virgin and regenerated GAC. The differences in surface chemistry of the GAC samples can clearly be seen from the Boehm titration results. Acidic group especially carboxylic functional groups have a growing tendency on GAC's surface after DBD process, and the lactonic and basic groups drop dramatically.

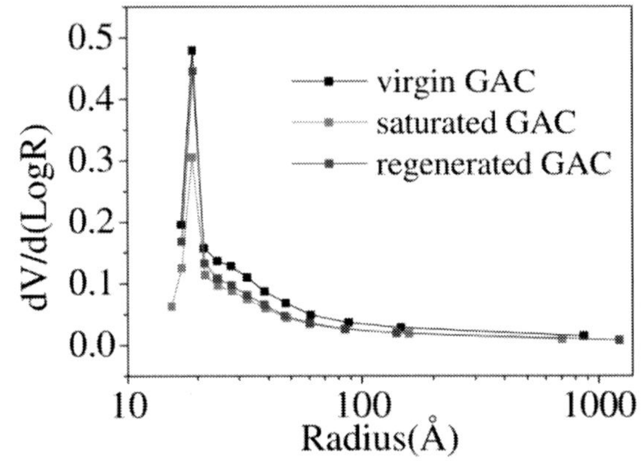

Figure 7. Pore size distributions by BJH method.

Table 3. Comparison of acid–base properties

sample	Groups (mmol g^{-1})				
	phenolic	lactonic	carboxylic	basic	acidic
virgin	0.26	0.12	0.45	0.7	0.83
regenerated	0.23	0.08	0.69	0.41	1

3.3. Different factors' contribution to GAC regeneration

DBD process is a combination of radical field and energy field. There are three important factors relate to GAC regeneration: O_3, free radicals, and thermal effect. Parallel experiments were carried out for the assessment of these three major factors' contribution separately. We set a blank test to eliminate the evaporation; and a parallel experiment without heating and discharging, only provide the same O_3 concentration as comparative DBD process to the reactor; and another parallel experiment without discharging and O_3 aeration, and only heat the reactor to get the same temperature as comparative DBD process. The higher regeneration efficiency of the comparative DBD process (9 kV driving voltage, 9 kHz frequency, 0.06 L min^{-1} air, 1.5 g GAC) over the summation of RE% of these parallel experiments can approximately character the contribution of free radicals. Figure 8 shows that the regeneration of GAC mainly attributes to high active species generated by DBD (64%) and thermal effect (25%), while O_3 has minor effect (8%).

Figure 9 is the Emission spectrum of DBD with damped O_2 aeration. As it shows, the main free radicals generated in this process are: •OH, •O, and excited state of O_2. This results interpret that O_2-DBD get much higher regeneration efficiency than N_2-DBD, and the moisture in gas initiate the generation of •OH.

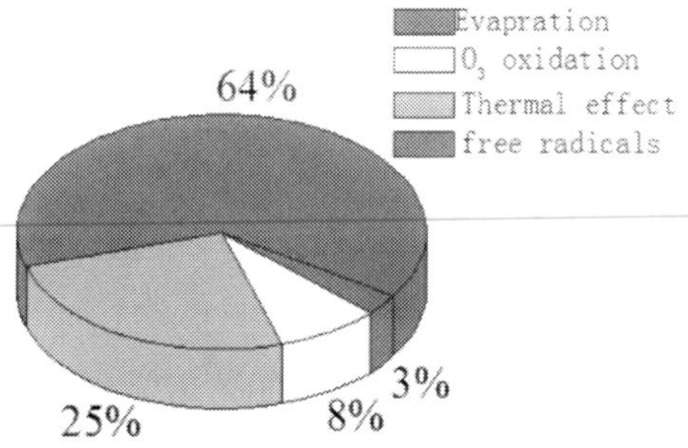

Figure 8. Different factors' contribution to GAC regeneration at DBD process.

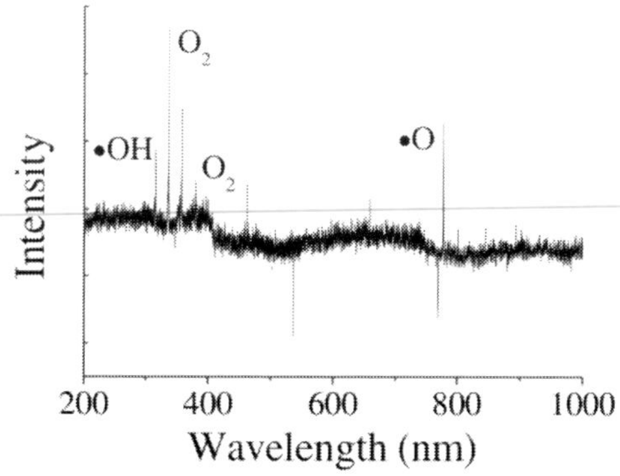

Figure 9. Emission spectrum of DBD.

4. Conclusions

The DBD process is a promising method for the regeneration of activated carbon. Although acidic groups especially carboxylic functional groups have a growing tendency on GAC's surface after DBD process, over 80% of regeneration efficiency can be achieved at best experimental condition, and the RE% remain nearly stable after 5 times of regeneration cycle. The regeneration of GAC by DBD method mainly attribute to high active species generated by discharge and thermal effect, while O_3 has minor effect.

Acknowledgments

The authors would like to acknowledge financial support for this work provided by MOST project of China (No.2008BAC32B06) and NSFC of China (No. 21076188, 20836008, 20976158, 20990221 and 21076189), the Key Innovation Team for Science and Technology of Zhejiang Province of China (2009R50047).

References

[1] Yenisoy-Karakas S, Aygun A, Gunes M and Tahtasakal E 2004 *Carbon* **42** 477
[2] Lim J L and Okada M 2005 *Ultrason. Sonochem.* **12** 277
[3] Foo K Y and Hameed B H 2012 *Chem. Eng. J.* **193** 404
[4] Morenocastilla C, Riverautrilla J, Joly J P, Lopezramon M V, Ferrogarcia M A and Carrascomarin F 1995 *Carbon* **33** 1417
[5] Guo Y and E Du 2012 *Energy Procedia* **17** 444
[6] Carratala-Abril J, Lillo-Rodenas M A, Linares-Solano A and Cazorla-Amoros D 2010 *Chem. Eng. Sci.* **65** 2190
[7] Charinpanitkul T and Tanthapanichakoon W 2011 *J. Ind. Eng. Chem.* **17** 570
[8] Duan X H, Srinivasakannan C, Qu W W, Xin W, Peng J H and Zhang L B.2012 *Chem. Eng. Process.* **53** 53
[9] Liu X T, Quan X, Bo L L, Chen S and Zhao, Y Z 2004 *Carbon* **42** 415
[10] Okawa K, Suzuki K, Takeshita T and Nakano K 2007 *Water Res.* **41** 1045

[11] Yap P S and Lim T T 2012 Water Res. **46** 3054
[12] Zhang H P 2002 *Chem. Eng. J.* **85** 81
[13] Alvarez P M, Beltran F J, Gomez-Serrano V, Jaramillo J and Rodriguez E M 2004 *Water Res.* **38** 2155
[14] Guo D S, Shi Q T, He B B and Yuan X Y 2011 *J. Hazard. Mater.* **186** 1788
[15] Aktas O and F Cecen 2007 *Adsorption* **13** 159
[16] Aktas O and Cecen F 2009 *Bioresource. Technol.* **100** 4604
[17] Hamdaoui O, Naffrechoux E, Suptil J and Fachinger C 2005 *Chem. Eng. J.* **106** 153
[18] Zhang G J, Wang S F and Liu Z Z 2003 *Environ. Eng. Sci.* **20** 57
[19] Qu G Z, Li J, Wu Y, Li G F and Li D 2009 *Chem. Eng. J.* **146** 168
[20] Qu G Z, Li J, Li G F, Wu Y and Lu N 2009 *Asia-Pac. J. Chem. Eng.* **4** 649
[21] Qu G Z, Lu N, Li J, Wu Y, Li G F and Li D 2009 *J. Hazard. Mater.* **172** 472
[22] Liu S and Wang R 2011 *J. Hazard. Mater.* **18** 99
[23] Rauthula M S and Srivastava V C 2011 *Chem. Eng. J.* **168** 35
[24] Wang S H, Zhang X J, Li Y, Liu Y, Chen C, Wang, Y, Zhu Z B and Zhao X S 2007 *China Water & Wastewater* **23** 1
[25] Borcia G, Anderson C A and Brown N M D 2005 *Plasma Sources Sci. T.* **14** 259
[26] Vertriest R, Morent R, Dewulf J, Leys C and Van Langenhove H 2003 *Plasma Sources Sci. T.* **12** 412
[27] Ognier S, Cavadias S and Amouroux J 2007 *Plasma Process Polym.* **4** *528*

Combination of ozone with nano TiO_2 supported by zeolite for the decontamination of raw water from Songhua River

Yi-xin Yang[1*], Jun Ma[2], Sheng-jun Wang[3], Jing Zhang[4]
1 School of Resources and Materials, Northeastern University at Qinhuangdao, Qinhuangdao, 066004, China
2 National Engineering Research Center of Urban Water Resources, Harbin Institute of Technology, Harbin, China
3 The 2nd Design Department, Beijing General Municipal Engineering Design & Research Institute, China
4 College of Architecture and Environment, Sichuan University, Chengdu, China

E-mail: yangyixininhit@gmail.com

Abstract. In the present paper, nanosized TiO_2 supported by zeolite was used in the ozonation of Songhua River as catalyst. The catalytic activity of TiO_2/Zeolite for ozonation was testified by nitrobenzene removal, which is one kind of representatively resistant organic contaminant. And the removal of some organic and inorganic compounds existing in the Songhua River by TiO_2/Zeolite catalyzed ozonation was investigated. GC-MS analysis showed that the Songhua River was obviously polluted by organic contaminants. Catalytic ozonation can remove about 30 kinds of organic matters from water and reduce the total organic peak area of feed water by 62.5%.

1. Introduction

Rapid population growth and economic development and the lack of effective environmental protection in China often lead to serious deterioration in the quality of water resources such as the Songhua River. Surface water contamination presents great difficulties and challenges to the capability and performance of conventional water treatment processes.

Ozone is widely used in drinking water treatment for many different purposes such as disinfection and algae control, taste, odour and colour control, oxidation of inorganic pollutants (iron, manganese), oxidation of organic micro- and macropollutants as well as for the improvement of coagulation [1]. It is now widely assumed that ozone reacts in aqueous solution on various organic and inorganic compounds, either by a direct reaction of molecular ozone or through a radical type reaction involving the hydroxyl radical induced by the ozone decomposition in water [2-4]. Since the oxidation potential of hydroxyl radical is much higher than that of the ozone molecule, radical oxidation is faster than that direct oxidation. Ozone decomposition is catalyzed by hydroxide ions and appropriate solid catalyst, which eventually produces free radicals, especially hydroxyl radicals [5-9].

In previous works, nanosized TiO_2 was chosen as a catalyst in the ozonation process and showed efficient activity for the degradation of nitrobenzene (NB), which is usually refractory to conventional chemical oxidation or toxic to micro-organisms of conventional biological oxidation [10]. In this work

* To whom any correspondence should be addressed.

the results of the investigation to the removal of some inorganic and organic compounds existing in water from the Songhua River by nano-TiO_2/Zeolite catalyzed ozonation processes are presented. These data provide a comprehensive evaluation of the efficacy of the nano-TiO_2/Zeolite catalyzed ozonation processes to improve the water quality for potable uses.

2. Materials and methods

2.1. Chemicals
The chemical reagents such as tetrabutylorthotitanate, absolute ethanol, sodium hydroxide and hydrochloric acid etc., purchased from Tianjin reagent Co. (China), were of analytical grade and used without further purification. Artificial zeolite pellets in a diameter of 1-2 mm were used as the supporting material and obtained from Shanghai reagent Co. (China).

2.2. Preparation and characterization of catalyst
Tetrabutylorthotitanate was used as a precursor of titania. Conforming to the sol-gel method, tetrabutylorthotitanate was added into the premixed solution of absolute ethanol, deionized water and hydrochloric acid at room temperature. The solution was mixed for 2 h to obtain titania sol. After the hydrolysis, the artificial zeolite pellets were slowly added into the titania sol and mixed for 24 h to obtain zeolite which carrying titania xerogel on the surface pores. Then the supported catalyst was calcined between 400 to 700 ^oC to control the TiO_2 crystallinity.

XRD studies of catalyst were performed by means of a D/max-r B diffractometer, at 40 kV, 30 mA (1200 W) and using nickel-filtered Cu kα radiation to determine crystal phase and estimate crystallite size. X-ray energy spectrum analysis was performed to test the TiO_2 content in S-570 electron microscope (Japan).

2.3. Experimental system

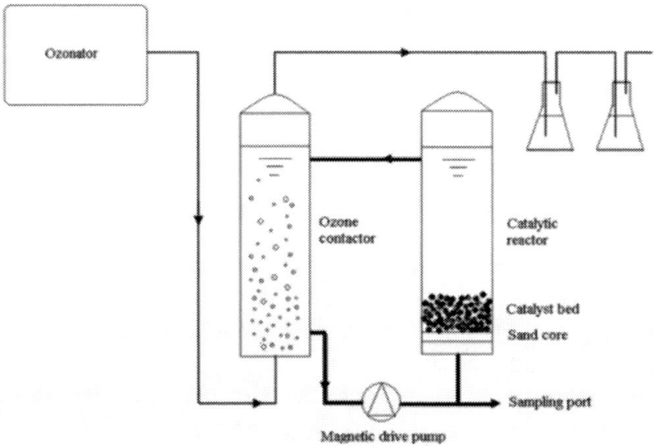

Figure 1. Experimental set-up.

Bench scale tests were performed using a two-stage flow-through system to facilitate the operation of all ozone-based oxidation processes, which consists of two series-wound cylindrical reactors made of pyrex glass with the dimensions of 80 mm in diameter and 500 mm in height (figure.1). 1.5 L test water was used for each experiment. Ozone was continuously introduced through a porous fritted diffuser that can produce fairly fine bubbles at a constant rate of 0.9 mg O_3 per minute. Ozone was produced from extra dry oxygen using the "corona discharge" method with a XFZ-581 Laboratory Ozonator. The system was operated by gravity feeding the water into the gas-liquid-contactor and then pumping the ozonated water through the catalyst bed. Surplus ozone in off-gas was trapped by two bubblers, containing 250 mL of a 2% KI solution. Unless otherwise stated, the catalyst dose was 10 g L^{-1}. Sample waters for nitrobenzene analysis were taken from the sampling port at an appropriate time interval. $Na_2S_2O_3$ was added to the sample immediately to quench possible further reaction by residual ozone. The temperature was maintained at 20±3 °C. Ozone decomposition experiments were performed using similar apparatus to that described above.

2.4. Feed water and analytical methods

Water samples used for this study were obtained from the Songhua River (Harbin, China), which is a water supply source for many northeastern cities of China. Since the traditional techniques such as coagulation and sedimentation are strictly connected with the transference of ozone from gas to solution phase, so water was pumped into the feed water tank for ozonation after the pretreatment of coagulation and sedimentation.

Water samples were analyzed for key water-quality parameters such as TOC, UV absorbance at 254 nm, NH_4^+ and NO_3^-. All analyses were performed following the Standard Methods. A TOC analyzer (Shimadzu TOC-VCPN, Japan) and a UV/VIS spectrophotometer (Shimadzu UV2550, Japan) were used in the analysis, respectively.

GC/MS apparatus (Agilent GC7890/MS5975) was used for qualitative analysis of organic matter. GC conditions are as follows: HP-5MS chromatographic column (30 m×250 μm×0.25 μm); column temperature is programming heated-up from 45 to 270 °C; detector temperature: 250 °C; injector temperature: 250 °C; carrier gas: N_2; split ratio: 20:1; shunt volume: 23.943 mL min^{-1}. MS conditions are as follows: ionization mode: electron bombardment; ion source temperature: 230 °C; scanning area: 45-450 m z^{-1}.

3. Results and discussion

3.1. Activity of nano TiO_2 supported by zeolite

The crystal structure and crystallite size of nano TiO_2 are important factors that determine catalytic activity, which is strongly influenced by heat treatment temperature. Table 1 showed the catalytic ozonation efficiency by the supported TiO_2/Zeolite catalyst calcined at different temperatures with nitrobenzene as indicator compound.

The supported TiO_2/Zeolite catalyst did increase the removal rate of nitrobenzene compared with single ozonation. And the heat treatment process enhanced the activity of catalyst by increasing the crystallinity of TiO_2 and the connection between TiO_2 and support. It is shown from table 1 that 600 °C is the best temperature for calcination. XRD analysis shows that TiO_2/Zeolite calcined at 600 °C were synthesized by a majority of rutile and the approximate crystallite size is 20-30 nm. X-ray energy spectrum analysis shows that the weight percentage of TiO_2 is about 8% as compared with the Zeolite support. Because of the best activity of the supported catalyst calcined at 600 °C, it was adopted for the latter research.

Table 1. Catalytic activity of TiO_2/Zeolite for nitrobenzene removal.

Process	O_3	O_3+C_1	O_3+C_2	O_3+C_3	O_3+C_4	O_3+C_5
NB removal, %	13.1	19.8	20.3	28.1	33.8	31.9

C_1: TiO_2/Zeolite calicined at 0 °C; C_2: TiO_2/Zeolite calicined at 400 °C
C_3: TiO_2/Zeolite calicined at 500 °C; C_4: TiO_2/Zeolite calicined at 600 °C
C_5: TiO_2/Zeolite calicined at 700 °C; O_3: 0.9mg·min^{-1}; Catalyst: 10g L^{-1}; Reaction time: 10 min; Temp: 20±3 °C; pH: 5.4~5.8

3.2. Catalytic ozonation of Songhua River for TOC and UV_{254} removal

The TOC and UV_{254} value of feed water approximately ranges from 7~13 mg L^{-1} and 0.08~0.13 cm^{-1}, which indicated that the Songhua River was obviously polluted by organic contaminants. The results are presented in table 2 and the data from the table clearly indicate a high catalytic activity of TiO_2/Zeolite in the process of TOC and UV_{254} due to ozonation.

Using ozonation alone with high ozone dosage, resulted in approximately 40% efficiency of TOC removal after 30 min ozonation time. About 15% TOC removal was achieved by adsorption of organic matter on catalyst surface. Ozonation in the presence of TiO_2/Zeolite gave as a result approximately 65% of TOC removal efficiency. This shows that catalytic ozonation increased 25% efficiency of TOC removal from water when compared with ozonation alone.

Despite the relatively low efficiency of dissolved organic carbon removal from water, ozonation alone resulted in the high efficiency of UV_{254} absorbance reduction of solution, 58% of UV_{254} absorbance reduction after 30 min contact time. Catalytic ozonation gave slightly better results: 73% of UV_{254} absorbance reduction after 30 min contact time. The high ability of ozone to lower UV_{254} absorbance of solution and its relatively low ability to degrade TOC can be explained by the fact that only a partial oxidation of organic matter, without organic matter conversion to carbon dioxide, takes place. This can be explained by the nature of natural organic matter, which is known to consist of aromatic and aliphatic unsaturated constituents. Because of the direct reaction of ozone molecules with natural organic matter, these components, which are responsible for the absorption of UV_{254} light, change into saturated, non-adsorbing UV_{254} light intermediates. In summary, the results showed that catalytic ozonation on TiO_2/Zeolite might be a successful tool for the removal of natural organic matter from water of high TOC value.

Table 2. Degradation of TOC and UV_{254} by TiO_2/Zeolite catalyzed ozonation.

Processes	O_3	C_4 adsorption	O_3+C_4
TOC removal, %	39.8	15.0	65.2
UV_{254} removal, %	57.9	9.0	72.5

C_4: TiO_2/Zeolite calicined at 600 °C; O_3: 19.8 mg min^{-1}; Catalyst: 100 g L^{-1}; Reaction time: 30 min; Temp: 20±3 °C; pH: 7.3~8.0.

3.3. Catalytic ozonation of Songhua River for ammonia and nitrate removal

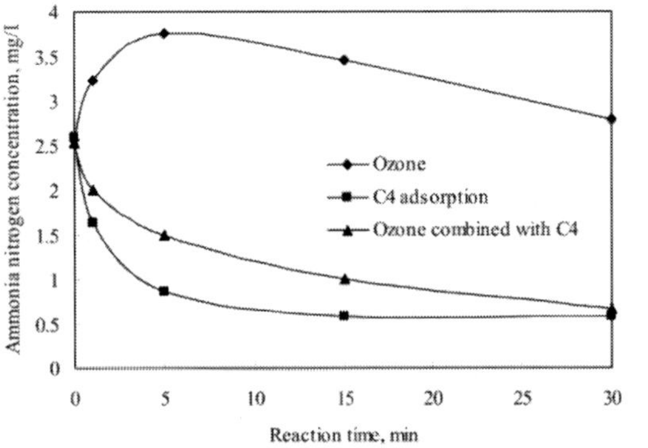

Figure 2. Variation of ammonia nitrogen concentration during catalytic ozonation.

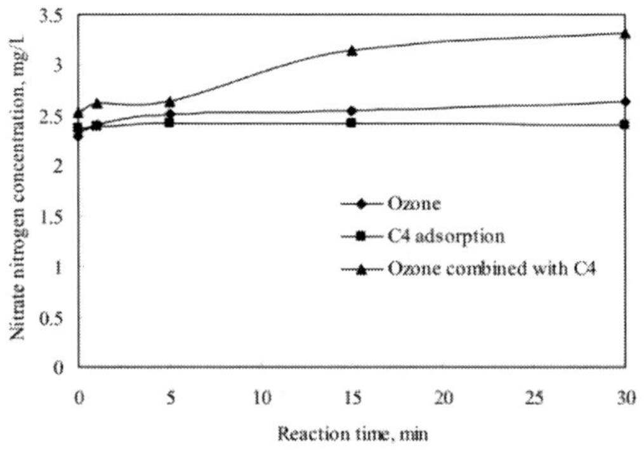

Figure 3. Variation of nitrate nitrogen concentration during catalytic ozonation.

Catalyst: TiO_2/Zeolite calicined at 600 °C; O_3: 19.8 mg min^{-1}; Catalyst: 100 g L^{-1}; Reaction time: 30 min; Temp: 20±3 °C; pH: 7.3~8.0

The concentration of ammonia nitrogen and nitrate nitrogen need to be controlled because of the possible eutrophication that may be caused by nitrogen. The results about the variation of ammonia nitrogen and nitrate nitrogen concentration during catalytic ozonation are shown in figure 2 and figure 3, respectively.

It is noticeable that single ozonation of the feed water caused the increase of ammonia nitrogen concentration at the first 5 min and then followed a decrease trend with reaction time. It is assumed that parts of organic nitrogen in the raw water were converted to inorganic nitrogen which is in the formation of ammonia by oxidizing action of ozone. That is the reason about the increase of ammonia concentration in the first stage of single ozonation process. And then ammonia was sequentially oxidized to nitrate or other form of inorganic nitrogen, thus lead to the following decrease of ammonia concentration as shown in figure 2.

Since zeolite has well adsorption ability and can do ionic exchange with ammonia in the solution, the ammonia nitrogen concentration decreased from 2.59 mg L^{-1} to 0.57 mg L^{-1} within 30 min of contacting time to TiO_2/Zeolite catalyst.

With the combination of ozone and TiO_2/Zeolite catalyst, ammonia was transferred from solution to the surface of zeolite support by adsorption effect and ion-exchange reaction on the one hand, on the other hand, organic nitrogen in the solution was oxidized by catalytic ozonation so caused the increase of ammonia concentration. As a result, the ammonia nitrogen concentration follows a slower decrease curve during catalytic ozonation process with comparison to adsorption process and then approach to the final value at about 30 min.

It can be seen from figure 3 that the nitrate concentration of the solution keeps unchanged during the adsorption process, but nitrate concentration increased from 2.3 to 2.6 mg L^{-1} in ozonation process and from 2.5 to 2.2 mg L^{-1} in catalytic ozonation process, respectively. That is to say, some other forms of nitrogen were converted into nitrate nitrogen, which indirectly testified the imagination above-mentioned.

3.4. GC-MS chromatographic analysis results about catalytic ozonation

Organic impurities in surface water supplies have always been one of major concerns in drinking water treatment. Sewage discharge without sufficient treatment has become the main source of organic impurities in surface water. The Songhua River was obviously polluted by sewage and industrial waste. GC-MS analysis showed that even the raw water from Songhua River was treated by coagulation and sedimentation, there still remained about 50 kinds of organic matters in the solution. It can be seen from figure 4 that the organic mainly are acid, ester, phthalate ester, heterocyclic hydrocarbon, and aliphatic hydrocarbon.

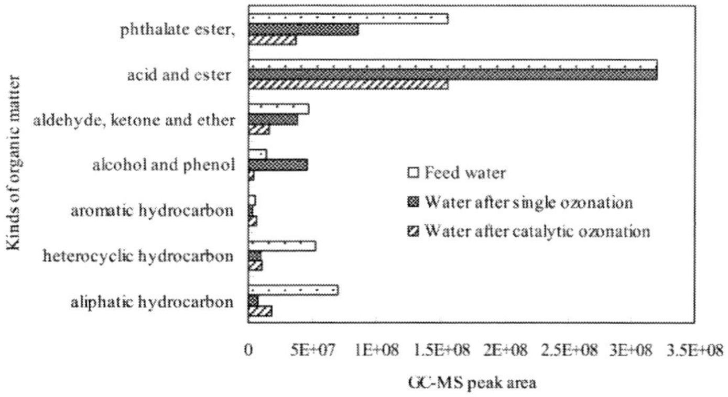

Figure 4. GC-MS chromatographic analysis results.

Catalyst: TiO_2/Zeolite calicined at 600 °C; O_3: 19.8 mg min^{-1}; Catalyst: 100 g L^{-1}; Reaction time: 30 min; Temp: 20±3 °C; pH: 7.3~8.0.

Both single ozonation and catalytic ozonation can decrease the organic content of the water. There were about 36 kinds of organic matters after the single ozonation process and the total organic peak area reduced by 23.5%. It is evident that catalytic ozonation process degraded the organic matter from water more efficiently. When the catalytic ozonation reaction carried 30 min, there still remained 20 kinds of organic matters which can be detected and the total organic peak area reduced by 62.5%.

4. Conclusions

Nanosized TiO_2 supported by zeolite was used in the ozonation process as catalyst. Nitrobenzene removal showed that TiO_2/Zeolite catalyst did increase the ozonation efficiency and 600 °C is the appropriate heat treatment temperature. Investigation about the TiO_2/Zeolite catalyzed ozonation of Songhua River was carried for TOC, UV_{254}, ammonia nitrogen and nitrate nitrogen removal. Catalytic ozonation gave as a result approximately 65% of TOC removal efficiency and 73% of UV_{254} removal efficiency. This shows that catalytic ozonation increased 25% efficiency of TOC removal and 15% of UV_{254} removal when compared with ozonation alone. The zeolite support shows good ability to remove ammonia nitrogen from water. It is assumed that some forms of organic nitrogen in the solution were oxidized into ammonia and then transferred into nitrate in catalytic ozonation process. GC-MS analysis results showed that about 30 kinds of organic matters were removed from water by catalytic ozonation and the total organic peak area was reduced by 62.5%. As a conclusion, TiO_2/Zeolite catalyzed ozonation is a promising deep treatment technique which can be used to control organic contaminant for drinking water.

Acknowledgments

Support from the Natural Natural Science Foundation of China (No. 51008052) and Hebei Provincial Natural Science Foundation of China (No. E2011501037) is greatly appreciated and acknowledged.

References

[1] Langlais B, Reckhow D A and Brink D R 1991 Ozone in Water Treatment: Application and Engineering (Lewis Publishers, Chelsea, Michigan, USA)
[2] Legube B, and Karpel Vel Leitner N 1999 *Catal. Today* **53** 61
[3] Ma J, and Graham Nigel J D 2000 *Water Res.* **34** 3822
[4] Beltra'n F J, Rivas F J and Montero-de-Espinosa R 2005 *Water Res.* **39** 3553
[5] Pines D S, and Reckhow D A 2002 *Environ. Sci.Technol.* **36** 4046
[6] Beltra'n F J, Rivas F J and Montero-de-Espinosa R 2003 *Ind. Eng. Chem. Res.* **42** 3210
[7] Beltra'n F J, Rivas F J and A'lvarez P 2002 *Ozone Sci. Eng.* **24** 227
[8] Fu H, Karpel Vel Leitner N and Legube B 2002 *New J. Chem.* **26** 1662
[9] Ping T S, Hu L W and Qing Z J 2002 *Ozone Sci. Eng.* **24** 117
[10] Yang Y X, Ma J and Zhang J 2009 *Ozone Sci. Eng.* **31** 45

Purification of water by bipolar pulsed discharge plasma combined with TiO_2 catalysis

Yongrui Zhang, Ruobing Zhang* (Member, IEEE), Wenchang Ma, Xian Zhang, Liming Wang, and Zhicheng Guan (Member, IEEE)
Graduate School at Shenzhen, Tsinghua University, Shenzhen 518055, China

E-mail: zhangrb@sz.tsinghua.edu.cn

Abstract. In the process of water treatment by bipolar pulsed discharge plasma, there are not only the chemical effects such as the cold plasma, but also the physical effects such as the optical radiation. The energy of the optical radiation can be used by photocatalyst. Therefore, the effect of the photocatalyst to the degradation of the organic pollutant was investigated using a packed bed reactor by bipolar pulsed discharge in the air-liquid-solid mixture. The nanoparticle TiO_2 photocatalyst was obtained using the sol-gel method and the typical dye solution Indigo Carmine was chosen as the degradation target to test the catalytic effect of the nanoparticle T_iO_2 photocatalyst. Experiment results proved that the degradation efficiency of the Indigo Carmine solution was increased by a certain extent with the T_iO_2 photocatalyst. It was totally decolorized within 3 minutes by bipolar pulsed discharge in the condition that the peak voltage was 30 kV and the air flow was 1.0 m^3 h^{-1}.

1. Introduction

The water treatment by pulsed discharge plasma is a new advanced oxidation technology which includes the features of various advanced oxidation treatment such as electrics, chemistry, optical radiation etc. [1]. Because of several obvious advantages such as mild operating conditions, high processing efficiency, lack of secondary pollution, and non-selectivity for the degradation of organic pollutants, this technology attracts more and more attention and has a broad application prospect [2-6].

Most of previous researches were concentrated on the development of power supply like the DC power and unipolar pulsed power [7-8], or on the development of discharge reactor such as the needle-plate electrode reactor and the coaxial electrode reactor etc. [9]. Besides, the effects of discharge plasma in air and in air-liquid mixture have been extensively studied [10-14]. In our previous studies, the bipolar pulsed discharge in air-liquid-solid mixture system had been proved to be very efficient in decoloration of the dye solution when the "solid" in the system comprises a quantity of glass pellets [15-16]. To make the most of the physical and chemical properties especially the light emission in the reactor to achieve greater removal of organic pollutants and decrease the energy consumption of the power supply, the research of the photocatalyst which could be applied in the discharge plasma is necessary.

In this experiment, the packed-bed dielectric barrier discharge reactor with coaxial electrode configuration was adopted, which takes use of the bipolar pulsed high voltage power supply. A typical dye wastewater Indigo Carmine was chosen to study the purification effect by

* To whom any correspondence should be addressed.

bipolar pulsed discharge plasma in the reactor with changed parameters like the peak voltage and air flow, especially to analyze the actual catalysis of the TiO_2 photocatalyst.

2. Experiment system and method

2.1 Experiment system

The experimental setup system is shown in figure 1, which consists of the bipolar pulsed power supply, discharge plasma reactor, the air and water supply system and electrical detection system.

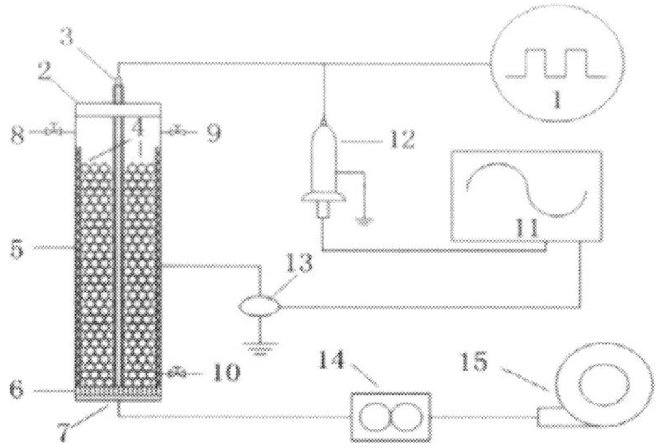

1. Bipolar pulsed power supply; 2.Reactor; 3.High voltage electrode ; 4.Packing layer ; 5.Ground electrode ; 6.Air plate; 7.Air inlet ; 8.Air outlet ; 9.Water inlet ; 10.Water outlet11.Oscilloscope ; 12.High-voltage probe ; 13.Current probe ; 14.Air flow meter; 15. Air pump

Figure1. Experimental setup system.

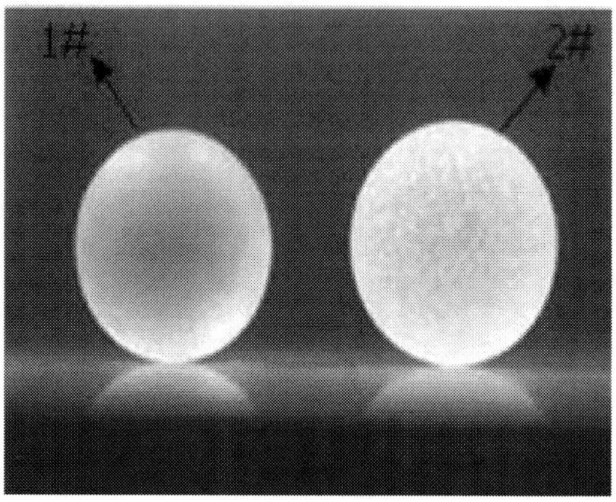

Figure 2. Photo of packing glass pellets.

2.2 Experimental method

The pulsed power frequency was 50 Hz and three different applied peak voltages (20 kV, 25 kV, 30 kV) were chosen. Three air flow rates of 0.5 $m^3 h^{-1}$, 0.75 $m^3 h^{-1}$ and 1.0 $m^3 h^{-1}$ were adjusted for the experiment. Our previous experiments had shown that when the diameter of glass pellets was 9 mm, the discharge characteristic in the reactor was better than the status with other diameters [17]. Therefore, the 9 mm's glass pellets were chosen as the packing layer and two kinds of glass pellets with different surface roughness (labeled 1# and 2# in figure 2) were selected to compare the adhering condition of the nano-TiO_2 film on the surface. The calculated surface roughness data acquired by the provided graphic and analysis tools of MicroXAM-3D was shown in table 1.

Table 1. Calculated surface roughness data

Surface morphological characteristics	Roughness Average, μm	Root Mean Square, μm	Peak-Peak, μm	Density of Summits,μm^{-2}
Data 1#	0.771	0.894	8.12	0.305
Data 2#	6.83	8.07	77.0	0.202

The sol-gel method was used to prepare the nanoparticle TiO_2 photocatalyst and its catalytic effect was obtained by measuring the absorbance of the target dye solution after being treated in the discharge plasma reactor using the spectrophotometry.

3. Preparation and characterization of photocatalyst

3.1 Catalytic Mechanism of TiO_2 Photocatalyst

When the discharge happened, the cold plasma along with the optical radiation was generated in the reactor. The electron-hole pair was produced when the TiO_2 film absorbed the light emission. On the surface of the glass pellets, a series of reactions were happened as the following equations showed.

$$TiO_2 + hv \longrightarrow TiO_2 + h^+ + e^- \tag{1}$$

$$H_2O + h^+ \longrightarrow \bullet OH + H^+ \tag{2}$$

$$OH^- + h^+ \longrightarrow \bullet OH \tag{3}$$

$$O_2 + e^- \longrightarrow \bullet O_2^- \tag{4}$$

$$O_2 + H^+ \longrightarrow \bullet HO_2 \tag{5}$$

$$H_2O + \bullet O_2^- \longrightarrow \bullet OOH + OH^- \tag{6}$$

$$2 \bullet OOH \longrightarrow H_2O_2 + O_2 \tag{7}$$

$$\bullet OOH + H_2O + e^- \longrightarrow H_2O_2 + OH^- \tag{8}$$

$$H_2O_2 + e^- \longrightarrow \bullet OH + OH^- \tag{9}$$

$$H_2O_2 + \bullet O_2^- \longrightarrow \bullet OH + OH^- \tag{10}$$

Through the reactions above, several very active species were produced like the $\bullet OH$, $\bullet O_2^-$ and $\bullet HO_2$, which could oxidize the organic pollutants into small molecules because of their strong oxidizability. This was the catalytic mechanism of TiO_2 photocatalyst.

3.2 Pretreatment of Glass Pellets

To make the glass pellets clean enough, they were pretreated as the following steps: clean sufficient number of glass pellets three times with deionized water and place them in ultrasonic cleaning machine for 30 minutes; soak the pellets respectively in 15% hydrochloric acid solution and 20% sodium hydroxide solution for one hour and wash them with deionized water; clean the pellets with ultrasonic cleaning machine for 10 minutes and put them in the drying oven for later use.

3.3 Preparation of TiO_2 Colloidal Sol

Exactly put 85.1 mL tetrabutyl titanate ($Ti(OBu)_4$) into 175 mL anhydrous ethanol (EtOH), then added 12.8 mL acetyl acetone (AcAc) and 2.25 mL 65% nitric acid(HNO_3)into the solution as the inhibitor to restrain the strong hydrolysis of $Ti(OBu)_4$ and had the mixture strongly stirred. After one hour, added a mixture of 9 mL water (H_2O) and 87.5 mL anhydrous ethanol with a burette. Stirred the solution strongly for another 2 hours and had the solution static stand for 24 hours at room temperature. After the steps above, a beaker of yellow uniform transparent TiO_2 sol was prepared, in which the concentration of titanium ion is about 0.65 mol L^{-1}. The molar ratio of the substances is $Ti(OBu)_4$:EtOH :H_2O:HNO_3:AcAc=1:18:2:0.2:0.5. The sol prepared by this method was very stable that could be kept in a closed environment at room temperature for about one year.

3.4 Preparation of TiO_2 Photocatalyst

Put the prepared glass pellets into a funnel of 12 cm in diameter and immersed the funnel into the prepared sol for 3 minutes so that the glass pellets were coated with sol film. Put the pellets in glass dishes and dried them in a drying cabinet at 110 ℃ for 30 minutes, then took them out to reach the room temperature. Repeat the steps above for three times to ensure that the sol film was thick enough. Roasted the glass pellets in an resistor furnace at 300 for 30 minutes and then at 550 ℃ for 2 hours. Took the glass pellets out and let them be natural cooling to room temperature. After all the steps, the glass pellets were coated with TiO_2 film on the surface.

3.5 Characterization of TiO_2 Photocatalyst

3.5.1 Surface Morphology of Glass Pellets

The glass pellets coated with TiO_2 film were shown in figure 3. The adhering condition of TiO_2 on the surface of glass pellets labeled 2# was better than that of labeled 1# because of the bigger roughness and the area and thickness of TiO_2 film on the surface of glass pellets coated twice were better than that coated once.

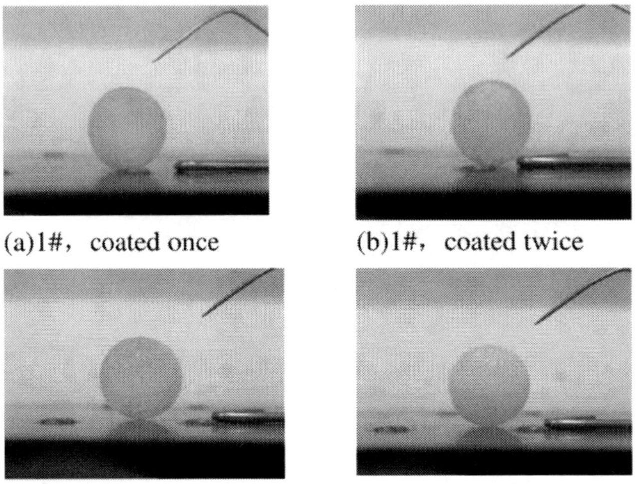

(a)1#, coated once (b)1#, coated twice
(c)2#, coated once (d)2#, coated twice

Figure 3. Packing glass pellets coated with TiO_2

To observe the micromorphology on the surface of the glass pellets when coated with TiO_2 film, the scanning electron microscope was used in the experiment. The SEM images of the glass pellets were shown in figure 4.

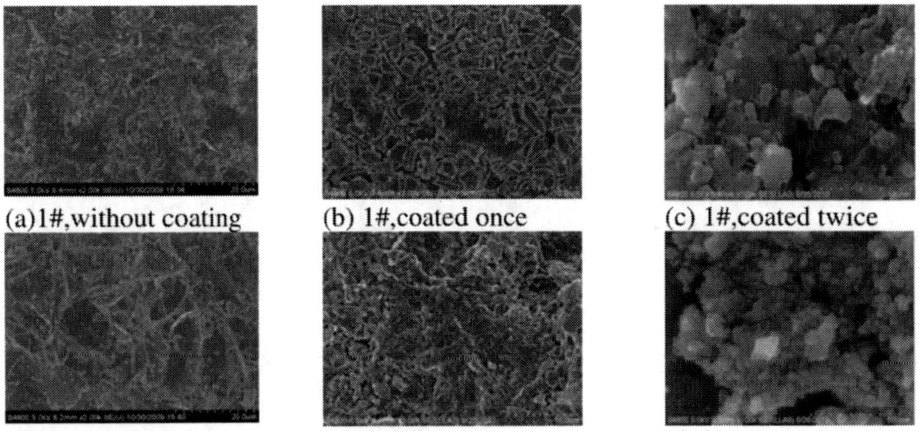

(a)1#,without coating (b) 1#,coated once (c) 1#,coated twice
(d)2#,without coating (e) 1#,coated once (f) 2#,coated twice

Figure 4. SEM images of glass pellets

It should be specially explained that the SEM images (c) and (f) of the glass pellets' surface shown in figure 4 were magnified for 100 thousand times. It can be seen from the images that the size of TiO_2 particles, which were aggregated by some smaller particles, was about tens to hundreds of nanometers.

3.5.2 Analysis by XRD

XRD is short for X-ray diffraction, which is used to obtain the information such as composition, internal atomic or molecular structure and morphology by analyzing the X-ray diffraction pattern of target material. The diffraction pattern of TiO_2 film prepared with sol-gel method was shown in figure 5.

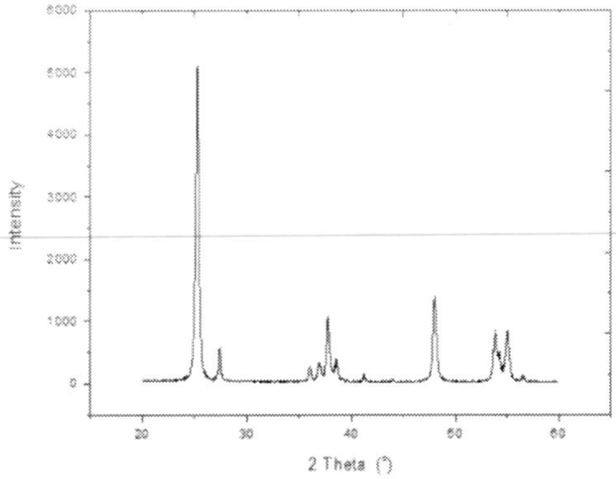

Figure 5. X-ray diffraction pattern of TiO_2 film

As shown in figure 5, there was a strong diffraction peak at the spot where $2\theta = 25.3^o$, which was the characteristic diffraction peak of anatase titanium dioxide's lattice plane; and there was another strong diffraction peak at the spot where $2\theta = 27.4^o$, which was the characteristic diffraction peak of rutile titanium dioxide's lattice plane. These results indicated that the TiO_2 film obtained was composed by two kinds of crystal grains: anatase titanium dioxide and rutile titanium dioxide.

The Scherrer equation can be used to calculate the average size of nano-TiO_2 particles through the X-ray diffraction pattern.

$$D = \frac{K\lambda}{B\cos\theta} \tag{11}$$

In equation (11), D represents the average size of nano-TiO_2 particles (nm); K represents the Scherrer constant, which is 0.89 in usual cases; λ represents the wavelength of the X-ray (nm); B represents the full width at half characteristic peak (rad); θ represents the Prague diffraction angle (°). Take all the parameters into equation (11):

$$D = \frac{0.89 \times 0.15418}{(25.44 - 25.10) \times \frac{\pi}{180} \times \cos\frac{25.3^o}{2}} \approx 23.7nm$$

The calculated value showed that the average diameter of the nano-TiO_2 particles was 23.7 nm.

4. Results

4.1 Comparison between Loaded TiO_2 and PowderTiO_2

The comparison of catalytic treatment effect between loaded TiO_2 photocatalyst and powder TiO_2 in the condition that the voltage was 25 kV and the air flow rate was 0.75 m^3 h^{-1} was shown in figure 6. The catalytic degradation effect of indigo solution by adding powder TiO_2 photocatalyst was almost the same with the effect by adding loaded TiO_2 photocatalyst. The method of adding powder TiO_2 was simple and convenient but had the serious disadvantage that the catalyst cannot be recycled which led to a secondary pollution to the environment. In contrast, although the loaded TiO_2 photocatalyst was prepared a little complexly, it overcame the very disadvantage which made it have a broader space for development. Figure 6 proved that the loaded TiO_2 photocatalyst on the surface of the packing glass pellets we studied in this experiment was very meaningful.

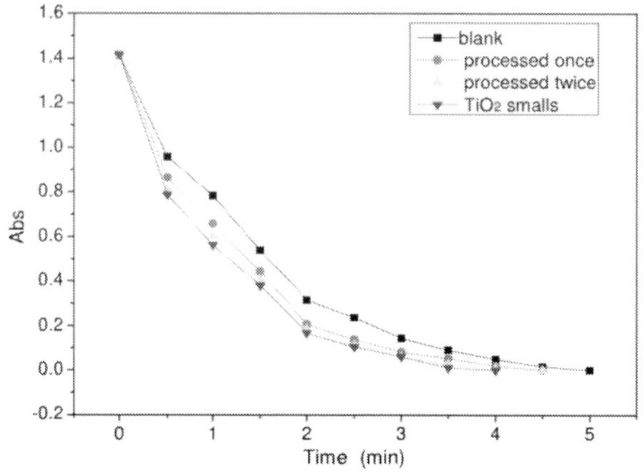

Figure 6. Comparison of catalytic treatment between loaded TiO_2 photocatalyst and powder TiO_2

4.2 Influence of Amount of TiO_2 Photocatalyst

The amount of TiO_2 photocatalyst loading on the surface of glass pellets was an important factor to the catalytic treatment. The glass pellets coated twice had more catalyst on the surface than that coated once. The curves that the time varying absorbance of Indigo solution measured in the condition that air flow rate was 0.5 $m^3 h^{-1}$ and voltage was 30 kV or 20 kV were shown in figure 7.

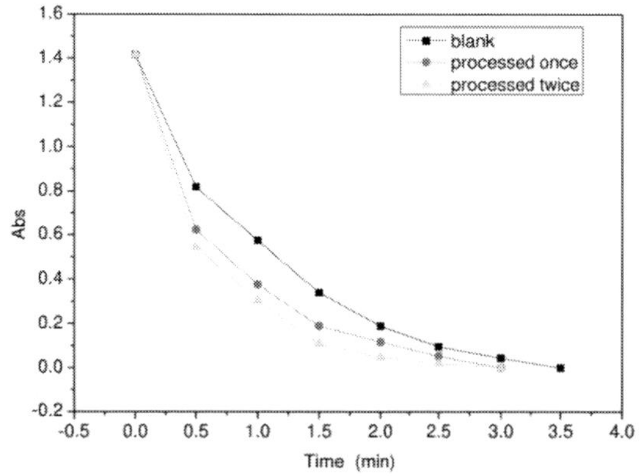

(a) The voltage was 30 kV

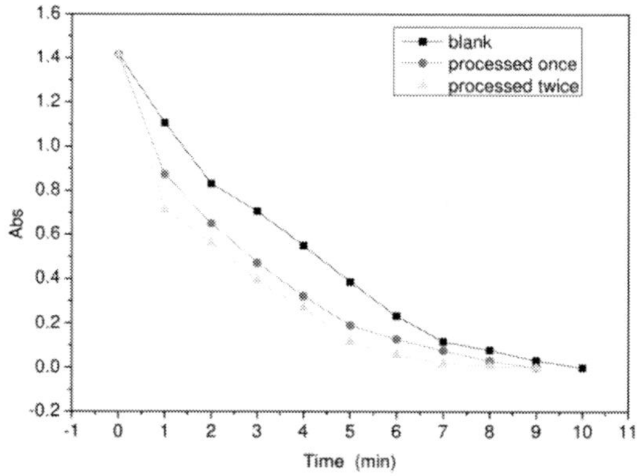

(b) The voltage was 20 kV

Figure 7. Influence of amount of TiO_2 photocatalyst

As shown in figure 7, in comparison with the condition without catalyst, the degradation rate of dye solution increased obviously when coated with TiO_2 photocatalyst and the catalytic effect was a little better when the glass pellets were coated twice than that coated once.

Table 2. Detailed data of degradation ratio

		Without catalyst	Coated once	Coated twice
0.5 $m^3 h^{-1}$	20 kV	41.26%	53.90%	60.38%
0.5 $m^3 h^{-1}$	30 kV	86.72%	91.78%	96.66%

Table 2 showed the degradation ratio of indigo solution after two minutes discharge treatment. When the voltage was 30 kV, the degradation ratio of indigo solution reached 96.66% in the condition that the glass pellets packed in the plasma reactor were coated twice, which suggested that the catalytic effect of the nano-TiO_2 photocatalyst was satisfying.

4.3 Influence of Applied Voltage

The voltage of the power supply was an important factor to the catalytic treatment and its influence was shown in figure 8 with the air flow rate of 1.0 $m^3\ h^{-1}$. As the voltage increased, the degradation rate of indigo solution was accelerated and the degradation time was shorten, which suggested that nano-TiO_2 photocatalyst achieved an obvious catalytic effect in the degradation of the dye solution. But the enhanced extent was different in different voltages. The enhanced extent of catalytic effect decreased a little along with the increase of voltage, this was because the discharge was very serious at a high voltage and many indigo molecules had already been resolved in the strong electric field. As a result, the number of indigo molecules that can reach the center of catalyst reduced.

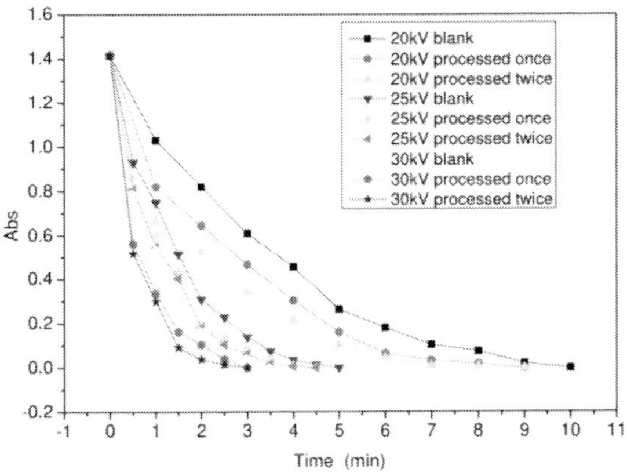

Figure 8. Influence of applied voltage

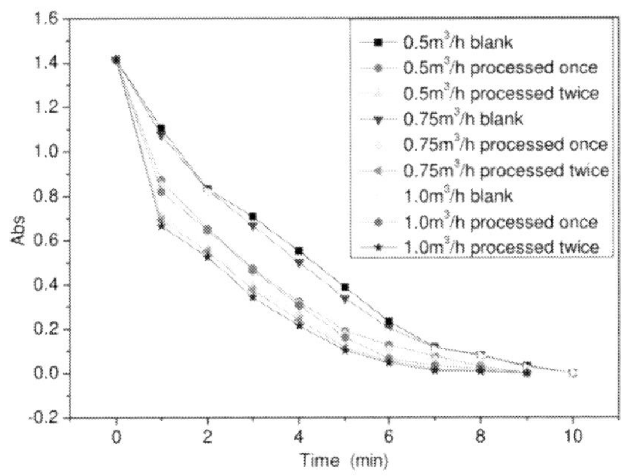

Figure 9. Influence of air flow rate

4.4 Influence of Air Flow Rate

The influence of air flow rate to the catalytic treatment effect of nano-TiO_2 photocatalyst in the condition of 20 kV voltage was shown in figure 9. The degradation effect of indigo solution enhanced along with the increase of air flow rate but the enhanced extent was not that much. The promotion of degradation effect of indigo solution by increasing the air flow rate was far less than that by adding TiO_2 photocatalyst. Possible reason was that when the air flow rate increased, the number of bubbles increased and the discharge enhanced in the solution. As a result, the intensity of ultraviolet light generated by the discharge promoted and the catalysis of TiO_2 photocatalyst enhanced by absorbing more ultraviolet light.

5. Discussion

The TiO_2 film on the surface of glass pellets prepared by the sol-gel method was composed by two kinds of crystal grains: anatase titanium dioxide and rutile titanium dioxide, and some anatase TiO_2 will change into rutile TiO_2 at 550 ℃. Overall, the adhering condition of TiO_2 on the surface of 2#glass pellets was better than 1# and the amount of TiO_2 film on the surface of glass pellets coated twice was larger than that coated once. From the X-ray diffraction analysis we could calculated that the average diameter of the nano-TiO_2 particles was 23.7 nm, which was the very original size. However, the particles observed by SEM about tens to hundreds of nanometers were aggregated by some particles with smaller size.

Based on the catalytic mechanism of TiO_2 photocatalyst and the surface morphology of the packing glass pellets by the SEM images, it seemed that the catalytic effect by adding powder TiO_2 should be better than the effect by using loaded TiO_2 photocatalyst. The result in our experiment showed that the two different patterns had almost the same catalytic effect. Possible explain was that the powder TiO_2 dispersed in the discharge plasma reactor lowered the optical transmissivity of the dye solution so that the strongest discharge zone, i.e. the surface of packing glass pellets, absorbed less optical radiation. This negative effect counteracted its original advantage. What should be noticed was that the method of adding powder TiO_2 was simple and convenient but had the serious disadvantage that the catalyst cannot be recycled which led to a secondary pollution to the environment. In contrast, although the loaded TiO_2 photocatalyst was prepared a little complexly, it overcame the very disadvantage which made it have a broader space for development.

6. Conclusion

The purification of dye solution by bipolar pulsed discharge plasma in the air-liquid-solid mixture combined with TiO_2 photocatalyst was investigated using a coaxial electrode configuration. The experiment results showed that the degradation effect was enhanced obviously by the loaded TiO_2 photocatalyst on the surface of packing glass pellets. The catalysis of namo-TiO_2 photocatalyst was influenced by voltage and air flow rate. The enhanced extent of catalytic effect to the indigo solution decreased a little along with the increase of voltage and the degradation effect of indigo solution enhanced along with the increase of air flow. The promotion of degradation effect of indigo solution by adding TiO_2 photocatalyst was far more than that by increasing the air flow rate but a little less than that by increasing the voltage.

References

[1] Locke B R, Sato M , Sunka P, et al.2006 *Ind Eng Chem.* **45** 882
[2] Lukes P, Clupek M 2008 *IEEE Trans. Plasma Sci.* **36** 1146
[3] Zhang R B 2007 Bipolar pulsed discharge in water-solid particle mixture and its application [D]. Beijing: Tsinghua University
[4] Malik M A , Ghaffar A and Malik S A 2001 *Plasma Sources Sci. Tech.***10** 82
[5] Sun B, Sato M and Clements J S 2000 *Environ. Sci. Technol.* **34** 509
[6] Zhang R B, Wu Y, Li J et al. 2004 *J. Adv. Oxidation. Technol.* **7** 172
[7] Jiang Y, Wen J P and Hu Z D 2004 *Chem. Eng. Prog.* **23** 256
[8] Arif M M, Ubaid-Ur R and Ghaffar A, et al. 2002 *Plasma Sources Sci. Technol.* **11** 236
[9] Malik M.A. et al. 2003 *Plasma Sources Sci. Technol.* **12** 26
[10] Hoeben W F L M, Veldhuizen E M van, Rutgers W R, et al. 1999 *J. Phys. D: Appl. Phys.***32** 133
[11] Birmingham J G, Hammerstro D J 2000 *IEEE Trans. Plasma Sci.* **28** 51
[12] Akiyama H. 2000 *IEEE Trans. Dielectr. Electr. Insul.* **7** 646
[13] Sunka P, Clupek M, Lukes P, et al. 1999 *Plasma Sources Sci. Technol.* **8** 258
[14] Anto T S, Takayuki O and Sato M 2002 *Thin Solid Films* **407** 174
[15] Zhang R B, Wu Y, Li G F, Wang N H and Li J 2004 *J. Environ. Sci.* **16** 808
[16] Zhang C, Zhang R B, Wang L M and Guan Z C 2010 *IEEE Trans. Plasma Sci.* **38** 100
[17] Chen X X, Zhang C and Zhang R B 2008 *High Voltage Technology (in Chinese).* **34** 401

Journal of Physics: Conference Series **418** (2013) 012126 doi:10.1088/1742-6596/418/1/012126

Determination of potassium and sodium ions with diaphragm glow discharge plasma in aqueous solution

Yongjun Liu[1*], Bing Sun[1], Lei Wang[2]
[1]College of Environmental Science & Engineering, Dalian Maritime University, Dalian 116026, People's Republic of China
[2]Water Resources & Environmental Institute, Xiamen University of Technology, Xiamen 361005, People's Republic of China

E-mail: lyjglow@sohu.com

Abstract. In the present work, a new apparatus of diaphragm glow discharge emission spectroscopy was described for the determination of potassium and sodium ions in aqueous solution. The discharge was formed in a pin hole on a dielectric diaphragm interposed between two submerged graphite electrodes. Effects of pH and applied voltage on the determination have been examined. It was found that decreasing the solution pH and increasing the applied voltage were favorable for the determination performance. Limits of detection for Na and K were 0.002 and 0.05 mg L^{-1} under the optimum conditions, respectively. It demonstrates that the diaphragm glow discharge emission spectroscopy is a promising technique in measurements of metal ions in aqueous solution, because no optical interferences from the electrodes were found.

1. Introduction

Recently, electrolyte as cathode glow discharge (ELCAD) has been extensively investigated as a novel analytical tool to determine the concentrations of trace metals in aqueous solutions [1-8]. In ELCAD, a rod anode is placed over the surface of an aqueous solution, and a sheath of plasma is sustained between the anode tip and the solution as a high DC voltage is applied to the system. During ELCAD, the metal ions in the solution were sputtered to the plasma zone and were excited by three body reactions. These excited atoms will spontaneously radiate the characteristic atomic lines. As the intensity of these lines was often proportional to the concentration of the metal ions in the solution, ELCAD can be used for the quantitatively analyses of metal elements in liquids [3-8]. The most obvious advantage of ELCAD is its instrumental simplicity and high sensitivity. However, this process uses tungsten or platinum as the working electrode, which interference the determination because of thermal radiation [9]. Previously, we designed a new diaphragm glow discharge plasma reactor in which the electrodes do not directly contact with the highly reactive plasma. The discharge mode used was termed as liquid diaphragm glow discharge (LDGD) [10]. Experimental results showed that the LDGD reactor could be operated in electrolytic solution with no direct electrode contact. Considering the problem of ELCAD, we attempted to use LDGD for the detection of aqueous Na and K.

[*] To whom any correspondence should be addressed.

Journal of Physics: Conference Series **418** (2013) 012126 doi:10.1088/1742-6596/418/1/012126

2. Experimental

Experimental apparatus consisting of a discharge reactor, a dc power supply and a series of measuring instruments were shown in figure1.

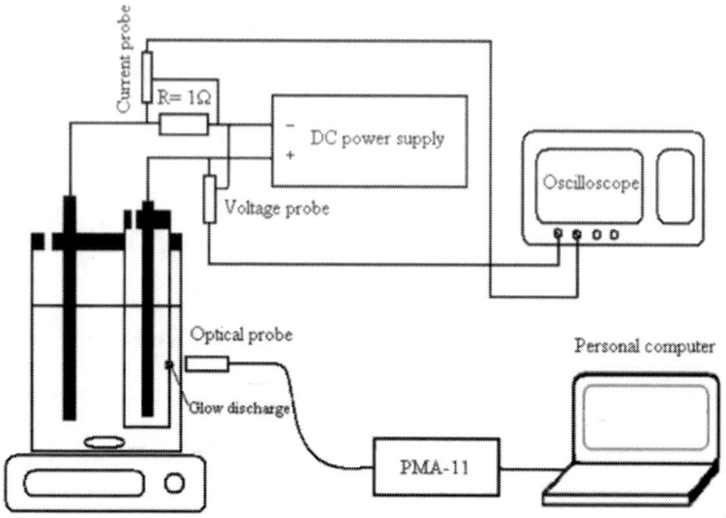

Figure 1. Experimental apparatus for the optical determination of Na and K.

The discharge reactor was a quartz cylinder with 60 mm inner diameter containing two electrodes. The anode was a graphite rod (diameter: 0.5 cm) put into a quartz tube with a pin hole in the tube, which was immersed in 100 ml of dilute sulfuric acid solution. The cathode was placed in the reactor and was separated from the anode by the quartz tube.

A dc voltage varied from 0 to 1000 V was applied across the anode and the cathode. The bulk solution in the reactor was maintained at 293±2 K by running cooling water in the outer water jacket, in order to avoid the variation in discharge characteristics. A quartz fiber probe was located 10 mm vertically from the pin hole of the quartz tube, where the optical intensity emitting from the discharge was measured using a photo-multichannel analyzer (PMA–11, Hamamatsu Photonics Co., Ltd) with a wavelength ranging from 200 to 900 nm, where the measurement was repeated 20 times and the average value was recorded.

3. Results and discussion

3.1 Emission spectra

When the voltage was applied to electrodes, emissions were observed by the PMA–11 when the applied voltage was over 700 V. The emission spectra between 200 and 900 nm in 0.04 M H_2SO_4 solution containing 200 mg L^{-1} K^+ and 100 mg L^{-1} Na^+ at applied voltage of 850 V were shown in figure 2.

As indicated in figure 2, there are several lines and bands present in the spectra. The bands in the wavelength from 250 to 340 nm are due to the emissions of excited OH radicals. The main transitions of the OH radicals generated in the discharge are from $A^2\Sigma^+ \rightarrow X^2\Pi$ [9]. The emission line at 656.3 nm were ascribed to the H_α (n=3 to n=2). Emission lines at 589 and 766.5 nm were attributed to the radiations of Na ($2p^63p \rightarrow 2p^63s$) and K ($3p^64p$-$3p^64s$) atoms [11].

H and OH are generated from the decomposition of H_2O. Although a high concentration of SO_4^{2-} ions are present in the solution, no emissions from S or SO were observed in the emission spectra. At the same time, no thermal emission was observed from the electrode [9]. The present investigation indicated that LDGD is a good tool for detection of aqueous metal ions because the discharge is simple and no emissions of the supporting electrolyte and of the electrodes were present.

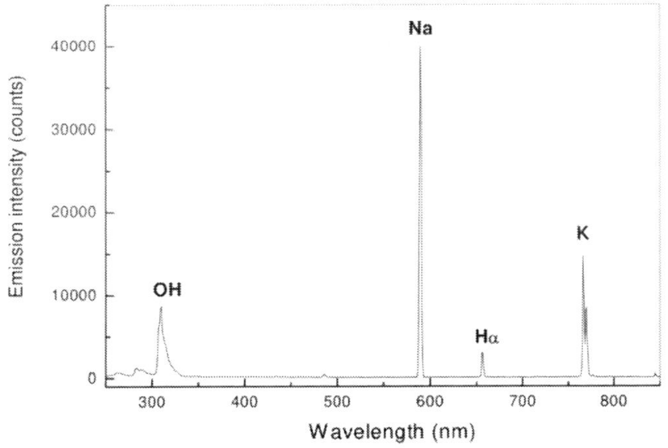

Figure 2. Emission spectra of LDGD ([H_2SO_4]: 0.04 M; [K^+]: 200 mg L^{-1}; [Na^+]: 100 mg L^{-1}; voltage: 850 V; current: 340 mA)

3.2 Effects of experimental variables on the determination

The emission intensities of the K (766.5 nm) and Na (589 nm) increased with the concentration of these ions contained in the solution. As the emission intensity seems to increase linearly with the concentration in solution, the emission intensity with the concentration was fitted with the following equation:

$$I = A + Sc \tag{1}$$

where I denote the emission intensity (counts) and c is the concentration of metal ion in the solution. A is the background emission intensity and S is the detection sensitivity. The results at applied voltage 850 V are shown in figure 3.

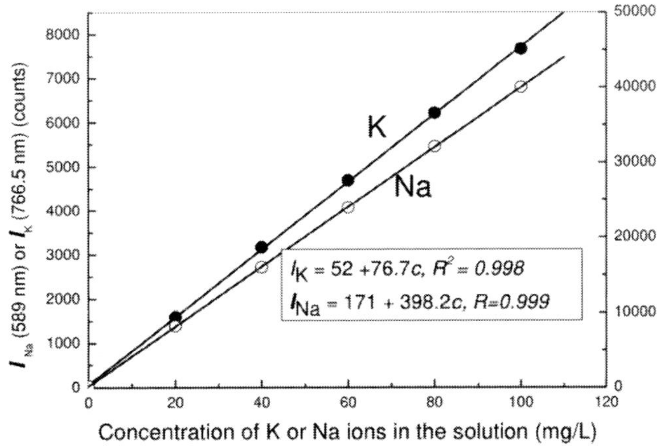

Figure 3. Calibration curve of the emission intensity versus concentration of K or Na ions in the solution ([H_2SO_4], 0.04 M; voltage, 850 V; current, 340 mA).

As shown in figure 3, straight lines with good correlation were obtained for both the K and Na ions.

The square regression coefficients R^2 were all close to 1 suggests that the emission intensity at 589 and 766.5 nm changed linearly to the concentration of sodium ion and potassium ion. So, the emission intensity at 589 and 766.5 nm can be used as the analytical signal for the determination of Na ion and K ion in the solution, respectively.

In order to ensure the maximum possible sensitivity and accuracy, effects of experimental variables on the determinations of Na and K were examined. The effects of applied voltage on the determinations of Na and K ions were shown in table 1.

Table 1. Effects of applied voltage on the determination of Na and K ions ($[H_2SO_4]$: 0.04 M)

Voltage and current (V/ mA)	Determination of $[Na^+]$ (mg L^{-1})			Determination of $[K^+]$ (mg L^{-1})		
	Calibration curve	R^2	Detection limit (mg L^{-1})	Calibration curve	R^2	Detection limit (mg L^{-1})
750/254	63 + 135 c	0.997	0.08	32 + 15.4c	0.995	0.23
800/310	137 + 253 c	0.998	0.01	37 + 38.3c	0.996	0.15
850/340	171 + 398.2 c	0.999	0.004	52 + 76.7c	0.998	0.09
900/410	183 + 431.4 c	0.998	0.002	69 + 94.3c	0.999	0.05
950/460	194 + 442.6 c	0.997	0.002	79 + 100.4c	0.997	0.06

As indicated in table 1, all the determination sensitivities increased with increasing applied voltage. For example, when the applied voltage increased from 750 to 950 V, the determination sensitivity of Na at 589 nm increased from 63 to 194 counts per mg L^{-1}, while the sensitivity of K at 766.5 nm increased from 32 to 79 counts per mg L^{-1}. The detection limits can be obtained at 900 V, where 0.002 mg L^{-1} for Na and 0.05 mg L^{-1} for K.

Cserfalvi et al. [2] observed that the emission intensity of the lines depend strongly on pH of the water sample in ELCAD. Since the plasma generation mechanism is similar to that of ELCAD, effects of pH on the detection sensitivity were shown in figure 4.

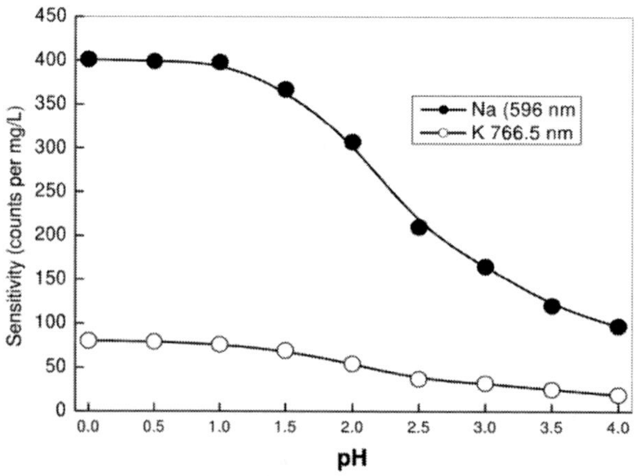

Figure 4. Effects of pH on the detection sensitivity of Na and K ions (Applied voltage, 850 V; current, 340 mA).

As indicated in figure 4, the determination sensitivity decreased with increasing pH. However, sensitivity did not increase further when the pH decreased to 1.0. Therefore, pH=1.0 maybe the optimum condition in the present investigations.

4. Conclusions

Decreasing the solution pH and increasing the applied voltage were favorable for the determination performance. Limits of detection for Na and K were 0.002 and 0.05 mg mg L^{-1} under conditions of 900 V and pH < 1. The diaphragm glow discharge emission spectroscopy was a promising technique in measurements of metal ions in aqueous solution, because no optical interferences from the electrodes were present and the good detection sensitivity.

Acknowledgement

This work was supported in part by the National Science Foundation of China (11005014, 20807008 and 51008262) and the Fundamental Research Funds for the Central Universities (2011JC016). The Natural Science Foundation of Fujian province, China (2010J05035), Open Foundation of the Key Laboratory of Marine Spill Oil Identification and Damage Assessment Technology (201016) and the Science-Technology Foundation of Education Bureau of Fujian Province, China (JA10247) are also acknowledged.

References

[1] Mezei P and Cserfalvi T 2007 *Appl. Spectrosc. Rev.* **42** 573
[2] Cserfalvi T, Mezei P and Apai P 1993 *J. Phys. D: Appl. Phys.* **26** 2184
[3] Park Y S, Ku S H, Hong S H, Kim H J and Piepmeier E H 1998 *Spectrochimica Acta B* **53** 1167
[4] Kim H J, Lee J H, Kim M Y, Cserfalvi T and Mezei P 2000 *Spectrochimica Acta B* **55** 823
[5] Davis W C and Marcus R K 2001 *J. Anal. At. Spectrom.* **16** 931
[6] Webb M R, Andrade F J and Hieftje G M 2007 *J. Anal. At. Spectrom.* **22** 766
[7] Webb M R, Andrade F J and Hieftje G M 2007 *Anal. Chem.* **79** 7899
[8] Zhu Z, Chan G, Ray S J, Zhang X and Hieftje G M 2008 *Anal. Chem.* **80** 7043
[9] Liu Y, Sun B, Wang L and Wang D 2012 *Plasma Chem. Plasma Process.* **32** 359
[10] Liu Y J and Jiang X Z 2005 *Environ. Sci. Technol.* **39** 8512
[11] http:// physics. nist. gov/ PhysRefData/ ASD/ lines_ form. htm

Experimental study on scale removal from special-shaped conduits through underwater electrical discharge

Y Cao*, Z Q Wang, G F Li, Y Wu, J J Zhou
Address: College of Electrical Engineering, Dalian University of Technology, No.2 Linggong Road, Ganjingzi District, Dalian City, China

E-mail: cayo@163.com

Abstract. Underwater electrical discharge technology is an innovative technique that can be used to enhance the stress intensity of water and improve the load addition. The technique enlarges the section area and compresses the surrounding water using a high-powered shock wave, which is induced by an underwater electrical discharge. This paper investigates the effectiveness of scale removal for special-shaped conduits employing underwater electrical discharge. Experimental results show that the pressure wave generated by underwater electrical discharge is capable of eliminating scale in special-shaped conduits. The data indicates that when the capacitance of the parallel-pulsed capacitors was 4 μF, the high pulsed power voltage was 33 kV and the primary discharge gap was 48 mm, the result of scale removal was remarkable. In laboratory tests, the scale of special equipment was removed to a great extent by this method. Because of its effectiveness and low cost, this method improves the practice and extends the lifetime of such equipment, and thus has potential application and economic value.

1. Introduction

Underwater electrical discharge is a process of instantaneous strong current pulse discharge in a liquid medium. Additionally, this discharge of high voltage and large current is a kind of high-energy-density transient conversion process. It has characteristics such as high efficiency, strong shockwave pressure and low energy expense. Underwater electrical discharge technology (UEDT) has thus been widely used in scientific research and industrial application, such as machinery manufacturing, mining, environmental engineering and biotechnology [1–5].

In production processes and related areas of the petroleum chemical industry, heat-exchange equipment, reaction devices and chemical containers have differently shaped pipes. After a long period of operation, scale commonly becomes a problem in pipes and containers. Because of scaling, devices need to be stopped, replaced or cleaned after a few months. The removal of scale requires a lot of manpower and materials each year. Currently, the main cleaning methods are chemical corrosion, mechanical processing and hydraulic injection. However, actual surveys show that these methods are not satisfactory. Therefore, there is an urgent need for an effective solution.

2. Underwater electrical discharge system and components

An experimental system was designed to systematically generate high-voltage pulses under reproducible conditions. The system comprises a dc power supply with capacitive energy storage and

* To whom any correspondence should be addressed.

a discharge source immersed in tap water at room temperature. A schematic diagram of the arrangement is shown in figure 1. The electrode system was located in the centre of a water tank with dimensions 800 mm (length) × 500 mm (width) × 500 mm (depth).

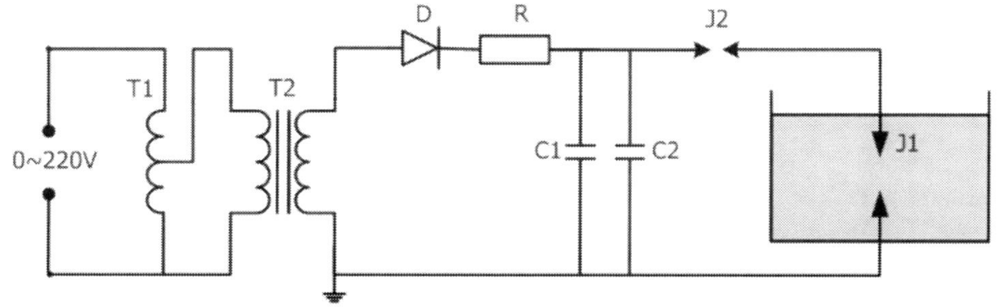

Figure 1. Schematic diagram of the experimental arrangement.

In figure 1, the coupling voltage regulator T1 (output voltage range of 0-280 V) receives external 220-V ac power, thus providing variable voltage for the high-voltage transformer T2. T2 and a silicon rectifier stack D constitute a high-voltage dc power supply (output voltage range of 0–50 kV); both T2 and D serve as a reserved capacitance charging power supply. The current-limiting resistance R limits the current charging loop and protects the high-voltage dc power supply. C1 and C2 each having capacitance of 2 μF withstand a voltage of 50 kV and provide total capacitance of 4 μF in a parallel manner. When the closure of J2 is triggered, the resultant short-circuit has a fast-fall-time step voltage. Owing to the charge stored in the capacitance, there is an electrical energy potential across the water-filled gap between the main electrodes J1. After a short delay time on the order of milliseconds, there is electrical breakdown of the water and the capacitor discharges through the plasma channel. The auxiliary discharge electrode J2 enhances the strength of the electrical discharge effect, and the gap J2 can be adjusted (adjustable range of 0–30 mm).

In this paper, the experimental subject was an ordinary steel tube with cement evenly spread on its inner wall. To study the mechanism of scale removal, the test model was constructed, simulating real conditions.

3. Experiment process, results and analysis

3.1. Experimental process

According to the structure of a "U-shaped" pipe and the actual working conditions of a sink, the experiment adopted indirect discharge. Indirect discharge is adopted (indirect discharge means external electrode discharges workpiece, instead of taking workpieces as discharge electrode). The main electrode is adjusted and moved both vertically and horizontally under water, 15 cm away from the upper outer wall of pipe fitting vertically. In the experiment, the main electrode discharge gap was set as Z = 48 mm. The tested common steel pipe fitting had an inner diameter of 25 mm, thickness of 2 mm, and length of 500 mm. To stimulate actual scale, the steel tube was bent in a U-shape, cut along its central axial, and uniformly coated with a 2-mm layer of cement along its inner walls. Because silicate cement is composed of silicates, its comprehensive mechanical and physical performance is similar to that of scale, but its adhesion to metal is stronger than that of scale. In this experiment, the applied cement is ordinary Portland cement, mode 300, and the proportion between cement and slurry is 1:3. After being mixed and tumbled with an appropriate quantity of water, the Portland cement was evenly smeared on the inner wall of the tube. The two halves of the pipe were then clamped and fixed.

3.2. Experiment results

The experimental results are given in tables 1 and 2. The tables show that for the same number of electrical discharges, electrical discharge energy increased and scale removal improved distinctly with an increase in the pulsed discharge voltage between the main electrodes; however, when the voltage of storage capacitances remained stable, even with a greater number of discharges, the result of scale removal was inferior to that for the pulsed voltage step-up. In the tables, single-discharge energy was calculated using equation (1). The removal rate was the weight percentage removed relative to the total amount of cement.

Table 1. Experiment data of scale removal with increasing pulsed voltage.

pulsed voltage (kV)	Single discharge energy (J)	Number of discharge	Removal rate
24	1152	20	7.1%
27	1458	20	34.2%
30	1800	20	68.5%
33	2178	20	97.3%
36	2592	20	98.8%

Table 2. Experiment data of scale removal with increasing number of discharges.

Pulsed voltage (kV)	Number of discharge	Removal rate
20	25	6.8%
20	30	16.0%
20	35	31.5%
20	50	66.7%
20	65	81.3%

Figure 2 shows clear dependency of the scaling removal rate on the pulsed voltage and the discharge frequency. The trends of the two lines verify that increasing the pulsed voltage achieves a better result.

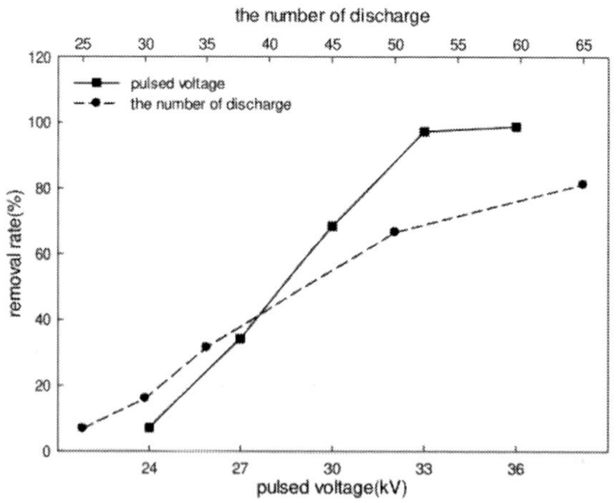

Figure 2. Descaling rate versus discharge voltage and number of discharges.

Figure 3 shows the part section of semi-tube that is daubed by cement, and figure 4 shows the result of inner wall descaling by increased pulsed discharge when its value is 36 kV. Although figure 4 shows that there is some residuum of the daubed cement not removed completely, its total descaling result is excellent compared to figure 3.

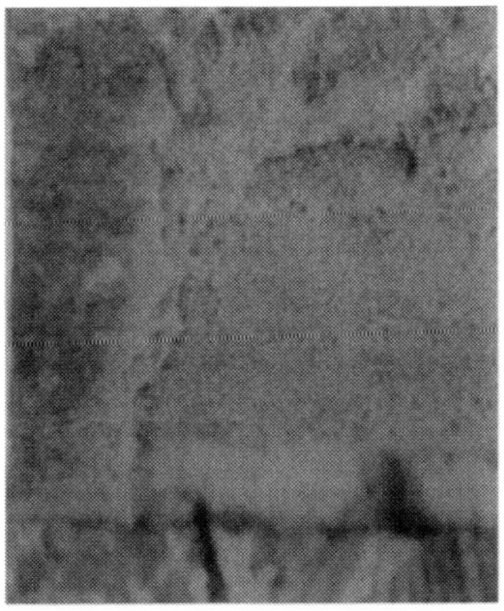

Figure 3. Photograph of inner wall before descaling.

Figure 4. Final result of descaling the inner wall through pulsed discharge.

3.3. Analysis of Experimental Results

When electrical breakdown occurred between the main discharges electrodes, strong pulsed current reached 4100 A and pulsed voltage reached 750 kV cm^{-1}. A shunting plasma channel with high temperature and high pressure developed on a nanosecond timescale. Therefore, the applied voltage and consequently high energy (several tens of kilojoules) were used to induce the electrical breakdown of water. In the first phase, the application of this long-duration high-voltage pulse between a pair of electrodes leads to the development of bubbles where gas discharge takes place. Moreover, the UEDT is associated with the emission of a powerful shock wave propagating radially into the water. The pressure waves generated have peak values in the range of 10^2–10^3 MPa [7].

The single-discharge energy W of the main electrodes is described by

$$W = CU^2 \cdot 2^{-1} \tag{1}$$

where C is the value of the tank capacitor and U is the charging voltage. The shock wave propagated along the radius of the spread. The peak pressure $P(r,\, t_r)$ can be approximated by [8]

$$P(r,t_r) = A\left[\frac{2}{n+1} + \frac{n-1}{n+1}\left(1 + \frac{n+1}{rC_0^2}G\right)^{1/2}\right]^{(2n/(n-1))} - B \tag{2}$$

According to equation 2, r is 48 mm, and the metal susceptive pressure strength is calculated as about 69.42 MPa.

Scale attached to the walls of the tube is mainly composed of inorganic salt deposits such as sour barium, magnesium silicate and calcium carbonate. These deposits attach to the tube inner wall through a strong adhesion force. Because of the elastic modulus, the vibration frequency and especially the acoustic impedance of the metal are different from those of scale, and the shock-wave propagation therefore produces different mechanical effects. Usually, owing to the small pipe diameter, the spherical surface pressure released from the channel of the discharge can be approximately considered as a planar pressure wave in the case of a long discharge channel distance. Theoretical analysis of a shock wave propagating in different media reveals that the acoustic impedance of scale is less than that of metal. Under the action of the shock wave, scale and the metal wall will be compressed, but when the wave reflected from the metal wall reaches the boundary between the scale and water surface, there is a new reflection wave. This reflected wave can be transformed to a stretching wave. Influenced by the stretching wave, descaling will occur along the direction from the metal wall to the scale. The tensile stress is synthesized by a wave front (reflection wave of scale and water) and wave tail (reflection wave of the metal wall). With the generation of spalled scale and a new boundary surface appearing repeatedly, scale constantly undergoes spalling and breaks down, until the majority of scaling is divorced from the inner tube wall.

4. Conclusion

Scaling is removed by UEDT through the effect of the stress wave function in different solid media stuck to a pipeline. The strong tensile stresses separate the scale layer from the conduit wall, and thus have an excellent cleaning effect. By increasing the discharge voltage or the number of discharges, the cleaning effect can be improved. Compared with other removal methods, UEDT is not affected by the structure and the forms of a metal tube, and the technology does not negatively affect the mechanical properties of the tube. Meanwhile, this method has characteristics of high efficiency, strong shock pressure and low energy cost, and is thus very effective in removing scale from pipe fittings such as "U-shape" and spiral pipes. Therefore, UEDT has large-scale application prospects in industrial pipeline cleaning.

References

[1] LeBlanc J and Shukla A 2010 *Compos. Struct.* **92** 2421
[2] Timoshkin I V, Mackersie J W and MacGregor S J 2004 *IEEE Trans. Plasma Sci.* **32** 2055
[3] Gao H, Jia Z, Mao Y, Guan Z and Wang L 2008 *IEEE Trans. Dielectr. Electr. Insul.* **15** 435
[4] Locke B R, Sato M, Sunka P, Hoffmann M R and Chang J S 2006 *Ind. Eng. Chem. Res.* **45** 882
[5] Tao J, Shih A J and Ni J 2008 *J. Manuf. Sci. E-t. Asme.* **130** 11002
[6] Gidalevich E, Boxman R L and Goldsmith S 2004 *J. Appl. Phys.* **10** 1509
[7] An W, Baumung K and Bluhm H 2007 *J. Appl. Phys.* **101** 53301
[8] Lu X, Pan Y, Liu K, Liu M and Zhang H 2002 *J. Appl. Phys.* **91** 24

Oxidation of ammonium sulfite by a multi-needle-to-plate gas phase pulsed corona discharge reactor

Hua Ren, Na Lu, Kefeng Shang, Jie Li, Yan Wu*
Institute of Electrostatic and Special Power, Dalian University of Technology, Dalian 116024, China

Email: wuyan@dlut.edu.cn

Abstract. The oxidation of ammonium sulfite in the ammonia-based flue gas desulfurization (FGD) process was investigated in a multi-needle-to-plate gas phase pulsed corona discharge reactor in this paper. The effect of several parameters, including capacitance and peak pulse voltage of discharge system, electrode gap and bubbling gas flow rate on the oxidation rate of ammonium sulfite was reviewed. The oxidation rate of ammonium sulfite could reach 47.2% at the capacitance, the peak pulse voltage, electrode gap and bubbling gas flow rate equal to 2 nF, -24.6 kV, 35 mm and 4 L min^{-1} within treatment time of 40 min The experimental results indicate that the gas phase pulsed discharge system with a multi-needle-to-plate electrode can oxide the ammonium sulfite. The oxidation rate increased with the applied capacitance and peak pulse voltage and decreased with the electrode gap. As the bubbling gas flow rate increased, the oxidation rate increased first and then tended to reach a stationary value. These results would be important for the process optimization of the $(NH_4)_2SO_3$ to $(NH_4)_2SO_4$ oxidation.

1. Introduction

Ammonia-based flue gas desulfurization (FGD) is a well known technique with high desulfurization efficiency [1], which flue gases are scrubbed with aqueous ammonia to reduce their sulfur oxide content [2]. In China, the ammonia-based wet FGD process has been increasingly used in recent years because of high efficiency, lower cost, less land occupied, rare fouling, etc. Ammonium sulfite as a byproduct of dusulfurization is easily decomposed into SO_2 and NH_3 which produce secondary pollution in the environment, while ammonium sulfate whose chemical property is stable can be used as a fertilizer or in fertilizers in agriculture. Thus, it is critical to oxidize ammonium sulfite to ammonium sulfate efficiently for ammonia flue gas desulfurization (FGD) process.

* To whom any correspondence should be addressed.

Over the past few decades, several oxidation technologies, such as air diffusion method, catalytic oxidation technology and advanced oxidation technology (AOT), have been utilized for the purpose of oxidizing ammonium sulfite generated in FGD process [1, 3, 4]. The oxidation of ammonium sulfite by using air diffusion method is usually limited by the gas-liquid mass transfer coefficient of oxygen and thus needs consuming more time. Catalytic oxidation technology could efficiently oxidize the concentrated ammonium sulfite [3] and lots of researches focus on the use of homogeneous catalyst, but the method can cause secondary pollution. Therefore, these methods could not meet the econo-technical norms in the oxidation of ammonium sulfite field.

Advanced oxidation technologies (AOTs) including UV photolysis, photocatalysis, sonochemistry, supercritical water oxidation, electrical discharge plasma technology and electron-beam irradiation, etc, has been widely used in wastewater disposal fields. In the application of electrical discharge plasma technology, glow discharge, dielectric barrier discharge and pulsed corona discharge are the three main discharge types used for the purpose of water disposal [5]. Most previous works have been investigated in pulsed corona discharge plasma systems [6]. The technique has a higher energy efficiency compared to other AOTs because of the very short pulse width (approximately 500–1000 ns) and the fast pulse rise time (nanosecond level) and can produce more various reactive chemical radicals such as ·OH, ·O, ·O_2, and molecular species such as H_2O_2,O_3 [7, 8]. All these active species have high oxidation potentials to oxidize the ammonium sulfite. Therefore, the pulsed corona discharge technology can be used as one promising candidate for the oxidation of ammonium sulfite with low operation costs. The ammonium sulfite oxidation reactions are a series of radical chain reactions. The net reactions can be written as:

$$2(NH_4)_2SO_3 + O_2 \xrightarrow{OH,O,O_3,\ HO_2,\ e,\text{etc}} 2(NH_4)_2SO_4 \quad (1)$$

$$2NH_4HSO_3 + O_2 \xrightarrow{OH,O,O_3,\ HO_2,\ e,\text{etc}} 2NH_4HSO_4 \quad (2)$$

In the pulsed high voltage discharge system, the structure of the discharge reactor (needle-to-wire, needle-to-plate, wire-to-plate and ring-to-cylinder), which was determined by the configuration of the discharge electrode and ground electrode, could influence the formation of active species and organic degradation [9, 10]. Multi-needle-to-plate discharge rector has been used for wastewater treatment for many years, demonstrating that it could induce an oxidation process for organic contaminants degradation in aqueous solution [6, 11]. In liquid discharge reactor, the effect of conductivity is critical to the formation of discharge in water. Compared with this reactor, gas phase reactor is not affected by conductivity.

In this paper, we have investigated the feasibility of using multi-anode pulse discharge reactor to oxide the ammonium sulfite. The experiment was carried out to determine the effect of such factors such as the capacitance, the peak pulse voltage, electrode gap and bubbling gas flow rate on the oxidation rate of ammonium sulfite. This paper could provide some experimental support for further improvement the oxidation of ammonium sulfite caused by the non-thermal plasma technology.

2. Experimental sections

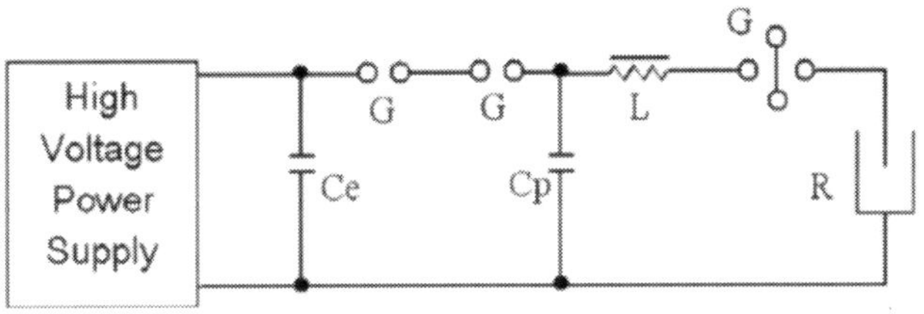

Figure 1. The sketch of the pulsed high voltage source.

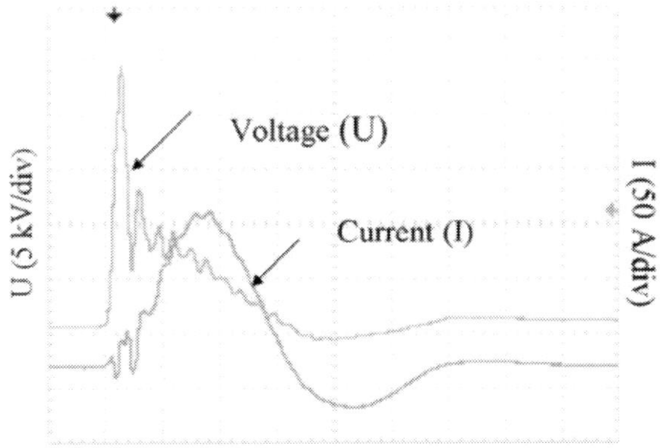

Figure 2. The waveforms of voltage and current input into the reactor by the pulsed power source.

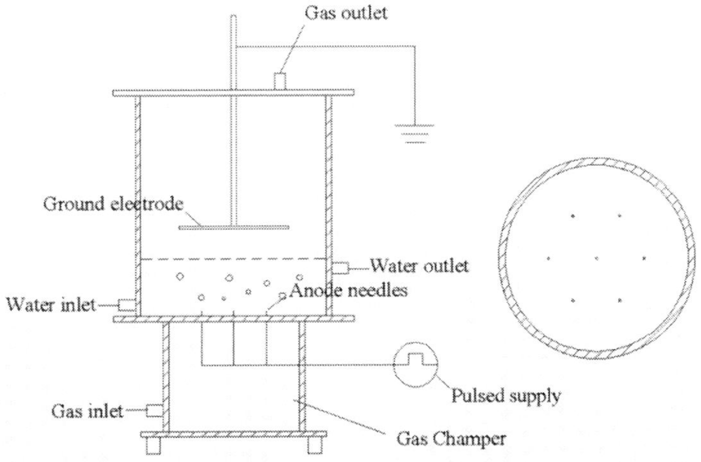

Figure 3. The schematic of pulsed high voltage discharge system.

The power supply was realized by making use of rotating spark gap to produce pulsed high voltage (figure 1). It consisted of a 0–50 kV adjustable DC power supply, a storage capacitor (*Ce*), an adjustable pulse-forming capacitance (*Cp*, 2 nF) and a rotating spark-gap switch (*G*). Typical voltage and current waveforms of input into the reactor by the pulsed power source are shown in figure 2. The rise time and width of the voltage was less than 100 ns and 500 ns, respectively.

The oxidation of ammonium sulfite was carried out in a multi-needle-to-plate gas phase pulsed corona discharge reactor (figure 3). This reactor used was a Plexigas^{TM} cylinder vessel, 11 cm in internal diameter and 12 cm in height. The ground plate electrode which consisted of a 90 mm diameter stainless-steel plate of 1.5 mm thickness was placed in gas phase. The high voltage needle electrode which produced positive streamer corona discharge at its needle tips was placed in water. The distance from needlepoint to plate was adjustable between 30 mm and 50 mm. The high voltage electrode, comprised of one needle in the center and the other six distributed uniformly around a circle of 15 mm radius, was secured within a resin disc. Each needle consisted of a No. #12 stainless-steel hypodermic pinhead. Silicone insulation encased the needle tips that protruded from the resin disc, with only 1 mm of length exposed in water. The air flowed into the gas chamber, through the seven needles and into the liquid chamber, forming gas bubbles.

The used ammonium sulfite solution was prepared via ammonium sulfite of analytical grade. The total volume of the treated ammonium sulfite solution was 400 mL, and its concentration C_0 was equal to 0.3 mol L^{-1}. For liquid circulation, ammonium sulfite solution in container was pumped into the discharge rector, and then was diffused into the discharge region. Subsequently, the solution flowed back into the container. The solution flow rate was kept at 225 mL min^{-1}. Sulfite ion was titrated in iodine solution by using dextrin solution as an indicator. Details of the method are reported in the national standard [12].

The input power W of a pulsed discharge system can be calculated from the following formula:

$$W = \frac{1}{2} C f V^2$$

Here C is the storage capacitance, f the pulse frequency and V the DC source voltage. If the energy in this reaction system is neglected, the input energy can be calculated by the above formula.

3. Results and discussion

3.1 Effect of capacitance

The influence of the capacitance on the oxidation rate of ammonium sulfite was displayed in figure 4 (a). It was obvious that increasing the capacitance benefited the oxidation of ammonium sulfite, due to a higher energy input introduced in the discharge. When the capacitance was improved from 1 to 2 nF, the oxidation rate of ammonium sulfite enhanced rapidly from 7.9% to 29.7%, while from 2 to 3 nF, the oxidation rate improved a few, only from 29.7% to 32.3%.

The energy efficiency was used as an indicator to evaluate the energy consumption of the system, which was shown in figure 4 (b). As the capacitance increased, the oxidation rate went up at first and then decreased. Thus, 2 nF was chosen for our experiments.

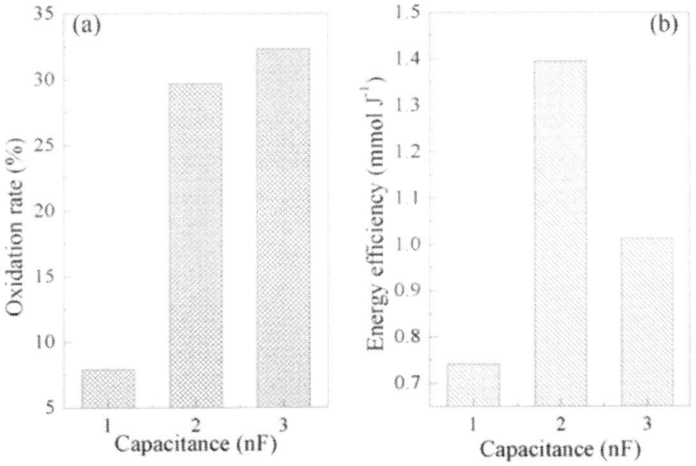

Figure 4. Effect of capacitance on the (a) oxidation rate of ammonium sulfite and (b) energy efficiency. (U=22.6 kV, f=50 Hz d=10 mm, C_0= 0.3 mol L^{-1}, Q_{air}=3 L min^{-1})

3.2 Effect of peak pulse voltage

Figure 5 showed typical results of the effect of peak voltage on the oxidation rate of ammonium sulfite. It could be seen that the oxidation rate increased with treatment time and the peak voltage within treatment time of 40 min, the oxidation rate increased from 29.7% to 48.3% as the peak voltage increased from 22.6 to 26.6 kV. Since, the electrical field strength becomes larger and the density of plasma active species and oxidant agents increases with the voltage [13, 14]. In consideration of the oxidation rate, the maximum discharge voltage reviewed in this paper was chosen as 24.6 kV.

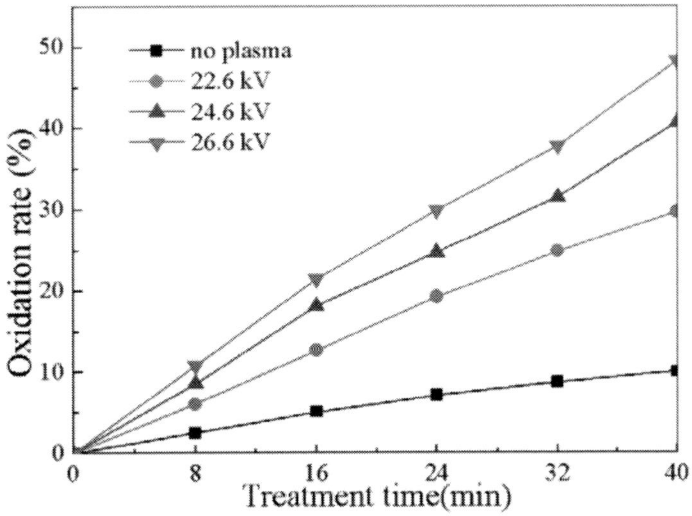

Figure 5. Effect of peak voltage on the oxidation rate of ammonium sulfite. (C=2 nF, f=50 Hz, d=10 mm, C_0= 0.3 mol L^{-1}, Q_{air}=3 L min^{-1})

3.3 Effect of the electrode gap

The effect of the electrode gap on the oxidation rate of ammonium sulfite was studied by varying gas layer heights (GH, i.e. the height from the gas–liquid interface to the surface of the top stainless-steel plate). At fixed LH (LH, i.e. the height from the tip of the needle to the gas–liquid interface, 25 mm), as shown in figure 6, the oxidation rate of ammonium sulfite decreased with the increase of GH at a treatment time of 40 min. In experiment, fewer streamers were observed at a wider GH. In fact, the increase of GH increased the electrode gap, and the resistance between two electrodes therefore increased. As a result, more energy was consumed to conquer the increased resistance and less energy was supplied to the plasma channels in the solution [15], resulting in the decrease of the oxidation rate.

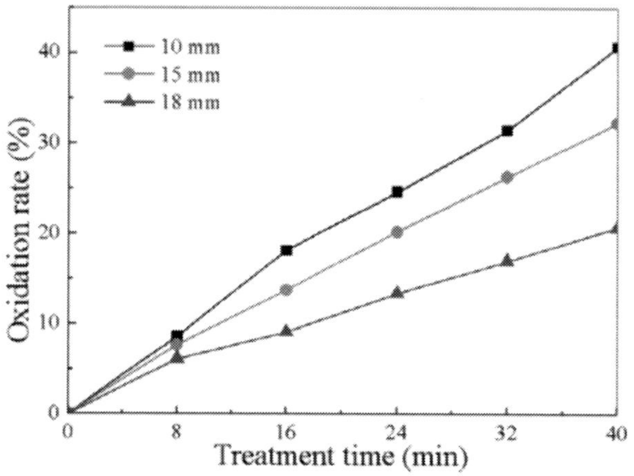

Figure 6. Effect of the electrode gap on the oxidation rate of ammonium sulfite.
(U=24.6 kV, C=2 nF, f=50 Hz , C_0= 0.3 mol L^{-1}, Qair=3 L min^{-1})

3.4 Effect of bubbling gas

Figure 7 showed the oxidation rate of ammonium sulfite at different flows of bubbling gas. The oxidation rate increased with the treatment time (figure 7 (a)). Figure 7 (b) showed the results at a treatment time of 50 min. It was concluded that the oxidation rate increased first and then tended to reach stationary values. Only 36.3% ammonium sulfite was oxidized at 40 min at the bubbling gas flow rate of 2 L min^{-1}, while more than 46% ammonium sulfite was oxidized at the rate equal to and higher than 4 L min^{-1} at the same discharge time. The reason for the increase of oxidation rate was that more bubbling gas could make much more ozone and •OH generated in the reactor during discharge, which could speed up the oxidation of ammonium sulfite. It was also seen from figure 7 that the oxidation rate at 4 L min^{-1} was very close to that at 5 and 6 L min^{-1}. Higher bubbling flux means more wastage energy. Therefore, bubbling flux was kept at 4 L min^{-1} in the experiment.

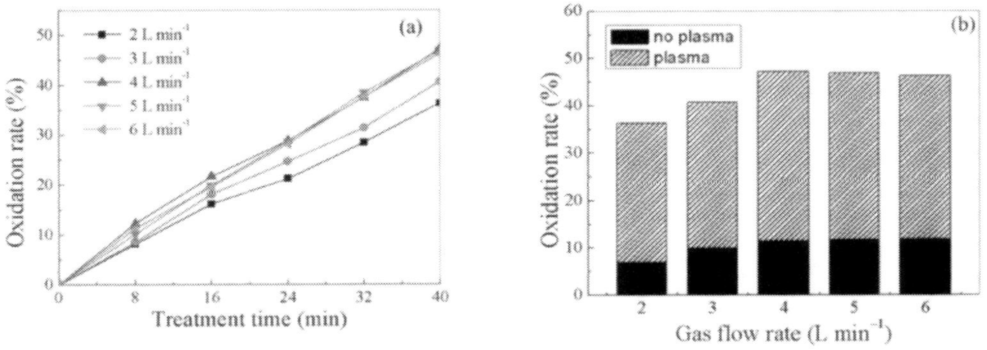

Figure 7. Effect of the bubbling gas on the oxidation rate of ammonium sulfite.
(U=24.6 kV, C=2 nF, f=50 Hz, d=10 mm, C_0= 0.3 mol L^{-1})

4. Conclusions

An experimental study on the oxidation of ammonium sulfite (initial concentration of 0.3 mol L^{-1}) in aqueous solutions was carried out in a multi-needle-to-plate gas phase pulsed corona discharge reactor. The main findings are as follows.

1) A multi-anode gas phase pulse discharge rector could realize the oxidation of ammonium sulfate. The oxidation rate of ammonium sulfite after 40 min treatment processing could reach 47.2% at the capacitance, the peak pulse voltage, electrode gap and bubbling gas flow rate equal to 2 nF, -24.6 kV, 35 mm and 4 L min^{-1}.

2) The increase of the capacitance and peak pulse voltage improved the oxidation of ammonium sulfite, while the increase of electrode gap inhibited the oxidation of ammonium sulfite. The oxidation rate of ammonium sulfite increased first and then tended to reach stationary values with the solution flow rate.

References

[1] Lourenco F and Bischoff L 1998 *Chem. Eng. World* **33** 75

[2] Slack A. V and Hollinden G. A 1975 *Sulfur dioxide removal from waste gases* (Noyes Data Corporation)

[3] Long Xiang-li, Li W, Xiao Wen-de and Yuan Wei-kang 2006 *J. Hazard. Mater.* **129** 260

[4] Hu X， Zhu T, Jiang X, Wang Y, Qiu R, Zhang H, Li R and Yan K 2008 *Chem. Eng. J.* **139** 469

[5] Malik M. A and Ghaffar A 2001 *Plasma Sources Sci. Technol.* **10** 82

[6] Wang H, Li J and Quan X 2006 *J. Electrostat.* **64** 416

[7] Elsawah M, Abdel Ghafar H H, Morgan N N, Hassaballa S, Samir A, Elakshar F F and Garamoon A A 2012 *IEEE Trans. Plasma Sci.* **40** 29

[8] Sunka P, Babicky V, Clupek M, Lukes P, Simek M, Schmidt J and Cernak M 1999 *Plasma Sources Sci. Technol.* **8** 258

[9] Sano N, Kawashima T, Fujikawa J, Fujimoto T, Kitai T and Kanki T 2002 *Ind. Eng. Chem. Res.* **41** 5906

[10] Sugiarto A T, Ohshima T, Sato M 2002 *Thin Solid Films* **407** 174
[11] Bian W, Ying X and Shi J. 2009 *J. Hazard. Mater.* **162** 906
[12] State environmental protection administration of China HJ/T 56-2000 Determination of sulphur dioxide from exhausted gas of stationary source Iodine titration method 2000
[13] Suarasan I, Ghizdavu L, Ghizdavu I, Budu S and Dascalescu L 2002 *J. Electrostat.* **54** 207
[14] Ge H, Zhang L, Yan L, Mi D and Zhu Y 2011 *J. Electrostat.* **69** 529
[15] Locke B R, Sato M, Sunka P, Hoffmann M R and Chang J S 2006 *Ind. Eng. Chem. Res.* **45** 882

Journal of Physics: Conference Series **418** (2013) 012129 doi:10.1088/1742-6596/418/1/012129

Study on the factors influencing phenol degradation in water by dielectric barrier discharge (DBD)

Lijuan Duan[1], Kefeng Shang[2], Na Lu[2], Jie Li[2*], Yan Wu[2]
1 School of Environmental Science and Technology, Dalian University of Technology, Dalian, China
2 Institutes of Electrostatics and Special Power, Dalian University of Technology, Dalian, China

E-mail: lijie@dlut.edu.cn

Abstract. Dielectric barrier discharge (DBD) is an effective method to produce active species including ozone for remediation of pollutants in water, gas, etc. In the present work, a DBD type surface discharge system was utilised to degrade phenol in solutions. The influence of experimental conditions including solution pH values, initial concentration of phenol and applied voltage on the degradation rate of phenol, the removal of COD and the byproducts was examined and discussed as well as the concentration change of ozone during phenol was examined. Experimental results reveal that high pH value advanced the degradation of phenol and COD removal, which maybe contributed to the enhanced generation of hydroxyl radicals under higher pH value; applied voltage influenced the transformation of phenol to byproducts which could be demonstrated by the different COD removal under different applied voltages. Under an optimization condition, the degradation rate of phenol reached 100% in 80 minutes and COD removal reached 68% in 120 minutes.

1. Introduction

Electrical discharge technologies have attracted many attentions for the degradation of organic compounds in aqueous solutions in the past decades [1]. As a high efficiency method, DBD is an effective method to removing biorefractory organic compounds in aqueous solutions. The most abundant oxidizing species produced by dielectric barrier discharge plasmas is ozone that is formed in the gas phase, although small amounts of other species such as hydrogen peroxide, hydroxyl radicals, and hydroperoxyl radicals are also produced in the water phase [2]. Therefore, in order to efficiently utilize ozone for removing the water contaminants, it is very important to study the factors affecting the ozonation efficiency.

Phenol has been widely used as an object pollutant to elucidate the complex discharge reaction mechanisms and to evaluate the performance of many reactors designs [3]. Moreover, a high content of phenols wastewater exhibiting high chemical oxygen demand (COD) is still less effective treated through conventional biological processes due to their poor biodegradability. Their toxic effects on human health are well documented, being related to severe illnesses and some serious human organ malfunctions.

* To whom any correspondence should be addressed.

In this study, we investigated the application of a DBD reactor submerged in water to degrade phenol. Meanwhile, several factors affecting the degradation efficiency of phenol were studied in detail, such as pH values, initial concentration of phenol and applied voltage.

2. Experimental

2.1. Experimental Apparatus

The DBD reactor was made with a quartz glass cylinder (inner diameter: 10 mm; outer diameter: 12.5 mm; length: 280 mm) and a stainless steel wire, in which the wire coiled inside of the cylinder was supplied with an AC high voltage power supply. The wastewater was grounded. The volume of the wastewater contained in the reactor vessel was 600 mL. The flow rate of dry air fed into the DBD reactor was 2 L min^{-1}.

The AC high voltage applied to the DBD reactor was varied from 6 to 10 kV (peak value) to change the discharge power, and eventually to change the concentration of ozone. For measurement of the voltage applied to the DBD reactor, a 1000:1 high voltage probe （Tektronix P6015A） and a digital oscilloscope （Tektronix TDS2014） were used.

2.2. Experimental Methods

The chemical oxygen demand (COD) was measured by potassium dichromate titration method.

The percentage removal of the parameters evaluated (degradation) was determined using the following equation (1):

$$\eta_{COD}\% = (m_0 - m)/m_0 \qquad (1)$$

where η: degradation efficiency of COD, m_0 and m are initial and final values of COD in the synthetic wastewater.

The concentration of ozone in the gas was analyzed by the iodometric method [4].

The determination of phenol concentrations was measured by HPLC. The degradation of phenol was determined by equation (2):

$$\eta\% = (C_0 - C)/C_0 \qquad (2)$$

η: degradation efficiency of phenol, where C_0 and C are the initial concentration of phenol and the concentration of phenol at any time in the synthetic wastewater.

3. Results and discussion

3.1. Effect of solution pH

Since pH is a key parameter for both ozone stability and the production of hydroxyl radicals in aqueous solution, it is important to examine the influence of pH on ozonation of phenol-contained water. Figure 1 shows the COD removal efficiency of phenol solution with an initial concentration of 80 mg L^{-1} and a different initial pH values at 3.0, 6.0, 9.0 and 11.0, respectively. It is obvious that the degradation efficiency of COD increased with discharge time. The degradation efficiency reached the high value of 54.32 % during 80 min in the solution with pH 9.0. For the solutions with pH 3.0 and 6.0, the degradation efficiencies were more and more close to 65 % in 120 min. However， the degradation efficiency decreased in the solution with pH 11.0, which only reached 12.60 %.

It is well known that ozone decomposes into hydroxyl radicals more easily in basic solution than neutral or acid circumstances because hydroxide ions may play a role of initiator of the chain reaction [5]. The variations of ozone at different pH can be observed in table1, the results show that the higher pH was, the lower the amount of ozone was, which maybe attributed to the production of hydroxyl radicals in basic solution through ozone decomposition. Because hydroxyl radicals were a less selective and more powerful, more phenol molecules can be oxidized in a basic pH solution since the concentration of hydroxyl radicals is much higher than acid solution. Therefore, the COD removal efficiency reached the high value at pH 9.0. However, the COD removal efficiency was close at pH=3, 6 and 9 after 80 min, especially in 120 min. As shown in figure 2, there is a tendency toward a decrease in pH with the ozonation time; the decrease in pH may be due to phenol molecules being oxidized by ozone, resulting in small organic molecular fragments, such as acetic acid [6], aldehydes, ketones and so on. pH was almost same in 120 min with different initial pH values at 3.0, 6.0 and 9.0. The main reactions taking place in the acid aqueous solution were those between ozone and phenol. So the COD removal efficiency was close at 120 min. However, the COD degradation efficiency decreased in pH 11.0. In the above experiments, the working gas was dry air, which includes CO_2. In basic solution condition, CO_2 can generate carbonate ions which are the suitable •OH scavengers. So phenol decomposition in the synergistic system of DBD was decreased.

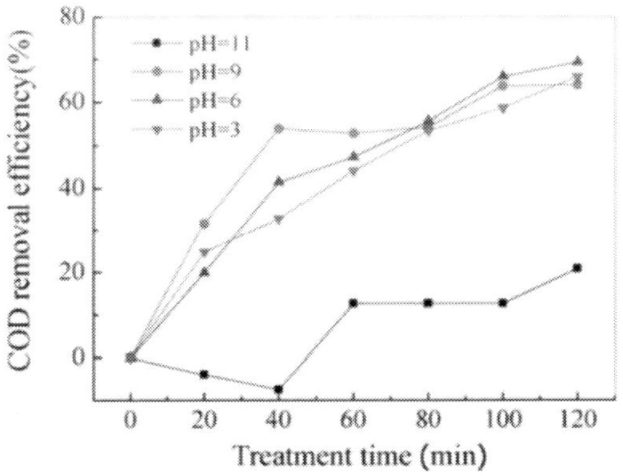

Figure 1. Effect of pH on COD removal efficiency (applied voltage 7 kV, initial concentration 80 mg L^{-1}, frequency 7 kHz)

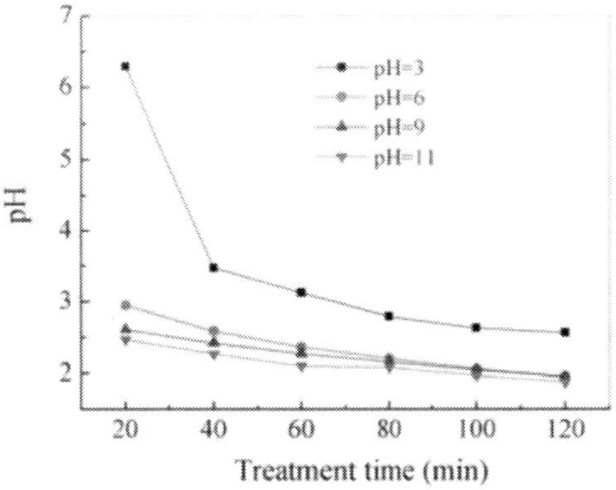

Figure 2. pH variations of the solutions (Initial concentration 80 mgL^{-1}, applied voltage 7 kV, frequency 7 kHz).

Table 1. Ozone variations of different pH.

	pH = 3	pH = 6	pH = 9	pH = 11
Ozone (mg^{-1})	0.459	0.445	0.364	0.315

3.2. Effect of initial concentration of phenol

The results of the COD removals at an initial concentration of 60, 80, 150, 200 mg L^{-1} are shown in figure 4. A maximum phenol conversion of 68 % attained at an initial concentration of 60 mg L^{-1}. As shown in figure 3, the removal rate of phenol reached to 100 % after 80 min discharge treatment, while the COD removals efficiency was 45 % which was much less than the removal rate of phenol. It may be due to phenol molecules being easily oxidized by ozone, resulting in small organic molecular fragments, which are not completely mineralized under the oxidative conditions described. So the degradation efficiency of pollutants through examination of the COD removal is more reasonable.

The initial concentration of phenol had the lowest effect on its degradation. An increase in the initial amount of the pollutant from 60 to 150 mg L^{-1} in solution led to a decrease in the degree of COD conversion only by 6.3 %, while a decrease of 26.3 % occurred when initial concentration changed from 60 to 200 mg L^{-1}. It may be there was plenty of ozone for removal of phenol below 150 mg L^{-1}. So the amount of ozone was a constant under certain conditions, the COD removals efficiency was also certain.

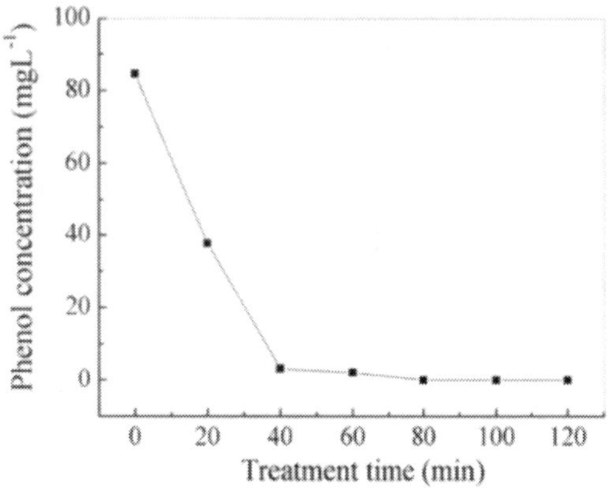

Figure 3. The concentration of phenol variations of the solutions.

(Initial concentration 80 mgL^{-1}, applied voltage 7kV, frequency 7 kHz).

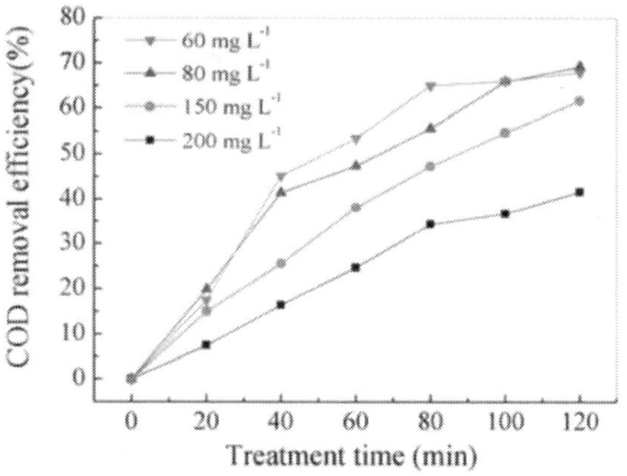

Figure 4. Effect of initial phenol concentration on COD removal efficiency (Applied voltage 7 kV, frequency 7 kHz)

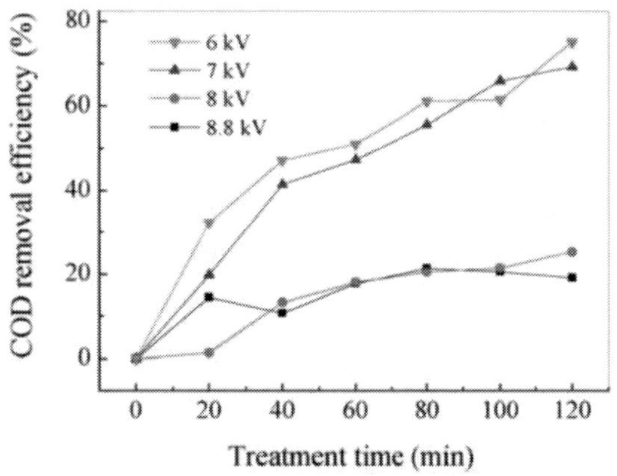

Figure 5. Effect of applied voltages on COD removal efficiency (initial concentration 80 mgL^{-1}, frequency 7 kHz)

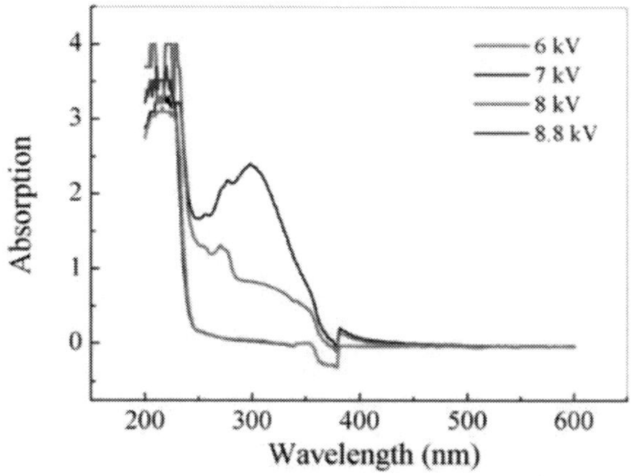

Figure 6. UV/visible spectra for different voltages (treatment time: 60 min).

3.3. Effect *of applied voltage*

The effects of applied voltage on COD decomposition were revealed in figure 5. It could be observed that the COD decomposition decreased with the applied voltage. At applied voltage of 7 kV, about 69 % of COD was removed, while approximately 75 %, 22 % and 19 % of phenol could be degraded at pulse voltages of 6, 8 and 8.8 kV under the same condition, respectively.

As shown in table 2, higher applied voltage would generate more ozone, which should result in increasing COD removal, but the result is not accord with it. This can be explained as follows: the degradation pathways of phenol in different voltages are diverse. Figure 6 shows the UV/visible spectra of the samples after being treated for 60 min at different applied voltages. These spectra of phenol degradation at the applied voltage of 7 kV were much different from that at the applied voltage of 8 kV, which indicates the by-products of the applied voltage at 7 kV are different from that at 8 kV. It may be that the by-products produced at higher voltage are more refractory. The main intermediates detected in the phenol oxidation were ring compounds and short chain acids. The ring intermediates mainly included catechol, hydroquinone and benzoquinone [7]. So the ring compounds are likely to the main by-products of higher applied voltage. In all cases, maximum absorbance occurred at around 210 nm, which is attributed to the presence of the byproducts produced from the phenol. In this study, the byproducts were not fully identified, but these spectra indicate that most of the benzene and naphthalene rings existing in the original phenol molecules were destroyed.

Table 2. Ozone variations of different applied voltage.

	6 kV	7 kV	8 kV	8.8 kV
Ozone（mg^{-1}）	0.2853	0.415	0.473	0.603

4. Conclusions

The aim of the present work was to investigate the factors affecting the COD degradation efficiency of phenol. The optimal conditions for the COD removal are as follows:

(1) The optimum pH for phenol degradation was about 6, although the degradation efficiency reached the high value in the solution with pH 9.0, but degradation efficiency at pH =6.0 were slightly less than at pH =9.0, especially the innate pH of phenol is about 6. So the degradation effect of phenol is good at pH =6.0, and the operation of experimental system is easy.

(2) The best range of initial concentration is from 80 mg L^{-1} to 150 mg L^{-1} in this reactor. The amount of ozone in this reactor is abounding within this limit.

(3) The highest value of COD degradation efficiency was 75 % at 6 kV. However, voltage instability may occur at 6 kV. The value of COD degradation efficiency was 69 % at 7 kV, so there was a high COD degradation efficiency at 7 kV.

Acknowledgments

This work is partially supported by the Fundamental Research Funds for the Central Universities [Grant NO. DUT12RC (3)12].

References

[1] Locke B R , Sato M, Sunka P, Hoffmann M R and Chang J S 2006 *Ind. Eng. Chem. Res* **45** 882
[2] Young S M, Jin-O. Jo and J. Christopher Whitehead 2007 *Chem. Eng. J.* **142** 56
[3] Aaron Ortiz-Gomez, Benito Serrano-Rosales and Hugo de Lasa 2008 *Chem.Eng. J.* **63** 520
[4] Bader H , Hoigné J 1981 *Water Res* **15** 449
[5] Staechlin J and Hoign´e J, *Environ Sci. Technol* 1982 **16** 676
[6] Hoeben W F L M , Veldhuizen E M V , Rutgers W R ,Cramers C A M G and Kroesen G M W 2000 *Plasma Sources Sci. Technol* **9** 361
[7] Wang H J, Li J, Quan X,Wu Y, Li G F and Wang F Z 2007 *J. Hazard. Mater* **141** 336

Journal of Physics: Conference Series **418** (2013) 012130 doi:10.1088/1742-6596/418/1/012130

Oxidation of ammonium sulfite in aqueous solutions using ozone technology

Yue Li[1], Kefeng Shang[1,2], Na Lu[1,2], Jie Li[1,2*], Yan Wu[1,2]
1 Key Laboratory of Industrial Ecology and Environmental Engineering (MOE), School of Environmental Science and Technology, Dalian University of Technology, Dalian, China
2 Institute of Electrostatics and Special Power, Dalian University of Technology, Dalian, China

E-mail: lijie@dlut.edu.cn

Abstract. How to deal with unstable ammonium sulfite, the byproduct of flue gas desulfuration by ammonia absorption methods, has been a difficult problem in recent years. Oxidation of ammonium sulfite in aqueous solutions using ozone produced by a surface discharge system was investigated in the paper. The oxidation efficiency of ammonium sulfite by ozone and traditional air aeration were compared, and the factors including ozone concentration, gas flow rate, initial concentration of ammonium sulfite solution and reaction temperature were discussed. The results show that the oxidation efficiency of ammonium sulfite by ozone technology reached nearly 100% under the optimum conditions, which had a significant increase compared with that by air aeration.

1. Introduction

Along with economic development and social progress as well as people's raised awareness of the environment protection, the researches for industrial flue gas desulphurization technology are increasingly emphasized [1]. In recent years, the technology of ammonia process of desulfurization, which uses ammonia liquid to remove SO_2 in flue gas, has been researched and applied. The byproduct produced in this process mainly is ammonium sulfite [2], which can be converted to ammonium sulphate after oxidation treatment. Since ammonium sulphate can be used as fertilizer, recycle of its by-products has high economic value [3]. However, how to convert ammonium sulfite into ammonium sulphate economically and efficiently is one of the keys to achieve industrialization of the ammonia process of desulfurization technology [4].

Ozone is a kind of strong oxidizer inferior to fluorine, which can oxidize a variety of compounds with fast reaction rate and low energy consumption. Meanwhile, it does not produce sludge. In developed countries, the ozone, as a kind of strong oxidizer and disinfectant, is widely applied in food, water treatment, chemical industry, medical service, catering storage, cultivation and other fields [5].

In this paper, the oxidation efficiency of ammonium sulfite by ozone technology and traditional air aeration were compared. In addition, the factors such as ozone concentration, gas flow rate, initial concentration of ammonium sulfite solution and reaction temperature were investigated and the corresponding reasons were analysed as well as the optimized experimental conditions were obtained.

---

[*] To whom any correspondence should be addressed.

2. Material and method

2.1 Systematic setup

A schematic diagram of the experimental setup is shown in figure 1.The setup mainly consists of an air pump, an ozone generator, a flow meter and a reactor. The ozone gas, which is produced by an ozone generator, flows into the ammonium sulfite solution placed in the reactor through an aerator and turns sulfite into sulphate. The reactor is an organic glass tube, with the size of 44 mm in height, 2.9 mm in thickness and the inner diameter of 50 mm. It has a capacity of 400 mL ammonium sulfite solution.

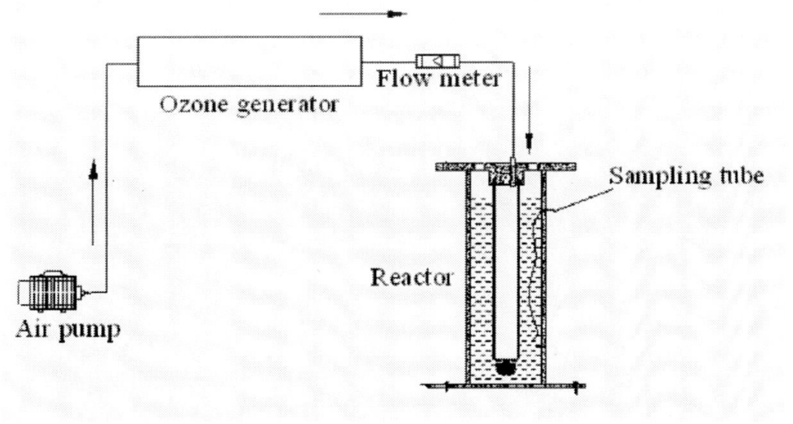

Figure 1. Schematic diagram of the experimental system.

2.2. Analysis of ozone

In this paper, the concentration of ozone was analysed by the iodometric method titrated by $Na_2S_2O_3$ standard solution [6].

O_3 reacted with KI as follows:

$$O_3+2KI+H_2O \rightarrow 2KOH+I_2+O_2 \quad (1)$$

2.3. Analysis of ammonium sulfite

In this work, ammonium sulfite was also analysed by the iodometric method [7].

$(NH_4)_2SO_3$ reacted with I_2 as follows:

$$(NH_4)_2SO_3+I_2+H_2O \rightarrow (NH_4)_2SO_4+2HI \quad (2)$$

3. Results and discussion

3.1. Comparison between ozone treatment and traditional air aeration

The ozone treatment was compared with the traditional air aeration in terms of oxidation efficiency, and the result was shown in figure 2. The gas flow rate was 3 L min^{-1}, and the initial concentration of ammonium sulfite was 0.2 mol L^{-1}.

It was obvious that the oxidation efficiency of ammonium sulfite, which was treated by ozone technology, was nearly 100% after 25 minutes, which was far higher than that (about 50%) by traditional air aeration. This can be explained by the stronger oxidbillity of O_3 than O_2, so O_3 can react with ammonium sulfite in a very short time.

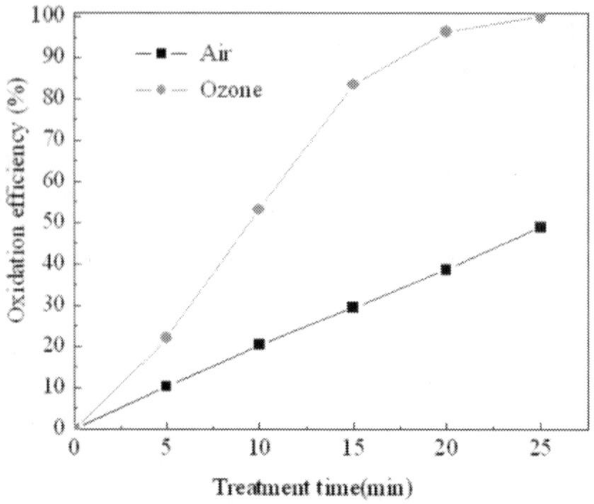

Figure 2. Efficiency of air aeration and ozone induced ammonium sulfite oxidation.

3.2. Influencing factors in oxidation

3.2.1. Ozone concentration

The effect of ozone concentration on the oxidation efficiency of ammonium sulfite was illustrated in figure 3, in which the ozone concentration was adjusted by changing the applied voltage connected with the ozone generator. The treatment time was 25 minutes, the gas flow rate was 3 L min^{-1}, and the initial concentration of ammonium sulfite was 0.2 mol L^{-1}.

It was found that the oxidation efficiency of ammonium sulfite had an apparent promotion with the increase of ozone concentration, which was close to 100% when the ozone concentration reached 702 ppm. It could be explained that the more effective collision probability between ozone and sulfite happened with the higher ozone concentration, which could enhance the oxidation of ammonium sulfite.

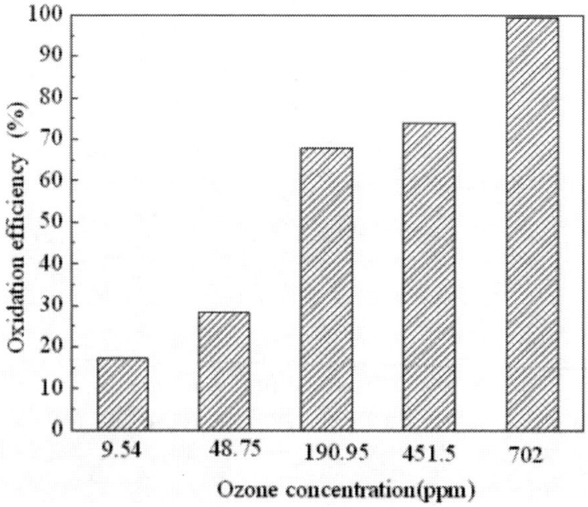

Figure 3. Oxidation efficiency of ammonium sulfite as functions of ozone concentration.

3.2.2. Gas flow rate

The influence of ozone flow rate on the oxidation efficiency of ammonium sulfite was shown in figure 4, in which the gas flow rate was adjusted by a flow mete in range of 1 to 3 L min^{-1}. The treatment time was 25 minutes, the ozone concentration was 702 ppm, and the initial concentration of ammonium sulfite was 0.2 mol L^{-1}.

It was obvious that the oxidation efficiency of ammonium sulfite had a significant increase as gas flow rate increased. Meanwhile, ammonium sulfite was totally oxidized when the flow rate reached 3 L mol^{-1}. The reason was that on condition of the same ozone concentration, a bigger flow rate could lead to a larger total quantity of ozone exposed to the solution.

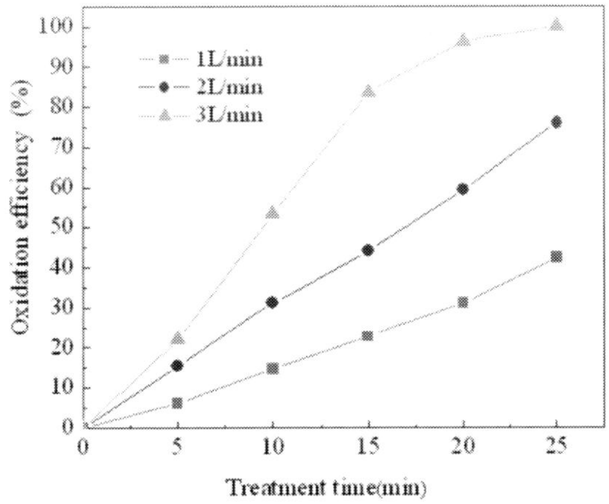

Figure 4. Oxidation efficiency of ammonium sulfite as functions of gas flow rate.

3.2.3. Initial concentration of ammonium sulfite

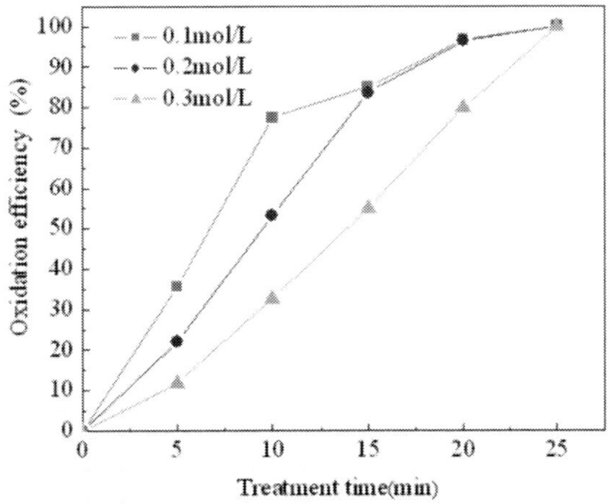

Figure 5. Oxidation efficiency of ammonium sulfite as functions of initial concentration.

Figure 5 showed the comparison of the reaction efficiency in different initial concentration solutions. The treatment time was 25 minutes, the ozone concentration was 702 ppm, and the gas flow rate was 3 L min^{-1}, the reaction temperature was 30 ℃. It could be seen from figure 5 that the oxidation efficiency of ammonium sulfite decreased with the increase of the initial concentration. This was because the density of ozone approximately remained constant at the given ozone flow rate. The higher initial concentration of pollutants would need more ozone, which might become insufficient for the decomposition of ammonium sulfite.

3.2.4. Reaction temperature

The effect of reaction temperature on the oxidation of ammonium sulfite was investigated, in which the temperature was controlled at certain value by a waterbath, and the result was shown in figure 6. The treatment time was 25 minutes, the ozone concentration was 702 ppm, and the gas flow rate was 3 L min^{-1}, the initial concentration of ammonium sulfite was 0.2 mol L^{-1}.

The figure illustrated that the increase of the reaction temperature facilitated the oxidation of ammonium sulfite. The reasons could be stated as the following three points. Firstly, high temperature of the solution could reduce the stability of ozone, which would benefit the oxidation reactions of ammonium sulfite. Secondly, the increase of temperature could enhance the transfer of particles, which improved the reaction rates between ammonium sulfite and oxidative species. Thirdly, the increasing chaos of the particles enlarged the equivalent resistance of solution and enhanced the effective voltage [8].

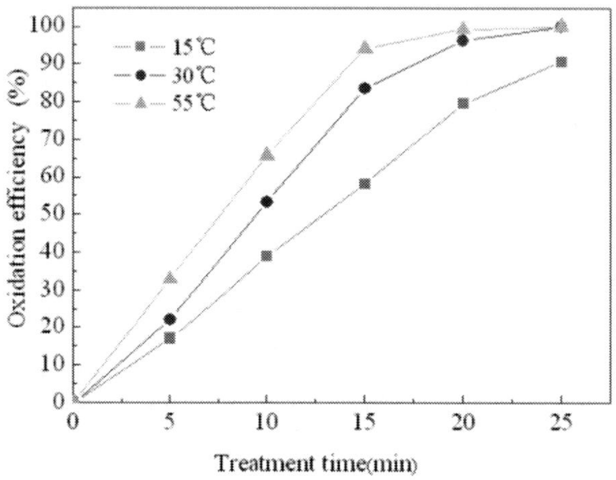

Figure 6. Oxidation efficiency of ammonium sulfite as functions of reaction temperature.

4. Conclusion

An ozone oxidation system was employed to oxidize ammonium sulfite in aqueous. The oxidation efficiency of ammonium sulfite by ozone and traditional air aeration was compared and the effect of experimental conditions including ozone concentration, reaction temperature etc on the oxidation of ammonium sulfite was examined. The results show that ozone oxidation presented a better performance; an increase in the ozone concentration, gas flow rate and reaction temperature advanced improved the oxidation efficiency of ammonium sulfite; Meanwhile, reduction of the initial concentration of ammonium sulfite would bring lower oxidation efficiency.

Acknowledgments

This work is partially supported by the Fundamental Research Funds for the Central Universities [Grant NO. DUT12RC (3)12].

References

[1] Wang Y J and Wang W Y 2011 *Boiler Manufactures (in Chinese)*. **3** 45
[2] Hikita H, Asai S and Tsuji T 1978 *J. Chem. Eng. Jpn.* **11** 236
[3] Kang H W, Zhou Y B and Li J P 2009 *Guangzhou Chemical Industry (in Chinese)*. **37** 132
[4] Wang C Y, Li H Y and Xue S K 2009 *Guangzhou Chemical Industry (in Chinese)*. **36** 139
[5] Li H and Yu Y Q 2009 *Guangzhou Chemical Industry (in Chinese)*. **37** 12
[6] Birdsall C M, Jenkins A C and Spandinger E 1952 *Anal. Chem.* **24** 662

[7] Nair S A, Yan K P, Safitri A, Pemen A J M, van Heesch E J M and Ptasinski K J 2004 *J. Electrostat.* **178** 261
[8] Yu Z M, Lu F, Ye Q Z, Liang F H and Li J 2009 *High Voltage Engineering (in Chinese)*. **35** 36

Influence of power supply on the generation of ozone and degradation of phenol in a surface discharge reactor

Yan Zhao, Kefeng Shang, Lijuan Duan, Yue Li, Jiutao An, Chunyang Zhang, Na Lu, Jie Li*, Yan Wu
Institute of Electrostatics and Special Power, Dalian University of Technology, Dalian, China

E-mail: lijie@dlut.edu.cn

Abstract. A surface Dielectric Barrier Discharge (DBD) reactor was utilized to degrade phenol in water. Different power supplies applied to the DBD reactor affect the discharge modes, the formation of chemically active species and thus the removal efficiency of pollutants. It is thus important to select an optimized power supply for the DBD reactor. In this paper, the influence of the types of power supplies including alternate current (AC) and bipolar pulsed power supply on the ozone generation in a surface discharge reactor was measured. It was found that compared with bipolar pulsed power supply, higher energy efficiency of O_3 generation was obtained when DBD reactor was supplied with 50Hz AC power supply. The highest O_3 generation was approximate 4 mg kJ^{-1}; moreover, COD removal efficiency of phenol wastewater reached 52.3% after 3 h treatment under an AC peak voltage of 2.6 kV.

1. Introduction

In recent years, degradation of biologically recalcitrant and toxic organic compounds in water has become the focus of environmental remediation efforts. As a new advanced oxidation process, non-thermal discharge plasma technology exhibits a great prospect in wastewater treatment. There are two kinds of processes (physical and chemical processes) during non-thermal discharge plasma treatment, which are thought to occur simultaneously in synergism for direct or indirect degradation of toxic organic compounds. The physical processes include strong electric field, ultraviolet light and shock waves etc. The chemical processes mainly include the formation of various active radicals, such as •OH, •H, •O, $•O_2$ and $•HO_2$, and active molecules, such as H_2O_2 and O_3 [1].

In the present study, a surface discharge type DBD reactor with a waste water volume of 40 L was designed by Dalian university of technology to purify the phenol wastewater. As we know, the types of power supply for the surface discharge type DBD reactor has an important effect on pollutants degradation efficiency and energy yield. The discharge characteristics of self-designed power supplies (50 Hz AC, 200 Hz AC and bipolar pulse power supplies) and their influences on the O_3 generation were studied. The power supply which is advantageous for O_3 generation was selected to degrade phenol in water.

* To whom any correspondence should be addressed.

2. Experimental

2.1. Surface discharge type DBD reactor

The schematic diagram of the surface discharge type DBD reactor was shown in figure 1. The reactor was a cylindrical stainless steel tube with an inner diameter of 280 mm and a height of 1000 mm. The high-voltage electrode was made of stainless steel spiral, which was tightly attached on the inside wall of a cylindrical quartz tube [2]. The diameter of the spiral was 1.5 mm and the thread pitch was about 15 mm. The inner diameter of the quartz tube was 40 mm and the length was 900 mm. The whole discharge length of quartz glass tube was 700 mm. The high-voltage electrode system was up-scaled by adding four parallel connections of quartz tubes. There was an aerator at the bottom of each quartz glass tube .The outside wall of the cylindrical stainless steel tube reactor was grounded.

In each experiment, 40 L phenol wastewater was treated. Air was injected into the four quartz glass tubes before discharge treatment, and then the high voltage electrode system was immersed into the wastewater.

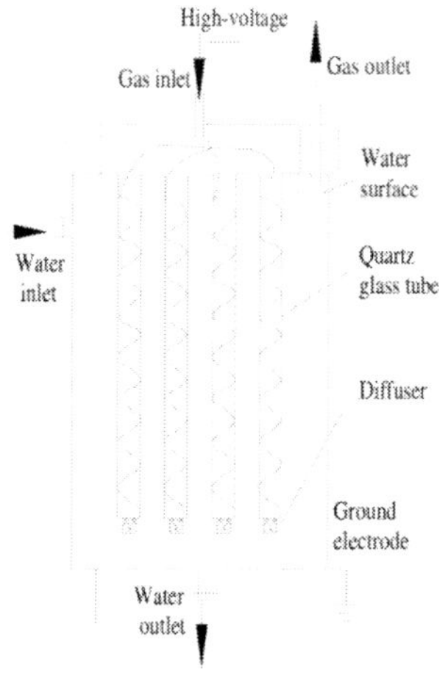

Figure 1. schematic diagram of the reactor

2.2. Methods and analyses

The discharge voltage and current were measured with a Tektronix TDS2012 digital oscilloscope equipped with a Tektronix (P6015A) high voltage probe and a Tektronix (A6021) current probe.

Ozone concentration was measured by the method of potassium iodide. The equation is showed following:

$$c\ (O_3) = A \times B \times 2400 / V_0 \tag{1}$$

Where c (O_3) is the concentration of O_3, A is the usage volume of $Na_2S_2O_3$, B is the concentration of $Na_2S_2O_3$ and V_0 is the volume of a gas extracted from reactor.

The COD removal efficiency (η) is defined as the equation (2).

$$\eta = \frac{m_{\mathrm{COD}_0} - m_{\mathrm{COD}_t}}{m_{\mathrm{COD}_0}} \times 100\% \tag{2}$$

Where $m_{\mathrm{COD}0}$ is the removed amount of COD (g), the $m_{\mathrm{COD}0}$ and $m_{\mathrm{COD}t}$ are the amount of COD at time 0 and t, respectively.

3. Results and Discussion

3.1. Effect of power supply type on discharge characteristics of reactor

Three power supplies (power frequency AC, high frequency AC, bipolar pulse power supply) were designed to investigate the effect of power supply types on their discharge characteristics. Figure 2-4 present the discharge characteristics of surface discharge reactor supplied by AC power supply (50Hz), AC power supply (200Hz) and bipolar pulse power supply, respectively.

The waveforms in figure 2-4 were obtained at the highest normal input, namely, when the applied voltage increased further, spark breakdown or other abnormal discharge would occur. From figure 2 and 3, the amplitude of current and voltage input by 50 Hz AC power supply was higher than 200 Hz AC power supply.

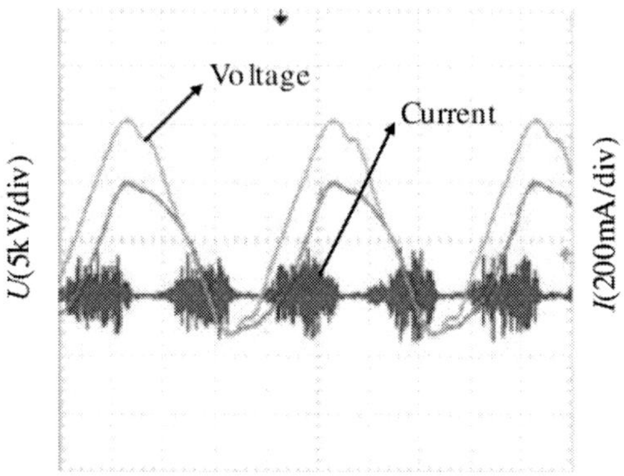

Figure 2. Voltage and current waveforms 50Hz AC power supply.

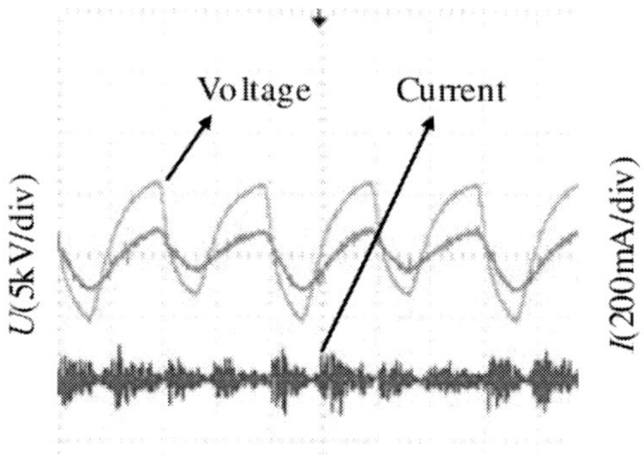

Figure 3. Voltage and current waveforms of 200Hz AC power supply.

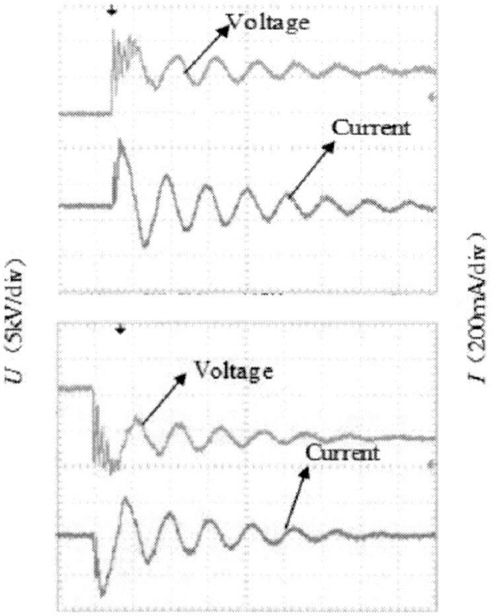

Figure 4. Voltage and current waveforms of bipolar pulse power supply.

The input power (P) of AC power supply was calculated by voltage-charge (V-Q) Lissajous method, and the input power (P) of pulse power supply was calculated by integration of the voltage (u) and current (i) according to equation (3). The input energy was calculated by discharge power (P) multiplied by time.

$$P = \int vi\mathrm{d}t \tag{3}$$

According to the calculation results of discharge power, higher input power was obtained by 50 Hz AC power supply, which may attribute to the fact that the breakdown voltage of 50Hz AC power supply was much higher than 200Hz AC power supply and bipolar pulse power supply in our experimental systems.

3.2. Effect of types of power supply on generation of ozone

Gas phase surface discharge plasma can generate a large number of reactive species (such as high energy electrons, ions, radicals and O_3) and ultraviolet light etc, which are directly related to the type of power supply for plasma reactor [3]. Figure 5 presented the effect of types of power supply on zone generation efficiency. The experimental results show that compared with 200 Hz AC and bipolar pulse power supply, the O_3 generation efficiency was higher when DBD reactor was supplied with 50Hz AC power supply. For example, when the air flow rate was $1 m^3 h^{-1}$, approximately 4 mg kJ^{-1} of O_3 concentration was obtained when 50Hz AC power supply was utilized, while only 1.5 and 1.6 mg kJ^{-1} O_3 was obtained when 200 Hz AC and bipolar pulse power supply were utilized, respectively. Therefore, 50Hz AC power supply was selected to supply reactor for phenol degradation in water.

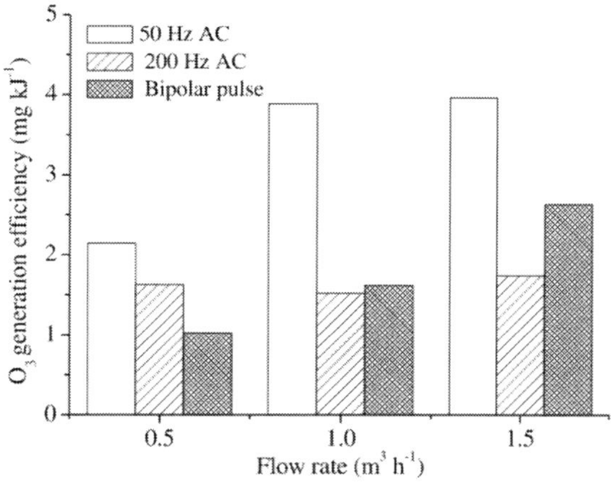

Figure 5. O_3 concentration can be obtained by different type of power supply.

3.3. Effect of discharge voltage on COD removal of phenol wastewater

Discharge voltage is an important parameter to evaluate the practical application of this technology in pollution control. The effect of discharge voltage on COD removal of phenol wastewater was evaluated and the result was shown in figure 6. The air flow rate was $1 m^3 h^{-1}$. It was found that COD removal efficiency increased with the discharge voltage. At discharge voltage of 26 kV, COD removal rate reached 52.8% within 3 h, while approximately 47.4% and 39.2% of COD could be removed at 22 kV and 18 kV within the same time, respectively. It is generally believed that more energetic electrons are produced at higher discharge voltage, leading to the accelerated formation of active species such as ozone [4].

Therefore, higher COD removal efficiency was achieved in higher discharge voltage, and the optimized discharge voltage was 26 kV in the present study. The experiments were all conducted under the condition of discharge voltage of 26 kV in the following studies.

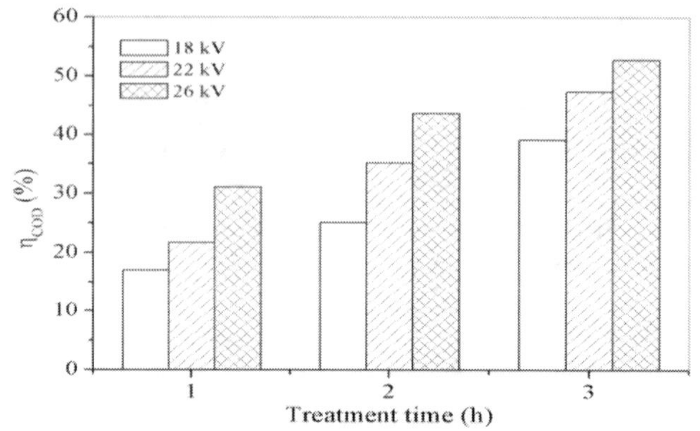

Figure 6. The COD removal efficiency at different discharge voltage.

3.4. Effect of air flow rate on COD removal of phenol wastewater

Figure 7 presented the effect of air flow rate on COD removal of phenol wastewater. As shown in figure 7, COD removal efficiency decreased with the increase of air flow rate [5]. When the air flow rate increased from 1 to 2 $m^3 h^{-1}$, COD removal efficiency decreased from 52.8% to 40.3% after 3 h of discharge treatment. It may be due to the fact that the residence time of ozone in aqueous was shortened and the utilization efficiency of ozone decreased with the increase of air flow rate.

Due to the pressure restriction of water, the air flow rate must be not less than 1 $m^3 h^{-1}$ (if the air flow rate decreased further, water maybe penetrate the aerator into the discharge tube) in the present system. The optimized air flow rate was 1 $m^3 h^{-1}$.

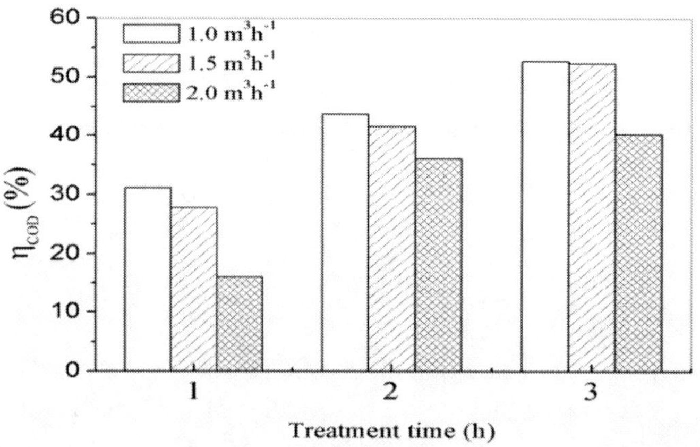

Figure 7. The COD removal efficiency at different air flow rate.

4. Conclusions

In this paper, a surface discharge type DBD reactor was utilized to degrade phenol in water, and the effect of different power supplies on the ozone generation was studied. The energy efficiency of ozone generation in surface DBD reactor with 50 Hz AC power supply was higher than 200 Hz AC power supply and bipolar pulse generator, respectively. Thus, 50 Hz AC power supply was selected to degrade phenol in water. The results show that COD removal efficiency of phenol wastewater increased with the discharge voltage and decreased with the air flow rate. COD removal efficiency of phenol wastewater reached nearly 53% at the air flow rate of 1 $m^3 h^{-1}$ and a peak voltage of 26 kV,

Acknowledgments

This work is partially supported by the Fundamental Research Funds for the Central Universities [Grant NO. DUT12RC(3)12].

References

[1] Hoigne J, Bader H 1983 *Water Res.* **17** 185
[2] Sunka P, Babicky V, Clupek M 1999 *Plasma Sources Sci. Technol.* **8** 258
[3] Gai K, Dong Y J 2005 *Plasma Sour Sci .Tech.* **14** 589
[4] Suarasan I, Ghizdavu L, Ghizdavu I 2002 *J. Electrostat.* **54** 207
[5] Zhang Y Z, Zheng J T, Qu X F 2008 *Chemosphere* **70** 1518

Journal of Physics: Conference Series **418** (2013) 012132 doi:10.1088/1742-6596/418/1/012132

Degradation of m-dihydroxybenzene by contact glow discharge electrolysis in aqueous

Ke Gai[1*], Huili Qi[1], Dongping Ma[1], Chunlin Wang[2]
1 Chemistry and Chemical Engineering, Longdong University, Lanzhou road, Qing yang, China
2 College of Life Sciences, Longdong University, Lanzhou road, Qing yang, China

E-mail: gaike310@sohu.com

Abstract. This paper reported the degradation of m-dihydroxybenzene aqueous solution with contact Glow Discharge Electrolysis. The rate of degradation in different conditions such as pH, H_2O_2, Fe^{2+}, methanol, and other affecting factors were studied. The results showed that there is faster removal rate when the solution is in a relatively higher acidity; H_2O_2 can improve the efficiency rate. Fe^{2+} can promote reaction, but radical elimination agent of methanol will decrease the rate of the reaction. On the basis of analyzing the ultraviolet (UV) spectra of the solution and the intermediate products from High Performance Liquid Chromatography-Mass Spectrum (HPLC-MS), reaction pathway was proposed.

1. Introduction

Contact glow discharge electrolysis (CGDE) has been studied since the 1950s [1-3]. CGDE is a novel type of electrochemical process in which plasma was directly produced by the electrical discharge against water. In CGDE, various active species such as hydroxyl radicals, hydrogen peroxide, as well as several other species was generated locally between the platinum wire anode electrode and the surface of liquid solution when the applied voltage was increased to a threshold value, normally 480V. In the reaction zone within the plasma around the anode, vapor H_2O molecules were ionized, activated and bombarded each other, and then the free OH radicals were produced through charge transfer, with a resultant of the production of free OH radicals and sometimes H atoms. On the other hand, in the liquid-phase reaction zone near the plasma-anolyte interface, some liquid H_2O molecules were broken up into H_2, H_2O_2 and O_2 as it is being bombarded by $H_2O^+_{gas}$ from the anodic plasma. When a mixture of the active species OH, H and $H_2O^+_{gas}$ diffuse across the primary zone into the bulk electrolyte, they will act as strong oxidizing agents and induce oxidation reaction in the solution. Therefore, CGDE

[*] To whom any correspondence should be addressed.

could be used to induce some unusual chemical changes in the solution and could be used in an advanced oxidation processes [4-8]. Recently, there have been several reports on the degradation of certain organic target compounds dissolved in water by use of CGDE [9-17]. In this work, degradation of m-dihydroxybenzene aqueous by CGDE has been studied.

2. Experiment

The experimental apparatus consists of a reactor and a high voltage power supply, which is illustrated in figure 1. The reaction vessel is made up of dual glass tubes connected by a sintered quartz glass disk at the bottom. The anode was a platinum wire of 0.5 mm diameter encased in a glass tube located in the bottom center of a glass cylinder reactor, with the needle tip protruding 1 mm from the glass tube. The cathode was a stainless steel stick of 5 cm inserted through the top of the reactor. The power supply was the WYJ-1000V2A direct current power unit which provides voltages of 0-1000 V and current output of 0-2 A. During the reaction course, the cell voltage was adjusted to 480 V and the current was about 80 mA. A glow spot was observed and stretched into the solution of nearly 3mm in height. In addition, the reaction cell was coated by a water jacket, in which the temperature was maintained at room temperature to ensure the long-time reaction.

M-dihydroxybenzene was dissolved in a potassium sulfate solution (2 $g \cdot L^{-1}$). The pH of the solution was adjusted with dilute sodium hydroxide and dilute sulfuric acid to the expected value. During the reaction, a small portion of the solution was periodically sampled out from the sampling port to measure the reduction of reactant and intermediate product immediately by a UV-V is spectrophotometer (Analytic Jena) and GC-MS and HPLC from Shimadzu Company respectively.

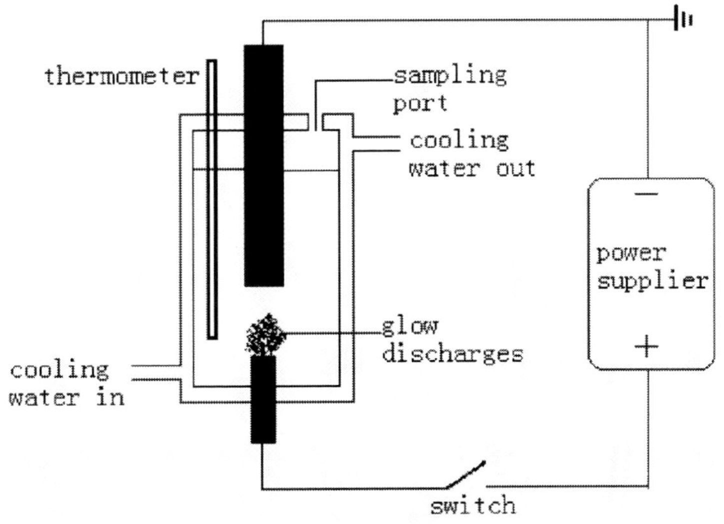

Figure 1. The schematic diagram of the experimental apparatus.

3. Results and discussion

3.1. Accelerated effect of ferrous ions

It is well known that the ferrous ion could react with hydrogen peroxide and produce hydroxyl radicals with powerful oxidizing abilities to degrade organic pollutants. Hickling et al found that the primary yield of oxidizing species was about four times larger than the case without ferrous ion in the solutions [18]. This can be achieved by Fenton's reaction, therefore, some ferrous sulfate was added into the solution and results were shown in figure 2. The initial concentration of m-dihydroxybenzene was 200 $mg{\cdot}L^{-1}$, pH value was 6.00 and applied voltage was 480 V. It can be seen that Fe^{2+} has an evident catalytic effect. Previous studies showed that the predominant products were hydrogen and hydrogen peroxide in the gas-phase and the solution when CGDE was carried out on an aqueous solution without organic substances. Generally, hydrogen peroxide was not able to cause the degradation of pollutants in CGDE. Therefore, the resultant hydroxyl radical from plasma might be responsible for the oxidation of m-dihydroxybenzene as well as its incompletely degraded products. The reasons as follows:

$$Fe^{2+} + H_2O_2 \rightarrow Fe^{3+} + HO\cdot + OH^- \quad (1)$$

$$Fe^{2+} + HO\cdot \rightarrow Fe^{3+} + OH^- \quad (2)$$

$$OH\cdot + H_2O_2 \rightarrow H_2O + HO_2\cdot \quad (3)$$

$$Fe^{3+} + H_2O_2 \rightarrow Fe^{2+} + HO_2\cdot + H^+ \quad (4)$$

$$Fe^{3+} + e^- \rightarrow Fe^{2+} \quad (5)$$

From the above reactions, it can be found that in the presence of Fe^{2+}, the H_2O_2 produced by the discharge was converted to HO· again. In other word, it accelerates the degradation.

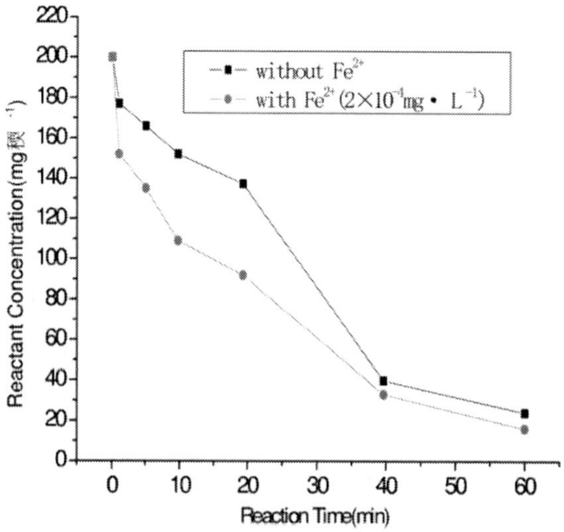

Figure 2. Effects of ferrous sulfate (U=480 V, T=293 K, c_0=200 ppm, V=200 mL, pH=6.00).

3.2. pH Effects

Special attention was paid to investigate the effects of solution pH on m-dihydroxybenzene degradation since the pH often plays an important role in the wastewater treatment. Figure3 shows the effects of pH on the degradation of m-dihydroxybenzene. Faster m-dihydroxybenzene removal rate can be achieved in a relatively higher acidity conditions. This maybe presumed that more ·OH was produced in CGDE under acidity conditions.

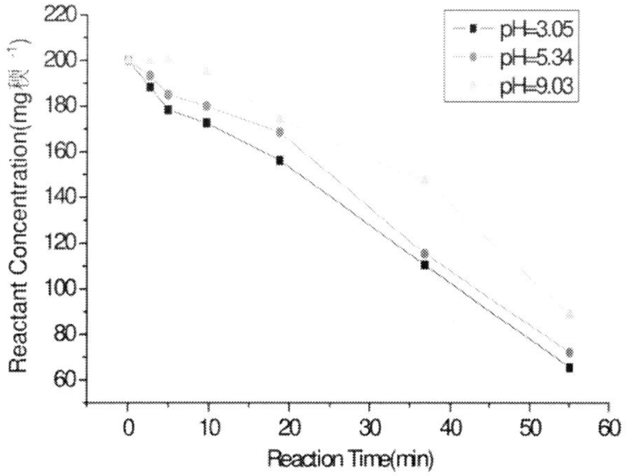

Figure 3. Effect of pH on the degradation (U=480 V, T=293 K, c_0=200 ppm, V=200 mL).

3.3. Effects of methanol and hydrogen peroxide

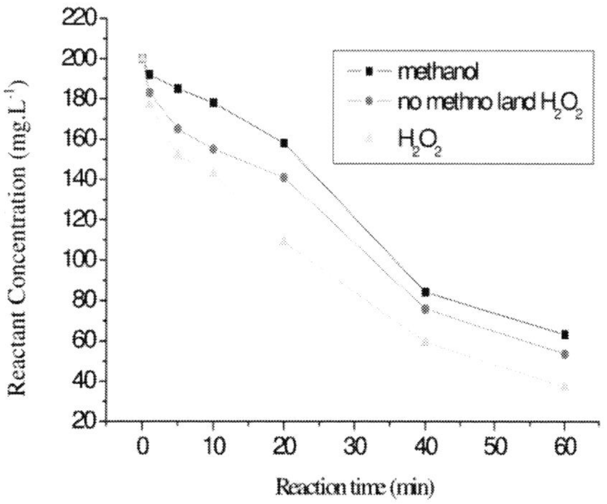

Figure 4. Effects of methanol and hydrogen peroxide (U=480 V, T=293 K, c_0=200 ppm, V=200 mL, pH=6.00).

In order to investigate the role of H_2O_2 and methanol in removing of m-dihydroxybenzene, both H_2O_2 and methanol were added into the solution. Here, 100 mg·L^{-1} H_2O_2 and 100 mg·L^{-1} methanol were added to 200 mL solutions of m-dihydroxybenzene, respectively, in order to study the degradation behavior. Figure 4 indicates the concentration change of m-dihydroxybenzene in the presence of H_2O_2 and methanol during 60 minutes of treatment. The results show that the removal efficiency of m-dihydroxybenzene in the presence of hydrogen peroxide is higher than without hydrogen peroxide, and the removal efficiency without hydrogen peroxide and methanol is higher than with methanol. It may be explained that the concentration of ·OH is increased when H_2O_2 was added into the m-dihydroxybenzene solution and hydroxyl radical is the most responsible oxidant in the degradation of m-dihydroxybenzene.

3.4. Byproduts formation for m-dihydroxybenzene degradation

Special attention was paid to analyze the intermediate in the process of degradation of m-dihydroxybenzene. On the basis of the detailed analysis of the HPLC and HPLC-MS, some intermediate such as benzoquinone, phenol, benzene were produced in the initiation of the discharge. As the electrolysis went on, some carboxylic acids such as formic, oxalic and malonic acids and so on were produced in the process of reaction [19]. All of the intermediate products were easily oxidized continuously under the present conditions. The ultimate products were inorganic carbon, which might exist as bicarbonate or carbon dioxide. Figure 5 gives the variation of intermediate products in degradation process.

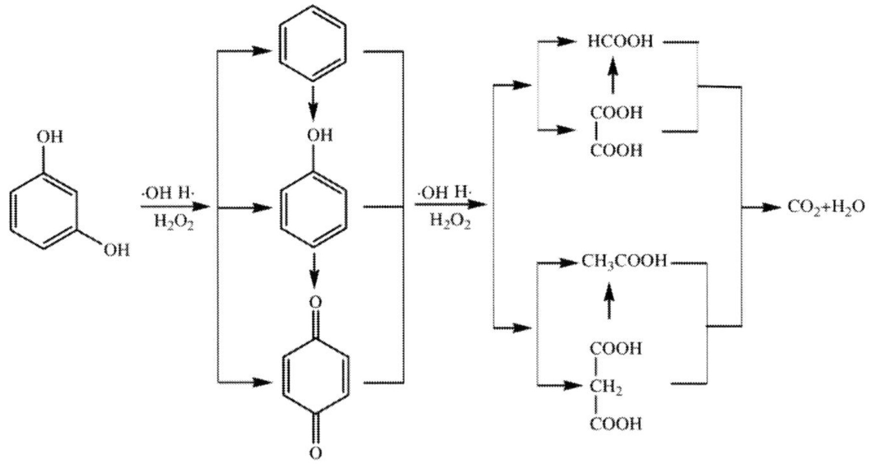

Figure 5. M-dihydroxybenzene degradation pathway in CGDE.

Previous studies have showed that the predominant products were hydrogen and hydrogen peroxide in the gas-phase and the solution respectively, when CGDE was carried out on an aqueous solution without organic substances. Under the experimental condition, the hydrogen peroxide was not able to cause the degradation of m-dihydroxybenzene, therefore, the resultant hydroxyl radical from plasma might be responsible for the oxidation of m-dihydroxybenzene as well as its incompletely degraded products. On the basis of the detailed analysis of the intermediate products and kinetically consideration, it was proposed that the successive attack of hydroxyl radicals on the aromatic rings of starting materials was presumed to be the key steps, especially in the initial stage of overall reactions.

4. Conclusions

From the above experiments, the following conclusions can be drawn:

i) m-dihydroxybenzene can be rapidly, exhaustively degraded by glow discharge electrolysis under water.

ii) The degradation can be greatly accelerated in the presence of Fe^{2+}.

iii) Faster m-dihydroxybenzene removal rate can be achieved in a relatively higher acidity conditions.

iv)The degradation can be inhibited in the presence of mechanol, and hydrogen peroxide can greatly improve the indole degradation rate.

These wonderful results suggested that glow discharge electrolysis was an efficient method for degradation of m-dihydroxybenzene, and further study is under way.

References

[1] Hickling A and Ingram M D 1963 *Trans. Faraday Soc.* **60** 783
[2] Denaro A R and Hickling A 1958 *J. Electrochem. Soc.*.**105** 265
[3] Dewhurst H A, Flagg J F and Watson P K 1959 *J. Electrochem. Soc.* **106** 366
[4] Gai K and Dong Y 2005 *Plasma Sources Sci. Technol.* **14** 589
[5] Amano R, Tomizawa S and Tezuka M 2004 *Electrochemistry.* **72** 836
[6] Tezuka M and Iwasaki 1998 *M Denki Kagaku.* **66** 662
[7] Malik M A, Ghaffar A and Malik S A 2001 *Plasma Sources Sci. Technol.* **10** 82.
[8] Ogata A, Miyamae K, Mizuno K, Kushiyama S and Tezuka M 2002 *Plasma Chemistry and Plasma Processing* **22** 537
[9] Tezuka M and, Iwasaki M 2001 *Thin Solid Films* **386** 204
[10] Tezuka M and Amano R 2006 *Water Research* **40** 1857
[11] Tezuka M and Iwasaki *M* 1997 *Denki Kagaku.* **65** 1057
[12] Tezuka M and Iwasaki M 1998 *Thin Solid Films.* **316** 123
[13] Sharma A K, Josephson G B, Camaioni D M and Goheen S C 2000 *Environ. Sci. Technol.* **34** 2267
[14] Sugiarto A T, Ito S, Ohshima T, Sato M and Skalny J 2003 *J. Electrostatics.* **58** 135
[15] Tezuka M and Iwasaki M 2002 *Thin Solid Films.* **407** 169
[16] Kobayashi K, Tomita Y and Sanmyo M 2000 *J. Phys. Chem. B.* **104** 6318
[17] Iniesta J, Michaud P-A, Panizza M, Cerisola G, Aldaz A and Comninellis C 2001 *Electrochem. Acta.* **46** 3573
[18] Hickling A and Ingram M D 1964 *J. Electroanal. Chem.* **8** 65
[19] Gai K 2006 *J. Chin. Chem. Soc.* **53** 627

Journal of Physics: Conference Series **418** (2013) 012133 doi:10.1088/1742-6596/418/1/012133

Analysis of phenol degradation in pulsed discharge plasma system based on Back-Propagation artificial neural network model

Chao-hua Liu[1], Hui-juan Wang[2, 3*], Cheng-wu Yi[2]
1 School of Management, Jiangsu University, Zhenjiang 212013, China
2 School of Environment, Jiangsu University, Zhenjiang 212013, China
3 Key Laboratory of Modern Agricultural Equipment and Technology, Ministry of Education & Jiangsu Province, Jiangsu University, Zhenjiang 212013, China

E-mail: hjwang@ujs.edu.cn

Abstract. Due to the advantages of Artificial Neural Network (ANN) for analyzing complex reaction system, the oxidation process of phenol in a pulsed discharge plasma system is simulated using an ANN model. Reaction factors including solution with pH values of 3.6, 5.4 and 9.8, and hydroxyl radicals (·OH) scavengers (Na_2CO_3 and n-butyl alcohol) are considered, and the changing trends of phenol degradation under various experimental conditions are simulated and predicted by the Back-Propagation (BP) neural network model. The obtained results show that the BP neural network model can effectively predict the degradation efficiency of phenol in the reaction system. According to the results, acidic solution is favourable for phenol oxidation and increase in the Na_2CO_3 and n-butyl alcohol addition will greatly restrain the phenol degradation. The restraining effect of scavengers on phenol degradation indicates that ·OH is one of most important active species for phenol oxidation in the pulsed discharge plasma system.

1. Introduction

As one of advanced oxidation technologies (AOTs), pulsed discharge process in aqueous solution can produce several oxidation processes, such as UV oxidation, electrical oxidation and chemical oxidation etc. Thus pulsed discharge plasma technology has been paid increasing attention for organic wastewater treatment since Sato et al. reported the decolorization effect of pulsed discharge plasma on dye wastewater [1]. Many conclusions have also been obtained, including factors analysis of the oxidation process [2-4], such as electrical parameters (pulse voltage, pulse frequency), reactor parameters (electrode geometry and electrode distance), gas-liquid conditions (initial solution pH, initial solution conductivity, gas bubbling varieties and gas bubbling rates etc.), and mechanism analysis of pollutants degradation in the pulsed discharge system [5].

For the mechanism of organic compounds degradation in the pulsed discharge plasma system, researchers have done many works. In their researches, hydroxyl radicals (•OH) is considered as one of most important species for organic pollutants degradation in the pulsed discharge plasma system, so it is very important to measure the change of •OH generation amount in the pulsed discharge plasma system for illustrating the important effect of •OH on pollutants degradation. However, •OH has the

* To whom any correspondence should be addressed.

characteristics of short existence time (10^{-6} s), higher reaction activity and lower concentration. Therefore, there is no an effective method to determine •OH concentration in a reaction system. However, the effect of •OH for organic compounds oxidation in the pulsed discharge system could be indicated in other ways, such as investigating the change of solution pH or •OH scavengers addition on the change of organic compounds degradation efficiency. On the other hand, pulsed discharge plasma for wastewater treatment is a complicated process and a variety of factors could control the oxidation process. So it is difficult to clarify the effect of various factors on organic compounds degradation in the pulsed discharge system.

Back-Propagation (BP) artificial neural network is a kind of multilayer feed forward neural network, including an input layer, several hidden layer(s) and an output layer [6]. The basic principle of BP is the input signals transmit forward. Errors of the hidden layers are estimated according to the results obtained from the hidden layers and then the error of the input layer is estimated. The process circulates constantly until the precision of the output layer is achieved. BP has functions of non-linear mapping, high degree fault-tolerance, self-organization and self-learning. So it has been applied in many fields, such as function approximation, pattern recognition, category message, and data compression etc. [7-8].

According to the advantages of BP and the characteristics of pulsed discharge plasma system, the BP artificial neural network was used to simulate the oxidation process of the pulsed discharge plasma to show the crucial effect of •OH in the oxidation process. In this paper, phenol was chosen as a target pollutant and then models of phenol degradation under different conditions of initial solution pH values, Na_2CO_3 addition and n-butyl alcohol addition were simulated. The research would provide some theory basis for explaining the mechanism of pulsed discharge plasma in wastewater treatment.

2. Experimental

2.1. Experimental system

The experimental apparatus consists of a pulsed high voltage power supply and a multi-needle-to-plate electrode geometry reactor, which is same as those of our previous work [9]. The reactor vessel is made from a Plexiglas TM cylinder (75 mm diameter and 100 mm length). The separation distance (d) between the needle anodes and the plate cathode is 15 mm. The seven-needle stainless-steel anode, comprised of one needle in the centre and the other six distributed uniformly around a circle of 7.5 mm radius, is secured within a resin disc. The ground plate electrode is a stainless-steel disc (40 mm diameter, 1.5 mm thickness) with eight 0.5 mm diameter finestras on it. In the investigation, the corona discharge voltage was 21 kV and pulse frequency was adjusted to 50 Hz. Gas was bubbled into the reactor through the finestras by pump and the air bubbling rate was controlled at 5 L min^{-1} in all experiments of the study.

2.2. Experimental method and analytical procedure

Phenol is selected as a target pollutant in the study. 250 mL phenol solution with 100 mg L^{-1} concentration is used in all experiments. The phenol solution was circulated through the reactor by a peristaltic pump. Solutions of 0.01mol/L KCl was employed to adjust solution conductivity. In the paper, the initial electrical conductivity of phenol solutions were adjusted to 100 μS cm^{-1} when n-butyl alcohol was used as radical scavenger and 200 μS cm^{-1} when sodium carbonate was used as radical scavenger. The initial pH value of 100 mg L^{-1} phenol solution, with 100 μS cm^{-1} solution conductivity was about 7.0. Solution pH was hardly affected by the addition of n-butyl alcohol with different concentration into experimental solutions. PH values of solutions were about 10.5 by adding Na_2CO_3 within 0.1–1.0 mmol L^{-1} concentration. In the study, the initial pH of phenol was controlled at 3.6, 5.4, and 9.8, respectively. The addition amount of Na_2CO_3 was 0.1, 0.5 and 1.0 mmol L^{-1}, respectively, while the n-butyl alcohol amount was 0.1, 0.5, 1.0 and 5.0 mmol L^{-1}, respectively. Concentration of phenol was analyzed by means of high performance liquid chromatography.

3. Theory and model analysis

3.1. BP artificial neural network

BP has a structure of three layers or more, including one input layer, several hidden layers, and an output layer. As a kind of multilayer feed forward neural network, the main characteristics of BP are the signals transmit forward and the errors transfer reversely. Through studying and training, the input signals transmit from the input layer to the hidden layers and then to the output layer. The output values are gotten from the functions of every layer. Figure 1 illustrates a structural drawing of a three layers structure of the BP. In the figure, X_1, X_2, $\cdots X_n$ are the input values, Y_1, Y_2, Y_m are the predictive values, and w_{ij} 和 w_{jk} are the weights. Through training, the projective relation between $X^n = (X_1,\ldots,X_n)^T$ and $Y^m = (Y_1,\ldots,Y_m)^T$ could be set up. The network set up through training has capacities of associative memory and prediction.

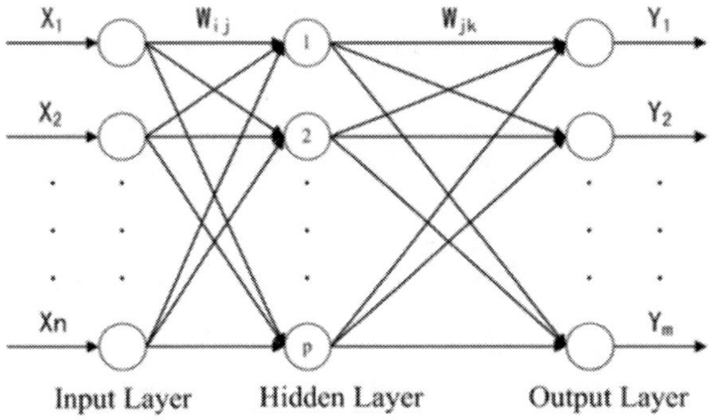

Figure 1. Topology of BP neural network.

3.2. Model and simulation

In the study, initial solution pH, Na_2CO_3 concentration and n-butyl alcohol concentration were taken as influencing factors and the degradation rates of phenol under different experimental conditions were reviewed. Then the optimization toolbox of Matlab was firstly used to determine the reviewing target in the study. The transfer function of the hidden layers nodes was selected as tansig, the transfer function of the output layer nodes was selected as purelin. After normalization, the samples should be training in the network and the initial weights matrix would be generated at random. During the training process, weights and threshold values of every layer could be adjusted and the iteration finished until achieving the given studying target. Precision of the model could be evaluated by comparing the actual values and predicted values, and then the changing trend of phenol degradation under different initial solution pH, Na_2CO_3 addition and n-butyl alcohol addition was simulated.

4. Results and Discussion

4.1. Effect of pH

According to the established BP neural network model, the effect of initial solution pH values for phenol degradation was first considered in the paper. The initial solution pH values were controlled at 3.6, 5.4 and 9.8, respectively. The simulated model of phenol degradation under different initial solution concentrations and discharge time are shown in figure 2.

From the figure 2, it can be seen that the degradation rates of phenol increase with the increase of pulse discharge time within the investigated time and solution pH. On the other hand, it also can be indicated from the figure that the acidic solution condition is favourable for phenol degradation. The degradation rates of phenol are lower in neutral and basic solution and change of the degradation rate is gently under both conditions of neutral and basic solution. Therefore, the solution pH should be controlled at the acidity for achieving higher degradation rates of phenol. The reason for the results is that •OH has higher oxidation effect under the condition of acidic solution.

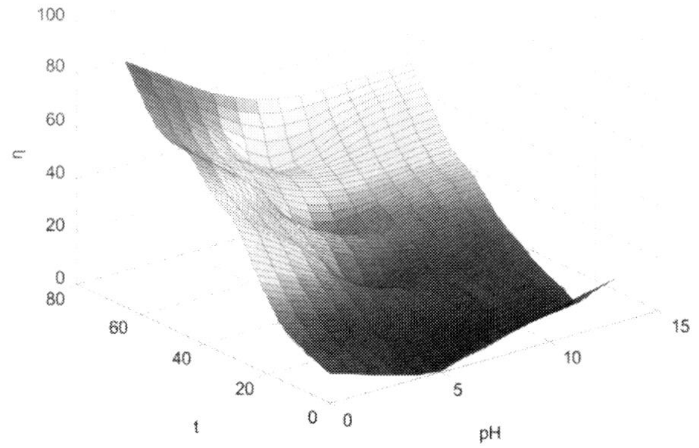

Figure 2. Degradation rate of phenol with different initial solution pH values and time.

4.2. Effect of Na_2CO_3 addition

The simulation results of phenol degradation under different concentration Na_2CO_3 addition and reaction time according the BP model are shown in figure 3.

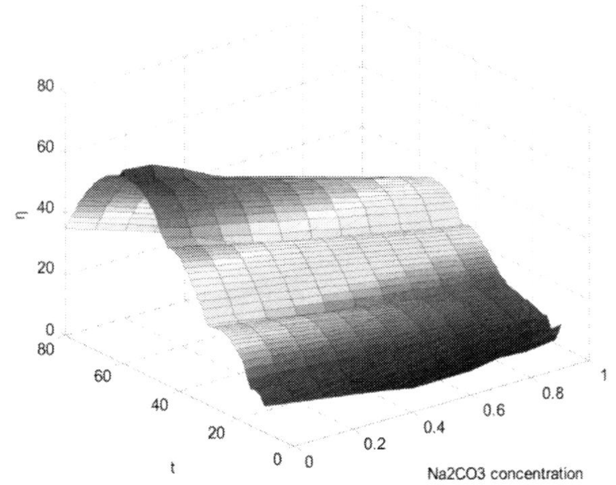

Figure 3. Degradation rate of phenol with different Na_2CO_3 concentration and time.

From the results of figure 3, it can be seen that the degradation rates of phenol decrease with the increase of Na_2CO_3 addition within the concentration scope of Na_2CO_3 addition. The prediction results indicate the collection of Na_2CO_3 for •OH, and also show the effect of •OH on phenol degradation in the pulsed discharge plasma system.

4.3. Effect of n-butyl alcohol addition

Using the BP model, the simulation results of phenol degradation under different concentrations of n-butyl alcohol addition are shown in figure 4.

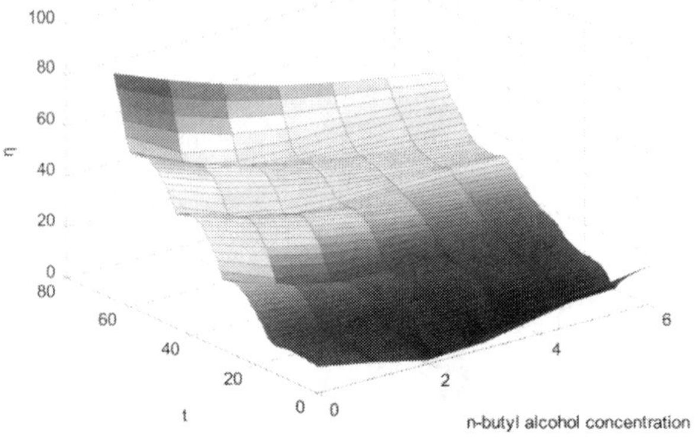

Figure 4. Degradation rate of phenol with different n-butyl alcohol concentration and time.

It can also be seen from figure 4 that with the increase of n-butyl alcohol concentration, the degradation rate of phenol decreases. When the addition concentration of n-butyl alcohol exceeds 2 mmol L^{-1}, the reduction of phenol degradation is little. The results indicate the collection effect of n-butyl alcohol for •OH and all the •OH produced during pulsed discharge process might react with the n-butyl alcohol under the condition of certain amount of n-butyl alcohol addition, phenol degradation under such a condition might due to other physical and chemical effect produced in the reaction system. The results confirm the effect of •OH for phenol degradation and also indicate the change of •OH amount under different n-butyl alcohol addition during pulsed discharge process.

5. Conclusion

In the paper, the BP artificial neural network model of phenol degradation in a pulsed discharge plasma system is set up to review the effect of pH values, Na_2CO_3 addition and n-butyl alcohol addition on phenol degradation. Phenol degradation by changing solution pH values and adding •OH scavengers can be predicted well by the established BP model in the research. From the research, •OH is confirmed to be one of most important species for phenol degradation in the pulsed discharge system.

Acknowledgments

The project sponsored by China Postdoctoral Science Foundation (20110491353), Jiangsu Postdoctoral Science Foundation (1102116C), and the foundation for Talents in Jiangsu University (Jiangsu, China, project No. 08JDG057). The authors would like to thank the financial support.

References

[1] Clements J S, Sato M and Davis R H 1987 *IEEE Trans. Ind. Appl.* **IA-23** 224
[2] Wang H J, Li J and Quan X 2006 *J. Electrost.* **64** 416
[3] Lei L C, Zhang Y, Zhang X W and Shen Y J 2008 *J. Electrost.* **66** 16
[4] Yin X L, Bian W J and Shi J W 2009 *J. Hazard. Mater.* **166** 1474
[5] Sun B, Sato M and Clements J S 2000 *Environ. Sci. Technol.* **34** 509
[6] Wang W Q, Gao N Y, Li L, Zhang K J and Rong W L 2011 *J. Central South University (Chinses)* **42** 260
[7] Li Y, Lei N. 2004 *Value Eng. (Chinese)* **6** 126
[8] Lu J, Huang G L and Li S Z 2009 *J. Shanghai Jiaotong University (Science)* **14** 482
[9] Wang H J, Li J, Quan X and Wu Y 2008 *Appl. Catal. B: Environ.* **83** 72

Degradation of methyl orange waste water by electrochemical oxidation method

Jingyi Tian[1], Kefeng Shang[2*], Xiangxin Xue[2], Lei Yang[3]
1 College of Resources and Materials, Northeastern University at Qinhuangdao, Qinhuangdao, P. R. China
2 Institute of Metallurgical Resources & Environmental Engineering, Key Metallurgical Resourc Recycling Science Laboratory of Liaoning Province, Northeastern University, Shenyang, P. R. China
3 Beijing Institute of Technology, Beijing, P. R. China

E-mail: shangkf2002@yahoo.com.cn

Abstract. Degradation of methyl orange (MO) waste water was conducted by electrochemical oxidation method with PbO_2/Ti electrode as anode. PbO_2/Ti electrode was fabricated by electrochemical deposition of PbO_2 on Ti foil. The micrograph and crystal structure of PbO_2 show that uniform coating of PbO_2 on titanium foil was obtained and the dominant crystal structure was β-PbO_2. Degradation experiments of MO solution indicate that the degradation rate increased with cell voltage and solution conductivity. In addition, air aeration also improved the degradation of MO solution; but an increase in cell voltage or input energy decreased the energy efficiency of MO removal. The energy efficiency reached over 0.1mg kJ^{-1} under a cell voltage lower than 15V, and the removal rate could reach 90%.

1. Introduction

Dyes have wide application in textile or cosmetic industries for dyeing fibres, and a considerable percent of the dye is lost during the dyeing process into the effluents [1].

The emission of dyes into surface water reduces its transparency and the dissolved oxygen [2], furthermore, certain dyes and its degradation products are carcinogenic and highly toxic [3], all of these influences pose a great threat to the water aquatic ecosystem. In recent years, besides microbial degradation process of dyes [3], some advanced oxidation processes (AOPs) for degradation of dyes waste water have been developed with more stringent environmental standards [4-7]. In AOPs, hydroxyl radicals are the important oxidative agents for dye pollutants, therefore, some practices including addition of hydrogen peroxide into reaction system for enhancing the production of •OH were often adopted.

Electrochemical oxidation (EO) is one of the effective technologies for degradation of dye waster water, and EO has obtained extensively attentions [5, 8, 9] since it holds the advantages such as the potential of complete oxidization of dyes to CO_2 and H_2O, operation at room temperature and atmospheric pressure, simple process, etc. Since electrode materials have important effect on the life of electrode and the production active species as well as PbO_2 electrode has a good anti-oxidation

[*] To whom any correspondence should be addressed.

performance. In this paper, we used EO to decolorize methyl orange (a type of acid dye) on Ti/PbO_2 electrode, and the factors influencing the degradation rate of methyl orange were studied.

2. Materials And Methods

2.1. Chemicals

All reagents used in experiments, including methyl orange ($C_{14}H_{14}N_3O_3SNa$), lead nitrate ($Pb(NO_3)_2$), sodium sulfate (Na_2SO_4), hydrochloric acid (HCl), were analytically pure, and the titanium substrate for preparation of PbO_2 electrode was ultra-pure with a purity of 99.99%.

2.2. Preparation of Ti/PbO_2 electrode

Ti/PbO_2 electrode was prepared by electrical deposition method. Titanium sheet with a dimension of 30 mm×30 mm ×0.2 mm was used as the substrate for electrical deposition of PbO_2. Prior to electrical deposition, the Ti substrate underwent five minutes of ultrasonic cleaning in acetone, deionised water in sequence for diminish the grease and acetone, and 30 min of etching in boiling hydrochloric acid with a volume concentration of 10%, and then the Ti substrate was ultrasonically cleaned in deionised water for another 5min.

Electrical deposition of PbO_2 on the clean Ti sheet was performed at a constant anodic current of 60 mA for about 60 min and then at 120 mA for another 30 min. The micro-morphology and crystal form were tested by SEM and XRD, respectively.

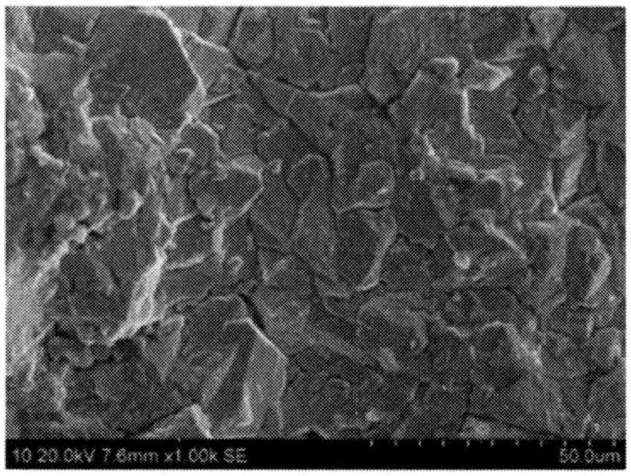

Figure 1. Micro-morphology of Ti/PbO_2 by SEM.

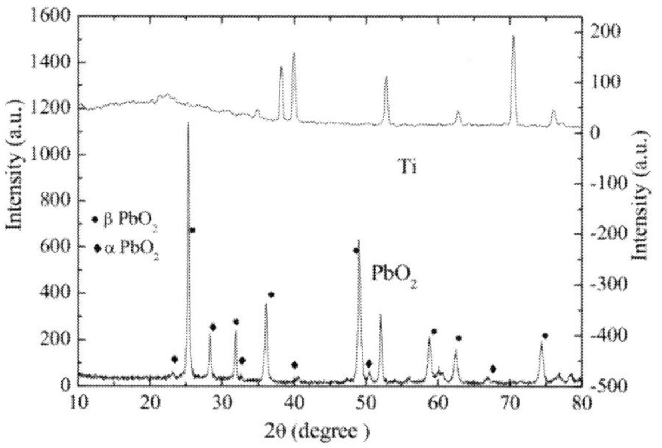

Figure 2. Crystal form of PbO_2 electrode.

Figure 1 shows micro-morphology of Ti/PbO_2 electrode. It can be seen that uniform PbO_2 thin film was formed on the surface of Ti sheet. Figure 2 shows the XRD pattern of Ti/PbO_2 electrode. The figure shows that no diffraction peaks of Ti appeared in the sample of PbO_2 electrode, which means that the Ti substrate was completely covered by PbO_2; also the sample hold some strong diffraction peaks at 25.3°, 36.2°, 49.1°, 58.8°, 74.4° and so on, these characteristic diffraction peaks at 25.3°, 36.2°, 49.1°, 58.8° corresponds to (110), (200), (211), and (310) reflections of β-PbO_2, respectively. Furthermore, the sample hold some weak diffraction peaks at 28.5, 23.4, etc, which correspond to that of α-PbO_2.

2.3. Experimental procedures

Electrochemical degradation of methyl orange was carried out in a 300 mL batch electrolytic cell. 200 mL methyl orange solution with a concentration of 2.5-10 mg/L was used as the simulated waste water, and 0.2 g-0.8 g Na_2SO_4 was added into the simulated waste water were used as supporting electrolyte, the conductivity of solution was gauged by a conductivity meter DDSJ-308A (China). The electrolytic cell was located on a magnetic stirrer to keep the electrolyte well mixed. The Ti/PbO_2 electrode was used as anode, and a pure Ti sheet, vertically and parallel to the anode, was used as the counter electrode in the electrolytic cell.

Potentiostatic electrolyses were carried out in the range of 10 V to 20 V. All the experiments were operated at room temperature. The pH value of methyl orange were measured by a pH meter PB-10 (Sartorius, Germany) , since the solution was not buffered, the pH value changed in the range of 6.2-6.7, which depended on the degradation level and the concentration of methyl orange. The concentration of methyl orange was determined by testing the absorbance of the samples at regular intervals of 15min or 30min. The absorbance of methyl orange were monitored at 466 nm by a UV-vis spectrophotometer UV-1800 (Shimadzu, Japan) The degradation rate of dye waste water was calculated according to the equation below:

$$\eta = \frac{C_0 - C}{C_0} \times 100\%$$

Where is the degradation rate of methyl orange; C_0 is the initial concentration of methyl orange, and C is the concentration of methyl orange at sampling after electrolysis.

In this paper, the initial concentration of MO was 2.5 mg/L, and the cell voltage was fixed at 15 V if not specified.

3. Results and Discussion

Electrochemical oxidation of methyl orange was conducted in an electrolysis cell with Ti/PbO_2 electrode as anode. The influence of air aeration, cell voltage, solution conductivity on degradation of MO was examined.

The influence of air aeration on the degradation rate of MO was shown in figure 3. The flow rate of air was about 0.3 L min^{-1}. Figure 3 indicates that air aeration improved the degradation rate of MO by 11%-25% comparing with that only under magnetic agitation. When air was bubbled into MO solution, the mass transfer of MO to electrode surface can be improved; in addition, oxygen content in solution increased by aeration, and more active species such as H_2O_2, O_3 could be produced and these species could effectively enhanced the decomposition of MO. So in our experiments, air with a flow rate of 0.3 L min^{-1} was always bubbled into the MO solution if not specified.

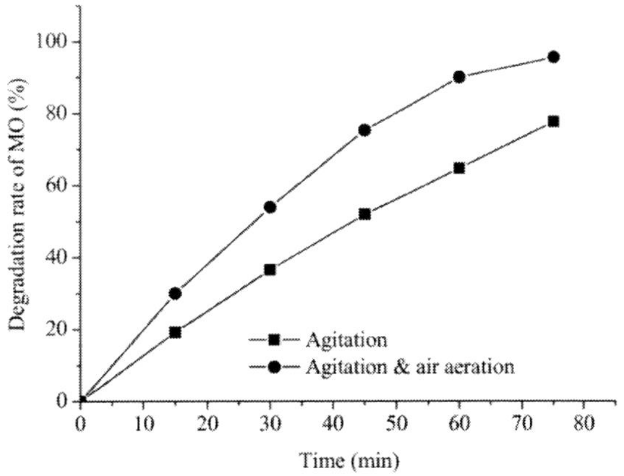

Figure3. Influence of aeration on degradation rate of MO.

Cell voltage had great influence on COD removal of azo dye, and COD removal increased with cell voltage since the increased production of ClO^- species [9]. In our experiments, the degradation of MO also increased with the cell voltage. Figure 4 indicates that the decoloarization of MO at electrolysis time of 60 min reached over 85% under the cell voltage beyond 15 V; but in the same interval, the degradation rate of MO reached 34% under the cell voltage of 10 V. When the cell voltage increased from 10 V to 15 V, and 20 V, the degradation rate increased with voltage, but the averaged power also increased from 0.3 W, to 1.2 W, and 2.7 W, that means a big increase in the energy consumption. In order to find a balancing point on the energy efficiency and the removal rate of MO, the energy efficiency of MO removal was also calculated in figure 5. The energy efficiency of MO removal was defined as the removal mass of MO per input energy (kJ). Figure 5 shows that energy efficiency of MO removal decreased with the input energy and the cell voltage. Considering the energy efficiency, the degradation of MO and the treatment time, electrolysis of MO at 15 V was selected in our experiments.

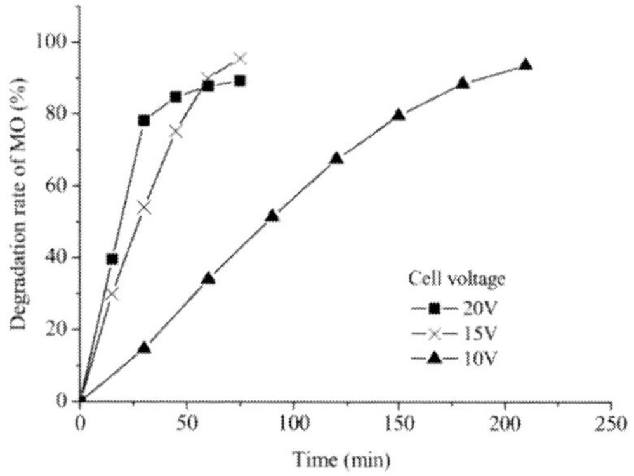

Figure 4. Influence of cell voltage on degradation rate of MO.

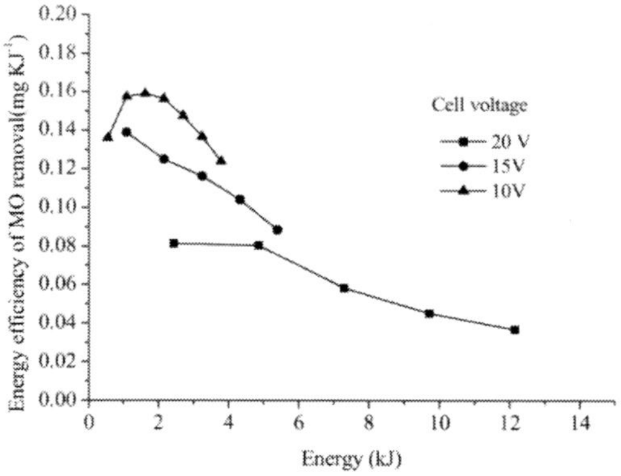

Figure 5. Removal efficiency of MO

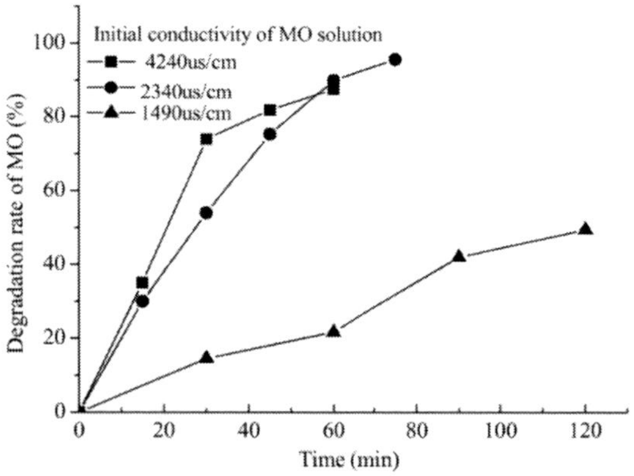

Figure 6. Influence of conductivity of solution on degradation rate of MO.

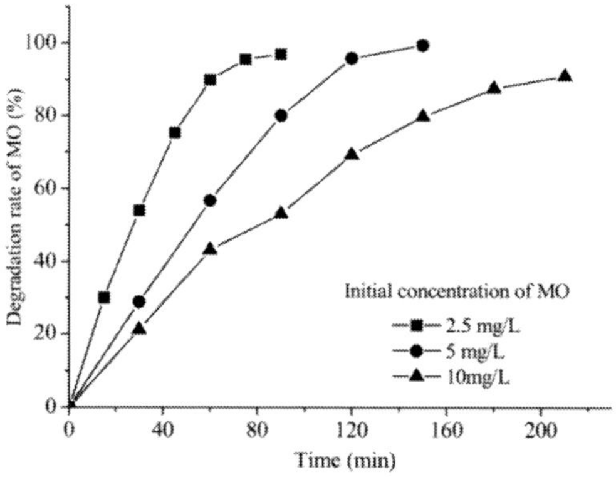

Figure 7. Influence of initial concentration of MO on degradation rate of MO.

Conductivity of MO solution influenced the current density of electrode, so influenced the energy input. The conductivity of solution was regulated by the addition quantity of Na_2SO_4. Figure 6 shows that the degradation of MO increased with the conductivity of solution. When the conductivity increased from 1490 μs cm^{-1} to 2340 μs cm^{-1}, and 4240 μs cm^{-1}, the current increased from 50 mA, to 80 mA, and to 140 mA, which means higher energy was inputted into the electrolytic cell and more active species for oxidation of MO were produced, so MO removal rate increased with the solution conductivity.

Degradation rate of MO under different concentration of MO was shown in figure 7. Figure 7 indicates that degradation rate of MO decreased with the increase of concentration of MO. When the initial concentration of MO was 2.5 mg L^{-1}, the degradation rate of MO reached 97% after 30 minutes, but the degradation rate decreased correspondingly to 80% and 53% when MO concentration increased to 5 mg L^{-1} and 10 mg L^{-1}, respectively. Because the generation of active species is the same under the same experimental conditions including cell voltage and solution conductivity, so the oxidation of MO has an upper limit, which maybe leads to the decrease in the removal rate of MO solution.

4. Conclusions

The results of electrochemical oxidation of MO solution indicate that an increase in cell voltage and solution conductivity within a certain range was advantageous to increase the degradation of methyl orange solution, but an increase in cell voltage or input power decreased the energy efficiency of MO removal. Compared with no air aeration in solution, air aeration could significantly improve the electrochemical oxidation of methyl orange by 11% and 25%. But the removal rate of MO solution decreased with the initial concentration of MO.

Acknowledgments

This work is supported by Natural Science Foundation of Hebei Province (Grant No. E2012501026). The authors thank for the financial supports.

Reference

[1] Gupta Vi K, Jain R, Nayak A, Agarwal S and Shrivastava M. 2011 *Mater. Sci. Eng. C* **31** 1062
[2] Khataee A R and Zarei M 2011 *Desalination* **273** 453
[3] Banat I M, Nigam P, Singh D and Marchant R 1996 *Bioresour. Technol.* **58** 217
[4] Khataee A R and Zarei M. 2011 *Desalination* **273** 453
[5] Parsa J B and Shojaat R 2007 *Physics and Chemistry of Liquids* **45** 479
[6] Brisset J L, Benstaali B, Moussa D, Fanmoe J and Njoyim-Tamungang E 2011 *Plasma Sources Sci. Technol.* **30** 1
[7] Papic S, Muzic M, Koprivanac N, Peternel I and Deanovic M. 2010 *Chemi. Biochem. Eng. Q.* **24** 9
[8] Parsa J B, Rezaei M and Soleymani A R 2009 *J. Hazard. Mater.* **168** 997
[9] Yusuf Y, Koparal S, Ogütveren A, Bakir U. 2011 *J Chem. Technol. Biotechnol.* **86** 261

Role of energy input model on the remediation of the p-Nitrophenol contaminated over-wet soil by pulsed corona discharge plasma

C H Wang[1], X Zhang [2], T C Wang [2], N Lu[2*], J Li [2], Y Wu[2]
1. East China Sea Fisheries Research Institute, Chinese Academy of Fisheries Sciences, Shanghai 200090, China.
2. Key Laboratory of Industrial Ecology and Environmental Engineering (MOE), Institute of Electrostatics and Special Power, Dalian University of Technology, Dalian 116024, China.

E-mail: luna@dlut.edu.cn

Abstract. Low-temperature plasma has exhibited high efficiency for fast remediation of organic-polluted soil with water content (less than 20%). In the present study, the feasibility of remediation of p-Nitrophenol (PNP) contaminated over-wet soil (water content of 100%) was studied using pulsed corona discharge plasma, which was generated in a needle-plate discharge reactor. Effect of energy input model, including pulse voltage and pulse frequency on PNP degradation, was studied. Experimental results showed that about 86.3% of PNP could be smoothly removed after 60 min discharge treatment. PNP degradation efficiency increased with an increase in pulse voltage or pulse frequency, due to the enhancement of energy input. Existence of water contributed to H_2O_2 generation and the amount of exhausted H_2O_2 increased with pulse voltage. This study is expected to provide an alternative method for remediation of contaminated soil containing much water by pulsed discharge plasma without drying pretreatment.

1. Introduction

Xenobiotic substances have been regarded as hazardous compounds in soil due to their stability and toxicity. As a kind of xenobiotic substances, nitrophenols are widely used in a number of industrials, among which p-Nitrophenol (PNP) is a very important member of nitrophenol family. PNP is released into soil by industries of pesticides, dyes and pharmaceuticals manufactures, and accumulates in soil as a result of hydrolysis of organophosphate pesticides parathion, methyl parathion and herbicides dinoseb and dinitrocresol [1-3]. The accumulation of PNP in soil can be accelerated through spreading of wastewaters and sewage sludge containing PNP and the PNP can also leach into the groundwater resources from soil. Because of its high toxicity, PNP has been listed as the 129 priority toxic pollutants by U.S. Environmental Protection Agency [4].

Traditional methods for remediation of organic-contaminated sites include physical [5], chemical and biological methods [6-7], which are often difficult to execute, not environmentally friendly and time consuming. Recently, advanced oxidation processes (AOPs) are receiving great emphasis because of their ability to rapidly oxidize refractory organic contaminants [8]. Pulsed discharge plasma,

[*] To whom any correspondence should be addressed.

one of the AOPs, has received the most extensive investigations [9]. Our previous studies have demonstrated that organic compounds in soil could be removed effectively by pulsed discharge plasma [10-11]. In this process, soli layer was placed onto the ground electrode and the discharge intensity was enhanced due to decrease of discharge gap. For successful plasma soil remediation, it is important to consider the discharge state that influences generation and activity of active chemical species, and it was found that the soil moisture played important role in determining the active species. On one hand, the existence of water molecular contributes to formation of hydroxyl radical (•OH); on the other hand, water molecular can absorb high-energy electrons to weaken the discharge intensity. The PNP contaminated soil, in which the water content is less than 20%, has been fast remediated by low-temperature plasma (LTP) [12-13]. However, no study is carried out on applying LPT method for treating PNP over-wet soil (water content of 100%), which has been existed due to PNP contaminated wastewaters spreading through surrounding soil. In order to save the drying process and realize fast remediate such contaminated soil, it is necessary to explore LTP method for direct degrading PNP in over-wet soil and it is believed that it may be powerful enough for pulsed discharge plasma to remove PTP in soil of water body.

Therefore, remediation of PNP contaminated over-wet soil was conducted in this study. Effects of operational conditions on PNP degradation were performed and the possibility of pulsed discharge plasma for over-wet soil remediation was evaluated. It is expected to provide an environmentally friendly method of pulsed discharge plasma for the remediation of organic pollutants contaminated over-wet soil.

1. Materials and methods

1.1. Materials

PNP (purity > 95%) was purchased from the Chemical Plant of Tianjin, China. All other organic and inorganic reagents used were analytical grade (Tianjin Fuyu Refinery Chemical Co., Ltd. and Shenyang Chemical Reagent Factory, China) and were used as purchased without further purification.

The PNP contaminated over-wet soil was artificially prepared by adding appropriate amount of PNP acetone solution into a given amount of soil. The soil was vigorously homogenized for several hours, and then placed in the fume hood until acetone evaporated entirely, which was confirmed by weight loss. This method gave a uniform PNP distribution in the soil at about 300 mg/kg. The characteristics of the soil could be seen in our previous study [10]. Prior to pulsed discharge plasma treatment, the moisture content of soil was adjusted to 100% with deionized water to simulate over-wet soil.

1.2. Treatment of contaminated over-wet soil

The schematic diagram of the experimental apparatus and a sectional view of the reactor were illustrated in figure 1 and figure 2, respectively. The reaction system consisted of a pulsed high-voltage power supply and a reactor vessel. High-voltage pulses were generated using the combination of a 0-50 kV adjustable DC power supply, a storage capacitor (Ce), an adjustable trim capacitance (Cp) and rotation spark gap switches (RSG1, RSG2). The pulse rise time and pulse width were less than 100 ns and 500 ns, respectively. The pulse frequency and adjustable trim capacitance Cp were 0-150 Hz adjustable and 200 pF, respectively. The high voltage electrode comprised of six stainless-steel needles and the distance of adjacent needle was 18 mm. The ground electrode was a stainless-steel plate. The distance between the high voltage electrode and the ground electrode was 10 mm. The peak pulse voltage and current were measured with a Tektronix TDS2014 digital oscilloscope equipped with a Tektronix P6015A high voltage probe and a Tektronix A6021 current probe, and then the electric power was calculated through the integral of pulse voltage and current under time.

In each experiment, PNP contaminated soil samples (approximately 2.0 g) were spread on the ground electrode, and then adjusted to 100% of water content. Air was used as carrier gas in this study, as shown in figure 1. Concentration of PNP in the over-wet soil was measured after desirable treatment time.

The extraction and analysis of PNP in over-wet soil were the same as previous study [13]. The energy yield (g kWh^{-1}) is defined as the mass of removed PNP divided by consumed energy.

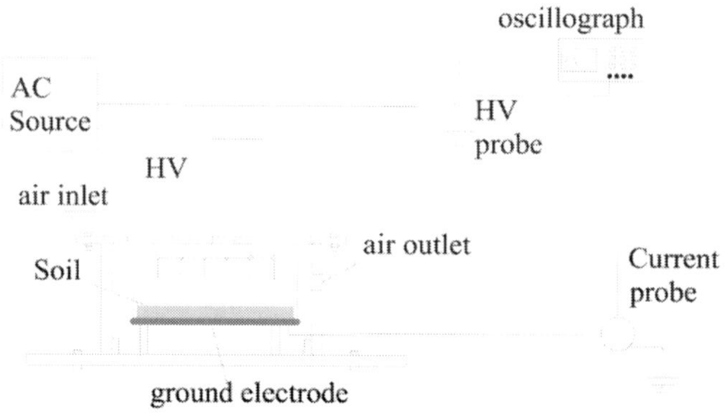

Figure 1. Schematic diagram of the experimental apparatus.

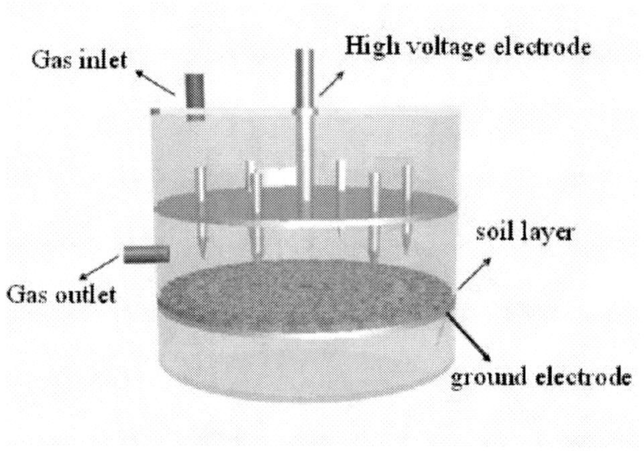

Figure 2. Sectional view of the reactor.

2. Results and discussion

2.1. Effect of pulsed discharge voltage on the discharge characteristics

The increased discharge voltage can make the corona range extended and air gap entirely breakdown, and then electrical filed and the energy input into the reactor will increase. The effect of different pulsed discharge voltage on energy input without the test regents was illustrated in figure 3.With the increase of pulsed discharge voltage from 17.6 to 21.6 kV, the energy input enhanced from 2.17 to 3.27 W under the condition of the pulsed frequency 70 Hz, the electrode 10 mm and gas atmosphere.

2.2. Effect of pulsed discharge voltage on the removal of PNP

Pulsed discharge can generate a large number of chemically active species, accompanied by some physical effects, which are directly related to the pulsed discharge voltage. Pulsed discharge voltage is an important parameter of this technology. The effect of pulsed discharge voltage on PNP degradation was illustrated in figure 4. Herein, the pulsed frequency was 70 Hz, and air flow rate was 0.4 L min^{-1}. Obviously, the increase of pulsed discharge voltage benefited PNP degradation. With the increase of pulsed discharge voltage from 17.6 to 21.6 kV, PNP degradation efficiency enhanced from 73.0% to 86.3% within 60 min of discharge treatment. On the other hand, with treatment proceeding, PNP degradation efficiency also improved deeply.

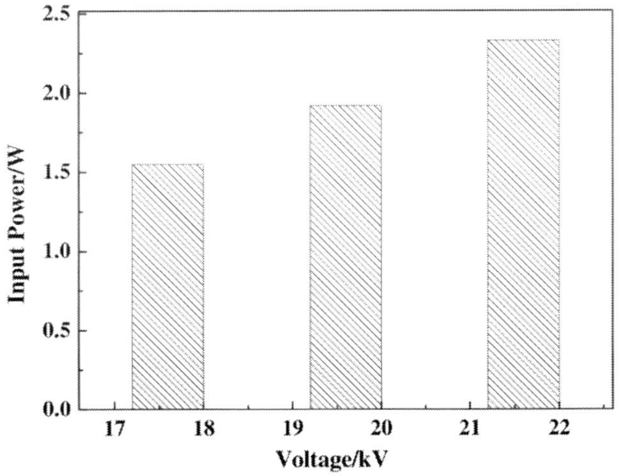

Figure 3. The input power of reactor with different applied voltages.

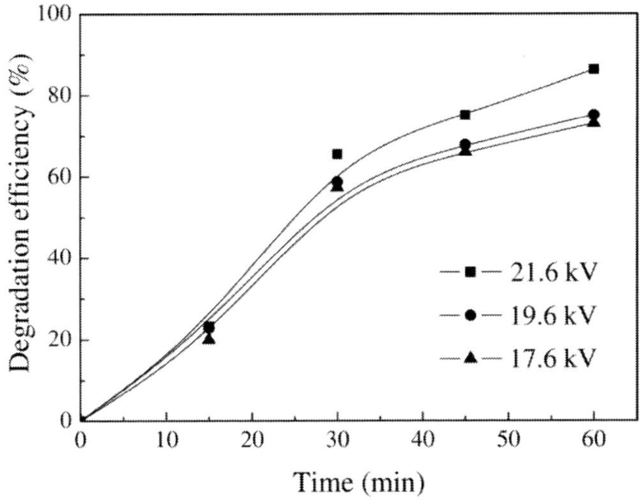

Figure 4. Effect of pulsed discharge voltage on PNP degradation.

With the increase of discharge voltage, more chemically active species would be generated. In the present study, H_2O_2 should be one of important oxidants because of high water content in soil, which can be generated accordingly to following equations [2].

$$H_2O + e \rightarrow H_2O^* + e$$

$$H_2O^* + H_2O \rightarrow H + OH + H_2O$$

$$H + H_2O^* \rightarrow H_2 + OH$$

$$OH + H_2O^* \rightarrow H_2O_2 + H$$

Accordingly, the H_2O_2 generation within 15 min of discharge treatment in different discharge voltages was examined and the result was shown in figure 5. On the one hand, more H_2O_2 was generated in higher discharge voltage. For example, only 0.11 mmol L^{-1} of H_2O_2 was generated in clean over-wet soil at 17.6 kV, while it was enhanced to 0.21 mmol L^{-1} at 21.6 kV. On the other hand, the concentration of H_2O_2 in clean over-wet soil was detected to be much higher than it in contaminated one, which suggested that H_2O_2 involved in PNP removal and played a key role in this discharge system.

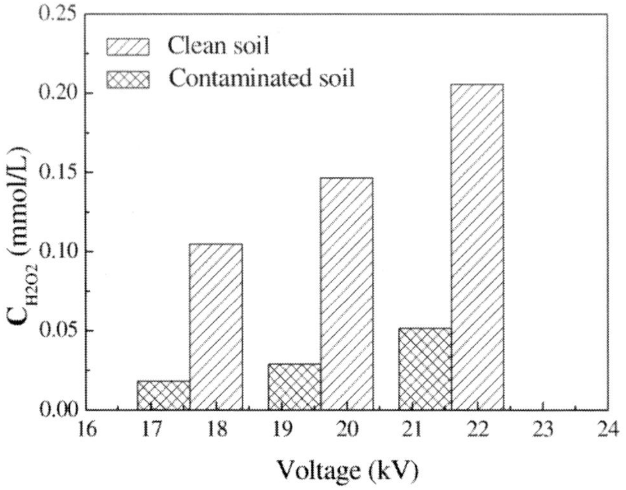

Figure 5. H_2O_2 generation in different discharge voltages.

2.3. Effect of pulse frequency on the removal of PNP

In pulsed discharge plasma processes, the energy injected into the reactor increases with the pulse frequency, which may benefit the removal of organic pollutants. The effect of pulse frequency on PNP removal was shown in figure 6. Herein, the air flow rate was 0.4 L min^{-1}. The increase of pulse frequency enhanced PNP removal. When the pulse frequency was 50 Hz, about 55.2% of PNP was removed within 60 min of discharge treatment, which was enhanced to 84.6% in 90 Hz.

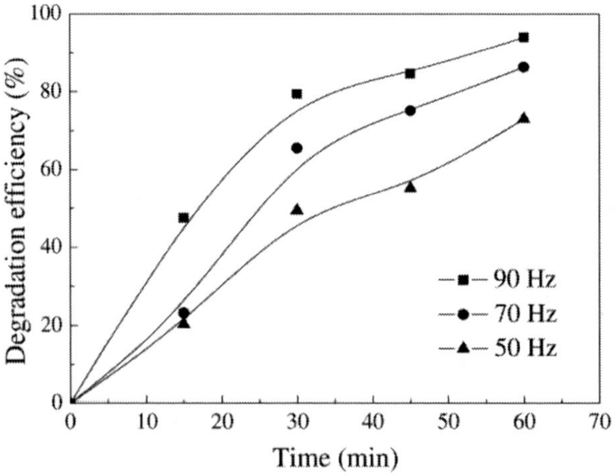

Figure 6. Effect of pulsed frequency of PNP degradation.

The energy yields of PNP degradation within 60 min of discharge treatment in different discharge voltages and pulse frequency were presented in table 1 and table 2, respectively. As could be seen, although increasing discharge voltage and pulse frequency was beneficial for PNP removal, its energy yield was deceased, e.g. the energy yield of discharge treatment in 90 Hz was lower than in 50 Hz. Considering the highest PNP degradation efficiency and comparative energy yield of 21.6 kV to 19.6 kV and 90 Hz to 70 Hz, the better optical experimental condition was conducted under 21.6 kV and 70 Hz.

Table 1. Energy yield of PNP degradation in different discharge voltages.

Discharge voltage (kV)	Degradation efficiency (%)	Energy yield (g kWh^{-1})
17.6	73.0	0.31
19.6	75.1	0.23
21.6	86.3	0.20

Table 2. Energy yield of PNP degradation in different pulsed frequencies.

Pulsed frequency (Hz)	Degradation efficiency (%)	Energy yield (g kWh^{-1})
50	55.2	0.20
70	75.1	0.23
90	84.6	0.19

3. Conclusions

The application of pulsed discharge plasma in removing PNP in over-wet soil was studied. Pulsed discharge plasma presented great potential for PNP contaminated over-wet soil remediation. Under the studied conditions, increasing pulsed discharge voltage or pulse frequency resulted in higher PNP degradation efficiency, due to the enhancement of energy input, and thus more active species would be generated. O-reagents played very important roles for PNP removal, and H_2O_2 also exhibited certain roles during the degradation process. This study is a fundamental research effort, trying to offer an alternative solution for organic pollutants contaminated over-wet soil remediation.

Acknowledgments

This work was supported by the National Natural Science Foundation of China (No. 41001188) and National Natural Science Foundation of China (No. 40901150).

References

[1] Shimazu M, Mulchandani A and Chen W 2001 *Biotech. Bioeng.* **76** 318
[2] Stevens T O, Crawford R L and Crawford D L 1991 *Biodegradation* **2** 1
[3] Schackmann A and Muller R 1991 *Applied Micro. Biotech.* **34** 809
[4] USEPA, July 2002, http://www.scorecard.org.
[5] Anthony E J andWang J S 2006 *Fuel* **85** 443
[6] Yap C L, Gan S Y and Ng H K 2011 *Chemosphere* **83** 1414
[7] Hu J, Nakamura J, Richardson S D and Aitken M D 2012 *Environ. Sci. Tech.* **46** 4607

[8] Fukushima M and Tatsumi K 2001 *Environ. Sci. Tech.* **35** 1771
[9] Joshi A A, Locke B R and Arce P 1995 *J. Hazard. Mater.* **41** 3
[10] Wang T C, Lu N, Li J and Wu Y 2010 *J. Hazard. Mater.* **180** 436
[11] Wang T C, Lu N, Li J and Wu Y 2010 *Environ. Sci. Tech.* **44** 3105
[12] Wang T C, Lu N, Li J and Wu Y 2011 *Environ. Sci. Tech.* **45** 9301
[13] Wang T C, Lu N, Li J, Wu Y and Su Y, 2011 *J. Hazard. Mater.* **195** 276

Journal of Physics: Conference Series **418** (2013) 012136 doi:10.1088/1742-6596/418/1/012136

Evaluation of matching between a pulsed-power and corona discharge reactor containing different thickness of soil

Na Lu[1*], Cheng Lou[1], Cuihua Wang[2], Jie Li[1], Kefeng Shang[1], Yan Wu[1]
[1]Key Laboratory of Industrial Ecology and Environmental Engineering (MOE), Institute of Electrostatics and Special Power, Dalian University of Technology, Dalian 116024, PR China
[2]East China Sea Fisheries Research Institute, Chinese Academy of Fisheries Sciences, Shanghai 200090, PR China.

E-mail: luna@dlut.edu.cn

Abstract. Soil contamination by organic compounds has become an issue of concern around the world. Currently, non-thermal plasma, especially pulsed corona discharge, has received a great attention in environmental protection field. As a result, the matching between a pulsed-power and corona discharge reactor containing different thickness of soil was a significant aspect in optimizing the pulse corona discharge. In this paper, some methods have been adopted to achieve the matching, including choosing a suitable capacity, adjusting the frequency, providing a suitable soil thickness and comparing the energy and energy utilization efficiency. The details of the matching and optimization discussed were based upon the theories of streamer formation and experimental results as well. The results indicated that energy injected into the reactor increased with the pulse forming capacity and pulse frequency. There existed an optimal energy utilization efficiency with the change of soil thickness and pulse frequency under the pulse forming capacity of 100 pF. The SED at pulse voltage of 19 kV and pulse frequency of 70 Hz was achieved 0.11 J g^{-1}_{soil} at the soil thickness of 3 mm, which was only 0.064 J g^{-1}_{soil} at the soil thickness of 9 mm; meanwhile, with the increase of pulse frequency from 50 Hz to 90 Hz, the SED increased from 0.075 J g^{-1}_{soil} to 0.146 J g^{-1}_{soil} at 19 kV and soil thickness of 3 mm. This study is expected to provide reference for the matching between a pulsed-power and reactor containing different thickness of soil for producing corona discharge.

1. Introduction

Atmospheric corona discharge plasmas has been widely used in environmental pollution control including water purification and gas conversion [1, 2] and various material modification occasions [3, 4]. Besides the capability for contaminated water and exhaust gas treatment, corona discharge plasma has given rise to interest in the application of soil remediation [5, 6]. Under the right conditions, the plasma can be efficient: a large amount of the energy is utilized to initiate the desired chemical reactions. However, a part of energy is dissipated by the process of plasma generation and only little information that is about the mechanisms involved in generating the plasma in an efficient way can be found. To obtain high energy transfer efficiencies from pulsed-power into plasma generation, the

[*] To whom any correspondence should be addressed.

matching between the power source and the corona plasma reactor is very important. Besides, the load impedance of reactor changes rapidly during the plasma generation [7] as the characteristic of plasma generation is transient. In a word, it's difficult for corona plasma system to establish a criterion.

In recent years, many studies have been launched researching the factors that influence the pulsed corona characteristics [8-10]. However, due to slowness of systematic researches, the optimum conditions have not been fully studied. The matching between pulsed-power and the reactor containing different thickness of soil, which serves all pulsed energy, is vital to realize the optimum pulsed corona characteristics. And the detailed discussion on the matching realization and the roles it plays in forming the pulsed corona plasma has not been presented.

In this study, from the point of view of energy and energy utilization efficiency, we investigated the matching between pulse forming capacity and the reactor containing different soil thickness, and the relations between frequency of pulsed-power and energy input. The effect of soil thickness on energy utilization efficiency was also examined.

2. Experimental

2.1. Pulsed high voltage source and reactor

The pulsed high voltage source was shown in figure 1, which consisted of a high voltage positive direct current (DC) source, a storage capacity C_e, a pulse forming capacity C_P and two rotary spark gap switches (RSG1, RSG2). Its principle was that the formed pulsed energy was injected into the reactor via rotary spark gap switch and transmission line after the forming capacity was discharged, and pulsed corona discharge was produced in the reactor at last. The pulsed high voltage source was with peak pulsed voltage in the range of 0-30 kV, repetition frequency of 0-150 Hz, pulse rise time of 30-100 ns, and pulse width of 200-1000 ns. The peak pulse voltage and current were measured with a Tektronix digital oscilloscope (TDS2012B, 100 MHz) equipped with a Tektronix P6015A high-voltage probe and a Tektronix A6021 current probe, and then the electric power was calculated through the integral of pulse voltage and current under time. Typical waveforms of the pulsed voltage and the discharge current were shown in figure 2.

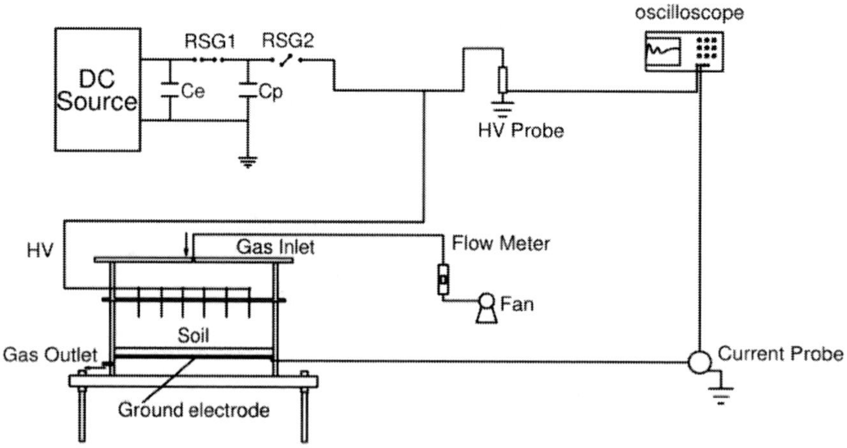

Figure 1. Schematic diagram of the experimental apparatus.

The reactor was a cylinder containing a point-to-plate geometry electrode system. The high voltage electrodes were six stainless steel needles (Ø 2 mm) that distributed uniformly around a circle of 20 mm radius. The six stainless steel needles were sealed in a Plexiglas™ cylinder (80 mm inner diameter and 37 mm length), herein named as part (a) of the reactor. The ground electrode was made of aluminium, which was embedded into a quartz glass cylinder (80 mm inner diameter and 22 mm length), herein named as part (b). The part (a) and part (b) were connected to form the reactor system. The distance of adjacent needle was 20 mm. The distance between the high voltage electrode and the ground electrode was 10 mm.

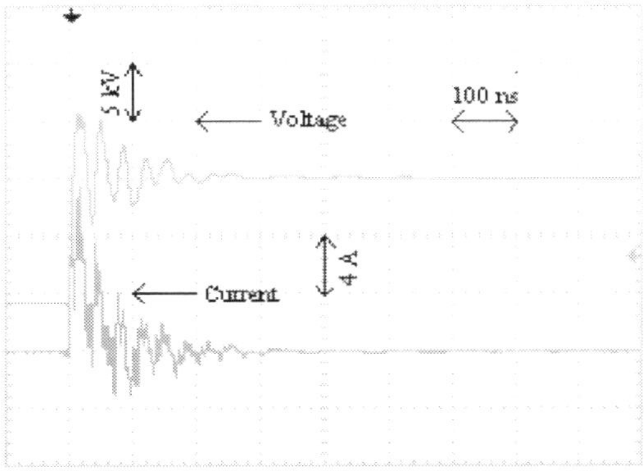

Figure 2. Typical waveforms of the pulsed voltage and the discharge current.

2.2. *Experimental methods*

In the present study, it achieved a relatively reasonable result by comparing four different forming capacities C_P that has been used. And in another experiment, soil samples were spread on the ground electrode with a thickness that ranged from 1 mm to 9 mm. The maximum pulsed high voltage was 22 kV because of the limit of experiment condition.

E_{pulse} (E_P) was the pulse energy delivered into the reactor and calculated by integration of the pulse voltage (V_P) times the discharge current (I_{cor}) over the pulse time [11]:

$$E_P = \int V_P \cdot I_{cor} dt \qquad (1)$$

In order to mutually compare the energy efficiency of different soil thickness, the specific energy density can be calculated by the following equation:

$$\mathrm{SED} = \frac{E_P f t}{m_{soil}} \qquad (2)$$

where SED was the specific energy density (J $g^{-1}{}_{soil}$), f was the pulse frequency, m_{soil} was the total mass of treated soil, and t was the discharge time.

3. Result and discussion

3.1. *Effects of forming capacity on the energy*

To investigate the effect of pulse forming capacity on the energy, it ranged from 50 pF to 200 pF and the frequency was fixed at 70 Hz. The energy input per discharge can be obtained by the integration of voltage and current measured at different pulse forming capacity. And variation of energy injected into the reactor with different pulse forming capacity was presented in figure 3. The energy that was injected into the reactor increased with the pulse forming capacity; however, the energy didn't increase too much when the pulse forming capacity was changed from 100 pF to 150 pF, as shown in figure 3. It was related to the energy transfer efficiency since the voltage level after corona extinction increased with the pulse forming capacity, and the residual energy in the pulse forming capacity also increased with the pulse forming capacity. Therefore, poorer energy transfer efficiency was obtained when the pulse forming capacity was larger [12].

However, it was believed that a too low capacity was not desirable by reason of all the energy delivered to the reactor would be used only to charge the reactor, while accordingly corona discharge cannot develop. So it was important for the reactor to choose a reasonable capacity. In the present study, the whole system got a relative appropriate situation when the pulse forming capacity was 100 pF.

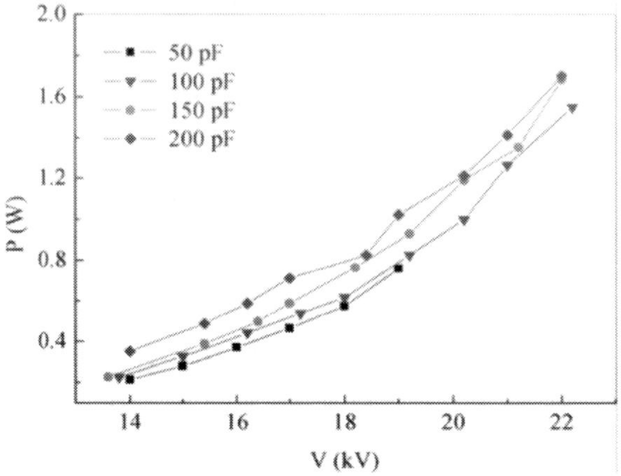

Figure 3. Effect of pulse forming capacity on the energy under pulse frequency of 70 Hz.

3.2. Effect of pulse frequency on the energy

Figure 4 showed the effect of pulse frequency on the energy. When the frequency was increased from 50 Hz to 110 Hz and the pulse forming capacity was fixed at 100 pF, it meant more energy was injected into the reactor at the same time. The energy delivered to the reactor can be divided into two parts: one delivered during the streamer propagation and the other delivered through the streamer channel after streamer propagation stops [9]. The energy delivered after streamer propagation was mainly used to sustain the low conductive streamer channel and to produce the secondary streamer, while the later part of the energy producing only slow electrons didn't contribute to energy utilization. Consequently, the major part of the energy producing energetic electrons and radicals was delivered during the streamer propagation. However, with the increase of frequency, the wastage of pulsed-power would also raise.

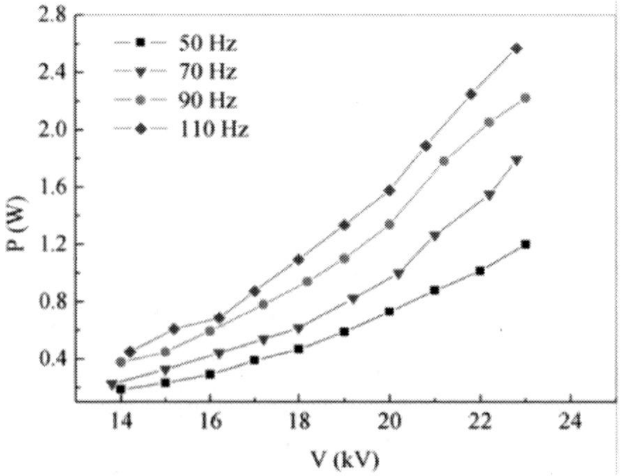

Figure 4. Effect of pulse frequency on the energy under the pulse forming capacity of 100 pF.

Therefore, it was noted that the pulse frequency related to the energy and wastage should correspond to energy utilization efficiency during the discharge process.

3.3. Existence of soil on the energy and energy utilization

In non-thermal plasma process, the energy injected into the reactor increased with the pulse frequency which may benefit the matching between a pulsed-power and reactor containing different thickness of soil. The evolution of energy input obtained at different soil thickness with the increase of pulse frequency was presented in figure 5. With the increase of pulse frequency from 50 Hz to 90 Hz, energy that was delivered into reactor without soil enhanced from 0.59 W to 1.1 W at the pulse voltage of 19 kV; while the energy input increased from 0.9 W to 1.76 W at the same condition, except the soil thickness of 3 mm. This phenomenon suggested that energy input conducted at the soil thickness of 3 mm was more than the one without soil since the reduction of distance between high electrode and ground electrode. The effects of pulse frequency on specific energy density (SED) were performed at the soil thickness of 3 mm as shown in figure 6. And a certain range of maximal SED existed along with the change of pulse frequency, which was obtained at frequency of 90 Hz in the experiments. It was important to note that the increase of pulse frequency was reasonable at a certain extent, because proper increase of pulse frequency can improve the energy input but the wastage of pulsed-power also enhanced if the pulse frequency was raised too much.

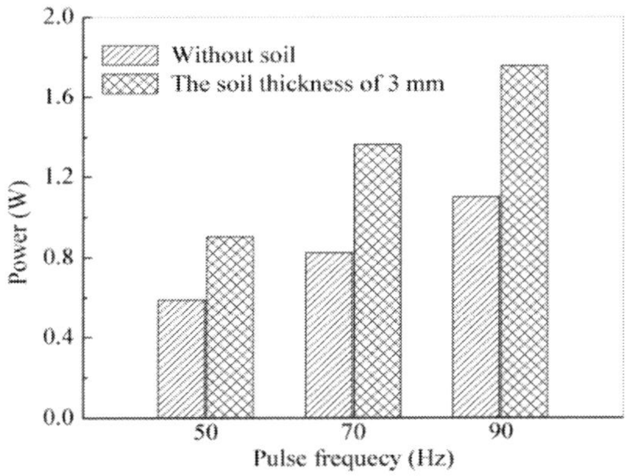

Figure 5. Effect of pulse frequency on the energy at the soil thickness of 3 mm.

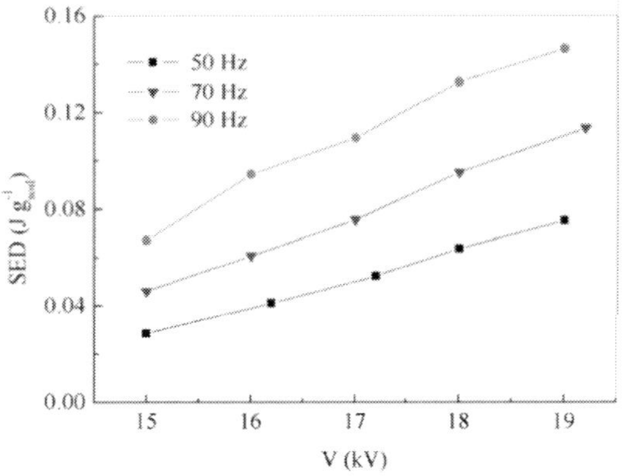

Figure 6. Effect of pulse frequency on the SED at the soil thickness of 3 mm.

3.4. Effect of soil thickness on the energy and energy utilization

The soil that was spread on the ground electrode with different thickness influenced the energy injected into the reactor and the SED. Increasing soil thickness could improve the energy delivered to reactor but the SED decreased at the same time. Hence, soil thickness was an important parameter to evaluate the matching between a pulsed-power and reactor. The effects of soil thickness on energy and SED under the pulse forming capacity of 100 pF were illustrated in figure 7 and figure 8, respectively. For example, when soil thickness increased from 3 mm to 9 mm, energy input enhanced from 1.36 W to 2.31 W at pulse voltage of 19 kV, while SED decreased from 0.114 J g^{-1}_{soil} to 0.064 J g^{-1}_{soil}. The increase of soil thickness benefited energy input due to the reduction of distance between high voltage electrode and ground electrode, which meant the decrease of corona inception voltage and breakdown voltage at all other things being equal. However, with the increase of soil thickness, the volume of soil was also gained; correspondingly, the energy delivered to per gram was decreased. Based on these results, it could be deduced that selecting an appropriate soil thickness that could get a relatively high energy and energy utilization efficiency was important.

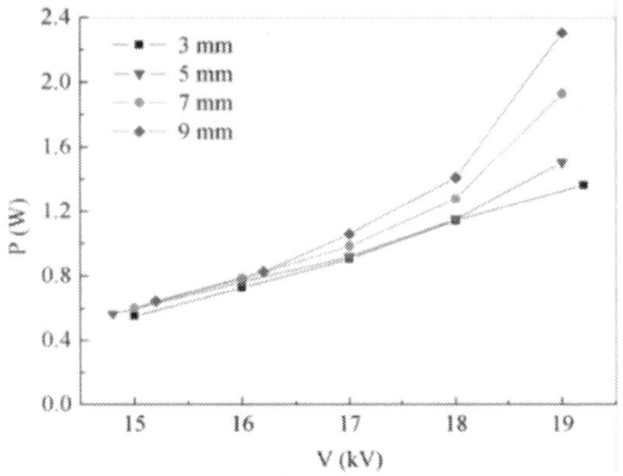

Figure 7. Effect of soil thickness on the energy under the pulse forming capacity of 100 pF.

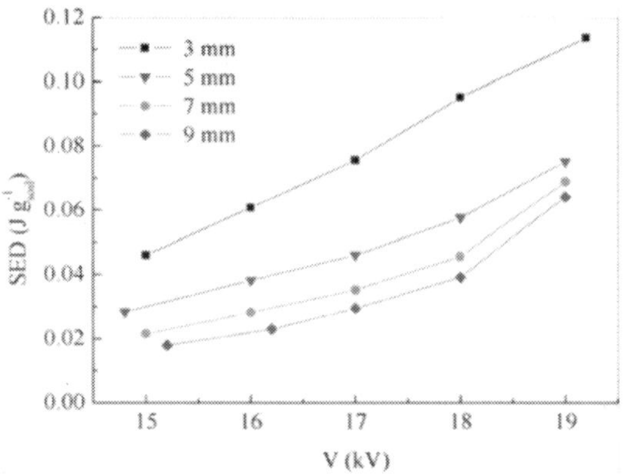

Figure 8. Effect of soil thickness on the SED under the pulse forming capacity of 100 pF.

4. Conclusion

The matching between a pulsed-power and reactor containing different thickness of soil for producing corona discharge was studied. Under the studied conditions, increasing pulse forming capacity or pulse frequency resulted in higher energy input, due to the enhancement of energy in per unit time. The investigation of soil thickness in the matching between pulsed-power and reactor indicated that increase of soil thickness benefited the energy input during the discharge process. The change of soil thickness could also affect the SED, and there existed a maximal SED, as a result of the energy input and the volume of soil.

This study was a fundamental research effort, trying to offer an alternative solution for the matching between a pulsed-power and reactor containing different soil thickness. The performance was showed for the matching between a pulsed-power and reactor containing different soil thickness, demonstrating that optimizing the pulse forming capacity, pulse frequency and soil thickness for the matching was a reasonable solution.

References

[1] Guo X H, LI J, Ye Z Q and Qi J 2001 *High Voltage Eng.* **27** 42.
[2] Luo Y H, Yang L J, Feng Y P, Yu L M 2001 *High Voltage Eng.* **27** 39.
[3] Vaidya R U, Castro R G, Peters M I, Gallegos D E and Petrovic J J 2002 *J. Therm. Spray Technol.* **11** 409.
[4] Sun Y, Li X Y, Bell T 1999 *J. Mater. Sci. Lett.* **34** 4793.
[5] Wang T C, Lu N, Li J, Wu Y 2010 *Environ. Sci. Technol.* **44** 3105.
[6] Wang T C, Lu N, Li J, Wu Y 2011 *Environ. Sci. Technol.* **45** 9301.
[7] Liu Z, van Heesch E J M, Pemen A J M and Yan K P 2008 *IEEE Trans. Plasma Sci.* **36** 243.
[8] Masuda S 1988 *Pure Appl. Chem.* **60** 727.
[9] Rea M, Yan K P 1995 *IEEE Trans. Ind. Appl.* **31** 507.
[10] Fujii T, Gobbo R and Rea M 1993 *IEEE Trans. Ind. Appl.* **29** 98.
[11] Hoeben W F L M, van Veldhuizen E M, Rutgers W R, Cramers C A M G and Kroesen G M W 2000 *Plasma Sources Sci. Technol.* **9** 361.
[12] Mok Y S, Ham S W and Nam I S 1998 *Plasma Chem. Plasma Process.* **18** 535.

Journal of Physics: Conference Series **418** (2013) 012137 doi:10.1088/1742-6596/418/1/012137

Electrostatic bio-manipulation for the modification of cellular functions

Masao Washizu[1, 2*]
[1]Department of Bioengineering, The University of Tokyo, 7-3-1 Hongo, Bunkyo-ku, Tokyo, 113-8656, Japan
[2]JST CREST

E-mail: washizu@washizu.t.u-tokyo.ac.jp

Abstract. The use of electrostatic field effects, including field-induced reversible-breakdown of the membrane and dielectrophoresis (DEP), in microfabricated structures are investigated. With the use of field constriction created by a micro-orifice whose diameter is smaller than the cells, controlled magnitude of pulsed voltage can be applied across the cell membrane regardless of the cell size, shape or orientation. As a result, the breakdown occurs reproducibly and with minimal invasiveness. The breakdown is used for two purposes, electroporation by which foreign substances can be fed into cells, and electrofusion which creates genetic and/or cytoplasmic mixture among two cells. When GFP plasmid is fed into MSC cell, the gene expression started within 2 hours, and finally observed in more than 50% of cells. For cell fusion, several ten percent fusion yield is achieved for most cell types, with the colony formation in several percents. Timing-controlled feeding foreign substances or mixing cellular contents, with high-yield and low-invasiveness, is expected to bring about a new technology for both genetic and epigenetic modifications of cellular functions, in such field as regenerative medicine.

1. Introduction

There are emerging needs for artificial control over cellular functions, for instance in reprogramming and/or differentiation control of ES (Embryonic Stem) or iPS (induced Pluripotent Stem) cells. Such processes require technologies for the introduction of foreign genes and substances into cells, followed by the evaluation of the cellular responses. In order to guarantee the purity and the traceability of the product, each step in these processes should be made in single-cell level. In this view, the author and his colleagues are focusing on electrostatic field-effects in micro-fabricated structures to realize such "cell surgery".

The field effects used are a) dielectrophoresis (DEP), an electrokinetic effect to move and position cells, and b) so called "reversible breakdown of the membrane", which uses an electrical impulse to create localized temporary disruption of cytoplasm membrane. During the breakdown, substances can diffuse into the cytoplasm (electroporation), or the membranes of two cells in contact can reconnect and merge into one (electrofusion). Microfabrication techniques are conveniently utilized here to make structures comparable to a cell to enable the manipulation in single-cell level, as well as to tailor the field pattern around the cell. In particular, with the use of the field constriction created by an orifice

* To whom any correspondence should be addressed.

smaller than a cell, the voltage drop can be concentrated to the cell membrane in contact with the orifice, so that it can be precisely controlled regardless of cell size, shape or orientation. Hence, reproducible and low-invasive breakdown is achieved.

In this paper, the principle of the field constriction is reviewed, and the experimental results of both electroporation and electrofusion are shown.

2. Reversible breakdown of the membrane based on field constriction

The conventional reversible breakdown method uses cell suspension under an uniform pulsed electric field. Modelling a biological cell as a spherical conducting particle with an insulating skin, the induced voltage V_m across the membrane is given by [1]

$$V_m = \frac{3}{2} a E_0 \left(1 - \exp[-t/\tau]\right) \cos\theta \quad (1)$$

$$\tau = a C_m (\rho_{in} + \rho_{out}/2) \quad (2)$$

where a is the cell radius, ρ_{in} and ρ_{out} are the resistivities of the inner and the outer medium, C_m is the membrane capacitance per unit area, E_0 is the strength of the uniform applied field, θ is the azimuthal angle measured from the direction of E_0.

The problem here is that V_m is size dependent. For an ensemble of cells with size distribution, large cells receive too much voltage and rupture, while the voltage is too small for small cells to induce effective breakdown. Likewise, for non-spherical cells, the membrane voltage depends on the shape and the orientation, which can make the yield even lower.

To solve the problem, we make use of the constriction field. As schematically illustrated in figure 1, a typical constriction field is produced by an insulating plate with an aperture, which is sandwiched by a pair of electrodes. In the figure is drawn the lines of force, as well as the equi-potential contours. It is seen that the potential drop occurs dominantly near the orifice, in this example case 80% drop occurs between z = 0 and 0.26 d. Therefore, if a cell, whose diameter is adequately larger than that of the orifice, is placed on the orifice, most potential drop will be impressed on the cell, and the membrane voltage will be virtually independent on the cell size.

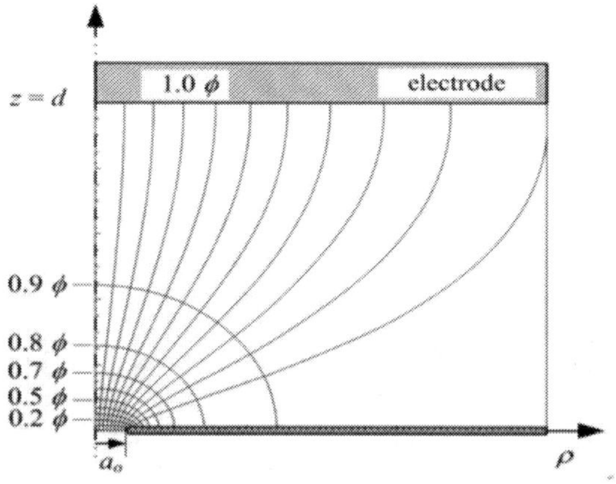

Figure 1. A typical constriction field.
An infinitely large insulator plate at z = 0, having an orifice with the radius a_o, is placed between infinitely large parallel plate electrodes at $z = d$ and z = $-d$, to which potential ± is applied. The equi-potential lines, as well as the lines of force are drawn for the case of $d = 10\ a_o$.

Figure 2 depicts such a field constriction geometry applied for electroporation and electrofusion. An insulator sheet with orifices, may it be single, in a linear array, or x-y 2-dimensional array, is used. In the case of electroporation, adhering cells may be cultured on the sheet, while non-adherent cells can be placed onto the orifices by aspiration or DEP. If the cells perfectly seals the orifice, all voltage applied to the electrodes V_{app} is impressed on the membrane extended by the orifice, hence $V_m = V_{app}$. Even if the sealing is imperfect, the field far from the orifice is weak, most voltage drop will take place in the cell, and $V_m \approx V_{app}$. Thus V_m can be controlled regardless of the cell size, shape or orientation, and reproducible reversibility in the membrane breakdown is guaranteed. Hence the foreign substances in the surrounding medium, say plasmids, can be brought into the cells with a high-yield.

For the case of electrofusion, the cells are positioned onto the orifices by DEP to form cell pairs, as delineated later. Field constriction in this case again plays a role of concentrating the applied voltage to the membrane on the orifice.

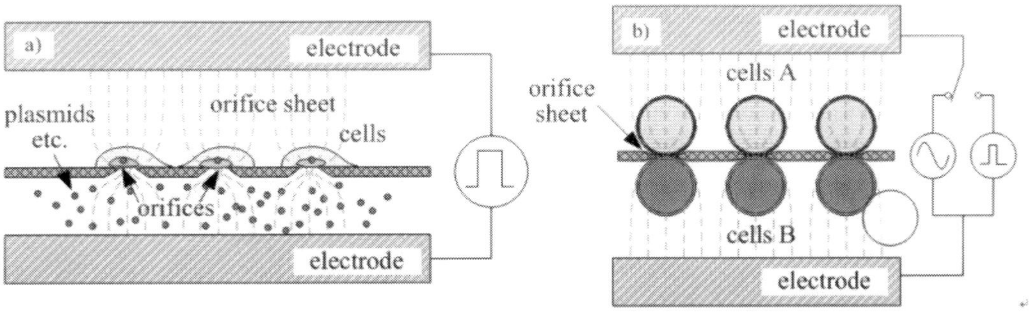

Figure 2. Electroporation (a) and electrofusion (b) based on field constriction at micro orifices.

Figure 3 shows a typical result obtained by the numerical analysis of V_m distribution on the cell pair sandwiching the orifice, assuming an axial symmetry, as well as the mirror symmetry. The method itself is delineated in ref. [2]. The calculated result is shown in b) in normalized form, i.e. as the membrane voltage V_m divided by the applied voltage V_{app}. It is seen that V_m / V_{app} is about 0.9 in the orifice ($\theta < 20°$ in this case), indicating 90% of V_{app} goes onto the membrane making contact with the other cell, while V_m outside the orifice (say $\theta > 50°$ in this case) is virtually zero. Hence a controlled voltage can be applied to the membrane around the contact point of two cells, and high-yield fusion can be expected. The fusion will be 1:1; even a third cell may be on the side as depicted in figure 2b) right bottom, this cell will not take place in fusion, because there is no V_m here.

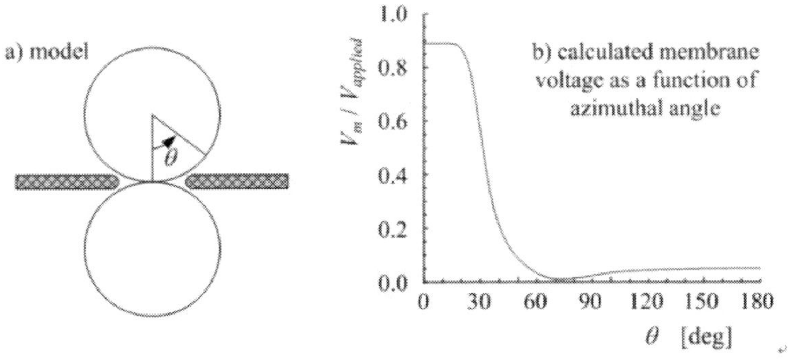

Figure 3. Calculated membrane voltage induced on a pair of cell at an orifice.

3. On-chip electroporation based on field constriction

Molecules smaller than protein, for example fluorescent dye, glutamate or GFP [3], have diffusion coefficients high enough that they can reproducibly be fed into cells by electroporation with a short (≈ ms) pulse. But in real biological applications, for example in generating iPS cells or its differentiation control, often required is the introduction of plasmid DNA of several ten kilo-base, which is far larger than proteins. In such cases, we make use of the electrophoretic effect induced by a pulse of several hundred ms in length.

To visualize how plasmids can be fed into a cell, we made a real-time observation of the process, fluorescence-labelling plasmids with Quantum dots (Q-dots) [4]. figure 4a) schematically shows the experimental setup. The device consists of a PDMS (silicone rubber) microfluidic chip having a vertical wall with an orifice at the center, and a pair of electrodes to apply voltage across the orifice. A cell is positioned on one side of the orifice by aspiration, Q-dot labeled plasmids fed on the other side, and an electrical pulse, with the cell side positive, is applied. The plasmids migrate along the field line due to electrophoresis, and move into the cell through the pores formed by electroporation. The nucleus itself has inherent pores which is used for material transport into and out from, so some of the flux penetrates into nucleus, by which the plasmids can be transported directly into nucleus. Figure 4 b) is the cell on the right of the orifice, where the nucleus is clearly visible. The orifice itself, and the part further left, is not visible due to halation of many Q-dots. When the pulse is applied, the influx of Q-dots into the cell is observed, some of them into the cytoplasm, and some into the nucleus. Figure 4c) is after the pulse, where two Q-dot labelled plasmids, indicated by arrows, are left in the nucleus.

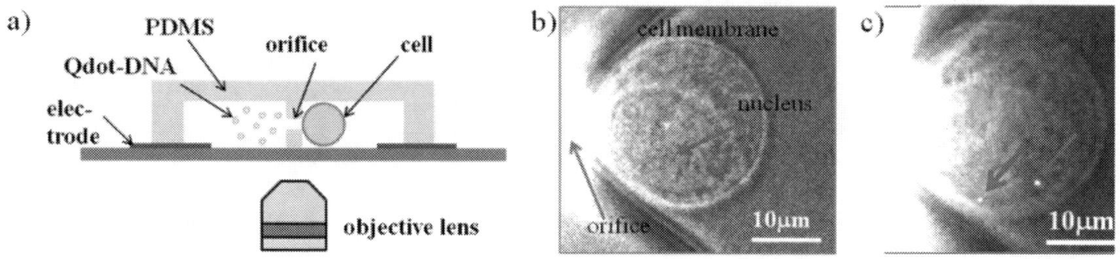

Figure 4. Fluorescence observation of electrophoretic plasmid transport into nucleus.
a) experimental setup; b) before pulse; c) plasmids fed into nucleus (fluorescence view).

The result suggests that plasmids can be directly transported into nucleus, if it is penetrated by the line of force. Therefore, we designed a high-density orifice array sheet, where the orifice pitch is made comparable to the nucleus diameter (when the cell is adhering on the surface and the nucleus is stretched).

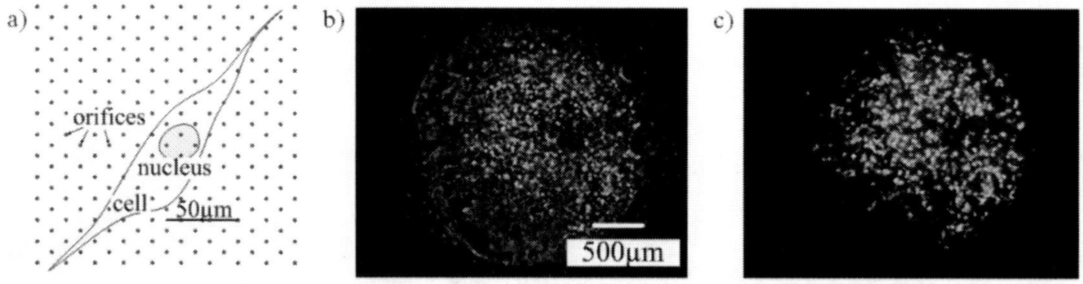

Figure 5. Electroporation of GFP plasmid into MSC (Mesenchymal Stem Cell)
a) the orifice array dense enough to have more than one orifice beneath nucleus (schematic); b) all alive cells on the orifice sheet stained by calcein; c) fluorescence from expressed GFP (after 5 hours).

Figure 5 shows the result of plasmid transfection. MSC (Mesenchymal Stem Cell), which is adherent, is used as the cell, to which GFP (Green Fluorescent Protein) plasmid is fed. MSC has a typical dimension as drawn in figure a), where the triangular lattice of 15μm satisfies the condition of "at least one orifice beneath nucleus". Figure 5b) is a fluorescence photo of alive cells as stained by calcein, and c) is the fluorescence from GFP taken after 5 hours of pulsing. The percentage of cells in figure c) divided by that of figure b) gives the transfection yield, which is roughly 50% in this case. In conventional electroporation, it takes about a cell cycle (10-20 hours) for the plasmid to go into nucleus before it is expressed. In contrast, in the case of this on-chip transfection, the plasmid is transported by electrophoresis into nuclei, and hence the expression starts as quick as 2 hours after pulsing.

4. On-chip electrofusion based on field constriction

Two types of on-chip electrofusion chips are developed, one being one dimensional orifice array to view the fusion process from a side (figure 2b) [5-7], and the other being two-dimensional orifice array for massively parallel fusion [8].

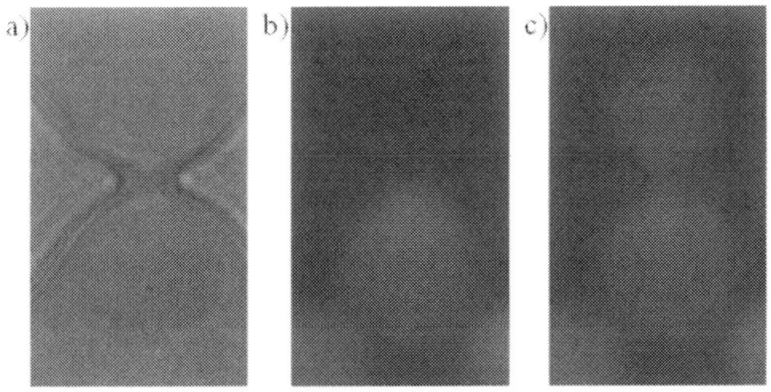

Figure 6. Electrofusion of Jurkat cells.
a) optical view; the orifice is at the center . b) fluorescence view before the pulse; cytoplasm of the cells below the orifice are fluorescence stained, while those above the orifice are not. c) fluorescence view after the pulse; the cytoplasm are mixed and the fusant takes a "snow-man" shape. Orifice diameter~10 μm.

Figure 6 shows the instance of fusion and how cytoplasm is mixed, where Jurkat is used as the cell. The cells whose cytoplasm is stained with calcein are fed into the channel below the orifice, while unstained cells into the channel above. The cells are attracted and positioned onto the orifice by DEP. As optically photographed in a), two cells are making contact at the orifice, and b) shows the same instance viewed by fluorescence, which shows the cells below the orifice are emitting fluorescence, while those above are not. Figure 6c) is just after the pulse, and the fluorescence dye in the lower cell has already been diffused into the upper cell to look like a "snow man". This cytoplasm mixing, at least as observed by the dye, takes only a few seconds. Also to note is that, while there are cells other than making a pair at the orifice, none of them takes part in fusion. This is as it should be, because the membrane voltage builds up only within the orifice. This guarantees that the fusion is exclusively 1:1, in contrast to unpredictable number of cells involved in fusion in conventional methods. The minimum extra gene brought in by 1:1 fusion is desirable in the survival of the fusant.

Time lapse observation of the fusant is made, a typical result is shown in figure 7. By the 1:1 fusion among diploid cells (L929, murine fibroblast), a tetraploid is formed, which in some case divided into two (figure 7 a), three (b) or four (c). Such multipolar divisions [9] have been reported in relation to oncogenesis. The fusion method developed here, enabling high-yield timing-controlled 1:1 fusion and its time-lapse observation, is expected to be a novel tool for such biological or biomedical investigations.

Massively-parallel 1:1 fusion can be achieved with the use of a two-dimensional orifice array. A 5×5 mm sheet with 50μm-pitch orifices, having 10^4 orifices altogether, is used in the experiment. Size-independence of V_m, a property of the field-constriction design, is effectively utilized in the fusion among different-sized cells.

Journal of Physics: Conference Series **418** (2013) 012137 doi:10.1088/1742-6596/418/1/012137

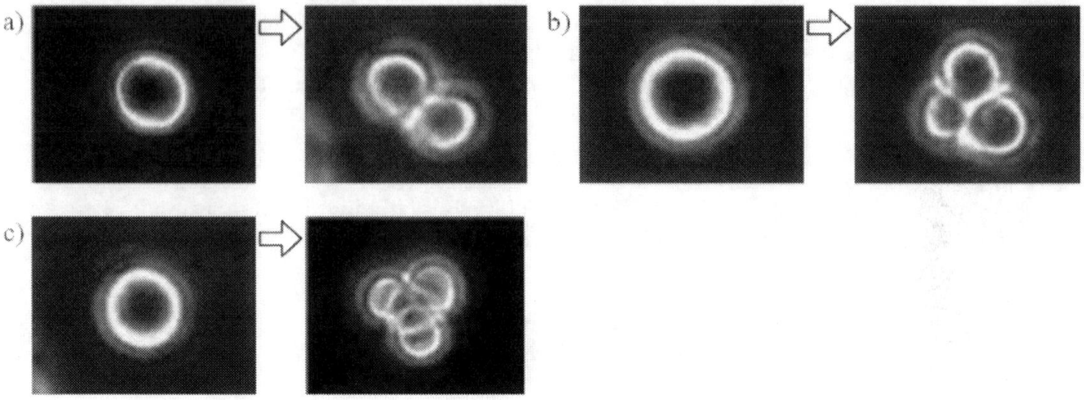

Figure 7. Real-time observation of multipolar division of the 1:1 fusant.

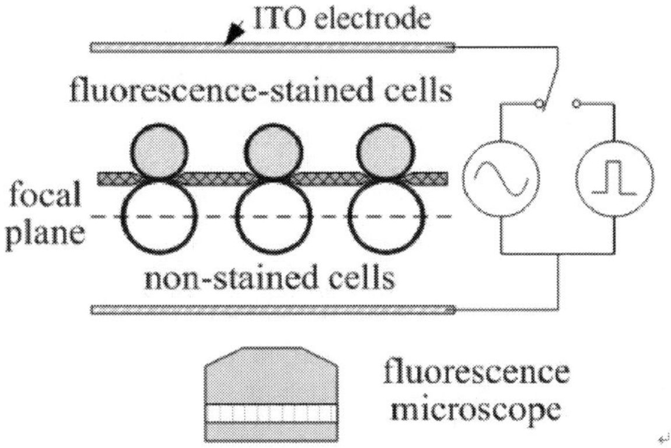

Figure 8. Observation of parallel fusion using x-y orifice array. Influx of fluorescence dye into lower cell is monitored with a microscope viewing the fluorescence from lower cells.

Figure 8 shows the experimental setup. A cell whose cytoplasm is stained with the fluorescence dye calcein is fed above the orifice sheet, while those unstained are fed below the sheet. They are positioned onto the orifices to form pairs by dielectrophoresis with the aid of gravity (flipping the chip over). The observation is made from the bottom by an inverted microscope, with the focal plane set at the lower cell. At this moment, as the fluorescence microscope is focusing on non-fluorescence cells, only out-of-focus view of the cells through the orifice sheet is observed. When the pulse is applied, the cytoplasm mixing, same as that viewed from side in figure 6, takes place and the lower cells become fluorescent, which can now be observed brighter and with a clear contour.

Figure 9 is an example of the experimental result, the fusion among K562 (human leukemia, ~24 μm) and HL60 (human leukemia, ~14 μm). In figure a), the fluorescent cells are out of focus and are hardly visible. When the pulse is applied, each cell pairs fuses, the dye transferred into the non-fluorescent cells on the lower side, and become clearly observable as in figure b). It is found that having a neat alignment is more difficult than the fusion itself. Even so, the fusion yield of ~70% is obtained.

One of the largest application of conventional cell fusion is the creation of antibodies by fusing B cell and myeloma cell. After the fusion operation, the cells are cultured in the selective medium, in which non-fused cells cannot survive, and only fusants can proliferate and form colonies. Whether or not, or how efficiently the fusants obtained by the on-chip electrofusion can form colonies is investigated. The fusion is performed with the same procedure as that in figure 9 using B cell (~7 μm^{ϕ}) and myeloma cell (~15 μm^{ϕ}), and after fusion, the device is disassembled, the orifice sheet taken out, and placed onto a layer of feeder cells in a dish for culturing.

Journal of Physics: Conference Series **418** (2013) 012137 doi:10.1088/1742-6596/418/1/012137

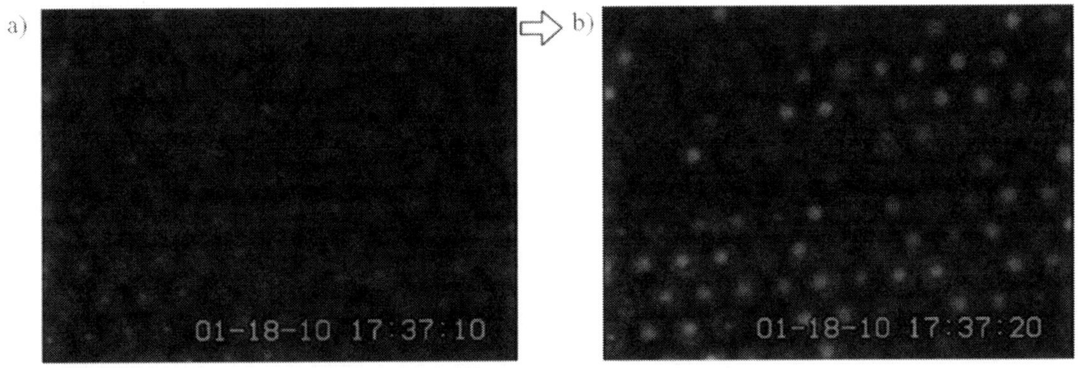

Figure 9. Electroporation of GFP plasmid into MSC (Mesenchymal Stem Cell). Orifice pitch = 50 μm.

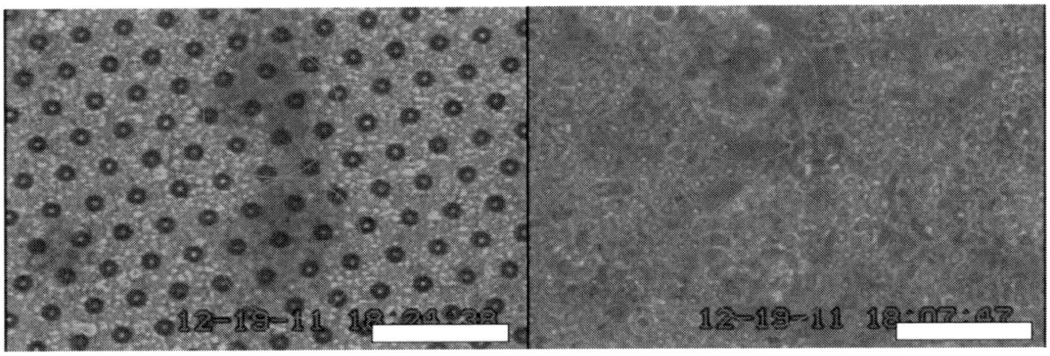

Figure 10. Electroporation of B cell and myeloma cell. Scale bar = 150 μm.

The result is shown in figure 10. The left and the right photo are of the different place in the same culture dish. In the left photo is seen the feeder cells almost uniformly covering the dish surface, and the array of orifices. Encircled are two colonies of cells, which appear no different from that obtained by the conventional fusion method. As the orifice sheet is smaller than the culture dish, and the colonies can float and move during culturing, colonies out of the orifice sheets are observed as well, as encircled in the right photo. About 100 colonies are obtained, using the sheet having 2000 orifices. This is very high yield compared with that of conventional methods, partly due to the yield of fusion event itself is high as seen in figure 9, and partly because the fusion is exclusively 1:1 and the fusant is tetraploid, which can better survive than hyperploids.

5. Conclusions

It has been shown that field tailoring using microfabricated structure can be a novel tool for the investigation, modification and utilization of cellular functions. By precisely controlling the membrane voltage, high reproducibility in the reversible breakdown is guaranteed. Foreign molecules including genetic materials, or the cytoplasmic components of other cells, can be brought into the target cell, and because all operations can be made in a microfluidic chip under a microscope, the cellular response can be evaluated in single-cell level. The technology developed here will find various applications in biological or biomedical field, in particular in regenerative medicine, where the creation, function control, evaluation, and quality control have an essential meaning.

Acknowledgments

The results shown in this paper were done under collaborations and discussions with the following people, Dr. Murat Gel, Dr. Hidehiro Oana, Dr. Osamu Kurosawa, Dr. Yuji Kimura, Prof. Takao Hamakubo, Dr. Hiroko Iwanari, Mr. Yasuhiro Nishigaichi, Mr. Shohei Suzuki of The University of Tokyo, Prof. Hidetoshi Kotera, Dr. Takashi Tada of Kyoto University, Dr. Boonchai Techaumnat of Chulalongkorn University, Thailand. The work was partly supported by JST CREST, NEDO "Dynamic Biology" and "Cell Array" projects, Ministry of Education Kakenhi, ASTEM "Nano-medicine" project and CNBI, U-Tokyo.

References

[1] Zimmermann U, Arnold W M 1988 J. *Electrostat.* **21**, 309
[2] Techaumnat B and Washizu M 2007 *J. Phys. D: Appl. Phys.* **40** 1831
[3] Kurosawa O, Oana H, Matsuoka S, Noma A, Kotera H, and Washizu M 2006 *Meas. Sci. Technol.* **17** 3127
[4] Kurosawa O, Sumita Y, Gel M, Oana H, Kotera H, Kato T, Toguchida J and Washizu M 2010 μ-TAS M49A (Groningen, Netherlands) p 217
[5] Techaumnat B, Tsuda K, Kurosawa O, Gel M, Oana H and Washizu M 2008 *IET Nanobiotechnol.* **2** 93
[6] Gel M, Suzuki S, Kimura Y, Kurosawa O, Techaumnat B, Oana H, and Washizu M, 2009 *IEEE Trans. Nanobiosci.* **8** 300
[7] Gel M, Kimura Y, Kurosawa O, Oana H, Kotera H and Washizu M 2010 *Biomicrofluidics* **4** 022808
[8] Kimura Y, Gel M, Techaumnat B, Oana H, Kotera H, Washizu M 2011 *Electrophoresis* **32** 2496
[9] Vitale, Senovilla L, Jemaa M, Michaud M, Galluzzi L and Kepp O et al. 2010 *The EMBO Journal* **29** 1272

Journal of Physics: Conference Series **418** (2013) 012138 doi:10.1088/1742-6596/418/1/012138

Transformation of soybean Gy3 gene into Artemisaarenaria mediated by corona discharge

Lu-meng Chao[1,2], Ri Na* [1,2], Dan Xue[1,2], Yongze Xu[1,2], Teng Liu[1,2]
1 School of Physical Science and Technology, Inner Mongolia University, Hohhot, China, 010021
2 Key Laboratory of Ion Beam Bioengineering, Inner Mongolia University, Hohhot, China 010021

E-mail: nari6363@sina.com

Abstract. In order to improve the protein content of desert plant, a method of genetic transformation mediated by corona discharge was established. Artemisia seeds were processed in corona electric field for 120 min at 12 kV, and then soaked in 0.1SSC media that contained Soybean Gy3 gene DNA to incubate for 12 h at 26 °C. Finally the seeds were inoculated on the differentiation medium. Polymerase Chain Reaction (PCR) and Reverse Transcription Polymerase Chain Reaction (RT-PCR) detection showed that the Soybean Gy3 gene had been successfully introduced into genomic DNA of the regenerated plants of Artemisaarenaria. The study provided a new way for corona discharge in plant genetic modification.

1. Introduction

In recent years, a lot of achievements about the sand plants growing in extreme climate on Northwest China's arid have been obtained. The responses of net photosynthetic rate (Pn), chlorophy II fluorescence parameters and trehalose content to drought stress were studied [1]. Water parameters of different parts of Artemisia ordosica and their roles in water transfer of SPAC system were studied by using the plant pressure-volume technique [2]. Leaf and young stem were used as the explant to study callus induction and proliferation in the desert plant Reaumuria soongorica [3]. But there are very few reports using transgenic methods to improve psammophytes protein content.

Artemisaarenaria are subshrub belonging to genus Artemisia of Compositae, which is grown in China's Junggar basin of xinjiang, irtysh river basin, Hexi corridor in gansu sand, inner Mongolia alashan desert, the qaidam basin in Qinghai province (China); Mongolia and southern parts of Europe, the caucasus and central Asia. Artemisaarenaria, with strong deep roots and low feeding value, is good at defending the breeze sand and have abilities of keeping water and soil to grow. Its seeds are edible, and can also extract oil for food, paints, or preparation of food conditioner. At the same time, they can be used as medicine, the efficacy of anti-inflammatory scattered swollen and Chest profit gas [4]. Currently, only the tissue culture, chemical composition analysis and its food value were studied towards Artemisaarenaria. The study of molecular biology, which will take an important role in improving crop drought-tolerant quality and is good for recovering the mechanism of Artemisaarenaria's drought-tolerant, however, has no reports in Artemisia transgenic. The

* To whom any correspondence should be addressed.

experimental results were found by Li Yu-hua [5]. Results show that the anthers of two types of strawberry cultivate Mingling and Minglei, which were treated with high-voltage electrostatic field before inoculation, could significantly increase the percentages of induction, differentiation and differentiated shoot clumps of strawberry antrher-derived callus, as well as increase the ratio of granular crowed callus. Song etc successfully observed the phenomenon that the power of the plasmid transfection, when E. coli (JM105) and scale ratio of DNA is 50~75 on low-frequency electric field's intensity is (5~20) $\times 10^3$ V·m^{-1} [6]. Liu Chang-jun [7] etc proved that low-intensity (2×10^3 V·m^{-1}) (Ensure a high survival rate of cells) and fast electromagnetic pulse (The pulse width of 10^{-7} s, the rise time is 1.2×10^{-9} s). Irradiated cells about 20-100 min by 300 Hz can lead to cell electroporation. Based on the above research, to better understand the biological effects of high-voltage corona field, expanding the application and study a new means of transgenic [8]. A new experimental device which can produce the high-voltage corona electric field is developed. The first time the pBI121 plasmid, which contains a copy of GUS gene and kanamycin resistance gene, was introduced into Artemisaarenaria by corona discharge, and regenerated plants were obtained.

2. Experimental materials and measuring method

2.1 Experimental material

Artemisaarenaria seeds are provided by NingXia BIG West Agriculture Seed Ltd. The soybean "Heinong 37" is provided by the Heilongjiang Province, Dragon Seed Industry. The pMD19-T is provided by the Takara Biotechnology (Dalian) Co. Ltd. PBI121 which contains GUS gene and kanamycin resistant gene is provided by Li Jing Shandong University.

2.2 Design of experimental device

The developed experimental device is shown in figure 1.

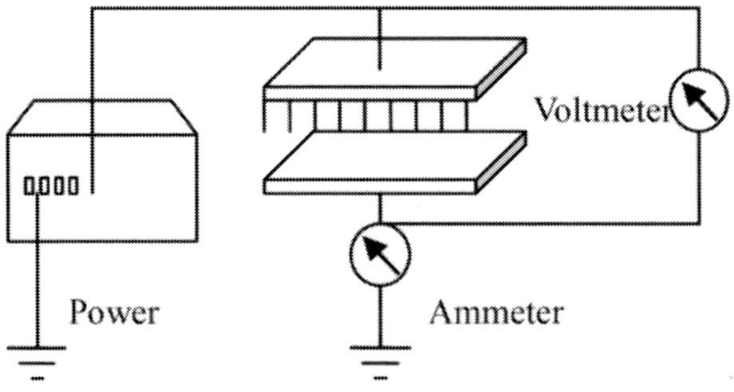

Figure 1. Sketch map of equipment for corona discharge

The voltage supplied by DC high voltage power can be continuously adjusted from 0 to 30 kV. An ammeter whose resolution is 10 μA and accuracy is 0.8% is used to detect the discharge current. The applied voltage is tested by Q-3V voltmeter. The determined optimum parameters are the following: the cone tip electrode is made by nickel-iron alloy. The needle tip's diameter is 0.05 mm, the needle tail's is 1.5 mm and the length of needle is 2.6 mm. There were 100 needles (n=100), the distance between the needle tips is 4 mm, and adjustable range is 4~32 mm, the distance between the adjacent needle plates is in the range of 0~60 mm.

2.3 measuring method

2.3.1 Preparation of plasmid pBI121 [9] According to literature reports [10], PBI121's registration number in gene sequences is M36686. The specific primer (RT-PCR) is designed by software Oligo6.0. The upstream primer is introduced into the XbaI restriction site. The gene sequences are: F1 (5′-tctaga AACACTCATCAGTCATCACC-3′). The downstream primer is introduced into the BamHI restriction site. The gene sequences are: F2 (5′-ggatcc CTAAGCCACAGCTCTCTTC-3′). The primers are synthesized by Dalian TaKaRa Biotechnology Company. The first strand of cDNA that is reversed transcription as a template, the target fragment is amplified after PCR reacted. Connecting the target fragment is amplified by PCR and pMD19-T vector. The positive clone expression vector that connect product is named pBI121/ Gy3.

Connecting the cDNA of the Gy3 subunit and pMD19-T carrier, the positive clones is sequenced by Dalian TaKaRa Biological Engineering Company. The results show that there are 1523bp in the gene fragment and reversing into the T vector. Then doing Homology comparison with corresponding sequence in Genbank (M36686), the results show that the homology of the corresponding sequence between the soybean Heinong No.37 Gy3 subunit gene nucleotide sequences and Gy3 subunit in GenBank is 99%. On pBI121 plasmid vector and the target fragment that connected product are identified by PCR. The target fragment has been successfully constructed into pBI121 plasmid of the vector. This phenomenon shows that the cDNA of Gy3 subunits expression vectors have been constructed successfully.

2.3.2 Transforming Of plasmid DNA into Artemisaarenaria by Needle - plate corona electric field Select Artemisia's vernalized seeds, stocke on a glass slide with double sided tape, and then process by Corona electric field (voltage 12 kV and time 120 min). After processing, the seeds are immediately soaked into 0.1 × SSC import media that containing 200 $\mu g \cdot ml^{-1}$ DNA. They are incubated for 12 h at 26 °C, and then removed and washed 3 times with 0.1 × SSC solution repeatedly. After that the seeds are sucked up the remnants of solution by sterile filter paper, and are inoculated on the differentiation medium. The above operations are performed in sterile ultra-clean workbench.

2.3.3 PCR detection of transgenic plant regeneration Genome of the leaf of Artemisaarenaria is extracted by Broad Tektronix genome mention kit, PCR is reacted by the specific primers which has designed, the PCR products are used as negative control. The cDNA sequence of Glycinin Gy3 subunit which has been cloned used as positive control. PCR is reacted by DNA of transgenic regenerated plant leaves which is extracted by Broad Tektronix genome mention kit.

2.3.4 RT-PCR detection of transgenic plant regeneration The leave has been detected by PCR, and the total RNA of leave were extracted. RT-PCR reaction was conduct by using the synthesized specific primers and the optimal reaction condition which have been found out. The cDNA sequence of Glycinin Gy3 subunit which has been cloned was used as positive control. The experiments were conduct according to plant RNA extraction kit and RT-PCR kit which purchased from Dalian TaKaRa Biotechnology Company.

3. Results and Analysis

3.1 Cultivation transgenic plants

The exogenous DNA was introduced into vernalized seeds of Artemisia by corona discharge. Then they were inoculated in differentiation medium at 26 °C, cultured for 15 days till began to sprout roots. They grew into small plants about 7 days, and it was shown in figure 2.

3.2 transformers were tested by PCR

Artemisia seeds were transformed and inoculated in differentiation medium, and then they grew into small plants. Differentiated Artemisia seedlings were detected by PCR and RT-PCR. According to the steps that it was provided by broad Tektronix genome miniprep kit, the genomic DNA was extracted from the leaves of Artemisaarenaria that were untransformed or transformed, PCR reaction was conducted by using the specific primers which has been designed. In the conducted electrophoresis experiment, PCR products about genomic DNA of Artemisaarenaria which have been transformed were used as the negative control, the cloned cDNA sequence of Glycinin Gy3 subunit were used as the positive control.

3.3 PCR experimental results of transformnts

Genomic DNA of Artemisaarenaria seedlings for PCR reaction, through the agarose gel electrophoresis, the PCR results were showed in figures 2 and 3.

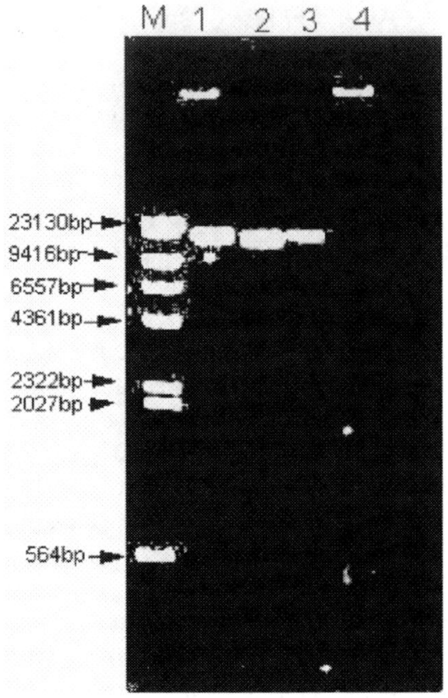

Figure 2. Genomic DNA M:Marker(λ-HindIII) 1,2,3,4: transformants.

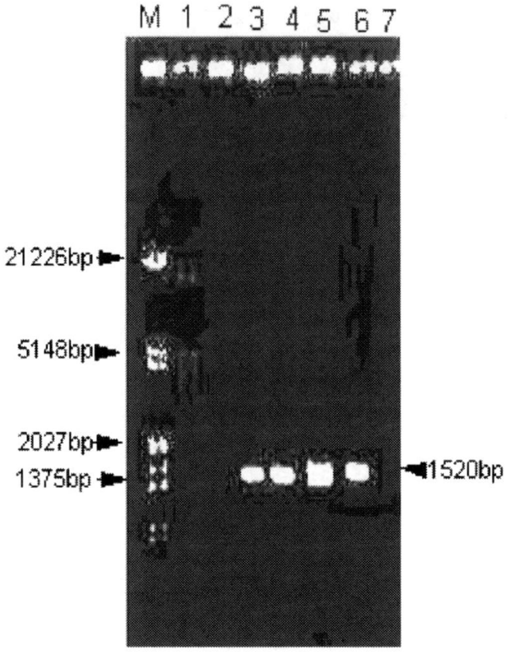

Figure 3. PCR Results M: Marker 1,2: negative control 3:positive control 4,5,6: transformants.

Figure 2 showed that the genomic DNA of Artemisaarenaria seedlings were extracted through the agarose gel electrophoresis. Figure 3 showed that PCR was reacted by specific primers heap of genomic DNA. It can be seen from figure 3, PCR test of transformants in lane 4, 5, 6 and positive control in lane 3 have the same size of 1520 bp fragments, which is cDNA sequence of Glycinin Gy3 subunit, but the negative control in lane 1, 2 can not amplify the expected fragment.

3.4 RT-PCR experimental results of transformants

The experimental were conduct according to plant RNA extraction kit and RT-PCR kit which purchased from Dalian TaKaRa Biotechnology Company. The total RNA was extracted from Artemisaarenaria seedlings which has been introduced the exogenous gene and not intrduced foreign gene, RT-PCR reaction was conduct by using the synthesized specific primers and the optimal reaction condition which have been found out, and then the extracted samples were identified by agarose gel electrophoresis. In the electrophoresis experiment that was conducted, RT-PCR products of the total RNA in Artemisaarenaria seedlings which have been transformed were used as the negative control, the cloned cDNA sequence of Glycinin Gy3 subunit were used as the positive control. The RT-PCR results were showed in figure 4.

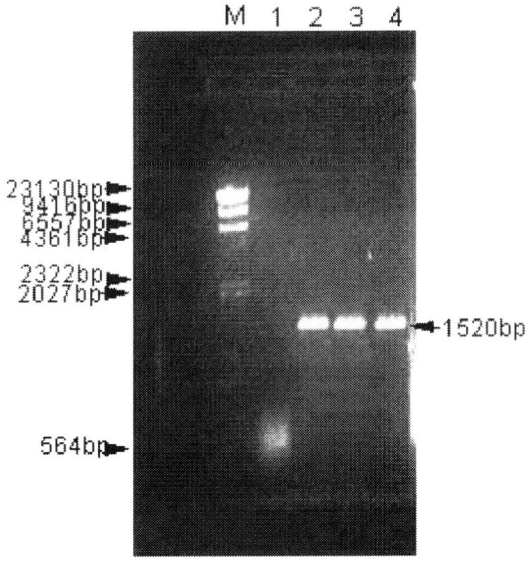

Figure 4 RT-PCR Results M: Marker(λ-Hind III) 1: negative control 2: positive control 3,4: transformants.

3.5 Determination of crude protein content in transgenic plants

The roots of the seedling and the above part were dried at 105 °C, 10 min to 30 min, and then dried at 80 °C 3 h to 10 h until the plant water was evaporated completely. The dried plant was crushed to determine the crude protein. The crude protein content was 3.50% in the control group, but it was higher in the transformed plants than in the control group, in which the average increases to 4.67%.

4. Conclusion

Soybean DNA was introduced into Artemisaarenaria seeds mediated by corona discharge, and then transformed seedlings were obtained. Experimental results showed that the indexes of physiological and biochemical of desert plant were affected by corona discharge. The method used was a genetically modified one that can obtain efficient expression of transgenic plants.

References

[1] Song W M, Zhou H Y, Jia R L, et al.2008 *J. Desert Res.* **28** 449
[2] Li X J, Tan H J, Zhang Z S and Li X R. 2007. *J. Desert Res.* **27** 448
[3] Li Y B, Zhao X and Tan H J. 2008 *J. Desert Res.* **28** 254
[4] Bai S N,Yong T W and Yun X F. 2000 *J. Packaging & Food Mach.* **18** 17
[5] Li Y H, Deng J G, Lei J J, et al. 1995 *J. Shenyang Agric. Univ.* **26** 136
[6] Song T Y 1990 *Biophys. J*.**58** 897
[7] Liu C J, Wang B Y and Zhang H 1999 *Chin. Sci. Bull.* **44** 1157
[8] Xu X, Na R, Yang J and Liu J J 2009 *J. Inner Mongolia Univ.* **40** 600
[9] Maniatis T, et al. 1982 *Mol. Cloning Lab. Manual*. (New York :Cold Spring Harber Laboratory) p 75
[10] Utsumi S, Kohno M, Mori T and Kito M 1987 *J. Agric. Food Chem.* **35** 210

Effect of corona electric field on the production of gamma-poly glutamic acid based on bacillus natto

Hong Qi [1,2], Ri Na[1,2 *] , Jiletu Xin [1,2], Ya Jie Xie [1,2] , Jiu Feng Guo [1,2]

[1]. School of Physical Science and Technology, Inner Mongolia University, Hohhot,China,010021
[2]. Key Laboratory of Ion Beam Bioengineering, Inner Mongolia University, Hohhot, China 010021

E-mail：nari6363@sina.com

Abstract. Bacillus Natto is an important strain for gamma-poly glutamic acid (γ-PGA) production. The mutagenesis of Bacillus Natto 20646 under corona electric field and the screening of high γ-PGA producing mutant were investigated. A new mutant bacillus natto Ndlz01 was isolated from Bacillus Natto 20646 after mutation in corona electric field at 9kV for 2min. The Ndlz01 exhibited genetic stability of high γ-PGA producing ability even after five generation cultures. When the bacterium was mutated in streamer discharge state at 9kV for 2min, its death rate was more than 90%. Compared with the yield of γ-PGA based on the original Bacillus Natto 20646, the γ-PGA yield of mutant bacillus natto Ndlz01 increased from 2.6 to 5.94 g/L, with an increase rate of 129.78%.

1. Introduction

The gamma-poly glutamic acid (γ-PGA) is a kind of new environment friendly biopolymer material, which consists of D-Glutamic and L-Glutamic connected by amino bond [1]. With the advantages of biodegradability, biocompatibility, edibility, especially the characteristic of non-toxic both to human and environment, the γ-PGA is being widely applied in several fields, such as medicine, food, cosmetics, sewage treatments and so on [2]. Among several common ways to synthesize γ-PGA, the biosynthesize is chosen in present paper, because this method processes gentle culture conditions, short production period and higher extraction efficiency. In addition, the producing process and the molecular weight are easily controlled and the way is environment friendly.

At present, UV, DES, NTG and microwave are the common methods to mutate the bacterium domesticly. Most of γ –PGA is produced by the mutant strains and its yield increased by in the range of 30% to 135.3%. In the present paper, the Bacillus natto is mutated in corona electric field with the aim to improve the yield of γ-PGA from Bacillus natto. Up to now, there is lack of information on the mutagenesis of Bacillus natto by corona electric field.

* To whom any correspondence should be addressed.

2. Materials and methods

2.1 Materials, reagents and apparatus

Bacillus natto20646 was chosen as the original bacterium, which was provided by Tianjin Institute of Industrial Microbiology (1.01034).

Chemical reagents of beef extract, peptone, agar, cane sugar, yeast extract, $MgSO_4$, $CaCl_2$, NaCl, $MnSO_4 \cdot H_2O$, K_2HPO_4, KH_2PO_4 and sodium glutamate are all analytical purity grade. KBr is spectroscopical purity grade.

The ultraviolet and visible spectrophotometer (HITACHI U-2900) was used to monitor the growth curve of bacteria. The Infrared spectrometer (Nicolet Awatar 330) was used to identify the γ-PGA. The PCR apparatus (ABI Veriti) was used to identify the mutant Bacillus natto.

2.2 Medium

The fluid nutriment medium with pH of 7.0 was prepared as follows: bean sprouts soup100 mL, cane sugar 5.0 g /L. The seed medium with pH of 7.5 was prepared as follows: beef extract 5 g /L, peptone 10 g /L, cane sugar 30 g /L, yeast extract 5 g /L, NaCl 5 g /L, $MnSO_4 \cdot H_2O$ 0.4 g /L, K_2HPO_4 2 g /L, KH_2PO_4 4 g /L, $MgSO_4$ 0.5 g /L and $CaCl_2$ 0.25 g /L. Fermentation medium with pH of 7.5 was prepared as follows: peptone 10 g /L, cane sugar 30 g /L, yeast extract 5 g /L, sodium glutamate 30 g /L, NaCl 5 g /L, $MnSO_4 \cdot H_2O$ 0.4 g /L, K_2HPO_4 2 g /L, KH_2PO_4 4 g /L, $MgSO_4$ 0.5 g /L and $CaCl_2$ 0.25 g /L. The solid-state medium with pH of 7.5 was prepared as follows: beef extract 10 g /L, peptone 10 g /L, cane sugar 30 g /L and agar 10 g /L NaCl 10 g /L.

2.3 Method

2.3.1 Cultivating condition. Dilution-plate method was used to cultivate the bacteria and each culture dish contains 25 mL solid-state medium. The culture dish was put up side down and cultivated at 37 ℃ in a constant-temperature incubator for 48h. Cultivating condition of seed was as follows: bacteria was inoculated into a 250 mL conical flask which contains 50 mL seed medium and inoculation quantity was 2 %, the bacteria was cultivated on a rotary shaker (200 rpm) at 37 ℃ for 24 h. Cultivating condition of fermentation was as follows: the bacteria was inoculated into a 250 mL conical flask which contains 50 mL fermentation medium and inoculation quantity was 2 %, the bacteria was cultivated on a rotary shaker (260 rpm) at 37 ℃ for 24 h.

2.3.2 Plate separation. The bacterial suspension was obtained by the bacillus natto seeds cultivated in seed medium for 24 h. After a diluting process, 20 μL of diluted suspension was taken and coated uniformly on the solid-state medium. After that, the solid-state medium was put into a constant-temperature incubator and cultured at 37 ℃ for 48 h. Then, the single colony was picked up and dissolved into the sterile water for further use.

2.3.3 Gram stain. Gram stain was carried out as the direction and examined by microscope.

2.3.4 Infrared Spectrum. The γ-PGA was composed of D-Glutamic and L-Glutamic by amino bond. The amino bond had a series of characteristic absorption peak, which was easy to be identified by infrared spectroscopy [3]. The sample was prepared by squashing method and characterized by infrared spectrometer in 4000~400 cm^{-1} at room temperature.

2.3.5 Growth curve. The cultivating time determines the density and vitality of the bacteria. Cultivated the bacillus natto seed in seed medium for 34 h, and then take 3 mL to measure its ABS value at OD_{660} every two hours, using the blank seed medium as the reference.

2.3.6 High voltage corona electric field. Pin-plate equipment was used to generate corona electric field, which was shown in Figure 1.

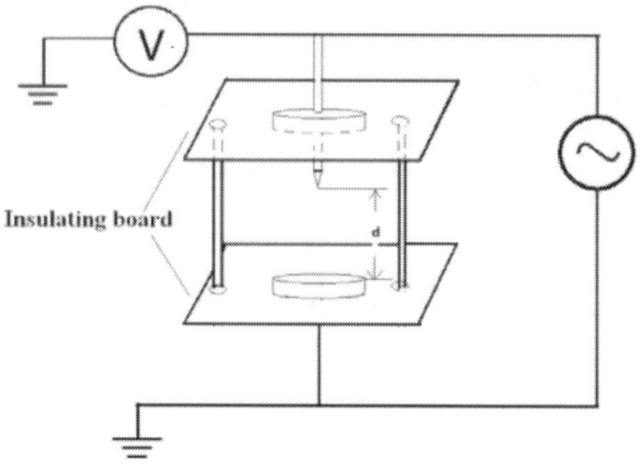

Figure 1. Schematic diagram of equipment for corona.

The bacillus natto 20646 was mutated in two corona electric fields, respectively. Both the critical state and the streamer discharge state were chosen. The critical voltage was 3.5 kV and the treatment time included 15 min, 20 min and 25 min. Because of the wide streamer discharge state, three different voltages (5, 7 and 9 kV) were chosen to mutate the original bacteria. The treatment time included 1, 2 and 3 min, respectively.

The bacillus natto seed was cultivated in seed medium for 12 h. When growing to exponential growth phrase, it was made into bacterial suspension. Firstly, 7 mL of bacterial suspension was put into culture dish with a diameter of 7 cm and shaken slightly to make it distribute uniformly in the culture dish. Secondarily, it was mutated in corona electric field according to the conditions mentioned above; finally, the mutated bacterial suspension was collected and diluted by ten-fold dilution method, and inoculated on the solid-state medium.

2.3.7 Primary screening. The mutant strains were coated on the solid medium immediately after mutation. If the yield of one strain was large enough, obvious pucker on intermediate protrusions of the colony would appear and obvious threadlike strains could be observed when it was drawn by inoculation needle. A high γ-PGA yield of the strain corresponded to obvious puckers and longer threadlike strains.

2.3.8 Freeze-drying way to secondary screening. After fermenting for 24 h, the fermented liquid was centrifuged and the supernatant was taken out. Then the supernatant was mixed with four volumes of absolute ethanol and held for 19 h to precipitate the γ-PGA at 4℃ under static condition. Finally, the sediment was obtained by centrifuging and eluting method and was frozen below -20 ℃. After freeze-drying, the sediment was weighed by electronic balance.

2.3.9 PCR identification. Because the 16SrDNA sequence (GenBank) of bacillus subtilis and bacillus natto were homologous, the primers were designed by choosing the conserved fragment in both ends of the 16SrDNA sequence, and PCR was amplified by using colony directly in 20 μL system.

2.3.10 Genetic stability of the high yield bacterium. Both the original bacterium and the mutated bacterium were inoculated in fermentation medium to cultivate for 24 h and were used to extract the γ-PGA. The yield of γ-PGA extracted from the original and that of the mutated bacterium was compared. In order to investigate the genetic stability of the bacterium, the bacterium was fermented further until the fifth generation.

Journal of Physics: Conference Series **418** (2013) 012139 doi:10.1088/1742-6596/418/1/012139

3. Results and discussion

3.1 Gram stain

The microscopic examination revealed that the bacterium was purple and it implied that the Bacillus natto was gram-positive bacterium, which was consistent with the report in the literature.

3.2 Infrared spectrum

According to the characteristic absorption peak of amino bond reported in Ref. [4], the Fourier Transform Infrared Spectroscopy shown in Figure 2 and Figure 3 suggested that the polymer obtained in the present experiment was γ-PGA.

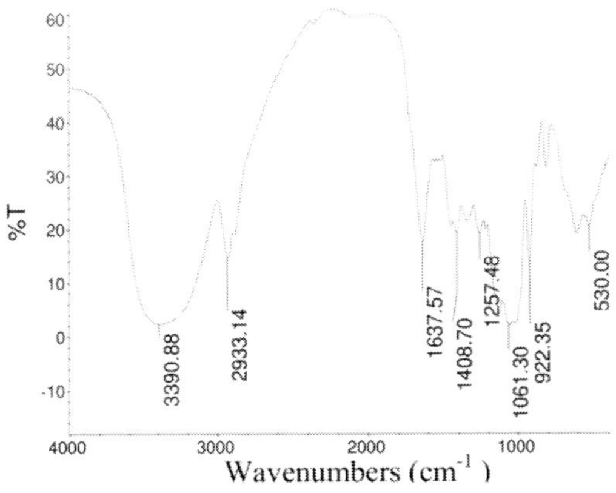

Figure 2. Fourier Transform Infrared Spectroscopy of original bacterium.

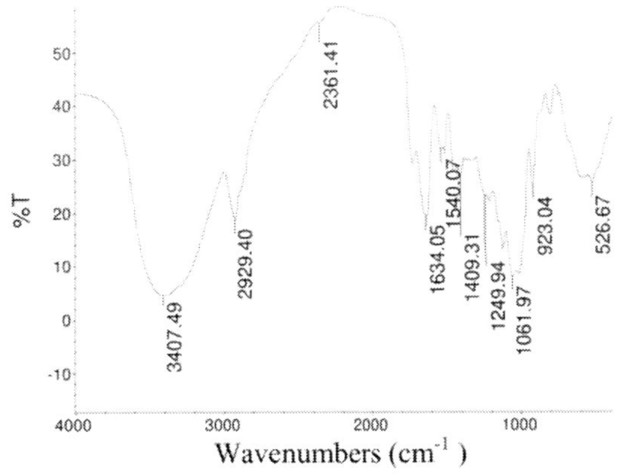

Figure 3. Fourier Transform Infrared Spectroscopy of Ndlz01 bacterium.

3.3 Measuring the growth curve

The growth curves of both the original bacterium and the mutated Ndlz01bacterium were shown in Figure 4. Fig.4 indicated that the lag phrase began at 4 h. The bacillus natto was in the exponential growth phrase between 4 to 10 h and it reached the stationary phrase after 10 h. During the first 18h, the Ndlz01 bacterium grew faster than the original bacterium.

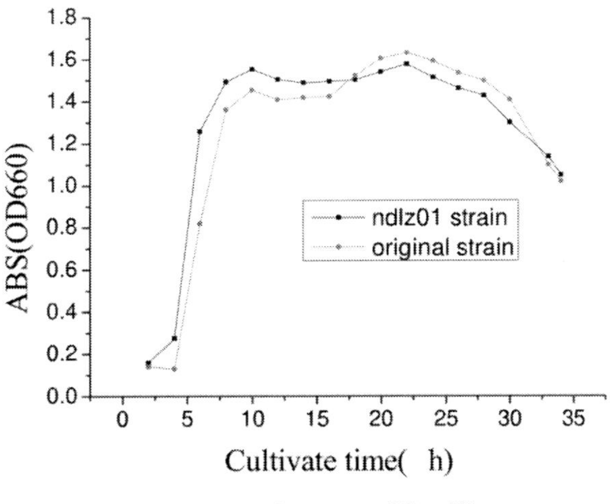

Figure 4 .The growth curve of Bacillus natto.

3.4 Mutagenesis *of bacteria in corona electric field*

After mutation, the death rate of the bacteria was calculated and the results were shown in Table 1. The data shows the death rate was 94.59% under the condition of 3.5kV for 25min, which was higher than 50%, but the mutagenesis time was too long. The death rate under the condition of 9kV for 2min reached 97.3%, which was higher than 50% and the time was acceptable.

Table1 The death rate of the bacterium under the different corona electric field.

Time	Voltage 3.5 kV	Time	Voltage		
			5 kV	7 kV	9 kV
15 min	no	1 min	no	no	70.27 %
20 min	13.5 %	2 min	59.46 %	5.4 %	97.3 %
25 min	94.59 %	3 min	nearly100 %	nearly100 %	nearly100 %

3.5 Primary screening

Compared the colony shape of the mutant bacterium with that of original bacterium, a lot of puckers were found on the edge of one culture dish. Drawn with an inoculation needle, obvious thread-like strains were observed. The corresponding mutation condition was under 9kV for 2min. This bacterium was named as Ndlz01 and reserved for secondary screening.

3.6 Freeze-drying way to secondary screening

After primary screening, the optimal bacterium named as Ndlz01 was isolated, fermented to synthesize γ-PGA. The yield of γ-PGA from Ndlz01 was compared with that from the original bacterium. The results were shown in table 2 (the γ-PGA was extracted from 50 mL fermented liquid). The table clearly showed that the yield of γ-PGA from Ndlz01 was much higher than that of original bacterium.

Table 2 The comparisons of γ-PGA yield from the original bacterium and the Ndlz01

bacterium	G1	G2	G3	G4	G5	Average
Original bacterium	117.5 mg	121.7 mg	144.4 mg	144.5 mg	122.5 mg	130.12 mg
ndlz01	260.2 mg	326.8 mg	332.5 mg	271.2 mg	295.2 mg	297.18 mg
Increment	21.45 %	168.53 %	130.26 %	87.68 %	140.98 %	129.78 %

3.7 The result of PCR identification

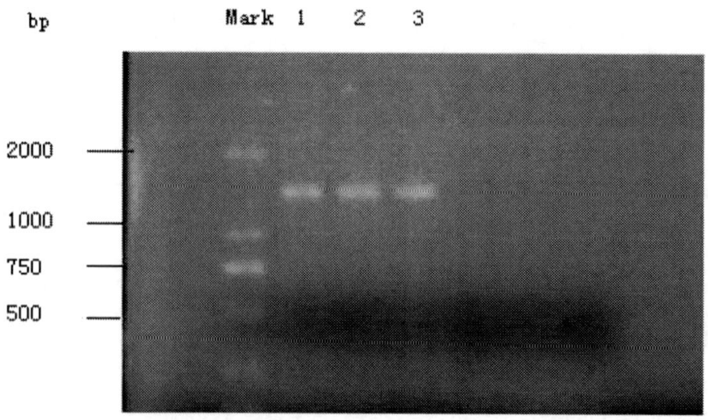

Figure 5. Electrophoretogram of the PCR production.

The PCR reaction used colony directly. The gel electrophoresis result of the PCR production was about 1450 bp (shown in Figure 5 for 1, 2, 3 lane), it was consistent with the expected result.

3.8 The genetic stability of the high yield bacterium

Table 2 gives the measuring results of the γ-PGA yield of original bacterium and that of Ndlz01 in each generation, and the calculated results of yield increment was given too. The result showed that the genetic of the Ndlz01 was stable.

4. Conclusion

The death rate of the bacterium increased with increasing voltage and treatment time. Under the optimal condition of 9 kV for 2 min, the death rate reached 97.3%. After mutation, the optimal bacterium was isolated and named as Ndlz01.The yield of γ-PGA from Ndlz01 increased from 2.6 g/L (130.12 mg per 50 mL fermented liquid) to 5.94 g/L (297.18 mg per 50 mL fermented liquid), the increase rate reached 129.78%. And what's more, Ndlz01 exhibited good genetic stability. These results indicated that the bacillus natto was sensitive to the corona electric field, and a positive effect corona electric field on mutation of bacillus natto existed.

The identification of mutant bacterium according to analysis of PCR revealed that the bacterium used was really the bacillus natto. The growth curves after 18 hours indicated that the growth rate of Ndlz01 was slower than that of the original bacterium, which may be attributed to the fact that the nutrients in the medium of Ndlz01 is less than that in the medium of original bacterium. So it's valuable to make further research to find out a way to keep the bacteria growing fast all the time, which contributes to improving the production.

References

[1] Shih I L and Van Y T 2001 *J. Bioresour technol.* **79** 207
[2] Wang H, Yang L P, Zhao X Y and Liu J J 2011 *Shandong Food Fermentation (in Chinese)* **163** 30
[3] Wan H G 2004 *J. Fine Chem.* **21** 492
[4] Wang Z X 1989 *The infared spectral analysis and identification of polymer.*(ChengDu: SiChuan University Publishing House) p 179

Journal of Physics: Conference Series **418** (2013) 012140 doi:10.1088/1742-6596/418/1/012140

Improvements in plant growth rate using underwater discharge

K Takaki[1,2*], J Takahata[1], S Watanabe[1], N Satta[3], O Yamada[4], T Fujio[4], Y Sasaki[4]
[1] Faculty of Engineering, Iwate University, 4-3-5 Ueda, Morioka, Iwate 020-8551, Japan
[2] Soft-Path Engineering Research Center (SPERC), Iwate University, 4-3-5 Ueda, Morioka, Iwate 020-8551, Japan
[3] Faculty of Agriculture, Iwate University, 3-18-8 Ueda, Morioka, Iwate 020-8550, Japan
[4] Iwate Agricultural Research Center, 20-1 Narita, Kitakami, Iwate 024-0003, Japan

E-mail: takaki@iwate-u.ac.jp

Abstract. The drainage water from plant pots was irradiated by plasma and then recycled to irrigate plants for improving the growth rate by supplying nutrients to plants and inactivating the bacteria in the bed-soil. *Brassica rapa var. perviridis* (Chinese cabbage; *Brassica campestris*) plants were cultivated in pots filled with artificial soil, which included the use of chicken droppings as a fertiliser. The water was recycled once per day from a drainage water pool and added to the bed-soil in the pots. A magnetic compression type pulsed power generator was used to produce underwater discharge with repetition rate of 250 pps. The plasma irradiation times were set as 10 and 20 minutes per day over 28 days of cultivation. The experimental results showed that the growth rate increased significantly with plasma irradiation into the drainage water. The growth rate increased with the plasma irradiation time. The nitrogen concentration of the leaves increased as a result of plasma irradiation based on chlorophyll content analysis. The bacteria in the drainage water were inactivated by the plasma irradiation.

1. Introduction

Hydroponics is widely used in the cultivation of many kinds of vegetable and fruit. In hydroponics, artificial nutrient solutions are used as fertiliser, circulated in the cultivation bed to improve the growth rate of the vegetables and fruits. The nutrient solutions used for plant growth in greenhouses are rich in electrolytes, such as K^+, NH_4^+, Ca_2^+, Mg^{2+}, NO_3^-, $H_2PO_4^-$ and SO_4^- [1]. Owing to this, the drainage nutrient solutions employed in greenhouses have relatively high nutrient ion concentrations, and therefore, they can be reused to save fertilisers and to prevent contamination of water sources with undesired ions, such as nitrate and dihydrogen phosphate [2].

One of the major problems encountered when recycling drainage solutions in soilless and/or artificial soil-based cultures is an exponential increase in pathogenic bacteria concentrations, such as agrobacterium, powdery mildew, and Fusarium wilt, among others [3, 4]. Furthermore, an excess

* To whom any correspondence should be addressed.

accumulation of Na^+, Cl^- and other nutrient ions in irrigation water is also a major problem [5]. In such cases, to reduce the incidence of pathogenic fungi and bacteria in the root environment, the pathogenic fungi and bacteria must be inactivated in the drainage solution. Additionally, to reduce the accumulated Na^+ and Cl^- concentrations, the initial ion concentrations in the artificial fertiliser must be decreased.

Discharge underwater is a promising candidate for solving such problems because discharge produces chemically active species, such as atomic oxygen (O), ozone (O_3) and hydroxyl radicals (OH), which work to inactivate the pathogenic fungi and bacteria [6-11]. Additionally, underwater discharge produces nitric acid, which works as a fertiliser [1, 6]. This paper describes the effect of underwater discharge on both the growth rates of plants and the associated bacterial activities.

2. Materials and methods

2.1. Materials

Figure 1 shows schematics of the setup used for plant cultivaion. The bed-pot was placed on a stand with a height of 10 cm. *Brassica rapa var. perviridis* (Chinese cabbage; Brassica campestris; Atariya Noen Co. LTD) was used as the specimen, cultivated in the bed-pots filled with artificial soil. The artificial soil was composed of expanded vermiculite (Gleen Plan Co. LTD) and Loamy soil (Gleen Plan Co. LTD) mixed at a rate of 1:2. Chicken droppings (1 g in weight) were placed on the soil surface as a fertiliser. The bed-pot was 9 cm in upper diameter, 7 cm in lower diameter and 7 cm in height. The bed-pot was filled with 93 g of artificial soil. Distilled water was supplied to the bed-soil on the first day of cultivation. A portion of the water was absorbed by the plants and the bed-soil, and the remainder was flown through a hole located at the centre of the lower side of the bed-pot. The drainage water was collected in a tank located under the bed-pot and then was recycled as the bed-soil water supply. The seedlings of *Brassica rapa var. perviridis* were cultivated for three days at 25 °C in a dark room after sowing. The three seedlings were planted in the bed-pot and then were cultivated for 28 days in a room near windows. The room temperature was approximately 20 C.

2.2. Plasma reactor

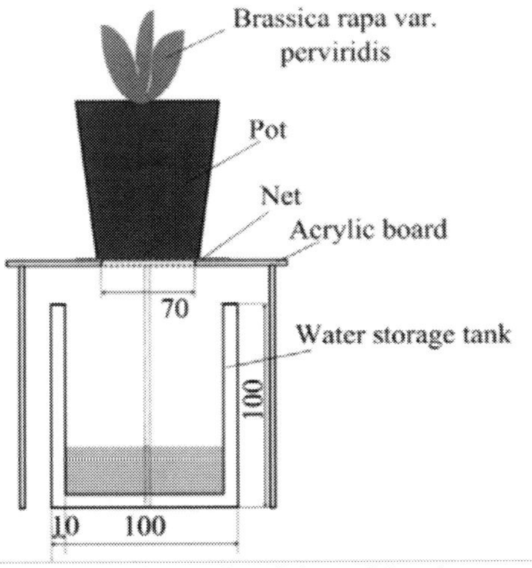

Figure 1. A schematic of the cultivation setup.

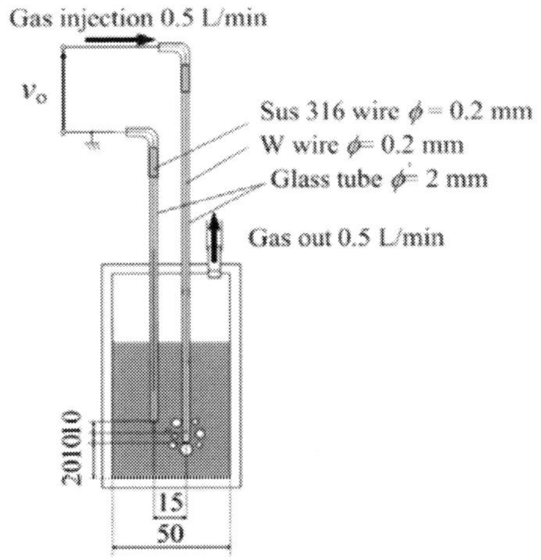

Figure 2. A schematic of the reactor.

A schematic of the plasma reactor used to treat the drainage water is shown in figure 2. The reactor was 7 cm in width, 7 cm in length and 12 cm in height, i.e., a total of 250 ml in volume. Air was injected into the water at an air flow rate of 0.5 L/min using a glass tube placed on the lid. A tungsten wire measuring 0.2 mm in diameter was located in the glass tube, used as a high-voltage electrode to

Journal of Physics: Conference Series **418** (2013) 012140 doi:10.1088/1742-6596/418/1/012140

generate the plasma within the drainage water. The drainage water was grounded by a SUS316 wire measuring 0.2 mm in diameter. The drainage water from the bed-pot emptied into the plasma reactor and then was treated by plasma irradiation, with the plasma produced in the glass tube and in the air bubble injected into the water through the glass tube. The plasma irradiation times were set as 10 and 20 minutes.

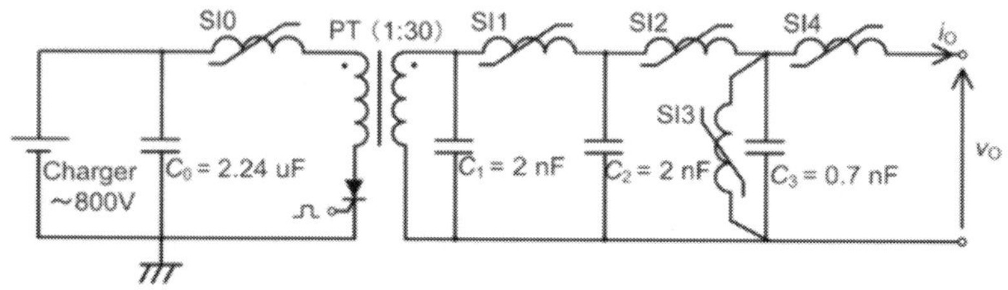

Figure 3. A diagram of the magnetic pulse power generator circuit.

The pulsed high voltage was supplied by a magnetic pulse compression circuit (Suematsu Electronics CO., LTD., MPC-3000S SP, Japan), as shown in figure 3, and was applied to the high-voltage wire electrode located in the glass tube. The capacitor C_0 was charged to the charging voltage by the charger. The energy stored in C_0 was transferred to C_1 through a pulse transformer, PT. The pulsed voltage was produced at the secondary side of the pulse transformer. The pulsed voltage was compressed by saturable inductors (SI_1, SI_2 and SI_4) and capacitors (C_2 and C_3). SI_3 was connected in parallel with C_3 to shorten the pulse width of the output voltage (v_O) [12].

Figure 4 shows typical waveforms of the output voltage (v_O) and the voltages on the capacitors C_1 (v_{C1}), C_2 (v_{C2}) and C_3 (v_{C3}) when the reactor was not connected to the circuit. The pulse compression procedure can be observed clearly. The pulse width of v_{C1} is 1.19 μs and is compressed to a v_O of 130 ns using magnetic compression processes. As can be seen in the figure, the output voltage has a peak value of 21 kV, and its rise time is 57 ns. The LC oscillation and the reflection due to the impedance mismatch cause the voltages to oscillate. The pulse repetition rate is maintained at 250 pps with a peak voltage of 30 kV in the experiment. The polarity of the applied voltage is fixed as positive. The energy in the reactor per pulse is calculated by integrating the electric power obtained by the output voltage and the current over time.

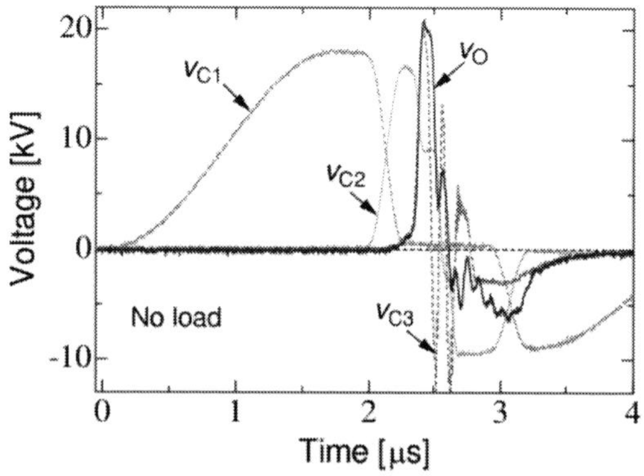

Figure 4. Typical output voltage waveforms of the magnetic pulse compression circuit without connection to the reactor.

Figure 5. Photographs of *Brassica rapa var. perviridis* cultivated for 28 days at (a) w/o plasma and with (b) 10 min. or (c) 20 min. of irradiation per day.

2.3. Analytical methods

The effect of plasma irradiation in the drainage water on the growth rate of *Brassica rapa var. perviridis* was evaluated by the amount of dried weight and the time history of the leaf length. The leaf lengths were monitored once per day during cultivation. The dried weight was obtained from cropped *Brassica rapa var. perviridis* after 28 days of cultivation. The cropped *Brassica rapa var. perviridis* was dried at 75 C for 48 hours using a constant-temperature box (Isuzu; Incubator SFR-113S). The weight was averaged from three samples from each pot. The number of bacteria in the drainage water was obtained by a colony counting method using a plate count agar. The bacteria were cultivated at 35 C for 48 hours. The nitrite and nitrate ions in the drainage water were measured with Ion Chromatography (IC) analysis. The nitrogen concentration of the leaves was measured as the chlorophyll content (Soil and Plant Analyser Development; SPAD) with a SPAD-502 meter.

3. Results

3.1. Growth rate

Figure 5 shows photographs of *Brassica rapa var. perviridis* at 28 days of cultivation for various plasma irradiation times. The plasma was produced in the drainage water using the plasma reactor mentioned in figure 4. The plasma was irradiated in the drainage water for 10 and 20 minutes each day. One group consisted of three cultivated pots of *Brassica rapa var. perviridis*. Another group (the control) was cultivated without plasma irradeation. All *Brassica rapa var. perviridis* plants grew by 28 days of cultivation. Figure 5 shows that the leaf size of the plants increased with plasma irradiation, changing with the irradiation time.

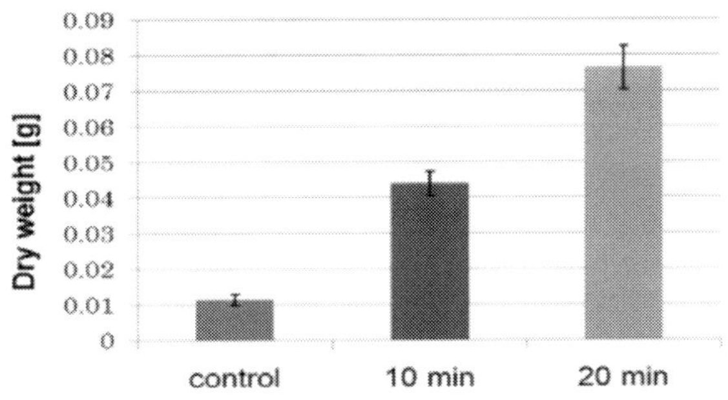

Figure 6. The dried weight of cropped *Brassica rapa var. perviridis* after 28 days of cultivation for various plasma irradiation times.

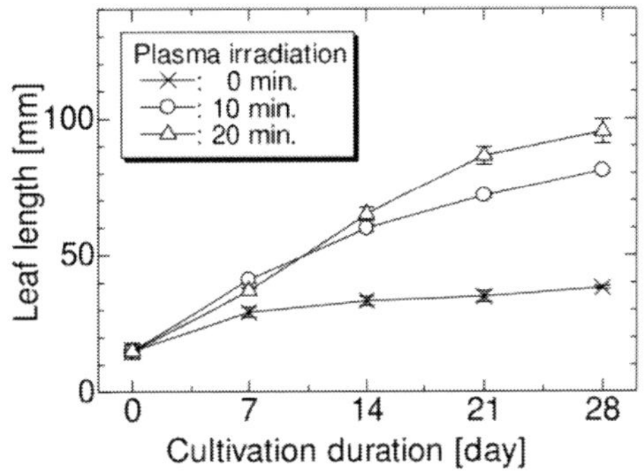

Figure 7. The time history in length of cultivated *Brassica rapa var. perviridis* leaf for various plasma irradiation times.

Figure 6 shows the dried weights of cropped *Brassica rapa var. perviridis* at 28 days after cultivation for various plasma irradiation times. The error bars show the standard diviations of the data. The dried weight of the plant without the plasma irradiation was an average weight of 0.011 g. The dried weight increased to 0.044 and 0.076 g as a result of 10 and 20 minutes of plasma irradiation, respectively. These values correspond to 3.9 and 6.6 times incremental increases in comparison to that of the control group. The statistical significance was confirmed by *t*-test with an accuracy of $P < 0.01$.

Figure 7 shows the time history in length of the cultivated *Brassica rapa var. perviridis* leaf for various plasma irradiation times. The leaf length in all cases was found to increase gradually with cultivation duration. The increase in the leaf length was saturated at 7 days after cultivation without plasma irradiation. The leaf length at 28 days of cultivation reached 38 mm without plasma irradiation. The leaf length at 28 days of cultivation increased to 80 and 95 mm as a result of 10 and 20 minutes of plasma irradiation, respectively. These values correspond to respective increases of 2.1 and 2.5 times in comparison to that of the control group.

3.2. Inactivation of bacteria

Figure 8 shows the time history of the bacteria count in the drainage water of cultivated *Brassica rapa var. perviridis* for various plasma irradiation times. The bacteria count was found to increase from 4.0 to 5.5 log-CFU at 28 days of cultivation without plasma irradiation. However, the bacteria count in the drainage water after 28 days of cultivation shows approximately the same value at day 0 of cultivation under the plasma irradiation condition. The bacteria count decreases by 2 log-CFU with the application of plasma irradiation into the drainage water. These results indicate that plasma irradiation of the drainage water is effective for inactivating bacteria not only in the drainage water but also in the bed-soil filled in the cultivation pots.

4. Discussion

The results of this study confirm that the growth rate of the plant *Brassica rapa var. perviridis* can be improved by plasma irradiation. The use of plasma in the water can supply many kinds of chemical species, such as OH, O, O_3, NO_2, NO_3, and H_2O_2, among others [11, 12]. Nitrate and nitrous species in the drainage water were measured to clarify the effect of plasma irradiation on the plant growth rate because nitrous nitrogen typically works as a fertiliser in plant cultivation.

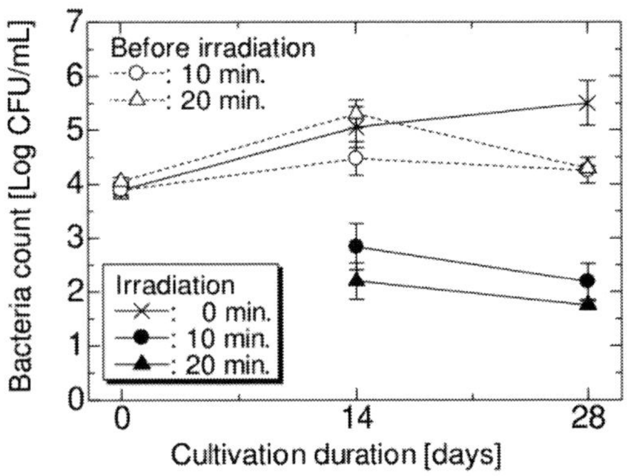

Figure 8. The time history of bacterial count in the drainage water of *Brassica rapa var. perviridis* cultivation for various plasma irradiation times.

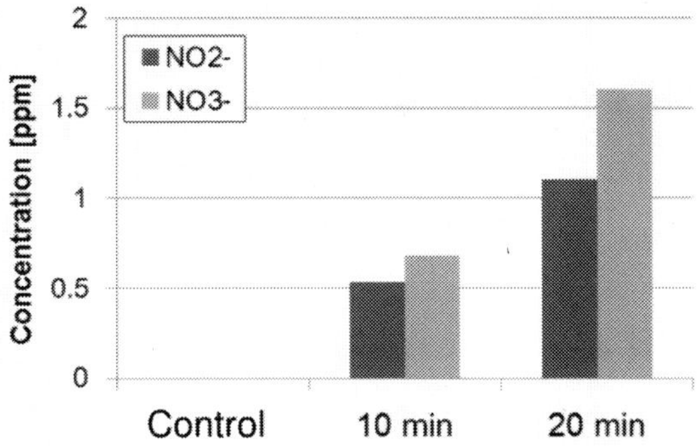

Figure 9. The concentration of nitrate nitrogen (NO_2^-) and nitrous nitrogen (NO_3^-) in the drainage water at 28 days of cultivation for various plasma irradiation times.

Figure 9 shows the concentrations of nitrate nitrogen (NO_2^-) and nitrous nitrogen (NO_3^-) in the drainage water at 28 days of cultivation for various plasma irradiation times. Nitrate nitrogen and nitrous nitrogen were not detected in the drainage without plasma irradiation treatment. Nitrate and nitrous nitrogen species were produced by plasma irradiation and increased with the irradiation time. The plasma causes the following reactions to occur in the air bubble [13]:

$$N_2 + O_2 + e^* \rightarrow 2NO + e \quad (R1),$$
$$2NO + O_2 \rightarrow 2NO_2 \quad (R2),$$
$$H_2O + e^* \rightarrow OH + H + e \quad (R3),$$
$$NO + OH \rightarrow HNO_2 \quad (R4),$$
$$NO_2 + OH \rightarrow HNO_3 \quad (R5).$$

The drainage water was acidified with HNO_2 and HNO_3 as HNO_2 (aq) and HNO_3 (aq) because the Henry constants of HNO_2 and HNO_3 are larger than that of the other species produced with the air plasma [14].

Figure 10 shows the nitrogen concentrations of the leaves of *Brassica rapa var. perviridis* at 28 days of cultivation for the various plasma irradiation times. The nitrogen concentrations were evaluated with the SPAD (Soil and Plant Analyser Development) value using a chlorophyll meter. The nitrogen concentration in the leaf was found to increase with the application of plasma irradiation into the drainage water. Generally, plants absorb NO_3^- (aq) as nutrients in the root, derived from the soil and the supplied water. NO_3^- (aq) in the drainage water is recycled because the water supplied to the bed-pot is produced by plasma irradiation, as shown in Fig. 9. The experimental results shown in Fig. 10 indicate that the produced NO_3^- (aq) ions are absorbed from the roots of *Brassica rapa var. perviridis* and are carried to the leaves.

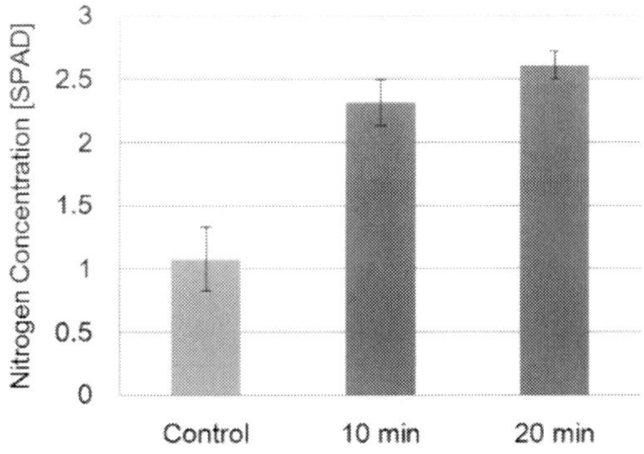

Figure 10. The nitrogen concentration of *Brassica rapa var. perviridis* leaves at 28 days of cultivation for various plasma irradiation times.

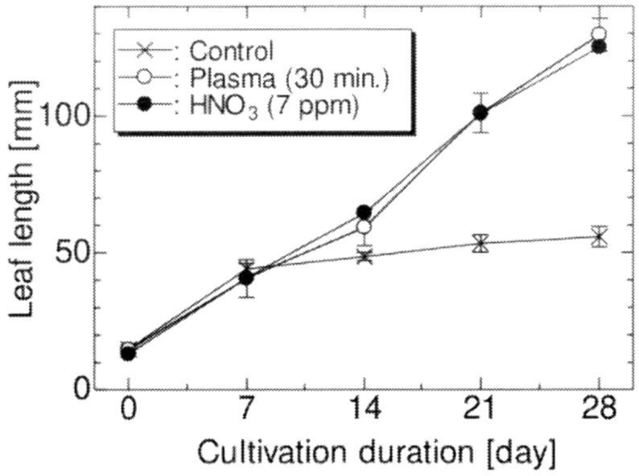

Figure 11. The time history in length of cultivated *Brassica rapa var. perviridis* leaves for various plasma irradiation times.

The experimental results shown in Fig. 9 and 10 strongly suggest that the increase in plant growth rate as a result of plasma irradiation of the water supplied to the bed-soil for plant cultivation is mainly caused by the air plasma-produced NO_3^- (aq) produced. To confirm the mechanism, the *Brassica rapa var. perviridis* was cultivated for 28 days under three different conditions, including 1) w/o plasma irradiation and w/o an additive (the control), 2) 30 minutes of plasma irradiation and w/o an additive, and 3) w/o plasma irradiation and with a HNO_3 addition as the additive but w/o plasma irradiation. The HNO_3 concentration in the drainage water was adjusted to 7 ppm, which was almost in agreement with that following 30 minutes of plasma irradiation. Figure 11 shows the time history in length of the cultivated *Brassica rapa var. perviridis* leaves under the three different conditions. The leaf length in all cases was found to increase gradually with cultivation duration. The increase in leaf length also was saturated at 7 days after cultivation without plasma irradiation or HNO_3 addition. The leaf growth rate was improved not only with plasma irradiation but also with HNO_3 addition. The growth rate at 7 ppm of HNO_3 addition was in good agreement with that at 30 minutes of plasma irradiation. This result clearly shows that NO_3^- (aq) production with the plasma is a dominant factor for increasing the plant growth rate.

5. Conclusions

The plasma was generated using a magnetic pulse compression pulsed power generator to improve the plant growth rate in a bed-pot based cultivation system. The experimental results showed that the growth rate increased significantly by the application of plasma irradiation to the drainage water. The growth rate of the plants increased with the plasma irradiation time. The bacteria in the drainage water were inactivated following plasma irradiation. The nitrate and nitrous nitrogen species were produced by the application of plasma irradiation to the drainage water. The produced nitrous nitrogen ions were absorbed by the plants, becoming a dominant factor for improvements in the plant growth rate by the plasma irradiation.

Acknowledgments

The authors are grateful to Drs. T. Fujiwara, S. Mukaigawa, and K. Takahashi of Iwate University for their many comments and discussions. The authors are also grateful to Mr. Y. Shida of Iwate University for his technical support. This work was supported by Sanriku Kikin H24.

References

[1] Mastalerz JW 1977 *The Greenhouse Environment* (New York, USA: John Wiley & Sons, Inc.)
[2] Savvas D and Manos G 1999 *J. Agricultural Eng. Research* **73** 29
[3] Mizukami K, Satoh K, Kanayama H, Itoh H, Tagashira H, Shimozuma M, Okamoto H and Takasaki S 2006 *IEEJ Transactions FM* **126** 688 [in Japanese]
[4] Ohshima T, Watanabe S, Fukagawa T and Sato M 2006 *J. Institute of Electrostatics Japan* **30** 236 [in Japanese]
[5] Koide S and Satta N 2004 *Biosystems Engineering* **87** 89
[6] Akiyama H 2000 *IEEE Trans. Diel. Elect. Insul.* **7** 646
[7] Sato M 2008 *Plasma Sources Sci. Technol.* **17** 024021
[8] Takahashi K, Sasaki Y, Mukaigawa S, Takaki K, Fujiwara T and Satta N 2010 *IEEE Trans. Plasma Sci.* **38** 2694
[9] Chang J S, Lawless P A and Yamamoto T 1991 *IEEE Trans. Plasma Sci.* **19** 1152
[10] Ebihara K, Takayama M, Ikegami T, Ogata K, Stryczewska H D, Gyoutoku Y and Sakai T 2006 *J. Adv. Oxid. Technol.* **9** 170
[11] Lukes P, Clupek M, Babicky V and Sunka P 2008 *Plasma Sources Sci. Technol.* **17** 024012
[12] Takahashi K, Takaki K and Satta N 2012 *J. Adv. Oxid. Technol.* **15** (in press)
[13] Bruggeman P and Leys C 2009 *J. Phys. D: Appl. Phys.* **42** 053001
[14] Daito S, Tochikubo F and Watanabe T 2000 *Jpn. J. Appl. Phys.* **39** 4914

Influence of strong electric field on MDA and SOD of rice under atmosphere pressure

Jianping Xiong *, **Shengyong Hu, Jikai Li, Songqing He, Lixin Feng**
Electrostatic Institute, Guangdong University of Petrochemical Technology, Maoming, Guangdong, 525000

E-mail: jianping422@163.com

Abstract. The content of MDA is measured by TBA method in the experiment. The results show that the MDA content of rice seedlings after being radiated in a strong electric field under atmosphere pressure decreases compared to that of those not being radiated while the SOD activity decreases. It indicates that radiated seeds' resistance against oxidative stress can be greatly enhanced. The mechanism and relation between them are analyzed in this paper.

1. Introduction

An adaption system can be formed in a plant and it is constrained by heredity of which reactive oxygen metabolism plays an important role. It is a primary reaction when the plant is stressed in adversity. The balance of generation and scavenging of intracellular free radicals is destroyed and thus the increase of free radicals causes damage to the cell. MDA is a decomposition product during the membrane lipid peroxidation. The reaction with proteins and nucleic acid modifies its nature to inhibit protein synthesis and it may react with enzyme to make the enzyme lose activity to be a metabolic molecule of error catalysis. Therefore, the content and MDA and permeability change of membrane can reflect the peroxidation degree of membrane lipid and meanwhile are important indexes indicating the damage of plasmalemma. MDA content in lamina of rice seeds in the test is basically decreasing as the SOD of lamina increases. As for rice seeds processed in a strong electric field, enzyme activity of SOD is increased, which indicates that rice seeds' function after radiation increases against oxidative stress. This is reflected form the test data of MDA content.

2. Measure MDA by TBA method

2.1. Principle

During membrane lipid peroxidation, the generated MDA content can reflect the damage of the plant in adversity. In acid and high temperature conditions, MDA may react with TBA to generate brown 3, 5, 5 trimethyl oxazole 2, 4 diketone of which the maximum absorption wavelength is 532 nm. During the test of MDA in the plant tissue, however, it is disturbed by many kinds of materials, mainly the soluble sugar of which the maximum wavelength of color reaction products with TBA is 450 nm and wavelength of 532 nm can be absorbed too. TBA can be reacted with other materials that may be absorbed at these wavelengths. In order to eliminate influence of the reaction, the absorbance at

* To whom any correspondence should be addressed.

wavelength of 600 nm should be measure as the MDA content is being measured, from which the absorbance difference at 532 nm, 600 nm and 450 nm can be used to calculate the MDA content.

2.2. Materials, instrument and reagent

2.2.1. Materials: Rice embryo leaves

2.2.2. Instrument:

Strong electric field radiometer (including high-frequency and high-voltage power supply and the discharge device with dielectric barrier, etc.), high-speed refrigerated centrifuge: TGL-16GR (Shanghai Anting Scientific Instrument Factory), T6 Xinshiji spectrophotometer, micro-syringe, 10 ml test tubes, water bath, electronic balance, mortar and pipettes (5 ml, 2 ml and 1 ml), etc.

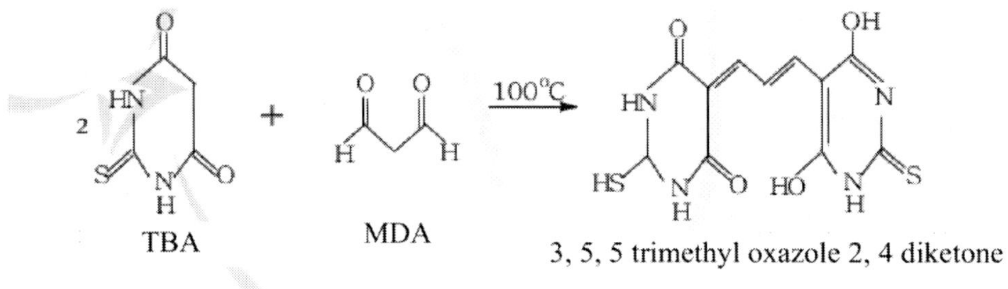

Figure 1. The production process of 3, 5, 5 trimethyl oxazole 2, 4 diketone.

2.2.3 Preparation of MDA reagent

(a) 0.05 mol L-1 pH 7.8 sodium phosphate buffer;

(b) 5 % trichloroacetic acid solution: 5 g trichloroacetic acid dissolved in distilled water at a constant volume of 100 ml;

(c) 0.5 % TBA solution: 0.5 g TBA dissolved in 5 % trichloroacetic acid at a constant volume of 100 ml.

2.3 Experimental steps and data

2.3.1. Strong electric field radiation:

(1) Select 6 groups of plump rice seeds in a uniform size, 100 seeds for each group; measure their dry weight and put them in an incubator, one group for reference and the other for test.

(2) Determine the spacing between counter electrodes of strong electric field radiation is 1 cm and apply a field intensity of 100 kV cm^{-1} for 5, 10, 15, 20 and 25 s.

(3) Add distilled water of same volume into each culture dish and place them in the incubator for soilless cultivation.

2.3.2 Extraction of MDA

(1) Weight 0.5 g sample and add 2 ml precool 0.05 mol L^{-1} pH 7.8 phosphate buffers;

(2) Add a little bit quartz sand ground into homogenate in an ice-bathed mortar;

(3) Transfer it in a centrifuge tube of 5 ml scale (including the fluid after the mortar is washed with buffer);

(4) Centrifuge it for 10 min at 4500 revolutions min^{-1} and exact the supernatant, i.e., MDA extract.

See table 1 below for experimental data:

Table 1. MDA extract of experimental groups and reference groups.

	CK group	Experimental group 2 (D2)	Experimental group 3 (D3)	Experimental group 4 (D4)	Experimental group 5 (D5)	Experimental group 6 (D6)
Vt(ml)	3.4	3.6	3.4	3.7	3.6	3.8
V1(ml)	4.0	4.0	4.0	4.0	4.0	4.0
V2(ml)	2.0	2.0	2.0	2.0	2.0	2.0

Vt: total volume of sample extract;
V1: total volume of reaction fluid of sample extract and TBA solution;
V2: volume of sample extract reacted with TBA; (the same below).

2.3.3. Determination of MDA content
(1) Place 2 ml extract in a graduated test tube;
(2) Add 3 ml solution of 0.5 % TBA in trichloroacetic acid;
(3) Heat it for 10 min in a boiling water bath;
(4) Centrifuge it for 10min at 4500 revolutions min-1 and exact the supernatant; determine the absorbance for 100 % transmittance with distilled water as the blank at a wavelength of 532 nm and 600 nm.

The experimental data are recorded in table 2 as below:

Table 2. Determination of Absorbance of Each Experimental Group and Reference Group.

	CK group	Experimental group 2 (D2)	Experimental group 3 (D3)	Experimental group 4 (D4)	Experimental group 5 (D5)	Experimental group 6 (D6)
A532 (abs)	0.140	0.154	0.109	0.153	0.159	0.160
k*A532	13.860	15.246	10.791	15.147	15.741	15.246
A600	0.057	0.067	0.034	0.067	0.076	0.076
k*A600	5.643	6.633	3.366	6.633	7.029	6.633
A450 (abs)	0.186	0.316	0.148	0.316	0.240	0.297
k*A450	18.414	31.284	14.652	31.284	21.879	31.284

A532 represents the absorbance at 532 nm and A600 represents the absorbance at 600 nm.

2.3.4. Determination of MDA content
(1) Place 2 ml extract in a graduated test tube and add 3 ml solution of 0.5 % TBA in trichloroacetic acid;
(2) Heat it for 10 min in a boiling water bath;
(3) Centrifuge it for 10 min at 4500 revolutions min^{-1};
(4) Exact the supernatant and determine the absorbance for 100 % transmittance with distilled water as the blank at a wavelength of 532 nm and 600 nm.

The following is the formula to calculate the MDA content:

$$C = 6.45(A532 - A600) - 0.56 * A450 \quad (1)$$

$$MDA\ content = \frac{C \times V_t \times V_1}{1000 \times V_2 \times W} \quad (2)$$

where 1000 is the coefficient for ml converted into L; W is the mass of fresh samples; A532 represents the absorbance at 532 nm and A600 represents the absorbance at 600 nm; Vt is the total volume of sample extract; V1 is the total volume of reaction fluid of sample extract and TBA solution; V2 is the volume of sample extract reacted with TBA; see table 3 for measured content of each group.

Table 3. MDA content of each experimental group and reference group (umol g^{-1}).

	CK group	Experimental group 2 (D2)	Experimental group 3 (D3)	Experimental group 4 (D4)	Experimental group 5 (D5)	Experimental group 6 (D6)
A532	0.140	0.154	0.109	0.153	0.159	0.160
A600	0.057	0.067	0.034	0.067	0.076	0.076
A450	0.186	0.316	0.148	0.316	0.240	0.297
C(umolL^{-1})	0.431	0.384	0.401	0.378	0.444	0.375
Vt(ml)	3.400	3.600	3.400	3.700	3.600	3.800
V1(ml)	4.000	4.000	4.000	4.000	4.000	4.000
V2(ml)	2.000	2.000	2.000	2.000	2.000	2.000
W(g)	0.500	0.500	0.500	0.500	0.500	0.500
MDA content (umol g^{-1})	0.005864	0.005532	0.005452	0.005591	0.005774	0.00571

2.3.5. Time effect curve

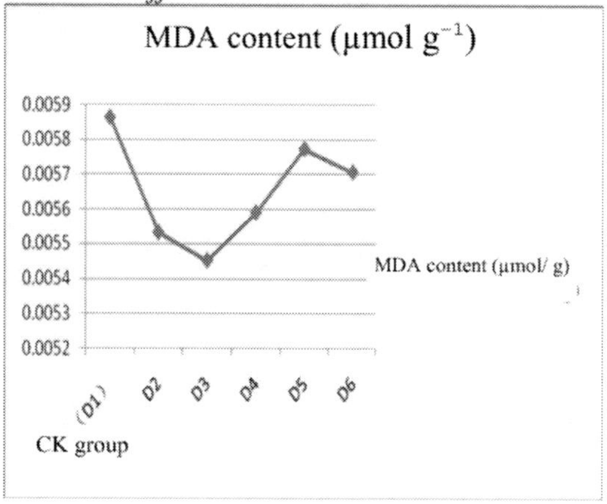

Figure 2. The curve of MDA content of each group

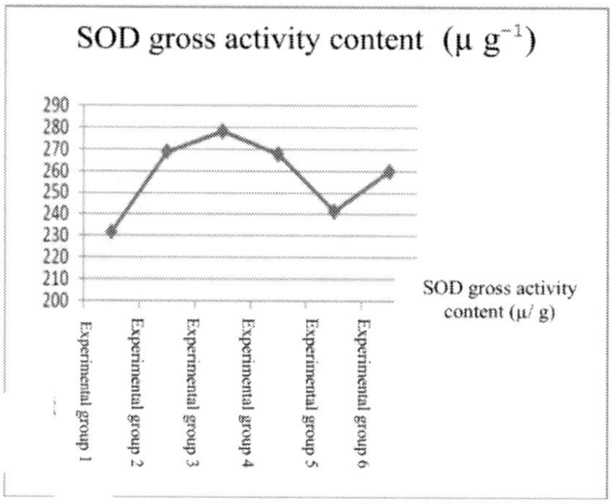

Figure 3. SOD gross activity content curve.

3. Measure SOD by NBT method

3.1. Preparation of SOD reagent

(1) 130 mmol L^{-1} methionine (Met) solution: 1.9399 g Met dissolved in phosphate buffer at a constant volume of 100 ml;

(2) 750 μmol L^{-1} NBT solution: 0.06133 g NBT dissolved in phosphate buffer at a constant volume of 100 ml;

(3) 100 μmol L^{-1} EDTA-Na_2 solution: 0.03721 g LEDTA-Na_2 dissolved in phosphate buffer at a constant volume of 1000 ml;

(4) 20 μmol L^{-1} riboflavin: 0.00753 g riboflavin at a constant volume of 1000 ml, kept in darkness.

3.2. Collection of SOD extract

(1) Take 0.5 g rice leaves in the pre-cool mortar;

(2) Add 1 ml pre-cool phosphate buffer ground into homogenate;

(3) Add buffer to 5 ml;

(4) Take 2 ml solution for centrifugation for 20 min at 1000 rpm;

(5) Collect the supernatant, i.e., SOD extract.

3.3. Color reaction

(1) Prepare four 5 ml finger-type tubes with good transparency, two for test and two for reference;
(2) Place solution: 1.5 ml phosphate buffer of 0.05 mol L^{-1}, 0.3 ml Met solution of 130 mmol L^{-1}, 0.3 ml NBT solution of 750 μmol L^{-1}, 0.3 ml EDTA—Na_2 solution of 00 μmol L^{-1}, 0.3 ml riboflavin of 20 μmol L^{-1}, 0.05 ml SOD solution of 10 μmol L^{-1} and 20 μmol L^{-1};
(3) Replace the SOD solution with buffer in two reference tubes;
(4) Add 0.25 distilled water to a total volume of 3.0 ml, mix it even and place one in darkness; place other tubes under 4000 Lx sunlight for 20 min reaction; after the reaction is completed, determine absorbance at 560 nm for other tubes.

3.4. Transmittance measuring of each group

(1) Replace the SOD solution with buffer in two reference tubes;
(2) Place 3 ml above solution into the finger-type tubes, place one of the in darkness and other three in positions receiving same light intensity;
(3) Use fluorescent tubes of 4000 Lx (15 W can be used) at 25-30 ℃ for 15-20 min illumination until the color changes;
(4) With the unlighted tubes as the blank and tubes without SOD solution as the reference group, place a spectrophotometer for colorimetric determination at a wavelength of 560 nm.

Record the transmittance as indicated in table 4 below:

Table 4. Transmittance by colorimetric method for each group at a wavelength of 560 nm.

	CK group	Experimental group 2 (D2)	Experimental group 3 (D3)	Experimental group 4 (D4)	Experimental group 5 (D5)	Experimental group 6 (D6)
Ack(abs)	0.260	0.241	0.340	0.307	0.213	0.274
k*Ack	25.74	23.859	33.660	30.393	21.087	27.126
Ae1(abs)	0.043	0.023	0.028	0.036	0.056	0.042
k*Ae1	4.257	2.277	2.772	3.564	5.544	4.158
Ae2(abs)	0.080	0.038	0.053	0.044	0.044	0.055
k*Ae2	7.920	3.762	5.247	4.356	4.356	5.445
Ae(abs)	0.0615	0.0305	0.0405	0.040	0.050	0.0485

where $Ae = \frac{(Ae1 + Ae2)}{2}$ is the average value of light absorption of the sample tube, Ack is the light absorption value of the reference tube, and Ae1 and Ae2 are light absorption values, respectively.

3.5. Measuring SOD gross activity of each group:

With unlighted reference tubes as the blank, measure the light absorption of each tube. As known that SOD activity unit restraining 50 % of NBT photochemical reduction is an enzymatic activity unit, calculate the SOD activity according to the equation below:

$$SOD \;\; gross \;\; activity = \frac{(A_{CK} - A_E) \times V}{A_{CK} \times 0.5 \times W \times V_t} \tag{3}$$

where the SOD gross activity is in a unit of fresh weight enzyme per gram; Ack is the light absorption value of the reference tube; and Ae is the light absorption value; V is the total volume of sample solution cm^{-3}; Vt is the sample consumption cm^{-3};W the weight of sample (g).

See table 5 for SOD gross activity of each group:

Table 5. SOD gross activity of each experimental group.

	CK group	Experimental group 2 (D2)	Experimental group 3 (D3)	Experimental group 4 (D4)	Experimental group 5 (D5)	Experimental group 6 (D6)
Ack (abs)	0.26	0.241	0.34	0.317	0.213	0.274
Ae (abs)	0.0615	0.0305	0.0405	0.04	0.05	0.0485
V (ml)	3.8	3.85	3.95	3.85	3.95	3.95
Vs (ml)	0.05	0.05	0.05	0.05	0.05	0.05
W (g)	0.5	0.5	0.5	0.5	0.5	0.5
SOD gross activity	232.0923	269.0207	278.3588	267.8697	241.8216	260.0657

3.6. Time effect curve

Based on table 5, the SOD gross activity time effect curve of each experimental group and reference group is graphed as the right figure.

4. Conclusions

(1) As can be seen in table 1, 2 and 4 and MDA content curve (left), a same electric field of 100 kV cm^{-1} is used to radiate the rice seedlings and the MDA content in rice seedlings is lower than that of reference, which indicates that time effect occurs for rice seeds in electric field radiation under atmosphere pressure. The MDA content decreases as radiation time extends, for example, group 1, 2 and 3. However, the MDA content, on the contrary, increases as the radiation time extends further, for example, group 4, 5 and 6, while the MDA content is still lower than that of reference group, which indicates that the change of MDA content is not monotonic as the radiation time extends. A threshold value can be found-10s radiation time in group 3. The MDA content is a key index reflecting the level of lipid peroxidation. MDA content in Rice seed after radiation in a strong electric field decreases; lipid peroxidation is weakened; stress resistance is enhanced. These promote the seed germination.

(2) As also indicated in table 4, 5 and the right experimental curve, the SOD gross activity of rice seedlings radiated under atmospheric pressure in a strong electric field is higher than that of reference group and is getting higher as the radiation time extends, for example, group 1, 2 and 3. SOD gross activity of experimental group 3 is the highest (by about 20 %). However, as the radiation time extends to a certain extent, the SOD gross activity decreases on the contrary but is still higher than that of reference group, for example, group 4, 5 and 6. While the activation of SOD eases the lipid peroxidation caused by active oxygen, which protects the membrane and makes it adapt better in the environment and is conducive to seed germination.

(3) After the rice seeds are radiated in a strong electric field, MDA content in rice seedlings decreases and SOD activity increases, which are two aspects of biological effect of a strong electric field, and the effect differs for different radiation time. As shown from research results, for rice seeds after being radiated in a strong electric field, the O^{2-} in seedling cells is comparatively lower to ensure the membrane system keeps stable.

(4) The mechanism is that the strong electric field polarizes the ordered arrangement of polar molecules such as cellular protein, carbohydrates and lipids as well as metal ions to cause change of metalloenzyme conformation. While MDA is a product of cell membrane lipid peroxidation and its generation will aggravate the damage of membrane. The decrease of MDA content of the radiated seedlings indicates that the membrane damage is reduced and oxidation resistance enhanced. The SOD, however, is a kind of protective enzymes clearing superoxide anion free radicals O^{2-}. Commonly, the generation and scavenging of free radicals in a cell is in a dynamic balance and the growth of SOD activity after radiation in a strong electric field is conducive to scavenging oxygen radicals, which eases peroxidation of membrane lipid. It can make the O^{2-} in the seedling cells maintain at a low level, thus to ensure stability of membrane system. Therefore, after having been radiated in a strong electric field, seeds' resistance against oxidative stress can be greatly enhanced. The researches present that

the SOD activity directly reflects the balance state of generation and scavenging of O2– in a cell and indirectly influences the level of MDA.

References

[1] Song T Y 1994 *In: First East Asian Symp. on Biophys.* (Japan) 39
[2] AnileevaID,KostinaLN,VaulinaEN. 1983 Experiments with airdried seeds of Arabidopsisthaliana (L.) Heynh.And Crepis capillaries(L.)wallr.Aboard salyut 6. Adv Space Res, 3(8):129-133.
[3] Xiong J P 2010 *Nat. Sci. J. of Hainan Uni.* **28** 222
[4] Qin P, Liu FH, Liang X N 2002 *Heilongjiang Agric. Sci.* **1** 31
[5] Zhang Z L 1990 *Higher Educ. Press* (Beijing) 27
[6] Li H S 1990 *Higher Educ. Press* (Beijing) 68

Quantitative and economical assessment of effectiveness of electrostatic pesticide spraying

Ryo Nishimura[1*], Satoko Fujita[1], Shota Michihara[1], Takashi Masuoka[2], Toshihiro Kimura[2], Shinji Yatsuzuka[2] and Shinobu Anaguchi[2]
1 Graduate School of Engineering, Tottori University, 4-101 Koyama-minami, Tottori 680-8552, Japan
2 Arimitsu Industry Co.,Ltd., 1-3-7 Fukae-kita, Tousei-ku, Osaka 537-0001, Japan

E-mail: ryo@ele.tottori-u.ac.jp

Abstract. Electrostatic pesticide spraying (EPS) improves the adhesion characteristics of the pesticide solution to agricultural crops. If the adhesion characteristics are improved, the requisite amount of the pesticide to be sprayed can be reduced in comparison with the conventional spraying method that uses non-charged pesticide. In this research, disease (rust) control experiments were carried out to substantiate the effectiveness of the EPS from a statistical point of view. We sprayed pesticide to potted Japanese pear trees under calm condition. The numbers of the rust lesions on the pear leaves were counted at fixed intervals after spraying to investigate the difference of the results of the disease control. The t-tests were carried out for the populations of the various spraying times and applied voltages. It was statistically-derived that EPS can reduce the amounts of pesticide to be sprayed by 50 % in comparison with the non-EPS method. It is also estimated from the results that about 55,000 kL year^{-1} of pesticides can be reduced for the Japanese pear cultivation in Tottori prefecture. Also, this means that the expense of the pear cultivation can be reduced by about 240 million yen (3 million USD) every year in Tottori prefecture by introducing EPS.

1. Introduction

Electrostatic pesticide spraying (EPS) improves the adhesion characteristics of the pesticide solution to agricultural crops. If the adhesion characteristics are improved, the requisite amount of the pesticide to be sprayed can be reduced in comparison with the conventional spraying method that uses non-charged pesticide. Some agricultural implement and sprayer manufacturers sell various kinds of EPS nozzles. In order to investigate the effectiveness of EPS, in comparison with the conventional spraying method, Yamane et al. [1, 2] investigated the adhesion of the sprayed water to crops such as cabbages and melons using pieces of water-sensitive paper attached to the target. Also, they compared the numbers of the insect pests that inhabit the crops and insect damages of the crops between EPS and the conventional spraying method.

In this research, disease (rust) control experiments were carried out to substantiate the effectiveness of the EPS from a statistical point of view. The experiments were carried out for the Japanese pears (*Pyrus pyrifolia*), one of the signature agricultural products of Tottori prefecture, to estimate the cost-cutting in the disease control for the pear cultivation.

---

* To whom any correspondence should be addressed.

2. Experiments

2.1. Spraying nozzle and test trees
In this research, we sprayed the pesticide solution that controls rust to potted Japanese pear trees (test trees) by using a battery-driven sprayer that is usually sold at Do-It-Yourself stores. The average values of width D and height H are 58 cm and 87 cm, respectively. Each of the trees was cultivated in a pot of which height is 31 cm. A grounding rod was put in the soil of the pot when the spraying experiments were carried out. The nominal spraying rate and pressure of the sprayer are 269 cc min^{-1} (4.48 cm^3 s^{-1}) and 2.7 atm (2735 hPa), respectively. Figure 1 shows the distribution of the diameter of the droplets from spraying nozzle. 55% of the droplets have the diameter of less than 0.1 mm. The pesticide droplets were charged by contacting charging. The voltages applied to the nozzle were -40 kV and 0 kV. The sprayer was fixed on the insulator during the experiments.

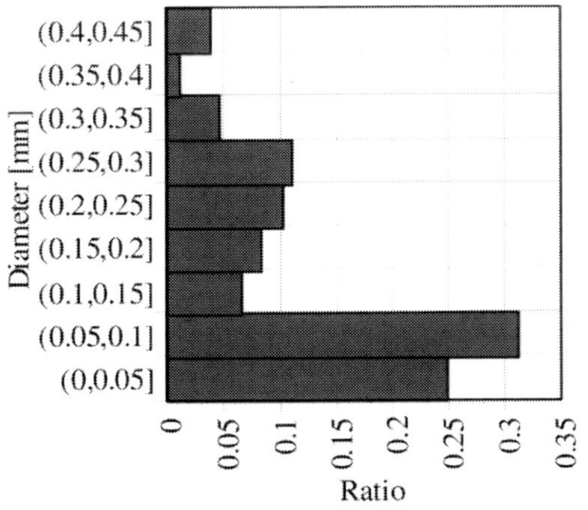

Figure 1. Distribution of the diameter of the droplets from spraying nozzle.

The spraying pressure of the sprayers usually used in agricultural field is 1 ~ 1.5 MPa. The spraying rate and pressure of the sprayer used in this research were rather smaller than those of the field-use sprayers. Also, the EPS nozzles sold in Japan for such sprayers are induction-charging type. In this research, as described below, the pesticide was sprayed to a single test tree for a trial spraying. The amounts of pesticide to be sprayed were controlled by changing the spraying time. If a sprayer with large spraying rate is used, it is concerned that too much pesticide is sprayed even for short spraying time and the amounts of sprayed pesticide will vary widely for small fluctuation of the spraying time. Because of these concerns, a battery-driven sprayer with not-so-large spraying rate was used for the research. The contacting charging method was adopted because an induction-charging electrode for the small sprayer was not able to be built.

2.2. Experimental conditions and procedures
Figure 2 show the arrangement of the spraying nozzle and a test tree. The nozzle height from the ground was fixed at 75 cm. The distance between the nozzle orifice and a pear trunk is 50 cm or 80 cm. The experimental condition was shown in table 1. The pesticide of the specified concentration was sprayed as follows:
(i) A grounded plate, which blocks the flow of the pesticide, was placed between the nozzle and a test tree.
(ii) The sprayer was activated.
(iii) The charging voltage was applied to the nozzle.
(iv) When the pesticide was stably sprayed and the nozzle voltage reached the setup value, the grounded plate was removed.

(v) After the pesticide was sprayed for a set time (4 s, 8 s, 12 s, 16 s and 20 s), the grounded plate was placed again between the nozzle and the test tree.

Table 1. Experimental conditions.

Date	Apr. 26, 2010
Temperature	27 ℃
Humidity	24%
Wind	Calm

The numbers of the rust lesions on the leaves were counted after 5, 11 and 17 days after spraying. For this number counting, 150 leaves were selected randomly from a single test tree. Fifty of them were selected permitting overlaps from the area lower than 65 cm from the bottom of the pot, others 50 were from 65 ~ 85 cm and the others 50 were from upper than 85 cm.

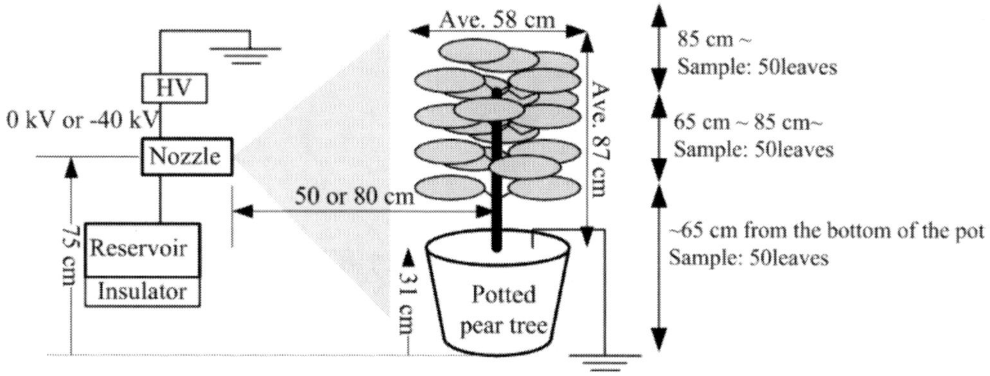

Figure 2. Arrangement of the spraying nozzle and a test tree.

2.3. Rust

Rust on pear leaves is one of fungi developed by Basidiomycetes (*Gymnosporagnium asaticum*). As shown in figures 3, this disease develops yellow lesions on pear leaves. These lesions are not curable and become holes. The leaves with a lot of lesions will defoliate. Rust was one of the important diseases of Japanese pears. Recently, some pesticides that work wonders have been developed and easily control the disease [3]. This disease develops after a rainfall in late of April or beginning of May in Japan. In this paper, this disease is adopted as the experimental object because the developing time is easily predictable.

(a) Adaxial side (b) Abaxial side

Figure 3. Rust lesions on Japanese pear leaves.

3. Results

3.1. Experimental results

Average lesion number of a test tree N_{ave} was calculated by the following equation.

$$N_{ave} = \frac{N_{lesion}}{N_{leaf}} \tag{1}$$

where N_{lesion} is the number of the counted lesions of the tree and N_{leaf} is the number of the sample leaves.

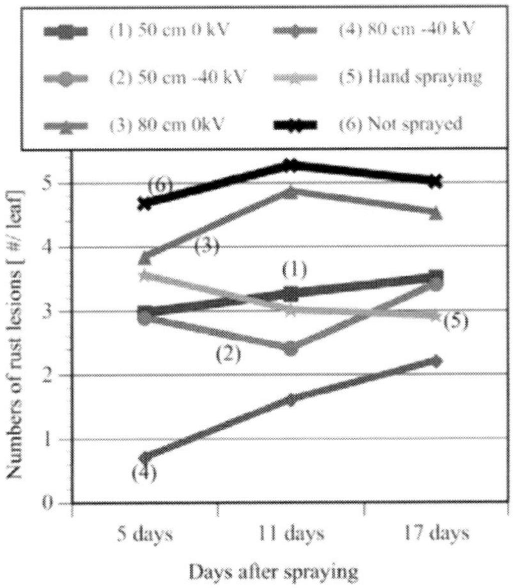

Figure 4. Time variation of the numbers of rust lesions after spraying (spraying time = 8 s).

Figure 4 shows the time variation of the average numbers of the rust lesions on a leaf for various spraying conditions for the case that the spraying time is 8 s. Comparing the changes of the average lesion numbers from 5 days to 11days after spraying (Time (i)) with those from 11 days to 17 days (Time (ii)), the numbers in Time (ii) change at a slower rate overall than those in Time (i) and most of them vary within one during Time (ii). From the above consideration, it is assumed that the lesion numbers converged at 17 days after the spraying and, in this paper, the effects of the disease control will be discussed using the data of "17 days after the spraying". The previous research [4] also shows that the numbers of the rust lesions converged at around 16 days after pesticide spraying. Though the rust lesions are not curable, the lesion numbers for some spraying conditions decrease with time in figure 3. One of the reasons of this phenomenon is that more than one closely-developed lesions became larger with time and united as a large single lesion and was counted as a "one" lesion.

3.2. Data analyses using t-tests

As described in section 2.2, 150 leaves were selected permitting overlaps from a single test tree. The lesion numbers of each leaf was recorded. So, a raw data for statistical analyses consists of 150 numbers. T-tests [5, 6] were carried out for the samples of the various spraying times and applied voltages to determine if a significant difference between two samples (e.g. -40 kV and 0 kV) exists. In t-tests, the researchers' hypothesis and the hypothesis that opposes "the researchers' hypothesis" are called the alternate hypothesis and the null hypothesis, respectively. The tests are carried out calculating the test statistic based on an assumption that the null hypothesis is correct. However, if the test statistics belongs to the rejection region, the null hypothesis is considered to be erroneous and rejected. The probability that null hypothesis is rejected is called significance level α. In this research, Welch's test is adopted and α is set at 0.05 (5 %), a commonly-used probability. The Welch's test is a kind of t-tests used when two samples are assumed to have different dispersions.

3.2.1. Relation between spraying distance and disease control

All the raw data, except for "Hand spraying" and "Not sprayed", are categorized according to the spraying distance (50 cm and 80 cm) into two samples without considering the nozzle voltage and the spraying time. Table 2 shows the result of a significance test for the spraying distance. This table was automatically obtained by using one of the analytical tools of Microsoft Excel [7]. The alternate hypothesis for this tests is that "the number of lesions was changed (increased or decreased) by changing the spraying distance from 50 cm to 80 cm" and the two-tailed test should be adopted. If the alternate hypothesis is that "the number of lesions was increased (or decreased) by changing the spraying distance", the one-tailed test would be adopted. If $P(T \leq t) < \alpha$ or $t_{boundary} < |t|$, the null hypothesis is rejected [7]. We can see that $P(T \leq t) > \alpha$ and $t_{boundary} > |t|$ for two-tailed test and no significant difference between the two samples is obtained. In the following discussions, for the same nozzle voltage and the same spraying time, the data for the two spraying distances will be merged into a single sample.

Table 2. Significance test for spraying distances.

Spraying distance	50 cm		80 cm
Average lesion number	3.375		3.559
Dispersion	10.96		13.70
t		-1.212	
$P(T \leq t)$ for one-tailed test		0.1128	
$t_{boundary}$ for one-tailed test		1.646	
$P(T \leq t)$ for two-tailed test		0.2256	
$t_{boundary}$ for two-tailed test		1.961	

3.2.2. Relation between nozzle voltage and disease control

For various spraying time, except for "Hand spraying" and "Not sprayed", the raw data are categorized according to nozzle voltage (-40 kV and 0 kV), without considering the spraying distances. The Welch's tests are carried out for two samples for each spraying time. The alternate hypothesis for these tests is that "the number of lesions was decreased by adopting EPS" and the one-tailed tests are carried out.

Table 3. Significance test for nozzle voltage.

Spraying time	20 s		16 s		12 s		8 s		4 s	
Nozzle voltage	0 kV	-40 kV	0 kV	-40 kV	0 kV	-40 kV	0 kV	-40 kV	0 kV	-40 kV
Average lesion number	3.443	3.763	3.286	3.505	2.825	2.621	4.219	3.204	4.258	3.496
Dispersion	9.785	13.61	7.826	11.37	12.99	7.505	11.79	11.23	22.51	11.68
t	-0.9767		-0.7086		0.6767		3.003		1.960	
$P(T \leq t)$	$0.1646 > \alpha$		$0.2395 > \alpha$		$0.2495 > \alpha$		$0.001422 < \alpha$		$0.025 < \alpha$	
$t_{boundary}$	$1.649 > \|t\|$		$1.649 > \|t\|$		$1.648 > \|t\|$		$1.649 < \|t\|$		$1.649 < \|t$	

As shown in table 3, for the spraying times of 12 s ~ 20 s, the null hypothesis is not rejected. This means that the significant differences between EPS and non-EPS for these spraying times are not observed. On the other hand, for the spraying times of 4 s and 8 s, the significant differences between the two spraying methods are observed. For the spraying time of 12 s, though the significant difference between EPS and non-EPS is not observed, the average lesion number was decreased by adopting EPS. These results mean that even for the non-EPS case, some amounts of pesticide, which gave little difference in the disease control in comparison with the EPS case, adhered to the test trees when the spraying time is longer than 12 s. For the spraying time shorter than 12 s, non-EPS could not control the disease well in comparison with the EPS. This difference was significant for the spraying time of 4 s and 8 s.

4. Discussions

4.1. Estimation of pesticide reduction adopting EPS

Figure 5 shows the average numbers of the rust lesions for various spraying time obtained 17 days after spraying. Spraying excessive amount of pesticide usually does not decrease the control ability. Because of this, the lesion number will monotonically decrease as the spraying time and ideally be zero for a long enough spraying time if the pesticide is sprayed evenly all over a test tree. In this research, it is concerned that the pesticide did not reach some leaves even for the EPS cases because the positions of the spraying nozzle and the test trees were fixed. For such spraying conditions, the average lesion number will not decrease to zero for a long spraying time. From figure 5, it seems that the average lesion number converged at the spraying time of about 12 s and the longer spraying times did not improve the control ability for non-EPS cases. For EPS cases on the other hand, the average lesion number did not decrease so much as the spraying time and little difference was in the average lesion numbers between the spraying times of 4 s and 20 s is obtained. This means that it is expected that the convergence average lesion number was obtained even for the 4-s-spraying. In other words, the disease was sufficiently controlled by the 4-s-spraying for the adopted experimental conditions. For the non-EPS cases, the average lesion number too rapidly decreased at the spraying times of 12 s. It is concerned that some unknown factors affected this rapid decrease. So, we assume that the average lesion numbers converged and the disease was sufficiently controlled by the 16-s-spraying for the non-EPS cases. We carried out the Welch's tests for the sample of 16-s-spraying for the non-EPS case and those of 4- and 8-s-spraying for the EPS cases. The alternate hypothesis was that "the number of lesions was decreased or increased by adopting EPS" and the two-tailed tests were adopted. The results (table 4) show that there are not significant differences between the non-EPS-16-s-spraying case and the both of EPS-4-s- and 8-s-spraying cases. These results mean that the amounts of pesticides to be sprayed can potentially be reduced by 50 % ~ 75 % by adopting EPS because the amounts of sprayed pesticides are proportional to the spraying time.

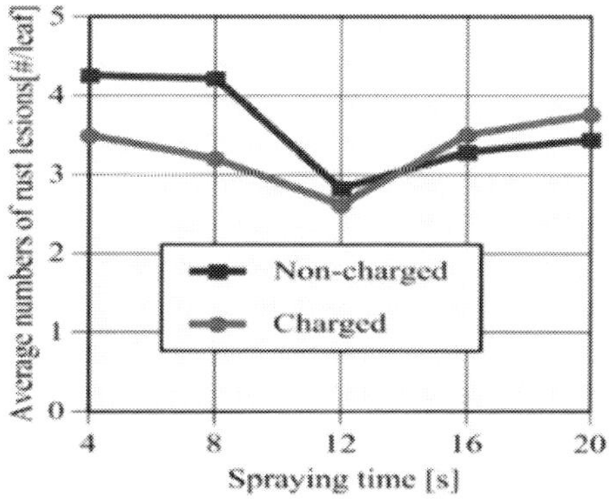

Figure 5. Average numbers of rust lesions for various spraying time (17days after spraying).

Table 4. EPS can reduce the amounts of pesticide to be sprayed.

	(a) Reduced by 50 %		(b) Reduced by 75 %	
	0 kV, 16 s	-40 kV, 8 s	0 kV, 16 s	-40 kV, 4 s
Average lesion number	3.286	3.204	3.286	3.496
Dispersion	7.826	11.23	7.826	11.68
t	0.2570		-0.6952	
$P(T \leq t)$	$0.7973 > \alpha$		$0.4873 > \alpha$	
$t_{boundary}$	$1.967 > \lvert t \rvert$		$1.966 > \lvert t \rvert$	

4.2. Estimation of pesticide reduction adopting EPS

Assuming that the adopting EPS can reduces 50 % of pesticides to be sprayed, the reduction of the pesticide cost in the pear cultivation in Tottori prefecture is estimated as follows.

The area for the pear cultivation in Field Science Center (FSC), a Tottori university's experiment farm, is 960 m^2. The total amounts and costs of all pesticides to be sprayed without adopting EPS in this area are 8,600 L $year^{-1}$ and 37,600 yen $year^{-1}$ (470 USD $year^{-1}$), respectively. The area for the pear cultivation, including for young trees, in Tottori prefecture is 1,230 ha [8]. Assuming that the planting density of the pear trees in all the pear farms in Tottori prefecture is the same of that in FSC and required amounts of pesticides are proportional to the planting area, the estimated total amounts and costs of all pesticides to be sprayed for the pear cultivation in Tottori without adopting EPS are $(8{,}600[\mathrm{L\ y^{-1}}] \cdot 1{,}230[\mathrm{ha}] \cdot 10{,}000[\mathrm{m^2\ ha^{-1}}]) \cdot (960[\mathrm{m^2}])^{-1}$ = 110,000 kL y^{-1} and 480 million yen y^{-1} (6 million USD y^{-1}), respectively. By adopting EPS, it is estimated that 55,000 kL y^{-1} (240 million yen y^{-1} or 3 million USD y^{-1}) of the pesticides can be reduced in the pear cultivation in the entire Tottori prefecture.

5. Conclusions

The spraying experiments described in this paper were carried out when calm. Because of this, the pesticide reduction discussed above can be applied for the real pear cultivation if the pesticides are sprayed only when calm or in greenhouses. However, the result that the amounts of pesticides to be sprayed can potentially be reduced by about 50 % means that excess pesticides are sprayed. Considering this result, EPS is effective not only for the cost saving but also for the environmental protection and food safety.

References

[1] Yamane S, Miyazaki M and Ohmura K 2010 *J. Jpn. Soc. Agric. Mach.* **316** 54

[2] Yamane S, Miyazaki M, Saito H, Ohmura K, Aono M and Ohsuka R 2010 *J. Jpn. Soc. Agric. Mach.* **321** 570

[3] Sakagami Y and Kudo A 2003 *A Handb. of Dis. and Insect Pests of Fruit Trees Vol. 2 Pears, Grape, Persimmon, Chestnut and Fig (Revised Edition)* (Tokyo: Japan Plant Protection Association) (in Japanese)

[4] Michihara S, Nishimura R, Ishihara N, Masuoka T, Kimura T, Yatsuzuka S and Anaguchi S *Proc.2010 Spring-time Meeting of Inst. Electrostat. Japan* (Tokyo: IESJ) p 35 (in Japanese)

[5] Katatani N and Matsutou T 2003 *Kankyo Toukeigaku Nyuumon* (Tokyo: Ohm-sya) (in Japanese)

[6] http://www.aoni.waseda.jp/abek/document/t-test.html

[7] http://www1.tcue.ac.jp/home1/abek/htdocs/stat/Excel/t-test/t-test.html

[8] Ministry of Agriculture, Forestry and Fisheries of Japan 2008FY *57th Annu. Rep. of Stat. on Agric., For. and Fish.*

Journal of Physics: Conference Series **418** (2013) 012143 doi:10.1088/1742-6596/418/1/012143

Combined electrohydrodynamic (EHD) and vacuum freeze drying of shrimp

Yucai Hu*, **Qiang Huang, Yaxiang Bai**
School of Science, Dalian Ocean University, Heishijiao Street, Dalian, 116023, China

E-mail: hyc@dlou.edu.cn

Abstract. To improve the drying qualities of shrimp, a combination of electrohydrodynamic (EHD) and vacuum freeze drying (FD) is examined. The drying rate, the shrinkage, the rehydration ratio, and the sensory properties including the color and trimness of the dried products under different drying methods (including combination drying of EHD and FD, EHD drying and FD drying) are measured. Compared with FD and EHD drying alone, the combined process consumes less drying time, and the product processed by combined drying displays lower shrinkage, higher rehydration rate and better sensory qualities.

1. Introduction

Shrimp and fish represent one major source of protein to human beings in the world. Their preservation is an important issue because they are perishable products. Drying has been proved to be an efficient and cheap method for food preservation [1-4]. It has been used to preserve fish and shrimp for a long time in most areas of the world.

At present, the conventional drying processes of shrimp include solar drying and hot air drying. The latter often causes heat damage and adversely affects texture, color, flavor, and the nutritional value of dried products [5]. Although solar drying does not need special equipment and technology, and neither does it incur high expenses, its drying conditions are difficult to control. As a result, the product quality is often poor [6].

To improve their preservation quality and consumer acceptability, new drying technology for shrimp is needed. It is widely known that although FD is an excellent drying method from quality standpoint, its energy consumption is excessive compared with other methods of drying. It is, therefore, desirable to evaluate new drying technologies to achieve better product quality and more reasonable cost.

As mentioned by Mujumdar, agricultural and food products should be dried by efficient industry equipment so how to select a proper dehydration method should be considered firstly [7]. Consideration should be given to many factors before selecting a drying process. These factors are the type of product to be dried, the product's susceptibility to heat, pretreatments required, and capital and processing cost. There is no one best technique for all products [8]. For shrimp, the drying method should have properties of low temperature and high efficiency. There have been reports about heat

* To whom any correspondence should be addressed.

pumps and solar tunnels used to dry fish [9, 10]. Hybrid drying methods are also important according to recent research results [11].

Combination drying technology can combine advantages of various drying methods, and avoid the disadvantages of a single drying method. Until now, many reports about combination drying have been developed. For example, Fito used five different combination drying methods (hot air-microwave, osmosis-hot air, vacuum osmosis-hot air, and vacuum osmosis-hot air-microwave) to dry banana and found that vacuum osmosis-hot air-microwave could improve its rehydration capability [12]. Bai et al. reported that a combination of freeze drying and electrohydrodynamic drying (EHD) could reduce drying time about 15% compared with freeze drying [13]. Xu et al also reported that convective air and vacuum freeze drying could get good product quality and relative short drying time during drying bamboo shoots [14]. FD technology also could be combined with heat pump or other drying methods; even atmospheric freeze drying (AFD) was developed to reduce power consumption [15].

From these research findings, FD could be combined with other drying methods conveniently for its own advantages. At present, there are no reports on drying shrimp using combination drying of EHD-FD (EHFD). The objective of this article is to examine the effect of three drying methods on rehydration and shrinkage of shrimp, and furthermore, introduce a new combination drying technology to improve dry shrimp quality and reduce drying time.

2. Materials and Methods

2.1. Preparation and drying of shrimps

Shrimps were procured from the local market. Before drying, shrimps were peeled and washed with tap water and were dipped in NaCl solution (8% w/v) at 90-99 ℃ for 10 min. They were then taken out and drained.

The samples were divided into three groups at random with similar weight in each group. Initial weights of the samples were measured by a digital balance. One group was subjected to EHD drying at 18±1°C. One group underwent FD drying, and the other one for EFD drying. The samples were dried by EHD for 3h, and then dried to the final water content (12±1%) by FD. All samples were dehydrated until they reached the final moisture content (12±1%).

2.2. Determination of moisture

Moisture content was determined by the oven method. At regular time intervals during the drying processes samples were taken out and dried in the oven for 7-8 h at 105°C until constant weight. Weighing was performed on a digital balance, and then moisture content (w.b.) was calculated.

2.3. Shrinkage measurement

Shrinkage determination was made by manually measuring the sizes of dried shrimp samples obtained from different drying processes using a vernier caliper. The equivalent spherical diameter of each sample was then determined and compared. Ten samples were taken for measurement for each experimental condition. Shrinkage was calculated using the following equation:

$$\text{Shrinkage}=\frac{Z_0 - Z_f}{Z_0}, \tag{1}$$

where Z_0 and Z_f are the volume of each group of shrimps at the beginning and at the end of each drying experiment, respectively.

2.4. Rehydration ratio measurement

Rehydration ratio was determined by soaking 10.00 ± 0.01 g of dried shrimps on a dry weight basis in 100 ml distilled water at ambient conditions for up to 24 h. The soaked shrimps were blotted with a paper towel every 2 h to remove excess water. They were then weighed and placed back into the soaking water. Water absorption value was expressed in percentage and calculated as grams of water absorbed per 100 g of dried shrimps through the following formula.

$$\text{Percentage water absorption} = \frac{m_g - m_0}{m_0} \times 100 \quad (2)$$

where m_0 and m_g are the weights of dry and soaked shrimps, respectively.

2.5. Sensory evaluation

A sensory evaluation of dried samples was carried out by a panel of six untrained judges. The panelists were asked to indicate their preference for each sample, based on the quality attributes of color, appearance, texture, aroma/flavor, and overall acceptability. A balanced 10-point hedonic rating was employed for all the attributes evaluated where 9-10 denoted "like very much" (7-8 "like," 5-6 "neutral," 3-4 "dislike") and 1-2 indicated "dislike very much." The judges were asked to give their remarks about each of the samples.

3. Results and discussion

3.1. Influence of different drying conditions on drying rate

As shown in figure 1, the drying rate of the FD technique was the slowest, followed by EHFD; EHD drying exhibited the fastest drying rate. The average drying rate for FD, EHFD, and EHD drying were 2.15, 2.85, and 3.25 g h^{-1}, respectively.

3.2. Influence of different drying conditions on shrinkage

During drying, foods undergo volume changes either by shrinkage (due to the removal of water from the drying shrimp) or by expansion (due to gas generation or pore formation). Different drying techniques thus provide different degrees of volume changes. Normally, the muscle fibers shrink as they lose moisture during the drying process. Figure 2 shows the results of shrinkage of shrimp that underwent different drying methods. Shrinkage was found to be lower for EHFD than for EHD. The final shrinkage values for FD and EHFD drying were 7.32% and 17.35%, respectively, while it was 23.40% for EHD drying. This shows that EHFD significantly reduces shrinkage.

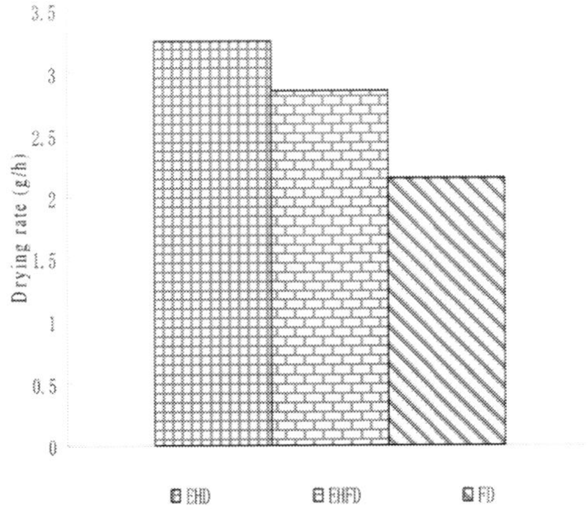

Figure 1. Effect of different drying methods on average drying rate of shrimps.

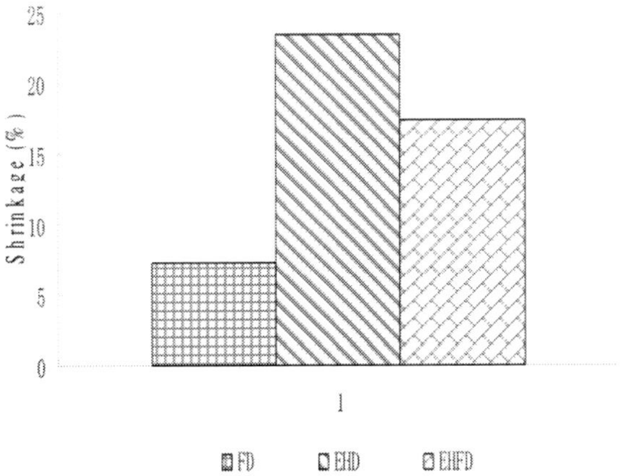

Figure 2. Effect of different drying methods on shrinkage ratio of shrimps.

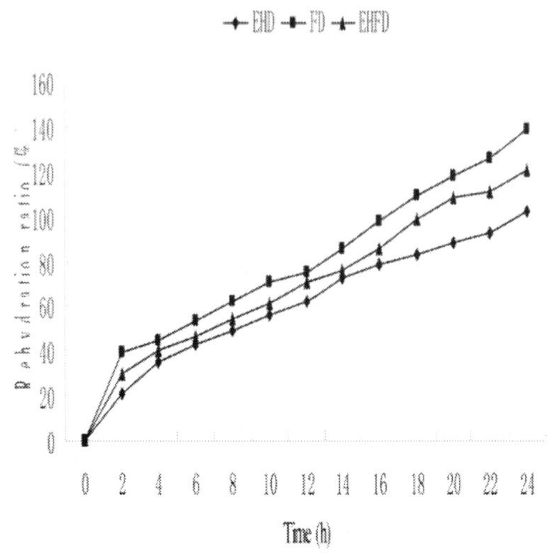

Figure 3. Effect of different drying methods on rehydration ratio of shrimps.

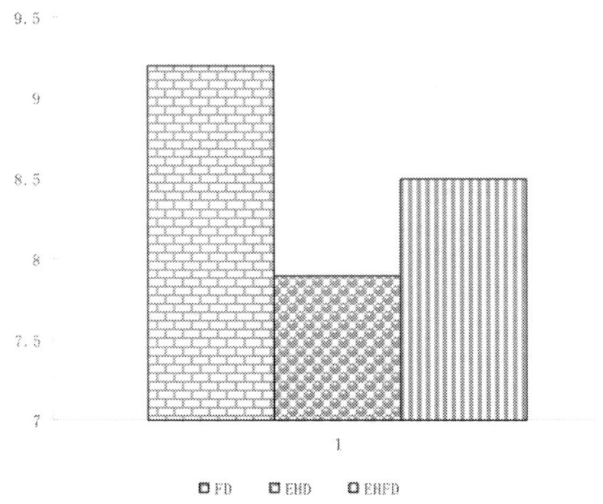

Figure 4. Effect of different drying methods on sensory qualities of shrimps.

3.3. Influence of different drying conditions on rehydration ratio

The rehydration characteristics of the dried product were used as a quality index because they could indicate the physical and chemical changes of samples during drying. The effect of the three drying methods on rehydration ratio is shown in figure 3. It shows that compared with EHD, the shrimp dried by EHFD had a good rehydration capability. The lower temperature used in FD drying might be the factor enabling EHFD -treated shrimp to hold more water compared with EHD-samples. The rapid evaporation of moisture in the shrimp induced a porous structure, which was in favor of the rehydration of products.

3.4. Influence of different drying conditions on sensory properties

As shown in figure 4. It can be seen that the total score of shrimp dried by FD was the highest followed by EHFD and EHD. FD could lead to the best quality, and EHFD could achieve a better quality.

4. Conclusions

Different drying methods such as EHD, FD, and EHFD had different effects on the quality and drying rate of the dehydrated shrimp. FD could lead to the best quality, but had the lowest drying rate. EHD shrimp had poor quality, and had a short drying time. Combination drying of EHFD could lead to better quality and drying efficiency.

References

[1] Ratti C 2001 *J. Food Eng.* **49** 311
[2] Bai Y X, Sun B 2011 *J. Food Process Pres.* **35** 891
[3] Duan X, Zhang M, Mujumdar A S, Wang S J 2010 *J. Food Eng.* **96** 491
[4] Bai Y X, Yang G X, Hu Y C, Qu M J 2012 *Aquat. Food Prod. Tech* **21** 238
[5] Barbara G 1998 *Food Manufact.* **73** 30
[6] Wall R, Bindu J, Howard J J 2001 *J. Applied Ecol.* **38** 339
[7] Mujumdar A S 2001 *Drying Technol.* **19** 1217
[8] Mujumdar A S 2006 3rd Ed (Boca: Raton/ FL, CRC Press) p 552

Journal of Physics: Conference Series **418** (2013) 012143 doi:10.1088/1742-6596/418/1/012143

[9] Chua K J, Chou S K, Shen D L 2002 *Drying Technol.* **20** 1579
[10] Bala B K, Mondol M R A 2001 *Drying Technol.* **19** 427
[11] Mujumdar A S 2004 *Drying Technol.* **22** 1
[12] Fito P J 2001 *Alimentaria* **320** 113
[13] Bai Y X, Yang Y X, Huang Q *Drying Technol.*, DOI:10.1080/07373937.2012.663435
[14] Xu Y Y, Zhang M, Tu D Y 2005 *Int J Food Sci Tech ,* **40** 589
[15] Claussen I C, Ustad T S 2007 *Drying Technol.* **25** 947

Application of high voltage electric field (HVEF) drying technology in potato chips

Yaxiang Bai[*]**, Hua Shi, Yaxin Yang**
School of Science, Dalian Ocean University, Heishijiao Street, Dalian, China

E-mail: Byx0671@163.com

Abstract. In order to improve the drying efficiency and qualities of vegetable by high voltage electric field (HVEF), potato chips as a representative of vegetable was dried using a high voltage electric drying systems at 20℃. The shrinkage rate, water absorption and rehydration ratio of dried potato chips were measured. The results indicated that the drying rate of potato chips was significantly improved in the high voltage electric drying systems. The shrinkage rate of potato chips dried by high voltage electric field was 1.1% lower than that by oven drying method. And the rehydration rate of high voltage electric field was 24.6% higher than that by oven drying method. High voltage electric field drying is very advantageous and can be used as a substitute for traditional drying method.

1. Introduction

Drying is defined as a process of moisture removal due to simultaneous heat and mass transfer. It is also a classical method of food preservation, which provides longer shelf-life, lighter weight for transportation and smaller space for storage [1]. It has been used for a long time to preserve fruits, vegetables, fish and shrimp etc in many areas of the world [2, 3]. Dried vegetables and fruits products and dried yoghurt has been used in food industry as an ingredient or protein enrichment like confectionery or bakery industry [4, 5]. Besides, dried shrimp and dried shark fin are some of the more popular items that have gained wide acceptance in the US and European markets.

At present, the conventional drying processes of vegetables and fruit are based on convective, radiative, conductive, dielectric heat, superheated steam, and lyophilization, which have been thoroughly discussed. These energy-intensive methods require heat to be applied externally or removed from the drying material while the liquid component in it undergoes a phase change. Osmotic drying does not involve a phase change, and de-watering from cellular materials occurs by applying a hydrostatic pressure different across the semipermeable cell membranes. The fluidized bed drying consists of passing hot air through a bed of the material converted into a fluidized state. In acoustic drying, mass transfer enhancement is brought about by ultrasonic vibrations which also produce heat. Since high temperature dryings usually degrades color, flavor, texture, and nutrients of food [6], non-thermal drying is of importance in industries where heat-sensitive materials are manufactured.

However, there is growing interest in the application of non–thermal processing of food and similar materials. High voltage electric field (HVEF) method is a novel non-thermal technique of drying and is being developed recently [7-10]. Since corona discharge can be produced at room temperature and

[*] To whom any correspondence should be addressed.

atmospheric pressure, the technique is particularly attractive for low-temperature applications. Compared with hot air drying systems, high voltage electric field drying systems offer lower food production costs as well as superior quality in terms of physiochemical properties such as color, shrinkage, flavor, and nutrient content. Compared with convective and freeze drying, High voltage electric field drying systems have a simpler design, which consumes less energy [11, 12]. Several earlier studies have demonstrated the effectiveness of high voltage electric field drying. Some have even been applied to food preservation with great success. For example, these include the drying of cookie dough, potato slabs, apple slices and spinach, as reported by a recent review by Bajgai et al. [7]. Thus, it shows great potential for bulk and industrial drying applications.

At present, there are few reports available on high voltage electric field drying potato chips. The objectives of this research were to explore a method for potato chips drying. The effects of high voltage electric field and oven drying on potato chips' quality in terms of shrinkage ratio, water absorption and rehydration ratio were investigated and compared.

2. Experimental setup and procedure

2.1. Experimental setup

The experimental setup for high voltage electric field drying is shown in figure 1.

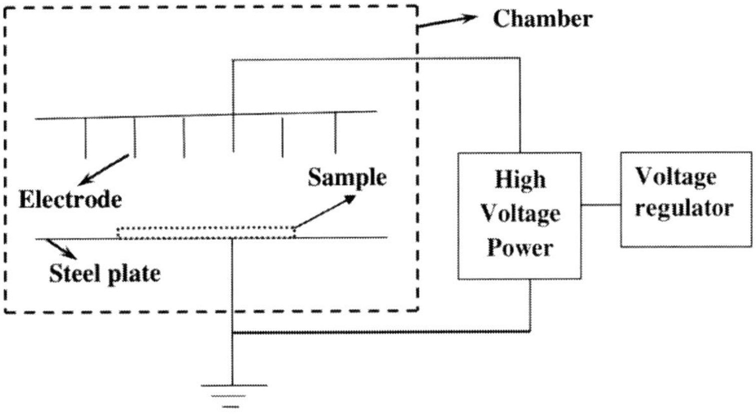

Figure 1. Diagram of high voltage electric field drying.

It consisted of a vertically mounted electrode with multiple pointed needles projected to a fixed horizontal grounded metallic plate, on which the potato chips to be dried are placed. The distance between the high-voltage electrodes and the grounded electrode was 9 cm. The sharp points of 25 needles were fixedly connected to a DC high-voltage power source. To set the desired high-voltage parameters for high voltage electric field drying, the power source was connected to a voltage regulator, with the voltage adjustable range from 0 to 60 kV by a controller. The grounded plate electrode was a 40×35 cm rectangular stainless steel plate. The apparatus was placed in a stainless steel chamber, the temperature and relative humidity of which were controlled. All the samples were spread in single layer on the grounded plate electrode.

2.2. Experimental materails

Fresh potato was purchased from the local market near Dalian Ocean University, China. It was washed with tap water and the rectangle slices of 4 cm in length, 4 cm in width and 4 mm in thickness were cut using fine edged knife. The slices were blanched for 3 min in boiling water and followed by cooling in tap water. Water left on the surface of the slices was soaked by blotting paper. The initial weight of the slices was taken and they were subjected for drying. High voltage electric field drying was carried out at 20 °C and 36% relative humidity. Slices were also dried using oven at 60 °C and ambient at 20 °C. Moisture content of potato slice was determined at one hour intervals using a digital electronic balance. Potato slices also were dried in an oven at 80 °C until constant weight and the equilibrium moisture content was determined.

2.3. Shrinkage measurement

The Shrinkage determination was made by manually measuring the sizes (length, width and thickness) of dried potato slices samples obtained from different drying processes (High voltage electric field, oven and ambient air drying) using a vernier caliper. The equivalent spherical diameter of each sample was then determined and compared. Ten samples were taken for measurement for each experimental condition. All experiments were performed twice. Shrinkage was calculated using the following equation:

$$\text{Shrinkage} = \frac{Z_0 - Z_f}{Z_0} \times 100\% \quad (1)$$

Where Z_0 and Z_f are the potato slices volume of each group at the beginning and at the end of each drying experiment, respectively.

2.4. Water absorption measurement

The water absorption was determined by soaking 10.00 g of dried potato slices on a dry weight basis in 100 ml distilled water at ambient conditions for up to 1 h. The soaked potato slices were blotted with a paper towel every 10 min to remove excess water. They were then weighed and placed back into the soaking water. Water absorption value was expressed in percentage and calculated as grams of water absorbed per 100 g of dried potato slices through the following formula.

$$\text{Percentage water absorption} = \frac{m_g - m_0}{m_0} \times 100\% \quad (2)$$

Where m_0 and m_g are the weights of dry and soaked potato slices, respectively.

2.5. Rehydration ratio measurement

The rehydration potential of potato slices dried through high voltage electric field, oven, and ambient air was evaluated by immersing them in hot water at 100 °C. The samples were drained, and their weights were measured by a digital electronic balance at 10 min interval for 1 h. The incremental mass due to absorbed water was then calculated as the degree of rehydration per 100 g of dried potato slices expressed as follows:

$$\text{Percentage rehydratim ratio} = \frac{m_g - m_0}{m_0} \times 100, \quad (3)$$

Where m_0 and m_g are the weights of dry and boiled potato slices, respectively.

3. Results and discussion

3.1. Influence of different drying conditions on drying rate

The drying rate curves of potato slices treated by high voltage electric field, oven, and ambient air drying are shown in figure 2. The drying rate of potato chips was significantly improved in the high voltage electric drying systems. The average rates of high voltage electric field, oven, and ambient air drying from the first to the fifth hour were 15.2, 12.1, and 2.7 mg min^{-1}, respectively. High voltage electric field drying as a novel non-thermal technique can dry potato slices samples at a low temperature because of its unique drying mechanism. In the process, the needle electrode produces corona wind (resembling a round jet) that impinges and removes moisture from the surface of potato slices.

3.2. Influence of different drying conditions on shrinkage

During drying, foods undergo volume changes either by shrinkage (due to the removal of water from the drying potato slices) or by expansion (due to gas generation or pore formation) [13]. Different drying techniques thus provide different degrees of volume changes. Normally, the fibers shrink as they lose moisture during the drying process.

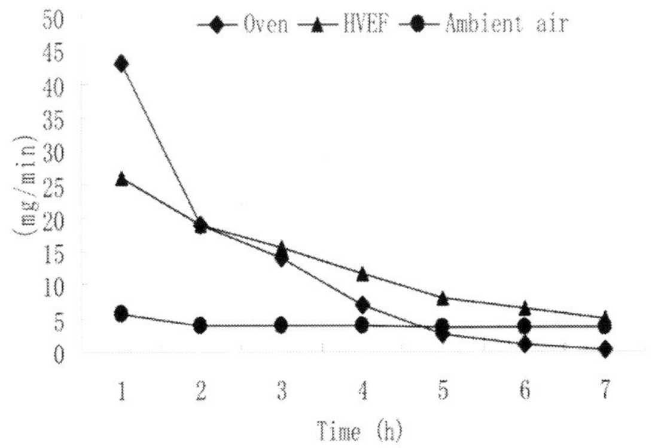

Figure 2. Drying rate of high voltage electric field, oven and ambient air drying for potato slices.

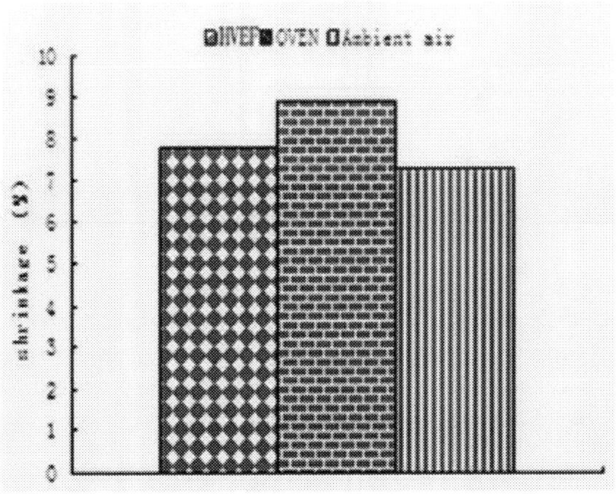

Figure 3. Effect of different drying methods on the shrinkage ratio of dehydrated potato slices.

Figure 3 shows the results of shrinkage of potato slices that underwent different drying methods. Shrinkage was found to be lower for high voltage electric field than for oven. The final shrinkage values for high voltage electric field and ambient air drying were 7.8% and 7.3%, respectively, while it was 8.9% for oven drying. This result resembles the shrinkage in radish slices [14]. Lesser shrinkage in potato slices dried under high voltage electric field was observed. This shows that high voltage electric field significantly reduces shrinkage. Oven drying resulted in extensive shrinkage, which might be associated with cellular collapse probably due to heat damage of the wall and cell membrane [15].

3.3. Influence of different drying conditions on water absorption

The water absorption curves of potato slices treated by high voltage electric field, oven, and ambient air drying are shown in figure 4. According to the results, high voltage electric field-dried potato slices absorbed more water compared with oven-dried samples when they were soaked in water at 20 °C for 1 h. The water absorbed by the high voltage electric field-dried sample was 268 g/100 g, which is 21 g/100 g more than the water absorbed by oven-dried samples on a dry weight basis.

3.4. Influence of different drying conditions on rehydration rate

Figure 5 illustrates the rehydration behavior of dried potato slices samples obtained from different drying methods at a rehydrating temperature of 100 °C. It was found that high voltage electric field-dried potato slices absorbed more water compared with oven-dried samples when they were soaked in water at 100 °C. The water absorbed by the high voltage electric field-dried sample was 494.6 g/100 g, which is 24.6 g/100 g more than the water absorbed by oven -dried potato slices, on a dry weight basis. High voltage electric field-dried potato slices also showed better rehydration than oven-treated samples. The high temperature used in oven drying might be the factor enabling oven-treated potato slices to hold less water compared with high voltage electric field and ambient air-dried samples.

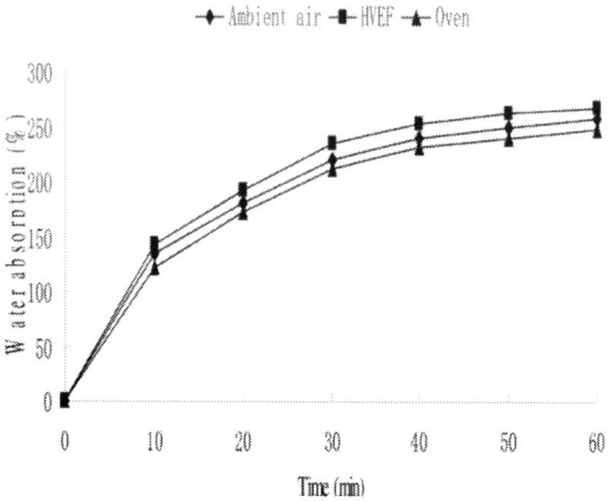

Figure 4. Effect of different drying methods on the water absorption of dehydrated potato slices.

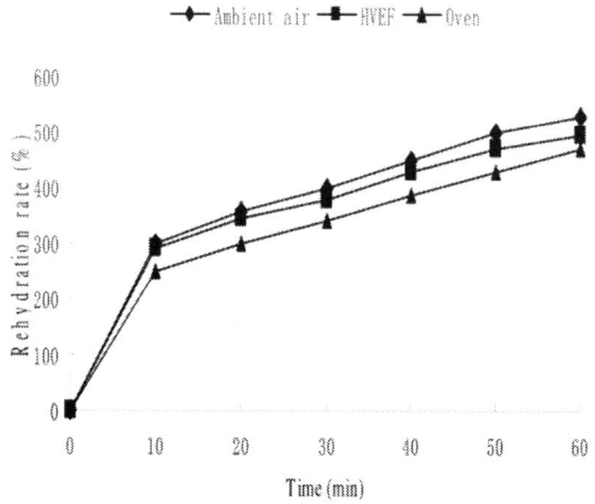

Figure 5. Effect of different drying methods on the rehydration rate of dehydrated potato slices.

4. Conclusions

High voltage electric field drying as a novel technique can dry potato slices samples at a low temperature because of its unique drying mechanism. Thus, the shrinkage rate of potato slices dried by high voltage electric field is lower, and their water absorption and rehydration rate are higher compared with oven drying. Furthermore, as high voltage electric field is a non-thermal technique, no heat energy is lost and energy can be saved significantly. Drying temperature had significant influences on the drying rate, shrinkage, water absorption, and rehydration rate of dried potato slices samples. The drying rate and shrinkage of dried potato slices samples increased with drying temperature, while their water absorption and rehydration ratio decreased with drying temperature.

References

[1] Ertekin C and Yaldiz O 2004 *J. Food Eng.* **63** 349
[2] Pabis S and Jaros M 2002 *Biosyst. Eng.* **81** 201
[3] Duan X, Zhang M and Li X L 2008 *Drying Technol.* **26** 413
[4] Leonid A B, Vladimir P G, Andrew V B, Alexander M L,Valeriy L and VladimirA K 2006 *J. Food Eng.* **74** 410
[5] Hayaloglu A A, Karabulut I, Alpaslan M and Kelbaliyev G 2007 *J. Food Eng.* **78** 109
[6] Wall R, Bindu J and Howard J J 2001 *J. Applied Ecol.* **38** 339
[7] Bajgai T R, Vijaya Raghavan G S and Hashinaga F 2006 *Drying Technol.* **24** 905
[8] Lai F C and Wong D S 2003 Drying Technol. **21** 1291
[9] Bai Y X, Li X J, Sun Y and Shi H 2011 *Int. J. Appl. Electrom.* **36** 217
[10] Ramachandran M R and Lai F C 2008 *Drying Technol.* **28** 1477
[11] Bai Y X and Sun B 2011 *J. Food Process Pres.* **35** 891
[12] Alemrajabi A A, Rezaee F, Mirhosseini M and Esehaghbeygi A 2012 *Drying Technol.* **30** 88
[13] Namsanguan Y, Tia W, Devahastin S and Soponronnaritl S 2004 *Drying Technol.* **22** 759
[14] Bajgai T R and Hashinaga F 2001 *Drying Technol.* **19** 2291
[15] Valle J M, Del A V and Leon H 1998 *Food Res. Int.* **31** 557

Journal of Physics: Conference Series **418** (2013) 012145 doi:10.1088/1742-6596/418/1/012145

The dry process of ZnO film deposition by atmospheric pressure plasma

K Ting[1*], J Y Kao[1], Y S Hsieh[1], C C Chen[1], C C Chang[2] and C J Wu[2]
1 Graduate School of Engineering, Lunghwa University of Science and Technology, Taiwan
2 Mechanical and Systems Research Lab., Industrial Technology Research Institute, Taiwan

E-mail: kuenting@mail.lhu.edu.tw

Abstract. An atmospheric pressure plasma system has been developed to prepare zinc oxide (ZnO) thin film on the silicon substrate. The precursor of dry powder of Zinc acetylacetonate hydrate was directly sublimated as vapor and carried by Argon into the downstream of Nitrogen plasma jet. The films deposition process was carried out at low temperature and the substrate temperature was not over 300 ^{o}C. The examinations of thin film sample by X-ray diffraction (XRD) showed the major peak at 34° was the hexagonal wurtzite structure of ZnO. The surface morphologies and thickness of the films were evaluated by scanning electron microscopy (SEM). The mean square surface roughness of the ZnO film was only 4.063 nm and a smooth surface of ZnO thin film was obtained.

1. Introduction

Over the past few years, several methods were employed for the deposition of ZnO thin film on different substrates, including spray pyrolysis [1-3], sol–gel technique [4, 5], chemical vapor deposition [6, 7], pulsed laser deposition [8], sputtering [9], laser ablation [10] and chemical bath deposition [11]. Sputtering is preferred among these techniques due to its relative high deposition rates, good film properties and process stability [12]. However these processes must be conducted under vacuum in batch process chambers and complicated costly.

Recently, several alternative processes operated at atmospheric pressure were developed. Various types of electrical discharge can be used to generate non-thermal plasmas at atmospheric pressure: dielectric barrier discharge (glow discharge), corona discharge, RF discharge, microwave discharge and arc discharge at atmospheric pressure. Few of studies were restricted as toxic liquid or $ZnCl_2$ precursor [13-18].

This paper presented the simple process under atmospheric pressure for the deposition of ZnO thin films on silicon wafer. Dry powder Zinc acetylacetonate hydrate was used as the precursor carrying by Argon and sprayed into downstream of the Nitrogen plasma jet to deposit thin film. The temperature of the substrate was not over 300 ^{o}C. The power and frequency were maintained at 500 W and 25 kHz, respectively. XRD Rigaku 2200 was used to characterize the diffraction patterns of ZnO thin films. Major peak at 34° of the hexagonal wurtzite structure ZnO were clearly observed. Surface morphology and thickness of the film was evaluated by scanning electron microscopy (SEM) JEOL JSM-6500F.

[*] To whom any correspondence should be addressed.

2. Experimental Details

The ZnO films were deposited on 5×5 cm silcon wafer by atmospheric pressure plasma jet (APPJ) of CVD system. The schematic of the setup was shown in figure 1. N_2 gas was selected as the main gas to produce plasma and passes through the cannular region between two diodes with 2.5 slm. In this process, dry powder water Zinc acetylacetonate hydrate (Zn $(C_5H_7O_2)_2$) was used as the Zinc precursor and was put into oil bath for the vaporization using heating oven. The vapour of the Zinc precursor mixed with gas Argon with 900 sccm flow rate were carried to the downstream of the nitrogen plasma jet and sprayed to perform thin film deposition. Oxygen was used as to control the oxygen contents of ZnO films and flow rate was fixed at 30 sccm. The frequency and power of power suppliers were fixed at 25 kHz and 500 W, respectively. The parameters in the process were listed in Table 1.

Rigaku2200X X-ray was used to characterize as-deposited films. The crystallinity was examined by X-ray diffraction analysis. Thickness and surface morphology of the films were evaluated by scanning electron microscopy (SEM) and atomic force microscopy (AFM), respectively.

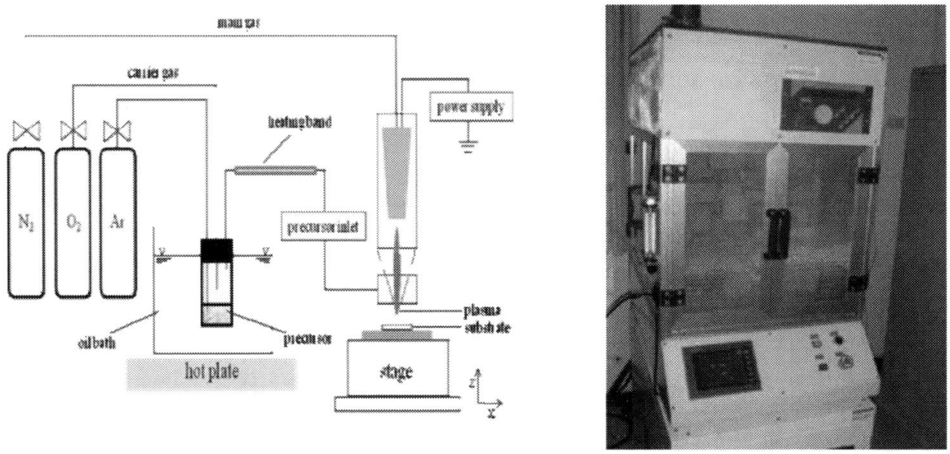

Figure 1. Atmospheric Pressure Plasma Jet.

Table 1. Parameters used in the deposition process of ZnO film.

Substrate	Silicon wafer, P-type
Substrate temperature	350 °C
Source	Zinc acetylacetonate hydrate, 98 %
Source Temperature	150 °C
Main gas (N_2)	2.5 slm
Ar carrier gas flow rate	900 sccm
O_2 carrier gas flow rate	30 sccm
Gap	20 mm
Sweep speed	4 mm·s^{-1}
Deposition time	5 min

3. Results and Discussions

The crystal structure of the ZnO film deposited by this APPJ was first examined by Rigaku2200X XRD, and figure 2 showed the diffraction pattern. Three major peaks corresponding to 32 °, 34 ° and 36 ° were significant present. The strong peak observed at 34 ° can be attribute to the (002) plane of the hexagonal wurtzite structure ZnO [19].

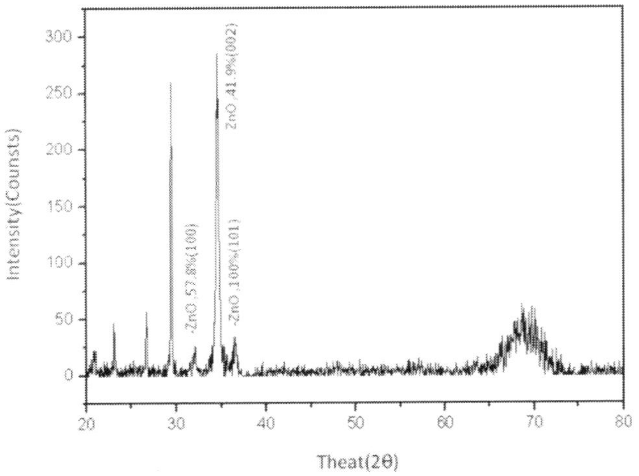

Figure 2. The XRD pattern of the ZnO deposited film by using the APPJ.

Figure 3 showed the SEM morphologies for ZnO film. The cross section SEM image for the deposited ZnO thin film in figure 4 showed a well defined compact and uniform. EDS analyzed to identify the Zn element on silicon substrate shown in figure 5. The deposition rate 150 nm/s was demonstrated in figure 6.

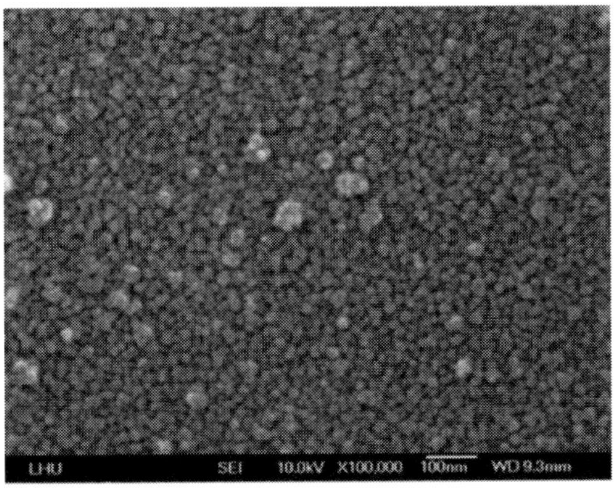

Figure 3. SEM morphologies for ZnO film.

Figure 4. Cross section SEM image for ZnO thin film.

AFM 3D and 2D images of the ZnO film of 1×1 μm^2 sample were shown in figure 7 and 8 respectively. The mean square surface roughness for a 1×1 μm^2 square section of the ZnO film was only 4.063 nm. The surface morphology of the Zno film in this proposed process was smooth.

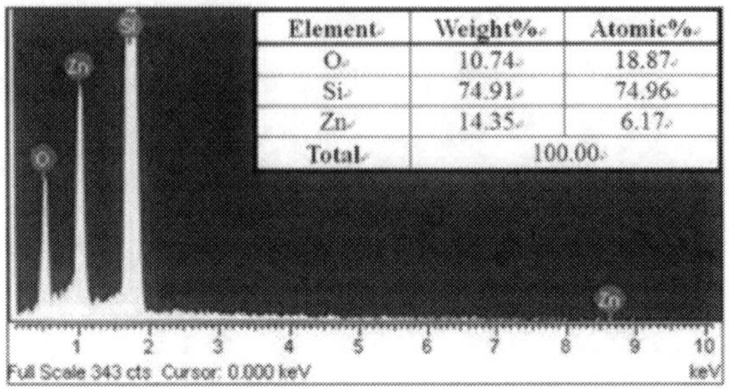

Figure 5. EDS analysis of Zn element.

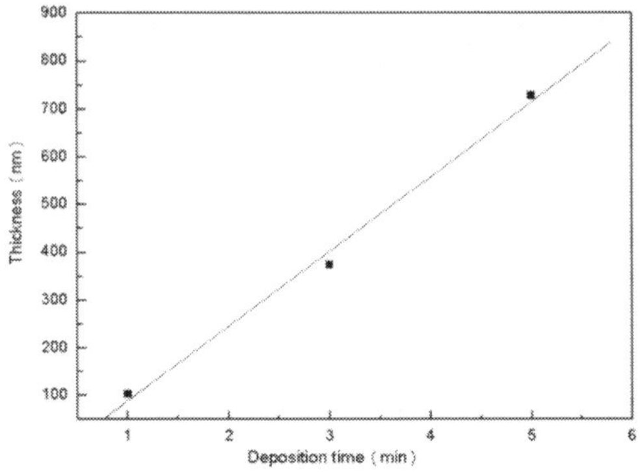

Figure 6. Deposition rate for ZnO thin film.

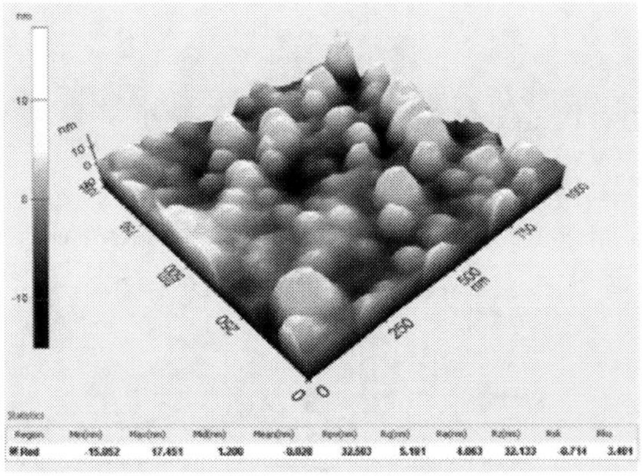

Figure 7. 3D surface morphology of ZnO film by AFM.

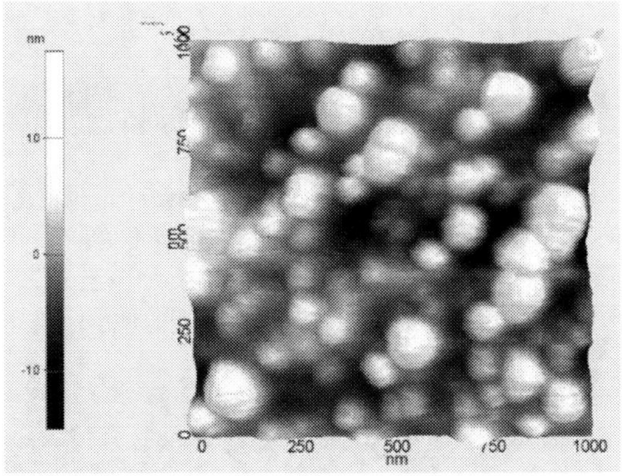

Figure 8. 2D surface morphology of ZnO film by AFM.

4. Conclusions

Thin films of ZnO were successfully deposited on silicon substrates by atmospheric pressure plasma jet in vapor phase using a dry powder Zn precursor. The X-ray diffraction pattern confirmed that the growth of hexagonal ZnO was formed. Thus, this present process provided a potential simple, inexpensive technique to grow ZnO film at a high growth rate with approximate 0.5 μm h^{-1}.

Acknowledgments

The authors would like to thank the National Science Council Taiwan for supporting this project (NSC-99-2815-C-262-009-E).

References

[1] Suchea M, Christoulakis S, Moschovis K Katsarakis N and Kiriakidis G 2006 *Thin Solid Films* **515** 551
[2] Perednis D, Ludwig J and Gauckler J 2005 *J. Electrochem.* **14** 103
[3] Sivalingam D, Gopalakrishnan J B, Krishnan U M, Madanagurusamy S and Rayappan J BB 2011 *Phys. E* **43** 1804
[4] Trinchi A, Li YX, Wlodarski W, Kaciulis S, Pandolfi L, Russo SP, Duplessis J and Viticoli S 2003 *Sens. Actuators A: Phys.* **108** 263.
[5] Sahay PP and Nath RK 2008 *Sens Actuators B: Chem.* **134** 654
[6] Liang HW, Lu YM, Shen DZ, Li BH, Zhang ZZ, Shan CX, Zhang JY, Fan XW and Du GT 2006 *Solid State Commun.* **137** 182

[7] Lupan O, Chow L and Chai G 2009 *Sens. Actuators B: Chem.* **141** 511
[8] Kim JP, Lee SA, Bae JS,Park SK, Choi UC and Cho CR 2008 *Thin Solid Films* **516** 5223
[9] Jayaraj MK, Antony A and Manoj R 2002 *Bull. Mater. Sci.* **25** 227
[10] Hu WS, Liu ZG, Guo XL, Lin C, Zhu SN and Feng D 1995 *Mater. Lett.* **25** 5
[11] Hari P, Baumer M, Tennyson WD and Bumm LA 2008 *J. Non-Cryst. Solids* **354** 2843
[12] Norton DP, Heo YW, Ivill MP, Ip K, Pearton SJ, Chisholm MF and Steiner T 2004 *Mater. Today* 34
[13] Hsu YW, Li HC, Yang YJ and Hsu CC 2011 *Thin Solid Films* **519** 3095
[14] Takahashi N, Kaiya K, Omichi K, Nakamura T, Okamoto S and Yamamoto H 2000 *J. Cryst. Growth* **209** 822
[15] Barakin MD, Gonzalez II E, Ladwig AM and Hicks RF 2007 *Sol Eneg Mat Sol C* **91** 924
[16] Ahn K, Jeong YS, Lee HU, Jeong SY, Ahn HS, Kim HS, Yoon SG and Cho CR 2010 *Thin Solid Films* **518** 4066
[17] Ito Y, Sakai O and Tachibana K 2010 *Thin Solid Films* **518** 3513
[18] Lua GJ, Kawaharamurab T, Nishinakab H, Kamadab Y, Ohshimab T and Fujita S 2007 *J. Cryst. Growth* **299** 1
[19] JCPDS- The International Centre for Diffraction Data, File No. **36** 1451

Hydrogen production by methane reforming based on micro-gap discharge

N N Liu, M D Bai*, M X Wang, K Y Liu
Dalian Maritime University, Environmental Science and Engineering College, Dalian 116026, China

E-mail: mindong-bai@163.com

Abstract. Based on micro-gap strong ionization discharge, this paper presents a study of hydrogen production by methane reforming at room temperature and atmospheric pressure without catalyst. Influence rules of conversion of methane and production of hydrogen were studied by changing discharge power and feed gas flow rate. Results show that when the discharge power was about 341 W, the discharge gap was 0.47 mm and the flow rate of feed gas was 100 mL min^{-1}, the conversion of methane and yield of hydrogen reached optimization. The conversion rate of methane and the highest yield of hydrogen were 68.14 % and 51.34 %, respectively.

1. Introduction

Methane and natural gas reforming are widely applied in industry to obtain hydrogen. Hydrogen is definitely a clean energy source, and fuel-cell technology has been investigated extensively and vigorously to practically utilize fuel cell vehicles and cogeneration system. Conventional technologies of hydrogen production, i.e. coal gasification, hydrocarbon reforming and water electrolysis, are too expensive or not applicable for specific applications, due to technical reasons. Thus, new methods are under development, like steam reforming [1], biological and plasma methods. Recently, developed micro-gap discharge operated at atmospheric pressure [2-7] seems to have a high potential for hydrogen production via hydrocarbon reforming.

In this paper, a method for production of hydrogen via methane reforming by the use of an atmospheric pressure micro-gap discharge reactor is presented.

2. Experiment setup and method

The plasma reaction chamber of hydrogen produced by methane was shown in figure 1. The plasma reactor is rectangular, and its dimensions are 260 mm (long) × 130 mm (wide) × 35 mm (thick). The thinner dielectric layer of α-Al_2O_3 is on the both sides of discharge electrodes, the dimensions of the dielectric layer are 175 mm (long) × 80 mm (wide) × 0.64 mm (thick), and the discharge gap is 0.47 mm. The dielectric constant and the electric insulation intensity are 10 and 350 kV cm^{-1}, respectively. The high voltage electrode is made from silver, grounding electrodes on both sides of which are made from aluminium. With this thinner dielectric layer, the high energy electrons can be obtained

* To whom any correspondence should be addressed.

further to form the strong electric-field discharge. A YDKS-WK power supply was used in the experiment, whose frequency was 9.85 kHz.

In this study, the energy efficiency of the methane conversion and the production of compounds were evaluated using the specific input energy (*SIE*) which could be determined as:

$$SIE = \frac{P_0}{v} \tag{1}$$

Where P_0 was discharge power and v was flow rate of feed gas. The discharge power in micro-gap was measured and calculated using the method of charge-voltage graphics [7]. The generation rate r could be expressed as:

$$r = \frac{yield \times v}{22.4(\mathrm{L \cdot mol^{-1}})} \tag{2}$$

The electric field intensity E_g could be calculated by

$$E_g = \frac{V_{pp}\varepsilon_d}{2l_d\varepsilon_g + l_g\varepsilon_d} \tag{3}$$

Where V_{pp} was the voltage peak value, ε_d was the dielectric constant. l_g is the gap of discharge, l_d is the dielectric thickness. The electron power is in direct proportion to the square of E_g. The electron energy is restricted by the electric field intensity of critical breakdown, the gas density or the gas pressure.

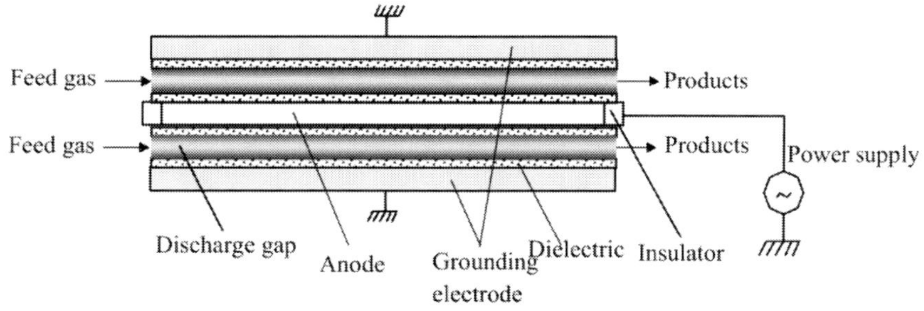

Figure 1. Structural diagram of plasma reaction chamber.

3. Experiment result and discussions

Table 1. Influence of discharge power for hydrogen generation and methane conversion.

Discharge power (W)	H_2 yield (v/v, %)	Methane conversion (v/v, %)	H_2 Generation rate (μmol min^{-1})	Eg (kV mm^{-1})	SIE(kJ L^{-1})
143.8	20.44	30.0	1825	11.37	43.13
153.4	25.13	34.2	2244	15.49	46.01
226.0	29.20	40.0	2607	17.56	67.81
297.8	30.65	41.9	2737	19.73	88.28
352.7	32.16	45.1	2871	20.70	105.81

The experiments were carried out under the conditions of the room temperature, the 105 Pa pressure, without any catalyst. The mixed gases of CH_4/O_2=95/5 (v/v) were introduced into the reactor at the flow rate of 200 mL min^{-1}. Table 1 shows the effects of discharge power for hydrogen generation and methane conversion. The SIE was 43.13 kJ L^{-1} under the 143.8 W discharge power. According to equation (1) and (3) the SIE and electric field intensity increased with the improvement of discharge power, the electron density of discharge gap discharge and electron collision frequency increased accordingly. Therefore an increase in discharge power can help to improve the chances of methane and electron collisions, the methane conversion rates and the hydrogen yield.

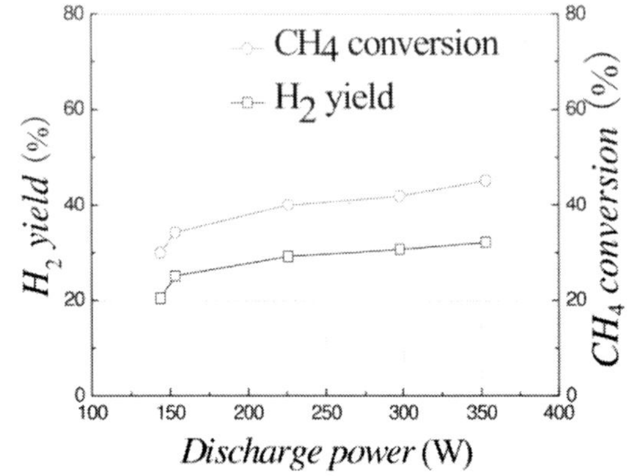

Figure 2. Influence of discharge power for H_2 yield and conversion of CH_4.

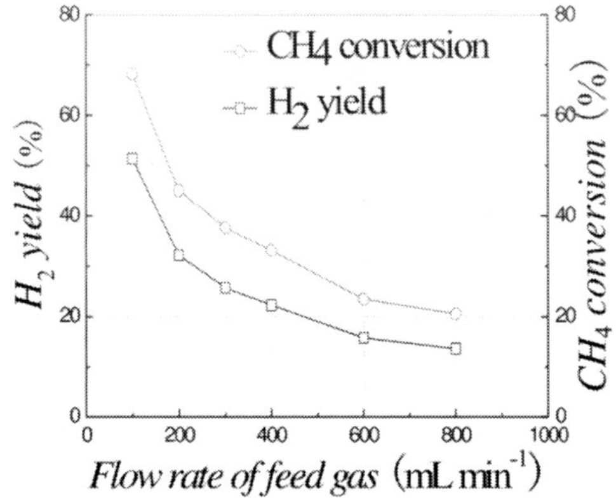

Figure 3. Influence of flow rate of feed gas power for H_2 yield and conversion of CH_4.

As shown in figure 2 and figure 3, both discharge power and flow rate of feed gas had great effects on the experimental results. The yield of H_2 was 51.34 % (v/v) and the conversion of CH_4 was 68.14 % (v/v) when the flow rate of feed gas was 100 mL min^{-1}. Here excitation voltage and discharge power were 2.8 kV and 341 W respectively. In figure3, when the excitation voltage was unchanged, the change of electric field intensity and discharge power were not obvious with the improvement of flow rate of feed gas, while the SIE decreased obviously. Hence, the yield of H_2 and the conversion of CH_4 decreased.

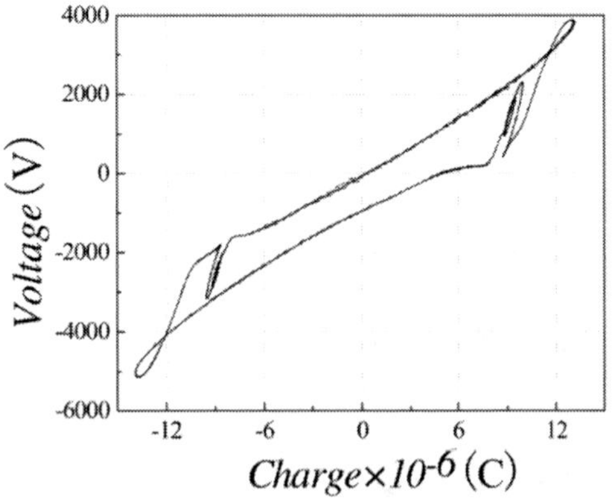

Figure 4. Voltage-charge Lissajous graphics.

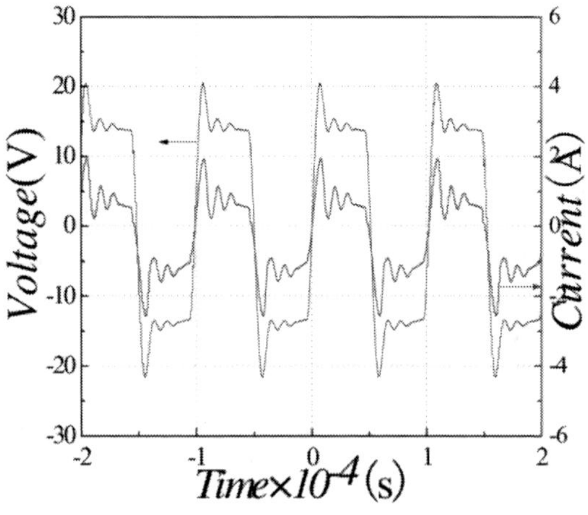

Figure 5. Voltage&Current oscillograms.

Figure 4 and figure 5 show the voltage-charge Lissajous graphics and the Voltage & Current waveform which were used for measurement and calculation of discharge power. It can be inferred that YDKS-WK power supply provides a high degree of electric field intensity and power, since a large number of CH_4 molecules were ionized and dissociated to produce H_2.

4. Conclusions

This paper presents a method of methane conversion to hydrogen. Influence rules of conversion of methane and production of hydrogen were studied by changing discharge power and feed gas flow rate. As shown in results, with the improvement of discharge power, electric field intensity increases, hence, the conversion rate of methane and hydrogen yield increase. As the feed gas flow rate increases, specific input energy diminished, and the conversion rate of methane and hydrogen concentration reduce. Under optimal conditions conversion rate of methane and hydrogen yield reached 68.14 % and 51.34 % respectively.

Acknowledgments

This work is supported by The National High Technology Research and Development Program of China under Grant No 2012AA062609, The International Science and Technology Cooperation Project of China under Grant No 2010DFA61470, The National Science Foundation for Distinguished Young Scholars of China under Grant No 61025001, The Fundamental Research Funds for the Central Universities under Grant No 2012QN067 and 2012TD061.

References

[1] Liguras D K, Kondarides D I and Verykios X E 2003 *Appl. catal. B: Environ.* **43** 345
[2] Shigeru K, Kohei U, Yasushi S and Kaoru F 2003 *Fuel* **82** 1377
[3] Krause M, Eliasson B., Kogelschatz U. and Wokaun A 2001 *Phys. Chem.* **3** 249
[4] Matsumoto H, Tanabe S, Okitsu K., Hayashi Y. and Suib S L. 2001 *J. Phys. Chem. A.* **105** 5304
[5] Eliasson B., Liu C. J. and Kogelschatz U. 2000 *Ind. Eng. Chem. Res.* **39** 1221
[6] Kabashima H., Einaga H. and Futamura S. 2003 *IEEE Trans. Ind. Appl.* **39** 340
[7] Zhang Z. T., Xian Y. Z. and Bai M. D. 2003 *J. Phys. (in Chinese)* **32** 458

Optimizing the formulation of cyclosporine A electret patch and the controlled release of drug

H Y Liu[1], P Wang[1], Y Y Liang[2], X Guo[2], J Jiang[2*] and L L Cui[1*]
1. Department of Inorganic Chemistry, School of Pharmacy, Second Military Medical University, China
2. Department of Physics and Mathematics, College of Basic Medical Sciences, Second Military Medical University, China

E-mail: cuilili39@hotmail.com, jjiang@smmu.edu.cn

Abstract. The polypropylene (PP) film coated with/without aluminum electrode were charged with the gird voltages of -500 V, -1000 V and -2000 V to prepare the electrets and produce electric field for control of drug release. The model drug of cyclosporine A (CsA) was loaded on a patch and ethyl oleate was used as the chemical enhancer in the manufacturing process. The formulation of the CsA drug patch enhanced by chemical was optimized, and the in vitro release behaviours of drug in the patches were studied to explore the enhancing effect of the external electrostatic field on the CsA release from the patch. Besides, the piezoelectric d_{33} coefficient was also determined to study the polarization of the drug in the patch under the action of the internal electrostatic field of the electret. The results indicate that the electrostatic field produced by the electret could polarize the drug in patch and enhance the release of CsA from the patch, and the effect depended on the electrode coating condition and charging voltage of the electret.

1. Introduction

Transdermal drug delivery systems are of great interest to researchers for its advantages over the oral, intravenous or intramuscular administration, such as avoiding hepatic first-pass effects, continual delivery of drug, circumvent the problem of rapid drug elimination and patient compliance etc. [1]. However, the stratum corneum (SC), the outmost layer of the skin, provides uniquely impressive resistance to molecular transport both into and from the body. The SC is a lipophilic barrier that is particularly impermeable to hydrophilic drugs [2, 3]. Therefore, only small neutral molecules could be delivered through this route. Peptides and proteins which have large molecular weight and radii are often charged. Thus, their passive transdermal delivery is very difficult [4]. Therefore, it is very important to develop an approach which can achieve a higher release rate of drug from transdermal formulation to the local site of the skin or into the systemic circulation. Now, some commonly used approaches include chemical approaches (using chemical enhancers), physical approaches (eg. iontophoresis, electroporation, electret etc.) [5-7].

Electret is a functional material which can provide stable electrostatic field and microcurrent. The external and internal electrostatic field produced by electret could result in a large amount of ducts in

[*] To whom any correspondence should be addressed.

SC for drug penetrating or the polarization of the drug, which could increase the drug release from the formulation. Our previous studies indicated that electret could increase transdermal delivery of lidocaine and meloxicam [8, 9]. The internal electric field of the electret transdermal system could induce the polarization of the drug and effectively control the release of the model drug. The electret transdermal system may have a better application for its easy manufacture and use, low cost and no need of auxiliary power.

Cyclosporine A (1202 Da) is a cyclic nonpolar oligopeptide which was composed of 11 amino acid residues. It is a 3rd generation immunosuppressive agent used in organ transplantation. Its oral administration might associate with gastrointestinal tract and hepatic first-pass effect to result in the low bioavailability, slow and incomplete absorption [10]. In order to increase its bioavailability, some new preparations such as microemulsion, suppository, nano particles and transdermal system have been developed. In this study, we used the polypropylene (PP) electret with or without aluminum electrode coated to prepare the drug electret patch. The drug release behaviour and piezoelectric modulus d_{33} of the patch were studied to optimize the formula of the preparation and to evaluate the release behaviour of the CsA PP electret patches and the effect of both the external and internal electrostatic field of the PP electret on controlled drug release.

2. Materials and Methods

2.1. Preparation of electret

The PP film, 6×6 cm^2 in area and 13 μm in thickness, with the aluminium electrode coated on one side or with no aluminium electrode coated on both sides were corona charged for 5 min under constant voltage and 77% relative humidity to prepare the electrets. The corona charging system in this study was manufactured by Dalian University of Technology, China. The point voltage used was -25kV and the grid voltages were -500V, -1000V and -2000V, respectively. The effective surface potentials of all the PP electrets were measured by compensation method using a surface potentiometer (ESR102A, Beijing Huajinghui Technology Co. Ltd. China).

2.2. Preparation of CsA patch

0.25g of Eudragit 100 (E100, pressure sensitive adhesive, PSA) was dissolved in ethanol to form a clear solution. Then CsA and 50% of tributyl citrate (plasticizer, w/w of total polymer) were dissolved in the adhesive solution and mixed until a clear solution was obtained. The solution was cast on the backing side of the PP film and was allowed for evaporation of the solvent at room temperature. After 12 h, the dried CsA patch was obtained and was used as control.

2.3. Preparation of chemical enhancer loaded CsA patch

Ethyl oleate was used as chemical enhancer. The chemical enhancer loaded CsA patches containing different amount of E100, plasticizer and chemical enhancer were prepared using the same method as described above. 9 combinations were possible (table 1). The variables were optimized by taking cumulative drug release amount (Q, μg cm^{-2}) as the dependent factor.

2.4. Preparation of chemical enhancer loaded CsA electret patch

The chemical enhancer loaded CsA electret patch (drug or CsA electret patch) was prepared by covering the charged face of the electret on the back side of the drug patch loaded with chemical enhancer.

2.5. In vitro drug release studies

In vitro drug release studies were performed by using vertical Franz diffusion cells with a receptor chamber volume of 6.75 ml and the effective permeation area of 2.8 cm^2. The patch was sandwiched between the receptor and donor chambers with the drug loaded PSA layer facing the receptor cell. The receptor chamber was filled by phosphate buffer solution (PBS) with pH 7.4. The solution was stirred with magnetic bar and maintained at 32±0.5°C. Samples (0.3 ml) were withdrawn at predetermined time intervals and the receptor phase was replenished with an equal volume of fresh PBS at each sample withdrawn.

Table 1. Formulation and drug release evaluation of the chemical enhancer loaded CsA patches

Patch code	E100 (g)	Tributyl citrate (%, w/w)	Ethyl oleate (%, w/w)	Q^a (μg cm^{-2})
P1	0.20	50	8	30.798
P2	0.20	55	10	27.007
P3	0.20	60	12	31.842
P4	0.25	50	10	49.699
P5	0.25	55	12	29.986
P6	0.25	60	8	29.964
P7	0.30	50	12	30.490
P8	0.30	55	8	31.010
P9	0.30	60	10	32.558

[a] Cumulative released amount at 28th hour.

2.6. Data analysis

The cumulative amount of CsA released from the patch was analyzed by using high performance liquid chromatograph (HPLC, Agilent Tech, USA) system. A 250 mm×4.6 mm DiamonsilTM ODS column (Dikma Tech Inc, China) was used. The UV detection was set at 214 nm. The mobile phase in the ratio of 340:100:60 of acetonitrile, methanol and water respectively was used and was delivered at a flow rate of 1.0 ml/min.

2.7. Determination of piezoelectric d_{33} coefficient

The piezoelectric d_{33} coefficient of CsA patch, chemical enhancer loaded CsA patch and CsA electret patch were determined by using a piezo-d_{33} meter (ZJ-3A, Chinese Academy of Science Institute of Acoustics, China).

2.8. Statistical analysis

Data from the in vitro release studies and piezoelectric modulus determination were treated for statistical analysis using SPSS13.0 analytical software for single factor analysis of variance (ANOVA) and *t*-tests. *P* below 0.05 was considered to be significant.

3. Results and discussion

3.1. Optimization of patch formula

The drug release characteristics of all nine patches were studied. The patch containing 0.25g PSA, 50% plasticizer and 10% ethyl oleate (P4) presented the best drug release rate. The drug release behaviour followed the Higuchi model with the relative coefficient of 0.998:

$$Q = 8.4267\sqrt{t} + 4.6227 \qquad (1)$$

P4 had the highest cumulative release amount among the nine patches (table 1). The relative release amount of CsA from the patch was about 21% at 28th hour. So P4 was used to prepare the drug electret patch for further studies.

3.2. In vitro drug release studies of electrode coated drug electret patches

Figure 1 is the in vitro drug release characteristics of control patch and three electrode coated drug electret patches. It indicates that the drug release behaviour of all the patches followed the Higuchi model. The cumulative released drug amounts of CsA from -500 V, -1000 V and -2000 V electrode coated drug electret patches were 1.286-fold, 1.368-fold ($P<0.05$) and 1.086-fold compared with that of the control patch. The enhancing effect was dependent on the charging voltage of the electret. And the -1000 V electrode coated electret had the most effective drug release enhancement.

Figure 2 was the piezoelectric d_{33} coefficient of the control patch and electrode coated drug electret patches. From the figure we can see that all three electrode coated drug electret patches had the certain d_{33} values but the control patch had a d_{33} value of zero. This suggested that the drug in electret patch was polarized under the action of internal electric field of the electret. -1000V electret could induce more drug polarization than those of the -500V electret and -2000V electret, which was coincided with the results of the drug release experiment. The results also indicate that there was an optimized surface potential range for electrode coated electret to polarize the drug and to increase the drug release amount under the action of the unidirectional electrostatic field of the electret.

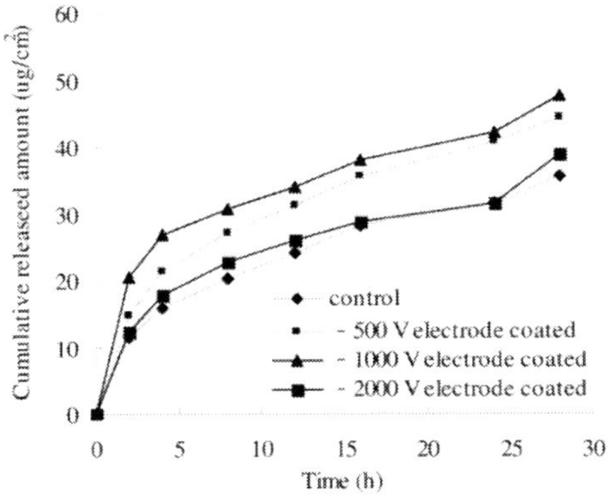

Figure 1. In vitro drug release for CsA patch and electrode coated drug electret patches.

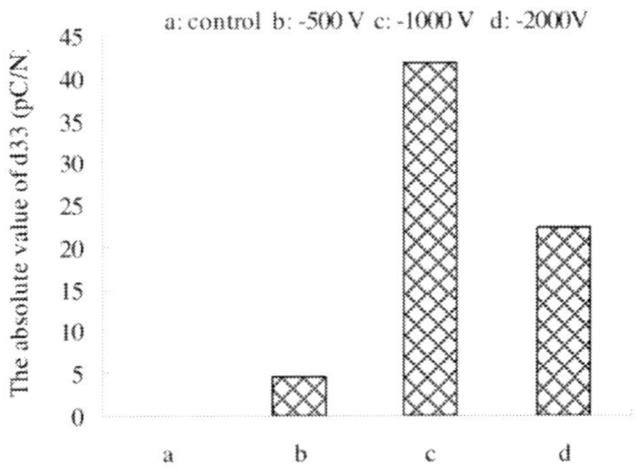

Figure 2. The piezoelectric d_{33} coefficient of control patch and electrode coated drug electret patches.

3.3. In vitro drug release studies of non-electrode coated drug electret patches

Figure 3 is the in vitro drug release characteristics of control patch and three drug electret patches without electrode coated. The drug release behaviour of all the patches also followed the Higuchi model. -500 V and -1000 V drug electret patches showed almost the same release characteristics. Although the cumulative released amounts of CsA from both two patches at 28th h were respectively 1.301-fold and 1.313-fold compared with that of the control patch, they were of no significance. However, -2000 V drug electret patch had a far more cumulative released amount of CsA. It was 1.509-fold compared with that of the control patch and was of significance.

From figure 4 we can see that when the electret was corona charged in the rage of -500 V to -1000 V, the polarization degree of the drug in the patch was increased greatly. However, when the charging voltage of the electret was in the range of -1000V to -2000V, the drug polarization degree of both the patches were almost the same.

Comparing figure 2 and figure 4, figure 1 and figure 3, we can see that the electret induced drug polarization behaviour and drug release behaviour for both the electrode coated and non-coated drug electret patches were different. The results were due to the different distribution rule of space electric field of the electrode coated electret and non-electrode coated electret. Therefore, the drug release behaviour was also different. The detailed mechanism needs further study. Anyway, the results give us the information that it is possible to control the drug release by reasonably choosing the electret with or without electrode or changing the charging voltage of the electret.

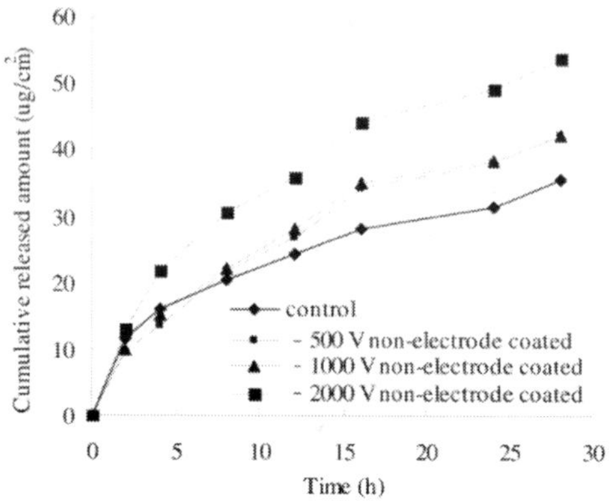

Figure 3. In vitro drug release for CsA patch and non-electrode coated drug electret patches.

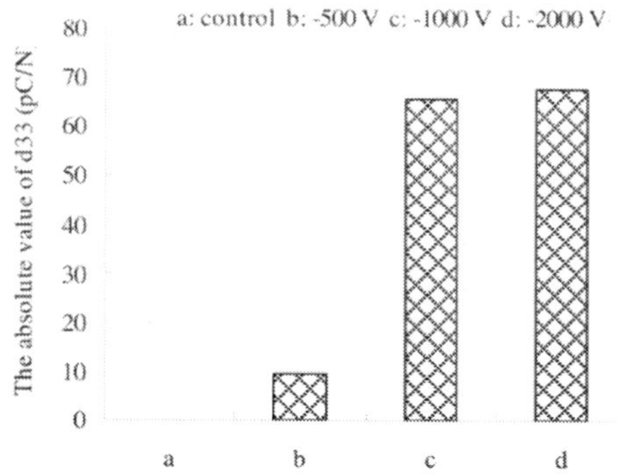

Figure 4. The piezoelectric d_{33} coefficient of control patch and non-electrode coated drug electret patches.

3.4. Combination effect of electret with chemical enhancer

Figure 5 is the in vitro drug release characteristics of control patch, chemical enhancer loaded drug patch, as well as chemical enhancer loaded -1000 V and -2000 V drug electret patches without electrode coated. 10% ethyl oleate could effectively increase the drug release from the patches. The cumulative released amounts of the drug at 28th h for the chemical enhancer loaded drug patch was 1.39-fold (P<0.05) compared with that of the control. On the other hand, the cumulative released amounts of the drug at 28th h for -2000 V and -1000 V drug electret patches without electrode coated were 1.15-fold (P<0.05) increased and 1.083-fold (P<0.05) decreased compared with that of the chemical enhancer loaded patch. Therefore, the reasonable combination of electret and chemical enhancer could effectively regulate the release rate and released amount of the drug from the patch.

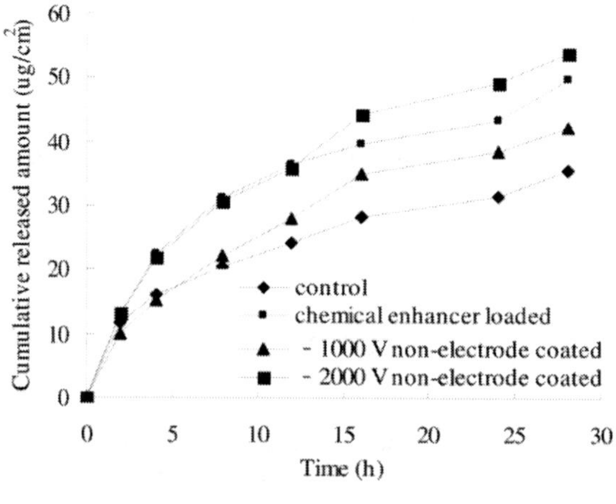

Figure 3. In vitro drug release for CsA patch, chemical enhancer loaded drug patch and non-electrode coated drug electret patches.

4. Conclusions

The electrostatic field of the electret could induce the polarization of the drug in the patch. The polarization degree of the drug depended on the charging voltage and the electrode coating condition of the electret. A combination of negative PP electret and 10% ethyl oleate could be an effective strategy for controlled release of CsA from the patch. The drug electret patch could be a potent formulation for study of transdermal delivery and preparation of transdermal formulation of polypeptide drugs.

Acknowledgments

The authors thank the National Nature Science Foundation of China (Grant No. 50977089) and Natural Science Foundation of Shanghai (Grant No. 10ZR1437600) for financial support provided.

References

[1] Amnuaikit C, Ikeuchi I, Ogawara K, Higaki K and Kimura T 2005 *Int. J. Pharm.* **289** 167
[2] Bouwstra J A, Honeywell-Nguyen P L, Gooris G S and Ponec M 2003 *Prog. Lipid Res.* **42** 1
[3] Proksch E, brandner J M and Jensen J 2008 *Exper. Dermatol.* **17** 1063
[4] Yerramsetty K M, Neely B J, Madihally S V and Gasem K A M 2010 *Int. J. Pharm.* **388** 13
[5] Cui L L, Hou X M, Jiang J, Li G D and Liang Y Y 2008 *J. Phys.: Conf. Ser* . **142** 012015
[6] Dhote V, Bhatnagar P, Mishra P K, Mahajan S C and Mishra D K 2012 *Sci. Pharm.* **80** 1
[7] Paudel K S, Milewski M, Swadley C L, Brogden N K, Ghosh P and Stinchcomb A L 2010 *Ther. Deliv.* **1** 109
[8] Cui L L, Hou X M, Jiang J, Li G D and Liang Y Y 2008 *J. Phys.: Conf. Ser* .**142** 012015
[9] Han D, Cui L L, Jiang J, Li G D, Song M H and Wang X P 2005 *Proc. Int. Symp. On Electret* (Bahia:Salvador-Brazil) 60
[10] Lopes L B, Collett J H and Bentley M V L B 2005 *Euro. J. Pharm. Biopharm.* **60** 25

Journal of Physics: Conference Series **418** (2013) 012148 doi:10.1088/1742-6596/418/1/012148

Study on the electrostatic and piezoelectric properties of positive polypropylene electret cyclosporine A patch

X Guo[1#], Y Y Liang[1#], H Y Liu[2], L L Cui[2*], J Jiang[1*]
1. Department of Physics and Mathematics, College of Basic Medical Sciences, Second Military Medical University, Shanghai, China
2. Department of Inorganic Chemistry, School of Pharmacy, Second Military Medical University

E-mail: jjiang@smmu.edu.cn, cuilili39@hotmail.com

Abstract. Corona charged electrets at voltages of +500 V, +1000 V and +1500 V were prepared for manufacturing polypropylene (PP) electret blank patches and PP electret drug patches. The stability of external electrostatic field of the electret patch and the polarization of the drug in patch under the internal electrostatic field of the electret were studied. The results indicate that all the electret drug patches had good charge storage stabilities. However, the non-electrode coated electret drug patch had better stability in the external electrostatic field than that of the electrode coated electret drug patch. The higher the charging voltage of the electret, the faster the surface potential of the electret drug patch decayed, and the worse the stability of the external electrostatic field. All the electrets used in this study could result in the polarization of the model drug in patch. The piezoelectric properties of non-electrode coated electret drug patch increased with the charging voltage of the electret. However, excessively higher charging voltage could result in the decreased polarization of the drug in patch. Both the stability of the external electrostatic field of electret and the polarization of drug were the key factors for controlled drug release and skin permeation.

1. Introduction

Transdermal drug delivery is a method to apply the drug directly to the skin to reach the systemic circulation. The commonly referred transdermal drug delivery system is a patch. It contains a small pad that has the desired drug, and the drug can be delivered continuously when the patch is applied to the skin. Now there has been some successful transdermal delivery system for delivering of fentanyl, lidocaine, testosterone and nitroglycerin for therapeutic purposes [1]. Although transdermal drug delivery has many advantages over the traditional way for drug administration, the natural barrier function of the skin, mainly the stratum corneum, makes it difficult for most of the drugs to penetrate through the skin to reach a therapeutic concentration in the blood. Therefore, the development of some strategies to enhance the drug transdermal delivery is of great interest to researchers. Nowadays, several methods including chemical enhancer or physical methods have been developed, and the physical methods include iontophoresis, electrophoresis, ultrasound, electret [2-6]. However, some of the physical methods like iontophoresis and electrophoresis need external power and conductive electrode, which make them inconvenient to be used.

[*] To whom any correspondence should be addressed.
[#] Both of the authors contributed equally to this work.

As a new type of physical penetration method, electret can produce stable and strong enough electrostatic field and microcurrent to open up the skin by changing the lipid structure in stratum corneum and produce many new channels in the skin [7]. Electret method is expected to become an ideal enhancement technology for drug absorption due to its advantages such as simple preparation, easily to use and safety, etc.

In this experiment, we used cyclosporine A as model drug and electrode coated or non-coated polypropylene (PP) film as electret material to prepare the electret cyclosporine A patch by using pharmaceutical method and electret experimental techniques. The stability of the external electrostatic field of the patch was studied by measuring the surface potential decay of the patch. The polarization of the drug in patch under the action of the internal electrostatic field of the electret was studied by measuring the piezoelectric d_{33} coefficient of the patch. The stability of the external electrostatic field of the electret and the polarization of the drug were very important in controlled drug release and skin permeation.

2. Materials and Methods

2.1. Preparation of electrets

The PP film with a thickness of 13 μm (Toray Industries Inc., Japan) was an aluminium coated on one side by using the vacuum resistance evaporation coater machine (Beijing techno. imp. & exp. Co. Ltd, China). The electrets of electrode coated and non-electrode coated were prepared by corona charging technology under the grid voltages of +500 V, +1000 V, +1500 V and tip voltage of 10 kV using a corona charging system (Dalian University of Technology, China). The electrode side of the coated electret was grounded. The charging time was 7 min.

2.2. Preparation of patches

The blank patch was prepared by dissolving the Eudragit E100 (pressure sensitive adhesive) in ethanol to form a clear solution. Then the tributyl citrate (plasticizer) was added into the solution. The solution was coated on the PP film with the area of 6 cm× 6 cm to prepare the patch. The patch was allowed for solvent evaporation at room temperature for 12 h to be used as the blank patch.

The ethanol solution of Eudragit E100 and tributyl citrate was prepared as described above. Then a drug coating solution formed by dissolving the cyclosporine A and 10 % ethyl oleate (chemical enhancer) into the ethanol solution was coated on the PP film with the same size as the blank patch to prepare the cyclosporine A patch (drug patch). The solvent evaporation procedure was the same as that in blank patch preparation.

The charged face of the electrode coated or non-coated electret was covered on the back layer of the blank/drug patches to obtain the all kinds of electret patches.

2.3. The surface potential, piezoelectric coefficient measurements

The vibrating capacitor electrometer (ESR102A, Beijing HuaJingHui Technology Co., Ltd, China) was used to measure the surface potential of the patches. The quasi-static d_{33} meter ((ZJ-3A, Chinese Academy of Science Institute of Acoustics, China) was used to measure the piezoelectric d_{33} coefficient.

3. Results and discussion

3.1. Stabilities of the electrostatic fields of the patches

The normalized decay curves of the surface potential on electrode coated or non-coated +500 V electret blank patches are shown in figure 1. It indicates that the surface potentials of both the patches decayed exponentially, and the external electrostatic field of the patches exhibited better stability. The surface potentials of the electrode coated and non-coated PP electret blank patches at 48 h were 433 V and 372.5 V, which were 86.6 % and 74.5 % of their initial values, respectively. After 100 hours, both of the patches could still remain 81.9 % and 73.9 % of their initial surface potential. Compared with the non-electrode coated +500 V PP electret blank patch, the electrode-coated patch had more stable external electrostatic field.

Figure 2 is the normalized decay curves of surface potential on +500 V PP electret cyclosporine A patches. The surface potentials of the electrode-coated and non-coated +500 V electret drug patches were 386.5 V and 356.0 V, respectively, which were 77.3 % and 71.2 % of the initial values accordingly. After 100 hours, both of the patches could still remain 67.7 % and 66.96 % of the initial surface potential, which suggested that both the electret drug patches had good charge storage stabilities and could produce a stable external electrostatic field. Compared with the non-electrode coated +500 V PP electret drug patch, the electrode coated patch had more stable external electrostatic field.

Comparing figure 1 with figure 2, it can be seen that the surface potential decay behaviour for all the patches are the same. However, the patches made from electrode-coated electret had higher absolute and normalized value of the surface potential than those of the non-electrode coated electret patches. Therefore, both the electrode-coated electret blank patch and electret drug patch had more stable external electrostatic field.

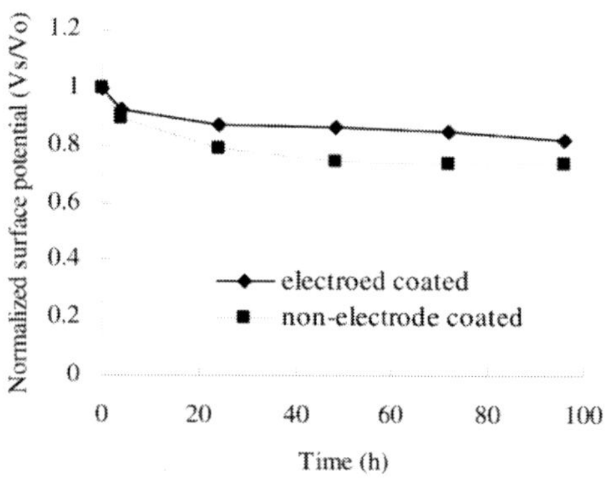

Figure 1. Normalized surface potential decay of electrode coated and non-electrode coated +500 V PP electret blank patches (n=6).

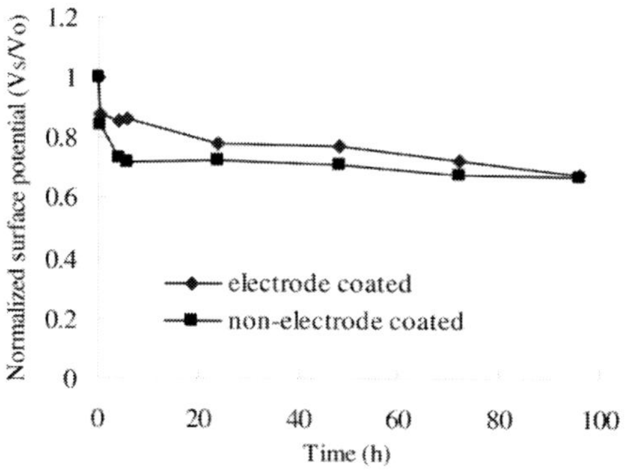

Figure 2. Normalized surface potential decay of electrode coated and non-electrode coated +500 V PP electret drug patches (n=6).

The normalized surface potential decay curves of +500, +1000 and +1500 V electrode-coated and non-coated PP electret cyclosporine A patches are shown in figure 3 and 4, respectively. From the figures we can see that when the electret was corona charged with the same grid voltage, the electrode-coated PP electret cylosporine A patch had higher surface potential value and stronger external electrostatic field than those of the non-electrode coated electret drug patch. For the electrode-coated +500 V, +1000 V and +1500 V PP electret drug patches, the surface potentials at 48 h were 77.3, 66 and 59 % of the initial surface potentials, respectively. For the non-electrode coated +500 V, +1000 V and +1500 V PP electret drug patches, the surface potentials at 48 h were 71.2, 49.4 and 41.2 % of the initial values, respectively. Therefore, the electrode coated PP electret drug patch had more stable external electrostatic field than that of the non-electrode coated electret drug patch.

Figure 3 and 4 also indicate that the higher the charging voltage of the electret, the faster was the surface potential decay of the patch and the worse was the stability of the external electrostatic field. The explanation for this is as follows: with the increase of the charging voltage, the charge density injected into the PP film was increased. Then a stronger self-electric field was produced to induce more charge migration from deep trap to shallow trap. Charges in shallow trap were easily affected by the outside and detrapped. Therefore, the penetrated surface potential decayed more quickly, resulting in a more unstable external electrostatic field.

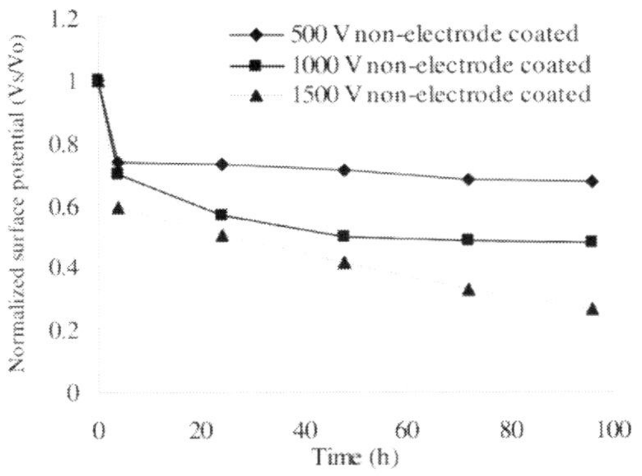

Figure 3. Normalized surface charge decays of +500 V, +1000 V and +1500 V electrode coated PP electret drug patches (n=10).

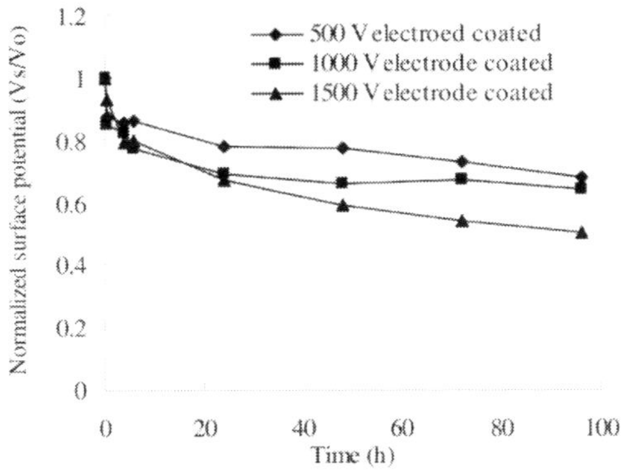

Figure 4. Normalized surface charge decays of +500 V, +1000 V and +1500 V non-electrode coated PP electret drug patches (n=6).

The electrostatic field of the elelctret patch consists of two parts: the external electrostatic field and internal electrostatic field. Our previous studies indicate that the higher external electrostatic field could open up more skin structure and result in enhanced drug penetration through the skin [8-9]. The function of the internal electrostatic field is to induce polarization of the drug.

3.2. The piezoelectric properties of positive electret cyclosporine A patches

Studies have shown that the piezoelectric property of the materials can be affected by means of the chemical treatment, process improvement, and polarization [10-13]. To study the piezoelectric properties of electret cyclosporine A patch and to explore the polarization of the model drug in patch, the electret was removed from the electret cyclosporine A patch after the electret covered on the back side of the patch for 100 h and then the piezoelectric d_{33} coefficient of the patch was measured. The piezoelectric d_{33} coefficients of the blank patch and electret treated cylosporine A patches are shown in figure 5 and 6.

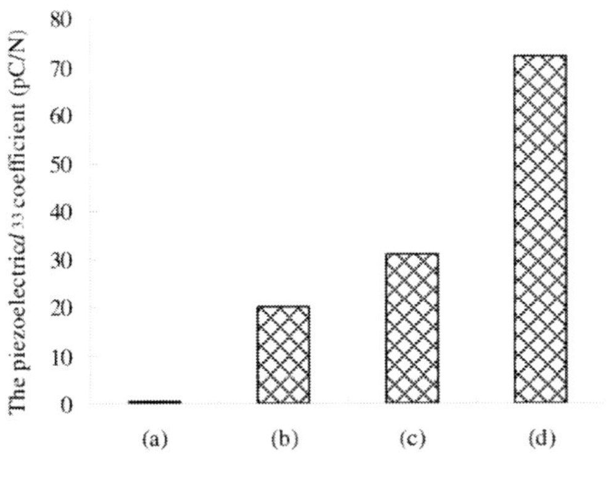

Figure 5. The piezoelectric d_{33} coefficient of drug patch after treated by non-electrode coated PP electret for 100 h (n=6).

(a) Blank patch; (b) +500 V electret treated drug patch; (c) +1000 V electret treated drug patch; (d) +1500 V electret treated drug patch.

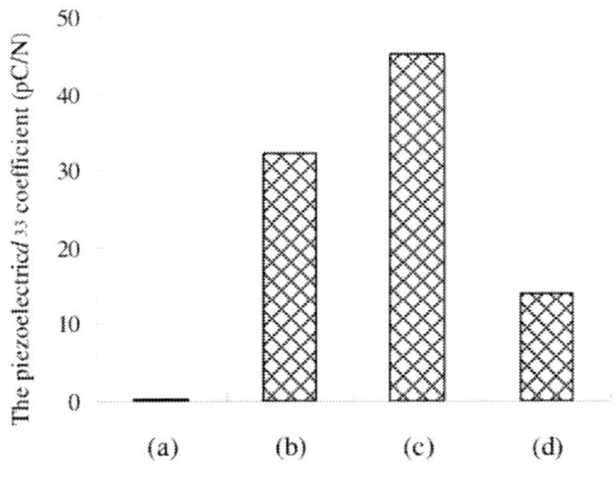

Figure 6. The piezoelectric d_{33} coefficient of drug patch after treated by electrode coated PP electret for 100 h (n=5).

(a) Blank patch; (b) +500 V electret treated drug patch; (c) +1000 V electret treated drug patch; (d) +1500 V electret treated drug patch.

Journal of Physics: Conference Series **418** (2013) 012148 doi:10.1088/1742-6596/418/1/012148

Compared with the blank patches, all the positive PP electrets used in this study could induce the polarization of the model drug in patch. However, the drug polarization properties were different when the patches were covered by electrode-coated electret or non-electrode coated electret. For the non-electrode coated patch, the absolute value of d_{33} increased with the charging voltage. For electrode-coated patch, the absolute value of d_{33} increased accordingly when the charging voltage increased from +500 V to +1000 V. However, when the charging voltage reached to +1500 V, d_{33} decreased significantly, suggesting that an excessive higher charging voltage would reduce the polarization of the drug in patch. Therefore, the piezoelectric properties of cyclosporin A patch were dependent not only on the electrode coating of the electret, but also on the magnitude of the charging voltage.

4. Conclusions

The stability of the external electrostatic field and polarization of drug in electret patch were affected by electrode coating and charging voltage of the electret. Since the electrostatic and piezoelectric properties of electret cyclosporine A patch would influence the controlled release and percutaneous absorption of model drug in patch. Therefore, an appropriate selection of electret type and manufacture coefficient is very important to achieve the regulation effect of electret on drug formulations when the electret transdermal drug delivery system is studied.

Acknowledgments

The authors thank the National Nature Science Foundation of China (Grant No. 50977089) for financial support.

References

[1] Prausnitz M R and Langer R 2008 *Nat Biotechnol.* **26** 1261

[2] Escobar-Chávez J J. & Merino, 2010 *Current Technologies to Increase the Transdermal Delivery of Drugs.*(Bussum, Bentham Science Publishers) p 120

[3] Zheng J M 2006 *Percutaneous Administration of the New Formulation* (Beijing, People's Health Publishing House) p 17

[4] Mark R P and Robert L 2008 *Nature Biotec.* **26** 1261

[5] Ibrahim A S and Li S K 2010 *Int. J. Pharm.* **383** 89

[6] Murthy N S, Boguda VA and Payasada K 2008 *Biol. Pharm. Bull.* **31** 99

[7] Jiang J, Liang Y Y, Dong F J, Liu H Y, Tu Y and Cui L L 2012 *J. Electrosta.* **70** 258

[8] Jiang J, Dong F J, Liang Y Y, Liu H Y and Cui L L 2012 *IEEE T. Dielect. El. In.* August, accepted

[9] Cui L L, Liang Y Y, Dong F J, Liu H Y and Jiang J 2011 *Proc. 14th Intern. Symp. on Electrets* (Montpellier: Electrets) p 193.

[10] Zhang X Q, Hillenbrand J and Sessler G M 2004 *Appl. Phys. Lett.* **85** 1226

[11] Tang M M, A Z L, Yang Q, Xia Z F and Zhang X Q 2006 *Piezoelectectrics & Acoustooptics. (in Chinese)* **30** 109

[12] Cao G X ,Zhang X Q, Sun Z L. Wang X W, Lou K X and Xia Z F 2010 *Acta Physica Sinica. (in Chinese)* **59** 6514

[13] Zhang X Q, Hillenbrand J and Sessler G M, 2006 *Appl. Phys. A: Mater. Sci. Proc.* **84** 139

Journal of Physics: Conference Series **418** (2013) 012149 doi:10.1088/1742-6596/418/1/012149

Enhancing effect of negative polypropylene electret on in vitro transdermal delivery of cyclosporine A solution and its synergistic effect with ethyl oleate

L L Cui[1#], L Ma[2#], Y Y Liang[2], H Y Liu[1], X Guo[2], J Jiang[2*]
1. Department of Inorganic Chemistry, School of Pharmacy, Second Military Medical University, Shanghai, China
2. Department of Physics and Mathematics, College of Basic Medical Sciences, Second Military Medical University

E-mail: jjiang@smmu.edu.cn

Abstract. In this study, the corona charged electrets at voltages of -500 V, -1000 V and -2000 V were made from polypropylene (PP) film. The cyclosporine A (CsA) and 10% ethyl oleate were chosen as the model drug and chemical enhancer, respectively. The charge storage stability of the electrets and the in vitro transdermal behaviour of the model drug in solution under different conditions were studied. The results indicate that the external electrostatic field of the negative PP electrets could penetrate through the rat skin and enhance the transdermal delivery of cyclosporine A. A synergistic effect on enhancing the transdermal delivery of cyclosporine A was observed by combining different surface potential negative PP electrets with 10% ethyl oleate, and the amount of transdermal delivery of CsA was greatly increased comparing with only application of electrets. Therefore, the combination application of electret and chemical enhancer could be a feasible strategy in enhancing transdermal delivery of small peptide drugs or some large molecular drugs.

1. Introduction

Skin is considered as a promising route for delivery of therapeutic molecules into the body due to its several advantages such as avoiding of the potential hepatic first pass effect, constant drug delivery, self-administration, non-invasive and improving patient compliance, and so on [1]. However, the stratum corneum (SC), the outmost layer of the skin, serves as a main barrier for transdermal delivery due to its rigid lamellar structure. Therefore, only small, potent and moderately lipophilic molecules could be effectively delivered via this route. However, the peptides, proteins and drugs which are often charged and highly hydrophilic in nature are limited to achieve the transdermal systemic delivery [2].

To overcome the barrier function of the SC and to increase the flux through the skin, several enhancing approaches have been investigated, such as chemical enhancers, iontophoresis, electroporation, electret, thermal energy, skin metabolism inhibitors, microneedles, etc [3, 4]. Among these approaches, chemical enhancers promote drug permeation through skin by increasing drug partitioning into the SC domain, the drug diffusivity in the SC domain or the combination of both [5]. Iontophoresis uses continuous current to drive peptide, proteins or charged drug molecules to

[*] To whom any correspondence should be addressed.
[#] Both of the authors contributed equally to this work.

cross the skin [5]. However, it has to use external power and electrodes, and erythema may present under the electrode application site [6]. Electret is a dielectric material whose external electrostatic field could penetrate through the skin and facilitate the drug to permeate through the skin by alteration of the skin structure [7, 8]. Besides, it has been proved that electret could increase the flux for both of the small molecules and some macromolecules with the molecular weight lower than 1000 Da [9].

Cyclosporine A (CsA) is a 3rd generation immunosuppressive agent used in organ transplantation. It is a cyclic nonpolar oligopeptide composed of 11 amino acid residues [10]. Its systemic administration might associate with narrow therapeutic index and side effects [11]. So topic delivery is desirable for treating skin disorders. However, this approach has been reported as ineffective in most case due to its large molecular weight (1202 Da). Therefore, we used the electrostatic field produced by PP electret to function as physical enhancer and 10% ethyl oleate as chemical enhancer, and the enhancing effects of single use of the electret or 10% ethyl oleate and the combinational use of both the strategies on transdermal delivery of CsA were studied.

2. Materials and Methods

2.1. Materials

The PP film with a thickness of 13 μm was obtained from Toray (Toray Industry Co. Japan). The cyclosporine A was purchased from Yuanye Biotechnology Co. Ltd. (Shanghai, China). The ethyl oleate was purchased from Feixiang Chemical Plant (Shanghai, China).

2.2. Preparation of electret

The double bare surface PP film was corona charged at constant voltage for 5 min using a corona charging system (Dalian University of Technology, China) with the point voltages of -10 kV and grid voltages of -500 V, -1000 V and -2000 V, respectively. The effective surface potentials of the electrets were measured by compensation method using a surface potentiometer (ESR102A, Beijing Huajinghui Technology Co. Ltd., China).

2.3. Preparation of skin

Male Sprague-Dawley (SD) rats with weight of 200±10 g were selected for transdermal permeation study. The study was conducted with the approval of Institutional Animal Ethical Committee of Second Military Medical University. The rat was anesthetized by intraperitoneal injection of 20% urethane (0.5 ml/kg body weight). The hair from the abdominal region was carefully shaved using an animal hair clipper before the day of the experiment. After a period of 24 h, the rat was sacrificed and the shaved region was incised and adherent fat and other visceral tissue were removed from the undersurface. Then the skin was washed with phosphate buffer solution (PBS), wrapped in aluminium foil and stored in deep freezer for use.

2.4. In vitro permeation experiment

A vertical Franz diffusion cells (receptor volume=6.75 ml, effective permeation area=2.8 cm^2) were used to perform the in vitro permeation experiment. A piece of the excised skin was sandwiched between the donor cell and receptor cell with the SC facing the donor chamber. The donor cell was filled with saturated CsA solution, saturated CsA solution and 10% ethyl oleate in combination, saturated CsA solution covered by differently corona charged PP electret which was 1 mm away from the surface of the solution, or a combination of saturated CsA solution and 10% ethyl oleate with electret 1 mm away from the surface of the solution. Phosphate buffer solution (pH 7.4) was used in the receptor and stirred with magnetic bar. The cells were maintained at 32±0.5°C. Samples (0.3 ml) were withdrawn at regular time interval up to 24 h and were replaced with fresh PBS.

2.5. Data analysis

The cumulative amount (Q, μg cm^{-2}) of CsA that permeated across the skin was calculated based on the equation:

$$Q = (c_n + V/V_0 \sum_{i=n-1} c_i) \times V/A \quad (1)$$

Where V is the volume of the receptor cell; A is the effective permeation area; V_0 is the sampling volume; c_n and c_i are the concentrations of CsA in receptor cell at n h and i h, respectively.

The concentration of CsA was analysed by high performance liquid chromatograph (HPLC, Agilent Tech, USA) with a 250 mm×4.6 mm DiamonsilTM ODS column (Dikma Tech Inc, China) and UV detection at 214 nm. The mobile phase consisted of acetonitrile, methanol and water in the ratio of 340:100:60 and delivered at a flow rate of 1.0 ml/min. The inter-day and intra-day S.D. of the HPLC analysis were below 5%.

2.6. Statistical analysis

Data were treated for statistical analysis using SPSS13.0 analytical software for single factor analysis of variance (ANOVA) and *t*-tests. *P* below 0.05 was considered to be significant.

3. Results and discussion

3.1. Stability of the electrostatic field penetrated through the rat skin

Figure 1 is the time courses of the effective surface potentials of negative PP electrets penetrated through the rat skin. It indicates that all the external electrostatic field of the electrets used in this study could effectively penetrate through the rat skin. A positive correlation between the penetrated effective surface potential and the initial surface potential of the electret was presented. The higher the initial surface potential of the electret, the larger the penetrated effective surface potential was. Although a quicker decay of the penetrated effective surface potential was observed when the electret had a higher initial surface potential, its absolute value of the remaining effective surface potential was still higher than those electrets with lower initial surface potentials. The penetrated electrostatic field of all the negative electrets could keep strong enough and relatively stable. Therefore, the PP electrets could be used in drug transdermal enhancement experiment.

3.2. Enhancing effect of electret for transdermal delivery of cyclosporine A

Figure 2 shows the percutaneous cumulative amounts of CsA from the saturated solution (control) and the saturated solution plus -500 V, -1000 V or -2000 V PP electrets. An 8 h lag time was observed for all the samples, which could be attributed to the higher molecular weight of the CsA. The natural barrier function of the SC to CsA was significantly increased, and the time for the drug to cross the SC and diffuse, the transfer as well the as metabolism in various layers of the skin was prolonged. Therefore, a delaying effect up to 8 h was presented and even could not be shortened after the application of negative PP electrets. Therefore, electret with the surface potentials in the range of -500 V to -2000 V could not effectively reduce the lag time of the CsA through the rat skin.

However, an enhancing effect of the electret on CsA cross the rat skin was observed within the time interval of 8 h to 24 h. The cumulative amounts of the drug at 24 h after application of -500 V, -1000 V and -2000 V PP electrets were 7.52±0.70 μg cm^{-2}, 9.28±1.87 μg cm^{-2} and 8.86±4.57 μg cm^{-2} respectively, which were 1.2-fold, 1.48-fold and 1.42-fold with respect to the control. The highest enhancing effect on permeation was obtained when the PP electret surface potential was -1000 V, suggesting a "potential window" was existed for electret transdermal enhancing effect. And this result has been proved by our previous studies [12]. Our scanning electron microscopy (SEM) study indicates that among the three negative electrets, -1000 V PP electret had the most significant effect on loosening the SC and disengaging the compact SC lipid structure, which were in favour of drug transdermal delivery.

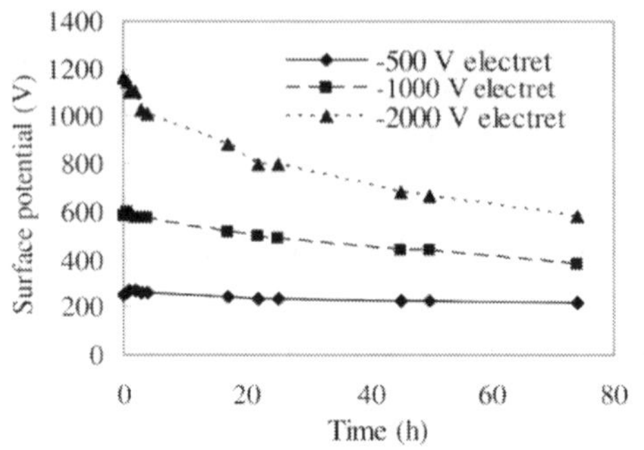

Figure 1. Skin penetrated surface potential decay of negative PP electrets.

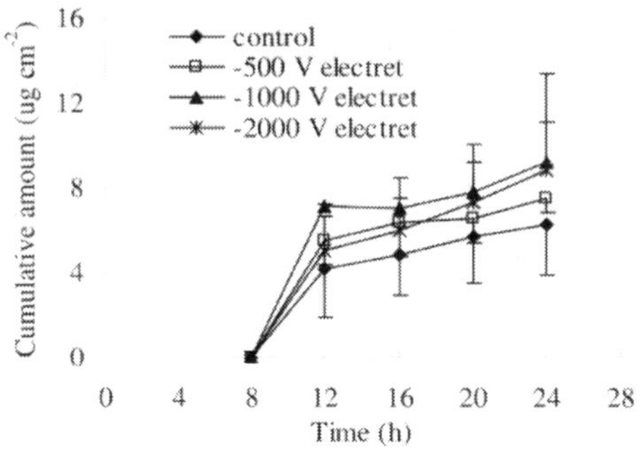

Figure 2. Time course of cumulative amount of CsA through rat skin.

3.3. Synergistic effect of electret and ethyl oleate for transdermal delivery of CsA

Figure 3 shows the cumulative amounts of CsA from the saturated solution, the saturated solution plus 10% ethyl oleate, and the saturated solution plus 10% ethyl oleate and negative PP electret. It indicates that ethyl oleate could significantly increase the cumulative amount within 24 h as compared with the control. Besides, the lag time for CsA penetrating through the rat skin was shortened. The cumulative amounts of the drug at 8 h and 24 h were 18.94±21.72 μg cm^{-2} and 89.69 μg cm^{-2}, respectively, which were 18.94-fold and 14.30-fold with respect to those of the control. Ethyl oleate, a kind of fatty acids, is proved to be an effective enhancer for transdermal drug delivery [13]. The action mechanism was that Ethyl oleate as a separate phase within the bilayer lipids can disturb the intercellular lipid packing and thus facilitating the permeation of the drug penetrating through the skin.

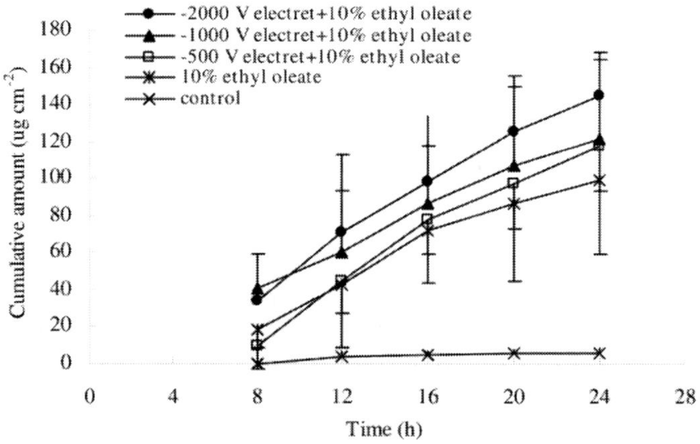

Figure 3. Time course of cumulative amount of CsA through rat skin.

Figure 3 also indicates that when -500 V, -1000 V or -2000 V PP electret was used in combination with 10% ethyl oleat, the cumulative amounts of CsA at 24 h were respectively 109.86±45.12 μg cm^{-2}, 127.15±38.63 μg cm^{-2} and 142.51±42.71 μg cm^{-2}, which were 17.50-fold, 20.25-fold and 22.69-fold increases with respect to the control, and 1.1-fold, 1.28-fold and 1.43-fold with respect to single use of 10% ethyl oleate. Therefore, the combinational use of both the electret and chemical enhancer had the synergistic effect on CsA transdermal delivery. The higher the corona voltage of the PP electret, the more the synergistic effect was. The mechanism could be explained as follows: electret could not only regulate the electric polarized state of the skin, but also change the structure of the SC to form many reversible ducts. The ethyl oleate diffused to SC and various layers of the skin via these ducts, further disturbed the arrangement of the lipid bilayers of the SC and increased the fluidity of the lipids. Therefore, more drugs could dissolve and trap in the skin, and then penetrate through various layers of the skin into subcutaneous tissue and further into blood circulation under the action of natural diffusion and electrostatic field of the PP electret.

4. Conclusions

Negative electret alone or in combination with ethyl oleate could increase the in vitro transdermal delivery of CsA. -1000 V electret had the best enhancing effect when the electret was used alone. However, -2000 V electret exhibited the best synergistic enhancement when it was in combination with the ethyl oleate. These findings suggeste that the electrostatic field produced by the PP electret could be a potential physical factor to enhance the transdermal delivery of macromolecule, especially used in combination with chemical enhancers.

Acknowledgments

The authors thank the National Nature Science Foundation of China (Grant No. 50977089) and Natural Science Foundation of Shanghai (Grant No. 10ZR1437600) for financial support provided.

References

[1] Prausnitz M R and Langer R 2008 *Nat. Biotechnol.* **26** 1261
[2] Kalluri H and Banga A K 2011 *AAPS. Pham. Sci. Tech.* **12** 431
[3] Dhote V, Bhatnagar P, Mishra P K, Mahajan S C and Mishra D K 2012 *Sci. Pharm.* **80** 1
[4] Cui L L, Hou X M, Jiang J, Li G D and Liang Y Y 2008 *J Phys.: Conf. Ser.* **142** 012015
[5] Paudel K S, Milewski M, Swadley C L, Brogden N K, Ghosh P and Stinchcomb A L 2010 *Ther. Deliv.* **1** 109
[6] Singh J, Gross M, Sage B, Davis H T and Maibach H I 2001 *Food Chem. Toxico.l* **39** 1079
[7] Jiang J, Liang Y Y, Dong F J, Liu H Y and Cui L L 2012 *J. Electrostat.* **70** 258
[8] Jiang J, Dong F J, Liang Y Y, Liu H Y and Cui L L 2012 *IEEE T. Dielec.t El. In.* To be piblished.
[9] Murthy N S, Boguda V A and Payasada K. 2008 *Bio. Pharm. Bull.* **31** 99
[10] Nakarani M, Patel P, Patel J, Patel P, Murthy R S R and Vaghani S S 2010 *Sci. Pharm.* **78** 345
[11] Lopes L B, Collett J H and Bentley M V L B 2005 *Euro. J. Pharm. Biopharm.* **60** 25
[12] Cui L L, Liang Y Y, Dong f J, ma L, Tu y, Liu H Y and Jiang J 2011 *J. Phys.: Conf. Ser.* **301** 012027
[13] Williams A C and Barry B W 2004 *Adv. Drug. Deliver. Rev.* **56** 603